MINING SCIENCE AND TECHNOLOGY '99

PROCEEDINGS OF THE '99 INTERNATIONAL SYMPOSIUM ON MINING SCIENCE AND TECHNOLOGY / BEIJING / CHINA / 29-31 AUGUST 1999

Mining Science and Technology '99

Edited by
Heping Xie
China University of Mining and Technology, China

Tad S.Golosinski
University of Missouri-Rolla, USA

A.A.BALKEMA / ROTTERDAM / BROOKFIELD / 1999

The texts of the various papers in this volume were set individually by typists under the supervision of each of the authors concerned.

Published by
A.A. Balkema, P.O. Box 1675, 3000 BR Rotterdam, Netherlands
Fax: +31.10.413.5947; E-mail: balkema@balkema.nl; Internet site: www.balkema.nl
A.A. Balkema Publishers, Old Post Road, Brookfield, VT 05036-9704, USA
Fax: 802.276.3837; E-mail: info@ashgate.com

ISBN 90 5809 067 1
© 1999 A.A. Balkema, Rotterdam
Printed in the Netherlands

Mining Science and Technology' 99, Xie & Golosinski (eds) © 1999 Balkema, Rotterdam, ISBN 90 5809 067 1

Table of contents

2   *Geology, geophysics and methane drainage*

3 Geomechanics

4 *Mine construction and tunneling*

5   *Coal processing and clean coal technology*

7 *Mine control, automation and mechanization*

8   Mine economics, management and administration

Mining Science and Technology' 99, Xie & Golosinski (eds) © 1999 Balkema, Rotterdam, ISBN 90 5809 067 1

Preface

During the last decade the situation of world's mining industry has been changing rapidly. The pace of change has been especially intensive during the most recent years. Among the many reasons for this change the most significant are: Development of new technologies directly applicable to mining, rapid growth of the demand for fuels and minerals in many developing countries such as China, structural readjustment of the countries that constituted the former Soviet Union, low mineral commodity prices and the like. All these changes bring to the forefront the need to produce fuels and minerals more efficiently, thus at lower overall cost, and to provide the market with specialized, high quality products. The related problems are being addressed by both the mining industries worldwide and by scientists involved in mining-related research and development.

The 1999 International Symposium on Mining Science and Technology ('99ISMST) gathers mining scientists and engineers from all over the world and provides the most up-to-date overview of the recent changes and developments related to mining worldwide. Major topics to be covered are: Mine Operations and Safety; Geology and Coal Bed Methane Drainage; Rock Mechanics and Ground Control; Mine Construction and Tunnelling; Mineral and Coal Processing, and Clean Coal Technology; Mine Subsidence, Reclamation and Environment; Mine Mechanization and Automation; Mine Economics, Management and Administration. The Symposium emphasizes China and her mining industry for two reasons. First, the mining industry of China is undergoing a major structural change and faces a multitude of new challenges as a result. Second, many innovative mining technologies and techniques were developed there facilitated by availability of advanced research and development capability. Much of this work is described and discussed in this book. The Symposium also deals with other mining industries worldwide. To briefly illustrate the scope of the Symposium it describes the problems faced by the dimensional stone mining in developed Italy and very different problems faced by the copper mining in third-world Peru. It also deals in detail with many problems of American mining, still the largest in the world, as well as with those of mining in Russia, Australia, and other mining countries of the world.

China University of Mining and Technology (CUMT) organized the first International Symposium on Mining Science and Technology in 1985. Three Symposia in this series have been held succcessfully so far. Founded in 1909, CUMT is a key Chinese University that features mining science and technology. Additionally it has expertise in and deals with all sciences, including engineering science, and with liberal arts, including economics and management. The University's main campus is located in the beautiful south suburbs of Xuzhou City, Jiangsu Province. Two other campuses are in Beijing. The importance of CUMT was officially confirmed by its listing as one of the select universities for the National '211 project'. Tens of thousands of CUMT graduates, high quality scientists, engineers and technologists, have made their impact on both Chinese and world mining industries.

CUMT is the lead organizer and sponsor of the ISMST in 1999. Five main co-organizers/co-sponsors of the Symposium are the University of Missouri-Rolla, USA; the National Natural Science Foundation of China (NSFC), China; the University of Nottingham, UK; Yanzhou Mining Group Corporation Ltd., China; and Datong Coal Mining Administration, China. The academic institutions represent the three countries with the commanding position in world's mining, NSFC coordinates development of science nationwide in China and the last two are the key players in Chinese mining industry. In their work on organization of the Symposium the five enjoyed cooperation and help of many other mining-related educational, research and industrial institutions worldwide. While these are too numerous to list here, three that must be mentioned are the Wroclaw University of Technology, Poland; the University of Leeds, UK; and Western Australian School of Mines, Curtin University of Technology, Australia.

The organizers wish to thank all the parties involved for their support and cooperation, all the authors for their preparation and delivery of the papers and the China University of Mining and Technology for providing the venue for holding this world class event.

Heping Xie
Professor and President
China University of Mining and Technology

Tad S.Golosinski
Professor
University of Missouri-Rolla

Mining Science and Technology' 99, Xie & Golosinski (eds) © 1999 Balkema, Rotterdam, ISBN 90 5809 067 1

Organization

SYMPOSIUM ORGANIZING COMMITTEE

Chair

Prof. Heping Xie, President of CUMT, China

Co-chair

Prof. Tad S.Golosinski, University of Missouri-Rolla, USA
Prof. Yuehan Wang, Vice-president of CUMT, China
Prof. Bryan Denby, University of Nottingham, UK

Members

Prof. Huizheng Pan, Vice-director of China Coal Mining Society, China
Prof. Monika Hardygora, Wroclaw University of Technology, Poland
Prof. Jingche Zhao, President of Yanzhou Coal Mining Group Ltd., China
Prof. Lev Punchkov, Moscow State Mining University, Russia
Prof. Xiaoxi Zhu, Director General of Datong Coal Administration, China
Prof. Minghong He, Natural Scientific Foundation of China, China
Prof. Kezhuang Feng, Vice-President of CUMT, China
Prof. F.L.Wilke, Technical University of Berlin, Germany
Prof. David Spottiswood, Curtin University of Technology, Australia
Prof. P.A.Dowd, University of Leeds, UK
Prof. Yuzhuo Zhang, Director of China Coal Research Institute, China

Secretary-general

Prof. Yuehan Wang, Vice-president of CUMT

Deputy Secretaries-general

Prof. Wenjin Ke, Vice-president of CUMT
Prof. Suping Peng, CUMT
Dr Yaodong Jiang, CUMT

Mining Science and Technology' 99, Xie & Golosinski (eds)© 1999 Balkema, Rotterdam, ISBN 90 5809 067 1

The National Natural Science Foundation of China (NSFC)

The National Natural Science Foundation of China (NSFC) was established in 1986 as a major consequence of the reform of science and technology system in China. The objective of the foundation is to promote and financially support both basic and applied basic research and to gradually institute a science funding system in China.

NSFC has enjoyed great support from the government since its founding. Its annual budget has been rapidly increasing, a big leap from 80 million yuan RMB in 1986 to over 570 million yuan RMB in 1996. The Government input a total of 1.66 billion-yuan RMB at an average annual rate of increase of 29.8 percent in the period of the Eighth Five-Year Plan (1991-1995). The Foundation has now become the biggest funding agency for basic research in China.

In the past ten years NSFC has created the categories of General Program, Key Program, Major Program, and the National Science Fund for Distinguished Young Scholars, as well as several other types of grant in accordance with the requirements of scientific research. Up to 1996, it has supported more than 33,000 projects in the General Program, 300 key projects, and 122 major projects, and granted the National Science Fund for Distinguished Young Scholars between 1994-96 to 214 young scientists, who came from universities of the State Education Commission, research institutes of the Chinese Academy of Sciences, and universities and institutions under the ministries and provinces. As a result, NSFC has made significant contributions to the acquirement of scientific achievements, the fostering of talented persons and the promotion of prosperity of science in China.

NSFC attaches great importance to international scientific cooperation and exchange. It has signed collaborative agreements and memoranda of understanding with science funding organizations in 31 countries, and in regions of Hong Kong and Macao. Great efforts have been made to maintain these as stable and flexible channels of communication and to create other venues of international cooperation and exchange for Chinese scientists.

In accordance with the state strategy of revitalizing the Chinese nation through science and education, NSFC will continue to support basic and applied basic research. It will coordinate this research with the national goals to assure that these contribute to the economic development and social progress of the country.

Mining Science and Technology' 99, Xie & Golosinski (eds) © 1999 Balkema, Rotterdam, ISBN 90 5809 067 1

Yangzhou Coal Mining Company Limited (YZC Ltd.)

Yanzhou Coal Mining Company Limited (YZC Ltd.) is the leading producer of coal from its underground mines in Shandong Province, P. R. China. The Company currently owns and operates seven mines. It is the most profitable coal mining company in China and one of the country's largest coal exporters. In 1997, YZC Ltd. produced 19.42 million tonnes of raw coal, of which 12.975 million tons were excavated by the Company's modern longwall caving mining methods. Of this 4.88 million tonnes of clean coal were exported, mainly to Japan. The proven coal reserve of Yanzhou coal field is 4 billion tons of prime quality, low-sulphur coal, capable of yielding a product with a controlled ash content as low as 6%.

YZC Ltd. was listed on The Stock Exchange of Hong Kong on 1st April 1998 and on the New York Stock Exchange on 31st March 1998. The latter was a combined offering of Company's H shares ("H Shares") and American Depositary Receipts (ADSs), each representing 50 H Shares.

Located in the heart of one of China's fastest growing regions, YZC Ltd. will continue to enjoy a higher profit margin than most of its domestic competitors due to low transportation costs. In recent years YZC Ltd. has made great strides in the development of longwall top-coal caving technology (LTCT), and set a world record of 5.01 Mta coal production from single longwall operations. With advanced mining technology, a highly experienced management team and access to international capital markets, YZC Ltd. will remain one of the most competitive coal mining enterprises in this new era.

Mining Science and Technology' 99, Xie & Golosinski (eds) © 1999 Balkema, Rotterdam, ISBN 90 5809 067 1

Datong Coal Mining Administration (DCMA)

Located in Shanxi Province, Datong Coal Mining Administration (DCMA) is one of the largest modern coal enterprises in China. Datong Coal Mining Administration is mainly engaged in underground mining of high quality coal with low ash (below 10%), low sulfur (below 0.9%) and high heating value. Coal from its mines is sold to electronic power plants, gas plants, metallurgical plants, chemical factories and other customers located in 27 domestic provinces or regions. It is also exported to Japan and Korea.

Currently Datong Coal Mining Administration operates 15 mines and employs 130,000 employees. The largest producer in China, the Datong Coal Mining Administration produced 37.06 million tonnes of coal in 1997, and realized the income of RMB 47.5 billion. About 97% of this coal was extracted by machines. Longwall mining accounts for 90% of total production..

About 1 billion tons of quality coal has been produced since Datong Coal Mining Administration was established fifty years ago. In 1995 Datong Coal Mining Administration was listed as the 37th largest industrial enterprises in China according to their income and taxes paid to Chinese government, the 71st according to the sales value, and the 77th according to the capital value.

Datong Coal Mining Administration, an enterprise with glorious history, is working hard for a brighter future Shanxi Province and China.

1 Mine operations, ventilation and safety

Mining Science and Technology'99, Xie & Golosinski (eds) © 1999 Balkema, Rotterdam, ISBN 90 5809 067 1

Developing coal mining technology for the 21st century

Michael Kelly
Commonwealth Scientific and Industrial Organisation (CSIRO), Kenmore, Qld, Australia

ABSTRACT: In July 1997, the CSIRO commenced implementation of a triennium plan for coal exploration and mining research. The plan has a total funding of AUD$31 million and is focused on project initiatives in the six broad areas of resource assessment, operations improvement, gas control and utilisation, automation, safety and material characterisation for fines control.
The triennium is already more than halfway completed. This paper summarises the achievements so far, the likely outcomes at the end of the triennium and some future research directions.

1 INTRODUCTION.

The CSIRO is a statutory agency of the Australian Government that provides benefit to Australia's industry and economy, environment and social system and support to Australia's international objectives through the provision of science and technology. It employs a total of 6,800 people in 22 industry sectors and is supported from both Government and private funding.

One of these areas is the Energy Sector, in which Coal Exploration and Mining is the largest component. Within this component, the CSIRO commenced implementation of a triennium plan in July 1997. The plan has a total funding of AUD$31 million and is focused on 18 specific project initiatives in six broad target areas. The coal industry is Australia's largest export industry with 1998 exports reaching a record of 168 million tonnes, an increase of 15% on 1997.

The six overall target areas were:

- To develop processes and tools that will allow mining companies to better assess current and future mining areas.
- To extend and optimise established open cut & highwall mines, improve the viability of established underground mining, and provide new mining techniques for current and future mines.
- To remove gas from mine leases for productivity and safety enhancement, for commercial utilisation technologies, and also to reduce greenhouse effects.
- To increase automation of mobile equipment in mining based on novel sensors for navigation, robotics and communications.
- To improve mine safety including safety management systems and technologies to control hazards; new extraction methods to avoid hazardous conditions; and adequate emergency response capability.
- To predict material variability and its impact during mining and processing, and develop methods to control product from pit to port through effective breakage and reduction of fines.

By March 1999, the component had 54 active projects in these areas employing 63 people within CSIRO and collaborating with more than double that number from industry and consultant groups. These projects are well advanced to achieve most of the objectives for the current triennium. In addition strategic planning for the next triennium from July 2000 has commenced, and it is interesting to reflect upon the new technology required to take the coal industry into the next century.

2 CURRENT TRIENNIUM OUTCOMES.

Before looking at the future, it is worthwhile to look at the current state of technology development by examining the outcomes of some of the present suite of coal mining projects within CSIRO.

2.1 Geological/geotechnical assessment.

Many coal mining projects in Australia have suffered as a result of inadequate or incorrect geotechnical assessment. In many cases this has stemmed from poor communication through project stages as well as insufficient understanding of consequence of risk. There has been a raft of projects initiated to overcome these issues with a view to reducing the financial risk of new projects.

The foundation of all geotechnical assessment is a comprehensive understanding of the geological environment. This has been aided with the use of 3D block models to assess sedimentology from a mine to regional basis (Esterle and Fielding, 1996). The models have now been extended to include regional structural regimes that can be applied directly to mines down to individual panel scale.

Communication has been greatly enhanced with advances in 3D visualisation (Le Blanc Smith & Caris 1998) which allow many disparate data sets to be displayed in an interactive 3D (and 4D) package. These can be transferred and updated right from pre-feasibility stage through to rehabilitation.

Advances in geophysical tools and assessment of results have been highlighted by microseismic monitoring of mining (Luo & Hatherly 1998), interpretation of full waveform sonic logging (Hatherly et al 1997) and improved logging systems. As an example, the SIROLOG neutron gamma logging system is ideally suited for exploration and reserve quality estimation in new coal developments (Charbucinski et al. 1986). The tool can be calibrated to give depth and thickness of the seams and the lithology of associated host rocks, as well as in situ quantitative estimates of BTU, ash, iron in ash, density and both silica and alumina content. All of these tools have an input into 3D geotechnical block models that are under development.

Improvement in computer hardware is allowing the realistic 3D computational modelling of rock mass deformations. Rock failure and fluid flow algorithms are being developed for particle codes as well as conventional FE/FD models. These require extensive knowledge of coal (Medhurst & Brown 1998) and rock failure criteria.

These advances, being developed under the umbrella of a framework of assessment protocols and criteria, will ensure a significant reduction of risk for future coal projects.

2.2 Highwall Mining

Highwall mining utilises the asset of the existing seam access from the final highwall prior to rehabilitation activities. During the triennium, CSIRO has developed several tools that have enhanced this access and allowed operating companies to maximise the existing resource.

Geotechnical design tools for pillar (Duncan Fama 1995) and unsupported span (Shen & Duncan Fama 1997) in highwall mining have allowed for a penetration depth of more than 500 metres and various highwall geometries.

Equipment advances for directional control now result in an accuracy of better than 1m cross-track in 500 m depth (Reid, D.C. et al 1997). A spin off CSIRO company, Cutting Edge Technology, is now producing high technology, steerable augers for both highwall and underground applications.

2.3 Longwall Geomechanics.

Longwall geomechanical research being carried out by CSIRO and Strata Control Technology has resulted in a better understanding of rock failure mechanisms around longwall extraction (Kelly & Gale 1999). Shear, rather than tensile failure has been the predominant failure mechanism in the Australian environments monitored. Failure has occurred further ahead of the retreating face than predicted by conventional longwall geomechanics theory. In some cases rock breakage and/or slippage has been detected several hundred metres ahead of the face position with demonstrated influences of minor geological discontinuities. Major structural features have been shown to have a dominant control on failure mechanisms. Layout geometry, the previous goafing mechanics and pore water pressure have also been shown to influence failure. Validating technologies of microseismic monitoring and new face monitoring techniques have assisted the development of predictive 2D computational modelling tools. The demonstrated 3D consequences of failure has led to the future direction of the research to further investigate these effects.

2.4 Gas Assessment and Alleviation

Applied research into outburst mechanisms (Choi & Wold 1996) hydrofracturing (Jeffrey & Settari 1998), reservoir modelling (Wold et al 1999), tight radius drilling (Meyer & Howarth 1998) and goaf gas control have all had direct benefits for mining companies. The use of Computational Fluid Dynamics for goaf gas modelling has been particularly beneficial with Dartbrook mine increasing its gas capture by 250% partly as a result of this technology. The reservoir model SIMED has been transferred to the consultant group Geogas for direct use as its preferred reservoir model.

2.5 Roadway development

Research into underground roadway development has varied from industry applications of a systems approach to minesites (Kelly 1997) to the development of new equipment. The highlight of this has been the commencement of a five-year project in 1998 to develop a fully automated conveyor/bolting module that sits between a remotely controlled miner and coal haulage units. The JCOAL/CSIRO initiative involves manufacturers from both Japan and Australia and is also sponsored by Australian mining groups. The automated feed/drilling component will utilise a new self-drilling bolt that has been under development by BHP for several years.

2.6 Dragline Automation

A prototype automated swing control system has been successfully commissioned on a production dragline at the Meandu mine. This system is expected to increase the dragline productivity by 4%, translated Australian wide to an annual saving of $280M/year. (Winstanley 1998) Other automation technology also being developed under the Centre for Mining Technology and Equipment for underground mobile machines can be readily translated for coal mining applications. New safety systems are also being developed for large surface vehicles that utilise new hazard proximity sensors.

2.7 Emergency response

The development of the LAMPS (Location And Monitoring for Personal Safety) system is well progressed (Einicke et al. 1998). The system, comprising of control & monitoring, network beacons and personal transponders will provide a wireless capability to report the location and health of underground staff, even after an explosion or other underground incident.

The technology developed for the "Numbat" remote controlled emergency response vehicle is now being utilised to construct a remote emergency vehicle capable of carrying up to 12 men. The vehicle will be able to be used to recover injured or trapped personnel after an explosion without putting a mines rescue team at risk.

New inertisation equipment coupled with computational fluid dynamics simulation will be an effective response to issues of spontaneous combustion.

2.8 Coal fines minimisation

An integrated tool set is being developed for assessing, monitoring and alleviating the impact of size degradation and fines generation during the mining process (Esterle et al 1998). Trials conducted have indicated that plant yields may be increased in the order of 3-5% by assessing coal structure and managing of energy inputs during the complete mining cycle.

3 FUTURE STRATEGIES

The project summaries above, although not a complete description of the group's work are a representation its current direction and expertise. It has been an onerous task to review this direction and set a new suite of targets for the next triennium beginning in July 2000.

The first priority for future work is, of course, to complete the current tasks as described. The impact of these successfully implemented into the Australian industry will result in many hundreds of millions of dollars in reduced costs and opportunity benefits. Researchers tend to be good at creating new ideas rather than their implementation, so there continues to be a major challenge in coordinating such a group to set realistic implementation strategies. As such all projects must have an implementation component, preferably with manufacturing and/or mining company involvement from the beginning.

The second priority is to set ongoing strategies to facilitate "Developing Coal Mining Technology for the 21st Century". The passage of the new century is an opportunity to set some new directions and to let go of some of the paradigms of our past. For the future of the Australian Coal mining industry, the group has determined seven key research and development areas listed below that either are new or should have an increased focus:

3.1 Geotechnical Assessment

With new mining areas being targeted, the need for integrated geotechnical assessment that extends from the geological environment to predicted mining performance and risk assessment is greater than before. Companies need to be assured of return on investment and further tool development and integration of assessment techniques is essential.

3.2 Automation

Automation of longwalls, mobile and development equipment are clear priorities for the future. Automation of longwalls is required to promote consistent performance and the removal of persons from hazardous face areas. Repetitive operations like coal haulage vehicles are relatively easy to automate, but technology to protect and exclude personnel requires development. The ultimate in roadway development systems will be a fully automated TBM type machine and there are many steps of technical development required to achieve this outcome.

3.3 Virtual mining

Controlling mine operations within a virtual mining environment is not something for the future, it is an imperative for here and now. Ninety per cent of the technology required has been developed and it only requires cooperation from manufacturers and the vision from mining companies to realise its potential. The paybacks from consistency, safety and increased quality flexibility of operations are enormous. One only needs to look at the developments in coal processing to realise the economic potential of "hands off" mining.

3.4 Thick and multiseam mining

Australia has billions of tonnes of quality coal that can be extracted via thick seam and multiseam operations. It is imperative that new extractive methods are developed that maximises the recovery from this resource *and* enhances company profitability. Methods of longwall/sublevel caving and thick seam hydraulic mining are two potential methods requiring further development. Both of these will be greatly enhanced by virtual reality technology.

3.5 Rehabilitation/subsidence

Increasing environmental awareness and standards, and the legacy of the past combine to demand action on surface rehabilitation. This requires increased resources applied to mine site rehabilitation that focuses on development of cost effective ecosystems.

Future thick seam extraction requires detailed understanding of comprehensive subsidence mechanisms, focusing on key areas of impact on farming, hydrology and surface structures to enable mining without devastating surface effects. Failure to develop this technology will essentially sterilize large tracts of quality coal reserves.

3.6 Greenhouse gas issues

Coal mining, because of its inherent emission of methane and carbon dioxide in addition to its energy intensive nature, has unique greenhouse issues that the industry must address that should not be lost in the general carbon cycle arguments. Utilisation of methane emissions from mine atmospheres impacts on the world stage from all coal mining areas. CSIRO can make a significant contribution to this issue for the East Asian region.

3.7 In situ gasification

Australia has large coal resources that may be suitable for in situ gasification. This technology has the potential for competitive energy production with less environmental impact both in terms of surface disturbance and production of greenhouse gases. Access, mining processes and environmental impact can all be improved through improved technology and "virtual mining".

4 CONCLUSIONS

The CSIRO has developed a world class coal mining research capability that will continue to improve the competitiveness of the Australian coal industry into the future. Its capability has been further enhanced by strong collaborative links with industry, consultant groups and other research agencies. Technology transfer and implementation mechanisms are an essential part of this capability.

Dramatic changes in the coal market require an even more dramatic review of future research and development requirements. The pace of computer and information technology improvements creates an opportunity for revolutionary developments in mining practice. Operators who can implement these practices will be at a great advantage in the market place.

Obligations to promote best practice in safety and environmental outcomes also are strategic drivers. Applications extend past Australia's shoreline and allow mutual benefit for Australia and her Asian neighbours through cooperative development programs, sale of technology and improved environmental and safety outcomes.

The quality of Australia's coal reserves and the regional energy requirements present an outlook of at least a further hundred year business for the Australian coal industry. This massive resource demands a commensurate response for research and development strategies to maximise its utilisation within environmental imperatives and return to the community.

5 REFERENCES

Charbucinski, J., Youl, S.F., Eisler, P.L. & Borsaru, M., 1986. Prompt neutron-gamma logging for coal ash in water filled holes. *Geophysics* 51,5: 1110-1118.

Choi, S.K., & Wold, M.B., 1996. A Mechanistic Approach in the Numerical Modelling of the Cavity Completion Method of Coalbed Methane Stimulation. *2nd North American Rock Mechanics Symposium*, Montreal Vol.2, 1903-1910.

Duncan Fama, M.E., Craig, M.S. & Trueman, R., 1995. Two- and Three-Dimensional Elasto-Plastic Analysis for Coal Pillar Design and its Application to Highwall Mining. *Int. J. Rock Mech. Min. Sci & Geomech.* Abstr. 32, 3, 215-225.

Einicke, G., Dekker, D. & Gladwin, M., 1998. Location and Monitoring for Personal Safety (LAMPS*), Proc. Qld. Mining Industry Health and Safety Conf.*, Yeppoon, Australia, 191 – 195.

Esterle, J.S. & Fielding, C.R., 1996. Sedimentological analyses of goafing interval for longwall mining at Goonyella Mine, Bowen Basin. *Geology in Longwall Mining*. Coalfield Geological Council of NSW, Sydney, 27-34. Conference Publications.

Esterle, J.S., O'Brien, G., Sexty, G., Thornton, D., Kojovic, T., Djordjevic, N., Kolatschek, Y., Firth, B. & Clarkson, C.J., 1998. Assessing coal fines and their impact on plant performance. *XIII International Coal Preparation Congress*, Brisbane, 173-182.

Hatherly, P., Fallon, G.N., Fullaghar, P.K. & Zhou, B., 1997. Geotechnical characterisation of the rock mass with sonic logging. 4th Intl Symp. on Mine Mechanisation and Automation, July 1997, Brisbane. A9:15-24.

Jeffrey, R. G. & Settari, A., 1998. An instrumented hydraulic fracture experiment in coal. SPE Paper 39908. *1998 SPE Rocky Mountain Regional/Low-Permeability Reservoirs Symposium and Exhibition*, Denver, CO.

Kelly, M.S. & Gale, W., 1999. Ground Behaviour about Longwall Faces and its Effect on Mining. *ACARP Report C5017.*

Kelly, M.S., 1997. Coordination of Roadway Development Strategy. *ACARP Report C5013.*

LeBlanc Smith G. & Caris C., 1998. Internet Virtual Reality Tools: A New Paradigm Unfolds for Exploration and Mining. *Geological Society of Australia, Abstracts No. 49, 14th AGC.* Townsville. 264.

Luo, X. & Hatherly, P., 1998. Application of microseismic monitoring to characterise geomechanical conditions in longwall mining. *Exploration Geophysics*, 29, 489-493.

Medhurst T.P. & Brown E.T., 1998. A study of the mechanical behaviour of coal for pillar design. *Int. j. rock mech. min. sci. & geomech abstr.* 35: 1087-1105.

Meyer, T. & Howarth, D., 1998. Tight Radius Drilling. *Annual Report, Cooperative Research Centre for Mining Technology & Equipment*, Brisbane. 22.

Reid, D.C., Hainsworth D.W. & McPhee, R.J., 1997. Lateral Guidance of Highwall Mining Machinery Using Inertial Navigation, *Proceedings of the 4th Intl Symp on Mine Mechanisation and Automation*, Brisbane, Australia, July 6-9 B6:1-10.

Shen, B. & Duncan Fama, M.E. 1997. A laminated span failure model for highwall mining span stability assessment. *Computer Methods and advances in Geomechanics,* Yuan (ed.), Balkema, Rotterdam. Vol 2, 1587-1591.

Wold, M.B., Choi, S.K., George, S.C., Wood, J.H. & Williams, D.J., 1999. Coal mining beneath a gorge; induced fracturing and the release of reservoired gases. *Paper accepted for 9th ISRM Congress*, Paris.

Winstanley, G., 1998. Dragline Automation. *Annual Report, Cooperative Research Centre for Mining Technology & Equipment*, Brisbane. 30.

Mining Science and Technology' 99, Xie & Golosinski (eds) © 1999 Balkema, Rotterdam, ISBN 90 5809 067 1

The structure model and its application to roof rock pressure control in gateways of a deep mine

Song Zhenqi, Jiang Jinquan & Sun Xiaoming
Shandong Institute of Mining and Technology, People's Republic of China

Zhao Jingche
Yanzhou Mining Corporates Group, Shandong, People's Republic of China

ABSTRACT: The key of the roof rock pressure control in gateway lies first in having thorough knowledge of the dynamic change of the surrounding rock movement and stress fields as the face advances so as to establish the dynamic face-structure mechanics model in accordance with the objective reality. Then, on this basis, by correctly selecting the location of gateways and their excavating time, one can set up the corresponding dynamic gateway-structure mechanics model so as to solve problems concerning the gateway support design techniques.

1 INTRODUCTION

The yearly coal output in China is nearly 1.14 billion tons, most from underground mining, the length of gateways needed to be excavated and maintained is over 6 million meters annually. As mining depth increases, the costs of excavation and maintenance steadily increase as gateways have gradually deteriorated. Therefore, developing and perfecting the gateway rock pressure control theory and support technique have already been the key problems in achieving high output and high efficiency of coal production.

The aspects that needed to be dealt with in developing gateway rock pressure control theory at present is:

1. As mining depth increases, the roof rock pressure increases (including strata pressure and residual tectonic stress), the stability of surrounding rock deteriorates, the possibility of serious accidents, such as impact rock pressure, gas and coal out bursts, increases. Improper layout of gateway will result serious maintenance problem even accidents.

2. Since lack of reliable theory and scientific means, gateway support design actually still remains at the stage of statistical empirical decision-making.

Aimed at the above, the basic tasks of studying roof rock pressure control theory in gateways are:

1. On the basis of comprehensive understanding the surrounding rock movement and the stress field changes as the face advances, the dynamic face-structure mechanics model should be established and further perfected by determining related parameters at the lab and in-situ with simple and easy means. 2. Based on the above mechanical model study, the computer decision system for gateway design could be developed according to the concrete coal seam bedding condition and supporting techniques.

2 THE DETERMINATION OF THE FACE-STRUCTURE MODEL AND ITS PARAMETER USED IN THE PRESSURE CONTROL OF GATEWAY

Under normal mining conditions, the face-structure model in which the face wall is broken ahead of the faceline affected by the abutment pressure is shown in Figure 1.

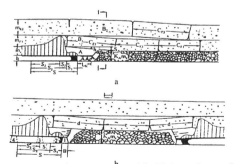

Figure 1 The face structure model with internal stress field
a--advancing direction; b--perpendicular to advancing direction

[1] Supported by the major project of the National Scientific Foundation of China

The strata extent required to be controlled in the structure model generally consists of two parts: "immediate roof" composed of the strata which have already caved in gob (Figure 1a) and "main roof" composed of "the transmission rock beams" whose movement affects obviously the coalface roof behavior (Figure 1b).

The thickness of the immediate roof m_z can be obtained by the following formula:

$$m_z = \frac{h - S_A}{K_A - 1} \qquad (1)$$

Where, h = mining height; S_A = the value of roof sagging at the contact point where the rock beam contacts the waste in gob,

$$S_A = \frac{C_1 \Delta h_A}{L_K} \qquad (2)$$

K_A = the expansion factor of the caved roof at the contact point.

For the inactive face, when the thickness of each layer of overlying strata is known, total thickness m_z of the strata (i.e. immediate roof) can be deduced layer by layer in ascending order according to the conditions of instability and the cave of strata.

$$m_z = \sum_{i=1}^{n} m_i \qquad (3)$$

Where, m_i = the thickness of each caved stratum; n = the number of caved strata counted from coal seam in ascending order.

Of which, the thickness m_n of the nth stratum and the thickness m_{n+1} of the $(n+1)$th stratum will satisfy the following conditions:

$$\begin{cases} m_n \leq \dfrac{C_n \mathrm{tg}(\theta - \alpha)}{4} + S \\ m_{n+1} > \dfrac{C_n \mathrm{tg}(\theta - \alpha)}{4} + S \end{cases} \quad \begin{array}{l}\text{stability-losing by}\\ \text{rock beam sliding}\end{array} \quad (4)$$

or

$$\begin{cases} m_n \leq h - \sum_{i=1}^{n} m_i (K_A - 1) \\ m_{n+1} > h - \sum_{i=1}^{n} m_i (K_A - 1) \end{cases} \quad \begin{array}{l}\text{stability-losing by}\\ \text{rock beam rotating}\end{array} \quad (5)$$

where, h = the mining height; C_n, C_{n+1} = the first breaking span of nth and $(n+1)$th stratum respectively; ψ, θ are the internal friction angle and the rock fracture angle of the related strata respectively; S = the value of the roof sagging before the stability-losing of the structure.

During the extraction of thick seam by slice mining, the total thickness of the immediate roof m_{zn} for nth slice at the lower part of the seam can be deduced by the following formula:

$$m_{zn} = \frac{h_n - S_A}{K_A - 1} + m_{n-1}\left(1 - \frac{K_A - K_C}{K_A - 1}\right) \qquad (6)$$

Where, h_n = the mining height of nth slice; K_C = the expansion factor of the false roof at the location of nth slice face; m_{n-1} is the caved height during the extraction of $(n-1)$th slice.

Mining practice has demonstrated that the total thickness of the main roof is about 4~6 times of mining height for the coalface covered by the general strata. The number of rock beams involved in the main roof, in general, is not over 3. The thickness of "the transmission beams" is sum of the thickness of the simultaneously moved strata in the main roof. The conditions of simultaneous movement of two (up and down) layers of strata can be expressed by the following formula:

$$E_n m_n^2 \geq K^4 E_c m_c^2 \qquad (7)$$

where, m_n, m_c = the thickness of the lower stratum (supporting stratum) and the upper stratum (stratum moved with the lower one) respectively; E_n, E_c = the elastic modulus of the lower and upper strata respectively; K = coefficient of proportionality, it can be 1.15~1.25 according to the specific condition.

The first weighting span C_0 and the periodic weighting span C_1 for each "rock beam" of the main roof can be expressed by the following formula:

$$C_0 = \sqrt{\frac{2m_n^2 [\sigma_n]}{(m_n + m_c)\gamma}} \qquad (8)$$

$$C_i = -\frac{1}{2}C_{i-1} + \frac{1}{2}\sqrt{C_{i-1}^2 + \frac{4m_n^2 [\sigma_n]}{3\gamma(m_n + m_c)}} \qquad (9)$$

where, $[\sigma_n]$ = the tensile strength of the main roof rock; γ = the unit weight of the main roof rock; C_i, C_{i-1} = the periodic weighting span for the current weighting and last weighting respectively.

In production site, the number and moving span of the main roof rock beams can be deduced by the face support resistance or the roof convergence versus the face advancing distance gained from the field observation.

The research has demonstrated that the fracture pattern of the main roof under the condition of the solid coal supporting state (i.e. before mining) at the both ends of the face as the face advances is shown is Figure 2. The value of the lateral fracture span (in the faceline direction) of the main roof can be derived by using the plate limit analysis method:

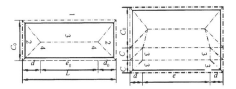

Figure 2 The fracture pattern of the main roof under the condition of the solid coal supporting state

The first fracture is:

$$d_0 = \frac{C_0}{2} \cdot \frac{C_0}{L} \left[\sqrt{1 + 3\left(\frac{L}{C_0}\right)^2} - 1 \right] \qquad (10)$$

The periodic fracture is:

$$d = \frac{2C}{17} \left[\sqrt{\left(10\frac{C}{L}\right)^2 + 102} - 10\frac{C}{L} \right] \qquad (11)$$

where, L = the length of the coalface; C_0 and C = the span of the first weighting and periodic weighting of the main roof respectively (in the face advancing direction).

When the main roof weighting comes, the immediate roof and the lower part of rock beams fracture and the fractured line goes deep in front of the faceline, the "abutment pressure" (or the corresponding reaction force) on the coal seam or strata, caused by the action of the overlying strata can be obviously divided into two portions, which is an important characteristics of the structure model under this specific face conditions. Of which the coal mass has been broken within "the internal stress field" (portion S_1 in Figure 1), the magnitude of the pressure applied on it is determined by the applied force arisen from the movements of the immediate roof and the main roof. Obviously, when the weighting of the main roof has completed and it got into the stable state, it is feasible that bearing resistance of that portion of coal mass is only to balance the action force of the immediate roof. The abutment pressure distributed on the coal mass in "the external stress field" (portion S_2 in Figure 1) is determined by the overall action force of the overlying strata, its magnitude and distribution extent are slightly affected by the immediate roof and the main roof.

The experiment research has verified that under the condition of "the internal stress field" occurred at

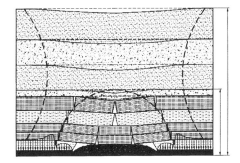

Figure 3 Movement of the overlying strata and distri-bution of the abutment pressure

deeper mining depth, the abutment pressure on the face wall reaches the maximum value before supercritical mining as the face advances. The distance L_h that the face advanced is about 1.5 times the mining depth (H) (i.e. L_h=1.5H) as the abutment reaches maximum value. At the time, the height of the arch connected by the boundaries of the caved and fracture-broken strata is nearly the half of the face advanced span, i.e. h_L=0.5L_h. The extent of the moved strata is delimited by the moving boundary and arch-forming boundary of each stratum, as shown in Figure 3.

By this time, the abutment distributed in front of coal wall around the face will be composed of the following two parts:

1. The action force F_{1max} which is transmitted by the fractured "rock beams" in the arch.

Under the obvious movement of the "rock beams", the action force will be mainly supported by the coal mass in "internal stress field". The magnitude of the action force after the rock movements stabilizing can be expressed approximately by the following formula:

$$F_{1max} = \int_0^{\delta_1} \sigma_{1x} dx = \frac{3}{8} \gamma H C_E \qquad (12)$$

where, C_E = the mean span of the fractured rock beams in the arch (i.e. the mean of the weighting spans).

If the compressed amount y_0 at the edge of the coal wall and the coal mass stiffness G_0 at the end of the internal stress field are known, and the maximum abutment stress of the internal stress field, $\sigma_{1max} = G_0 y_0 / 4$, can be derived, we can get

$$\int_0^{\delta_1} \sigma_{1x} dx = G_0 y_0 S_1 / b = \sigma_{1max} S_1 / 2$$

From this, the expression of the extent of the internal stress field can be obtained as follows:

$$S_1 = (3\gamma H C_E)/(G_0 y_0) \qquad (13)$$

2. The action force (i.e. the pressure on arch abutments) F_{2max} which is transmitted by the moving strata outside the arch through "the arch axis".

The action force is a main source of the stress in "the external stress field"; its magnitude can be expressed approximately as follows:

$$F_{2max} = 0.37\gamma H^2 \qquad (14)$$

If σ_{2max} used to express the max. bearing stress in external stress field, it is easy to obtain the approximate expression of the extent of external stress field (S_2):

$$S_2 = 0.74\gamma H^2/\sigma_{2max} \qquad (15)$$

When obtained the extent of the internal stress field and external stress field S_1 and S_2, then the extent ($S=S_1+S_2$) affected by the face abutment pressure is determined.

After the model of structure mechanics are known, the position of gateways prepared for the vicinal working face, the time of extraction and preparation and the correct maintenance method of gateways are not difficult to determine.

If the face belongs to the structure model with interior stress field and the floor heave does not happen, the proper designing procedure is as follows.

The gateway maintained along the gob edge (position 1 in Figure 1): at this time, the minimum resistance of the support must balance with the load from the immediate roof, the yield value of the support must be adapted to total sagging value caused by the lower rock beam contacting and compressing the gob waste. The minimum value Δh_{min} can be approximately derived from the following formula:

$$\Delta h_{min} = (S_1 + B)\left[h - m_z(K_A - 1)\right]/b \qquad (16)$$

where, B = the width of the gateway; d = the broken width along the side direction for the lower rock plate of the main roof, in general, it approaches the roof weighting interval along the face advancing direction, C_1.

The gateway maintained by the coal pillars (position 4 in Figure 1): the gateway should be driven in the position near the virgin stress zone, the supporting pattern and the supporting resistance should be designed in accordance with the ground pressure in the relevant zone.

The gateway driven in interior stress field (position 2 in Figure 1): the gateway must be driven at that time when the main roof movements have basically stopped and the main roof is in stable state. The fact in which the coal mass in the interior stress field has been broken should consider for determining the supporting pattern. The supporting resistance required should not be less than the force that balances with the load from the immediate roof. The supporting retraction value should be adapted to the possible maximum roof sagging of the gateway arisen from the interior stress field during the face advancing to the position where the main roof weighting occurs.

Under the conditions in which the structure with the interior stress field occurs, layout of the gateway in the high stress zone (position 3 in Figure 1) should be absolutely avoided. In this position, the surrounding rocks will bear the stress which is several times greater than the virgin stress (γH). The gateway is not only difficult to maintain, but also often causes serious accidents such as impact ground pressure and coal and gas burst during driving the gateway.

If a strong tectonic stress exists within the initial stress field, especially when both roof and floor composed by the soft and lower strength strata, it is necessary to avoid pre-excavating and maintaining gateways in the initial stress field. The best choice is to adjust the development roadway layout, realizing the layout and maintenance of gateways in the internal stress field where the overlying strata movement and abutment pressure has tended to be stable. By the time, since the overlying strata have broken, the tectonic residual stress will be fully released.

It can be seen from above analysis that realizing the excavation and maintenance of gateways in stable internal stress field is essential measurements in deep mining, especially in preventing accidents and improving roadway maintenance under the conditions of the soft roof and floor or tectonic stress existing. Therefore, when conducting the development roadway layout and deciding excavation sequence, the plan of one-winged extraction layout, the coal extraction and heading faces advance in same direction should be employed. In order to realize the excavation and maintenance of roadways in stable internal stress field, the panel length along strike should be enlarged as much as possible.

It must be pointed out that since the extent of the internal stress field enlarges as the mining depth increases, it gives better conditions to realize the excavation and maintenance of roadways in internal stress field. Therefore, it is not definite to confirm simply that the impact rock pressure and burst out of the gas and coal must increase in the gateway excavation.

3 THE ROCK PRESSURE AND SUPPORTING FORCE OF GATEWAY DRIVIVG IN INTERNAL STRESS FILED

Under the condition of the face structure in which the internal stress field exists, the roadway structure mechanics model of the roadway driving along next goaf is shown in Figure 4.

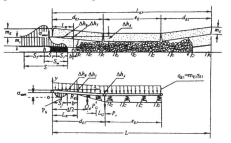

Figure 4 The structure mechanics model of the roadway driving along next goaf

According to the static equilibrium condition, the expressions of the roadway support resistance (R_E) and the action force on coal pillars under the action of the rock beam can be derived as follows:

$$R_E = \frac{1}{b}\left[d_E\left(\frac{Q_d}{2}+P_e\right)-P_G(d-L_G)-P_s\left(b+\frac{L_k}{2}\right)\right]$$
(17)

$$P_s = \frac{1}{L_k+2b}\left[d\left(\frac{Q_d}{2}+P_e\right)-P_G(d-L_G)-R_Eb\right]$$
(18)

By simplifying the above equations, one can get the supporting resistance needed when the coal pillars do not bear pressure (P_S=0) and rib heaving of the roadway do not occur as follows:

$$R_E = A+\left(B-C\Delta h_k^2\right)$$
(19)

where, Δh_K = the compressed amount of coal pillars at the both ends of the working face (i.e. convergence). The related constants respectively are:

$$B = m_E\gamma_E d_E\left(L_E-d_E\right)/(2b)$$
(20)

$$C = d_E^3 E_G\left(L_E-d_E\right)/4b\left(b-l_K\right)^2$$
(21)

where, m_E, γ_E and d_E = the thickness and unit weight of the rock beam, and the lateral span after the rock beam fractured respectively; L_E = the span of the rock beam; b is the width of the roadway; L_K = the width of the coal pillars; E_G = the resistance-adding factor of the stiffness of waste foundation.

When the sagging of several combination rock beams occur simultaneously, then one gets:

$$R_{En} = A+\left(B_n-C_n\Delta h_K^2\right)$$
(22)

here

$$B_n = dE_1\sum_1^n m_{Ei}\gamma_{Ei}d_{Ei}\left(L_{Ei}-d_{Ei}\right)\bigg/(2b)$$
(23)

$$C_n = \frac{B_n}{\Delta h_K^2} = \frac{d_{E1}^2 B_n}{\left(b+l_K\right)^2\Delta h_{An}^2}$$

$$= \frac{d_{E1}^3 E_G\sum_1^n m_{Ei}\gamma_{Ei}\left(L_{Ei}-d_{Ei}\right)\left(L_{Ei}-2d_{Ei}\right)}{4b\left(b+l_K\right)^2\sum_1^n m_{Ei}\gamma_{Ei}\left(L_{Ei}-2d_{Ei}\right)}$$
(24)

If the waste foundation is fully compressed, then

$$R_{EH} = B_H - C_H\Delta h_K^2$$
(25)

here:

$$B_H = H\gamma d\left(L-d\right)/(2b)$$

$$C_H = \gamma Hd^3\left(L_E-d\right)\bigg/\left[2b\left(b+L_K\right)h^2\right]$$

Where, H = the mining depth; h = the mining height.

Under the condition of the roadway driving along next goaf, if the supports work with "the given deformation" plan, then the pressure P_S on the coal pillars applied by the main roof rock beam sagging can be, in the same way, derived as:

$$P_s = B - C_n\Delta h_K^2$$
(26)

of which, when applied by single rock beam sagging:

$$B = \frac{M_E\gamma_E d_E\left(L_E-d_E\right)}{L_K+2b}$$
(27)

$$C = \frac{d_E^3 E_G\left(L_E-d_E\right)}{2\left(L_K+2b\right)\left(b+L_K\right)^2}$$
(28)

When applied by multi-rock beams sagging.

$$B_n = \left[\sum_1^n m_{Ei}\gamma_{Ei}d_{Ei}\left(l_{Ei}-d_{Ei}\right)\right]\bigg/\left(L_K+2b\right)$$
(29)

13

$$C_n = \frac{d_{E1}^3 E_G \sum_{1}^{n} m_{Ei} \gamma_{Ei} \left(L_{Ei} - d_{Ei}\right)\left(L_{E1} - 2d_{E1}\right)}{2\left(L_K + 2b\right)\left(b + L_K\right)^2 \sum_{1}^{n} m_{Ei} \gamma_{Ei} \left(L_{Ei} - 2d_{Ei}\right)} \quad (30)$$

Under the condition of the supercritical mining:

$$B = \gamma H d\left(L - d\right)/\left(L_K + 2b\right) \quad (31)$$

$$C = \frac{\gamma H d^3 \left(L - d\right)}{\left(L_K + 2b\right)\left(b + L_K\right)^2 h^2} \quad (32)$$

After knowing the pressure P_S on coal pillar, the maximum stress value σ_{max} in coal pillars can be approximately derived by the following equation:

$$\sigma_{max} = 2P_s/S_1 = 2P_s/L_K \quad (33)$$

It can be clearly seen from the analysis of above equations that under the condition of using roadway driving along next goaf, regardless of the support resistance or the pressure borne by coal pillars all are related to the actual compressed value Δh_g of coal pillars when the roadway excavates. Under the condition of the values of the resistance and pressure are very small, it is very difficult for the support resistance alone to support the roof or for the coal pillars to bear the roof. Therefore, the key to reduce the roadway pressure manifestation is to correctly choose the time of roadway excavation.

Summarizing the above analysis, the basic principles of the technical design of the roadway driving along next goaf are:

1. The gateway excavation should be conducted under the conditions that the structure sagging is stable, i.e. the stress in internal stress field reaches the minimum value. It is wrong idea that no attention is to be paid to the time of the roadway excavation and the roadway maintenance relies only upon the small coal pillars.

2. The key for realizing the roadway excavation in stable stress field is to enlarge the advance distance of the working face and employ the mining-district roadway layout with one way mining and simultaneous working in two raises.

3. Under the condition of realizing the roadway excavation in internal stress field in which the structure is tending stability basically, if the immediate roof is in good condition, Bolting, guniting and grouting can be used to reinforce coalwall. If the immediate roof is not so good, the chain mesh of roof-bolt and guniting, or frame metal support should be used.

4. When the roadway is forced to be pre-excavated, then the supports (such as the arch support, etc.), adopting the surrounding rock deformation should be selected according to "the given deformation",

until the surrounding strata become stable, the supports, such as bolts and guniting can be used instead of the previous one. Under this circumstances the design of the initial cross section of the roadways must be done according to the predicted amount of the surrounding strata deformation.

When employing "big coal pillars" for roadway retaining, the roadway-structure mechanics model and its dynamic feature of the structure depends upon the magnitude and composition of stresses in initial stress field, the thickness and strength of coal seams and the condition of roof and floor. The basic principles for control design of the roof behavior are:

1. The width of the previously remained coal pillars must ensure that the roadway is in the initial stress field. The excavation and maintenance of roadways in high stress zone should be absolutely avoided. Therefore, making clear understanding of the extent of the abutment pressure distribution, especially the location of the high stress zone is a prerequisite for the layout of the roadway remained by coal pillars to enter into the reliable application.

2. Under the condition of the small thickness and lower strength of coal seams, the roadway must be driven along the roof, otherwise it is very difficult to keep stability of surrounding rocks at the roadway ribs.

3. When there is the residual stress in initial stress field, the use of the layout of roadway remained with big coal pillars should be avoided, and the roadway excavation in the unreleased initial stress field should be also avoided as far as possible.

Mining Science and Technology' 99, Xie & Golosinski (eds) © 1999 Balkema, Rotterdam, ISBN 90 5809 067 1

Control of coal output at the Hambach opencast mine

L. Kulik
Rheinbraun AG, Cologne, Germany

R. Durchholz
Rheinbraun Engineering und Wasser GmbH, Cologne, Germany

ABSTRACT: In order to ensure the reliability of supplies, both quantitatively and qualitatively, to the power plants and upgrading plants, a complex system to control coal mining is nowadays essential in the opencast mines of Rheinbraun AG. This system is based both on accurate daily scheduling for the bucket wheel excavators that are used in the mining process and also on accurate management and control of the bunker supplies. Furthermore, an evaluation of records and reports is also necessary in order to subsequently monitor the development of the coal output. Taking Hambach opencast mine as an example, the means employed for this purpose and their correlation will be discussed. Scheduling already includes a qualitative classification of the coal face prior to the mining process. The correlation of quantity and quality and the fact that block working sequences are specified in advance result in accurate daily scheduling of excavator operations.

1 INTRODUCTION

In the Rhenish lignite mining area, Rheinbraun AG is currently operating four opencast mines with a yearly output of some 100 Mt of lignite. 85 % is used to generate electricity in the power plants of RWE Energie AG, and the remaining 15 % serves to produce upgraded products in the three Rheinbraun-owned upgrading plants. The three opencast mines Hambach, Bergheim and Garzweiler are connected with the upgrading plants and power plants via a company-owned mine railway system, the North-South railway and the Hambach railway. Of secondary importance are connections via belt conveyors between the Garzweiler opencast mine and the Frimmersdorf power plant as well as between the Bergheim opencast mine and the Niederaussem power plant and the Fortuna-Nord upgrading plant. With the Weisweiler power plant, the Inden opencast mine constitutes an isolated operation and is thus excluded from a special control system of the coal flows (Figure 1.).

The different deposit conditions prevailing in the opencast mines connected to the mine railway system call for higher-level control of the coal flows in order to meet the quantity and quality requirements of the consumers and guarantee the sales planning of the opencast mines. This task of coal distribution is performed by the general mine management of the Garzweiler opencast mine.

With the nearing exhaustion of the Bergheim opencast mine, it is only the Hambach and Garzweiler opencast mines that remain as coal suppliers for the consumers along the North-South railway system (Table 1.).

Already on the suppliers' side, the thus growing demand for higher-level co-ordination requires reliable processes for quantity and quality control, which are available in the opencast mines and described in the following, taking the Hambach mine as an example.

2 HAMBACH OPENCAST MINE

The Hambach opencast mine developed in 1978 extends over a mine field of 85 km² with a lignite content of 2.5 Gt. The coal comes from the Garzweiler and Frimmersdorf seams and occurs with a thickness of up to 70 m as one seam, but also with interburden (Figure 2.).

Due to the dipping of the deposit averaging 3° to the north-east, the present depth of 330 m below surface will drop down to 450 m.

15

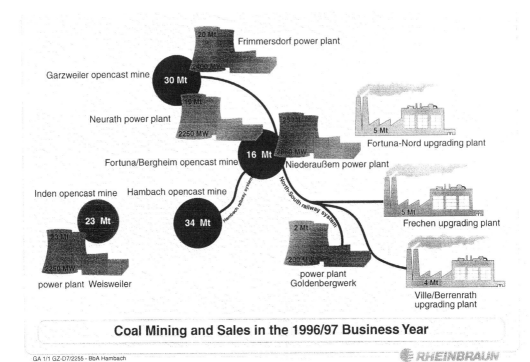

Figure 1. Coal Mining and Sales in the 1996/97 Business Year

Table 1. Long-term mining capacity in the rhenish lignite mining area

SUPPLIER		CONSUMER
Opencast mine	Total mining capacity	
Garzweiler	35 - 45 mill. t	4 lignite-fired pwer plants
Hambach	45 - 55 mill. t	3 upgrading plants
Inden	20 - 25 mill. t	1 lignite-fired power plant

Long-Term Mining Capacity in the Rhenish Lignite Mining Area

GA 1/1 GZ-D7/2260a

The current quality of the run-of-mine coal is specified below. The figures in brackets are the average values:

Ash	1.6 to 5 %	(3.5 %)
Iron	1500 to 5000 ppm	(2800 ppm)
Potassium	50 to 500 ppm	(150 ppm)
Sodium	200 to 1000 ppm	(450 ppm)
Sulphur	0.15 to 0.45 %	(0.25 %)
Heating value	9000 to 10,000 kJ/kg	(9500 kJ/kg)

For overburden and coal mining and overburden dumping, continuous equipment (bucket-wheel excavator and spreader) with daily capacities of 240,000 m³ + t is used. More than 80 % of the coal is extracted on the bottom-most bench. The excavator working above only compensates for lacking coal quantities, thus acting as a quantity buffer. The coal mined in the opencast mine is transported via the tripper car to a two-compartment stockpile with a theoretical capacity of 2 x 400,000 t in order to be stored there. Depending upon the consumers' requirements, the coal can be reclaimed by means of two slewable bucket-wheel reclaimers each having a daily capacity of 120,000 t and supplied by two loading facilities of the same capacity to the trains (Figure 3.). The possibility of dispensing with the stockpile and conveying the material directly to the loading system is only utilised in exceptional cases since this approach permits consumer-oriented quality supply only to a limited extent. What is more, the bucket-wheel excavators working in the opencast mines will have to face considerable capacity restrictions.

Since October 1997, the Bergheim opencast mine has been part of Hambach's general mine management. Since the Bergheim mine will be exhausted in some years and thus lose more and more importance as far as coal mining is concerned, this lecture will not elaborate on it.

3 QUALITY REQUIREMENTS

When dealing with the quality issue, it is first of all the individual upgrading plants offering various product ranges that have to be dealt with. In addition to the upper limits of ash content, viz. 2.5 %, and sulphur content, viz. 0.24 %, every upgraded product makes further requirements that are to be met. Besides the existence of large wood-rich portions, gel and fusite stripes are to be considered as well. Attention has also to be paid to the iron content. In view of the upgrading plants' requirements, the Hambach opencast mine has to provide six different types of coal to be used for upgrading, only one type of which is suitable for all three plants.

As far as the steam coal used for electricity generation is concerned, six parameters are decisive for combustion reasons, viz.

- ash
- iron
- potassium
- sodium
- sulphur
- heating value.

According to the experience gained by the power plants from lignite-based electricity generation, classes were formed for the individual parameters. After being weighted correspondingly, the similar parameters of potassium and sodium are grouped together (Table 2.).

Thanks to these classes, deposit-specific variations in the coal qualities can be compensated by mixing of coals from various opencast mines so that the power plants can be supplied with a constant coal quality that ensures trouble-free boiler operation.

Together with the class system, the large number of parameters result in plenty of steam coal types. In the Hambach opencast mine, with special importance being attributed to the slagging-relevant components iron, potassium and sodium, these types are grouped as follows:

- Standard steam coal = KK[1]
- critical steam coal due to its iron content (> 4000 ppm) = KKK[2]
- critical steam coal due to its sodium / potassium content (> 1100 ppm) = US[3]
- critical steam coal due to its iron and sodium/potassium contents = KKK/US[4].

The great number of quality requirements and the necessity of co-ordinating the coals from the different opencast mines call for the use of sophisticated instruments to control the coal output.

4 INSTRUMENTS OF CONTROL

The control of lignite output in the Hambach opencast mine is based on the combined components *materials planning* (including deposit and operating conditions of the bucket-wheel

---

[1] KK is a German abbreviation standing for standard steam coal.
[2] KKK is a German abbreviation standing for critical steam coal due to its iron content.
[3] US is a German abbreviation standing for critical steam coal due to its sodium/potassium content.
[4] KKK/US is a German abbreviation standing for critical steam coal due to its iron and sodium/potassium contents.

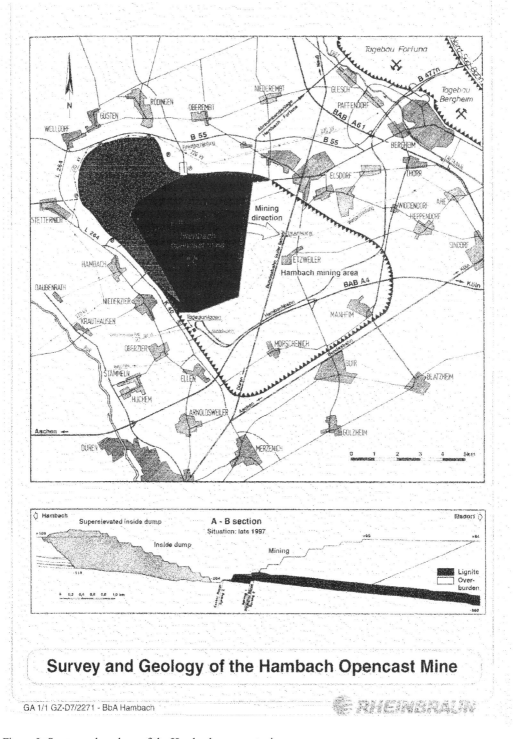

Figure 2. Survey and geology of the Hambach opencast mine

excavators), *stockpile management* and *historical evaluation*. The following will deal with the individual components and their interplay.

4.1 *Materials Planning*

Materials planning in respect of the continuous equipment (bucket-wheel excavator and spreader) use covering a period of up to six weeks--referred to as *small-scale planning*--is carried out by the mining department. In this context, it is in particular the two coal benches that call for very detailed consideration.

First, this consideration is based on the operation schedules of the planning department with data on overburden and total coal amounts.

Secondly, computerised deposit-specific data are provided for each operation schedule. These include all quality parameters with physical data as requested by the consumers.

Materials planning first adjusts the planned quantities shown in the operation schedule to the conditions occurring at the face and determines--in 100 m conveyor segments--the actual quantities for the individual mining blocks. In addition to this blockwise quantity determination, the coal quality has to be determined as well. This determination is based on the deposit data file developed from borehole logging and continuously updated by means of geological opencast mine surveys and face samples. To optimise deposit identification, this data file serves to derive hypothetical quality columns for 50 m conveyor segments. At the face, these columns can also be identified by the occurrence of marked layers, such as dark horizons or interburden (Figure 4.). The quality columns are divided into 1 m sections and include all necessary quality parameters.

The hypothetical columns now serve as a basis that, already before the start of actual mining, allows by means of the IKOLA programme (German abbreviation standing for interactive determination of the coal quality based on deposit-specific data) the mining block to be divided into excavator slices using the PC, with account being taken of the equipment geometry, and the respective qualities for the individual slices to be determined automatically. For quality determination, up to three columns over the width of a mining block are used. Besides the quality per excavator slice, the thickness and the output to be expected are displayed as well (Figure 5.).

As far as the division into excavator slices is concerned, it is necessary to take account of multifarious requirements, as e.g. the maximum output of lignite to be upgraded or the quality of the occurring steam coal in the other opencast mines and the thus imperative provision of blended coal meeting the quality requirements. The quality data on the blockwise preplanned slices are then stored and linked with the mining blocks. As a result of the division of the 3 km long coal bench into 50 m long conveyor segments and the different travelling planes of the bucket-wheel excavator, more than 200 different quality sections per bench are obtained which in the case of block division can increase even more.

The following and most important step includes the task of arranging the mining blocks provided with quantity and quality data in a time sequence such that, as the first priority, the quantitative and qualitative requirements of the consumers are always met. It is in particular the deposit-related varying distribution of the upgrading coal that calls for special attention. What is more, empty running of the coal mining excavators on the bench is to be minimised, and high and deep cuts on the bottom-most bench have to be performed in a steady way to ensure the safe development of the inner dump. Depending upon the conditions prevailing at the face, daily output figures are to be specified for materials planning, and maintenance and repair measures as well as belt conveyor relocation are to be included in coal planning with maximum accuracy in terms of the time schedule. At the same time, the coal quantities specified by coal distribution management are considered. The result of this materials planning process is the *coal flow* (Table 3.).

With exact data for the respective day, the *coal flow* represents the mining output to be expected for steam coal and upgrading coal. The development of the stockpile levels as a function of the planned sales quantities is shown as well.

The simultaneous representation of the steam coal qualities to be expected then permits the higher-level co-ordination of coal mining with the Garzweiler opencast mine. Any required interventions in the excavator operation are thus recognised early in time. Monitoring and control of operations nowadays call for daily updating. All steps from the combination of masses with the coal qualities all the way down to the representation of the qualities every week are nowadays linked in an Excel-based programme and permit a check of the operation variants in a short time.

4.2 *Stockpile Management*

The stockpile management, which is based on materials planning, forms the linkage between the bucket-wheel excavators operating at the face and the consumers. In addition to quantity buffering, the stockpile above all assumes the task of coal stockpiling and supply in compliance with the quality requirements. In the Hambach opencast mine, the stockpile management is divided into the processes of *storage* and *discharge*.

Table 2. Quality characterisation of the lignite

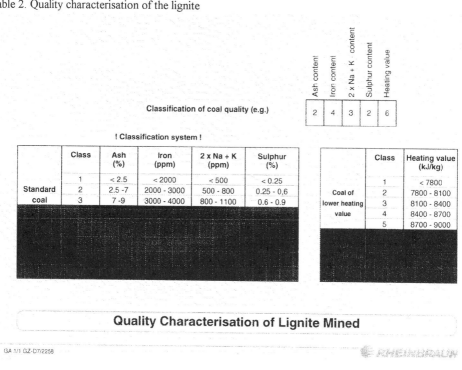

Classification of coal quality (e.g.)

	Ash content	Iron content	2 x Na + K - content	Sulphur content	Heating value
	2	4	3	2	6

! Classification system !

	Class	Ash (%)	Iron (ppm)	2 x Na + K (ppm)	Sulphur (%)
Standard coal	1	< 2.5	< 2000	< 500	< 0.25
	2	2.5 -7	2000 - 3000	500 - 800	0.25 - 0,6
	3	7 -9	3000 - 4000	800 - 1100	0.6 - 0.9

	Class	Heating value (kJ/kg)
Coal of lower heating value	1	< 7800
	2	7800 - 8100
	3	8100 - 8400
	4	8400 - 8700
	5	8700 - 9000

Quality Characterisation of Lignite Mined

GA 1/1 GZ-D7/2258
RHEINBRAUN

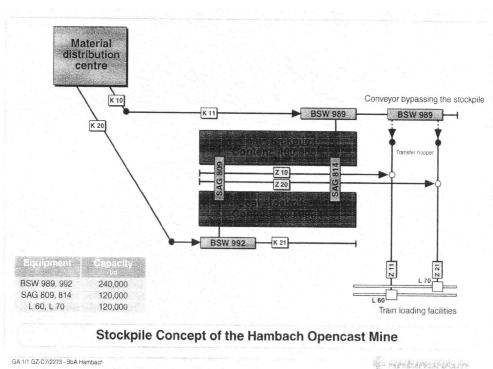

Stockpile Concept of the Hambach Opencast Mine

GA 1/1 GZ-D7/2273 - BbA Hambach
RHEINBRAUN

Figure 3. Stockpile concept of the Hambach opencast mine

20

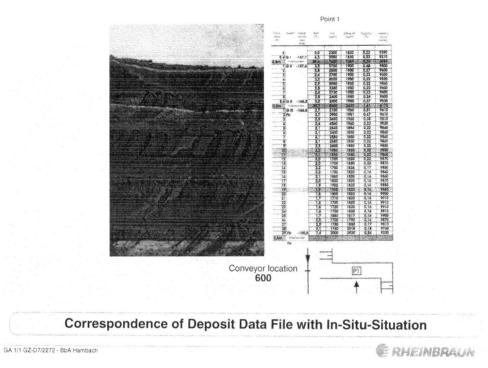

Figure 4. Correspondence of deposit date file with in-situ-situation

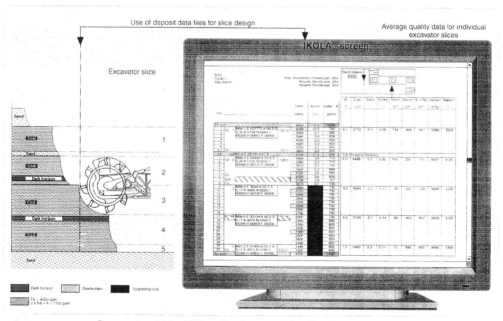

Figure 5. Computer-aided slice design and quality determination

Table 3. „Coal flow" as a result of materials planning

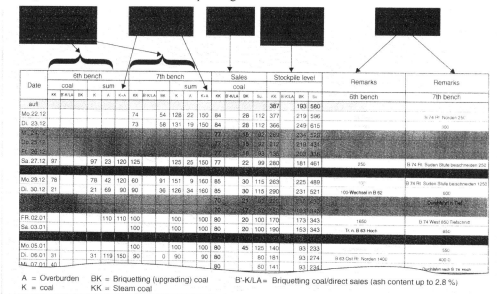

Date	6th bench coal KK	B'K/LA	BK	6th sum K	A	K+A	7th bench coal KK	B'K/LA	BK	7th sum K	A	K+A	Sales coal KK	B'K/LA	BK	Su	Stockpile KK	B'K/LA	BK	Su	Remarks 6th bench	Remarks 7th bench
aufl																	387		193	580		
Mo.22.12							74		54	128	22	150	84		28	112	377		219	596		B 74 Rt Norden 250
Di. 23.12							73		58	131	19	150	84		28	112	366		249	615		300
Mo.24.12													77		15	92	289		234	523		
Do.25.12													77		15	92	212		219	431		
Fr.26.12													77		16	93	136		203	338		
Sa.27.12	97			97	23	120	125			125	25	150	77		22	99	280		181	461	250	B 74 Rt Suden Stufe beischneiden 250
Mo.29.12	78			78	42	120	60		91	151	9	160	85		30	115	263		225	489		B 74 Rt Suden Stufe beischneiden 1250
Di. 30.12	21			21	69	90	90		36	126	34	160	85		30	115	290		231	521	100-Wechsel in B 62	600
													70		21		220		216	430		Durchfahrt in Tief
													70		17	150	150		193	343		
FR.02.01			110	110	100		100			100	100	100	80		20	100	170		173	343	1650	B 74 West 850 Tiefschnitt
Sa.03.01				100			100			100	100	100	80		20	100	190		153	343	Tr. n. B 63 Hoch	850
Mo.05.01				100			100			100	100	100	80		45	125	140		93	233		550
Di. 06.01	31			31	119	150	90		0	90	90	90	80		80	181	181		93	274	B 63 Ost Rt Norden 1400	400-0
Mi. 07.01	40												80		80	141	141		93	234		Durchfahrt nach B 74 Hoch

A = Overburden   BK = Briquetting (upgrading) coal   B'-K/LA = Briquetting coal/direct sales (ash content up to 2.8 %)
K = coal   KK = Steam coal

„Coal Flow" as a Result of Materials Planning

GA 1/1 GZ 7/2263 - BbA Hambach   RHEINBRAUN

Field 7   Field 6   Field 5   Field 4   Field 3
Sector 1   Sector 2   Sector 1   Sector 2   Sector 1   Sector 2   Sector 1
K 11
Z 10
Z 20
AG 814

Diagrammatic View of Stockpile Operation

GA 1/1 GZ-D7/2253 - BbA Hambach   RHEINBRAUN

Figure 6. Diagrammatic view of stockpile operation

22

4.2.1 Storage

Storage is based on the division of the mining block into excavator slices, defined in advance by the IKOLA programme. Here, the preplanned division is once again checked by the competent in-situ supervising personnel of the mining operation and adjusted, if necessary. The defined qualities of the individual slices--referred to as batches--are now allotted to free stockpile sections. Before the mining process, the supervisor also determines the discharging section of the stockpile. In this context, it is to be seen to it that--on the one hand--the coal-mining excavators in the opencast mines are operated with the requested high utilisation in terms of time and--on the other hand--collision of the stockpile equipment is excluded. This fundamental requirement and the multitude of qualities cause the useful stockpile content to decrease from 800,000 t to 600,000 t.

The in-situ definitions are passed on by the supervisory personnel to the mine control centre. Thus, the latter is capable of directing the mining output from the bucket-wheel excavator--depending upon the quality mined--to the allotted stockpile section. For this purpose, the mine control centre is in constant contact with the excavator operator in the opencast mine and the operators of the tripper car and the reclaiming equipment in the stockpile.

For the accurate monitoring of the coal flows and the allocation of qualities, each stockpile compartment is divided into individual fields. Each field in turn consists of two sectors, each of which is 50 m long and contains 25,000 t of coal (Figure 6.).

Only the fields at the stockpile ends have other contents. To guarantee unambiguous identification of the sector boundaries, large boards are installed; at the same time, the operator of the tripper car receives an audible alarm signal when crossing a sector boundary.

As a rule, the coal is dumped according to the cone shell technique so that the individual coal batches are placed in line. In the Hambach opencast mine, only such coal batches are dumped together in one sector that have similar qualities. The upgrading coal is clearly separated from the steam coal by means of *foot* dumping.

When storage operations are completed in one sector, the supervisory personnel of mining operation determines the available coal quality and enters the result in the *stockpile pattern* (Table 4.). In addition to the quality data, the *stockpile pattern* also includes the quantities stored and the individual filling levels in the sectors. The horizontal division of the sectors into four slices takes account of coal mining in slices during discharge.

Total storage with quality determination by means of IKOLA is performed by the supervisory personnel that also inspects the mining face on the coal benches in the opencast mine. If the in-situ situation differs from the deposit data determined, the geology department is informed of these deviations which will then include them in the deposit data file. If coal for upgrading purposes has to be mined in the boundary area towards steam coal qualities, daily co-ordinations are made with the consumers. In isolated cases, special samples are taken which are analysed in Rheinbraun's laboratory.

4.2.2 Discharge

Every week, the quantities and qualities to be discharged are co-ordinated between the opencast mines and the coal distribution management. As explained above, the supervisory personnel of the mining operation uses this basis in order to determine for every shift the zones to be discharged which are also made known to the coal distribution management.

As the implementing body of the coal distribution management, the coal control centre co-ordinates these specifications with those of the Garzweiler opencast mine. Then, the coal control centre informs the mine control centre of the Hambach opencast mine on the train requirements in respect of quantity and quality. The reclaimers in the stockpile are used accordingly. For this purpose, the mine control centre has direct access via the PC to the topical *stockpile pattern*. Discharge normally takes place in slices along one sector. This method guarantees good mixing and thus a uniform quality already at that point. After the completion of the loading operation, the quality derived from the *stockpile pattern* data is fed into the mine control computer. The link-up with the computer of the train operation permits further monitoring of the coal flow.

The mass flows are updated by the mine control centre in the *stockpile pattern*. Hence, the discharging operation is concentrated on the mine control centre since it is also in constant contact with the coal distribution management.

The provision of a uniform steam coal quality interlinked with the Garzweiler opencast mine and the supply of the upgrading plants are today only guaranteed if the two reclaimers and the train loading facilities work on a continuous basis. As a result, maintenance and repair measures have to be co-ordinated, with account being taken of the mining operations concerned.

4.3 Historical Evaluation

Today, the reliable supply of the consumers calls for the most accurate determination possible of the coal quality. Possible deviations of the actual quality from the deposit data, which may result from complaints

Table 4. Section of the sockpile pattern

Hambach opencast mine																	Supervision, mine control centre	Rossaint		Date:	21.12.1997
Mining department, conveyor operation, stationary																	Supervision, mining department	Faßbender		Time:	04:44
																				Shift:	3

Field	9		8		7		6		5		4		3		2		1	\ Stockpile 1		
Tripper car	S1	S2	S1	S2	S1	S2	S1	S2	S1	S2	S1	S2	S1	S2	S1	S2			
Quantity, 'BK'																	Sum 'BK'		
Quantity, 'KK'																	Sum 'KK'		
																	Sum		

Field	9		8		7		6		5		4		3		2		1		
	S1	S2	S1	S2	S1	S2	S1	S2	S1	S2	S1	S2	S1	S2	S1	S2		
AG 809 BK												5,5	5,5				BK 809	11.0
KK							8,3	4,1	6,9								KK 809	19,3
AG 814 BK																	BK 814	
KK																	KK 814	
																	Sum 'BK'	11.0
																	Sum 'KK'	19,3
																	Sum	30,3

Field	9		8		7		6		5		4		3		2		1		
	S1	S2	S1	S2	S1	S2	S1	S2	S1	S2	S1	S2	S1	S2	S1	S2		
Stockpile level	924517														BK			
															812318			
	US	US	US	KUS	KUS	KUS	US	US				BK	BK	BK	BK	BK		
	923517	923518	923627	925618	924618	924417	923418	911628				911618	812518	811518	811518	812518		
Quantity, 'BK'												19,5	25,0	25,0	25,0	30,0	Sum 'BK'	124.5
Quantity, 'KK'	20.0	8.4	25.0	25.0	25.0	25.0	25.0	7.0									Sum 'KK'	160,4
																	Sum	284,9

German abbreviations standing for:
'BK' = briquetting (upgrading) coal
'KK' = steam coal
'BSW' = tripper car
'BU' = mine control centre

'US' = critical steam coal according to the sodium/potassium content
'KUS' = critical steam coal according to the iron and potassium/sodium contents
'BbA' = mining department

Section of the "Stockpile Pattern"

GA 1/1 GZ-D7/2262

RHEINBRAUN

---

of the consumers, have therefore immediately to be attributed to the corresponding deposit section. This section is to be checked once again, and the updated deposit data are to be taken into account for subsequent applications. Hence, already the mining location is to be determined accurately and a continuous data flow to be guaranteed up to the consumer.

To this end, it is already during mining of an excavator slice that in addition to quantity and quality, the supervisory personnel of the mining operation records a large number of further data, such as the number of the excavator, the belt conveyor, the mining position, the excavator's travelling track, the excavator slice and thickness as well as the date. What is more, data entry of the storage zone in the stockpile allows the coal flow to be definitely monitored up to that point.

The monitoring of the discharged quantities and qualities permits the daily updating of the coal flow and the early recognition of necessary intervention in the mining sequence. Besides these data, the IKOLA evaluation is printed out and filed for each mining block. As the results from special samples, mean value determinations obtained from the storage of different batches in one sector are filed for twelve weeks.

SUMMARY

Today, the reliable supply of the power plants and the upgrading plants located along the North-South railway system in quantity and quality terms calls for sophisticated control of lignite mining. The basis for this is both materials planning of the bucket-wheel excavators taking part in this process so that they exactly meet the daily requirements, and accurate stockpile management. Furthermore, a historical evaluation for the subsequent monitoring of coal extraction is necessary.

The Hambach opencast mine was taken as an example to show the instruments available for this purpose and their linkage. Materials planning already considers a quality-related classification of the coal face, which precedes the actual mining process. The combination of quantities and qualities as well as the determination of block sequences result in a speci-fication for the excavator operation that exactly meets the daily requirements.

Mining Science and Technology' 99, Xie & Golosinski (eds) © 1999 Balkema, Rotterdam, ISBN 90 5809 067 1

Technology development of coal surface mining in Yugoslavia

R.D.Simic & V.J.Kecojevic
Department of Surface Mining, University of Belgrade, Yugoslavia

M.R.Gomilanovic
Ministry of Industrial, Energy and Mining, Republic of Montenegro, Yugoslavia

ABSTRACT: The objective of this paper is to present the state of art in Yugoslav surface mining. Particular attention has been paid to a graphically designed program covering researches in this field of science. The program encompasses a database, application of commercial softwares in Yugoslav surface mining, program packets and models resulting from detailed studies carried out by Yugoslav scientists, as well as the concept on making the systems for supervising and managing the surface exploitation processes in real time.

1 INTRODUCTION

The territory of what is today Yugoslavia is known for its abundant and varied mineral wealth, partly being active and partly under investigation. For the last twenty years the reserves of coal, petroleum, natural gas, copper, lead and zinc, bauxite, magnesite, asbestos, barite, feldspars and other non-metallic raw materials, technical and decorative stones as well as other mineral materials have considerably increased and new deposits of nickel, phosphate, zeolite, wollastonite, tungsten etc. have been established. It can be concluded, on the grounds of rough estimations that the degree of exploration for non-ferrous metals ranges approximately to 40%, for oil below 35% and for coal even up to 70%. Thus, the very exploration of Yugoslav mineral deposits confirms, on one hand, the already available coals reserves to be the most stable source of primary and other forms of energy, and on the other, the production of aluminum, copper, lead and zinc to increase with requirements. Formation of a significant base for obtaining nickel and some other rare metals should be expected, as well. As to non-metallic mineral materials, the opinion prevails that a more intense exploration might result in production to meet domestic requests.

The deposits of metallic minerals are various in size, content, way of bedding, exploitation conditions etc. Consequently, the applied exploitation technologies, different working effects and other specific parameters vary from mine to mine.

Copper, iron and nickel ores are mostly exploited in open pits, bauxite in open pits and underground mines, lead and zinc are mainly recovered from underground mines, while antimony and other ores are obtained by underground winning only. Mechanization of great unit capacity, meaning also great annual production, is applied in open pits, primarily of copper and iron, while mines with underground exploitation are usually of medium and small capacity. And this results, first and foremost, from geological characteristics of environment and recovery of deposits.

The achieved level of mining technics and technology is based on the long-lasting experience acquired in mineral materials production, production of explosives and other reproductive materials and mining equipment in Yugoslavia.

Opening and development of numerous underground mines, particularly those of lead, zinc and coal, as well as open pits of coal, copper, bauxite, non-metals and other mineral materials, was followed almost simultaneously by foundation of powerful scientific-researching, design and education centers, greatly contributing to both successful opening and development of certain mines and formation of professional and scientific personnel, as well as providing high current mining production and modern technic-technological solutions.

This paper gives a review of Yugoslav production capacities covering open pits production of mineral materials and global research program in the same field.

2. A REVIEW OF YUGOSLAV OPENCAST EXPLOITATION

The opencast mining provides more that 90% of all mineral raw materials under exploitation in Yugoslavia. High technical and technological levels characterize this modern opencast exploitation.

Considered from production capacity aspect, the opencast exploitation of coal, copper ores and non-metallic minerals as well as bauxite ores in Yugoslavia can boast of high achievements.

The opencast coal exploitation is carried out in five coal basins: Kolubara, Kosovo, Kostolac, Kovin and Pljevlja.

The present production in the Kolubara basin, ranging approximately from 25 to 29 million of tons per a year, is performed at the open pits "Field B", "Field D", and "Tamnava-East Field" while the production at the "Tamnava-West Field" open pit is about to start.

Coal production at the "Field B" is planned to stop in 2.000 and at the "Tamnava-East field in 2005. After 2.000 two new opencast mines are going to be opened: first the "South-Field" and then the "Radljevo".

The annual production of the Kostolac basin ranging approximately from 8 to 9 million of tons, is carried out at the "]irikovac" and the "Drmno" opencast mines. The production of the "]irikovac" opencast mine will stop in 2005, while the production of the "Drmno" mine will increase.

At the territory of Vojvodina, the opening and experimental exploitation started at the "Kovin" opencast mine by submerged way of mining, being applied for the first time both in our country and abroad. If techno-economic justification proves this exploitation method, the optimum capacity and dynamics of a plant will be defined.

The very production of coal in the Kostolac basin, ranging at the time being from 6 to 9 million of tons per annum, is carried out at the "Bela}evac" and the "Dobro Selo" opencast mines, the closure of which is expected to be in 2.005.Within the period up to 2010, and even after it, opening of new opencast mines are expected: "Sibovac" having capacity of 17 million of tons, "Sibovac-East" having capacity of 12 million of tons and "Kru{evac" with the capacity of 14 million of tons per a year.

All the above mentioned basins operate within the Electric Power Supply of Serbia, and have a status of public enterprises.

The essential coal reserves in Monte Negro are located in the Pljevlja coal basin. The Spatial Plan of this Republic, based on the mentioned reserves, estimates the mining capacities of lignite production at 2,4 million of tons. The facilities for lignite production and combustion are nowadays built in the vicinity of the Pljevlja town and located at places richest in quantities of the established balance lignite reserves. The present installed capacities for coal production amount to approximately 2 to 3 million of tons of coal per a year, the established reserves of which, including all other reserves (Mao~e, Otilovi}i, Mataruge etc.) provide exploitation life for more than 100 years. This fact indicates long-lasting coal exploitation at these areas. The Pljevlja coal basin operates within the framework of the Electric Power Supply of Monte Negro.

Serbia is a well-known producer of copper ore with annual production capacity of 24 million of tons of ore, and the same capacity covering flotation and separation. Production of copper ore is carried out in the Bor Pit and in the Bor opencast mines, Veliki Krivelj, Cerovo I and Majdanpek.

The Nik{i} bauxite basin is the main bauxite ore producer in Monte Negro. The reserves of red bauxite ore are distributed almost over 1/3 of the territory of Monte Negro. The largest deposits are located on the area of the Nik{i} community where a number of deposits are exploited by surface and pit mining. Annual production capacity of red bauxite ore amounts to 750.000 tons. Processing capacities for this ore were built in Podgorica providing 600.000 t/y of bauxite and 100.000 tons of aluminum.

Serbia produces annually about 100.000 tons of bauxite and the main producer is the "Kosovo" mine located in Klina, while ferronickel is produced at the "^ikatovo" and "Glavica" opencast mines in Glogovac with annual capacity of 1 million tons of ore. The ironworks have annual capacity of 11.000 tons of nickel, namely 58.000 tons of ferronickel.

The most important producers of non-metallic minerals in Serbia are: Magnohrom-Kraljevo - magnesites; Rgotina - quartz sand; Jelen do - limestone; Granit pe{~ar - granite; Slavkovac - dacite; Keramika, Mladenovac - clay; cement raw materials, Beo~in, Popovac, Kosjeri}, \eneral Jankovi}; brick raw material Potisje, Kanji`a, Be~ej, Kikinda, "Kubu{nica", Aran|elovac, "]ele Kula", Ni{; technical stone-Rakovac; asbestos Korla}e etc.

Reserves of decorative stones are found in several places in Monte Negro, in the area of Danilovgrad, Kola{in, Andrijevica, Ulcinj, Bar communities, etc., but exploitation is performed only at the area of Danilovgrad community. The present production capacity of blocks is about 5.000m per a year, and cut plates of approximately 50.000 m per a year. Having in mind the already established reserves and the increasing interest of consumers, this production has a promising future that certainly should be used in the period to come. The Saltworks in Ulcinj produces salt from seawater both through natural and industrial processing. This Saltworks, based on solid grounds, is expected to develop into a great producer of salt and salt-based products. Monte Negro disposes of enormous reserves of good quality technical-building stone, distributed throughout the whole Republic. Exploration works were carried out only in several deposits and considerable reserves were established there. Nowadays, 12 quarries are under exploitation.

Domestic mechanical engineering industries manufacture a wide assortment of products for opencast mines, such as:
- heavy dump trucks (Radoje Dakic - Podgorica and FVK - Kraljevo),
- wheel and track bulldozers (Radoje Dakic - Podgorica and 14. Oktobar - Krusevac),
- universal excavators (Radoje Dakic - Podgorica and IMT - Belgrade),
- stable and mobile crushers (14. Oktobar - Krusevac),
- dewatering pumps, flotation pumps and separation pumps (Jastrebac - Nis),
- mobile drillholes (Geomasina - Zemun),
- loaders, graders, excavators and scrapers (Radoje Dakic - Podgorica and 14. Oktobar - Krusevac),
- plant engines of 2600 kW and more; shovel excavators (Sever - Subotica),
- bucket wheel excavators (Kolubara metal - Veroci, Gosa - Smederevska Palanka, Ivo Lola Ribar - Zeleznik),
- spreaders (Kolubara metal - Veroci, Gosa - Smederevska Palnaka),
- self-propelled transporters (14. Oktobar - Krusevac),
- apparatus and machines for waste dumps (14. Oktobar - Krusevac),
- rail transporters (Kolubara metal - Veroci, Gosa - Smederevska Palanka, Kolubara metal - Vreoci),
- dewatering pumps of great capacities (Jastrebac - Nis),
- rubber belts for transporters (Balkan - Suva Reka),
- autocrane (Ivo Lola Ribar - Zeleznik, MIN - Nis);
- stationary and mobile cranes (Gosa - Smederevska Palanka and 14. Oktobar - Krusevac).

3. SCIENTIFIC RESEARCH IN YUGOSLAV OPENCAST MINING

Explorations carried out in Yugoslav opencast mining have proved to be at high scientific level. The most recent equipment has been engaged in opencast mines, while scientific and professional personnel follows the world's trends in this field.

The main tasks of scientific-investigation study include the following topics: formation of data bases, usage of commercial and own program packages and models and mastering the management process in real time. A global algorithm of research is given in Fig. 1 where are particularly listed the commercial softwares that have already found application in Yugoslav opencast exploitation, as well as

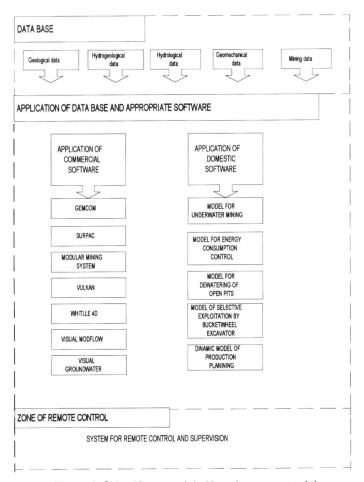

Figure 1. Scientific research in Yugoslav opencast mining

program packets and models resulting from own investigations.

A detailed explanation of the mentioned own investigations will be the subject of oral presentations at the Conference.

4. CONCLUSION

The Yugoslav opencast mining, based on great experience, represents an independent, multidisciplinary and intellectually developed technical field, capable of surmounting the ever increasing problems the opencast mining is fated with.

The technics and technology applied in opencast mines throughout Yugoslavia has followed the modern world's trends and the achieved capacities and usage of equipment for surface exploitation of copper, bauxite, coal and non-metallic minerals, may keep in step with the most developed mining countries.

Diversity of long-lasting surface exploitation in Yugoslavia resulted in precious experience in exploration, in opening and development of capacities, in assembly, in running up the equipment and in opening of designed capacities. Both design and scientific institutes in Yugoslavia may provide successful solutions covering introduction and application of modern equipment, construction and running up the new mining machines, as well as improvement of technological processes and elaboration of new solutions. Institutes and design organizations base their knowledge on great experience acquitted during elaboration of numerous projects, studies and investment programs not only for the requirements of Yugoslav opencast mines but also for important facilities abroad.

Particular attention in Yugoslav opencast mining has been given to scientific researches, strongly supporting the research program presented in this paper.

And finally, it should certainly be mentioned that the future of opencast mining in Yugoslavia is very certain. Production capacities of coal, copper, bauxite and non-metallic mineral raw materials will be increased. The already existing capacities and enlargement of the same will create new spaces for opencast exploitation, that reconciled with modern world's trends should also be, in the period to come, the leading mining technology and a catalyst of knowledge for the entire technological development.

Mining Science and Technology'99, Xie & Golosinski (eds) © 1999 Balkema, Rotterdam, ISBN 90 5809 067 1

High performance longwall extraction in large depth

H.C.W. Knissel & H. Mischo
Technical University of Clausthal, Institute for Mining, Clausthal-Zellerfeld, Germany

ABSTRACT: Stagnation of coal prices to a low level on the international coal trading markets reduces profits and pressured the German hardcoal companies to reduce the mining costs by increasing of the face output and at the same time by reducing the numbers of working points. This concentration could be achieved by the insertion of high performance longwall operations.

The success of longwall operations is dependent upon key points like Geology, equipment and the mine layout. This paper discusses the definition of high performance longwall extraction in large depths and introduces the standard parameters of the equipment and the typical coefficients of these longwall operations as well.

1. INTRODUCTION:

The economic conditions for the German hardcoal mining industry have changed significantly in recent years. The German hardcoal industry had until the early 1990s a secure selling market. The „Jahrhundertvertrag" guaranteed the purchase of German hardcoal by heavy industry and steel and energy providers. In the last few years the development of the European Union and a sweeping liberalization of the energy markets lead to heavy European and international rivalry. This increasing competition, caused by the import of inexpensive hardcoal from overseas, pressured the German hardcoal industry to react. It was necessary to concentrate the extraction on the most profitable collieries and to excavate only the most suitable parts of the deposits. Until recently it was required to mine the entire deposit. Although the number the mining operations decreased from 147 in 1990 to 64 in 1997 a 56 % decrease, the coal output decreased by only 32 % to 47 mil. tons. This concentration will continue. Additionally, a reduction of the costs of production and an increase in productivity required thedevelopment of new and innovative techniques and face equipment. The goal of these developments is to compensate for the competitive disadvantage due to the great depth of the German hardcoal deposits with the most modern technology and mining methods.

Figure 1: German hardcoal basins

Hard coal mining in Germany is based on a century-old tradition. Figure 1 shows the German hardcoal basins.

Even 500 years ago only the out-crop of the

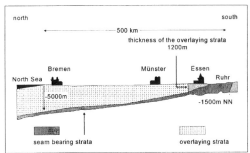

Figure 2: Sinking of the seam-bearing strata in the Ruhr-area in the northern direction

coal seams was mined. Since the mid-19th century, the Industrial Revolution and the resulting need for energy promoted the recovery of the deposits and the extraction in ever-increasing depths. Since 1920 the average mining depth has increased from 330 m to 648 m in 1959 and lies today at 1,006 m (1997). Some longwall operations have reached depths of around 1,450 m.

An additional problem is the increasing coverage of the remaining hardcoal deposits. Figure 2 shows the sinking of the seam-bearing strata in the Ruhr-area in the northern direction.

The new working areas and hardcoal mines, which open up the deposits in great depth, are attached without exception to the existing old coal mines. The main problem is to improve these existing mines, originally designed for much smaller working point capacity, to handle high performance longwall operations and to transport the entire conveyance discharge through the old mine to the coal preparation plant.

2. HIGH PERFORMANCE LONGWALL OPERATIONS

In Germany the term "high performance longwall operation" is not clearly defined. Usually we describe it as the longwall which extract a net amount of over 16,000 t high quality coal per day. That means 30,000 t run-of-mine coal per day. These operations can be roughly described by the following simple guidelines (1).

- Net production (effective delivery):
  Over 16,000 t high quality coal from great depths > 1,000 m
- Gross Production (total delivery):
  Up to 1.6 times the net production,
- 30,000 t run-of-mine coal/d

- Face length:
  Up to 480 m (600 m are planned)
- Power consumption:
  Up to 4,500 kW at the longwall
- Working length:
  Several kilometers
- Area increment of face advance:
  Over 16 m²/min up to 25 m²/min
- Supporting Performance:
  At least 1.2 times the area increment of face advance, up to 30 m²/min
- Productivity in the longwall:
  Over 200 t v.F./MS (European record: 452 t v.F./MS in Ensdorf Colliery)
- Goaf treatment :
  Roof-fall exploitation
- Development design:
  With parallel headings and inclines

In order to integrate the high performance longwall operations in the existing collieries, it was necessary to carry out adaptation measures. The first measure was to optimize the development design and orientation of the new working areas and attached mines.

2.1 Demands on the mine layout

The great depth, in which the German hardcoal deposits lie, has a large influence on the mine layout. Each roadway must, therefore, be lined with extensive and expensive sliding roadway arch supports regardless if it is a main transport drift or only a short-term parallel gate. The construction costs reach up to DM 15,000 ($US 9,000) per meter roadway. Attempts to replace the sliding roadway arch supports with bolted supports or to develop the roadway in a rectangular rather than arch profile, did not achieve all desired results to stabilize the roadway or reduce the development costs.

The mine layout connected to the high performance longwall must fulfill the following requirements:

- Suitable infrastructure for the efficient transport of material to the longwall
  - Short traveling time for the workers for a long effective working time
  - Fast transport of workers with passenger lifts and belt riding
  - Transport of large cross-sections, for example transport of complete shield units
  - Efficient material transport, maximal 3 h

from storage depot over the surface to the longwall

- Suitable infrastructure for product extraction:
  • Sufficient dimensional analysis of the belt conveyor
  • If possible, inclines to avoid vertical conveyance
  • If necessary, storage bunkers for homogenization of the conveyance discharge
- Adequate ventilation area
  • Control of the climatic conditions, for example formation and mining product temperature, waste heat of the mining machinery
  • Control of the gas emission

- Adequate energy supply
  • Electrical energy
  • Compressed air and hydraulics
  • Water for cooling water, dust consolidation and nozzle reception
  • Cooling capacity and air conditioning

Since the above-mentioned requirements were, in many cases, taken into consideration in the planning and development of new fields and connecting mines, they are capable of economical and trouble-free high performance longwall operations today.

2.2   Connection of the high performance longwall to the mining layout

The layout of the fields and the connection of the longwall operations to the mining layout is intimately associated with the development of the mining layout itself. The following requirements should, therefore, be considered (1):

- Development of the connecting mine and the working areas in the coal seam with inclines even in great depth
  • Delivery with belt conveyors: no junctions from horizontal to vertical haulage
  • No junctions from horizontal to vertical haulage in transport and carriage roads

- Parallel headings should be directly connected with a main delivery or transport road if possible

  • Linear product extraction from the longwall
  • Faster material transport to the longwall, reduction of the transport time
  • Short travelling time, extension of the effective working time
  • Short ventilation circuits

Under certain circumstances a homogenization of the conveyance discharge may be necessary. In particular for the continuous operation of the small belt conveyors in the old parts of the mines and a continuous preparation in particular, a homogenous conveyance discharge is necessary.

2.3   Design of the longwall

In order to operate high performance longwalls under the difficult geologic and climatic conditions prevalent in depths greater than 1,000 meters, a special longwall design is needed. These longwalls are generally cut as retreating faces to the main haulage road. This has the advantage of obtaining information about the seam, for example coal gas content, before starting the excavation.

With gas rich longwalls, in particular, a preliminary degassing can be undertaken before the extraction begins. Due to the high overburden pressure at great depths, a significant expenditure is necessary to prepare and maintain the parallel extraction and transport roads. To limit this expenditure the parallel gates are abandoned after passage of the face. There are no coordination problems between the longwall operation and the drifting as well, and there is no additional material handling in the parallel mining heading with building and support material. An additional argument for retreating longwalls is the shortening of the conveyance distance and the constant optimization of the conveyance over the running time of the working panel.

Until the early 1990s face lengths of only up to 270 m were technically possible and allowed at great depths. Today the new high performance longwalls are designed as double longwall systems with two 350 m face lengths or as single longwalls with up to 480 m face lengths. Greater face lengths are not practical at this time because the layout of the face conveyor reaches its limits. At present the length of the face conveyor and the resulting vibrations create significant problems for durability. The maximal possible power consumption of the face conveyor limits the loading rate and, therewith, the total length of the face conveyor length of the escape way exceeds the regulated length. Another significant problem is the air cooling.

The maximal seam pitch in the longwall is generally reported to be 40 gon. Steeper deposit sections are not suitable for economically profitable recovery using standing high performance longwall techniques. A gently declining mining direction through the strike has proven to be useful in increasing the stability of the wall and avoiding the flaking of the seam through the tipped coal face.

2.4   Safety Regulations

Due to the slow escape speeds in the longwall, a long face increases the duration of a possible escape over the acceptable and allowed limits. In order to confront this problem several measures were implemented. Care was taken when selecting and constructing the individualshield supports in order to attain comfortable and sufficiently wide gangway. In addition it became necessary to equip the face workers with the most modern filter self-rescuers. These filter self-rescuers guarantee lower inhalation resistance at significantly reduced temperatures of the breathable air ($< 65°C$). The organization of the escape routes was also reorganized. Medical examinations show that regular short pauses regulated by specially-trained escape leaders strongly reduce the physical burden of the individual miner during the escape without significantly increasing the escape time. Moreover, it may be necessary to employ selected, physically fit miners at steep faces. When calculating the length of the escape ways, according to the German hardcoal mining regulations, only the speed traveled by foot is taken into account. Passenger lifts, which would increase the escape speed and would usually be used, are not included in the calculation (2).

The introduction of high performance longwall operations and the planning of overly long faces also creates new problems for explosion protection. The German hardcoal mining industry requires the erection of explosion water barriers with 200 l water/m² roadway cross-section at 400 m intervals to extinguish the beginnings of methane gas explosions. These requirements are clearly exceeded by the introduced great face lengths with a distance of up to 120 m from the face end to the nearest barrier. Two different methods to meet the latent danger of a methane explosion were developed. The entrance of fresh air through the lower section of the shield column and the abandoned belt road is drastically reduced by consequent sealing of the gate end using grouting

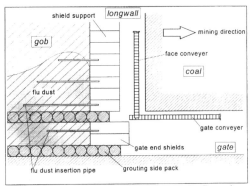

Figure 3: Grouting side packs and flue dust insertion piping for the prevention of air leakage in the abandoned workings

side packs and piping of flue dust against this grouting side pack. The goal is to reduce air leakage and to avoid the formation of an explosive gas mixture behind the shield column. Figure 3 shows the arrangement of the grouting side packs and flue dust in the abandoned workings.

In order to limit the starting length of a methane explosion, a mobile explosion water barrier (Saar-Ex 2000) was developed. This system is based on the active Tremonia barriers which, sensor-controlled, produce and distribute a fine water mist throughout the roadway cross section before the explosion wave can continue. This Saar-Ex 2000 explosion water barrier reduces the distance fromthe face end to the first barrier to a constant 30 m.

2.5   New developments of longwall techniques

The desired high face output could not be achieved using the formerly applied face equipment. It was, therefore, necessary to modernize and, if need be, redevelop the individual components of the face equipment for the demands of a high performance longwall. (1)

These demands to achieve a high face output are summarized as follows:

- High coal output of the longwall machine with a power consumption up to 500 kW per drum
- Application of point attack bits with a necessary high bit cutting depth of 8-10 cm even by lower drum rotational speeds; this is necessary for an effective reduction of the dust production
- Effective pic track flushing to avoid Hot-Spots and to consolidate the dust

34

- Large drum cutting depth, up to 1,000 mm
- Optimized drum loading capacity by the use of cowls and Globoid-drums
- High winning speed, speed over 13 m/min needs a powerful wheel-rackatrack haulage system
- High technical availability

The shearer loaders SL from the company Eickhoff which fulfill the above-mentioned criteria, are most commonly used in German high performance longwall operations. It was then necessary to switch from the previously used 1 kV-technology in the longwall, present in most mines, to a 3 or 5 kV power supply. Figure 4 shows a shearer loader SL.

The armored flexible face conveyors in the longwall area were equipped to reach high chain speed and manage large loaded cross sections in order to handle the expected increase in tonnage.

Much effort has been placed in the face support in order to continue the development of the longwall technique. The newly developed two-leg lemniscate powered face supports are used today without exception with the IFS (immediate forward support). The yield support resistances are suited for the high demands of great depths and amount up to 5,700 kN (yield load density of 600 kN/m²). An additional demand on the shield support was the necessity for high supporting performance of up to 30 m²/min in connection with short cycle times and quick roof support through the use of extensible canopies. The large cutting depth of the drum shearer requires a maximum advance distance for the shearer and the support of 1,200 mm. The operating range of this support should make heights of 1.80 to 4 m possible.

In order to optimize the longwall face move, the transport dimensions and weight were limited to make complete transport under ground possible. Within the framework of the new developments in longwall techniques, the system width was increased from the customary spacing of 1.50 m to 1.75 m for the longwall conveyor as well as the shield support. This enlargement of the system width primarily serves the purpose of minimizing the number of possible trouble sources in the longwall.

An analysis in the late-eighties and nineties showed that great expenditure was required to control the face-end zone and the belt entry. It was not possible to use the common face support in the parallel mining roadways due to the lining of the mining parallel headings with TH-sliding roadway arches. Moreover, several meters between the parallel mining roadway and the first shield had to

Figure 4: Shearer loader SL from Eickhoff

be built up conventionally using single legs and strike beams. Only after the development of the new face end shields and prop drawer shields for the parallel mining roadways was it possible to use high performance longwall technology to support the face end zone and the belt entry with modern powered supports (3).

To control the longwall face-end it was necessary to develop new power support systems as seen above. There was, however, no need for new side discharge technology. The direct side discharges used in the Ruhr area and the free side discharges with discharge pan used in the Saar area were able to handle even the great output of high performance longwalls.

The DSK (Deutsche Steinkohle AG) has realized the above-mentioned concept in different high performance longwall operations. One of these operations is described below.

3. HIGH PERFORMANCE LONGWALL EXTRACTION "LONGWALL 2000" AT THE ENSDORF COLLIERY

The Ensdorf colliery was one of the first to introduce high performance longwall operations to the German hardcoal mining industry in 1995 under the concept "Longwall 2000". The goal of this trial was to install a longwall system that could guarantee the daily output of 12,000 t of the Ensdorf mine out of a single longwall. Based on the positive results, the idea was then taken over by the other mines in the Saar deposit.

The Ensdorf colliery mines the northern section of the Saar hardcoal deposit on the Schwalbach coal seam (Coal seam 930) and the Wahlschied coal seam (Coal seam 950). The underlying Grangeleisen coal seam (Coal seam 970) is developed as a reserve. The mining done in this colliery, formed from the formerly independent Griesborn, Schwalbach and Ensdorf mines, concentrated on the southern deposits in shallow depths up to the 1950s. The mine concessions Ostfeld and Nordfeld were also mined later. Because the minable deposits were limited here, the Dilsburg field with the north shaft for material transport, man haulage and downcast ventilation shaft as well as the south upcast ventilation shaft were connected to the mining layout in the 1970s with the main haulage road on the fourteenth floor. Today the Primsmulde field is joined with the Dilsburg field as an additional reserve.

As can be seen in Figure 5, the Dilsburg field is developed with centered inclines and cross-cuts. The parallel mining headings are directly connected with the inclining main haulage roads. These transport requirements were ideally met using a suitable infrastructure for material transport and haulage. Both are processed over the Nordschacht. Using a powerful shaft haulage layout it was possible to transport complete shield units and machinery up to 35 tons, or up to 160 persons per haul. The travelling from the 20th level to the face entrance proceeds by belt riding over particular haulage and level belt conveyors under ground. Figure 6 shows the vertical sections of the mine. The bed inclination to the north is clearly recognizable.

The material transport is processed over the 18th level. The pieces to be transported are transferred directly from the cage to the transport site using locomotive haulage. At this point the material is transferred to the flat top track cable ways which carries it to the longwall. These track cable ways can also be used for face moves and transport up to 3 complete shield units at a time. Ideally the material transport from the storage depot on the surface to the longwall can be completed in 2 hours. All roadways in the working panel must have a cross-section of 23.5 m² to guarantee the necessary transport and ventilation diameters (4).

The product transport from the parallel gate continues on the conveyor belts in the inclines. Using additional conveyor belts, the raw coal is raised to the 14th level and carried over a lateral road on the 14th level to the old Duhamel Colliery. There it is raised through the Barbara drift over a length of 3.500 m and a vertical interval of 650 m to the surface. The material is then processed by the Duhamel shaft plant. Due to the generous design of the conveyor belts, at least 1,400 mm wide and a speed of up to 3.5 m/s, and a number of interposed field bunkers to homogenize the conveyance

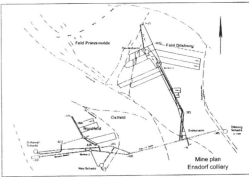

Figure 5: Mine plan of the Ensdorf colliery

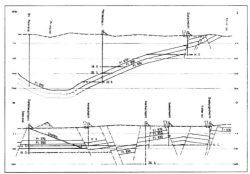

Figure 6: Vertical section of the Ensdorf colliery

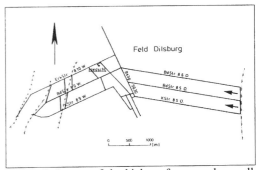

Figure 7: Layout of the high performance longwall in the Ensdorf colliery

discharge, up to 3,200 t/h can be hauled to the surface. The total length of the conveyor belts connected in series amounts to up to 17 km. The use of conveyor belts to the surface enabled higher output than the discontinuous skip hoisting.

Since the energy supply of the Dilsburg field comes through the Nordschacht, it was no problem to hang up the additional 5 kV lines needed for the high performance longwalls in the shaft.

When planning the layout of the high performance longwalls, the double longwall systems tried and proved in Ensdorf were used. Figure 7 shows the layout for the first double longwall using the high performance longwall technique. It ran from the beginning of 1996 to the middle of 1997.

When using the double longwall system, two longwall operations in neighboring panels are run simultaneously, using the same center gate. Both longwalls are layed out as retreating faces to the coal seam. The upper longwall hauls over the center gate, which also serves for ventilation and the lower longwall hauls over lower gate. The face length of the individual longwalls is 310 m, the panel length around 2,850 m. The direction of face advance is slightly inclined out of the drift into the dip in order to prevent the running out of the coal from the wall face. Fresh air flows in through the lower and the center gate. The upper gate serves as a return air road.

The Eickhoff SL 500 is used as the standard shearer loader. Figure 8 shows this shearer with labeled transport units.

A total performance of 1,240 kW was installed in this shearer loader. Each drum has 500 kW (5kV) and each of the two winches has 60 kW (1kV) to its disposal. With a maximal speed of 13 m/min during the extraction and a cutting depth of up to 1,000 mm for both Globoid drums, up to 39 m^3 coal per minute can be extracted from a 3 m thick coal seam.

To be able to extract these amount, a armored face chain conveyor HB 1000 280 V with a engagement width of 1,000 mm and a maximal drive power of 615 kW was installed in the longwall. The chain speed of the 34 x 102 double center chain can be increased to up to 1.61 m/s. In this case the speed had previously been restricted to 0.93 m/s. It was necessary to develop new racks to fulfill the input requirements of the wheel rackatrack system. The transfer to the 1,000 mm gate chain conveyor width with a conveyor belt speed of 1.5 m/s proceeds by the means of unbound side discharge. The conveyor hauls on a 1,200 mm wide conveyor belt, with a speed of 3.2 m/s, and a haulage capacity of 2,200 t/h, the limits of the capacity of the longwall system. Through the use of overhanged loading sections of the stage loader and belt storage, the shortening of the belts was no longer coupled with the face advance.

The newly developed Saartech-Ecker Shield 16 - 18/40 was introduced for face support. This shield support with its small transport height of 1,650 mm, large operating range from 1,800 to 4,000 mm, and large support resistance of

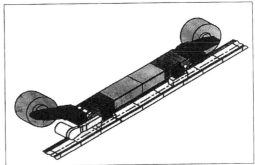

Figure 8: Shearer loader Eickhoff SL 500.

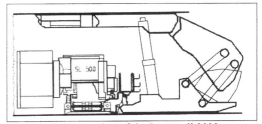

Figure 9: Cross-section of the Longwall 2000

Figure 10: Longwall 2000

5,700 kN (507 kN on the front tip of the canopy and a load density of 574 kN/m²) corresponded exactly to the demands of a face support at great depths. Based on the modern concept of a carriage road shifted into the shield, a large support advance, and an earlier roof support through the use of forepole canopy, this shield type has proved to be useful for other new longwalls in the Saar area and worldwide. Figure 9 shows a cross-section of the longwall system. Clearly identifiable are the generous dimensions of the longwall as well as the operator friendly and safe location of the carriage road in the shield.

Figure 10 takes a look into the longwall during the extraction. The size of this high performance longwall equipment is clearly seen when compared with the shearer operator.

The shield column could be extended into the top as well as into the bottom road using the new gate shields first used in this double longwall. The personnel and time intensive conventional single leg support of the belt-entry could be forgone.

For the continuous performance trials with these high performance longwalls in 1997, each individual longwall ran a daily output of over 18,500 t which corresponds to a gross output of about 27,000 t. The goal of guaranteeing a daily output of 12,000 t from a longwall was achieved. In the meantime, a record daily output of 23,700 t was set inEurope by the new panel 8.6 West on the Ensdorf colliery.

4. OUTLOOK

In recent years, the German hardcoal industry has put great effort in attempting to provide a secure future for German coal in spite of significant competition from inexpensive coal importers. In particular, the introduction of the most modern extraction technology has allowed for reduction of production costs almost to the levels of the world markets. Due to the great depths of the deposits, special demands set put on the layout of the longwall operations and the development of newer equipment. The positive results from the introduction of the first high performance longwall operations in Germany have lead to the standard use of this technology for the extraction of deposits at very great depths. The constant development of the installed technology and the adaptation of the winning method to the particular demands of these depths increases the possibility for the German hardcoal industry to offer hardcoal to the German economy at competitive prices into the future.

REFERENCES

(1)   Martens, Per Nicolai Universitätsprofessor Dr.-Ing. Dipl.-Wirt.Ing. et al.: *Proceedings of the "Internationales Kolloquium Hochleistungs-Strebbetriebe – High-Performance Longwall Extraction"*, 1. Auflage, Aachen, Verlag der Augustinus Buchhandlung, 1997

(2)   Oberbergamt für das Saarland und das Land Rheinland/Pfalz: *Richtlinien des Oberbergamtes für das Saarland und das Land Rheinland/Pfalz für die Ermittelung zulässiger Fluchtweglängen im Steinkohlenbergbau unter Tage (Fluchtweg-Richtlinien)* Verfügung vom 12. November 1979 – I 4800/13/79, Saarbrücken, Oberbergamt für das Saarland und das Land Rheinland/Pfalz, 1979

(3)   Hunfeld, Hanns-Hermann Dipl.-Ing. und Geißler, Hans-Joachim Dipl.-Ing.:*Vom Raubschild zum Raub-Rück-Tand*em, bergbau 4/97, p. 165 - 166

(4)   Clarner, Peter Dipl.-Ing. und Wahlmann, Harald Dipl.-Ing.: *Strebsystem 2000 des Bergwerks Ensdorf - Überzogene Technik oder zukunftsweisende wirtschaftliche Notwendigkeit* bergbau 7/97, p. 310 – 318

Mining Science and Technology' 99, Xie & Golosinski (eds) © 1999 Balkema, Rotterdam, ISBN 90 5809 067 1

Research on roadway layout for ensuring safety in longwall top-coal caving system*

Jingli Zhao & Qiang Fu
University of Mining and Technology, Beijing, People's Republic of China

ABSTRACT: For the sustainable development in coal industry, this paper presents a new roadway layout system of stagger arrangement in thick seam according to the theory and practice of mining system design to enhance the safety in operation of longwall top-coal caving face. This system solves the problems of the spontaneous combustion, the methane accumulation, and the difficulty of the extraction and maintenance of roadway. It raises the recovery ratio of coal mining and promotes the research on the sustainable development of longwall top-coal caving technology.

1. CURRENT SITUATION AND MAIN PROBLEMS OF LONGWALL TOP-COAL CAVING TECHNOLOGY

1.1 Current situation

The development of the longwall top-coal caving technology is a very important way for realizing high productivity and efficiency in coal mining. As far as the productivity of single face is concerned, the longwall top-coal caving has become the main coal mining method with high productivity and efficiency in underground mining in China. In July 1990, the first achievement of monthly productivity 140kt in longwall caving system was made in face No.8603 of Yangquan Coal Mining Administration (CMA). In 1993, a record of monthly productivity 310kt, yearly 2.5Mt and efficiency 100t/man was created in Wang Zhuang coal mine of Lu-an CMA. In 1996, batches of high productivity faces were presented with the longwall top-coal caving technology coming to maturity. 8.5 out of the 11 mining teams with yearly productivity over 2Mt applied the longwall top-coal caving technology. In 1997, all of the four with yearly productivity over 3Mt were longwall top-coal caving teams, including the No.2 team of Yanzhou Dongtan mine with yearly productivity 4.1Mt and efficiency 208t/man. In 1998, a new breakthrough of monthly productivity 500kt in that mine was made, which matched for the advanced level of the world. In recent years, it was the top-coal caving method to create new records of

high productivity and efficiency, and promote the development of coal industry especially in the thick seam mining in China.

Since 1993, the productivity of longwall top-coal caving faces has increased by 30% or so per year. In 1996, the productivity from the longwall caving faces is 8-10%of the national running coal mines. Its applying scope has also been enlarged constantly, especially in some coal seams difficult to mining, "three-soft" seams and unstable or thinner thick-seams.

At the same time, care must be taken to the problems of recovery ratio, spontaneous combustion, methane accumulation and dust concentration etc., which have restricted the further raise of mining benefit and application of the method in a long run.

1.2 Main problems

The four problems existed in the longwall top-coal caving method can be summed to two groups. One is the problem of safety, including spontaneous combustion, disasters of methane and dust, where spontaneous combustion and methane accumulation directly threaten to the normal operation in longwall top-coal face. The other is recovery ratio. Furthermore, there are always problems of extraction and maintenance of coal roadway driving along the floor.

Most of the thick seams in China are liable to spontaneous combustion which is of rare occurrence in the first layer of the thick seam with slicing

---

* National nature science key project (No. 59734090)

mining, but of frequent occurrence in the second or the following due to the repeated uncovering of gob. The Longwall top-coal caving technology realizes the whole seam mining, simplifies the layout of roadway and avoids the repeated uncovering. Owing to the great amount of the loss of coal, spontaneous combustion has still taken place frequently in the longwall top-coal caving district for the over ten years. According to the statistics in 89 longwall top-coal caving faces, the threat of spontaneous combustion has existed in over half of them, even causing the stopping of the mining operation or the accidents. Since the roadway of the long-wall top-coal caving face (including open-off cut) drives along the bottom where roof and two-rib of the roadway are all in coal and are easy to fall, the caving cavity is easy to induce spontaneous combustion. Statistics shows the times of spontaneous combustion taking place in roadways exceeds 2/3 of the total (Wu 1997). · If these roadways are extracted along roof cr under metal mesh in gob, the serious roof fall cavity will not come into being and the fire accident can be avoided effectively. According to the 13-year statistics of the longwall top-coal caving mining, the frequency of spontaneous combustion on the side of section road adjacent to the upper section is only next to that on the roadways (shown in Fig.1). Since the two-end support in most top-coal caving faces can not draw out the coal, the top-coal caves after the face advances. Thus, there is a lot of loose coal near the section roadways even some of the coal in the roof fall cavity was of occurrence in spontaneous combustion before. It is easy for leakage air passing through the new extracted roadways to ignite the coal again. If this part of top-coal can be extracted, the size of section pillar can be reduced to a minimum and it is not broken, so that the favorable condition for the ignition will be eliminated. If the combustion in roadways and on the one side of section roadway can be controlled effectively, the difficult situation of fire prevention in long-wall top-coal caving system will be changed basically.

The other problem is to discharge methane. The typical section roadway in longwall top-coal caving face is extracted along the bottom layer of coal seam. As methane emission goes up, the space above the supports will accumulate methane after top-coal caves. Since the main airflow in the face flows from the cutting space to return airway, it is impossible to dilute the local accumulated methane (shown in fig.2). When the roof falls, the methane above the supports is pressed to face as a result. The methane concentration in the face, especially in the upper corner will exceed the allowed limit, which is

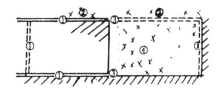

1--- Serious roof fall cavity and rib fall area of air-intake and air-return roadways, open-off cut area and terminal line; 2---Gob nearby no-pillar mining face; 3---Upper and lower corners; 4---Gob behind the face
Fig.1 Location and sequence of spontaneous combustion

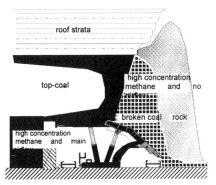

1---Roof strata; 2---Top-coal; 3---Broken coal; 4---Refuse; 5---High concentration methane and main airflow; 6---High concentration methane and no airflow;
Fig.2 Methane discharge of top-coal caving face

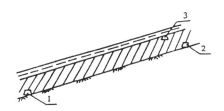

1---Air-intake roadway; 2---Air-return roadway; 3---Special methane discharge roadway;
Fig.3 Location of special methane discharge roadway

a threat to the safe operation. To solve the problem, a special methane discharge roadway is extracted in some mines, which forms a complex ventilation system with one air-intake and two air-return roadways (shown in Fig.3).

To keep high recovery ratio is related to the sustainable development of coal industry. It is also very important for the popularization of the top-coal caving method. The loss of coal in section pillar and in upper section roadway is still a long-term problem. In addition, the caving of the top-coal in the two ends of the face can not be done, as the supports there can not draw out coal. The problem can not be solved by improving the structure of those supports because the conveyor head in the lower end hinders coal drawn out and there is no conveyor to transfer caved coal in the upper end.

The practical situation that the recovery ratio is difficult to raise greatly comes into conflict with the regulation for recovery ratio which should be over 75% in the top-coal caving face. To solve the caving problem of the ends of the face and the loss in the section pillar becomes an important project at present. In the slicing mining, although the recovery ratio is high, the loss of coal in the section pillar is greater, so that it is a common problem existed in the long-wall top-coal caving and the slicing mining.

The section roadway of the long-wall top-coal caving face drives in the bottom of the seam. The extraction is often affected due to the caving of top-coal, and the maintenance is also difficult. If the roadway can be arranged along the roof, its extraction and maintenance will be easier.

In order to solve above difficult problems comprehensively, it is an important way to seek for more reasonable operation system in the thick seam.

2. RESEARCH ON THE STAGGER ARRANGEMENT SYSTEM IN THICK SEAM

2.1 Roadway layout system

According to the theory and practice of mining system design, this paper presents a new roadway layout system with stagger arrangement (shown in Fig.4). It can be used to solve above ticklish problems in thick-seam with longwall top-coal caving. In the system, the intake and return airways, usually driving along the lower layer of the thick seam, are arranged in different layers in the thick-seam under the roof or the metal mesh in the gob respectively to avoid the occurrence of serious roof fall cavity and spontaneous combustion. At the same time, the extractions of the top-coal above the two-end and section roadways eliminate the main factor of the combustion in nearby gob. Furthermore, arranging return airway in the upper layer of seam can discharge the accumulated methane in the upper corner of the face effectively. The scheme can also raise recovery ratio. The practice of extraction under gob and in uphill of face can be realized by synthesis application of available techniques.

To contrast easily, the roadway layout of typical longwall top-coal caving face and the coal loss in the two-end of the face and the section pillar are shown

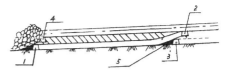

1---Intake airway; 2---Return airway; 3--- Intake airway of next section; 4--- Return airway of upper section; 5---Residual triangle pillar;
Fig.4 Stagger arrangement system of roadways in thick seam

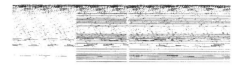

1---Intake airway; 2---Return airway; 3--- Intake airway of next section; 4--- Return airway of upper section; 5--- Loss of top-coal and section pillar; 6---Methane discharge roadway along the roof.
Fig.5 Roadway layout in typical longwall top-coal caving system

in Fig.5. The contrast of the two roadway layout forms is analyzed as follows.

2.2 Reduce the coal loss of the two-end and section pillars; decrease the possibility of spontaneous combustion

Fig.4 indicates, in the new system, the top-coal above the No.1 section roadway and transformation supports has been mined by the upper section. In the system shown in Fig.5, the top-coal above No.1, No.4 roadways and the transformation supports caves after that of the coal face caving. It becomes loose and easy to leak air. In addition, the little pillar is also broken under roof pressure and is easy to leak air. As a result, spontaneous combustion is easy to occur. In the system shown in Fig.4, not only the top-coal of the two-end is extracted but also section pillar becomes triangle one (shown in Fig.5). Thus, the loss of coal is reduced greatly. As far as coal loss is concerned, only the triangle pillar is lost. The following calculation indicates the difference of coal loss between the two systems.

Suppose the thickness of a coal seam is 6m, the cutting height 2.5m, the caving height 3.5m, the width and height of section roadway 3.0m and 2.5m respectively. Suppose there are two transformation supports, with a total width of 3m, which can not draw out the coal. In the system shown in Fig.4, the triangle pillar comes into being with the raising of conveyor gradually. Each trough is raised about 3 degree each time. After the angle reaches to 15 degree, 3 troughs remains at this degree and then about 3 degree for each trough is reduced each time until it becomes horizontal. When the trough is raised about 2.5m, the triangle pillar on the left of the No.1 and No.2 roadways shown in the Fig.4 is come into being, which is the loss of coal. The upper side of the pillar is composed of broken lines with the length of 1.5m(the length of trough). Its cross-area is 14.3m^2. In the system shown in Fig.5, suppose the little pillar of no-pillar mining is 4.0 wide and the other parameters are all the same as above. The cross-area of the coal mass lost is 66.0m^2. In other words, the system shown in Fig.4 can recovered 80% of the top-coal lost in the two-end and section pillar in the system shown in Fig.5.

In the system shown in Fig.4, the No.2 roadway is along the roof, which avoids the occurrence of the serious roof fall cavity. The No.1 Roadway is under the gob. The probability of spontaneous combustion in these two roadways is decreased greatly.

As far as spontaneous combustion on the side of roadway is concerned, in the system shown in Fig.4, the residual coal is a little. Its triangle pillar is in the stress-relieved area and has no direct link with the seam roof. Accordingly, the pressure acting on it is small, the pillar is not easy to break and leak air and the probability of spontaneous combustion is very low.

Obviously, in the system shown in Fig.5, the loss of coal is large. The little pillar up to the roof is higher than the No.1 and No.2 roadways, so that it is easy to break and leak air. The more the loose coal is, the greater the air leakage is, and the easier the spontaneous combustion takes place. Moreover, in the system shown in Fig.5, the top of roadway is easy to cave, which induces spontaneous combustion. After caving of the fired area in top-coal, spontaneous combustion is easy to occur again.

2.3 Discharging methane

Fig.4 indicates, in the system, the layout of the No.2 return airway along the roof can eliminate the local methane accumulation in face effectively, which acts as the No.6 roadway shown in Fig.5. For the face with low methane emission, the roadway can be saved, which simplifies the productive system and reduces the driving expenses of roadway.

2.4 Roadway driving and maintenance

The system shown in Fig.4 is simple, reasonable and saves one roof roadway. The driving and the maintenance of the No.2 roadway extracted along the roof are better than that in the system shown in

Fig.5. In addition, one side of the triangle pillar is maiden coal with great width and in the stress-relieved area, so that the maintenance is easy and it is not easy to leak air. The No.1 roadway can be driven under the metal mesh or the regenerated roof according to the concrete situation.

3 CONCLUSIONS

The superiority of the stagger arrangement of roadway is prominent. The layout is simple and practical. It not only keeps the high productivity and efficiency of long-wall top-coal caving technology, but also helps to solve the problems of spontaneous combustion frequently occurring in the serious roof fall cavity of the roadway and the nearby gob. Combining the new layout with current techniques, an overall fire prevention effect will be achieved. It is also in accord with the law of methane moving to arrange the return airway along the roof, which increases the safe reliability greatly.

The roadway layout raises recovery ratio obviously. It is possible to make recovery ratio reach even exceed the relevant national stipulation. In a long run, the economical and social benefits of raising recovery ratio will be an important yardstick and a development direction of the mining method. It has a great future to be widely applied in the longwall top-coal caving faces.

With the end of the shortage of coal, the coal market nowadays is saturated. How to raise recovery ratio at the time of high productivity has become a common knowledge gradually. The foundation of the operation system with high recovery ratio and reliability is significant for the sustainable development and popularization of the longwall top-coal caving technology.

REFERENCES

Wu Jian & Zhao Jingli, *1997.4,* Several opinions on present application of longwall top-coal mining method. *Coal:* 16.

Xu Yongqi, chief editor, 1990, *Atlas of mining methods in China:* 295. Publishing House of China University of Mining & Technology.

Mining Science and Technology' 99, Xie & Golosinski (eds) © 1999 Balkema, Rotterdam, ISBN 90 5809 067 1

Acceleration of massive roof caving in a longwall gob using a hydraulic fracturing

Kikuo Matsui & Hideki Shimada
Kyushu University, Fukuoka, Japan

Herryal Z. Anwar
R&D Center for Geotechnology-Indonesian Institute of Sciences, Bandung, Indonesia

ABSTRACT : Periodic weighting and windblast which are caused by related to the dynamic behavior of overburden in a longwall panel significantly affect longwall mining. This study proposes a new loosening system using hydraulic fracturing and moistening, and also discusses strength deterioration in hard-to-collapse roof rocks due to water in order to solve the problems caused by the effects of roof rocks behavior in a longwall panel.

1 INTRODUCTION

In certain underground coal mines where the roof consists of hard-to-collapse rocks, or strong and massive rocks, the roof rocks do not always cave in regularly as longwall extraction proceeds, but roof overhanging does occur, leading to extensive areas where the roof is unsupported in the gob. Here, the dynamic effects of roof rock behavior present difficulties in a highly mechanized longwall face. The major problems are periodic weighting (Peng et al., 1984), rock burst and coal bump (Sugawara et al, 1987) and windblast (Matsui et al., 1998).

Conventional techniques using longhole blasting have been used, in order to prevent these dynamic effects in faces with hard-to-collapse roofs. However, rock loosening using this method is the following disadvantages:
1) Rock loosening with this method is uneven: near the hole the rock is overfractured, while at a given distance it is destroyed by isolated cracks.
2) The effective range of the blast is small; 5-25 radii of the explosive charge.
3) Drilling operation is hard and dangerous.

This paper proposes a new hydraulic fracturing technique for rock loosening in order to control hard-to-collapse roofs in longwall panels, and discusses the characteristics of the strength deterioration of hard and strong sandstone roofs due to water by means of laboratory tests.

2 ROOF STRATA BEHAVIOR

The problem of dynamic behavior has already been identified as being related to the presence of hard-to-collapse strata or strong and massive strata in the immediate roof (Frith, 1996).

The massive overlying strata cantilever over the powered shield supports into the gob and pivot about a point that is located ahead of the longwall face. This can lead to fracturing of the strata ahead of the faceline, resulting in face falls and cavities developing to considerable heights above the face area (Matsui et al., 1997). No amount of increased powered support capacity will prevent this problem from occurring. When the immediate roof consists of more friable or weaker strata such as shale, coal, mudstone, it will readily cave in behind the powered supports. At the same time, if the thickness of the immediate roof is large enough, the caved roof rocks bulk up sufficiently to fill the gob and stabilize the dynamic behavior resulting from the movement of the overlying strata. However, when the thickness of the immediate roof is not enough thick, there is insufficient bulking, resulting in large, continuous voids in the gob extending back from the faceline. Any subsequent failure and movement of the main roof would have the potential to create severe problems on the face such as periodic weighting, face instability, and windblast.

Coal outburst is also related to the existence of the strong and massive main roof (Sugawara et al., 1987).

3 HYDRAULIC FRACTURING TECHNIQUE

3.1 *Hydraulic fracturing system*

Hydraulic loosening is performed through boreholes by hydraulic moistening and fracturing. When water is pumped into a section of the borehole that is isolated by packers, some water penetrates into the surrounding strata. As the water pressure increases and reaches the maximum stress level that the strata can sustain, a crack is formed. If pumping is continued, the crack will extend.

However, with the hydraulic moistening of the roof, uniform and adequate saturation of the strata is quite difficult because of the relatively low permeability of the constituent strata. Permeability is reduced with the increasing depth of mining operations.

Hydraulic fracturing occurs in random segments of the hole, mainly through existing natural cracks. Water opens them even more and develops only these cracks, starting to move through those randomly oriented cracks that have with the widest openings.

Using present methods, it is impossible to control the rock loosening process, leading to uneven changes in the mechanical properties of the rocks.

The main structural difference between easy- and hard-to-collapse rocks is the stratification of easily collapsible rocks, with relatively weak contact between layers. The hard-to-collapse rocks, even though sedimentary, are massive and monolithic and have no distinct cleavages or bedding planes. If the hydraulic fracturing technique is used effectively and controllably, the instability of the longwall face which is caused by the dynamic effects of roof rocks can be limited by inducing artificially directed cracks in the hard-to-collapse rock mass and dividing it into many layers.

3.2 *Hydraulic fracturing mechanism*

A new method for loosening hard-to-collapse roof rocks is based on the hydraulic fracturing technique that has been used in petroleum or rock engineering such as the extraction of geothermal energy from hot dry rocks or in-situ stress measurement. In general, two types of fractures are created by hydraulic fracturing, i.e., vertical or longitudinal and horizontal or transverse. Creating vertical fractures has primarily been studied, however, forming horizontal fractures is useful and can effectively control the caving of hard-to-collapse roofs in

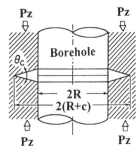

Figure 1. An artificial circumferential crack model of a borehole.

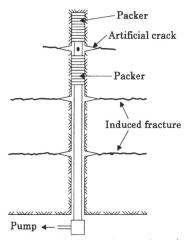

Figure 2. Hydraulic fracturing and moistening system.

longwall panels. Forced horizontal fractures can be formed through one bore hole, creating two, three, or more layers along the bedding planes, depending on the number of artificial cracks which are created on the internal walls of the borehole before fracturing.

Using the theory of elasticity or fracture mechanics can treat a condition for creation of the horizontal fracture. With the elastic theory, the conditions for the propagation of this horizontal fracture can be met if the internal water pressure equals the vertical stress plus the tensile strength of the rock. However, the propagation conditions depend on the existence of numerous natural cracks in the rock mass. Therefore, controllable crack initiation needs the treatment of cracks using fracture mechanics. Figure 1 shows an artificial circumferential crack model of a borehole in order to create a horizontal fracture plane along the bedding plane in the hard-to-collapse roof (Keer et al., 1977). A borehole is drilled under the magnitude of initial vertical stress P_z and an artificial circumferential crack is formed on the internal walls of the borehole. Water is pumped into a section of the borehole that is isolated by packers. As the water pressure increases, a fracture begins at the tip of the artificial crack. In order to simplify the analytical treatment, let the angle of the artificial crack edge $\theta_c = 0$. Then, breakdown water pressure P_c is estimated by the following equation:

$$P_c = P_z + K_{IC}/(k_c c^{1/2}) \qquad (1)$$

where K_{IC} = fracture toughness; c = crack length and k_c = a constant depending on the ratio of crack length c to borehole radius R. Once formed, the fracture will continue to propagate as long as the pressure is greater than the stress normal to the plane of the fracture, or the vertical stress.

Figure 2 shows a hydraulic fracturing and moistening system. First, a long borehole is drilled into the roof vertically along the setup entry, the headentry, or tailentry. Next, one, two, or more artificial circumferential cracks are made on the internal walls of a borehole at a given depth with slot cutting equipment or a water jet. Then, the packers are installed and, after that, hydraulic fracturing begins. Finally, the hard-to-collapse roof rocks are loosened by both fracturing and moistening.

4 MECHANICAL PROPERTIES OF COAL MEASURES ROCKS

4.1 Uniaxial strength and Brazilian tensile strength

Coal measures in Japan are Paleogene and consist primarily of sandstones, shales, sandyshales, mudstones, tuffs, conglomerates and coals.

Figure 3 shows the relationship between uniaxial compressive strength, (UCS) and Brazilian tensile strength, (BTS), of the Coal Measures rocks at Miike and Ikeshima Collieries. These data were obtained under dry and wet (saturated) conditions.

These figures show that the strength properties cover a wide rage and decrease under wet conditions. Some shale exhibits a slaking phenomenon when it comes in contact with water.

Figure 4 shows the mechanical property changes of Miike sandstone under different contents of water. The properties of the rock decrease with an increase in the water content.

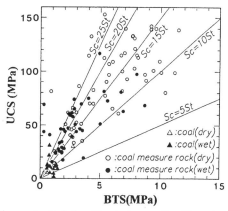

Figure 3. Relationship between Brazilian tensile strength and uniaxial compressive strength of the Coal Measures rocks at Miike and Ikeshima Collieries.

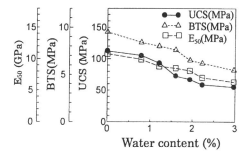

Figure 4. Change in mechanical properties under different water contents.

4.2 Fracture toughness of strong sandstone

Fracture toughness tests were performed according to the ISRM suggested methods (ISRM, 1988).

The fracture toughness of Miike strong sandstone was determined by using chevron bend and short rod tests under dry and wet conditions. From these two tests, the anisotropy of fracture toughness can be estimated. Level II testing was done in all fracture toughness testing.

The sandstones, which are Miike-1 and –2, were taken from the roofs of the Upper Seam and the Second Upper Seam in the West-80 area at Miike Colliery, Japan. These rocks were fine-grained sandstones. The physical and mechanical properties, including fracture toughness, obtained from laboratory tests are listed in Table 1. Other rocks are also listed for reference.

Table 1. Physical and mechanical properties of rocks.

Sample	Effective porocity (%)	Water content (%)	UCS (MPa)	BTS (MPa)	E_{50} (GPa)	K_{CB} (MPa·m$^{0.5}$)	K^C_{CB} (MPa·m$^{0.5}$)	K_{SR} (MPa·m$^{0.5}$)	K^C_{SR} (MPa·m$^{0.5}$)
Sandstone (Miike-1)	9.30	0.0 (dry)	118.3	8.2	8.2	1.56	2.77	1.43	1.79
		3.75 (saturated)	37.8	1.1	5.1	0.63	0.82	0.23	0.51
Sandstone (Miike-2)	9.67	0.0 (dry)	126.1	8.3	6.5	1.27	1.15	1.42	2.20
		3.92 (saturated)	47.2	2.1	4.1	0.35	0.62	0.33	0.56
Granite (Tokuyama)	1.65	0.0 (dry)	218.9	10.9	23.7	2.08	2.35	1.75	2.50
		0.62 (saturated)	194.1	9.1	22.0	1.95	2.18	1.79	2.40
Andesite (Karatsu)	12.05	0.0 (dry)	116.9	8.7	16.3	1.49	2.58	1.78	2.63
		4.59 (saturated)	79.1	4.9	11.1	1.10	2.07	1.30	2.13

Note : Sandstone Miike 1 and 2 were taken from the roof of the Upper Seam and the 2nd Upper Seam, respectively.
UCS : Uniaxial compressive strength.
BTS : Brazilian tensile strength.
E_{50} : Tangent Young's modulus at 50% peak stress.
K_{CB} : Fracture toughness from chevron bend test.
K^C_{CB} : Corrected fracture toughness of K_{CB}.
K_{SR} : Fracture toughness from short rod test.
K^C_{SR} : Corrected fracture toughness of K_{SR}.

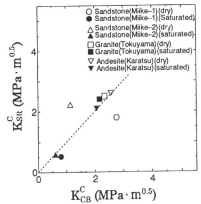

Figure 5. Relationship between corrected fracture toughness K^C_{CB} and K^C_{SR}.

Figure 5 shows the relationships between K^C_{CB} and K^C_{SR}. K^C_{CB} and K^C_{SR} are the corrected fracture toughness and are always greater than the initial values of K_{CB} and K_{SR}. Miike sandstone-1 and 2 show some anisotropy but not clear. When saturated, the rocks decreased in fracture toughness, especially Miike sandstone. Using equation (1) and these data, an effective hydraulic fracturing system can be designed.

CONCLUSION

Dynamic roof rock behavior can be controlled by the acceleration of roof caving with hydraulic fracturing and the moistening technique. In the near future, the hydraulic fracturing system that is proposed in this study would have practical applications.

ACKNOWLEDGMENT

The authors would like to thank the staff of Miike Colliery for their cooperation in providing relevant information in the preparation for this paper.
Any opinions stated in this paper are those of the authors themselves and are not necessarily those of the colliery.

REFERENCES

Frith, R., 1996, Development and Demonstration of A Longwall Monitoring System for Operational Decision Making, *ACARP Project Report No. C4017.*

ISRM, 1988, Suggested Method for Determining the Fracture Toughness of Rock, *Int. J. Rock Mech. Min. Sci. & Geomech. Abstr.*, Vol. 25, No. 2, pp. 73-96.

Keer, L.M., Luk. V. K. and Freedman, J. M., 1977, Circumferential Edge Crack in a Cylindrical Cavity, *Journal of Applied Mechanics, Transactions of the ASME*, June, pp. 250-254.

Matsui, K. and Shimada, H., 1997, Roof Instability of Longwall Face at Ikeshima Colliery, *Proc. 16th Int. Conference on Ground Control in Mining*, Morgantown, WV, USA, pp. 92-97.

Matsui, K., Shimada, H. and Anwar, H., 1997, Control of Hard-to-Collapse Massive Roofs in Longwall Faces Using a Hydraulic Fracturing Technique, *Proc. 17th Int. Conference on Ground Control in Mining*, Morgantown, WV, USA, pp. 79-87.

Peng, S.S. and Chaing, H.S.,1984, *Longwall Mining*, John Wiley & Sons, New York, 1984, pp. 17-49.

Sugawara, K., Okamura, H., Obara, U. and Kimura, O., 1987, Transversal Coal Outburst in Miike Coal Mine, *Proc. Int. Symposium on Coal Mining and Safety*, Seoul, Korea, pp. 175-184.

Mining Science and Technology' 99, Xie & Golosinski (eds)© 1999 Balkema, Rotterdam, ISBN 90 5809 067 1

Graphical representation and modelling of an airflow in ventilation network

F. Rosiek, M. Sikora, J. Urbański & J. Wach
Wrocław University of Technology, Poland

ABSTRACT: For many years a computer system called AutoWENT has been used in LGOM mines, as AutoCAD R12 application. The basic system role is modeling of airflow in a ventilation network and creating, editing and updating a graphical representation of the ventilation network. Besides of this, the system contain many others programs which are very helpful, for instance, in creating a mathematical model of a ventilation network or pre-processing measurements data necessary for the model. From AutoWENT one can run programs calculating air distribution in ventilation network as well as visualizes obtained results on spatial diagram that is regular AutoCAD drawing. Recently the AutoWENT system has been rewritten, and now runs in Windows (Win32) environment. The new version of the program comprises a simple CAD subsystem. The system is very useful in managing, planning and developing a ventilation network.

1 INTRODUCTION

Mathematical methods of analysis of airflow in a ventilation network have been known from many years. However these methods could not be practically applied till computer technology became available to mine companies at a reasonable cost. Computer analysis can provide fast access to all necessary information about airflow within the mine network. However, before airflow calculations can be carried out a numerical model of the network should be created. This model should contain data describing the network structure and parameters related to all its components.

For ventilation purposes it is necessary for mines to maintain up to date ventilation network schemes. It is necessary for such schemes to be in accordance with state regulations. Creating, modifying and updating such schemes is a time consuming and cumbersome job. In order to make this work easy the mine ventilation services should posses proper tools for drawing, updating and modifying ventilation schemes.

In polish mines different calculation programs have been used. In LGOM mines (i.e. Legnica and Głogów Copper Basin) an application called AutoWENT has been used for many years. Apart from its main role, which is the modelling of airflow, the system can perform many other calculations. For example pre-processing data which

has been measured and is necessary for the model. This function is important in the case of LGOM mines, because these mines have large airway cross-sections and exploitation systems with parallel excavations in one airway, which produce serious problems in determining aerodynamic resistance of air branches.

The system is very convenient for a user because creation of a mathematical model is performed simultaneously with drawing a space diagram of a ventilation network so that coherency between the diagram and the data is kept all the time. Under control of the AutoWENT application the user can modify both data and a space diagram, as well as run calculations and display calculation results on a scheme.

The versions of the AutoWENT that were created before 1998 were ADS applications, and worked within AutoCAD for DOS environment. They were made up of programs written in C, AutoLISP and Pascal languages. Recently the AutoWENT system has been rewritten, and now is available as a single executable running in Windows (Win32) environment. This version of the program comprises a simple CAD subsystem, which makes the whole application independent of any other CAD systems available on the software market. The CAD subsystem has the ability to create and modify two dimensional (2D) vector drawings in a similar manner as AutoCAD does. It can insert and edit simple geometrical figures and texts. It uses layers,

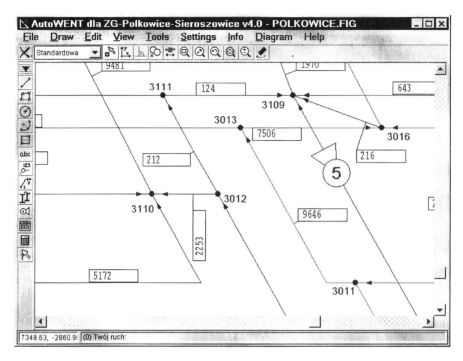

Figure 1. Application window and a part of ventilation network space diagram created by the AutoWENT

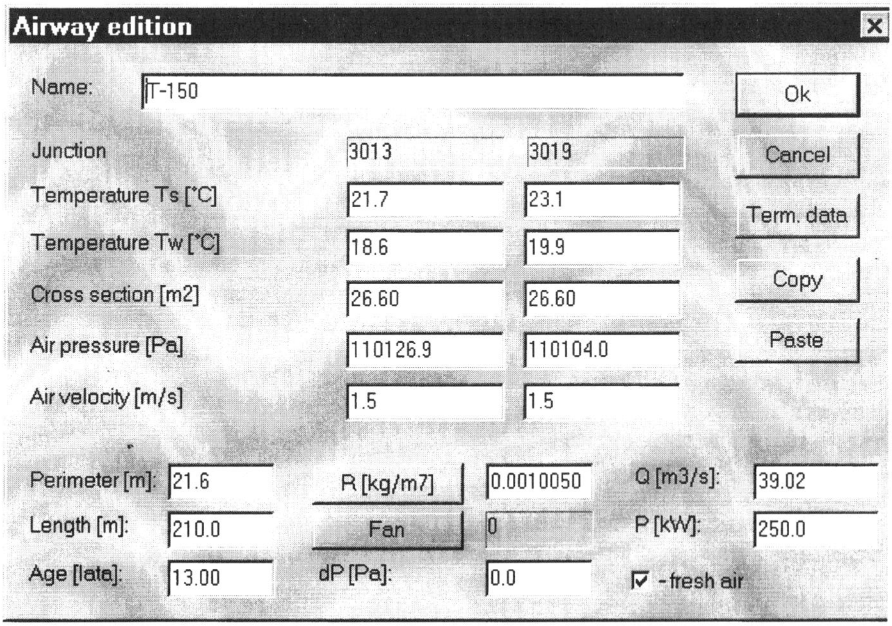

Airway edition

Name:	T-150	Ok
Junction	3013	3019	Cancel
Temperature Ts [°C]	21.7	23.1	Term. data
Temperature Tw [°C]	18.6	19.9	
Cross section [m2]	26.60	26.60	Copy
Air pressure [Pa]	110126.9	110104.0	Paste
Air velocity [m/s]	1.5	1.5	

Perimeter [m]:	21.6	R [kg/m7]	0.0010050	Q [m3/s]:	39.02
Length [m]:	210.0	Fan	0	P [kW]:	250.0
Age [lata]:	13.00	dP [Pa]:	0.0	☑ - fresh air	

Figure 2. The dialog window for an airway data editing

blocks and provides many useful commands that make drawing easy. However, the main aim of the system is the performance of airflow calculations, and the assistance in creating, modifying and updating all kind of diagrams of the ventilation network.

2 THE GRAPHICAL REPRESENTATION OF A VENTILATION NETWORK

The basic graphical representation of a ventilation network in the AutoWENT system is a space diagram. The system contains tools necessary to create and edit the drawings representing such a diagram. The appropriate commands are located in the application menus and toolbars. They allow the user to insert into the drawing all components of the diagram, for example junctions, airways, fans, shafts, stoppings, doors, regulators etc. and edit both their graphical form as well as numerical data associated with them. The main components of the drawing are labelled. Airway labels are framed and represent selected physical parameters of the airflow corresponding to that airway. The ventilation network diagram is drawn on many layers and these makes changing its appearance easy. Figure 1 shows the application window and an example of ventilation space diagram. Among other commands the AutoWENT system provides search commands that let a user easily find interesting junctions or airways either by name or by number. It is a very important function particularly in the case of large graphs, which contains many airways and junctions.

3 ENTERING AND UPDATING NUMERICAL DATA

In the AutoWENT system entering and editing data of junctions, airways, stoppings and fans is easy and can be realised with a couple of clicks of the mouse. For instance to edit an airway data one should select an appropriate command from the menu or the toolbar and then click the mouse when its cursor covers a line representing the airway. This displays the appropriate dialog window, which allows a user to enter and edit necessary data. The dialog window of an airway data edition is shown on Figure 2. In the similar manner a user can recall proper dialog window to edit data of junctions, stoppings and fans. A set of data describing a single junction contains junction location i.e. depth and geological co-ordinates as well as side airflow parameters (temperature and volume airflow rate). Stoppings are described by two parameters namely its

aerodynamic resistance and type. In case of fans their pressure-volume characteristics are determined by a set of polynomial coefficients of power of three. A fan data contains also pressure ranges and volume values for which its characteristic has been verified experimentally. Data and drawing of a diagram are saved together in a single file. In a version for AutoCAD they were stored in two separate files associated by name. Each time, when numerical data was altered or graph was changed, the program would check the coherency of the data and structure of the ventilation network before the airflow calculations were carried out.

4 CALCULATIONS AND RESULTS VISUALISATION

Airflow calculations are necessary for an effective airflow management in a mine ventilation network. It applies particularly to the large networks that have complex topological structure. Adjusting airflow pattern to excavation evolution requires knowledge about future investments in advance. In order to adjust airflow pattern suitable software performing airflow calculations is necessary. On the other hand, to run the calculations a proper numerical model of the ventilation network is to be created. If numerical data describing junctions, airways, stoppings and fans are available when the space diagram of a ventilation network is drawn, then the mathematical model and the diagram can be created simultaneously. This was described in the previous section.

Airflow calculations can be performed in two variants (namely with and without temperature prognosis) chosen in appropriate dialog window. Selected calculation results are immediately visible on the screen in the airway frames. They can also be printed as a report, which can be formatted according to user request. According to the user decision, numbers displayed in the airway frames can represent volumetric airflow rate, air velocity or temperatures at inlet and outlet of an airway. The airway frames can display two sets of results, firstly "current" results and secondly "prognosis" results. It makes the analysis of airflow convenient because all changes resulting from network structure modifications or operating with airflow controls such as stoppings, fans, etc. are clearly visible. This helps to make a correct decision and choose the best from all variants, which have been considered.

If numerical data describing the network components are not available when the space diagram has been drawn, then measurements have to be performed in mine excavations in order to determine the resistance of the network's airways. The measurement results have to be pre-processed and this is performed in two stages. In the first stage volumetric airflow rates are adjusted so that I Kirchhoff's law is fulfilled at each junction. In the second stage corrections are applied to potential fall in order to fulfil II Kirchhoff's law in each network loop. It should be emphasised that potential fall tuning is made with additional assumption, i.e. constant airflow potential at chosen network junctions. Once the pre-processing has been finished a user can continue the airflow calculations.

5 SUMMARY

AutoWENT system both in AutoCAD and Windows version is a helpful tool. Its main objective is to create and modify space diagram and numerical model of a ventilation network. The system ensures coherency between the space diagram and its numerical model. The system provides software tools for drawing, editing and displaying a space diagram, conducting airflow calculations (also with temperature prognosis) and visualising its results. The program is a valuable tool suitable for airflow management in a mine ventilation network. The main disadvantage of older AutoWENT versions is that they are available only to licensed AutoCAD users. The newest version requires only licensed version of Windows 9x/NT operating system.

REFERENCES

Hartman H. L., 1982 „Mine ventilation and air conditioning", New York, Wiley,
Voss J., 1981. „Grubenklima", Essen, Verlag Glückauf GmbH,
Rosiek F., Sikora M., Urbański J. 1993, „Determining resistance of ventilation side branches for the purpose of constructing a digital model of mine ventilation network", Katowice, Przegląd Górniczy Vol. 49 No 6

Mining Science and Technology' 99, Xie & Golosinski (eds) © 1999 Balkema, Rotterdam, ISBN 90 5809 067 1

Research on the E-type ventilation system in the working face with the fully mechanized top coal caving in the Luan Coal Mining Administration

Qinfang Duan
Luan Coal Mining Administration, Shanxi, People's Republic of China

ABSTRACT: In the Luan Coal Mining Administration, the problem of local accumulation of methane at the upper corner of the goaf is solved successfully by changing the conventional U-type ventilation into E-type ventilation. According to the in-situ measurement and the physical experiments, the advantages of E-type ventilation are analyzed and an example for using E-type ventilation is introduced in this paper.

In the past several years, the top-coal caving method is widely practiced in the Luan Coal Mining Administration. Because of the effects of top coal caving on the emission of methane in coal seam, the spatial distribution of methane in top-coal caving face is very different from that of the general sliced longwall mining face. The factors caused the increasing of the amount of methane emission in a fully mechanized top-coal caving face include the following 4 aspects. (1) The mining intensity of top-coal caving increases obviously. As a result, the absolute methane emission increases. (2) The remaining coal in the goaf increases. Therefore the amount of methane released from remaining coal in the goaf also increases. The released methane enters into the coal face and accumulates at the upper corner of the goaf. (3) The coal wall, the caved zone, the fractured zone and the cracked zone of top-coal increase the free face for releasing of methane. As a result, the amount of methane emission increases. (4) Compared to the slice mining face, the prepared gateways decrease obviously. Therefore the methane in the coal seam can not be released before coal mining and it induces the increasing of the methane emission during coal mining. In past several years, the method for methane control in the fully mechanized top-coal caving face has been being a key point of safe production in Luan Coal Mining Administration. The problem of the local accumulation of methane at the upper corner of the goaf is resolved successfully by a series of new technologies, of which the most important one is to change the conventional U-type ventilation system into the E-type ventilation system in working faces.

1 INTRODUCTION TO E-TYPE VENTIATION

In the coal face of fully mechanized top-coal caving, an additional roadway called the methane emission roadway with a relatively small area, which is specially used for emission of methane, is excavated at the upper part of coal seam closed to the roof. The methane emission roadway is parallel to the return airway. In general, the distance between the return airway and the additional roadway is 10-12m. In the process of coal mining and caving, part of air is forced to enter the methane roadway through the cracks in top-coal and forms an return airflow path. This type of ventilation is called E-type ventilation.

The following example shows the effects of E-type ventilation in No.5111 coal face of the Wangzhuang Coal Mine in the Luan Coal Mining Administration.

From 11 to 18 March 1995, the methane concentration was monitored at the measurement station (The distance ranges from 250m to 280m away from the coal face) in the methane emission roadway in the Wangzhuang Coal Mine. The maximum concentration was 1.24% and the minimum was 0.64%, the average was 0.98%. Moreover, the methane distribution in the coal mining face was also monitored. 2394 measurement data show that the methane concentration in the area closed to the coal wall was minimum and the methane concentration in the area closed to the coal caving windows of supports was maximum (as shown in Fig.1).

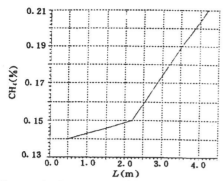

Figure 1 The distribution of methane ratio of coal face in the direction of coal face advancing

Four measurement stations are installed in the return airway in No.5111 coal face whose distance ranges from 10m to 270m away from the coal face. The measurement results show that the average maximum concentration of methane in 4 measurement stations was 0.36%, while the average minimum was 0.28% and the average was 0.33%. On the basis of measurement results, the absolute methane emission and the relative methane emission can be estimated to be $4.2m^3/min$ and $2.72m^3/t$ respectively. The absolute methane emission in the methane emission roadway is estimated to be $1.3m^3/min$.

Therefore, based on the data mentioned above, it can be estimated that the total absolute methane emission and relative methane emission are $5.5m^3/min$ and $3.56m^3/t$ respectively.

The measurement data show that the average ratio of methane at the upper corner of the goaf was 0.89%, the maximum was 0.95% and the minimum was 0.63. It is indicated that the problem of the local accumulation of methane at the upper corner of the goaf can be solved successfully by using the E-type ventilation.

2 PHYSICAL MODELS AND ANALYSIS OF THE U-TYPE AND E-TYPE VENTILATION

At the present research, the physical models are employed to investigate the distribution of methane in the fully mechanized top-coal caving face of U-type and E-type ventilation. The physical models are composed according to the in-situ conditions of No. 5111 coal face. Four layers of the measurement points are installed in the goaf. The vertical distance of first layer is 19mm away from the floor of coal seam. The spacing of other layers is 65mm. On the basis of measurement data of each layer, the contour of methane concentration can be given.

2.1 Distribution of methane concentration of the U-type ventilation in coal face

In the process of experiment, the methane is injected into the model from different points of the model. The basic parameters of experiment are given by: injected volume 1500L/h, area of air intake gateway $1.5945 \times 10^{-3} m^2$, velocity of air 14.35m/s, rate of flow $2.288 \times 10^{-2} m^3/s$, area of air return roadway $1.6 \times 10^{-3} m^2$, velocity of air 16.84m/s, and rate of flow $2.6944 \times 10^{-3} m^3/s$. Fig.2 shows the distribution and the contour of methane concentration of the second layer.

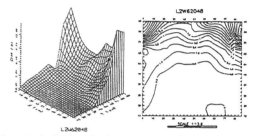

Figure 2 Distribution of methane on the second layer

2.2 Distribution of methane ratio of E-type ventilation in coal face

In order to investigate the effects of the position of the methane emission roadway on the distribution of methane in the coal face, the methane emission roadway is excavated in different position of the top coal closed to the roof of physical models. The present result shows the distribution of methane in coal face with the methane emission roadway in the place No.3 which distance is 42m distant from return airflow in-situ. The basic parameters of the experiment are shown by: injected volume 1000L/h, area of air intake roadway $1.5945 \times 10^{-3} m^2$, velocity of air 15.65m/s, rate of flow $2.495 \times 10^{-2} m^3/s$, area of air return roadway $0.016m^2$, velocity of air 14.52m/s, rate of flow $0.02322m^3/s$, area of the methane emission roadway $7.068 \times 10^{-4} m^2$, velocity of air 10.77m/s, and rate of flow $7.61 \times 10^{-3} m^3/s$.

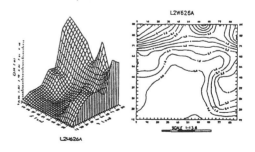

Figure3 Distribution of methane on the second layer

According to the experimental data, the distribution and the contour of methane concentration of the second layer in coal face with a methane emission roadway are shown in Fig.3.

2.3 Comparison between U-type and E-type ventilation

Based on the experimental data, the characteristics of spatial distribution of methane and air velocity in the coal face with the U-type ventilation can be understood. The characteristics are concluded by the following six aspects. (1) the distribution of air velocity in the area closed to the coal wall: The air velocities are relatively high at the ends of the air intake roadway and the air return roadway, while in the middle of coal face, they are relatively low. (2) the distribution of methane concentration in the area closed to coal wall: The methane concentration increases gradually from the end of air intake roadway to the end of air return roadway. (3) the distribution of air velocity in the passageway of supports: The velocity of air is maximum at the end of air intake roadway, minimum in the middle of coal face, and medium at the end of air return roadway. (4) the distribution of methane concentration in the passageway of supports: The methane concentration increases gently from the end of air intake roadway to the place where is 87m distant from the end of air intake roadway, while it increases rapidly from the place where is 87m from the end of the air intake roadway to the air return roadway. (5) the distribution of air velocity in the area closed to the top-coal caving windows of supports: The air velocity increases gently from the end of air intake roadway to the center of the coal face, and it increases rapidly from the center of the coal face to the end of the air return roadway. (6) the distribution of methane concentration in the area closed to the top-coal caving windows of supports: The methane concentration almost doesn't change from the end of air intake roadway to the center of coal face, however, the methane concentration increases rapidly from the center of the coal face to the end of the return airway. The average methane concentration at the place, where is 6m distant from the end of the air return roadway, is 1.05%.

The characteristics of the spatial distribution of methane and air velocity in the coal face with E-type ventilation can be expressed by the following six aspects. (1) the distribution of air velocity in the area closed to the coal wall: The air velocity fluctuates with a small amplitude along the coal face. (2) the distribution of methane concentration in the area closed to coal wall: The methane concentration increases gradually from the end of air intake roadway to the end of air return roadway. However, the maximum and minimum methane concentration are smaller than those of U-type ventilation. (3) the distribution of air velocity in the passageway of supports: The velocity of air decreases gradually along the coal face. (4) the distribution of methane concentration in the passageway of supports: The methane concentration increases gently from the end of the intake airway to the end of the return airway. But the maximum methane concentration is 50% smaller than that of U-type ventilation, while the minimum is the same as that of U-type ventilation. (5) the distribution of air velocity in the area closed to the top-coal caving windows of supports: The air velocities at the end of air intake/return roadway are larger than that at the central part of coal face. And the velocity at the end of return airway is larger than that at the end of intake airway. (6) the distribution of methane concentration in the area closed to the top-coal caving windows of supports: The methane concentration increases gradually from the end of intake airway to the end of return airway. Compared with U-type ventilation, the maximum methane concentration decreases by 60%.

The experimental results show that the leakage of air is related to the flow rate of air and the position of the methane emission roadway. In the coal face of U-type ventilation, the more the air is emitted from the goaf, the greater the methane concentration is at the upper corner of the goaf. The opposite situation occurs in the coal face of E-type ventilation. The more the air is emitted from the goaf, the smaller the methane concentration is at the upper corner of goaf.

When the location of the methane emission roadway is determined, the methane concentration at the upper corner of the goaf can be controlled by adjusting airflow rate. The methane concentration at the upper corner of the goaf decreases if the flow rate of air increases, while the methane concentration at the upper corner of the goaf increases if the flow rate of air decreases. It has been shown that the problem of the local accumulation of methane at the upper corner of the goaf can be solved successfully by using the E-type ventilation.

The in-situ research and experimental results indicate that the rational location of the methane emission roadway is 4-23m distant from the air return roadway. In other words, the ratio of the distance of the methane emission roadway from the return airway to the length of the coal face should be in the ratio of 1:7.74-1:44.5. The flow rate of air in the methane emission roadway and the flow rate in the return airway should be in the ratio of 1:5-1:8.

In addition, it should be pointed out that the E-type ventilation is suited to the fully mechanized top-coal caving face in which the absolute methane emission ranges from 5 to 15m^3/min, and the local methane concentration at the coal caving windows of supports and the upper corner of the goaf going beyond the limit.

Table 1 Methane ratio and flow rate in No.1408 fully mechanized top coal caving face of Zhangcun Mine

| Date | In air return roadway | | In methane emission roadway | | Methane ratio at the upper corner of the goaf (%) |
	flow rate(m^3/min)	methane ratio(%)	flow rate (m^3/min)	methane ratio(%)	
92.12.06	563	0.29	179	0.30	0.43
92.12.14	458	0.22	228	0.68	0.28
92.12.25	440	0.30	210	0.62	0.18
93.01.09	475	0.34	227	1.35	0.34
93.01.14	531	0.29	163	1.26	0.37

3   AN EXAMPLE FOR USING THE E-TYPE VENTIATION

The local accumulation of the methane at the upper corner of the goaf often occurs in the fully mechanized top-coal caving face No. 1408 of the Zhangcun Coal Mine. In order to solve this problem, the airflow rate is increased. In the initial stage of excavation of coal face No. 1408, the airflow rate is 470m^3/min, but the methane concentration often goes beyond 1%, and the maximum accounts for 4%. The methane concentration in the return airway is about 0.35%, the maximum accounts for 0.6%. With the increasing of the methane emitted from the goaf, the airflow rate also increases. However, when the flow rate increases to 650-750m^3/min, the methane concentration in the return airway is still as high as 0.6-0.7%, and the methane concentration at the end of the return airway accounts for 2% sometimes. Therefore, the methane can not be diluted effectively by increasing the airflow rate.

In order to solve this problem, the E-type ventilation is practiced in coal face No.1408 of Zhangcun Coal Mine. The methane emission roadway is 12m distant from the air return roadway.

After the E-type ventilation is practiced in the coal face No.1408, the methane concentration at the upper corner of the goaf is smaller than 0.5%. And the methane concentration in the air return roadway is smaller than 0.4%(as shown in Fig.1). It has been indicated that the E-type ventilation is an effective way to solve the problem of the local accumulation of methane.

In addition, the average methane concentration between the supports decreases from 1.124% to 0.26%, while the average concentration at the roof of supports decreases from 2.3% to 0.54%. To sum up, the advantages of E-type ventilation are (1) the airflow resistance can be decreased by adding the methane emission roadway; (2) the methane in the coal face and the methane in the goaf can be controlled by using the E-type ventilation. Compared to the U-type ventilation, the E-type ventilation can changes the flow direction of methane, $i.e.$, the methane emitted form the goaf can not enter into the coal face. Therefore, the local accumulation of methane at the upper corner of the goaf, between the supports and at the roof of supports can be controlled.

4 CONCLUSIONS

The experimental results and the coal mining practice in the Wangzhuang Coal Mine and the Zhangcun Coal Mine have shown that the E-type ventilation can be used successfully to solve the problem of the local accumulation of methane at the upper corner of the goaf. At present, the E-type ventilation has been widely used in the Luan Coal Mine Administration.

Mining Science and Technology' 99, Xie & Golosinski (eds) © 1999 Balkema, Rotterdam, ISBN 90 5809 067 1

Study on the remote control system for mine doors by using computer

Deming Wang, Yuejun Wang & Jianzhi Fang
China University of Mining and Technology, XuZhou, People's Republic of China

Qingguo Bao, Shaoju Yin & Benxu Xie
Chaili Coal Mine of Zaozhuang Mining Bureau, Shandong, People's Republic of China

ABSTRACT: The control system for the local airflow reversing or short circuit established in a coal mine, will plays an important role for fire fighting and the reduction of fire losses. The remote control system for mine doors adopts a telephone network to carry the control signals and control the doors installed underground. The mine door controller is developed by a single-chip microcomputer. It not only can do the logical analysis and judgement, but also has a strong anti-disturb ability and sufficient safety reliability. The system is successfully applied in a coal mine in China and the paper presents the case study.

1 FOREWARD

The exogenous mine fire seriously threatens the mine safety production with its bursting and destruction. While a mine fire happens, if not tackling timely and properly, it will claim a heavy loss of life and property. As the space in the mine is restricted, a great deal smoke generated by combustion will rapidly spread to other roadways and working faces. A system for local airflow reversing or short-circuit of should be established for meeting the need of fire fighting, because the system can not only control the direction of the smoke and the ranges of the contaminated area, but also can create the safe conditions for fire fighting and the evacuation of miners. Accordingly, an airflow control system is very important for fire fighting and reduction of the fire losses.

At present, the most doors used in the airflow control system in mines are manual doors, so that it is necessary to appoint special persons to open or close the doors during a mine fire. However, the operation of manual doors is generally very difficult and danger, because the distance may be very long and need a long time to reach the location of the doors, or the temperature of smoke in that area may be too high to access for miners. Sometimes the fire control may be failure and causes a big disaster. So the manual doors can not meet the need of the rapid

rescue action during a mine fire. To enhance the ability of disaster-resistant in a mine and meet the need of fire fighting by using modern technology, the authors have developed a new remote control system for mine doors by using computer.

The system adopts a telephone network which carries the control signals to control the doors installed underground. The mine door controller is developed by a Single Chip Microcomputer (SCM). It not only do the logical analysis and judgement, but also has a strong anti-disturb ability and sufficient reliability. Telephone lines provide communications, so that it is not necessary to lay the special communication cables and can reduce the costs in the system installation.

2. THE COMPOSITION OF THE SYSTEM AND ITS WORK PRINCIPLE

The remote control system is shown in Fig. 1. It consists of the main control system (computer), telephone network, the mine door controller, the electronic motor and doors.

In order to change the digital signals into the simulating signals for remote distance transmission, the digital signals should be modulated into the the simulating signals and be transported to the controller in the underground by telephone network. Therefore, a modem must be installed in the serial interface of the main control system on the surface. The mine door controller in the underground is

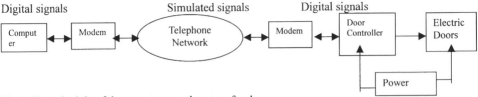

Fig.1. The principle of the remote control system for doors

based on a SCM and can analyze and detect to the signals from a remote distance. The control signals for switching the mine doors can be sent to the electrical-motors. As the signals, which are sent to the SCM, should be the digital ones, the control simulating signals, which are taken out from the remote distance (the surface), must be restored to digital signals again. The modem of the door controller is installed in an explosion-proof control box in the underground. The digital signals modulated by the modem are processed by the SCM in the local station in underground and are sent out to control the electrical-motors which implements the switching of the mine doors.

3 DESIGN AND MANUFACTURE OF THE SYSTEM

3.1 Software design of the control system
The software of the main control system in the surface is written by language C++. The program has the function of communication with the door controller. In the software, the keys of the program design are the initiation of the modem and the serial interface, the connection of the computer and modem, the sending or receiving of the messages after receiving the signal of the successful connection of the computer and the modem. The flow chart of the signal communication between computer and modem is shown in Fig.2.

3.2 The development of door controller
The SCM shares the advantages such as high liability, easily enhanced functions and powerful control actions. It can be widely used in the industry monitoring and controlling. The door controller of the system adopts the 80C552 of the SCM, which communicates with the main computer on the surface by a serial interface and controls the switches of electric-motor door by P4 interface.
The connecting block diagram of the door controller is shown in Fig. 3. The door controller is used as a terminal of the telephone network and is allocated a telephone number. In order to avoid receiving a

wrong-dialing and other disturbing pulse, the full security and reliability have been strengthened to ensure the door to act properly. A password is set in the controlling software of the main computer on the surface. Only after a right password is input, one can enter into the controlling software. The commands, which are sent out by the computer on the surface to the underground controller, have special data formats. At first, the command of the door beforehand acting must be sent out, then the command of the door acting. The door controller does some judgements to the signals received. After receiving the command of the door beforehand acting, it cancels the safeguard and performs the functions according to the door acting command.

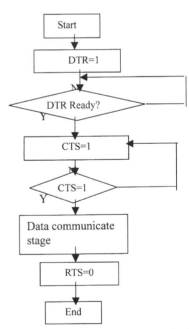

Fig.2　Flow Chart of the Signal Communication between the computer and modem

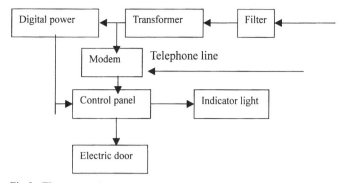

Fig.3   The connecting block diagram of the door controller

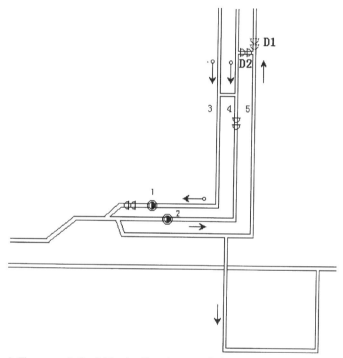

1. The upcast shaft.   2. The Auxiliary downcast shaft   3.The out-take raise roadway of the southern wing
4. The outtake raise roadway   5. The intake raise roadway

Fig4. the airflow control lay-out of the southern district in the second Level of Chaili coal mine

3.3 *Electric-motor operating door system*
In the design, the MZF-2 type the infrared automatic door, which is in the type of flameproof and intrinsically safe, is used as the power mechanism of the system. The door control panel is accordingly modified as follows:
(1) Canceling the automatic control of the infrared sensors and changing into receiving the switch control of the underground controller switches.

(2)Opening the door while receiving the high electricity level and closing the door while receiving the low electricity level.

4 APPLICATION

Chaili Coal Mine in China has driven a pair of rise roadway (No. 4, 5 as shown in Fig.4) in the southern district of the second level in order to solve airflow rate deficiency. Thus part of the airflow in the southern district of the second level comes from the inlet shaft of the adjacent mine. Two mines shares a common part of intake airway, namely the auxiliary shaft, the horizon inset, the main intake airway. If fire breaks out in the common intake part of the auxiliary shaft in the adjacent mine, smoke will spread into the southern district of the second level in Chaili Coal Mine through the intake raise roadway. The whole working area will be directly invaded by the smoke, which endangers greatly the mine safety. Therefore, it is much more important to establish the fire smoke control system.

One pair of the remote control automatic doors (usually closed), named as D2, is installed in the transverse between the intake raise roadway and the outtake raise roadway, and another pair of the same doors (usually opened), named as D1, is installed in the overall intake. Once fire breaks out in the adjacent mine, the doors of D2 will be opened and the doors D1 will be closed immediately to shortcut smoke by operating the computer in the control center of Chaili Coal Mine. Thus the smoke will flow out to the surface through the air return shaft , which avoids the smoke flowing into the southern district of the second level in the mine. Therefore the region directly invaded by smoke and the loss caused by the fire will be effectively lessened.

The remote control system for doors by using computer has been working perfectly since it was installed in the Chaili Coal Mine in September, 1997. It has been shown that the system is of great importance in safeguarding the mine operation.

Mining Science and Technology' 99, Xie & Golosinski (eds) © 1999 Balkema, Rotterdam, ISBN 90 5809 067 1

Study on optimization of mine ventilation network

B. Li, K. Uchino & M. Inoue
Department of Earth Resources Engineering, Kyushu University, Japan

ABSTRACT: This paper considers that the most economical ventilation network control is obtained by the least change of resistance, whether increasing resistance or reducing resistance performs the control. Based on this idea, a new method for the optimization of mine ventilation network is presented and its usefulness is illustrated by an actual example from the cost of electric power and the cost of construction. Then, it is discussed to judge that the control of ventilation network is optimal in what way, and a judgement method is given from the economical standpoint.

1 INTRODUCTION

In general the optimization of ventilation network signifies either that of consumption energy or that of total cost. This paper describes a new method for mine ventilation network analysis from the standpoint of total cost. In most of other studies on this problem fictitious model mine are used to verify their theories, which sometimes leads to the lark of persuasiveness. In this paper, therefore, the method proposed by the author is applied to an actual coal mine in China and the usefulness of this method is illustrated. Also, the optimal control of ventilation network is discussed from economical standpoint as well.

2 PRINCIPLE AND MATHEMATICAL MODEL

The optimization of ventilation network must be the optimization of total cost. The total cost includes the electric power cost for ventilation and the construction cost to control the airflow. Based on this idea, various factors which affect the total cost to control airflow distribution are discussed and it is concluded that the most economical ventilation control is obtained when the change of resistance is the least, whether the control is performed by increasing resistance or reducing resistance. Namely, the optimization of ventilation network is a problem that calculates the least change of resistance.

However, if we take the quantity of change of the resistance as the objective function of optimization, the calculation of the mathematical model is very difficult. Generally, as the smaller the change in air quantity, the smaller the resistance change in the ventilation network, the mathematical model of mine ventilation network optimization can be expressed as

$$\text{minimize} \quad J = \sum_{j=1}^{n}(qb_j - q_j)^2 \tag{1}$$

$$\text{subject to} \quad \sum_{j=1}^{n}a_{ij}(r_j + s_j)q_j^2 - hf_j - hn_j = 0 \tag{2}$$

$$(i = 1, 2, \cdots, m)$$

$$Lq_j \le q_j \le Uq_j \tag{3}$$

$$Ls_j \le s_j \le Us_j \tag{4}$$

$$Lh_j \le hf_j \le Uh_j \tag{5}$$

where n is the number of airways; j the airway number; r_j the resistance factor; s_j the change quantity of resistance; q_j the air quantity; a_{ij} the element of the fundamental mesh matrix; i the mesh number; m the number of fundamental meshes; hf the fan characteristic curve; hn the natural ventilation pressure; Lq_j, Ls_j and Lh_j the lower limits of air quantity, the change quantity of resistance and fan pressure; Uq_j, Us_j and Uh_j the upper limits of the variables respectively.

The computer program has been written to calculate the model.

3 AN EXAMPLE OF COAL MINE

3.1 *An outline of the mine*

There is a coal mine producing 1.5million tons annually. The ventilation network of the mine is shown in Figure1 and equipped with a main fan and fifteen regulators. Eighteen airways need to be ventilated in certain required air quantities. The specifications of the main fan and related data are as follows: the diameter is 2.8m, blade angle 45 degrees, revolutions 580rpm, efficiency 0.67, air quantity 130m³/s and fan pressure 1476Pa. Table1 and 2 show actually provided air quantities and required air quantities at working places and measured resistance of all airways.

In the mine, the air temperature of the long wall face (airway 8-10) is about 30℃, and actual supplied quantity (10.3m³) is insufficient. The measures for improving the mine ventilation have been planned to supply enough air quantity to the coal working face 8-10 and the heading face (airway 9-11). For this purpose it is planned to excavate two new airways. One of them is an inclined airway 32-A-33 (Fig.1. A broken line,

length is about 700m) and the other a horizontal airway 37-B-12 (length is about 300m).

Table1 Actual air quantity distribution and required air quantity

Airway Number	Actual Air Quantity m³/s	Required Air Quantity	
		Lower Limit	Upper Limit
8-10	10.81	12	15
9-11	8.44	6	8
19-12	1.88	2.5	5
20-13	9.77	5	7.5
20-13	2.90	4	6
18-15	9.23	10	13
18-14	5.34	4	6
5-16	5.18	2.5	5
21-27	1.86	2.5	5
22-27	21.82	2.5	5
23-31	4.83	5	7.5
23-28	1.52	4	6
31-29	8.09	10	13
24-29	5.80	4	6
25-30	16.42	5	7.5
25-6	8.25	4	6

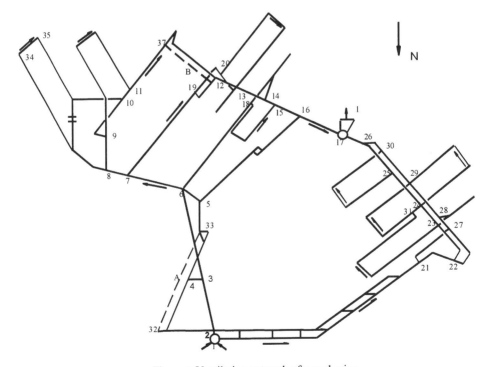

Figure 1 Ventilation network of a coal mine

Table2 Measured resistance of the ventilation network

Airway Number	Resistance of Airway	Resistance of Regulator	Airway Number	Resistance of Airway	Resistance of Regulator	Airway Number	Resistance of Airway	Resistance of Regulator
17-1	0.0140		2-3	0.1675		6-18	0.2134	
8-10	1.11		2-4	0.1568		7-19	0.4504	
9-11	0.533	1.105	4-3	1.0573		19-20	0.0293	
19-12	0.6234	41.018	4-5	0.1047		2-21	0.0547	
20-13	1.0653	1.0303	3-6	0.1078		2-22	0.0323	
20-13	0.5223	23.1017	5-6	0.0189		22-23	0.0279	
18-15	1.0567	4.7645	6-7	0.0459		23-24	0.0338	
18-14	0.4231	13.6270	7-8	0.0032		24-25	0.0373	
5-16	0.1056	21.7940	8-9	0.0464		27-28	0.0155	
21-27	77.0448		9-10	0.0817	0.7303	28-29	0.0533	
22-27	0.1066	0.1872	10-11	0.0117		29-30	0.0515	
23-31	1.5233	0.6299	11-12	0.1106		30-26	0.0479	
23-28	0.2305	43.0955	12-13	0.0517		26-17	0.0538	
31-29	1.0270		13-14	0.0248		24-31	0.0567	0.4728
24-29	0.8053	2.000	14-15	0.037		1-2	0.0141	
25-30	0.5734		15-16	0.0091				
25-26	0.0325	4.4962	16-17	0.0316				

* Resistance: Ns^2/m^8

3.2 Results of the optimization

On the condition that the resistance of airways 2-4 and 11-12 can be reduced, and other airways are can be increased, the optimization of this ventilation network was performed by the proposed method. In the calculation, the following conditions were also considered.

1. Because the air temperatures of working face 8-10 and heading face 9-11 are high, the upper limits of 15 and 8m³/s shown in Table1 are chosen as required air quantity respectively to cool down the airways.

2. Required air quantities of all airways are increased by 10% of the quantities given in Table 1 to consider the air leakage.

3. According to the actual measurement, the natural ventilation pressure of the mine is 100Pa, and with the same direction as that created by the main fan. To guarantee that the main fan can supply enough air quantity to the ventilation network even if the natural ventilation pressure is smaller than 100Pa, the natural ventilation pressure has not been considered in this calculation.

As the results of the optimization, the airflow distribution and the resistance changes in all airways are shown in Table3. The air quantity of main fan is 91.3m³/s, and the pressure is 1164Pa. According to the results of calculation, the characteristic curve of main fan has to be adjusted so that the revolution is reduced from 580rpm to 396rpm and the blade angle from 45degrees to 30degrees. In this case, the efficiency of main fan is 0.848.

3.3 Comparison of the actual data and the results of optimization

1. *Distribution of airflow.* In the actual data of the mine ventilation, the air quantity of main fan is 130m³/s; the air quantity of eight airways is smaller, and airways 9-10, 22-27 and 25-30 are larger than required, respectively, as shown in Table1. However, according to the results of optimization shown in Table3, the air quantity of the main fan is 91.3m³/s, and the air quantities of all working places have converged which the range given in Table1. Especially, the air quantity of working place 8-10 is increased to 16.5m³/s and is larger than the actual by 5.7m³/s. Meanwhile, the air quantity in the airway 9-10 is decreased from the actual 11.7m³/s to 0.001m³/s and that in the airway 22-27 is decreased from 21.8m³/s to 2.8m³/s. Namely, the two regulators are quite important to raise the ventilation efficiency. The results of calculation clearly show that there is no working place any more where the air quantity is insufficient or excessive.

2. *Excavating new airways.* The mine has planned to excavate two new airways 32-A-33 and 37-B-12 to supply more air to the working places 8-10 and 9-11. However, according to the results of optimization, the resistances of airways 2-4 and 11-12 do not have to be changed.

To discuss the economical effects of excavation of new parallel airways, the optimization of the ventilation network was executed again, on the assumption that the airways 32-A-33 and 37-B-12 with the resistance is 0.1464 and 0.0639N·s²/m⁸, respectively. The results of this calculation show that the air quantity of main fan is 91.3m³/s and the pressure 1123.5Pa. Compare these results with those shown in Table3, the pressure of the main fan is reduced by only 40.5Pa and the electric power consumption reduced by only 3.7Kw. As shown in Table3, even if the two new airways are not excavated, the main fan is able to supply enough air to the working places 8-10 and 9-11. Moreover a large amount of money is needed to excavate them. Therefore, it is unnecessary to excavate the new airways 32-A-33 and 37-B-12.

3. *Electric power consumption.* In the actual ventilation, the air quantity of main fan is 130m³/s and pressure 1476Pa and efficiency 0.617. If we assume the efficiency of electric motor is 0.91 and transmission efficiency between the main fan and the electric motor is 0.95, the electric power consumption for the ventilation is 3.15×10^6kw·h/year. If the control shown in Table3 is performed, the electric power consumption for the ventilation is 1.27×10^6kw·h/year and is about 40% of the actual ventilation.

Table 3 Result of the optimization

Airway Number	Resistance of Regulator Ns²/m⁸	Air Quantity m³/s
8-10		16.498
9-11	3.362	8.801
19-12	39.842	2.750
20-13	10.287	5.503
20-13	17.216	4.402
18-15	3.373	11.004
18-14	23.843	4.401
5-16	82.419	2.75
21-27	93.724	2.751
22-27	89.265	2.751
23-31	16.077	5.504
23-28	33.443	4.402
31-29		11.005
24-29	32.454	4.402
25-30	21.675	5.503
25-26	37.164	4.403
9-10	2.9×10^8	0.001
24-31	17.131	5.500
Air quantity of main fan		91.3 m³/s
Pressure of main fan		1164Pa

4 VERIFICATION OF OPTIMAL CONTROL OF MINE VENTILATION NETWORK

Verification of the optimum of ventilation network control is important not only for the optimization of ventilation network but also for the usual management of the ventilation. A method to judge whether or not the optimal control of ventilation network is obtained from the economical point of view is discussed.

4.1 *Adjustment of resistance*

Firstly, network control by increasing resistance is discussed.

A chain of airways from the intake airshaft to the return airshaft is called a path. Clearly, the pressure drop along the path equals the pressure created by the main fan. For convenience of discussion, it is assumed that Q is the optimal airflow distribution and S is not the optimal change in quantity of resistance. Here, $Q = (q_1, q_2, \cdots, q_n)$ represents the airflow distribution and $S = (s_1, s_2, \cdots, s_n)$ the adjustment value of resistance. Because S is not the optimal value of resistance, it is possible to reduce S further. When the airflow distribution is kept invariable, if S is reduced, the pressure of main fan decreases. Because the main fan and the path have the same pressure, when the pressure of main fan is reduced, the pressure of all paths also decreases by the same quantity. Namely, it means that all path have airways where S can be reduced further. If it is impossible to reduce S further in a path, the pressure of main fan cannot be reduced, either and S is optimal. Clearly, when s_j contained in a path is zero, it is impossible to decrease S any further. Therefore, if S is optimal, it has to satisfy the condition as follows: there exist some paths where the resistance changes for adjustment of all airways contained in the path is zero in the ventilation network.

Next, the case in which network control is performed by reducing resistance is discussed.

As described above, the most economical control is obtained by the least change of resistance, whether resistance is increased or reduced to perform the control. Therefore, it can be similarly demonstrated that the same conclusions as in the case of increasing resistance can be obtained, in this case as well.

A ventilation network where the multiple fans are operated can be considered as the multiple sub-ventilation networks. The aforementioned conclusions can be applied to each sub-ventilation networks as well.

4.2 *Airflow distribution*

If a set of airways can be cut out the network is divided into two parts, the set is called a cut set of the ventilation network. Here, a cut set containing all the airways where specific air quantities are needed is defined as an objective cut set (objective cut set is abbreviated as OCS). The airflow distribution of ventilation network is determined by the air quantities of OCS. Therefore, if the air quantities of OCS are optimal, also the airflow distribution of ventilation network is optimal. If the resistances of airways contained in OCS can be changed, it is possible to minimize their air quantities by changing respective resistance. Obviously, the minimum air quantity are Lq_j respectively (Lq_j is the lower limit of required air quantity). If there are multiple airways in OCS, the resistances of which cannot be changed, it is possible to minimize their air quantities by adjusting the main fan. As the air quantity of the main fan is adjusted, the air quantities of all airways in the network change by the same ratio. Therefore, when the air quantity of any airway is equal to Lq_j, the air quantity of main fan becomes the least.

As discussed above, if the airflow distribution of the ventilation network is optimal, it has to satisfy the conditions as follows: in airways contained in OCS, if its resistance can be changed, its air quantity is equal to its Lq_j. If there are multiple airways in OCS, where the resistance cannot be changed, one of the airways at least is equal to its Lq_j.

4.3 Number and of airways where the resistance is changed

According to the basic equations of the airflow control

$$\sum_{k=1}^{n} a_{ij}(r_j + s_j)q_j^2 - hf_j - hn_j = 0 \qquad (6)$$

when every mesh has only one airway where the resistance is able to change, S has only one solution. Moreover, if the control of main fan and the meshes where the resistances do not need to change are considered, the number of airways where the resistances can be changed, can be expressed as

$$R_e \le m - f \qquad (7)$$

where, R_e is the number of airways where the resistances are changed.

As stated above, if the control of ventilation network is optimal, there are some paths where the resistances are not changed in the network. Because the number of the paths is equal to that of meshes, the same conclusions with equation (7) also can be obtained from the discussion in 4.1.

This proposed method is applied to the verification of the optimization results of the ventilation network as shown in Figure1. From the calculation results shown in Table3, the following conclusions are drawn.

1. The path 1-2-3-6-7-8-10-11-12-13-14-15-16-17-Fan does not contain airways where the resistances are changed.

2. The air quantities of all airways shown in Table1 are equal to respective Lq_j.

3. The number of airways where the resistances have been changed is smaller than that of the fundamental meshes.

Therefore, we can say that the calculation results of the actual example shown in Figure1 are optimal control.

5 CONCLUSIONS

In this paper a new method for optimization of mine ventilation was proposed and verified by applying it to an actual problem in a Chinese coal mine. This method of optimization can be utilized for technical guideline for rational design of mine ventilation and for the optimization of ventilation system in actual mines in terms of both technology and economy. Finally, a method for verification of the optimization of ventilation network is proposed.

REFERENCES

X.Wu, E.Topuz & M.Karfakis 1992, Optimization of Ventilation Control DeviceLocations and Sizes in Underground Mine Ventilation System. Proceedings of 5th US Mine Ventilation Symposium: 391-399.

Changhong Huang & Y.J.Wang 1993, Mine Ventilation Network Optimization Using the Generalized Reduced Gradient Method. Proceedings of 6th US Mine Ventilation Symposium: 153-161.

Bingrui Li, K. UCHINO & M. INOUE 1995, Optimization of Ventilation Network by the Control of Resistance. Journal of the Mining and Materials Processing Institute of Japan, 11(12): 829-834.

Bingrui Li, M. INOUE & K. UCHINO 1997, Studies of a Mine Ventilation System Optimization Based on Actual Mine Data. Journal of the Mining and Materials Processing Institute of Japan, 113(9): 677-682.

Bingrui Li, K. UCHINO & and M. INOUE 1998. Optimization of mine ventilation network on the idea of total cost. Proceedings of the International Mining Tech, 98 Symposium: 348-354. Chongqing, China.

Mining Science and Technology' 99, Xie & Golosinski (eds) © 1999 Balkema, Rotterdam, ISBN 90 5809 067 1

Coal mine gob area methane flow control

N.O. Kaledina
Moscow State Mining University, Russia

ABSTRACT: The aim of the given work is the increase of safety of underground coal mining at simultaneous decrease of atmosphere methane pollution by way the expense of extraction and utilising of conditional methane on the base of using an optimum ventilation regime. The criterion is offered for the methane control, which may be used at any control method and also takes into account the development of new technologies of methane utilisation.

The most widespread methane control methods are ventilation and degasification. The main purpose of these methods usually was struggle with methane. But methane has a value as a fuel and chemical recourse. So most methane volumes have to be output by degassing and utilized.

The main quantity of gas is thrown out from mines in atmosphere with a ventilation flow, but in many countries up to 30 % of general issue makes methane from degasification systems and some of this volume of gas is used for industrial purposes. In coal-produced countries on the average of 20 % is given by a coal cycle, in which the maximum contribution is given by underground production - about 86.

With increasing of depth the total methane emission of mines and contribution of gobs in their gas balance are growing. The increasing of intensity of ventilation so as degasification increases methane issue into earth atmosphere, if methane is not utilised. Methane utilising is possible only at maintenance of stable outputs and high concentration of gas in air-gas mixture, which are reached only at combined methods of control, including gob area degassing control.

Efficiency of gob area degassing much depends on an aerodynamic regime or on intensity of leakage: the increase of methane volume in ventilation flow reduces gas volume and concentration in degassing holes. The regime of air leakage plays also major role in process of spontaneous combustion in gob areas, since the leakage velocity defines a temperature regime in a gob. Thus increased gas emitting reduces concentration of oxygen $_2$, that is necessary for coal oxidation. Thus, there is the certain contradiction between measures of struggle with gas and with endogenous fires.

Therefore the gob areas gas monitoring and control by aerodynamic and gasdynamic (degassing) methods gets the major meaning as from the point of view of mining safety, so as of environment.

Gas emitting from the gob areas in coal mine represents extremely complex process of active gas diffusion in air leakage through porous matter. This process is defined by a set of the diverse factors and only approximately can be described by a system of differential equations. Their decisions by both of an analytical and numerical methods does not give accuracy acceptable to practice because of complexity of the description boundary conditions, distribution of intensity of gas emitting sources and structural characteristics of filtering environment.

So it is expediently to proceed from complex multiparametrical models to dynamic models of integrated character, using the most essential interrelations between major factors and parameters of described processes. It simplifies the description of the complex phenomena without loss of necessary accuracy, reduces volume of the analysed information and facilitates acceptance of the control decisions.

In Moscow State Mining University by the physical model there were investigated as stationary, so as non-stationary gasdynamic processes, for different types of ventilation scheme, as under degassing of gob areas, so as at its absence.

For the description of laws of air leakage in gob area it is most expediently to use the two-partial law of resistance. In such form the aerodynamic resistance coefficients are depending on the gob area pore characteristics. Variation of the coefficients are described by empirical relations, containing coefficients which are not depending on ventilation regime, but depending on character of strata collapse in gob areas [Puchkov L.A.,1993].

As it was shown by analysis of a ratio of appropriate forces in real conditions, having place in mines, the main criteria of similarity under physical modelling is Reynolds number, what is applied to pore environments, can be described by following:

$$Re = \frac{uk}{vl}. \tag{1}$$

where u – air velocity, v - dynamic viscosity coefficient,; k, l – coefficients of air-penetrance and unevenness, depending on the type of surrounded rocks and the working face advance rate.

On the base of the received descriptions of Reynolds number fields for main types of ventilation schemes an average-integrated Reynolds number (Re*) is accepted. It is a complex integrated parameter, taking into account all major factors, determining specific character of gob area aerodynamics:

$$Re^* = \frac{1}{(X - x_0) \cdot L} \int_{x_0}^{X} \int_{0}^{L} Re(x, y) \, dxdy , \tag{2}$$

where X, x_0, L – geometrical dimensions of gob area; x, y – co-ordinates.

The ventilation regimes, determining methane distribution in system "workings - gob area - degassing holes", is characterised by integrated Reynolds criterion, which can be determined on data of mine atmosphere monitoring on the base of established laws of leakage-flow through a waste zone.

It is established, that the formation of a methane concentration field is defined, mainly, by intensity of filtering regime, spatial structure of leakage-flow (i.e. ventilation scheme) and intensity of gas-emitting source. For any type of the scheme a zone of high gas concentration has place, which is a main source of gas emitting in the workings. The high concentration zone (HCZ) localisation depends on distribution of emitting source on the length of gob area determined by character of rocks displacement and yielding by coal seam extraction, ventilation regime and scheme. Variation of maximum methane concentration in dependence on integrated Reynolds number (Re*) is described by exponential functions.

Thus ventilation regime and scheme are the main factors, influencing on efficiency of gob area degassing. The relation between methane debit and aerodynamic regime of the gob area can be described by following form:

$$I_d = I_0 \exp(-A Re^{\cdot c}), \tag{3}$$

where A - coefficient, depending on ventilation scheme and geometric sizes of gob.

Researches have shown, that for extractive unit there is some range of ventilation regimes, the most rational from the point of view at combined ventilation and degasification, ensuring a safe gas conditions in the workings and high efficiency of degassing. There will be optimum the least intensive of ventilation regime providing the permissible methane concentration in outletting flow [Puchkov L.A., Kaledina N.O., 1995]:

$$0{,}01 Q_u (Re^\cdot) = I_f + I_{ga} - I_d(Re^\cdot), \tag{4}$$

where I_f - working face gas-emitting, m^3/min; I_{ga} - gob area gas-emitting, m^3/min; I_d - methane, extracted by degassing holes, m^3/min; Q_u - inletting airflow, m^3/min.

The meanings I_f and I_{ga} change in accordance with advance rate and even during a shift with characteristic rhythms. So they have to be counted in dependence on data of the mine atmosphere monitoring on the base of established lows.

So the algorithms of mine degasification systems control based on a principle of supreme of total methane debit by means of optimum distribution of vacuum on holes and pipelines, has to be added by a method of aerodynamic influence on gas-emitting from gob area.

The strategy of control by a ventilation mode should take into account not only gas danger, so as the mine ventilation regime is, as a whole, defined by a set of the diverse factors, reflecting characteristic, individual peculiarity for each object for technology and mining process organisation. The optimum mine ventilation regime has to be chose, as a whole, in view of all set of the limiting factors.

The functioning of coal mine's ventilation systems can be presented by multi-levels hierarchical model of air-gas flow aggregate. The model's levels conform to the technologic hierarchy levels. For every level the criteria of optimal gas control has to supply maximum output of methane by degassing, because of this mode allowed to get methane-air mixture, suitable to utilisation (with high gas concentration). Besides the optimal criterion should provide necessity of mine methane usage, but also opportunity of development of new technologies of its utilisation.

Such criterion can be presented as fallow:

$$\frac{I_d}{I_v} \rightarrow \max \quad , \tag{5}$$

where I_d, I_v – methane flow of degassing and ventilation system.

The coal mining experience and the numerous scientific researches prove a necessity of co-ordinating of gas-emitting control by using of air- and gasdynamic methods, besides, the ventilation and degasification are necessary to be considered as uniform system.

The conclusions

1. The range of optimum mine ventilation regimes is considerably defined by air-gasdynamic processes in gob area of extractive units, the authentic description of which is carried out on the base of established laws of distribution of air-leakage and methane concentration in volume of gob area depending on ventilation regime.
2. Usage of established laws of an aerodynamic regime influence on gob area degasafication parameters lets to increase degasafication efficiency and to supply teleological control of gas flows in ventilation-degasification mine system.
3. The functioning of coal mine's ventilation systems can be presented by multi-levels hierarchical model conforming to the technologic hierarchy of mine. For every level the criterion of optimal gas control has to supply maximum output of methane-air mixture, suitable to utilisation.
4. The criterion, offered for the methane control, is universal, because it may be used at any control method and also takes into account the development of new technologies of methane utilisation.
5. Obtained results may be used by mine ventilation and degasification systems constructing.

REFERENCES:

1. L.A.Puchkov. (1993).Underground mining gob area aerodynamics. Moscow State Mining University, Moscow, Russia.
2. L.A.Puchkov & N.O.Kaledina. (1995). Methane dynamics in coal mine gob area. Moscow State Mining University, Moscow, Russia.

Mining Science and Technology' 99, Xie & Golosinski (eds)© 1999 Balkema, Rotterdam, ISBN 90 5809 067 1

Behavior of airflow and methane at heading faces with auxiliary ventilation system

Shinji Tomita, Masahiro Inoue & Kenichi Uchino
Department of Mining Engineering, Kyushu University, Japan

ABSTRACT: The airflows and methane gas concentrations at a heading face with forcing, exhausting and combined systems of ventilation were examined in a reduced scale model gallery. Methane was emitted at the face. Methane concentration was measured and its accumulation was visualized by Titanium tetrachloride (TiCl₄). The location of the duct end gives significant effects on the airflow. The optimal airflow rate through a forcing and exhausting duct and duct end locations to reduce hazard of methane accumulation were investigated for the combined system of ventilation. Methane accumulation was investigated in another reduced scale model. Water was used instead of air and very fine bubbles generated by electrolysis were employed as tracer in this model. The behavior of bubbles in water is quite similar to that of methane in the air. The airflow and methane accumulation were also examined when a large machine such as a road heading machine is placed near the heading face.

1 INRTODUCTION

Auxiliary ventilation at heading faces is very important in coal mines for controlling gases, dust and heat which are emitted at increasing rates today as a result of enhanced driving rates and deepening working levels. There are several researches about the airflow at heading faces (e.g. Wesely, 1984; Shuttleworth, 1963; Uchino & Inoue, 1997).

However, because of the complexity of airflow in the face area it is difficult to find the ventilation method that is technologically optimum for heading faces of various conditions. From this point of view, the airflow and methane concentration at heading faces with auxiliary ventilation were investigated using a visualization technique by laser light in two reduced scale model galleries.

Firstly, methane accumulations by the use of real methane in the air were investigated. Secondly, they were examined by using water and bubbles generated by electrolysis instead of air and methane.

2 EXPERIMENT ON METHANE ACCUMULATION BY REAL METHANE

Airflow and methane accumulation in a heading face were investigated in a one-fifteenth-scale model gallery with rectangular cross section 40 cm in width

and 20 cm in height. Methane was mixed with TiCl₄ and was emitted at the face. TiCl₄ reacts with the moisture and generates white smoke, by which methane accumulation can be visualized by the aid of light. Methane concentration was also measured.

Fig.1 shows the distribution of methane when there is no flow in the model. Brighter regions in the figure show higher concentration of methane. Methane generated at the face moves up along the face first, then advances along the roof. It is clearly seen that methane layer is formed below the roof.

Fig.1 Distribution of methane at the heading face. (No flow. Methane flow rate: 5 l/min)

Fig.2 (a) and (b) show the distribution of methane when the forcing ventilation system is employed. The distance from the face to the forcing duct end is 50 cm (7.5 m in actual size) and the duct position is in the center of the roof. The figure (a) shows the contour of methane concentration and the figure (b) the visualized concentration of methane on a vertical section near the sidewall. Methane concentration is high near the roof and the floor at the face. A low concentration region is observed between the two high concentration regions. Brighter regions in the visualized picture coincide well with the distribution of methane concentration.

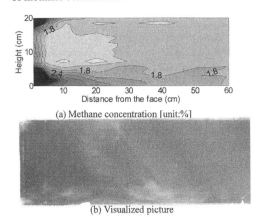

(a) Methane concentration [unit:%]

(a) Forcing velocity is 1 m/s

(b) Forcing velocity is 2 m/s

Fig.3 Distribution of methane at the heading face with forcing ventilation(Vertical section near the sidewall; Methane flow rate: 3 l/min; Duct position: corner of the roof.)

(b) Visualized picture

Fig.2 Distribution of methane at the heading face with forcing ventilation (Vertical section near the sidewall; Flow velocity: 3 m/s; Methane flow rate: 5 l/min; Duct position: center of the roof)

Fig.3 (a) and (b) show the distribution of methane gas concentration when the forcing duct is set on the corner of the roof and the distance from the face to the duct end is 47 cm (7 m in actual size). The figure (a) shows the distribution of the concentration when forcing velocity is 1 m/s. The air jet doesn't reach the face in this condition, and the methane concentration is high near the roof at the face. In this case methane is difficult to be discharged from the face because airflow velocity is very slow and stagnated region is formed. The figure (b) shows the distribution of the concentration when forcing velocity is 2 m/s. The air jet reaches the face and return flow is formed in this condition. Methane concentration is relatively high at the corner of the face and under the forcing duct, but methane is diluted better near the roof as compared to the figure (a).

When the forcing velocity was 4 m/s, the methane concentration was lower than the figure (b) but the distribution pattern was similar to it. The distribution of methane seems to significantly depend on whether forcing jet reaches the face. It must be noted that methane accumulation is often observed near the floor.

The overlap ventilation system, that is, forcing system combined with exhausting system, is going to be used considerably in Japanese coal mines to control both the dust and methane gas at the same time. However, there is almost no published work about the optimization of the overlap system.

Fig.4 shows the distribution of methane when the forcing overlap system is employed. The distance from the face to the exhausting duct end are (a) 20 cm and (b) 33.3cm (3 and 5 m in actual size). Flow velocities are 2 m/s in the forcing duct and 4 m/s in

the exhausting duct. Methane is diluted in the figure (b) better than in the figure (a). In the figure (a), some methane is not sucked into the exhausting duct and flows away to outby side (right side of the figure), because the distance from the face to the exhausting duct is too short. The figure (c) shows the distribution when all the conditions are the same as that of (b) except the flow velocity from the exhausting duct is decreased from 4 m/s to 3 m/s. In this case, the effect to diffuse methane decreases apparently. When the flow velocity from the exhausting duct is increased 6 m/s, the effect to diffuse methane does not vary very much.

(a) Distance from the face to the exhausting duct end is 20 cm and exhausting flow velocity is 4 m/s.

(b) Distance from the face to the exhausting duct end is 33.3 cm and exhausting flow velocity is 4 m/s.

(c) Distance from the face to the exhausting duct end is 33.3 cm and exhausting flow velocity is 3 m/s.

Fig.4 Distribution of methane at the heading face with forcing overlap system (The distance from the face to the forcing duct end is 33.3cm and visualized section is the vertical plane through the exhausting duct.)

3 FLOW VISUALIZATION BY BUBBLES

Visualization by Titanium tetrachloride ($TiCl_4$) is useful to examine the distribution of methane concentration. However, this experiment is very dangerous, because methane gas is explosive and $TiCl_4$ generates poisonous gas by reacting with moisture. Therefore, a combination of laser light and minute bubbles is used to simulate and visualize the behavior of methane at the heading face in a reduced scale model.

It is a one-thirtieth-scale model with rectangular cross section 20 cm in width and 10 cm in height. Water was used instead of air and minute bubbles generated by electrolysis as tracer instead of methane. Bubbles are safe and they hardly pollute water in the gallery. Bubbles were generated at the face. Brighter regions in visualized pictures show higher concentration of bubbles. Flow rate, the distance between the duct end and the face with forcing, exhausting, or combined ventilation system were examined. The diameters of the forcing and exhausting duct were 3 cm and 2 cm, respectively. Ducts were set on the two corners of the roof.

Fig.5 shows the distribution of bubbles when there is no flow in the model. Bubbles generated at the heading face move upward along the face first, then go to outby side along the roof. A layer of bubbles was formed below the roof just as methane does. The behavior of bubbles in water is quite similar to that of methane in the air (Fig.1).

Fig.5 Distribution of bubbles generated by electrolysis at the heading face (No Flow)

Fig.6 shows the distribution of bubbles when the exhausting system is employed. The figure (a) and (b) show the distribution when the distance from the face to the duct end is 10 cm and 16.7 cm (3, 5 m in actual size). The exhausting system is very effective in controlling coal dust but a region ventilated directly by the flow which comes from behind the exhaust duct end is limited within quite a short distance ahead of the duct end. Consequently the concentration of bubbles is very high near the face as shown in Fig.6. The bubbles are not observed behind the duct end.

Fig.7 shows the distribution of bubbles when the forcing overlap system is employed. The distances from the face to the exhausting duct end are (a) 10 cm and (b) 16.7 cm (3 and 5 m in actual size), the

distance between the face to the forcing duct being kept 23.3cm (7m in actual size). Flow velocities are 1.4 m/s (flow quantity is 60 l/min) in the forcing duct and 5.3 m/s (flow quantity is 100 l /min) in the exhausting duct. The effect of exhausting bubbles was greater with the face to the exhausting duct of 16.7 cm than with 10 cm. The whole distribution of bubbles is similar to that of methane (Fig.5 (a) and (b)) but the contrast between bright and dark regions is clearer in visualization picture by bubbles. It is not easy to control methane and $TiCl_4$ experimentally, but generation of bubbles by electrolysis is safe and effective for visualization.

(a) Distance from the face to the duct end is 10 cm.

(b) Distance from the face to the duct end is 16.7 cm.

Fig.6 Distribution of bubbles at the heading face with exhausting ventilation. (Vertical plane through the exhausting duct; Duct position: corner of the roof)

(a) Distance from the face to the exhausting duct end is 10 cm.

(b) Distance from the face to the exhausting duct end is 16.7 cm

Fig.7 Distribution of bubbles generated by electrolysis at the heading face (Visualized section is the vertical plane through the exhausting duct.)

At heading faces in actual mines, there are some obstacles such as road heading machines. Fig.8 shows the distribution of bubbles in this case when the forcing system is utilized. The distance from the face to the forcing duct end is 23.3cm (7 m in actual size) and flow velocity is 1.4 m/s (flow quantity is 60 l /min). The figure (a) shows the case that there is no obstacle and the figure (b) shows the influence of a road heading machine placed at the face. In this case, the concentration of bubbles near the face and the roof are higher when the machine is placed at the face although the concentration are lower near the floor than when the machine is not placed. Even if the forcing duct end is brought close to the face, the concentration of bubbles near the face is not lower.

(a) The case of no obstacle

(b) The case of existing a road heading machine

Fig.8 Distribution of bubbles generated by electrolysis at the heading face.(Visualized section is the vertical plane through the center of the model gallery)

It is not so easy to recognize clearly the difference in concentration of bubbles from the figures shown in Fig.7, etc. However, these differences can be easily represented if pictures are analyzed by a computer. Fig.9 shows the same distribution of bubbles as Fig.7. The minimum brightness in the figure is set to be 0 (painted black) and the maximum one is set to be 255 (painted white). It is easy to recognize that the brightness below the roof and below the exhausting duct is higher and in the figure (a).

(a) Distance from the face to the exhausting duct end is 10 cm.

(b) Distance from the face to the exhausting duct end is 16.7 cm

Fig.9 Brightness contours about Fig.7

4 CONCLUSION

Airflow and concentration of methane gas at the heading faces with auxiliary ventilation were investigated. The results are summarized as follows:
(1) Methane accumulation at the heading face was visualized by using $TiCl_4$ as a tracer. Visualized results were similar to the distribution of methane concentration. When the overlap system was utilized, methane accumulation was affected not only by the locations of the ventilation duct but also by the flow rates through the ducts.
(2) Visualization by bubbles generated by electrolysis was also conducted. The distribution of bubbles shows good coincidence with that of methane. It is difficult to diffuse bubbles near the face by the exhausting system. When a road heading machine was placed at the face, bubbles accumulated near the roof and the face.

REFERENCES

Sheila, E. H. Shuttleworth, 1963, "Ventilation at the Face of a Heading, Studies in the Laboratory and Underground," Int. J. Rock Mech. Mining Sci., Vol. 1, pp. 79-92.
Wesely, R., 1984, "Airflow at Heading Faces with Forcing Auxiliary Ventilation," 3rd Int. Mine Vent. Cong., England, pp. 73-82.
Uchino, K., and Inoue M., 1997, "Auxiliary Ventilation at Heading Faces by a Fan," 6th Int. Mine Vent. Cong., Pittsburgh, pp. 493-496

Mining Science and Technology'99, Xie & Golosinski (eds) © 1999 Balkema, Rotterdam, ISBN 90 5809 067 1

An expert system for real time predicting the concentration of combustion product in mine ventilation network

A.R.Green, Z.X.Tan & J.Cross
University of New South Wales, Sydney, N.S.W., Australia

ABSTRACT: Expert systems have applications in many areas. An expert system for predicting the concentration of combustion products in mine ventilation networks has been developed to take advantage of expert systems over conventional simulation programs. With this expert system, a variety of simulation and prediction scenarios can be evaluated. It is then possible to see for example which junctions have critical concentrations at particular times. The validity of the application written in Prolog has been verified by comparing the junction concentration results with those obtained from a widely used Fortran program. Compared to the Fortran based application, the Prolog application is superior in its readability, ease of implementation and its user-friendly interface.

1 INTRODUCTION

Mine fires are a serious hazard in the mining industry leading to the death of miners from the toxic combustion products. Therefore, it is extremely important to predict the path and the concentration of the combustion products for fire emergency planning.

At present, a simulation program for this purpose developed by the Michigan Technological University in 1980s (MTU) is being widely used. In the MTU program (Greuer 1981), the simulation is carried out by generating, advancing and deleting the contaminated waves (contaminated air segments with homogenous concentrations). Waves are initially generated by the fire sources, and then they are advanced in each time interval simultaneously. When the waves meet at junctions, they are deleted and the new waves are generated and then are advanced into the outgoing airways. During the mixing process, the mass conservation law is applied to determine the concentrations of the junction and the newly generated waves. The major feature of the MTU program is that it can be used to predict the concentration distribution resulting from the multiple fire sources in a mine ventilation network. However, it is not applicable in the case of a fire source varying with time because the number of waves generated by the fire becomes infinite. To overcome the limitation existing with the conventional simulation program, a new model and the associated expert system have been developed.

2 MATHEMATICAL MODEL

Fig.1 shows junction X and the airways connecting with it in a ventilation network. According to the mass conservation law, the concentration of junction X at time T can be established as:

$$C(X, T) \quad \frac{\Sigma\, C(N_i,\, T)Q_i}{\Sigma Q_i} \tag{1}$$

Where:
Q_i —— Volume airflow rate of airway N_i
$C(N_i, T)$ —— Concentration of airway N_i at time T

The above formula implies that the concentration of a junction at a given time in the ventilation network is the volume weighted average of the concentrations of the airways entering the junction at their ends at the same time.

The concentration of junction X at time T can not be obtained directly from formula (1) because the concentrations of the airways at their finishing junctions at time T are unknown. To continue, the concentrations of those airways are required. If the longitudinal diffusion in an airway can be ignored, then it can be inferred that the concentration of the air current at an airway end at time T is same as that of the air current at its beginning at the time T_i. If there is no fire source in the airway, then the concentration of the airway at its beginning will be equal to the concentration at its starting junction at the same time. Therefore, we have the following equation:

$$C(N_i, T) = C(X_i, T_i) \quad (2)$$
$$(T_i = T - L_i S_i Q_i)$$

Where:

L_i —— Length of airway N_i
S_i —— Cross sectional area of airway N_i
Q_i —— Airflow rate of airway N_i
X_i ——Starting junction of airway N_i
$C(X_i, T_i)$ ——Concentration of junction X_i at time T_i

Combining (1) and (2) , we have

$$C(X, T) = \frac{\Sigma\, C(X_i, T_i) Q_i}{\Sigma Q_i} \quad (3)$$

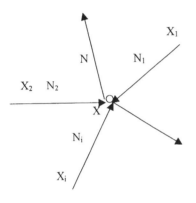

Fig.1 Airflow mixing at a junction

If there is a fire at the airway beginning at time T_i, then the concentration of the airway at its beginning at time T_i will be equal to the concentration behind the fire at the same time. This can be expressed by:

$$C(N_i, T) = \frac{C(X_i, T_i)(Q_i - Q_a) + M(T_i)}{Q_i} \quad (4)$$

Where:

$M(T_i)$ —— Contaminant production rate
Q_a —— Additional gas entering the airway

Combine (1) and (4), we have:

$$C(X, T) = \frac{\Sigma\, C(X_i, T_i)(Q_i - Q_a) + M(T_i)}{\Sigma Q_i} \quad (5)$$

The initial condition and boundary conditions are:
- $C(X_i, T_i) = C_0$ (C_0 is the background value of the product of combustion), if $T_i < 0$.
- Boundary condition
  $C(X_i, T_i) = \phi$ (ϕ is the contaminant concentration in the atmosphere) if X_i is a surface junction.

3 IMPLEMENTATION

3.1 *Introduction to expert system*

Expert system is an intelligent computer program that uses knowledge and inference procedures to solve problems that are difficult enough to require significant human expertise for their solutions (Joseph & Gary, 1994).

A typical expert system comprises three basic components, that is knowledge base, inference engine and user interface (See Fig.2).

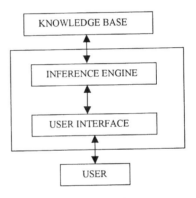

Fig.2 Components of Expert System

The knowledge base comprises the knowledge that is specific to the domain of application, including the things such as simple facts about the domain, rules that describe relations or phenomena in the domain, and possibly also methods, heuristics and ideas for solving problems in this domain.
The inference engine is the component which can scan facts and rules, and provide answers to the queries from the user.
The user interface is the means by which the user communicates with the expert system. It provides a mechanism for the input of information concerning a problem, the examination and modification of the knowledge base and the output of recommendations and reasoning by the system.

3.2 *Knowledge base building*

In order for the built-in inference engine in Prolog to predict the concentration of the combustion product, a knowledge base must be built. In our case, the knowledge base consists of the facts related to the mine ventilation network and the fire sources, and the rules generated from the mathematical model. The details of the representation of these facts and rules in Prolog are introduced below.

3.2.1 *Ventilation network*

A mine ventilation network consists of a series of airways. The information associated with each airway includes airway number, starting junction, finishing junction and the associated attributes. The attributes of an airway which have to be represented in the network are airflow rate, airway cross-sectional area and airway length. Therefore, each airway can be represented by: airway(AN,JS,JF,Q,S,L). The arguments in the predicate are airway number, starting junction, finishing junction, airflow rate, and cross-sectional area and airway length respectively.

3.2.2 *Fire*

The knowledge related to a fire includes fire identification number, airway containing the fire, location of the fire, starting time, ending time, fire type and contaminant production characteristics. Predicate fire/9 has been defined to represent the fire specified. Because the contaminant production characteristics vary with the type of fire, the meaning of the arguments associated with the production characteristics are different. For all types of fires, the meanings of the first six arguments are same. The last two arguments vary from the type of the fire. If the fire is of a contaminant type, they are Q_f (Volume flow rate of the gas current carrying the contaminant.) and C_f(contaminant concentration in the gas current carrying the contaminant). If it is oxygen rich, they are MINO2 (Oxygen concentration in the fumes leaving the fire) and a redundant variable. If it is fuel rich, they are SMPO2 (Contaminant production rate per cubic feet of oxygen delivered) and a redundant variable. The redundant variable is used here only for balancing the structure of predicate fire/9. The general formats of three fire types are :

- Contaminant type
fire(FN.AN.FIRE.DIST.ST.ET.TYPE.Qf.Cf).
- Oxygen rich type
fire(FN.AN.FIRE.DIST.ST.ET.TYPE.MINO2._).
- Fuel rich type
fire(FN.AN.FIRE.DIST.ST.ET.TYPE.SMPO2._).

3.2.3 *Concentration calculation rules*

The major rules generated from the model are stated as follows:
Rule1 initial condition rule
If the calculation time T<0, then the concentration of this junction is equal to the background value before the start of the fire.

Rule 2 boundary condition rule
If the junction for calculation is a surface junction, then the concentration of this junction is equal to the concentration of the atmosphere.

Rule 3 mass conservation rule
If neither rule 1 and rule 2 is true, the junction concentration at a given time T is the weighted average of the concentration of the airways entering the junction at their ends at the same time.

Rule4 rule to calculate the weighted average of the concentration of the entering airways at their ends at the given time
If the sum of the products of the concentrations at the ends of these airways and their airflow rates is SUMCQ,
and the sum of the airflow rates of these airways is SUMQ,
then the weighted average of the concentrations at all the entering airways ends at the given time is SUMCQ/SUMQ.

Rule 5 rule to determine SUMCQ
If the concentration of one of the entering airways at its end is C,
and the airflow rate of this airway is Q,
and the sum of the products of the concentrations and the corresponding airflow rates for the rest of the entering airways is SUMCQ1,
then SUMCQ is SUMCQ1+C*Q.

Rule 6 rule to determine SUMQ
If the airflow rate of one of the entering airways at its end is Q,
and the sum of the airflow rates of the rest of these airways is SUMQ1
then SUMQ is SUMQ1+ Q .

Rule 7 rule to determine the concentration of an airway at its end for case 1
If the airway is not the airway on fire,
and $T_i =T- L_i/(Q_i /S_i)$,
then the concentration of an airway at its end at time T will be equal to the concentration of the stating junction of the airway at time T_i.

Rule 8 rule to determine the concentration of an airway at its end for case 2
If the airway is the airway on fire,

and $T_i = T - L_i/(Q_i/S_i)$,
then the concentration of an airway at its end at time T will be equal to the concentration behind of the fire source at time T_i.

Rule 9 rule to determine the concentration behind the fire source
If the staring junction concentration of the airway on fire at the time T_i is C_j,
and the contaminant production rate is M ,
and the additional gas entering the airway is Q_a,
and the airflow rate of the airway is Q,
then the concentration behind of the fire can be calculated with the formula $C_j*(Q-Q_a)/Q + M/Q$.

Rule 10 rule to determine the M and Q for case 1
If the calculation time T_i is larger than the staring time ST,
and less than the ending time ET,
and the fire is of contaminant type,
then M is $Q_f*C_f/100$ and Q_a is Q_f.

Rule 11 rule to determine the M and Q for case 2
If the time Ti is larger than the staring time ST,
and less than the ending time ET,
and the fire is oxygen rich,
and the concentration of the fumes leaving the fire is O2MIN,
and the airflow of the airway on fire is Q ,
and the concentration of the starting junction of the airway on fire at the time T_i is C_j,
then M is $(0.21-C_j-O2MIN/100)*Q$ and Q_a is 0.

Rule 12 rule to determine the M and Q for case 3
If the time T_i is larger than the staring time ST,
and less than the ending time ET,
and the fire is fuel rich ,
and the airflow of the airway on fire is Q ,
and the concentration of the starting junction of the airway on fire at the time T_i is C_j ,
then $M = (0.21-C_j)*Q*SMPO2$ and Q_a is 0.

Rule 13 rule to determine the M and Q for case 4
If $T_i < ST$ or $T_i > ET$,
then M is 0 and Q_a is 0.

3.3 Inference process

The operation of the inference engine of Prolog can be understood as a recursive cycle of unification (pattern matching) and sub-goal evaluation

Triggered by a query, the inference engine will descend as deep as necessary into the structure program to find facts that validate the query, and then return having proved or failed to prove the query. When a user types in a query to the system, the query is activated. The inference engine searches

through the currently asserted rules for the first rule whose head unifies with the query. For a query to unify with the head of a rule, both must have the predicate name and the same number of arguments, and all arguments of both must be unified.

The inference process for the junction concentration predication is illustrated in Figure 3. As shown in Figure 3, the inference process starts from the user query, then determine which rule of the first three rules are satisfied. If rule 1 or rule 2 is satisfied, then the inference process will stop and the concentration value is returned. If rule 3 is satisfied, and then rule 4 is invoked. Consequently, rule 5 and rule 6 will be activated by rule 4. The activation of rule 5 will cause the activation of rule 7 and rule 8. Rule 7 will invoke one of the first three rues. Because of the activation of rule 8, rule 9 will be invoked. One of the Rule 10, rule 11, rule 12 and rule 13 will be invoked from the activation of rule 9 as well as one of the first three rules. This process will repeat until all the inference flows meet the initial condition rule or the boundary condition rule.

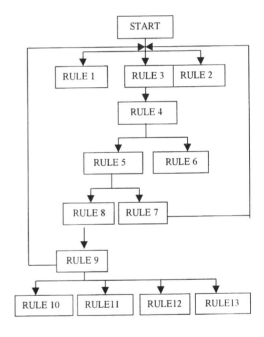

Figure3 Demonstration of the Inference Process

3.4 User interface

Prolog comes with an editor for building the knowledge base and a console window for user to query the system and output the results of the queries. It seems to be not necessary to build a separate interface for the input and output of a query. However with this built-in interface, a user needs to know some basics of Prolog and also the structures of the predicates defined in the program to make the correct queries. In order for users without any knowledge to Prolog and the program to use the system properly, a user friendly interface has been created with the aide of the dialog editor provided with Win-Prolog (Brian,1997). The details regarding the interface are introduced as follows.

3.4.1 *Main menu*

In order for user to select the desired query, a main menu window is created. There are 10 button controls contained in this main menu window. Each button's caption is named after the corresponding query, that is junction_con, junction_simu, junction_distr, junction_critical, network_distr, airway_calc, airway_distr, airway_simu, airway_critical and network_simu.

3.4.2 *Input dialog windows*

Once a query is selected, it is necessary to provide some immediate input parameters to evaluate the query. A series of input dialog windows have been designed to accept the input data from user during the querying time. The common format of the dialogs is a dialog window together with some controls. The types of controls include edit control, static control and button controls. The edit control is used to receive the input data needed for carrying on the selected query. The static controls are used to label the corresponding control. The button controls are used to accept or reject the input data. One common feature of the input dialogs is that all of them contains two button controls named as 'ok' and 'cancel' which are used to accept or reject the input parameters.

3.4.3 *Output dialog window*

To display the output data of each query, an output dialog window is designed. This dialog window consists of a main dialog window and some controls (child windows). One of the controls is the edit control for displaying the output data. This edit control is designed with both vertical and horizontal scroll bars. Others are three button controls which are labeled as 'repeat', 'main', and 'exit'. The 'repeat' button is designed to bring the previous query input dialog back in order for user to carry out the same query with different input parameters. The main' button is used to bring the main menu window back in order for user to select a different query. The 'exit' button is utilized to quit from the current consulting session and exit from the system.

4 VALIDATION

To check the validity of the expert system, it has been applied to the ventilation network shown in Figure 4 and Table 1.

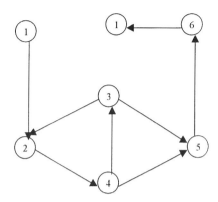

Figure 4 A Simplified Mine Ventilation Network

It is assumed that a beginning smouldering fire in airway 1 produces 1000 ft^3 /min of fumes with 100% concentration for 60 minutes. Using the query junction_critical, the junctions with the critical condition and their concentrations can be determined for the time 10, 20, 30, 40, 50, and 60 minutes after the fire starts. The results from this query and those obtained from the MTU program are shown in Table2.

Table 1 Airflow rates and airway dimensions

Airway	JS	JF	Q(ft^3/min)	A(ft^2)	LA(ft)
1	1	2	100000	300	2000
2	3	2	25000	180	1000
3	2	4	125000	180	1400
4	4	3	35000	36	500
5	3	5	10000	180	760
6	4	5	90000	180	900
7	5	6	100000	300	2000
8	6	1	100000	75	20

As shown in the Table 2, the concentration predication results from the two programs are almost identical except some minor differences for some junctions after 50 minutes. The existence of the minor differences between the two sets of the results can be explained by the fact that a calculation error is introduced in the MTU program because the waves with concentration levels lower than the specified criteria have been ignored in the MTU program (Greuer,1981).

Table 2 Junction concentration Results

Time	Junction	Concentration(%)	
		Expert system	MTU Program
	2	0.8000	0.8000
	3	0.8000	0.8000
10	4	0.8000	0.8000
	5	0.7200	0.7200
	2	0.9600	0.9600
	3	0.9600	0.9600
	4	0.9600	0.9600
2 0	5	0.8640	0.8640
	6	0.7200	0.7200
	2	0.9920	0.9920
	3	0.9920	0.9920
	4	0.9920	0.9920
30	5	0.9728	0.9728
	6	0.9440	0.9440
	2	0.9984	0.9984
	3	0.9984	0.9984
	4	0.9984	0.9984
40	5	0.9946	0.9946
	6	0.9888	0.9888
	2	0.9997	0.9997
	3	0.9997	0.9997
50	4	0.9997	0.9997
	5	0.9989	0.9989
	6	0.9978	0.9978
	2	0.9999	0.9999
	3	0.9999	0.9997
60	4	0.9999	0.9997
	5	0.9998	0.9996
	6	0.9996	0.9989

- No approximation procedures are introduced in the calculation process; therefore this leads to a higher predicating accuracy.
- The use of the expert system is guided by a main menu and a series of dialog windows. Therefore it is user-friendlier.
- The solution process to the concentration predication problem is a backward tracking inference process while this is just how Prolog inference engine works. Therefore it can be implemented in much easier way in Prolog
- The data related to the problem has been represented by facts and rules in Prolog. The meanings of the facts and rules are self-evident.
- The expert system can be used to provide answers to as many as 10 different queries related to the concentration predication problem.

6 REFERENCES

Brian D Steel. *LPA WIN-PROLOG 3.5-Win32 Programming Guide*. Logic Programming Associates Ltd, London, 1997. pp. 114-125.

Greuer, R.E. ,1981. *Real Time Pre-calculation of the Distribution of Combustion Products and Other Contaminants in the Ventilation System of Mines*. United States Department of the Interior, Bureau of Mines, OFR 22-82.

Joseph Giarratano, Gary Riley. *Expert Systems-Principles and Programming*. Boston, 1994. pp. 1-5.

5 CONCLUSION

To sum up, based on a new mathematical model, an expert system for predicating the concentration of product of combustion in mine ventilation network has been developed and verified against the conventional program in the same problem domain. Compared with the conventional program, the expert system has the following features:
- The expert system is not only applicable to the constant fire sources but also the fire sources varying with time continuously.

Mining Science and Technology' 99, Xie & Golosinski (eds) © 1999 Balkema, Rotterdam, ISBN 90 5809 067 1

Coal mine ventilation practices in the United States

Jerry C.Tien
University of Missouri-Rolla, Mo., USA

ABSTRACT: Today, over 40% of the U.S. coal is mined using underground mining method. Mining methods used underground has also been slowly chaning over the years too, which has presented challenges in ventilation. This paper presents the current status of ventilation practices in U.S. underground coal mines. It also discusses current regulations as pertaining to underground coal and the proposed regulations on diesel-powered equipment to enhance the health and safety for underground coal miners.

1 INTRODUCTION

According to the Mine Safety and Health Administration (MSHA), there were 2,644 coal mining operations in the United States in 1997, of which, 971 (or 36.7%) are underground mines (Anon., 1998). During that year, U.S. coal production was at an all time high of 1.0886 billion tons and underground mining accounted for about 39%, or 410.4 million tons, of total U.S. coal production (Anon., 1998 and Tien, 1998).

In the U.S., two major methods are used in mining underground coal: longwall and continuous mining, although conventional and shortwall also continue to play a minor role (accounting for 8.64% for conventional and less than 0.1% for shortwall in 1996). In the past two decades, longwall mining has helped revolutionize underground coal mine operations, with its share of total U.S. underground production increasing from 10 percent to 48 percent, surpassing continuous mining tonnage in 1994 and the trend has held true since then (Tien, 1998).

The face of longwall mining in the U.S. has changed dramatically over the years as well. Ten years ago, 25 U.S. coal companies operating 92 longwall faces; today, 22 companies operating a total of 62 longwall faces with an average shift production of 3,139 tons of clean coal in 1997 (Fiscor, 1999), and the figure for 1998 is expected to go up significantly.

2 REGULATIONS THAT AFFECT COAL MINING

In the U.S., health and safety regulations and enforcement functions reside both at the federal and state levels. Federal regulations are specified in the Code of Federal Regulations (CFR) which is divided into 50 titles, with Title 30 covering mineral resources. Health standards are set forth in Part 70, while ventilation regulations for coal mines are described in Subpart D of Part 75. MSHA, administered by the Department of Labor, is charged with enforcing these regulations (Tien, 1996a).

Since 1982, Part 75 has gone through a vigorous reviewing process by way of public hearings where industry, government, labor union, academia, and those who are involved in mine ventilation participated and provided recommendations. Final version was released in the spring of 1996 (Tien, 1997). As new findings and proven experience became available, these regulations, upon permission from governing authorities, have been modified in the field.

These regulations provide a minimum quality and quantity requirements expected for a coal mine ventilation system, such as the installation of mine fan(s); the minimum oxygen (19.5%), maximum requirements for carbon monoxide (0.5%) and other pollutants (H_2, H_2S, H_2S_2, CH_4, diesel particulate matters, etc.); minimum air quantity requirements at the last open crosscut be 9,000 cubic feet per minute (cfm) and the working face be 3,000 cfm; minimum air velocity requirement at all working faces be 60 feet per minute, (fpm) and maximum air velocity of 250 fpm in trolley-haulage entries. In addition, each mechanical section must be ventilated by a separate split of intake air; the various specifications for ventilation controls (doors, stoppings, overcasts, seals, bleeder systems, etc.); mandated the separation of airways for belt-haulage entries, etc.

Since its early introduction to U.S. coal mines in the early 1970s, the number of diesel-powered equipment operating in underground coal mines has increased from approximately 150 in 1974 to over 2,900 units operating in 173 mines in 1995. MSHA projects that the number of diesel units operating in underground coal mines could increase to approximately 4,000 in 250 underground coal mines by the year 2000 (Anon., 1998). This increasing popuparity of diesel equipment has also brought concerns about the health and safety of workers exposed to diesel emissions. As a result, two proposed regulations on diesel engines for coal mines were released in the past few years. One (released on October 25, 1996) established approval, exhaust gas monitoring, and safety requirements for the use of diesel-powered equipment in underground coal mines and has become effective since April 25, 1997 (Anon., 1999). The other (released on April 5, 1998), which set standard for diesel particupate matters, is still going through the phasing in period. Both regulations sginificantly strengthened the equipment emission and approcal standards and will greatly enhance the heath and safety of the coal workers in the U.S. (Tien, 1999).

3 MAJOR MINE VENTILATION SYSTEMS

MSHA estimates that between 93-95% of U.S. coal mines use an exhausting system where the fan(s) are located on top of the return airshaft(s), with blowing systems accounting for 3-4 pct, and the remainder using a combined (push-pull) system. There are pros and cons to each system depending on particular mining and environmental conditions (Tien, 1978). In complying with the federal mandate that air traveling in coal haulage entries is not to be used in working faces, a so-called "neutral" airway is created which adds another dimension to the ventilation system. All entries are driven in coal seams so coal is produced even during mine development, although at a slower advanc-

ing rate, and ventilation system typically make use of multiple entries.

3.1 Room-and-Pillar Ventilation System

Both conventional and continuous mining are room-and-pillar systems which have been the predominant method used in the U.S. In 1994, coal mined using longwall system exceeded that of the room-and-pillar method for the first time, although it is still used during development in longwall mining.

The room-and-pillar method will usually yield between 50 to 55% extraction ratio. If a pillar-recovery process is applied where the continuous miner mines into the pillars when mining reaches the end of the panel and the direction of the mining is reversed, up to 5% of additional coal can be extracted in the pillar recovering process as the roof is allowed to systematically collapse. Special permission for ventilation plan must be obtained for a pillar recovery operation.

To design a ventilation system in the multiple entries found in a room-and-pillar workings, either a unidirectional flow or bidirectional flow system can be used for airflow distribution. In the former system, the air in adjacent openings flows in the same direction and is entirely fresh or exhaust air; whereas in a bidirectional flow system, the air in adjacent openings flows in opposite direction; the latter system is also referred to as a "fish-tail" arrangement (Figure 1).

In either system, stoppings must be erected in connecting crosscuts to maintain separate flow. Between the two systems, the bidirectional arrangement would normally have a worse leakage problem and require additional stopping line. For these two reasons, the unidirectional system has been more popular in the U.S., especially in main entries. To reduce air leakage, parallel mains separated by long barrier pillars are often used. They driven next each other connected only once every 10 to 15 crosscuts to facilitate mining (Tien, 1978). Traditionally, stoppings made of concrete blocks or cinder blocks are used for air distribution, although metal stoppings have gained increasing popularity over the years because of its easy of use and re-usability (Tien, 1996).

In the face area where coal is being mined, methane and dusts have become increasingly important in recent years because of the increased coal production. Methane is controlled by providing adequate amount of air airflow and frequent methane checks while dust

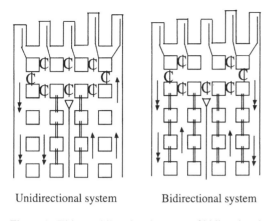

Unidirectional system   Bidirectional system

Figure 1. Either unidirectional system of bidirectional
system can be used for air distribution.

is controlled primarily through the use of a machine-mounted dust scrubber which has become a standard equipment on continuous miners (Figure 2).

Surfactants have also been used in spray water to facilitate dust suppression (Tien and Kim, 1997). Figure 3 shows a flooded-bed scrubber on a remotely-controlled continuous miner and how it is used in combination with a blowing curtain in a heading. Typically, 7,000 to 8,000 cfm of fresh air is provided at the end of a "blowing curtain" where it is further directed toward the face. Contaminated air is discharged at the end of the continuous miner and carried away by return air current without exposing workers in the fresh air current. It is essential that a blowing curtain be used along and ad-

equate amount of fresh air must reach the entrance of the curtain in order for this system to be effective, or return air will be recirculated back to the heading.

3.2 Longwall Ventilation System

Because of superior productivity and safety, longwall mining has been increasing in popularity in the past two decades in the U.S. The average panel width for a U.S. longwall is 875-ft, although the longest has reached 1,200-ft in length. Aver panel length is 8,235-ft with the longest being at 17,600-ft.

The basic practice for longwall ventilation is simple: course the intake air through the headgate panel entries, across the face and down the tailgate entries, while causing a portion of the air traveling through the gov and bleeder entries (Fuller, 1989 and Figure 3).

In practice, problems often found in controlling respirable dust and methane along the longwall face area and behind gobs, and to maintain effective bleeder system. Coalbed and strata methane released during the mining process is oftentimes removed using de-gasification drilling in advance of mining and dilution during mining. Although the law (§75.325(c)(1)) requires that at least 30,000 cfm (14.2 m^3/sec) must be provided at the working face of each longwall, there have been between 40,000 to 60,000 cfm (18.9 to 28.3 m^3/sec) of fresh air reaching the end of headgate entries for most longwall faces depending on methane content, with one face in West

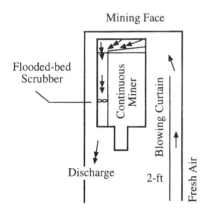

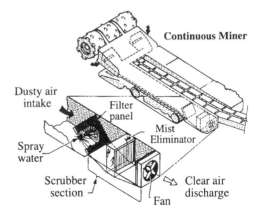

Figure 2.   Diagram showing a continuous miner in a heading (left) and a flooded-bed scrubber
on a continuous miner (right).

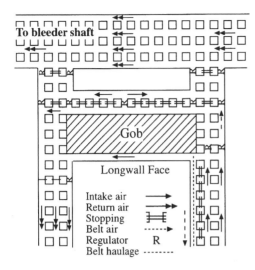

To bleeder shaft

Gob

Longwall Face

Intake air ⟶
Return air ⟶⟶
Stopping ⊨E
Belt air ------⟶
Regulator   R
Belt haulage --------

Figure 3. A typical longwall ventilation system used in the U.S.

Virginia having over 100,000 cfm (47.2 m³/sec) at the face because of high methane concentration (Anon., 1999).

In the U.S., a minimum of two entries are required on a longwall panel, although three is the most commonly used number of entries for operational reasons. The law also mandates that there should be least one entry be kept open at all times in the bleeder entries; most bleeder system has at least two, but oftentimes three or four are used in case of roof falls that block the bleeder entries.

4 SUMMARY

Although U.S. underground coal mines are blessed with relatively horizontal coal seam, consistent coal thickness, and low methane emission, however, challenges abound. Two areas that brought about the most concerns are the increasing coal dust concentrations as a result of the increased coal production (more specifically, increased longwall activities), and exposure to diesel pollutants because of the increased use of diesel equipment in underground coal mines.

Even with all technological advances in dust control in the past two decades, there are still as much as 15 to 20% longwall operations are out of compliance with the 2 mg/m³ respirable dust standard, and there have been discussions to strengthen current respirable dust standard to 1 mg/m³, which will certainly present mine operators with even more challenges.

The recently released diesel eimssion standard for coal mines have also presented operators with challenges, To be in compliance with new stardards, additional air quantity will have to be provided which will greatly increase operational cost.

Looking ahead, the following is a list of issues that warrant attention in the next millennium: use of controlled recirculation, use of booster fans underground, better dust control technology,more sophisticated mine ventilation planning, and computerization and automation for mine ventilation (Tien, 1999).

REFERENCES

Anon. (1999) "Code of Federal Regulations," Mine Safety and Health Administration (MSHA) web page (http://www.msha.gov), Department of Labor, as of February 15, 1999.

Fiscor, S. (1999) "U.S. Longwall Census," *Coal Age,* Feb., pp. 30-35.

Fuller, J.L. (1989) "An Overview of Longwall Ventilation System Design," SME annual Meeting, Las Vegas, NV, *Preprint #89-53,* 13 pp.

Tien, J.C. (1978) "Pros and Cons of Underground Ventilation System," *Coal Mining and Processing,* Jun., pp. 110-113.

Tien, J.C. (1996) "Air Leakage Costs," Coal Age, Sep., pp. 88-96 & 121.

Tien, J.C. (1996a) "U.S. Mine Ventilation Regulations," *Proc. '96 Int'l Sympo. on Min. Sci. and Tech.,* Guo, Y.G. and Golosinski, T.S., eds., Xuzhou, Jiangsu, China, pp. 163-170.

Tien, J.C. (1997) "Safety Underground: Part I, The Regulation Maze," *Rock Products,* Apr., pp. 46-48; Part II, Rules in the Making, May, pp. 38-40; Part III, Standards for Air Flow and Quality, June, pp. 33-36.

Tien, J.C. and Kim, J. (1997) "Respirable Coal Dust Suppression Using Surfactants," *Appl. Occup. Eniron. Hyg.,* Vol. 12, No. 12, Dec., pp. 957-963.

Tien, J.C. (1998) "U.S. Coal Industry - Past, Present, and Challenges," *Proc. Asian-Pacific Workshop on Coal Min. Tech.,* Tokyo, Japan, Sep. 8-9, 11 pp.

Tien, J.C. (1999) "Chapter 15: Diesel Underground," *Practical Mine Ventilation Engineering,* Intertec Publishing Company, Chicago, IL, pp. 323-335.

Mining Science and Technology '99, Xie & Golosinski (eds) © 1999 Balkema, Rotterdam, ISBN 90 5809 067 1

The application of the qualitative and quantitative synthetic analysis method to decision making in mine fire fighting[1]

Xinquan Zhou & Hongqing Zhu
China University of Mining and Technology, Beijing, People's Republic of China

ABSTRACT: The airflow control during mine fire is a most important task in fire fighting. The qualitative and quantitative synthetic analysis method is applied to define safe airflow control plan for miner evacuation and fire fighting. Using the qualitative analysis method to locate the airflow state regulators, and the quantitative analysis method, shows the effect of the airflow control plan quantitatively. The characteristic zone method is employed to solve the difficulty in the estimation of fire intensity in practical fire fighting.

1.PREFACE

The mutual interaction of thermal draft caused by mine fire and mechanical pressure destroys the former balance between ventilation power and resistance of normal mine ventilation system, and causes disorder of airflow state. It not only changes airflow distribution greatly, but also causes airflow reversal in some airways. The reversal airflow with poisonous and harmful gases enters the entries area of mine, and enlarges the risk region. The reversal of airflow becomes the major factor to claim more loss in lives and property, and add more difficulties in fire fighting. The airflow state control, aiming at the regulation of fume flowing path, is the most important task in fire fighting.

However, it has been proved by fire fighting practice for years that it's difficult for decision makers to make decision correctly and timely, because sometimes, they don't know the state and the influence of fire. Therefore, the decision made from experiences maybe cause the big problem at the complex mine fire cases. For a long time, the decision-makers have wished to get the help by solving the following problems(Zhou 1996):

(1)Obtaining the reliable information in time in relation to the dynamical state change of mine ventilation system during mine fire such as the changes of airflow quantity, direction, temperature, fume spreading areas and it's concentration.

(2)Obtaining the help of some relevant technologies to decide the airflow control measures for miner evacuation and fire fighting safely.

(3)Implementing the airflow control measures timely and safely.

For above three problems, some works have been done as follows(Zhou, 1994, 1996):

(1)Estimating the location of fire source with the locating technology, and analyzing how to optimize the sensor location of mine environmental monitoring system.

(2)Applying the non-steady state airflow simulation technology to help ventilation engineers to know the dynamical state variation of the airflow distribution during mine fire .

(3)Helping decision makers during mine fire for the state control by applying qualitative control method such as Budryk's method or applying the modern control theory to develop the dynamic quantitative optimizing control technology.

(4)Applying electronic technology to carry out the long distance control and automation of the regulators to implement the control measures.

There are, however, some problems left in fire fighting.

(1)It's difficult to estimate the combustion characteristic of fire source in a practical mine. The heat released from fire source is a fundamental basis to calculate the variation of airflow state and prerequisite condition for the correct application of airflow state simulation technology. This problem will impede the correct application and popularization of mine fire simulation technology.

(2)In recent years, the rapid development of airflow state simulation technology and computer software plays an important role to reveal the dynamic state variation of airflow distribution and threat to miners(Greuer 1983,1988). However, there are still some problems to apply the state simulation technology to solve airflow control problem, such as how to determine the optimal number, size and location of the regulators, how to prevent fume from entering the miner evacuation path, while miners escape to surface by going through the path.

(3)Although the airflow qualitative control technologies, such as Budryk's method, have played

[1] The Project Supported by National Science Foundation of China (No.59874030)

an important role in mine fire fighting since 50's, it cannot meet the complicated dynamic change of mine fire by its nature of steady state control on the local network. It is very difficult to determine some airflow state parameters in the airflow direction discriminant during mine fire, and is difficult for ventilation engineers to apply in situ. Although Budryk's method has been developed for the following years, the discriminant becomes more and more complicate to give more difficulties to mine ventilation engineers to master the discriminant. On the other hand, it's more difficult to determine the airflow state parameters in the discriminant during mine fire.

The dynamic state quantitative optimal control technology developed in recent years solves successfully the problem of airflow state dynamic control(Zhou 1996). The optimal control technology requires the dynamic variation of the state regulators, and it needs the support of ventilation system automation control. Under the present condition, it's difficult to get wide application for its expensive cost. Therefore, a new kind of airflow control method should be developed to meet the requirement of mine fire fighting in situ. There are two main contents discussed in this paper, the qualitative and quantitative synthetic analysis method for the airflow state control and the characteristic zone method for the estimation of fire intensity.

2. THE APPLICATION OF THE QUALIATIVE AND QUANTITATIVE SYNTHETICAL ANALYSIS METHOD TO THE DECISION-MAKING OF MINE FIRE FIGHTING

Presently, mine fire simulation technology reveals the dynamic changes of airflow state, the location and period of airflow reversal, and the fume influence area. On the basis of above information, decision makers can select the evacuation and fire fighting routes which could not be entered by fume during fire fighting. However, it may be impossible to select miner evacuation path by averting from the fume spreading area passively during a serious mine fire. It's necessary to control airflow state and to regulate the fume flowing path so as to ensure miners evacuation and fire fighting safely.

For one specific mine fire, there might be several control plans for us to select. The qualitative and quantitative synthetic analysis method is that applies essentially experiential and qualitative method to determine the location and the number of airflow state regulators, and then applies airflow dynamic simulation technology to check the airflow state control effect. The procedures are as follows:

(1)Determining the location and the resistance of state regulators during mine fire taking place in a specific region in mine.

(2)Applying fire simulation software to simulate the change of airflow distribution on the mutual interaction of thermal draft, mechanical pressure and added regulators during fire. It can help decision-makers to know about fire influence on whole ventilation system.

(3)Applying the qualitative analysis method to determine the airflow control plans for the certain mine fire to protect miner evacuation or fire fighting path. The control plan includes the number, location and resistance of regulators.

(4)Applying dynamic state simulation technology to simulate the state control result of the control plans determined by step (3) one by one to check the plentitude and necessity of plans.

(5)Selecting one or two plans to act as the optimal control plan after comparing the state control effects of the selected plans.

(6)Repeating step (1) to (5) and selecting the optimal control plans one by one for each easy-burning area of mine and inputting into computer as an important part of the prevention and fire fighting plan which should be used as a reference for mine fire fighting decision-making.

(7)Selecting a suitable airflow state control plan from computer and using as a reference for mine fire fighting decision-making in the light of the location of fire source and the combustion characteristic when a mine fire takes place.

On the base of non-steady airflow simulation technology, the above method shows the practical dynamic effect of the airflow state control measure to be determined by qualitative method, and the dynamic variation of airflow and fume distribution. It can be applied conveniently and meet the present technical and equipment level in coal mines

There are following merits of the qualitative and quantitative synthetic analysis method.

(1)It shares the merits of the qualitative method, such as easy to be applied and use friendly. The resistance of regulator is fixed, which need not match for the state automation control system.

(2)It shares the merits of the non-steady state simulation technology, which can check practical effect of the airflow control plan and optimize the number, location and the resistance of regulators, but also can show the dynamic variation of the state parameters such as, the time and concentration of fume invading airways. The decision makers can select the escaping path by using the period before the airways are invaded by fume.

However, the above method has still some demerits. It is difficult to estimate the intensity of fire source, which impedes the application of airflow dynamic simulation technology. Surely, it becomes the bottleneck in the application of dynamic state simulation technology. The problem is solved by applying the characteristic zone method which will be introduced as follows:

3. THE CHARACTERISTIC ZONE METHOD OF FIRE SOURCE

There is an important difference between airflow state control during mine fire and general airflow state regulation during normal operation. The former mainly aims at controlling the airflow direction to protect the escaping route not to be polluted by fume. The latter, however, aims at controlling airflow rate for safe operation. The realization of the airflow direction control is easier than that of the airflow rate control. Thus, the airflow direction control needed during mine fire gets larger room for state control. In other words, the influence of fire on airflow direction in airways is same if the intensity of fire estimated varies in same characteristic zone with the intensity of realistic fire. The so-called characteristic zone means that each fire intensity in that zone has same influence on the airflow direction in ventilation system. The airflow state control plan based on the intensity estimation of fire source will still show a good effect for airflow state direction control. This method breaks the bottleneck in the application of dynamic state simulation of mine fire.

Fig.1 shows a simple ventilation network. Fresh airflow passes branch 1, then enters respectively through branches 2 and 3, and joins together in branch 4. Branch 5 connects to another working face. In order to ensure the airflow quantity in branch 2, a regulating air door is set in branch 3. If fire occurs in branch 2, fumes will pass only through branch 4.

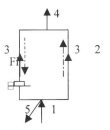

Fig.1 Illustration of Characteristic Zone

With the increase of thermal draft, the airflow rate in branch 3 reduces. When the intensity of fire source keeping normal direction increases to a certain value, the airflow direction in airways may reverse. Accordingly, the intensity of fire source under the certain value is defined as the first characteristic zone, and for different intensity in this zone, the airflow direction in every branch is same. In the first zone, the airflow direction needn't be controlled because fume flows directly out from return airway 4. If the intensity of fire source is over that certain value, the second zone should be defined and in that zone, the airflow in branch 3 reverses. Fume will back to the end junction of branch 1 and invade working face in branch 2 as well as pass through branch 5 to endanger another working area. Therefore , a regulator should be set on branch 2 to decrease airflow quantity and the regulator in branch 3 should be open in order to increase the quantity of branch 3, and avoid the reversal of airflow in branch 3. When airflow state dynamic simulation is involved, if the estimated fire intensity and real fire intensity share the same characteristic zone, the control plan, which is obtained according to the

estimated fire intensity, can achieve a satisfactory result of airflow control in that real fire fighting. The plan, according to the qualitative and quantitative synthetic analysis method, is feasible if the characteristic zone is estimated correctly and the estimated fire intensity shares the same characteristic zone with the real fire intensity. Consequently, the dot-to-dot relation between the estimated and the real fire intensity, which is very difficult to meet, is turned to the dot-to-region relation between estimating and the real fire intensity, which is much easier to achieve.

In addition, it should be noticed that the airflow control plans for two adjacent zones generally can be combined. As an example in Fig 1, for fire fighting plan, it don't need to control airflow because fume enters directly into return airways in the case of less fire intensity in branch 2. In order to prevent that the increscent intensity causes airflow reversal, the airflow control measure, which is corresponding to the characteristic zone with greater fire intensity is applied in advance. It means that the airflow control measure in the second zone for the developing stage of fire, to set a regulator in branch 2 and open the regulator in branch 3, is achieved in advance in the first zone for the early stage of mine fire. It can prevent the airflow reversal while the fire intensity spreads into the second characteristic zone. Because it is impossible to change the airflow control measure frequently during fire fighting, the above characteristic of fire airflow control plays an important role to ensure the airflow control plan meeting the dynamic change of fire source intensity. Therefore, the estimated error between the estimated fire intensity and the real fire intensity doesn't influence greatly the correct selection of fire airflow control plans.

Consequently, the characteristic zone method solves following problems.

(1)If the estimated fire intensity and the real fire intensity share same zone, the airflow control plan, got from the simulation, can be regarded as a real fire control plan. It solves the difficulty of estimation of the fire intensity when applying fire state simulation technology.

(2)When mine fire precautionary disposal plan is developed, it needn't try various airflow control plans in relation to the different fire intensity for a certain fire, but need only to select a fire source intensity from each characteristic zone to obtain the corresponding airflow control plan.

4 DETERMINING THE CHARACTERISTIC ZONE OF THE FIRE SOURCE

A possible fire intensity region is determined firstly for a certain location of fire, and then this region is divided into the different zones. The whole region, which is between the possible maximum and minimum fire intensity, should be determined by analyzing the species, amount and distribution of the fuel, the airflow rate of the airway and the environmental condition.

It's key point to divide the fire characteristic zones by finding the critical point of the zone. There are two methods to determine the critical point.

(1)Sequential Search Method

It has been known that the airflow state reversal caused by mine fire is similar in the same fire characteristic zone. Applying the airflow state simulation method, the different fire intensities are selected one by one to simulate airflow state starting with the lower limit of the fire intensity. When the state of airflow reversal is changed, the fire intensity characteristic will fall into another zone. The searching interval of finding critical point in the region can be determined according to the actual situation. When airflow reversal changes frequently with the variation of fire intensity, the selected interval should be smaller in case larger error in searching the zone critical point, and vice versa. All zone critical points will be found by searching from lower limit to upper limit of the fire intensity region.

(2)Two-end Approach Search Method

When finding the critical point of the zone, it can search beginning with the upper limit and the lower limit synchronously as follows.

Simulating by using the upper limit and the lower limit of the fire intensity characteristic respectively, there is no critical point in this zone if the simulation result shows that there is same airflow reversal. If there is the critical point, the simulation will go on by selecting the middle point between the upper and lower limit and there will be three kinds of simulation result as follows:

Firstly, the simulation result in the middle point shares same airflow reversal result with the upper limit. It means that there is no the critical point between the middle point and the upper limit, and the critical point is in the region between the middle point and the lower limit.

Secondly, the simulation result in the middle point shares same airflow reversal result with the lower limit. It means that there is no the critical point between the middle point and the lower limit, and the critical point is in the region between the middle point and the upper limit.

Thirdly, the simulation results in the middle point differ from the airflow reversal result in the upper and lower limit. It means that there are the critical points between the middle point and the upper or the lower limit respectively.

Repeating above steps to reduce the searching scope of the critical point, until finding all critical points. The fire source characteristic zones are determined too.

5 SELECTING PRINCIPLE OF THE AIRFLOW CONTROL PLAN DURING FIRE

By combining the fire characteristic zone method with the qualitative and quantitative synthetic analysis method, the better decision making of airflow control can be achieved during mine fire fighting. At that time, the optimal state control plan should be selected from various plans which can achieve same airflow control result. There are the following principles which should be considered when selecting the plan.

(1)The reliability of the airflow control effect of the selected plan. One should select the reliable control plan if the implement of the plans has similar difficulty.

(2)Difficulty of installing the regulator. It's not safe to install the regulators in return airways full of fume, so that the remote control regulator should be preinstalled in that position.

(3)The number of the airflow control regulators. The number of air doors being regulated should be reduced as small as possible.

(4)The airflow quantity in airway on fire must exceed a certain value to prevent explosion of combustible gases especially in high methane mine.

(5)The airflow control plan should have common suitability for different fire location.

(6)The airflow control plan gives smaller influence on other mine section.

In order to testify the feasibility of the methods mentioned above, they have been applied to Xian Dewang coal mine and XingTai coal mine to make the mine fire prevention and fighting plan. The practical application has proved the feasibility of the method.

6.CONCLUSION

The qualitative and quantitative synthetic analysis method has been proved to be a reliable method for airflow state control during a mine fire. It matches the present level of technology and equipment in mines around the world. The characteristic zone method solves the problem in estimating the intensity of a practical mine fire.

The combination of the qualitative and quantitative synthetic analysis method and the characteristic zone method plays an important role in setting up a better decision making system for mine fire fighting in situ.

REFERENCES

Greuer R.E.,1983, "A Study of Pre-calculation of effect of fires on Ventilation System of Mines", USBM Contract J0285002, OFR 1984,(NTISPB84-159979), pp293.

Greuer,R.E.,1988,"Computer Models of Underground Mine Ventilation and Fire", Recent Development in Metal and Nonmetal Mine Fire Protection , USBM IC. 9206.

Zhou,Xinquan & Wu bin, 1996, The Theory and Practice of Mine Fighting, China Publish House of Coal industry. P406.

Zhou,Xinquan & Greuer R.E,1994. "Specialized FORTRAN Computer Programming and Analysis Services to Upgrade Capability of MFIRE Program", USBM, Contract N.O P0241054,Bureau of mines, USA.

Mining Science and Technology' 99, Xie & Golosinski (eds)© 1999 Balkema, Rotterdam, ISBN 90 5809 067 1

The evaluation of fire hazard in underground coal mines

A. Strumiński & B. Madeja-Strumińska
Institute of Mining Engineering, Technical University of Wrocław, Poland

ABSTRACT: Two methods of testing the evolution of fire hazard in the underground coal mines are presented in the paper. The first concerns the fire hazard evaluation in cavings and is based on the courses of average speeds of the chemical reactions between oxygen and coal in gobs. In the second method the amount of oxidised coal in the mining sections and from the time changes of the coal oxidation intensity we can conclude about the spontaneous fire hazard in the tested mine region.

1 INTRODUCTION

The rise of underground fire is connected with presence in the given area of the mine flammable material and the temperature high enough to initiate the burning process of this material. The flammable material ignition can occur particularly as a result of external heat source action with high enough temperature and long enough period of action as well as a result of exothermic physical and chemical changes leading to the self-ignition of flammable material. Therefore in the underground mining we can distinguish two types of fire i.e. exogenous and spontaneous fires [1,5].

The exogenous fires i.e. risen due to external reasons are caused usually by open flame, defective work of electrical and mechanical equipment, gas explosions and wrong running of blasting works. The exogenous fires may rise in every mine and in principle in each place. Usually they arise unexpectedly, without long worning symptoms. In coal mines, however, the most frequent are spontaneous fires i.e. caused by the coal self-ignition. They are mostly located in gobs, fissures of the fractured coal, accumulations of muck from the caved workings as well as in separated coal in the roof over the mine working support.

In this paper we concern only the evaluation of hazard caused by the spontaneous fires.

2 FIRE HAZARD EVOLUTION IN CAVINGS

Simultaneous occurrence of disintegrated coal inclined to low-temperature oxidation, air inflow to the aggregated coal as well as accumulation of the heat emitted during the coal oxidation is necessary to cause the simultaneous fire [1,5,6]. Such conditions are very common in face cavings in Polish coal mines.

The fire hazard is big when over the coal bed, strong roof rocks occur which break only in huge blocks, and then, the residues of coal bed left in cobs are crushed and the air inflow is easier due to leaky caving. The similar great risk of the fire occurrence in the gobs is when coal from upper beds is introduced into the caved rocks or when unmined fenders, ribs and coal pillars are left in the gobs.

When we evaluate the fire hazard in the gobs in Polish coal mines, we use different methods now [1,3÷6]. Generally, according to the mine regulations, we use Graham factor, carbon oxygen amount rate and carbon oxygen increase index.

In all methods of fire hazard evaluation in gobs used until now we assume that the carbon monoxide is the earliest product of coal oxidation in the process of self-ignition [1,5]. In the earliest phase of the coal self-heating process, however, in the site of fire focus we state the oxygen loss in the mine air, and only after that carbon monoxide occurs. Therefore it is reasonable to take bigger note on studying the coal oxidation process in the mine gobs. Among the other things we discuss one part of this problem in the paper.

The process of coal combustion is complicated [4÷6], multistage and includes in turn: heating (for example self-heating) when the water is evaporated, volatile parts emission and organic compounds reconstruction, combustion of volatile parts in gas phase, changes of mineral components of fuel and coke combustion. During the coal combustion, ash and nitrogen form so called ballast.

From the theoretical point of view the fire hazard evaluation in, for example, gobs isolated from the open drives, could be based on the speed of chemical reaction between oxygen and coal located in gobs.

The real (temporary) speed of chemical reaction is defined by the differential equation [2]

$$\vartheta = -\frac{dc}{d\tau} = \frac{dr}{d\tau} = k(T)C^n \qquad (1)$$

where $\dfrac{dc}{d\tau}$ is the reduction of substrate concentration in the time unit, $\dfrac{dr}{d\tau}$ the increase of product concentration, $k(T)$ – constant reaction speed in T temperature, n – summary reaction grade.

As it follows from the formula (1) in order to determine the speed of any chemical reaction it is necessary to now the reaction grade n. In practice, due to complexity of the coal oxidation process in cavings, determination of the reaction grade is difficult. Therefore we suggest using, in such case, AB index based on the average speed of reaction [2] between oxygen and coal in cavings, which is defined by the following formula:

$$AB = \frac{r_{O_2} - r_{O_2}^{\bullet}}{\tau_2 - \tau_1} \qquad (2)$$

where r_{O_2}, $r_{O_2}^{\bullet}$ are oxygen molar concentrations at the initial τ_1 and final τ_2 time of the time range $\Delta\tau = \tau_2 - \tau_1$ respectively.

The average speed of the oxygen reaction with coal can be different. The highest reaction speed is during the advanced combustion process, and the lowest when the flammable material oxidation is at the initial stage. Therefore knowing average speeds of oxygen – coal reaction in cavings we can observe there the fire hazard evolution. We assume at the same time, that the oxygen loss in cavings occurs mostly as a result of carbon oxidation process and we do not record the oxygen displace by methane or carbon monoxide. If there is accumulation of methane or carbon monoxide in cav-

ings, we can assume that the coal oxidation is impeded and the coal self-ignition will not occur.

The presented method of the fire hazard evaluation in gobs concerns the earliest stage of coal oxidation period (self-heating) i.e. the period when we do not record the carbon monoxide presence. If we already state the CO presence, we can use indexes other to AB, like Graham, Morris, Trickett, Bystroń [6] index, to evaluate the fire hazard. Using the indexes, which need measurements of carbon dioxide, as Young, Willet index, may lead to wrong conclusions in cases of natural CO_2 emission in cavings.

3 FIRE HAZARD EVALUATION IN EXPLOITATION AREAS

In the exploitation areas air composition in normal conditions (non emergency conditions) show generally small deviations from the atmosphere composition. During the under-ground fire, however, in the mine air, gases which do nor occur in the atmosphere are present [1,4,5]. They are carbon monoxide, methane, hydrogen and aliphatic hydrocarbons. A consi-derable amount of the carbon dioxide is produced during the fire.

The outside symptoms of the coal self-heating in the initial phase manifest themselves in composition changes of air flowing through the fire focus. In particular we can state oxygen amount decrease in the air, its higher humidity as well as the increase of the carbon dioxide. At the same time small amounts of carbon monoxide are formed. At the final stage of self-heating coal temperature is 100°C or higher, and in the air flowing out of the region so called "fire smells" (aromatic hydrocarbons) arise.

When the air flows through the mine workings located in the certain exploitation area, the oxygen loss is caused mainly by different substances oxidation, especially coal. Therefore we can assume that carbonate substance oxidation in the exploitation panels is a process of slow coal combustion without light effect.

In the initial stage of coal self-heating, as it was said before, beside of oxygen loss in the mine air, we can record increase of carbon dioxide concentration, but in general carbon monoxide does not occur. Thus during this period we may say about the full combustion, because at that time in mine air tare are not flammable components, except the possible natural methane. In that case we can assume that at this stage coal combustion runs according to reaction $C + O_2 = CO_2$, it means that to combust 12 kg of coal we need 32 kg of oxygen, and thus for 1 kg of oxygen fall 12/32 kg of coal and for m_{O_2} kg of oxygen - $m_{O_2} \cdot 12/32$ kg of coal [4].

If though the mining region flows the air stream $\dot{V}$ (m^3/s), and the oxygen concentration in volume fraction on the region inlet and outlet is c_{O_2} and $\overset{*}{c}_{O_2}$ respectively, the oxygen stream used in the oxidation process is:

$$\dot{V}_{O_2} = \frac{c_{O_2} - \overset{*}{c}_{O_2}}{100} \dot{V} \tag{3}$$

and the stream of the oxygen mass is:

$$\dot{m}_{O_2} = \rho_{O_2} \dot{V}_{O_2} \tag{4}$$

where ρ_{O_2} is oxygen mass density in kg/m^3.

In relation to this the amount of coal being combusted (oxidised) in the time unit can be calculated from:

$$\dot{m}_C = \frac{12}{32} \rho_{O_2} \frac{c_{O_2} - \overset{*}{c}_{O_2}}{100} \dot{V} \tag{5}$$

Evaluating the fire hazard state in the mining region where coal is self-heated, for examples in the fractured unmined coal, muck accumulated due to cavings occurred in the workings or uninsulated gobs, we assume that at the big amounts of combusted (oxidised) coal, fire hazard in the mining panel in big, and for small amounts is small.

Studying the changes of oxidised coal mass in the mining panel in the course of time, we can define the tendency of fire hazard changes i.e. its evolution. It allows, therefore, to estimate the increase or decrease of the fire hazard in the mining region.

The presented method of fire hazard estimation concerns the earliest period of the coal self-heating i.e. period where there are not any flammable gases, especially carbon monoxide, in the mine air. If the process of coal self-heating is advanced and in the air flowing from the mining region, carbon monoxide is present, then the coal combustion into the

CO_2, we have also coal combustion into CO according to the reaction $C + 0.5O = CO$. In that case from combustion of the 12 kg of coal at the presence of 16 kg of oxygen, 28 kg of carbon monoxide is formed. Thus for the 1 kg of carbon monoxide fall 16/28 kg of oxygen, and for m_{O_2} kg of carbon monoxide fall $m_{O_2} \cdot 16/28$ kg of oxygen. The steam of $\dot{m}_{CO}$ carbon monoxide in the air flowing through the mining region can be calculated from the equation:

$$\dot{m}_{CO} = \rho_{CO} \frac{c_{CO} - \overset{*}{c}_{CO}}{100} \dot{V} \tag{6}$$

where c_{CO} and $\overset{*}{c}_{CO}$ – carbon monoxide concentration in the volume fraction at inlet and outlet from the mining region respectively, ρ_{CO} is density of the carbon monoxide mass.

The stream of the oxygen mass $\overset{*}{m}_{O_2}$ (kg/s) which reacts with coal forming carbon monoxide is described by the following relationship:

$$\overset{*}{m}_{O_2} = \frac{16}{28} \rho_{CO} \frac{c_{CO} - \overset{*}{c}_{CO}}{100} \dot{V} \tag{7}$$

When the summary oxygen loss in the mining region is $\Delta O_2 = r_{O_2} - \overset{*}{r}_{O_2}$,, then without taking into consideration the oxygen emitted form the fuel during the oxidation process, we can calculate the stream of the oxygen loss $\overset{**}{m}_{O_2}$ (kg/s) connected with the carbon dioxide formation as follows:

$$\overset{**}{m}_{O_2} = \left(\rho_{O_2} \frac{r_{O_2} - \overset{*}{r}_{O_2}}{100} - \frac{16}{28} \rho_{CO} \frac{c_{CO} - \overset{*}{c}_{CO}}{100} \right) \dot{V} \tag{8}$$

The amount of the coal combusted into the carbon monoxide $\overset{*}{m}_C$ (kg/s) and carbon dioxide $\overset{**}{m}_C$ (kg/s) we calculate from the formulas:

$$\overset{*}{m}_C = \frac{12}{16} \overset{*}{m}_{O_2} \tag{9}$$

and

$$\dot{m}_C^{\bullet\bullet} = \frac{12}{32}\dot{m}_{O_2}^{\bullet\bullet} \tag{10}$$

Thus $\dot{m}_c$ (kg/s) of coal is combusted in total of

$$\dot{m}_C = \dot{m}_C^{\bullet} + \dot{m}_C^{\bullet\bullet} \tag{11}$$

Taking into consideration that in the combusted coal is p percent of the pure carbon, we can calculate an approximate mass of the combusted material in the time unit

$$\dot{m}_s = 100\frac{\dot{m}_C}{p} \tag{12}$$

Studying the changes of oxidised coal amount in the course of time in the mining panel, we can define the tendency of fire hazard changes in this region.

The presented method of the fire hazard evaluation can relate to one part of the mining region or to single workings as well.

4 EXAMPLES OF FIRE HAZARD EVALUATION IN COAL MINES

Methods of the fire hazard estimation in the underground hard coal mines presented in the paper we exemplify with two examples from the mining practice.

Example 1

In N mine in gobs of 414 longwall, mined towards the field boundaries using roof caving method, a fire originated [6]. After the fire had been dammed-in, gas samples were taken regularly from behind the T_2 dam (fig. 1) for the chemical analysis. The oxygen concentrations in the dammed fire field, starting from the fifth day after closing the dams are presented in the table 1. Percentage and molar concentrations of oxygen and average speeds of chemical reactions causing the oxygen loss i.e. average values of the AB index, are showed in the table too.

It results from these calculation, that big 10^3 AB index was recorded between 6 and 7 day and between 9 and 10 day, what means that at that time fire activity increased. As it turned out, in dammed-in cavings methane explosion probably occurred,

and the methane burned out, what was suggested by miner observations. It results from data analysis, however, that 10^3 AB index was reduced in course of time, so we can assume that fire gradually went out. Negative values of 10^3 AB index show about disturbances of chemical reaction of combustion process after periodical feeding with nitrogen to the fire fields from special tanks.

This example shows that using AB index (average reaction speeds) we can analyse fire hazard evolution in cavings fairly well, and even conclude about the methane or other fire gases explosion, about unsealing the dammed gobs or fire fields etc.

Table 1. Specification of data for AB indexes value calculation in 414 longwall cavings in N mine

Time τ, days	O_2 concentration		Average index 10^3 AB, mol/($m^3 \cdot$ day)
	r_{O_2}, %	r_{O_2}, mol/m^3	
5	20.20	9.017	
			35.78
6	20.12	8.981	
			2606.90
7	14.28	6.374	
			1504.37
8	10.91	4.870	
			549.07
9	6.68	4.321	
			2517.75
10	4.04	1.803	
			-1013.18
11	6.35	2.834	
			803.52
13	4.55	2.031	
			669.60
14	3.05	1.361	
			245.52
15	2.50	1.116	
			-848.16
17	6.30	2.812	
			301.32
25	0.90	0.401	
			-15.19
62	2.12	0.946	
			128.57
67	0.68	0.303	

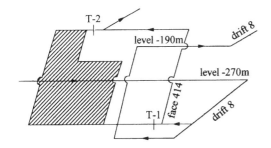

Fig. 1. 414 face cavings in N mine

Example 2

This example concerns G-VI mining section in A mine [7] where coal was extracted in bed no 507. In this region the bed was 3.8 m. thick and its dip was 4°. Up-dip longwall transverse method with roof caving was used there (fig. 2). 1300 tonnes of coal was mined per day from the face no 1001. Tests on coal inclination to self-ignition showed that coal in G-VI region has a strong inclination to self-ignition. The amount of air flowing through the face

Table 2. Parameters and calculations for fire hazard estimation in G-VI section, 507 bed, mine A

Date	Measuring station	O_2 %	$c_{O_2} - c_{O_2}^{*}$ %	$\dot{V}$ m^3/s	$\dot{V}_{O_2}$ m^3/s	$\dot{m}_{O_2}$ kg/s	$\dot{m}_C$ kg/h	$\dot{m}_s$ kg/h
01/08/97	S_1	20.71	0.11	6.33	0.007	0.01000	13.5	20.149
	S_2	20.60						
05/08/97	S_1	20.71	0.15	6.67	0.010	0.01429	19.296	28.8
	S_2	20.56						
08/08/97	S_1	20.72	0.19	6.83	0.013	0.01858	25.092	37.451
	S_2	20.53						
12/08/97	S_1	20.60	0.23	7.60	0.017	0.02429	32.796	48.949
	S_2	20.37						
15/08/97	S_1	20.66	0.29	6.92	0.020	0.02858	38.592	57.6
	S_2	20.37						
18/08/97	S_1	20.71	0.26	7.00	0.018	0.02572	34.740	51.851
	S_2	20.45						
21/08/97	S_1	20.60	0.30	6.96	0.021	0.03001	40.5	60.448
	S_2	20.30						
24/08/97	S_1	20.64	0.38	6.85	0.026	0.03715	50.148	74.848
	S_2	20.26						
28/08/97	S_1	20.61	0.32	6.94	0.022	0.03144	42.444	63.349
	S_2	20.29						
30/08/97	S_1	20.67	0.37	6.21	0.023	0.03287	44.388	66.251
	S_2	20.30						
02/09/97	S_1	20.70	0.36	7.12	0.026	0.03715	50.148	74.848
	S_2	20.34						
06/09/97	S_1	20.75	0.30	8.02	0.024	0.03430	46.296	69.099
	S_2	20.45						
13/09/97	S_1	20.62	0.39	6.90	0.027	0.03858	52.092	77.749
	S_2	20.23						
16/09/97	S_1	20.75	0.31	8.30	0.026	0.03715	50.148	74.848
	S_2	20.44						
20/09/97	S_1	20.72	0.37	8.00	0.030	0.04287	57.852	86.346
	S_2	20.35						
23/09/97	S_1	20.67	0.46	6.80	0.031	0.04430	59.796	89.248
	S_2	20.21						

was between 380 and 500 m³/min. Measuring stations for early spontaneous fire detection were located on transport incline I (station S_1, fig.2) and in ventilation incline II (station S_2). Detailed data - about chemical composition of air samples taken at these measuring stations as well as results of calculations made using formulas 3 to 5, 9 and 11 are presented in Table 2. They show that the amount of oxidised (combusted) air was regularly increasing from 20.149 kg/h to 89.248 kg/h. It was assumed that at the end of testing period in G-VI region, intensive coal self-heating occurred. Therefore preventive actions like gobs sealing on the caving line or gypsum and anhydrite wash (milk) forcing into transport incline walls on sections where coal was fractured, were taken up. These actions allowed to finish the face exploitation with no damage and to fill the mining region.

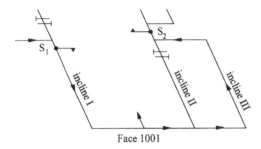

Fig. 2. G-VI section in bed no 507, coal mine A

5 CONCLUSIONS

The methods of fire hazard evaluation in coal mines, presented in the paper, are innovative with regard to this problem. They need further tests and practical, industrial verification. But their usefulness for evaluation of spontaneous fire hazard evolution, especially in cavings and working mining panels can be now stated.

REFERENCES

[1] Bystroń H., Jaroń S., Kołodziejczyk B., Markefka P., Strumiński A.: *Miner handbook* (in Polish). Wyd. Śląsk, Katowice

[2] Całus H.: *The basics of chemical calculation* (in Polish). Wyd. Naukowo-Techniczne, Warszawa 1971

[3] Madeja Strumińska B., Nędza Z., Strumiński A.: *Evaluation of spontaneous fire hazard in longwall cavings* (in Polish). Wrocław 1994

[4] Madeja Strumińska B., Strumiński A.: *Mining aerothermodynamics* (in Polish). Wyd. Śląsk, Katowice 1998

[5] Strumiński A.: *Fighting fires in underground mines* (in Polish), Wyd. Śląsk, Katowice 1996

[6] Strumiński A.: Madeja Strumińska B.: *Evaluation and liquidation of spontaneous fire hazard in coal mines* (in Polish). Wyd. DWE, Wrocław 1997

[7] Strumiński A.: Madeja Strumińska B.: *Fire evolution assessment in dammed-in fire fields*, (in Polish). International Conference on Mining Rescue, 90 Anniversary of Mining Rescue in Poland 1901-1997, Bytom, Lubliniec-Kokotek, 25-27 Novem-ber 1997, Central Mining Rescue Station in Bytom

Mining Science and Technology' 99, Xie & Golosinski (eds) © 1999 Balkema, Rotterdam, ISBN 90 5809 067 1

Study on 3-D dynamic monitoring of spontaneous combustion of coal seam

J.M.Zhang, S.N.Ning, S.L.Wang & H.Y.Guan
China University of Mining and Technology, Beijing, People's Republic of China

Z.Vekerdy & J.L.van Genderen
International Institute for Aerospace Survey and Earth Sciences (ITC), USA

Abstract: In China heavy loss of lives and coal resources result from spontaneous coal combustion. The project, 'Coal fire Fighting and Monitoring System in China', was undertaken to counteract this. It was developed in Rujigou coalfield, the most representative coalfield in the north of China. Performed by BRSC (China), EARS, ITC, and NRTG (Netherlands), the project was focused on a dynamic monitoring and operational ability of fire extinguishing thus to provide technically feasible system for environmental restoration and management. The study combines detective approaches, data management and analysis methods with the monitoring system and uses 3-D logical concepts and interpreting methods. GIS and 3D-visualization software support data management and dynamically show visual effects of the fire control strategy.

1. INTRODUCTION

1.1 *Previous study*

Coal fire, in general, is detected by using remote sensing method, because the thermal anomalies appeared when the heat released from fire is transferred through the surrounding rocks to surface. Based on this thermal features, Point model was presented to interpret it (Genderen, 1996). Some applications at Xinjiang, Ningxia and other places in the north of China demonstrate that the thermal IR is an effective method for monitoring SCCF. Another case study of coal fire in Jharia coalfield of India (Saraf, 1997) proved its effectiveness too. According to detecting the residual magnetism in rocks after coal fire, The magnetic detection (Zhang X.S, 1986 and Wang Z.C, 1996) that can indicate the area effected by SCCF is proved to be another effective approach. All results from above works that use remote sensing approaches are displayed with plane and profile images.

1.2. *3-D dynamic monitoring*

Recently, with the enhancement of coal fire hazards, their impacts have been intensively expanded. In order to treasure resources, reduce the SCCF impacts on the environment and make ecological circumstance favorable to human being, fire fighting projects have been planned and carried out by the government. Dynamic monitoring technique has been applied to provide the continuous information of SCCF. However, the 2-D and profile images that are acquired with the previous methods can not describe the real 3-D SCCF situation. It is difficult for

fire fighter and decision-maker to find the effective methods and make reasonable design for fire fighting. It is very urgent to provide the situation of coal fire in 3-D, which would give the realistic results of resources lose and potential impacts on the surrounding environment.

2. THE FUNDAMENTAL FEATURES OF COAL FIRE

2.1 *Development of coal fire*

The SCCF dynamic process that correspond with four zones spatially can be divided into four stages (See Fig.1). At the first stage, coal seams are affected by the external geologic action and periodic strong radiation from sun for a long time and it is named the oxidation zone because of suffering from weathering and oxidization. The following stage is characterized with temperature continuously rising of coal due to the oxidization happened in coal seam. It is named as the volatile zone and its temperature may increase to the critical value of spontaneous combustion eventually. If the temperature goes up to the critical temperature, the combustion will start at the area where the surrounding condition is favorable. Coal fire can spread over intensively when heat transfer is easy and there is enough air for maintaining combustion. If there are carbonaceous materials within the surrounding rocks, combustion will extend into rocks. The burning zone is formed with the feature of the high temperature. Coal fire will extinguish gradually, when carbonaceous material and coal are burn out. Coal is turned to ash and surrounding rock is turned to the burnt rock. These rocks can be classed by its burnt degrees such as baked rock, melted rock. The last stage is named extinguishing zone, refer to the places

where temperature goes down and the burnt rocks exist.
The observation of the spread of the coal fire shows that it starts at lower or middle part of coal seams and extends both upwards and downwards progressively. Underground coal fire spreads upward easily due to open cracks and fissures in the surrounding rock. Whereas, surface coal fire extends downwards where coal seam exists and is slower

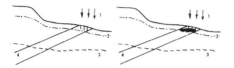

Weathering & oxidation stage   Spontaneous combustion stage

1. Radiation, 2. Temperature variation surface. 3. Underground water surface, 4. Coal seam, 5. Burning point

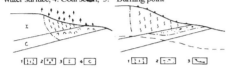

Burning stage.   Extinguishing stage

1. Burnt rocks and cracks zone. .   1. Crack zone. 2. Collapse area

2. Gases   3. Surrounding rocks   3. Surface slide area.

4. Coal seam

Fig. 1 the cultivated diagram of the SCCF

than underground fire. There are many factors related to SCCF, such as periodic solar radiation, human-induced heat and air supply. In the special geological and weather environment with rich coal resources, favorable mining condition in the northwest of China, the coal fire is easy to take place and causes some unneglectful impacts on the surrounding environment.

2.2 *Physical parameters for monitoring in 3D*

Along with fire developing, thermal anomalies induce physical actions on the surrounding rock. These actions change their physical properties of the surrounding rock, which include thermal stress, conductivity, magnetism and radiation etc. The geophysical field is constituted with these changing parameters. For 3D monitoring, these parameters used in measurement should contain the information about SCCF in a certain depth. As above analysis, if the above approaches are used for monitoring, heat and magnetism are two parameters to indicate the situation of SCCF in 3D. Therefore, temperature and magnetism fields are selected for studying the 3D situation of SCCF. If the imaging method are used, the deep information of SCCF would be extracted to generate 3D images for describing the spatial distribution of coal fire. These images would be favorable to make the various estimations with more reality by offering the real-time visual images.

3 THERMAL IMAGING - VOLUME EVALUATION

With recent development of sensors, there are greater progress in imaging method for extracting the deep information, such as EM and seismic imaging. These techniques are widely utilized to detect the structure of the shallow strata in earth, inner status of human body and quality of structure. Based on the study on the physical parameters, the information of inner structure could be deduced. The methods can extract the deep information of the objects because there are relationships between the detected objects and the physical fields. With the help of imaging methods, the object could be realized for the purpose of understanding inner structures.

3.1 *Thermal imaging-obtaining depth information*

In computer vision analysis, it is difficult to analyze the

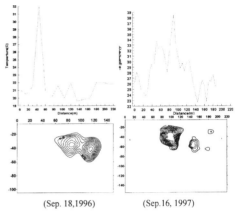

(Sep. 18,1996)   (Sep.16, 1997)

Fig.2 The thermal imaging section of Dafeng area, Rujigou coalfield, Ningxia, China

information content of image for each pixel. Indeed, the value depends partly upon the lighting conditions and contains some noise. However, it is more important for signal analyzing to the local variations of the image intensity, which is related to signal sources. When the local signal is analyzed, the region of the analysis scale, in which the relative resolution is computed, must adapted to the size of the objects analyzed. This size defines a correspondent resolution for measuring the local variations of the image. Generally, the structures we want to recognize have greatly different sizes. Hence, it is impossible to define an optimal resolution for analyzing images in prior. In order to identify the objects at various sizes, Mallat (1989) developed the pattern-matching algorithms, which can process the image at different resolutions. For this purpose, one can reorganize the image information by decomposing it into a set of details appeared at different resolution. Following this idea, wavelet methods were created for multi-resolution signal decomposition. When the idea of wavelet was introduced into decomposition of the thermal signal analysis, we get:

$$W_{\phi_{a,b}}(T(x)) = \frac{1}{\sqrt{a}} \int_{-\infty}^{+\infty} T(x)\phi(\frac{x-b}{a})dx \quad (1)$$

Where $T(x)$ is an analyzed thermal signal, $\varphi(x)$ is filter function at the scale a and the position b, $W_{\varphi_{a,b}}$ represents the filtered result at scale a and position b. If the scale is adjusted, its frequency will vary continuously and the different deep information of each imaging position could be resulted. Under the potential field, if frequency is varied with sampling length, the relative deep information could be obtained with the adjustment of scale.

Upon the thermal model analysis of SCCF, the model of point heat source is used for setting up kernel function for filter. The filter could figure the information about the position and the state of heat source, that is the direct indicator to the situation of coal fire.

3.2 Volume evaluation

The method of volume evaluation is that all states of coal fires are considered in three dimensions. From thermal imaging, a series of 3-D images (or volume image) could be found for the state analysis of SCCF. Above approaches for image processing can be used for quantitative study. For convenience, we assume that $M(x,y,z)$ represents spatial position, $T_i(M,a,b,c,\cdots)$ as the function related to SCCF situation at fixed time. $T_{t_i}(M,a,b,c,\cdots)(i=1,2\ldots)$ presents the status of SCCF at different time. Supposing that the different algorithms are symbolized as operator (Yang, 1997), 3D image can be given by following equation:

$$T_{t_2}(M,a,b,c,\cdots) = T_{t_1}(M,a,b,c,\cdots) \otimes \Delta T_{t_1}(M,a,b,c,\cdots)$$

Where $\otimes$ is an operator used for analysis, and Δ means the state variation during the period of $t_1 \square t_2$. Therefore,

$$\Delta T_{t_2}(M,a,b,c,\cdots) = T_{t_2}(M,a,b,c,\cdots) \otimes T_{t_1}(M,a,b,c,\cdots)$$

Where $\Delta T_{t_2}(M,a,b,c,\cdots)$ is volume image deduced by operator between the images obtained at different time. The consequence would take different forms such as residual image or accumulated image, according to different operator selected upon the images. They can present different state of SCCF in 3D format.

Rujigou coalfield, located at the north part of Ningxia Hui autonomous region of China, is selected for the imaging experiment. Temperature measurement was finished respectively in 1996 and 1997 for dynamic monitoring. The graphs in Fig.2 show the temperature anomalies over coal fire. After the graphs are re-sampled based on possible resolution of satellite image, sectional images at different times are constructed using the imaging method. Each image indicates the position of the heat sources. The difference between images at separate duration reveals the movement of heat source.

The example offers the possibilities to present 3-D image

for describing the spatial process of coal fire. Based on imaging process, the above operational detective methods for monitoring coal fire can be used for acquiring remote sensing and ground geophysical data. Imaging method will be put on the extraction of coal fire information. According to the dynamic features, the analysis method of image sequences (Thomas, 1983, S.N. Ning and S.L Wang, 1995), which is available to understanding spatial spreading of underground coal fire, has been provided. Developing tendency and spreading direction of coal fire has been estimated with the dynamic image analysis. It can forecast how the coal fire will evolve 3-D image, which is identical to the spatial development of coal fire, and presents the visual results and spatial extents. With the help of the volume images, various estimations could be made, such as the loss of coal resources, the affected region, and the different kind of harmful gases that are contributed to atmosphere from underground coal fire. The evaluating results will be more correct and realistic.

4 3-D SYSTEM DESIGN

The monitoring system consists of effective detective methods, data processing, analytical models and fire fighting-orientated software with which the functions of data management, analysis and 3-D visualization of product will be provided. According to the imaging idea of 3D-figure extraction, the volume analysis can be introduced into the system design. With its development, new functions are added in, such as environment assessment, the history study and the future forecasting of SCCF, aided-design for the layout of fire fighting, information issue and allocation of resources for fire fighting etc.

4.1 Software support

The system structure designing is based on the process features of SCCF, the detecting approaches, the data analysis and the extraction of deep information, the requirement of time-series analysis, the dynamic display and output. DBMS, ILWIS and Visual Basic support the current development of the system for programming. 3D visual software is used for data management, analysis, function development, and 3D visualization.

4.2 System structure and function

For convenience, the present software of GIS and the image processing serve to the development of the 3-D system. Fig.3 shows the system structure, containing data management, 2D and 3D data analysis, visualization of 3D images. The last output will be visualized to end-user in different levels. The functions is as follows:

- Management of data base
- Image processing
- Digital imaging
- Tool package (fire detection, hazard evaluation, loss evaluation, allocation of fire region, priority analysis,

risk analysis for mining, auxiliary design, support for decision-making)
- Study on the development history of coal fire
- Information assistance (data management, process of SCCF and affected parameters, estimation methods, detective tools, image processing for SCCF).

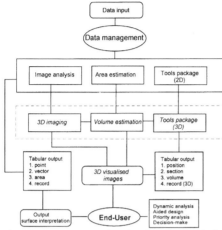

Fig. 3 the system structure based on the volume analysis of SCCF

4.3 data-flow process

Following the way indicated in Fig.4, data could be classified into original, analytical and resulted data. Data standardization, geo-coding and various corrections are finished before it is analyzed. Imaging methods will be used for the extraction of depth information. The analysis of volume images would be operated for constructing the 3D images to describe SCCF. The visualized result is presented to user under the support of 3D visualized software.

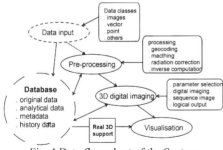

Fig. 4 Data flow-chart of the System

5. CONCLUSIONS AND FUTURE WORK

The developing process of SCCF can be divided into four stages, featured with weathering and oxidation, spontaneous heating and temperature increase, intense burning and gradually extinguishing. With SCCF's development, thermal anomalies and thermal residue

magnetism are the direct indicators, containing the information of SCCF in 3-D. 3D-volume image, available to describe SCCF in three dimensions, could be generated using imaging approaches. It provides the possibility of the volume evaluation on the development and influences according to volume alteration of SCCF, with reality, real-time and visualization. It supplies the end-user powerful support on the data management and decision-making.
The further development will be based on 3D features of SCCF and focus on:
- The processing methods of volume image
- The approaches of risk analysis on mining operation and resource utilization
- Fire fighting priority
- Aided design and process control for coal fire fighting
- PC-based 3D visualization support

6 ACKNOWLEDGEMENT

The authors would like to thank the members of BRSC (China), ITC and EARS (the Netherlands) for their support.

REFERENCES:

Genderen, J.L. van, Cassells, C.J.S., and Zhang, X.M., 1996, The synergistic use of remotely sense data for the detection of underground coal fires. *International Archives of Photogrammetry and Remote Sensing.* Vol. XXXI, Part B7, Vienna, 9-19 July .

Ning S.N. Wang S.L. 1995. *Image processing of remote sensing and application.* Seismic Press, Beijing.

Saraf A.k., Prakash A., Sengupta S. and Gupta R.P. Landsat TM , 1995, Data for Estimating Ground Temperature and Depth of Subsurface *Coal Fire in Jharia Coal field, India. International Journal of Remote sensing,* 16(12): 2111-2124.

Stephanie G. Mallat. 1989. A theory for multi-resolution signal decomposition: *The wavelet representation. Transaction on Pattern Analysis and Machine Intelligence* IEEE, 11(7): 674.

Wang Z.C, 1996, Detection of fire boundary using geophysical methods. Study on geophysical features of rock for exploration, pp237-246, *Coal* Press of China.

Yang C.T. and Chen Z.Y., 1997, The algorithm of 3D positioning for residual objects, *Signal Processing.* Vol. 13, No.2, pp177-181.

Thomas S. Huang T. 1983. Image Sequence Processing and Dynamic Scene Analysis. NATOASI Series. Series F: *Computer and Systems Sciences* No.2,

J.M. Zhang S.N. Ning Y. Cao 1998.The environmental impacts of spontaneous combustion of coal seam and related fire in northern China and administration study. *Disaster Reduction of China.* 8(1) pp55-66.

Mining Science and Technology' 99, Xie & Golosinski (eds)© 1999 Balkema, Rotterdam, ISBN 90 5809 067 1

Research on dust diffusion and distribution during coal-cutting at the fully mechanized face

Haiqiao Wang & Shiliang Shi
Xiangtan Polytechnic University, Hunan, People's Republic of China

Hong Xie & Xiusheng Chu
Xingtai Mining Industry Group Limited, Hebei, People's Republic of China

ABSTRACT: Dust diffusion features and its concentration distribution are analyzed during coal-cutting at the fully mechanized working face based on the theory of turbulence hydrodynamics in this paper. The dust diffusion during coal-cutting is a synthesized diffusing process, which consists of the motion along the main airflow and the transverse diffusion caused by the turbulent ventilation. The dust diffusion equation during coal-cutting at the working face is derived from the mass conservation law, and then the math model of the dust distribution is established. This paper presents a practical sample analyzed with the math model. The dust distribution around the driver is calculated, and the theory analysis results match the practical measuring data approximately. At last, the relation among the dust concentration around the driver, the amount of dust-producing and the air velocity at the working face is analyzed, and the reasonable range of the air speed is deduced. The above research achievements show the biggish practical significance to the dust control at the fully mechanized working face in coal mine.

1. INTRODUCTION

There is a great deal dust generated in the coal mining operation, especially in the process of the coal-cutting, coal-transporting, support-moving and conveyor-moving. Because of the continuous operation of the coal mining machine, the dust is greatly generated, and over 60~80% of the dust at the working face is generated in its coal-cutting process. At the fully mechanized working face, the dust mainly consists of the coal dust, and the dust concentration is about 60~4000 mg/m^3, and the most of which is the harmful respiration dust. The dust generated at the working face can not only cause the explosion disaster, but also pollute the operating environment, and is harmful to miner health. The machine driver, who works in the environment with high concentration of dust, would be greatly damaged by the respiration dust. Therefore, research on the law of dust diffusion during the coal-cutting process is important for dust prevention at the fully mechanized face.

2. THE DIFFUSION CHARACTERISTICS OF DUST DURING COAL-CUTTING PROCESS IN FULLY MECHANIZED MINING FACE

The contaminants(e.g. dust) in airflow will keep the flow properties of airflow(i.e., momentum, energy and temperature, etc.). The process, in which the substances are transported from one location to another, is called the transportation process. There are several factors related to the transference, one of which is diffusion. The diffusion is a process that the substances contained in airflow move from higher density to low one, and it can be induced by the molecule motion, so called molecule diffusion. In turbulent flow, the diffusion also occurs because of the turbulent motion of the mass groups , which is called turbulent diffusion. Therefore, the turbulent diffusion is with the feature of the turbulent airflow at the working face.

The turbulent flow is with a complex motion that consists of several types of the whorl flow, and shows the characteristic of the diffusion and the irregularity. The turbulent motion is not the intrinsic characteristic of the fluid such as viscidity, but a special type of flowing motion. Therefore, its features are closely related to the boundary condition of the fluid. In mine ventilation, it is mainly affected by the air velocity and the friction coefficient in airway. There are different kinds of the turbulent flow due to the different boundary condition.

The speed of the turbulent diffusion is about 10^5~10^6 times as larger as that of the molecule diffusion, so that the molecule diffusion is ignored in analyzing the disturbance of air motion. At the fully mechanized mining working face, because of the action of the friction resistance at the face, the distribution of the air velocity is not even in the cross section of the face. The motion of the dust, called synthesized diffusion, is caused by the turbulent diffusion and the dispersion.

When the mining machine cuts the coal continuously, it can generate a great deal of dust by breaking off the coal with the bite of the drum. The dust is mixed with and carried by the airflow. This flowing motion appears the feature of horizontal transportation along the main current of the airflow and of the vertical transferring motion by the influence of the concentration grads of the contaminant. In the turbulent airflow, because of the pulse diffusion of air, the dust is diffused along the cross section of the face. At the working face, the motion of airborne dust is also the synthesized action with both the horizontal transportation and the vertical turbulent diffusion process. If only considering the turbulent diffusion generated by the cutting drum against the wind, the performance of the turbulent diffusion could be observed in the ventilation simulation tunnel when the dust is released at one side of the tunnel, the size of which being 550×600×6000mm. The experiment result is described in Fig.1 when the air velocity is 0.72 m/s.

Fig.1 The photo of the dust diffusion in simulating way vs. air speed 0.72 m/s

3. CALCULATION OF THE DIFFUSION CONCENTRATION OF DUST AT DRIVER'S LOCATION

At the fully mechanized mining face, if the mining machine cuts coal continuously and the airflow is in the turbulent state, the cutting drum is considered a continuous and even dust source. The effect of the dust on the density of the airflow and the second-rising dust on the concentration in air current are ignored. Therefore, the motion of the dust in airflow is the synthesized diffusion.

The dust diffusion during the process of the coal-cutting appears mainly as the transfer flow along the main air current and the crossing diffusion due to the

turbulent ventilation, which is a synthesized diffusion process. The turbulent diffusion makes the dust radial motion mixed with air, and the longitudinal diffusion of the fluid affects the spreading of the dust along the main current. The former is caused by the speed pulse of the turbulent flow, and the latter is resulted from the changes of the turbulent speed at the cross section of the working face. If the machine cuts coal continuously, the process of the synthesized diffusion could be simplified into the two-dimension transfer diffusion of the equal-intensity and continuous dust source. If starting to add dust continuously at a certain location at t=0 (t—time), the diffusion equation is as follows according to the mass conversation law:

$$\frac{dc}{dt} + u\frac{dc}{dx} = D(\frac{d^2c}{dx^2} + \frac{d^2c}{dy^2}) \qquad (1)$$

where, c is average dust concentration, mg/s^3; m is average air speed at the face, m/s; t is time of the dust diffusion, s; D is synthesized diffusing coefficient of dust, m^2/ s;

$D = D_X + K$

where D_X is longitudinal diffusing coefficient due to the uneven velocity distribution along the main

$$c(x,y) = \frac{G}{u\sqrt{4\pi xD/u4}}\exp(-\frac{y^2u}{4\pi D}) \qquad (2)$$

airflow at the face, m^2/s; K is radial diffusion coefficient due to the grads of the dust concentration and the turbulent flow pulse, m^2/s.

The resolution of the equation (1) is :
where, G is output of the dust per unit of time in the dust resource at the face, mg/s.

If the coordinate of the dust source (x, y) equals to (0,0), the equation (2) is the distribution of the dust concentration with the two-dimension diffusion at any coordinate point.

According to the model of the transformation of contamination, the equation of the longitudinal turbulent diffusion coefficient and the radial diffusion coefficient can be derived from the theory of the momentum transfer resemble principle and the distributing function of the turbulent airflow speed, which is described as follows:

$$D_X = 65.47ru\sqrt{a} \qquad (3)$$

$$K = 0.0076ru\sqrt{a}R_e^{-0.04} \qquad (4)$$

Where, r is hydraulic radius of the cross section of the face, m; α is friction resistance coefficient of the working face, N.s^2/m^4; R_e is Renault number.

4. PRACTICAL EXAMPLE

The working face No.1326 is equipped with a double-drum machine in Gequan mine of Xingtai

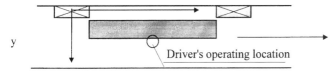

y

Driver's operating location

Fig.2 The layout of the mining machine and the working face

Mining Industry Group Ltd., and the model of the machine is MPX-240W with 2.6m long and two 1.4m-diameter drums. The roof is supported with ZYQ1800/14/32 double-pillar- hydraulic shield support. The mining height is 3m, the average section area 5.85m², and the perimeter of the section is 9.9m. The average airflow rate is 400 m³/min., and the average airflow velocity is 1.2m/s. The friction coefficient of the face is 0.033 N.s²/m⁴. The layout of the working face is illustrated in Fig.2.

Supposing:

hydraulic radium $r = S / U = 5.85 / 9.90 = 0.591$; Renault number $R_e = ud / v$, where, the kinematic viscosity coefficient $v = 14.4 \times 10^{-6} \text{m}^2/\text{s}$, equivalent diameter $d = 4r$, i.e., R_e is $164166u$,

therefore, from equation(3,4), obtaining:

$$D_X = 65.4 \times 0.591 \times \sqrt{0.033} \times u = 7.029u$$

$$K = 0.076 \times 0.591 \times \sqrt{0.033} \times u \times (164166 \times u)^{-0.04}$$

$$= 0.0046u^{0.96}$$

The equation is obtained as below:

$$D = D_X + K = 0.0046u^{0.96} + 7.029u$$

When the mining machine is operated, the effect of the dust on the driver is mainly determined by the location of the drum against the airflow. The drum against the airflow is considered as the linear dust source along the main air current. Because of the dust source near the coal wall, the strength of the dust generated from the source should be doubled when the calculation of the dust concentration distribution is carried on by employing the principle of the wall reflection. For example, if the strength of the dust generated at the drum against airflow is G_0, the strength of dust generated after dust being reflected by the wall surface is $G=2G_0$. Supposing the drum against the wind cutting the roof coal, the dust source is a linear and even source along the face, and the strength of generated dust in the linear source is the total dust quantity divided by the drum diameter, i.e.

Supposing that the X coordinate represents the airflow direction, and the Y coordinate is perpendicular to the direction of airflow, and the coordinates of the dust source are $(0, 0)$, the dust concentration in any location of the face can be calculated with equation (5):

$$c(x,y) = \frac{2G_0/1.4}{u\sqrt{4\pi xD/u}} \exp(-\frac{y^2u}{4xD}) \qquad (5)$$

where, (x, y) is the distance of the location of the driver to dust source.

Supposing that the distance of the location of the driver to the dust source $x=3.0$m and $y=1.0$m, the relation, among the dust concentration, the generated dust quantity of the drum ,and the air velocity at the

$$G = G'/1.4 = 2G_0/1.4$$

$$c(3,1) = \frac{G_0}{4.3\sqrt{(0.0046u^{0.96} + 7.029u)u}} \bullet$$

$$\exp(-\frac{u}{12(0.0046u^{0.96} + 7.029u)}) \qquad (6)$$

location of driver is as follows:

According to equation (6), if the dust quantity of the cutting drum is 400, 300, 268 and 150 mg/s respectively, the curves describing the relation of the dust concentration vs. the air speed are described in Fig.3.

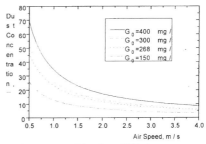

Fig.3 The curves of the dust diffusion concentration vs. the air speed

When the average velocity of airflow at the face is 1 m/s and the dust quantity generated at the drum is 300 mg/s, the dust concentration at the driver's location is 25.98mg/m³. When the dust quantity of the drum against airflow is 150 mg/s, the dust concentration at the driver's working place is 12.99mg/m³.

In Gequan Coal Mine, when the average air velocity at the No.1326 working face is 1.2m/s, and the dust quantity of the drum against airflow is 268 mg/s, the dust concentration at the driver's location is measured as 21.08 mg/m³. The calculating concentration from equation (6) is 19.11 mg/m³, and the difference rate between them is only 9.3%.

In Fig.3, the diffusion dust concentration at driver's location decreases when the air speed in the face decreases, and the decreasing rate of the dust concentration also reduces with the air velocity. When the air speed at the face is changed in the region of 0.5~2.5m/s, the decreasing rate of the dust concentration gets the maximum. As the air speed is over 2.5m/s, the function of diluting the dust is enhanced because of the increase of air speed. At the same time, the performance of the turbulent flow in airflow increases too, so that the decreasing rate of the dust concentration at the driver's working place is reduced.

According to the practical working condition in Gequan Coal Mine, when the average air speed is 1.2m/s and the dust quantity by the cutting drum against airflow is 268mg/s, the distribution of the dust concentration is calculated with equation (5) and illustrated in Fig.4.

In Fig.4, the dust concentration at the face decreases continuously when the distance from the driver to the dust source increases, and reaches a stable state gradually. The reason is that the dust diffusion from the source mixed with airflow induces the even distribution of the dust concentration.

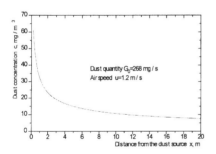

Fig.4 The distribution of the dust concentration vs. the distance between the driver and the dust source

5 CONCLUSION

5.1 The dust concentration at the driver's working place is affected by the dust quantity generated at the cutting drum and the speed of airflow at the fully mechanized working face. The more the dust quantity is generated at the drum, the greater the dust concentration diffused to the driver's location is. With the application of the water injection into the coal seam and the inside and outside water spray around the cutting drum, the dust quantity generated at the cutting drum can be reduced during the coal-cutting process, and the dust concentration is also decreased too.

5.2 Increasing the air speed at the working face is in favor of reducing the dust concentration at the driver's location. Nevertheless, if the air velocity is increased too much, the second-rising dust is formed, and the dust concentration goes up too. When the dust diameter is 10 μm for coal dust, the region of the air speed inducing the second-rising dust is from 1.6m/s to 2.7m/s(Wang 1994). We can see from Fig.4 that there is no obvious effect for reducing the dust concentration at the driver's location while the air speed is over 2.5m/s.

5.3 In Fig.4, it shows that the dust concentration at the face decreases when the distance from the dust source increases, and tends to be stable gradually. Actually, with the distance increasing, the bigger dust particles will fall down to the floor, and the dust staying in the air is the airborne dust, therefore, the practical dust concentration measured is less than that of the calculated one shown in Fig.4.

5.4 From the view point of ventilation and dust prevention, the best way to protect the driver from the dust is to apply the method of the parallel ventilation and the replacement ventilation at the mining face, so that the dust diffusion process can be stopped at the working face.

5.5 The application of the dust preventing schemes affects the variation of the dust concentration at the working face. With the application of the inside water spray in the cutting drum, the coal dust is partly grasped by the water fog, and the dust concentration at the fully mechanized working face is cut down.

REFERENCES

Yu Changzha, 1992, *Introduction to Environment Hydrodynamics*. Beijing:Tsinghua university publishing house.
Wang Yingmin. 1994, *Aerodynamics And Ventilation System in Mine* . Beijing: Metallurgy Industry Publishing House.

Mining Science and Technology' 99, Xie & Golosinski (eds)© 1999 Balkema, Rotterdam, ISBN 90 5809 067 1

New technologies for dust control in the longwall faces with top-coal caving

Gui Fu, Xuexi Chen & Zhiping Lei
University of Mining and Technology, Beijing, People's Republic of China

ABSTRACT: Some of the newly developed methods and technologies for dust depressing in the longwall mining faces with top coal caving are discussed. The water wettability research results can be also applied in coal processing, and the dust extraction technology may be useful in other dust-producing sources such as boiler chimney, etc.

In recent years the longwall mining with top-coal caving(LMTC) has been widely used in the coal mines in China. Now more than 70 mining faces using this method are operating in China, about 70 million tons of coal are produced in those working faces annually. According to the recent information LMTC will be more widely used both in China and in the world. But dust control in LMTC faces has been a problem which draws intensive attention, much relative research work has been done in recent years and will be done in the future. Some newly developed technologies for dust control in LMTC faces are discussed in this paper.

1 RESEARCH ON THE WETTABILITY OF COAL SEAM

The dust depressing efficiency is affected in two ways by the wettability of coal. Firstly, when depressing dust by spaying, good wettability produces high efficiency. Secondly, much water, which is a key factor for the effect of dust depressing, can be injected into coal seam if the coal has good water wettability. The wettability of coal can be described by the contact angle( ) between water and coal. But the contact angle is very difficult to be measured accurately. Some valuable results and the general changing trend(shown in Fig. 1), that describes the relations between the water wettability and the attributes of coal, have been achieved by the authors of this paper with the experimental samples from seven coal mines of China by means of rapid photography. And the following equations show the relations between the coal-water contact angle and the vitrinite and innertinite content in coal.

$$\Phi = 96.40 - 0.88J \quad (\mathrm{abs}(R) = 0.83)$$
$$\Phi = 19.5 + 0.82S \quad (\mathrm{abs}(R) = 0.87)$$

Where, Φ is the contact angle between coal and water, degrees; J is the vitrinite content in coal, %; S is the inertinite content in coal, %; R is linear coefficient.

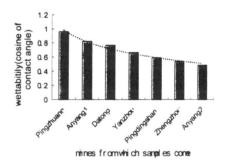

Fig. 1. Wettability of coal

2 STARTING POINT OF THE CRACKS GENERATED IN FRONT OF LMTC FACES

What is important for dust depressing by using water injection is the generated cracks and the original pores in the coal seam or coal body. They act on the penetrating feature of injected water extremely actively. Because of the large-area downfall of rock and top coal caving behind LMTC faces, a lot of micro-cracks are generated at the same time in front of the LMTC faces. By means of the observing at the mine sites, the authors have got the starting point of the cracks generated. Those cracks can be made good use of to achieve favorable effect in water injection and dust depressing. Figure 2 shows the process of rock and coal downfall behind the LMTC mining faces, and the starting point of generated cracks can be seen in Figure 3. The LMTC mining face discussed here is typical of ones in western Henan Province(one of the main coal production area) in central China.

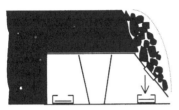

Fig. 2. Process of top coal caving

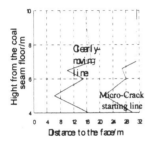

Fig. 3. Starting point of the generated cracks

3 LAYOUT OF BORE HOLES OF WATER INJECTION FOR DUST DEPRESSING

According to the observing results in situ, there is a directly proportional relation between the water injection period(T) for the coal, in front of faces, to be wet and the dust depressing efficiency(C). If the period is converted to the distance(L) from water injection hole to mining face, the proportional

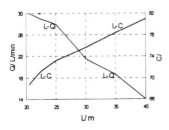

Fig. 4. Suitable point for water injection holes

relation will not change. But an indirectly proportional relation exists between the amount of water (Q) injected into the hole and the distance (L) from injection hole to LMTC face, so that the suitable position(M) of the water injection hole can be drawn out by means of above relations. In fact, this method for laying out the water injection hole is being applied in Chaohua Mine and Micun Mine in central China.

4 APPLICATION AND RESEARCH ON DUST EXTRACTOR(DE) IN LMTC FACES

The extraction technology for dust depressing has been widely used in China and other countries, for example the dust extraction drum(ED) manufactured by the Hydra Tools in England. Here the application of another way of extraction in the LMTC faces is discussed. An extractor for dust depressing in LMTC faces is designed and assembled (Fig 5). The extractor which is equipped under the shield of the hydraulic support can generate the airflow at the rate of more than 26m^3/min with the pressure 12MPa, and good dust depressing effect has been achieved in laboratory. Within the device the particles of the water spray goes at a speed of 30 to 40m/s, the diameter of the water spray particles is 20 to 30um in average, and the concentration of particles can also reaches a required state.

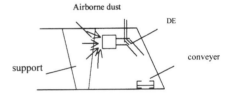

Fig. 5. Application of dust extractor and its assembling

The key part within the dust extractor mentioned above is the spray nozzle with a special structure. The nozzle consists of a shell and a screw rotator (Fig 6). The special structure of the nozzle makes the extractor inhale large amount of air, and makes the particles of the water spray to be fine and more suitable for dust depressing compared with other kinds of dust extractors.

Generally, the ratio B between the amount of air inhaled by the dust extractor and the water consumed by the extractor(nozzle) is used to evaluate the efficiency of the extractor. The greater the value of B is, the more efficient the dust depressing effect of the extractor is.

Some important parameters closely related to the value of B will be discussed in the following. The angle Φ_1, Φ_2, by which the water sprays out, diameter D(mm) and length L(mm) are 4 parameters

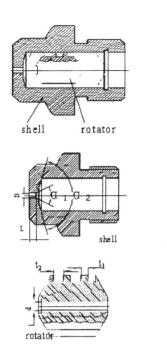

shell   rotator

shell

rotator

Fig. 6. Structure of the nozzle

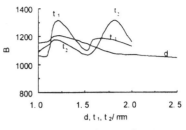

(a) relations between B and t_1, t_2

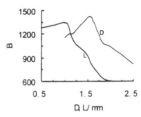

(b) relations between B and D, L

Fig. 7. Relations between B and other 5 parameters

for the shell. The diameter d, tooth width t_1, pitch t_2 are 3 parameters for the screw rotator. It is confirmed in the experiments that $_1 = _2 = 120^0$ is optimal for the shell. 6 shells and 13 screw rotators are selected in order to get the best values of other 5 parameters by experiments for the structure design of the nozzle. Fig 7 shows the results got in the experiments. It is can be concluded that the values in table 1 are the best ones for the design of the shell and rotator according to the curves related to B in Fig 7.

Table 1 the optimal parameters of the nozzle

$_1$	$_2$	D	L	d	t_1	t_2
120^0	120^0	1.5	1.0	1.3	1.2	1.8

5 nozzles with the optimal parameters mentioned above are manufactured. The value of B is 1600 at a pressure of 12 MPa (all of the research experiments are finished under 12 MPa pressure). Then LPDA(Laser Particle Dynamics Analyzer) test system is applied to analyze the properties of water particles sprayed by the five nozzles fixed in the dust extractor. The results are shown in Fig 8, and in that figure V_p, D_p, and S are the velocity, diameter and the distance to the nuzzle of water particles along the nozzle axis respectively. The velocity and the diameter of the water particles along nozzle axis generated by those five nozzles is 55m/s and 35um

in average respectively. Furthermore, very good dust depressing effects can be achieved when the velocity and diameter of particles of water spray is above 20-30m/s and is between 20-50um respectively according to related research results in China. So it is can be said that the dust extractor equipped with a special nozzle discussed in the paper will produce a very good dust depressing efficiency. In fact, the field tests have shown that more than 60% dust particles can be controlled in

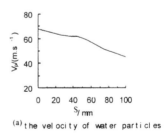

(a) the velocity of water particles

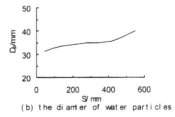

(b) the diamter of water particles

Fig. 8. Velocity and diameter of water particles

the air flowing around the conveyor in Fig 5, and more than 80% dust in the air going through the extractor can be controlled. And 26m³/min of air is inhaled with 16L/min of water consumed. Although the amount of water consumed by the extractor is great, the dust extractor it can be turned off when the top coal caving operations stopped.

5 CONCLUSIONS

This paper discusses the several dust supression techniques that are either used or tested in LMTC faces in China. In fact, many other technologies developed these years are also effective, such as dust depressing by second-negative-pressure, spraying controlled by picks on mining machine, improved nozzle layout, etc.

REFERENCES

Wu, Jian, *et al*, 1994,Safety problems in Fully-Mechanized Top-Coal Caving Longwall Faces, *J. of China Univ. of Mining & Tech.* V(4),2:20~25
Fu, G., *et al*, 1997, Study on Dust-Control Water Injection in Fully-Mechanized Top-Coal Caving Longwall Faces. *J. of China Univ. of Mining & Tech.*V(7),2:56~59
Clarke, R. D. C., Oct 1987, *The Extraction Drum as a Ventilation Device*, Colliery Guardian.

Mining Science and Technology '99, Xie & Golosinski (eds) © 1999 Balkema, Rotterdam, ISBN 90 5809 067 1

Dust suppression at the fully mechanized top-coal caving face[1]

Jingang Guo
Luan Coal Mining Administration, Shanxi, People's Republic of China

Xiexing Miao
China University of Mining and Technology, Xuzhou, People's Republic of China

ABSTRACT: Dust is one of the greatest hazards that affects the safety and working environment of fully mechanized top-coal caving faces. Depending on the source of dust generated in the process of coal extraction, several dust suppression techniques were developed in Zhangcun Coal Mine of Luan Coal Mining Administration.

1. INTRODUCTION

Because the mining intensity of a fully mechanized top-coal caving face is nearly twice higher than that of a fully mechanized slice mining face, its dust intensity is larger than that of a slice mining face. The distribution of dust intensity in the operations of a fully mechanized top-coal caving face is determined by the operation types. The main dust sources in a fully mechanized top-coal caving face include cutting and loading of the coal mining machine, periodic advancing of supports, caving of top-coal, falling of face wall and breaking of top coal caused by the mining pressure. To improve the working environment is very important for higher product capacity and economic efficiency. Based on in-situ measurements of dust intensity, several dust suppression technique is practiced in Luan Coal Mining Area, China. The principles and effectiveness of these dust suppression technique of a fully mechanized top-coal caving face used in Zhangcun Coal Mine of Luan Mining Area are introduced in this paper.

2. AUTOMATIC WATER SPRAYING DURING ADVANCING OF SUPPORTS AND CAVING OF TOP COAL

The two work operations of top-coal caving and supports advancing are main dust sources in a fully mechanized top-coal caving face. The floating dust caused by advancing and caving, including not only coal dust but also silicon dust, enters the work space along with the air flow. The dust, especially the

breathing dust that is difficult to fall down, causes a dust hazard to the coal face. In order to suppress dust diffusion in advancing and caving, the automatic spray measure is practiced in Zhangcun Coal Mine. The operation principle of the automatic spray device is shown in Fig.1.

A one-way hydraulic valve is linked with the high pressure emulsion circuit of five actions including raising and dropping the front prop of supports, raising and dropping the back prop of support and pulling the supports. During raising of the front prop, liquid moves into the below cavity of the front prop and the water circuit of valve 1 is connected (Fig.1). The four nozzles between supports begin to spray. When raising of the front props stops, the hydraulic circuit stops supplying liquid. As a result, the water circuit is cut off, *i.e.* spraying stops. When dropping the front prop or pulling the support, the case is the same as in raising the front prop. During caving top-coal and dropping the back prop, liquid moves into the top cavity of the back prop and the water circuit of valve 2 is connected. The three nozzles for spraying in caving take their work. When dropping of the back pillar stops, the hydraulic circuit stops supplying liquid. And the water circuit is cut off so that the spraying stops. When raising the back prop and closing the drawing window, the case is the same as in dropping the back prop. In addition, a reversible valve 6 is paralleled to the water circuit of spraying between supports and can take the work besides spraying during caving. Therefore automatic spraying can be realized in pulling the supports (including three actions of dropping, pulling, and raising the supports) and drawing (including caving

---
[1] The project is supported by National Science Found of China (59734090)

Table 1. Effectiveness of dust suppression during advancing and caving.

pump pressure (MPa)	flow rate of water (L/min)		dust content in advancing (mg/m³)		dust content in caving (mg/m³)	
			at driver's position	15 meter behind the drum	at driver's position	15 meter behind the drum
3.5	34 in advancing	61 in caving	30.9	71.6	49.3	105

and closing the caving windows). The layout of nozzles of supports is shown in Fig.2. The spraying system between supports consists of four nozzles vertical to the coal wall and two nozzles fixed on the back-timber and the front-timber. The normal lines of the nozzles are vertical to the back-timber of the support.

The three nozzles for coal cutting are fixed on the front timber of the supports (as shown in Fig.2). The normal line of the nozzles to the front-timber is in an angle of 45 degree. The nozzles aim at the caving windows. The five hydraulic control one-way valves of the spray system are fixed under the shield and connected with the hydraulic circuit. The two-way

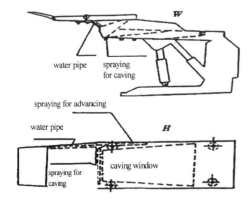

Figure 1. The operation principles of automatic spray in advancng of supports and caving

Figure 2. Spraying system of supports

valve is connected in the water circuit in the back-timber. The water supplying valve is connected in the main water circuit.

The effectiveness of dust suppression in advancing of supports and caving of top-coal is shown in Table 1. According to in-situ measurement, in case of no spray, the average dust content at the driver's position is as high as 78mg/m³ during advancing of the supports. It is 137 mg/m³ during caving of top-coal. While in case of automatic spray, the dust content at the driver's position is decreased 69% during advancing of supports and 64% during caving of top coal.

3. TECHNIQUE IMPROVEMENT OF SPRAYS BETWEEN SUPPORTS

The dust content during advancing of supports in a fully mechanized top-coal caving face is as high as 500-600 mg/m³. The rising dust in coal face greatly pollutes the environment and affects the safe production. In order to solve this problem, two types of spraying instruments between supports are developed and practiced in Zhangcun Coal Mine.

3.1 *Layout of spray instruments between supports in the fully mechanized top-coal caving face No. 1403 (type A)* The high pressure rubber pipe with the diameter of 32mm is used as the main pipeline, while the rubber pipe with the diameter of 13 mm is used as the branch pipeline. A set of spray instrument is installed in every three supports. And a set of spray instrument, which is formed by small spray polls with three nozzles, is controlled by a stop-valve. The small nozzles are fixed on the front-timbers of supports. The angle between nozzles is 30 degree so that the water can cover the face. The plane of nozzles of the spray polls to the direction of the air current is in an angle of 45 degree, which increases the velocity of water spraying and improves the spraying effect.

3.2 *Layout of spray instruments between supports in the fully mechanized top-coal caving face No. 1302 (type B)*

The high pressure rubber-pipe with the diameter of

Table2. Dust suppression effectiveness of spraying instruments between supports.

Coal face	Pump pressure (Mpa)	Water flow (L/min)	Dust content in advancing without spraying the average	Dust content in advancing with spraying the average
No. 1103	3.5	71	620	213
No. 1302	3.5	32	473.5	193

32 mm is used as the main pipeline, while the rubber-pipe with the diameter of 10 mm is used as the branch pipeline. A nozzle is installed in a set of supports. Every two spray nozzles is controlled by one valve. The nozzles are installed in the timber of supports.

Every four nozzles can cover a section of face during advancing of supports. After the use of spray instruments between supports in coal face, the coal dust can be suppressed. As a result, the amount of dust in advancing of supports is efficiently decreased. The effect of dust suppression is shown in Table 2.

According to Table 2, the dust content in the coal faces is decreased greatly. The dust content decreases 65.5% in the coal face No.1403 and 59.2% in the coal face No.1302. Therefore, the working environment of the coal face is improved obviously.

4. WIDE USAGE OF HIGH EFFECTIVE DUST SUPRESSION MEDIUM IN FULLY MECHANIZED TOP COAL CAVING FACE

In order to improve the quality of water and the effect of dust suppression, the filters with 60meshes are installed in water supply pipe and before the spray pump separately. In addition, a special mixed pump is employed to add the high effective dust suppression medium to the water with the ratio of 1% in the water flow. After pressurized by the spray pump, it can be transported to the coal cutting machine, the spray system between supports, transfer points, water curtain and other dust suppression systems.

Comprehensive dust suppression measures were practiced in Zhangcun Coal Mine. To some extent, coal dust and rock dust appear to be hydrophobic. On the other hand, it is usually difficult for dust to be wetted rapidly and fully because of the great surface tension of water. It causes the efficiency of dust suppression to be decreased to some degree. Research results of many countries and our investigations indicate that to add the dust suppression medium to water can greatly improve the wetting ability of coal body and coal dust.

In general, the hydrophic rate of coal dust ranges from 0.5 to 2.5mm/min. In other words, the coal dust is hydrophobic dust and can not be wetted by water rapidly and fully. However, the property of the water with dust suppression medium is different from that of general water. The dust suppression medium contains hydrophilous group and hydrophobic group that are different from each other in property. When dust suppression medium is added to water, hydrophilous group is wetted and dissolved by water, while hydrophobic group is emitted in the air. So the tight interface absorbed layer is formed on the surface of water. The existence of the interface absorbed layer causes the surface tension of water to be decreased. In addition, the hydrophobic group facing to the air can absorb the dust. As soon as the dust contacts the water with dust suppression medium, a parcel surrounding the dust, in which the hydrophobic group faces to the dust particle while the hydrophilous group faces to water, will be formed. It makes the dust can be wetted fully and fall down. When the water with dust suppression medium is injected into the coal seam of a fully mechanized top-coal caving face, its wet capacity increases so that it can much easily infiltrate along with the cracks in the coal body and wets the coal seam. A special mixed pump i.e. pump of adding the dust suppression medium made in China is employed to suppress the dust in Zhangcun Coal Mine. It is paralleled in the dust suppression pipe net of a fully mechanized top-coal caving face. The high efficient dust suppression medium called R-89 is added to water by the ratio of 1%. With experiments of many times, the instrument of intake water and the connecting way of pumps are improved. Four pumps adding dust suppression medium are installed successfully and take their work.

According to the in situ measurements, it has been indicated that the special mixed pump can be used to suppress the dust with a better effect in coal face, especially in a fully mechanized top-coal caving face. After the use of dust suppression medium, the dust content decreases about 60% (as shown in Table 3).

5. THE SEALING TECHNIQUE OF CRUSHING MACHINE FOR DUST SUPPRESSION

The amount of lump coal mined with the fully mechanized top-coal caving is relatively large, which

Table3. Effectiveness of dust suppression medium.

coal face		cutting	driver's position	advancing	else	average
No.1403	with dust suppression medium	10	116	180	11	80.6
	without dust suppression medium	87	382	302	58	207.5
No.1332	with dust suppression medium	12	110	192	20	83.5
	without dust suppression medium	140	460	240	96	235.3

Table4. Effectiveness of dust suppression by sealing of crushing machine

coal face	water flow at nozzles (L/min)	dust content before sealing (mg/m^3)	dust content after sealing (mg/m^3)
No 1403	40L/min	210	7
No 1302	40L/min	188	7

results in the amount of dust in the transfer place and crushing machine increase greatly. Dust rises everywhere and endangers the safe production.

For many years, systemic research on sealing technique of crushing machine has been carried out. According to Table 4, the average dust content of the top-coal caving face decreases from 200mg/m^3 to 7mg/m^3. It has been shown that this technological innovation improves successfully the working environment of coal face.

6. CONCLUSIONS

In the practice of preventing coal dust hazard in the fully mechanized top-coal caving face of Zhangcun Coal Mine in Luan Mining Area, a series of comprehensive measures are taken according to the distributions of coal dust resource and coal dust content. The dust content caused by advancing of supports and caving of top coal decreases over 60% by means of innovation of spraying system and automatic spraying. The average dust content in coal face decreases 62.8% by using high-effective dust suppression medium. With the sealing technology of crushing machine for dust suppression, the dust content in crushing decreases from 200mg/m^3 to 7mg/m^3.

Mining Science and Technology' 99, Xie & Golosinski (eds) © 1999 Balkema, Rotterdam, ISBN 90 5809 067 1

Composition and dispersity of soaring coal dust of mines

L. Ya. Kizilshtein & Yu. I. Kholodkov
Rostov State University, Rostov-on-Don, Russia

ABSTRACT: The soaring dust of coal mines is the subject of study of this paper in connection with the anthracosis of miners. It has been shown that 50%-90% of coal particles belong to the fraction lesser than 5μm which causes most of lung anthracosis. The dust under study contained free state pyrite particles. The presented data describe particles' morphology in relation with the kind of macerals. The authors think, that the composition of the soaring dust may, to a considerable degree, initiate the blast danger.

1 RESULTS AND DISCUSSION

The coal dust in a mine air bears a professional threat to the health of miners, provocating a specific kind of pneumoconiosoanthracosis. It has been established that within the metamorphic ranks the coal dust, originating in the process of working off the anthracite beds, causes most of lung pathologies.

Among the coal macerals the fusenite is more anthracosically harmful than vitrinite. The pathologenesis of the dust increases with the increase of quartz particles in its composition. Polycarbonic acids are formed in oxidic-hydrolic decay of the dust in the moist medium of lungs. According to the experimental data, the latter is the most important factor of the dust pathogenesis (Dinkelis at al., 1981; Kukharenko at al., 1982).

The higher is the rank of metamorphism of coal, i.e. it contains more aromatic rings and is more condensed, the more intensively occurs the formation of polycarbonic acids. Polycarbonic acids of high rank coals also contain more aromatic acids than the coals of low rank metamorphism, which increases their anthracosic hazardous effect (Dinkelis at al., 1981, Kukharenko at al., 1981).

The observations over the distribution of dust professional deseases have shown that the anthracosic hazard of the coal dust depends on its concentration in a mine air, the exposition time, the composition and dispersity.

The authors have studied the fractional and petrographic composition of the soaring mine dust, accumulated on the filters of the anti-dust respirators during the work of one shift, while working off the rock coal beds and the anthracites in the Donbass. The dust has been extracted from the dried filters by simple shaking and the polished section briquettes, made of this dust, have been studied under the microscope reflection light.

The accepted way of dust samples selection permitted to characterise the particles' composition, which may penetrate into the lungs of miners, provided there was no special protection. Besides, we studied the composition of macerals of every sample and measured the maximum and minimum diametres of about 300 specks of the dust and then calculated their effective diametre. The data presented in Table 1 characterise the particle size distribution in accordance with their classes of largeness.

Table 1. Size of dust particles, μm.

Sample	< 5	5 - 25	26 – 45
1.	52.0	42.7	5.3
2.	68.3	28.0	3.7
3.	90.9	8.8	0.3
4.	77.2	18.2	4.6

The Table shows, that in all cases 50 % - 90 % of coal particles belong to the fraction lesser than 5 mm. The dust of such kind of dispersity is believed to penetrate into the alveolar part of lungs bringing by such the greatest harm. Some scientists (Pryadilova et al, 1978; Borisenkova et al., 1983), however, indicate that the large dispersed dust may turn no less pathogenic than the smaller one. Variation of dust dispersity may be explained not only by the specific character of coal but also by difference in mechanisms and efficiency of

Soaring coal dust of mines explanations are given in the text.
Fig. 1 - x 280; 2, 3 - x 230; 4, 5 - x 250; 6 - x 300; 7, 8 - x 230. Fig. 1-6 - white - pyrite; Fig. 7, 8 - white - inertinite.

dust protection measures. One may see it from the comparison of samples, obtained in working off the same bed by different mechanisms. For instance, during the operation of a mining combine "Donbass-1g" the watering of the coal-face was low and the content of small specks of dust reaches its maximum (sample 3). Relationship between dust formation and mining technologies used has been shown in some special research studies (Vorontsova et al, 1982).

Comparing the compositions of coal bed macerals and the dust formed, during their working off we may conclude that there is no regular and sufficient changes in the composition of dust and in content of the organic microcomponents, macerals. At the same time there is 1.5 – 2.5 times increase in the concentration of pyrite and more than 3 times of quartz and to a lesser degree – carbonates i.e. all mineral admixtures, which are easily separated from coal during its desintegration and thinning. The content of clay, which is usually deeply incorporated into coal, remains practically unchanged.

Probably together with the ability of mineral particles to incorporate with coal, the enrichment of coal dust by minerals may be explained by fragility, which leads to the finer dispersity of mineral grains, compared with the coal.

Some peculiarities in morphology of dust particles have been revealed during the observation under the microscope. The macerals of vitrinite and semi-vitrinite groups form particles of different morphology (Photos 1 – 8). Inertinites, when being crushed, give exclusively oblong, sharp fragments (Photos 7, 8).

High pathogenesis of inertinite particles is well known (Kukharenko, Kilkeev, 1982). The authors put a stress upon the fact that the inertinite particles, at least in many cases, have the oblong form of pointed debrises capable to harm the lung tissue far more, than the others do.

Pyrite is found in dust either in the form of separate grains preserving their characteristic forms (globules, crystals, Photo 4) or as irregular fragments (Photos 1, 2), formed in crushing of bigger fragments. Most of pyrite is densely incorporated into organic macerals (Photos 2, 3, 5, 6). Being oxidized in lungs it forms the sulphuric acid with a pathologic effect similar to the "acid rains".

The data concerning the sulphur content in some coal beds of the Eastern Donbass are given in Table 2. They relate to pyrite in the soaring dust of the bed working off zone. Besides, it is shown how much pyrite is present in separate grains not bound with organic macerals.

The most interesting fact is the presence of separate grains of pyrite in the air which, because of their high density (about 5000 kg m^{-3}), should have settled down.

Considerable amount of pyrite in the mine dust probably increases its blasting ability, so far as the sulphide dust has much more perceptibility to inflame than methane and coal particles. The inflammability of sulphide dust occurs at the temperature 450 – 550 °C (methane – at 650 – 750°C, soaring coal dust – at 750 – 800°C) (Komarov, Kilkeev, 1969). Blast danger concentration of sulphide dust is 0.08 – 0.1 kg m^{-3}. Sulphides are distributed in coal seams of the Donetsk and some other basins quite irregularly. The concentration of pyrite in separate coal beds reaches 50% – 60 % and even more. Pyrite is found as the finest (disperse) impregnations (the size of separate grains lies within the limits from 1–2 to 50 μm). Therefore, it is easy to conclude, that the destruction of coal during coal shooting and the operation of mine mechanisms may lead to blast dangerous concentration of sulphides in local spaces of a mine atmosphere. Though the power of possible blasts is not high, they nevertheless bear a great danger due to the presence of methane and SO$_2$ in coal dust.

So far there was no complete understanding about the effect of sulphides, present in the dust, to respiratory organs. Meanwhile the subtle dispersity of sulphide particles, determining their high surface activity, in moist medium of lungs, leads to quick oxidation with formation of chemically active products of decay,

Table 2. Pyrite in the soaring dust.

Bed	Sulphur in bed, %	Pyrite in dust, %	Pyrite grains in dust, %
l$_3$	1.4	2.8	0.9
l$_4$	1.3	2.7	0.4
l$_6$	1.6	1.1	0.4
l$_7$	1.6	1.8	0.6
m$_3$	1.4	1.4	0.2
m$_3$	-	1.4	0.4
m$_9$	3.2	2.7	0.5
m$_9$	3.4	2.2	0.6

including the sulphuric acids. Such kind of ethiologic aspect of pneumoconiosis deserves a serious attention.

CONCLUSIONS

1. Soaring dust, present in coal mines and formed due to the working off coal beds is the cause of lung pathologies of miners.

2. The degree of hazardous effect of coal dust is determined by the coal rank, largeness of particles and composition of macerals.

3. In studied mines of the Donets basin 50 % - 90 % of particles in the soaring coal dust belong to the fraction lesser than 5mm.

4. Free state pyrite, reveald in the dust together with organic macerals, may oxidize in the moist conditions of lungs initiating deseases, similar to "acidic rains".

REFERENCES

Borisenkova, et. al. Results of joint, with the COMECON member states, studies on unification of approaches to the coal dust requirements (regulation). // Hygiene and Sanitation magazine. 1983. No.5, (Russian).

Dinkelis S.S., Derr E.A., Kukharenko T.A. The ethiological role of hard coal dust in the development of pneumoconiosoanthracosis. //Zdravookhranenie Kazakhstana (Public Health of Kazakhstan), 1981. No.3, (Russian).

Komarov V.B., Kilkeev Sh.Kh. Mine ventilation. M., "Nedra", 1969, (Russian).

Kukharenko T.A., Kilkeev Sh.Kh., The problem of anthracosis and oxidation-hydraulic decay of hard coal in organism. // Ugol (Coal), 1982, No.8 (677), (Russian).

Pryadilova N.V., Bykhovskaya I.A. History and contemporary state of relations between particle sizes and pathogenity of dust.// Hygiene of labour and professional deseases. 1978. No 3, (Russian).

Vorontsova E.I. et al. Labour conditions and peculiarity of dust pathology of respiratory organs of mines of the Donbass anthrocite mines.// Struggle with the silicosis. M.: Medizdat. 1982, (Russian).

Mining Science and Technology '99, Xie & Golosinski (eds)© 1999 Balkema, Rotterdam, ISBN 90 5809 067 1

Study and application of the o-shaped circle theory for relieved methane drainage*

Jialin Xu & Minggao Qian
China University of Mining and Technology, Xuzhou, People's Republic of China

ABSTRACT: Experiments and discrete simulation results have revealed that as the excavated area is large enough, mining-induced fractures are repressed and closed in the middle zone of goaf , but the fractures can still exist around the goaf for a long time, so that an O-shaped circle with developed fracture is formed around the goaf with the width about 30m. The experiment results have shown that the O-shaped circle is the main flowing passage of relieved methane, and the relieved methane drainage holes should be laid in the O-shaped circle. Guided by the O-shaped circle theory about relieved methane drainage, the industrial experiment to drain methane from long away overlying coal seam through surface well has been completed successfully. The plan to drain the relieved methane in the goaf of LuLin colliery has been put forward and applied.

1 INTRODUCTION

Methane is a harmful gas to threaten mine safety and also a kind of clean and efficient energy. Methane drainage is an important measure to reduce the mine methane emission rate and prevent methane explosion and coal-methane outburst. Methane drainage methods include the un-relieved methane drainage before coal mining and relieved methane drainage after coal excavation. The mining-induced fractures cause methane to be relieved and become the flow passage for the relieved methane. The flow pattern of the relieved methane is related with the mining-induced fracture distribution. The relieved methane drainage is adopted by almost all gassy mines and its key point is the drainage holes layout. In order to improve the relieved methane drainage rate and reduce the drilling works, the layout of the relieved methane drainage holes should be optimized by using the distribution law of the mining-induced fractures.

2 DISTRIBUTION CHARACTERISTICS OF MIINIG-INDUCED FRACTURES IN THE OVERLYING STRATA

2.1 *Physical model experiments*

In order to study the distribution characteristics of the mining-induced fractures in the overlying strata after coal seam is excavated, 5 model experiments of equivalent material have been done in TaoYuan Colliery. The first long wall mining face No.1018 excavates No.10 coal seam with the mining height 2.5m, the face length 180m, and the mining depth 500m. An plane-stress model with dimensions $2500 \times 200 \times 1600$ mm^3 is adopted. The strata displacement of the models is measured and the fractures distribution of the models is photographed.

The mining-induced fractures in the overlying strata can be classified into two groups as follows: one is the fracture occurred with the bed separation while roof caving, and it can be developed between the bedding planes along the whole overlying strata; the other is cross-breaking fracture, and it becomes the flow passage for the released gas to flow down into the longwall mining face and its goaf from near overlying coal seams. The experiment results indicate that the cross-breaking fracture is only developed up to 30m from No.10 seam. The indexes of bed-separation ratio and fracture density are adopted to assess quantitatively the development of the mining-induced fractures. The bed-separation ratio (mm/m, or ‰) expresses the height of bed-separation in the unit thickness of rock. The bed-separation ratio is got by measuring the displacement difference between two strata(in the physical model). The fracture density is got by image analysis technique. An image of fracture distribution obtained by scanning the photographs of the physical model is

* Project supported by Coal Mining Science Foundation of China(No.97 Min.10102) and by National Science Found of China (59734090)

measured and counted by a special computer program FIMAGE(Xu, 1998). The image of fracture is divided into pixels and the fractures density expressed by the pixels numbers. The more the pixels numbers are, the more the fractures are developed. The fracture density distribution of the models is illustrated in Fig.1. The bed-separation ratio contour in the goaf, as the face is advanced 250m along strike, is illustrated in Fig.2.

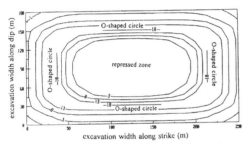

Fig.1 the distribution of the fracture density by image analysis

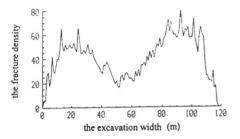

Fig.2 the contour lines of the bed-separation ratio in the goaf

Fig.1 and Fig.2 indicate that as the excavated area is big enough, the mining-induced fractures are repressed and closed in the middle zone of the goaf, but the fractures can still exist around the goaf for a long time, so that an developed fracture zone is formed around the goaf. The zone is named as o-shaped circle and its periphery width is about 30m.

2.2 Discrete element simulation

The simulation condition of the discrete element method is same as the physical simulation model, but the roof is only simulated up to 30m above No.10 seam. The elements selected in the computer simulation model are based on the size of the broken blocks in the layers shown in the physical simulation model. The fracture distribution simulation results, as excavation length of working face is up to 140m, is illustrated in Fig.3. Fig.3 indicates that a developed bed-separation zone exists in two boundaries of the goaf, so that an o-shaped circle of bed-separation is formed around the goaf. The bed-

separations mainly occur between the overlying strata with large broken blocks and the lower strata with smaller broken blocks. The discrete element simulation results verify the o-shaped circle distribution characteristics.

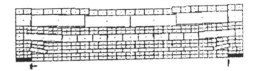

Fig.3 Discrete element simulation results of fracture distribution

2.3 *The strata movement mechanism of mining-induced fracture distribution*

The mining-induced fractures are formed due to the strata movement and breaking after excavation. The overlying rock mass is the rock strata with different thickness and strength. The broken blocks of hard strata form the structure of voussoir beam. Based on the analysis of the mechanics model of the voussoir beam(Qian,1995) and the s-shaped curve characteristics of the subsidence profile of strata(the subsidence profile is convex in the forefield and then turn to concave curve), the formula about the subsidence curve of the voussoir beam has been shown as follows:

$$w = w_0(1 - 1/(1 + e^{(x-0.5l)/0.25.l})) \qquad (1)$$

where w is the subsidence of the voussoir beam
w_0 is the maximum subsidence of the voussoir beam
x is the distance to the excavation boundary;
l is the length of broken block in the voussoir beam.

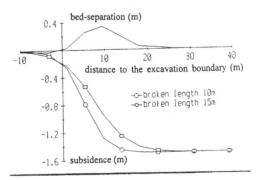

Fig.4 influence of the length of broken blocks on the subsidence and forming of bed-separation zone

The formula (1) indicates that when the mining thickness and the distance between the voussoir beam and the coal seam are defined, the subsidence curve of the voussoir beam depends on the broken length of the voussoir beam and the distance to the mining boundary. If there are two adjacent strata with the same $w_0=1.5m$ and the length of the broken blocks in the upper or the lower strata are 15m or 10m respectively, by using the formula (1), the subsidence curve of the two strata and the bed-separation distribution curve between the two strata can be illustrated in Fig.4. Fig.4 indicates that in the goaf the subsidence of the longer blocks is less than that of the shorter blocks, and their subsidence tends to be same when the distance to the mining boundary becomes longer. The difference of the subsidence of the two strata causes the bed-separation occurrence. The bed-separation mainly occurs near the mining boundary, so that the o-shaped circle of the bed-separation is formed around the goaf.

3 TEST AND APPLICATION OF THE O-SHAPED CIRCLE THEORY OF RELIEVED METHANE DRAINAGE

3.1 *The o-shaped circle theory about relieved methane drainage*

The flow of the relieved methane is with two steps: firstly, the methane flow from coal mass into the surrounding fractures by diffusing, then the methane flow through the fractures to gas drainage holes with permeating manner. The mining-induced fractures become the passage for the methane migration. Obviously, if the drainage holes is laid in the area with developed fractures, the methane can easily flow to the drainage holes. Based on the mining-induced fracture distribution characteristics, the o-shaped circle theory about the relieved methane drainage has been put forward as follows: the o-shaped circle of the mining-induced fractures is like a "methane river bed". The relieved methane from the surrounding rock and coal migrates continuously to and gather in the "methane river". Accordingly, the relieving methane drainage holes should be laid in the o-shaped circle to keep the period of the effective drainage longer, the drainage area bigger and methane drainage rate higher. In order to keep the drainage holes laid in the o-shaped circle, the horizontal distance of the bottom of drainage hole to the return airway should be calculated by the following formula:

$$S=[H-(B+Hctg\phi)tg\alpha]\sin\alpha$$
$$+(B+Hctg\phi)/\cos\alpha \qquad (2)$$

where S is the horizontal distance of the bottom of the drainage hole to the return airway, m.
H is the vertical distance of the bottom of drainage hole to the excavated coal seam, m, different types of drainage hole types have different value of H.

B is the distance of the drainage hole to the outer boundary of the o-shaped circle, m, in general, B stands for the value at 0-30m.
ϕ is the dip angle of the coal seam,(°).
α is the intersection angle between the coal seam and the line connecting the outer boundary of the o-shaped circle and the mining boundary (°).

The numerical simulation result (Ding,1996) about the relieved methane permeating in the goaf testify the o-shaped circle theory. The tracer analysis results for gas migration in the goaf using the gas SF_6 also testify the o-shaped circle theory about relieved methane drainage.

3.2 *The application of the o-shaped circle theory about relieved methane drainage*

The o-shaped circle theory about the relieved methane drainage has been applied in TaoYuan colliery and LuLin colliery of Huaibei Coal Mining Bureau in AnHui province.
TaoYuan Colliery is a gassy mine. Being conducted by the o-shaped circle theory, the industrial experiment to drain methane from long away overlying coal seam through surface gas well has been completed in its first longwall face No.1018. The No.1018 face excavates No.10 seam with the excavated height 2.5m and the working face is 150m long along dip. The methane releasing amount of the middle group coal seams (No.7,8,9) is about 4.1~5.9m³/t, and the gas pressure 0.72~1.03 MPa, and the permeability coefficient is 0.03m³/Mpa².d. The middle group coal seams are not mined because of its little thickness, and the thickness of 7, 8, 9 coal seams is 0.9m-0.3m and 0.3m respectively. The distances of middle group coal seams to No.10 seam are about 84-150m, so that the released methane from the coal seams in the middle group can't flow down to the goaf and the working face. Based on the o-shaped circle theory about the relieved methane drainage, the layout plan of the surface gas well 94-w₁ which is used to drain methane from the middle group coal seams has been put forward. The surface gas well 94-w₁ has kept draining for 15 months, and the maximum drainage rate is up to 1008m3/d, the average drainage rate up to 521m³/d. The gas drainage ratio is about 64.1□in the drainage radius 100m. The successful drainage experiment of the 94-w₁ surface gas well proved that the long away overlying coal seam could be relaxed completely due to the bed-separation development in the overlying strata. Its permeability increases greatly, and its relieved gas can be drained out through the O-shaped circle. It has provided a new way for many coal mines in China to exploit the methane resource.
Lu Ling Colliery is a gassy coal mine. The permeability of No. 8 seam is low, and its methane drainage rate is only about 15 before excavation. As

the first flat of No.8 seam is excavated, the released methane from No.8 seam and the adjacent seams(7,9) migrate into the goaf. The methane concentration often exceeds the safe limit and the mining operation has to stop, so that the high productivity and the safety can't be guaranteed. To drain the relieved methane in the goaf is an essential way to solve the problem. Based on the o-shaped circle theory about relieved methane drainage, the plan of the floor cross-measure drainage holes, the plan of horizontal drainage tunnel in the roof and the plan of the surface gas well have been put forward to drain the relieved methane in the goaf of the first layer of No.8 seam. An good effect has been achieved and the surface gas well has drained a lot of methane which has been used. The maximum draining rate is up to $2300m^3/d$ and good economic results has been got.

4 CONCLUSIONS

1 The mining-induced fractures are repressed and closed in the middle zone of the goaf, but an O-shaped circle with developed fracture around the goaf exist for a long period.

2 The o-shaped circle of the mining-induced fractures is like a "methane river bed", the releasing methane migrate continuously to and gather in the "methane river". The drainage holes should be laid in the o-shaped circle to keep methane drainage rate higher. To keep the drainage holes to be laid in the o-shaped circle, the location of the drainage holes bottom should be calculated by the formula (2).

3 The o-shaped circle theory about relieved methane drainage has been applied successfully in HuaiBei Coal Mining Bureau.

REFERENCES

Xu J.L. & Qian M.G. 1998,The study on the quantitative analysis of the experiment results of mining-induced fractures, in Chinese, *Journal of Liao Ling University of Engineering and Technology*, P37~39.

Qian M.G. & Liao X.X. 1995,The analysis on the structure and mechanic characteristic of the overlying strata of working face, in Chinese, *Journal of rock mechanics and engineering*,P97~106;

Ding D.X. 1996, *Mine Atmosphere and three-dimensional how of methane in Chinese*, The Publishing House of China University of Mining & Technology, P217~222.

Wang J. Z. Liang D. & Zhang Q.F. 1995, The tracing analysis on the methane migration in the semi-seal goaf, in Chinese, *Journal of Fu Xin Mining Institute*, P1-3,14(2).

Mining Science and Technology'99, Xie & Golosinski (eds) © 1999 Balkema, Rotterdam, ISBN 90 5809 067 1

Boundary element method to calculate the heat dissipation of the surrounding rock in airway

Yueping Qin & Haizheng Dang
China University of Mining and Technology, Beijing, People's Republic of China

ABSTRACT: The boundary element method is used to calculate the heat dissipation of the surrounding rock in airway. With the basic solution of two-dimensional unsteady heat transfer differential equation, the boundary integral equation is established. By using the similarity theory, this boundary integral equation is made into a dimensionless form. With the discrete boundary integral equation, the linear equations are set up and a computer program is developed. By analysis the result of calculation, a series of curves are given to show the unsteady heat transfer criterion change with Biot Number and Fourier Number.

1 INTRODUCTION

The heat dissipation of the surrounding rock is the main source of heat in airway, and it plays an important role in the prediction of the air temperature in airway, in the choice of cooling measures and in the calculation of cooling quantity need in the working face. Finite element and finite difference method are often used to solve this problem. These methods need to presume an outside boundary, although the temperature field is semi-infinite in fact. Therefore, these mathematical models deviate from the practice. The boundary element method is another numerical method to calculate heat transfer and temperature field. It is divided into the direct method and the indirect method. In this paper the direct method is used, which based on the Green second formula and the basic solution of infinite mediator.

2 THE ANALYSIS OF THE BOUNDARY ELEMENT METHOD ON THE PROBLEM OF TWO-DIMENSION HEAT CONDUCTION

When the functions "U" and "V" is the second order continuously differentiable in the whole region "Ω" that includes boundary "Γ", then the Green second formula in the two-dimensional problem can be expressed as:

$$\iint_{\Omega} (U\nabla^2 V - V\nabla^2 U)d\Omega = \int_{\Gamma}(U\frac{\partial V}{\partial n} - V\frac{\partial U}{\partial n})d\Gamma \quad (1)$$

where Ω is the whole temperature field; Γ the boundary of temperature field; Δ^2 lapacian.

With this formula, the integral in the region can be exchanged with the integral on boundary.

For an airway, we can consider approximately that:

(1) The surrounding rock of the airway is isotropic;

(2) The temperature in a certain section of the airway doesn't changed with time;

(3) The section of the airway is circular.

In this way, the temperature field of the surrounding rock of the airway can be considered as the unsteady two-dimensional temperature field. Its heat conduction equation is expressed as follows:

$$a(\frac{\partial^2 t}{\partial x^2} + \frac{\partial^2 t}{\partial y^2}) = \frac{\partial t}{\partial \tau} \quad (2)$$

$$q = -\lambda\frac{\partial t}{\partial n}\Big|_{r=r_0} = a(t_w - t_f) \quad (3)$$

$$t = t_0, r - \infty \quad (4)$$

where n is exterior normal of boundary; q heat-flux density on boundary; t the temperature in rock, ^{O}C; t_0 virgin rock temperature, ^{O}C; t_w temperature of the rock on boundary, ^{O}C; t_f air temperature in airway, ^{O}C; the coefficient of heat conduction of the surrounding rock; convection coefficient of air; a heat diffusion of rock; r_0 radius of airway.

In isotropic mediator, the basic solution related to time of two-dimensional heat conduction equation can be expressed as:

$$t^*\left(x,\tau,x',\tau'\right)=\frac{1}{4\pi a\left(\tau-\tau\right)}\exp\left[\frac{-r^2}{4a\left(\tau-\tau\right)}\right]\quad(5)$$

The basic solution means the temperature field produced by an instantaneous point heat source setting up on point x' at the time τ. r is the distance between the point x and x'. The basic solution of heat-flux density related to time expressed as:

$$q^*\left(x,\tau,x',\tau\right)=\frac{\partial t^*}{\partial n}=\frac{-r}{2a\left(\tau-\tau'\right)}\frac{\partial r}{\partial n}t^*\quad(6)$$

The boundary integral equation of the solution of heat conduction equation is:

$$ct=\iint_{\Omega}t_0 t_0^*\,d\Omega(x)+a\int_{t_0}\int\left(t^*\frac{\partial}{\partial n}-t\frac{\partial^*}{\partial n}\right)d\Gamma(x)d\tau\quad(7)$$

where

$$t_0=t(x,\tau_0),\ t_0^*=t^*(x,\tau;x,\tau_0)$$

$$c(x)=\iint_{\Omega}\Delta(x-x')d\Omega(x')\quad(8)$$

$\Delta(x)$ =δ-function.

3 THE BOUNDARY ELEMENT METHOD TO CALCULATE THE HEAT QUANTITY OF THE SURROUNDING ROCK IN AIRWAY

3.1 The dimensionless differential equations and integral equations

In order to get general conclusion, the authors introduces some similarity Number and dimensionless variables:

$$\theta=\frac{t-t_f}{t_0-t_f};\ Bi=\frac{r_0\alpha}{\lambda};\ F_0=\frac{a\tau}{r_0^2};\ R=\frac{r}{r_0};$$

$$X=\frac{x}{r_0};\ Y=\frac{y}{r_0};\ N=\frac{n}{r_0};$$

Where Bi is Biot Number, Fo Fourier Number, R Dimensionless radius, (X,Y) Dimensionless coordinate, N Dimensionless exterior normal of boundary.

Then the temperature field in the surrounding rock can be expressed as:

$$\begin{cases}\dfrac{\partial^2\theta}{\partial X^2}+\dfrac{\partial^2\theta}{\partial Y^2}=\dfrac{\partial\theta}{\partial Fo}\\[2mm]-\dfrac{\partial\theta}{\partial N}\Big|_{R=1}=Bi\,\theta_w=Bi\,\theta\big|_{R=1}\\[2mm]\theta=\theta_0,Fo=0\end{cases}\quad(9)$$

Unsteady heat transfer coefficient of any one section of airway k_r is defined as when the air temperature is 1 degree less than the virgin rock temperature, the heat dissipation of the unit area of the wall of the airway to air flow in unit time. Then heat-flux density from rock to air is

$q=k_r(t_0-t_f)$.

One defines the unsteady heat transfer criterion $k_{u\Box}$as $k_{u\tau}=\dfrac{k_r r_0}{\lambda}$, then $k_{u\tau}=Bi\theta_w$, The heat-flux density and heat quantity on boundary can be calculated by

$$q=\frac{k_{u\tau}\lambda}{r_0}(t_0-t_f),\ Q=2\pi k_{u\tau}\lambda(t_0-t_f).$$

If θ_w or $k_{u\tau}$ has been known, q and Q can be worked out from above formula.

From formula (9), it can be known that $k_{u\tau}$ and θ_w are the function of Bi and Fo, i.e.

$$\theta_W=f_1(Bi,Fo)\ \Box\ k_{u\tau}=f_2(Bi,Fo)$$

Formula (7) is changed into dimensionless form:

$$c\theta=\iint_{\Omega}\theta_0\theta_0^*\,d\Omega(X)$$

$$+\int_0^{Fo}\int\left(\theta^*\frac{\partial\theta}{\partial N}-\theta Q^*\right)d\Gamma(X)dFo'\quad(10)$$

$$\theta^*(X,Fo;X,Fo)=\frac{1}{4\pi(Fo-Fo)}Exp\left[\frac{-U^2}{4(Fo-Fo)}\right]$$

$$Q^*(X,Fo;X,Fo)=\frac{\partial\theta^*}{\partial N}=\frac{-U}{2(Fo-Fo)}\frac{\partial U}{\partial N}\theta^*$$

where U is the distance between the point X and X'; Then formula (10) can be changed into:

$$\frac{1}{2}\theta_w=P-2\int_0^\pi\int_0^{Fo}\frac{\theta}{4\pi(Fo-Fo')}\left[Bi+\frac{\sin^2\frac{\beta}{2}}{(Fo-Fo')}\right]$$

$$\times Exp\left[\frac{-\sin^2\frac{\beta}{2}}{(Fo-Fo)}\right]dFo'\,d\beta\quad(11)$$

where

$$P=\iint_{\Omega}\theta_0\theta_0^*\,d\Omega=\iint_{\Omega}\theta_0^*\,d\Omega\quad(12)$$

β' is the included angle between vector U and N; and β is the included angle between segment OX and OX', as shown in Figure 1.

Formula (11) is the boundary integral equation of

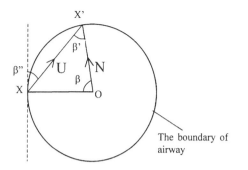

Figure 1. The vectors and their included angles

The boundary of airway

dimensionless basic solution related to time of the surrounding rock in airway.

3.2. The discretization of the integral equation

3.2.1. Discretization of time

The dimensionless period [0,Fo] is divided into n time intervals, and in the interval S,

$$Fo_S = Fo_S - Fo_{S-1}; \quad \theta = \theta_{s-1} + (\theta_s - \theta_{s-1})\eta$$

$$Fo' = Fo_{S-1} + i \div Fo_S \eta; \quad dFo' = i \div Fo_S d\eta,$$

where θ_s is the dimensionless temperature when $Fo=Fo_s$. η varies from 0 to 1.

Because only the temperature on boundary is calculated, in the following of the paper, $\Box_k$ means the temperature on boundary when Fo equals to Fo_k. Define that :

$$x = \frac{\sin^2 \beta}{F_{OK} - Fo_{S-1} - i \div Fo_S \eta}$$

Then formula (12) can be changed to:

$$\frac{1}{2}\theta_k = P - 4\sum_{s=1}^{k} \int_{\frac{\pi}{2}}^{} \int_{x_{s-1}}^{x_s} \frac{\theta_{s-1} + \frac{(\theta_s - \theta_{s-1})}{F_{OS}}(F_{OK} - Fo_{S-1} - \frac{\sin^2 \beta}{x})}{4\pi x}$$
$$\times (Bi+x)Exp(-x)dxd\beta \qquad (13)$$

3.2.2. The discretization of the boundary

Because formula (13) can't be integrated, β should be dispersed. the region $[0,\pi/2]$ is divided into J spans, and in span J, $[\beta_{j-1}, \beta_j]$, $\beta=(\beta_{j-1}+\beta_j)/2$. formula (13) can be expressed approximately as:

$$\frac{1}{2}\theta_k = P - 4\sum_{s=1}^{k}\sum_{j=1}^{J} \int_{x_{s-1,j}}^{x_{s,j}} \frac{\theta_{s-1} + \frac{(\theta_s - \theta_{s-1})}{F_{OS}}(F_{OK} - Fo_{s-1} - \frac{\sin^2 \frac{\beta_{j-1}+\beta_j}{2}}{x})}{4\pi x}$$
$$\times \Delta\beta_j(Bi+x)Exp(-x)dx$$

Integrating above formula:

$$\frac{1}{2}\theta_K = P - \frac{1}{\pi}\sum_{s=1}^{K}\sum_{j=1}^{J}\left[W(s,j)\theta_{s-1} + V(s,j)\theta_s\right] \qquad (14)$$

$$W\pounds s,j) = \{(Bi - GBi - HBi + H)\left[E_1(x_{s-1,j}) - E_1(x_{s,j})\right]$$
$$- HBi\left[\frac{Exp(-x_{s,j})}{x_{s,j}} - \frac{Exp(-x_{s-1,j})}{x_{s-1,j}}\right]\}(\beta_j - \beta_{j-1})$$
$$- (1-G)[Exp(-x_{s,j}) - Exp(-x_{s-1,j})] \qquad (15)$$

$$V\pounds s,j) = \{(GBi + HBi - H)\left[E_1(x_{s-1,j}) - E_1(x_{s,j})\right]$$
$$+ HBi\left[\frac{Exp(-x_{s,j})}{x_{s,j}} - \frac{Exp(-x_{s-1,j})}{x_{s-1,j}}\right]\}(\beta_j - \beta_{j-1})$$
$$- G[Exp(-x_{s,j}) - Exp(-x_{s-1,j})] \qquad (16)$$

$$x_{s,j} = \frac{\sin^2(\frac{\beta_{j-1}+\beta_j}{2})}{Fo_K - Fo_s}; \quad x_{s-1,j} = \frac{\sin^2(\frac{\beta_{j-1}+\beta_j}{2})}{Fo_K - Fo_{s-1}}$$

$$H = \frac{\sin^2(\frac{\beta_{j-1}+\beta_j}{2})}{i \div Fo_s}; \quad G = \frac{Fo_K - Fo_{s-1}}{i \div Fo_s} \qquad (17)$$

$$E_1(x) = -C_0 - \ln(x) + \sum_{n=1}^{\infty}(-1)^{n-1}\frac{x^n}{nn!} \qquad (18)$$

Formula (18) is an exponential integral function, and Co is Euler's constant.

3.2.3. The calculation of integral in region-P

As denoted in Figure 1, one divides the region Ω into part I and part II, by an imaginary tangent line that goes through the point X.

β" is included angle of vector **U** and imaginary tangent line. In region I, β varies from $\pi/2$ to $3\pi/2$,

$$P = \frac{1}{2} + \frac{1}{\pi}\int_0^{\frac{\pi}{2}}Exp\left[\frac{-\cos^2\beta}{Fo_K}\right]d\beta$$

and in region II, β varies from $-\pi/2$ to $\pi/2$, then

With the same discretization method mentioned above, the P can be expressed as:

$$P = \frac{1}{2} + \frac{1}{\pi}\sum_{j=1}^{J}Exp[-\frac{\sin^2(\frac{\beta_{j-1}+\beta_j}{2})}{Fo_K}]\beta_j \qquad (19)$$

from formulas (14) and (19), one obtains:

$$\frac{1}{2}\theta_K = \frac{1}{2} + \frac{1}{\pi}\sum_{j=1}^{J}Exp[-\frac{\sin^2(\frac{\beta_{j-1}+\beta_j}{2})}{Fo_K}]\beta_j$$

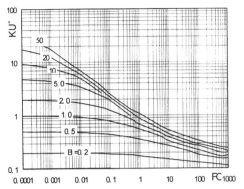

Figure 2. The variation curves in which the non-steady transfer criterion vary with the Biot and Fourier's numbers

REFERENCES

Yugeng Zeng, Jingsheng Liu & Xueyang Zhang 1991. *Finite element method and boundary element method.* Book concern of Xi'an Electronic science and technology university.

Qin Yueping, Dang Haizheng 1998. Dimensionless Analysis on Heat Dissipation of Rock surrounding in Coal Face. *Journal of China Coal Society.*Vol.23 No.1 62-66.

[i] Supported by National Natural Science Foundation of China (N0. 59704003)

$$-\frac{1}{\pi}\sum_{s=1}^{K}\sum_{j=1}^{J}\left[W(s,j)\theta_{s-1}+V(s,j)\theta_{s}\right] \qquad (20)$$

Using the formula (20), one can calculate θ_k corresponding to any Biot Number Bi and Fourier Number $Fo_k(k=1,2,...,n)$. The authors develop a FORTRAN computer program calculating $\theta_{w,k}$ based on the theory mentioned above, and draw a series of variation curves that K_{ur} follows Bi and Fo, in which Bi is between 0.1 and 50, and Fo is between 0.001 and 1000, as shown in Figure 2.

For a real airway, one can calculate its Bi and Fo at first, then run the computer program, and get the parameters such as θ, K_{ur}, q and Q of this airway, or can get the K_{ur} by Figure 2, and use K_{ur} to calculate q and Q.

4. CONCLUSIONS

By introducing some similarity modules based on the similarity theory, the boundary element equation is changed into dimensionless form, and the calculation is made simpler and speedier.

When calculating the heat dissipation of the surrounding rock in airway, the boundary element method needn't presume an outside boundary. It reduces a dimension from the integral equation, and decreases the calculation work.

A series of variation curves, in which the unsteady heat transfer criterion vary with Biot Number and Fourier Number, make the calculation of heat dissipation of the surrounding rock in airway more convenient.

Mining Science and Technology'99, Xie & Golosinski (eds)© 1999 Balkema, Rotterdam, ISBN 90 5809 067 1

Surface mining technology optimization based on artificial neural network and fuzzy logic decision

Y. D. Zhang, X. C. Li & C. S. Ji
China University of Mining and Technology, Xuzhou, People's Republic of China

ABSTRACT: A system of combining the artificial neural network (ANN) and fuzzy logic has been developed for selection of surface mining technology system in this paper. The ANN system is applied to determine the weight value of the various factors that will be used to evaluate the feasible mining technology. The fuzzy logic is applied to optimize the feasible mining technology alternatives that have been obtained from the expert knowledge base. This system has been applied to a Chinese surface coal mine.

1 INTRODUCTION

The application of ANN in the mineral industry has obtained rapid progress since ANN has been used in early 1980's, such as evaluation of resources, underground mining, surface mining and mine safety etc.

The selection of surface mining technology is an important part of the mine design. Generally, the mining technology system will be in service in a long term even over the total life of the mine.

Surface mining technology system can be divided into four categories: continuous technology, discontinuous technology, semi-continuous technology and combined technology. The selection of different mining technology systems is influenced by a series of natural, social and economical factors. For a large mine, the mining technology system for top soil stripping, hard rock stripping and mineral mining may also be different.

The main factors, which should usually be considered in mining technology selection, are as follows:
(1). Geological conditions of the deposit. Including mineral reserves, characters of mineral and overburden, natural occurrence of the deposit, topography of the mining area, hydrogeological conditions, etc.
(2). Climate conditions of the mining area. Such as: annual rainfall, wind conditions, frozen period and

frozen depth, etc.
(3). Mine capacity. Both mining capacity and stripping capacities are involved for surface mining.

Here, we built an expert knowledge base to give the feasible mining technology alternatives, which includes the factors and the expressions of the technology alternatives, as well as the rules those are applied to determine the possible mining technology alternatives.

2 WEIGHT VALUE DETERMINATION OF EVALUATION FACTORS BY ANN

For optimizing the feasible mining technology alternatives obtained from the expert knowledge base, ten decision-making factors are adopted to evaluate the mining technology alternatives. They are: total investment; unit investment; unit operation cost; mine capacity; number of operating personnel; labour productivity; capital construction volume; capital construction period; mining period from putting into operation until reaching designed mine capacity; stripping ratio in early operation stage.

A three layer ANN with BP is developed to determine the weight values of the ten decision-making factors. The input units include the natural factors such as the occurrence conditions, the structure of the mineral seams, the mineral reserves, the mineral quality and the local climate. The mining technology system for top soil stripping, hard rock stripping and mineral mining are also taken as the input units. The output units include the ten decision-making factors.

After training the network by practical data from a number of surface mines, the weight values for the input units and the output units of the network will be obtained. In order to obtain the actual weight values of the decision-making factors, four formulas are applied to analyze the values between the input units and the output units. The formulas are shown as the following:

$$r_{ji} = \sum_{k=1}^{k=p} w_{jk}(1-e^{-x}) / (1+e^{-x}) \qquad (2.1)$$

$$x = w_{ki} \qquad (2.2)$$

$$F_{ij} = r_{ji} / \sum_{j=1}^{j=n} \sum_{i=1}^{i=m} r_{ji} \qquad (2.3)$$

$$R_{ji} = \left| (1-e^{-y}) / (1+e^{-y}) \right| \qquad (2.4)$$

$$y = r_{ji} \qquad (2.5)$$

$$S_{ji} = R_{ji} / \sum_{j=1}^{j=n} R_{ji} \qquad (2.6)$$

Where:

r_{ji} is the correlation marked coefficient between unit j and unit I

F_{ij} is the relative effect coefficient between unit j and unit i

R_{ji} is the correlation index between unit j and unit I

S_{ji} is the absolute effect coefficient between unit j and unit i

i is the input units of ANN, $i=1\ldots m$

j is the output units of ANN, $j=1\ldots n$

k is the hidden units of ANN, $k=1\ldots p$

w_{ki} is the weight values between the input units and the hidden units

w_{jk} is the weight values between the output units and the hidden units

Here, the S_{ji} stands for the weight values of various decision-making factors to the natural factors and the technology systems. As ANN outputs, the weight values of the decision-making factors are as follows.

Total investment	0.13
Unit investment	0.10
Unit operation cost	0.10
Mine capacity	0.14
Number of operating personnel	0.05
Labor productivity	0.10
Capital construction volume	0.13
Stripping ratio in early operation stage	0.05
Capital construction period	0.07
Mining period for reaching mine capacity	0.13

3 MINING TECHNOLOGY SYSTEM OPTIMIZATION BY USE OF FUZZY LOGIC DECISION

Fuzzy logic is applied to optimize the feasible technology alternatives that come from the expert knowledge base. The two steps fuzzy logic is adopted here.

Before optimization, the decision-making factors are summarized into three categories: the economic target, the production target and the time target. Adding the weight values of all subsidiary factors of each target forms the weight values of the three targets. All the weight values of the targets and each decision-making factors are listed in the following:

A = (A1,A2,A3)=(0.33,0.47,0.20)
A1 = (0.4,0.3,0.3)
A2 = (0.3,0.1,0.21,0.29,0.1)
A3 = (0.35,0.65)

Where: A stands for the three categories of targets, A1 stands for the economic target , A2 stands for the production target and A3 stands for the time target

The first step is to evaluate the single decision-making factors. There are three matrixes, R1, R2 and R3, which are obtained by evaluating the single decision-making factors of the technology alternatives by experts. The evaluation results are composed of four grades: excellent, good, middle and poor. The ratio of the numbers of excellent to all the numbers of the four grades will form the grade of membership of excellent in the matrixes, similarly we can obtain the grade of membership of the good, middle and poor.

Then we will obtain the first evaluation result:

$B_i = A_i \cdot R_i$ (i=1,2,3) $\qquad$ (3.1)

where: R_i is the result of the evaluation to i^{th} target by experts.

The second step is to evaluate the three category targets. The evaluation result can be obtained from the following expression:

$D = A \cdot [B1,B2,B3]^T = (d1,d2,d3,d4)$ $\qquad$ (3.2)

where: D is the final evaluation result, d1-d4 are the grade of membership of excellent, good, middle and poor accordingly.

Comparing the result data of excellent, good, middle and poor, the maximum one is the final evaluation result of the technology alternatives. For example, if the data of excellent is maximal, then the final evaluation result of the technology alternatives belongs to excellent. We can also obtain the weighted average score from each evaluation grade as another evaluation index.

4 CASE STUDY

Anjialing surface mine in Pingshuo Coal Mining District is taken as an example.

By inputting the actual conditions of Anjialing, we get the feasible technology alternatives from the expert knowledge base as follows.

Top soil stripping technology: shovel-truck, scraper mining

Hard rock stripping technology: shovel-truck technology ; semi-continuous technology; shovel-truck-railway technology

Mineral mining technology: shovel-truck technology; semi-continuous technology

For example, in order to optimize the hard rock stripping technology, the original scoring for two feasible alternatives has been obtained from the field technician first, then by optimizing feasible alternatives with fuzzy logic, the results can be obtained as follows:

$D_1=(d_1, d_2, d_3, d_4) = (0.3940, 0.4415, 0.1644, 0)$
$D_2=(d_1, d_2, d_3, d_4) = (0.2998, 0.4921, 0.2079, 0)$

We can also obtain the average score of two alternatives by another evaluation index as follows:

$D\square_1=90*d_1+80*d_2+70*d_3+60*d_4=82.29$
$D\square_2=90*d_1+80*d_2+70*d_3+60*d_4=79.82$

Here, D_1 or $D\square_1$ is the evaluating result of the shovel-truck technology and D_2 or $D\square_2$ of the shovel-truck and semi-continuous combination technology, so we can obtain the best alternative, shovel-truck, from the optimization results.

REFERENCES

Zhang, R., Ren, T.X., Yu, R., Zhang, Y., Li, K., 1998. Application of in-house CAD Software in Surface Mine Design and Planning. 27th APCOM Symposium Proceedings. IMM: 91-98. London.

Li, X.C., Zhang, Y.D., Zhang, R.X., 1998, Artificial Neural Network System with Hierarchic Structure to Evaluate Resource Condition of Surface Mining Areas. 3rd Regional APCOM Proceedings. Kalgoorlie: 137-140. Western Australia.

Ji, C.S., 1996. Fuzzy Evaluation in Mining Equipment Selection. 26th APCOM Symposium Proceedings. SMME: 447-449. Littleton.

Mining Science and Technology'99, Xie & Golosinski (eds)© 1999 Balkema, Rotterdam, ISBN 90 5809 067 1

Safety information management system for Jinchuan mine No. 2

Jiaqian Yuan, Tongyou Liu & Qianli Zhao
Jinchuan Non-ferrous Metal Corporation Jointing, Gansu, People's Republic of China

Qian Gao
University of Science and Technology of Beijing, People's Republic of China

ABSTRACT: In order to enhance the capability of the fault prediction in underground mining operation, a safety expert system, which possesses the function of managing safety information, analyzing accidents and predicting faults, has been developed for Jinchuan mine No.2. Safety Information Management System (SIMS) is one of the sub-systems. This paper introduces the structure and functions of SIMS.

1 INTRODUCTION

Jinchuan nickel mine is situated in Jinchang city, Gansu province, China. The mine owns the largest nickel and copper deposits in China and it is with one of the most abundant ore resources in the world. With more than three hundred and twenty four million tons ore, there are 18 kinds of metals such as copper , cobalt , gold , silver , platinum and palladium.

That area has undergone more than one time of geological construction actions so that the engineering geological conditions in mine become very complex. Faults and joints in the rock masses are developed and the horizontal stress is high. More than 75% of the developed roadways are located in the fractured or jointed rock masses, so that the driving and the mining operation are very difficult. There are many hidden dangers in mining (Liu 1996, Liu, Tian & Shugao 1996).

Due to the complex geological conditions in Jinchuan mine, the stability of the rock strata is poor. By the action of the high structural stress, there are many faults during mining operation. An integrated system combining the mining and geological conditions in Jinchuan mine, which is used to manage safety information, carry out the fault tree analysis and predicte the faults, has been studied and developed. Safety Information Management System (SIMS) is one of the sub-systems.

2 THE STRUCTURE OF SMIS

The undercut-and-fill stoping is adopted in Jinchuan mine No.2. The safety production is influenced by many factors such as geological structures, quality of filling body and mining design etc.. The system analysis theory is used for the development of accident prediction system, which combine digging, supporting, lifting, transiting, mining, filling and machine & electricity working procedures as a macro-system. Because of the mutuality of the above working procedures, the factors related to the safety production are classified into two groups. One is the natural factors and the engineering factors which involves the geological structures, ore and rock quality, engineering and hydrological conditions, mining sequences, the types and parameters of the operation support etc.. The other is the management and economical factors, which covers the operation procedure organization, safety production management, safety measures, equipment expenses, procedure running expenses and maintenance costs etc..

From the viewpoint of the information science, the information management includes two tasks: obtaining the information and using the information. It is necessary to capture the geological, mining and operation data, and the factors related to the safety production in time. On the basis of observation and measurement in-situ, some original data can be obtained. The aim of the data and information analysis is to obtain and apply the knowledge to predict the hidden dangers. It is the aim that SIMS has been studied and developed. The structures of the system is shown in Fig.1.

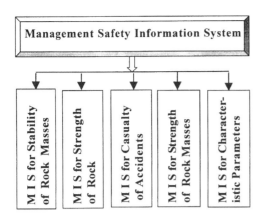

Figure 1 The structure of SMIS

Figure 1. SIMS contains 5 sub-systems, that is
(1)The Casualty of Employee Information Management System (CEIMS)
□The Stability of Rock mass Information Management System (SRIMS)
□ The Rock Strength Information Management System (RSIMS)
□The Rock Mass Strength Information Management System (RMSIMS)
□The Rock Characteristics Information Management System (RCIMS)

3 THE FUNCTIONS OF SIMS

The system is able to obtain , manage and store the safety information which is used to predict the hidden faults in the mining production as shown in Fig.2.

For several decades in the construction and mining operation of the mine, a lot of valuable data and information, such as geological and mining operation data, figures mining design parameters and producing management practice, have been accumulated. A lot of research results have been obtained. The researches not only had solved the problems in mining production in Jinchuan mine No.2 , but also will provide valuable data and information to predict faults for the future mining of depth ore. It is necessary to collect and analyze these data so that the data will be applied.

Considering the aim of the safety production management and fault prediction, there are two classes of information collected in SIMS, namely the direct safety information and the indirect safety information.

3.1 *Direct Safety Information Management*

The direct safety information is namely the records and the analysis on the casualty events, which are related to death, grievous bodily harm and flesh wound. In order to manage the direct safety information, CEIMS has been developed which possesses following functions:
□ Information input
Man-machine dialogue is adopted for data input in order to use the system conveniently. When input information, activate input menu and show inputting dialogue window in the screen, and then input necessary data only in the corresponding station. A record is inputed by filling a table sheet.

□ Data modification
In order to modify the information, one should search for and revise the records that should be modified. The system provides several ways as follows:
A. Modifying all records: When one selects the revising way, all records in the database will be shown from first record to the last record.
B. Based on the revising conditions: When choosing the revising way, one inputs the keywords of revising condition(s), and the system will find out these records which satisfies the searching of the condition(s) and show them in the screen.
C. Based on the code number of record: When one chooses the modifying way, the system will search for and show the record corresponding to the input code number. When a record is input, the system will give a chance to decide if you continue to input or quit the system.

□ Data deleting
Data deleting is similar to data modification. Firstly one searches for these records needed to be deleted, the system will add a deleting remark for the record.

□ Statistic analysis of Data
The module possesses multifunction to meet the demands of the different information and the multi-statistic analysis. For the direct safety information , the functions are as follows:
□ Statistic of the casualty event which involves the event times, the time of flesh wound or GBH or casualty, losing working days and economic loss etc..
□ Statistic of casualty rates which involve GBH rate per thousand men, the death rate per thousand men, the injured rate per million man-hour and the death rate per million ton ore.
□ Single item contrasting statistic: Based on user appointed time scope, the system is capable of performing the following several item statistics:

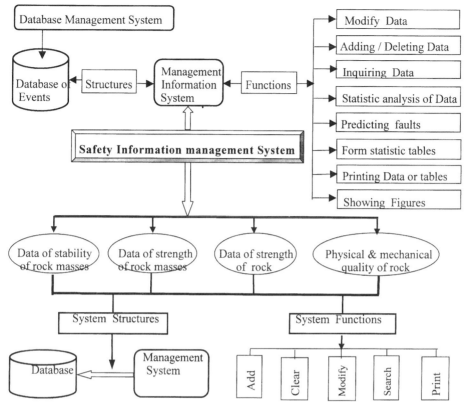

Figure 2. Functions of the Safety information management system for Jinchuan mine No.2

Such as unit, age, type of work, period of service and type of accidents.

☐ Multi-item cross statistic: This function provides the following groups cross statistic, that is unit - type of accidents, unit – type of works and type of work – type of accidents.

☐ Inquiring data
This module provides 5 inquiring ways: the number, location and type of accident , type of work and unit. Each way contains all information or some information of inquiry.

☐ Making and showing figures
This module provides the functions to make and show figures which contain both round and square figures involving casualty happening in any year.

☐ Faults Prediction
With the help of regressive method, this function can be used to predict the possible accident times in next year, based on the time of the casualty, the death toll, GBH and the injured number. The system is able to display and / or print the predicting results.

☐ Data, tables and figures output
Based on the content and data , the module can fulfil the following functions:
☐ Printing any records in the data-base.
☐ Printing the statistic analysis results.
☐ Printing the report tables of the casualty event including the yearly or monthly statistic report tables.

3.2 *Indirect Safety Information Management*

The indirect safety information is mainly related to the geologic and mining information. Geologic information involves geologic construction, original stress, faults and joint groups etc.. Mining information contains the mining method, the stability of rock masses and the types and parameters of supports etc.. The above information is very valuable for predicting the operation faults and making more mining profit, so that it is also necessary to obtain the most of the information in mining design and safety production management. The indirect safety information will be used as follows:

□ Providing input information for fault system prediction. In order to predict mining faults, the fault prediction system (FPS) had been developed which is related to a normal inference expert system(NIES) and neural network expert system (NNES). Then when making use of the system to predict faults, some information need to be input.

□ Determining mechanical parameters of rock masses for evaluation the rock masses stability. For very complicated geologic conditions, the theoretical or numerical analysis is needed to evaluate the stability of the surrounding rock. The determination of the calculated parameters of rock masses is very difficult. In order to determine these parameters accurately, the safety information is needed .

For managing the indirect information, four sub-IMS had been developed as follows:

□Stability of Rock mass Information Management System (SRIMS)

□Rock Strength Information Management System (RSIMS)

□ Rock Mass Strength Information Management System (RMSIMS)

(4) Rock Characteristics Information Management System (RCIMS)

SRIMS is related to the following data:

□ Types of rock strata, geological characters and rock mass structures.

□ Shapes of cross section of roadway and types of ground pressure.

□ Types of supports.

RSIMS contains the following data:

□ Mechanical parameters of intact rock which is related to compressive strength and shearing strength of intact rock.

□ Characteristic parameters of deformation of intact rock which is related to elastic modulus and Poisson's ratio of intact rock.

RMSIMS contains the following data:

□ Mechanical parameters of rock masses such as the compressive strength and shearing strength are related to the mining production.

□Characteristic parameters of deformation of rock mass, which are related to its elastic modulus and Poisson's ratio

RCIMS contains the unit weight, rock color, and percentage of porosity and percentage of water absorption.

The 4 sub-systems mentioned above possess the management functions of data which can inquire, search for, modify, add and delete data when performing management.

4 INSTALLATION AND APPLICATION OF SIMS

□ SIMS installation

SIMS contains 5 databases and they are put in different sub-systems. It is very simple to install SMIS. By inputting an install order , the system will automatically copy all software to hard disk and put into different subsystem respectively.

□ Software environment of SIMS

Operation system □□□6.2; Chinese platform: UCDOS 6.0; Database management system: FoxPro 2.6.

□ Application of SIMS

□ After starting computer, at first enters sub-system "jcmis".

□Enter your using information management system. For example, if you hope to perform casualty information management, at first, enter the sub-system "sgdj"

5 CONCLUSIONS

As a part of the integrated system of the safety information management and the fault prediction for Jinchuan No.2, SIMS possesses an important function. This system has been used in Jinchuan mine for safety information management. The application of the system not only provides the important experience and information for forecasting the production faults, but also heightens the modernizing management level.

REFERENCES

Liu Tongyou etc. 1996, *The problems of the control on the surrounding rock masses of the stope in Jinchuan nickel mine. Evolvement of Rock mechanics & Engineering in Gansu*, Lanzhou University Press.

Liu Tongyou , Gao Qian & Zhao Qianli: 1997, *Research Report on Safety Expert System for Jinchuan Mine No.2 , Jinchuan Non-ferrous Metal Corporation* , University of Science & Technology of Beijing.

Liu Tongyou , Gao Qian & Zhao Qianli . System Analysis and Integrated Management for Underground Mining: *Geology* Press, 1998.

Mining Science and Technology' 99, Xie & Golosinski (eds) © 1999 Balkema, Rotterdam, ISBN 90 5809 067 1

Safety in Japanese coal mining industry and analysis of the recent trend

K.Uchino & M.Inoue
Department of Earth Resources Engineering, Kyushu University, Japan

ABSTRACT: This paper reviews the safety in Japanese coal mining industry and the influences of related technologies and social factors on the accidents are discussed. Although the various safety records have been improved significantly over the last fifty years, they still remain at unsatisfactory levels as compared to those of other industries. Aiming to reduce the accidents further, a new study project that focused on the human factors in mine accidents was started in 1989. The detailed results of the study are analyzed. The results show that there are still many technological problems and their solutions related to the accidents in the coal mining industry. Further efforts from both aspects of human factors and technologies must be made to reduce the injuries.

1 INTRODUCTION

Major coal mines in Japan are underground. In 1940 the annual coal production in Japan recorded 56.3 million tons, which is the maximum in the history. However, because of the shortage of labor, machinery and materials the safety level in the coal mining industry gradually lowered toward 1945 and the production decreased to 20 million tons. The coal mining industry was reconstructed rapidly by the government policy and attained a production of 55.4 million tons in 1961, the second peak of annual output. After that, the production has decreased consistently until today because of cheap imported oil at first and the foreign coal later as well. Current annual coal production is about 3.6 million tons from two major coal mines.

2 CHANGES OF SAFETY PERFORMANCE

Figure 1 shows the coal production, productivity and injury frequency rate*(MITI 1995). Figure 2 shows the number of coal mines, workers and fatalities. The trend of mine accidents from 1946 until today can be divided into five periods (UCHINO 1993).

(1) 1946-1955
Each period can be characterized by technological and social changes and events.
a) The government policy to reconstruct coal mining industry and steel industry prior to others as a base for the whole industry.
b) Introduction of steel support and mechanization of coal face.

The coal production and productivity increased and injury frequency decreased as shown in Figure 1. However the number of fatalities did not decreased as the injury frequency, shows in Figure 2.

(2) 1956-1965
a) Transition of principal energy source from coal to oil.
b) Rapid rationalizations of coal mining industry.
c) Many labour disputes.
d) Mechanization of coal face and introduction of many new technologies.
e) Shift to worse mining conditions.
f) Injury frequency increases.

Figure 3 shows the trend of injury frequency categorized by the causes of accidents and Figure 4 the variation of coal mining method. Injury frequencies increased in all categories except surface. Materials handling became the second cause of injuries and 196 serious disasters** occurred in the period. The coal policy of the government in this period set the annual coal production at 55 million tons.

(3) 1966-1975
a) Increase of oil import.
b) First oil shock.
c) Introduction of shield support.

The government policy did not make quantitative target of coal production in the beginning of this period. However, 20 million tones target was made in 1973.

(4) 1976-1985

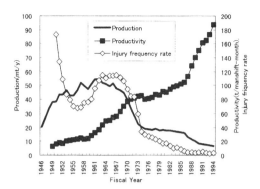

Figure 1. Coal production, productivity and injury frequency rate.

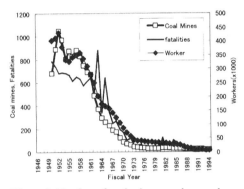

Figure 2. Number of coal mines, workers and fatalities.

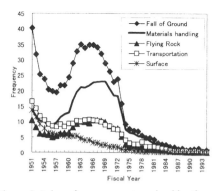

Figure 3. Injury frequency categorized by the cause of accident.

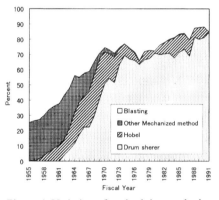

Figure 4. Variation of coal mining method.

a) Government coal policy to maintain the annual production at 20 million tones.
b) No significnat change of technologies.
c) Prevalence of longwall mining system equipped with drum shearer and shield support.
  (5) 1986-1996
a) An enhanced exchange rate of the yen.
b) New government coal policy that the annual coal production should be 10 million tons.

3 INVESTIGATIONS FOCUSED ON HUMAN FACTORS

No serious accident has occurred since 1988. However injury frequency still remained at a higher level compared to those of other industries. There was no need to report the minor accidents to the mine inspectorate untill then. In order to collect the enough and detailed information about the background of the accidents it was decided to collect the data of all accidents which caused any medical treatment even if there was no lost time of working. A new and detailed form of the report of injury was designed and requested to report all accidents. The report contains following items
1. Date, time and location of accident
2. Accident type.
3. Description of injury.
4. General situation of injury.
5. Information of sufferer; age, experience, education, character.
6. Information of person(s) concerned ; age, experience, education, character.
7. Cause of accident; unsafe conditions, unsafe behavior.
8. Behavior of the worker and motion of materials.
9. Environmental conditions of the place where the accidents occurred.
10. Standard and method of the job which was being done by the sufferer.

  Figure 5 shows the variation of injury rate

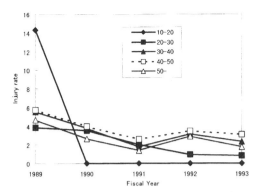

Figure 5 Variation of injury rate (number of injures / number or workers) classified by the age of workers.

Table 1. Injuries from 1989 to 1995.

Mine	Fatal	Serious	Major	Minor	No lost time	Period
A	1	14	1	2	68	1989-1992
As	0	4	0	0	20	1989-1992
S	2	22	2	0	297	1989-1992
T	2	36	0	1	48	1989-1992
	0	10	4	0	23	1995
I	0	27	2	12	377	1990-1993
	0	4	1	0	65	1994
	2	2	0	3	61	1995
M	4	173	12	4	344	1990-1993
	1	16	3	3	101	1994-1995

Table 2 Number of injured workers who had injured for the previous three years.

number	mine M6	mine M7	mine I6
0	51	43	27
1	8	13	15
2	4	5	13
3	0	0	4
Total	63	61	59

classified by the age of workers. Table 1 shows the injuries that occurred from 1989 to 1995. It was found that many minor and no lost time injuries occurred during the period. Table 2 shows the number of workers suffered for the previous three years.

Interesting information about the accidents in this period were found in the analysis of the reports as follows (CMRC 1997).

1) Injury rate of the workers is high whose age are 30's to 40's and experience are more than 10 years.

2) Almost all the accidents occurred when the sufferers were familiar with both the working place and the job.

3) The jobs which were being done by the sufferers were in good progress when the accidents occurred.

4) The probability that the same worker meets with an accident again is several times higher than that expected by simple analysis of the statistics.

These facts mean that there might be many accidents caused by the problem related to human factors. However, no apparent relationship between human characters and accidents was observed from the results of the investigation.

4 INVESTIGATIONS BY MINE'S RECORD

Investigations described above were finished in 1995. Accident data are still continuously recorded briefly and investigated in individual mines. Although it is brief the records are valuable because the accidents were described in the context of human behaviour.

Table 3 shows the causal breakdown of all accidents occurred from 1991 to 1998 in a mine. The grouping is different from ordinary one because the causes of accidents are classified from the viewpoint of its prevention.

Fall of ground: Almost all accidents of this kind occurred when the ground rock stress was changed artificially by such as excavation. These accidents decreased with the increase of usage of road heading machines that give less damage on the ground rock as compared to blasting. Introduction of a new machine to assist the construction of a steel support at a heading face also contributes to reducing the accidents. A gap width between the roofs of shield support at a long wall face was minimized as possible to prevent the roof fall through them.

Roof control on shield support: Although it is dangerous, it was necessary to work with the roof rock on the shield supports at a long wall face when the roof condition was very bad. Improvements of the shield support itself contributed to decreasing the accidents.

Machinery in motion such as a drum shearer, shield support, conveyer, road heading machine cause the accidents even if it moves slowly. Accidents by a flying piece by a drum cutter occured seven times and are included in this category.

Fixing and weight: This means the failure of fixing of heavy materials. Various materials became the cause of this kind of accidents.

Tension, pressure, strike: A rope is the major cause of this kind of accidents. Flying pieces by strike caused 9 injuries on eyes, however they are categorized into eye.

Careless mistake: Various materials have the potentials to be the cause of this accidents, especially, hook, knife and pin causes 12, 5, and 4

Table 3 Causal breakdown of all accidents occurred from 1991 to 1998 in a mine.

	1991	1992	1993	1994	1995	1996	1997	1998	Total
Fall of ground	17	21	20	14	7	9	3	2	93
Roof control on shield support	7	5	3	2	4	1	2	1	25
Machinery in motion	4	6	8	4	4	2	1	1	30
Fixing and weight	7	11	2	7	7	2	4	1	41
Tension, pressure. strike	6	4	6	6	4	1	3	0	30
Careless mistake	14	20	12	9	12	5	3	1	76
Communication	7	8	4	10	9	4	2	6	50
Slip and fall	9	14	10	6	7	5	4	3	58
Inappropriate facility	3	17	7	7	3	1	0	0	38
(Buttery)		(10)	(2)	(3)					
Injury of eye	1	8	10	5	7	4	1	1	37
Others	0	1	0	0	4	0	0	0	5
Total	75	115	82	70	68	34	23	16	483

*exclude November and December in 1998

injuries respectively.

Communication; Fourteen accidents were brought about by other persons as a result of insufficient communication. Thirteen accidents occurred in heavy lifting work and five in unloading.

Inappropriate facility; Buttery was the major cause of the accidents. However it becomes zero because they changes the buttery to Ni-Cd type without liquid.

Eye: Most accidents occurred when dusty works were being done. Employers were continuously recommended to use safety glasses.

5 CONCLUSION

Investigations focused on human factors were started on an assumption that the causes of many accidents might be attributed to human factors. Detailed reports supported this assumption to some extent. However, it was also found that there are still technological causes and their solutions. Both efforts related to human factors and technologies are important to increase the safety performance in coal mining industry.

*Injury frequency rate is the number of injured persons per million worked hours.

Serious disaster is defined that causes more than three fatalities or five major injuries*.

-Injury is defined as one that causes lost time of three days or more.

-Serious injury is defined as one that heavily injured or more than 14 days lost time.

-Major injury is defined as one that causes lost time more than 3 days and less than 14 days

-Minor injury is defines as one that cause less than 3 days.

REFERENCES

Ministry of International Trade and Industry (MITI), 1995, Annual report on mining safety.

Uchino, K. 1993, Safety and mechanization in Japanese coal mining industries, United Nations Interregional symposium on safety and mechanization in underground coal mining, Omuta, Japan.

The coal mining research centre, Japan (CMRC), 1997, Report on mine accidents research focused on human factors,

Mining Science and Technology' 99, Xie & Golosinski (eds) © 1999 Balkema, Rotterdam, ISBN 90 5809 067 1

Experimental study on the Electromagnetic Radiation (EMR) during the fracture of coal or rock

Xue-qiu He, En-yuan Wang & Zhentang Liu
China University of Mining and Technology, Xuzhou, People's Republic of China

ABSTRACT: Electromagnetic radiation (EMR) during coal or rock fracture is measured and analyzed by both laboratory and site experimental study. The results show that EMR truly exists during the fracture of coal or rock (containing or not containing gas), it follows Hurst statistical distribution, and it is enhanced during the process of breaking. The EMR strength and frequency are correlated to coal or rock fracture process, and the signal of EMR is more sensitive than that of AE. Based on these properties a new method for coal and methane outbursts prediction is proposed - the so called EMR method. It makes the prediction of coal and methane outbursts easier, with continuing measurement and without touching the coal surface.

1. INTRODUCTION

Coal and methane outbursts and rock outbursts are the violent disasters in coal mining, especially in coal mines at great depth. How to monitor the dynamic process of coal or rock, which contains gas, is a very important technology in modern coal mining, especially in the prediction of coal and methane outburst, rock outburst, and roof fall and abutment pressure displacement. Although the prediction technology has been developed for a long time, for example, acoustic emission monitoring technology, etc, most of these methods are complex in application and the sensors must be coupling with coal or rock wall during the process of monitoring.

Recent research of Authors reveals, however, that electromagnetic radiation (EMR) takes place during the deformation and fracture of coal or rock (and/or containing gas)(He, X.Q., 1994, 1995). This also accords with EMR researchers in the field of earthquakes (Gress,G.O., 1987). The research of EMR can be widely used to assess the stress condition of rock, to reveal the mechanism of deformation and fracture, to predict the catastrophic phenomena of rock or coal. These catastrophes are caused by the stress change, and it follows the "Rheology Hypothesis"(He, X.Q., 1992). This kind of stress change must lead to some characteristic change of EMR, which can be taken as the omen of rock destabilization. It is obvious that the researches and applications of EMR are important, especially to the safety of coal mining.

2. EXPERIMENTAL STUDY

In the experiment two types of coal sample are tested. One is molded under the pressure of 135Mpa in a steel model with coal powder (the particle sizes are less than 0.4 mm); the other is molded with original coal. The sizes of tested coal specimen are □50×100mm. Totally, 22 different coal specimen are tested under the load of axial compressive stress.

The experiment system consists of EMR sensors, AE sensors, signal amplify system, load system, electromagnetic wave shield system, coal specimen vacuuming system and gas adsorption system. The sketch of laboratory experiment system is shown in fig 1.

On the basis of the test results, a general concept of EMR during the fracture of coal is built up, and it is found out that the frequency is spread at a very wide range. Every micro-fracture or displacement in the coal emits a different frequency signal of EMR. And some times it is coupling with the AE of coal fracture, even it isn't homologous to the AE. The general characteristics of EMR and AE during the whole process of coal fracture are shown in Fig 2.

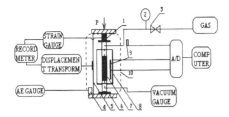

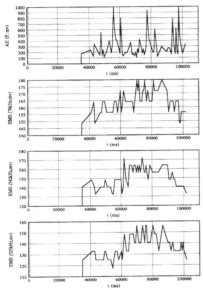

a Time sequence of AE and EMR with different frequency

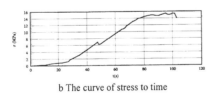

b The curve of stress to time

Fig 2 The general characteristics of AE and EMR during the fracture of original coal specimen (from Xuzhou Coal Mining Bureau, Jiangsu, China).

1–load sensor□2–pressure gauge□3–valve□4–AE sensors□5–displacement sensor□6–isolator□7–coal specimen□8–pressure vessel□9–antennae□10–shield net

Figure 1. Design of the experiment

It also is found out that EMR is more sensitive than AE during the fracturing process of coal. And the frequency of EMR signals widely distributes, and some times it doesn't synchronize with AE. According to the curves of stress and strain versus

time, the EMR signal number direct proportion to the load in some way. ᴗral signals of

3.   THEORETICAL ANALYSIS

In order to confirm the reliability of using the signal of EMR to predict the fracture or outburst of coal or rock, the method of R/S statistical analysis is used. H. E. Hurst put forward the *R/S* (re-scaled range analysis) method in 1965. For a temporal signals of EMR $\{x(t), t=1,2,...N\}$, their mean value is given by formula (1).

$$\langle X \rangle_k = \frac{1}{k} \sum_{t=1}^{k} x(t), \ k=1,2...N \qquad (1)$$

Then we can get the cumulative, maximum and standard deviations as formula (2)-(4).

$$X(n,k) = \sum_{i=1}^{n} (x(i) - \langle X \rangle_k) \ \square \ 1 \leq n \leq k \qquad (2)$$

$$R(k) = \max_{1 \leq n \leq k} X(n,k) - \min_{1 \leq n \leq k} X(n,k) \qquad (3)$$

$$S(k) = \sqrt{\frac{1}{k} \sum_{t=1}^{k} (x(t) - \langle x \rangle_k)^2} \qquad (4)$$

Hurst found out the following relation:

$$R(k)/S(k) \propto k^H \qquad (5)$$

Where H called Hurst index.
   The physical meanings of R/S is that, if the series $\{x(t), t=1,2,..., N\}$ is an independent random series with limited variances, $H=1/2$; if $H>1/2$, the signals versus time series are dependent with the increasing tendency of the past signals. Upon the suitable rearrangement of equation (5), we can obtain:
$$H = d\log(R(k)/S(k))/d\log(k) \qquad (6)$$
   The equation (6) shows that *Hurst* index H is the tangent of the curve of $\log(R(k)/S(k))$ to $\log(k)$.
H can be obtained by regression analysis of the experimental data using equation (6). Fig.3 shows the statistical results of the pulse number of EMR within the period of deformation of coal specimen from Xuzhou Coal Mining Bureau. In Fig.3, 10k, 2.6M, 814k and 138k represent the EMR acceptance frequency of 10kHz□2.6MHz□814kHz□138kHz respectively. *R* represents the regression coefficient. The calculating results of each frequency band are given in Table 1.
In Fig.4, a□b□c represents the statistical results of

EMR pulse number with equal time span during deformation of coal specimen respectively from Huainan Coal Mining Bureau, with the acceptance frequency of 542KHz ,81KHz and 10KHz. The calculating results of each frequency band are given in Table 2.

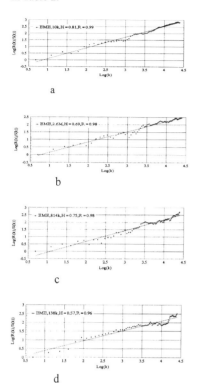

a

b

c

d

Fig.3 the statistical rule of EMR of raw coal specimen (from Xuzhou Coal Mining Bureau)

Table1 Results of R/S statistical analysis of EMR of raw coal specimen (from Xuzhou Coal Mining Bureau)

f of EMR	3kHz	10k Hz	81k Hz	2.6M Hz	814k Hz	542k Hz	138k Hz
H	0.71	0.81	0.72	0.69	0.75	0.72	0.57
R	0.95	0.99	0.97	0.98	0.98	0.97	0.96

From the results, it can be found out that the frequency band of EMR during deformation and fracture of coal is wide, and that EMR of each band conforms to Hurst statistical distribution, with Hurst index H between 0.5~1.0□ and regression coefficient R is above 0.95. The results also show that EMR exhibits gradually enhance tendency

during the increase of the deformation. The R/S statistical distribution of EMR provides a new tool for the prediction of catastrophic dynamic phenomena of coal or rock.

The Hurst index of EMR in each frequency band is different from each other. The leading frequency band is changing during the process.

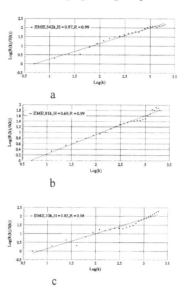

a

b

c

Fig.4 the statistical rule of EMR of molded coal (No.2, from Huainan Coal Mining Bureau)

Table 2 The results of R/S statistical analysis of EMR of molded coal specimen (from Huainan Coal Mining Bureau)

f of EMR	wide	138kHz	542kHz	81kHz	10kHz
H	0.70	0.54	0.88	0.69	0.86
R	0.96	0.97	0.99	0.99	0.98

The above results show that it is not very ideal to predict the catastrophic phenomena of coal or rock if just using spot frequency as a criterion, therefor the wide frequency band is recommended.

4. APPLICATIONS

On the basis of laboratory experiment and general analysis, an equipment of monitoring EMR with wide frequency band during digging or mining in coal mine is developed and used in the prediction of rock outburst, coal and methane outburst. Some of the results measured in 8th Coal Mine, Pindingsan Coal Mining Bureau, Henan Province, are shown as fig 5.

From the site experiment results, we can find that the amplitude and the pulse number of EMR before and after explosion changes greatly. Especially all the EMR pulses number oversteps the record limit of the equipment just 23 minutes after the explosion, as shown in fig 3---b.

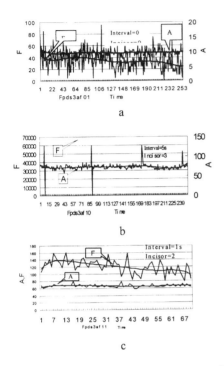

a

b

c

In the above figures, A and F mean separately the amplitude and frequency of EMR measured in the site.
a EMR monitoring results before the explosion in the coal head of a downgrade roadway.
b EMR monitoring results, 23 minutes after the explosion in the coal head of a downgrade roadway.
c EMR monitoring results 41 minutes after the explosion in the coal head of a downgrade roadway.

Fig 5 Site EMR monitoring results during an explosion in the outburst prone coal head of a downgrade roadway, 8th Coal Mine, Pingdingsan Coal Mining Bureau.

The number of EMR pulses and the amplitude then decline with time. Some parallel experiments between outburst prone and non-outburst prone heads are carried out, and in the same time other outburst prediction methods are also used. Finally, it is proved that the bigger the EMR pulse number and the EMR amplitude are, the larger the outburst prone is.

5.   CONCLUSIONS

Through the experiment study both in laboratory and field it can be found out that the signal of EMR is more sensitive than that of AE during the damaging or fracturing of coal or rock. By the EMR monitoring in the course of coal or rock fracture, more fracture information could be obtained. And this technology can be used in the prediction of coal and methane outburst or rock outburst.

However, more attentions should be paid to the future research such as, how to get the outburst prediction guideline by more site experiments, how to ascertain the orientation of EMR, etc.

ACKNOWLEDGEMENT

The authors would like to express thanks to the "CHINA NATIONAL SCIENCE FOUNDATION COMMITTEE", "THE FOUNDATION COMMITTEE OF SPAN CENTURY PERSON WITH ABILITY OF CHINA" and "FOUNDATION COMMITTEE OF DOCTORAL SUBJECT RESEARCH " for their financial support.

REFERENCES

Cress, G. O., et al. 1987. Sources of Electromagnetic Radiation from Fracture of Rock Samples in Laboratory. Geophys.Res.Lett.. Vol.14, pp. 331-334.

He, X.Q. & Zhou, S. N. 1992. The Rheological Fracture Properties and Outburst Mechanism of Coal Containing Gas. Proc. 11th Intl. Conf. on Ground Control in Mining. Aziz, ed., Wollongong, Australia, pp575-579.

He, X. Q. & Liu, M. J.. 1994. Laboratory Research on Electromagnetic Emission from Fracture of Coal and Rock Containing Porous Gas. Proc. Intl. Symp. on new Develop. in Rock Mech. and Eng. X. H. Xu, ed., Shengyang, China, pp567-572.

He, X. Q. 1995. Fracture electromagnetic dynamics of coal or rock containing gas. China University of Mining & Technology Press, Xuzhou, China.

Mining Science and Technology'99, Xie & Golosinski (eds)© 1999 Balkema, Rotterdam, ISBN 90 5809 067 1

Extinguishing spontaneous combustion in coal seam with heat-resistant, rich-water gel

Jingcai Xu, Jun Deng, Hu Wen, Xingming Guo & Xinghai Zhang
Department of Mining Engineering, Xi'an Mining Institute, People's Republic of China

ABSTRACT: The injection technique of heat-resistant, rich-water gel (HRRWG) is successfully employed to extinguish spontaneous combustion in coal seam. HRRWG is gelatinated with primary material (PM), accelerant, hardener and water in proportion. Its properties prevent and extinguish spontaneous combustion. Based on previous applications of this technique a simple technology for coal fire extinguishing was developed. The technique of HRRWG has been employed in 26 nation-owned coal mines in 16 Mining Administrations and other 8 local coal mines in China. It had helped to extinguish 67 fires by the end of 1998. Now, the technique of fire extinguishing with HRRWG is one of the primary means to fight spontaneous combustion in coal seam.

1 INSTRUCTION

About 56% of coal mines are liable to spontaneous combustion in China. In recent years, many mine fires, caused by spontaneous combustion, results in the loss of property and the closure of mine. Coal seam self-fires, caused by spontaneous combustion, results in the loss of property and the closure of the mine. Coal seam self-ignition originates from the exothermic reaction of coal and oxygen. If the concentration of oxygen is more than 3%, the oxidation heat will be released from and accumulated in coal. Coal seam fires hide usually in high places. Therefore in order to fight fire in coal seam effectively, the extinguishing agent must be able to: 1) block porosity and stop air-leakage in coal, 2) absorb heat released from coal and adjacent rock, 3) destroy the active particles in coal and retard coal oxidation.

2 CHARACTERISTICS OF HRRWG

HRRWG is gelatinated with primary material (sodium-silicate), coagulant (ammonium bicarbonate, sodium bicarbonate, or ammonium chloride), hardener (clay or soot) and solvent (water) in proportion. The table 1 contrasts HHRWG with other analogous techniques in the world. The weight of water is more than 90% in the HRRWG. Before the materials gelatinating, the gel is liquid which can infiltrate into the cracks of loose coal. But

after gelatinated, it can block the porosity, stop the air-leakage, and then retard coal oxidation. The gel comprising hardener can fill the caving area in the gob or the empty place in the roof-coal of gates. The primary material, the coagulant and their products of reaction are all retarders. The gel can slow down coal oxidation and exothermicity. HRRWG is not easy to vapor (fig.1), just shrivel slowly above $1000°$, so that the safety reliability for extinguishing is high enough. Under the normal conditions of underground coal mine (relative humidity:> 90% ;temperature: <28°), the gel's life-span is longer than 13 months. The time of gelation can be controlled between several seconds to several hours by adjusted coagulant. Because the gelation is endothermic reaction and the gel's specific heat is greater than water, it can absorb heat vastly inside coal and bedrock, and then reduce the temperature of coal. The materials of gelation are inexpensive and get easily.

3 EQUIPMENT AND TECHNOLOGICAL PROCESS

According to the characteristics of fire in different locality, such as roof-coal of gate, gob, the area adjacent to gob, the back of powered supports, stable-hole, and so on, a series of technological process and the equipment for fire direct extinguishing directly have been developed.

Table 1. HRRWG compared to other analogous techniques in world

Items	Oversea			china	HRRWG
				$Na_2O\cdot SiO_2$	$Na_2O\cdot SiO_2$
PM	Bentonite	Clay	$Na_2O\cdot SiO_2$	$Na_2O\cdot SiO_2$	$Na_2O\cdot SiO_2$
Accelerant	No	No	$(NH_4)_2SO_4$	$(NH_4)_2SO_4$	NH_4HCO_3 or$NaHCO_3$
Hardener	No	No	No	No	Clay or soot
PM dosage	30%	50%	10%	10%	6%□10%
Accelerant dosage	No	No	8%□10%	8%□10%	2%□3%
Extinguishing directly	bad	bad	No report	No report	Good
Gelatinating speed	inadjustable	inadjustable	adjustable	adjustable	Adjustable
Stop airleakage and absorb heat	less	less	better	better	Best
Retarder	Bad	Bad	better	better	Best
Series equipment			No	No	Ok
Flux equipment	Ok	Ok	No	No	Ok
Causticity	No	No	Intensive	Intensive	Weak
Available longevity	Longer	longest	Long	long	Longest
Cost	Expensive	Cheap	Expensive	Expensive	Inexpensive

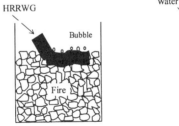

Fig.1. Contrasting experiments on the extinguishing properties of HRRWG and water

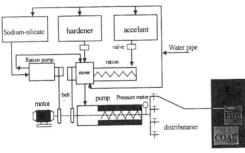

Fig.2 Moveable technological process of HRRWG infusion

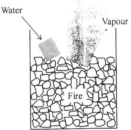

Fig.3. Technological process of HRRWG infusion

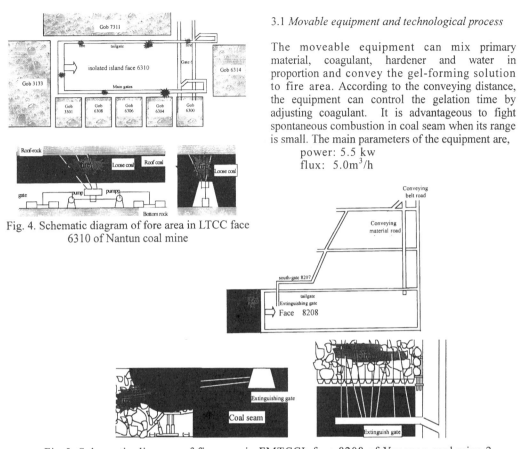

Fig. 4. Schematic diagram of fore area in LTCC face 6310 of Nantun coal mine

3.1 *Movable equipment and technological process*

The moveable equipment can mix primary material, coagulant, hardener and water in proportion and convey the gel-forming solution to fire area. According to the conveying distance, the equipment can control the gelation time by adjusting coagulant. It is advantageous to fight spontaneous combustion in coal seam when its range is small. The main parameters of the equipment are,

power: 5.5 kw
flux: 5.0m^3/h

Fig.5. Schematic diagram of fire area in FMTCCL face 8208 of Yanquan coal mine 2

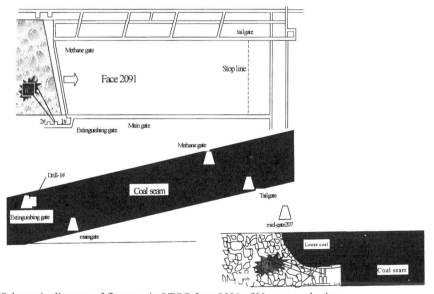

Fig.6. Schematic diagram of fire area in LTCC face 2091 of Nantun coal mine

pressure: 1.8 MPa

hardener diameter: <5mm

size(length×width×height):1600mm×56
0mm×1100mm

weight: 230kg

Fig. 2 shows the technological process.

3.2 *Immovable flux technological process of HRRWG infusion*

Based on the grouting system, the gel-forming solution is conveyed to fire area. In the technological process, primary material, hardener and water are mixed in proportion on the ground, only coagulant is added in the vicinity of the spontaneous combustion region in underground mine(Fig.3). Gel infusion flux is about $30\sim100m^3/h$. The technological process is enough to treat the spontaneous combustion spreading large area which occurs on the back of power supports, gob, roof-coal of gate, and so on.

4. APPLICATION OF HRRWG IN COAL MINES

The technique of HRRWG has been popularized in 26 nation-owned mines of 16 Mining Administration Bureaus(MAB), such as Yangquan, Yanzhou, Yima, Zaozhuang, Datong, Datun, Jingyuan, Hami, and other 8 local coal mines in China. 67 fires had been extinguished by using the technique by the end of 1998. Several representative examples of fire extinguishing are illustrated as follows.

4.1 *Roof-coal fire in gate near gob (see fig. 4)*

The spontaneous combustion of the roof-coal occurred in the tailgate in the fully mechanized top-coal caving longwall (FMTCCL) face 6310 of Nantun mine, Yanzhou MAB, early September, 1993. Many measures (water infusion, grouting, and so on) were adopted, but the fire still spread along the gate rapidly. The length of the fire area was more than 40m, and the other six fires were occurred in gates with the length of 1800m in September. Finally, HRRWG was introduced into the mine. The first fire was completely extinguished by infusing $70m^3$ HRRWG in three days, and other fires were extinguished quickly.

4.2 *Fires on the back of power supports in gassy mine (see fig.5)*

The remained coal fire was result from gas ignition in the gob of FMTCCL face 8208 in Yanquan No 2 coal mine, October 1997, and the face was sealed. The fire extinguished completely by infusing $800m^3$ HRRWG, and the face was unsealed successfully.

4.3 *Gob fire in the operating workface*

The gob of FMTCCL face 2091 occurred self-ignition in Xiasijie coal mine of Tongchuan MAB, October 1998. The working face would have been closed down, where the concentration of CO was more than 0.25%. But after adopting flux technological process and infusing $500m^3$ HRRWG with the mining operation continuing, the concentration of CO decreased to 0.006%. The technique ensured face 2091 normal operation.

5. CONCLUSIONS

The Technique of HRRWG is employed to extinguish spontaneous combustion successfully, which was not solved in before. It possess following features.

(1)Extinguishing rate is fast.

(2)Reliability is good.

(3)The time of reopening the seal is shorten.

(4)The possibolity of re-ignition of fire area is small.

Now the technique is one of the leading means to fight the spontaneous combustion in China.

REFERENCES

Singh, R.V.K.; Tripathi,D.D. 1996, Fire fighting expertise in Indian coal mines , *Journal of Mines, Metals & Fuels*, vol. 44 no.6-7 Jun-July, pp210-212.

Evwseev V.S.; Yu.A. 1985,*Application of coal gel-forming solution in prevention of coal self-ignition* ,ugol n9,sep. pp16-18.

Highton, W. (NCB,UK); 1982, Cooper,J.M. Spontaneous combustion: Its early detention and thelatest means of combustion it in the western area, *Min Eng (London)*, V142 n 250, July PP43-48.

Miron ,Y. ,1995, *Gel sealants for the mitigation of spontaneous heatings in coal mines*, U.S. Bureau of mines ,report of Investigation, No 718, PP34-38.

Mining Science and Technology' 99, Xie & Golosinski (eds)© 1999 Balkema, Rotterdam, ISBN 90 5809 067 1

The 3-D visual simulator of top coal caving and its application

Zaikang Lin, Jian Yang, Xuefeng Yan & Weizhong Tao
China University of Mining and Technology, Xuzhou, People's Republic of China

ABSTRACT: The basic principle and method of 3-D visual simulator of longwall top-coal caving(LTC) technology in FMF are introduced in this paper. Blocks-substitution theory is applied in the program of simulator generated by software VC++5 and OpenGL. The main parameters of LTC, such as seam pitch, control distance, control height breaking angle of top coal, control type etc., are input to the simulator and the reasonable results are obtained.

1 INTRODUCTION

The study of computer simulation techniques began from the 1980's in mining industry in China. Some universities and research institutes made simulation models, such as "Production simulator of coal face", "Production system simulator of mining district", "The major production system simulator of mine", "Substitution plan for mining" and "Dynamic program of mining area", etc. These models have been used in some mining areas and mines, which have made contribution to the normal production and management of mines.

The computer simulation techniques have been focused on the development of 3-D visual simulator software under Windows interface since 1990's. The development of 3-D visual simulator of longwall top-coal caving (LTC) technology was one of them in mining industry. The main purpose of developing and studying lies in that how the main parameters of LTC, such as seam pitch, control distance, control height, breaking angle of top coal, control type, etc., are input to the simulator and then the reasonable results can be obtained through the simulator.

2 BLOCKS-SUBSTITUTION THEORY

In 1968, A Canadian scholar, David Jolley, put forward mine caving simulation method, simulated model as illustrated in Figure 1.
It divided mining rock into model blocks with equal size and shape. The random up and down substitution movements of the model blocks simulate the flowing process of the fallen rocks.

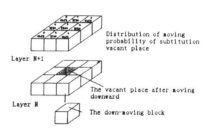

Figure 1 D.Jolley Simulation Model

The substitution among model blocks is realized by the vacant space with random movement to the opposite direction. To be more specific, there is a vacant space in the leaking hole once a block released. That is equal to putting a vacant space in the leaking hole, and this vacant space is substituted randomly by the neighboring block in the first layer above it on the given probability. After the substituted block moves down, its former space becomes vacant, so the vacant space is moved to the first layer from the leaking hole. The vacant space, that is moved to the first layer, is then moved to the second layer, and in this order the vacant space moves randomly toward above and at last it goes into the rock. The program follows the vacant space and records the change of the moving blocks and at the same time calculates the number of released model blocks and rock model blocks. In this way it simulates the whole process of mine caving.

In fact, there are two ways for later research work. One is to study the problem of numbering probability in D.Jolley model blocks, and the other is to study model blocks substitution model. In other research work, besides D.jolley's mine substitution models, there are six-angle substitution model and seven-substitution model.

3 SIMULATION METHOD AND MAKING SIMULATOR

This program is developed by Visual C++5.0(VC5) and OpenGL

The program aims to create a dynamic representation of the top coal caving under the condition of different breaking angle of top coal, breaking angle of roof rock, mining height, control height, control distance, control type through simulation calculation. And it reveals in the ways of 3-D the change of top coal in the process of coal caving. At last the program will give a group of reasonable parameters to direct the production. In addition, the study of 3-D visual simulation techniques and inter 3-D promotes the development of the simulation techniques in coal mining.

According to the theory of drawing ellipsoid-sphere coal in order to realize continuous dynamically changing of ellipsoid-sphere coal, the key point is to calculate the motion of particles on surface. The program takes continuous equal-amount drawing method to get a picture by calculating the position of released ellipsoid, released funnel and particles on the surface of ellipsoid. If the memory of the computer is big enough, 80 pictures can be made every second, and continuity effect of animated cartoon is reached.

OpenGL must be initiated before it is render, which includes tinting equipment, color index, substance definition, light definition and variant setting of OpenGL state etc. The setting of some parameters of the program can also be initiated here. The following work is to put in the parameter setting of the theory of drawing ellipsoid-sphere coal, which includes rock flowing parameters □,□, the length of the long axis of released ellipsoid H and other parameters. These parameters can decide the size and shape of released object, released funnel and flexible ellipsoid.

To reach a more genuine effect, simulation of top coal caving must have the pictures of broken top coal, mined hollow area, incompletely broken front coal wall, working face top coal support and working face conveyer, etc. The supports in the picture should have two states, i.e., control opening being opened or control board being opened and closed. In addition, the coal flow on conveyer in the picture is dynamic, that is, dynamic coal flow is revealed with the change of control openings position and control quantity. The move of coal flow can be realized by dynamically changing quadrilateral or particle system.

There are many ways and functions of broken top coal object, which have three categories: parameter setting function, blocks-substitution calculation function and 3-D drawing function. Blocks-substitution calculation function is the key point to realize simulation. There are many parameter-setting functions and each parameter can be changed. Substitution calculation is based on D.Jolley model, calculation in substitution upwardly from control opening. drawing function only draws the edge blocks of broken top coal.

From the analysis of the above the simulation flow-chat is in Figure 2.

4 THE APPLICATION OF 3-D VISUAL SIMULATOR OF TOP COAL CAVING

4.1 *The study of reasonable control parameter and control type of top coal caving*

It needs some suitable conditions to apply top coal caving, such as coal thickness, hardness of coal, seam pitch, position of coal caving opening, the size of coal caving opening, control distance, breaking angle of top coal etc. The selection of suitable conditions and reasonable parameters can make the techniques of top coal caving apply and spread well.

4.2 *The design of simulation program and the choose of parameter*

In the LTC technology, work face seam pitch, work face control height, breaking angle of top coal, control distance, control type are very important parameters. So these parameters can be calculated as separated numbers and the general rule can be gotten through calculating the parameters. The assumptions and parameters in the programming are adopted as the following:

1. Coal is soft, all top coal can be crushed and doesn't stick to the top.

2. All the blocks can be regarded as even divided particles and they don't interfere with each other.

3. The other basic conditions remain unchanged, and have little influence on the result of simulation.

4. Only calculate one time of control work.

Provides simulation's constant parameter is: the coal cutting height is 2.8m, the length of work face is 120m, the number of support of control is 60, the length of top beam is 3.4m, the size of control opening is 1.0*0.5m.

Numbering the studied parameters:

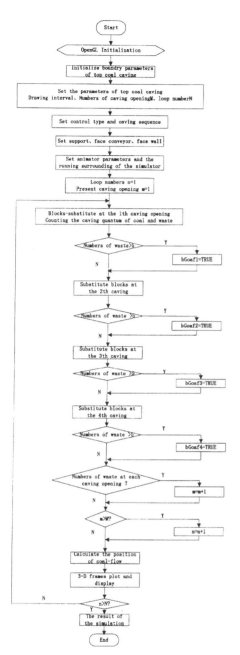

Figure 2 The Flow-Chart of Simulator

1. Seam pitch: 12^0, 25^0
2. Control distance: 0.6m, 1.2m, 1.8m
3. Control height: 3.2m, 5.0m, 10.0m
4. Breaking angle of top coal:60^0, 90^0,120^0
Controlling types are:

1. multi-turn, division, orderly equal-amount control
2. single-turn, multi-openingorderly, unequal-amount control
3. single-turn, intermittent orderly equal-amount control
4. single-turn, intermittent multi-opening control

There are 54 plans of the 4 controlling types after above parameters grouping.

4.3 *Simulation results and its analysis*

After the running of the 54 plans and 4 kinds of control type, the simualted results are obtained, some results are listed in Table 1.

Table 1 Particles results table

Pl-an	Multi-turn Division Orderly Equal-amount		Single-turn Multi-opening Orderly Unequal-amount		Multi-turn Intermittent Orderly Equal-amount		Single-turn Intermittent Multi-opening	
	Coal	Waste	Coal	Waste	Coal	Waste	Coal	Wast
1	8863	481	8912	514	9252	520	9022	594
2	11270	570	11855	560	11629	507	11562	670
3	9987	493	10163	498	10074	522	10768	1052
4	9252	500	9624	449	10243	653	9624	584
5	13663	489	14572	499	14641	511	14048	608

4.3.1. The influence of control type active on caving quantity

From the data of simulation, Mode 3, i.e., "multi-turn, interval, sequence, identical caving", is a better option than other Modes.

Figure 3 shows the caving quantity of coal and waste of the 4 control types. They are "multi-turn, interval, sequence, identical caving","single-turn, multi-opening, orderly, unequal-amount", "multi-turn, intermittent, orderly, equal-amount", "single-turn, intermittent, multi-opening". When the same quantity of waste are drawn, Mode 3 can draw more coal than other modes. It is said that if Mode 3 is used, the waste intermingle in more slowly, the inter face between waste and coal can maintain stabilization more easily and come down more equably.

4.3.2. The influence of caving interval

The influence of caving interval on caving effect is very salient, different caving interval must be carefully selected in different conditions.

When the control height increase, a longer caving interval should be used, otherwise, a shorter caving interval should be used, in this way a higher recovery can be reached.

4.3.3. The influence of breaking angle of top coal

From the result of the simulator we can draw a conclusion that the influence of breaking angle of top coal to the caving quantity is distinct. When the breaking angle of top coal changes from 60^0 to 120^0 the caving quantity of coal go up rapidly (maybe it is more than 2 times) while the quantity of waste keep the same.

4.3.4. The influence of seam pitch

The influence of seam pitch active on the caving quantity is salient too. It is apparent that the caving quantity will increase with the seam pitch. Because the long axis of caving ellipsoid will become longer with the seam pitch increasing, and the caving quantity of the top coal will enhance. The larger the seam pitch is, The more the top coal is drawn while the same quantity of waste is drawn.

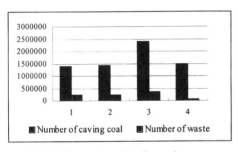

Fig 3 Caving number of the four caving ways

4.3.5. The influence of top coal height

The results of the simulator indicate that top caving cannot get a content result unless the top coal height fit the bill. Figure 3 shows that the caving quantum of top coal is primitively same when the caving interval is 1.2m or 1.8m, if the height of top coal is lower than 6m, the difference of quantity between 1.2m and 1.8m is small. When the height of top coal higher than 6m, the caving quantity will slow down and close to stabilization with the increase of top coal height under the condition that the caving interval is 1.2m. When the caving interval is 1.8m, the caving quantity will preserve rectilinear increasing. It is say that the caving interval should be enhanced rationally with the increase of top coal height. When the caving interval is not so high, a long caving interval will lead to lose more top coal, the recovery ratio is low. When the top coal height is not higher than 5.5m, "caving every two cuts", namely the caving interval is 1.2m, is reasonable.

When the top coal height is higher, the caving interval can be increased properly in order to enhance the recovery ratio of top coal multi-caving interval may be adopted.

5 CONCLUSIONS

1. The research shows in this paper that the 3-D visual simulator of top coal caving can be programmed based on the blocks-substitution theory. The software VC++5 and OpenGL are used to make the simulator. The 3-D visual on-time simulation and 3-D alternation are also realized. The continuous dynamic procedures of simulation is realized by make use of continuous creation of frames.

2. The simulator can represent the dynamic process of top coal caving in the new software introduced in this paper, and recovery ratio can be obtained from the software when various parameters are input such as control type, boundary parameters and top coal parameters.

REFERENCES

[1] Wu Jian 1991. The theory and practice of top coal caving in China. Coal Mining Journal. 16(3),1-11

[2] Zhang Dingli 1994. The rock constracture and mining presure of top coal caving face.Mining Presure and Roof Control. 1994(4)13-17

Mining Science and Technology '99, Xie & Golosinski (eds) © 1999 Balkema, Rotterdam, ISBN 90 5809 067 1

A blasting survey in limestone quarries

A. Bortolussi
Mineral Science Study Centre, National Research Council, Cagliari, Italy

R. Ciccu, S. Forte & B. Grosso
Department of Geoengineering and Environmental Technologies, University of Cagliari, Italy

ABSTRACT: A survey of a number of Italian quarries involved in the production of raw materials for cement manufacturing has been carried out by collecting the blasting data under different conditions regarding the kind of material, the structural features of the rock and the location of the quarry.

The data have been properly processed in order to put into light the relationship existing between the various blasting parameters and variables. Through the study of statistical correlation a straight linear function has been found to exist between the specific consumption of explosive and a global parameter accounting for the rock properties, the blasting geometry and the type and loading conditions of the explosive used, in which all the pertinent variables are incorporated as factors to a fractional power. The influence of individual issues according to the model is discussed and the outcome is explained on scientific grounds.

The reliability of the model appears so good that it can be proposed as a predictive tool for blasting design.

1 INTRODUCTION

The world demand of building materials, boosted by the growth of construction activity, has today reached considerable levels, especially in developed countries.

Consequently the scientific interest in the sector has been raised with the goal of finding suitable solutions to the problems related to the industrial production of such materials, including engineering, operations management, economic evaluation and technological improvement.

This is especially true for the quarrying activity which became a very complex concern, due to the increase in the size of the producing units, in presence of severe environmental constraints.

In the case of excavation methods based on the use of explosives, the accurate design of drilling and blasting is of a capital importance, since the economic profitability of the enterprise strongly depends upon the successful outcome of that operation.

The present trend in surface mining is toward larger blast volumes, entailing an increase in hole diameter, bench height and size of the drilling grid.

Consequently specific drilling is considerably reduced while the ratio between burden and spacing is kept within practical limits in order to control the product size and the side effects of blasting.

Although surface blasting has been the object of a great number of studies and investigations (Langefors & Kinlstrom 1963, Seguiti 1969, Berta 1983), no reliable theoretical model capable of describing the rock-explosive interaction is yet available, with the consequence that blasting results are difficult to predict, due to the complexity of the system as well as to the great variability of the conditions encountered.

As a matter of fact, some of the influencing parameters are not always quantitatively definable in advance, especially those related to the characteristics of the rock.

In spite of this, different models valid for particular conditions have been proposed in the literature. In practice, the optimum blasting plan for a given quarry is always the result of a series of field trials leading to the construction of empirical relationships between variables and parameters.

The present investigation is founded on the statistical analysis of data collected at a number of

Italian quarries involved in the production of limestone and marl for cement factories. In these quarries the blasting plan has been refined by the daily experience and it can be considered as optimal for each particular site conditions.

Gathered information consisted of data regarding the blasting geometry, the explosive loading parameters, the characteristics the rock and the results obtained with the typical rounds, including the size distribution of the product.

In order to evaluate the efficiency of the blasting plan adopted, field data have been elaborated by calculating the relevant parameters of the blasting process, which have always been found to fall within the range suggested by the technical literature for similar conditions.

In the subsequent stage of the investigation, field data have been used to build a comprehensive mathematical model, based on the "best fitting" approach, which can be used for prediction.

This correlation model, which takes into account all pertinent aspects of the blast, also enables to put into light the sensitivity of the blasting outcome to each of them, considered individually.

Although the reliability of the relationships found is only proven for the set of quarries examined in the present study, it is believed that their applicability can be extended to other quarries of the same kind with a good level of confidence.

Moreover, the validity of the model can be generalised by adding new data to be collected at other surface mining operations into the statistical process, although at the expense of accuracy.

On the other hand, the methodology followed can be applied to other instances of rock blasting, in order to define suitable specific models having the same mathematical structure but different constants.

2 ANALYSIS OF INDUSTRIAL DATA

As previously mentioned, the information collected in the field concerns the various parameters and variables of bench geometry, drilling pattern, explosive used and loading configuration, firing sequence, rock mass excavated as well as the blasting results obtained.

The parameters taken into consideration can be distinguished into "measured", i.e. directly obtained from the blasting plan, and "calculated", through the elaboration of crude data.

2.1 Measured parameters

Bench basting geometry has been characterised by defining the conventional parameters reported in table 1.

Table 1. Parameters of bench blasting geometry.

Parameter	Symbol	Units
Hole diameter	Φ	mm
Bench height	K	m
Hole depth	H	m
Theoretical burden	V_t	m
Burden	V	m
Spacing	E	m
Subdrilling	U	m
Hole inclination	i°	$^\circ$

Table 2. Loading configuration parameters.

Parameter	Symbol	Units
Top stemming	B_s	m
Intermediate stemming	B_i	m
Total charge	Q	kg
Total charge height	L_c	m
Column charge load	Q_c	kg
Column charge height	L_{cc}	m
Bottom charge load	Q_b	kg
Bottom charge height	L_{cb}	m

Table 3. Characteristics of explosives.

Parameter	Symbol	Units
Bulk density	δ_c	kg/dm^3
Charge diameter	Φ_c	mm
Detonation velocity	v_d	m/s
Specific energy	ε_{ev}	MJ/kg
Explosive impedance	$I_e = \delta_c v_d$	$10^3 kg/m^2 s$

Table 4. Rock-related parameters.

Parameter	Symbol	Units
Volumic mass	ρ_r	kg/dm^3
Compressive strength	σ_{cr}	MPa
P-wave velocity	c	m/s
Work Index (after Bond)	W_i	kJ/kg
Rock impedance	$I_r = \rho_r c$	$10^3 kg/m^2 s$
Blastability constant	s	

Charging configuration parameters are reported in table 2. The parameters characterising the explosives used for the column and bottom charge are reported in table 3, whereas the parameters defining the rocks are given in table 4.

The results of bench blasting are defined by considering the size of the broken material, in particular its top dimension (D_{max} [m]) and the values D_{80} [m] and D_{50} [m] of the size distribution.

2.2 Calculated parameters

The crude data of bench blasting collected in the quarries have been elaborated by calculating the conventional parameters commonly used for evaluating the efficiency of the blast, such as:

- Blasthole productivity [m³/hole]: ratio between the total blasted volume and the number of blastholes in the round;
- Specific drilling **S** [m/m³]: drilled metres per cubic metre of rock in situ;
- Specific charge **q** [g/m³] or [g/t]: explosive consumption per bank cubic metre or ton of rock;
- Decoupling ratio Φ_c/Φ : ratio between the diameters of explosive charge and of drillhole;
- Bottom charge concentration l_b [kg/m]: weight of explosive per metre of bottom charge;
- Column charge concentration l_c [kg/m]: weight of explosive per metre of column charge.

For all the quarries taken into consideration the measured and calculated parameters have been found to fall within the range suggested in the literature for similar situations, confirming that the blasting plans adopted are generally adequate.

3 CORRELATION MODEL

The various issues affecting the process of bench blasting are different in origin and often the numerical values assigned to them are the result of a qualitative evaluation. Moreover, the system is greatly complex and the significance of the many variables and parameters is not univocal. In these conditions predictive models are not easy to build, whereas a deterministic approach to the problem may not be realistic.

The solution proposed in the present work consists in a comprehensive mathematical model in which all the relevant variables and parameters characterising the different issues involved in the process are taken into consideration. The model enables to evaluate their relative influence on

blasting results, under the hypothesis that collected data are obtained from optimal blasting plans.

Specific charge, i.e. the amount of explosive consumed for cubic metre of rock has been considered as the most significant parameter of blasting, from both the technical and economic point of view.

In fact the knowledge of specific charge for each field of application of explosives (production blasting at surface or underground, tunnelling, trenching, presplitting, etc.) provides an immediate evaluation of the efficiency of the blast, when compared with the corresponding published values obtained by the experience.

Therefore, in the construction of our model it has been assumed that specific charge could be expressed as a linear function of a suitable global parameter accounting for the various relevant issues of the blasting plan.

Accordingly, the general form of the relationship searched is the following:

$$q = k_1 P + k_2 \qquad (1)$$

where k_1 and k_2 are constants typical of the field of application and P is the global parameter.

It has also been assumed that the various issues are represented in parameter P as factors to a fractional power to be determined through a statistical analysis of available data. The number and type of such factors and their mathematical form have been evaluated by considering each issue and its theoretical influence on specific charge.

The model has been built trough a trial-and-error procedure based on best fitting of field data collected in the quarries. According to the results obtained from such analysis, the following issues have eventually been taken into consideration:

Rock related parameters
- Work index: W_i
- Volumic mass: ρ_r
- Blastability constant: s
- Size of the product: D_{80}

Variables of bench blasting geometry
- Hole diameter: Φ
- Specific drilling: S
- Burden/spacing ratio: V/E
- Bench height: K
- Hole inclination: i°

Explosive-related parameters
- Specific energy: ε_{ev}

Energy transmission efficiency indexes
- Decoupling ratio: Φ_c/Φ
- Impedance ratio: I_r/I_e

The exponent appearing in each factor of the expression of P has been determined by maximising the correlation coefficient of the linear regression analysis of the field data.

The equation obtained with a high correlation coefficient ($r^2 = 0.956$) is the following:

$q = 0.4215\,P + 200.78$ [g/m^3]   where:

$P = [(W_i\rho_r)^5\,s^{-2}\,D_{80}^{-1.4}]\,[\Phi^{-2}S^{1.7}(V/E)^{0.5}K^{-1.5}(\sin i°)^{0.5}]$
$[\varepsilon_{ev}^{-2}(\Phi_c/\Phi)^{-2}(I_r/\,I_e)^{0.7}]$

Correlation resulted to be even better ($r^2 = 0.994$) if the anomalous point representing a quarry with a very low specific charge (181 g/m^3) is excluded, as justified by the fact that in this particular case the quantity of explosive is limited by environmental restrictions (vibrations) and excavation is carried out according to a combined method of blasting and drag scraper.

It is interesting to observe that the global parameter P was not affected at all and the equation became:

$q = 0.405\,P + 207.6$ [g/m^3]

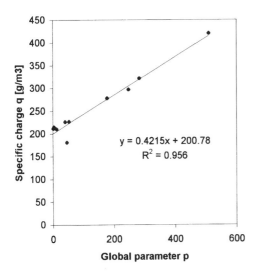

Figure 1. Correlation line of specific charge as function of global parameter P for all the quarries examined.

The correlation line of specific charge as a function of the global parameter P is shown in Figure 1.

4 DISCUSSION

The mathematical expression found for parameter P enables to evaluate the influence on the blasting process of the various issues.

Among rock-related parameters, the work index referred to the unit volume $W_i\rho_r$ has a great influence on specific charge, which increases considerably with it, thus confirming the blasting theory based on energy balance (Berta 1985). In fact the work index, commonly used in the design of mineral comminution machines, represents the specific energy required to crush the material from a given feed size to a given product size. The higher that value, the larger is expected to be the amount of explosive needed to fragment the rock to a given size D_{80}.

The influence of the work index is in some way balanced by the blastability constant s which describes the structural features of the rock. A homogeneous compact rock with no fractures (s < 0.4) requires much more explosive than a very jointed and fissured rock (s > 0.6) for achieving the same product size.

The size D_{80} of the blasted material, obtained from the size distribution curve at 80% undersize cumulative frequency, is the conventional parameter for assessing the quality of the blast product. Of course smaller quantities of explosives are required the coarser the fragments to be obtained.

Concerning bench blasting geometry parameters, hole diameter and bench height appear in the formula with a negative exponent, indicating their favourable influence (the higher their value the lower specific charge will be), while specific drilling and burden/spacing ratio seem to have a negative influence on specific charge, all the other conditions being the same, as it is well known from the blasting practice. A decrease in hole inclination i° with respect to the horizontal plane is always advantageous provided that drilling accuracy is maintained.

Explosive is characterised by its specific energy by mass, which has a positive influence on specific charge, decreasing when a more powerful explosive is used. This is a clear physical effect (more energy contained in a smaller mass).

Regarding the parameters of energy transmission efficiency, the value of decoupling ratio should be kept high and impedance ratio as small as possible (close to 1) in order to reduce specific charge. Bulk density of the explosive loaded into the blastholes and detonation velocity are incorporated in this latter parameter.

The indications given by the model are in perfect agreement with the well known thumb rules for the safe handling and best use of explosives.

The model here proposed can be adopted either for survey purposes or as a predictive tool.

In the first case, the information defining a blasting plan under evaluation is introduced into the model as input data for calculating the theoretical specific charge, to which the real specific consumption is compared. If the difference between the two values is within acceptable limits, the blasting plan can be considered adequate. However if that difference is statistically significant, this means that one or more parameters of the blasting plan are likely to be out of their optimal range, and consequently some technical or/and economic drawbacks are suffered, calling for suitable remedial measures. The advantage of such survey derives from the need to define clearly all aspects involved in the blasting process.

In the second case, the model can be applied as the starting point of a predictive procedure according to which each one of the variables of the system is calculated taking into account the characteristics of the rock (fixed data), while the other variables are set at their average value. This kind of procedure can provide a very important information on the combined influence of the various issues, indicating, through a specific sensitivity analysis, their optimal range of variation.

In a next step of the research in course, a larger set of quarries of the same kind will be considered in order to improve the reliability of the model. Further on, the model will be applied to rocks having different characteristics aiming at extending its applicability to all instances of surface mining operations.

Finally the research will be addressed to other field of application of explosive blasting, including underground excavation.

5 CONCLUSIONS

The proposed model linking specific charge to the relevant parameters and variables of bench blasting in surface quarries shows a statistical validity confirmed by a very high correlation coefficient.

The influence of the different issues can be quantitatively defined through the analysis of the factors appearing in the mathematical expression of the global parameter used in the linear correlation.

The model represents a guideline for the survey of blasting plans applied in active quarries, as well as a useful predictive tool for blast design.

ACKNOWLEDGEMENTS

The work has been carried out according to the research programme of CNR (National Research Council) and MURST (Ministry for University and Scientific and Technological Research). The co-operation of Italcementi S.p.A. is highly appreciated.

REFERENCES

Langefors, U. & Kinlstrom, B. 1963. *The modern technique of rock blasting*. Stockholm: Almquist & Wiksell.

Seguiti, T. 1969. *Le mine nei lavori minerari e civili*. Roma: Ed. "L'Industria Mineraria".

Berta, G. 1983 *L'esplosivo e la roccia*. Milano: Italesplosivi

Berta, G. 1985 *L'esplosivo strumento di lavoro*. Milano: Italesplosivi

Mining Science and Technology' 99, Xie & Golosinski (eds) © 1999 Balkema, Rotterdam, ISBN 90 5809 067 1

Numerical simulations of blast waves from spherical charge

S. Kubota, H. Shimada & K. Matsui
Kyushu University, Fukuoka, Japan

ABSTRACT : The spherical TNT charge exploded in air has been numerically simulated by a 2-D hydrodynamic code in which the JWL equation of state is employed for describing the expansion of detonation products and the air is assumed as the perfect gas. The simulation on the peak overpressure and positive impulse are compared with the corresponding experimental data by Swisdak. It is found that this code can be effectively applied to the calculation on the problem of blast waves.

1 INTRODUCTION

In the chemical plant for manufacturing the explosives or in the magazine for storing the explosives, if the accidental explosion of such energetic materials happens, it would bring out the disastrous effects to the nearby areas by the blast wave or the broken fragments. Although the considerable safety distance is now set for the prevention of the disasters to the houses and apartments nearby, however, with the increase of the number of the built houses, it requires to degrade the safety distance by putting the some shades near the explosion source. For this purpose, in order to assess the affects of the explosion to neighboring area under various circumstances, the authors developed a numerical hydrodynamic code to apply to the explosion problems and compared the results with those from the existed variety of computer codes so that the capability of the code was able to be examined. In this paper, a typical blast problem, a spherical TNT charge exploded in air, has numerically been analyzed and the questions raised in the numerical simulation on the explosion phenomenon are under investigation.

2 NUMERICAL PROCEDURE

The computer code for the numerical analysis was made on the basis of the ALE (Arbitrary Lagrangian-Eulerian Technique(Hirt et al. 1974;Amesden et al. 1980)) difference scheme. The governing equations included are 2-D inviscid flow equations of mass, motion and energy conservation and these are expressed in the following:

Table 1. The characteristics of TNT

Explosive	$\rho_0(kg/m^3)$	$D(m/s)$	$Pcj(GPa)$
TNT	1630	6930	21.0

$$\frac{\partial \rho}{\partial t} + \frac{1}{r}\frac{\partial \rho u}{\partial x} + \frac{\partial \rho v}{\partial y} = 0 \tag{1}$$

$$\frac{\partial \rho u}{\partial t} + \frac{1}{r}\frac{\partial r\rho u^2}{\partial x} + \frac{\partial \rho uv}{\partial y} = -\frac{\partial(P+q)}{\partial x} \tag{2}$$

$$\frac{\partial \rho v}{\partial t} + \frac{1}{r}\frac{\partial r\rho uv}{\partial x} + \frac{\partial \rho v^2}{\partial y} = -\frac{\partial(P+q)}{\partial y} \tag{3}$$

$$\frac{\partial \rho e}{\partial t} + \frac{1}{r}\frac{\partial r\rho eu}{\partial x} + \frac{\partial \rho ev}{\partial y} =$$
$$-(P+q)\left(\frac{1}{r}\frac{\partial ru}{\partial x} + \frac{\partial v}{\partial y}\right) \tag{4}$$

where ρ is density, u and v are the particle velocities in the x- and y-directions, respectively, P is pressure, e is specific internal energy and q is an artificial viscous pressure. TNT explosive was used for the explosive and the characteristics of TNT is shown in Table 1. The equations of state for materials are necessary. For the explosive, the JWL equation of state(Lee et al. 1973) was used. The expression is given in the below,

$$P = A(1 - \frac{\omega}{R_1 V})\,exp(-R_1 V)$$
$$+B(1 - \frac{\omega}{R_2 V})\,exp(-R_2 V) + \frac{\omega\rho_0 e}{V} \tag{5}$$

Table 2. JWL parameter of TNT

$A(GPa)$	$B(GPa)$	$C(GPa)$
373.8	3.747	0.734
R_1	R_2	ω
4.15	0.90	0.35

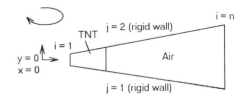

Figure 1. Calculation field.

The isentropic JWL equation of state for the detonation products is

$$P_s = A\,exp(R_1V) + B\,exp(R_2V) + CV^{-(\omega+1)} \quad (6)$$

where $V = v/v_0 = \rho_0/\rho$, v, ρ are the specific volume and density of detonation products, respectively, e is the specific energy. The subscript 0 denotes the state of the unexploded explosive. A, B, R_1, R_2, ω are so-called JWL parameters determined by the test of the expansion of cylinder(Hornberg 1986). The JWL parameters for the TNT in the calculation uses the published ones of TNT with the density of $1630kg/m^3$. Table.2 shows these parameters in the equation of state. The air is assumed as the perfect gas $(P = \rho(\gamma - 1)e)$ with the initial state of the standard atmosphere (pressure : $101.325kPa$, density : $1.20458kg/m^3$).

Owing to the computer code being 2-dimensinal, so the calculation field forms the cells as the wedge shape as shown in Fig. 1. Along the axial direction (y direction) only 1 cell is taken into account. If $j = 1$ and $j = 2$ are assumed as the rigid walls, so, the 1-dimensional spherical symmetry problem can be solved by the 2-dimensional axisymmetric code. The explosion conditions are specified with use of Fig. 2. The figure shows the calculation field near the central part of the explosive charge. i, t, V are the cell point along the radial direction, the time step and the volume of the cell, respectively. Because the detonation wave travels at the detonation velocity, the time step Δt can be defined as the ratio of the interval of the cell Δr to the detonation velocity D, i.e., $\Delta t = \Delta r/D$. At the beginning of the calculation, all the cells are inputted the values possessed by the solid explosive at the atmospheric state. When t=n (here n=1), the i=1 cell at this time has the values of pressure and specific energy corresponding to the original explosive with the density of $\rho_0 = 1630kg/m^3$. As the explosive at i= 1 cell becomes into the detonation products, the products will be thought to expand obeying the JWL equation. However, the cells below i=2 are the unexploded explosive and in the calculation they may be treated as the solid explosive at the atmospheric state. Moreover, the particle velocities at the cell points below i=3 are put into zero. After finishing the calculation at this stage, the velocity at i=2 cell point appears,

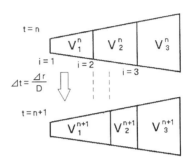

Figure 2. Calculation model for detonation process.

the value of the volume V_2^n is calculated to be V_2^{n+1} by the compressing the detonation products to this volume under the conservation of mass of the cell. The corresponding state can be calculated from eq.(2). The calculated values is acted as the initial conditions for the calculation at $t = n + 1$. The detonation calculation is made by repeating the above procedures.

As the calculating problem of TNT charge exploded in air, there are numerous experimental studies and numerical analysis, and a lot of data and knowledge have been obtained(Brode 1955 ; Brode 1959 ; Kinney 1962 ; Swisdak 1975) . The used experimental results for the comparison with this calculation are those summarized by Swisdak(Swisdak 1975) in 1975. As a comparison, the numerical analysis is made to one corresponding case of TNT charge with the mass of 156kg.

3 RESULTS AND DISCUSSION

Fig. 3 shows the blast energy varying with the time after the explosion of the charge. At the instant of the finish of the explosion, the internal energy accounts for about 94% in total energy, whereas the kinetic energy accounts only for 6% approximately. Brode also showed the same result in his calculation of blast wave by considering the detonation process of the charge(Brode 1959). From the result, it is seen that at the positions

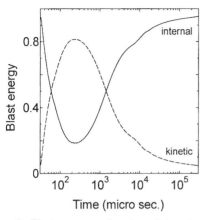

Figure 3. Blast energy as function of time for TNT blast(kinetic+internal=1).

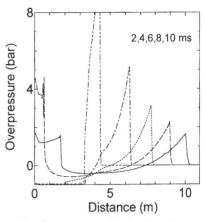

Figure 5. Overpressure distributions the main and second shock at different time.

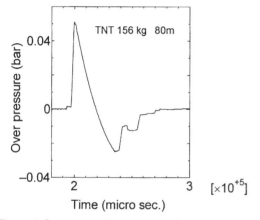

Figure 4. Overpressure history at 80m from the center of the blast.

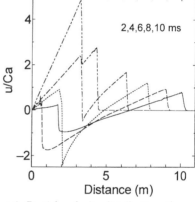

Figure 6. Particle velocity distributions the main and second shock at different time.

near the explosion source, the detonation stage is not important. As long as the blast energy can be assessed, it is all right. In practice, in the calculation made by the assumption that the charge is exploded in equal-volume explosion instead of the actual detonation process, the behavior of the blast wave at the distant positions do not show a distinctive result. And then, in the extremely short time of $200 \sim 300\mu sec$, the maximum kinetic energy arrives, even accounting for 80% of the total energy. The various cell length, varying from $10cm \sim 1m$ near the explosion source, are performed in the calculation, the arrival time of the maximum kinetic energy falls almost the same range. Only when the cell use the larger length, the time has the small increased tendency. Moreover, as the time passes by till 3ms, the varied parts of the curvature in two curves can be con-

firmed. The variation of curvature between the time of $2ms \sim 3ms$ is due to the occurrence of the flow toward the central part. Further, between the time of $8ms \sim 9ms$, the observable variation of the curvature comes from the rise of the pressure once more. The reason for this can be deduced by the occurrence of the second shock wave propagating from the center to the outward. Fig 4 shows the pressure profile of the blast wave at the distance of 80m away from the explosion source. In a fairly far distance from the explosion source, the pressure waveforms are clear and it shows that a stable computation is proceeding. Fig.5 and Fig. 6 give the diagrams of the pressure and particle velocity distributions from the explosion source center before and after the appearance of the second shock wave. It can be seen that the pressures at the central part rise again. The vertical axis in Fig. 6 is the scaled particle velocity by the sound speed at

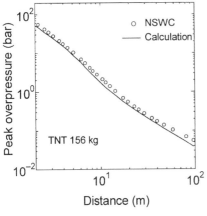

(a) Peak overpressure

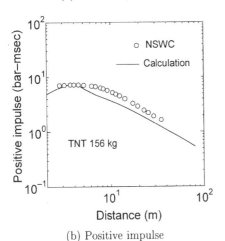

(b) Positive impulse

Figure 7. The comparison on the blast wave parameters between the experimental data and numerical calculation (experimental data coming from the published ones by Swisdak.

atmospheric state. According to the information from both figures, the over expansion of detonation products due to the inertia causes the sharp decrease of pressure at the central part and then the flow runs back. When it arrives at the central part, the pressure is quickly improved and the second shock is formed. This is the mechanism for the occurrence of the second shock. The above results demonstrate that the numerical calculation can indeed provide the useful knowledge on the understanding of the problem of blast wave. Fig.7 presents the comparison between the parameters of blast wave from the numerical calculation and the experimental data by NSWC. It can be seen that the peak overpressures in both cases have a good agreement. In any distance range, the over-

pressure has the value of 20% or less. The calculated result shows a lower value. Furthermore, at the range of 10m and 100m, the difference of 30% extent is confirmed. On the positive phase impulse, the two cases quite agree with each other until 5m distance. However, in numerical calculation, an obvious bending point appears and from this point the difference in two cases starts to become remarkable.

4 CONCLUSIONS

The numerical analysis was conducted on the problem of TNT charge exploded in air by a 2-dimensional hydrodynamic computing code. In the code a simple computing technique for the detonation of explosive was utilized. From it the initial blast energy has been assessed. The calculation shows that the internal energy, account for 94% total energy turns into the blast energy. Moreover, the numerically calculated results were compared with the corresponding experimental data in air blasting by NSWC and a good consistency was achieved, while on the positive impulse it is found that the two cases indicate a tendency of disagreement.

References

Hirt, C.W., Amsden, A.A. and Cook, J.L.,1974, An Arbitrary Lagrangian-Euerlerian Computing Method for All Flow Speeds, Journal of Computational Physics 14, 227-253

Amesden,A.A., Ruppel,H.M. and Hirt,C.W.,1980, LA-8095,UC-32.

Hornberg, H., 1986, Determination of Fume State Parameters from Expansion Measurements of Metal Tubes, Propellants, Explosives, Pyrotechnics, 11, 23-31

Lee, E.L., Finger, M. and Collins, W.,1973, JWL Equation of State Coefficients for High Explosives, Lawrence Livermore National Laboratory, Rept-UCID-16189

Brode, H.L.,1955, Numerical Solutions of Spherical Blast Waves, Journal of Applied Physics, 26, 6, 766-775

Brode, H.L.,1959, Blast Wave from a Spherical Charge, Physics of Fluids, 2, 217-229

Kinney, G.F.,1962, Explosive Shock in Air, Macmillan, New York

Swisdak,Jr.,M.M.,1975, NSWC/WOL/TR 75-116

Mining Science and Technology' 99, Xie & Golosinski (eds)© 1999 Balkema, Rotterdam, ISBN 90 5809 067 1

Recent advances in automated 3D measurement of blasting

Youzhi Wei & Xiaoliang Wu
CSIRO Exploration and Mining, Floreat Park Laboratories, Perth, W.A., Australia

ABSTRACT: This paper gives a brief introduction of recent advances in the development of the VirtuoBlast technology – a new, automated and highly interactive 3D data acquisition system which is being developed by CSIRO Exploration & Mining. The VirtuoBlast technology weaves together three advanced technologies – close-range digital terrestrial photogrammetry, oblique or super high-speed stereo imaging, and temporal-spatial image matching technique. The VirtuoBlast technology is specifically developed for automated 3D measurement of blasting through rapid collection, processing, analysis and visualisation of 3D data related to blasting. Some results from the field trials have demonstrated the potential of the technology to perform pre-blast 3D rock surface mapping, post-blast 3D performance measurement and temporal-spatial modelling of the dynamic blasting process.

1 MOTIVATING THE PROBLEM

Blasting, as a basic process in mining, has the most substantial impact in determining the efficiency of a mining operation. The output and performance from a blast affects every down-stream operation (loading, crushing and milling). Therefore, savings to the mining operation through optimised blast design and better blast control could be substantial (Duvall & Atchison 1957, Langefors & Kihlstrom 1978, Grant & Little 1995).

Over the past three decades, significant progress has been made in the development of new technologies for blasting applications. These include the advancement of modern instrumentation for monitoring, increasingly sophisticated computer models for blast design and blast performance prediction, and more versatile explosives and initiation systems. Blasting today is moving more and more towards a science (Fourney 1993, McKenzie 1993).

However, as blasting is extremely complex, precise control of blasting in rock remains a major challenge for blasting engineers in the mining and explosive industries. Some of the interrelated challenges are discussed in the following.

1. The mechanisms of rock fragmentation by explosive loading are not well understood (Fourney 1993, Starfield 1967, Barker & Fourney 1978, Holmberg & Persson 1979). A new and more reliable approach for the realistic physical modelling of blasting is required (Wei & Wang 1988, Fourney *et al* 1993).

2. Blast performance modelling and prediction are far from satisfactory as all available blast models encapsulate only elements of the explosive-rock interaction physics. Often these models involve either endless fuzzy factors or influencing parameters that are difficult to obtain or oversimplify the problem (Preece *et al* 1993, Paine *et al* 1987, Cunningham 1983, McKenzie *et al* 1995). Of all the blast models developed so far, the particle based method, coupled with other necessary modules, is perhaps the most promising approach to tackle the dynamic fracture and flow of jointed rock under explosive loading (Song & Kim 1995, Muehlhaus *et al* 1997).

3. Geological characteristics (joints, faults, bedding planes, weak zones *etc.*) are not yet systematically included in the blast design (Chakraborty *et al* 1994). Although it is well recognised by practitioners that rock properties have more influence on the blasting results than

the explosive properties, the significance of geological characteristics in blasting has not been sufficiently investigated (Aler *et al* 1996). The successful design of a blast, either on or within a discontinuous rock mass, is largely dependent on our ability to reliably characterise discontinuity and rock mass geometry (Maynard 1990, Lewandowski *et al* 1996). This requires new and innovative tools and systems for safe, detailed and efficient geological data acquisition and processing, and blast design (Wei & Wu 1998).

4. More efficient blast instrumentation and objective performance measurement techniques are required to meet the increasing demands of mining contractors and operators. Depending on applications, the most critical blasting performance measure in most cases is fragmentation (Franklin & Maerz 1995, Morrison 1995). Fragmentation measurements are commonly performed using the labour intensive and time consuming method of hand digitising photographs of a muckpile. There are a variety of such systems based on photographic or image analysis methods currently available for the measurement of particle sizes (Franklin & Maerz 1995, Morrison 1995, Ord & Cheung 1991, Poniewierski *et al*. 1995). A big problem with these techniques, however, is that the particle size distribution measured out of a small portion of the muckpile surface is not representative of the fragmentation of the whole muckpile. In most cases, such as in open pit bench blasting, the bigger size rock fragments tend to hide underneath the surface within the fragmented rock mass. In addition, there are other technical issues associated with the photographic or image analysis methods (Chiappetta *et al* 1987, Hendricks *et al* 1990, Kennedy 1994, Stephenson & Fuller 1995). In addition, blast performance measurements are important for further development of blast models, in particular for use in calibrating and testing fragmentation prediction models (theoretical, curve-fitting, empirical or hybrid). Such models require good quality data. Thus, an efficient 'seeing through rock muckpile' type of technology is required for the measurement of fragmentation.

A fundamental requirement for advancing the science and technology of blasting, is ready access to relevant, high quality and timely data, in particular spatial data. Such data must be in a form that can be visualised instantaneously and used to assist with blast design, modelling, monitoring and performance measurement. Important data includes that which quantifies the effects of rock structure, geological information from 3D bench and front surface mapping, 3D rock mass movement immediately after detonation, fragmentation, muckpile shape and swell, diggability and flyrock. This requires new and innovative tools for safe, detailed and efficient spatial data acquisition, and automated processing, reconstruction and visualisation of blasting process.

This paper outlines recent advances in the development of automated 3D measurement systems for blasting applications. While the research is at present ongoing within CSIRO Exploration and Mining, the results to date have shown that the new technology is promising.

2 CSIRO VIRTUOBLAST TECHNOLOGY

The VirtuoBlast technology developed by CSIRO Exploration and Mining weaves together three advanced technologies — close-range digital terrestrial photogrammetry, oblique or super high-speed stereo imaging and temporal-spatial image matching techniques.

Application of the VirtuoBlast technology to blast measurement and monitoring involves the following three steps:

1. field data acquisition (which usually includes control points survey and photography);
2. image processing (which involves image orientation, matching and the generation of a digital surface model (DSM) and 3D image); and
3. 3D visualisation and interpretation of results.

A process, which has been developed specifically for blasting application, is shown in Figure 1.
Blast measurement includes the following three stages:

1. pre–blast 3D surface mapping;
2. temporal-spatial modelling of dynamic blasting process; and
3. post–blast 3D performance measurement.

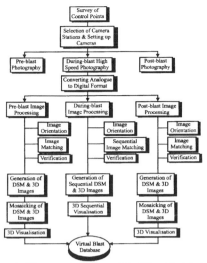

Figure 1. The procedural chart of the VirtuoBlast technology for automated 3D measurement of blasting.

3   3D DATA PROCESSING

The 3D measurement of blasting process involves several steps: field image capturing, image orientation, image processing (image matching), 3D surface model generation and 3D visualisation.

The most important step in 3D data processing is image matching. Image matching identifies the corresponding points between images. The absolute 3D coordinates of the matched points are calculated using their image coordinates and the image orientation parameters. Good image matching is critical to the accuracy and efficiency of the 3D surface reconstruction.

An integrated image matching approach has been attempted which incorporates multiple-points least squares matching, relaxation matching and an array algebra technique (Wu 1996) for pre- and post-blast mapping applications. In those cases, camera positions are relatively close to the object surface, and hence large image distortion and surface undulation often occurs. The preliminary results have shown that the proposed image matching technique produces accurate and reliable matching results for the efficient pre-blast surface mapping and post-blast muckpile mapping (Wei & Wu 1998).

A temporal-spatial image matching technique has been developed for the modelling of dynamic blasting process. Dynamic blasting process often involves large image distortion and surface undulation. In addition, the surface or object is rapidly deforming while moving. Thus this adds another dimension 'time' to the traditional 'static' image matching process (Wu & Wei 1998).

With all image matching techniques available, mismatching of points is relatively common. This affects the accuracy of the final 3D points. Therefore, manual checking and editing are often required. Manual editing is conducted with the aid of stereo glasses such as StereoView™.

4   3D VISUALISATION OF BLASTING

The visualisation of blasting process (pre-, during and post-blast) is achieved by draping the mosaicked 3D images over the mosaicked DSM which is corresponding to the 3D images. This allows the details of the blasting targets and process to be examined and analysed in a virtual environment. The spatial coordinates of a feature (open pit bench, rock face, heave, flying rock, muckpile, dust *etc.*) can be retrieved by pointing to the feature. In addition, certain parameters related to the features can be calculated.

The photorealistic 3D images of the dynamic blasting process can also be reconstructed and viewed in a virtual environment.

5   CASE STUDIES

CASE 1 Pre–Blast 3D Surface Mapping

A field trial of the VirtuoBlast technology for pre-blast mapping has been undertaken at a major open pit in Western Australia. The mapped area includes the free surface and bench prior to the execution of the production blast. The results are presented in Figure 2. Figure 2(a) is the rock surface to be mapped, whilst Figure 2(b) is the reconstructed DSM presented as contour lines and Figure 2(c) is the 3D images (also known as orthoimage) superimposed with contour lines.

Progress has been made in the development of virtual reality for future blast design. Figure 3 shows a completed reconstructed 3D photorealistic image of the Eureka open pit in Western Australia (Wei & Wu 1998). Figures 3(b) and 3(c) show an enlarged section of the reconstructed open pit (Figure 3(a)) where production blast can be designed in a virtual environment. This process and the virtual environment allow any area or section of the open pit to be scheduled and designed for production blast purpose.

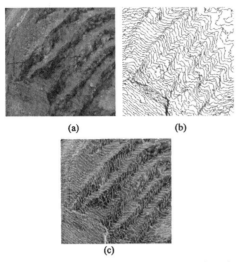
Figure 2. Pre-blast surface mapping: (a) A section of rock surface; (b) The corresponding DSM of the rock surface presented as contour lines; and (c) The reconstructed 3D image.

(a)

(b)   (c)

Figure 3. A completed reconstructed 3D photorealistic open pit: (a) A view of the digital photorealistic 3D open pit; (b) An enlarged 3D image section of the open pit; and (c) The enlarged 3D image's wireframe.

CASE 2 Temporal-Spatial Modelling of Dynamic Blasting Process

A field trial of production blast monitoring using the VirtuoBlast technology was undertaken in August 1996 at the super pit of the Kalgoorlie Consolidated Gold Mine in Western Australia. Because of the difficulties in getting two high-speed cameras on site at the time of the trial, two SONY™ digital video cameras were instead employed to demonstrate the feasibility of the proposed techniques.

Several production blasts were recorded and the results were downloaded onto a 4mm tape for image processing. The spacing between the two cameras was approximately 100m and the distance between the cameras and the object was about 800m. The cameras were set up according to field procedures specifically developed for this purpose. Altogether more than one hundred pairs of images where taken by the two video cameras and processed. For demonstration purpose, only selected results are presented here.

Figure 4 shows the results of reconstructed sequential 3D images of the dynamic blasting process. The four pictures in the right column in Figure 4 are the reconstructed DSM extracted from the temporal-spatial model of the blasting event, which are superimposed with the 3D images over four different time intervals after denotation of the explosives. The four pictures on the left column in Figure 4 are the corresponding DSM presented only as contour lines.

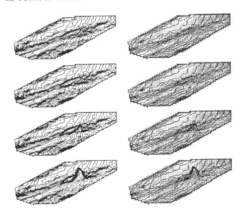
Figure 4. The reconstructed 3D images of the dynamic blasting process. The four pictures in the left column are the extracted DSMs from the temporal-spatial model presented as contour lines. The pictures on the right column are the corresponding 3D images of these DSMs.

CASE 3 Post-Blast 3D Performance Measurement

Post-blast mapping is mainly concerned with quantification of muckpile and rock fragmentation.

Immediately after blasting, mapping of the muckpile and rock fragments can commence. As the targets or objects are not moving, static mapping techniques apply.

A field trial of the VirtuoBlast technology for post-blast mapping was undertaken at a quarry site in Western Australia. Typical muckpile and rock fragments were photographed according to field procedures developed specifically for this purpose.

A reconstructed muckpile model is shown in Figure 5. Figure 5(a) is the DSM presented as wireframe and pseudo colours. Figure 5(b) is the 3D image (superposition of orthoimage with DSM) of the muckpile. With the accurate 3D data, it is possible to work out the muckpile shape and volume (or tonnage) *etc*.

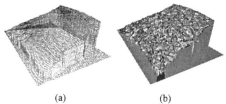

(a)   (b)

Figure 5. A reconstructed muckpile model from a stereo image pair: (a) The DSM of the muckpile presented as wireframe and pseudo colours; and (b) The corresponding 3D image.

Several pairs of images were taken at close range from selected portions of the muckpile. Typical results are presented in Figure 6. Figure 6(a) is the reconstructed DSM of the rock fragments presented as contour lines. Figure 6(b) shows the 3D rock fragment images. As can be seen clearly in Figure 6(b), the rock fragments are well reconstructed. The DSM and 3D image provide very accurate 3D data and texture details for further analysis and quantification of rock fragmentation and/or size distribution.

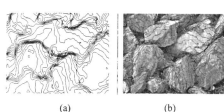

(a)   (b)

Figure 6. A reconstructed detailed rock fragment model from a stereo image pair: (a) The DSM of the rock fragments presented as contour lines; and (b) The orthoimage superimposed with contour lines.

6   CONCLUDING REMARKS

Demonstrated applications are showing an exciting future of the VirtuoBlast technology for quantitative 3D measurement of blasting.

The VirtuoBlast technology can also be applied to measure the movement of rock face and flying rock fragments during production blast. The measurements are important for further development of blast models, in particular for calibrating and testing our mathematical models which predict the results of specific blast designs - the shape of the muckpile, the size distribution and the distance and direction of fly rock and dust. From this research, new tools will be developed to enable mining companies and blasting engineers to better measure their blast performance, diagnose their blasting problems and reduce their costs.

The research is presently ongoing with focus on improving the efficiency of image processing, utilising higher resolution digital cameras, integration with other complementary techniques (such as 3D laser scanning) for efficient data acquisition and further field trails of the technologies under a variety of operating conditions.

ACKNOWLEDGEMENTS

The authors wish to thank Dr Bruce Hobbs for his encouragement and support of this important strategic research. Appreciation is also extended to Kalgoorlie Consolidated Gold Mines Pty Ltd for providing partial financial support during this project, for allowing access to the super pit to obtain the required stereo images and for allowing publication of the images and data related to the production blast operations. The Jasper Mining Ltd is appreciated for permission to publish the images of the Eureka pit. Thanks are given to our colleague Philip Soole for reviewing this paper.

REFERENCES

Aler J., Mouza J. D. and Arnould M., 1996. Measurement of the fragmentation efficiency of rock mass blasting and its mining applications. *Int. J. Rock Mech. Min. Sci. & Geomech. Abstr.* 33, 125-139.

Barker D. B. and Fourney W. L., 1968. Photoelastic investigation of fragmentation mechanisms – Part II. *University of Maryland Report.*

Chakraborty A. K., Jethwa J. L. and Paithankar A.G., 1994. Effects of joint orientation and rock mass quality on tunnel blasting. *Engineering Geology* 37, 247-262.

Chiappetta R. F. et al., 1987. Analytical high-speed photography to evaluate air decks, stemming retention and gas confinement in presplitting, reclamation and gross motion applications. In *Proc. 2nd Int. Symp. Rock Fragmentation by Blasting.* (Editors by Fourney and Dick), 257-301, Keystone, Colorado.

Cunningham C. V. B., 1983. The Kuz-Ram model for prediction of fragmentation from blasting. In *Proc. Int. Symp. on Rock Fragmentation by Blasting.* 439-453, Lulea, Sweden.

Duvall W. I. and Atchison T. C., 1957. Rock breakage by explosives. *Rep. Invest. – U.S. Bur. Mines* RI-5356.

Fourney W. L., 1993. Mechanisms of rock fragmentation by blasting. *Comprehensive Rock Engineering* 4 (Edited by J. A. Hudson), 39-69, Pergamon Press, England.

Fourney W. L., Dick R. D., Wang X. J. and Wei Y. Z., 1993. Fragmentation mechanism in crater blasting. *Int. J. Rock Mech. Min. Sci. & Geomech. Abstr.* 30, 413-429.

Franklin J. A., Maerz N. H. and Santamarina J. C., 1995. Developments in blast fragmentation measurement. In *Proc. 11th Annual Symp. on Explosives and Blasting Research.* 258-268, Nashville, Tennessee.

Grant J. R. and Little T. N., 1995. Blast driven mine optimisation. In *Proc. EXPLO '95 Conf.* Brisbane, Australia.

Hendricks C., Peck J. and Scoble M. J., 1990. Integrated drill and shovel performance monitoring towards blast optimisation. In *Proc. 3rd Int. Symp. Rock Fragmentation by Blasting.* 9-20, Melbourne, Australia.

Holmberg R. and Persson P. A., 1979. Design of tunnel perimeter blasthole patterns to prevent rock damage. In *Proc. Tunnelling '79.* (Edited by M. J. Jones), Institution of Mining and Metallurgy: London.

Kennedy P. S. G., 1994. Blast management. In *Proc. Open Pit Blasting Workshop 94.* Curtin University of Technology, Australia.

Langefors U. and Kihlstrom B., 1978. *The modern technique of rock blasting.* 3rd Edition, Halstead Press, New York.

Lewandowski T., Mai V. K. L. and Danell R., 1996. Influence of discontinuities on presplitting effectiveness. *Rock Fragmentation by Blasting.* 217-225, Balkema, Rotterdam.

Maynard B. C., 1990. A blast design model using the inherent fragmentation of a rock mass. *CIM Bulletin* 83, 71-77.

McKenzie C. K., 1993. Methods of improving blasting operations. *Comprehensive Rock Engineering* 4 (Edited by J. A. Hudson), 71-94, Pergamon Press, England.

McKenzie C. K., Scherpenisse C. R. and Jones J. P., 1995. Application of computer assisted modelling to final wall blast design. In *Proc. EXPLO '95 Conf.* 285-292, Brisbane, Australia.

Morrison D. M., 1995. Fragmentation – the future. In *Proc. EXPLO '95 Conf.* 11-18, Brisbane, Australia.

Muehlhaus H. B., Sakaguchi H. and Wei Y. Z., 1997. Particle based modelling of dynamic fracture in jointed rock. *Computer Methods and Advances in Geomechanics.* (Edited by Yuan), Balkema, Rotterdam.

Ord A. and Cheung L., 1991. Image analysis techniques for determining the fractal dimensions of rock joint and fragmentation size distributions. In *Proc. 7th Int. Conf. on Computer Methods and Advances in Geomechanics.* 89-91, Cairns, Australia.

Paine G. G., Harries G. H. and Cunningham C. V. B., 1987. ICI's computer blasting model 'SABLEX' field calibration and applications. In *Proc. 13th Explosives and Blasting Techniques Conf.* 33-48, Society of Explosives Engineers.

Poniewierski L, Cheung L. C. C. and Maconochie A. P., 1995. SIROFRAG - A new technique for post-blast rock fragmentation size distribution measurement. In *Proc. EXPLO '95 Conf.* Brisbane, Australia.

Preece D. S., Burchell S. L. and Scovira D. S., 1993. Coupled explosive gas flow and rock motion modelling with comparison to bench blast field data. *Rock Fragmentation by Blasting.* (Edited by Rossmanith), Balkema, Rotterdam.

Starfield A. M., 1967. Strain wave theory in rock blasting. In *Proc. 8th U.S. Symp. Rock Mech.* (Edited by C. Fairhurst), 538-548, University of Minnesota.

Song J. and Kim K., 1995. Numerical simulation of the blasting induced disturbed rock zone using the dynamic lattice network model. In *Proc. 2nd Int. Conf. on Mechanics of Jointed and Faulted Rock.* (Edited by Rossmanith), 755-761, Balkema, Rotterdam.

Stephenson B. and Fuller T., 1995. The development and application of laser surveying systems. In *Proc. EXPLO '95 Conf.* Brisbane, Australia.

Wei Y. Z. and Wang S. R., 1988. Superdynamic Moire method loaded explosively and its application in the experiment of explosion and fracture mechanics. In *Proc. 18th Int. congr. on high speed photography and photonics.* Bellingham, Washington.

Wei Y. Z. and Wu X., 1998. Reconstruction and simulation of 3D open pits: a critical path towards practical virtual mine. *Mineral Resources Engineering Journal* 7, No 3.

Wu X., 1996. Multi-point least squares matching with array relaxation under variable weight models. In *Proc. 18th ISPRS.* Vienna, Austria.

Wu X. and Wei Y. Z., 1998. Temporal-spatial modelling of fast moving and deforming 3D objects. *SPIE Proc.* 3545, 401-404.

Mining Science and Technology' 99, Xie & Golosinski (eds) © 1999 Balkema, Rotterdam, ISBN 90 5809 067 1

Coal mine planning with due consideration of strata control and environmental protection

J. Te Kook & W. Keune
Deutsche Montan Technologie, GmbH, Essen, Germany

ABSTRACT: The presented planning system
- is based on many years of experience in hard coal mining with winning coal on different seams in one of the most complicated deposits in the world,
- applies computer technology that handles the vast amounts of required data to describe a coal mine,
- reduces the planning times by approx. 50%.

The system is based on standard hard and software such as Oracle™ or MSOffice™. It runs on ordinary PCs and laptops to ensure world-wide availability. Special software such as the finite difference method is also used, if required, as is the physical modelling technique.

1 INTRODUCTION

Considering that a mechanised mine needs roughly US $ 100 to 150 million of investment, planning is very important for minimising the risks. A distinction can generally be made between seven types of risks:
- Political risks
- Social risks
- Geographical risks
- Legal risks
- Risk of Deposits
- Technical risks
- Financial risks

Taxation, customs and fees are not fixed for a long-term project, instead they may be decided upon within political processes.

Foreign investors usually provide their own management. Religious or ethnic conflicts may cause problems.

Mining activities often take place in geologically active formations. High floods or lack of water are definitely geographical risks.

Uncertainties in legislation on mineral ownership and changing of legislation on mining and the environment are further risk factors.

Although the first four items are of great importance, this paper will focus on the three other risks mentioned above.

2 COMPUTER AIDED MINE PLANNING AND DESIGN

To minimise the further risks, a large amount of data has to be evaluated. The aim of planning is to combine the deposit, mine infrastructure and capital input in a cost-optimised way. To reach this target, different scenarios need to be developed. For that purpose, these computer aided mine planning systems are very important.

For developing scenarios, different soft- and hardware tools are required. Fortunately most tasks can be solved using standard software like spreadsheet or CAD-programs. Special tasks such as strata control or subsidence prevention require special software.

All engineering disciplines involved in mine planning need to retrieve data from a unique source. Working within one framework makes great demands on communication [1]*.

2.1 Deposit

The assessment of the deposit requires information about e.g., the quantity and quality of coal, the amount of waste, thickness, depth, dipping of the seams and tectonical structures.

The result of the exploration illustrates the amount of data. Figure 1 describes a small deposit. It covers an area of roughly 4 km². The deposit is explored by about 100 drillings ranging up to 250 m in depth. This means 15 km core description and additional data depicting the surface and infrastructure has to be handled.

Unfortunately, the coal seams were not named during the sedimentation period. Therefore geologists have to identify the coal seams by using their experience with key strata. The data needs to be interpreted to derive a 3-dimensional model of the deposit. Therefore data sections and cross sections are conventionally generated using graphical programmes. After correlation of coal deposits ISO-patches and maps depicting floor-contours are to be created.

Computer-based planning is able to perform an analytical approach. Identification of coal seams from drillings works are as follows:

- Based on the data an inquiry is generated straight away. For instance, to find the top seam, all data describing coal at the upper position within each drilling is separated.
- From this data all drillings outside the outcrop are removed.
- To take the technical performance into account a minimum thickness is defined.
- A length-dipping ratio is calculated to find tectonical structures that distinguish between extractable parts of the deposit.

The results of analytical and conventional identification of the coal seams are compared and show that more than 95% of identified resources are identical. Remaining uncertainties have to be clarified geologically. Using the analytical approach, the time needed to derive a model of the deposit from the exploration results can be reduced by more than 50%.

---

* [N°] see paragraph literature

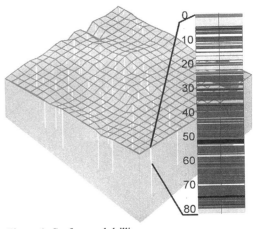

Figure 1: Surface and drillings

Due to the fact that this approach is based on a full 3D-model of the underground and surface, cross-sections, volumes, quality distribution, basically all types of ISO-lines may be generated. This means that the depiction of the deposit e.g. in maps, sections and 3D-views and all dependent data is fully included.

2.2 Technique

After fixing the deposit, engineering will proceed in mine layout. Conventional planning takes border conditions into account such as dipping, thickness, characteristics of seam and host rock etc. This is done by interpreting maps, sections and other bases containing the necessary information. The results are shafts, roadways and panels as shown in Figure 2. Analytically one divides that task into two subtasks.

The 1st step reduces the deposit geometrically to the mineable area. That area has been derived from the planning system, regarding the border conditions. Due to the fact that one has to mix two ISO-planes (e.g. seam -thickness and -level) the computer will decrease the time to fix that area dramatically.

The 2nd step covers the technical approach to fit the goal to extract the deposit into the frame, using the suitable extraction method, the machinery availablilty, transportation and so on.

A key criterion in underground mining is the control of the strata. The aim is to maximise the exploitation of the deposit regarding the stability of all excavations. Regarding roadways, the support and maintenance expenditure are to be minimised.

The German mining industry has more than thirty years of experience in strata control. Due to the average extraction depth nowadays of roughly 1000 m German mining investigation began in the early sixties with R&D programmes to develop a system of strata control.

Strata control is indispensable in the planning of underground mines. In Indian deposits with shallow depths and thick sandstones in the roof, strata control has the same importance as in deep mining.

The system of strata control includes the following elements:

- Measurements and observations in galleries and faces, estimation of the behaviour of support and strata, verification of the planning and developing the prediction methods.

- Rock pressure calculations and geologic mapping for optimising the mine layout.

- Physical and numerical modelling for the behaviour of support and strata in faces and galleries research.

- Testing rigs for investigation and improvement of gallery and face support.

- Test benches for analysing the properties of rocks and building materials [2].

The prediction tool of the system uses geometrical data and the representation tool of the mine planning system.

After preparation of a mine layout, including panel positions and time schedule one has to take into account that mining will influence the environment. For instance a river as shown in Figure 2 can be influenced by subsidence. Normally a waste dump will cover a certain area on the surface. It is needed for dumping extracted rock material. Generally the mine operator has to reclaim the affected area before leaving it [4].

Therefore the engineering workforce has to predict all influences of mining on the surface and to prepare an environmental impact plan (EIP). On the one hand measures have to be taken to prevent damage. On the other hand subsidence might help keeping e.g. a harbour open. The deposit of mud may make the floor rise and thus turn the harbour inoperable. Sub-

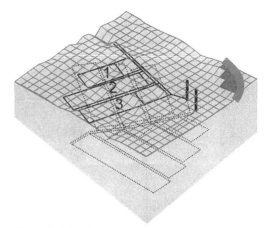

Figure 2: Mine layout

sidence can decrease the floor. The EIP is based on an exact prediction of all those impacts. For that reason a numerical prediction system for ground movement is one tool of the workbench the above described mine planning system represents. This system uses a stochastic approach to calculate ground movement. After calibration it is applicable under all mining conditions [3].

2.3 Finance

The success of a mine largely depends on low costs and high revenues. This requests the control of costs right from the start of the planning process.

Conventional finance and technical planning run more or less parallel. There is no online data exchange. Instead it is only applied at fixed points, e.g. for the selection of new equipment. Planning is -in this case- restricted to fewer planning scenarios, because it is too time consuming to consider a cash flow analysis for each modification introduced, e.g. in case of a new pressure or subsidence prediction.

The mine planning system provides methods and actual data to each member of the planning team. Therefore an engineer calculating subsidence is able to estimate the influence of modifications e.g. in panel length or time sequence on cost .

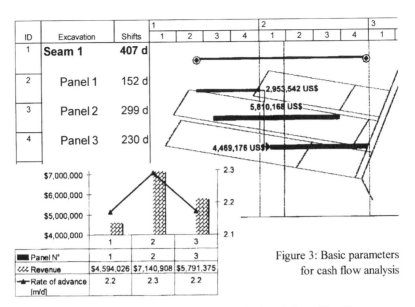

ID	Excavation	Shifts	1				2				3
			1	2	3	4	1	2	3	4	1
1	**Seam 1**	**407 d**									
2	Panel 1	152 d									
3	Panel 2	299 d									
4	Panel 3	230 d									

Panel N°	1	2	3
Revenue	$4,594,026	$7,140,908	$5,791,375
Rate of advance [m/d]	2.2	2.3	2.2

Figure 3: Basic parameters
for cash flow analysis

Figure 3 shows the mentioned multi-use of the same data. This figure combines the mining sequence with the mine layout and the rate of advance in relation to expenses and revenues. 90 % of all information shown has been generated with standard software. The data on which the figure is based, comes from the data base management system.

3   CONCLUSION

Under consideration that the removal of an error while planning charges a split part of the same while mining, planning becomes of higher importance in modern mining. In fact planning means exchange and processing of data between different mining disciplines. The ventilation system and requirements depend on the mine layout. But an optimal ventilation system may increase the costs beyond certain limits or make extraction impossible. The same data as mentioned before is used to estimate the length of transportation routes. Every change in the layout is automatically provided to all mine engineering disciplines and can be thus considered.

Five years of computer based planning taught us that the planning system may provide tools to accelerate the work, to avoid mistakes and to develop different scenarios easily. But successful planning mainly depends on the use which is made of those tools. On the one hand knowledge of mining operations and mine planning issues is of importance, on the other hand the knowledge of handling a very complex system guarantees success.

4   LITERATURE

[1] Keune W.: A mine information system based on an OMT model of the coal mine and deposits. in: R.V. Ramani, (Hrsg.): 26th International Symposium Application of Computers and Oparations Research in the Mineral Industry. Society for Mining, Metallurgy and Exploration Inc. Colorado, USA 1996, S. 347-353.

[2] te Kook J. and W. Keune: The German system of Strata Control and a Mine Information System to calculate subsidence and earth movement. 2nd seminar on ground control in mining. S.K. Sarkar (ed). Oxford and IBH Publishing Co. PVT. Ltd. New Delhi, Kalkutta 1996. S. 341-356.

[3] Keune W. and V.P. Talwar. Subsidence prediction techniques including modelling. Journal of Mines, Metals and Fuels. Vol XLV, Calcutta 1997. P. 49 – 56.

[4] Keune W. et al: Enlargement of a deep mine information system (IS) with a model of the surface. An application for Bobovdol in Bulgaria. In: Minning in the 3rd Millenium. Xth International Congress of the International Society for Mine Surveying. Congress Proceedings. Promaco Convetions Ltd. Freemantle (WA) 1997. P. 649 – 658.

Mining Science and Technology' 99, Xie & Golosinski (eds) © 1999 Balkema, Rotterdam, ISBN 90 5809 067 1

The pattern of overlying strata movement in longwall face

Dingli Zhang
Northern Jiao Tong University, Beijing, People's Republic of China

Yuehan Wang
China University of Mining and Technology, Xuzhou, People's Republic of China

ABSTRACT. This paper discusses the process and characteristic of overlying strata movement in longwall face. Based on field measurement, physical modeling and theoretical analysis, basic pattern of overlying strata movement and influence of different stratum combination on strata movement are put forwarded, bed separation type of overlying strata and its control are further analyzed. The study in this paper will be useful to estimate and control surface subsidence.

1 INTRODUCTION

The purpose of overlying strata control in longwall face includes two aspects: one hand to decrease the influence of surface subsidence and minimize its destroy degree on surface construction; on the other hand to effectively control the influence of strata movement on ground behavior in the face. The object of the above study is overlying strata. However, due to distinctness of their study purposes, each study emphasizes on different aspects: the ground pressure and the mining subsidence induced by limited strata broken nearby face or the movement and destroy of upside stratum.

In recent years, research of stratified rock mechanics has been receiving more attention worldwide [1]. Many scholars try to establish a general mechanical model of overlying strata movement, explaining surface subsidence and ground behavior around face, and significant achievements have been obtained [2-4].

However, movement and failure process of overlying strata are a quite complicated. This paper divides movement of overlying strata into several patterns, and analyses bed separation types and their conditions. As a result the process of strata movement can be rationally described.

2 BASIC CHARACTERISTICS OF OVERLYING STRATA MOVEMENT IN THE FACE

With mining of coal seam, overlying strata move successively from bottom to top. But due to unlikeness

of rock strength, layer thickness and the extent of fissure development, span of strata movement and caving are different. As face advance, the harder and thicker strata in the overburden may form certain structure stratum while the weaker and thinner strata will inevitably load on the structure stratum. As a rule, the most bottom stratum in a set of strata becomes key stratum. As a sign of key stratum, overlying strata in the face may be divided into some strata group, and deformation of every stratum in the same strata group is well coordinated each other. The stratum that its caving can affect surface becomes a dominant key stratum which is the emphases of surface subsidence control.

Owing to grouping movement of overlying stratum, bed separation generally appear at the dividing line that is between upper harder stratum and lower weaker stratum, and bed separation is caused by unequal deformation of both strata group. The magnitude of bed separation and its continuance period depend on the interaction among upper structure stratum and lower structure strata and mining conditions.

The movement process of overlying strata in the face obtained from physical modeling is shown in Fig.1. When face advance 85m, the harder stratum, locating 28m above the coal seam, separates from lower parts of stratum (Fig.1a). When face advance 120m, the harder stratum rupture and collapse in quite large-scale strata until second harder stratum that is 80m from above the seam, and a new bed separation be formed (Fig.1b). When face advance 180m, the second harder stratum rupture, and instability (Fig.1c) as well as surface subsidence is produced.

(a)

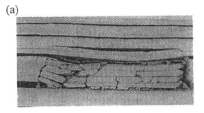

(b)

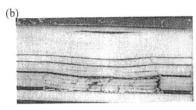

(c)

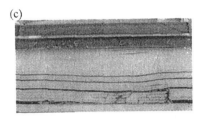

Fig.1 Movement procedure of overlying strata in longwall face

Transfer of overlying strata movement in most cases includes two aspects, i.e. the force transferring and the displacement (deformation) transferring. Transferring of force is achieved from top to bottom, while transferring of displacement is achieved from bottom to top. Both transfers are suspended in the positions where bed separation exists, but with rupture and instability of overlying stratum occurring separation interspace immediately disappear.

3 BASIC PATTERNS OF OVERLYING STRATA MOVEMENT AND ITS MECHANICS ANALYSES

Grouping movement of overlying strata in face may be divided into two situations:

□ Owing to cohesive force of strata, different strata in the same strata combination moved lower all together.

□ Lower stratum in the same strata combination become key stratum, upper strata load on key stratum while moving all together.

This paper focuses on the second situation, i.e. stability of key stratum will thoroughly be analyzed.

Movement of overlying strata in the face and its influence on surface subsidence mostly depend on destroy characteristic and instability of key strata, especially dominant key stratum. Accordingly, key strata successively carry out translation from continuum

medium to pseudo-continuum medium, then to non-continuum (damage) medium.

Therefore, movement of overlying key strata may be divided into three patterns:

□ Continuum beam founded on elasticity foundation——continuum medium beam.

□ Fractured soft beam founded on elasticity foundation——pseudo-continuum medium beam.

□ Block structure beam founded on elasticity foundation and waste rock——non-continuum medium beam.

Due to the difference of support foundation and loading stratum condition, the above three structure patterns in different positions will possess different equilibrium conditions. Considering control of surface subsidence, state and stability of dominant key stratum are of importance. Moreover, on account of location of dominant key stratum, its structure may be one of the above three patterns.

3.1 *Recognizing of key strata*

In analysis of overlying strata structure in the face, key strata should firstly be recognized. According to reference [5], loading of stratum n on the first stratum can be expressed as:

$$(q_n)_1 = \frac{E_1 h_1^3 \left(\gamma_1 h_1 + \gamma_2 h_2 + \cdots + \gamma_n h_n\right)}{E_1 h_1^3 + E_2 h_2^3 + \cdots + E_n h_n^3} = \frac{E_1 h_1^3 \sum_{i=1}^{n} \gamma_i h_i}{\sum_{i=1}^{n} E_i h_i^3} \quad (1)$$

where: $E_1 \pounds E_2 \cdots E_n$——modulus of elasticity of stratum, n is stratum number;

$h_1 \pounds h_2 \cdots h_n$——thickness of stratum;

$\gamma_1 \pounds \gamma_2 \cdots \gamma_n$——density of stratum.

When $(q_{n+1})_1 < (q_n)_1$, i.e. stratum $n+1$ don't affect on first stratum, temporally first stratum is a key stratum while stratum n+1 will become another key stratum.

As a rule, there might be several key strata in overlying strata, obviously, stability of dominant key stratum will directly affect surface subsidence, it is key issue of surface subsidence control. Due to most hypogynous key stratum affect directly ground behavior at face, thus, it is emphases of ground pressure control.

3.2 *Rupture laws of continuum medium beam*

According to Fig.1, it is obvious that key stratum ahead first ruptured may be considered as continuum medium beam founded on Winkler foundation, i.e. vertical force acted on elasticity foundation can be expressed as follows:

$$p = -ky \quad (2)$$

where: k——foundation coefficient, it may be expressed as follows:

$$k = \frac{1}{\sum_{i=1}^{n} \frac{h_i}{E_i}}$$

166

where: y——compress amount of Winkler foundation;

p——collect degree of force acted on Winkler foundation.

Using mechanics model of Winkler foundation, it is possible to calculate deformation, ruptured span and fracture position of key stratum ahead first broken, in succession, and influence of key stratum on overlying strata and surface subsidence might also be determined.

3.3 Rupture laws of fractured soft beam

According to viewpoint of fracture mechanics, because of crack existing in rock material, and mining process will produce more crevices, therefore, rupture of rock is a progressive damaging process. This progressive damaging process will lead final macro-rupture. In fact, influence of mining on upper key strata is smaller when mining depth is greater, no crevice is penetrated in some conditions, they may be treated as fractured soft beam, and damage degree depends on its state of "stress—strain", i.e. damage parameter $D = 1 - \exp\left(-\dfrac{\varepsilon}{\varepsilon_{\max}}\right)$, it can be described as follows:

$$\sigma = E(1 - D)\varepsilon \quad (3)$$

Due to no macro-rupture and penetrated crevices in rock beam, macro-continuity is preserved while rock deforming. But owing to reducing of intensity and rigidity, its deformation amount is larger than that of continuum medium beam, however, it is generally less than the deformation amount of block structure beam.

3.4 Rupture laws of block structure beam

Rock beam structure is very similar to the block structure beam after strata is ruptured, or called "voussoir beam" structure. Generally there are one or more key strata in overlying strata, but dominant key stratum is key issue of surface subsidence control. Therefore, a random key stratum is shown as in Fig.2, its stability after ruptured is analyzed.

Fig.2 Block structure of key stratum

In view of symmetry of influence extension ahead plenary mining, only half of influence extension will be considered, force analysis of structure is shown as fig.3.

According to Fig.3, relationship between displacement of middle part and end part in rock block is expressed as follows:

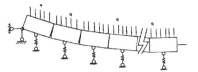

Fig.3 Mechanical analysis of block beam structure

$$\Delta_1 = \frac{1}{2} w_1$$

$$\Delta_2 = \frac{1}{2}(w_2 + w_1)$$

$$\Delta_3 = \frac{1}{2}(w_3 + w_2) \qquad (4)$$

$$\cdots$$

$$\Delta_{n-1} = \frac{1}{2}(w_{n-1} + w_{n-2})$$

$$\Delta_n = \frac{1}{2}(w_n + w_{n-1})$$

If convergence of key stratum touching waste rock is

$$m = M - \Sigma h \cdot (k_p - 1) \qquad (5)$$

then relationship of convergence between middle part and both ends subsidence can be written as follows:

$$\Delta'_1 = \frac{1}{2} w_1 - m$$

$$\Delta'_2 = \frac{1}{2}(w_2 + w_1) - m$$

$$\Delta'_3 = \frac{1}{2}(w_3 + w_2) - m \qquad (6)$$

$$\cdots$$

$$\Delta'_{n-1} = \frac{1}{2}(w_{n-1} + w_{n-2}) - m$$

$$\Delta'_n = \frac{1}{2}(w_n + w_{n-1}) - m$$

When the amount of block structure is even number $(2n)$, n block will be analyzed by symmetry, its mechanics equilibrium condition is matrix (7). When the amount of block is odd, its analysis is similar.

Further calculating of matrix (7), $w_1 \square w_n$ may be gained, namely, and the state of key stratum behind ruptured may be determined. Obviously, every structure stratum possess different equilibrium state. The equilibrium state of structure stratum is also influenced by loading of upper strata, support foundation of lower strata, length and thickness of ruptured rock block. But state of dominant key stratum decides appearance of surface subsidence in certain extent.

As face advance, state of lower structure strata is firstly changed, therefore it change support foundation of upper strata, in succession, state of upper strata is accordingly changed so as to achieve new equilibrium of structure strata. This is the basic characteristic of stress and deformation transfer among different stratum.

$$
\begin{bmatrix}
l_1 & \frac{1}{2}l_1 & 0 & 0 & 0 & 0 & 0 & 0 & 0 & 0 & 0 & -(h-w_1) \\
0 & \frac{1}{2}l_1 & l_1 & 0 & 0 & 0 & 0 & 0 & 0 & 0 & 0 & (h-w_1) \\
0 & 0 & l_2 & \frac{1}{2}l_2 & 0 & 0 & 0 & 0 & 0 & 0 & 0 & (w_2-w_1) \\
0 & 0 & 0 & \frac{1}{2}l_2 & l_2 & 0 & 0 & 0 & 0 & 0 & 0 & -(w_2-w_1) \\
\cdots \\
0 & 0 & 0 & 0 & 0 & l_i & \frac{1}{2}l_i & 0 & 0 & 0 & 0 & (w_i-w_{i-1}) \\
0 & 0 & 0 & 0 & 0 & 0 & \frac{1}{2}l_i & l_i & 0 & 0 & 0 & -(w_i-w_{i-1}) \\
\cdots \\
0 & 0 & 0 & 0 & 0 & 0 & 0 & 0 & l_{n-1} & \frac{1}{2}l_{n-1} & 0 & (w_{n-1}-w_{n-2}) \\
0 & 0 & 0 & 0 & 0 & 0 & 0 & 0 & 0 & \frac{1}{2}l_{n-1} & 0 & -(w_{n-1}-w_{n-2}) \\
0 & 0 & 0 & 0 & 0 & 0 & 0 & 0 & 0 & 0 & 1 & 0
\end{bmatrix}
\begin{bmatrix}
R_{0-1} \\ R_1 \\ R_{1-2} \\ R_2 \\ R_{2-3} \\ \cdots \\ R_y \\ R_{i-(i+1)} \\ \cdots \\ R_{(n-2)-(n-1)} \\ R_{n-1} \\ R_n \\ T
\end{bmatrix}
=
\begin{bmatrix}
\frac{1}{2}ql_1^2 \\ \frac{1}{2}ql_1^2 \\ \frac{1}{2}ql_2^2 \\ \frac{1}{2}ql_2^2 \\ \cdots \\ \frac{1}{2}ql_i^2 \\ \frac{1}{2}ql_i^2 \\ \cdots \\ \frac{1}{2}ql_{n-1}^2 \\ \frac{1}{2}ql_{n-1}^2 \\ ql_n
\end{bmatrix}
\quad (7)
$$

State of key strata notably influences on deformation of overlying strata, moreover in a certain extent, the dominant key stratum decides state of surface subsidence. When key stratum is in pattern 1, less influence on overlying strata is produced. When key strata is in pattern 2, due to damage of rock beam and reducing of rigidity, influence on overlying strata increases. When key strata is in pattern 3, deformation of key stratum rapidly increase, especially for dominant key stratum, more large-scale surface subsidence will appear. In case of structure instability, sidestep subsidence will also come out in the surface.

In fact, movement of key strata generally comes through above 3 patterns in proper order, and deformation and fracture gradually develop. Furthermore, the combination of different stratum structure patterns and its mechanics and mining conditions will also influence on the amount, coefficient and speed of subsidence. When sub-level caving mining is adopted, subsidence velocity of surface obviously increases, but accumulative total of surface subsidence comparatively decreases.

5 BED SEPARATION TYPE OF OVERLYING STRATA AND ITS CONTROL CHARACTERISTIC

Bed separation of overlying strata may be divided into 2 types:

☐ Bed separation above caving zone, i.e. bed separation between caving zone and fractured zone when caving interspace is not fully penetrated.

☐ Bed separation within fractured zone, namely bed separation among stratum group, this is caused by non-harmonize deformation between both strata group.

To counter different bed separation type, different measure of decreasing surface subsidence will be applied so as to get the better result. Bed separation filling is considered as an effective measure, it has been applied in many mines in China in recent years, and an obvious effect has been got in practice.

6 CONCLUSION

In view of grouping movement of overlying strata, strata movement and surface subsidence are controlled by certain key strata, among which, the stratum directly influencing surface subsidence is considered as the dominant key stratum, it is the emphases of controlling surface subsidence. After analysis of strata movement process, especially analysis on the state and stability of block beam structure, three movement patterns are put forwarded, and emphatically analyses state and stability of block beam structure, so it is possible to calculate and further control surface subsidence.

REFERENCES

1. Yuzhuo Zhang, Liliang Chen. The conditions of bed separation in longwall mining face, Journal of China Coal Socity, 1996

2. Minggao Qian, Xiexing Miao. Theoretical analysis on the structural form and stability of overlying strata in longwall mining. China Journal of Rock Mechanics and Engineering, 1995, 14(2):97—106

3. Fenghai Ma, Xueli Fan, Yongjia Wang. The model of rock stratum movement as huge system complex medium and its engineering application. China Journal of Rock Mechanics and Engineering, 1997, 16(6):536—543

4. Jingche Zhao, Manchao He. Sustainable mining theory and stratagem study on coal resources under buildings. (subject research report)1997

5. Minggao Qian, Tingcheng Liu. Ground pressure and its control. Coal industry publish house, 1992

Mining Science and Technology' 99, Xie & Golosinski (eds) © 1999 Balkema, Rotterdam, ISBN 90 5809 067 1

Longwall top coal caving method – Highly efficient mining method used in Yanzhou coalfield

Boyun Xu
Yanzhou Mining Group Corporation Limited, People's Republic of China

ABSTRACT: The top coal caving method was initiated in the 1940's. In early 1990's this method had been successfully introduced to longwall coal mines of China. It assures high production, safety and low production cost. Yanzhou Mining Group Corporation Ltd. has pioneered introduction of this innovate coal mining technique, and uses extensively in thick seams. Many experimental practices, regarding modified mining system, caving supports, operation processes and mining parameters, have been developed to make this mining method much more efficient and sophisticated.

1. INTRODUCTION

With the increase of more intensive competition in the global coal market, the current challenge is therefore to innovate extraction techniques in order to reduce producing costs. Longwall top coal caving method used extensively in China in recent years has been shown to have significant safety, very high efficiency and cost benefits for extracting thick underground seams.

Longwall top coal caving method (LTCC), which is called caving method for short, was initiated in former Soviet Union, France, Romania, Czechoslovakia, and Jugoslavia in the 1940', but the output of coal face was limited.

The caving method was firstly introduced in China in 1982, simultaneously, necessary steps were taken to start manufacturing of supports and equipment within the county. The first experienced mining activity was conducted in the thick gentle slope coal seam in Puhe Coalmine in China in June 6th, 1984.

Yanzhou Mining Group Corporation Limited (YMGC) had conducted extensive investigations and studies regarding the rock and coal seam properties, strata and support behavior before 1984. The first caving face was put into practical use at 5306 face at Xinglongzhuang coal mine in Yanzhou Coal Field in July 1992. Today the slicing method is replaced gradually by the caving. This innovation has made the caving method play a dominant role in Yanzhou coalfield, and so far there are six caving faces with the highest productivity of five-million-ton-per-year per face.

The YMGC's seven mines totalled 1924 million tons, of which 1296 Mt was to be recoverable by the caving method in 1997. This coal mining method has been proven to be highly efficient for the thick gentle slope coal seam in Yanzhou coalfield (Fig. 1 shows raw coal production from year 1992 to 1997.

2. DESCRIPTION OF SEAMS

The Company's primary reserves of coal are contained in the No 3 seam, which lies at depths of between 200 m and 1000 m below the surface and varies in thickness from about 2.0 m to 13.5 m. 42 per cent of the total reserves is very suitable for caving method. The No 3 seam is a particularly high quality coal seam producing a low sulphur coking coal capable of yielding a product with a controlled ash content of between 6 per cent and 12 per cent. The No 3 seam is continuous throughout the mining area with only generally shallow gradients. Additional reserves exist in other coal seams, but these are thin and slightly higher sulphur content.

Thousand Tons

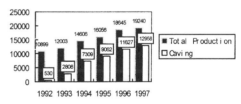

Fig. 1 The raw coal production from year 1992 to 1997

3. LONGWALL TOP COAL CAVING SYSTEM

All the coal mines in Yanzhou coalfield operate a system of fully mechanized retreat longwall mining, involving the removal of a rectangular shape panel of coal which typically measures approximately 150~220 *m* in width by 1000~2500 *m* in length. The extraction panels are firstly created by driving a pair of parallel gateways which serve only one coalface. Between the adjacent panels there is no pillar left, i.e., one of the roadway of an adjacent panel is developed in coal seam just nearby the consolidated goaf of another worked out panel, which is called no-pillar mining technology in China.

There are two variations on the mechanized longwall system in No 3 Seam, namely "slicing" and "caving". The slicing method, around which the mines were originally designed (Fig. 2), is used to extract the full section of a thick seam in a sequence of two or three slices of up to 3.0 m in height, commencing in the uppermost portion of the seam. The method involves standard longwall practice, but modifications to facilitate the spreading of wire mesh along the face, which is trapped against the roof by the hydraulic face supports as they advance. The mesh is left behind in the goaf to form a definite boundary between broken waste rock, and the remaining coal in the floor of the seam. After consolidation of the goaf, a new panel is established on a lower portion of the seam beneath the previously extracted panel.

The caving method (Fig. 3) has been extensively adopted in more recent years, and involves the extraction of a single panel in the lower portion of the thick seam in which the top coal, comprising initially of coal body, is allowed to cave in the normal manner, without the use of mesh. The roof supports are of a modified design incorporating a system of hydraulically operated tail-canopies at the rear of the support which can be moved up and down to allow broken coal in the goaf area to spill onto a second armoured face conveyor. This process is allowed to continue until all retrievable coal behind the face has been collected and waste rock appears. The principal data for longwall top coal caving coalfaces in No 3 seam are as follows.

Coalface length	*150/220 m*
Drvage length	*1000/2500 m*
Shearing height	*2.8/3 m*
Caving height	*3/7 m*
Protodyakonov scale of hardness	*f = 2/3*
Seam pitch	*3/15°*
Seam thickness	*8.43, m*

The operation steps in the process of longwall top coal caving are slightly different from standard longwall working, the steps in sequence are detailed as follows:
- Shearing coal in the front of the coalface
- Pushing the front conveyor
- Setting the support forward
- Opening the tail-canopy of support to allow broken coal to spill onto the rear conveyor
- Pulling the rear conveyor

The face advance rates vary from $100 \sim 260$ m per month. In general, on a caving face, only $2 \sim 4$ cuts are made each shift to allow time for the caving process to take place behind the face line and the subsequent collection of coal on the rear conveyor.

The longwall caving system has some advantages and disadvantages when compared with the slicing method. The caving system is a much more efficient and economical use of resources for the tons of coal produced, since it enables extraction of a greater seam height of coal within a production panel area with one set of gate drivages than would be the case using the slicing method. However, collection of the coal through the rear doors of the supports depends on many factors which are may outside the control of the support operators. It is clear that coal is left in the goaf. However some effective measures, such as optimization of the caving process, have been taken to improve the recovery of coal. Another efficient practice is to drill boreholes in the top coal at 15 m intervals along the coalface and two gateways to compute theoretical tons, and compare this with readings obtained from belt weighting machines so as to supervise the coal recovery and promote the control of support operators. The caving method normally results in a somewhat lower recovery percentage, about 80 85 per cent, comparing with 97 per cent recovery by slicing method. However the caving method has facilitated the control of support operators because of several years practice, the coal recovery percentage is increasing year by year, and has thoroughly met the need of regulations (more than 75 percent) promulgated by the Ministry of Coal Industry.

Apart from above advantages by the caving method, minimum roads are developed and minimum equipment is occupied in the same panel, achieving the result of the lowest production costs. The slicing method, in contrast, ensures a high recovery rate, but requires a new pair of gate drivages and a new set of equipment for each slice. In addition, it requires the infrastructure roads, to which the gateroads are linked, to be maintained for a long periods of time. Faults with throws of 5 m can be taken by caving faces whereas slicing faces are limited to only 3 m.

4. RATIO OF SHEARING HEIGHT TO CAVING HEIGHT

It is essential to evaluate the caving behavior of top coal before planning the caving panel. The ratio of shearing to caving height is generally taken as an assessment index to determine whether the caving method is chosen to utilize or not. Following formula is the calculating method.

$$\Delta = \frac{h_1}{h_2}$$

Where h_1 ---- caving height, m
h_2 ---- shearing height, m
and $h_1 + h_2 =$ total thickness of seam

According to numerous laboratory experiments and practical surveys, the ratio $\square$ can be coined to quantify the choice of the caving method. If the ratio is between 1:1 and 1:3, the top coal may be classified as very easily cavable. Slight difficulty in caving is experienced when the ratio increases above 1:3 or less than 1:1. For No 3 seam the ratio is around 1:1$\square$2, belonging to a very easily cavable seam.

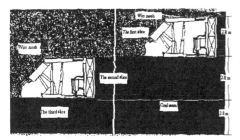

Fig. 2

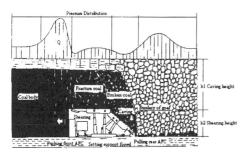

Fig. 3

5. TOP COAL CAVING INTERVALS

Circularly caving interval is a vital parameter which can be used to guide the process management of caving activities and affects the recovery of a seam. In general the higher the caving height of top coal and the harder the hardness of coal, the larger the caving interval should be, vice versa. The value of L is determined by the experimental operations and the empirical formula. In order to cater for the management of caving process the value of a caving interval must be the integer times of one cut depth (0.6 m).

The empirical relationship connecting the caving interval L with these parameters is as follows:

$$L=(0.15\square0.21)(h-0.7)$$

where h-- total thickness of seam, m

Collection of the coal through the rear doors of the supports depends on many factors, but the value of L and the control of the support operators are considered as the principle of the caving process.

In the case of No 3 seam, two kinds of caving interval are selected, i.e., 0.6m and 1.2m whilst the proportion of caving and shearing height are respectively 1:1 and 1:2. In the caving procedure, there are 2$\square$3 support operators, who are arranged one after another along the coalface in about five supports distance, control the valves of the tail-canopies of the supports to allow the top broken coal spill into an armoured conveyor at the rear of the supports and shut down the doors while rock waste in goaf occurs. However, both the shearing and caving operations are in parallel within the limit capacity of the stage loader

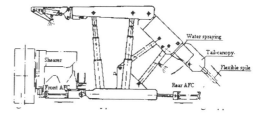

Fig. 4

in which coal flow from both front and rear conveyors is loaded.

6. EQUIPMENT FOR CAVING FACES

The caving faces use a range of equipment types and manufacture. Most of the Double Ended Ranging Drum Shearers ("DERDS") are of the AM 500 type, manufactured in China under license from Anderson Longwall International (UK). The hydraulically chock shield supports are generally of a 4-leg type specifically designed and manufactured in China. The AFCs, being used both in the front of the face and at the rear of the supports, are generally of Chinese manufacture, and vary according to their application, having different specifications depending upon duty.

Of all the equipment used in caving faces, only the supports have been modified. There are mainly three types of supports named "high-lying", "medium-lying" and "low-lying" top coal caving supports, of which the "low-lying" top coal caving support is dominantly used in China due to its significant advantages of higher recovery and lower dust emission. This kind of supports has a rear open canopy with a flexible spile, which is different from the standard hydraulic longwall supports.

Face-end supports have another characteristic on caving duty. Two face-end supports are respectively installed at the intake and the return of the face ends, and between these supports the general top coal caving supports are displayed.

6.1 General Top Coal Caving Supports

Fig 4 shows a typical chock shield caving supports used in Yanzhou coalfield.

The properties of the ZEP5600-1.7/3.2 chock-shield-type caving support used extensively in Yanzhou coalfield are as follows:

Supporting height	*1700 ~ 3200, mm*
Supporting width	*1482 ~ 1598, mm*
Setting load	*4986 kN*
Working resistance	*5600 kN*
Supporting strength	*0.8 MPa*
Supporting area	*8.45 m²*
Swinging angle of the front canopy	*+15 ~ -13*
Swinging angle of the tail canopy	*+0 ~ -58*

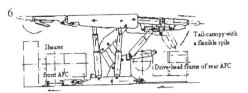

Fig. 5 Illustration of a face-end chock-shield-type caving supports

Longwall face-end chock shield supports are specially designed for the larger space at the longwall face-ends, where the conveyor drive devices are located both at the rear of the supports and in the front of the face. The top coal upon the face-end supports can be partly fallen onto the rear conveyor through the most adjacent caving support. Fig 5 illustrates the characteristics of this type of supports.

The properties of the ZTF5400-2.2/3.2 chock-shield-type face-end caving support used extensively in Yanzhou coalfield are as follows:

Supporting height	*2200/3200, mm*
Supporting width	*1482/1598, mm*
Setting load	*4850 kN*
Working resistance	*5400 kN*
Supporting strength	*0.695 MPa*
Supporting area	*7.87 m²*
Swinging angle of the front canopy	*+15 / -3*
Swinging angle of the tail canopy	*+0/-58*

7. STRATA AND SUPPORT BEHAVIOR ON CAVING FACES

Strata control and support behaviors are very concerned during full-scale trial. Support properties designed for caving operations are initially on the base of the standard support properties obtained from the slicing practice and further modifications have been made.

Of the standard longwall slicing faces in Yanzhou coalfield, obviously first weight and periodic weights were observed. The first weight occurred at a span varying from 38.6 m to 64.7 m whereas subsequent periodic weights occurred at an average interval from 19.6 m to 26.5 m. Contrasting to slicing method, at the longwall top coal caving faces the first weight occurred accompanying about 80 per cent of the relief of the safe valves, with first weight intervals from 45 m to 55.3m, and the subsequent periodic weights were surveyed varying from 13.2 m to 16 m. In the case of caving operation, the phenomenon of weighting are not much more obvious than the standard longwall coal faces, and during the period of weighting the available support resistance was as high as 78.7 per cent of the rated support resistance. In a word, the supports designed for caving faces are suitable for the caving extraction under the strata behaviors in Yanzhou coalfield and there is no support damage and roof rock or coal degradation.

8. SPONTANEOUS COMBUSTION PROBLEM

The No 3 seam is prone to spontaneous combustion. The incubation period for No 3 seam is 3□6 months. There are some practices during the caving process could encourage the onset of this condition, including the leaving of unrecovered coal in the goaf and the broken coal left on the face-end supports between the adjacent panels. However, the condition has not presented a major problem, which attributes to following measures.

- Continuous monitoring of the atmosphere in the underground working and goaf
- No pillar left between adjacent panels
- Large-scale ventilation balance to prevent the air leakage
- Rapid slaving of a face equipment following completion of a longwall panel
- The use of nitrogen injection for extinguishing fires

9. SUMMARY

Longwall top coal caving method is a modified longwall mining technique, which has been practiced by YMGC in recent years, and has been proven to be highly efficient and lower in production costs whilst comparing with the slicing method. This method has extensively adopted for thick coal seams in China.

The caving system is almost the same as the standard longwall mining, but the needed roads can be significantly reduced comparing with the slicing method for mining the same thick seam. The process of caving method consists of a series of subsequent steps, including shearing along the coalface, pushing front conveyor, setting supports forward, manipulating the rear doors to have top coal caving and pushing the rear conveyor.

The ratio of shearing height to caving height is a major index for assessing the caving properties of thick seam. In order to facilitate the complicate management of caving process and achieve the higher coal recovery, 0.6 m caving interval is determined for the ratio of 1:1, and 1.2 m for 1:2.

The specially designed supports are an utmost characteristic at caving coalfaces. The state of strata pressure collected from extracting practice shows that the supports for the caving faces in Yanzhou coalfield work on suitable situation.

Spontaneous combustion must be concerned as a potential risk because of the leaving of unrecoverable coal in goaf, and effective prevention measures must be taken.

2 Geology, geophysics and methane drainage

Mining Science and Technology'99, Xie & Golosinski (eds) © 1999 Balkema, Rotterdam, ISBN 90 5809 067 1

A high resolution 3D-seismic technique for distinguishing geological structure in the complex coalfield

Suping Peng & Yi Liao
China University of Mining and Technology, Beijing, People's Republic of China

Bingguang Zhang & Maoye Sun
Huainan Coal Administration, Anhui, People's Republic of China

ABSTRACT: Processing the high resolution seismic data and comparison with the data obtained from mining workings allows definition of the seam variations, the small-scale geologic structure and the texture of coal. For the described examples, nine faults with ≥ 3-m drop height, three faults with ≥ 1-m drop height and four structural loose zones located in about 680-m depth are found in 1.2 kilometers square area of the research district. It provides a useful large-scale experimental measure for the seismic exploration.

1 INTRODUCTION

With rapid development of the electric technology, the large-scale digital seismic exploration equipment has been developed to hundreds' channels since 1980s. As a result, the 3D seismic exploration technology is carried forward with the development of the multi-channel's seismograph and the large-scale computer technology. In the end of 1980s, about one hundred's oil or gas areas had been explored by the 3D seismic technology. In China, it was begun to experiment with 3D seismic technology in Jianghan Oil Field in the end of 1970s, which has an obvious geological reflection in research of the complex structure, lithologic and oil-bearing(or gas-bearing) changes in lateral. The result shows it is useful for looking for the covered oil field and for determining of the drilling position.

It is first used the 3D seismic exploration technology to the exploration of the Ruhr Coalfield and then about 70% seismic exploration is used by the 3D seismic exploration in Germany. The 3D seismic exploration is coordinated with core-drilling to use for dividing the coal-mining area, determining the pit shaft position and arrangement of the tunnel and the coal-mining faces. In China, the first experiment for 3D seismic exploration was conducted in 7.3 kilometers squares area in the Northern Yiminhe Mine District in 1978 to 1979, where the geologic structure is more complicated. In

1987, the data was analyzed again by the Bureau of Coal Geology, PRC, and the faults with about 20-m drop height were found out. In the end of 1980s the second 3D seismic exploration was undertaken in the Tankou exploration area, China, and the faults with 10 to 15-m drop height were made a through investigation. Since 1993, it has been studied on high resolution 3D-seismic exploration on the first coal-mining area in the coal mine district of Huainan coalfield, China. A 3D seismic data with high resolution, high signal/noise ratio and high density is obtained. It is found out that a set of faults buried in about 300-700m depth, are with ≤5-m drop height and the ≤ 10-m space position-fixing error.

The geological security system for the large-scale underground coal mine have a more high requirement, though the faults with ≥5-m drop height can be found out by the 3D seismic exploration. On the other hand, the results obtained by the 3D seismic exploration are frequently verified during the coal-mining process, and other much more useful information of 3D seismic technology that could be used for mining engineering is needed. Therefore, it is tried to analyze the high resolution seismic exploration data and the geological data from Huainan coalfield of China as follows:

1). further study the seismic reflect characteristics of the small-scale fault and the influence of the thickness changes of the thin coal seam to the seismic wave;

Table 1   The amplitude data base of the 1221 coal-mining face, Panji No.3 Coal Mine.

	Amp42	Amp62	Amp82	Amp102	Amp122	Amp142
1	10012.346	15946.028	16893.381	17390.201	3940.208	10124.469
2	11838.913	16477.252	20589.828	20927.658	7498.431	8664.864
3	12514.37	16106.793	22437.436	27146.385	11010.023	7777.715
4	14025.126	20273.654	25129.924	33981.082	15963.42	6243.9
5	15042.793	25810.668	30047.625	37745.02	20671.012	3073.436
6	15850.233	30102	35622.055	38009.59	24719.172	3251.209
7	17275.434	34846.797	39593.922	41695.793	29820.207	3379.739
8	18733.762	39208.254	42241.41	42780.988	32794.988	6134.559
9	18649.324	38658.801	43933.871	42546.41	32938.133	9577.455
10	18723.084	39001.559	44955.051	40863.406	33798.184	9815.056

2). build up a geophysical model for distinguishing the coal seam and the small-scale fault;

3). build up a simple, audio-visual method for mining engineering technician to use the 3D seismic exploration data conveniently during the coal-mining process.

2. LITHOLOGIC STUDY OF THE 3D SEISMIC STRATIGRAPGY

The 3D seismic exploration data is useful for the lithologic study in the following four points:

1). amplitude changes of the reflect wave for the 3D data process accord more with objective reality;

2). the velocity data are more detail;

3). the rate of the space samples in the 3d seismic is higher than that in the 2D situation. Therefore, the space changes of the wave amplitude in the 3D seismic data is easy to analyze and track;

4). different section demonstrations and mutual interpreting could improve the differentiate. Especially, the diagram along with the amplitude plane of the bed could directly show the sedimentary characteristics, such as the trend of coal thickness changes, the fold and the fault etc..

In the process of the 3D seismic data in Huainan coalfield,, the mutual interpreting system is used. Table 1 is a part of the amplitude data base in the 1552(3)coal-mining face at Panji No.3 Coal Mine

It is obtained from Tracking the maximum amplitude value of the coal seam, and cooperated to the time from the mutual interruting result, the seismic amplitude distribution of the seam 13-1 at the Panji No.3 Coal Mine can be grasped (Figure 1)

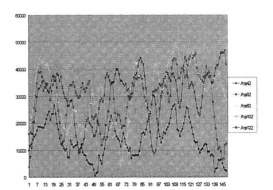

Figure 1 Seismic amplitude distribution of Seam 13-1

It is shown that the changes of the seismic amplitude are generally big, the absolute value of which is from 0 to 5000. Generally, the low position of the seismic amplitude is corresponded to the small-scale fault, the broken belt or the thickness changing belt of the coal seam. For the small-scale fault and the broken belt, the low value extent of the seismic amplitude is small, though the low value extent of the seismic amplitude is big at the thickness changing area. The abnormal amplitude area caused by the small-scale fault or the broken belt can be carried on space smooth.   Then, the smoothed comprehensive amplitude value can be used as a parameter to apply to scale the coal seam.

From the mutual interpreting result, the comprehensive amplitude value can be collected. Using the time data reflected from the coal seam and the depth-correct of drilling, the isogram map of the coal roof (or bottom) can be traced(Figure 2). The

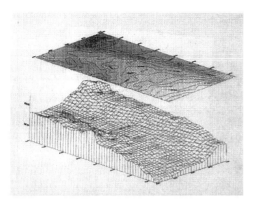

Figure 2 Isogram and 3D characteristics of the seam 13-1

small-scale fault can be distinguished from the map, because the time value is the maximum value of the reflected wave of the real coal seam and isn't the maximum value of the phase (Figure 3).

3. DISPLAY OF 3D SEISMIC DATA AND RESOLUTION FOR THE SMALL-SCALE FAULT

The seismic exploration of coalfields has been mainly used to solve the problems about the undulation and structure of coal seams. The appliance of high resolution 3D seismic exploration has reached a new level to solved this problem. The

high resolution 3D seismic exploration can identify the 5-m faults, 3-m broken points in vertical orientation , and has successfully identified two tunnels that has a distance of 50m in horizontal orientation in the Huainan coalfield. Coal-mining with high-production and high-efficiency requires higher demands for geological security system. It has become an important research objective for geophysicists to make good use of the 3D seismic exploration information and to improve the capability of 3D seismic exploration for solving the geological problems in mining. To do that, it is very important to change the display method of profile first, which had been used in the seismic exploration in coalfields for a long time. Although there have been some automatic identification methods for little structures of coal seams at present, the most reliable and immediate interpretation method for seismic information is manual interpretation.

The traditional seismic profile use the wave shape and variable area as the display mode, so it is difficult to show the large dynamic ranges received by the seismic wave at present. In order to fully use the seismic information to interpret, we use large scales and selective contour sections(Figure 4). The large scale display can fully show the identifiability of the stratum, while the selective contours can show the seismic information in the large dynamic ranges

In the past, the seismic profile only gave out the variations of the seismic waves, and the tracing

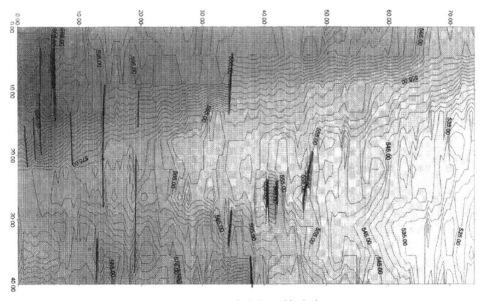

Figure 3 The small-scale fault showed in the isogram map

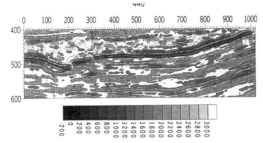

Figure 4 High resolution selective contours section line 62

methods of powerful phase-amplitude and wave groups are used to interpret the seismic profile. However, the change of small-scale faults and structures of coal seams is not enough to lead to obvious changes of the powerful phase-amplitude wave shapes. For the reason, it is very difficult to find out these little reflections by using the displays and the analysis of time section. Before the seismic waves is displayed it have been made some processing to strengthen the display of little wave shapes, especial the wave tails. Different parameters are selected in accordance with different aims. In order to improve the sensitivity for small-scale structures, the coal seams texture and the variation of coal thickness, the display mode of the instantaneous amplitude variation have been selected. The wave resistance and instantaneous frequent are used as parameters to interpret lithologic character. $\tilde{x}(t)$ can be obtained by Hilbert alternation from a real continuous seismic signal record $x(t)$. Also, an analytic signal can been formed as follows:

$$q(t) = x(t) + \tilde{x}(t)$$

Here, give a definition as follow:

$$e(t) = |q(t)| = \sqrt{x^2(t) + \tilde{x}^2(t)}$$

here, e(t) is the envelope of a real signal of $x(t)$;

$$\theta(t) = arctg \frac{\tilde{x}(t)}{x(t)}$$

$\theta i(t)$ is the instantaneous phase of the $x(t)$;

$$\mu(t) = \frac{d}{dt}\theta(t) = \frac{d}{dt} arctg \frac{\tilde{x}(t)}{x(t)}$$

$\mu(t)$ is the instantaneous frequency of the $x(t)$.

By comparing the processed 3D seismic datum with the site tunnel exploitation effects of a mine field in Huainan, we found that the section, which is got in the display mode of instantaneous amplitude variation, has a high sensitivity to the variation of the coal thickness, small-scale structures and coal seams texture. In a range of 2.5 km^2 3D seismic exploration, the seismic data in the four tunnels position had been processed, and 32 small-scale faults and 132 meters loosens coal bands are confirmed in total.

This study will provide a new way for high-resolution seismic exploration interpretation in coal fields and makes full use of the 3D information. Its application to mining will certainly produce a considerable economical benefit for the high-production and high-efficiency working faces.

REFERENCES

Ma Zhaitian, 1989, 3D seismic exploration methods, Petroleum Industry Press, Beijing.

Hao Jun et al., 1992, 3D seismic exploration technology, Petroleum Industry Press, Beijing.

Gao Luzeng, 1986, An attempt to use seismic dynamic information to find small-scale faults, *Petroleum Geophysical Exploration,* Vol. 24(4).

Mining Science and Technology'99, Xie & Golosinski (eds) © 1999 Balkema, Rotterdam, ISBN 90 5809 067 1

Scale effects in the destructive testing of borehole cores

R.Goodchild
Veritas DGC Limited, UK

L.Wade
Department of Mining and Mineral Engineering, University of Leeds, UK

ABSTRACT: This paper demonstrates that the uniaxial compressive strength (UCS) values of rock samples from two boreholes in Leicestershire, UK are subject to significant scale effect. The magnitude of these scale effects is shown to be lithologically dependent, as is the Representative Element Volume (REV) - the size of sample above which there will be no further scale effect. The results using the author's methodology are compared with published data and are found to be in close agreement. However, it is calculated that, for each of three lithotypes, the UCS of the bulk rock mass varies only between 73.7 and 76.6% of that of a 25mm diameter and 50mm long cylinder of the same material. It is therefore concluded that, for the purposes of a wider research programme studying correlations with geophysical values, no correction for this phenomenon is required as only 25mm diameter cylinders are being used in that work.

1 BACKGROUND

A research programme into correlations between the geophysical and geomechanical properties of rock has been conducted by the Department of Mining & Mineral Engineering of the University of Leeds. The geophysical data took the form of wireline logs from boreholes and thus represent the properties of the in-situ rock mass. The geomechanical properties, on the other hand, were obtained from laboratory tests on small samples taken from the cores recovered from those same boreholes. The two sets of data were, however, not directly comparable.

As the size of a sample increases, so does the likelihood that it will contain a plane or zone of weakness. This phenomenon, known as the scale effect, had to be accounted for in the research. The technique for doing so was that of Representative Elemental Volume (REV). This is defined as "the theoretical specimen size above which no significant changes in the geomechanical property of interest will occur" (Pinto da Cunha, 1990).

The literature clearly shows that the scale effect phenomenon exists but is lithologically dependent. The magnitude of the influence of specimen size has been shown to differ greatly from one rock type to the next for all geomechanical properties. The representative elemental volume is also shown to differ significantly between rock types. As a consequence of this the scale effect should be considered to be site specific and any corrections made for its influence should be conducted with this in mind. This paper therefore concentrates purely on uniaxial compressive strength (UCS) of samples of carboniferous period rocks, obtained from 2 boreholes within 2 kilometres of each other in Leicestershire, UK.

2 METHOD OF INVESTIGATION

Peng (1993) stated that there is no difference between the geomechanical properties of an underground coal pillar and those of a laboratory sized intact coal specimen, provided that the laboratory specimen is larger than the cleat spacing in the coal. Extending this theory to a larger scale, for the purposes of this project it was assumed that the REV for the rocks in this case will be of a size proportional to their mean joint spacing.

Based on this assumption, an average REV value was calculated for each of the 3 major rock types encountered in the boreholes (sandstone, siltstone & mudstone). The REV was calculated by assuming it to be in the form of a large cylindrical specimen of length to diameter ratio 2:1 (the same form as the intact test specimens), with its length equal to the joint spacing, whereby:

$$REV = \pi.r^2.J$$

where J = mean joint spacing and r = radius of specimen.

Because the length of the specimen is twice its diameter, the radius is one quarter of J. Hence:

$$REV = (\pi/16).J^3$$

The mean joint spacing (J) was calculated for each of the major rock categories by dividing the total length of core of each lithological type by the number of major fractures present in it. Fractures caused by removal of the rock from the wireline core barrel, or by damage during storage and transit, were disregarded and only those originally present in the rock mass were included. This distinction was aided by consideration of the unprocessed sonic channel in the geophysical data, where fractures and discontinuities in the borehole wall show up as spikes in the log.

The mean joint spacing, the calculated REV and the equivalent diameter of an REV-sized 2:1 right cylinder for each major rock type are summarised in Table 1.

Table 1: Mean REV dimensions.

	J (mm)	REV (m³)	Equivalent diameter (mm)
Sandstone	320	6.458 x10⁻³	160.2
Siltstone	273	4.013 x10⁻³	136.7
Mudstone	181	1.172 x10⁻³	90.7

Having calculated the REV for each rock type, UCS tests were conducted on 25mm, 54mm and 85mm core cylinders. Three samples of each lithotype were chosen for these tests and their geometry permitted each sample to provide two cylinders of 85mm diameter, three of 54mm and eight of 25mm. The mean results of these tests for each lithotype are summarised as Table 2.

Table 2: Mean UCS measurements

	UCS at 25mm (MPa)	UCS at 54mm (MPa)	UCS at 85mm (MPa)
Sandstone	40.03	38.11	32.64
Siltstone	48.90	45.29	39.09
Mudstone	56.55	55.28	42.25

As the equivalent diameters emanating from the REV analysis all exceeded 85mm it would have been better if some larger cylinders could have been tested, but the borehole core diameter of 85mm set an upper limit. Consequently a slight extrapolation had to be performed to predict the UCS at the REV.

This was achieved in a two stage process. Firstly, a curve was fitted to the regression of UCS against sample diameter for each lithotype. In all cases, logarithmic curves were found to produce the best fit, namely:

$$UCS_{sandstone} = 58.9168 - 5.6625\ln(d) \quad R = 0.91$$

$$UCS_{siltstone} = 74.1858 - 7.663\ln(d) \quad R = 0.96$$

$$UCS_{mudstone} = 92.6035 - 10.62\ln(d) \quad R = 0.83$$

The correlation coefficients (R) for sandstone and siltstone are significantly high, though the lower figure for mudstone indicates that there is more scatter of the data about the regression line.

The representative diameter of the REV-sized cylinder for each lithotype was then substituted back into the relevant equation. The predictions of UCS at the REV obtained are given in Table 3.

Table 3: Predicted UCS values at REV

Lithotype	Representative diameter (mm)	UCS at representative diameter (MPa)
Sandstone	160.2	30.17
Siltstone	136.7	36.50
Mudstone	90.7	44.73

3 COMPARISON WITH THE LITERATURE

Hoek & Brown (1980) collated data from the work of 8 different authors, covering a wide variety of rocks which had been tested for UCS using cylinders ranging in diameter from 10-200mm. They were able to derive the following empirical relationship between the UCS of a rock specimen of any diameter (d) and (UCS_{50}), the UCS of a 50mm diameter cylinder of the same rock.

$$UCS_d = UCS_{50}(50/d)^{0.18}$$

It can be assumed that the UCS test results from our 54mm cylinders would not be too different from those for 50mm cylinders. The data was therefore put into the Hoek & Brown equation to estimate UCS values for each lithotype at the REV. These are compared in Table 4 with the predictions derived from our own regression equations.

Table 4: Comparison of UCS at REV predicted by different methods

Lithotype	UCS at REV from Hoek & Brown	UCS at REV from current research
Sandstone	30.90	30.17
Siltstone	37.79	36.50
Mudstone	49.66	44.73

As can be seen, the UCS predictions at REV derived using the two approaches are very similar indeed, which gave the researchers confidence to proceed with the analysis.

4 ANALYSIS OF RESULTS

Having deduced a theoretical UCS value for each lithotype at its REV, and having shown these values to accord well with the literature, the next step was to calculate scaling factors between intact sample UCS values and those at the REV. When undertaking this analysis, it was realised that the raw data values for intact cylinders summarised earlier all exhibited minor deviations from the regression lines. It was therefore necessary to undertake this analysis using UCS values predicted by the regressions for 25mm cylinders. Then it became simply a matter of dividing the curve-fitted UCS at 25mm for each lithotype by the respective UCS predictions at REV. The scaling factors derived are shown in Table 5.

Table 5: Scale factors for down-rating UCS of 25mm cylinders to rock mass values

Lithotype	Fitted UCS at 25mm diameter (MPa)	Fitted UCS at representative diameter (MPa)	Scale factor
Sandstone	40.69	30.17	0.741
Siltstone	49.52	36.50	0.737
Mudstone	58.42	44.73	0.766

The scale correction factors calculated for sandstone, siltstone and mudstone are seen to be virtually identical. The range exhibited is less than 4%; a discrepancy which is considered to be within the accuracy of testing. This shows that, despite the fact that the magnitude of the scale effect differs between the rock types, the effect of down-rating from REV to 25mm diameter core specimens is virtually uniform. On average, an intact 25mm diameter specimen of any of the 3 rock types would be expected to exhibit a UCS value approximately a third larger than would a bulk specimen. For the main research purpose of correlating the geomechanical and geophysical properties of rocks it can be concluded that, as long as all of the UCS testing is carried out on 25mm diameter cylinders, no size correction is necessary for the rocks encountered in the Leicestershire boreholes. This would not be the case though if a mixture of cylinder sizes or sample origins had been used.

5 SUMMARY

The results presented in this paper show that the uniaxial compressive strength values of the rocks from two Leicestershire boreholes are subject to a significant scale effect. Analysis of the nature of this scale effect however shows that, provided the results from 25mm diameter core specimens are used exclusively for the purposes of correlation with geophysical values, no correction for this phenomenon is required.

It is shown that the scale effect is influenced by lithology, with sandstone and siltstone values being affected in a similar manner, but with mudstone samples showing a slightly smaller rate of strength reduction with size.

The diameter and quantity of the borehole core available from the boreholes limited the size and number of test specimens used for this analysis. The amount of homogenous rock available was restricted by the need to use the available borehole core for other testing purposes. It would have been preferable to conduct this investigation with a larger number of test specimens with a larger range of diameters. This may well have increased the accuracy of the procedure and would also have made it possible to establish the magnitude of the REV's through experimental methods, rather than having to rely upon predictions based on fracture spacings.

REFERENCES

Hoek, E. & Brown E.T 1980. *Underground excavations in rock*. London: IMM.

Peng, S.S. 1993 Strength of laboratory sized coal specimens vs. underground coal pillars. *Mining Engineering*: Feb. 157-158.

Pinto da Cunha, A. 1990. Scale effects in rock mechanics. *Proc. 1st Int. Workshop on scale effects in rock mechanics.* Rotterdam: Balkema. 3-27.

Mining Science and Technology' 99, Xie & Golosinski (eds) © 1999 Balkema, Rotterdam, ISBN 90 5809 067 1

Theory and methodology of a new, GIS based coal mine information system

Shanjun Mao

Institute of Remote Sensing and GIS of Peking University, Beijing, People's Republic of China

ABSTRACT: The limitations of present coal mine management & information system, and analysis of advantages and disadvantages of the GIS lead to development of a new system. The proposed system integrates structuralized triangulated irregular network (TIN) and GIS as well as a network model. Through establishing topologic structure relative to point, line and area, we have set up an algorithm that allows to dynamically modify a coal mine TIN which involves reverse faults. The related software on unified systems of TIN and GIS has been developed using Visual C^{++} language in Windows environment.

1. COAL MINE INFORMATION MANAGEMENT

The coal mine deposit is a three dimensional geologic entity. All the processes of coal mine production need to deal with three dimensional space, and a variety of maps and technical data are used in resource exploration to mine production in order to determine the correct spatial location of mine and corresponding parameters. Obviously, a coal mine information system should be a spatial information system.

The spatial information system (SIS) is also called the geographic information system (GIS). It is a computer system of collecting, storing, processing, analysing, converting and outputing spatial data, and is also an outcome of interdisciplinary development of computer science, technology of remote sensing and aerial survey, computer graphics, computational mathematics, graph theory, geography, and geology .

With increasing development of computer and communication technologies, the study and application of GIS have prevailed many fields of natural science and applied technology, such as geography, geology, environmental monitoring, land utilization, city planning, and traffic safety. Presently, GIS has receive more and more attentions from governmental authorities and industries. GIS is different from CAD in that the latter is used for displaying, processing and plotting spatial entities rather than for processing non-graphic information, and it is not concerned with the topological relationships of graphic units (such as point, line, area) when storing

and processing graphics. Unlike the CAD, the GIS can be used not only to display and plot various engineering maps, but also to provide decision-making services.

The process of coal mine production is completely different from geographic and geological research, environmental monitoring, city planning conducted on the earth surface. Firstly, its work space is a three dimensional geologic entity varying from gray to white, where lots of graphics need dynamic modification. Secondly, it has its own unique graphic generation and data processing methods. Finally, mining geological conditions have restricted mining methods into two categories, namely, underground mining and open pit mining. In China, the underground mining method is most commonly used, which is more complicated than the open pit mining, so corresponding MIS should process not only the entities of coal seams, but also the snarled roadway system. In order to improve the automation level of processing mine graphics and other information of the MIS to maximum extent, special data model and data structure must be designed. The current GIS is hard to be adequately applied in coal mines because the system design of current GIS does not take coal mines into account, especially those underground coal mine graphic entities such as point, line, area and their correlations. Therefore, it is indispensable to conduct research on coal mine GIS, to put forward the data model and data processing method of coal mine GIS, and to design specialized coal mine MIS.

2. DATA MODEL OF COAL MINE SPATIAL MANAGEMENT INFORMATION SYSTEM

2.1. *Unified dynamic data model of all essential factor's structuralized TIN model and GIS*

Two conceptions need to be introduced before discussing the model.

1) "All essential factor" means all parameters relative to a single triangle, which includes the number of the triangle itself, three end members and their numbers, three edges and the number of adjacent triangle, and the identifier of relationships between point, line, area (the area contained in the triangle) and objects of point, line, and area in GIS.

2) Two implications of the structuralization: 1) Establishing the topological data structure of triangle in order to express the correlation between entities of points, lines, and areas (the area contained in the triangle) in TIN; and 2) Setting record order to hide the relationships among these point, line and triangle unit, in order to reach the aims of saving storage space.

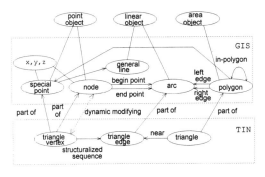

Fig.1 The unified data model of TIN and GIS.

Fig.1 gives the all essential factor's structuralized unified data model of TIN and GIS. From the graph, we can see that the model abstracts spatial data type into point, linear, and area objects. These three classes of graphic objects are expressed with special points, nodes, general line segments, arcs and polygons. It is worthy to note that, in order to express the measured strata-controlling points such as boreholes and underground coal discovery points, we add in an original data type "special points". "Special points" possess not only spatial geometric locations and topological relationships, but also more information on attributes. In addition, "special points" can also be used to express those completely isolated points such as end points of the sign separators. "General line segments" refer to those linear graphic objects such as exploration lines, separators among signs.

There are relationships of "part of" or inclusion between TIN and GIS, by which the unified data model is essentially achieved. Because the TIN is based on the vector graphs of the GIS, the spatial area contained in a single triangle and its points and lines can be a part of those point, linear, area objects in the GIS. The unification of the TIN and the GIS can be achieved in the data structure of the TIN and the GIS, provided the relevant identifier of point and vertex, line and edge, area and triangle are identical. In Fig.1, the dashed line between the "triangle vertex" of the TIN and the "nodes" of the GIS represents the dynamic modification relationships in the unified data model. The change of the spatial location and relationships of the nodes in the GIS will lead to the variation of triangle vertices, thus generating a new triangle network. In the same way, the variation of the triangle vertices will bring on the changes of locations of part nodes in the plane graphics, with which part of the regenerated linear, area graphics object are changed in this way.

2.2. *Network model*

Fig.2b shows the prototype of the coal mine GIS network model. In this model, nodes which are in contact each other, such as different development heading, preparatory roadway or extraction opening in Fig 2a), will be connected each other with pointers. The main features of the coal mine GIS network model are as follows:

1) each child node can have several parent nodes;

2) child nodes of the same parent node can be inter-related.

The network data model can be used to solve several problems as follows:

1) Because the spatial distribution and relationships of roadways are analogous to the graphics-expressing mode of the network model in the database (Fig 2a), it will be ideal to express the practical spatial distribution status and relationships with the aid of pointer.

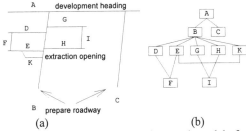

(a)   (b)

Fig.2 Spatial distribution and network model of roadway

2) Because of the existence of the correlation among roadway graphics entities, when generating coal mining engineering plane map automatically, it will be quite easy to solve problems like intersection of space and plane of roadways and non-linear variation of the coal mine roadway's width in different scale;

3) Dynamic generation of roadway paths such as those of ventilation and transportation lay a firm technical foundation for visualization of dispatcher management. Because in the data structure which is based on the network model, pointers are used to express the network relationships of the graphics entities, we can get network roadway paths in relation to practical application quickly and in realtime, through operating pointers.

3. BRIEF INTRODUCTION TO MAJOR ALGORITHM

3.1. *Surface spline function*

Surface spline function is an effective method to process geologic surface fitting and interpolation. The main merits of this function are as follows:

1) Interpolation result are all located on the same smooth surface;

2) Interpolation result of known data points are identical to original data;

3) When fitting surface, original data points (such as boreholes) need not to be arranged regularly, and natural border conditions without differential coefficient information of borders will be adopted.

3.2. *Automatic generation and dynamic modification of the TIN model*

Because geological data of coal mines have a very complicated spatial pattern (particularly the existence of reverse faults), various present triangulation algorithms according to the Delaunay triangulation rules are not suitable for the construction of the coal mine TIN model.

Through studying original data and correlation among various map border entities, it has been found that, no matter how complicated are data points, attributes, border shapes of the map area, we can solve various problems encountered during the process of constructing complicated geological TIN models, as long as we establish all kinds of topological relationships among original points, lines, areas and utilize the discrimnant model of positive and negative area related to original map borders.

4. BRIEF INTRODUCTION TO FUNCTION ABOUT COAL MINE SPATIAL MANAGEMENT INFORMATION SYSTEM

According to the data model of coal mine spatial information system and relevant algorithm, under the Microsoft Windows, we have developed 2D-MGIS (two dimensional coal mine spatial management information system) successfully by using the Visual C++, Powerbuilder, and SQL Server. This system has been applied successfully in some modernized coal mines in China such as the Nantun and Dongtan coal mines of the Yanzhou Coal mine Administration, the Bayi and Fucun coal mines of the Zaozhuang Coal mine Administration, the Xiqu coal mine of the Xishan Coal mine Administration, the DongSheng coal mine of the Huaneng refined coal Inc. It has promoted the modern management of coal mines to a higher level.

Based on the unified data structure (including graphics and attribute database), the 2D-MGIS is further divided into following subsystems:

1) plane graphic subsystem;
2) section graphic subsystem;
3) coal mining engineering plane map subsystem;
4) heading stope graphic subsystem;
5) dispatcher management information subsystem;
6) mining plan and arrangement management information system;
7) ventilation management information system;
8) legend library system.

5. CONCLUSIONS

1) The unified data structure of the TIN and GIS have been established, which not only benefits the inquiring of coal mine spatial objects, the analyzing of the spatial variables, and dynamic modification of graphics, but also provides the powerful theory and method to the automatic generating of the work face graphics and the predicting of section graphics in any directions.

2) The network model and its data structure have been developed, which lay a theoretical foundation for generation of mining engineering map as well as dynamic and relative inquiry of the ventilation and transport information.

3) The automatic generation algorithm of the TIN based on the topological relationships among points, lines, areas and the surface spline function algorithm provide a technical guarantee for processing arbitrary complicated geological entities.

4) Successful application of 2D-MGIS in coal mine adequately demonstrates the rationality and

correctness of the proposed data model and relevant algorithms.

REFERENCES

Anderson, D. 1996. Mine mapping, modeling and visualization systems. *E/Mj*, 197(7).

Darmody, R. G. 1995. Modeling agricultural impacts of longwall mine aunsidemce: a GIS approach. *International Journal of Surface Mining, Reclamation and Environment* 9(2): 63-68.

Kufoniyi, O. & T. Bouloucos 1994. Flexible intergration of terrain objects and DTM in vectos GGIS. *Proceedings of Integration, Automation and Intelligence in Photogrammetry. Remote Sensing and GGIS.*

Agatha, Y., T. Teresa, M. Adams & E. Lynn Usery 1996. A spatial data model design for feature-based geographical information systems. *INT. J. Geographical Information Systems* 10(5): 643-539.

Goodchild, M. 1992. Geographical information science. *INT. J. Geographical Information Systems* 6(1): 31-45

Mao, S. & Y. Xu 1996. The theory and technical way of achieving dynamical management of coal mines informations. *Proceeding of 3rd international Symposium on Mining Technology and Science*: Xuzhou, China.

Mining Science and Technology'99, Xie & Golosinski (eds) © 1999 Balkema, Rotterdam, ISBN 90 5809 067 1

The changes of structures and macerals of hard lignite during non-binder briquetting

J. Wang
China University of Mining and Technology, Xuzhou, People's Republic of China

ABSTRACT: In this paper a new technique is developed to convert hard lignite into briquets. Coal samples with Late Jurassic age were collected from the Zhalanoer mine. In a specially manufactured briquetting machine, the low quality lignite has been converted into a clean fuel with high strength and low moisture content, and the heating values increased by 1000 Kcal/kg. Huminites of briquets become soft and deformed during briquetting process. Importantly, the maceral groups gave rise to strong fluorescence during heating under certain pressure. Many new macerals were also produced, such as bituminite, terpenite and fluorinite, all of which may act as a part of binders. It is i implied in this paper that the mechanics of briquetting follows the rules of creep mechanics rather than that of the Hooke's law.

1. INTRODUCTION

China has abundant lignite resources, with a reserve up to 120 billion ton, and these lignites are mainly distributed in the Inner Mongolia of North China, and the Northeast and Southwest of China. Many of them can be worked by open pit mining.

It is known that lignite is a series of coals with different maturities, ranging from peat to bituminous coal. In China, according to maturity, moisture content and geological age, lignite is divided into soft lignite (young lignite) and hard lignite (old lignite). Lignite generally contains considerable amounts of moisture and oxygen functional groups, but is low in caloric values, thus it is a poor quality of energy resource. Therefore, it is necessary to develop an efficient technique to convert poor lignite to a new fuel. The study on converting lignite into coking coal has been undertaken in our lab since several years ago. This present study is mainly concerned with converting hard lignite into the product with high strength, high caloric value and low moisture content by briquetting.

In spite of various upgrade and utilization techniques of lignite, there are still two problems to be solved: 1) how to develop a new converting method which is of low cost, but high efficiency; and 2) how to improve the properties of coal in order to suit the converting process.

In general, there are two different techniques in the briquetting of coal, one with binders and another without binders (non-binders). The briquetting process of soft lignite (brown coal) has already been developed and has been used in large scale commercial production in the early of this century. However, non-binder briquetting of the hard lignites is still a problem in the world. In this study, samples of hard lignite are collected from mines of the Zalainuoer, Longkou and Kalimantan coalfields. These samples can be converted into high strength bricks, and the quality can be upgraded into level of bituminous coal in properties. After briquetting, the macerals and structures of coal can be changed and some new constituents can be formed. That is why the poor lignite becomes a new fuel with high quality.

The present paper will focus on the study of changes in structures and maceral compositions of lignite after process of non-binder briquetting.

2. EXPERIMENTAL PROCESS.

2.1. *Briquetting procedure*

The samples are collected mainly from the Zhalanoer coal mine, late Jurassic Hailear Basin. Some of them are obtained from the Longkou and Kalimamtan coal mines, with Tertiary age.

The briquetting process is carried out in a specially manufactured automated press, 50 g of each natural coal sample is loaded in mould, with grain sizes ranging from 0.5 to 3.0mm, and then heated at temperatures from 100 to 350°C, under pressures from 800 to 2000 kg/cm^2.

2.2 Briquetting technological process

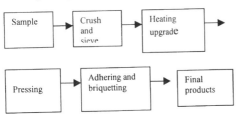

According to the highly gelification degree, porous and creep characteristics of hard lignite, the samples loaded in mould are heated and pressed simultaneously. It is very important to control the temperature and pressure properly to produce satisfied changes of maceral of coal.

3. RESULTS AND DISCUSSIONS

3.1. Evaluations and tests for briquet quality

Proximate and ultimate analyses of briquets are shown in Table 1. The chemical properties of the lignite from the Zhalanoer are as follows: M_{ad} 22.2%, A_{ad} 6.34%, V_{daf} 46.2%, CV_{ad} 21.63mj/kg. Compressive strength is shown in Table 2. The tests are carried out in a hydraulic press.

Table 1.   Proximate and ultimate analyses of briquetted products

No.of briquet	Mad (%)	Aad (%)	Vad (%)	CVad (Mj/kg)	Cad (%)	Had (%)
Z4-5	5.20	4.80	37.2	25.94	72.75	4.50
B131	5.70	7.60	34.3	26.00	75.65	4.70
C13	5.70	3.46	38.1	28.90	68.70	3.76

To date, there are no uniform standards to evaluate the quality of briquets with normal testing methods. In this paper methods of chemistry, physics, mechanics and coal petrology are used integratedly.

Incremental Drop Test (IDT) is also conducted . In average, the IDT resluts of briquets of the Zhalanoer mine are as follows: IDT ranging from 2m + 96% > 13mm to 1.25m + 100% > 13mm.

Water Resistance Tests (WRT) are as follows :
   Z13-2: WRT=99.70%

Z4-6: WRT=99.80%

Table 2.   Compressive strength*

No.of Briquets	Size (mm)	P (Mpa)	S (kg/cm^2)
Z1-2	1	11.5	149
Z4-3	1	11.0	141
B92	3	9.5	122
B102	3	8.0	104
ZT2-2	3	12.5	159
ZT3-1	3	18.0	231
C1-1	2	6.5	85
C1-2	2	8.0	103

* Samples lableled Z are briquets converted from lignite of the Zhalanoer mine, B and C are from the Longkou, Kalimantan coalfields, respectively.

The micropore texture has been tested with the Micromeritics Poresizer. From Fig.1 & Fig.2, it can be seen that micro and intermediate pores are less developed than the large pores for the unconverted lignites. But in the briquets converted from hard lignites, micro and intermediate pores become more developed than the large pores, as shown in Fig.3 & Fig.4.

3.2. Influences on briquet quality

3.2.1. Changes on the property of coal

Lignite, especially hard lignite can not be soften and deformed during heating. We know that the different lignitea have different briquetibility. The briquetibility influences not only the briquetting technique and process, but also the quality of the product. It becomes the key problem to recognize and control the property of coal to be able to fit the non-binder briquetting process. In this study huminites of briquets have become soften and deformed. There are many new fluoresent macerals appeared, such as fluorinites, resinites and bituminites etc. It is clear that many high fluoresent macerals act as binders in the briquetting process.

3.2.2. Critical temperature

When lignite samples are heated and pressed in the briquetting machine, they don't fit the rule: the higher the temperature, the stronger the strength of the briquet. In certain range of pressure when temperature is raised continuously, the strength of the brequet does not increase. There is a critical temperature existed, which may represent the activate energy. There is a close relationship between heating time and temperature. It is shown in Fig.5 that the temperature only rise a little during a long period of heating. In order to obtain the high strength briquets, it is necessary to raise heating temperature to exceed the critical temperature.

3.2.3. *Rheologic mechanics*

In this paper we suggests a new idea that the mechanical property of hard lignite doesn't obey Hooke's law, but follows the creep mechanics rule of rheological system. This rule controls all briquetting techniques and pressing processes. The core of rheologic mechanics is that everything can flow. In view of the briquetting process, the property of briquets proved to follow creep mechanics. Any sustaining parameters of creep stage with its action determined by the property of coal, load bearing and heating temperature. The creep process of coal may be decelerated or accelerated.

Total deformation = instantaneous deformation + time deformation + unloading deformation. This is shown in Fig.6. There are three stages when describing variation of deformation with time, that is, attenuation deformation stage, stable flowing deformation stage and sharp deformation stage. To obtain a high strength briquet, the second stage needs to be longer so that the broken time of third stage can be reduced. If it is over loaded, the lasting time of the second stage will become shorter, as a result, the broken of third stage appeared more quickly.

4. CONCLUSIONS

1) The testing data and spectra of briquetted lignites have correctly indicated the changes of parameters of lignite. This is why the hard lignites can be converted into high strength and higher caloric value briquet without adding any binders.

2) The briquetting technological and pressing processes of lignites follow the rule of creep mechanics rather than elastic mechanics.

REFERENCES

Crockett, L. A. 1973. Briquetting of brown coal in Victoria, Australia. *Proc. Inst. for Briquetting and Agglomeration 13th Biennial Conf.*: 237-250.

Elliott, M. A. 1991. Chemistry of Coal Utilization.

Gains, A. F. et al. 1976. Pyrolysis of lignite. *Fuel 52* : 129-137.

Gossens, W. et al. 1972. Hot briquetting of bituminous coal by the ancit process and test in the blast furnaces. *Stahl end Eison*: 1039-1044.

Gowrisan K. S. et al. 1990. Lignite utilization. *Fuel, Science & Tech.* 9(2-4): 75-84.

Lin, Q. & J. M. Guet 1990. Characterization of coals and macerals by X-ray diffraction. *Fuel 9*: 821-825.

Waldemar, I. F. & J. M. Randall 1988. Mercury porosimetry of coals. *Fuel 67*: 1516-1520.

Mining Science and Technology'99, Xie & Golosinski (eds) © 1999 Balkema, Rotterdam, ISBN 90 5809 067 1

The lithofacies-palaeogeography of the Huabei huge coal basin

Huanjie Liu, Yinghai Guo & Shuxun Sang
China University of Mining and Technology, Xuzhou, People's Republic of China

ABSTRACT: The characteristics of the epeiric sea, tidal deposits, event deposits, platform and barrier island systems and their composite systems of the coal-bearing sequences are discussed in this paper. There are two coal-forming environments namely peat swamps and peat flats. Sea level changes and sequence stratigraphy are also discussed. All these are essential to understand the Late Palaeozoic Huabei huge coal basin.

1. INTRODUCTION

The Huabei coal basin is located in the north of China and is a huge Late Palaeozoic cratonic basin. The ages of coal-bearing strata are Carboniferous and Permian, and are currently distributed in areas up to $80 \times 10^4 \ km^2$ (Shang 1997) where abundant coal and coalbed methane resources are preserved. It is an important base for coal and future coalbed methane industry in China.

As the mining intensity, machanization level and environmental protection requirement are being raised in the highly industrialized Huabei region, higher requirements have been put forward to geologists, that is, increasing the energy resources reserve and accurately finding out the accumulating regularities of these resources. The lithofacies palaeogeography researches are the foundations for re-understanding the Late Palaeozoic Huabei huge coal basin, particularly for exploring the new coal-accumulating regularities and coalbed methane-enriching regularities and enlarging the new resources of coal and coalbed methane. Great progresses have been made in the researches on the lithofacies palaeogeography in the Huabei huge coal basin in recent years.

2. PROGRESS

2.1. *Tidal deposits and characters of the epeiric sea origin of coal-bearing sequences*

Analyses of lithology, lithofacies and depositional models show that the coal-bearing sequences in the Huabei plate were formed in epeiric environments A lot of tidal deposits are recognized, which constitute important parts of the coal-bearing sequences. The barrier coastal zones of the epeiric sea were the main palaeogeographical setting for forming these Late Palaeozoic coal-bearing sequences and the tidal currents provided important hydrodynamic controls to the coal basin. The surrounding rocks of the coal seams were usually deposits of the mud flat, sand flat or mixed flat environments. The common sand bodies of the coal-bearing sequences were usually formed in tidal influenced environments such as tidal delta, tidal bars, tidal channels and barrier islands. Peat flats were the important coal-forming environments.

2.2. *Platform, barrier island systems and their composite systems*

The platform systems of the coal-bearing sequences are indicated by the coal-bearing carbonate deposits formed in the carbonate platform environments. The barrier island systems of the coal-bearing sequences are indicated by the coal-bearing clastic deposits which formed in the barrier island –lagoon environments. The platform-barrier island composite systems of the coal-bearing sequences are shown by the coal-bearing interbedded carbonate and clastic deposits which were formed in the platform-barrier island composite conditions (Fig. 1).

The main deposits of the delta and fluvial as well as alluvial systems are distributed in the northern and marginal parts of the Huabei huge coal basin. But in the central and southern parts of the basin there are the deposits of platform, barrier island systems, platform-barrier island as well as barrier island-delta composite systems. The palaeogeographic characters of the epeiric sea are shown by the development of the tidal deposits and various types of the coastal sand bodies. The coal-forming environments and the distribution regularities of coal seams and sand bodies in the Benxi, Taiyuan and part of Shanxi formations are controlled by the platform, barrier island systems and their composite systems in the central and southern parts of the Huabei huge coal basin.

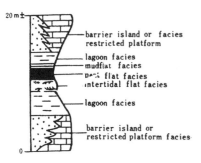

Figure 1. Vertical sequence of platform and barrier island composite system.

2.3. *Two different types of coal-forming environments*

There are two different types of coal-forming environments in the Late Paleazoic Huabei huge coal basin, each of which has its own distinct features of coal distribution and quality. One is peat swamps in continental facies fluvial, lacustrine and delta systems. Another is peat flats in marine facies and distributed at the barrier island, carbonate platform systems and their composite systems as well as part of the delta plain in the delta system.

Peat flats are the coal-forming environments of the tidal flat origin. Mangrove-like tidal plants could grow abundantly in the tropical and subtropical intertidal and supertidal flats as well as parts of shallow water subtidal zones and tidal channels. These plants could form peat deposits over a large area under suitable conditions. Such a direct coal-forming environment in the tidal flats is called "peat flats" (Liu, 1988). Peat flats are different from swamp environments of continental facies. The tidal characters of the peat flats include the features of their natural landscapes, their hydrodynamic, physical and chemical conditions, features of coal-forming plants and organic fossils and their environmental combinations.

The common types of peat flats include the lagoon peat flats, back barrier peat flats, tidal delta peat flats, platform tidal flats, estuary tidal flats, and lower delta plain tidal flats. The coal-accumulating features in peat flats show that the coal seams are distributed extensively and can be traced in a vast

area, but they are variable in thickness and splited frequently, with the relatively high sulphur contents.

Due to spatio-temporal evolution of depositional environments, peat flats and peat swamps may alternate vertically and laterally. The relative importance between the peat swamps and peat flats will depend on their lithofacies palaeogeographical locations and the influences of the palaeogeographical evolution due to the sea level changes. During sea level fluctuations, peat flats and peat swamps will have spatio-temporal alternations, and consequently coal seams were formed over a large area in the epeiric sea setting of the Huabei huge coal basin .

2.4 *Event deposits in coal-bearing sequences*

Storm and volcanism as the main geological events have been recorded in the Late Palaeozoic coal-bearing sequences. The Late Palaeozoic Huabei huge coal basin was formed in epeiric sea setting with the palaeolatitude of the equator hurricane zone. The palaeogeographic and palaeoclimatic conditions favored the development of many horizons of storm event deposits including carbonate and clastic tempestites. The main genetic types include the back flow tempestites, the stirring carbonate tempestites and the proximal sandstone tempestites. The back flow tempestites are further divided into distal, proximal and proximal shallow water carbonate tempestites (Fig. 2). A close relationship exists between the storm event and coal accumulation. The roof rocks of the coal seams are usually tempestites and the coal seams were eroded by storm flows and then became thinner, even pinch-out.

Several layers of pyroclastic rocks have been found in coal-bearing sequences which recorded the multiple volcanic events. Subaqueous pyroclastic gravity flow and a large number of whole body brachiopod fossils have been found in the coal-bearing Taiyuan formation of the Jungar coal field, Inner Mongolia. This represents the catastrophic events induced by volcanism.

Event deposits and their distribution regularities are the important signs for reconstructing the lithofacies palaeogeography of the coal basin and also constitute the important bases for subdivision and sub-correlation of the strata and coal seams.

2.5. *Sequence stratigraphic models and basin evolution*

The sequence stratigraphic frameworks have been reconstructed for the Late Palaeozoic Huabei huge coal basin, which include totally 6 sequence boundaries, 5 third-order sequences and 26 parasequences

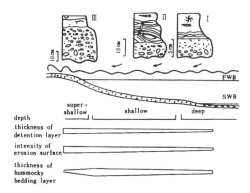

Figure 2. The depositional model of backflow-type tempestite.

(Fig. 3). The parasequences can be grouped into 4 types: bottom type, platform-barrier island composite type, clastic coastal type, and fluvial-lacustrine type. Three sequence stratigraphic models are established in Late Palaeozoic Huabei huge coal basin, including the epeiric sea type, transitional type and lake type. One second-order sea level cycle was formed from the Late Carboniferous (C_2^{1-1}) to the Late Permian (P_3^2), with a period of 80 Ma. There are 4 third-order sea level cycles and 26 fourth-order sea level cycles and the period of each fourth-order sea level cycle is 3.1 Ma (Fig. 4).

The processes of development and evolution of the coal basin can be divided into five phases, that is, basin formation phase, basin development phase, basin zenith phase, basin withering phase and basin transform phase. The coal seams were formed at the turning stage from fall to rise of fourth-order sea level and were developed at the top of the parasequence. The thick coal seams with extensive distribution were developed mainly at the top of the high systems tract in third-order sequences. The coal accumulation in the whole basin was controlled by the fourth-order sea level changes coupled with third-order sea level change, rather than only by fourth-order sea level changes. The coal accumulation centers were located around the shoreline and migrated with the sea level changes.

3. SIGNIFICANCE

1. The epeiric sea characters of the coal basin have been proposed, and the five phases of basin evolution, including basin formation, basin development, basin zenith, basin withering and basin transformation, have been divided. All these can explain the extensive distribution of the coal-bearing sequences and the coal seams in the Late Palaeozoic Huabei huge coal basin.

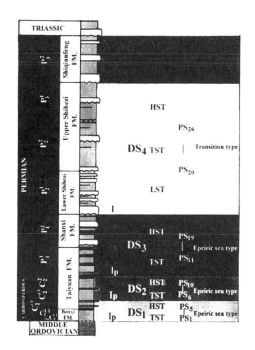

Figure 3. Sequence stratigraphic framework of the Permo-Carboniferous in Huabei coal-bearing basin.

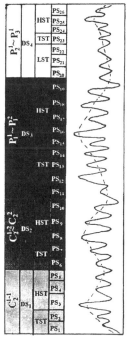

Figure 4. Sea-level change types and cycles.

2. Two coal-forming environments, peat swamps and peat flats, have been discussed and the development of tidal deposits, platform, barrier island systems and their composite systems as well as the spatio-temporal combination of the delta, fluvial, lacustrine and alluvial systems have been outlined. All of these have shown diversities and regularities of both coal-forming environments and coal qualities for the Late Palaeozoic Huabei huge coal basin.

3. Regional subdivision and correlation of the strata and coal seams are well supported by the regularity of sea level changes, development of sequence stratigraphic framework, storm and volcanic event deposits in the Late Palaeozoic Huabei huge coal basin.

4. The progresses on lithofacies palaeogeographic researches have established the basis for us to re-understand the Late Palaeozoic Huabei huge coal basin, to explore the coal-accumulating regularities and coalbed methane-enriching regularities and to enlarge the new resources of coal and coalbed methane.

4.CONCLUSIONS

1. The characters of the epeiric sea, tidal deposits, event deposits, platform, barrier island systems and their composite systems are the important features of the Late Palaeozoic Huabei huge coal basin.

2. There were two different types of the coal-forming environments, peat swamps and peat flats which could alternate and transform each other in space and time. As a consequence, coal seams were formed over a large area of the Late Palaeozoic Huabei huge coal basin.

3. The regularity of sea level changes and the development of sequence stratigraphic framework can be used to study the regional coal-accumulation and correlation.

REFERENCES

Liu, H. et al. 1987. The features of the barrier island systems of the epeiric sea and their event deposits of coal-bearing formations in Carboniferous of North China. *Acta Sedimentologica Sinica* 5(3): 73-80.

Chen, Z. et al. 1993. *The depositional environments and coal-accumulating regularities of the Late Palaeozoic coal-bearing measures in North China.* Press of China University of Geosciences.

Liu, H. 1988. Tidal deposits and coal-bearing formations. *Acta Sedimentologica Sinica* 6(2): 42-49.

Shang, G. 1977. *Late Palaeozoic coal geology of North China platform.* Shanxi Science and Technology Press.

Liu, H. et al. 1992. Progress in coal-forming theory of marine facies. *Acta Sedimentologica Sinica* 10(3): 47-56.

Cheng, B. 1992. *Late Palaeozoic sedimentary environments and coal accumulation.* Shanxi Science and Technology Press.

Liu, H. et al. 1997. *Comparative sedimentary research on coal-forming environments-Mangrove tidal flats and mangrove peats in the Hainan island at the South China Sea.* China University of Mining and Technology Press.

Mining Science and Technology' 99, Xie & Golosinski (eds) © 1999 Balkema, Rotterdam, ISBN 90 5809 067 1

Environmental-geological impact of karst coal mines in North China

Hanhu Liu
China University of Mining and Technology, Xuzhou, People's Republic of China

Yinchao Yang & Jingbo Guo
Xing Tai Mining Corporation, People's Republic of China

ABSTRACT: The age of coal series in North China is Permo-Carboniferous. Several thin limestone beds are intercalated in the coal bearing strata, below which is thick Ordovician limestones. So the hydrogeological conditions are complicated. In Karst coal mining areas, a series of environmental geological problems have arised during mining, such as Karst water inrush, rocks and soils brought from water inrush, contradiction of drainage and supply, deterioration of water quality, surface subsidence and surface collapse etc. In this paper above problems are addressed and corresponding countermeasures are proposed.

1. INTRODUCTION

The age of coal-bearing strata in North China is Permo-Carboniferous. There are several thin limestones in the coal measures, below which are thick Ordovician limestones. These features have been met in about 30 coalfields in the Hebei, Henan, Shandong, Jiangsu, Anhui, Shanxi, Liaoning and Jilin Provinces, where hydrogeological conditions are complicated. In the process of coal production in Karst areas, a series of environmental geological problems have occurred and presently the situation is getting more serious. Therefore, we should pay more attention to Karst limestones in the coal mine area.

2. ENVIRONMENTAL GEOLOGICAL PROBLEMS

2.1. *Karst water inrush*

Karst water often rushes into coal mines, with water yields up to several times to hundreds of times more than normal drainage capacity. China has the most serious water hazard of mining in the world. According to the statistics (Wang & Liu, 1993), there had been 218 cases of submergence and 769 cases of water inrush between 1955 and 1985. In the latest decade, the influence of the Ordovician Karst water

on Permo-carboniferous coal mines in North China becomes more serious. Several huge Karst water inrushes, with the yield more than 120 cubic meter per minute, occured and resulted in a great deal of economic loss. For example, in 1984, the Ordovician water rushed into the No. 2171 face of the Fangez-huang Coal Mine through the Karst collapse columns. The maximum yield is 2053 cubic meter per minute. The economic loss reached several hundred millions Yuan. In 1993, the Ordovician water also rushed into the Guojiazhuang Coal Mine with maximum yield of 549 cubic meter per minute. In 1997, the Ordovician water flooded to the Zhangji Coal Mine with maximum yield of 400 cubic meter per minute and caused a loss of 3 hundred million Yuan.

2.2. *Rocks and soils brought from water inrush*

In general, more or less rocks and soils can be brought out from Karst water inrush. For example, in the Doulishan Coal Mine and the Meitanba Coal Mine of Hunan province, when tunnels were being excavated through the Maokou limestone, 55 ton muds were brought out, 1 ton of rocks was pushed 10 meters away. And some workers were injured. In 1993, the Ordovician water rushed into the Guojiaz-huang Coal Mine and the mine was submerged for 6 hours. After the mine was recovered by grouting, 16

518 cubic meters of rocks and soils were left in the mine. According to author's on-site investigation (Liu, 1997), rocks and soils are mainly coal, siltstone, limestone and loose sands. The components are given in Table1.

These outbust rocks and soils have a very bad influence on the production recovery and environment.

Table 1. Constituents of outburst materials in 200m of the north roadway in the Guojia Zhuang Coal Mine in 1993

Type of materials	Volume (m3)	Percentage(%)	Size (mm)
Coal	6500	39. 4	0. 1~810
Siltstone	5000	30. 3	0. 01~310
Limestone	2010	12. 2	05~550
Sandstone and silicified wood	180	1. 0	0. 1~31. 0
Loose sand	2740	16. 6	0. 01~2. 0
Caco$_3$	88	0. 5	0. 1

2.3. *Contradiction between Karst water drainage and supply*

Because of deep drawdown and long time drainage of Karst water during mining, a series of contradictions have emerged. For example, the ground water level will be dropped a great deal, the ground water resource will be exhausted, the spring will become dried up and the natural landscape will be ruined. In one of coal mines in the Hebei province, water drainaging have resulted in many problems, water supply wells were abandoned, spring became dried up and landscape and environments were destroyed. Cases of spring becoming dried up include the Dalu spring in Jiaozuo, the Dahuo spring in Xingtai, the Mingjing spring the Doulishan.

Large-scale pumping test was carried out in the Handan Coal Mines. Totally 9 groups of pumping wells were used with yield of 1.6 cubic meter per second. The drawdown in the center of influenced cone was 2.8 meter and the influenced area reached 10 km square. As a result, the water table of supply well was dropped by 1.5 meter with yield decreasing from 17 000 to 3 000 cubic meters per day. About 100 thousand residents suffered the shortage of water supply at the time.

2.4. *Deterioration of karst water quality*

Because of draining of mine water and coal separa-

tion waste water, the quality of surface and ground water gets deterioration. The content of total minerals, total hardness, iron, manganese, and sulfate have increased resulting in decreased PH value. For example, the total hardness of mine water of the Jincheng Coal Mining Bureau is from 30 to 76 and arsenic content exceeded 2 to 8 times more than the national drinking standard.

2.5. *Surface subsidence and collapse*

After coal is excavated, surface subsidence will take place. For example, the area of subsidence in 9 mines of the Tangshan area reached 29,260,000 m^2 with watery depression of 5,620,000 m^2, and 12 meters water depth. In the Huaibei area, the subsidence reached 36,000,000 m^2, the average of subsidence per 10,000 ton of coal produced was 2440 m^2. In the future, the subsidence areas will be 85,320,000 m^2, some peasants won't have land to cultivate (Li, 1997).

Surface collapse caused by drainage is also common in North China. For example, in the Fangezhuang Coal Mine, collapse holes were formed after Karst water inrush. In some Karst water supply fields, intensive pumping also caused surface collapse.

3. COUNTERMEASURES

Primary environments can be greatly destroyed during coal production. We must study inherent relationships between coal production and environmental changes. At the same time, a series of countermeasures should be taken.

3.1. *Countermeasures for mine water prevention and cure*

For the situation of relatively more complicated hydrogeological conditions and Karst water inrush, , we should check up hydrogeological conditions first, then take following countermeasures: dewatering, mining with pressure, and underground grouting. There have been some successful examples in North China. In the Zhangcun Coal Mine, the Ordovician Karst water threatened the excavation of the lower coal measures. In 1974, the relevant Departments offered proposals of dewatering the Ordovician Karst water. After scientific demonstration, in 1982, the method of mining with pressure was adopted to avoid producing environmental geological problems. By the end of 1997, a total of 19 working faces had been excavated successfully and overall coal yield reached 3,175,000 tons with no Karst water inrush and no environmental geological problems happened. This is a typical model of mining with pressure in North China

3. 2. *Combination of drainage and supply*

Considering the factors of mine drainage, water supply and environmental protection, a mathmatics model should be constructed. For example, in a mine of Hebei province, surrounding with aquiclude, a separating hydrogeological unit is formed. Cambrian and Ordovician Karst water is the main source of mine water and also industrial water supply. Mine drainage would lead to water supplying wells to be abandoned and spring to be exhausted. Also, it will destroy landscape and ecological environments.

Therefore, considering the factor of mine drainage, water supply and environmental protection, a policy-making model of ground water resource should be constructed.

Objective functions:

$$\max Q(i) = \sum_{j=1}^{m1} Q1(i,j) - \sum_{j=1}^{m2} Q2(i,j)$$

$$\pounds^{..} i = 1,2,3,\cdots,n)$$

$$s(k,i) = \sum_{j=1}^{m1} Q1(i,j)\beta1(k,j,i) - \sum_{j=1}^{m2} Q2(i,j)\beta2(k,j,i)$$

$$(i = 1,2,3,\cdots,n)$$

Restrictive conditions:

$$s(k,i) \leq S_k^a \quad k \in G1$$

$$s(k,i) \geq S_k^b \quad k \in G2$$

$$s(k,i) \leq 0 \quad k \in G3$$

$$\sum_{j=1}^{m1} Q1(i,j) \geq Qa$$

$$\sum_{j=1}^{m2} Q2(i,j) \geq Qd$$

$$\sum_{j=1}^{m3} Q3(i,j) \geq Qp$$

$$Q(i,j), Q1(i,j), Q2(i,j), Q3(i,j) \geq 0$$

Optimum solution is sought using pure shape method. The result is given in Table 2.

Table 2 The result of ground water management ($\times10^4 m^3/d$)

Mine demand yield	Water resource		Optimum yield
	Karst water reservoir		
4. 08	2. 40	2. 74	5. 121

If above-policy making plan is carried out, the purpose of mine safety, water supply and environment protection will be satisfied.

3. 3. *Mine water treatment*

China is poor in water resources. Mine water treatment can compensate to some extent the shortage of water resources. Also, it is an important way for preventing and curing mine water pollution. Mine water after first grade treatment can be used to irritate farmland. After second grade treatment it can be used as cooling water in power plants and chemical factories.

3. 4. *Prevention and cure of subsidence and collapse*

When preventing and curing surface collapse, the influencing factor should be considered. The surface collapse located in industry squares, electrical substations, transport lines and constructions need to be refilled.

4. CONCLUSIONS

Hydrogeological conditions of Karst coal mine in North China are complicated, which has induced a series of environmental geological problems. The problems are Karst water inrush, rocks and soils brought from water inrush, contradiction of drainage and supply, deterioration of water quality, surface subsidence and collapse etc. In order to guarantee continuous developing of coal production, above environmental geological problems should be studied, and scientific, lawful, administrative and economical countermeasures should be taken.

REFERENCES

Wang, Z. & Liu, H. 1993. *The Excavation upon Karst Water*. The Publishing House of Coal Industry.

Liu, H. 1997. Research on the characteristics of soils brought from water inrush in guojiazhuang coal mine. *Journal of China University of Mining & Technology*: No 2.

Li, Y. 1997. Environment problem resulted from resource extraction and countermeasures. *Journal of China University of Mining & Technology*: No1.

Mining Science and Technology' 99, Xie & Golosinski (eds) © 1999 Balkema, Rotterdam, ISBN 90 5809 067 1

Study of fractal interpolation on fault surface

Hongquan Sun
China University of Mining and Technology, Xuzhou, People's Republic of China

Heping Xie
China University of Mining and Technology, Beijing Campus, People's Republic of China

ABSTRACT: In this paper, the mathematical principle of fractal interpolation surfaces is discussed and the calculation formula of fractal interpolation surfaces is given. The improved method of fractal interpolation surfaces using the ways of partition of local domains and vertical scaling factor is proposed. The roughness of the fault surface is simulated and the fractal dimension of the fault surface is obtained.

1. INTRODUCTION

The accidents of slope instability and roof falling in mining engineering and civil engineering occur frequently. The occurrences of the accidents are closely related to the influence of faults and joints in rocks. The roughness and surface shape of the faults and the joints affect the slope stability and the roof cave-in directly. For many years , the researchers in the field of rock mechanics, geology and mine have been paying much attention to the study of the shape and the surface roughness of the faults and joints.

However, faults and joints are in the different layers underground and it is difficult to obtain the data of the roughness of the fault surfaces . So, it is urgent to set up mathematical models(Benoit B.Mandelbrot 1982, Michael Barnsley 1986,1988, Peter R. Massopust 1994, Kenneth Falconer 1990) that the real fault shapes can be interpolated approximately for analyzing, simulating and predicting the influences of the roughness of faults and joints on the accidents in mining engineering and civil engineering.

H. Xie studied the influences of the roughness of faults on mining subsidence in detail and found out that the fault surfaces have fine fractal properties of statistical self-similarity. The rough shape of rock fracture surfaces is related to the rock characteristics and tectonic stress features. The conclusion is that the roughness of fault surface influences the mining subsidence directly.

In this paper, based on the theories and methods of fractal interpolation surfaces and using the altitude data of the fault surface measured by physical exploration, the roughness shape of fault surface is simulated and the fractal dimension of the fault surface is obtained. According to the principles of multivariate statistics, the attitude of fault is obtained.

2. THE PRINCIPLES OF FRACTAL INTERPOLATION SURFACE ON RECTANGLE AREA

Let $I = [a , b]$, $J = [c , d]$; and domain $D = I \times J = \{ (x , y) : a \leq x \leq b , c \leq y \leq d \}$. Subdivide D into the subintervals:

$$\begin{cases} a = x_0 < x_1 < \cdots < x_N = b \\ c = y_0 < y_1 < \cdots < y_M = d \end{cases} \quad (2.1)$$

Given a set of data on the grid: $(x_i , y_j , z_{i,j})$, $i = 0, 1, ..., N, j = 0, 1, ..., M$. We would like to find a interpolation function $f : D \rightarrow R$, so that $:f (x_i , y_j) = z_{i,j}$, $i = 0, 1, ..., N, j = 0, 1, ..., M$.

We will discuss on the three dimensional domain (Heping Xie and Hongquan Sun 1997) $K = D \times [h_1, h_2]$ $(-\infty < h_1 < h_2 + \infty)$. Then, for (c_1 , d_1 , e_1), $(c_2 , d_2 , e_2) \in K$, take $d ((c_1 , d_1 , e_1), (c_2 , d_2 , e_2)) = \max \{ | c_1 - c_2 | , | d_1 - d_2 | , | e_1 - e_2 | \}$.

Let $I_n = [x_{n-1}, x_n]$, $J_m = [y_{m-1}, y_m]$, $D_{n, m} = I_n \times J_m$, $n \in \{1, 2, ..., N \}$, $m \in \{1, 2, ..., M \}$. Let $\Phi_n : I \rightarrow I_n$, $\Psi_m : J \rightarrow J_m$ be contraction mappings and satisfy:

$$\begin{cases} \Phi_n(x_0) = x_{n-1} , \quad \Phi_n(x_N) = x_n \\ \Psi_m(y_0) = y_{m-1} , \quad \Psi_m(y_M) = y_m \\ | \Phi_n(c_1) - \Phi_n(c_2) | < k_1 | c_1 - c_2 | \\ | \Psi_m(d_1) - \Psi_m(d_2) | < k_2 | d_1 - d_2 | \end{cases} \quad (2.2)$$

where: c_1, $c_2 \in I$, d_1, $d_2 \in J$, $0 \le k_1 < 1$, $0 \le k_2 < 1$

Let $L_{n,m}: D \to R^2$ be a contraction mapping: $L_{n,m}(x, y) = (\Phi_n(x), \Psi_m(y))$; $F_{n,m}: K \to [h_1, h_2]$ is continuous, so that

$$\begin{cases} F_{n,m}(x_0, y_0, z_{0,0}) = z_{n-1, m-1} \\ F_{n,m}(x_N, y_0, z_{N,0}) = z_{n, m-1} \\ F_{n,m}(x_0, y_M, z_{0,M}) = z_{n-1, m} \\ F_{n,m}(x_N, y_M, z_{N,M}) = z_{n,m} \end{cases} \quad (2.3)$$

According to the theory of iterated function system, we obtain:

$$\Phi_n(x) = x_{n-1} + \frac{x_n - x_{n-1}}{x_N - x_0}(x - x_0) \quad n \in \{1, 2, \cdots, N\} \quad (2.4)$$

$$\Psi_m(y) = y_{m-1} + \frac{y_m - y_{m-1}}{y_M - y_0}(y - y_0) \quad m \in \{1, 2, \cdots, M\} \quad (2.5)$$

Let

$$F_{n,m}(x, y, z) = e_{n,m}x + f_{n,m}y + g_{n,m}xy + s_{n,m}z + k_{n,m} \quad (2.6)$$
$$n \in \{1, 2, \cdots, N\}, m \in \{1, 2, \cdots, M\}$$

$G_{n,m}(x, y, z) =$

$$\begin{cases} F_{n,m}(x, y, z) + kx\dfrac{x - x_{N-1}}{x_N - x_{N-1}}, n \in \{1, 2, \cdots, N-1\} \\ \qquad\qquad\qquad x \in [x_{N-1}, x_N] \\ \qquad\qquad\qquad m \in \{1, 2, \cdots, M\} \\[4pt] F_{n+1,m}(x, y, z) - kx\dfrac{x_1 - x}{x_1 - x_0}, n \in \{1, 2, \cdots, N-1\} \\ \qquad\qquad\qquad x \in [x_0, x_1] \\ \qquad\qquad\qquad m \in \{1, 2, \cdots, M\} \\[4pt] F_{n,m}(x, y, z) + ky\dfrac{y - y_{M-1}}{y_M - y_{M-1}}, n \in \{1, 2, \cdots, N\} \\ \qquad\qquad\qquad y \in [y_{M-1}, y_M] \\ \qquad\qquad\qquad m \in \{1, 2, \cdots, M-1\} \\[4pt] F_{n,m+1}(x, y, z) - ky\dfrac{y_1 - y}{y_1 - y_0}, n \in \{1, 2, \cdots, N\} \\ \qquad\qquad\qquad y \in [y_0, y_1] \\ \qquad\qquad\qquad m \in \{1, 2, \cdots, M-1\} \\[4pt] F_{n,m}(x, y, z), \qquad\qquad \text{Others} \end{cases} \quad (2.7)$$

where :

$$kx = (F_{n+1,m}(x_0, y, z) - F_{n,m}(x_N, y, z))/2 \quad (2.8)$$

$$ky = (F_{n,m+1}(x, y_0, z) - F_{n,m}(x, y_M, z))/2 \quad (2.9)$$

$s_{n,m}$ ($n \in \{1, 2, \ldots, N\}$, $m \in \{1, 2, \ldots, M\}$) is a freedom parameter, and $|s_{n,m}| < 1$ is called vertical scaling factor. From equation (2.7), we may define an iterated function system (IFS) $W_{n,m}(x, y, z)$ on the domain K:

$$W_{n,m}(x, y, z) = (\Phi_n(x), \Psi_m(y), G_{n,m}(x, y, z)),$$
$$n \in \{1, 2, \ldots, N\}, m \in \{1, 2, \ldots, M\}. \quad (2.10)$$

For such defined IFS, there is a unique attractor $G = \{(x, y, f(x, y)) : (x, y) \in D\}$ which is a graph of a continuous function f, so that:

$$f(x_i, y_j) = z_{i,j}; \quad i = 0, 1, \ldots, N, \quad j = 0, 1, \ldots, M \quad (2.11)$$

3. FRACTAL INTERPOLATION OF THE FAULT SURFACE

3.1 *Elevation data of the fault surface*

28 data were obtained with quake prospecting on a fault surface in a coal field in south China. For the simplicity, the coordinate of X and Y are moved into origin.

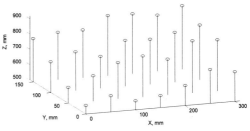

Figure 1 Space distribution of data on the fault surface

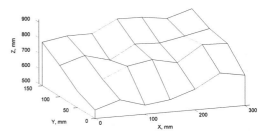

Figure 2 Mash of data on the fault surface

Figure 1 gives the space distribution of the 28 point data. The mash shape of the data is shown in Figure 2. The bases of these two figures are 500m. The roundlets show the locations of the data on the fault. The elevation data of fault surface include two kinds of information: attribute and roughness. To study the roughness of fault surfaces, we first divide the elevation data of the fault into the roughness and the attribute of the fault, then analyze the attribute analyses of the fault surface.

3.2 *Attribute analyses of the fault surface*

The attribute of fault surface including trend and the inclination in the study of the fault surface is an important index for researching faults.

The analyses of the attribute of the fault are based on the principle of the trend surface analyses. We use the practical data obtained from the fault surfaces to fit one order trend surface .

$$\hat{z} = b_0 + b_1 x + b_2 y \qquad (3.1)$$

Where x and y are coordinates, $\hat{z}$ trend value of the fault surface, and b_0, b_1 and b_2 are coefficients of the trend surface, which can be obtained by using the least square method. The intersect line of trend surface (3.1) and the plane z=0 is the direction of trend surface. The angle between the one order trend surface and the plane XY indicates the obliquity of the fault surface. On the practical data, we obtained the trend surface equation :

$$\hat{z} = 556.7590 + 0.5454\, x + 1.4678\, y \qquad (3.2)$$

The shape of the trend surface is shown in Figure 3.

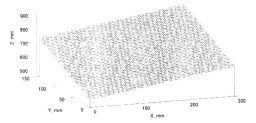

Figure 3 The one order trend surface of the fault

According to the analyses of the trend of the fault, let the z value of the one order trend surface equation (3.2) be zero. The equation of trend line can be obtained:

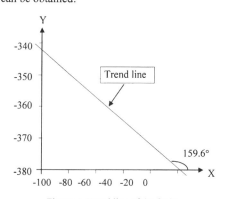

Figure 4 Trend line of the fault

$$y = -0.3716x - 379.3153$$
(3.3)

Suppose the angle of trend line and X axis is θ, then :

$$\theta = 180 - tg^{-1}(0.3716) = 180 - 20.4 = 159.6° \qquad (3.4)$$

Let the angle between trend line and Y axis be α, then:

$$\alpha = \theta - 90 = 69.6 ° \qquad (3.5)$$

So, the fault trend is NWW69.6 °. The trend line of the fault on XY plane is shown in figure 4.

The angle of the one order trend surface and XY plane is used to express the obliquity of the fault.

Based on the calculation formula of the angle of the two planes, we can obtain the angle of the one trend surface and XY plane

Let P_1 be the one order trend surface and P_2 be XY plane, that is:

$$P_1: \ z = 556.75090 + 0.5454\ x + 1.4678\ y$$
$$P_2: \ z = 0 \qquad (3.6)$$

then the angle of P_1 and P_2, φ, can be expressed as :

$$\cos\varphi = \frac{1}{\sqrt{0.5454^2 + 1.4678^2 + 1}} = \frac{1}{1.8579} = 0.5382 \qquad (3.7)$$

We have:

$$\varphi = 57.4° \qquad (3.8)$$

So, the obliquity of the fault is 57.4°.

3.3 Improved self-affine fractal interpolation

3.3.1 The partition of local field

The rock fracture surface possesses fractal characters, but not strictly self-similarity. The researche indicates that the variable $z(x)$ reflecting roughness of the rock fracture surface is a regional variable, that is it includes both pertinence and randomicity. Using the variagram theory of geostatistics(Hongquan Sun 1990) , we put forward the method of the local field partition. The expression of the spherical model of variagram is:

$$\gamma(h) = \begin{cases} 0 & h = 0 \\ c_0 + c\left(\dfrac{3}{2}\cdot\dfrac{h}{a} - \dfrac{1}{2}\cdot\left(\dfrac{h}{a}\right)^3\right) & 0 < h \leq a \\ c_0 + c & h > a \end{cases} \qquad (3.9)$$

where a is range, c_0 nugget and $c_0 + c$ sill.

The physical meaning of range a is that if the distance between two points is less than a, the variation of these two points is related to the distance between them. In the process of fractal interpolation, the range a is used as the basis of the partition of the local field. The calculation formulas of the fractal interpolation are used in the local fields.

3.3.2 *The way of selecting of vertical scaling factor*

We give the way of selecting vertical scaling factor as follows (Hongquan Sun 1998):

1. Using the practical data (interpolating data) (x_i, y_i, z_i) ($i = 1, 2, ..., n$) and the least square method, we can construct an one order trend surface equation :

$$\hat{z} = b_0 + b_1 x + b_2 y \tag{3.10}$$

2. According to the one order trend surface equation , we calculate the trend value on each interpolating datum point:

$$\hat{z}_i = b_0 + b_1 x_i + b_2 y_i \ (i = 1, 2, ..., n) \tag{3.11}$$

3. Using the practical data to detract the trend value on each point, we obtain the deviation value on the corresponding points :

$$e_i = z_i - \hat{z}_i \ (i = 1, 2, ..., n) \tag{3.12}$$

4. The deviation values can be used as the vertical scaling factors.

3.4 *Fractal interpolation of the fault*

Based on the practical data shown in figure 1 and figure 2, the roughness of fractal surface is obtained using the method of fractal interpolation surface discussed above. The result of the interpolation is shown in figure 4.

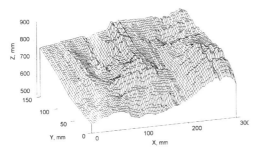

Figure 4 Fractal interpolated fault surface

From Figure 4, we can see that there are obviously scraggy local areas on the fractal interpolation fault surface. It gives us an intuitionistic roughness and is a virtue of the fractal interpolation. With the traditional methods, however, this result can't be obtained .

According to the method of calculating box dimension, we can obtain that the dimension of fractal interpolated fault surface is 2.1993.

According to the principles of precision analyses (Hongquan Sun 1998), the dimension precision and the deviation precision of the fractal interpolated fault surface can be calculated. As the practical data number is 28 and the interpolated point number is 4186. So the information content is $h = 28/4186 = 0.67\%$. We then obtain that the dimension precision and the deviation precision are 97.95% and 92.06% respectively.

REFERENCES

Benoit B. Mandelbrot 1982. The Fractal Geometry of Nature. W. H. Freeman and Company, San Francisco: 361~366.

Michael Barnsley 1988. Fractals Everywhere. Academic Press Orlando. FL, 172 ~ 247.

Peter R. Massopust 1994. Fractal Functions, Fractal Surfaces, and Wavelets. Academic Press: 135~355.

Michael F. Barnsley 1986. Fractal Functions and Interpolations. Constructive Approximation 2. 303~329.

Kenneth Falconer 1990. Fractal Geometry. Mathematical Foundations and Applications, New York: 92~100.

Heping Xie and Hongquan Sun 1997. The Study on Bivariate Fractal Interpolation Functions and Creation of Fractal Interpolated Surfaces Fractals. Vol.5, No.4: 625 ~ 634.

Hongquan Sun 1990. Geostatistics and its applications (in Chinese) China University of Mining and Technology Press.

Hongquan Sun and Heping Xie 1998. Fractal interpolation surface and its dimension estimation (in Chinese) The Journal of China University of Mining and Technology. Vol. 2.

Hongquan Sun , Heping Xie and Guanming Yu 1997. The method of creating fractal interpolation surface (in Chinese). Journal of Fuxin Mining Institute. Vol. 6.

Hongquan Sun 1998. The study on the theory of fractal interpolation surface and the interpolation of rock fracture surfaces. Doctoral Thesis. China University of Mining and Technology.

Mining Science and Technology' 99, Xie & Golosinski (eds) © 1999 Balkema, Rotterdam, ISBN 90 5809 067 1

Influence of stratigraphic facies variations on the roof stability by physical modeling study

Zhaoping Meng, Suping Peng, Hongliang Qu, Haiyun Zhao & Liping Luo
China University of Mining and Technology, Beijing, People's Republic of China

ABSTRACT: Controlled by depositional environment, the combined structure of roof rock layers behaves as vertical cyclicality, thickening or thinning up to pinchout discontinuously in horizon. Thereby, the roof stability is dominantly dependent on it. In order to investigate how the roof sedimentary combined structure and thickness of sandstone influence on the roof stability during and after coal mining, a physical modeling study is conducted in this paper. It is shown that with stronger sandstone of roof strata, roof has a longer distance of initial weighting and periodic weighting, higher abutment stress on the coal workings, larger caving and fracture zone. It is quite different in case of roof rock-mass without any framework sandstone. While in the transition zone of nipped sandstone, roof rock-mass is broken and with poor stability.

1. INTRODUCTION

The geological models with homogeneous, continuous, and equal thickness of rock seam, on which traditional theories of underground strata and pressure control were established in 1950s, are different from the real conditions. In fact, the lithological features of roof rock mass above the coal seam are variable and discontinuous in 3-dimension space, including thickening, thinning, or even pinching out in lateral. Different depositional roof types will certainly affect the roof stability in some degree [1], so that the study of mechanics and movement of the roof strata during mining will get better understanding of the roof control. Since 1988, The authors have undertaken several research project concerning the geomechanical modeling of the roof strata under discontinuous condition [2][3], geological prediction techniques of the roof stability [4]. In this paper, based on the field investigations of Xinji coal mine, Huainan coalfield of China, the influence of stratigraphic facies variation on the deformations, failure and underground pressure redistribution of roof strata during mining are studied in terms of physical modeling.

2. PHYSICAL MODELING

2.1. GEOLOGICAL CONDITIONS

In Xinji coal mine there are two main coal seam 13-1 and 11-2 in upper Shihezi formation. The coal seam 13-1 is about 6.2 to 10 meters thickness, buried under 293 meters from the surface; It is suitable to be mined out using Longwall top-coal caving technology. The immediate roof is gray-black mudstone or sandy mudstone with plentiful joints & fractures, block-fractures structure. Its thickness varies in lateral, in some areas even scoured away. The main roof is gray or gray-white, fine or medium quartz sandstone with poor joints & fractures, and is in contact with underlying coal seam some where, its thickness also varies in horizon direction (fig.1).

2.2. MODEL DESIGN

The physical model of plane stress was designed with dimension of 2.5m×0.2m×1.8m in length, width and height respectively. The range between 806 and 703 borehole was chosen as prototype of sedimentary facies variation. The real situation is 375 meters in length, 185 meter in height, while the height of 13-1 coal seam is 8.5 meters (fig.2). In the

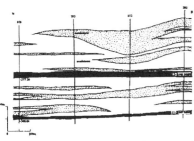

Fig.1 The sedimentary section in the southern center mining area

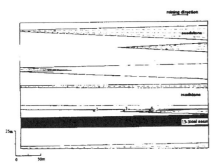

Fig.2 A physical model of sedimentary facies variations

model, the first layer of sandstone in the roof pinches out from the right to the middle of the model. And the thickness of the second layer of sandstone is stable, about 2.09 to 3.61 meters in the prototype, which is 26.05 to 29.87 meters far away from the 13-1 roof, called 13-1 main roof sandstone.

3. ANALYSIS OF RESULTS

3.1. FAILURE OF THE ROOF ROCK MASS

Controlled by depositional environment, the combined structure of roof rock layers indicates vertical cyclicality, thickening or thinning up to pinchout discontinuously in horizon. Thereby, the behavior of periodic weighting distance is different even for the same model. It is shown that the roof strata displacement is mainly controlled by the "framework" sandstone layers, which have high strength, stable thickness and compact cementation. In the experiment, the first layer of sandstone in the roof is 1.19 to 5.00 meters far away from the 13-1 roof. And its thickness is unstable, which is about 0 to 4.66 meters and pinches from the right side to the middle of the prototype. The middle part of the

model is a transition zone which characterized by the pinch-out sandstone. When mining from the left to the right of the model, the transition zone is controlled the failure and the underground pressure of the roof rock mass, where fractures are maturated and caving rock is broken. The left part of the model includes a thick mudstone and sandy mudstone with a thin coal seam as immediate roof. Caving mudstone filled the goaf immediately and no obvious concentration of stress and periodic weighting. In the right part of the model, periodic weighting distance is 18.0 meters long; and in the transition zones, i.e., the middle part of the model, the periodic weighting distance is 9.0 meters long. All of these proved that the difference of combined structures in roof generates different characters of deformation during mining. For example, in the left part of the model, the breaking angle is 50.5 to 66.0 degrees, because of no first sandstone layer, while in the right part of the model, breaking angle is 66.0 up to 78.0 degree. It was also found that the caving and breaking angle is 66.0 degree at the top of starting cut and 68.0 degree at the top of the stopping line in the modeling study. With proceeding of mining, splits of roof rock layers develop from bottom to top, especially in combined rock structures with bottom-weak and top-hard structure. In the left part of the model, the roof deformed continuously and quickly, because of its weak lithological characters, and deformed greatly in vertical. However, the caving and breaking were limited, no more than 73.5 meters, because the goaf were filled gradually. On the contrary, the roof in the right part of the model deformed discontinuously, generated shear, sliding between layers and breaking along the depositional structure faces, which were called splits failure. So the height of the caving and fracture zone was high up to 88.5 meters.

3.2. MOVEMENT OF THE ROOF STRATA

The roof strata in the right part of the model subside slowly during mining, which can be proved from the subsidence curve. In this example, the subsidence curve is gentle dip and asymmetrical with obvious deviation (fig.3-5). This phenomenon can be explained with the relations between high-strength sandstone and low-strength mudstone. In the roof rocks, comprised of high-strength sandstone framework, the sandstone displacement make its upper rock mass or its surroundings move with it too. On the other hand, the low-strength weak rocks (mudstone, sandy mudstone) played a role of follower, which move with the framework sandstone rocks. As a result, the roof strata begin to move a

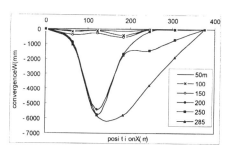

Fig.3

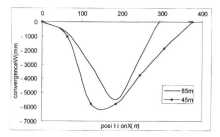

Fig.4

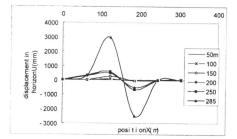

Fig.5

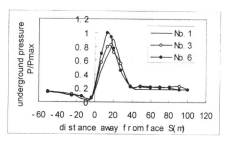

Fig.6 Underground pressure of the roof rocks while coal mining

than that of the lower roof strata and the convergence of the roof rock strata is the non-coordination.

3.3. DISTRIBUTION OF UNDERGROUND PRESSURE

In order to determine the influence of framework sandstone roof on the distribution of underground pressure, several pressure sensors were set at the top of the first fine sandstone layer above the coal seam 13-1.

While mining, the underground pressure of roof strata began to appear (see Figure 6). In fig.6, the No.1 sensor located in the roof strata without sandstone, the No.3 located in the transition zone of nipped sandstone, and the No.6 located in the fine sandstone roof. In the model, the left half is comprised of thick mudstone and sandy mudstone, with coal seam as the immediate roof. From fig.6, we found that the roof strata fell, during mining, filled the mining space. The periodic weighting were not obvious and did not concentrate on a certain zone, so that the abutment stress infront of working face were low. While in the transition zone of nipped sandstone, the thinner sandstone and the surrounding mudstone layer fell as mining. The periodic weighting and abutment stress ahead of face decreased to minimum, and the position of maximum abutment stress was about 17.25 meters away from working face, The reason for that is mining fractures of the roof developed and increased under the action of abutment stress. While in the right half of the model, as main roof was sandstone, the distance of periodic weighting increased to 18 meter, and the abutment stress reached its relative peak value, which located in 12.75 meters front of workings. From the above analysis, we can conclude that the abutment stress and its position are greatly different for the different sedimentary combinations. The higher the abutment stress, The nearer the peak

small displacement, 10 meters to 20 meters ahead of working face during mining. Away some distances from the working face, the roof rock mass, located between working face and the back of the face, presents a vertical movement which increase with the distance. This can be referred in figure 5. In the figure, the curve before transition point represents a tensile deformation zone generated by boom bending zone, and the residual part of the curve images a compression deformation zone generated by breaking rocks. From fig.5, it can be found that the convergence and subsidence factor decreased from bottom to top. Splits of roof rock layers develop with proceeding of mining, because there are a difference in combined structures of sedimentary rocks mass. Moreover, we concluded that the movement of the upper roof strata is smaller value to the working face when the roof is of high-strength framework sandstone. While for the low-strength mudstone roof, especially in transition zone of sandstone pinchout, the abutment stress decrease and the position of the peak value transited to far away rock strata, because the mudstone can not bear more underground stress.

4. CONCLUSIONS

(1) The roof stability depends strongly on the mechanical properties and distribution of the underground pressure during and after mining. Controlled by depositional environment, the combined structure of roof rock layers behave as vertical cyclicality, thickening or thinning up to pinchout discontinuously in lateral, the stability of roof strata significantly depends on the sedimentary facies. Certainly, the height of the caving and fracture zone, the distance of main roof breaking, and other dynamical phenomena associate dominantly with the sedimentary combinations tightly.

(2) In case of one or more layers of framework sandstone, the distance of initial and periodic weighting is larger, and the height of caving and fracture zone is higher. Without framework sandstone, the distance of the initial and periodic weighting decrease, and the height of caving and fracture zone reduce, underground pressure drops too. While in transition zone of sandstone pinch-out, the roof strata break and become difficult to be hold and supported.

(3)Compared with the lower strength mudstone roof or transition zone of sandstone pinchout, The roof strata of the higher strength framework sandstone will experience larger abutment stress and the peak value near the working face. The lower strength mudstone or transition zone undergoes less abutment stress and the peak value migrate to the inside of the surround of rock, because it can not bear pressure as same as the sandstone.

AKNOWLEDGEMENT

The authors would like to express thanks to the "CHINA NATIONAL SCIENCE FOUNDATION COMMITTEE" for their financial support (Projects: 49872053, 59774003).

REFERENCES

1 Suping Peng. Strata control of roof stability in mining workings. The proceedings of the international congress on mining science, applied geology and mining technology, St.Petersburg, Russia, ISBN 3-230-19519-3. 1993.85-91

2.Suping Peng, Li Yangbing. Wedge-shaped sandstone roof strata and roof stablity, examples from seam 13-1 in huainan coalfield, china. In: Zhu Deren,eds. Proceedings of the international mining tech(96 Symposium):Groundwater hazard control and coalbed methane development and application techniques. Xian: CCMRIC, 1996. 473-479

3.Suping Peng. Mining-induced stress redistribution and roof hazardous characteristics of the channel-filled sandstone roof : A physical modeling study. In: M.Aubertin et al. eds. Rock Mechanics (vol.1). A. A. A. Balkema, 1996. 423-429

4 Suping Peng, Y.P.Chugh. Geological moldering techniques of longwall mining roof stability: A case study. In: Zhu Deren,eds. Proceedings of rock mechanics and strata control in mining and geotechnical engineering. Beijing: 1995.232-240.

Mining Science and Technology' 99, Xie & Golosinski (eds) © 1999 Balkema, Rotterdam, ISBN 90 5809 067 1

The composite geophysical methods in monitoring spontaneous combustion of the coal seam

Feng Yang, Shunian Ning, Hongling Wang, Xuejun Zhao & Yizong Ling
China University of Mining and Technology, Beijing, People's Republic of China

ABSTRACT: In this paper, the application of geophysical methods to monitoring the spontaneous combustion of the coal seams is discussed, some specific formulas on temperature and rock's properties are suggested, and relative cases are also enumerated.

1. INTRODUCTION

The coal resource is the important resource on which the life of human beings depends. More attention needs to be widely paid to the process from prospecting and mining to using of the coal in the world. Now, the spontaneous combustion of the coal seam has become a serious problem threatening to the coal resource. The coal spontaneous combustion wastes above 100 million yuan every year in China, and at the same time, releases large amount of the harmful gases into the air. Because of this, the spontaneous combustion of the coal seam is treated as an important problem to be solved in many countries where coal is their main energy resource. Many modern geophysical prospecting methods, such as electrical surveying, electromagnetic surveying, nuclear surveying, seismic surveying, and remote sensing, are highly developed and are being used in dynamically monitoring the coal spontaneous combustion. In this paper, applications of these geophysical methods in monitoring coal spontaneous combustion are addressed.

2. THEORETICAL BASIS OF SOME GEOPHSICAL METHODS APPLIED IN MONITORING THE COAL SPONTANEOUS COMBUSTION

The process of the spontaneous combustion of the coal seam can be divided into three stages, including the coal oxidation in the seam, the coal burning, and coal disappearing. When the temperature is as high as 70°C to 80°C, the oxidation in the coal seam will be accelerated and the spontaneous combustion of the coal will take place. The temperature in the coal seam will go up with the combustion of the coal, and it can reach several thousands Celsius Degrees. Under such a high temperature, the surrounding rocks of the coal seam will be changed into the burnt rocks and their original structures will be destroyed. After combustion, the quantity of heat emitted from the coal seam will decrease and the temperature will gradually go down back to normal. Because the volume of the combustion residues is smaller than that of original coal, the top part of the coal seam will collapse, forming relatively incompact rock's structure. Here, we will discuss the variation of the geophysical properties of the coal seam and surrounding rocks during and after spontaneous combustion.

2.1. *The influence of the temperature on the rock's electrical properties*

The electrical resistance of the rock has close relationships with the porosities. The studies indicate that there is a relationship between the rock's electrical resistance and the water content in the rock:

$$\lg \rho = k \lg w + b \tag{1}$$

Where ρ represents the rock resistivity, and w represents water percentage in the rock.

The further study on the relationships between the coal's electrical resistance and temperature indicates that the following formula can be used to describe the changes of resistance with temperature from brown coal to anthracite:

$$\rho = kT^a \tag{2}$$

where T is the Kelvin.

In this formula, if we know the distribution of the underground strata, we can simulate the temperature of the underground strata and estimate stage of the coal spontaneous combustion. Furthermore, under the influence of thermic stress, the rock will break up at the 200~300oC, and the rock's fissure will increase accordingly. If the fissure is full of water, the rock's electrical resistance will decrease remarkably. Survey conducted in some areas has indicated that the resistance can decrease by half.

The high temperature will also affect another electrical property of the rock, the dielectric capacitivity (ε). The relationship between the temperature and the ε is expressed as:

$$\varepsilon = a + \frac{b}{T} \tag{3}$$

Although the ε is affected by the temperature, when the temperature is enough high, the ε changes little with the temperature. Figure 1 illustrates the variations of the temperature and frequency with ε.

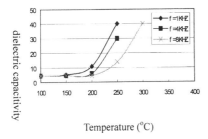

Fig1 The relationship of the temperature with the ε

2.2 The influence of the temperature on the magnetic properties of rocks

Characteristics of the crystal materials will change with the temperature, thus the physical properties of the materials, including the magnetism of the crystal material, will also change. The formula of the susceptibility of the paramagnetic material with the temperature is:

$$\kappa_{\text{ЦM}} = \frac{C}{T} \tag{4}$$

where C is the courier constant, and T is the Kelvin.

The relationship between the ferromagnetic and the temperature is complicated, but it is not the main basis for the geophysical surveying. If a coal seam contains abundant pyrites and siderites, when the coal seam is burning, the pyrites and the siderites will be transformed into the tetroxide with very strong magnetism. This is the main basis for mapping the boundary of the spontaneous combustion of the coal seam.

2.3. The influence of the temperature on the movement of the radioactive gas (niton gas)

It is generally believed that the niton moves from deep of the earth to the surface. The increasing temperature resulted from coal combustion will speed up this movement, and will lead to decreasing of the solubility of the niton. These two effects will produce the difference of the niton concentration between the combustion zone and non-combustion zone. Figure 2 shows the relationship between the solubility of the niton gas and the temperature.

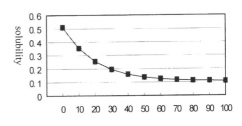

Fig2 The effect of the temperature to the solubility

2.4 The influence of the temperature on the wavefield properties of the rocks

The influence of the temperature on elasticity properties of the rocks depends on the cementation of the rock. In spite of this, the decreasing of the p-wave of all kinds of rocks can be detected when the temperature is over 200°C. The decreasing gradients can be up to 1-3m/s/°C. If the temperature of the rock increases by 100°C, the change in velocity of the p-wave will be reflected in the seismic results.

3. APPLICATION OF THE GEOPHYSICAL METHODS IN MONITORING COAL SPONTANEOUS COMBUSTION

Although there are few examples of applying compositive methods to monitor the coal burning, the above mentioned methods have been successfully applied in solving the coal spontaneous combustion and mapping the boundary of the burned coal seam.

At the beginning of the coal spontaneous combustion and during coal burning, the radioactive gases, remote sensing, and electrical surveying methods can be used to monitor the situation of the coal combustion. The radioactive gas method was applied for mapping the boundary of the coal burning in the Shigejie mine of the Shanxi Province in 1990. The surveying data was processed, and the area with the niton anomaly was mapped. The area with latent danger and combustion was distinguished. According to the surveying report, the concentration of the CO and CH4 released from incomplete burning of coals decrease largely after processed with surveying results, which confirms that the area mapped with the radioactive gas method is just the coal-burning area. The radioactive gas method has also been successfully used to inspect the high temperature area in the Liuzhuang mine, where the coal was gasified in the underground. This indicates that the radioactive gas method is an effective method for dynamic monitoring of the spontaneous combustion of coal seams.

As discussed previously, the electrical method can be used to monitor the whole process of the coal spontaneous combustion. Nevertheless, the precision of this method is limited, so at present there is no precedent for applying this method to map the burning area and the latent danger area. However, this method has been widely used to map the burned area. The coal spontaneous combustion is commonly seen in the Dahuangshan mine of the Xinjiang Uygur Autonomous Region. Because the rock destroyed by coal burning is full of water, it is necessary to map the area with distribution of burned coal in order to prevent water from breaking into the mining area. To do so, the transient electromagnetic method was used. The magnetic method survey is suitable for the survey in the area with distribution of abundant pyrites. This method was applied in the Huojitu mine at the boundary of the Shanxi province and the Neimenggu Autonomous Region, with the satisfied surveying results.

The remote sensing methods is one of the most effective methods. The first application of the remote sensing in inspecting the coal burning was recorded in the early 1960's. Now many new techniques have been adopted in the remote sensing, which provide more accurate monitoring to the coal combustion. Much development in remote sensing has been achieved..

The seismic surveying is the most difficult method to be used to dynamically monitor the coal spontaneous combustion, because its cost is very high and its theory for monitoring the coal combustion is not mature. There is no report on the application of this method in mapping the area with coal combustion, by contrast, this method proved to be very successful in mapping the empty mined area. After the coal combustion, the wavefield properties are same as those of the empty mined area. So, we are confident that this method can be used to map the boundary of the coal combustion area. As discussed before, the influence of temperature on the wavefiled can be detected with the seismic surveying method, but practical constructing techniques and methods need to be studied in detail.

4. CONCLUSION AND PROSPECT

Making a comprehension view on very fast development of the geophysical inspecting technology, on the increasingly perfect theories, and on the experiences that we have obtained in the past, we can realize the dynamic monitoring on the spontaneous combustion of the coal seam. With continuous improvements in the precision of the equipment and the technology of the construction, this monitoring will be more precise and effective. All of these will contribute to the human beings for them to make full use of the coal resources and to protect their environments.

REFERENCES

Yang, L., L. Yu 1997. Detecting the niton--- the new method for inspect the high area in underground gasified of the coal. *The Mining Science and Technology* l25(12).

Wei, S. & X. Wang 1997. The study on detecting niton to inspect the underground burning area. *The Mining Engineer* No.2.

Neuer G, Kochendörrfer R, Gern F, 1996. high temptreaure behaviour of the spectral and total emissivity of CMC materials. *High Temperature - High Pressure* 1995/1996 volumes 27/28.

Mining Science and Technology' 99, Xie & Golosinski (eds) © 1999 Balkema, Rotterdam, ISBN 90 5809 067 1

Deformation of coal measures and the factors that control it in China

Daiyong Cao & Wentai Gao
China University of Mining and Technology, Beijing, People's Republic of China

ABSTRACT: This paper studies the major controlling factors of coal measures deformation and coalfield tectonic framework in China, presents the concept of the coal measures deformational sub-region, and summarizes the deformational features of five coal bearing areas in China. The deformational sub-regions in China fall into two basic types: 1) the concentric rings association developed on the stable platform, and 2) the parallel strips association developed on the tectonically active belt. Controlled by the geodynamic framework of continental lithosphere, the deformation and distribution of coal measures in China can be divided into three parts: the eastern reformation zone, the western compression zone, and the middle transition zone.

1. INTRODUCTION

The significant characteristics of China's coal geology include various patterns of coal basins, remarkable deformation of coal measures and complicated structural styles of coalfields. These special coal geological conditions have seriously influenced on the value and difficulty of coal exploitation.

Five coal-accumulating areas have been divided by Chinese geologists, which include the North China, South China, Northwest China, Northeast China, and Yunan-Xizang coal-accumulating areas (Wang X. et al, 1992). This division also has a tectonic significance, reflecting the general control of geotectonic framework on coal measures formation and deformation. In view of coal resource evaluation and exploitation, we propose a concept of coal measures deformational sub-region to emphasize the reformation and distribution of coal measures. Based on this concept, the main structural characteristics of five coal-accumulating areas in China are discussed.

2. CONTROLLING FACTORS OF COAL MEASURES DEFORMATION

2.1. *Geodynamic environment*

China is a complex continent which consists of many tectonically stable blocks and active belts with weak basement and deformed cover (Ren, et al,

1990; Ma, 1992). Coal basins developed in this matching continent are easily deformed. Distribution of coal measures shows deformational zoning, which is different from coal measures in cratonic basins of the North America and East Europe.

2.2. *Tectonic evolution*

Since coal measures were formed, they have been subjected to influences of every episode of crustal movements and tectonic events in geologic history. One of the essential regularities is that the earlier were coal measures formed, the more complicatedly were coal measures deformed. The first valuable coal measures in China was formed in the Early Carboniferous, since then four tectonic cycles of Hereynian, Indosinian, Yanshan and Himalayan occurred one after another (Huang, et al, 1983). The spatio-temporal differentiation of tectonism results in different deformation of coal measures in different coal-accumulating periods and areas.

2.3. *Deep structure and basement properties.*

Deep structural framework and basement tectonic properties affect both the tectonic activity of coal basins and the deformation pattern of coal measures. Generally, coal measures formed in intraplate or platform with a stable basement and cover combination are weakly reformed and well preserved. On the contrary, coal measures formed in plate margin and orogenic belt with an active basement and deformed cover combination are extensively deformed.

2.4. *Tectonic stress fields*

Tectonic stress is the essential cause for coal measures deformation. The complicated regional tectonic stress background has produced complicated deformation of crustal strata including coal measures. For example, different coal measures in the same area has undergone different periods of tectonic stress fields, and the same coal measures in different area can be affected by very different stresses.

2.5. *Lithological associations of coal measures*

The lithological associations of coal measures are the material base for deformation. Lithological associations of coal measures are of layering and cycling, showing interbedding of hard and soft rocks. Because coal measures contain many weak layers such as coal and shale, they are highly sensitive to stress and are easy to be deformed. The special detachment styles of coal measures, such as thrust and nappe, gravity gliding, and extensional structures, are closely related to the special lithological associations.

3. COAL MEASURES DEFORMATION

3.1. *North China coal-accumulating area*

The North China coal-accumulating area refers to the area of North China and the south part of Northeast China, which is equal to the North China platform or the main part of North China plate. This area is well known for its rich coal resources of Permo-Carboniferous and Jurassic ages, accounting for 50.47% of the total coal reserves in China.

The North China platform is surrounded by tectonic belts. Tectonic activities between the North China palaeo-plate and its adjacent palaeo-plates have resulted in multi-periods of coal measures deformation since Late Palaeozoic. After deformation, the coal measures have lost their original whole and continuous state, and have been separated into coal-bearing blocks in different scales and different patterns. The coal measures deformation in the North China coal-accumulating area has a remarkable feature of zonation and shows an asymmetrical rings association. This association can be divided into compressional external ring, weakly compressional middle ring and extensional internal ring. The external ring is characterized by strongly deformed coal measures. A thrust and

nappe belt up to 3000 km long, along the margin of the North China coal-accumulating area, has been distinguished (Wang G. et al, 1992). The coal measures deformation in the middle ring is weaker than that in the external ring, with the broad folds and normal faults being its main structural style. Internal ring refers to the North China plain, where the essential tectonic framework consists of fault blocks.

3.2. *South China coal-accumulating area*

The South China coal-accumulating area lies to the south of Qinling-Dabie mountains. The Late Permain coal measures are distributed all over the area, and the Late Triassic and Early Carboniferous coal measures are distributed in some regions of the area. Coal in this area account for 6.8% of the total coal reserves in whole China.

This coal-accumulating area is equal to South China palaeo-plate, with basement comprising the Yangtze platform and the Southeast China Caledonian geosynclinal system. The coal measures deformation in the Yangtze platform shows approximately concentric rings association with deformation intensity decreasing from the external towards the internal. The Southeast China geosynclinal system has undergone multi-periods of tectonisms since the Late Palaeozoic, resulting in the complicated deformation of coal measures. In whole South China coal-accumulating area, both intensity of coal measures deformation and strength of magmatic activity increase from plate interior to plate margin, that is, from the Yangtze block to the coastland of Southeast China. A series of large-scale uplifts and depressions trended NE-NNE were alternately arranged. The thrust nappe and gliding nappe spread from uplift to depression, with obvious parallel strip association of deformational sub-regions.

3.3. *Northwest China coal-accumulating area*

The Northwest China coal-accumulating area refers to the vast domain to the west of the Huolanshan-Liupanshan mountains and the north of the Kunlun-Qinling mountains. This area is rich in coal resources, especially the Early and Middle Jurassic coal measures, which amounts to 35.52% of the total coal reserves in China.

The Northwest China coal-accumulating area spreads over several geotectonic elements including the Tianshan-Xingmeng geosyncline system, the Tarim platform, and the Qinling-Qilian-Kunkun geosyncline system. The prototypes of the Mesozoic coal basins are mostly extensional, which were inverted into composite foreland basins in the Late Mesozoic and Cenozoic. One of the common features of these basins is that thrust faults were developed along the basin margins (Cao et al, 1997). Coal basins within the Tianshan geosyncline system have a narrow ribbon shape, paralleling to the trend of orogenics, whereas coal measures in the Tarim, Junggar, and Qaidam basins were deformed in forms of concentric rings.

3.4. *Northeast China coal-accumulating area*

The Northeast China coal-accumulating area includes the Northeast China and the east part of Inner Mongolia. This area is well known for its Early Cretaceous terrestrial coal measures with coals accounting for 7.07% of total coal reserves in China. Coal measures in this area are preserved in three groups of basins from west to east: the West Daxinganlin mountain basin group, the Songliao basin group, and the east basin group.

The coal basins in Northeast China have undergone geodynamic environments of alternation of weak compression and extension. Inherited activities of faults control the deformation of coal measures, producing both half grabens and graben-horst associations. The effect of tectonic activity in plate margin decreases from east to west. In the eastern coal basins, some compressional structures, such as reverse faults and inclined folds, are developed, while the western coal basin group have preserved the previous extensional framework.

3.5. *Yunnan-Xizang (Tibet) coal-accumulating area*

The Yunnan-Xizang (Tibet) coal-accumulating area takes the Qinghai-Xizang plateau as its main part. The Palaeozoic and Palaeogene coal measures were developed in this area, with coals amounting to only 0.14% of reserves in whole China.

The Yunnan-Xizang coal-accumulating area belongs to the Tethys domain, consisting of several continental blocks (terranes) and suture belts of the Eurasia and Gondwanaland (Huang et al, 1987). The Palaeozoic coal measures developed in the Changdu block, the south margin of the Tarim - South China plate, have undergone a complicated history of deformation. This deformation resulted from the multi-periods of matching of terranes and collision of plates, with thrusts and linear folds being main structural style of coalfields. In the west Yunnan province, strike-slip faults are dominated, which are responsible for many small intermontane basins filled with weakly deformed Neogene coal measures.

4. CONCLUSIONS AND DISCUSSION

The China continent is a complex continent consisting of many tectonically stable blocks and active zones. In view of coal basin evolution and coal measures distribution, we can identify two types of basic tectonic coal-accumulating elements.

1) The craton or quasicraton type. This type means the platform or the main part of palaeo-continental plate with a stable basement. It is characterized by steady and continuous coal-accumulation within broad depressions, inherited tectonic evolution of coal basins, weak to middle reformation of coal measures, and concentric rings association of deformational sub-regions. The intensity of coal measures deformation decreases from basin margin to interior. In the basin interior, coal measures are well preserved and huge exploiting coalfields are distributed, such as the Ordos coal basin in North China, and the Sichuan coal basin in Southwest China.

(2) The tectonically active belt type. This type is equal to geosyncline or palaeo-continenal margin. It has an active basement and is characterized by narrow coal-bearing depressions or fault basins, various scales of coal accumulation with tectono-sedimentary differentiation, and strong coal measures reformation. The deformational sub-regions have a parallel strips association.

Three periods of geodynamic systems have been superimposed on the China continent since Palaeozoic, namely, Palaeo-Asiatic, Marginal Pacific, and Tethys-Himalaya geodynamic systems (Ren et al, 1990). The differentiation and polycycle of tectonic evolution lead the deformational sub-region association to have a complicated but regular pattern. Controlled by the continental lithosphere textural framework, the distribution of coal measures deformational sub-regions in China can be divided into 1) the eastern zone, 2) the western zone, and 3) the middle zone (Figure 1).

The eastern zone is situated to the east of the Daxinganlin-Taihangshan-Wuyishan mountains, and is characterized by various types of coal measures deformation. Compressional background dominates the region to the south of the Qinling-Dabie mountains; while extensional background dominates the North China and Northeast China areas. Basic pattern of deformational sub-regions in this zone is the NE-SW trending parallel strip association with deformation intensity decreasing from east to west.

The western zone is situated to the west of the Helanshan-Longmenshan mountains, and is characterized by compressional framework of coalfields, and the NW-SE trending coal measures deformational sub-regions. The sub-region patterns are transformed from the parallel strips association in Yunnan-Xizang coal-accumulating area to the multi-centered concentric rings association in the Northwest China coal-accumulating area.

The middle transitional zone between the eastern and western zones is characterized by the stable basement and inherited tectonic evolution of coal

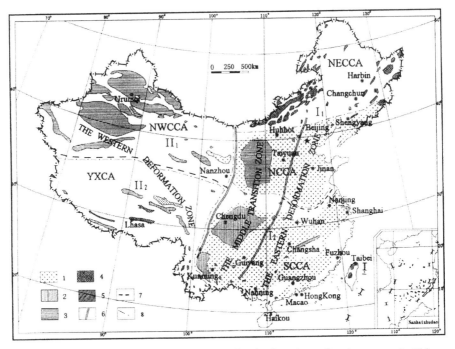

Figure 1. Schematic diagram showing deformational sub-regions of coal measures in China

Geological ages of coal measures: 1- Permo-Carboniferous, 2- Late Triassic, 3- Early-Middle Jurassic, 4- Early Cretaceous, 5- Tertiary; Geological borders: 6- first order deformational sub-region, 7- second order deformational sub-region, 8- coal-accumulating area. I_1- North and Northeast China extensional sub-region, I_2- South China superimposed deformation sub-region, II_1- Northwest China positive inverted sub-region, II_2- Yunnan-Xizang compressional sub-region; NCCA- North China coal-accumulating area, SCCA- South China coal-accumulating area, NECCA- Northeast China coal-accumulating area, NWCCA- Northwest China coal-accumulating area, YXCA- Yunnan-Xizang coal-accumulating area

basins. This zone is represented by the Ordos and Sichuan coal basins, where the deformation and distribution of coal measures show a typical platform-type concentric rings association.

REFERENCES

Cao, D., P. Zhang, K. Jin & G. Qian 1997. Tectonic evolution and inversion of the Turpan-Hami basin, Northwestern China. *Scientia Geologica Sinica.* 6: 407-412.

Huang, J., J. Ren, C. Jiang, Z. Zhang & D. Qin 1983. *The geotectonic evolution of China.* Beijing: Science Press (in Chinese).

Huang, J. & B. Chen 1987. *The evolution of the Tethys in China and adjacent regions.* Beijing: Geological Publishing House.

Ma, W. 1991. Analysis of regional structures. Beijing: Geological Publishing House (in Chinese).

Ren, J., T. Chen, B. Niu, Z. Liu & F. Liu 1990. *The tectonic evolution and metallogeny of the continental lithosphere in the eastern part of China and adjacent regions.* Beijing: Science Press (in Chinese).

Wang, G., D. Cao, B. Jiang, Z. Xu, D. Liu & S. Yan 1992. *Thrust nappe, extensional gliding nappe and gravity gliding structures in the southern part of North China.* Xuzhou: CUMT Press(in Chinese).

Wang, X., L. Zhu & J. Wang 1992. The formation and distribution of coalfields in China. Beijing: Science Press (in Chinese).

Mining Science and Technology' 99, Xie & Golosinski (eds) © 1999 Balkema, Rotterdam, ISBN 90 5809 067 1

Decision-making system for mine water inflow control and treatment in the North China coal bearing basin

Qiang Wu, Donglin Dong, Shuwen Li, Jintao Liu, Guoying Pan & Guimin Sun
China University of Mining and Technology, Beijing, People's Republic of China

Weidong Sun
Economic College of Shijiazhuang, People's Republic of China

ABSTRACT: Main characteristics of the mine-hydrogeological conceptual model in the North China coal bearing basin is a stereo water-filled geological structure formed by various inner- and outer-boundaries which hydraulically link up multi-layered aquifers. Finite-elemental numerical and optimal management models in "quasi-three-dimensions" have been used to describe the conceptual model in stereo scale, and to provide quantitative base for working out the decision-making system of mine water preventing and curing. The decision-making system can be further divided into two sub-systems. One is the preventing sub-system, with which some preventing engineering measures can be established. Another is the curing sub-system, with which concrete curing decision-making schemes can be scientifically worked out, based on analyses of mine-hydrogeological conditions and application of the numerical and optimal models.

1. INTRODUCTION

The direct water-filled aquifers in most coal mines in the North China coal-accumulating basin are mainly thin carbonate rocks or thick sandy rocks, with low permeability. It is believed that these aquifers can not form any water hazards for the safe mining of upper coal layers, providing these aquifers are in closed and independent states, and have no hydraulic connections with each other.

However, in fact, with increasing depth of mining and mining of base coal layers, quite a lot of water-bursting hazards below the coal layers would take place. The situation is now becoming more and more serious.

For example, water-bursting accidents took place in the Renlou mine of the Huaibei bureau of coal mine in 1996, and in the Zhangji mine of the Xuzhou bureau of coal mine in 1997. Maximum water-bursting inflows for the accidents are both over 60 m^3/min. Therefore, it is obvious that the above-mentioned hypothesis does not conform to the actual situation. That is to say, the direct water-filled aquifers are not independent closed systems, rather, they are hydraulically linked up together with other hard water-abundant aquifers by various inner and outer boundaries.

So, a stereo water-filled structure is one of the main mine - hydrogeological geological characteristics for the underlying water-bursting coal layers in the North China coal-accumulating basin.

2 MINE-HYDROGEOLOGICAL GEOLOGICAL CONCEPTUAL MODEL AND MATHEMATIC MODEL

In view of a profile, the North China coal-accumulating basin is of characteristics of multi-layer aquifer structures with closed hydraulic connections between them. The aquifers include porous, karstic and fissured ones.

For a hydraulic plane, all water-filled aquifers are considered to be of the natures of irregular spatial distribution, inhomogeneity, anisotropy, and complicated boundary position and condition. The occurrence medium of groundwater in a bedrock aquifer is thought to be a fissured network. The hydraulic types of the porous aquifers have obvious seasonal changes.

The hydraulic boundary can be divided into tow types: outer boundary and inner boundary.

The outer boundary means boundaries around each aquifer's periphery, which includes large-scale faults and surface water-bodies.

The permeable channel-way inside the aquifers is named as the "inner hydraulic boundary". Based on analysis of several hundreds of cases of serious water-bursting and inundation mine accidents in the basin, it can be found that almost all these incidents are related to various types of inner boundaries, and are caused by poor understanding to these inner boundaries for their hydraulic nature, concrete space position and shape. Therefore, systematically comprehensive studies on these inner boundaries are essential.

Studies on the inner boundaries should include division of their types, stereo simulation for their space positions, and assessment on their vertical permeability and concrete flux.

2.1. *Inner boundary of point karst collapse column*

Different scales of karst caves in the very thick carbonate rock aquifer have been found in the coal mining areas. These caves will collapse and can be filled with the brecciform rocks under the gravity action of overlying rocks and land surface buildings. A collapsed karst cave, from a plain view , looks like a point column. Therefore, it is called an inner boundary of point karst collapse column.

2.2. *Inner boundary of linear fault (fissure)*

The North China coal-accumulating basin as a geotectonic unit, belongs to the Zhongchao metaplatform. Major structural forms in the basin are faults and gentle folds due to its relatively long term stability. The permeable faults are generally located in the dense zone the convergent end of faults, or the intersection points of faults. The dislocation of permeable faults will destroy the integrity of both aquifers and aquifuges. This will not only shorten the distances between two neighboring aquifers situating in the two blocks of the fault, but also greatly reduce the impermeable strength of aquifuges.

2.3 *Inner boundary of narrow banded buried outcrop*

The thin-bedded limestone and thick-bedded sandstone aquifers in coal-bearing series and the very thick-bedded carbonate rock aquifers of the Middle Ordovician are often unconformably covered by porous deposits in most coal mines in the basin. The groundwater exchanging extent between the upper porous aquifer and the bottom bedrock ones in buried outcrop depends mainly on the following two factors. One is the permeability of weathered zone of the bedrock. Another is existence of a relatively thick aquitard between these aquifers.

2.4 *Inner boundary of planar fissure network*

Major water-filled aquifers in the coal-bearing series in the North China are the thick-bedded sandstones. Aquifuges between the two aquifers consist of fine sandstone, siltstone and mudstone. The brittle aquifuge releases stress in the form of fracturing under tectonic forces. This makes the aquifuge to be developed with a large number of mini-fissures, and to form a planar fissure network. Groundwater will flow through the network from a higher pressure aquifer to a lower one.

Past mathematical models for describing the stereo hydrogeological structure and predicting mine-water inflow have some similar characteristics, foe instance, all of them only consider the direct water-filled aquifer underlying coal-measures. Obviously, these models distort the stereo hydrogeological conceptual model in the basin.

After analyzing the regional mine-hydrogeological features in the profile, plane and boundary, this paper firstly apply a quasi-three-dimensional model and an optimal management model to the coal mines of the basin. These mathematical models describe a real stereo water-filled geological structure. They not only the hydraulic connections with multi-layer water-filled aquifers in shallow position, but also consider mine-water exchange between each inner boundary in quantity. At the same time, these models also overlay also groundwater leakage of the planar fissure network. Besides reflecting water-filled characteristics in the basin, the quasi-three-dimensional model has another important function, that is, it can easily provide scientific bases for working out decision-making system for the mine-water preventing and curing. It can be said that the model can be used not only to predict water inflows for different aquifers, exploitation levels and underground engineering tunnels, but also to determine quantitatively natures of different kinds of hydraulic boundaries which link up multi-aquifers. The natures include simulation of their extending positions in space, determination of their vertical leakage parameters, and prediction of the leakage inflow.

3. DECISION-MAKING SYSTEM FOR PREVENTING AND CURING WATER

Developed countries usually have advanced drainage equipment and technology, so they do not pay much attention to the studies on theories and methods of mine-water preventing and curing. In Germany, for example, pumping and dewatering equipment is commonly used in water curing, whereas the water-blocking is only regarded as a supplementary measure. It is generally thought that the water-blocking cannot replace the dewatering in practice, even though in the mines of the karst areas.

In China, the theory and practice of mine water preventing and curing are much better than in other countries. Many measures of the water preventing and curing have been utilized in practice, including pumping, dewatering, and water-blocking. In some cases, all these measures have to be used together. The dewatering measure includes the land surface dewatering and the underground tunnel dewatering. These two dewatering measures can be utilized together for some special coal mines. Another very efficient measure for preventing and curing water is a trinity combination of water-drainage, water-supply and ecological environmental protection. However, which measures should be chosen for mines with different mine-hydrogeological conditions? This is a very difficult question. Traditionally, if the leakage passage has little influence on total mine water inflows, according to qualitative analysis, the dewatering measure should be used, otherwise, the water-blocking measure should be adopted first. Yet, quantitative evaluation on the influence of the leakage passage is another question which can not be easily solved.

We believe that the decision-making system of the mine water preventing and curing in the North China coal basin should consider two aspects, one is preventing subsystem, the other is curing subsystem.

3.1. Preventing subsystem

The subsystem is mainly formed by establishing preventing measures and strengthening anti-water-hazard abilities. Enough drainage ability should be ensured, pumping-rooms should be enlarged, diving pumps should be utilized, and electricity supply should be guaranteed. It is necessary to establish independent water preventing systems for different mining districts, to improve water preventing walls and other sorts of anti-water enginerings, to setup groundwater dynamics measurement network system, and to raise measuring precision and dynamic analysis levels.

3.2. Curing subsystem

On the basis of analyzing mine-hydrogeological conditions, utilizing groundwater quasi-three-dimensional model and optimal model as evaluation tools, we can calculate a ratio between the leakage flow of the inner boundary and total mine water inflows and extra water resource situation of the whole groundwater system. Therefore the curing subsystem can scientifically work out a concrete curing water scheme. The steps includes:

Firstly, comprehensively analyzing mine-hydrogeological conditions, such as parameter distribution of water-filled aquifers, characteristics of recharge, flow and discharge of the groundwater system, hydrogeological boundaries, especially for the inner boundaries, all of these are a basis of the decision-making for the curing water schemes.

Secondly, utilizing the groundwater quasi-three-dimensional finite-element model to identify the stereo water-filled hydrogeological structure, for example, simulating concrete extending locations of the inner boundaries and determining their hydraulic parameters.

Thirdly, predicting both the total mine water inflows for a mining level and the leakage flows of the special inner boundaries based on the identified models. The leakage flows are predicted by both a real model and an ideal model. The ideal model is designed to be the same as the real one, except for without inner boundaries. Obviously, differences in the mine water inflows between these two models are considered to be the vertical leakage flows of the inner boundary.

Finally, after these tree steps, decision-making schemes of the water curing can be worked out as follows.

A. for coal mine districts with abundant extra water resource: (a). If the total mine water inflows are larger and the leakage flows of the inner boundaries are more than 40% of the total inflows, broad sense combination measure among water drainage, water supply and eco-environment protection should be adopted first as long as economic benefits of the combination are reasonable. Otherwise, the inner boundaries should be blocked.

(b) If the total mine water inflows are relatively smaller, no matter how large are the leakage flows of the inner boundaries, a dewatering measure should be recommended.

(c) If the total mine water inflows are larger and the leakage flows of the inner boundaries are less than 40% of the total inflows, broad sense combination measure among water drainage, water supply and eco-environment protection should be also utilized first providing the benefits are also reasonable. Otherwise, narrow sense combination among them is recommended to be used.

B. for coal mine districts with poor extra water resource:

(a) If the total mine water inflows are larger and the leakage flows of the inner boundaries are more than 40% of the total inflows, the inner boundaries should be blocked.

(b) If the total mine water inflows are smaller, no matter how large the leakage flows of the inner boundaries are, the dewatering measure should be directly recommended.

(c) If the total mine water inflows are larger and the leakage flows of the inner boundaries are less than 40% of the total inflows, narrow sense combination measure should be applied.

4 CASE STUDY

The Yanmazhuang coal mine in the Jiaozuo basin is one of the mines that has the most serious water-bursting problems from the footwall of the coal layers in North China. Three main underlying aquifers that threaten safe exploitation of coal layers are the thick carbonate rocks of the Middle Ordovician and the No.2 and No.8 thin carbonate layers of the Carboniferous. Inner boundaries in this mine are very complicated and critical.

Major causes for most serious water-bursting accidents happened in the past are unknown yet, although much attention need to paid to hydraulic natures and spatial locations of the inner boundaries. It can be said that one of critical factors for solving the serious water-bursting problems is to find out the inner boundaries. According to analyses of lithology, structure, groundwater flow and field hydraulic connection experiment in the mine, the F_{10}, F_3, F_5 and F_7 faults are regarded to be four water-inducing inner boundaries, in which The F_5 and the middle part of F_3 are considered to be the most serious water-inducing inner boundaries.

In the light of above analyses, the stereo water-filled structure is a major feature of mine-hydrogeological conceptual model in the Yanmazhuang coal mine. This structure is formed by three karst aquifers which are hydraulically linked up together by the four linear fault inner boundaries.

Field drainage data obtained in Dec. 24, 1985 in the mine are used to test the model. Each aquifer is discretized into 190 node points, 327 units. There are altogether 570 node points and 981 units for the three aquifers. According to results of model identification, simulated area can be divided into 12 homogeneous sub-areas.

Total mine water inflow for the mining level 2 in the Yanmazhuang mine is predicted to be 135 m^3/min by utilizing quasi-three-dimensional finite-element numeric model. The leakage flow for the southern two inner boundaries is calculated to be 65 m^3/min, it is about 48% of the total inflow for the mining level 2.

In the Yanmazhuang coal mine, there are abundant extra water resources, and the leakage flows of the inner boundaries is more than 40% of total mine water inflows. According to the decision-making subsystem of water curing, the scheme A(a) should be chosen, i.e. the broad sense combination measure among water drainage, water supply and eco-environment protection should be firstly utilized, otherwise, these two water-inducing inner boundaries in the southern part of the mine should be recommended to be blocked.

The practical verification indicates that the above curing water the scheme A(a) of water curing in the mine is feasible, and it is in accordance with field-measured data. In the light of assessment of financial situation in the mine, the scheme saves a drainage cost of 240×104 yuan/year after blocking the two inner boundaries in the southern part of the mine, and gains considerable economic and social benefits

5. MAJOR CONCLUSIONS

1. A stereo water-filled structure is formed by multi-layer aquifers which are hydraulically linked up together by various inner boundaries. This structure is a major mine hydrogeological characteristic for the North China coal-accummulating basin.

2. The quasi-three-dimensional finite-element model can be used to clearly describe the stereo water-filled structure, and it in reality reflects the main features of mine hydrogeological conceptual model.

3. The decision-making system for preventing and curing mine water can be divided into two subsystems, namely, water preventing subsystem and water curing subsystem.

REFERENCES

Wu, Q. & K. Tian 1991. The study on a quasi-three-dimensional finite-element model for a structure of multi-layer aquifers. *Proceedings of the International Conference on Modeling in Groundwater Resources.*

Mining Science and Technology'99, Xie & Golosinski (eds)© 1999 Balkema, Rotterdam, ISBN 90 5809 067 1

The Permian high-frequency sequence stratigraphy in western Shandong

Zengxue Li, Jiuchuan Wei & Meilian Han
Shandong Institute of Mining and Technology, Tai'an, People's Republic of China

ABSTRACT: The Permian strata in the western Shandong are composed of epicontinental and terrestrial deposits. There are four kinds of depositional systems: tidal flat, barrier-lagoon, river-controlled shallow water delta and lacustrine composite depositional systems. The high-frequency sequences are identified on the basis of recognition of the event surfaces caused by high-frequency sea level changes, and events represented by peat swamping, flooding, extensive erosion and exposure pedogenesis. The third-, fourth- and fifth-order sequences have been identified in the Permian of the western Shandong coalfields. The cyclic changes of the accommodation space caused by episodic tectonic subsidence in the continental lacustrine basin play principal controls on the development of these sequences. The periodicity of the spreading and withering of lake water is closely related to the Milankovitch periods.

1. INTRODUCTION

The fourth-order and even higher frequency cycles have been identified in the carbonate and siliciclastic successions. Goldhammer (1987,1990) investigated the fourth-order cycles superimposed on the third-order cycles. Mitchum et al (1991) discussed the stratigraphical records of high-frequency sea level cycles in the siliciclastic sediments and believe that the climatic changes caused by the cyclic changes of orbital factors such as eccentricity, obliquity and precession.

In the basin dominated by marine or paralic sediments, high-frequency sequences are the cyclic stratigraphic sequences produced by high-frequency sea levels, which occur as the fourth-order (0.1~0.2Ma), fifth-order (0.01~0.02Ma) (Mitchum et al, 1991) or even shorter cycles (or shorter period). These cycles are usually represented by parasequences or high-frequency sequences, which can further stack into the sequence sets through progradation, aggradation or retrogradation.

The sea level changes in the Early Permian epicontinental basin in the western Shandong were of high-frequency and of the characteristics of the composite sea level changes. During the Middle and Late Permian, the influence of sea level changes became weakening and was replaced by influence of the climatic changes caused by the Milankovitch events and the sedimentary accommodation changes caused by the basin tectonic events.

2. EPOSITIONAL COMBINATION AND FACIES

There were some different opinions on positions of the Carboniferous and the Permian boundary in the Shandong Province, even in North China. This boundary in North China was not correlated with that of other countries. In North China, this boundary was put on the top of the *Pseudoschwagerina* zone, while it is set at the bottom of the zone abroad. The L_{11} limestone is the lowest layer of marine limestones at the bottom of the Taiyuan Formation, directly overlying the No.17 coal bed. This limestone bed can be traced in Shandong and it's adjacent areas. Fossils such as *Quasifusulina, Rugosofusulina, Streptognathodus elongatus,* and *S. Gracilis* have been found in the L_{11} limestone, and this limestone can be regarded as the marker bed of the correlation. According to the characteristics of the fossil zone and it's distribution in the successions around the L_{11} limestone, we put the Carboniferous-Permian boundary at the bottom of the L_{11} (that is also the top of the No 17 Coal). This subdivision coincides perfectly with the international subdivision.

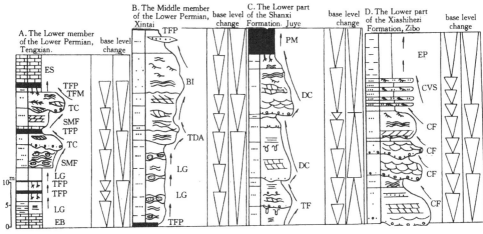

Fig.1 Typical combination of depositional facies in the Western Shandong coalfields

Four depositional systems have been identified in the Permian of the western Shandong, which include tidal flat, barrier-lagoon, river-controlled shallow water delta and fluvial-lacustrine composited depositional systems. The former two are developed in the Taiyuan Formation and the latter two in the Shanxi and Xiashihezi Formations.

The tidal flat depositional system was the relatively important system in the Early Permian epicontinental basin paralic coal-bearing successions. Gradual development of swamp on the tidal flat, and repeated occurrence of this event are significant characteristics of the coal measures in the epicontinental basin. The tidal flat depositional system is composed of mud flat (MF), sand and mud mixed flat (SMF), sand flat (SF), tidal channel (TC), tidal ridge (TR), tidal flat marsh (TFM) and tidal flat peat (TFP) environments. In vertical, the stacking of different kinds of depositional combinations often form more complicated depositional successions. For example, the mud mixed flat, mud flat, tidal channel and tidal flat peat swamp deposits can superimpose each other to form the mixed flat-dominated depositional succession. This kind of succession often contains several layers of peats with laterally extensive distribution, though they are thin-bedded. Another important depositional combination is composed of tidal flat (mud flat or mixed flat), tidal channel, tidal ridge, lagoon (or limited estuary bay) and carbonate platform deposits. The distinct feature of these two kinds of successions is that the tidal channel and the tidal ridge deposits are the framework facies. The weak progradation of the tidal flat produces depositional combinations with upward

fining successions and upward-increasing bioturbation. The tidal channel depositional combination often produces lenticular sandbodies, while tidal sand ridges often form sheet-like sandbodies.

The barrier-lagoon depositional system was developed in the southeast margin of the epicontinental basin of the western Shandong. The deposits of this system can be found in the mid-lower part of the Taiyuan Formation. The main genetic facies are barrier island combinations (BI) (including nearshore, foreshore, back shore and barrier flat), tidal channel (TC), tidal coast tidal flat (CTF), lagoon (LG), estuary bay (EB) and tidal flat marsh (TFM), tidal flat peat (TFP). In vertical, the deposits of the barrier-lagoon depositional system show upward coarsening then fining sequences with occasional marine erosional surfaces in the middle (Fig. 1B). In the study area, there are four basic units in the tidal delta depositional system: tidal flat (mainly mixed flat), finger tidal channel, sheet-like tidal channel, and tidal flat swamps. In the barrier-lagoon depositional system, the coastal tidal flat zones gradually prograde into the lagoon, and ultimately covering the whole lagoon. This is the process by which the lagoon is filled and developed into the tidal flat environments.

The fluvial dominated shallow water deltaic depositional system is developed on the top of the infilling succession of the epicontinental basin, which represents the latest stage of the evolution of the epicontinental basin. The shallow water deltaic depositional system in the Shanxi stage of the Permian is well developed in the west Shandong area. Due to strong progradation, the peat marsh development and the peat accumulation were closely associated with the developing of the shallow water delta. The rapid development of the delta, the progradation towards the basin and the overall sea level falling altogether caused the basinwide marine regression. The dominant factors influencing the development of the depositional systems were not only the high-frequency sea level changes but also the tectonic movement and the sediment supply. The vertical succession of the shallow water deltaic system is usually composed of two units, that is, the lower upward-coarsening unit and the upper upward-fining unit. The lower units are poorly developed, with thin delta front and prodelta sediments (Fig. 1C). The section is dominated by deltaic plain deposits. The main depositional facies include distributary channel (DC) (including the abandoned distributary channel), overbank (OB) (including crevasse splay (CVS) and crevasse delta (CD)), interdistributary depression (ID), interdistributary bay (IB), subaqueous distributary channel (SDC), and tidal flat and associated tidal channel (TF&TC) deposits, with less developed mouth bar (MB), sheet sand (SS), distal bar (DB) and prodelta (PD) deposits. During progradation of the delta, the deltaic plain deposits were developed and the distributary channels constituted their framework. There were two depositional processes in the interdistributary depressions: crevasse splay and overbank deposition, both of which filled the interdistributary depressions and caused swamps to be developed.

The Middle Permian Xiashihezi and the early Late Permian Shangshihezi Formations are dominated by deposits of the fluvial-lacustrine composited systems with various scales. The wide erosion events caused by the floodings can be used for stratigraphic subcorrelation and facies analyses. The well-developed fluvial and lacustrine deposits and their vertical superposition formed upward-fining depositional successions (Fig. 1D). The sedimentary facies include channel lag (LGD), overbank deposit (OB), channel filling (CF), crevasse splay (CVS), flood plain (FP), concave flood plain (CFP), lake deposit (shallow lake (SL), deep lake (DL)), levee (LV), and backswamp (BS) deposits. In the study area, the river channels in the Permian fluvial-lacustrine depositional system were highly meandering. The flooding plain and the flooding lake were developed as well, which were responsible for the thick sections of the muddy rocks. However, peat swamps were not well developed due to the influence of the climate.

3. ANALYSIS OF THE PERMIAN SEQUENCES

3.1. *Hierarchy and boundaries of the sequences*

There are seven orders of sequences: the first is the filling sequence of the basin; the second is the supersequence or tectonic sequence; the third is the sequence (the basic sequence); the fourth is the parasequence set; the fifth is the parasequence, and the six is the elemental sequence or microsequence. The third- to fifth-order sequences will be discussed here. The boundaries of the third-order sequence are the event surface caused by the regional sea level changes, such as the surfaces of the regional transgression and regression, the transforming surfaces of the regional tectonic stress field, and the regional unconformities. The boundaries of the fourth-order sequences are those transforming surfaces of the systems tracts, the surfaces across which the depositional system changes. The fifth-order sequences are formed by the high-frequency sea level changes, and can be especially identified with the transgressive surfaces, the surfaces of the large scale peat swamping and the exposed pedogenesis. In the western Shandong coalfields, the transgressive surfaces, the peat swamping surfaces, the exposed pedogenesis surfaces are well-developed in the Permian coal measures, whereas the flooding event surfaces, the large scale erosional surfaces and the surfaces of the abandoned lake are present in the noncoaly measures. These surfaces are the ideal boundary separating the fourth- or fifth-order sequences.

3.2. *The framework of the high-frequency sequences*

The Permo-Carboniferous successions preserved in the western Shandong coalfields can be divided into five third-order (sequence I-sequence V), eleven fourth-order and the twenty-eight fifth-order sequences, among which the fourth- and the fifth- order sequences are of high-frequency (Fig. 2). Sequence I belongs to the Carboniferous, and sequence II through sequence V belong to the Permian. Sequence II and III are the main parts of the infilling successions of the epicontinental basin.

Sequence IV and V are the infilling sequences of the terrestrial fluvial-lacustrine basin. The integrated processes of the sea level changes, tectonic subsidence of the basin, sediment supply, and climatic changes lead to the change of the depositional base level of the basin, and the latter, in turn, controls the filling of the basin.

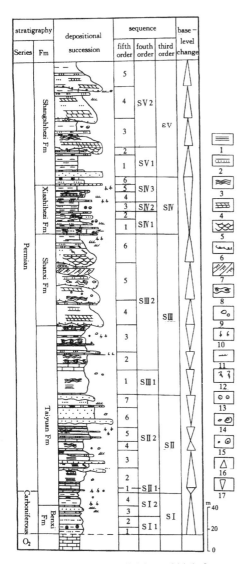

The formation of the fourth- and the fifth- order sequences in the sequence II and sequence III was controlled mainly by sea level changes. During this stage, the sedimentary base of the epicontinental basin in the western Shandong was gentle and the sediment supply was not high, hence the sea level rise, especially the rapid rise, would result in the large scale transgression.

By the time of deposition of sequence IV and V, the overall environments of the basin had become fully non-marine. The falling of the sea level and the rapid progradation of large amounts of clastic sediments caused large-scaled regression in the basin. The Milankovitch Periods lead to the periodical climatic changes, forming the rhythm in deposition. During this stage, the episodic tectonic subsidence in the fluvial-lacustrine depressions in the western Shandong caused the cyclic change of the accommodation. The sediment supply, together with the change of the climate, controls the architectural feature of the depositional sequence. The lake level change is not an uni-directional advancing or retreating of the water, but the spreading or withering of the water body on the whole. Such spreading or withering is of periodic changes, which may have genetic relationships with the Milankovitch periods.

This investigation is funded by the Coal Science Foundation of China. (No.97 Geology 10507).

Fig.2 Sketch map of the division of high-frequency sequences in the Western Shandong coalfilds. 1-horizontal bedding; 2-parallel bedding; 3-wavy bedding; 4-tabular cross bedding; 5-trough cross bedding; 6-erosional surface; 7-ripple bedding; 8-lenticular bedding; 9-mud gravel; 10-fossil plant root; 11-plant fragments; 12-bioturbation structure; 13-oolitic structure; 14-sideritic concretion; 15-pyrite concretion; 16-base level rise; 17-base level fall; 1, 2, ...- the fifth-order sequences; SII$_1$-SV$_2$ -the fourth order sequences; SII-SV - the third order sequences

REFERENCES

Brett, C. E., W. M. Goodman & S. T. Loduca 1990. Sequence, cycles and basin dynamics in the Silurian of the Appalachian foreland basin. *Sedimentary Geology* 69: 191-224.

Li, Z., J. Wei & S. Li 1997. The characteristics of sequence stratigraphy in the epicontinental basin. In: *Proc. 30th Intl. Congr.* 8: 141-151, VSP BV. The International Science Publishers.

Mitchum, R. M. & J. C. Van Wagoner 1991. High-frequence sequences and their stacking patterns: sequence stratigraphy evidence of high-frequency eustatic cycles. *Sedimentary Geology* 70: 131-160.

Mining Science and Technology' 99, Xie & Golosinski (eds)© 1999 Balkema, Rotterdam, ISBN 90 5809 067 1

Characteristics of the basal coal seam tonstein in the Taiyuan formation, western Shandong

Zuozhen Han, Jifeng Yu, Fengjie Yang, Xiuying Wang & Zhen Han
Shandong Institute of Mining and Technology, Tai'an, People's Republic of China

ABSTRACT: The Taiyuan formation is the main coal-bearing series in western Shandong and comprises a set of paralic coal-bearing series consisting mainly of clastic rocks with intercalation of marine carbonate rocks, with thickness of 135–180 m, and containing 2–4 mineable coal seams. Pyroclastic rocks exists in this formation, including some typical tonstein. These tonsteins can have industrial uses. In this paper, the origin, geological implications and utilization prospects of the tonsteins in the basal coal seams of the Taiyuan formation will be studied with the methods of mineralogy, petrology and geochemistry.

1. OCCURRENCE OF THE TONSTEIN

There is a layer of high quality tonstein at the bottom of the Taiyuan formation in western Shandong, which exists either as the coal seam floor or as the seam parting. In the Zibo, Jidong, Feicheng coal fields, this tonstein is preserved as the floor or parting of the No.10 coal seam, while in the Xinwen coal field, it occurs as the floor of the No.16 coal seam and in the Laiwu coal field as the floor and parting of the No.19 coal seam. Although these coal seams have different numbers in different localities, they were formed simultaneously, so was the tonstein layer. The thickness of the tonstein layer is generally 50cm, ranging between 10cm and 100cm in some places.

2. MINERALOGY AND PETROLOGY OF THE TONSTEIN

2.1. Macroscopic characteristics and textuural types

This tonstein is normally light brown colored, compacted massive, and fragile, and is easily weathered into pieces. It belongs to the family of hard claystones, The normal textures of this tonstein include cryptocrystalline texture, loosely mottled texture, mottled texture, crystalline texture, and cryptocrystalline pellet texture.

1. Cryptocrystalline texture: The tonstein with this texture usually shows compact and porodic, and contains little or no clasts but common uniform cryptocrystalline clay minerals. Under microscope, it shows light brown or brown in color and the micro-

lamellae can be observed. Under high-powered microscope, indistinct fine crystals can also be seen..

2. Loosely mottled texture: The characteristics observed with naked eyes are similar to those of cryptocrystalline texture. However, under microscope, a few vermicular kaolinite microcrystalline aggregates and quartz clasts can be seen, which occur in the form of phenocryst scattered within the matrix, with phenocrysts accounting for less than 10 percent.

3. Mottled texture: In this texture, the mineral phenocrysts are scattered in compact and porodic matrix, which can be recognized as aggregates of kaolinite, quartz and feldspar crystal clasts. The phenocrysts account for 10□30 percent of the whole rock.

4. Crystalline texture: This texture shows sand-like, and under microscope most clay minerals occur as fine to coarse crystal aggregates, in which kaolinite is their main constituents and are commonly vermicular kaolinite or kaolinites in the pseudomorph of biotites and feldspars. In addition, quartz, and feldspar crystals and accessory minerals are also commonly seen. Cryptocrystalline matrix accounts for less than 50 percent in total.

5. Cryptocrystalline pellet texture: In this texture, a large number of clay mineral cryptocrystals are accumulated into ellipsoid-like oriented pellets, with matrix between the pellets being organic matter and clay minerals.

2.2. Mineral compositions under the microscope and SEM

The tonstein studied here is mainly composed of cryptocrystalline kaolinite. Under the SEM, it can be seen that the cryptocrystals are mainly irregularly

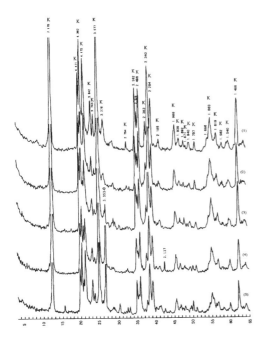

Figure 1 X-ray diffraction patterns of the basal coal seam tonstein from The Taiyuan formation, western Shandong. (1) Jidong (2) Laiwu (3) Xinwen (4) Feicheng (5) Zibo

arranged kaolinite microcrystalline aggregates. The mottles within the cryptocrystalline matrix are vermiform kaolinite crystal aggregates, which is the result of regular dislocation and accumulation of hexaplaty kaolinite lamellae along the 001 direction. There are also some amounts of other minerals such as quartz, feldspar, biotite and zircon, all of which have the features of typical pyroclastic crystals.

2.3. Characteristics of the X-ray diffraction of crystalline powder

A D/Max-R$_A$ X-ray diffraction instrument is used to analyze the tonstein samples (Fig 1). Experimental conditions are set as follows: CuKα radiation, Ni filtering, voltage 40KV, electric current 50mA. It is known from Fig.1 that the tonstein is mainly composed of kaolinites, with very little other minerals. Referring to standard diffraction pattern of kaolinites, our samples have following characteristics:

1. There are complete diffraction peaks of kaolinite, with oxy-symmetrical peak shape and clear fission, indicating that the major mineral components are those well-ordered kaolinites.

2. Strong interarea (00L) reflection is observed, comprising normally 5 to 6 diffraction peaks split up between 001 and 002. This suggests that kaolinite is fine-crystallized

3. Two groups of acute, symmetrical and threefold diffraction peaks are split out in the realm of 34 degree to 40 degree. All these (1,2,3) characteristics illustrate that kaolinites are highly ordered or well crystallized.

4. There are also the diffraction peaks of quartz (4.26Å, 3.35Å) and pyrites in the diffraction spectrum, indicating that there is a little quartz, pyrites in the sample.

2.4. Characteristics of infrared absorption spectrum

The Nicolet 170SX infrared absorption instrument made in the United States is also used in analysis of the sample. The results are shown in Fig. 2, from which we can see:

1. Two strong and two weak absorption peaks occur in the high-frequency realm from 4000 to 3000cm^{-1}, showing that the sample is composed of kaolinite. The strong peak at 3693 cm^{-1} is caused by the stretching vibration of the external hydroxyl of kaolinite octohedron lamellae, while the peak at 3620cm^{-1} is caused by the stretching vibration of its internal hydroxyl. The two weak peaks at 3668cm^{-1} and 3652cm^{-1} are caused by hydrogen bonds. The two strong peaks indicate that the hydroxyl radicels in packets of kaolinite are relatively perfect and well-ordered.

2. Nearly horizontal line with no peaks in the intermediate frequency realm from 1200 to 3000 cm-1 argues that there is no interlayer water existing between the packets of kaolinite in the sample and that the kaolinite is well crystallized.

3. Nine peaks with various strength appear in the low frequency realm from 1200 to 400 cm^{-1}, and they are mainly caused by the bending and stretching vibration of Si-O and Al-O bonds. These peaks show that the kaolinite has a high degree of order.

3. GEOCHEMISTRY OF THE TONSTEIN

The chemical compositions of this tonstein in different coal fields of western Shandong are shown in Table 1. Compared with the theoretical values of kaolinite, our tonstein sammples have following features:

1. The content of SiO$_2$ is slightly higher than the theoretical value of kaolinite, and this is caused by a small amount of quartz and chalcedony contained in the samples.

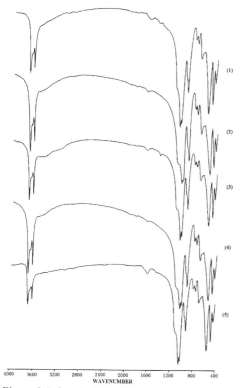

Figure 2 Infrared absorption spectrums of the basal coal seam tonstein from the Taiyuan formation, western Shandong. (1) Jidong (2) Laiwu (3) Xinwen (4) Feicheng (5) Zibo

2. The content of Al_2O_3 is slightly lower than the theoretical value of kaolinite. In general, the higher contents of Al_2O_3 in claystone than the theoretical value of kaolinite are caused by high-aluminum minerals while the lower values are caused by non-clay minerals in the rock, such as quartz and organic matters.

3. The content of Fe_2O_3 is relatively low. This is caused by existence of a small amount of pyrites in the claystone.

4. The content of K_2O+Na_2O is also relatively low, generally less than 0.7 percent, and the content of K_2O is higher than that of Na_2O. This is caused by existence of a small amount of illite in the claystone.

5. The content of $CaO+MgO$ is relatively low, generally less than 0.6 percent. This is caused by existence of a small amount of calcite in the claystone.

6. The loss of ignition (L.O.I) is generally higher than the theoretical value of kaolinite, which is possibly caused by the organic matter in the bulk.

In essence the chemical compositions of our samples are similar to the theoretical values of kaolinite.

4. ORIGINS OF THE TONSTEIN

There are several point of views as to the origins of claystones in the form of coal partings, such as, they can be allochthonous deposits, colloidal deposits, biochemical deposits or results of alteration of volcanic ashes. From above analyses, it is believed that this layer of tonstein originated from alterations of the volcanic ejecta (volcanic ash) falling into the peat-forming swamps. The alterations include leaching of the alkali metals and alkaline-earth metals from the ash, and desilication of the ash, under the processes of the organic acid. This can be supported by following evidences:

1. This layer of claystone is extensively distributed over the western Shandong, even all over the North China, the thickness of which is small, only several centimeters. The extensive distribution can be reasonably explained with the falling volcanic ejecta origin.

2. The mineral components of the claystone are simple, mainly kaolinite. The combination of non-clay minerals is stable, mainly volcanogenic high-temperature quartz, glassy feldspar, rod-shaped euhedral zircon, and biotite, with limited quantities. In non-coaly facies the tonstein is replaced by tuff.

Table 1 The chemical compositions of the basal coal seam tonstein from the Taiyuan formation, in different coal fields of western Shandong

Fields Comp.	Jidong	Laiwu	Xinwen	Fei-cheng	Zibo
SiO_2(%)	40.84	45.00	44.62	47.65	44.68
Al_2O_3(%)	34.10	38.92	36.88	33.07	37.86
Fe_2O_3(%)	0.69	0.47	1.32	0.82	0.82
TiO_2(%)	0.85	0.72	0.41	0.34	0.62
MgO(%)	0.62	0.20	0.42	1.10	0.08
CaO(%)	0.22	0.28	0.15	0.74	0.18
Na_2O(%)	0.18	0.00	0.27	0.12	0.18
K_2O(%)	0.48	0.06	0.28	0.65	0.24
L.O.I(%)	21.97	14.40	13.41	14.90	14.41
SiO_2/Al_2O_3 (mol)	2.03	1.96	2.05	2.45	2.01
TiO_2/Al_2O_3	0.02	0.02	0.01	0.01	0.02

3. The differences in contents of TiO_2 and ratio of TiO_2 and Al_2O_3 in different rocks reflect the differences in their precursory rocks. The content of TiO_2 in the claystone is normally less than 0.8 percent with the average of 0.59 percent, and the TiO_2/Al_2O_3 ratio usually ranges from 0.01 to 0.02. However, the content of TiO_2 in the surrounding mudstone is higher than 1.0 percent with the TiO_2/Al_2O_3 ratio from 0.03 to 0.06. Senkayi et al. (1984) believe that the TiO_2/Al_2O_3 ratios over 0.08 indicate provenance of basic volcanic ash while those below 0.02 indicate provenance of acid volcanic ash. Therefore, the

225

layer of claystone discussed here originates from the intermediate acidity volcanic ash.

4. The result of rare-earth element analyses shows that the average LREE/HREE ratio of the claystone is 10.4, while Eu/Sm 0.09, $\sum$REE 40-118ppm. The North American shales are the typical shales with sedimentary origins and their LREE/HREE values averaging 3.51, Eu/Sm averaging 0.20, $\sum$REE being 187.6. By comparison, the claystone studied here is relatively rich in the light rare-earth elements with evident Eu negative anomaly. These characteristics coincide with those of the intermediate acid volcanic rocks.

5. There are several layers of extensively distributed pyroclastic rocks in the Taiyuan formation of western Shandong. Normal sedimentary rocks are also found to contain many pyroclasts. The pyroclastic materials are mainly crystal clasts of qaurtz and feldspar, with minor lithic clasts and rare vitric clasts. These features also reflect that the original magma providing the pyroclastic materials was slightly acid.

5. IMPLICATIONS OF THE TONSTEIN

5.1. *Coal-seam correlation*

The volcanic event layer is of isochronic nature and can be used in stratigraphic correlation not only within one coalfield but also between different coalfields over a large area. As the depositional environments of the Permo-Carboniferous period in different areas are various, the number of marine beds in these areas are not consistent, thus resulting in certain confusions in stratigraphic correlations. However, two layers of tonsteins of volcanic ash origin (one at the bottom of the Taiyuan formation and another in the middle part of the Taiyuan formation) can be traced over the whole western Shandong area. These tonstein layers, assisted with marine bands, can provide a reasonable scheme for coal seam correlation between and within the coal fields in western Shandong.

5.2 *Relations to coal accumulation*

Volcanic events have twofold influences on the forming of coal seams, including both favorable and unfavorable influences. They can cause the growth of plants to break, which is unfavorable for coal accumulation but only lasts for a rather short time in the chronic geological history. However, as they happen paroxymally and fall into a large area within a short time, their deposits can cover the peat bed and prevent organic matter in peats from being oxidized, thus favorable for coal accumulation. The facts that carbonized plant materials have been found in the Tertiary basalts in the Jilin Province of China and that coal has been mined from the Quaternary volcanic rocks in the Java island of Indonesia illustrate that volcanic events are not so detrimental to the coal accumulation as some people imagine. Furthermore, volcanic events could even make special contributions to coal accumulation. For instance□ the fine-grained ash is rich in nutrients which are essential to the growth of plants. Flourishing plants are growing over a 300 years old volcano island in Alaska, showing the feature of temperate rain forest due to the adjustment of oceanic climates and the rich mineral nutrition in the volcanic rocks, although the island is situated at the high-latitude belt.

5.3. *Sedimentary environments*

It is generally believed that the Permo-Carboniferous depositional setting in North China was an epicontinental sea, and coal was formed in peat flat environments. With the research going on, more and more event deposits, such as storm deposits and volcanic deposits, are being continuously recognized in the Late Palaeozoic coal measures in North China. These event deposits and their distribution regularities can be used in the reconstructing of ancient sedimentary environments. The fact that pyroclastic rocks are widely distributed, with grain sizes ranging from fine, medium to fine gravel in various localities, suggesting that volcanism not only took place at the plate edge but also within the depositional basin or at its margin, and that some storm events might have something to do with volcanic eruptions.

REFERENCES

Addison, R. et al. 1983. Volcanogenic, tonstein from Tertiary coal measure, East Kalimantan, Indonesia. *Int. J. Coal Geol.* 3: 1-30.

Linda, M. & R. Smith 1990. Mineral assemblages of volcanic and detrital partings in tertiary coal beds, Kenal Peninsula, Alaska. *Clays and Clay Minerals* 27: 19-130.

Liu, Y. 1984. *Element Geochemistery.* Beijing: Science Press.

Senkayi, A. L. et al. 1984. Mineralogy and genetic relationships of tonstein, bentonite, and lignitic strata in the Eocene Yegua formation of east-central Texas. *Clays and Clay Minerals* 32 (4): 259-271.

Xu, X. et al. 1993. Volcanic deposits of the Carboniferous coal-bearing series in western Shandong province. *Acta Sedimentologica Sinica* 11(4).

Mining Science and Technology'99, Xie & Golosinski (eds) © 1999 Balkema, Rotterdam, ISBN 90 5809 067 1

The statistics of the discontinuities in rock mass in coal mines and their application

Shaoxiang Hu & Weiming Wang
Shandong Institute of Mining and Technology, Tai'an, People's Republic of China

Xianwei Li & Yongshuang Zhang
China University of Mining and Technology, Beijing, People's Republic of China

ABSTRACT: This paper presents studies or rock mass characteristics in coal mines. The statistical approach to defining the trace spacings in the condition of multiple sets of discontinuities coexisting in the same rock mass is presented. Window statistic method of trace length is introduced and the way to determine the maximum movable domains of the dangerous rock in the roadways is presented. Selection of the optimum axial directions using the statistical parameters is introduced, which provide scientific basis for the support design of roadways.

1. INTRODUCTION

The investigations in underground coal mines show that there often exist minor faults with the same attitudes in the vicinity of the major faults, and smaller scaled joints, in turn, occur near the minor faults. The shorter the distance to the faults, the denser the joints. This phenomenon reveals that the minor structures with the same scale in the same structural district have strong geometric self-relativity and special statistical feature. Theoretical analysis and engineering practices show that the statistical feature of the attitudes of the minor structures may approximately represent that of the discontinuities in the same structural block. The predominant directions of the minor structures obtained by polar projection methods and rose charts can approximately represent the dominant directions of the discontinuities, by which the ascription of every discontinuity can be determined. Measuring the attitudes of this type of structures and complementing sample survey data of the densities and trace lengths of discontinuities in a certain surrounding rock surfaces, we can establish the statistical model of discontinuities.

As an application of the engineering example: we analyzed the floor bed of No.2 coal seam in a mine. In the coal mine area, there are 20 large-scaled faults (H>30m), 49 mid-scaled faults (H=5~30m) and 839 small-scaled faults among which 490 faults are of 0.25~0.5m. The mine district was divided into eight structural zones according to the structures

distribution. Four dominant directions were founded in each zone.

2. THE STATISTICS OF TRACE SPACING OF DISCONTINUITIES

The trace of discontinuities is referred to the intersecting lines of the discontinuities with the exposure surfaces of rock mass. There are many research results about the statistics of the traces of the discontinuities[1][3], while most of them are aimed at one single group of discontinuities. The discontinuities in rock mass often occur in several groups, forming intersecting trace net in exposure surfaces. In order to obtain the space of each group, several measure lines are needed. In the case two groups of discontinuities as shown in Fig. 1, two lines are used. The intersecting angle between the first line (L_1) and the first group of discontinuities (choose the acute angle) is θ_{11}, and the intersecting angle of L_1 with the second group of discontinuities is θ_{21} (also choose the acute angle). If the dips and dip angles have been measured, the intersecting angles of these two groups of discontinuities may be calculated:

$$\alpha_{12} = \arccos(\hat{n}_1, \hat{n}_2) \tag{1}$$

The α_{12} can also be treated as the intersecting angle of the two group of traces. Thus:

$$\theta_{21} = 180 - (\alpha_{12} + \theta_{11}) \tag{2}$$

where: $n_i = (\sin\alpha_i \sin\beta_i, \sin\alpha_i \cos\beta_i, \cos\alpha_i)$ (3)

i=1, 2

$\alpha_i \beta_i$ are the dip and dip angle of the i group discontinuities respectively.

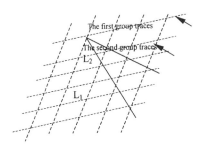

Fig.1 The arrangement of measuring lines

The total intersecting times of the two groups of traces with the first line are:

$$K_1 = \frac{l}{E(X_1)}\sin\theta_{11} + \frac{l}{E(x_2)}\sin[180 - (\alpha_{12} + \theta_{11})] + 1$$

(4)

where E(x₁) and E(x□) are the average spacing of the two group of traces which remain unknown, other parameters are all known. Therefore, the second line should be take into account. The intersecting times of the two groups of traces with the second line can also be calculated in the same way.

If there are m groups of discontinuity traces in the same exposure surfaces, m lines should be arranged accordingly, thus there are:

$$[L]\{E(x)\} = \{K\}$$ (5)

thus $$\{E(x)\} = [L]^{-1}\{K\}$$ (6)

where $\{E(x)\}$- the matrix of the spacings of traces,

$[L]$- the matrix of the trace directions,

$\{K\}$- the intersecting times of the traces with lines

For example: there are four groups of discontinuities in the surrounding rock of a roadway. The dip β and dip angle α are measured as (88,62), (133,59), (98,73), (119,68) respectively. In order to obtain the spacing distribution of the four groups of discontinuities, four survey lines are arranged. The length of each line is 4 meters. The intersecting angles of the four lines with the first group of discontinuities are $\theta_{11} = 20°; \theta_{21} = 30°; \theta_{31} = 45°; \theta_{41} = 90°$ respectively. The intersecting times of each line with discontinuities are $K_1 = 6; K_2 = 8; K_3 = 12, K_4 = 24$

respectively. From the formula (6□, we have:

$E(x_1) = 0.552$m,

$E(x_2) = 1.120$m,

$E(x_3) = 0.35$m,

$E(x_4) = 1.381$m

3. THE STATISTICS OF THE TRACE LENGTH

The common statistical methods are scanline survey statistics and statistical window methods. Paper [4] gives the detailed description to this method, where the distribution functions and averages are presented. The main defects of the scanline method lie in that the various density functions of trace length and their relations must obtained first, in site workload is quite heavy and difficult in operation. The window method does not need to know the distribution function of the average trace length. The intersecting point numbers of the traces with the statistical window and the dip angle distribution function of the traces and window length can be accounted for the average trace lengths. Following is the basic method and formula of the statistical window.

The statistical window method need a rectangle scope with a length a and width b in the exposure surface. The trace numbers and the intersecting relation of traces with the rectangle in the $a \times b$ area should be counted. This rectangle is called "statistical window".

The intersecting ways of traces with the statistical window can be grouped into three types. (1) containing type: the both ends of the trace are covered in the statistical window; (2) penetrating type: one end of the trace is in the window; (3) intersecting type: the both ends of the trace are out of the window. The procedure is that: record the dimension of the window, measure the attitudes, trace length, spacings of the same group and the types of the discontinuities. Group the discontinuities (parallel discontinuities are classified into one group) and coordinates the numbers of each group of traces, i. e., the number of containing type N_1, the intersecting type N_2, penetrating type N_3 and the total number N. $N = \sum_{i=1}^{3} N_i$. In addition, the intersecting angle (acute angle, θ_A) of the trace with window length is also need to be recorded.

The average trace length can be estimated as:

$$\bar{l} = \frac{ab(1 + \frac{N_3}{N} + \frac{N_1}{N})}{(1 - \frac{N_3}{N} + \frac{N_1}{N})(aB + bA)}$$

(7)

where: $A = E(\cos\theta_A); B = E(\sin\theta_A);$ other symbols are as above.

4. THE APPLICATION OF THE STATISTICAL PARAMETERS OF THE DISCONTINUITIES

According to the Shi's theory, the geometric conditions of moveable rock blocks in the roadway are:

$$\pm \hat{a} \notin JP \qquad (8)$$

Where $\hat{a}$ -the axis vector of the roadway; JP-the joint pyramid cut by discontinuities.

Combined with stress analysis, the potential dangerous rock in the roadway can be determined. Before excavation, the maximum moveable area must be calculated in order to provide data for the support and controlling the strike of the roadway.

The maximum moveable area may be regarded as the outer enveloping lines. The dangerous rock is a piece of multiple-edged pyramid. The vector of each edge is $\hat{I}_{ij} = (\hat{n}_i \times \hat{n}_j),$ (n_i/n_j) are the

outer normal vector $(i.j = 1,2,\cdots m, i \neq j)$. The projections in the sections are $\hat{I}_i{}'(i = 1,2\cdots m)$, the outer normal of I_i' are $\hat{n}(\eta_i) = \hat{I}_i' \times \hat{a}$. There are surely at least 2 edges in the m projection edges intersect with the sections of roadway, the normal vectors of the intersecting points are

$\hat{n}(\eta_1) = \hat{I}_1' \times \hat{a}$ and $\hat{n}(\eta_m) = I_m' \times \hat{a}$ respectively. The two edges that intersect with the section of roadway are named "boundary edges". The relations of the normal vector of the "boundary edges" with other edge vectors are as following:

$$\hat{n}(\eta_1) \cdot I_1 = 0, \ \hat{n}(\eta_1) \cdot I_N < 0 \qquad (i = 2,3,\cdots m)$$
$$\hat{n}(\eta_m) \cdot I_m = 0, \hat{n}(\eta_m) \cdot I_i < 0 \qquad (i = 2,3,\cdots m_{-1}) \Bigg\} (9)$$

From the formula (9), we can differentiate the boundary edges and determine the maximum movable area (Fig. 2). Using the statistical parameters of the discontinuities, we can also simulate the joint trace net of the unexcavated roadway surfaces and establish the prediction models, providing credible basis for the stable analysis of the rock surrounding roadways.

Fig. 2 The maximum movable area of the dangerous rock

For example, a semi-circular arch roadway, span 4 meters, height 3.4 meters, the axis direction $\beta = 259°$, axis dip angle $\alpha = 89.7°$. There are four groups of discontinuities.

Table1 shows that the maximum dangerous rock in the roadway is the No.1100 block. The maximum movable volume is 2.7036m³ with the weight of 67.5893kN. It has a sliding trend along the single plane p₄, with a net sliding force of 45.17KN. It is the dangerous rock that must be reinforced in the supporting design of the roadway.

In order to obtain the optimum axial direction, in the permission of engineering, the maximum movable area is obtained after adjusting the axes of the roadway (table 2&3).

Comparing table 1 with table 3, when the axes circum-gyrates 9 degree southward, the volumes and sliding forces 61.5949kN from the original

Table 1 The maximum movable area of the dangerous rock in a roadway (β = 259°)

The number of rock	Volume (m³)	Net sliding force (kN)	Sliding patterns of the dangerous rock	Coordinate of the move direction		
				X	Y	Z
1000	0.3096	3.7881	Sliding along single plane p₂	0.3767	-0.3512	-0.8572
1101	0.8136	16.1539	Sliding along single plane p₃	0.2902	-0.0356	-0.9563
1100	2.7036	45.7146	Sliding along single plane p₄	0.3417	-0.1894	-0.9205
1010	2.7036	1.6509	Sliding along two planes p₂, p₄	0.1908	-0.5257	-.8290
1110	8.1189	10.9222	Sliding along two planes p₃, p₄	0.2318	-0.3697	-0.8998

Table 2 The maximum movable area of the dangerous rock in a roadway ($\beta = 250°$)

The number of rock	Volume, m³	Weight, kN	Net sliding force , kN
1000	☐0.3688	8.4211	4.0672
1101	☐1.0788	26.9688	21.4197
1100	☐3.6428	91.0697	61.5949
1010	☐3.6428	91.0697	2.2777
1110	☐9.6903	242.258	13.0362

Table 3 The maximum movable area of the dangerous rock in a roadway ($\beta = 265°$)

The number of rock	Volume, m³	Weight, kN	Net sliding force, kN
1000	0.3019	7.5480	3.6455
1101	0.8046	20.1144	15.0757
1100	2.2019	55.0474	37.2312
1010	2.2019	55.0474	1.3768
1110	7.3213	183.3023	9.8492

45.7140kN. When the ax circum-gyrates 6 degrees of every group significantly increase. The sliding force of the No.1100 dangerous rock block increases up to north, the sliding forces of every dangerous rock block decrease distinctly (table 1&3). The dangerous rocks with the sliding forces more than 10kN become two groups from the original three groups. The occurring frequency of the dangerous rock decreases. The weight of the largest dangerous rock reduces from 45.7140 to 37.2312 kN. Anyhow, the supporting load increases when the roadway axes turn toward south and decreases when turn toward the north.

REFERENCES

[1] L. A. Hudson & S. D. Priest. 1976. Discontinuities and Rock Mass Geometry, *Int. J. Rock Mechanics and Mining Sciences, Abstr.* Vol.16, pp339~362

[2] Wallis, P. F. and King, M. S. 1980. Discontinuity Spacings in a Crystalline Rock, *Int. J. Rock Mech. Min, Sci, &Geomech. Abstr,* Vol.D,PP63~66.

[3] Prist, S. D. and Hudson, J. A. 1976. Discontinuity Spacings in Rock, Int. J. Rock Mech, Min. Sci. and Geo mech. Abstr. Vol13. pp135~148.

[4] Priest, S. D. and J. A. Hudson 1981. Estimation of Discontinuity Spacing and Trace Length Using Scanling Surveys. *Int. J. Rock Mech. Min. Sci & Geomech. Abstr* Vol.18. pp183~197.

[5] Biedong,Pan et al. 1989. Statistical model of rock mass and its application. *New Advances of Rock Mechanics*, Northeast Technical University Press.

Mining Science and Technology' 99, Xie & Golosinski (eds)© 1999 Balkema, Rotterdam, ISBN 90 5809 067 1

Coalbed methane drainage technology in Henan Province

Xianbo Su & Youyi Tang
Jiaozuo Institute of Technology, People's Republic of China

Jianhai Sheng
Coal Geological Bureau of Henan Province, Zhengzhou, People's Republic of China

ABSTRACT: Development of coal bed methane drainage technology in Henan Province started in early 70's Considering geological conditions, three methods for drainage of coalbed methane in Henan Province are proposed in this paper. The first is roof stimulation that is suitable for structural coal. The second is acid treatment that is suitable for Anthracite in which cleats are filled with calcite. And the third is conventional development that is suitable for primary and fissure structural coal.

1. INTRODUCTION

There are three advantages to develop coalbed methane: 1) coalbed methane is a kind of new clean energy resources; 2) gas hazard of coal mining can be avoided to some extent, 3) the quality of atmosphere can be improved. As we know, the hazard of coal gas is generally concentrated in the area with structural coal distribution, and we also have not found outburst of coal and coal gas in the area with primary structural coals. At present, coalbed methane development projects are all concentrated in the area with primary or fissure structural coals, and there have not been any kinds of development technologies which are applicable to the areas with clastic coal and mylonitic coal distribution In Henan province, structural coals are distributed widely, which account for a large proportion. How to develop coalbed methane in these areas becomes an interesting research focus.

The development of coalbed methane in Henan province began in the Jiaozuo mining district. In the seventh "five-year plan", the new theory of coalbed methane geology was introduced. In the eighth "five-year plan", the North-China Petroleum Geology Bureau undertook the research project of the China National Planing Committee: "Geology Assessment of Coalbed Methane and Exploration/ Development Technology". In this project the coalbed methane resources in Henan province have been entirely evaluated, and the experiment of exploration/development was performed in the Anyang coal district. At the beginning of 1990's the exploration/development of coalbed methane came into a climax. Some companies performed experiments of

exploration and development in the Xinggong, Pindingshang, Anyang, Hebei and Jiaozuo areas. Because of the complicated geological conditions and improper technology, the results of the experiments were unsatisfactory. Based on practices and achievements of the past research projects, the authors will present three development methods, as summarized in Table 1.

Table 1. Stimulation technologies and their applications

Stimulation technology		Application
Roof stimulation		Structural coal reservoir
Acid treatment		Anthracite, the cleats are filled with calcite
Conventional Stimulation	Hydraulic Fracturing	Primary or fissure coal reservoir
	Open-hole cavity	High pressure and high permeability reservoir

2. ROOF STIMULATION

In Henan Province, the major mineable coalbed is No.II$_1$ coal in the Lower Permian Shanxi formation, which are commonly reworked into structural coals structural processes. Clastic and mylonitic coals account for more than 70% of the total No.II coal. Moreover, gas contents of these structural coals are usually as high as 10m^3/t. Because this kind of coal is reworked seriously, cleats do not exist and permeability is very low. A new technology must be adopted to replace the conventional technology for developing coalbed methane from this coal reservoir. According to the principle of mining "liberating coalbeds" that was used to prevent outbursts of coal and coal gas, we suggest a roof stimulation technology which will be discussed here.

Firstly, for multiple seamed coal mines, the coalbeds without danger or with least danger of gas outbursts will be mined firstly. These coalbeds are called "liberating bed". The unmined coalbeds with obvious danger of gas outbursts are called "liberated bed", which are influenced by liberating beds. Secondly, for single thick coalbed with danger of gas outbursts, the upper part of the bed will be mined firstly, which is named as liberating subbed and the lower unmined part of the bed is named as liberated subbed. When the liberating bed is mined, the gas existed in the liberated one will largely escape, and the reservoir pressure will be largely lowered, and then the danger of gas outbursts will be obviously reduced or fully relieved. It has been confirmed in the practice that the mining of upper part subbed may reduce 60~70% of total gas content in the lower subbed. In view of that coalbed methane can migrate out from the coalbed by mining the liberating bed, we can also stimulate the roof of the coalbed through surface drilling, so that permeability of the coalbed can be improved. In this way coalbed methane in structural coal reservoir with low permeability can be effectively extracted.

According to coal reservoir characteristics, roof stimulation can be further divided into three types. The first one is designed for clastic and mylonitic coal with low permeability ($n \times 10^{-3}$ md), such as II_1 coal in the Xinggong area. In this case, the surface well is completed when drilling to the coalbed roof, not into coalbed. A portion of coalbed roof can be regarded as "pseudoreservoir" which will be stimulated by hydraulic fracturing or open-hole cavity. Coalbed methane is then defused into this pseudoreservoir from coal reservoir, and serve for final extraction. The thickness of pseudoreservoir is not constant with different lithologies. The second stimulation is designed for the coalbed such as V_{9-10} coal seam in the Pingdingshan. district with a subbed of primary structural coal or fissure coal in the top. In this case, the well is completed when drilling into this subbed. This subbed and a portion of coalbed roof is regarded as pseudoreservoir to be stimulated. Also, the coalbed methane diffused from the coal reservoir can be developed. The determination of thickness of "Pseudoreservoir" is of the same principle as that of the first type. See Figure 1. Also, the coalbed methane diffused from the coal reservoir can be developed. The determination of thickness of "Pseudoreservoir" is of the same principle as that of the first type. See Figure 1. The third one is designed for multiple coal seams, such as VI_{15} and VI_{16-17} coal seams of the No.8 coal mine in Pingdingshan

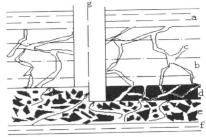

Figure 1. Mechanism of roof stimulation for single coal seam. a-mudstone, b-pseudoreservoir, siltstone, c-joints or fractures, d-primary or fissure coal subbed, e-structural coal, f-mudstone, g-well

coal district. The interval between these two coal seams is only 7-10m thick and is composed of siltstones with common joints and fractures. The siltstones are stimulated into the pseudoreservoir for the two overlying and underlying coal seams. Coalbed methane from these two coal reservoirs will be developed. See Figure 2.

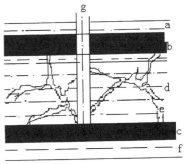

Figure 2. Mechanism of roof stimulation for mult-coal seams. a-roof, siltstone, b-upper coal seam, c-lower coal seam, d-pseudoreservoir, siltstone, e-joints or fractures, f-floor, mudstone, g-well

It is possible to use roof stimulation to extract coalbed methane from the structural coal reservoir. Because of low permeability, the coalbed methane in this reservoir can not be developed by conventional technologies. The mechanism of stimulation is that permeability of pseudoreservoir and connectivity to coal will be obviously increased after hydraulic fracturing and open-hole cavity. In the course of pressure reducing, coalbed methane constantly desorbs, diffuses and migrates into pseudoreservoir, where coalbed methane is finally developed. Although the permeability of structural coal reservoir is very low, it is possible that coalbed methane moves to roof pseudoreservoir. The liberating mining practice has indicated that more than 60% coalbed methane will diffuse or migrate into the pseudoreservoir. However, attentions must be paid to the proper selection of stimulating technology because joints are commonly developed in the roof rocks of the coalbed. The effective method to increase the aperture and length of these joints is the hydraulic fracturing. The use of the open-hole cavity completion method has special requirements to pressure and permeability of pseduoreservoir, so caution must be paid when selecting this technology.

3. ACID TREATMENT

Acid treatment is a common stimulating technology in the development of conventional oil and gas. A kind of acid (such as hydrochloric acid, sulfuric acid, hydrofluoric acid or organic acid) is injected into coal reservoir when under the formation fracturing pressure and the acid can dissolve some minerals in the cleats or fractures of coal reservoir, and then permeability will be increased. In some districts of anthracite (such as the Jiaozuo), the gas content is high, primary structural coal or fissure coal exists in same parts of the area, as a consequence, the developing potential of coalbed methane is very great.. When coal reservoir is investigated carefully, it will be found that the cleats in anthracites are not fully closed (cleat density is 0 to 2 cleats per 5cm) and these cleats are usually filled with carbonate. To deal with this situation, we proposes the acid treatment technology. This technology possesses more advantages than hydraulic fracturing. Hydrochloric acid and sulfuric acid are used to dissolve carbonates in cleats and to make these cleats open. After acid treatment, the permeability of coal reservoir will be increased and the connection between coal reservoir and surrounding aquifers will be avoided. When hydrochloric acid is used, the chemical reaction is expressed as follows:

$$CaCO_3 + 2HCl \rightarrow CaCl_2 + CO_2 \uparrow + H_2O$$

The type and concentration of acid, and the treatment pressure and temperature are determined by some special requirements.

4. CONVENTIONAL STIMULATION TECHNOLOGY

Primary and fissure structural coals are distributed in some districts in Henan Province. This kind of reservoir can be stimulated by conventional development technology - hydraulic fracturing or dynamic open-hole cavity completion - in coalbed methane development. Hydraulic fracturing is widely adopted to stimulate the reservoir in Henan Province and has increased producibility to some extent. The adoption

of dynamic open-hole cavity completion is limited by the coal reservoir conditions. The activity of coalbed methane development in the San Juan basin shows that this method is suitable for those reservoirs with high pressure and high permeability. The permeability is usually above 10 md and at least is not below 5 md. This reservoir must be matched with underground water dynamic conditions. It is difficult to find this kind of reservoir in Henan Province, so it must be cautious to adopt this technology.

5. DISCUSSION

The success of coalbed methane development mainly depends on technology. Different geological backgrounds resulted in different coal reservoirs. The development technology must be consistent with the type of coal reservoir. The detailed description of coal reservoir is a critical work. Only by understanding the property of reservoir systematically can the proper development technology be selected.

The roof stimulation is feasible in Henan province. This technology is probably the best method for developing coalbed methane from structural coal reservoirs, but more experiments and studies need to be performed.

The acid treatment is only suitable for the coal reservoir with the cleats filled with calcite. Although acid treatment is very common in the development of conventional oil and gas, there is no precedent in the development of coalbed methane. Therefore, experiments and research are needed.

ACKNOWLEDGEMENTS

This research was funded by National Natural Science Foundation of China (NSFC) with project No.4970207.

REFERENCES

Jiaozuo Mining Institute, 1990. *Gas Geology*. Coal Industry Publishing House. 100-116.

Tyler, R., A. R. Scott & W. R. Kaiser et al. 1996. Geologic and hydrologic controls critical to coalbed methane producibility and resource assessment: Williams Fork Formation, Northwest Colorado. *Topical Report of Gas Research Institute*: 319-334.

Mining Science and Technology'99, Xie & Golosinski (eds) © 1999 Balkema, Rotterdam, ISBN 90 5809 067 1

Optimal sample spacing for gas content measurement in a coal seam: A statistical approach

S. Xue
CSIRO Exploration and Mining, Brisbane, Qld, Australia

Z. Jiao
Queensland University of Technology, Brisbane, Qld, Australia

ABSTRACT: This paper studies the optimisation of sampling space for gas content measurements in a coal seam. The methodology developed is based upon establishing the sample variograms and block krigging by using exploration data on gas content and determining the influence area of each sample by using ranges of the variograms. By optimising sampling configurations, sample spacing has increased from between 70 to 170 m to 300 m, hence to minimise the effect of sampling practice on mine production.

1 INTRODUCTION

In order to minimise the risk of coal and gas outbursts in underground coal mines working on gassy coal seams, it is a common practice to monitor and measure gas contents prior to and during roadway developments and longwall operations. The practice consists of drilling an in-seam hole, taking a core and measuring its gas content and comparing this with a threshold value. It is time-consuming, expensive and disruptive to normal mine operation.

The mine studied has a moderate gas content between 6 m^3/t to 9 m^3/t and selective gas pre-drainage is practiced. During roadway development, a coal sample has to be taken every 100m ahead of the development face to measure its gas content. In most cases, the measured gas content is well below the threshold value for safe mining (8 m^3/t in this case). This study was undertaken to statistically determine the optimal configuration and spacing for coal sampling, the outcome of which might reduce costs and down-time for the mine by reducing the number of cores required to characterise seam in terms of its gas content.

2 METHODOLOGY

The following steps were taken to accomplish the objectives.

(1) Exploration of data set

- Examine available data of gas contents and sampling locations.

- Eliminate data that may not be appropriate in a sense that the measurement of gas content may not be accurate due to the leakage of a canister. Data taken from areas where geological structures such as dykes, faults and shear zone exist was also eliminated.

(2) Establish sample variograms of data set

- Omnidirectional variogram
- West-East variogram
- South-North variogram
- North-East variogram
- North-West variogram

(3) Establish block krigging system based on the variograms established

(4) Calculate and analyse the error variances of the mean gas contents of blocks of 200×170 m^2

(5) Determine the area of influence of each sample with ranges of variograms

3 ANALYSIS & DISCUSSION

3.1 *Data summary*

The data set consists of 158 gas content values of coal samples taken at spacings of 100, 170, or 200 m. These values and their corresponding geographical locations are closely examined against any obvious error or abnormality. Statistical summaries of the data are given as follows:

Minimum value:	2.25
Maximum value:	11.11
First quantile:	5.72
Third quantile:	8.37
Median:	7.12
Mean:	6.99
Standard deviation:	1.98
Coefficient of variation:	0.28

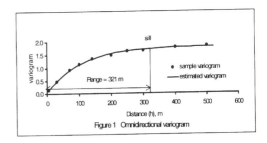

Figure 1 Omnidirectional variogram

3.2 Calculation of variograms

One of the methods in assessing the geological variability is to analyse the continuity of data. In geostatistics, continuity is usually described by variograms. A variogram in a specific direction is a measurement of the similarity of values a projected distance. A variogram has two elements: distance and direction. There are two kinds of variograms: directional and omnidirectional. Directional variogram is a variogram along a specified direction while omnidirectional variogram is the average variogram in all possible directions. Generally the variogram value will increase as the separation distance between two sample points increases. Eventually, however, an increase in the separation distance no longer causes a corresponding increase in the average squared difference between pairs of values and the variogram reaches a plateau. The distance at which the variogram reaches this plateau is called the range. In geostatistics the range is usually used to define the influence area of a sample. The value of a variogram at the range is called the sill of the variogram.

Five variograms (Figs 1 to 5) are calculated for the data set. These consist of four directional variograms along West-East, South-North, North-East, and North-West, and an omnidirectional variogram. East-West direction corresponds to the direction of the longwall while South-North corresponds to the direction of the longwall face.

Most of the samples are regularly spaced at 100, 170 or 200 m, so lag increments chosen in calculating the variograms are the same as the regular sample spacing, i.e. 100, 170, and 200 m. Smaller lag increments are used for the areas where samples were densely spaced.

3.3 Modelling of variograms

All these sample variograms show similar shapes with very small values near their origins. The similarity in shapes indicates the homogeneity of data in all directions while the small values of variograms near the origin suggest a good statistical continuity of data. This coincides with the contour

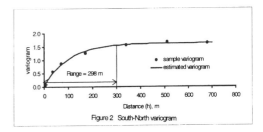

Figure 2 South-North variogram

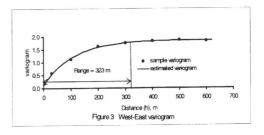

Figure 3 West-East variogram

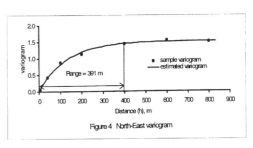

Figure 4 North-East variogram

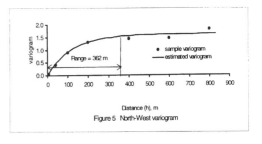

Figure 5 North-West variogram

plot of gas content (Figure 6), in which no obvious minimum or maximum direction of continuity is shown.

Sample variograms also indicate that they are the transition type meaning that these variograms possess a sill. So exponential models are chosen to fit these variograms (Figs 1 to 5). The exponential models chosen to fit the sample variograms are of the form:

$$r(h) = C_0 + C_1(1 - \exp((-3 \times h)/a))$$

where

C_0	nugget
a	range
$C_0 + C_1$	sill

For each directional and the omnidirectional variogram the nugget effect, sill and range are simultaneously estimated. The results are given below.

Nugget effect	0.01 - 0.1
Sill	1.6 - 1.9
Range	300 – 400 m

R^2 of the fit is above 95%, indicating the exponential models are indeed fitted very well with the sample variograms.

3.4 Ordinary krigging system

For a given location the accuracy of an estimate of gas content is associated with the distribution and spacing of sample points (chosen to estimate the given points). A measurement of such accuracy is the krigging variance. In order to calculate the krigging variance of an estimate of gas content an ordinary krigging system needs to be established. Establishment requires a variogram model. In our case, the omnidirectional variogram model was chosen to build the ordinary krigging system because of the homogeneity of the data. The variogram model r(h), which is a function of h, is given below:

$$r(h) = 0, \qquad (h = 0)$$
$$r(h) = C_0 + C_1(1 - \exp((-3 \times h)/a)) \qquad (h > 0)$$

Where

h	distance in metre
C_0	0.059
C_1	1.763
a	321

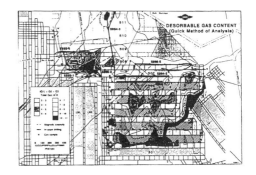

Figure 6 Spatial distribution of desorbable gas content

In geostatistics a covariance model is often used instead of a variogram model in krigging processes for computation efficiency (Davis, 1986). In this case, the corresponding covariance model of the omnidirectional variogram model was used in the krigging process. The covariance model c(h), which is also a function of h, is given below:

$$c(h) = C_0 + C_1, \qquad\qquad h = 0$$
$$c(h) = C_1(1 - \exp((-3 \times h)/a)), \qquad h > 0$$

Where
h is distance in metre
C_0 is 0.059, C_1 is 1.763, a is 321

An ordinary krigging system is established based on the specified coviance model.

3.5 Ordinary krigging

In this section the krigging variances of an estimate of gas content for a given location are calculated using samples arranged in three different ways shown below.

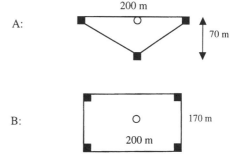

237

C:

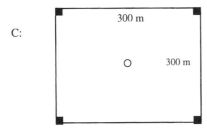

■   Sampling points

O   Point where gas content is estimated

The ordinary krigging system was used to calculate the variances of the three types of sampling configurations. Standard error (S.E.), the square root of variance, is often used in practice instead of variance. S.E. for each type of the configuration are calculated and shown below.

Type	S.E.
A	1.08
B	1.19
C	1.35

4 SUMMARY

- Coal sample spacing in any direction can be up to 300m.
- The influence area of a sample is a circle centred at sample location with a radius of about 300m.
- Standard errors calculated for the three types of sampling arrangements indicate that there is little to choose among sampling configurations as far as the accuracy of an estimate of gas content for a given location is concerned.

5 CONCLUSION

If the area is geologically undisturbed, the results of this study suggest that a sample spacing of 300 m is adequate to encapsulate the spatial variability in gas contents.

REFERENCE

Davis J. C., 1986 *Statistics and Data Analysis in Geology (2ⁿᵈ Edition)*, John Wiley & Sons, New York.

Mining Science and Technology' 99, Xie & Golosinski (eds)© 1999 Balkema, Rotterdam, ISBN 90 5809 067 1

High resolution sequence stratigraphy of the Late Permian coal measures in southwestern China

Longyi Shao, Liming Hao, Lijun Yang, Pengfei Zhang & Baolin Tian
China University of Mining and Technology, Beijing, People's Republic of China

ABSTRACT: The Late Permian coal measures in southwestern China consist of a series of coal seams and sedimentary rocks formed in environments ranging from non-marine, paralic, to fully marine. In this paper, depositional environments of these coal measures have been discussed and a further analysis is made on high resolution sequence stratigraphy of typical paralic coal measures in western Guizhou. In each of the 17 fourth-order sequences, coal seams at base represent a transgressive systems tract and overlying limestone and siliciclastic successions represent a highstand systems tract, and the base of the limestone bed overlying coal seams represents a fourth-order maximum flooding surface. These fourth-order sequences have stacked into three third-order composite sequences in which lowstand, transgressive and highstand sequence sets can be distinguished based on stacking patterns of fourth-order sequences. These sequence sets can also be recognized through tracing landward distribution of coal seams and limestone beds which are believed to represent third-order marine flooding zones.

1. INTRODUCTION

Sequence stratigraphic analyses have been conducted for the Late Palaeozoic coal measures in which coal-bearing cyclical successions are well developed (e.g. Hamilton et al., 1994; Flint et al., 1995; and Aitken et al., 1995, Howell and Aitken, 1996 eds.). Laterally extensive coal seams are generally considered to be restricted to the TST in high frequency (fourth-order) unconformity bounded sequences (Flint et al., 1995). It is also believed that those extensively distributed thick coal seams are deposits of the third-order maximum flooding zones (Aitken et al., 1995). These models are mostly derived from the studies on coal-bearing foreland basins where tectonic subsidence and sediment supply might play more important controls to changes in the accommodation space which is essential for coal accumulation. They need to be further tested in coal-bearing cratonic basins where the eustatic sea level changes are important controls to coal accumulation.

The southwestern China was a huge cratonic basin during the Late Palaeozoic (Liu et al., 1993), where the Late Permian coals were extensively deposited. The coal measures were formed in various environments ranging from non-marine alluvial plain to transitional coastal plain and fully marine (CNACG, 1996). In this paper, depositional environments of these coal measures are discussed and an attempt is made to analyze the high resolution sequence stratigraphy of a typical paralic coal measures in western Guizhou.

2. GEOLOGICAL SETTING

The research area includes Guizhou, southern Sichuan, and eastern Yunnan, which were tectonically a part of the cratonic basin within the Late Palaeozoic South China Plate (Liu *et al.*, 1993). For lithostratigraphy, the Upper Permian in these areas are subdivided into the Longtan Formation (Wujiaping Formation, Middle and Lower Members of the Xuanwei Formation) and the Changxing Formation (Wangjiazhai Formation, Dalong Formation, and the Upper Member of Xuanwei Formation). The Longtan Formation is further subdivided into the Lower Member and the Upper Member. For chronostratigraphy, two stages are defined for the Late Permian: Longtanian and Changxingian.

Each of these Formations has its own unique lithological associations. The Xuanwei Formation in eastern Yunnan is mainly composed of non-marine clastic rocks. The Longtan and Changxing Formations in western Guizhou and southern Sichuan consist of paralic clastic rocks intercalated with limestones. The Wujiaping and Changxing Formations in eastern and southeastern Guizhou, eastern part of southern Sichuan, and the southern part of eastern Yunnan consist of marine carbonate and siliceous rocks.

A wide range of palaeogeographical units occurred in southwestern China during the Late Permian (Figure 1). Non-marine alluvial plain dominated by fluvial channel and alluvial fan was developed in eastern Yunnan. Transitional paralic plain dominated by delta-tidal flat system and lagoon-tidal flat system was developed in western Guizhou and southern Sichuan. Marine carbonate platforms were developed in eastern and southern Guizhou, eastern part of southern Sichuan, and southern part of eastern Yunnan, and central Guangxi. Deep water faulted basins was developed in the vicinity of Ziyun and Luodian Counties of southern Guizhou.

3. DEPOSITIONAL ENVIRONMENTS OF COAL MEASURES

Sedimentary characteristics of the Late Permian coal measures used for environmental interpretation include lithology, sedimentary structure, palaeocurrent, clay mineral combination, geophysical well logging, geochemistry analyses. It is considered here that Late Permian coal measures in research areas were mainly formed in following system: (1) braided river and braided delta system, (2) alluvial plain fluvial system, (3) delta-tidal flat system, (4) lagoon-tidal flat system, and (5) carbonate platform system.

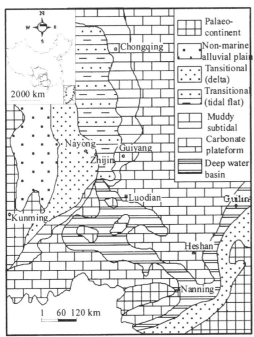

Figure 1. Palaeogeographical outlines of the Late Permian in southwestern China

The Xuanwei Formation in eastern Yunnan is Characterized by clastic coal measures including massive conglomerates, cross bedded sandstones, laminated siltstones, mudstones, and up to 20 coal seams. These rocks are interpreted as formed in braided river and braid delta system and alluvial plain fluvial system. The Longtan and Changxing Formations in western Guizhou and the Longtan and Xingwen Formations in southern Sichuan consist of cross-bedded sandstones, laminated siltstones, mudstones, and up to 36 coal seams, with the intercalation of marine bioclastic limestones in eastern part of this belt. Bidirectional cross beddings and tidal laminations are particularly developed in sandstone and siltstones. These deposits were formed in delta-tidal flat system and lagoon-tidal flat system. The delta-tidal flat system is believed to be one of the most important sites for coal accumulation in western Guizhou. This system can be further divided into fluvial-dominated upper delta plain, fluvial and tide co-effect transitional delta plain, and tide-dominated lower delta plain and tidal plain environments. The Wujiaping Formation in eastern and southern Guizhou and the Heshan Formation in Guangxi are characterized by coal-bearing carbonate sequences in which thick coal seams (up to 7 seams with each up to 4m thick) are directly interbedded with bioclastic limestones (Shao et al., 1998). The shallow marine fauna in limestones, such as Brachiopods, Mollusks, Echinoderms, Sponges, Bryozoans, Foraminifers, and calcareous algae, indicate a marine carbonate platform setting. The coals in carbonate sequences show various evidences of marine influence such as marine fossils in coal and the mudstone partings, high organic sulphur contents in coal (6-9%), high MgO and CaO level, and the occurrence of dolomite and calcite minerals in coals (Shao et al., 1998).

4. SEQUENCE STRATIGRAPHIC ANALYSES OF COAL MEASURES IN WESTERN GUIZHOU

The Late Permian coal measures in western Guizhou have been developed with most abundant coal resources in southwestern China, and a paralic setting for coal deposition has been generally considered (CNACG, 1996). The coal measures typically consist of interbedded clastic rocks, limestones, and coals, which are interpreted as formed in the transitional marine and non-marine environments. High resolution sequence stratigraphy has proved to be a useful tool in characterization of oil and gas pools (Howell and Aitken, 1996). The development of typical marine limestone marker beds and the high frequency coal-bearing cyclicities of the Late Permian in research area allow us to reconstruct their stratigraphic framework with methods of high resolution of sequence stratigraphy.

The coal measures in the Zhijin-Nayong region of western Guizhou are about 300m thick and contain about 17 limestone marker beds and more than 35 coal seams. They were formed in tidal influenced lower delta plain and tidal plain environments. The precursor mires were also developed on these environments, and peat formation occurred during water table rise related to relative sea level rise.

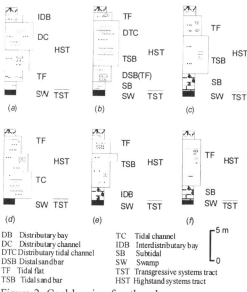

(a) (b) (c)

(d) (e) (f)

DB	Distributary bay	TC	Tidal channel
DC	Distributary channel	IDB	Interdistributary bay
DTC	Distributary tidal channel	SB	Subtidal
DSB	Distal sand bar	SW	Swamp
TF	Tidal flat	TST	Transgressive systems tract
TSB	Tidal sand bar	HST	Highstand systems tract

Figure 2. Coal-bearing fourth-order sequences present in the Late Permian in the Zhijin-Nayong region, western Guizhou Province. a – tidal dominated lower delta plain; b – subaqueous delta plain; c – tidal sand bar; d – tidal channel; e – interdistributary bay; f – tidal flat

A total of 6 types of coal-bearing sequences (fourth-order) have been recognized, with each characterized by a rooted paleosol (seat earth) at the top and the coal and marine beds at the bottom (Figure 2). Within each sequence, coal seams represented deposits of a transgressive systems tract and marine band and overlying siliciclastic sediments are attributed to a highstand systems tract. The fourth-order maximum flooding surface is put at the bottom of the limestone bed overlying coal seams. In the Zhijin-Nayong region, there are totally 17 fourth-order sequences, namely, S1 to S17 (Figure 3).

In the high resolution sequence stratigraphic framework proposed by Van Wagoner et al. (1990), the fourth-order sequences stack each other in special ways to form third-order composite sequences (Mitchum *et al.*, 1991) in which third-order lowstand sequence set, transgressive sequence set, and high sequence set can be further divided. In western Guizhou, the stacking styles of these sequence sets

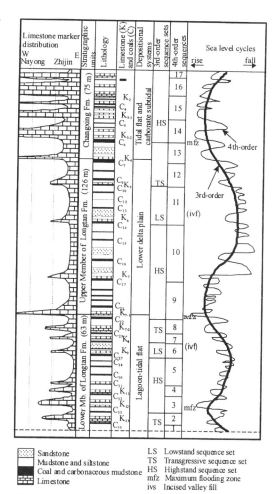

| Sandstone |
| Mudstone and siltstone |
| Coal and carbonaceous mudstone |
| Limestone |

LS Lowstand sequence set
TS Transgressive sequence set
HS Highstand sequence set
mfz Maximum flooding zone
ivs Incised valley fill

Figure 3. Late Permian sequence stratigraphic framework in western Guizhou.

can be recognized through tracing landward distribution of coal beds and limestone bands which represent deposits of third-order flooding zones. The third-order maximum flooding zones are represented by those couplets of marine band and coal seams with maximum distribution towards land to the west. During maximum flooding period, rapid increase in accommodation space, resulting from rapid sea level rise, would easily cause peat mires to be rapidly drown and to be replaced by deposition of subtidal carbonates. In this case, thick coal seams are not necessarily deposits of the third-order maximum flooding zones. This is different from the model proposed by Aitken et al. (1995) who suggested that only those thickest coal seams were deposited during maximum marine flooding period. Accordingly, the third-order sequence boundaries are indicated by minimum distribution of the coal beds and limestone bands which usually accompanied by thick channel sandbodies.

In the research area, three third-order composite sequences are those from S1 to S5, from S6 to S10, and from S11 to S17, respectively (Figure 3). Each composite sequence can be subdivided into three sequence sets: the transgressive sequence set (TS) characterized by the retrogradation sequences, the early highstand sequence set (EHS) characterized by the progradation to aggradation sequences, and the late highstand sequence set (LHS) characterized by the progradation to aggradation sequences.

5. EPISODIC COAL ACCUMULATION

The "episodic coal accumulation" has been proposed to represent the large regional (even global) coal accumulation across different subenvironments and facies belts or even different basins, induced by regional or global sea level (base level) changes (Shao, 1992). This theory emphasizes that coal accumulation in coastal plain settings occurred during the sea level rise episode and also emphasizes the synchronizing of one episode of coal accumulation. In southwestern China, the migration of the coal-accumulating center during the Late Permian was controlled by the third sea level changes (corresponding to composite sequence) and the horizon of the single coal seam was controlled by the episodic coal accumulation which were related to fourth-order sea level changes. While coal accumulation is controlled by the sea level changes, the coal seam . thickness variation is still closely related to the environments prior to coal deposition.

6. CONCLUSIONS

1. Late Permian coal measures in southwestern China were formed in five different systems: 1) braided river and braid delta system, 2) alluvial plain fluvial system, 3) delta-tidal flat system, 4) lagoon-tidal flat system, and 5) carbonate platform system. Coal accumulation was stronger in the delta-tidal flat system than in other systems.

2. Detailed analyses in the Zhijin-Nayong region of western Guizhou lead to recognition of three third composite sequences and 17 fourth-order sequences. Coal seams and overlying marine bands are deposits of third-order flooding zones. Third-order maximum flooding zones and sequence boundaries can be recognized through tracing landward distribution of the marine band and coal seams. Within a fourth-order sequence, coal seams represent deposits of transgressive systems tract and overlying sediments represent deposits of highstand systems tracts.

3. "Episodic coal accumulation" is proposed to

represent large scale coal accumulation which is responsible for laterally extensive and thick coal seams. The migration of the coal-accumulating center is controlled by the third sea level changes and the horizon of the single coal seam is controlled by the episodic coal accumulation which is related to fourth-order sea level changes.

This research was supported by the NSFC with project no. 497721129.

REFERENCES

Aitken, J. F. & S. S. Flint 1995. The application of high resolution sequence stratigraphy to fluvial systems: a case study from the Upper Carboniferous Breathitt Group, eastern Kentucky, USA. *Sedimentology.* 42: 3-30.

China National Administration of Coal Geology (CNACG) 1996. *Sedimentary environments and coal accumulation of Late Permian coal formations in western Guizhou, southern Sichuan and eastern Yunnan.* Chongqing University Press, 277pp. (in Chinese)

Flint, S. S., J. F. Aitken & G. Hampson 1995. Application of sequence stratigraphy to coal-bearing coastal plain successions: implications for the UK coal measures. In M. K. G. Whateley & D. A. Spears (eds.), *European Coal Geology.* Geological Society London, Special Publication, 82: 1-16.

Hamilton, D. S. & N. Z. Tadros 1994. Utility of coal seams as genetic stratigraphic sequence boundaries in nonmarine basins: an example from the Gunnedah basin, Australia. *AAPG Bull* 78:267-286.

Howell, J. A. & J. F. Aitken 1996 (*eds.*). *High Resolution Sequence Stratigraphy: Innovations and Applications.* Geological Society London, Special Publication. 104: 371pp.

Mitchum, R. M. & J. C. Van Wagoner 1991. High-frequency sequences and their stacking patterns: sequence stratigraphic evidence of high-frequency eustatic cycles. *Sedimentary Geology* 70: 131-160.

Shao L., P. Zhang, D. Ren & J. Lei 1998. The Late Permian coal-bearing carbonate sequences in South China: coal accumulation on carbonate platforms. *Int. J. Coal Geol.,* 37: 235-257.

Shao L., P. Zhang, Q. Liu & M. Zheng, 1992. Lower Carboniferous Ceshui Formation in central Hunan, South China: Depositional sequences and episodic coal-accumulation. *Geological Review,* 38: 52-59. (in Chinese with English abstract).

Van Wagoner, J. C., R. M. Mitchum, K. M. Campion & V. D. Rahamanian 1990. Siliciclastic sequence stratigraphy in well logs, cores and outcrops: concepts for high-resolution correlation of time and facies. *AAPG Methods in Exploration Series,* No.7: 55p.

Mining Science and Technology' 99, Xie & Golosinski (eds) © 1999 Balkema, Rotterdam, ISBN 90 5809 067 1

Geological seeking for potential CBM-accumulation zones and districts in China

Yong Qin
China University of Mining and Technology, Xuzhou, People's Republic of China

Jianping Ye & Dayang Lin
China National Administration of Coal Geology, People's Republic of China

ABSTRACT: New system is proposed for geological definition of potential coalbed methane districts. It is called "Progressive Key Factor Optimum Seeking" and comprises two parts: progressive seeking and quantitative ordering. This system has been successfully applied to search for the potential CBM-accumulating zones and districts in China. With this new system, the limitations of the traditional method "comprehensive evaluating standards" are alleviated, and the accuracy and reliability of the search results are improved.

1. INTRODUCTION

In the previous work for the potential coalbed methane (CBM) zone and district seeking, the "comprehensive evaluating standards" method was generally used, and the parameters tended to be more and more complex (Zhang, 1995; Liu *et al.*, 1998; Sun *et al*, 1998). Although this method reflects the increasingly deepening of the understanding of CBM geological controls, it also increases the difficulty and fuzziness in practical operation, resulting in decreasing reliability. Meantime, the sufficient attention is not paid still to the quantitative optimum seeking and the different standards need to be established for the evaluation units with different scales.

How could the qualitative and quantitative optimum seeking be realized based upon sophisticated geological analyses? What are the key CBM controlling factors or elements? Which factors or elements can be easily digitized for quantitative evaluation and which are hardly digitized at present and are only suitable for qualitative analysis? With what clues can the qualitative *versus* quantitative and "higher level units *versus* lower level units" be effectively jointed together in the optimum seeking. These technological keys must be envisaged in order to achieve successful exploration and development to the CBM resources in China.

2. SEEKING FOR POTENTIAL CBM ZONES AND DISTRICTS

2.1 *Thoughts to the optimum seeking*

Based upon the above considerations, we have sys-

tematically analyzed the CBM geological conditions and previous experiences in CBM exploration and development in China, and propose a new optimum seeking system, called "Progressive Key Factor Optimum Seeking" (shortly PKFOS system). In this system, the twofold characteristics of the CBM geological conditions, i.e. the accuracy and fuzziness, are sufficiently considered, and two fundamental methods of the factor progressive screening (geological risk analysis) and the quantitative ordering are simultaneously adopted in the optimum seeking for the potential CBM zones and districts in China (Figure 1).

The scientific thought of the AHP Hierarchy process in the decision-making (Wang *et al*, 1990) was used to establish the hierarchical structure model with progressive control function surrounding the final objective during carrying out the KFPOS system. The model includes three fundamental levels: objective level, that is, the optimum seeking for the potential zones and districts; criterion level, including major geological risk factors or elements; schedule level, that is, the seeking targets, consisting of the 30 CBM-accumulating zones and 115 districts in China. The objective determines the criterion and the criterion controls the potential of the zones and districts. Following this principle, the decisive system for the optimum seeking for the potential CBM zones and districts has been scientifically set up.

2.2. *Progressive Key Factor Optimum Seeking*

Geological analyses have indicated that there are five elements which could be key factors with unique veto power to the CBM potential, which are coalbed area, CBM content, CBM abundance, coal-

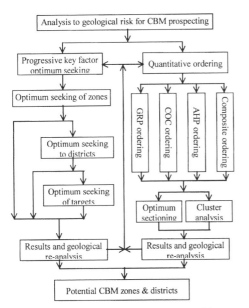

Figure 1. A framework of optimum seeking system for potential CBM zones and districts in China

Table 1 Progressive correlation of key element association for optimum seeking of potential CBM zones and districts

Scale	key element					Geologic setting
	area	CBM abundance	CBM content	Coalbed permeability	Ratio of CDP to CRP	
Zone						
District						
Target						

bed permeability, and critical desorption pressure (CDP) to coal reservoir pressure (CRP) ratio. These elements could be interrelated but cannot substitute each other. The CBM resource abundance involves such elements as coalbed thickness, absorption and coal composition. The permeability is controlled integratedly by coalbed pore-fissure system, reservoir pressure system and coal compositions. The unit area and the CBM content are the basic measures for the scale and gas-bearing characteristics of the seeking targets.

The relative importance of a single key element varies with the scale of the seeking targets, and through comparison the key element association can be set up (Table 1). According to the key element association, the units without obvious potential could be rejected through screening. The procedures of the optimum seeking are unfolded scale by scale. The units rejected upon the zone seeking can't be considered in principle as the candidates for district seeking, and the those rejected upon the district seeking can't be considered in principle as the candidates for the target seeking. This essential principle is also applicable to the quantitative unit ordering by computer.

The optimum seeking for the CBM districts follows procedures shown in Figure 2. As for the CBM-accumulation zones, the correlations of the geotectonic and sedimentary settings of the coal-bearing formation with the CBM accumulation should be focused. As far as the CBM districts concerned, the attention should be paid to the tectonic frameworks and the matching of the key structure-forming periods with the main CBM-forming periods in the district. As for the CBM target, more concerns should be paid to the secondary structure, the CBM reservoir to cap rock association, and the hydrogeological conditions.

2.3. Quantitative ordering of CBM-accumulating zones and districts

There is the overlapping of some single key elements among the units sought preliminarily by the progressive analysis of key geological risk factors. Furthermore, some non-key elements and their associations also control the CBM prospects of the seeking units to some extent. In this case, the simple or qualitative seeking would result in both difficulties in operation and some false conclusions. Therefore, the quantitative methods need to be introduced into the seeking procedure in order to guarantee the accuracy and reliability of the seeking results.

Method of geological risk probability (GRP) If there are n main risk elements in a seeking unit i, this unit has the risk probability Pij for the CBM prospecting as follows:

$$P_{ij}=\Sigma p_{ij}=(\Sigma f_{ij}Q_j)/f_{j,max} \quad (i=1, 2, ..., n)$$

In which: P_{ij} represents relative risk probability of a single element j; f_{ij} represents the absolute value of the risk element j in the unit i; Qj represents the power weight of the element j. The maximum of the element j exists in all the units and the P_{ij} value ranges between 0 to 1. Apparently, the higher the P_{ij}, the lower the potential of the seeking units. Then, the optimum sectioning is used to process the ordering values so that the units can be classified into a number of the risk groups according to similarities of their probabilities. The results are beneficial to the evaluating of the CBM prospecting risk and to further comparisons with the results from qualitative geological risk analysis.

Method of comprehensive ordering coefficient (COC) The principle of this method was suggested by Wu (1994). The coalbed methane resources are

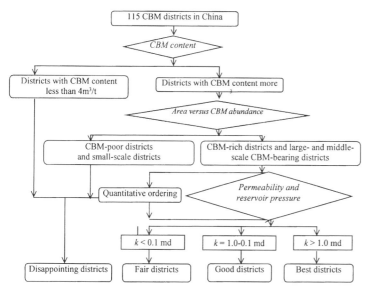

Figure 2. Procedure of progressive optimum seeking for CBM districts in China

Table 2. Vectors of relative power weight of CBM geological risk factors or elements in China

Vectors of relative power weight $w^{(i)}$	Geometry factor			Coal reservoir factor					Cap rock factor			Resources factor			Geological setting factor			
	B101	B102	B103	B201	B202	B203	B204	B205	B301	B302	B303	B401	B402	B403	B501	B502	B503	B504
Relative to the factors in the above level	0.481	0.114	0.405	0.045	0.063	0.491	0.127	0.274	0.105	0.258	0.637	0.455	0.090	0.455	0.462	0.351	0.140	0.047
Relative to total goal	0.019	0.005	0.016	0.024	0.034	0.266	0.069	0.148	0.008	0.020	0.049	0.113	0.022	0.114	0.043	0.033	0.013	0.004

Note: B101, seam area; B102, seam stability; B103, seam thickness; B201, coal rank; B202,CBM saturation; B302, reservoir permeability; B204, reservoir absorptivity; B205, reservoir pressure; B301, cap lithotype; B302, cap rock thickness; B303,cap rock physics; B401, CBM resources abundance; B402, CBM resources; B403, CBM content; B501, hydrogeological condition; B502, tectonic condition; B503, geothermal history; B504, sedimentary environment

different from the conventional oil and gas, and their geological controls are also different. In view of this, we use the X-axis to represent the *Resource Coefficient* which is the probability sum of the gas content, resources amount resource abundance and theoretical saturation of CBM, and use the Y-axis to represent the *Safety Coefficient* whose value equals to 1-R_i. Here, the R_i is the probability sum of other main risk elements. The equation of the comprehensive ordering coefficient R_i is given as follows:

$$R_i = \{(1\text{-}\log Q_i/\log Q_{max})^2 + (1\text{-}G_i/G_{max})^2\}^{-1/2}$$

in which: the $\log Q_i/\log Q_{max}$ represents the normalized resources coefficient; the G_i/G_{max} represents the normalized safety coefficient. The values of R_i vary between 0 to 1. The lower the R_i of the seeking unit, the lower its prospecting risk, and the higher its resource potential.

Analytical Hierarchy Process (AHP) The hierarchies in the decision-making system and the hierarchical structure model can be established. At the same time, a comparison matrix should be set up so that the relative importance of all elements in a same level relative to a single criterion in the above level needs to be compared one by one. Based on the matrix, the relative power weights of the compared elements to this criterion and the resultant power weights or ordering vectors of all the elements in each level relative to the total goal can be calculated (Table 2), and the final ordering results of the seeking units could be obtained. Compared with previous two methods, the AHP is logically more strict and has more reliable results.

Composite method of AHP and GRP In the progressive hierarchical structure model, the number of the levels is unlimited but the number of the elements allocated to each factor should be generally no more than nine, otherwise, the comparisons between these elements will become difficult (Wang et al, 1990). When the methods of geological risk probability are used, the numbers of the seeking

Table 3. Results of quantitative ordering for potential of CBM-accumulating districts in China

Category	Grade of potential of CBM districts			
	Best	Good	Fair	Disappointing
Confirmed	Yangquan-Shouyang Hancheng Jincheng	Liliu-Sanjiao Huaibei Anyang-Hebi	Tiefa, Pingdingshan, Hongyang, Jiaozuo, Kailuan, Dacheng, Fengfeng, Luan	Fengcheng, Fuxin, Shenbei, Wupu, Huodong, Huainan, Sanjiaobei, Xinggong, Nantong, Lianshao, Chunlei, Xishan
potential		Liupanshui Enhong Baiyanghe	Guishan, Hegang, Shuanyashan, Jixi, Zhongliangshan, Tianfu, Furong, Songzao, Guxu, Laojunmiao, Heshun-Zuoquan, Zhina, Aiweiergou, Shizhuishan, Hulushitai	Seventy-one districts, including Fushun, Junlian, Qinyang, Pingxiang, Yuxian, Yanlong, Xiangning, Huozhou and so on

units and the elements in each unit can be unlimited, and the ordering results will be identical to those made with the AHP if the power weights of the elements are reasonably set up. In view of this fact, we suggest a new method of the quantitative ordering, that is the Composite Method of AHP and GRP, in which the advantages of both AHP and FRP are adopted. In this composite method, the relative power weight can be scientifically obtained by AHP, and the risk probability can be further calculated by GRP. This probability vector is considered as the reference for ordering the seeking units. Thus, under the prerequisite that the reliability of results must be satisfied, not only can the artificial interfere to the setting of power weight in GRP be avoided, but also the setting and analysis of the judging matrix in GRP can be simplified. Using the composite method, the accuracy and reliability of the quantitative ordering results can be effectively enhanced.

3. EXAMPLE OF OPTIMUM SEEKING FOR CBM ZONES AND DISTRICTS IN CHINA

As shown in Table 3, the 115 districts in China are firstly classified into two categories after current CBM data have been analyzed. The first category is called the "confirmed" districts, which are supported with the borehole testing data such as the permeability, and includes four grades of the CBM potential, i.e., best, good, fair and poor. The second category is named as the "potential" districts, which are not supported with any borehole testing data, and comprises only three grades, i.e., good, fair and poor.

It is obvious that the best and good grades of districts in the confirmed category benefit the development of the CBM resources, and the good and fair grades of districts in the potential category might be valuable for the further testing for exploration and development of the CBM. In the confirmed category, six districts with best to good potential, such as the Yangquan-Shouyang and Hancheng, are considered to have priority over all other others for the CBM development in China, and the CBM devel-

opment tests have been or are being operated in these districts. Eight districts, including the Tiefa, Pingdingshan, and Hongyang, have the fair CBM potential, where the small-scale well pattern can be arranged after the more detailed research is conducted on the CBM geological conditions. In the potential category, the CBM potential of three districts, Liupanshui, Enhong and Baiyanghe, seems to be good, and can provide for the CBM exploration and development test in near future. Fourteen districts such as Guishan and Hegang are of the fair potential, and are suggested as the selected sites in the middle to long term.

REFFERENCES

Liu, H., Y. Qin, S. Sang et al. 1998. Geology of Coalbed Methane in northern Shanxi: CUMT Publishing House. Xuzhou

Sun, M., S. Huang, Xinmin et al. 1998. Handbook of CBM development and Utilization: Coal Industry Press, Beijing.

Wang, L. & S. Xu 1990. Introduction to Analytic Hierarchy Process: China People's University Press, Beijing.

Wu, S. 1994. Introduction to Evaluation for Petroleum Resources Geology: Petroleum Industry Press, Beijing.

Zhang, Q. 1996. An outlook of coalbed methane development in China. Proceedings of the International Conference on Coalbed Methane Development and Utilization: Beijing: 6-9.

Mining Science and Technology'99, Xie & Golosinski (eds) © 1999 Balkema, Rotterdam, ISBN 90 5809 067 1

Gas-bearing properties of coal reservoirs in China

Shuxun Sang & Yong Qin
China University of Mining and Technology, Xuzhou, People's Republic of China

Jianping Ye, Shuhen Tang & Aiguo Wang
The First Exploration Bureau of China National Administration for Coal Geology, People's Republic of China

ABSTRACT: Gas-bearing properties constitute important basis for geological selection of target areas for coalbed gas exploration. The South China region is characterized by highest gas contents and methane density of coalbed. Northwest China region is characterized by highest gas intensity. Northeast China region is characterized by highest gas saturability, and the North China region is characterized by balanced and constant values of all four parameters. As far as temporal distribution is concerned, the gas saturability, gas content, and gas qualities are highest in the Late Permian and Late Triassic coal seams and are second highest in the Late Carboniferous and Early Permian coal seams. Gas –bearing intensity is much higher in the Early and Middle Jurassic coals then in other coals, and decreases towards the Late Carboniferous, Early Permian to Early Cretaceous coals in order. Tectonics, coal ranks, coal burial depth, coal petrology, lithology of seals, and hydrogeological conditions provide direct and major controls to gas-bearing properties.

1. INTRODUCTION

Gas-bearing properties of coal seams are key parameters for carrying on geological selection for target areas of coalbed gas exploration, and constitute the foundation for evaluating coalbed gas resources. In coal geology exploration over last 5 decades, test wells for coalbed gas have been operated in most coal-bearing areas of China, from which gas contents and compositions have been measured and a large number of coal samples and data on gas bearing properties of coal seams have been obtained. The exploitation of coal resources in China has reached a very high extent, with large and small mines covering all main coal-bearing areas. A lot of data on coal mine gas outburst quantity and measured gas data by mine drills have been obtained in the coal mining process. In the recent decade, over 100 wells for data acquisition and development experiments have been operated in China, and more accurate data on gas-bearing properties of main coal seams have been obtained for some mine areas. On the basis of above data, a great number of statistics and analyses have been made on gas content, gas composition, gas saturability and gas intensity of major coal seams for main mine areas in China, and attempts are made to detect regularities of gas bearing properties, regional strata distribution of coalbed gas, and the changing regularities.

2. FUNDAMENTAL ELEMENTS OF GAS-BEARING PROPERTIES

Gas-bearing is the inherent characteristic of coalbeds, because coal seams and coalbed gases are paragenous closely. However, only when gases contained in coal seams meet certain conditions, are they of energy significance. Gas-bearing properties can be described with such four fundamental elements as gas content, gas quality, gas intensity and gas saturability, of which gas content is the most important and it influences three other elements to some extent.

1. Gas content, defined as gas volume per unit of weight of in-situ coal, is often expressed as m^3/t or ml/g on the basis of ash and water free or water free. Most of our data on coalbed gas contents are measured with the Desorption Method and few of them are obtained with the Collection Method. In addition, a part of gas content data are from transfers of relative outburst quantities of coal mine gas; only a few data are from coalbed gas test wells drilled recently and obtained with the Direct Method (Qian et al., 1996; Ning et al., 1995; Mavor, 1995; Li et al., 1990).

2. Gas quality refers to relative composition of gases in coal seams, particularly to relative percentage of methane (including heavy hydrocarbon gas sometimes), or simply, methane density. In China the measurement of gas compositions are mainly

undertaken with the gas chromatography.

3. Gas intensity, also called resource abundance, is expressed as total gas volume contained in coal seams per unit of evaluated areas. This parameter has close relationship with gas content and coalbed thickness, and is shown in a unit of 100 million m^3/km^2. In other words, it equals to the ratio of the total resources of coalbed gas and the area of the coal-bearing areas. In this paper, the basic unit for calculating is a mining area and the depth for calculation is from bottom of gas weathering zone to 2000m deep.

4. Gas saturability is subdivided into measured saturability and theoretic saturability, the former represents the degree of coalbed gas content reaching saturation relative to the measured reservoir pressure, and the latter refers to the degree of coalbed gas content reaching theoretic saturation. The measured saturability is the same or larger than theoretic saturability. Gas saturability is an important parameter affecting extraction of coalbed gas (Wang et al., 1996). When the gas contents are constant, the higher the coal rank, the lower the saturability, and the more difficult the development. It needs to point out that the theoretic saturability is only valuable for regional comparison, with its variation reflecting relative difficulties of coal reservoir development when evaluating coalbed gas resources. Although the theoretic saturability is helpful to the selection of coalbed gas target areas and macroscopic decision, its significance in directing development engineering of coalbed gas is limited.

Apart from the above four fundamental elements, gas-bearing gradient is also an important parameter for describing gas-bearing nature of coal seams. It is used to represent the increasing rate of gas contents with burial depth of coal seams, and is normally expressed as gas content variation when the coalbed burial depth is increased by 100m. In the mining areas where coalbed burial depth is the main gas-controlling factor, there is a relatively remarkable gas gradient, with values being generally within 0.5~3.0 m3/t·100m, and its regional variation are relatively stable also. Nevertheless, in those where the burial depth is not important controlling factor, it is difficult to determine if a gas gradient exists. That is why gas gradient is not considered to be a fundamental gas-bearing element in our research.

3. REGIONAL DISTRIBUTION OF GAS-BEARING PROPERTIES OF COAL SEAMS

3.1. General features

A large number of statistics have been done on every element of gas-bearing properties for main coal seams of 115 typical mining areas in China. The results are as follows: within the depth of coal exploration (<1000m in general), gas contents of coal seams range from 4.0 to 27.1 m^3/t, with the average of 9.76 m^3/t; below weathering zone and within coal exploration depth, methane density of coalbed gas generally varies from 83.3% to 97.0%, with its average being 90.6%, the average nitrogen density is 8.0%, the average carbon dioxide density is 2.0% and there is a little heavy hydrocarbon gas; from lower limit of the methane weathering zone to 2000m deep, gas intensity ranges from 0.03×10^8 to 8.77×10^8 m^3/km^2, with the average of 1.34×10^8 m^3/km^2; within coal exploration depth, theoretic gas saturability varies from 26% to 76%, with the average being 41%.

3.2. Regional distribution of gas-bearing properties

In China, coalbed gas is distributed in four regions, that is, the North China, South China, Northeast and Northwest gas-accumulating regions. The statistic results of average gas-bearing properties of above 4 regions are shown in Table 1. In general, gas content of coal seams is highest in the South China, higher in the North China and the Northeast China, but low in the Northwest China. The methane density values of coalbed gas of four regions are close to each other, with descendent order of the South China, North China, Northwest, to Northeast gas-accumulating regions. Gas intensities differ from each other in these regions largely, with the highest in the Northwest, higher in the Northeast and South China, and lower in the South China. Gas saturability is highest in the Northeast, second highest in the South China, high in the North China, but remarkably low in the Northwest.

It needs to be noted that, in the North China, advantages of the four elements of gas-bearing properties are of best equilibrium with the smallest regional variations. The main coal seams are regionally stable, and are extensively distributed. Besides, coalbed gas resources in the North China take up 70% of total $14 \times 10^{12} m^3$ in whole country. At present, exploration and development of coalbed gas are mainly concentrated in the North China. In the South China, since gas intensity is low and coal seams are thin and unstable, development extent of coalbed gas is limited, although the resources are rich and exceed 25% of the total. In the Northeast and Northwest, gas intensity is high due to existence of multiple, thick coal seams, but development potential of coalbed gas is also influenced by lower gas content, poor gas quality, lower resource amounts (less than 5% in total), and relatively limited distribution (Han et al., 1980, Zhang et al., 1991, Lin, 1998).

Table 1. The average gas-bearing properties of coal seams of gas-accumulating areas in China

Gas-accumulating regions	North China	South China	Northeast	Northwest
Gas content, m³/t	9.34	9.66	9.50	5.40
Gas intensity, 10⁸m³/km²	1.13	0.86	1.16	3.75
Methane density, %	90.5	91.8	89.6	89.9
Gas saturability, %	41.8	51.6	52.7	29.5

Table 2. The average gas-bearing properties of main coal seams of the major periods in China

Periods / Elements	C3-P1	P2	T3	JI-2	K1	E*
Gas content, m³/t	9.4	10.9	11.5	6.0	6.4	12.0
Methane density, %	90.5	91.4	92.7	89.9	89.2	>90.0
Gas intensity, 10⁸ m³/km²	1.24	0.75	0.97	5.00	1.11	3.72
Gas saturability, %	42.6	49.6	54.4	33.7	47.7	52.8

* Data from the Fushun and Shenbei mining areas with limited significance

4. GAS-BEARING PROPERTIES OF COAL SEAMS AND COAL-ACCUMULATING PERIODS

Main gas-bearing coal seams were formed in 6 geological periods in China, that is, Late Carboniferous-Early Permian (in North China), Late Permian (in South China), Late Triassic (in South China), Early and Middle Jurassic (in Northwest and Northeast), Early Cretaceous, and Tertiary (in Northeast) (Han et al., 1980). The average gas-bearing properties in 6 coal-accumulating periods, based on statistics of main coal seams in major coal-bearing basins and coal-bearing areas, are shown in Table 2.

In Table 2, except for the Tertiary period, which the averages of gas bearing property data are obtained from a few mine areas in the Northeast and are not representative, all other periods have better representative values of the average gas bearing properties because the data are from main mining areas of the whole country. Variations of gas-bearing properties with time also show some regularity. Gas content, gas quality and gas saturability of coal seams are highest in the Late Permian and the Late Tertiary, second highest in the Late Carboniferous-Early Permian coal seams, high in the Early Cretaceous coal seams, and lower in the Early-Middle Jurassic coal seams. However, gas intensity is highest in the Early-Middle Jurassic coal seams, high in the Late Carboniferous-Early Permian and Early Cretaceous coal seams, and lower in the Late Permian and Late Tertiary coal seams. In view of this, coal seams of the Late Carboniferous-Early Permian are the most important ones for gas-bearing, with better elements of gas bearing properties; coal seams in other periods are difficult for assessing their relative importance, because their elements of gas-bearing properties are less harmonious.

It is worthy to note that coal seams of the Early-Middle Jurassic and Early Cretaceous need to be paid more attention, because they have relatively high gas intensity and concentrated resources of coalbed gas, although their gas content is lower.

5. CONTROLLING FACTORS OF GAS-BEARING PROPERTIES

Gas-bearing properties of coal seams are controlled by three basic geological controlling systems, that is, gas generation system, reservoir physical property system, and sealing and preserving system. In the gas generation system, gas controlling geological factors includes the abundance, type (coal petrological components) and maturity (coal rank) of organic matter, and the history and gas generation rate of coal seams. In the reservoir physical property system, gas-controlling factors include coal rank, moisture in coal, coal petrology, and geothermal history. In the sealing and preserving system, gas controlling factors involves burial depth of coal seams, tectonic and stress fields, lithology of cap rock, and hydrological features (Sang et al., 1997). In general, most direct and major gas controlling geological factors are coal ranks, burial depth, petroloy, tectonic setting, lithology of cap rocks, and hydrology.

Further research shows that, tectonic setting, coal rank and coalification history altogether result in differences of gas-bearing properties between each gas-accumulating area; tectonic zonation and hydrological conditions determine distribution of gas-bearing properties within one gas-accumulating area and variation of gas-bearing properties among mining areas; burial depth, petrographic feature and cap lithology are major factors controlling variation of gas-bearing properties within a mining area. Gas-bearing properties of coal seams result from comprehensive interactions of above gas-controlling geological factors of different levels, and the importance of these factors normally vary from area to area.

REFERENCES

Qian, K., Q. Zhao & Z. Wang 1996. *Theory on exploration and development and technology on experiments and tests of coalbed methane.* Beijing: Oil Industry Press

Ning, D., B. Tang & Y. Liu 1994. Coal seam methane content measuring technology in China. In M, Chen & K. L. Ancell (eds), *Coalbed Methane-Conf. Pro. UN Int. Conf. CBM Dev. Uti.*, Beijing, 17-21 Oct. 1995:240-246

Mavor, M. J., T. J. Pratt & C. R. Nelson 1995. Quantitive evaluation of coal seam gas content estimate accuracy. *Unconventional Gas- Pro. Int. Unc. Gas Symp.*, Tuscaloosa, Alabama 14-20 May 1995:379-388

Li, M. & W. Zhang 1990. *Shallow coal-related gas of main coal fields in China.* Beijing: Science Press

Wang, S., Z. Chen & M. Zhang 1996. Research on coalbed methane reservoir characteristics and exploration target area in Panji Mining area of Huainan. *China Coalbed Methane* 2:41-43

Han D. & Q. Yang 1980. *China coalfield geology* (Volume 2). Beijing: Coal Industry Press

Zhang, X., S. Zhang, L. Zhong et al. 1991. *Coalbed Methane in China.* Xian: Shannxi Press of Science and Technology

Lin, D. 1998. Project of "Evaluation of coalbed methane resource in China" passes the preliminary examination by the experts. *Scientific and technical Dynamics on Coal Geology* 5:6-7

Sang, S., H. Liu & G. Li 1997. Generation and enrichment of coalbed methane I. Gas yield in effective stage and concentration of coalbed methane. *Coal Geol. & Exp.* 25 (6): 14-17

Mining Science and Technology' 99, Xie & Golosinski (eds)© 1999 Balkema, Rotterdam, ISBN 90 5809 067 1

Regional distribution of Carboniferous-Permian coalbed gas in China

Zimin Zhang, Junmin Sun, Dongsheng Yuan & Ruilin Zhang
Jiaozuo Institute of Technology, People's Republic of China

ABSTRACT: On the basis of drawing the Geological Map of Coalbed Gas in China with scale at 1:2,000,000, the regional distribution features of Carboniferous-Permian coalbed gas in China was studied using the theory of plate tectonics. During the stable development of the North China and South China plate, the Carboniferous-Permian coals were extensively deposited. Due to the high coalification degree with geothermal and later telemagmatic metamorphism in neotectonic epoch, the forming conditions of the gas from Carboniferous-Permian coals are superior to those from other age coals .This has important influence on the distribution of high gas grade coalfields. During the active development of the plates, the compressive deformation belts, folds and thrusting nappes caused by plate movement also affects the distribution of coalbed gas. More importantly, the composite compressive belts caused by the old plate and modern plate movement are dangerous districts for coal-gas outburst and unfavorable areas for gas exploitation. On the contrary, in the pull-apart basins, the gas content is very low due to the release of coalbed gas.

1. REGIONAL DISTRIBUTION FEATURES OF CARBONIFEROUS-PERMIAN COALBED GAS

China is abundant in coal and coalbed gas resources. It is estimated that resources of coalbed gas is as high as $3.268 \times 10^{12} m^3$ trapped in coalbeds which are lower than 2000m in stratigraphic depth. As is known, Carboniferous-Permian is the predominate age for coal accumulation in China, with coal resources exceeding 2,200 billion tons, accounting for 35.5% of all the coal resources in China. Futhermore, the coalbed gas of this age is up to $1.7037426 \times 10^{12} m^3$, comprising 52.1% of the total. Recently, great achievements have been obtained in coalbed gas exploration in China, and favorable areas for coalbed gas exploitation have been identified, most of which are Carboniferous-Permian coal mining districts. For example, there are 9 first class favorable areas explored in China, while 7 of them belong to Carboniferous-Permian coal mining districts, namely Hedong, Huainan, Huaibei, Xishan, Hancheng, Jincheng and Tunliu coalfields; There are also 7 Carboniferous-Permian coalfields in 9 second class favorable areas, including Liupanshui, Yangquan, Jiaozuo, Pingdingshan, Hebi, Kailuan and Songzao coal mining districts; Two Carboniferous-Permian coalfields, Nantong and Zhongliangshan, were determined as third class favorable areas.

Coal-gas outburst is one of most dangerous disasters in coal mining. Unfortunately, China lies first in the world in terms of the number of coal-gas outburst. The statistics shows that more than 11,500 times of outburst occurred in coal mines in China during last 5 decades, while the outburst from Carboniferous-Permian coalbeds exceeds 8000 times occupying 70% of the total.

During Carboniferous-Permian, coal accumulation predominately occurs in north and south China. Coal resources in these regions are 1,8778 billion tons and 3249 billion tons, accounting for 85.3% and 14.7% respectively. Among the 825 high gas grade coal mines, there are 537 Carboniferous-Permian coal mines including 131 mines in North China and 406 in South China. Among the 274 outburst style coal mines, there are 207 Carboniferous-Permian coal mines accounting for 75.5% of the total. These include 53 mines in North China and 207 mines in South China. In other geotectonic units, Carboniferous-Permian coals are not abundant, and all the coal mines are lower in gas grade.

Eight-eight distribution belts of coalbed gas, including 37 higher gas grade belts and 51 lower gas grade belts, have been divided in the geological map of coalbed gas of China with scale at 1:2000000. Among the high gas grade belts , there are 19 belts in which the gas is originated from Carboniferous-Permian coal measures, accounting for 51.4% of the total. These 19 belts include 8 in North China and 11 in South China. All the above-mentioned high gas grade or outburst style coal mines are situated in the high gas grade belts.

2. EFFECT OF GEOTECTONIC EVOLUTION ON CARBONIFEROUS-PERMIAN COALBED GAS

2.1. Gas formation

In the late palaeozoic period, Carboniferous-Permian coals were formed extensively in the North China carton basin. Due to the geothermal and later telemagmatic metamorphism caused by plate movement, most of the Carboniferous-Permian coals are of bituminous coal with middle to high rank or anthracite. Significant amounts of coalbed gas were formed during coalification.

In South China , Yangtze old land combined with the mobile belt of South China in late palaeozoic and formed the intergrated old land . However, it is still very active locally, and the area of carton basin is much less than that in North China. Coal accumulation occurred in early Carboniferous, late period of early Permian and late Permian, but the intensity of coal accumulation is much less than that in North China . Most of the coalbeds are thinner in thickness and the coal resources is only one sixth of that in North China . Whereas, tectonically, from Devonian to middle Triassic, most areas of South China were characterized by pull-apart and subsidence. The coal basin is very deep and the covering rocks are thick, so the degree of geothermal metamorphism is higher than that in North China. Additionally, during the movement of plates, magmatic intrusion is frequent and extensive, causing the intensive telemagmatic metamorphism of coalbeds. So the Carboniferous-Permian coals are mainly high rank bituminous coal and anthracite, resources of anthracite exceeds 50% of the total. Moreover, in some areas such as Longyan, Shangjing, Tianhushan in southwestern Fujian province, Siwangzhang, Meixian in eastern Guangdong province, Anfu and Lianhua in central Jiangxi province, the anthracite with superior metamorphis degree was formed.

2.2. Preservation and outburst of coalbed gas

2.2.1. Effect of the tectonic depression and uplift
In North China, after Carboniferous-Permian coal-bearing formation was deposited, the north edge of North China landmass was jointed with Siberia plate and the landmass was uplifted. The south edge of the landmass was combined with Yangtze plate. Since then the sea water has disappeared from North China and the basement presented elevation or subsidence movement with different degree. The Indosinian orogeny marks the neotectonic cycle of plate movement, it caused the uplift of the following

districts: Jiaoliao , Luhuai and Shanxi province. Except for ordos depression, the late Triassic deposits was commonly absent all over the North China. More importantly, due to the early uplift of Jiaoliao, Luhuai and the north edge of the landmass, early to middle Triassic deposits were also absent in these districts, thus caused the longer weathering and erosion of coal deposits. In coalfields in Shandong province, northern Jiangsu and Anhui province as well as eastern Henan province, the depth of the erosion zone of coalbed gas generally exceeds 500~600m. According to the statistics data of 71 state-owned coal mines, there are 68 mines with low gas grade. Typically, in Yongxia coalfield, all the 4 mines are low in gas grade in spite of mining anthracite. On the contrary, In Qinshui coalfield in the west of North China and Hedong coalfield in the east of Ordos basin, as the depression was deep and the coalbed gas was well-preserved, the content of coalbed gas is up to $25{\sim}32.74m^3$ / ton. The gas resources of these coalfields are 4,837.4 billion m^3 and 1,640.7 billion m^3 respectively. Recently, these coalfields have been regarded as the first class favorable zone in coalbed gas exploration of China.

In South China, however, during most periods from Devonian to middle Triassic, the basin was depressed and the shallow marine or offshore depositional environments lasted continuously. The covering rocks is thick and the size of them is normally very fine .The Carboniferous-Permian coalbed gas is well preserved except for the coal-bearing formation characterized by carbonate deposits such as Heshan Formation in the center of Guizhou province, Wujiaping Formation widely distributed in southwestern Hubei province, northwestern Hunan province and eastern Guizhou province. Generally speaking, the conditions for preservation of coalbed gas in South China is superior to that in North China .

2.2.2. Effect of geodynamics
(1) North China
During Indosinian movement, the coal basin was compressed in S-N trend and also compressed by Kula-Pacific plate striking NW. Because of the longer consolidation of the rigidity basement, the interior of the basin was still stable in spite of the intensive compression from the south and north edge , and the coalbeds in this region were not damaged seriously. For example, in Qingshui coalfield, the density of faults ranges from 2 to 3 /km², and the structure coal is rarely found except within the small area 50 cm near the faults . On the contrary, in Pingdingshan coalfield near the fault zone in the south edge, the density of faults are up to 200~ 300 faults /km², and the coalbeds were damaged strongly, structure coals were well-developed with

the thickness up to 1.5m .

During Yanshan movement, the North China landmass was active due to the underthrust of the Pacific plate. The intensive compression and twist toward the left during early to middle stage of Yanshan epoch caused the formation of Helan mountain, Taihang mountain and Jiaoliao uplift. The coalbeds in these areas were commonly strongly damaged by compression and shear, with the well-development of structure coals. For example, the coalfields near Taihang uplift such as Jiaozuo, Hebi , and Anyang mining districts are all high in gas grade. From late Yanshan movement to early Himalayan movement, namely after late Cretaceous, with the close of the Tethys, India plate collided strongly with Eurasian plate and moved northward continuously. At the same time, the Siberia plate compressed the landmass in meridional direction. The east edge of the mainland dispersed towards the Pacific. The tectonic zones trending NE, NNE were formed. The system of epicontinental ditches, arcs and depressions started to appear. The landmass was

Figure 2 Geodynamic sketch in Indosinian periods in China
(from Yuqi Cheng,1994 [a])

① landmass ② folded zone ③ basin ④ pull-apart zone ⑤ thrusting nappe zone
⑥ faulted zone ⑦ movement trend of plate or landmass

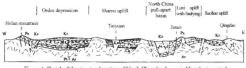

Figure 1 Sketch of the structural section of North China landmass in Himalayia periods
(From Yuqi Cheng,1994 [a])

characterized by the extension and subsidence. Original compression was transformed into extension. The famous pull-apart basin in Xialiaohe-Bohai-North -China plain was formed in the east of Taihang uplift zone(Figure 1). With the uplift of Ordos basin, a range of pull-apart basins such as Hetao garden and Yinchuan graben, etc, were formed around the Ordos basin. Additionally, the well-known Fenwei garden was also formed due to rift faulting during the uplift of Shanxi upwelling area. The above-mentioned tectonic movement caused the release of coalbed gases in a large amount. This is one of the reasons that the number of outburst-prone coal mines in South China is more than that in North China , even though the coal resources in north China is 5 times more than that in south China. On the contrary, Pingdingshan coalfield and Huainan coalfield located in the southern North China are high gas grade or outburst-prone mining districts, because they were influenced by the intensive deformation belt of eastern Qinlin and Dabie mountain, which was formed under the same tectonic stress field from Indosinian to Himalayan movement.

(2) South China

The geotectonic environment of Carboniferous-Permian coal-bearing basin in South China is characterized by the compound and union of all the four tectonic systems of China. So the compression is more intensive in this region during each geotectonic movement. In Indosinian orogeny, the basin was compressed surroundingly by Tarism-North China plate, Xizhang-Yunnan plate and Indosinian landmass. The southeast of the basin was influenced by the underthrust of Paleopacific crust toward the main land(Figure 2). The coalfields in eastern Yunnan province, southern Sichuan province, western Guizhou province and northern Guangxi province, located between Hengduan mountain and Fanjing mountain, were compressed by Tythys and Palaeocathaysian structural systems in both west and east sides, and also were compressed by Indosinian landmass and Yangtze landmass in south and north sides, respectively. In this region, the N-S and W-E structural zones, compartmentized by NE and NW tectonic belts, developed. This is one of the reasons that the coal mines with high gas grade or outburst-prone are common in these districts. According to the statistical data, among the 99 state-owned mines, there are 74 mines with high gas grade, accounting for 75% of the total. There are 24 outburst style mines with up to 1469 times of coal-gas outburst . For example, in Liuzhi coal mining district, all the 7 mines are high in gas grade, the average gushing amount of gas is up to $57m^3/t$, the gas content exceeds 20~30 m^3/t. Wuling-Xuefeng mountain and Ningzhen mountain are situated in the folded belt of foreland. Affected by this, the coalfields including Huayingshan, Tianfu, Zhongliangshan, Nantong and Songzao etc. in eastern Sichuan province are all mining districts with high gas grade. There are more than 20 outburst style mines which have occurred coal-gas outburst more than 1500 times. The most intensive outburst occurred in Sanhui mine in Tianfu coalfield, the extrusive rocks is as much as 12870 tons, which is also the biggest outburst in China. Between the Cathaysia and Jiangnan old land, a huge "S" or inverse "S" type folded belt, namely Hunan-Guangxi folded belt, was formed from Leping-Pingxiang in the central Jiangxi province, through Hunan province to eastern Guangxi province. In Lianshao and Baisha coalfields affected by this belt, the high gas grade mines comprise more than 70% of the total. There are more than 80 outburst style mines in which there have been more than 4000 times of coal-gas outburst. In addition, because North China landmass depressed striking south, Qingling orogenic zone thrusted also towards south but were blocked by Yangtze landmass, Tancheng-lujiang fault strike-slipped and whirled towards left and Dabie landmass moved southward , a rotation to NW-SE of the tectonic belt occurred. Consequently, the folded belts of the western Nandaba mountain were curved into an arc and formed the well-known thrust fold belt, namely the Huaiyang arc. In those districts such as Huangshi coalfield in Hubei province, Wutong and Xuanjing coalfields in southern Anhui province, Changguang coalfield in Zhejiang province, etc, although the Carboniferous-Permian coalbeds are not thick, coal mines with high gas grade or outburst proneness comprise more than 50% of the total.

During Yanshan movement, South China land in the southeast of the Eurasian plate was not only compressed by North China landmass, but also influenced by the subduction and transformation of Tethys plate and Kula-Pacific plate. During the early and middle Jurassic, Zhuguangyunkai uplift with NNE or NE trend and the well-known Nanling granite belt were formed between central Hunan province and Wuyi moutain. From late Jurassic to early Cretaceous, the Wuyi-Daiyun doming comprised of granite and the volcanic arc along Zhejiang, Fujian and Guangdong province were formed. A range of overthrust faults and nappe structures developed in Jiangsu, Zhejiang and Anhui province. From late Yanshan to Himalayas movement, with the close of the Tethys, the India plate depressed northward and west of South China was compressed again. Consequently, the Ailao-Longmen mountain over thrust belt was formed.

Comprehensively, the Carboniferous-Permian coal-bearing formation in South China has undergone a long compressive deformation caused by plate movements. The structure coals are commonly developed and the permeability of coal beds are abnormally lower. The condition for the preservation of coalbed gas is superior to that in North China.

3. CONCLUSIONS

Geotectonic evolution has had great influence on the formation, preservation, outburst and exploitation of coalbed gas. The proneness of coal and gas outburst depends upon the following factors: high gas content; seriously damaged structure coals; concentration area of tectonic stress, which are all related to compression caused by plate movement.

The Carboniferous-Permian coal measures formed during the stable development of old plates have undergone intensive deformation during each geotectonic cycle. Therefore, most of Carboniferous-Permian coal mines are high in gas grade or of outburst style. By comparsion, Carboniferous-Permian coals in South China are damaged more strongly than those in North China, and the outburst of coal and gas is more serious in this region. Moreover, as the primary coal structure is not well-preserved, the exploitation of coalbed gas is more limited in South China. Much attention should be paid to some coalfields in North China, where coalbeds are not seriously damaged. Additionally, in the tectonic units characteristized by extension, coal-gas outbursts have rarely occurred and it is unsuitable for gas exploitation due to the lower content of coalbed gas.

REFERENCES

Sun Maoyuan, Huang Shengchu etc. A handbook on development and utilization of CBM, Coal Industry Publishing House. Beijing: 1998.

China General Company of State-owned Coal Mines: Introduction of the Geological Map of Coalbed Gas in China, scaled at 1:2,000,000. Xian Map Publishing House , 1992.

China General Company of State-owned Coal Mines: Geological Map of Coalbed Gas in China, scaled at 1:2,000,000. Xian Map Publishing House , 1992.

Cheng Yuqi, Regional Geology of China, Geological Press , Beijing: 1994.

Yang Qi, Development of Coal Geology , Scientific Press Beijing: 1997.

Zhang Zimin, Some questions about coalbed gas in China , In: The Articles of China Science & Technology . Science & Technology Literature Press. 1997.

Mining Science and Technology' 99, Xie & Golosinski (eds) © 1999 Balkema, Rotterdam, ISBN 90 5809 067 1

New pressure-transient analysis of fractured well for coalbed methane

Wei Chen, Yong-Gang Duan, Qi-Shen Li & Chen Huang
Southwest Petroleum Institute, People's Republic of China

Yue Li
Xinjiang Petroleum Institute, People's Republic of China

ABSTRACT: Coalbed methane is mostly adsorbed in the matrix of coal seam. The process for extracting coalbed methane from matrix of coal seam to the well is pressure reducing-desorption-diffusion-flowing, which is much more complicated than that of extracting conventional gas. Most coalbed methane reservoirs in China are primarily low permeable to very low permeable and do not produce water. In China, recovery of coalbed methane can only be achieved through fracturing, so the testing of fractured well for coalbed methane is very important. In this paper, the single-phase flow of coalbed methane in coal matrix and cleats is studied using non-equilibrium isotherm adsorption model. The effect fracture face damage skin factor on pressure-transient performance and the effects of six parameters on pressure-transient performance are also considered in the new mode.

1. INTRODUCTION

Coalbed methane reservoir is a typical natural fracture system consisting of cleats and matrix, in which cleats are flow paths and matrix constitutes the storage space. Conventional natural gas is stored in the interconnected pore space of sandy reservoirs as free gas. The volume of gas in the conventional reservoirs is a function of pressure, temperature and porosity. Except for a little free gas of coalbed methane which is stored within the natural porosity of the coal (joints and fractures) or within the formation water, most coalbed methane is preserved as adsorbed layers on the internal surfaces of micropore spaces in the matrix of coal. Fine micropores of coal matrix usually have very high storage capacity. Extraction of methane from coals is a three-stage process, which includes (1) diffusion from the matrix to cleat, (2) desorption from the internal surface of matrix-cleats, and then (3) flowing through the cleat system to the well. The first and second stages follow the Fick's law, the third stage obeys the Darcy's law, and the relationship between the adsorption volume of coalbed methane and the pressure is expressed with the Langmuir's equation. As a result of these characteristics, numerical model used to describe flow for coalbed methane becomes more complicated.

The most coalbed methane reservoirs in China are characterized by low permeability or very low permeability and don't produce water. Coalbed methane development and recovery in China is different from the Black Warrior Basin and Appalachian Basin in America. Therefore, the wells for development of coalbed methane in China have to be fractured first. As a result, the testing of fractured well for coalbed methane development is very important. According to characteristics of coalbed methane development in China, a finite-conduct fractured well model of the single-phase flowing of coalbed methane has been established in this paper, which uses non-equilibrium isothermal adsorption model and Darcy's law. The new type curves are obtained with the boundary element method. This new model considers the effects of fracture face damage skin factor on pressure-transient performance.

2. MATHEMATICAL MODEL

2.1. *Flow Equation of Coalbed Methane*

It is assumed that coal seam is an uniformly distributed double-porosity system, with matrix being sphere and the coalbed methane gas diffusions into fracture being in balance state, that methane

flow in fractures follows the Darcy's law, that coalbed methane production is targeted to the adsorption gas. Then, under the single-phase radial flowing conditions, the pseudo and transient state mathematical models are described as follows:
For the pseudo-state model,

$$\frac{1}{r_D}\frac{\partial}{\partial r_D}\left(r_D\frac{\partial P_D}{\partial r_D}\right) = \omega\frac{\partial P_D}{\partial t_D} - (1-\omega)\frac{\partial V_D}{\partial t_D} \quad (1)$$

$$\frac{\partial V_D}{\partial t_D} = \frac{1}{\lambda}\left(V_{ED} - V_D\right) \quad (2)$$

for the transient state model,

$$\frac{1}{r_D}\frac{\partial}{\partial r_D}\left(r_D\frac{\partial P_D}{\partial r_D}\right) = \omega\frac{\partial P_D}{\partial t_D} - \frac{(1-\omega)}{\lambda}\frac{\partial V_D}{\partial r_{mD}}\bigg|_{r_{mD}=1} \quad (3)$$

$$\frac{1}{r_{mD}^2}\frac{\partial}{\partial r_{mD}}\left(r_{mD}^2\frac{\partial V_D}{\partial r_{mD}}\right) = \lambda\frac{\partial V_D}{\partial t_D} \quad (4)$$

$$V_D\big|_{r_{mD}=1} = V_{ED} \quad (5)$$

The dimensionless reference length is defined as r_w and other dimensionless variables are defined as:
Pseudo-pressure,

$$\psi = \frac{\mu_i Z_i}{P_i}\int_0^P \frac{P}{\mu Z}dP \quad (6)$$

Dimensionless Pseudo-pressure,

$$P_D = \frac{\psi_i - \psi(r,t)}{1.842\times10^{-3}q_{sc}B_{gi}\mu_{gi}}\times kh \quad (7)$$

Dimensionless Pseudo-time,

$$t_D = \frac{3.6k}{\sigma\cdot r_w^2}\times t \quad (8)$$

Comprehensive compress coefficient,

$$\sigma = \phi\mu C_g + \frac{KhP_{sc}TZ_i}{1.842\times10^{-3}q_{sc}B_{gi}T_{sc}P_i} \quad (9)$$

Dimensionless fracture and storage ratio,

$$\omega = \frac{\phi\mu C_g}{\sigma} \quad (10)$$

Interporosity flow coefficient,

$$\lambda = \frac{3.6k\tau}{\sigma\cdot r_w^2} \quad (11)$$

$$\tau = \frac{R_m^2}{6\pi^2 D} \quad (12)$$

Dimensionless radius,

$$r_D = \frac{r}{r_w} \quad (13)$$

Dimensionless radius of matrix,

$$r_{mD} = \frac{r_m}{R_m} \quad (14)$$

Dimensionless concentration of coalbed methane,

$$V_D = V - V_i \quad (15)$$

Dimensionless coalbed methane gas concentration of balance-state,

$$V_{ED} = V_E - V_i \quad (16)$$

Langmuir equations for pseudo-pressure are as follows:

$$V = \frac{V_L\psi}{\psi_L + \psi} \quad (17)$$

and

$$V_{ED} = \frac{V_L\psi_L(\psi - \psi_i)}{(\psi_L + \psi)(\psi_L + \psi_i)} \quad (18)$$

Defined as adsorption coefficient:

$$\alpha = \frac{1.842\times10^{-3}q_{sc}B_i\mu_i}{Kh}\times\frac{\psi_L}{(\psi_L + \psi)(\psi_L + \psi_i)} \quad (19)$$

Under assumption of adsorption coefficient α=Const, and $V_{ED} = -\alpha\cdot P_D$, through the Laplace transfer for equations (1)-(5), the fracture equation can be expressed as follows:

$$\frac{1}{r_D}\frac{\partial}{\partial r_D}\left(r_D\frac{\partial\widetilde{P}_D}{\partial r_D}\right) = sf(s)\widetilde{P}_D \quad (20)$$

Considering two cases for pseudo-state and transient state, the expressions are as follows:
For pseudosteady function,

$$f(s) = \omega + \frac{\alpha(1-\omega)}{s\lambda + 1} \quad (21)$$

For transient state function,

$$f(s) = \omega + \frac{(1-\omega)}{s\lambda}\cdot\alpha\cdot\left[\sqrt{\lambda s}\cdot\coth(\sqrt{\lambda s}) - 1\right] \quad (22)$$

Line-source solution of Eq. 20 for constant rate:

$$\widetilde{P}_D(r_D,s) = \frac{1}{s}\times K_0(r_D\sqrt{s\cdot f(s)}) \quad (23)$$

2.2. Fractured Coalbed Methane Well Model

The flow within the hydraulic fractures can be considered as linear because the fracture width is much smaller than fracture length and fracture height. It is assumed that the flow into the well takes place via the hydraulic fractures and the flow from the reservoirs into the hydraulic fractures only occur within the reservoir fracture network. In his model, pressure distribution in an uniform-flux fracture well is expressed as:

256

$$\widetilde{P}_D(x_D,y_D,s)=\frac{1}{2s}\int_{-1}^{+1}K_0\left(\sqrt{(x_D-\alpha)^2+y_D^2}\cdot\sqrt{s\cdot f(s)}\right)d\alpha \qquad (24)$$

In Eq. 24, the pressure distribution of infinite-conductivity vertical fracture is obtained using an uniform-flux rate solution for $x_D=0.732$, $y_D=0$.

For the finite-conductivity fracture well, the flux $q_D(x_D, t_D)$ varies with time and space in the fracture direction, with its pressure distribution as follows:

$$\widetilde{P}_D(x_D,y_D,s)=\frac{1}{2s}\int_{-1}^{-1}\widetilde{q}_D(\alpha,s)K_0\left(\sqrt{(x_D-\alpha)^2+y_D^2}\cdot\sqrt{s\cdot f(s)}\right)d\alpha \qquad (25)$$

In Eq. 25, the $\widetilde{q}_D(x_D,s)$ is the Laplace transfer of $q_D(x_D, t_D)$ and the $q_D(x_D, t_D)$ is symmetric distribution, and meets the conditions of:

$$\int_0^1\widetilde{q}_D(\alpha,s)d\alpha=\frac{1}{s} \qquad (26)$$

Considering fracture face skin factor effect of fracture damage, relationship between fracture pressure $\widetilde{P}_{fD}$ and formation pressure $\widetilde{P}_D$ can be expressed as follows:

$$\widetilde{P}_{fD}(x_D,s)=\widetilde{P}_D(x_D,y_D=0,s)+\widetilde{q}_D(x_D,s)\cdot S_f \qquad (27)$$

If we ignore the compressibility of fluid in fracture, the finite conductivity fracture flow Equation can be expressed as follows:

$$\frac{\partial^2\widetilde{P}_{fD}}{\partial x_D^2}+\frac{2}{C_{fD}}\frac{\partial\widetilde{P}_D}{\partial y_D}\bigg|_{y_D=0}=0 \quad 0<x_D<1 \qquad (28)$$

$$\widetilde{q}_D(x_D,s)=-\frac{2}{\pi}\frac{\partial\widetilde{P}_D}{\partial y_D}\bigg|_{y_D=0} \qquad (29)$$

$$\frac{\partial\widetilde{P}_{fD}}{\partial x_D}\bigg|_{x_D=0}=-\frac{\pi}{sC_{fD}} \qquad (30)$$

Combined with boundary condition Eq. 29 and Eq. 30, the integral Eq. 28 can be changed as follows:

$$\widetilde{P}_{wD}-\widetilde{P}_{fD}(x_D,s)=\frac{\pi}{sC_{fD}}\left[x_D+\frac{2s}{\pi}\int_0^{x_D}\int_0^{x'}q_D(x',s)dx'dx\right] \qquad (31)$$

According to Eq.25 and Eq.27, the Eq. 31, can be modified as follows:

$$\widetilde{P}_{wD}-\frac{1}{2}\int_0^1\widetilde{q}_D(x',s)\left[K_0(|x_D-x'|\sqrt{sf(s)})+K_0(|x_D+x'|\sqrt{sf(s)})\right]dx'$$

$$-S_f\cdot\widetilde{q}_D(x_D,s)+\frac{\pi}{C_{fD}}\int_0^{x_D}\int_0^{x'}q_D(x',s)dx'dx=\frac{\pi x_D}{sC_{fD}} \qquad (32)$$

This equation can be solved if we seperate the fracture (half-length) into n segments of uniform flux, assume that length of each segment is equal, and consider x_{Dj} to be located at the jth segment. The Eq. 32 can be substituted by:

$$\widetilde{P}_{wD}-\frac{1}{2}\sum_{i=1}^n\widetilde{q}_{Di}\int_{x_{Di}}^{x_{Di+1}}\left[K_0(|x_{Dj}-x'|\sqrt{sf(s)})+K_0(|x_{Dj}+x'|\sqrt{sf(s)})\right]dx'$$

$$-S_f\cdot\widetilde{q}_{Dj}+\frac{\pi}{C_{fD}}\left\{\sum_{i=1}^{j-1}\widetilde{q}_{Di}\left[\frac{\Delta x^2}{2}+\Delta x\cdot(x_{Dj}-i\cdot\Delta x)\right]+\frac{\Delta x^2}{8}\cdot\widetilde{q}_{Dj}\right\}=\frac{\pi x_{Dj}}{sC_{fD}} \qquad (33)$$

If we write this equation for every fracture segments, we obtain n equations with $(n+1)$ unknowns variables ($q_{fDi}(s)$, $i=1,...,n$ and $P_{wD}(s)$). One additional equation is accordingly resulted if we assume that the flow rate entering the fracture is equal to the well flow rate; that is

$$\Delta x\sum_{i=1}^n\widetilde{q}_{D,i}=\frac{1}{s} \qquad (34)$$

$$\widetilde{P}_{wfD}=\frac{\widetilde{P}_{wD}}{1+s^2C_D\widetilde{P}_{wD}} \qquad (35)$$

3. CHARACTERISTICS OF TYPE CURVES

The Eq. 35 for solution of the well pressure as contains the well storage coefficient C_D, fracture face skin factor S_f, adsorption coefficient α, fracture storage ω, interporosity flow coefficient λ, and fracture half length X_f These six parameters have influences on pressure-transient performances, and these perfomences is of four stages:

Stage 1. This is the early stage of the perfomence, when the transient pressure is controlled by well storage and skin factor. The effect of fracture face skin factor on transient pressure shows a well storage hump when fracture face skin factor is larger than zero ($S_f>0$).

Stage 2. This is a stage of linear flow towards fractures, with the pressure-transient curves having a characteristic of 1/2 slope line in pressure and pressure deviation curves.

Stage 3. This is a stage of interporosity flow between twofold media. The pressure-transient performance shows features of pseudo or transient states and the adsorption coefficient α affects interporosity concave depth and lasting time.

Stage 4. This is a final flow stage, with a characteristc of pseudo-radial flow.

4. CONCLUSIONS

1. The single-phase flow of coalbed methane in coal

matrix and cleats have been studied and a new evaluation model for finite conductivity fractured well of coalbed methane has been established with non-equilibrium isotherm adsorption model.

2. The new model has give considerations to the effect fracture face damage skin factor, adsorption coefficient α, dimensionless fracture storage ω, and interporosity flow coefficient λ on pressure-transient performance.

3. This model can be used not only for the calculation of theoretical curves but also for fractured well test design and interpretation.

4. Obtained formation parameters will be applied to coalbed methane reservoir development.

NOMENCLATURE

B	Gas formation volume factor
C_D	Dimensionless well storage, $C/(2\pi\phi\mu hC_t r_w^2)$
C_{fD}	Dimensionless fracture conductivity
D	Fick's coefficient for coalbed methane
$f(s)$	Transient flow function
K_0	Modified Bessel function of second kind of order zero
P	Pressure, MPa
P_D	Dimensionless Pseudo-pressure
t	Time, hour
t_D	Dimensionless Pseudo -time
R_m	Radius of sphere shape in matrix, m
r_D	Dimensionless radius, r/r_w
r_{mD}	Dimensionless radius of matrix, r_m/r_w
r_w	Well radius, m
s	Laplace variable
V_D	Dimensionless concentration of coalbed methane
V_{ED}	Dimensionless coalbed methane gas concentration of balance state
Z	Gas super compressibility factor
X_f	Fracture half length
x,y	Space coordinates
x_D, y_D	Dimensionless space coordinates, $x_D = x/X_f$, $y_D = y/X_f$
α	Adsorption coefficient
σ	Comprehensive compress coefficient, MPa^{-1}
μ	Gas viscosity, mPa.s
ω	Fracture storage ratio
λ	Interporosity flow coefficient
τ	Adsorption time of coalbed methane
ψ	Pseudo -pressure of gas

SUBSCRIPTS

D	Dimensionless properties
f	Fracture
g	Gas properties
i	Initial, also identify fracture segments
m	Matrix
sc	Standard condition

ACKNOWLEDGEMENTS

The authors thank Dr. Kang Yi-li for his helps in writing this paper.

REFERENCES

Cinco-Ley H., Meng H. Z.: "Pressure Transient Analysis of Wells With Finite Conductivity Vertical Fractures in Double Porosity Reservoirs", paper SPE18172 presented at the 1988 63rd Annual Technical Conference and Exhibition, Houston, TX, Oct. 2-5.

Rushing J. A., Blasingame T. A., B. D. Poe Jr., Brimhall R. M., and Lee W. J.: "Analysis of Slug Test Data From Hydraulically Fractures Coalbed Methane Wells", paper SPE21492 presented at the 1991 SPE Gas Technology Symposium, Houston, TX, Jun. 23-25.

Rodriguez F., Cinco-Ley H., and Samaniego-V F.: "Evaluation of Fracture Asymmetry of Finite-Conductivity Fractured Wells", paper SPE20583 presented at the 1990 65th Annual Technical Conference and Exhibition, New Orleans, LA, Sept. 23-26.

Wong D.W., Harrington A.G. and Cinco-Ley H.: "Application of the Pressure Derivative Function in the Transient Testing of Fractured Wells", SPE Formation Evaluation, Oct. 1986, P470-480.

Ertekin T. and Sung W.: "Pressure Transient Analysis of Coal Seams in the Pressure of Multi-Mechanisitic Flow and Sorption Phenomena", paper SPE19102 presented at the 1989 SPE Gas Technology Symposium, Dallas, TX, Jun. 7-9.

Walls J., Nur A. and Dvorkin J.: "A Slug Test Method in Reservoirs with Pressure Sensitive Permeability", Proceedings of the 1991 Coalbed Methane Symposium, the university of Alabama/Tuscaloosa, May. 13-16.

Koening.R.A, Dean A.K. and Lupton G.: "A Two-Phase Well Testing Tool to Measure Relative Permeability of Coal", Proceedings of the 1993 Coalbed Methane Symposium, the University of Alabama/Tuscaloosa, May. 17-21.

Olivier P.Houze, Roland N.Horne and Henry J.Ramey Jr.: " Pressure-Transient Response of an Infinite-Conductivity Vertical Fracture in a Reservoir with Double-Porosity Behavior", SPE Formation Evaluation, Sept. 1988, P510-518.

258

Mining Science and Technology' 99, Xie & Golosinski (eds)© 1999 Balkema, Rotterdam, ISBN 90 5809 067 1

Simulation of methane gas distribution by computational fluid dynamics

S. Nakayama
Kyushu Kyoritsu University, Fukuoka, Japan

B. Y. Kim & Y. D. Jo
Korea Institute of Geology, Mining and Materials (KIGAM), Taejon, Korea

ABSTRACT: In general, heading faces are located deep underground and isolated from main ventilation networks. Workers at heading faces are exposed to air contaminated mainly by dust and gas. Although forced auxiliary ventilation is widely employed at heading faces, it is not easy to design the optimum ventilation method. In particular the accumulation of explosive methane gas at a heading face has the potential of leading to a serious disaster. To tackle this problem, it is essential to figure out the exact nature of methane gas flow and movements in this space. This paper presents the behavior of methane gas at heading faces so as to provide information for designing adequate ventilation systems. Newly developed three-dimensional gas flow and movement software LASAR98, which is based on Computational Fluid Dynamics (CFD), was used to analyze methane gas behavior.

1 INTRODUCTION

To dilute and exhaust contaminated air, mechanical ventilation is required, but historically it has not been easy to design a reasonable ventilation method due to the complex air flows at a heading face. Recently CFD has proven as an effective tool in the calculation of data for the research of mine ventilation.

CFD software LASAR95 was developed for the analysis of three-dimensional air flow in a mine(Nakayama et al. 1995). Three-dimensional air flow in a 1:1 scale model of a gallery was predicted by LASAR 95 and compared with the results of measurement sampling (Nakayama et al. 1996, Uchino & Inoue 1996, Kim 1996).

Since it's inception this software program has been revised as LASAR98, and enhanced so as to be able to predict not only air flow but also individual and specific gas type movements and flow behaviors. The behavior of NOx gas exhausted by diesel equipment at heading faces was analyzed by this software (Kim et al. 1997). Methane gas concentrations were determined using this software and compared with those obtained using a reduced scale model and with those measured on site in an active coal mine (Nakayama et al. 1998, Ichinose et al. 1998).

This paper proposes a way of determining the behavior of methane gas movements at heading faces in a scientific manner. The detailed gas con-

centration patterns of methane gas were determined using LASAR98 and compared with the results of measurement sampling.

2 SIMULATION PROGRAM

The newly developed LASAR98 CFD software is based on the k-e turbulent model and SIMPLE method which are commonly accepted world wide. This program is coded in FORTRAN and can simulate turbulent flow, heat transfer, gas and dust movements on any computer which uses Windows 95 or 98. LASAR98 employs the Hybrid method for estimating the balance of convection and diffusion currents, and TDMA (Tri-Diagonal-Matrix Algorithm) for solving algebraic equations.

3 FUNDAMENTAL GAS DISTRIBUTION

To obtain the distribution of methane gas at a heading face, simulations were carried out using a simple model as described by figure 1. A circular duct of 1.0m in diameter was set in the corner of the ceiling with the duct outlet positioned 10m from the face end. The air volume discharged from the duct outlet was set at $300 m^3/min$. A 100% methane gas concentration being emitted from the face end at a rate of $3 m^3/min$ was chosen and input into the program. The buoyant force was set in the N.S. equation so as to take into account the difference in density between air(1.2039 kg/m^3) and methane gas (0.6682

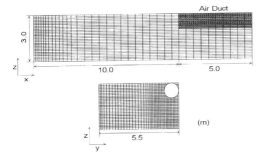

Figure 1. Division of calculations for a simple model

kg/m³).

Figure 2a and 2b show respectively the vector plots and the percentage of gas volume concentrations (%) as displayed over a horizontal plane sliced along the longitudinal axis 5cm from and parallel to the ceiling, as viewed from the top. Air flows from the duct and progresses toward the face end where it's direction is reversed and it then moves toward the free end, as can be seen in Figure 2a.

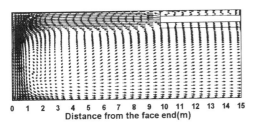

(a) Vector plots

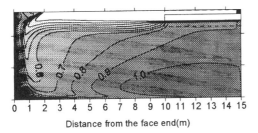

(b) Methane gas concentrations

Figure 2. Vector plots and methane gas concentrations in horizontal section 5cm from the ceiling

In Figure 2b, a high gas concentration area is formed in the corner of the gallery where the face end meets the left wall, as can be seen at the bottom

left corner of the diagram. A large quantity of methane gas stagnates in this area as a result of reduced air flow due to it's relative location from the air duct. Another high concentration area, of over 0.9%, is shown in right half of this figure, where the return flow of the mixed gases is highest.

Figure 3a and 3b show respectively vector plots and gas concentrations through a vertical section of the gallery taken longitudinally along the center line of the air duct. In Figure 3a, the air jet from the duct advances toward the face end and widens as a result of an exchange of momentum with the surrounding air. It can be seen that high methane gas concentrations stagnate in the spaces along where the face end meets the ceiling and floor, as shown in Figure 3b. Although the methane gas concentration beneath the air duct is over 0.8%, due to a reduction of speed of the return flow, it is believed that the methane gas around the face is diluted enough to be exhausted effectively.

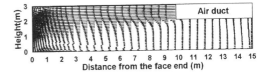

(a) Vector plots

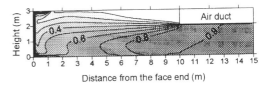

(b) Methane gas concentrations

Figure 3. Vector plots and methane gas concentrations in longitudinal section through the center of air duct

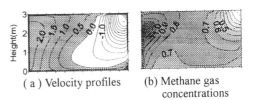

(a) Velocity profiles   (b) Methane gas concentrations

Figure 4. Velocity profiles and methane gas concentrations 4.75m from and parallel to the face end

Figure 4a shows a velocity profile, through the vertical axis, 4.75m from and parallel to the face end. Figure 4b shows the gas concentrations in this section. The air duct can be seen in the top right corner of the diagram. The gray area shows the return flow from the face end in Figure 4a. It can be seen that the high concentrations areas of methane gas shown in Figure 4b coincide with the areas of high return flow in Figure 4a. This indicates that there is a close relationship between the pattern of return flow and gas concentrations.

4 RESULT COMPARISON

4.1 A reduced scale model

For the practical application of the LASAR98 program, the methane gas concentrations calculated by this software were compared with those measured using a reduced scale model.

Figure 5 shows the division of calculated areas for a reduced scale model, which is 20 cm high, 40cm wide, and 1.5 m long. A circular duct of 6 cm in diameter was affixed longitudinal along the center of the ceiling. The velocity of air jet at the outlet was 1.0 m/s. 100% methane gas was emitted from the face end at the rate of 5 l/min (Ichinose et al. 1998).

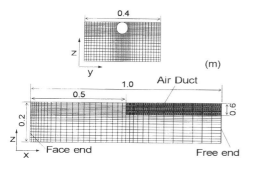

Figure 5. Division of calculated area for reduced scale model

The results of the experiment and simulation are displayed in Figure 6a and 6b respectively. These figures show gas concentration as described through a vertical slice of the gallery taken longitudinally along the centerline of the air duct, as seen from the side. Although the concentrations in the simulation are lower than those of the experiment, similarity in results has been observed. It is shown that the lowest gas concentrations formed in front of the duct outlet, and the highest gas concentrations formed along the corner where the face end meets the floor and in the space underneath the air duct itself.

Figure 7 shows the relationship between the concentration of methane gas through the longitudinal axis and the distance from the face end. Although the concentrations for the computer simulation were lower than those of the experiment, the results of both indicate a high degree of correlation.

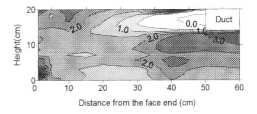

(a) Experimentation (Ichinose et al. 1998)

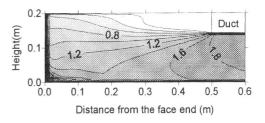

(b) CFD simulation

Figure 6. Methane gas profiles along longitudinal section through the center axis of air duct

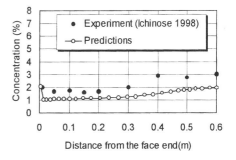

Figure 7. Methane gas concentration along the center axis of a reduced scale model

4.2 An active coal mine

The heading face which measured is one of these in Miike coal mine, which is 2.8m high, 5.4m wide, and 15m long. The effective cross section area of this roadway is 12.7m². A circular duct of 75 cm in diameter is hung longitudinally 40 cm from the ceiling and off to one side of the center line, with the duct outlet positioned 6.8m from the face end. The velocity of the air jet at the outlet was 9.8 m/s.

Table 1 shows the volumes of outflow air and outlet and at the free end of the gallery. The volume of methane gas emitted into the gallery, from the

261

Table 1. Air volume and methane gas concentrations at an on site heading face.

Outlet of air duct		Free end of the gallery
Air Volume (m³/min)	Concentration of CH₄(%)	Concentration of CH₄ (%)
266	0.08	0.14

gallery surfaces, calculated from this data, is 0.16 m³/min (Ichinose et al. 1998).

Figure 8a and 8b show respectively the concentrations of methane gas 6.6m from the face end as measured by actual measurements and as determined by CFD simulation. In both figures it can be seen that lower levels of methane gas concentrations are located immediately in front of the air duct outlet in right side of this figure, and high levels are located in left side.

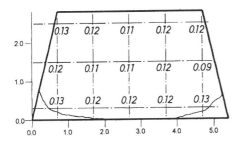

(a) Measurement sampling

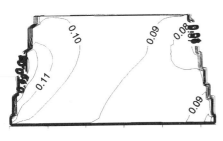

(b) CFD simulation

Figure 8. Methane gas concentrations 6.6m from and parallel to the face end

5 CONCLUSION

To optimize the working environment at heading faces, it is essential to figure out the three-dimensional behavior of methane gas. Comparison of results obtained by experiments and CFD simulations showed that CFD provide a cost effective and practical solution to get the important information needed to be used in the control of air flow and gas behavior at heading faces.

The following characteristics of air movements and gas distribution patterns, at heading faces employing auxiliary ventilation, as determined by CFD simulation, were observed;

1) Return flow of exhausted gases is directly related to the concentration of the methane gas being exhausted. Specifically high concentration areas of methane gas coincide with high velocity areas of the return flow.
2) Problem areas of high methane gas concentrations were confirmed to be located transversely in and along the corner space where the face end meets the ceiling, in and along the corner space where the face end meets the floor, and in the space underneath the air duct.

REFERENCES

Wesely R. 1984. Airflow at Heading Faces with Forcing Auxiliary Ventilation. *Proc. 3rd Intl. Mine Vent. Cong.*, England, pp.73-82.

Launder B. E. & Spalding, D. B. 1974. The Numerical Computation of Turbulent Flows. Computer Methods in Applied Mechanics and Engineering, Vol. 3, pp.269-289.

Nakayama S., Uchino K. & Inoue M. 1995. Analysis of Ventilation Air flow at Heading Face by Computational Fluid Dynamics. *Journal of the Mining and Materials Processing Institute of Japan*, No.4, pp.27-32.

Nakayama S., Uchino K. & Inoue M. 1996. 3-Dimensional Flow Measurement at Heading Face and Application of CFD. *Journal of the Mining and Materials Processing Institute of Japan*, No.9, pp.56-62.

Uchino, K. & Inoue M. 1996. Auxiliary Ventilation at Heading Faces by a Fan. *Proc. 6th Intl. Mine Vent. Cong.*, Pittsburg, pp.493-496.

Kim B. Y., Jo Y. D. & Nakayama S. 1997, 3D flow Simulation for Optimizing Environment of Tunneling Faces. Proc. *1st Asian Rock Mech. Symp.*, Seoul, pp.49-54.

Nakayama S., Uchino K. & Inoue M. 1998. Simulation of Methane Gas Distribution at a Heading Face. *Journal of the Mining and Materials Processing Institute of Japan*, No.4, pp.17-23.

Ichinose M., Nakayama S. & Uchino K. & Inoue M. 1998. In-situ Measurement and simulation by CFD of Methane Gas Distribution at Heading faces. *Journal of the Mining and Materials Processing Institute of Japan*, No. 11, pp.17-23.

Mining Science and Technology '99, Xie & Golosinski (eds) © 1999 Balkema, Rotterdam, ISBN 90 5809 067 1

Oil derived from the Jurassic coal measures in the Junggar and Turpan-Hami basins

Kuili Jin, Suping Yao, Hui Wei & Duohu Hao
Beijing Graduate School, China University of Mining and Technology, People's Republic of China

ABSTRACT: In this paper, the Early and Middle Jurassic terrestrial oil-related coal measures in the Junggar and Turpan-Hami basins, NW China, are subject detailed studies of their sedimentology, organic petrology and geochemistry properties. The study concluded that: 1) the main stage for oil generation occurred earl and the oil is mainly generated from fluorescent desmocollinite (desmocollinite B), bituminite, cutinite and suberinite of the source rock; 2) the oil-generating models of some individual macerals and oil expulsion experiment of vitrain sample are given; 3) the new organic petrological oil-source correlation techniques using laser-induced fluorescence, CLSM (Confocal Laser Scanning Microscope) and TEM (Transmission Electron Microscope) parameters for oil-source correlation are adopted; and 4) the sedimentary organic facies have been established with sedimentary, organic petrological and organic geochemical parameters, and it is suggested that the running water swamp/marsh organic facies is best responsible for coal measures-related oil.

1. GEOLOGICAL SETTING

The important Junggar and Turpan-Hami basins (Figure 5) are located in XinJiang Autonomous Region, NW China and occupy an area of 130000 km^2 and 50000 km^2 respectively. Many Chinese researchers documented that both basins used to be a great unified flooding basin formed after collision of the surrounding plates during the Late Carboniferous to Early Permian. There had been marine transgression and volcanic events that occurred in its early history. At the beginning of Late Permian, the basin was filled with nonmarine sediments and was separated into two basins after the Late Mesozoic collision.

The Lower and Middle Jurassic coal-bearing strata (Figure 5) vary in thickness from less than 1000m up to 2500m. These strata consist of mudstones, sandstones and more than forty coal seams, with a thickness of 100m or more. The coal seams in the Badaowan and Xishanyao Formations were deposited in a lacustrine-swamp system. It is difficult to determine the sedimentary cycles of nonmarine coal measures. We use the lake level as the indicator, then there are three sedimentary cycles from Badaowan Fm to Xishanyao Fm.

2. OIL-GENERATING MACERALS

The main macerals of coal and carbonaceous mudstone for liquid hydrocarbon generation are fluorescent desmocollinite (desmocollinite B), bituminite, cutinite and suberinite, which have generated liquid hydrocarbon in low rank coal/low maturity level carbonaceous mudstone (VRr=0.4%-0.6%).

3. OIL-SOURCE CORRELATION

At first, we examine oil samples from these basins with the Confocal Laser Scanning Microscope (CLSM) and the Transmission Electron Microscope (TEM) (Yao et. al., 1997), and discover some vitrodetrinites under CLSM. The reflectance of these vitrodetrinites in the oil is similar to that of coal or carbonaceous mudstone in coal measures of the two basins. In addition, a lot of submicro-macerals and Jurassic microfossils were discovered under TEM. However, the conventional oil produced in the basins is short of these evidences.

In order to verify oil-source correlation, some indicators of the different oils and aromatic fractions from source rocks are used, which include fluorescence spectrum, fluorescence lifetime fingerprint,

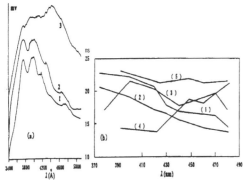

Figure 1. Oil-source correlations used in the Junngar-Turpan-Hami basins. (a) oil-source rock spectra: (1)carbonaceous mudstone; (2) coal-generating oil; (3) conventional oil. (b) oil-oil lifetime fingerprints: (1), (2) coal-generating oil; (3) Junngar's conventional oil; (4), (5) Tarim's conventional oil.

standard compound and maturity. The laser-induced fluorescence method used here was first proposed by Jin et. al. (1996).

Here, we only select the fluorescence spectrum and the fluorescence lifetime fingerprint as parameters for correlation. Figure 1(a) shows the spectrum of the coal measures-related oil (Curve 2) compared with that of the carbonaceous mudstone (Curve 1). Nevertheless, Figure 1(b) shows that the lifetime fingerprints of the same coal measures-related oils (Curve 1 and 2) from different wells are quite different from those of the Tarim's or Junngar's conventional oils (Curve 4 and 5).

4. SIMULATION OF HYDROCARBON GENERATION FOR INDIVIDUAL MACERALS

To further study the hydrocarbon generation, it is necessary for us to use the Rock-Eval to define hydrocarbon potential (S_1+S_2) for individual macerals (Table 1), and to use the pressure vessel together with quartz tube methods to define the hydrocarbon generating process for individual macerals. In addition, the PY-GC is used to analyze the residues after the Rock-Eval analysis, and the Micro-FT-IR is used to analyze the residues after the pressure vessel process. The results are given in Figure 2 and 3.

5. OIL-EXPULSION EXPERIMENTS

In order to confirm that "oil can be expelled from coal", some further experiments are designed. The mercury pressure porosimetry and SEM/TEM analyses are used to compare the coal macropore volume of the extracted vitrain samples with that of the unextracted. The results indicate that the total pore volume increases after extracting. In the SEM image, the intergranular/intragranular macro-pores of macrinites have been found. All of these suggest that oil can be expelled from coal.

Table 1. The hydrocarbon potential (S_1+S_2) of macerals

Maceral	VRr(%)	S_1(mg/g)	S_2(mg/g)	S_1+S_2 (mg/g)
Cutinite	0.5	9.91	358.23	368.14
Desmocollinite B	0.48	6.08	245.74	251.82
Suberinite	0.44	3.90	140.8	144.7
Mineral-bituminous groundmass	0.51	0.63	9.03	9.66
Bituminite	0.51	9.10	4.39	448.1

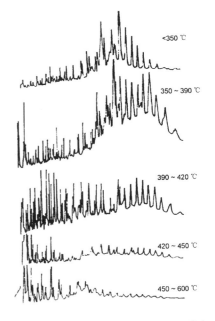

Figure 2. Pyrolysis gas chromatograms of desmocollinite B

264

In addition, we designed an oil-expulsion experiment as follows:

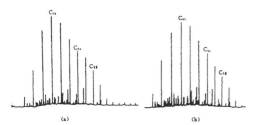

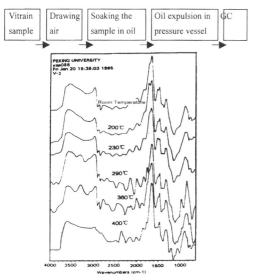

Figure 3. Micro-FT-IR spectra of desmocollinite B

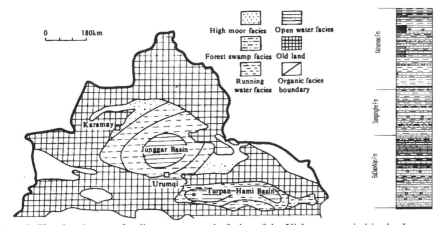

Figure 4 The GC spectrum of press-in oil (a) compared with that of the expelled (b)

The pressure vessel is also used here to study oil expulsion. The expulsion experiment conditions are: time, 72 hrs.; temperature, below 210 °C; and pressure, 18 atm. After thorough comparison of GC spectra (Figure 4) of the oil soaked in and the oil expulsed, we find out that both oils are basically identical.

6. SEDIMENTARY ORGANIC FACIES

The concept of organic facies was first proposed by Rogers. Then Jones defined it as a mappable subdivision of a designed stratigraphic unit. Huc suggested to use organic facies to improve quantitative petroleum evaluation. Based upon these ideas, we emphasize sedimentary parameters including maceral and palynofacies analysis, in addition to lithofacies study. The revised organic facies can be put onto the paleographic map to quantitatively predict and evaluate the coal measures-related oil resources. Therefore, we call the revised organic facies as sedimentary organic facies and use the sedimentological, organic petrological and organic geochemical parameters to define the sedimentary organic facies. Four sedimentary organic facies of coal and carbonaceous mudstone have been distinguished, namely, high moor, forest swamp/marsh, running water swamp/marsh and open water facies (Table 2 and Figure 5). The terms of the running water facies originates from Haymoba (1940) and others are from Teichmüller (1982).

Figure 5. The sketch map of sedimentary organic facies of the Xishanyao period in the Junggar and Turpan-Hami basins, Sinjiang Autonomous Region, N.W. China

Table 2. The classification of the sedimentary organic facies

Type		High moor facies	Forest swamp/marsh facies	Running water marsh Facies	Open water facies
Facies marks	No.	1	2	3	4
Organic petrologic character	V+I %	>90	70-90	40-70	<40
	E %	0-10	10-30	30-60	>60
	Main maceral	Fusinite	Telocollinite	Desmocollinite	Alginite, Cutinite
	Microlithotype	Fusite	Vitrite, Clarite	Clarite	Durite
Sedimentary character	V/I	<1	>1	>1	>1
	GI	1-2	2-50	0-50	2-10
	TPI	0-2	2-6	0-2	2-10
Organic geochemical character	H/C	<0.95	0.95-1.15	1.15-1.4	>1.4
	HI(mg/g.COT)	<125	125-250	250-400	>400
	S_1+S_2(mg/g)	<50	50-200	200-300	>300
	Organic matter type	III	IIB	IIA	I
Jones' organic facies		D□CD	C□BC	BC□B	B□AB□A

In addition to sedimentological and organic petrological parameters of facies samples, the organic geochemical parameters can also be obtained from maceral analyses. For example, the S_1+S_2 values can be obtained from the maceral analysis and the statistic analyses on known data (Table 2), although they are usually measured with the Rock-Eval (Table 1). The running water swamp/marsh facies is believed to be the best one for coal measures-related oil generation (Figure 5). Allochthonous /hypautochthonous transportation and disintegration within the running water swamp/marsh would cause accumulation of hydrogen-rich macerals, such as fluorescent desmocollinite, cutinite, bituminite and suberinite.

7. CONCLUSIONS

1. The simulation of individual macerals and the hydrocarbon generating phenomenon of coal and carbonaceous mudstone have been studied. The results confirm that macerals as fluorescent desmocollinite, cutinite, bituminite, suberinite play an important role in oil generation from low rank coal or low maturity carbonaceous mudstone. The oil generating models of some macerals are estabished.
2. A new organic petrological oil-source correlation technique has been adopted, which uses the macerals observed under the CLSM/TEM, such as vitrodetrinite/submicro-macerals, together with the laser-induced fluorescence parameters.

3. The mercury pressure porosimetry study shows that the total pore volume increases after extracting. Moreover, oil-expulsion experiment shows that oil can be expelled in the basins.
4. Sedimentary, organic petrological and organic geochemical parameters are used to subdivide sedimentary organic facies. The running water swamp/marsh facies is believed to be the best for the generation of coal measures-related oil.

REFERENCES

Haymoba, C. H. 1940. The genesis classification for coal of suburb-basin of Moscow. *USSR Mineral Resource Institute Bull.* 159 (in Russian)

Huc, A. Y. 1990. Understanding organic facies: a key to improve quantitative petroleum evaluation of sedimentary basins. In: A. Y. Huc ed. *Organic Facies.* AAPG, Tulsa, Oklahoma 1-11.

K. Jin, , J. Fang, Y. Guo, C. Zhao, & N. Qiu 1996. The use of new laser-induced fluorescence indicators in determining thermal maturation of vitrinite-free source rock and oil-source correlation . *Abstracts of 12th annual meeting of the Society for Organic Petrology*, Woodlands, Texas, 1996: 12: 18-20

Teichmüller, M. 1982. Origin of macerals. In: E. Stach, M.-TH. Mackowsky, M. Teichmüller et al., eds. *Stach's Textbook of Coal Petrology.* Gebrüder Borntraeger, Berlin, Stuttgart: 285-290.

Mining Science and Technology' 99, Xie & Golosinski (eds)© 1999 Balkema, Rotterdam, ISBN 90 5809 067 1

New method of coal production and utilization: Underground coal gasification

Liu Shuqin & Yu Li

China University of Mining and Technology, Beijing, People's Republic of China

ABSTRACT: A new underground coal gasification technique called "long tunnel, large section, two-stage" is introduced in this paper. Its features are summarized and compared with other techniques, and its economic and environmental aspects also evaluated.

1 INTRODUCTION

China is the largest country of coal production and consumption in the world. In the long period of coming time, coal will continue to occupy the leading position in the primary energy resource construction, which led to aggravation of environmental pollution. Coal production and utilization has become the main influence factor of air pollution, acid rain, climate change, and health. China will face more and more environmental problems with the development of economics and the increase of coal consumption. Statistics shows that total 4 billion tons of coal every year (i.e, 3 tons each person) is needed if attaining the present living standards of South Korea and Taiwan Province of China. Therefore, in order to accelerate coordinated and sustainable development of energy resource and environment, the new clean and harmless technology of coal production and utilization have to be found out.

Underground coal gasification (UCG) is such a process in which coal underground is burned under control to produce combustible gas through the action of heat and chemistry. It combines the techniques of shaft-building, coal-mining and gasification, changes physical mining into chemical mining, and large, cumbersome equipment is eliminated, So it boasts merits such as more safety, less investment, high efficiency, and less pollution. Early in 1888, the world-renowned chemist Mendeleev put forward the assumption of UCG: instead of directly extracting raw coal, the purpose of coal-mining should be to extract energy components of raw coal. It is also referred in UN conference of world coal prospect in 1979: developing UCG is one of the research direction of world coal-mining, through which we can solve a series of technical and environmental problems in traditional coal-mining methods. Up to now, a lot of scientific research project of UCG have been conducted worldwide and some significant progresses have been made. No-shaft techniques are often used in UCG worldwide, the problems arising from it are small gasifier, low heat value, unstability, high costs, this kind of techniques are also difficult to be industrialized. The sketch map of the old UCG technique of Former Soviet is shown as figure1.

Based on the study of different old UCG techniques worldwide, A new UCG technique of "long tunnel, large section, two–stage" has been developed and introduced in this paper, which will be hopeful to solve the real situation in China that there will be more and more abandoned coal resources.

2. NEW TECHNIQUE OF UCG

2.1. *Principle & technique*

The principle of the new UCG technique is the same as that of the ground coal gasification, but

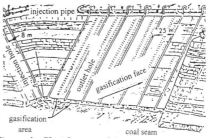

Figure 1   Sketch map of the old UCG technique of Former Soviet

techniques are different. The three essential component parts of the underground coal gasifiers are an inlet hole, a vent hole and a gasification tunnel. The whole process is conducted through three action zones, i.e, an oxidation zone, a reducing zone, and a dry distillation zone. The fire is made in gasificatin tunnel near by the inlet hole and the gasification agent (O_2 or H_2O)is blown in from the inlet hole. In the oxidation zone, O_2 meets the very hot coal and they are burning and reacting, then a great deal of heat is produced and its temperature will be up to what the gasifier need to react (about $1000°$). In the reducing zone, CO_2 and $H_2O(g)$ meet the very hot coal. CO_2 is reduced into CO and H_2O is decomposed into H_2 and CO. In the dry distillation zone, because of the action of the high temperature airstream, the coal bed will decompose and the dry distillation gas forms. In the tunnel near by the vent hole, the excessive $H_2O(g)$ and CO react, then H_2 is formed. After reaction in the three zones, gas (main ingredients: H_2, CO, CH_4) forms. The output of the combustible gas depends on the temperature in the reducing zone and the dry distillation zone.

"Long tunnel, large section, two-stage" underground gasification is a new technique of UCG. Two–stage means supplying air and water vapor circularly. In the first stage, air is blown in the gasifier and helps burning. Heat is stored up in the coal seam and the blast gas is produced. In the second stage water vapor is blown in and the underground water gas forms, the influence of N_2 on the gas calorie is dispelled and the high temperature oxidation zone in the first stage becomes a reducing zone. In the whole gasification tunnel, the dry distillation gas and underground water gas is produced, so high calorie gas can be obtained, in which H_2 and CO hold big content proportion. Gas calorie can be further raised due to the catalysis of some metallic oxides on methanation reaction. The

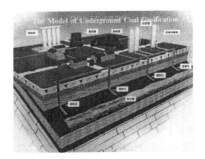

Figure 2. Sketch map of the new technique of UCG

crux of the two-stage gasification is that much coal is burned and much enough heat is supplied. The long gasification tunnel will get the heat airstream to have long enough time and area to be transferred into the coal bed. The large section makes the amount of the burning coal much big , the heat stored up great, and supplies enough reaction heat for the second stage. Therefore, the two-stage gasification method must be combined with the long tunnel and the large section gasifier. Figure 2 is the sketch map of new technique of UCG , and comparison between the new and the old technique is shown in table 1.

2.2 . Results

The new technique is a breakthrough of the traditionals. It is shown by semi-industrial trial that high calorie gas can be stably produced using the above technique. Comparison of the underground gas ingredients of different countries is shown in table 2, and gas ingredients and heat value in the second stage of Xuzhou Xinghe semi-industrial trial are shown in table 3.

Table 1. Comparison of gasifier structure and technology between the new and the old

Item	Old technique	New technique
Drilling holes for injection and output	More than ten holes with short diameter(150mm); big flowing resistance and high electricity consumption; function of hole can not be interchanged; holes contacting with burning area are not easy to be protected and have short life.	Less holes (5-6) with larger diameter (377mm); small flowing resistance and low electricity consumption; function of hole can be interchanged; holes not contacting with burning area are easy to be protected and have long life.
Tunnel	Small section(A < $0.1m^2$);short tunnel; big flowing resistance; two months to attain the expected gas output; bad airstream regularity.	Large section (A>$3.5m^2$);long tunnel(>200m); large amount of burning coal; only one day to attain the expected gas output; good airstream regularity.
Technology	One stage; gasification agent include air, air plus water vapor; gas heat value is less than $10.5MJ/m^3$; imperfect technology.	Two stage; the second stage is the sum of water gas and dry distillation gas with high content of methane and hydrogen, heat value is about 12-$15MJ/m^3$; gasification is combined with other assistance measures.

Table 2. Comparison of UCG of different countries

	Time (year)	Gas ingredient (%)					Heat value
		CO_2	CO	H_2	CH_4	N_2	MJ/m^3
U S A	1948	6.0	0.5	0.9	0.4	79.5	1.88
	1952	11.7	7.1	7.6	2.1	70.9	2.68
	1948	44.0	1.9	25.1	10.1	16.1	8.42
	1979	15.0	8.0	12.4	2.9	49.0	4.31
U K	1950	15.5	4.9	7.9	1.0	70.7	2.05
Italy	1979	19.7	4.5	15.6	2.2	57.8	3.43
Belguim	1979	36.1	18.5	36.1	5.4	0.0	8.55
	1979	13.4	36.2	31.2	3.0	2.0	9.67
	1979	19.3	53.3	17.6	0.7	0.0	9.27
France	1955	19.5	4.0	15.0	4.5	57.0	4.08
Former Soviet	1952	12.1	15.9	14.8	1.8	54.1	4.18
	1956	19.5	7.1	14.1	1.5	55.9	3.50
	1964	5.6	28.7	18.4	2.1	44.9	6.49
China	1958	15.8	7.3	15.9	1.2	58.1	3.81
	1958	8.3	15.0	10.7	0.7	62.6	3.50
	1958	10.3	21.3	16.0	2.8	49.2	5.53
	1994	12.6	11.9	63.6	10.2	1.7	13.68
	1996	10.5	24.8	53.1	7.2	4.4	12.74

Table 3. Gas ingredients and heat value by two-stage Technique in Xuzhou Xinghe semi-industrial trial

1994		Gas ingredient (%)					Heat value	Flow
Month	Data	H_2	CO	CH_4	CO_2	N_2	MJ/m^3	m^3/h
11	9	58.29	8.59	9.28	19.63	4.21	12.22	1920
11	10	53.03	18.59	15.08	13.16	0.13	15.12	1900
11	11	53.38	10.35	14.32	16.38	3.59	14.45	1400
11	12	57.10	11.66	14.89	13.81	2.50	14.70	1500
11	28	62.07	14.43	10.13	11.07	2.30	13.78	1650
11	30	54.25	15.72	10.65	15.26	4.12	13.14	1810
12	3	67.94	14.97	6.30	10.69	0.00	12.96	1450
12	6	70.51	12.42	5.23	10.83	1.00	12.66	1720
12	7	64.07	11.31	9.94	11.13	3.55	13.57	1900
12	8	60.42	16.57	9.54	12.52	0.95	13.61	1550
12	9	72.95	6.07	8.30	12.17	0.51	13.39	1520
12	10	69.16	10.32	9.10	10.94	0.48	13.76	1750
12	11	66.14	10.75	11.98	11.10	0.03	14.58	1580
12	12	64.63	12.47	9.65	11.70	1.55	13.69	1850
12	13	71.58	6.16	10.00	11.92	0.27	13.77	1600
12	14	67.56	9.47	9.50	12.92	0.53	13.60	1830

2.3 . *Creativity*

Compared with traditional techniques, the new technique possesses following creativity:

1. Production of water gas with high heat value can be realized by no-man, no-equipment, long-wall (more than 200m) gasification face.
2. The temperature of gasifier can be raised rapidly and then capability of heat storage is increased due to long tunnel and large section in the new technique. At the same time, methane content in dry distillation gas is increased and the gas heat value is improved which are beneficial to stability of production.
3. Long tunnel, large section can also reduce flowing resistance, decrease electricity consumption of air supply, and lower the cost of production.
4. Multi-hole gasifier can be used for supplying air forward, so the efficiency of air supply is improved. Drilling holes with multi-function can be interchanged through which monitoring on the spot can be conducted.
5. Gas leaking can be avoided by pressuring combined with pumping, which ensures safety of production .
6. The whole seam can be gasified from low level to high level through direct-adverse blasting.
7. Ground surface over the gasifier is not destroyed by gasification accompanied by filling.
8. Some metallic oxides such as Al_2O_3, CaO, MgO, Fe_2O_3, Fe_3O_4 in the coal residue can catalyze the water gas reaction and methanation reaction under the condition of long tunnel, which will increase the content of methane in the gas.
9. Compared with natural gas, underground water gas possesses high H/C ratio while its cost of production is lower. It is expected to be used as the substitution for natural gas. As indicated in table 4.

3. ECONOMIC & ENVIRONMENTAL BENEFIT

According to statistics, there are about 30 billion tons of abandoned coal resources and 31.5 billion tons of oil shale in china. Underground gasification can regenerate present fixed assets, save investment

Table 4. H:C mole ratio of different fuels.

	Coal	Oil	Natural gas	Xuzhou water gas
H:C	0.86:1	1.76:1	3.71:1	4.76:1

Table 5. Comparison of underground gasification & ground gasification

Items	Ground gasification	Underground gasification
Investment for civil construction（Yuan/m^3）	350—450	120—150
Cost （Yuan/m^3）	0.4—0.6	0.15—0.25
Producing technology	Coal preparation and selecting	Underground coal resource
Environmental protection	Much ash discharge	Less ash,less pollution

for heavy equipment , and recover abandoned coal resources and oil shale which can not be mined by traditional mining methods. Dirty resources can be converted into clean and cheap resources — underground water gas or hydrogen.

Cheap water gas can be obtained using two-stage technique, technic economic analysis of ground and underground gasification is shown in table 5. The heat value of water gas is higher than 12.60MJ/m^3 and hydrogen content higher than 60%, which make it possible to be used for extracting pure hydrogen (with purity 99.9%, cost 0.5 Yuan/m^3).

Ashes, oxides, radioactive materials, and waste rock after gasification are left underground, which will eliminate the pileup of waste on the ground, reduce the cavity space, make surface subsidence avoidable. Moreover, Gas can be concentratedly purified and decarbonized and obtain cheap hydrogen fuel .

4. CONCLUSION

The new UCG technique of "Long tunnel, large section, two-stage" is feasible in both principle and technique, and it enjoys good economic and environmental benefit, therefore, is surely of profound significance to clean production and utilization of coal in the future.

REFERENCES:

Yu Li , 1990. The past and future of underground coal gasification. *Kuang Ye Yi Cong*. 4:1-10.
Yu Li ,Yu Xuedong, 1998. The new technique of underground coal gasification. *coal*. 1.
Liang Jie & Yu Li, 1995. The theory and practice of producing hydrogen by underground coal gasification. *Ke Ji Dao Bao*. 8:50-52.

Mining Science and Technology'99, Xie & Golosinski (eds)© 1999 Balkema, Rotterdam, ISBN 90 5809 067 1

Trial study on underground coal gasification of abandoned coal resource

Jie Liang, Shuqin Liu & Li Yu
China University of Mining and Technology, Beijing, People's Republic of China

ABSTRACT: The field trial of underground coal gasification(UCG) at Liuzhuang mine is introduced in this paper. The long tunnel, large section, advancing gasifier is designed in the trial according to the character of abandoned coal seam. The design rule and structure of the gasifier are studied and the stability of air successive gasification process is analyzed. In addition, the moving velocity and the length of the gasification face are measured by radon technology. The moving velocity measured is about 0.204□0.487m/d. The trial has confirmed that UCG of abandoned coal resource is feasible.

1. INTRODUCTION

Coal is the main energy resource in China and accounts for about 70% in the primary energy resource, while only about 50% can be recovered by underground mining. According to statistics, there are 297 abandoned mines and 30 billion tons of abandoned coal resource in China. When bringing economic benefit the tradition coal mining also relate to a serious of negative effects on environment, such as surface subsiding, lost of underground water, and discharging of sulfur dioxide to air. In order to resolve above problems, clean coal technology must be developed. Therefore, UCG in abandoned mine has been studied by the UCG Research Centre at China University of Mining & Technology. The semi-industry trial of UCG at 2nd shaft of Xinhe coal mine in Xuzhou was first finished in 1994. Then the trial of UCG at Liuzhuang mine in Tangshan has been conducted since 1996. In the trial the long tunnel, large section, advancing gasifier was designed, and the air successive gasification process and its moving velocity of gasification face were researched.

2. THE ADVANCING GASIFIER STRUCTURE AND ITS SECURITY MEASURES

A distinguish between UCG and ground gasification is that the material layer of UCG can not move, so the movement of gasification face must be controlled to realize successive gasification. The best method of controlling the face is adjusting position of air injecting and gas vent with the movement of gasification face (Liang,1997).

The gasification tunnel in traditional no-shaft gasifier is linked by special technology, such as fire linkage, electricity linkage, and drilling in direction. It has problems of short tunnel, little diameter, big flowing resistance, and low stability. The points of air inlet and gas vent are fixed, which has three shortages: (1) It is difficult for oxygen to reach fire face, so combustion velocity is low. (2) Oxygen make a detour to the gas beside vent hole, and second combustion is produced, thus heat value of the gas is greatly lowered. (3) Gasification rate of coal seam is low. In order to overcome these drawbacks, the long tunnel, large section, advancing gasifier was designed in the trial (shown in figure 1). The gasifier mainly consists of a gasification tunnel, two auxiliary holes and two auxiliary tunnels. The gasification tunnel is a man-made coal tunnel, which can overcome the shortages of no-shaft gasification tunnel.

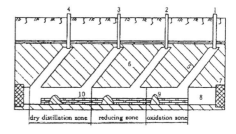

Figure.1 The UCG gasifier of the trial at Liuzhuang mine 1-inlet hole   2,3-auxiliary hole   4-vent hole 5-inclined connecting drift   6-gasifiable coal seam   7-sealed wall   8-gasification tunnel   9-coal pile   10-auxiliary tunnel

The length of the gasification tunnel is shown as:

$$L \geq L_1 + L_2 + U_s \tau \qquad (1)$$

Where L is the length of gasification tunnel(m), L_1 is the length of oxidation zone(m), L_2 is the length of reducing zone(m), U_s is the moving velocity of gasification face(m/day), $\Box$ is the single gasification time(day). The results of numerical simulation and laboratory experiment indicate that the length of gasification tunnel depends on the length of oxidation zone. When the concentration of oxygen is little or equals to zero, oxidation zone ends.In order to study the distribution of oxygen diffusion in the free tunnel . The assumptions of oxygen with one dimension constant flowing and two dimensions diffusion are adopted, so the eqaution of oxygen diffusion is shown as:

$$U\frac{\partial C_{O_2}}{\partial X} = \left[\frac{\partial^2 C_{O_2}}{\partial Y^2} + \frac{\partial^2 C_{O_2}}{\partial X^2}\right] \qquad (2)$$

Oxygen concentration distributed along axis direction (x direction) is shown as:

$$C_{O_2} = C_{O_2}^0 \exp(-0.02b^2\sqrt{\lambda}d^2 x) \qquad (3)$$

Where the term U is the flowing velocity(m/s), D is the diffusion coefficient(m²/s), C^0_{O2} is the oxygen concentration of inlet point(%), $\Box$ is the resistance coefficient of gasification tunnel, d is the hydraulic diameter(m), b is decided by

$$btg(bh) = K/D \qquad (4)$$

$$d = \frac{2hw}{h+w} \qquad (5)$$

Where the term K is the reaction rate of oxygen and carbon (mol/m³·sec),while h and w are the height and the width of the gasification tunnel in m.

The calculation results from formula (2) to (5) are shown in figure 2.

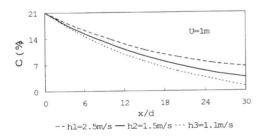

Figure 2 Oxygen distribution along axis direction

Figure.2 shows that the length of oxidation zone is proportional to flowing velocity of gas and the diameter(section) of the gasification tunnel.

Table 1. The main parameters of the gasifiers

Item	9^s	12^s
Length of gasification tunnel(m)	120	220
Thickness of coal seam(m)	2.0-3.5	4-6.5
Inclined Length of coal seam(m)	65-75	80-96
Angle of coal seam(Deg)	38-55	40-65
Gasification reserve(ton)	33000	112000

In the trial two gasifiers, 9^s and 12^s ,were built. The main parameters are shown in table 1.

The auxiliary holes are multifunction holes arranged between inlet hole and vent hole. It can be used to inject air, water vapor or to discharge gas. The trial results indicate that the distance between different auxiliary hole, injection hole or vent hole must be less than the sum of oxidation zone length and reducing zone length, which can ensure the gasification face of adjacent holes to connect.

The auxiliary tunnel is a auxiliary installation of air injection to prevent the gasification tunnel from blocking up. It is a little tunnel support by bricks.

Gasifier must be operated in the condition of seal abandoned coal underground. Otherwise, the combustible gases will leak to other areas of the mine, which will lead to accident. Therefore, in the trial of liuzhuang mine, the sealed separating bands with total length of 1019.4m were built. The tunnels connected with the gasifiers are also isolated by building sealing wall (Yang,1998).

3. AIR SUCCESSIVE GASIFICATION PROCESS AND THE STABILITY ANALYSIS

3.1 Successive air gasification

The coal of 9^s seam was fired by the first group of flame lighter at 23/22 and the second group lighter was ignited at 23/35 on May 17, 1996. Air supplying was adjusted to the optimal value to increase the temperature of gasifier. With the temperature increase of the gasifier, the gas was gradually producing, the heat value of gas also became stable.

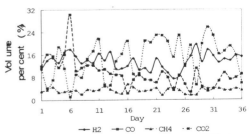

Figure 3 The curve of daily average change of (9^s) air gas ingredient

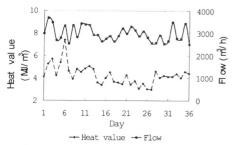

Figure 4 The curve of daily average change of
(9ˢ) air gas heat value and flow

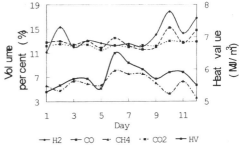

Figure 7 The curve of daily average change of (12ˢ) air gas
ingredient and heat value

The time from firing to gas producing is 81 hours. The trial of air successive production was conducted about one month from May 21 to June 20. The gas ingredient, heat value, and flow of the trial are shown in figure 3 and figure 4.

The coal of 12ˢ seam was fired at 19/28 on June 12,1996 ,and gas heat value reached 3.71MJ/m³ after 85 hours, the increase tendency of gas heat value is very similar to 9ˢ .The trial of air successive gasification of 12ˢ coal was conducted from June 15 to July 10. The change of ingredient, heat value, and flow of gas are shown in figure 5 and figure 6.

The gasification trial at Liuzhuang mine is still in operating. If calculated according to coal reserves, gasification of 9ˢ coal seam is coming to the end, while gasification of 12ˢ coal seam can be lasting another two years. Air gas has been steady and

successively produced in the lasting two years. The daily average change of air gas ingredient, heat value of 12ˢ gasifier from December 10 to December 21 in 1998 are shown in figure 7 .

3.2 Stability analysis

The unstability coefficient of gasification parameters can be expressed as follows:

$$\eta = \frac{2(M - N)}{M + N}$$

in which, η is Unstability coefficient in %; M is the maximum valve of parameters in measuring cycle. N is the minimum valve of parameters in measuring cycle.

According to above formula, the stabiliy of air successive gasification process can be analyzed. Unstatitity coefficient of gas heat value of 9ˢ gasifier is about 45%, and unstability coefficient of flow is about 30%.

The Unstability coefficient of heat value of 12ˢ gasifier is 53% in the earlier stage of firing and about 21% in the later stage. Unstability coefficient of flow is about 21%.

We can see from the above, unstability exists to some extent in the process of air gas production. Especially in the beginning period after firing, the variation of heat value is large because the temperature field for gasification reaction has not been developed. Heat value becomes more stable with the increase of gasifier temperature . The trial shows that heat value of air gas is always higher than 3.34MJ/m³, which can be used for burning of industrial boiler. The reason for stables heat value of gas is that the gas possesses stable methane content , which indicate the large production of dry distillation gas in the long tunnel underground gasifier. Therefore, the long tunnel, large section, advancing underground gasifier is favorable to the stability of gasification process. The variation of gas flow is due to the change of tunnel resistance influenced by coal collapse on the space of gasification reaction. Thus in order to keep stable

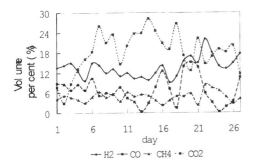

Figure5 The curve of daily average change of (12ˢ) air gas
ingredient

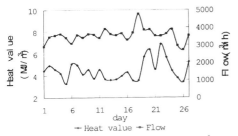

Figure 6 The curve of daily average change of (12ˢ)air gas
heat value and flow

gas flow, adjustable fan must be used to change pressure of air supplying with the variation of resistance in the gasification tunnel.

4. MEASURING OF THE LENGTH AND MOVING VELOCITY OF THE GASIFICATION FACE

During the trial at Liuzhuang Mine, the technology of radon measuring is applied to measure the moving velocity of gasification face. Radon(^{2}22Rn) is the only gaseous stage produced by radioactive elements during decaying. It has the character of vertically moving from underground to the surface when stimulated. Its movement has close relationship with temperature. The emanation power increases with the increasing of temperature. Therefore, the density of radon at the surface will indicate the relative temperature underground. Figure 8 is the distribution of radon density along the top of the tunnel at 9^s gasifier after 88 days operation. Figure 9 is the distribution of radon density along the top of the tunnel at 12^s gasifier after 64 days operation. The distribution of high temperature zone is determined based on the above results.

It can be seen from figure 8 that the first abnormal point of radon at the 9^s gasifier is 17m away from the ignition point, reaching as far as 42m. It means that the high temperature zone has moved for 17m forward, at least 42m in length(including oxidation and reduction zone). It can be seen form figure 9 that the first abnormal point of radon at the 12^s gasifier is 29m away from the ignition point, reaching as far as 38m .It shows that the high temperature zone has moved 29m forward, at least 38m in length (including oxidation and reduction zone). Since the coal seam at the 9^s is thicker than that at the 12^s ,it can be said that the thicker the coal seam is ,the shorter the high temperature zone is.

The moving speed of radon varies in different materials (Wu,1994). Usually, this speed is 0.3-1.9m/h within solid. Taking the average value as 1.1m/h and considering the depth of coal seam, it can be deferred that the emergence of radon abnormal point is 4.5 days later than the emergence

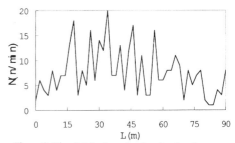

Figure 9 The distribution of radon density along the top of the tunnel at 12^s gasifier

of high temperature zone. Therefore, based on the location of high temperature zone and gasification time, the moving velocity is calculated to be 0.204m/d at the 9^s, and 0.487m/d at the 12^s. The moving speed at the 9^s is lower than that at the 12^s, which is mainly because two-stage UCG technique is applied at the 9^s for most time while air successive gasification at the 12^s. Therefore, the moving speed of gasification face in two-stage UCG is lower than that in air continuous gasification.

5. CONCLUSIONS

Based on the above study, the following conclusions can be obtained:

(1) Abandoned coal resource can be recovered by UCG ,which can increase the utilization ratio of coal resource.

(2) The construction of the long tunnel, large section, advancing underground gasifier is relative easier and the investment for it is lower. The successive and steady UCG can be realized. Security measures such as separation bands and sealed wall can ensure the stable operating of gasification process.

(3) Heat value obtained by air successive air gasification is about 4.18 MJ/m^3 .The moving speed of gasification face is from 0.204 to 0.487m/d.

(4) UCG can reduce the effect on environment during coal mining, transportion and using ,and supply clean energy resource and chemical unstripped gases, so it is an important research direction of clean coal technology in China.

REFERENCES

Liang Jie. 1997. Study on the stability and controlling technology in steep incline seam (*dissertation*) CUMT, China

Yang Lanhe et al.1998. Underground coal gasification. CUMT *Journal* 23(3):254~258.

Wu Jianming. 1994. Measuring radon in nature fire zone. Taiyan:Shanxi College of M & Technology.

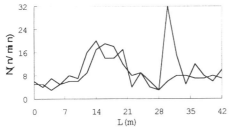

Figure 8 The distribution of radon density along the top of the tunnel at 9^s gasifier .

Mining Science and Technology'99, Xie & Golosinski (eds) © 1999 Balkema, Rotterdam, ISBN 90 5809 067 1

Formation and processing technology of kaolinite rocks in China

Xiaojie Yang, Qinfu Liu & Jingyan Cheng
China University of Mining and Technology, Beijing, People's Republic of China

ABSTRACT: Kaolinite rocks in China are a precious non-metallic mineral resource. They are thick-bedded, wide-spread and of good quality.. This paper presents the formation, development and utilization of kaolinite rocks in China.

1. INTRODUCTION

Kaolinite rocks, occurring in Permo-Carboniferous Coal measures in North China, is a kind of kaolinitic clay material, which is different from that in Germany, France, Russia, U.S.A. and other European-American Countries, featuring thick-bedded, wide-spread and of good quality. The thickness of kaolinite rocks in North China is 70 cm or so, but that in the above-mentioned countries only 3 cm. From Hunjiang County, Jilin Province to Shandan County, Gansu Province, from Huainan and Huaibei coal field to Inner Mongolia coalfield, kaolinite rocks deposits are scattered over vast areas on the North China Platform with great proved reserves of 1.6 billion tons.

It is of great interest and high significance to study the mineralization mechanism of kaolinite rocks for its processing and utilization. By means of mineralogy, petrology, geochemistry and organic chemistry, formation of kaolinite rocks has been systematically deliberated and the origin model of it has been proposed. On basis of this, the processing technology of kaolinite rocks has been put forward, which had been successfully employed in China.

2. FORMATION OF KAOLINITIC ROCKS

2.1. *Mineralogy*

Results of XRD, IR, DTA and SEM studies show that the content of kaolinite mineral in kaolinite rocks is in general more than 70% and up to 90%-100% in some good quality rocks. Based on the intensity feature of the (001), (001), (003), (004), (020), (060) peaks and their ratios in the XRD spectrum, it is found that some coarse crystals, which have been called vermicular kaolinite, are indeed kaolinite/dickite transitional minerals. Other clay minerals include I/S mixed-layer clay mineral, NH_4-illite, boehmite, diaspore, chlorite, etc. The accessory minerals observed include beta-quartz, detrital quartz, feldspar, zircon, toarmaline, apatite, hematite, ilmenite, rutile, calcite and siderite.

The Hinkley index (*HI*) varies with different kind and origin of kaolinite (Table 1), being the highest for the soft kaolin clay, the lowest for the terrigenous kaolinite.

2.2. *Petrology*

A number of microtextures were identified in the kaolinite rocks by using polarized light microscope, which include intraclast, pisolite, oolite, pellet, vermicular crystals, pseudomorphous crystal of feldspar and biotier, matrix and organic matter. The pisolite and oolite are mostly found in "flint clay type", whereas vermicular and pseudomorphous crystals are in "tonstein type".

Several types of structures were recognized in kaolin tonsteins, including graded bedding, small cross bedding, wavy and horizontal lamination, interbedding of organic matter and clay, micro-washing structure, and seepage flow tube structure. The small cross bedding and micro-washing structures represent the influence of current water in low energy environment.

Table 1. The HI for different type and origin of kaolinite

Rock types	Occurrence	HI	Ordered degree	Origin
Soft kaolin clay	associated with weathered coals	1.33-1.65 1.46	Well ordered	organic colloid chemical crystallization in surficial environment
Cryptocrystalline kaolinite rock	in the roof, parting, floor of coals	1.04-1.45 1.2	ordered	organic colloid chemical crystallization in swamps
Oolite, pelletoid, and cryptocrystalline kaolinite rock	not adjacent to coals	0.25-0.99 0.73	disordered	derived from terrigenous clay
Crystalline kaolinite rock	in the roof, parting, and floor of coals	0.78-1.22 0.96	disordered	pseudomorph after feldspar and biotite by alteration, or from diagenetic recrystallization
Intraclastic kaolinite rock	in the roof, parting, floor of coals, or not adjacent to coals	0.97-1.10 1.02	disordered	derived form terrigenous clay
Graupen kaolinite rock	in the roof, parting, floor of coals, or not adjacent to coals	0.95-1.13 1.05	disordered	derived from terrigenous clay
Sandy kaolinite rock	in the roof, parting, floor of coals, or not adjacent to coals	0.45-0.65 0.5	disordered	derived from terrigenous clay

2.3. Geochemistry

Based on the neutron activation analysis, the rare elements and REE are used to indicate the ancestor minerals of forming kaolinite.

The average contents of the elements of Mo, W, Zr, Hf, Th, Ag and Sb in kaolinite rocks are higher than those in the earth's crust, basalt and granite, which is related to the absorption of these elements by organic matter and clay minerals. The ratios of Ti/Al, Cr/Al, Zr/Al, Ni/Al and Zr/Hf indicate that the mother rocks for forming-kaolinite are mainly connected with acid rocks, but locally with neutral and basic rocks.

Results of REE analysis show that tonsteins from the alteration of acid volcanic ash are rich in LREE and the ratios of LREE/HREE range from 8.25 to 18.25, being higher than those of the kaolinite derived from terrigenous clay.

2.4. Organic matter and microorganism

The study on the features of the organic petrology and organic geochemistry of kaolinite rocks, shows that there are four types of organism and organic matter in kaolin tonsteins, Terrestrial plant dedris,

Amorphous organic matter, Fulvic acid and Ultra microorganisms, which take an important role in the formation of kaolin tonsteins including forming the acid environment by organic acids, accelerating the solution of aluminosilicate minerals, prompting the transformation of layered minerals and secreting strong organic acid dissolving the surrounded minerals.

2.5. Formation of kaolinite rocks

More and More tonsteins have been reported by many geologists all over the world in past decades. They are generally considered as being volcanic origin. At present, there is a tendency among geologists to explain all the kaolin tonsteins in coal measures by volcanic origin, however it is hard to imagine that some kaolinite rocks bedding up to 1000mm thick can be formed in this way. We therefore think that the process of forming tonstein is complicated, involving many factors.

Based on the mineralogical, petrologic and geochemical evidences and the occurrence of kaolinite rocks, we found out that the formation of tonsteins depends on two important factors: one is the original kaolinite-forming materials and the other the transformation and crystallization of kaolinite in swamp environment at diagenetic stage. Two types of original materials can be transformed into kaolinite in swamp environment and at diagenetic stage:one is the terrigenous clay and aluminosilicate minerals such as feldspar and biotite, and the other the volcanic ashes fallen into swamps. All the original materials, which entered the swamp basin, could be transformed and crystallized to form kaolinite with the leaching off of alkaline earth ions and silicon under the influence of organic acids.

3. PROCESSING OF KAOLINITE ROCKS

3.1. *Introduction*

Kaolinite rocks, with the great reserves and good quality, is the precious non-metallic minerals resource of both China and the world, attracting common concerns of scientists and enterprises all over the world.

For the last ten years, we have been studying and utilizing kaolinite rocks. Because of features of the raw minerals and the complicacy of the mineralization mechanism, the processing of the kaolinite rocks is very difficult. Especially the calcination technology differs from the calcination of the concentrates of the ordinary kaolin, during which the calcination is only for the modification of the structure not for the complete removal of the organic matters. Recently we have successfully developed the new calcination technology, the fluidized-bed moving calcination, instead of the conventional static calcination, which has been put into use in a kaolinite rocks processing plant of China Coal Non-metallic Minerals Co. Ltd. (CCAM)

3.2. *Calcination of kaolinite rocks*

Calcination technology is key issue of the processing of kaolinite rocks. Though calcination organic matter in kaolinite rocks is completely removed and the structure of kaolinite is also changed, resulting in the high whiteness and the multi- function of the final products, which decide the application value of the special material. The conventional calcination technology applies the static calciner, which is introduced from the kaolin processing field in U.K. or the ceramic indurstry in China. During calcination, the powder material in the calciner keeps the static state, leading to the lack

of the oxygen-supply and the uneven of the calcination temperature of the powder at different place so that the calcination products show the characteristics of the unstable quality and the low whiteness.

The fluidized-bed moving calcination technology developed recently by authors includes the systems of the coal gas heating resource, the feeding, the calcination, the heated air flow, the discharging and the cooling. This new technology, featuring the moving calcination of the powders, overcomes the shortcomings of the static calcination, resulting in the final products being characterized by the even and stable quality and the high whiteness, and the production capacity has also raised from 0.1 T/H to 0.5 T/H significantly.

3.3. *Processing technology of kaolinite rocks*

After many field tests and considerable study, it is found out that the rational and efficient processing technology of kaolinite rocks is firstly crushing and grinding the raw minerals to the particle sizes of - 600 mesh or -1250 mesh, then calcining the black powder to remove the organic matter and modify the crystal structure by the advanced moving calcination technology, and finally microgrinding the white powder to the particle size of-2 microns. The flow-chart of technology is shown in Figure 1. Now, CCNM, the largest kaolinite rocks processing and utilization company in China, has been adopting the technology to process kaolinite rocks into various high quality products. The quality of new products can match for the products of Engelhard Co. in U.S., the worldly famous kaolin company, and have been applied to such industrial fields as papermaking, coatings, rubber, plastics, color masterbatch, adhesives and others.

4. CONCLUSIONS

Kaolinite rocks in China is the important composition of the kaolin resources of the world. Especially at the time when the want of good quality kaolin resources in the world is more and more serious, it is of great significance to study and utilize kaolinite rocks in China for keeping the continuance of prosperity of the worldly kaolin indurstry.

Formation of kaolinite rocks is related to two important factors including the original kaolinite-forming materials and the transformation and crystallization of kaolinite in swamp environment at

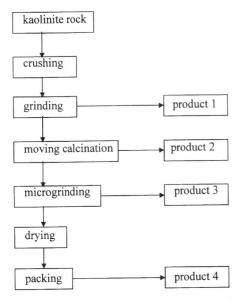

```
kaolinite rock
      │
      ▼
  crushing
      │
      ▼
  grinding ──────────► product 1
      │
      ▼
moving calcination ──► product 2
      │
      ▼
 microgrinding ───────► product 3
      │
      ▼
   drying
      │
      ▼
  packing ───────────► product 4
```

Figure 1.   Flow diagram of the processing of kaolinite rocks

diagenetic stage. The complicacy of the mineralization mechanism of kaolin tonsteins results in the difficulty of the processing technology of it. By the processing technology, crushing-grinding-moving calcination-microgrinding, kaolinite rocks has been successfully processed into the superior quality products with high-whiteness and super-fine grains. This new technology is very useful for industrial applications such as papermaking, paints, plastics, rubber and so on.

REFERENCES

Linares J, Huertas F. Kaolinite: Synthesis at room temperature.Science, 1971; 171:986 - 897.

Staub J R, Cohen A D. kaolinite-enrichment beneath coals:a modern analog, Sunggedy swamp, South Carolina. Journal of Sedimentary Petrology, 1978; 48 (1):203   210.

Tim P. Mossbauer spectra of soil kaolin from south-western Australia. Clays and Clay Miner, 1995, 40: 341.

Williamson I A. Tonstein-Their nature, origin and uses. Mining Magzine, 1970; 22 (2): 120- 125, 22 (3): 203 - 211.

Yang Xiaojie, Liu Qinfu. Study on the processing technology of Kaolin in Coal Measures. Coal Processing and Comprehensive Utilization, 1997, No.1.

Yang Xiaojie, Chenkaihui and liu Qinfu. Occurrence state of Fe in the structure of kaolinite. Chinese Science Bulletin, 1997,No.9.

Zhang Pengfei, Liu Qinfu and Yang Xiaojie. Microscopic characteristics and origin of pellitoid kaolinite rock in Huainan coalfield, Scientia Geologica Sinica, 1994, No.1.

3 Geomechanics

Mining Science and Technology' 99, Xie & Golosinski (eds) © 1999 Balkema, Rotterdam, ISBN 90 5809 067 1

Fractal evolution of a crack network in overburden rock strata due to coal mining

Heping Xie & Hongwei Zhou
China University of Mining and Technology, Beijing Campus, People's Republic of China

Guangming Yu & Lun Yang
Liaoning Technical University, Fuxin, People's Republic of China

ABSTRACT: The distribution of cracks caused by underground coal mining strongly affects stability of overburden rocks, fluid flow in fractured rocks and ground surface subsidence. Understanding of the distribution and evolution of crack network is useful for predicting surface subsidence. In this paper, plane stress physical models are employed to investigate the spatial distribution of a crack network. By using fractal geometry, the statistical self-similarity of a spatial distribution of crack networks is proposed. Additionally, the evolution of a crack network with an increase of the mining length of advance is investigated. It is shown that, (1) the spatial distribution of a crack network displays fractal behavior, so, the fractal dimension can be used as an alternative indicator to describe quantitatively the complexity of the crack network, (2) the fractal dimension of the crack network increases with an increase of mining length.

1 INTRODUCTION

In general, underground coal mining causes deformation and fracture of rocks surrounding the coal mining face. The fractured and deformed rocks will form a new structure in which exist numerous intersecting cracks. To some extents, the stability and mechanical behavior of rocks as well as fluid flow in rocks are controlled by this kind of crack network. For example, the network of cracks in overburden rocks due to coal mining is the main factor which influences the rock movement and subsidence at ground surface. Quantitative measurement of spatial distribution of crack network will help us to estimate the surface displacements. Additionally, underground coal mining causes crack network in floor strata. This crack network induces potentially dangerous water burst when coal is extracted above seams containing pressurized water, because the pressurized water bursts easily into the goaf and the coal mining face though the crack network. In order to avoid water burst when extracting coal from such seams, the spatial distribution of cracks should be investigated. So, an investigation of the spatial distribution of mining cracks is important for assessing the engineering stability of rocks, for investigating fluid flow in fractured rocks during underground coal mining, and for predicting surface subsidence (Liu, 1995).

With the advance of a coal mining face, the height

of crack network increases and the area of crack network enlarges. In other words, the process of advancing of coal mining face is also the process of evolution of complexity of spatial distribution of the crack network, so there would exist quantitative relation between them. However, because of the extremely complex spatial distribution of cracks in rocks, it is almost impossible to describe quantitatively the spatial distribution by using conventional parameters such as average spacing and average length within the framework of classical methods.

Fortunately, fractal geometry introduced by B. B. Madelbrot(1982) provides a very effective tool to quantitatively describe an extremely irregular object and discontinuous phenomena existing widely in nature. During the past two decades, extensive effort has been directed toward quantitative descriptions of complex and irregular objects and phenomena by using fractal geometry. For example, much attention has been devoted to applications of fractal geometry to earth science, particularly to descriptions of micro-fractures, cracks, pores and faults in earth (Tucotte, 1986; Taksyuki, 1989; Xie, 1988~1997).

As mentioned above, a crack network usually occurs in the overburden rocks. However, it is almost impossible to measure the distribution of cracks in the field. In this case, physical model will provide a good way to carry out research. In this paper, the two dimensional physical models are employed to study the distribution and evolution of cracks in overburden rocks. The degree of complexity for evolution of crack network is described quantitatively by fractal dimension. Research results indicate that the crack network exhibits the property of self-similar fractal and the fractal dimension can be used as an alternative indicator of complex degree of crack network in overburden rocks.

2 DESCRIPTION OF EXPERIMENTS

As mentioned above, cracks in overburden rocks will develop gradually in the process of advancing of a underground coal mining face. In this paper, two dimensional physical models are used to simulate this process, especially to visualize the spatial distribution and evolution process of cracks in overburden rocks.

2.1 Principle for design of physical models

The physical model is a kind of experimental method based on theorems of simulation. In order to insure that the physical phenomena displayed by physical models more accurately reproduce the processes of displacement and failure of the natural strata encountered during underground coal extraction, physical models must comply with three theorems of simulation. The physical models should follow the first theorem and the second theorem strictly and the third theorem approximately.

In the present research, the ratio of length of physical models is chosen to be $\alpha_l = l_m : l_p = 1:100$, where l_p and l_m present the length of natural strata and model, respectively. In addition, according to the second theorem of simulation, the physical models must comply with the following conditions of simulation:

(1) ratio of density $\alpha_r = r_m / r_n = 3:5$;
(2) ratio of velocity $\alpha_v = v_m / v_n = \sqrt{\alpha_l} = 1:100$;
(3) ratio of gravitational acceleration $\alpha_g = 1:100$;
(4) ratio of displacement $\alpha_s = 1:100$;
(5) ratio of strength, elastic module and cohesion $\alpha_R = \alpha_E = \alpha_C = 3:500$;
(6) ratio of fraction angle $\alpha_\varphi = \varphi_m / \varphi_p = 1:1$;
(7) ratio of applied force $\alpha_f = f_m / f_p = 0.6 \times 10^{-6}$
where r is the density, v denotes velocity, s denotes the displacement, R is the strength, E is elastic modulus, C is cohesion, φ is the fractional angle, f means applied force, the subscripts p and m represent the natural strata and the model, respectively.

2.2 Conditions of the natural strata

It is well known that the spatial distribution of a crack network is influenced by many factors such as the kind of overburden rocks, depth and thickness of coal seam, the mining length and advancing velocity of the coal mining face. In order to investigate the distribution and evolution of a crack network, some of the factors must be simplified. In the present research, only one kind of rock is simulated. The depth and thickness of the coal seam are 50m and 1.6m, respectively, the dip angle of the coal seam is 0°, the maximum mining length is 70m. The average density of overburden rocks is 0.025MN/m³; the average density of coal seam is equal to 0.014MN/m³; the average uniaxial strength of overburden rocks is 40MPa; the average uniaxial strength of coal seam is 20MPa.

2.3 Size and components of the physical models

According to the condition of the natural strata and three theorems of simulation, the parameter and component of physical models can be selected. In the present research, sandstone and coal are simulated. The physical models are composed of quartz sand, mica, dense spar, lime, gypsum and borate in appropriate proportions (as shown in Table 1).

Table 1 Volumetric proportions of simulated materials.

Kind of Materials	Proportion of matrix to cementing agent	quartz sand : dense spar : mica	lime: gypsum
Rock	4:1	4:2:2	5:5
Coal	6:1	6:1:1	3:7

Physical models with a length of 2m (corresponding to 200m in natural strata) and a height of 0.6m were used to simulate the coal mining process (as shown in Fig.1). The pictures depicted in Fig.2a, Fig.2d, Fig.2g and Fig.2j show the state of overburden rocks with coal mining length L of 24cm, 40cm, 60cm, and 70cm, respectively.

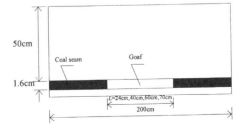

Figure1 Layout of physical model

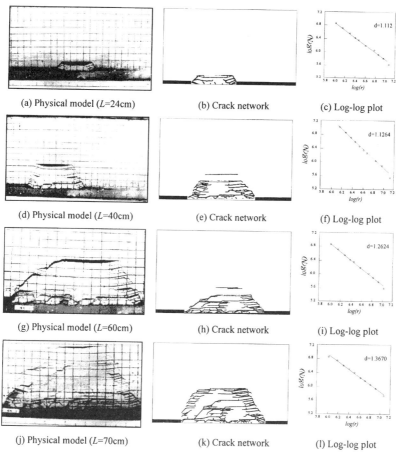

(a) Physical model (*L*=24cm)　　　(b) Crack network　　　(c) Log-log plot

(d) Physical model (*L*=40cm)　　　(e) Crack network　　　(f) Log-log plot

(g) Physical model (*L*=60cm)　　　(h) Crack network　　　(i) Log-log plot

(j) Physical model (*L*=70cm)　　　(k) Crack network　　　(l) Log-log plot

Figure 2 Evolution of crack network in mining overburden rocks during coal mining

3 EXPERIMENTAL RESULTS AND THEIR ANALYSIS

3.1 *Experimental results*

After extraction of coal seam, the roof behind the advancing coal mining face bends and moves under gravity. The overburden rocks will fracture, resulting in separation between layers, failure and fall of the entire upper strata if its internal tension stress exceeds the strength of rock (Fig.2). The fracture and separation in roof strata form so-called crack network. It is an extremely complex mechanical procedure correlated with mechanical properties of the overburden strata, geological condition, the velocity of the advancing coal mining face etc..

In Fig.2, as the coal face advances (*i.e.*, increasing of the mining length *L*), the crack network in overburden rocks extends to a larger area. In other words, each coal mining step causes a new crack network continuously. In this process, the crack network produced in the current mining step overlaps the network produced in the former mining step, and a new composite network is formed. At the same time, the cracks caused by the former mining step undergo a process of extending, closing or opening. As a result, the crack network in overburden rocks becomes more and more complex.

3.2 *Estimation of the fractal dimension of a crack network*

A fractal dimension can be used as an indicator of the complexity of a formation, extension and spatial distribution of a crack network. In order to describe the degree of complexity of crack networks, we should choose a method to quantitatively estimate the portion that the plane is occupied with crack network. The box-covering method is the most direct and simplest one. First, we choose a square grid and a box size adequate to cover the crack network. Then count the number of boxes needed to cover the crack network. Decreasing the box size (*i.e.*, scale) increase the number of boxes for covering the crack network. By repeating this process, if the distribution of crack network in overburden rocks appears to be self-similar, the following relation can be obtained:

$$N(r) \sim r^{-D} \qquad (1)$$

where $N(r)$ is the number needed to cover the crack network, r is the measurement scale, D is the fractal dimension. $N(r)$ and r will fall on straight line in log-log plot, the slope of the straight line is equal to $-D$.

In this paper, a special program developed by Chen (1995) is employed to estimate the fractal dimension of a crack network. First, scan the pictures of experimental results (Fig.2 a, d, g, j); second, digitize these pictures and form pictures containing only cracks (Fig.2 b, e, h, k). Then the program can count automatically the numbers of boxes in different scales and export the calculated results, *i.e.*, the fractal dimension of spatial distribution (Fig.2 c, f, i, l and Table 2). The fractal dimension is equal to minus slope of the straight line of log-log plot.

Table 2 Increasing of fractal dimension and subsidence with increasing of mining length

Mining length L (m)	24	40	60	70
Fractal dimension	1.1120	1.1264	1.2624	1.3870
Correlation coefficient	0.9999	0.9999	0.9999	0.9999

3.3 Evolution of fractal dimension with the mining length

Fig.2 c, f, i, l show that there exists a good linear relationship between $N(r)$ and r in log-log plot. It is indicated that the spatial distribution of cracks displays fractal behavior.

According to Table 2, the regression relation between fractal dimension D and mining length L can be given by:

$$D = 0.000172432L^2 - 0.0102654L + 1.2596 \qquad (2)$$

where D is the fractal dimension of spatial distribution of a crack network in mining overburden rocks, and L is the mining length.

Fig.2c, f, i, l and Eq.(2) show that as the coal mining length L is increased, the spatial distribution of crack network will become more complex, and the fractal dimension of spatial distribution will increase.

4 CONCLUSIONS

In this paper, plane stress physical models are employed to investigate the spatial distribution of cracks in mining overburden rocks during underground coal mining. The research results indicate that

(1) the spatial distribution of a crack network in overburden rocks appears to be statistically self-similar. The self-similar fractal dimension can be used to describe quantitatively the complexity of crack network caused in overburden rocks by underground coal mining;

(2) the statistically self-similar fractal dimension of crack network in mining rocks can be used as a quantitative indicator of degree of development of cracks in the process of coal mining. The greater the mining length, the greater the fractal dimension. The regression relation between fractal dimension and the mining length can be given by Eq.(2).

REFERENCES

Chen, J. P. 1995. Numerical simulation of fractal distribution of cracks in mining rocks. *J. of Engineering Geology.* 3(3): 34-42. (in Chinese)

Liu, T.Q. 1995. Influence of mining activities on mine rockmass and control engineering. *J. of China Coal Society.* 20(1): 1-5. (in Chinese)

Mandelbrot, B. B. 1982. *The fractal geometry of nature.* San Francisco : W.H.Freeman.

Turcotte, D.L. 1986. Fractal and fragmentation. *J. Geophys Res.* 91:1921-1926.

Taksyuki, H. 1989. Fractal dimension of fault system in Japan: Fractal structure in rock fractal fracture geometry at various scales. *Pageoph.* V.B1:157-170.

Xie, H. & Chen, Z.D. 1988. Fractal Geometry and Fracture of Rock. *Acta Mechanica Sinica.* 4(3): 255-264.

Xie, H. 1989. Studies on Fractal Models of the Microfracture of Marble. *Chinese Sci. Bulletin.* 34(15):1292-1296.

Xie, H. 1993. *Fractals in Rock Mechanics.* Netherlands: A. A. Balkema Publishers.

Xie, H. 1995. Effects of fractal crack. *Theory. & Appl. Fracture Mech.* 23: 235-244.

Xie, H. & Xie, W. & Zhao, P. 1996. Photoelastic study on the mechanical properties of fractal rock joints. *Fractals.* 4(4): 521-531.

Xie, H. Wang, J.A. and Xie, W. 1997. Fractal effects of surface roughness on the mechanical behavior of rock joints. *Chaos, Solitons & Fractals.* 8(2): 221-252.

Mining Science and Technology'99, Xie & Golosinski (eds) © 1999 Balkema, Rotterdam, ISBN 90 5809 067 1

Continuous improvement in geotechnical design and practice

T. Li & D.J. Finn
St. Ives Gold, WMC, Kombalda, W.A., Australia

E. Villaescusa
Western Australian School of Mines, W.A., Australia

ABSTRACT: Geotechnical design and practice has been recognised as a key issue to be addressed by the mining industry in Western Australia in order to eliminate the unacceptably high number of fatalities and to reduce injuries and accidents. A sharp increase in the geotechnical efforts committed by the underground mines has been experienced. Consequently, the incident rates at the mine sites have been significantly reduced. In this paper, several key features of the efforts to improve geotechnical design and practice are discussed. A number of successful initiatives are presented to highlight the progress that has been made. The geotechnical challenges facing the underground mines are also discussed.

1 INTRODUCTION

Rockfall related fatalities accounted for almost 50% of all fatalities in Western Australian underground mines in the last several years. The lack of sound geotechnical practice has been identified by the companies and the mine inspectorate as the key issue that must be addressed by the industry to reduce the accidents. The mining industry has responded over the last a few years.

The cases in this paper are mainly drawn from the underground mines at St Ives Gold, WMC Resources (WMC). St Ives Gold operated three underground mines and two open pits in 1998, producing about 3.1 million tonnes of ore grading 4.5g/t. St Ives Gold has now one underground mine and four open pits.

A wide range of geotechnical programs have been implemented by the management and the geotechnical professionals. These programs are aimed at improving the geotechnical design and practice on a continuous basis.

2 GEOTECHNICAL EFFORTS AND EFFECTS

Adequate geotechnical input into mine design and operational practices has been recognised as an integral part of the overall drive to improve safety and efficiency in underground mines. There has been a sharp increase in the geotechnical efforts undertaken by the mining industry, especially in the last few years since the release of Regulation 10.28 in 1995 (Lang 1999).

WMC set up a corporate wide Elimination of Fatalities Taskforce (EOFT) in mid 1996, with the objective to develop and implement standards and procedures that will eventually lead to fatality free operations (Harvey 1999). By mid 1997, the first of the 20 standards: Underground Ground Control Standard (UGCS) was finalised and implemented.

One measurement of the increased geotechnical effort is the increasing number of geotechnical professionals, in particular the site based geotechnical engineers. For instance, St Ives Gold has increased the geotechnical staff from 2 part-time in 1996 to 4 full time at present. WMC has increased its site based geotechnical staff from 4 in 1995 to 16 in 1998. Another measurement of the increased geotechnical effort is the increased geotechnical training courses conducted by educational institutions and the mines.

Most importantly, a number of geotechnical programs are being initiated and implemented at the mine sites. The geotechnical programs can be summarised as the following: the systematic collection of geotechnical data and the processing of this data, the use of geotechnical modelling and design parameters, the optimisation of ground support/reinforcement methods and procedures, the implementation of monitoring and testing, the back-analysis of stope and pillar performance and geotechnical training.

Rockfall related incident rates and severity have been reduced significantly, largely due to these measures. Figure 1 illustrates the trend of the rockfall related incident rates at St Ives Gold. The actual consequences of the incidents were recorded as fatalities, lost time injuries (LTI), medical treated injuries (MTI), and minor injuries (MI). The data shows that the severity of the potential consequences of the incidents has also decreased significantly.

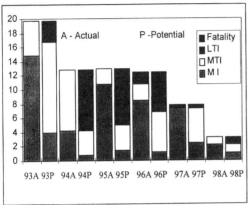

Figure 1. The trend of the rockfall related incident rates at St Ives Gold.

The decrease observed in the incident rates since 1997 coincided with the appointment of two full-time geotechnical engineers; and the implementation of the EOFT Underground Ground Control Standard (UGCS), in particular the 20 key elements of the UGCS (Harvey 1999).

Decreasing incident rates and high fatalities seem to be a trend across WA underground mines. One possible explanation of this seemingly unmatching trend is that in the past, mines might not have had adequate geotechnical input to the design and operation at an early stage. While this may not have caused problems in the past, the inadequacy or the lack of geotechnical input may become apparent later in the mine life.

Some geotechnical measures can have immediate impact on an operation, such as changes in development heading design and procedures, and alterations to the ground support methods and procedures. Some others may have an impact long after they have been implemented.

More critical issues are the understanding of the geotechnical conditions at the current and future stages of mining, the awareness of the potential consequences of not following sound geotechnical approaches, and consciously striving for quality implementation of sound geotechnical measures.

It is this cultural change combined with technical excellence that can minimise incidents, eliminate fatalities and sustain excellent performance. This fundamental change is taking place industry wide in WA. The experience in North America suggests that it could be several years after the systematic efforts have been directed to addressing the fatalities and accidents before significant results would be forthcoming.

Figure 2 is a graph of the fatal injuries caused by rockfalls in Western Australia and Ontario from 1980 to 1997 (Lang 1998). The underground workforces were approximately 4,000 in WA and 8,000 in Ontario in 1997.

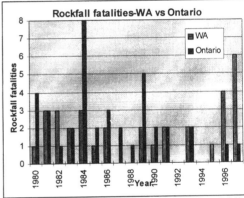

Figure 2 Graph of fatal injuries caused by rockfalls in Western Australia and Ontario from 1980 to 1997 (adapted from Lang 1998)

The fatalities in Ontario underground mines have remained low since 1990. However, this was only achieved after major geotechnical efforts were initiated and being implemented since 1985 following four fatalities in a single rockburst accident in 1984. These efforts included the establishment of three mining geomechanics chairs (Professorship) in three universities, the formation of a rockburst research program and the significant increase in the number of the mine site based geotechnical personnel.

Figure 3 is a graph of the number of fatalities due to all causes in USA coal mines (Tattersall 1993). There was a devastating fatal accident in 1970 resulting in the loss of 78 lives. US Federal Coal

Mine Health and Safety Act took effect in that year. The graph illustrates that it took a decade to halve the fatalities since 1970 (Tattersall 1993).

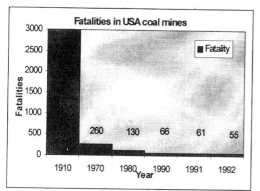

Figure 3 Graph of all fatalities in USA coal mines (adapted from Tattersall 1993)

The experience of both the Canadian and USA coal mines indicate that a systematic and continued effort is required to achieve the desired safety and efficiency.

3 MINE SITE BASED GEOTECHNICAL MANAGEMENT SYSTEM

A geotechnical group was established at St Ives Gold in late 1997 in an effort to provide improved services to the operations. Geotechnical programs consisting of long term and short term projects and goals were developed and approved. The core of the program is a geotechnical management system that features a number of points including:

• Adequate geotechnical resources, especially at mine sites.
• Systematic geotechnical data collection, processing and application.
• Geotechnical design and evaluation methods and procedures.
• Geotechnical training for the geotechnical engineers and all the other personnel.

The thrust of this approach is to deliver quality and timely geotechnical services for the short and long term needs of the mines. The short term geotechnical routine practices and the long term projects are integrated.

3.1 Geotechnical data collection and processing

Up to few years ago, geotechnical data collection had not been a routine practice in our mines. Efforts by the geotechnical engineers and geologists in the last two years has seen significant progress towards systematic data collection and processing. At present, a geotechnical database has been set up in the WMC Geobase. Geotechnical data, mainly core logging, is stored in the Geobase through the Data Entry System installed on standard PCs.

The development and implementation of geotechnical data processing make the whole process from data collection to geotechnical information streamlined. The 3D distribution of geotechnical domains, and the interpreted and interpolated geotechnical conditions within each domain are delineated following standard methods and procedures.

3.2 Optimisation of development and ground support methods and practices

The Junction Mine at St Ives Gold offers a good example on the continuous improvement in the design of development headings and ground support standards. Before 1997, the standard decline design at Junction Mine was 6m by 6m square profile, while the ore drives were mined to shanty profile at full width of the orebody, which can range from 5m to 15m wide. Figure 4 illustrates the effects of development sizes and shapes on the rockfall potential.

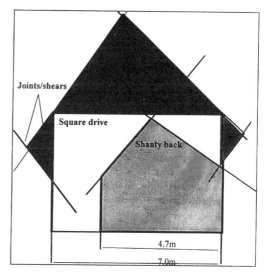

Figure 4. Effect of development profile and size on rockfall potential.

The geological discontinuities which can be present in and around the orebody have the potential to define large wedges at the back of the ore drives.

The wider the drive. the larger the potential wedge. The square shaped drive backs tend to be more prone to wedge formation. The shanty back shapes are likely to minimize the potential for large wedges.

Two major changes in development heading design were introduced in mid 1997, reducing the sizes of the drives and declines, and adopting the arched back profile for declines and modifying a shanty back profile for ore drives. The size of the decline was reduced from 6m by 6m flat back to 5.5m by 5.8m arched back. The size of the ore drives was reduced from full orebody width to 4.7m wide by 5m high with shanty backs (Figure 4). For the drives at the high stress deeper mining blocks, the shanty back profiles have been modified to semi-arched profiles to reduce stress effects.

These simple changes have resulted in several benefits. Development rates are higher, as the development cycle time is reduced due to the reduction in drilling, charging, mucking, and ground support. Ground support/reinforcement is more effective as the load distribution over the support elements is more even. The risk of large wedge rockfalls has been removed significantly, and there has not been a single large wedge rockfall since the implementation of these geometrical changes.

The improvement in development size and profiles were accompanied by perimeter blast and the optimisation in ground support/reinforcement methods and procedures. Extensive trials and tests were performed by the site geotechnical engineer over a wide range of support/reinforcement elements and schemes (Finn, et al 1999, Villaescusa & Wright 1997).

3.3 Stress measurement and monitoring

The two key parameters that measure rock mass behaviour are stresses and deformation. The Hollow Inclusion (HI) cell stress measurement techniques have been the major method for in situ stress determination. A renewed interest in other alternative stress measurement methods has emerged due to various reasons. The determination of in situ stresses by the Acoustic Emission (AE) and Deformation Rate Analysis (DRA) methods using a piece of diamond drill core, for example, has been compared well with that using the HI cell method (Seto & Villaescusa 1999). The benefit of the AE and DRA methods is that they only require oriented diamond drill cores and therefore the in situ stresses could be determined well before any

access to the underground excavations is available.

Monitoring of stress changes and deformation due to mining has also been implemented at an increasing number of mines (Brown 1993). The increased effort in geotechnical monitoring is probably better demonstrated in the case of mining induced seismicity monitoring. The information derived from the monitoring is used to assess the rock mass behaviour and allow the geotechnical engineers and mine managers to make informed decisions on mine design, planning, and operational issues.

Conventional observational methods and simple devices, such as extensometers and crack monitors, also provide reliable information about the likely rock mass behaviour.

3.4 Stability performance measurement and monitoring

St Ives Gold has used the Cavity Monitoring System (CMS) survey method to determine open stope voids, along with a number of mines in Australia. The CMS survey, for the first time in the history of underground mining, is able to quantify the performance of stopes and pillars. The amount of overbreak and underbreak, dilution, and ore losses are determined to a reasonable accuracy. Analysis of the CMS survey results are used to highlight the potential problems in stope design, to assess the reinforcement effectiveness, and also to optimise the drilling and blasting practices.

4 BACK-ANALYSIS TO IMPROVE THE GEOTECHNICAL DESIGN

Precedence of stope stability performance has been the practical design method for stope walls and pillars. In recent years, methods based on back-analysis of the stope walls and pillars performance have been developed and used extensively (Potvin et al 1989, Nickson 1992, Villaescusa et al, 1997). These back-analysis based methods typically incorporate geological conditions, stope geometries, and stress analysis to quantify the impact of geotechnical parameters and provide guidelines for the acceptable range of design parameters, such as exposed hangingwall span dimensions.

4.1 Modified Stability Graph

The Modified Stability Graph method was developed based on a large set of data from Canadian open stope mines (Potvin et al 1989).

This method has been extensively used in WA by mine geotechnical engineers and consultants for the determination of stope dimensions. St Ives Gold has been using this method in the last two years. Since early last year, an attempt was made by the geotechnical engineer at Junction to develop a local stability chart using similar principles and methodology such as those used in the Modified Stability Graph method. Extensive back-analyses of stope performance have been carried out, which would improve the geotechnical design.

4.2 Hangingwall Stability Rating

Hangingwall Stability Rating (HSR) is an empirical model of hangingwall stability at the Lead Mine in Mount Isa (Villaescusa et al 1997). The method is based on the quantified stope hangingwall performance, using observations and CMS surveys to establish a stability chart which links the factors controlling ground behaviour and excavation geometry for each of the orebodies at Mount Isa. The HSR method, assumes that the geological discontinuities, induced stresses, blast damage, and excavation geometry are the main factors controlling hangingwall stability (Villaescusa et al 1997).

The method has been calibrated and used as a predictive tool in mine planning to optimise new mine block designs(Villaescusa et al 1995, Harris & Li 1995). There is potential that this method can be used in the tabular orebodies to determine the optimum stope dimension.

The continued improvement in back-analysis methods and the application of these methods will gradually lead to better understanding of the stope performance and high quality stope design and sequencing.

5 GEOTECHNICAL TRAINING

The geotechnical training is an essential component for improved geotechnical design and practice. Although the level of understanding of geotechnical design and practice is different for different groups of personnel, there is a recognition that a geotechnical culture needs to be established to maximise the impact of geotechnical input.

A geotechnical culture can be simply defined as an environment where all personnel are consciously pursuing the understanding of geotechnical conditions, value the geotechnical input and play their respective roles for the implementation of geotechnical programs. This cultural change can only be brought about by a systematic and structured approach to conducting geotechnical programs and in particular, geotechnical training.

The approach pursued by St Ives Gold has both formal and informal components. The formal geotechnical training consists of selected training courses for different personnel.

The informal training has been conducted by the site geotechnical engineers by engaging the relevant personnel in geotechnical data collection, ground support/reinforcement trials and monitoring, and geotechnical design.

6 GEOTECHNICAL CHALLENGES

Two major geotechnical challenges facing the industry have been identified (Brown 1994). These relate to technical and education/training challenges. These are still the issues that the WA mining industry has to address today in order to ensure safe and economical mining operations.

A number of technical challenges, such as deeper mines and remnant mining, are currently being experienced. Whilst more deposits are hosted in more complex geological and geotechnical settings as exploration targets new areas and geological structures.

Economic conditions dictate that the current mining methods and sequences be designed to extract orebodies cheaper and quicker. These methods and sequences must be also justified to be viable through geotechnical evaluation.

Industry leaders have realised that mining is no longer being accepted and tolerated as a high risk activity (Morgan 1997, Carter 1998). Licences to operate could be at stake as community sentiments change. The values of mining houses have also shifted from the sole profit generating for the shareholders to the well-beings of all stakeholders and the society at large.

All of the above changes suggest an increased demand for quality geotechnical input to bring about safer and more efficient operations. The technical efforts should be accompanied by efforts in training and cultural changes.

7 CONCLUSIONS

An increased effort on geotechnical design and practice has been experienced in Western Australia. A systematic approach to sound geotechnical design and practice has demonstrated that a positive impact can be achieved to reduce the number of injuries and fatalities. There are a number of geotechnicl challenges facing the mining industry and the geotechnical professionals. The magnitude of these challenges requires a continuous improvement on all aspects of geotechnical endeavours.

ACKNOWLEDGEMENTS:

The permission by WMC management to use the materials contained in this presentation is appreciated. The authors are grateful to Mr Luke Tonkin and Mr Peter Crooks of St Ives Gold for reviewing this paper. The second author acknowledges the financial support from the Australian Centre for Geomechanics and Curtin University of Technology for his position as Professor of Mining Geomechanics at WASM. The Australian Centre for Geomechanics has received funding for this position from the Government of Western Australia, Centres of Excellence Program.

REFERENCES:

Brown, E. T. 1993. Geotechnical monitoring in surface and underground mining - An overview. *Proc. Geotechnical Instrumentation and Monitoring in Open Pit and Underground Mining.* Szwedzicki (Ed.) Balkema, Rotterdam. pp3-11.

Brown, E. T. 1994. Australian mining geomechanics - development, achievements and challenges. Short Course Note: Acceptable Risks and Practical Decisions in Rock Engineering Hoek, E. (Ed.).

Carter, R. J. 1998. AusIMM and professional competence in occupational health and safety. AusIMM/Chamber of Minerals and Energy Seminar, Perth, June 1998.

Harris, A. and T., Li. 1995. Mining of the southern 1900 Orebody at Mount Isa. *Proc. 6th Underground Operators' Conf.* Kalgoorlie.

Harvey, S. 1999. Elimination of fatalities taskforce – underground rockfalls project. International Symposium on Rock Support and Reinforcement Practice in Mining, (E. Villaescusa, C.R. Windsor & A. Thompson, Eds),Kalgoorlie.

Lang, A. 1998. How does this relate to Western Australia? *ACG Workshop notes: Mine Seismicity and Rockburst Risk Management in Underground Mines.* Perth. 1998.

Lang, A. 1999. Geotechnical mining regulations in Western Australia. International Symposium on Rock Support and Reinforcement Practice in Mining, (E. Villaescusa, C.R. Windsor & A. Thompson, Eds),Kalgoorlie.

Morgan, H. M. 1997. WMC's number 1 safety objective: Elimination of Fatalities. *Presentation to the Queensland Mining Industry Health and Safety Conference*, Yeppoon, 10 September.

Nickson, S. D. 1992. *Cable Support Guidelines for Underground Hard Rock Mine Operations.* M.Appl.Sci. Thesis. The University of British Columbia.

Potvin, Y., M. Hudyma & H. Miller, 1989. Design Guidelines for Open Stope Support, *CIM Bulletin.* **82**. pp53-62.

Seto, M. and Villaescusa, E. 1999. In situ stress determination by acoustic emission techniques from McArthur River Mine Cores. To be presented on 8th ANZ Geomechanics Conf. Hobart.

Villaescusa, E., Karunatillake, G. and Li, T. 1995. An integrated approach to the extraction of the Rio Grande silver/lead/zinc orebodies at Mount Isa. *Proc. 4th Int. Symp. On Mine Planning and Equipment Selection.* Calgary.pp 277-283.

Villaescusa, E. Tyler, D. and Scott, C. 1997. Predicting underground stability using a Hangingwall Stability Rating. *Proc. 1st Asian Rock Mechanics Symposium.* (H.K. Lee, H.S. Yang and S.K. Chung, Eds.). Seoul. pp171-176.

Villaescusa, E. and J. Wright, 1997. Permanent excavation support using cement grouted split set bolts. *The AusIMM proceedings.* pp65-69.

Mining Science and Technology '99, Xie & Golosinski (eds) © 1999 Balkema, Rotterdam, ISBN 90 5809 067 1

Dynamic mechanics model of mining subsidence

Yuehan Wang, Kazhong Deng, Dingli Zhang, Kan Wu & Guangli Guo
China University of Mining and Technology, Xuzhou, People's Republic of China

ABSTRACT: Different mechanics models are used to simulate the strata movement in different phases of subsidence. This paper proposes the differential equations to model strata movement in different phases, defines the boundary conditions for each model and relations between different models. The approach assumes that rockmass changes from continuous medium to discontinuous medium. This assumption makes the model f strata movement more practicable.

1. INTRODUCTION

Up to now, the existing prediction methods of mining subsidence can be divided into two types: empirical and theoretical. Theoretical method is mostly based on static mechanics, in which the fact that the range of broken rockmass increases with face advance length is not considered. Comparing to real situation, thus, the results of strata and ground movement calculated have much bigger error. The method can not be used to predict dynamic strata and ground movement, too. In this paper, considering the feature of the broken range of mining rockmass increased with the working face advanced, different models are used to predicted the strata movement in different phases, which makes the results much closer to practice.

2. DYNAMIC MOVEMENT PROCEDURE OF MINING ROCKMASS

The study on field and similar material test shows that the movement procedure of overburden rockmass duo to underground mining can be expressed as follows: when face advance length is less than length of roof initial caving, rockmass do not fall with face advancing. Roof first falls when face advance length is bigger than the length of roof initial caving, but fallen rockmass are not enough to fill the goaf. Thus,

$$h \le \frac{m}{k-1} \qquad (1)$$

Where h = the height of caved rockmass; m = working thickness; k = broken rock bulking coefficient.

With the face's advancing, the rockmass in front of it can support the weight of overburden rockmass and form overhanging beam. Overburden movement can be considered as deflection of overhanging beam in front of the face and the beam on elastic foundation behind. With face advancing again, overhanging beam ahead of face breaks and falls. Rock beam falls and caving range develops upward if length of beam is beyond the allowable length of beam breaking. If the height of caving rockmass accord with the formula (2), the caved rockmass will fully fill up goaf.

$$h \ge \frac{m}{k-1} \qquad (2)$$

Strata movement is circled like this: overhanging beam falls with the advancing of the face ahead of it → free beam over caving area breaks and falls → the beam's position goes up, gradually until the ground reaches critical mining.

Based on above analysis, the dynamic movement procedure of mining rockmass can be divided into four phases:

(1) When roof rockmass don't cave, strata movement is not severe and can be considered as a beam deflection, in which the middle of beam is not propped and two sides of beam are located on elastic foundation.

(2) When roof rockmass has caved but not fully filled the goaf, strata movement can be regarded as a beam deflection, in which the middle of beam is still not propped and two sides of beam are located on elastic foundation. The length of beam is smaller than the length of face because the rockmass surrounding above face boundary don't fall.

(3) When caved rockmass has fully filled the goaf, strata movement can be thought as the deflection of beam on elastic foundation, in which the middle of beam is located on caved rock and the two sides of beam are located at coal seam.

(4) There is a middle state in the beam deflection mentioned above, in which beam has cracked and been broken, but it has some sustaining ability and can support the weight of overburden. Movement and deformation of beam can be thought as a deflection of building block beam or cracked beam. Beam property changes from continuous medium to discontinuous medium;

(5) The movement and deformation of beam progress from the bottom to the top until get to surface. The movement of beam is group movement.

According to the above discussion, mining rockmass can be regarded as three kinds of media: 1) a continuous medium before caving; 2)a quasi–continuous medium when beam cracks yet has not broken; 3) when beam has fully broken, but it still has some sustaining ability, it can be thought as discontinuous medium and its movement and deformation can be gained by building block method (Voussoir beam); 4) caved rock can be considered as elastic foundation of upper rock beam.

The above analysis shows that dynamic mechanics procedure of mining rockmass can be characterized chiefly as follows:

(1) The rock beam property changes continually, from continuous medium (before cracking) to quasi-continuous medium (after cracking) to discontinuous medium (after breaking) to crushing medium or bloc medium (after caving);

(2) The position and length of beam changes continually. The length of beam equal to the distance of advancing face before roof rock caves first. The beam position moves up and the length of beam is less than the advancing distance after roof rock caves. The length of beam can be calculated by the following formula:

Where S = half of beam length; L = half of

$$S = L - h \cot \psi \qquad (3)$$

advancing face length; h = caving height; ψ =caving limit angle.

(3) Strata movement model is different in different phases.

(4) The elastic basement is changed in strata movement procedure. Along with the caving are the enlarged beam rises and the elastic basement is enlarged. The elastic basement is not static because the caved rock lumps are compressing.

(5) With face advancing beam moves up and the load acting on beam changes continuously.

Owing to the complexity and dynamic characteristic of strata movement, strata movement can not be analyzed by single beam model while different models should be taken into account in different conditions. The calculations of strata and ground movement can be undertaken by linking these different models together.

3. DYNAMIC MECHANICS MODEL ON MINING SUBSIDENCE

3.1 *Basic Hypothesis*

(1) The movement and deformation of rockmass can be calculated by elastic theory before rock cracking.

(2) The strata movement in longwall mining can be considered as the beam deflection and can be calculated by beam theory.

(3) Overburden movement is mainly controlled by the strata with bigger stiffness, while the strata with smaller stiffness will follow it in the movement, thus, strata movement can be considered as a movement of several compassed beam.

(4) Suppose that in-situ stress in rock beam is

$$q = \sum_{i=1}^{n} r_i h_i \qquad (4)$$

Where, r_i = the gravity density of layer i rock beam; h_i = the thickness layer i rock beam.

Owing to the stratification of strata and the difference of property in different layers, practical stress acting on a rock beam can be calculated as follows:

$$q(x) / n = \frac{E_1 h_1^3 \sum_{i=1}^{n} r_i h_i}{\sum_{i=1}^{n} E_i h_i} \qquad (5)$$

When

$$q(x) / n > q(x) / (n+1)$$

calculated as formula (4).
Where, E_i = elastic modulus of layer i rockmass; n = the layer quantity of strata acting on beam.

(5) Suppose that elastic foundation accord with Winkler's hypothesis.

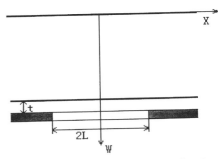

Figure 1. The model sketch of before roof caving

3.2 Dynamic mechanics model

3.2.1 Mining subsidence model before roof caving

Since roof has not fallen, mining subsidence model can be considered as the deflection of beam on elastic foundation (Figure 1), and its differential equation is as follows:

$$E_2 J_2 \frac{d^4 W_2}{dx^4} = q_2, \qquad (-L \le x \le L) \qquad (6)$$

$$E_2 J_2 \frac{d^4 W_2}{dx^4} = q_2 - \varphi_2(x) \qquad (7)$$

$$J_2 = \frac{Et^3}{12} \qquad (8)$$

Where, E_2, J_2 = the elastic modulus and moment of beam inertia; W_2 = deflection of beam; q_2 = the load acting on beam; φ_2 = counter stress of foundation. According to Winkler hypothesis, thus:

$$\varphi_2(x) = E_3 \frac{W_2}{m + \sum\limits_{i=1}^{n} h_i} \qquad (9)$$

$$E_3 = \frac{\sum\limits_{i=1}^{n} E_i h_i}{\sum\limits_{i=1}^{n} h_i} \qquad (10)$$

Where, E_3 = the average modulus of coal and rock under floor; m = working thickness; E_i, h_i = the elastic modulus and thickness of layer i rockmass under floor.
Boundary condition:

$$when \quad x = 0, \qquad \frac{dW_2}{dx} = 0, \frac{d^3 W_2}{dx^3} = 0 \qquad (11)$$

$$When \quad X \to \pm\infty, \quad W_2 \to 0 \qquad (12)$$

3.2.2 Mining subsidence model after roof caving

After roof rock caving, there are two kinds of situations, one is that caving rock has not filled goaf, the other is that caving rock has fully filled goaf. At the same time, rock beam moves up where roof rock caves, moreover, the face continues advancing and rocks ahead of advancing don't cave and form overhanging beam. Strata movement can be considered as the deflection of overhanging beam ahead of work face advancing and deflection of free beam (h≤m/(k-1)) and beam on elastic foundation (h≥m/(k-1)) (see Figure 2) after face advancing.

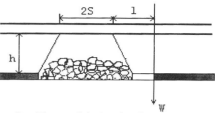

Figure 2. The model sketch of beam on elastic foundation

The differential equation of strata movement is as follows:
(1) For overhanging beam:
The deflection of overhanging beam can be calculated layer by layer. The touch and slip condition is satisfied between layers. For the lowest layer, there is:

$$E_1 J_1 \frac{d^4 W_1}{dx^4} = q_1, (-l \le x \le 0, \ -l_1 \le -l - 2s) \qquad (13)$$

$$E_1 J_1 \frac{d^4 W_1}{dx^4} = q_1 - \varphi_2(x), (0 < x < \infty, -\infty < x < -l \qquad (14)$$

Where, E_1, J_1 = the elastic modulus and moment of

$$l_1 = l + 2s + hctg\psi \qquad (15)$$

overhanging beam inertia; q_1 = load acting on overhanging beam; S = half length of free beam; φ_2 = counter stress of foundation, which is calculated by (9).

Boundary condition:

When $x \to \infty, W_2 \to 0$

$$When \quad -l \le x \le 0, \qquad \frac{d^2 W_1}{dx^2} = \frac{q_1(l + x)^2}{2E_1 J_1} \qquad (16)$$

$$\frac{d^3 W_1}{dx^3} = \frac{q_1(l + x)}{E_1 J_1} \qquad (17)$$

When $x=0$, W_1 is introduced by formulas (13) and (14) should be equal.
The boundary condition of left side of beam is similar to that of its right side.
For second layer beam, touch and slip condition lead to:

$$W_1^1(x) = w_1^2(x) = w_1^3(x) = \cdots W_1^i(x) \qquad (18)$$

Where W^i= subsidence of the layer i beam.

(2) For free beam:

$$E_2 J_2 \frac{d^4 W_2}{dx^4} = q_2 - \varphi_2(x) \qquad (-l_1 \le x < -l) \qquad (19)$$

$$W_2(x) = W_1(x), (-l \le x < \infty, -\infty < x \le -l_1) \qquad (20)$$

Where W_2, W_1 = subsidence of free beam and overhanging beam; $\varphi_1(x)$ = counter stress of caved rockmass foundation; q_2 = load acting on free beam.

For foundation counter stress, when caved rockmass don't fill the goaf, $\varphi_1(x) = 0$. When caved rockmass has filled up the goaf, considered as the simplest case, supposed that φ_1 accord with the Winkler's hypothesis is considered, thus,

$$\varphi_1(x) = \begin{cases} 0 & h < \dfrac{m}{k-1} \\ K_1 W_2 & h \geq \dfrac{m}{k-1} \end{cases} \quad (21)$$

Where K_1=foundation coefficient of caved rockmass.
Boundary condition:
When $x=-l$ or $x=-l_1$ □

$$W_2 = W_1, \qquad \frac{dW_2}{dx} = \frac{dW_1}{dx} \quad (22)$$

3.2.3 *Calculation of rock beam subsidence when there are fractures in beam.*

When there are cracks in beam and beam hasn't broken, the above method should be used in the calculation of strata movement, but the efficient stiffness of rock beam decreases because of beam cracking in beam. Rock modulus is its nature coefficient. Suppose that it doesn't change and that the decrease of rockmass stiffness is the result of the decrease of beam effective thickness, thus, moment of beam inertia decreases. Moment of beam inertia is before cracking:

$$J_2 = \frac{t^3}{12} \quad (23)$$

After cracking:

$$J_2' = \frac{(t-a)^3}{12} \quad (24)$$

Where, t = thickness of beam; a =depth of crack.
When rock beam has cracked but hasn't broken, the movement of cracking beam can be calculated by the above method through substituting J_2' for J_2.
When rock beam fully break and don't fall, rock can not be considered as continuous medium and can not be analyzed by continuous mechanics method and should be calculated by discontinuous mechanics method. In this paper, building block beam method is adopted so that the movement and deformation of discontinuous medium rock can be obtained. This method will be discussed in details in another paper.
Calculation model of subsidence has been put forward above; thus, slope; curvature, horizontal movement and horizontal strain can be calculated by the following equations:

$$i(x) = \frac{dw(x)}{dx} \quad (25)$$

$$k(x) = \frac{dW^2(x)}{dx^2} \quad (26)$$

$$U(x) = B \cdot i(x) \quad (27)$$

$$\varepsilon(x) = B \cdot k(x) \quad (28)$$

Where $i(x)$, $k(x)$, $u(x)$, $\varepsilon(x)$ = slope, curvature, horizontal movement and horizontal strain; B = coefficient of horizontal movement, constant.

Up to now, we have set up the dynamic mechanic model of mining subsidence, which is much closer to field observations, however, further research in needed in order to get better understanding of mining subsidence.

4 CONCLUSION

The dynamic movement procedure of mining rockmass can be divided into four phases. 1) continuous changes of the position and length of rock beam with face advancing; 2) continuous changes of property of rock beam; 3) change of strata movement model; 4) change of elastic foundation; 5) change of load acting on beam;
Based on study on dynamic property of mining rockmass, a dynamic mechanic model of mining subsidence is set up and corresponding equations and boundary conditions are formed.

REFERENCES

Deng Kazhong, Zhou Ming et al. 1998. Study on laws of rockmass Breaking induced by mining. *Journal of China University Of Mining and Technology*. 27(3): 261- 264
Wang Yuehan, Deng Kazhong et al. 1998. The Study on the Character of Strata Subsidence During Repeat Mining. *Journal of China Coal Society*. 23(5). 470-475.

Mining Science and Technology' 99, Xie & Golosinski (eds)© 1999 Balkema, Rotterdam, ISBN 90 5809 067 1

Control of the difficult caving massive roof at the fully mechanized coal face

Jingquan Xie, Linsheng Xu, Yanfeng Yang & Zhanhai Zhang
Datong Coal Mining Administration, Shanxi, People's Republic of China

ABSTRACT: At Datong Coal Mining Administration in China, with the more extensively application of the fully mechanized mining at longwall face and the increase of the economic benefits of the enterprises, a long-term study and practice on the weakening and the supporting of the difficult-caving roof have being done to solve the problems of the difficult-caving of the massive roof jeopardized to the mining operation. The massive rock weakening is going to be an important branch in rock mass alternating engineering gradually. The tight roof weakening theory is going to be perfect, and a series of theories and techniques on the control of the difficult-caving massive roof have been formed. The theory and technology have been applied successfully in the control of the difficult-caving massive roof at longwall face in coal mine based on the study on the failure mechanism of the fracture rock mass structure, organic structure, fractal dimension and mechanic properties of coal and rock, etc.

1 THE MECHANIC CHARACTERISTICS AND MOVEMENT MECHANISM OF THE DIFFICULT-CAVING MASSIVE ROOF

Lots of studies on the mechanic characteristics and the movement mechanism of the difficult-caving massive roof have been made by use of the comprehensive geological method and rock mechanics. They can be identified as follows.

1.1 Rock mechanic characteristics

The most of the difficult-caving massive roof above coal seams consist of thick or especially thick sandstone, conglomerate, limestone and sand-shale, etc. Generally, its thickness is over 10m, e.g. the thickness of conglomerate in Yungang mine Dating is 37 m. The thickness of medium-grit grain stone in a mine of India is 50 m. The rock masses are usually with smaller fractures, less joints and bedding planes and the linear fracture ratio is below to 0.02. Sometimes, the horizontal stress occurring in coal seams is larger, e.g. the horizontal stress measured in underground is 1.6-2.4 times of the vertical stress in Yan Zishan and Xin Zhouyao Mines Datong (see Table.1).

Table 1. Comparative table of maximum horizontal stress and vertical stress

Mine name	Yan zishan	Yan zishan	Xin zhouyao
Depth (m)	153	245	362
Horizontal stress MPa	5.69	12.30	21.57
Vertical stress MPa	3.63	6.08	9.02
$\Delta x/rH$	1.57	2.02	2.39

The mechanical parameters of the difficult-caving massive roof vary in a wide range e.g. the tensile and compressive stress ratio of grit stone in the Mine of India is 1:5-1:10, the ratio in Datong China is 1:23-1:35. The compressive strength of medium-grit sandstone in Datong China is 35-94MPa. The strength in the Mine in India is 17-19MPa. It is only 20-50% the strength in Datong China. But the tensile strength in both of mines is almost same. The tensile strength of the Mine in India is 0.8-2.8MPa, and that of Datong in China is 1.4-3.9 MPa (see Table 2).

Table 2 Mechanical parameters of sandstone roof

	Compressive Strength	Tensile Strength	Stress Ratio	Location
Silt-fine stone	80-180	1.7-4.4	1:15-1:50	Datong China
Medium-grit stone	35-94	1.4-3.9	1:23-1:35	Datong China
Medium-grit stone	17-19	1.6-4.4	1:5-1:10	Indian Mine

- The mechanical parameters of the Mine in India were measured by the Institute of Dating Coal Mining Administration

The rock mechanical properties of the difficult-caving massive roof have close relationship with the mineral contents, pores, fractures and structures in rock, etc. In recent years, a new development has been made in the study of fractal dimension on the characters of the deposited structure of coal and rock and the physical and mechanical properties of the difficult-caving massive roof. The results indicate that the rock mechanical properties have self-similarity and the complexity of rock mineral contents, structures, pores and fractures, etc, which related to rock mechanical properties directly, can be described with fractal dimension, which the rock mechanical characteristics of the difficult-caving massive roof can be used to evaluate and classify comprehensively. By applying fuzzy mathematics, the proper fuzzy subset is established to make fuzzy cluster analysis, and produce fuzzy comprehensive evaluation. So the recognization on rock mechanical properties of the difficult-caving massive roof is more scientific and accurate. The relationship between the compressive strength and the fractal dimension of rock granularity and pore of the difficult-caving massive roof in Datong China is shown in Fig.1.

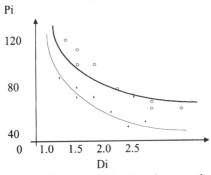

Fig1. The relationship Curve between the strength and fractal dimension of sandstone

1.2 The movement mechanism of the difficult-caving massive roof

There are 3 kinds of movements of the difficult-caving massive roof at fully mechanized mining face. The greatest danger to safety and production would be incurred if all of them appear simultaneously.

The shock load acts on the supports. In the gob area, the large area of roof rupturing produces a strong and wide spread deformation energy which makes the shock loading acting on support. Along with it, a large number of the hydraulic components are damaged, the pusher jack is deformed and even cracked. E.g. at a working face of Si Laogou colliery in Datong, the back column jacks of two supports were cracked while the safety valves could not be opened immediately, and the cracked jack (200-700 mm) was drawn to the wall for 1 m. The time for shock loading action was only 21.8ms and the impact pressure was even over 100MPa.

The roof is caved in large blocks behind the support. The roof beneath the lower crosscut forms a smaller inclined angle. The suspended roof strata with 3-5m, even up to 10m length is formed behind supports in the gob. Especially in the area behind the supports in the two ends of the face, the suspended length of such triangular roof stratum is much bigger. According to the measurement in Si Laogou colliery, the occurrence rate of the 5-10m suspended strata was up to 42.2%. Such roof was caved in large blocks which usually was about 10x3x2 cubic meters, sometimes up to or over 40x5x5 cubic meters. The weight of the most of them was over 100 tones; the biggest one was over 1000 tones. The horizontal thrust induced by the big rock caving on the supports pushes the supports advanced, damages the pusher jack of supports, bent the pans of the AFC line and even causes the fracture of the connected part of the canopy and the shield of the supports. At Shi laogou colliery of Datong in China, the large block caved is about the size of 75×8×2.5m with the volume 1500 cubic meters. Its caving advanced 17 supports to the coal wall for 0.8m and damaged 15 AFC conveyor pans

The strong periodic weighting takes place in working face. The expansion coefficient of the entire roof strata is only 1.08-1.1 after caving in Datong China. The gob can not be filled up with caved rock while the caving height is up to 20-25m at the working face with 3m in mining height. This would be more obviously if the caving thickness of the lower roof is small. Accordingly, the upper strata are also at suspension state in wide range,

Fig.2 damaged supports

even up to 70-90 thousand square meters. Normally, the roof periodic caving is in a large spacing between two times. The caving spacing is 30-80m, even over 100m. The bulk caving causes the large movement of roof, accompanied by big noise, gradually extended and intensified, the pillar spalled, the roof benched subsided, and caved along the coal wall. The load on the support increased in times, and the column of support subsides in great extent. On the fully mechanized mining face in the mine in India, the length of the face was 150m and the average thickness of seam was 2.65m The initial caving of roof took place when the advanced distance was 54m and mined-out area was 8100 square meters along with 8-10m suspend strata appeared behind supports. At that time, only 5-8m of lower roof strata was caved. The roof periodic weighting appeared for 3 times with great pressure when the working face advanced to 85.5,104 and 133m. During this period, there was no accident to occur, so that they had not paid any attention to prevent the damage. When the working face advanced to 182m, the fourth periodic weighting came and the gob area was up to 29000 square meters. Very large roof weighting appeared and the subsidence of entire roof strata was over 1.5m, The shearer and 101 supports were damaged seriously and covered with caved rock blocks. This accident made the mining operation stop all the time, and caused a large economical loss, only 117000 tones of coal being worked-out from the face. In Datong the very large roof weighting took place too and damaged the supports seriously (See Fig.2). All of this indicates that the rock masses alternating engineering is very important and can not be ignored in coal mining.

2 CONTROL ON THE DIFFICULT-CAVING MASSIVE ROOF

The practical and economical roof processing methods should be selected according to the type and the structure of roof, the rock properties and the characteristics of rock masses movement, to reduce the roof accidents and ensure the coal mining with safety and efficient. At present, some ways for roof control widely used in Datong are described as below.

2.1 *Induced caving by deep Hole Blasting in Gob*

The arrangement of blasting hole, parameters, the amount and type of powdered charge are decided comprehensively according to the working condition, roof suspended area in the gob, the roof processed height, the strength and the blast ability of rock, etc. Normally many groups of the vertical fan-shaped blasting holes are made along the face track or the road in the two ends of the face to the gob. The loading and blasting of every group of holes with millisecond delay detonator are made at same time. The amount of powdered charge detonated in a time is 1-4 tones in Datong mine China. The purpose is to alter the entire roof structure and ensure the entire roof in the gob caved in safe range. Normally the suspended roof area in the gob should be limited in 5000 square meters. The deep hole blasting caving not only makes part of roof caved directly, but also forms lots of fractures and free faces in the surrounding rock masses and weakens the tight roof.

2.2 *The Roof Weakening by High Pressure Water Injected*

The high-pressure water is injected into the entire roof strata through the roof infusion holes with the high-pressure pump to crack and weaken rock masses, and alter the physical, mechanical properties and the caving performance of rock masses. The trial results of industry tests indicate that the difficult-caving massive roof of sand stone and conglomerate can be weakened by high pressure water injecting to enlarge the original fissures into cracks and react with hydrophilic minerals of the rock. The reduction of compressive strength for the entire grit sandstone after immersed in water is 14-34%, 57% for the grit sandstone with micro structures, 43% for the medium grain sandstone with clay mineral colloidal matters and 19.7% for the sandstone with carbonate colloidal matters in Datong mine. Consequently, the roof weakening result is mainly related to the rock microstructure, mineral contents and hydro-physics.

The injecting holes can be arranged with the ways of single or double sides and multi-ways due to the condition of the roof strata to be weakened. The methods of high pressure water injecting can be selected according to the strata structure and the hydro-physics features of the rock in the roof, that is single stratum injecting, double or multi-strata injecting and double strata slice injecting, etc. For the application in different stope region, the gob injecting, the abutment area injecting, and the virgin rock stress region injecting etc., can be selected. The main parameters, the spacing of injecting holes and the quantity of water injected, can be decided comprehensively by the thickness of strata, rock structure, rock properties, caving performance, condition of production and regularity of strata behavior, etc. The methods mentioned above do not disturb the operation of coal mining and the processing result is obviously while all of the technological process is practiced in gate road.

The roof of the thick conglomerate and sandstone at Si Laogou Colliery had been processed with water injecting in the later stage of mining. The analysis shows that the suspended roof area was 62% of the reduction, the periodic weighting was reduced to 50%, the shock pressure impact was reduced to zero. The appeared rate of the suspended roof with the width larger than 5m, was reduced from 42.2% to 3.7%. The roof caving angle was increased from 30-45 to 70-80 degrees. And normally the fragmentation coefficient of caved rock was decreased and the expansion coefficient was increased after injecting. Therefore, the efficient of the rock filling in worked-out space was improved greatly.

2.3 *The selection on the parameters of the supports*

The efficient way to control the difficult-caving massive roof and ensure the production safety is to adopt the proper supports. The hydraulic supports for the difficult-caving massive roof should offer the enough resistance against the impact pressure. Its function is to buffer the great roof subsidence and the damaged shock load in time. So, the relief valve with large flow rate should be adopted to ensure the safety while the support is impacted by ground pressure. The parameters of the relief valve with large flow rate are established by the closed speed of roof and floor or by the pressure released as soon as the roof moving. The speed of liquid-release is up to 10000 litres/min for the relief valves with large flow rate used in Datong. The ability of resistance against the impact on support legs, which is over 50MPa, is increased obviously for 1.65 times by applying the liquid-filling legs.

The problem of the support, which would be damaged by the horizontal thrust induced by the large blocks caving in the gob, has been solved by the application of the vertical tail-bar short 4 links

compact hydraulic supports. The external part of the shield of TZ-720 4 legs chock-shield supports used in Datong is 819 or 1030mm shorter than Mitsui560 type from Japan and DT-550 type supports from U.K respectively. (see Fig.3).

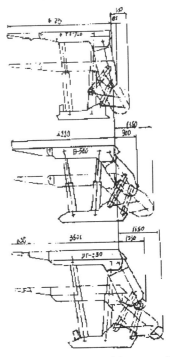

Fig.3 The external size of the waste shield

The support strength is a very important parameter related to the ability to deal with the strong actions of the roof strata. According to the practice in Datong, the parameters of the support strength can be adopted as below:

Coal seam thickness (m)	support strength (KN/m2)
less than 1.5	700-800
1.5-2.5	800-900
2.5-3.5	900-1100
more than 3.5	1100-1200

Many tests and field measurement have to be made, especially for the geological environment, the rock mechanical properties, the roof movement characteristics and regularities before the processing on the difficult-caving massive roof is put into practice. The movement of overburden strata will be observed with bore hole television in fully mechanized face. The intelligent computer controlled multi- parameters unit for mining pressure monitoring and data processing unit will observe the impact pressure and the support operation. For high-pressure water injecting, the measurement of the hydro-physics and microstructure features of the rock should be done. On the basis of the results of these studies, a single, two or three ways can be used properly to control and process the difficult-caving massive roof comprehensively, so that the satisfied results will be achieved successfully.

3 CONCLUSIONS

(1). The study on rock masses structure is very important in the controlling of the difficult-caving massive roof at the fully mechanized coal mining face. The rock masses of the difficult-caving massive roof are a natural geological body that has a certain entirety, but it is an integrated body. Its structure is neither a continuous bar nor a complete plate, and it is an uncontinuous, uneven and anisotropic body with weak planes. The rock masses structure is one of the main factors, which has influence on the roof deformation, breaking, caving and energy relieving.

(2).There are many methods of the study on the control of the difficult-caving massive roof. On the basis of the comprehensive study on rock mechanical properties combining the traditional geology method and the rock mechanical measurements all together, the versatile recognition of rock properties should be achieved through many methods, such as the fractal dimension method. The results in the control of the difficult-caving massive roof will be improved by deeper recognition of the features in the roof.

(3). In the course of the selection on the treatment methods of the difficult-caving massive roof at fully mechanized longwall face, not only study the mineral contents, rock structure, mechanical characteristics and original environment, etc., but also pay a great attention to the regularities of rock masses movement and breaking to ensure the efficiency of roof treatment methods in the practice.

(4).In the engineering of rock masses control, to select a proper and reasonable support method is very important. There is a close relationship between the roof control and space maintenance in working face, so that the working face need to be kept efficiently while the difficult-caving massive roof is processed. Only if the support pattern is selected properly, can the rock masses control engineering be carried out

successfully to achieve safe and effective production.

REFERENCES

Xu L.S.,Zhang Z.H.,1996. Study on relations between the fractal dimensions and mechanical properties of roof sandstone, *Journal of China Coal Society,* 245

Song Y.J., 1991.The application and prospect of rock mechanics in hard roof control of coal mine, *Proceedings of Chinese North Rock Mechanical and Engineering Conference,* 7 -19.

Xu L.S., Song Y.J., 1992.A study on application of comprehensive rock mechanical properties and geological characteristics in coal mine, *Proceedings of 11th International Conference on Ground Control in Mining,* 562-567.

Mining Science and Technology' 99, Xie & Golosinski (eds) © 1999 Balkema, Rotterdam, ISBN 90 5809 067 1

Experimental study of the breaking mechanism of coal and rock by impact

Quzhen Tian, Lixun Kang, Longwang Yue & Tianhu He
Department of Mining and Civil Engineering, Taiyuan University of Technology, People's Republic of China

ABSTRACT: The underground trials have shown that use of impact mining machine increases fraction of lump sized coal, reduces dust generation, slows gas release rate and reduces the energy used in the process of mining. The coal and rock impact mining equipment was developed by the authors to study the mechanism of coal breaking by impact. Propagation of impact induced stress in coal mass, and some interrelated parameters of impact and split have been analyzed in the paper. Selected mechanical and breaking characteristics of coal mass under impact are analyzed and influence of discontinuities in coal mass on effectiveness of breaking is analyzed.

1. INTRODUCTION

In underground longwall mining, a drum shearer is widely used as a coal-cutting machine. Because of higher linear velocity of the rotating drum, this type of shearer makes coal overbreakage. So dust is flying up in coal face, which makes the working environment become worse. This not only harms the workers` health, but also maybe leads to the dust explosion.

Compared with shearer, mechanism of a plough is planning breaking. This type of coal-cutting machine uses its haulage force to make plough body slide along face at a high speed. During the sliding coal is planned down. The plough is similar to a planing machine of mechanical processing. Because of the many confining factors such as coal seam conditions, energy consumption, coal lump rate, plough is less used in underground mining at present.

During a shearer milling breaking, the cutting speed of cut-in point depends on the rotating speed and drum diameter. Taking shearer AM-500 as an example, when a 2m-diameter drum is equipped, the minimum linear speed of cut-in point is 3.46m/s and the maximum may be up to 5.66m/s. Under such linear speed, coal is broken as it could be. The dust lifted up by the rotating drum flies up in the face by means of air current, which makes the working environment become seriously worse. Although the dust suppression measurements such as water spray,

coal seam infusion etc are adapted, the effectiveness of dust suppression is not ideal yet. Production cost also increases considerably because of those measurements.

In order to effectively improve lump-producing rate and reduce dust-producing rate, lots of technical methods such as using big bit, slowering rotating speed, reducing the number of cutting point, changing bit arrangement, increasing cutting web have been adapted worldwide. Though some effectiveness has been achieved, the desired results do not occur yet. Meanwhile, because the mechanism of milling breaking does not change, the power of shearer increases remarkably by adapting above measures.

2. IMPACT BREAKING OF COAL

In the coal seam near the face wall a series of compressive and tensile cracks are formed because of the gravity of overlying strata and abutment pressure induced by mining. These cracks are vertical to bedding plane and parallel to free surface[1]. The existence and influence of these cracks makes the mechanical strength of coal near wall lower remarkably. Experimental studies have indicated that coal mass as well as other rock mass whose tensile strength is far less than both of shear strength and compressive strength, and the character of mechanical strength is mainly brittleness. It is tensile stress or tensile stress derived from the interior of rock that causes the brittle rock mass breaking[1]. Because the tensile strength of coal and rock mass is the weakest, we should utilize this feature to generate positive tensile stress in coal and rock mass, and achieve purposes of breaking coal and rock mass. Under such concrete engineering background, the idea of impact and split coal breaking is put forward.

The coal mass near face wall is in an uniaxial compression condition under the action of abutment pressure .The cracks are not caused by compressive stress in coal mass directly but caused by the tensile stress derived from the compressive stress. Its fracture plane is vertical to the direction of the compressive stress and parallel to the derived tensile stress as shown in figure 1 :

The interrelation of each stress is:

$$\sigma_1 = k_1 \gamma H$$
$$\sigma_2 = k_2 \sigma_1 \qquad (1)$$

where k_1=the concentrated stress coefficient of the abutment pressure; γ=the average density of overlying strata, 10kN/m3; H=mining depth, m; k_2=the coefficient of lateral stress.

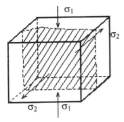

Fig.1 The stress analysis of the unit of coal mass near face wall.

If impact bit is parallel to the crack plane, when cutting bit impacts into coal mass the force imposed on the bit will superpose σ_2 at the same direction,

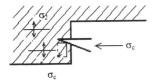

Fig.2 Principle of impact breaking

then the coal and rock mass will be split down. The purpose of increasing lump-producing rate would be completed. Figure 2 shows above principle:

3. IMPACT BREAING EXPERIMENT OF COAL AND ROCK MASS

3.1 *Experiment system*

Based on the principle of similarity, we use the similar material to simulate the real coal mass (uniaxial compressive strength: σ_d=34.5MPa, tensile strength σ_p=2.2MPa), the imitation ratio is 2.5, three dimensions of the test sample are 500*500*500mm. Some sensors are preset in the sample. The location and amount of the sensors depends on the request of the testing. The test sample is placed on the hydraulic impact breaking test bed developed by our own. In order to imitate front abutment pressure σ_1 of coal seam, we apply a side pressure on the sample's lateral surfaces. During the experiment, impact force, impact velocity and the split thickness of impact have been changed. The experimental instruments include dynamic strain gage, ray oscilloscope, impact experiment parameter measuring apparatus. The experiment system of coal mass is shown in figure 3

No.1 test system: The measuring system of impact velocity of cutting bit and the acceleration

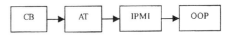

No.2 test system: The measuring system of impacting stress-strain:

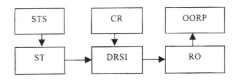

Fig.3 The block diagram of the test system

where CB-cutting bit; AT-acceleration transducer; IPMI-impact parameter measuring instruments; OOP-output of printer; STS-simulation test sample; CR-compensating resistance; OORP-output of recorder paper; ST-strain transducer; DRSI-dynamic resistance strain instrument; RO-ray oscilloscope;

3.2 *Experimental results and analysis*

The wing angle of the impact bit is related to the fracture angle of coal mass. Without changing the wing angle α , we can draw a conclusion from impact experiments that when the coal is homogeneous the fracture angle β is almost the same. So we can say that the fracture angle depends on the wing angle, the relationship between them as the following equation:

$$\beta=90+0.35\alpha-0.015\alpha^2 \qquad (2)$$

where α=the angle of cutting bit; β=the fracture

Fig.4 The relationship between wing angle, width of bit and fracture angle of coal and rock mass and crack development.

angle of coal mass. At the same time, the developing direction of the cracks is determined by width of the

$$\frac{\partial^2 \Phi}{\partial t^2}=c_1^2\nabla^2\Phi \qquad (3)$$

cutting bit blade. Figure 4 shows the relation:

During coal impact breaking, the impact load applied by impact bits induces hemisphere impact wave in coal. The center of the sphere impact wave is the impact point. In the sample, impact wave propagates in the hemisphere forms. Because of the geometrical increasing of the surface area in front of the spherical wave, the stress distribution in front of wave will vary quickly and the stress will decay inversely proportional to r which is the distance from wave source. The wave equation is:

where $\phi=\phi(r,t$☐stress function; c_1= wave velocity; r=the distance between wave front surface and wave source; t=time.

$$\frac{\partial^2 \Phi}{\partial t^2}=c_1^2(\frac{\partial^2 \Phi}{\partial r^2}+\frac{2\partial\Phi}{r\partial r})=c_1^2\frac{\partial^2}{r\partial r}(r,\Phi)$$
$$(4)$$

$$\Phi=\frac{g_1(r-c_1t)}{r}+\frac{g_2(r+c_1t)}{r}$$

Solved above equation, the following is obtained: where g_1 is the function of $(r-c_1t)$; g_2 is the function of $(r+c_1t)$.
When test sample is impacted, generating impact

wave pulse is virtually that the pressure at the wave source raises suddenly to P_0, then decays exponentially. The pulse expression is:

$$P=0 \qquad \Delta t<0☐☐$$
$$(5)$$
$$P=P_0e^{-\alpha t} \qquad \Delta t<0☐☐$$

where α is a decay coefficient.

The coal and rock mass is a kind of brittle material. Under the action of impact stress, its fracture is brittle one. The brittle fracture is caused by the cracks in the mass or by the cracks growing rapidly. The brittle fracture is the result of the sudden fracture without large deformation[2]. Just as mentioned above, because there are preexisting compressive and tensile cracks in the coal mass, and cracks can spread under conditions of lower than yield stress when the impact direction of the impact bit is parallel to the crack direction, the external load generated by the impact bit is symmetric to propagation plane of cracks. Just as figure 5 shows:

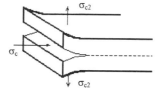

Fig.5 The pattern of crack propagation of the impact breaking.

Under the action of the impact stress, impact breaking cracks propagation is caused by tensile stress which is vertical to the crack plane and liquid pressure on the crack surface in coal mass.

$$\frac{dR}{dt}=T(\sigma_{c2}+P_l-\sigma_j)R \qquad (6)$$

where T=expansion coefficient; P_l=the liquid or gas pressure in the coal and rock mass;σ_j=the limited stress of the cracks expansion; R=the crack expansion.
According to the critical radius, the critical stress (limited stress) of the cracks expansion can be determined by:

$$\sigma_j = K\sqrt{\frac{\pi}{4R_l}} \qquad (7)$$

where K=the fracture toughness property, it reflects the resistance property to the crazes propagation.

In impact breaking of coal and rock mass , either σ or R reaches to the limited value, coal and rock mass will break. The breaking is mainly caused by σ_{c2} produced by impact stress σ_c.

4. IMPACT MINING EXPERIMENT

The impact mining experiment was carried on the face 2319 Fenhuangshan colliery in JinCheng, China. The width of the face is 130m, the seam thickness is 2.8m and the Protodyakonov coefficient of coal f is 3.5. The face is equipped with chock-shield supports and shearer. The impact mining mechanism is realized by just replacing new impacting device to original shearer's ranging arm and drum. After change, the original production system is almost unchanged. Figure 6 is the first impact mining machine experimented.

Fig.6. The first experimented impact mining machine.

The experiments have indicated that coal breaking mechanism of the impact operating mechanism is in accordance with the theoretical studies when coal is impacted. After the cutting bits impacts the coal mass, impact wave forms at the impact points. As the cutting bits impacting into fracture, split and tensile cracks form in the coal mass. It is very effective for increase of lump coal and reducing cutting specific energy when coal mass is under the action of the cracks and the impact wave. The lump-producing rate in the face equipped with impact shearer increases by 83%. Compared with that of drum shearer, 28% rises. Meanwhile, the dust concentration of face would decrease not less than 85%. At the same time, because of the increasing of lump coal, velocity of methane release is lowered.

5. CONCLUSIONS

1. During the impact breaking of coal and rock mass, the propagation law of impact stress wave in the coal and rock mass is in accordance with that of spherical wave.
2. The wing angle of impact cutting bit controls the fracture angle of coal and rock mass and their interrelation decides the dimension of lump coal.
3. The coal and rock mass is mainly broken by tensile stress caused by the cutting bit during impact.
4. The impact shearer has many advantages such as increasing lump-producing rate, reducing dust-producing rate, saving energy and slowing down velocity of methane release etc.

REFERENCES

Tian Quzhen 1998. No.2. The study on the mechanism of coal mining and breaking by impact. The journal of TaiYuan University of Technology.
Ma Xiaoqin 1992. Impact dynamics. Beijing University of Technology publishing press.

Mining Science and Technology' 99, Xie & Golosinski (eds) © 1999 Balkema, Rotterdam, ISBN 90 5809 067 1

Dip-oblique longwall mining along the strike of steeply inclined seams

Daojin Yin, Kelin Zhang, Ruiguo Zou, Guiwen Wu & Wen Hu
Guangwang Coal Mining Administration, People's Republic of China

ABSTRACT: It is very difficult to mine coal in thin and medium-thick steeply inclined seams under complicated geological conditions. Adoption of the dip-oblique mining method in Guangwang Coal Mining Administration improves safety and economics in this conditions. The paper describes the typical dip-oblique longwall method along strike, the face layout and selection of main parameters, observations on strata control and other aspects of this type of operation.

1. INTRODUCTION

Main mining seams in Guangwang CAM (Coal Mining Administration) of China are thin and medium thick. When overhand stope system or overhand-oblique mining method etc. were applied, roof death rate amounted to 4~7 men per million tons, recovery ratio was only 70% generally (less than 50% specially) with low working efficiency, but consumed timber came to 1000m³/kt (3000m³/kt in some seams).

The difficult situation mentioned above has forced us to reform the traditional mining methods of steeply inclined seam. A great breakthrough was achieved when adopting dip-oblique longwall mining along strike after 15 years' hard work. A lot of research projects were also completed concerning about the mechanism of strata movement and its control. These solved our problems of 'three-soft' and 'three-hard' and close-seam mining in steeply inclined seams. Now, single face output was increased by 30%; working efficiency was increased by 17.6%; consumed timber was decreased to less than 100m³/kt; death rate was decreased to 0.615 men per million tons. To the end of 1997, raw coal output amounted to 9970kt with the benefit of 118 million RMB Yuan, which made great contribution to safety and economic benefit in our Administration.

2. DIP-OBLIQUE LONGWALL MINING METHOD ALONG STRIKE

As shown in Fig.1, Sublevel height is 50~60m; working face is along the oblique with pseudo-angle 30~35°and length 100~120m; maximum roof-control distance (L) is synthetically decided

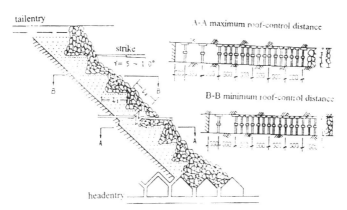

Fig.1. Face layout of dip-oblique longwall along strike in steeply inclined seam, face pseudo-angle 35, blasting mining; basic props (row x prop) 0.9x0.8 (mxm). close–standing props. 5 props per meter. L1=5m.L2=3~5m.

basic props(row×prop)0.9×0.8(m×m), close-standing props, 5 props per meter, L_1=5m, L_2=3~5m.

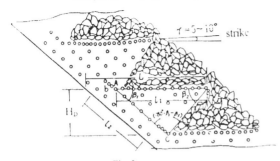

Fig. 2

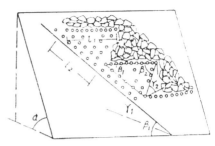

Fig. 3

by the state of roof and the necessities of ventilation, walk and chute, generally L=5~6m; strike close-standing length L_1=L-C-B, generally L_1=4~5m (in order to keep the stability of refuse, the angle between close-standing and strike is 5~10°); the relation of close-standing space (along the dip)H_D, oblique length L_2 and seam angle α, face pseudo-angle γ_1 and oblique angle β_1 is shown as the following (Fig.2 & Fig.3), where the angle between close-standing and strike is ignored.

The relation of natural grade γ_2 (generally 39~41°) of the refuse between the two strike close-standing and its oblique angle β_2 and seam angle is shown as the following:

$$\sin\beta_1 = \sin\gamma_1 / \sin\alpha_1 \quad (1)$$
$$\sin\beta_2 = \sin\gamma_2 / \sin\alpha \quad (2)$$
$$L_2 = L_1 \cdot [\sin\beta_2 / \sin(180 - \beta_1 - \beta_2)] \quad (3)$$
$$H_D = L_2 \sin\beta_1 \quad (4)$$

Formula (1)~(4) indicates seam angle is the main factor in the problem. When =45~90°, H_D=4~2m and L_2=5~3m. In practice, integer times of basic prop value are selected.

Row space and prop space of basic props are : row space × prop space =(0.9~1.0)m×0.8m; strike close-standing space is 5 props per meter generally.

3. MINING HEIGHT RESTRICTED BY SUPPORT CONDITION AND SEAM ANGLE

At first, mining height was chosen less than 2m in dip-oblique longwall mining along strike, which was in accord with the handbook of *Picture Set of China Mining Method* and *Mining Method of China*. Mining practice and research indicate that mining height is restricted by seam angle under the condition of individual hydraulic prop (or friction prop, timber prop).

3.1 Restriction decided by static load balance

The mechanics model of dip-oblique longwall mining can be demonstrated as Fig.4.

Assume dh along X-axis and its upper pressure is $P_\perp$; gravity component along X-axis is $\gamma m \sin\alpha dh$; friction force caused by roof and floor pressure is $2P_\perp f_1 dh$; friction force caused by gravity is $\gamma m f_1 \cos\alpha dh$. The mechanical balance equation can be expressed as:

$$(P_\perp + dP_\perp)m - P_\perp m + 2P_\perp f_1 \xi \cdot dh +$$
$$mf_1 \gamma \cdot \cos\alpha \, dh - \gamma \cdot m \sin\alpha \cdot dh = 0 \quad (5)$$

Then,

$$P_\perp = \frac{\gamma m(\sin\alpha - f_1\cos\alpha)}{2\xi f_1} \times (1 - e^{-2f_1\xi h/m})$$
$$(10^4 N / m^2) \quad (6)$$

When $h \to \infty$,

$$P_\perp = P_{\perp max} = \frac{\gamma m(\sin\alpha - f_1 \cos\alpha)}{2f_1\xi}$$
$$(10^4 N/m^2)$$

Where, f_1 - friction coefficient of rock and roof or floor, generally f_1=0.6;
ξ - average lateral pressure coefficient ;
h - rock height, m;
m - seam thickness, m;
γ - specific gravity of loose rock, 10^4N/m³;
- seam angle, degree;

Considering the difference of lateral confining pressure near the roof or floor and that in the middle, we assume the load change as linear from the middle to the two sides. Then, the load per meter along strike is:

$$P_{max} = \frac{\gamma m^2 (\sin\alpha - \cos\alpha)}{4 f_1 \xi} \times \left(\frac{1}{\xi_D} + \frac{1}{\xi_Z}\right)$$

$$(10^4 N/m^2) \quad (7)$$

Where,

$$\xi_D = \frac{1}{1 + 2f^2 + 2f\sqrt{(1+f^2)\cdot(f^2 - f_1^2)}}$$

$$\xi_Z = \frac{1}{1 + 2f^2 + 2f\sqrt{(1+f^2)}}$$

f - internal friction coefficient of loose rock, generally f=0.7.

3.2 Lateral thrust resistance of close-standing prop

The minimum lateral thrust resistance of each prop should be:

$$f_{k\,min} = f_0 \cdot P_c$$

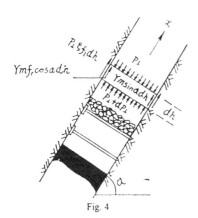

Fig. 4

If f_0=0.35 and P_c is the regular setting load or working resistance, then the lateral thrust resistance of each prop can be calculated.

3.3 Restriction relation of seam angle on mining height

The discriminating formula of close-standing prop being not thrust down is:

$$5 f_{k\,min} \rangle P_{max}$$

$$(8)$$

Calculating formula (7) and (8), we can reach the following conclusion: when same type of prop is applied, the greater the seam angle is, the less the applicable mining height should be, that is, seam angle restricts mining height and thus the applicable scope of dip-oblique longwall along strike is determined. In Wangcang mine of our

administration, 1Mt coal has been extracted safely by dip-oblique longwall along strike with =40~45° and mining height h=2.8~3.5m.

3.4 Pound resistance check

The pound force of close-standing prop can be calculated by the following formula:

$$F_c = \frac{m\sqrt{2gH_D}(\sin\alpha - f\cos\alpha)}{n\cdot t_c}$$

$$(9)$$

Where, m - mass of pound rock kg ;
  t_c - pounding time (generally 0.2~0.5s);
  n - number of props being pounded simultaneously;
  H_D- space of close-standing props, m ;
  F - friction coefficient of rock and floor, generally 0.4~0.7;

Generally, about 1.0m wide buffer zone is made behind close-standing props, which increases safety in production by t_c going up and F_c down.

4. STRATA CONTROL OBSERVATIONS

4.1 Behavior of basic prop Load

The behavior of basic prop load is:
the load of upper props > that of middle props> that of lower props;
in medium-thick seam, the load is between 100~80 KN/prop;
in thin seam, the load is between 60~80 KN/prop.

4.2 Main bearing of roof load

Close-standing props are the main bearing of roof load (about 60~40KN/prop). The average support intensity of close-standing props is 1.3~1.7 times that of basic ones (1.5~2.1 times when loading).

4.3 Lateral load of strike close-standing props

The lateral load of upper props < that of middle props< that of lower props;
in medium-thick seam, the lateral load is about 8~18Kn/prop;

4.4 Period loading

Period loading phenomena is the same as that of single longwall with similar loading distance, but the dynamic loading coefficient is different by section (dynamic loading coefficient of upper props> that of the lower); in medium-thick seam, dynamic loading coefficient of upper props is 2, that of lower props 1.5; in thin seam, dynamic loading coefficient of upper props is 1.2, that of lower props 1.2. The dynamic coefficient increases according as the mining height does.

5. CONDITIONS OF EMPLOYING DIP-OBLIQUE LONGWALL ALONG STRIKE

The seam is stable comparatively with dip angle 35^0, and the roof is medium stable. Mining height is restricted by seam angle. On the condition of individual hydraulic prop, when seam angle is less than 45°, mining height can amount to 3.0m; when seam angle is 50~60°, mining height can be 2.5~1.5m; when seam angle is 60~90°, mining height can be 1.2~0.8m.

6. CONCLUSIONS.

6.1 Conclusions

(1) Dip-oblique layout changes Mechanics State of roof and floor. The tensile and shear state of overhand-oblique or dip system (if there is crack, rotation will take place) is turned into compress state. Since the compress strength of rock is much greater than its tensile (or shear) strength (if there is crack, it will be close), the stability of rock and floor in dip-oblique system is increased greatly, which can prevent from thrusting down props or large scope floor.

(2) Sublevel strike close-standing props can hold back gob refuse, and come into being a continue refuse zone. They can improve the serious difference (section) of the upper and the middle and the lower caused by the violently rolling of gob refuse in steeply inclined seam. They can also buffer the pounding of large compress and reduce the intensity of period loading.

(3) High face recovery ratio. In overhand stope or overhand-oblique system etc., most of the extracted coal of the middle and the upper goes to gob and can not be taken back, generally the recovery ratio about 70%. In dip-oblique system, the coal goes back to chute by gravity and the little amount of loss coal on the floor is easy to be taken back, generally recovery ratio over 95%.

(4) Reducing the slope is convenient for walk and operation. It can prevent from falling down and improve safety in production.

6.2 Further development

(1) Mechanization in steeply inclined seam should first develop superior conventional mining and then fully mechanized mining.

(2) Steeply inclined close-seam applying dip-oblique layout has been successful in Guangwang CAM, realizing no or less loss of coal. It develops the technique of steeply inclined close-seam mining. This research achievement has been awarded the Second Scientific and Technological Progress Reward of Sichuan Province.

(3) According to support capacity to increase the mining height will significantly enhance the production capacity. If increasing height can not satisfy the permitted mining, condition, top-coal caving technology should be applied.

(4) When seam thickness is less than 1.3m, the support way of individual hydraulic prop can be changed to that of multi-cavity air-holder in order to improve the state of compound or broken roof. Now, three-cavity air-holder has been successfully used in seams of 1.0m mining height.

Mining Science and Technology' 99, Xie & Golosinski (eds)© 1999 Balkema, Rotterdam, ISBN 90 5809 067 1

Mechanical characteristics of the immediate roof in working face and working resistance of supports

Shenggen Cao, Minggao Qian, Changyou Liu & Jialin Xu
China University of Mining and Technology, Xuzhou, People's Republic of China

ABSTRACT: Based on numerical modeling and field observation, this paper analyzes displacement distribution, the P-Δl curve as well as support controlling effect on immediate roof in long wall top-coal caving(LTC) technology. FLAC and UDEC software is used for modeling. The relationship between support and surrounding rock of different immediate roof thickness is also studied in detail.

1.INTRODUCTION

With rapid development of long wall top-coal caving(LTC) technology, it is found out that support resistance of field observation in LTC is quite different from that predicted by traditional methods adopted in slice mining system. The reason is that immediate roof has been always regarded as rigid body in the traditional methods over a long period of time. However, recent research shows that the load acting upon the support is composed of two parts, one is from the weight of immediate roof while the other is from the given deformation pressure which depends on main roof rotation angle and mechanical characteristics of immediate roof.

As well known that the relation of P-Δl represents interacting mechanism of support and surrounding rock. In traditional long wall slice mining, P-Δl curve obtained from both field measurements and lab tests is similar to a hyperbola, which means that increasing working resistance, when it reaches a set value, has little influence on roof convergence. While this working resistance is below the set value, the influence on roof convergence increases greatly. Will the traditional P-Δl hyperbola curve still be suitable to LTC mining and will the P-Δl curve still be a hyperbola after the thickness of immediate roof has doubled and redoubled? In order to probe into these problems, numerical analyses method (FLAC and UDEC programs) are applied in this paper.

2 . DEFORMATION AND FAILURE CHARACTERISTICS OF IMMEDIATE ROOF

2.1 Numerical modeling by FLAC

Deformation and failure characteristics of immediate roof are analyzed by using FLAC program when the thickness of immediate roof is 2 times (3 m), 4 times (6 m) and 6

times (9 m) the mining height respectively. The analyzing model in FLAC is shown in Fig.1. The upper boundary load of model is equal to the rock weight of 300 m the mining depths. The bottom boundary is fixed, right and left boundaries are fixed in the direction of horizon, and the upper is the given deformation of main roof.

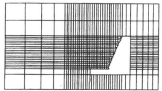

Fig. 1 Numerical model of FLAC

2.2 Displacement distribution feature in the immediate roof

Two layers of immediate roof are considered. The distance from upper layer to main roof is 1 m, and the lower layer to coal seam is also 1 m. Displacement distribution curves of the two layers are shown in Fig.2. (The upper set of curves represent those of upper layer, the lower set of curves represent those of lower layer).

From above, the following conclusions can be drawn:

(1) The vertical displacement increases gradually along the direction from coal rib to goaf. The feature of given deformation exerted to immediate roof is very clear.

(2) If the thickness of immediate roof is 2 times or 4 times the mining height, the displacement distribution of the whole immediate roof can be influenced by support working resistance. The displacements of each point will begin to decrease with the increasing of working resistance. On the contrary, it will begin to increase with the decreasing of working resistance. Meantime, main roof rotation angles will also influence the displacement from

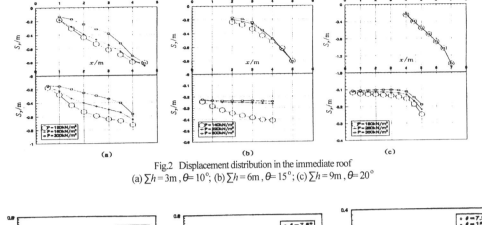

Fig.2 Displacement distribution in the immediate roof
(a) $\sum h = 3$m, $\theta = 10°$; (b) $\sum h = 6$m, $\theta = 15°$; (c) $\sum h = 9$m, $\theta = 20°$

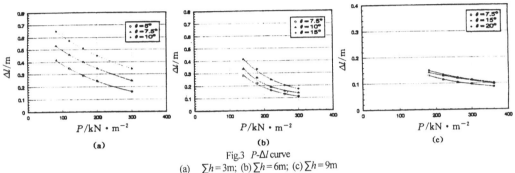

Fig.3 P-Δl curve
(a) $\sum h = 3$m; (b) $\sum h = 6$m; (c) $\sum h = 9$m

upper to lower layer.

(3) If the thickness of immediate roof is 6 times mining height, the upper layer displacement will be influenced only by main roof rotation angle, while the lower layer displacement will be mainly influenced by working resistance. This indicates neither working resistance nor main roof rotation angle can influence the displacement of the whole immediate roof. The working resistance can only influence the lower layer and the main roof rotation angle can only influence the upper layer. The main roof rotation deformation absorbed by the immediate roof will not act on the support. This is the difference in support and surrounding rock relationship between 6 times and 2, 4 times of mining height.

2.3 Relation between working resistance and roof convergence

By regressing the vertical displacement and working resistance obtained from numerical calculation, a set of curves can be gained shown in Fig. 3.

From above, we can see that the P-Δl curve is similar to a hyperbola while thickness of immediate roof is 2 or 4 times mining height. This coincides with the results of tests in the lab and field measurements. While it is 6 times of mining height , the hyperbola curve of P-Δl is not quite

clear, in other words, it is only a section of the whole hyperbola which is below the inflection point. In this section of curve, the working resistance has little influence on roof convergence.

The support's working load is composed of both the weight of immediate roof and given deformation pressure of main roof. While thickness of immediate roof is 6 times the mining height, the given deformation pressure can be fully absorbed by the immediate roof so that it cannot be transferred to the support. So the given deformation pressure may be considered as zero, then the support only needs to bear the weight of immediate roof. This is the reason why there is a difference of P-Δl curve when the height of immediate roof is different.

From the P-Δl curve, the following conclusion is also obtained. Main roof rotation angle can only change the absolute value of roof convergence rather than the whole curve shape in different heights of immediate roof. This means that the location and state of main roof cannot change the characteristics of P-Δl curve. P-Δl curve is the direct result of the interaction between support and immediate roof. So the regularity of overlying strata movement cannot be changed by the support's working resistance. This conclusion goes for all types of immediate roof.

310

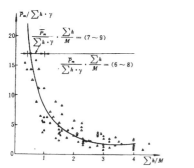

Fig.4 (left) Statistics curve between $\overline{P}_m$ / $\sum h \cdot \gamma$ & $\sum h / M$

4. RESISTNACE OF WORKING SUPPORT

4.1 *Statistics analyses of support's working resistance*

The load of support usually comes from the weight of immediate roof and periodic weighting after the main roof's breaking.

Immediate roof is regarded as the strata which cannot be compressed and self-balanced, the whole weight Q_1 must be undertaken by the support. If the length of working face is treated as a unit, then:

$Q_1 = \sum h \cdot L \cdot \gamma$

Where $\sum h$ is the height of immediate roof, L is face width, γ is unit weight of rock.

$\sum h$ is determined by the following formula :

$\sum h + M = K_0 \sum h$

$\sum h = M / (K_0 - 1)$

Generally $K_0 = 1.15 \sim 1.2$, so $\sum h = (2 \sim 4) M$

Where M is the mining height.

So the load of immediate roof is similar to the rock weight of $(2 \sim 4)$ times the mining height.

The dynamic coefficient η is usually used to estimate the load of main roof. Generally $\eta = 2$. So when taking the load of main roof into account, the load of support should be similar to the rock weight of $(4 \sim 8)$ times the mining height.

When $\sum h / M \square 1$, $\overline{P}_m = (6 \sim 8) M \gamma$

When $0.5M < \sum h < M$ $\overline{P}_m = (7 \sim 9) M \gamma$

When $\sum h < 0.5 M$, $\overline{P}_m$ depends on the weighting intervals and the load caused by unstability in rotation .

4.2 *Relation between support's working resistance and the stability of tip-to-face*

From above analyses, the support's working load does not increase with the increasing of the height of immediate roof. It is known that roof accidents take place mainly in the tip-to-face unsupported area after hydraulic supports are used. Reasonably determined support's working resistance can provide a key to dealing with roof flaking control and smooth top coal caving in LTC at the same time. The relation between support working resistance and roof flaking in tip-to-face area is analyzed by UDEC program in the next paragraph when the thickness of immediate roof is 6 times the mining height.

Under the conditions of different working resistance and tip-to-face distances, roof flaking in tip-to-face area is simulated by UDEC program. Five kinds of working resistance ($P = 800, 1200, 1600, 2800, 4400$ kN) and four different tip-to-face distances ($s = 0, 1, 2, 3$ m) are considered in calculating .

The calculating results of roof flaking in tip-to-face area are shown in Fig.5.

In China this has been confirmed by the results of field measurements at 71 mining faces as shown in Fig. 4.

The calculating results are as following:

(1) Whether roof flaking will occur in tip-to-face area mainly depends on working resistance P and tip-to-face distance support resistance. In Fig.5, roof flaking occurs, when $P = 800$ kN. But when $P = 4400$ kN, the tip-to-face blocks can be kept in temporary balance. This illustrates that increasing the support's working resistance can control roof flaking when tip-to-face distance is within a certain scope.

(2) When tip-to-face distance $s = 0$, no matter what P is, roof flaking will occur. While roof flaking in tip-to-face will never occur when $s = 3$ m.

(3) When $s = 1$ or 2 m, roof flaking is closely related to the

(4) The balancing state after roof flaking is similar to a falling arch whose height depends on support's working resistance. Take $s = 2$ m as an example, when $P = 4400$ kN,

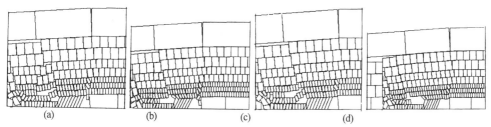

Fig.5 The roof flaking in tip-to-face area
(a) $P = 800$ kN, $s = 1$ m; (b) $P = 800$ kN, $s = 2$ m; (c) $P = 4400$ kN, $s = 1$ m; (d) $P = 4400$ kN, $s = 2$ m

roof flaking cannot occur; when P =2800 kN, the blocks of tip-to-face begin to fall but the falling arch height reaches minimum. When P = 800 kN, the height reaches maximum and the roof is in uncontrolling state.

5. CONCLUSIONS

(1) Working resistance of support and main roof rotation angle can influence the displacement distribution of the whole immediate roof if its thickness is 2 times or 4 times the mining height. If its thickness is 6 times the mining height, the upper layer displacement will be influenced only by main roof rotation angle while the lower only by working resistance.

(2) If thickness of immediate roof is 2 times or 4 times the mining height, the P-Δl curve is similar to a hyperbola which coincides with the results of tests in lab and field measurements. If its thickness is 6 times the mining height , the law of P-Δl curve is changed to gentleness, which means that support working resistance has little influence on roof convergence.

(3) In all types of immediate roof, the location and state of main roof cannot change the law of P-Δl. The P-Δl curve is the direct result of the interaction between support and immediate roof.

(4) Roof flaking in tip-to-face area mainly depends on working resistance and tip-to-face distance if thickness of immediate roof is 6 times the mining height. When the tip-to-face distance is within a certain scope, the falling arch height can be reduced and even the blocks in tip-to-face area can be kept in temporary balance by improving support's working resistance.

(5) The load acting upon mining face support comes from the weight of immediate roof and the given deformation pressure which depends on main roof rotation angle and mechanical characteristics of immediate roof. If the immediate roof is 6 times the mining height, the given deformation pressure can be partly or wholly absorbed by the immediate roof. The working resistance of sublevel caving hydraulic support may properly be improved in order to control roof flaking in tip-to-face area.

ACKNOWLEDGEMENT

The project is supported by National Science Found of China (59734090)

REFERENCES

Qian ,Minggao, etal. A further discussion on the theory of the strata behavior in longwall mining. *J. of China University of mining and Technology*, 1994 No.3, pp 1-9

Qian, Minggao etal. "S-R" stability for the voussoir beam and its application. *Ground Pressure and Strata Control*, 1994, No.3, pp 6-10

Wu , jian etal. Basic concept of determining resistance of the support with fully mechanized mining and top coal caving. *Ground Pressure and Strata Control*, 1995, No.3, pp 69-71

Qian, Minggao etal. Analyses of the key block in the structure of voussoir beam. *J. of China Coal Society*, 1996, No.4

Mining Science and Technology' 99, Xie & Golosinski (eds)© 1999 Balkema, Rotterdam, ISBN 90 5809 067 1

Theory research for the longwall top-coal caving technology

Jian Wu & Yong Zhang
China University of Mining and Technology, Beijing, People's Republic of China

ABSTRACT. For the last 15 years, the longwall top-coal caving (LTC) technology has been developing rapidly in China. But the theory research for LTC technology lag well behind the production. In this paper, the authors analyze and review the theory background of the LTC technology, outline the future research and development of LTC technology.

1. INTRODUCTION

A great development has been achieved on the thick-seam LTC technology in China during last 15 years. The practice has proven that LTC technology can achieve high output of longwall face and high economic benefit. However, comparing to traditional longwall slice mining system, some new problems are arising when LTC technology is introduced in production system, rock-strata control and safety technology.

(1). In most LTC faces, the working resistance of powered support doesn't increase rapidly as expected, in some case even decrease. How to explain the relationship between support and surrounding rock in LTC face? How to determine the direction and amount of load on the support in LTC face?

(2). How to explain that periodic caving is unremarkable in LTC face? How to establish new theory to describe the surrounding rock movement in LTC face?

(3). How to obtain both high recovery ratio and low dilution ratio in LTC face? How to understand the movement behavior and dynamic of the roof broken coal, rockmass and waste so as to upgrade recovery ratio? How to calculate the load amount and direction on the support in order to study the relationship between support and loose coal-waste movement?

(4). The safety technologies of LTC, such as fire prevention, dust control and anti-gas, have been greatly changed. Some safety technologies used in longwall slice mining face in the past need to be researched again.

So, solving these problems is an important duty for the mining engineers in China.

2. RESEARCH

Because most phenomenon of underground pressure are caused by mining influence, the research for LTC technology is the research for mining influence in fact.

2.1. *The research for top-coal damage and failure*

Comparing with immediate roof in slice mining face, top-coal in LTC face has the same space place but different mechanical property. Top-coal is a low strength and low elastic modulus medium because there are lots of inter-weakness in it. Within a big scope in front of coal wall, the inter-weakness can expand to be fracture under abutment, and top-coal has great change in mechanical properties. As a result, top-coal has a remarkable movement in front of the face wall, especially in softy coal seam (shown in Fig.1). The movement characteristics are:

(1) horizontal movement is dominant in the initial stage of top-coal movement,

(2) upper top-coal move priority to lower top-coal,

(3) vertical movement exceeds horizontal movement nearby the wall.

As a brittleness material, coal has a low tensile strength so that the coal failure type is brittle fracture due to tensile strength. The theoretical base for the research of top-coal movement is the fracture mechanics. Coal failure process has three stages: microcosmic damage stage, microcosmic fracture stage and macroscopic failure stage. The top-coal can be regarded as a damaged material. Damage stage of top-coal is developing with development of inter-weakness. The top-coal damage degree connects with top-coal movement:

$$\widetilde{E}_{lh} = (1 - \frac{\Delta S}{S_p})E \qquad (1)$$

where, l—the distance from the first moving point

L—the distance from the first moving point to failure point

ΔS—top-coal movement at the point which distance is l

S_p—top-coal movement at failure point

E—elastic modulus of coal

$\widetilde{E}_{lh}$—damaged elastic modulus of coal

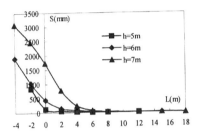

Fig.1 The relationship between
top-coal movement S and distance from wall L.

Varying with damage degree, top-coal mechanical properties have influence on surrounding rock. Because top-coal damage have great influence on support design, head face stability, top-coal caving level, gas emission and dust prevention. The research of top-coal damage mechanism and its influence on the production processes are the key problems to the LTC mining technology.

2.2. The research for the upper rock-strata balance structure

After coal extracted, roof rock collapse in goaf and fallen rock will support the upper rock stratum finally. The upper rock stratum form a certain balance structure in which two bearing-point are the coal in front of wall and compacted waste rock in goaf (shown in Fig.2). Coal damage status and waste rock compacting status have great influence on the parameters of upper rock-strata balance structure, and in turn the load of the cover rock-strata also has influence on damaged coal and compacted waste. That is to say, coal, waste, lower and upper rock-strata influence each other.

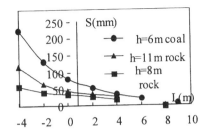

Fig.2 Upper rock-strata balance structure
load on top-coal and waste in goaf.

Firstly, the one-time mining thickness is very big in LTC face. The bigger one-time mining thickness is, the higher the height of roof rock which collapse to fill extracted space in goaf. The higher the height of upper rock-strata balance structure is, the bigger the span of upper rock-strata balance structure is. With the decreasing of upper rock-strata influence on LTC face, the periodic pressure phenomenon may relax. Secondly, top-coal damage in a large scope in front of wall decrease the load-carrying capacity of top-coal so as to extend abutment zone and move the abutment pressure peak ahead. So, the LTC face is within a decompression zone and far away from the abutment pressure peak. Thirdly, because bigger movement of top-coal occurs in front of wall while cover rock-strata have smaller movement in corresponding place, asynchro movement of top-coal and cover rock-strata cause top-coal apart from cover rock-strata (shown in Fig.3), in which rock load within upper rock-strata balance structure can't put pressure on supports through top-coal. Supports bear only weight of top-coal and part of collapsed waste so that working resistance decreases.

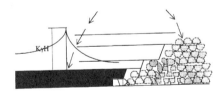

Fig.3 Top-coal apart from cover rock-strata.

Coal damage in a large scope and caving of top-coal cause violent movement of cover rock-strata, and deformation and failure of cover rock-strata change the boundary condition of top-coal. After every mining cycle, the relationship between top-coal and surround rock will adjust dynamically. The larger deformation theory of the nonlinear mechanics are used to establish a new mechanics model for upper rock-strata balance structure:

$$\sigma_c = \frac{k(1+2\mu)}{4(1-\mu)}(1-\cos\theta) \qquad (2)$$

where σ_c —tensile strength of rock of rock-layer forming balance structure

μ—Poisson's ratio

θ —average rotating angel of balancing rock-layer

E—elastic modulus of rock

k—cons.

2.3. The research for top-coal caving process

Caving process in LTC face dominates the efficiency, recovery ratio, coal quality. The possibility of caving, caving range and the movement behavior of top-coal, as well as sustainable mining are the important contents in the LTC research. Top-coal caving degree depends on top-coal breaking and damaging degree. On the other hand, the loose caving top-coal also has great influence on un-caved top-coal damage degree. The main research tasks of top-coal caving process are:

a. To get better understanding of movement behaviors of loosed top-coal under boundary restraining during caving process,

b. To study the distribution behaviors of float coal in goaf so as to decrease coal-loss and provide basis for fair prevention in goaf,

c. To seek for more effective ways to upgrade the recovery ratio and decrease refuse ratio,

d. To study the load behavior of caving top-coal on support in order to furnish reference for support design.

2.4. The research for roadway support

It is a precondition for normal production in LTC face to keep two roadways in good condition. It has been proven that the movement law and ground behaviors of roadway in LTC faces are different from those in slice mining faces. The abutment zone surrounding the both side roadways is bigger in LTC face than that in slice mining face, which leads a large deformation of roadways, and make roadway supports very difficult. The key problem is how to establish roadway support design principles. Moreover, whether to use pillarless mining technology in LTC face or not is anther important aspect in the future research.

2.5. The research for gas control

The practice shows that using LTC technology creates favorable conditions for gas control, but also bring some disadvantageous. Some LTC faces are in the high gassy seam in China. Change of mining system, change of abutment, and full development of crack have great influence on gas emission and gas or coal burst.

The research shows that outflow of gas in LTC face is relatively smaller than that in the slice mining face. But because of gas pressure decreasing caused by crack in top-coal, lots of absorbed gas turn into free gas so that absolute gas emission is much more than that of slice mining faces. Lots of gas enter the goaf and accumulate in the top-goaf to form a huge gas-storehouse. When cover rock-strata collapse, collapsed rock press gas to enter into face. The gas must be sucked out through some special measures in order to avoid it enter into faces.

Moreover, because the scope of mining influence in LTC face is bigger than that in slice mining face, gas from adjacent coal seam or rock-strata may enter the working face though rock crack so as to increase gas emission. It must be solved by some special methods.

2.6. The research for dust fall

Because the automatic mining technology hasn't been achieved in LTC faces, the workers will still have to work at the place with high dust concentration for a long time. How to reduce mining dust is another very important research project in China. Although with more dust sources in LTC face, dust stemming from support advance is in the main intake, dust become more serious. Besides the general dust prevention by water-cloud method, water injection is the main method used in China. So the main research contents are:

a. how to improve hydrophilicity of coal in order to upgrade efficiency of water injection?

b. how to establish the relationship between the efficiency of water injection and parameters of water injection, such as water pressure, injecting rate, injecting time and water amount of unit mass of coal, etc.

c. how to seek for the water moving behavior through crack of coal so as to upgrade efficiency of water injection by using top-coal crack developing laws under abutment.

Moreover, water injection, as a dust fall method, has a certain influence on top-coal caving level. So more researches should be conducted in the future.

2.7. The research for fire prevention

Most of the thick-seams in China are spontaneous combustion coal. With slice mining, the top slice has little chance of spontaneous combustion because the goaf is asphyxiated soon. While next low slices are mined out, spontaneous combustion is easier to occur because the goaf is exposed again. This condition is relieved in LTC faces. In LTC faces, with regular rate and without air leakage chance, spontaneous combustion will avoid generally except during the pre-cut and the time of withdrawing equipment. In LTC faces, the main spontaneous combustion comes from top-coal in roadways. With lots of developed cracks in top-coal over roadway, top-coal may collapse for un-reasonable and un-immediate support. Top-coal around caving zone may become potential trigger of spontaneous combustion.

So the research for spontaneous combustion prevention must be connected with the roadway support and the reasonable caving-process parameters.

3. CONCLUSIONS

As a new mining technology, the LTC technology involves many aspects such as production, safety, economic benefit, and resources protection, etc. The research for new technology and its theory base will be very useful for the sustainable development of LTC technology. LTC technology is a complicated system, in which strata stress and strata movements are dominant factors. Re-distribution of abutment and upper rock-strata balance structure are the main cause for the ground behaviors and key.issues for future research. More attentions are also paid to aspects such as safety, resources recovery ratio, and the sustainable mining.

REFERENCES:

Wu Jian, Theory researches and practices of top-coal caving technology in China, Coal Journal, 1991
Wu Jian, Three technology models of top-coal caving in flat dipping thickness seam, Coal Journal, 1994

Mining Science and Technology' 99, Xie & Golosinski (eds) © 1999 Balkema, Rotterdam, ISBN 90 5809 067 1

Modeling coal outbursts using non-linear instability simulation model

Guojing Zhao, Wei Jiang, Jifeng Zhu & Jinru Che
China University of Mining and Technology, Beijing, People's Republic of China

ABSTRACT: The outburst phenomenon of rock or coal outbursts has been studied by computing the buckling modes and analyzing bulging instabilities. The model takes into consideration both geometric non-linearity and elastic-plastic properties of rock mass. The simulation runs of the FEM based model are in good agreement with the actual outbursts reported in this paper.

1. INTRODUCTION

Rock or coal outbursts frequently arise in underground mining with increasing of the working depth. It is a disastrous collapse that outburst occurs suddenly in rock excavation or coal mining. The most important features of rock or coal outburst based on field observations are:

1. Outburst often happen in some special situation. That is: the ground stresses are in a higher level. Usually the coal seam is located deeper than $500 - 600m$ or large tectonic stresses exist. The coal seam and its roof and floor strata have a higher stiffness, and the Yong's module is about $8.0 \times 10^3 MPa - 1.0 \times 10^6 MPa$.

2. Very large displacements and deformations bulge to the empty space and result in local failure near the free surface of coal body when coal-body outburst happened.

3. Coal-body outburst is a sudden occurrence. It is a dynamical process essentially.

Because of the complexity in question, the researchers initially got some results from energy equilibrium to explain qualitatively the mechanism of rock outbursts, but it is difficulty to be applied in practical situations. The analysis models for outbursts are still on the static models so far, though it is a dynamic process. In the recent years, scholars have paid more attention to the theory of instability for rock or coal outburst. There are basically two kinds of mechanical model concerning the coal outburst in the technical reports: limiting equilibrium model and surface instability model. The first one emphasised that rock outburst is a process of losing the limiting static equilibrium (Lippmann 1978, 1987); and the second supposed that outburst is a surface instability problem (Vardoulakis 1984; Zhao & Li 1991) based on Biot's free surface instability theory (Biot 1965; Bazant 1991).

In this paper, based on the viewpoints above from practical observations, the rock outburst is regard as a bifurcation phenomena, which means that the original state of equilibrium is stable no longer and the rock body get a new equilibrium state (buckling mode or bulging instabilities) in which the large strains and displacements are inevitable, so that it must result finally in local failure of rock body. The solutions of bulging instabilities provides a new way for the study of outbursts.

Obviously, to establish a instability theory based on the framework of small deformation is difficult. For rock or coal body the large deformation state of bulging instabilities should be described by geometric non-linearity and elastic or plastic model. So the following hypothesis about rock or coal body were adopted :

1. Rock or coal body is in equilibrium and not in failure caused by reaching its strength when its critical state of stability not occur;

2. Rock body is homogeneously isotropic, and obeys the Mohr-Coulomb elastic-plastic properties.

By analysing the bulging or buckling modes of rock-body structures from a static state in which the initial stresses including the perturbation of mining excavation is considered, a unified non linear instability model for rock or coal body outburst has been established and the computational method using non-linear FEM to simulate outburst has been developed in this paper.

In section 2, a new instability theory of rock or coal outburst is established. Some practical examples of outburst in coal mining shown in section 3. It indicates that the computational results of the instability theory has a well consistency with practical situation. The results obtained in this paper could provide a theoretical basis and numerical ground for the prediction and treatment of rock or coal body outbursts in coal mining or other rock engineering.

2. INSTABILITY MODEL FOR OUTBURST

For three dimensional rock-body, the following formulations of instability model are based on the non linear displacement-strain relations and Mohr-Coulomb elastic-plastic constitutive equations.

2.1 Stress equilibrium equation

According to the momentum conservation low of solid mechanics, the equilibrium equations in Cartesian co-ordinate system, for the currant configuration of coal-body are

$$\frac{\partial}{\partial x_j}\sigma_{ij} + f_i = 0 \qquad (2.1)$$

where σ_{ij} is stands for the Cauchy stress tensor, and f_i the body forces. The equilibrium equations in terms of the Kirchhoff stress are

$$\frac{\partial}{\partial X_k}\left[S_{lk}\left(\delta_{il} + \frac{\partial u_i}{\partial X_l}\right)\right] + f_i = 0 \qquad (2.2)$$

where X_k is Lagrangian co-ordinates of coal body; u_i is the components of displacement, and S_{lk} is Kirchhoff stress tensor which is conjugate with the Green strain:

$$E_{kl} = \frac{1}{2}\left(u_{k,l} + u_{l,k} + u_{m,k}u_{m,l}\right) \qquad (2.3)$$

2.2 Constitutive equations

In large deformation theory, coal-body is considered as subelastic materials by which the constitutive equations in the increment form can be written as

$$\Delta S_{ij} = D_{ijkl}\Delta E_{kl} \qquad \square \qquad (2.4)$$

where ΔS_{ij} is the Kirchhoff stress increment, ΔE_{kl} is

Green strain increment, D_{ijkl} is the tangential stiffness tensor referred to Kirchhoff stress and Green strain. (2.1)-(2.4) constitute the mechanical model of coal- body as which has large deformation but the plasticity may be negligible. When the plastic properties need be considered, the inelastic effects must be counted in the tangential stiffness tensor. For this instance, the Mohr-Coulomb plasticity was advised to the model: A piecewise linear curve is adopted as the Mohr-Coulomb hardening property and the Mohr-Coulomb criterion is

$$s + \sigma_m - c\cos\phi = 0 \qquad (2.5)$$

where

$$s = \frac{1}{2}\left(\sigma_1 - \sigma_3\right), \quad \sigma_m = \frac{1}{2}\left(\sigma_1 + \sigma_3\right) \qquad (2.6)$$

and c and ϕ are the cohesion and the material angle of friction respectively.

2.3 Boundary conditions

By adding some suitable boundary conditions to the model discussed above, a boundary value problem applied to the outburst problems is put forward. The boundary conditions for coal-body are:

$$u_k = \bar{u}_k \quad on \ \Gamma_u \qquad (2.7)$$

$$\sigma_{ij}n_j = \bar{T}_k \quad on \ \Gamma_S \qquad (2.8)$$

2.4 The instability analysis

To solve outburst problems of various structures formed by rock or coal excavation, some initial state including the perturbation of mining excavation must be considered so that the buckling modes or bulging instabilities can be analysed. There are initial stresses including zero surface forces at free surfaces of the coal-body in the initial state which is solved by FEM counting the geometric non-linearity and elastic-plastic properties of coal-body discussed above .

The analysis of bulging instabilities will be performed in accordance with the initial state to solve eigen-modes. The minimum second order work principle on the bases of the Second Law of the Thermodynamics can be used as the criteria of outburst happening in three dimensional body under finite deformation (Bazant 1991; Zhao & Li 1998). The second order work of the deformed body is

$$\delta^2 W = \frac{1}{2}\int_V \delta\left(dS_{ij}\right)\delta\left(dE_{ij}^L\right)dV$$

$$+ \int_V \sigma_{ij}^0 \left[\delta \left(dE_{ij} \right) - \delta \left(dE_{ij}^L \right) \right] dV \qquad (2.9)$$

where σ_{ij}^0 is the initial stress, $E_{ij}^L = e_{ij}$ is the linear part of Green strain; and $\delta^2 W$, a scale of second order, represents the total energy of the structure in the increment of load step. The instability criteria are as the follows:

1. $\delta^2 W > 0$, the equilibrium state of the structure is stable;

2. $\delta^2 W < 0$, the equilibrium state of the structure is unstable. It implies that the outburst of coal-body happened owing to instability.

3. When the equilibrium state of the structure is critical, the second order work vanishes

$$\delta^2 W = 0 \qquad (2.10)$$

So (2.10) can be applied as a unified stability criterion under geometric non-linearity. If substituting e_{ij} and σ_{ij} for E_{ij} and S_{ij} respectively, one can obtain stability criterion for a special cases under the small deformation condition.

The mechanical model of the outburst for rock or coal body due to instability consists of the equations (2.1)-(2.8) and (2.10).

2.5 The FEM equations

Making the structure discretization with FEM, the eigen-value equation in matrix form instead of (2.10) can be formulated as:

$$\left(K_0 + \lambda_i K_\delta \right) v_i = 0 \qquad (2.11)$$

where K_0 is the stiffness matrix depending on the initial state; K_δ is the stiffness matrix caused by the load increment of the step of instability analysis, according to (2.9), it is relevant merely to the linear part of the increment of Green strain; λ_i and v_i are the bulging eigen-values and the corresponding eigen-modes respectively. Only the first eigen-pair (smallest eigen-value) has the meaning in practice. Finally, because the structure's load level corresponding to the computed eigen-value is lower than the failure level which is a strength condition caused the coal-body to collapse, the occurrence of outburst owing to instability was concluded.

3. OUTBURST EXAMPLES

3.1 Outburst in one side of a transport roadway

Some transport roadway happened outburst in one side which consists of coal pillar. The geometric and material properties of that coal pillar are given in Table 1.

Table 1. Geometric and material properties of coal pillar

Property	Value
Thick	2.4 m
Width	18 m
Young's modulus	9 GPa
Poisson's ratio	0.21
Uniaxial strength	32MPa
Maximum plastic strain	0.02

In the numerical simulation for this pillar by FEM, we use two rigid surfaces, the upper one represents the roof strata and the lower one the bottom, to exert the loads to the coal seam because of the higher stiffness of the roof strata and the bottom of the coal seam. The friction actions between the roof strata and the top of coal seam is considered.

The first step of the analysis is exerting pressure of about 30 Mpa on the upper surface to make a initial statically deformed state of the coal pillar. In Figure 1. the undeformed mesh consists of dashed lines and the deformed one consists of dark lines in which a small rotation representing the leaning subsidence of the roof has been performed. The distribution of the vertical stress component from the first step of analysis is shown in Figure 2. A area of stress concentration which is not reaching the compression strength in triaxial stress state appears at the position of about six metres from the left free surface.

Figure 1. The mesh and displacement field of the coal pillar

Figure 2. The mesh and displacement field of the coal pillar

The second step of the analysis is to compute the bulging load and the corresponding buckling modes representing approximately that the collapse style of coal pillar owing to instability. The diagram of the first buckling mode obtained in the second step has shown in Figure 3. It is obvious that the top part of coal-body near the left free surface (the side of larger

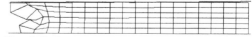

Figure 3. The first buckling mode of the coal pillar

subsidence) bulged to the empty space. This picture is well consistent with the scene of the collapse in field observations. From this example it is evident that the occurrence of coal-body outburst in the situation of the high level ground stresses and high stiffness of the roof and bottom is owing to bulging instability. Two important factors influence on the outburst can be concluded as:

1. Easy sliding between the roof strata and the top of the coal seam is necessity for outburst;
2. The position of the occurrence of outburst is sensitive to the leaning subsidence of the roof strata.

3.2 Outburst in a long-wall face

The possibility of outburst in coal seam of thickness of 10 m (one-time mining thickness) is shown in this example. Plane strain model which reflects the situation of the middle of work face is still adopted for conciseness. To consider the distance in front of the coal face as infinite, the edge that is 60 m away from the working face is constrained in horizontal direction. The material properties of coal-body are given in Table 2:

Table 2. Material properties of the coal face

Property	Value
Young's modulus	15 GPa
Poisson's ratio	0.25
Uniaxial strength	36MPa
Maximum plastic strain	0.02

Figure 4. The mesh and displacement field of the coal face

All of the remainder of the analysis model are similar to the former example. Making the first and second steps of analysis, the mesh of this example and the displacement field obtained by the first step with compression load of 35 Mpa is shown in Figure 4; and the graph of the first buckling mode is drawn in Figure 5.

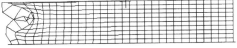

Figure 5. The first buckling mode of the coal face

From Figure 4. and Figure 5, we can see that coal-body outbursts owing to instabilities in the similar conditions are possible.

CONCLUSIONS

1. A mechanical model and non-linear FEM have been developed to analysis the rock outburst owing to instabilities in this paper.
2. The numerical simulations are conducted based on the data from practical measurements. The computational results indicated that the theory model is accordance with the scene of the collapse in field observations. It is evident that the theory model of instability are reasonable to simulate rock or coal body outburst.
3. The results in this paper provide a theoretical basis and a numerical ground for the prediction and treatment in coal mining or other rock engineering.
4. Plane strain examples given above may be going simplification too far for rock outburst in practice. Hence three dimensional models to matching the geometrical complexity in many situations to be simulated should proceed study further.

REFERENCES

Biot, M. A. 1965. *Mechanics of Incremental Deformation*, NewYork:
  John Wiley & Sons Inc.
Bazant Z P. 1991. *Stability of Structures*. New York Oxford: Oxford
  University Press.
Lippmann, H. 1978. The mechanics of translatory rock bursting.
  Advance in analysis of Geothechnical Instabilities. Univ. Waterloo
  SM Study No. 13: 25-63.
Lippmann, H. 1987. Mechanics of "bumps" in coal mines: A
  discussion of violent deformations in the sides of roadways in coal
  seams. *Appl.. Mech. Rev.* 40(8): 1033-1043.
Vardoulakis, I. 1984. Rock bursting as a surface instability
  phenomenon. *Int. J. Rock Mech. Min. Sci. Abstr.* 21(3): 137-144.
Zhao, G. J. & Li, Y. 1991. The finite element modelling of the coal
  and rock outburst in mining, *Proc. of The 2nd Int. Symp. Min. Tech.
  & Sic.* 2: 905□907. Xuzhou China.
Zhao, G. J. & Ding, J. H. 1998. Non-linear Numerical Simulation on
  Coal-Methane Outburst as a Coupling Instability Problem of Solid-
  Fluid Biphase Media. *J. CUMT.* 8(1): 103-107

Mining Science and Technology '99, Xie & Golosinski (eds) © 1999 Balkema, Rotterdam, ISBN 90 5809 067 1

Comprehensive analysis of coal mine roadway deformations in weak rock

Yaodong Jiang & Shiliang Lu
China University of Mining and Technology, Beijing, People's Republic of China

Changhai Liu & Xianhui Wang
Xiaokang Coal Colliery, Tiefa Coal Administration, People's Republic of China

ABSTRACT: The comprehensive understanding mechanism of large deformation of mine roadway in weak rocks will lead to design support system both more safely and economically. In this paper, the mechanisms related to the large deformation and failure of roadway excavated in soft rock mass are discussed, a quantitative relationship between the deformation of surrounding rocks and the support pressure of roadway is proposed on the basis of field investigations and laboratory tests.

1. INTRODUCTION

The large deformational behavior of coal mine roadway in weak rocks is becoming a crucial problem as mining progresses deeper and has been receiving great attention recently in the field of rock mechanics and mining engineering. Better understanding mechanism of large deformation of mine roadway in weak rocks will lead to design support system both more safely and economically. Although there are some proposals for definition of large deformation mechanism of roadway in literature worldwide, the problems in question have not been well understood yet. A wide variety of reports about unsuccessful roadway support in deep mines in week rock are a proof of this fact.

There are two basic types of instabilities of roadway: joint-controlled instabilities and stress-controlled instabilities. The jointed-controlled instabilities of roadways are due to the pre-existing discontinuous structures that divide the rock mass into discrete blocks; under the effect of gravity the blocks or wedges may fall or rotate towards the roadway at the crown and the walls. The joint-controlled instabilities are the only ones which occur in many underground tunnels or roadways at low depth or in hard rocks (Fairhurst, 1990, Panet, 1993), their deformations or movements can be analyzed by a simple graphic method using stenographic projection or by more sophisticated two or three dimensional distinct element methods (DEM or UDEC).

The stress-controlled instabilities of roadways, on the other hand, are induced by high initial stresses which exceed the strength of the surrounding rocks but the effect of discontinuities in the rock mass is secondary (Panet, 1993). The deformation behavior of the surrounding rocks in failure zone depends strongly on the rock properties. In hard rocks, field and laboratory investigations shows that macro-fracture or shear-bands will occur in the surrounding rocks, Vardoulakis (1988) explained this phenomenon as a bifurcation. Usually the homogeneous failure zone does not exist around the roadway, and deformation of roadway does tend to cease even for an unsupported roadway. The rocks around the roadway always retain some residual strength to resist tangential stresses. In very harder rocks, the spalling, popping and even rockbursting may happen due to a sudden energy release when the resulting stresses exceed the strength of the surrounding rocks.

In soft rocks the deformation behavior of the surrounding rocks in failure zone is quite different from that in hard rocks. A large deformation of the surrounding rocks undergoes slowly and gradually, and the large homogeneous failure zone continues to develop around the roadway as time lasts. For example at Xiaokang coal colliery in China, the mining depth is 600 meters, and the strength of rock surrounding roadway is about 14.5 Mpa. The deformation of roadway was 23.5 *mm* on the first day after roadway was excavated, and reached 385.7 *mm* in the first month, the deformation rate was still about 0.04 *mm* per day after 950 days excavation. The large deformation of roadway seriously influenced the normal production of the coal mine. In this paper, we focus our attention on the features and mechanism of large deformation of roadway in weak rocks and its support designs.

Figure1 The failure of roadway at Xiaokan coal mine, the total amount of deformation is over 4000*mm*.

2. FEATURES OF ROADWAY LARGE DEFORMATION IN WEAK ROCK

2.1 *Unconventional concepts about the instability of roadway*

A majority of large deformation of mine roadway in weak rock is dominantly contributed by the broken rock mass located in the failure zone surrounding the roadway. So first step in question is to study the failure mechanism of surrounding rock mass. However, when the rock mass surrounding a mine roadway is stressed to the level at witch failure initiates, the subsequent behavior of the rock mass is extremely complex and falls into the category of problems which sometimes are classed as "indeterminate". In other words, the process of fracture propagation and the deformation of the rock mass surrounding the roadway are interactive processes which cannot be represented by a simple set of equations. The challenge is also arising in this study about some traditional concepts. For example, rock strength is classically viewed as a strict material property, but this perception is relaxed here and failure is considered as a system property. This also leads to rethink about the traditional concept of "stability" or "instability". Conventionally "failure" means not "stability" or "instability", however anyone who has visited a deep level mine will be familiar with the sight of fractured rock surrounding the mine

roadways and yet these roadways are accessible and clearly have not failed. Different from traditional terms, stability should be judged here to be acceptable when the deformation of the rock mass is controlled and when the support elements are not overstressed.

2.2 *The features of roadway large deformation in weak rock*

Mine roadway fall into two categories: (a) main roadways which usually have a life of several years, and (b) service roadways for coal face which seldom have a life in excess of 12-24 months. Based on the extensive survey of coal mine roadway in China, the main deformational features of roadway in weak rocks are:

- Large excavating deformation;
- Long-time effects of deformation;
- Violent floor heave;
- Large borcken zones around the tunnel;
- Large re-deformation due to support failure, rehabilitation, water saturation and the like.

2.3 *Main factors related to roadway large deformation*

Many researchers have studied the major factors related to large deformation of road in weak rocks without mining influence during last decade. These factors are:

- The uniaxial stength of rocks, σ_c <25 Mpa;
- Competency factor, $S=\Delta H/\sigma_c$ =0.5;
- The tangential strain of the roadway wall, $\tau=u/R=0.5$;
- Rockmass quality degree, RQD<25%;
- Natural water content, w=25%;
- The thickness of broken zone, t=300;

3. THE MECHANISM OF ROADWAY LARGE DEFORMATION IN WEAK ROCKS

Coal mine roadways are always excavated in sedimentary rocks. The rocks surrounding roadway usually are stratified structure including a number of cracks and joints. The strength of surrounding rocks largely depends on the developing degrees of the cracks and joints.

In order to understand the interrelation between the structure of surrounding rockmass and the mechanism of roadway large deformation, we have conducted a large number of model studies using equivalent material on the roadway model test rig with tri-axial servo system designed by authors (Jiang, 1993). Figure 2 and figure 3 show the typical curves of floor heave verse loads, creep curves of floor heave for different rockmass structure in model tests respectively.

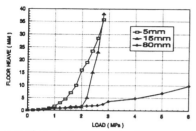

Figure 2 The typical curves of floor heave versus loads for different rockmass structure in model tests

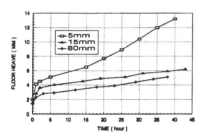

Figure 3 The creep curves of floor heave for different rockmass structure of floor in model tests

According to the field investigation and laboratory model test, a majority of large deformation of mine roadway in weak rock, which takes place slowly and gradually described as squeezing phenomenon, is dominantly contributed by the broken rock mass located in the failure zone surrounding the roadway. Although the broken & weak rocks in the failure zone still remain some strength and cohesion initially, both strength and cohesion will decrease with the increasing deformation, and they will eventually be exhausted if the support pressure is not sufficient enough. The progressive failure of the rocks around roadway results in a large increasing of rock volume due to rotating, relative sliding and deformation between the individual rock pieces within the zone. At this point the roadway becomes intrinsically unstable and brings about the large deformation of the roadway. As a result roadway finally loses its stability and will collapse spontaneously. The research shows, however, a small value of radial pressure at the roadway wall (or a residual cohesion) will suffice to ensure stability of the roadway at some value of deformation. This illustrates the value of roadway

support or reinforcement of the rock around the periphery (e. g. by frame support or rock bolting)

The further research proves that the large deformation of roadway in weak rocks features as bifurcation phenomena, which means at some critical state the deformation process does not follow its 'straight ahead' continuation but turns to an entirely different mode. Typical example of such a 'spontaneous' loss of homogeneity is shown in figure 4.

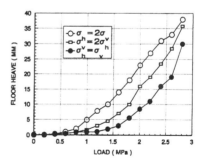

Figure 4 Bifurcation of large deformation of roadway in weak rocks

The features of roadway large deformation in weak rocks are also dependent on the ground stresses of far field. In the following laboratory tests, three models were designed to study the influences of ground stress state on floor heave.

(a) The vertical stress is higher than the horizontal stress ($\sigma_v = 2\sigma_h$)

(b) The horizontal stress is higher than the vertical stress ($\sigma_h = 2\sigma_v$)

(c) The hydrostatic pressure state ($\sigma_h = \sigma_v$)

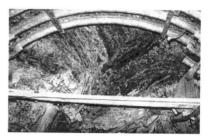

Figure 5 Curves of floor heave versus load in different ratio of σ_h over σ_v

All other parameters are the same. The curves of floor heave versus load for the three cases are shown in Figure 5. The test results, as shown in Figure 6,

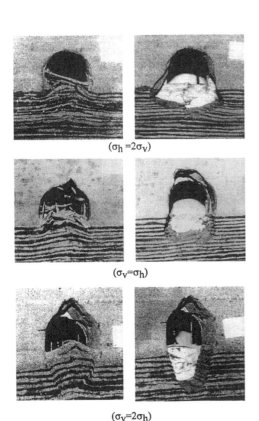

$(\sigma_h = 2\sigma_v)$

$(\sigma_v = \sigma_h)$

$(\sigma_v = 2\sigma_h)$

Figure 6 Failure mode of floor heave in model test

demonstrate that the floor heave is likely to occur under the three cases, while the failure zone of floor heave is different. When the horizontal stress is higher than the vertical one, the failure zone is a triangle. The vertical stress is higher, the failure zone is ladder-shaped. For the hydrostatic case, the failure zone is semicircle.

4. AN EMPIRICAL RELATIONSHIP BETWEEN CONVERGENCE OF ROADWAY AND SUPPORT PRESSURE

A number of observations and measurement of the relationship between deformation of roadway and support pressure in soft or extra-soft rocks have been carried out in China. Figure 7 is an empirical relationship between deformation of roadway and support pressure on the basis of field investigations and laboratory tests (Lu and Jiang, 1998). The results of surveying reveal that the primary function of support is to control the inward displacement of roadway and to prevent the progressive loosening of the broken rocks. The installation of rockbolts, shotcrete lining or steel

sets cannot prevent the failure of the surrounding rocks subjected to overstressing. The support pressure has significant influences on amount of roadway deformation in soft rock and does play a major role in controlling large roadway deformation.

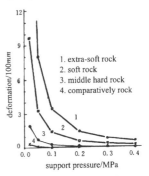

1. extra-soft rock
2. soft rock
3. middle hard rock
4. comparatively rock

support pressure/MPa

Figure 7 an empirical relationship between deformation of roadway and support pressure

5. CONCLUSIONS

A majority of large deformation of mine roadway in weak rock, which undergoes slowly and gradually, is dominantly contributed by the broken rock mass located in the failure zone. The instability of roadway excavated in weak rocks behaves as bifurcation, which also occurs in hard rocks.

The stability of roadway in weak rocks should be judged to be acceptable when the deformation of the surrounding rocks is controlled and when the support elements are not overstressed.

The support pressure has significant influences on amount of roadway deformation in soft rock and does play a major role in controlling large roadway deformation.

REFERENCES

Fairhurst, C, 1990. Deformation, yield, rupture and stability of excavations at depth in rock. *Rock at Great Depth* (Eds: Maury and Fourmaintraux), pp 1103-1114.

Jiang, Y D, 1993. The study of mechanism and control of floor heave. *PhD thesis*, China University of Mining and Technology.

Jiang, Y D, *et al*, 1995. Deformation, failure and stability in squeezing rocks, Proceedings of underground operators conference organized by the Australian Institute of Mining and Metallurgy, Kalgoorlie, Western Australia, November, PP.63-68.

Lu, S L and Jiang, Y D, 1998, The control of support resistance over strata around roadway in soft rock, Rock and Soil Mechanics, Vol.19, No.1, P1-6, (in Chinese)

Panet, M, 1993. Understanding deformation in tunnels, in *Comprehensive Rock Engineering* (Ed: J A Hudson), 1:663-690.

Mining Science and Technology'99, Xie & Golosinski (eds)© 1999 Balkema, Rotterdam, ISBN 90 5809 067 1

Modeling transient pressure at the bottom of coal seam gas well

Wei Jiang
China University of Mining and Technology, Beijing, People's Republic of China

Detang Lu
China University of Science and Technology, Hefei, People's Republic of China

ABSTRACT: Based on the non-balanced sorption model, this paper proposes mathematical modeling of coal seam gas in both quasi-steady flowing state and non-steady flowing state. It further reports on studies of the relationship between non-dimensional pressure and time at the well-bottom assuming the round closure and definite pressure with constant flow rate of coal seam gas.

1. INTRODUCTION

The reservoir of coal seam gas is a kind of double-pore media seams consisted of cleats and coal body. The coal seam gas exists in the coal body and the cleats in three ways: absorbed gas, free gas and dissolved gas. 90% of coal seam gas exists in the coal seams in the way of absorption. Because of the small pore diameter within the coal body, the main method of gas movement from coal body to cleats is diffusion. Generally, the diffusion of absorbed gas accords with Fick diffusion principle. The process that the coal seam gas flows from coal seams to shaft can be expressed as: with the continuous exploration of water in coal body, the seam pressure will decrease, and the gas absorbed by coal seam will be desorbed and diffuse into cleats. Then the gas in the cleats will seep into the shaft. For this reason, the movements of coal seam gas include two processes: diffusion and seepage flowing.

2. MATHEMATICAL MODEL AND PHYSICAL MODEL

The sate of gas absorbed in coal body can be mathematically expressed in two ways: the balanced absorption and non-balanced absorption. According to the non-balanced absorption model, this paper first supposes that the diffusion of gas in coal body is quasi-steady state and non-steady state Under this suppose, then seeks for the movement function of the coal seam gas. Furthermore, under some given conditions and the inversion of Laplace transforms,

the non-dimension real-space pressure function of the well-bottom is also obtained

Regarding the quasi-steady state flow, suppose coal seam gas is single phase, the cleat pressure satisfies the following equation according to the continuity equation and real gas principle.

$$\frac{1}{r_a}\frac{\partial}{\partial r_a}(r_a\frac{p_a}{\mu Z}\frac{\partial p_a}{\partial r_a}) = \frac{\Phi_a C_a p_a}{3.6KZ}\frac{\partial p_a}{\partial t} + \frac{p_{sc}T}{3.6KT_{sc}}\frac{\partial V}{\partial t} \quad (1)$$

In this equation, V is the concentration of the coal seam gas; subscript a means relating physical factors in the cleats; sc stands for the normal state physical factors.

Under the stable conditions, the gas concentration in the cleats and coal body can be achieved from following equation:

$$\frac{\partial V}{\partial t} = \frac{6D\pi^2}{R^2}(V_E - V) \quad (2)$$

In this equation, D is the diffusion coefficient of coal seam gas; R is the outer radius of the coal body; V_E is gas concentration at balance state.

Using Laplace transforms, the equation (1) and equation (2) can be simplified and the following equations can be obtained:

$$\frac{1}{r_D}\frac{\partial}{\partial r_D}(r_D\frac{\partial \overline{m}_D}{\partial r_D}) = \omega s \overline{m}_D - (1-\omega)s\overline{V}_D \quad (3)$$

$$s\overline{V}_D = \frac{1}{\lambda}(\overline{V}_{ED} - \overline{V}_D) \quad (4)$$

where s is Laplace variable.

Under the condition of quasi pressure, non-balanced unstable flow, considering coal body as a spherical media, the non-dimension equation can be described as:

$$\frac{1}{r_D}\frac{\partial}{\partial r_D}\left(r_D\frac{\partial m_D}{\partial r_D}\right)=\omega\frac{\partial m_D}{\partial t_D}-\frac{1-\omega}{\lambda}\frac{\partial V_D}{\partial r_{Di}}\Big|_{r_{Di}=1} \quad (5)$$

$$\frac{1}{r_{Di}^2}\frac{\partial}{\partial r_{Di}}\left(r_{Di}^2\frac{\partial V_D}{\partial r_D}\right)=\lambda\frac{\partial V_D}{\partial t_D} \quad (6)$$

$r_{Di}=r_i/R$ is non-dimension radial distance of coal body; $V_D=V-V_{ic}$ is non-dimension concentration of coal seam gas; r_i is the radial distance of coal body; R is the sphere radius; V and V_{ic} are the coal seam gas concentration under random pressure p and original pressure p_{ic} respectively; ω is reservoir space ratio; λ is non-dimension co-flowing coefficient, t_D is non-dimension time.

Applying Laplace transforms to equation (5) (6), the following equations can be achieved after simplification:

$$\frac{1}{r_D}\frac{\partial}{\partial r_D}\left(r_D\frac{\partial \overline{m}_D}{\partial r_D}\right)=f(s)\overline{m}_D \quad (7)$$

$$f(s)=\omega s+\frac{(1-\omega)}{\lambda}\alpha\left[\sqrt{\lambda s}\coth(\sqrt{\lambda s})-1\right]$$

Regarding non-balanced absorption model, no matter under quasi-stable state or non-stable state, a similar equation can be converted to equation (7) but $f(s)$ is different. If the shaft storage effect caused by working and hydraulic press crack is considered, the general result of equation (7) can be expressed by the first and second kind of Bessel function $I_0(x)$ and $K_0(x)$, under the condition of finite round seam, fixed pressure, closed round seam and infinite seam with fixed flow rate.

According to the inner boundary conditions and outer boundary conditions, the constants in the general solution can be determined and the final solution for the non-dimension well-bottom pressure in Laplace space can be obtained after simplification:

$$\overline{m}_{WD}=\left\{K_0\left(\sqrt{f(s)}\right)+MI_0\left(\sqrt{f(s)}\right)\right.$$
$$\left.+S\sqrt{f(s)}\left[K_1\left(\sqrt{f(s)}\right)-MI_1\left(\sqrt{f(s)}\right)\right]\right\}$$
$$\times\left\{C_D s\left[K_0\left(\sqrt{f(s)}\right)+MI_0\left(\sqrt{f(s)}\right)\right]\right.$$
$$\left.+(C_D sS+1)\sqrt{f(s)}\left[K_1\left(\sqrt{f(s)}\right)-MI_1\left(\sqrt{f(s)}\right)\right]\right\}^{-1} \quad (8)$$

where C_D is non-dimension shaft storage constant;

S is surface factor; $\overline{m}_{WD}$ is non-dimension quasi pressure in Laplace space; R_{eD} is non-dimension outer radius.

Applying Laplace inversion to equation (8), the curve of non-dimension well-bottom pressure vs. time under different time and different coefficients can be achieved. In order to simplify the surface factor and conditions of shaft storage, the combined coefficient $C_D\ e^{2S}$ can be applied. The $\overline{m}_{WD}\sim t_D/C_D$ and $\dfrac{d\ \overline{m}_{WD}}{d(t_D/C_D)}\sim\dfrac{t_D}{C_D}$ are obtained as log-log curve of quasi pressure and its derivative. The curve is showed in figure 1.

3. RESULT ANALYSIS

According to the above deduction, the log-log coordinate figure of non-dimension well-bottom quasi pressure and non-dimension time can be achieved. Four control parameters are included in figure 1: combined parameter ($C_D\ e^{2S}$), storage ratio (ω), co-flowing coefficient (λ), absorption parameter of coal seam gas (α). The shape of the curve is dominantly dependent on these four factors.

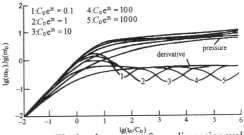

Figure 1. The log-log curve of non-dimension well-bottom pressure

Figure 1 is the log-log curve of non-dimension well-bottom pressure in coal seam gas well with quasi-stable flowing and its derivative under the condition of $\omega=0.5$, $\lambda=10^5$, $\alpha=10$ and different $C_D\ e^{2S}$. In this figure, the curve of quasi pressure comes together with the derivative curve when $t_D/C_D\to0$. A part of straight line with slope value 1 happened at the beginning of the curve. This means the existence of shaft storage. $C_D\ e^{2S}$ determines the persistence time of shaft storage. The greater the value, the longer the persistence time and the higher the curve. According to the figure, for all $C_D\ e^{2S}$ when $t_D/C_D\to\infty$, the curve of the well-bottom non-dimension pressure derivative becomes flat and inclines to straight line with the value of 1/2.

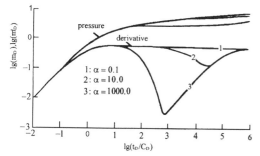

Figure 2. The log-log curve caused by different absorption factors (α) under quasi-stable flow

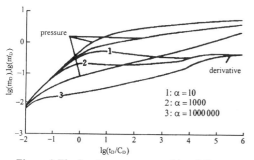

Figure 3 The log-log curve caused by different absorption factors (α) under unstable flow

Figure 2 and figure 3 show the affection of the log-log curve caused by different absorption factors (α) under the conditions of quasi-stable flow and unstable flow respectively. The parameters in these figures are $C_D e^{2S} = 1$, $\omega = 0.5$, $\lambda = 10^5$. The persistence time of the beginning straight line with deviation of 1/2 in the derivative curve is mainly affected by α. The greater α is, the earlier the derivative curve deviates the straight line and the greater the deviation extent is. According to the figures, the log-log curve of quasi-stable flow is different from the unstable flow's curve, and the quasi-stable flowing curve presents derivative curve of "V" shape.

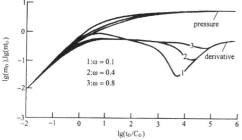

Figure 4. The log-log curve caused by different ω under quasi-stable flow

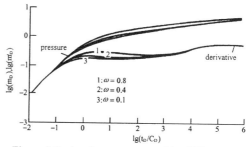

Figure 5 the log-log curve caused by different ω under unstable flow

Figure 4 and figure 5 show the affection of the log-log curve caused by different ω when $C_D e^{2S} = 1$, $\lambda = 10^5$, $\alpha = 10$ under the situations of quasi-stable flow and unstable flow respectively. From the figures, we can find the affection of ω is as same as that caused by α. However, ω's affection is a little bit smaller than α's affection. To the quasi-stable flow log-log curve, ω affects the depth of the "V" shape of the curve, the smaller the ω, the deeper the "V" shape. To the unstable flow log-log curve, the smaller the ω, the lower the derivative curve.

Figure 6 and figure 7 show the affection of the log-log curve caused by different co-flowing coefficient (λ) under the situations of quasi-stable

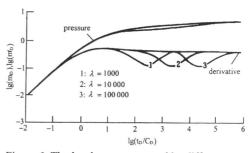

Figure 6. The log-log curve caused by different co-flowing coefficient (λ) under quasi-stable flow

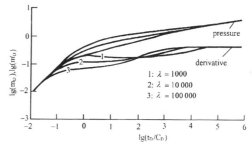

Figure 7. The log-log curve caused by different co-flowing coefficient (λ) under unstable flow

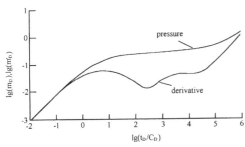

Figure 8. The quasi pressures and the derivative 's
log-log curve with the finite round closure

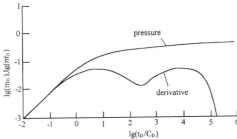

Figure 9. The quasi pressures and the derivative 's
log-log curve with the quasi-stable flow

flow and unstable flow respectively. The parameters
of the curve are $C_D e^{2S} = 1$, $\omega = 0.5$, $\alpha = 10$. When α
= 10, we can find in the figures that the main
affection caused by λ is the time which will lead the
derivative curve deviate or decline to the straight
line. To the quasi-stable flow log-log curve, the
bigger the λ, the later the "V" shape presents. To
unstable flow log-log curve, the bigger the λ, the
later that the derivative curve inclines to the straight
line with the value of 1/2.

The condition of the above discussions is the
specialty curve of infinite seam. To the finite seam,
the non-dimension log-log curve of well-bottom will
change when the boundary condition affects the
well-bottom pressure. Figure 8 and Figure 9 are the
quasi pressures and the derivative 's log-log curve
with the situation of the finite round closure as well
as the quasi-stable flow in the seam with definite
pressure. The parameters of the curves are $C_D e^{2S} = 1$,
$\omega = 0.5$, $\lambda = 10^5$, $\alpha = 10$, $R_{eD} = 150$. From the
figures, the curves of the pressure and derivative go
up under the condition of closed seam, and turn into
tangents with the slope value 45°. Regarding the
seam with definite pressure, the log-log curve
becomes flat, and the derivative turns to zero
quickly.

4. CONCLUSIONS

1. By replacing quasi-pressure in the Langmuir
equation, a new coal seam gas absorption model is
proposed.

2. Supposing absorption coefficient (α) is
constant, and its value is determined by m_{ic}..

3. Giving the quasi-stable and unstable flow
equations of the non-balanced absorption model in
Laplace space, the numerical result is achieved by
using the method of Laplace inversion.

REFRENCES

Anbarci, K. & Ertekin, T.A. Comprehensive study
of pressure transient analysis with sorption
phenomena for single-phase gas flow in coal
seams. *SPE*, 20568
King, G.R. & Ertekin, T.A. 1986. Numerical
simulation of the transient behavior of coal seam
degasification wells. *SPEFE*.
Stehfest, H. 1970. Numerical inversion of Laplace
transforms. *Communication of ACM*, 47-49
Langmuir, I. 1916. The constitution of fundamental
properties of solids and liquids. *Journal of
American Chemical Society*, 38
Lu, Detang. 1998. Bottomhole transient pressure
calculation of the coal seam gas well with
constant flow rate. *Natural Gas Industry*, Vol.18,
no. 3, pp34-39. (In Chinese)

Mining Science and Technology' 99, Xie & Golosinski (eds) © 1999 Balkema, Rotterdam, ISBN 90 5809 067 1

Study of the motion of overlying strata in top-coal caving system by finite deformation mechanics

Bao Li & Qiang Fu
China University of Mining and Technology, Beijing, People's Republic of China

Yuhuai Wang
North China Mining College, Beijing, People's Republic of China

ABSTRACT: The motion character of main moving strata, which plays a leading role to the whole overlying strata, is studied by means of traditional elastic and plastic theories as well as finite deformation mechanics. The concept of finite deformation beam is used to explain the mechanical behavior of longwall top-coal caving area and forecast the stress distribution in the process of mining. The method to discriminate the stability of finite deformation strata is also proposed in this paper.

1. INTRODUCTION

Coal is the main energy resource in China which remains about 70% of the total energy resources consumed until 2030[1]. The thick coal seam accounts for 44.1% of the total coal reserve. In the beginning of 1980s, longwall top-coal caving (LTC) technology was introduced to take place of slice mining technology in the mining of thich coal seam. As the mining space in LTC system is enlarged, the movement of overlying strata is increased and the ground pressure behavior also becomes different. Then, what is the motion character of overlying strata in LTC system? Deep and comprehensive studies on this subject has not been undertaken as yet.

2. ANALYSISES OF SUPPORT RESISTANCE AND FIELD TEST OF TOP-COAL AND ROOF IN LTC SYSTEM

2.1 Analysis of support resistance

The earliest observation on ground behavior, movement of top-coal and roof was launched successfully in China University of Mining and Technology (Beijing). The observation in Tonghua Daoqing mine was completed in 1987. The observation in Yangquan No.1 mine was taken in 1989. After these, many observations of ground behavior and movement of top-coal and roof were made in Micun, Chaohua, Wulan, Xingtai, Weijiadi and Xieqiao mine etc.

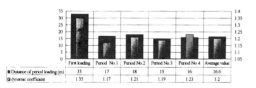

	First loading	Period No.1	Period No.2	Period No.3	Period No.4	Average value
■ Distance of period loading (m)	33	17	18	15	16	16.6
▨ dynamic coefficient	1.35	1.17	1.21	1.19	1.23	1.2

Fig.2.1 Distance of period load and layout of its dynamic load coefficient

The analysis results of support resistance indicate that most of the resistance strength in LTC system of thick and extra-thick seams is less than that of the first layer in slice mining system. The distribution of maximum resistance strength in Wulan 5321 LTC face is shown in Fig.2.1.

There is an obviously periodical change, but the maximum period load and distance are not very large, and the dynamic load coefficient is only 1.2 or so. The active point of composite force on support by top-coal and roof goes ahead. The increased area of abutment goes far from the face. Further analysis indicates that the main source of support load comes from the above loose coal and part of broken rock mass (shown in Fig.2.2).

Fig.2.2. Relation of top-coal and cover strata and support

2.2 Observation on the movement of top-coal and roof in different mining areas

The data of field test indicate that the moving quantity and the increased are different in similar position of cover strata of LTC system when the seam conditions are different. As the increasing of strength, the scope of abutment and the moving course and the quantity all become small (shown in Fig.2.3). Comparing the tested strata of Micun 10m high and Xieqiao 11.5m high and Yangquan11m high indicates that the moving speed in Micun is the biggest with the quantity of 3000mm.

Although the position in Xieqiao is a little higher and it is also in soft seam, but the first moving point is near the face as the seam thickness is small. The character of 11m position of mid-hard seam in Yangquan is very similar with that of 22m position in Xieqiao. So, the active of moving in mid-hard seam is far from that of soft seam.

2.3 Several opinions on movement of top-coal and roof in LTC system

1) The upper of top-coal moves earliest, and then the lower top-coal and immediate roof.

2) The positions of first moving point are different in different seams. There is no balance structure of rock biting in top-coal and the lower strata under the action of lateral force.

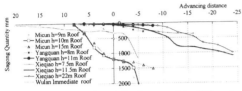

Fig.2.3 Comparison of roof displacement in different areas and different position

3) Top-coal less than mid-hard falls in front of the hinged point of support. Immediate roof falls as mining after the top-coal falls. The falling position of thick sandstone or limestone roof is behind that of soft strata.

4) The main roof, which can form boom structure in slice mining system, will be broken in front of the coal wall in LTC system. The distance between falling point and coal wall is different in different rock characters. The balance structure is changed to the hard and thick strata in higher position. So, concepts of immediate roof and main roof in LTC system are different with those of the traditional concepts in slice mining system. For example, the limestone roof was considered as main roof in $15^{\#}$ seam of Yangquan when slice mining method was applied. But it fell as mining and had no characters of main roof when LTC method was applied.

3. MODEL TESTS

Extensive simulation model tests indicate that the strata with high integral strength in thick and extra-thick seam of LTC system fall on the fashion of macro-integral bending (shown as the light-colored curve in Fig.3.1). The above comparatively low strength strata bend accordingly. Although the structure is shown as integral moving, it is actually a compound body of many irregular broken rocks (shown in fig.3.2).

Fig.3.1 Integral bending of cover strata in LTC system

Fig.3.2 Beam structure composed by irregular broken rock

Because of the early large falling and disintegrating of the lower coal and roof , the front supporting point of the structure goes ahead. The softer and thicker the seam is and the softer and more unstable the lower strata are, the higher the structure in overlying strata are and the larger the space is and the smaller the direct affect on mining area is.

4. ANALYSIS OF VARIATION OF POSITION AND FORM IN OVERLYING STRATA

4.1 Foundation of the analysis model of position and form in cover strata

Based on analysis of extensive field test data and results of simulation model, a deeply multi-angle research on the moving characters of overlying strata in LTC system is made by applying the traditional elastic and plastic analysis methods and the basic ideas of finite deformation mechanics and mathematics statistics. An analysis model of the motion form of overlying strata in LTC system is founded according to the typical concrete situation in Weijiadi mine (shown in Fig.4.1).

Fig.4.1 Motion form and relation of the definite deformation cover strata in LTC system

The deformation character of overlying strata in LTC system is different from that of slice mining. It is difficult to form the structure as the mining height and irregular falling height are all enlarged. The structure can only come into being in higher ,stronger and thicker strata, which is called definite deformation strata (shown in Fig.4.1).

The driving strata behave as gradually bending, even sagging and continually developing in macroscopic as the working face moves forward. The curvature is increased accordingly as mining space is enlarged. When strength in tensile face amounts to the limit, the weak face is open, the lump is turned and the integral bending of the beam is presented

4.2 Foundation of mathematics model of position and form in overlying strata

Based on the above analytical model of position and form in overlying strata and the results of field test data, a series of typical data is summarized by means of statistics and mechanics theories. These data are only the vertical deformation due to the limit of side in the axial direction. A sets of parameter equation, which can simultaneously describe the space state of overlying strata , is given by means of mathematics method.

$$\omega(x) = \left\{ \frac{s}{s_0} \cdot \sum_{k=0}^{6} \alpha_k \left[(x + 2 \cdot s) \frac{s_0}{s} \right]^k + \chi \right\} \cdot \eta \cdot 10^{-2} \quad (4.1)$$

Where, $\chi = s \cdot \left(1 - \frac{s}{s_0}\right) + 10^{-3} \cdot l(l_0 - l) \cdot \ln\left(\frac{l}{l_0}\right)$ is a modifying parameter of the deformation on all kinds of mining conditions; □ is the affecting coefficient of mining space, generally 1.8-2.0; S is the calculation fracture distance of the main structure which is related to the character (structure and strength) and size(thickness H_z etc.) of strata, the calculate total thickness H_s of following strata, the thickness H_j and falling character of immediate roof, seam conditions (thickness and strength, etc.), technical conditions(advancing speed V_g of face and caving distance) etc. It is a synthetic coefficient of the moving characters of overlying strata:

$$S = \psi(\bar{H}_m, \sigma_m, \Delta H_j, H_z, \sigma_z \cdots)$$

a_k — coefficient function of the equation, (x) ;
l — advancing speed of face on the time t , m;
n — numbers of overlying strata fracture : $l = n \cdot S$.

The variation law of position and form of the finite deformation beam is indicated in function (4.1) basically. The key problem is the value of the synthetic coefficient S. If the S is different, the curve described by function $f(x)$ is different too (shown in Fig.4.2. The bigger the S is, the larger the affected scope and the stronger the mechanics character of the beam are, and vice versa.

The linear relation in Fig.4.3 can be obtained by the analysis of S and the maximum quantity of $f(x)$:

$$s = 2.9131 \cdot \omega(x)_{max}$$

Assumed the calculate total thickness of the immediate roof under the finite deformation beam is δh_j, the coefficient of gob filling is K; the mining height in LTC system is h_m; the recovery ratio is μ; the expansion coefficient of loss coal is K_2; the not filling height under the beam is □h. Then:

$$\Delta h = \Sigma h + h_m - [K \cdot \Sigma h + h_m \cdot (1-\mu) \cdot K_2]$$

The maximum sagging amount $\omega(x)_{max}$ of the beam is equal to the not filling height of the gob, that is, $\omega(x)_{max} = \Delta h$, Thus:

$$s \approx 2.9131 \cdot \Delta h \qquad (4.2)$$

4.3 Application of the rational mechanics in the analysis of the overlying strata in LTC system [a]

Assumed micro-unit is 2m long (the beam is 35-140m long) according to the rule that micro-unit must be much smaller than the object. As the compressive strength of rock mass is very high, and the longitudinal compression is very small compared to the length of beam, the longitudinal deformation can be ignored. In order to

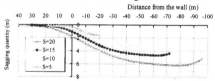

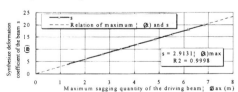

Fig.4.2 Moving character of the beam in different S

Fig.4.3 Linear relation of coefficient s and the maximum sagging quantity

simplify the study, the horizontal displacement of neutral stratum is restricted, that is, it can only rotate and move along the vertical direction (shown in Fig.4.4).

The micro-unit ▭1234 of the beam in Fig.4.4 can be changed to ▽1'2'3'4' after moving and rotating. The lines in the unit are also rotated besides the horizontal u and longitudinal displacement v of the points.

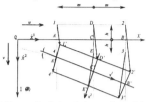

Fig.4.4 Dynamic finite deformation of the beam

Supposed fixed line Cartesian coordinates(X^1, X^2) (equal to that of $(X, (x))$). The co-moving coordinates (x^*, x^*) is also a fixed line Cartesian coordinates(equal to the fixed one). The co-moving coordinates is (x^*, x^*) after the deformation of micro-unit.

The lines of matrix X_c indicate the beginning coordinates of the four points $DBEA$. The coordinates of the neutral point C' is $(x_c, \omega(x_c))$ after the deformation. The coordinates of four points $D'B'E'A'$ after the deformation can be indicated in X_c.

Matrix U_c indicates the displacements of the points in the unit: $U_c = X_c - X_c$.

Considering the situation of large displacement, the rotation axis is perpendicular to the deformation plane(X^1, X^2). As $\Gamma_{jk} \equiv 0$, So

$$\hat{u}^i\Big|_j = \sqrt{g_{(ii)} / g_{(ii)}}$$

$$\hat{u}^i\Big|_j = \frac{1}{\sqrt{g_{(ii)}}} \frac{\partial u^i}{\partial x^j} = \frac{\partial u^i}{\partial s^j}$$

As $ds^j = \sqrt{g_{jj}} dx^j$ is the length of co-moving coordinate (x^j) after the deformation, applying the theory of "Stokes-Chen", then,

$$F = S + R \qquad (4.3)$$

The finite stress and rotation of ▽1'2'3'4' after deformation is shown in Fig.4.4. Applying the method of finite difference then:

$$\Theta = \arcsin\left(\frac{\omega(x_c - m) - \omega(x_c + m)}{2\sqrt{4m^2 + [\omega(x_c - m) - \omega(x_c + m)]^2}} + \frac{\sin(\theta_c)}{2}\right) \qquad (4.4)$$

$$\hat{S}_{c1}^1 = 1 - \cos\Theta \qquad (4.5)$$

$$\hat{S}_{c2}^2 = \cos\theta_c - \cos\Theta \qquad (4.6)$$

$$S_{c1}^2 = \frac{\omega(x_c - m) - \omega(x_c + m)}{2\sqrt{4m^2 + [\omega(x_c - m) - \omega(x_c + m)]^2}} - \frac{\sin\theta_c}{2} \qquad (4.7)$$

$$\hat{S}_{c2}^1 = \hat{S}_{c1}^2 \qquad (4.8)$$

where, $\theta_c = arctg(\omega'(x_c))$;□— germain rotation angle (suppose counter clockwise from x_1 to x_2 is positive),

The above functions represent the stress of neutral point C'. But we are concerning the degree of outside curved surface. As the outside curved surface is tensed and the tensile strength of rock blocks is very small, the stability of driving stratum (strata) is affected by the

331

deformation degree of outside curved surface. The stress of outside curved surface is studied as the following according to the above results. The geometric relation of outside layer and neutral layer is shown in Fig.4.5.

As the finite deformation mechanics is based on the measurement of deformation degree by the physics length on time of micro-unit, the stress of D' in Fig.4.5 is:

$$\hat{S}_{d_1}^{~1} = \frac{h \cdot \Delta\theta}{\dfrac{2 \cdot m}{1 - \hat{S}_{c_1}^{~1}} + h \cdot \Delta\theta} + \hat{S}_{c_1}^{~1} \tag{4.9}$$

The line stress of outside surface of the finite deformation beam (along the layer) is:

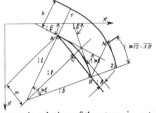

Fig.4.5 Geometric relation of the stress in outside curved surface and neutral layer.

$$\varepsilon = \varepsilon_d = S_{d_1}^{~1} \tag{4.10}$$

According to the relation of stress and strain of the large displacement finite deformation beam, strains of the neutral axis and outside of the beam are:

$$\sigma_c = \frac{2\mu \cdot (1 + \upsilon)}{8 \cdot (1 - 2\upsilon)} \hat{S}_{c_1}^{~1} \tag{4.11}$$

$$\sigma = \frac{2\mu \cdot (1 + \upsilon)}{8 \cdot (1 - 2\upsilon)} \varepsilon \tag{4.12}$$

$$\tau = G \cdot \gamma \tag{4.13}$$

The constants G, τ and γ in function (3.48) to (3.50) are respectively: $G = \dfrac{E_1}{2(1 + \upsilon)}$, $\mu = \dfrac{E_1}{4(1 + \upsilon)}$, $\gamma = \hat{S}_{c_1}^{~2}$.

4.4 Calculation and analysis of the position and form of the driving strata by finite deformation mechanics

According to functions (4.10-4.13) of axis stress, strain and angle strain and shear strain, the relation of stress and strain and geometric parameters (height, length, bending degree) of the beam can be calculated. In the calculation, Poisson's ratio $\mu = 0.1$, unidirection compress modulus $E_1 = 2 \cdot 10^3$ MPa, uniaxial tensile modulus $E_7 = 1000$ MPa.

The surface stress of different rock characters of the 8m high beam is calculated when the synthetic coefficient S is changed (shown in Fig.4.6)

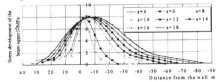

Fig.4.6 Relation of stress and coefficient S on the condition of same strata thickness

It can be seen that the maximum stress is nearly the same when different S is supposed on the condition of same beam height. This indicates that the maximum stress bearing by the beam is not changed on the condition of same rock character whatever the distance may be. The position of maximum stress goes ahead to the wall as S increases. The stress rapidly increases in front of the wall, and then slowly decreases on the gob after reaching the maximum. The phenomena represent that there is a great bending moment in front of the wall, and it amounts to the maximum above the face and then decreases gradually.

If the tensile strength is 5MPa, the fracture of beam in front of the wall is obviously (shown in the figure), Furthermore, the fracture positions are also different with different S. The bigger the S is, the further the fracture position will be.

As the law of configuration variation is decided by S basically, the curvature in same position is similar and the stress in upper surface is different and then the strain is also different whatever the thickness may be (shown in Fig.4.7).

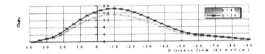

Fig.4.7 Comparison of stress in different thickness driving strata when s=20

5. CONCLUSIONS

The following conclusions can be preliminarily obtained by the analysis of overlying strata in LTC system:

1). The concept of finite deformation beam is supposed. It is the main cause of moving that affects the integral motion form of overlying strata in LTC system.

2). The model of finite deformation beam is founded preliminarily and the function f(x) of the configure variation of the beam is also put forward.

3). The mechanics model of finite deformation beam is founded to describe its strain, stress and germain integral rotation angle, and its mathematics equation is also given.

4). The opinion is proposed of linear relation of synthetic coefficient S and the maximum vertical displacement $H(x)_{max}$ of finite deformation beam, including the linear relation of S and not filling height h in the gob.

5). The idea is preliminarily given that the integral instability condition of the beam is the stress σ_c of neutral layer great than the limit strength.

A lot of studies on the ground behavior and movement of roof and top-coal are made in the paper. A new integral moving model, that is, finite deformation beam model, is put forward on the condition of limited base reference. The integral and continual character of the beam are stressed and analyzed quantitatively. A fundamental mechanics model of the stability condition is made preliminarily.

* The project is supported by National Science Found of China (59734090)

Mining Science and Technology' 99, Xie & Golosinski (eds) © 1999 Balkema, Rotterdam, ISBN 90 5809 067 1

Numerical simulation study of separations between strata

Hong Liu, Wei Jiang & Guojing Zhao
China University of Mining and Technology, Beijing, People's Republic of China

ABSTRACT: This paper presents a Discrete Strata Method that simulates the strata movement and separation between strata caused by mining. It regards the rock mass as stacked by discrete strata. A mixed finite element formula of the incremental mode is developed. New separations can be judged by the state stress between strata and then the separating process can be simulated dynamically by taking the continuous conditions away from the new separation. The numerical examples show that the Discrete Strata Method (DSM) is effective and feasible for the dynamical simulation of rock strata movement and deformation.

1. INTRODUCTION

In the process of strata movement caused by mining, each stratum will bend and fall at different degrees due to the different properties of strata. Therefore, the separation between strata takes place. The better understanding of the separation will have significant influence on, both theoretically and practically, the control of strata movement and subsidence, and the support design of the roadways. The separation is a developing process with mining. Numerical simulation is an important way for studying the developing process of the separation.

Because the separation is a problem that strata are changed from macro-continuum to discontinuum, it is difficult to simulate the separation by the general finite element method. According to the structural characteristics of the stratified rock mass, the rock mass can be regarded as stacked by discrete strata. The strata are joined together without separation by the continuous conditions of the displacement and the stress between strata. A mixed finite element formula of the incremental mode is developed by means of the Weighted Residual Method, which can analyze both the strata displacement and the stress between strata. New separations can be judged by the state stress between strata and then the separating process can be simulated dynamically by taking the continuous conditions away from the new separation. The numerical examples show that the Discrete Strata

Method (DSM) is effective and feasible for the dynamical simulation of rock strata movement and deformation.

2. COMPUTATIONAL MODEL & FORMULA

The stratified rock mass, as shown in Figure 1, can be considered to be stacked by discrete strata. The strata deform respectively, which may slip, or separate, but can not move into each other and the deformation within each stratum is continuous. Before the separation or slippage occurs, the displacement keeps continuous and so does the inter-stratum stress.

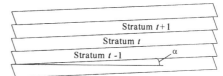

Figure 1 Stratified rock mass

Because of mining, rock strata move and deform. So the inter-stratum stress vary continuously. When the stress somewhere reaches the adhesive strength, the adhesion in this place is broken and the separation or slippage takes place between strata. At the interface of inter-stratum where the slippage, not separation, has occurred, the normal displacement and stress still remain continuous. Where the separation has occurred, the interface of inter-stratum becomes free surfaces.

Since mining, rock stratum movement and development of the separation are the dynamic processes, the incremental algorithm should be adopted. If all the displacement ($^{n}\mathbf{u}$), strain ($^{n}\varepsilon$) and stress ($^{n}\sigma$) after the nth step mining have been obtained, the displacement ($^{n+1}\mathbf{u}$) and stress ($^{n+1}\sigma$) after the next step mining should satisfy the following equation set:

In stratum:

Equilibrium equation:

$$A\left(^{n+1}\sigma\right) + {}^{n+1}\overline{\mathbf{f}} = 0 \qquad (1)$$

Relation of stress and strain:

$$\Delta\sigma = D\Delta\varepsilon \qquad (2)$$

On boundary: External force boundary S_σ:

$$\mathbf{n}\left(^{n+1}\sigma\right) - {}^{n+1}\overline{\mathbf{F}} = 0 \qquad (3)$$

Fixed displacement boundary S_u:

$$^{n+1}\mathbf{u} - {}^{n+1}\overline{\mathbf{u}} = 0 \qquad (4)$$

Interface of continuous displacement S_t^+:

$$\Delta\mathbf{u}_{t+1}^- - \Delta\mathbf{u}_t^+ = T(\Delta\mathbf{u}_{t+1}^- - \Delta\mathbf{u}_t^+) = 0 \qquad (5)$$

In the normal direction of slippage interface:

$$\Delta u_{t+1}'^- - \Delta u_t'^+ = 0 \qquad (6)$$

Here, $\mathbf{A}$ is the differential operator matrix; $\overline{\mathbf{f}}$ is the column matrix of body force; $\mathbf{D}$ is the elasto-plasticity matrix; $\mathbf{n}$ is the direction cosine matrix of boundary normal; $\overline{\mathbf{F}}$ is the column matrix of external surface force. For the matrixes above, the left superscript, n, is the mining step indicator and the right subscript, t, is the stratum indicator. The right superscript indicates the surface of stratum. The sign "+" means the upper surface of a stratum and the sign "-" means the reverse surface of stratum. $\Delta\mathbf{u}$ is the increment of displacement; $\Delta\mathbf{u}'$ is the increment of displacement at the normal and tangential. $\mathbf{T}$ is the transformation matrix from the general coordinate to the local coordinate of inter-stratum's normal and tangent. The exact form of each matrix can be found in Zienkiewicz (1977).

By means of the Weighted Residual Method, the integral equation that is equivalent to the above equation set can be obtained:

$$\sum_{t=1}^{m}\left[\int_{V_t} w_1\left(A\left(^{n+1}\sigma\right) + {}^{n+1}\overline{\mathbf{f}}\right)dV + \int_{S_t} w_2\left(\mathbf{n}\left(^{n+1}\sigma\right) - {}^{n+1}\mathbf{F}_t\right)dS\right]$$

$$+\sum_{t=1}^{m-1}\int_{S_t^+} v_t T\left(\Delta\mathbf{u}_{t+1}^- - \Delta\mathbf{u}_t^+\right)dS = 0 \qquad (7)$$

In the above integral equation, w_1, w_2 and v_t are arbitrary weighting functions; $\mathbf{F}_t$ is the surface force of stratum t; V_t means the space of stratum t; S_t

means the total surface of stratum t; S_t^+ means the interface that no slippage and separation have occurred between stratum t and stratum $t+1$; m is the number of strata.

Without losing generality, let $w_1 = -w_2$, and then integrate the first term of the first integration of the above equation by parts:

$$\sum_{t=1}^{m}\left[\int_{V_t}\left(Aw_2\left(^{n+1}\sigma\right) - w_2{}^{n+1}\overline{\mathbf{f}}\right)dV - \int_{S_u} w_2\mathbf{n}\left(^{n+1}\sigma\right)dS\right]$$

$$-\sum_{t=1}^{m}\left[\int_{S_t^+} w_2{}^{n+1}\mathbf{F}_t^+ dS + \int_{S_t^-} w_2{}^{n+1}\mathbf{F}_t^- dS\right]$$

$$+\sum_{t=1}^{m-1}\int_{S_t^+} v_t T\left(\Delta\mathbf{u}_{t+1}^- - \Delta\mathbf{u}_t^+\right)dS = 0 \qquad (8)$$

The inter-stratum stresses $^{n+1}\mathbf{F}_t^+$ and $^{n+1}\mathbf{F}_{t+1}^-$ are a pair of action and reaction force. They can be expressed by the $^{n+1}\lambda$, which is the normal and tangential stress of interface between strata.

$$^{n+1}\mathbf{F}_t^+ = -{}^{n+1}\mathbf{F}_{t+1}^- = T^{T}\,{}^{n+1}\lambda_t, \qquad t = 1,2,3,\cdots,m-1 \qquad (9)$$

Substitute the λ into equation (8):

$$\sum_{t=1}^{m}\left[\int_{V_t}\left(Aw_2\left(^{n+1}\sigma\right) - w_2{}^{n+1}\overline{\mathbf{f}}\right)dV - \int_{S_u} w_2\mathbf{n}\left(^{n+1}\sigma\right)dS\right]$$

$$+\sum_{t=1}^{m-1}\left(\int_{S_{t+1}^-} w_2 T^{T}\left(^{n+1}\lambda_t\right)dS - \int_{S_t^+} w_2 T^{T}\left(^{n+1}\lambda_t\right)dS\right)$$

$$-\int_{S_\sigma} w_2\left(^{n+1}\overline{\mathbf{F}}\right)dS$$

$$+\sum_{t=1}^{m-1}\int_{S_t^+} v_t T\left(\Delta\mathbf{u}_{t+1}^- - \Delta\mathbf{u}_t^+\right)dS = 0 \qquad (10)$$

Then introduce the formulas $^{n+1}\sigma = {}^{n}\sigma + \Delta\sigma$, $^{n+1}\overline{\mathbf{f}} = {}^{n}\overline{\mathbf{f}} + \Delta\overline{\mathbf{f}}$, $^{n+1}\overline{\mathbf{F}} = {}^{n}\overline{\mathbf{F}} + \Delta\overline{\mathbf{F}}$, $^{n+1}\lambda_t = {}^{n}\lambda_t + \Delta\lambda_t$ to the above equation:

$$\sum_{t=1}^{m}\left[\int_{V_t}\left(Aw_2\Delta\sigma - w_2\Delta\overline{\mathbf{f}}\right)dV - \int_{S_u} w_2\mathbf{n}\Delta\sigma dS\right]$$

$$+\sum_{t=1}^{m-1}\left(\int_{S_{t+1}^-} w_2 T^{T}\Delta\lambda_t dS - \int_{S_t^+} w_2 T^{T}\Delta\lambda_t dS\right)$$

$$-\int_{S_\sigma} w_2\Delta\overline{\mathbf{F}}dS + \sum_{t=1}^{m-1}\int_{S_t^+} v_t T\left(\Delta\mathbf{u}_{t+1}^- - \Delta\mathbf{u}_t^+\right)dS$$

$$+\sum_{t=1}^{m}\left[\int_{V_t}\left(Aw_2\left(^{n}\sigma\right) - w_2\left(^{n}\overline{\mathbf{f}}\right)\right)dV - \int_{S_u} w_2\mathbf{n}\left(^{n}\sigma\right)dS\right]$$

$$+\sum_{t=1}^{m-1}\left(\int_{S_{t+1}^-} w_2 T^{T}\left(^{n}\lambda_t\right)dS - \int_{S_t^+} w_2 T^{T}\left(^{n}\lambda_t\right)dS\right)$$

$$-\int_{S_\sigma} w_2\left(^{n}\overline{\mathbf{F}}\right)dS = 0 \qquad (11)$$

The total integral domain $V = \sum V_t$ is discretized into finite elements. Nodes and elements in all strata are numbered in a unified way. In the interface between strata, the nodes of the upper stratum and the beneath stratum must be arranged to form node pairs, which is the same to the elements. Thus the two nodes of a pair have different nodal number but same coordinates. Before the separation occurs, the two nodes will be joined together by constrain conditions. The increment of displacement and the inter-stratum stress are selected as the essential unknown functions, and are expressed approximately within an element in terms of interpolation function of their nodal value:

$$\Delta\mathbf{u}(x,y) = \sum_{i=1}^{n_e} N_i(x,y)\Delta\mathbf{u}_i = \mathbf{Na}^e \qquad (12)$$

$$\Delta\lambda_t = \sum_{k=1}^{n_s} \overline{N}_k \Delta\lambda_{tk} = \overline{\mathbf{N}}\mathbf{b}^e \qquad (13)$$

Here, n_e is the node number of an element; $\mathbf{a}^e$ is the column matrix of node displacement increment of an element; $\mathbf{N}$ is the matrix of displacement interpolation of an element; n_s is the node number on the interface of an element. $\mathbf{b}^e$ is the column matrix of inter-stratum stress increment of an element; $\overline{\mathbf{N}}$ is the matrix of inter-stratum stress interpolation of an element.

According to the Galerkin Method, the variations of unknown functions are chosen as the weighting function w_2 and v_t.

$$w_2 = \delta\Delta\mathbf{u} = N_j\delta a_j \qquad j=1,2,\cdots,n \qquad (14)$$

$$v_t = \delta\Delta\lambda_t = \overline{N}_k\delta b_k \qquad k=1,2,\cdots,n_c \qquad (15)$$

Here, n is the total node number obtained by discretizing the domain $V = \sum V_i$; δa_j is the displacement variations of the node j; n_c is the number of node pairs between stratum; δb_k is the variation of inter-stratum stress in the kth node pair. Because the approximate field functions are defined in elements, the formula (11) can be rewritten as the sum of element integral. Then formula (12), (13), (14), (15) and $\Delta\sigma = \mathbf{D}\Delta\varepsilon$, $\Delta\varepsilon = \mathbf{A}^T\Delta\mathbf{u}$, $\Delta\mathbf{u} = \mathbf{Na}^e$, $\mathbf{B} = \mathbf{A}^T\mathbf{N}$ are substituted into it. Considering the variation $\delta\mathbf{u}$ identically vanishes at the fixed displacement boundary S_u, the equation is obtained as follows:

$$\delta\mathbf{a}^T(\mathbf{Ka} - \mathbf{C}^T\mathbf{b} - \Delta\mathbf{P} - \mathbf{R}) - \delta\mathbf{b}^T\mathbf{Ca} = 0 \qquad (16)$$

In this equation:

$$\mathbf{K} = \sum_e \left(\mathbf{G}_a^{e^T} \int_{V_e} \mathbf{B}^T\mathbf{DB}dV\mathbf{G}_a^e \right) \qquad (17)$$

$$\mathbf{C} = \sum_{t=1}^{m-1}\left[\sum_{e_t^+} \mathbf{G}_b^{e^T} \int_{S_\lambda} \overline{\mathbf{N}}^T\mathbf{TN}dS\mathbf{G}_a^e \right.$$
$$\left. - \sum_{e_{t+1}^-} \mathbf{G}_b^{e^T} \int_{S_\lambda} \overline{\mathbf{N}}^T\mathbf{TN}dS\mathbf{G}_a^e \right] \qquad (18)$$

$$\Delta\mathbf{P} = \sum_e \mathbf{G}_a^{e^T} \int_{V_e} \mathbf{N}^T\Delta\overline{\mathbf{f}}dV + \sum_{e_\sigma} \mathbf{G}_a^{e^T} \int_{S_\sigma} \mathbf{N}^T\Delta\overline{\mathbf{F}}dS \qquad (19)$$

$$\mathbf{R} = \sum_e \mathbf{G}_a^{e^T} \int_{V_e} \mathbf{N}^T({}^n\overline{\mathbf{f}})dV + \sum_{e_\sigma} \mathbf{G}_a^{e^T} \int_{S_\sigma} \mathbf{N}^T({}^n\overline{\mathbf{F}})dS$$
$$- \sum_e \mathbf{G}_a^{e^T} \int_{V_e} \mathbf{B}^T({}^n\sigma)dV$$
$$- \sum_{t=1}^{m-1}\left[\sum_{e_{t+1}^-} \mathbf{G}_a^{e^T} \int_{S_\lambda} \mathbf{N}^T\mathbf{T}^T({}^n\lambda_t)dS \right.$$
$$\left. - \sum_{e_t^+} \mathbf{G}_a^{e^T} \int_{S_\lambda} \mathbf{N}^T\mathbf{T}^T({}^n\lambda_t)dS \right] \qquad (20)$$

In above formulas, $\mathbf{G}_a^e$ and $\mathbf{G}_b^e$ are transform matrixes by which the nodal displacement matrix and inter-stratum stress matrix of elements can be converted to the global displacement matrix and global inter-stratum stress matrix. Due to the arbitrary property of δa and δb, we can obtain:

$$\mathbf{Ka} - \mathbf{C}^T\mathbf{b} = \Delta\mathbf{P} + \mathbf{R} \qquad (21)$$
$$\mathbf{Ca} = 0 \qquad (22)$$

They can be united into a matrix equation:

$$\hat{\mathbf{K}}\hat{\mathbf{a}} = \begin{bmatrix} \mathbf{K} & -\mathbf{C}^T \\ -\mathbf{C} & 0 \end{bmatrix}\begin{bmatrix} \mathbf{a} \\ \mathbf{b} \end{bmatrix} = \begin{bmatrix} \Delta\mathbf{P} + \mathbf{R} \\ 0 \end{bmatrix} = \hat{\mathbf{P}} \qquad (23)$$

Here, $\mathbf{R}$ on the right side of equal is the residual force of the last step; $\Delta\mathbf{P}$ is load increment of the present step. Calculation of constraint matrix $\mathbf{C}$ is in the need of the present contact state of strata, which can be done by supposing the present state based on the last step. $\mathbf{K}$ is exactly the stiffness matrix of the general finite element method, and is symmetric and positive definite, so the $\hat{\mathbf{K}}$ is a symmetric matrix with zero diagonal element. The linear equation (23) can be solved with appropriate algorithm. Then $\mathbf{a}$ (increment of displacement in global coordinate) and $\mathbf{b}$ (increment of normal and tangential stress between strata) can be obtained. On the basis of displacement and stress of inter-stratum, a judgement is formed whether a new separation takes place at a certain place or old separation is closed. If so, the matrix C will be changed, then the next step of iterative computation will be performed until the matrix C need not be changed and the computation converges. Thus the strata movement and development of the separation between strata can be simulated dynamically.

3. EXAMPLE

In order to easily check the feasibility and validity of the algorithm, simulation of a simple example that

Figure 2 The deformation of strata after 56 meters have been dig out

simulates the movement of strata and several separations caused by mining is examined. Here the depth of coal seam is 80 meters underground and 2 meters thick with two layers of hard and thick rock 12 meters above it. So no separation is supposed to occur above these rocks. A plane strain model is set up along the mining direction, which is 160 meters long and 38 meters high with 8 layers of 160 elements numbered from the bottom to the top. The second layer is a coal seam and the rocks above the eighth layer is 50 meters thick represented by the pressure of .35MPa. First, the initial rock stress loaded on the boundary horizontally and vertically is simulated. From the second step on, the mining process is simulated, which starts from the seventh element on the left to the right with eight meters mining out each time. The increase in mining leads to the rise in the strata deformation. The slippage and separation occur between the third and the fourth layers (Interface 1) when 24 meters have been mined out and several separations take place

Table 1 Relation between the slippage/separations and the digging progress

Inter face		unit	mining progress				
			24m	32m	40m	48m	56m
5	slippage width	m			104	112	120
	separation width	m			0	0.5	0.5
	separation height	mm			0	0.1	0.1
4	slippage width	m		64	88	104	104
	separation width	m		8	24	28	48
	separation height	mm		38	112	311	722
3	slippage width	m		40	48	72	80
	separation width	m		8	24	36	48
	separation height	mm		27	40	115	82
2	slippage width	m		40	56	76	96
	separation width	m		16	24	32	40
	separation height	mm		33	66	137	210
1	slippage width	m	12	48	56	72	80
	separation width	m	8	16	24	32	44
	separation height	mm	1	3	30	40	55

(Interface 1, 2, 3 and 4) when 32 meters mined out, the largest of which is between the sixth and seventh layers with 38 mm wide and 8 meters long. When 40 meters are mined out, slippage takes place on all the interfaces with the characteristic of decreasing in size up to down. The roof fails when 56 meters have been mined out and the deformations are shown in Figure 2. The relation between the slippage or separations and the mining progress is shown in the Table 1.

The algorithm is proven by the example to be feasible and valid to calculate the strata movement caused by the mining, which especially can simulate the strata separation dynamically.

4. CONCLUSIONS

Compared with the Distinct Element Method which treats the unbroken rock mass as discrete block, the Discrete Stratum Method, which takes the stratified rock mass as the discrete strata adhering to one another by the stress between them, is more reasonable and realistic, and therefore is much more efficient. The DSM can not only deal with the discontinuous problem without a lot of contact check, but also have the advantage of domain decomposition algorithm, which can be processed in parallel. The algorithm put forward in this paper suggests that the displacement continuous constraint is put on the possible fracture and whether the fracture occurs or not can be judged by the constraint force. Along with the adaptive re-meshing technique, this algorithm may be developed into a general algorithm, which analyzes problems from continuity to discontinuity, and it may promote the development of the numerical modeling in rock mechanics.

REFERENCE

Zienkiewicz O C. 1977. *The Finite Element Method Third Edition.* McGraw-Hill, Inc.

Mining Science and Technology' 99, Xie & Golosinski (eds) © 1999 Balkema, Rotterdam, ISBN 90 5809 067 1

Influence of joint orientation on key block size distributions

Amarin Boontun
Mining Engineering Department, Chiang Mai University, Thailand

ABSTRACT: The numerical approach to probabilistic key block analysis is used to evaluate the influence of joint orientation on key block size distribution. A series of tests are conducted by variation of three joint sets. Since joint orientation is recognized by the dispersion of pole vectors (k-factor) so that they are varied. The other joint parameters are kept constant during the test. The relative angle between pole vectors (A) is tested at 90, 60, 30 and 0 degrees from each other. The concentration factor of pole vectors (k-factor) is varied for each different angle. The distribution of key block size is observed from each test. The results show that a unimodal-shaped distribution occurs when k-factor is reduced from 10,000 to 200. When joint plane orientations are random or k-factor is less than 200 the resulting key block size distribution fits a reverse J-shaped Weibull distribution. The fitted shape parameter of Weibull distribution reduces as k-factor is reduced. The angle between pole vectors has less influence on the key block size distribution than does the pole dispersion.

1 INTRODUCTION

Naturally, joints play an important role in dividing the rock mass into blocks. It has been known that the rock block size distribution is affected by the three main joint parameters that are joint orientation, joint length and joint spacing (Young et al. 1995, Young & Boontun, 1996). Since block theory was proposed by Goodman & Shi (1985), key block determinations are notable and widely used for engineering projects dealing with jointed rock mass. Most key block analysis provides the number and volume (size) of key block as a result. Joint parameters are expected to play a role on the size distribution of key block as well.

Both deterministic and probabilistic methods of block theory in rock analysis were advanced by a number of researchers. Each approach is different and uses different parameters to analyze the size of key block. An inaccuracy in prediction of key block size can be found if some significant parameters are omitted during analysis and may cause severe problem for the overall project. Therefore, it is useful to study the joint parameters that strongly affect the key block size distribution.

In this paper, the effect of joint orientation on key block size distribution is studied. The Discrete Region Key Block Analysis, a probabilistic numerical approach, is used as a tool to determine the size and number of key block. The key block size distribution is presented in the form of relative histogram for each test. The shape of key block size histogram is compared directly to evaluate the changes in key block size distributions.

2 JOINT SIMULATION

The Discrete Region Key Block Analysis (DRKBA) program developed by Stone (1994) is a numerical technique for solving the probabilistic key block analysis problems. The region around the excavation is represented by a three dimensional grid of disconnected discrete nodal points. The grid is assembled with standard finite element procedures, employing an eight-node cubic element with 28 connection bars. Each connection bar of the global connectivity matrix represents the space in the rock mass between two nodal points.

Joints are simulated as circular disc placed in the rock mass according to probabilistic distributions. The orientation of the joint normal (pole) vectors are simulated from a Watson bipolar distribution in which a concentration factor, k, is used to represent the concentration joint plane normal vector about some pole vector; u (Fisher et al. 1987). As the value of k-factor approaches zero the distribution approaches a random distribution defined on the unit sphere. As k-factor increases without bound, the distribution approaches a single point with orientation u. The joint length and spacing are characterized by Beta distribution; β (a,b,p,q). Therefore, the range of distribution can be easily controlled by using parameters a and b, and the shape of the distribution can be controlled using parameters p and q.

During each simulation, each connection bar in the matrix is checked to determine whether any joint in the rock mass passes between the two nodes associated with the connection bar. Removal of any such connection bars from the matrix isolates sub-structures within the matrix which correspond to individual rock blocks. These blocks are subsequently analyzed with the use of traditional block theory developed by Goodman & Shi (1985) to determine whether they are geometrically stable. The knowledge of node numbers within the independent matrix elemental block is enough to identify the shape, size or volume, and location of a rock block or a key block.

3 JOINT MODELLING AND TEST PROCEDURE

A finite element model was developed for a 250 feet × 250 feet × 250 feet cube rock mass consisting of 100 × 100 × 100 nodal points. The joint test model consisted of three joint sets with input parameters were tested on this rock model. The key block sizes were calculated on a vertical free face of this rock mass model as well.

The input parameters of each joint set were joint orientation, joint length, and joint spacing. Joint orientation was defined by the dispersion of pole vectors (k-factor) according to Watson bipolar distribution. Joint length and spacing were characterized by Beta distribution with uniform shape. The tests were conducted by varying the concentration factor of pole vectors (k-factor) and the angle between these joint pole vectors (A) while joint length and joint spacing were kept constant. The length and spacing of each joint set were fixed at 10,000 feet and 10 feet respectively.

The relative angle between joint pole vectors (A) of each joint set was identical, and was set at 90, 60, 30, and 0 degrees apart from each other. The concentration factor, k, was varied for each different angle (A) at k-factor values of 10,000, 700, 500, 300, 200, 100, 50, 20, and 5 as shown in Table 1. Each test was run for 100 simulations on each k-factor. The size or volume and the number of key block were calculated as a result.

4 RESULTS AND INTERPRETATIONS

4.1 Perpendicular joint sets: A=90 degrees

All test results are shown as histograms of key block

Table 1. A series of tests on variation of k-factor for each angle between pole vectors (A).

k-factor	A	(degrees)		
10,000	90	60	30	0
700	90	60	30	0
500	90	60	30	0
300	90	60	30	0
200	90	60	30	0
100	90	60	30	0
50	90	60	30	0
20	90	60	30	0
5	90	60	30	0

size (volume) and are used to fit for the distribution. Figure 1 shows the examples of histograms of key block size distributions for A equals 90 degrees with k-factor equals 10,000, 700, 300, 200, 20, and 5. The corespondent shape parameters of Weibull distribution for each k-factor are listed in Table 2. It was found that when k-factor is equal 10,000, discontinuous distributions are observed. Since the large k-factor value is essentially equivalent to a set of parallel joint plane which implies that less key blocks are generated. A unimodal shape can be seen as k-factor is gradually reduced from 700 to 200 and fitted shape parameters also reduced from 1.12 to 0.83.

When k-factor is less than 200 or joint plane is random, the resulting key block size distribution becomes slightly skews to the right and fits well with a reverse J-shaped Weibull distribution. The fitted shape parameter reduces to 0.66 when k-factor is reduced to 5. This means that more small key blocks are generated when joint plane is random.

The test results obviously indicate that the variation in a dispersion of pole vectors despite in a few degrees can alter the key block size distribution drastically.

4.2 Other joint sets: A=60 and 30 degrees

When the angles between pole vectors (A) are 60 and 30 degrees apart from each other, the key block size histograms from these sets are comparable to those from the perpendicular set. The example of histograms for A equals 60 degrees are shown in Figure 2. The discontinuous distributions are still observed when k-factor is equal 10,000. As k-factor is reduced from 700 to 200, the unimodal-shaped can be seen and fitted shaped parameter of Weibull distribution is reduced from 1.02 to 0.75 for A equals 60 degrees and from 0.89 to 0.75 for A equals 30 degrees.

When k-factor is reduced from 200 to 5 the histograms matched very well with the reverse J-shaped Weibull distribution and fixed with this shape with shape parameters of about 0.7-0.6 for A equals 60 degrees and about 0.6 for A equals 30 degrees

4.3 Parallel joint sets: A=0 degrees

Discontinuous key block sizes appear as results from parallel joint sets with k-factor ranging between 700 and 200. This event can be explained by k-factor value. When the k-factor value is between 700 and 200, joint planes are more random (not perfectly parallel) than higher k-factor value so that there are some intersections of these joint planes which generate some key blocks. The histograms of key block size closely match reverse J-shape Weibull distribution with about 0.6 shape parameter when k-factor is equal to 100 and became fixed at this shape when k-factor is reduced gradually to 5.

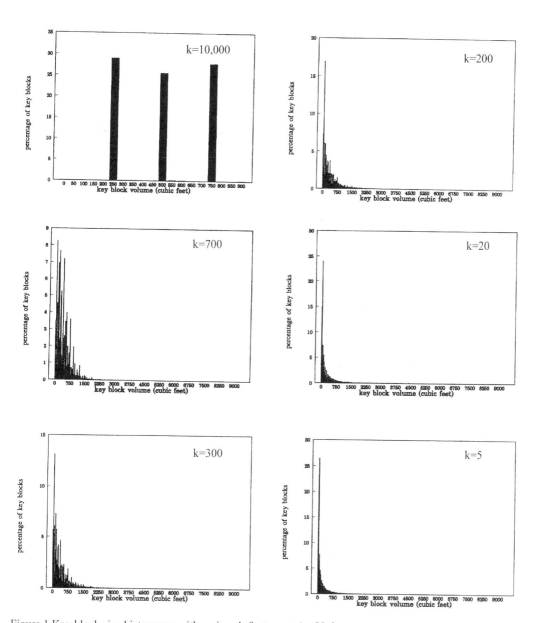

Figure 1 Key block size histograms with various k-factors at A= 90 degrees.

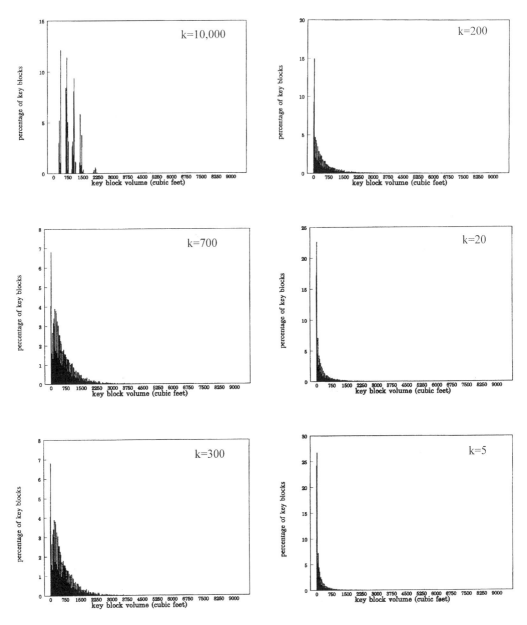

Figure 2 Key block size histograms with various k-factors at A=60 degrees.

5 CONCLUSIONS AND DISCUSSIONS

The following conclusions can be drawn based on the tests performed so far:

- The dispersion of pole vectors around the mean attitude of a joint set is a significant parameter influenced the key block size distribution. The key block size distribution is very sensitive to dispersion of joint plane pole vectors (k-factor) comparing to the relative angles among them. Variation in a dispersion of pole vectors can alter the key block size distribution drastically.

- The shape of key block size distribution alters when k-factor changes. The fitted shape parameter of Weibull distribution for key block size distribution decreases with decreasing k-factor.

Table 2 Fitted shape parameters of Weibull distribution for key block size distribution of different k-factors and A angles.

k-factor	(a) A=90	(b) A=60 degrees	(c) A=30 degrees	(d) A=0 degrees
10,000	N/A	N/A	N/A	N/A
700	1.12	1.02	0.89	N/A
500	1.02	0.92	0.83	N/A
300	0.89	0.83	0.76	N/A
200	0.83	0.75	0.75	N/A
100	0.74	0.68	0.63	0.59
50	0.69	0.65	0.58	0.59
20	0.66	0.63	0.57	0.57
5	0.66	0.63	0.60	0.60

- The unimodal shape of key block size distribution is observed when the concentration of joint pole vectors, k, is larger than 200. When joint plane is random or concentration factor of joint pole vectors is less than 200, key block size distribution fixes as a reverse J-shaped Weibull distribution with shape parameter of about 0.6.
- In any key block size analysis, the whole pole vector dispersion should be considered and applied through the probabilistic approach rather than deterministic analysis that employ only the mean attitude of poles in a joint set.

REFERENCES

Fisher, N.J., T.lewis & B.J.J.Embleton 1987. *Statistical Analysis of Spherical Data*. London: Cambridge University Press.

Goodman, R.E. & Gen-Hua Shi 1985. *Block Theory and Its Application to Rock Engineering*. Englewood Cliffs: Prentice Hall.

Stone, C.A. 1994. *A matrix Approach to Probabilistic Key Block Analysis*. Ph.D Dissertation. Michigan Technological University.

Young, D.S., Boontun, A. & Stone, C.A. 1995. Sensitivity Tests on Rock Block Size Distribution *Proceedings of the 35th U.S. Symposium on Rock Mechanics*: 849-853. Rotterdam: Balkema.

Young, D.S. & Boontun, A. 1996. Joint System Modeling for Rock Blocks. Proceeding of the 2nd *North American Rock Mechanics Symposium: NARM'96*: 1237-1244. Rotterdam: Balkema.

Mining Science and Technology' 99, Xie & Golosinski (eds) © 1999 Balkema, Rotterdam, ISBN 90 5809 067 1

Slake-durability behaviour of coal mine shales

Herryal Z. Anwar
R&D Centre for Geotechnology-Indonesian Institute of Sciences, Bandung, Indonesia, or Kyushu University, Fukuoka, Japan

H. Shimada, M. Ichinose & K. Matsui
Kyushu University, Fukuoka, Japan

ABSTRACT: In this paper, a quantitative discussion on the durability of coal mine shales is given on the basis of the results obtained by means of three different kinds of slaking tests. Vertical closure, the problem that arising due to slaking phenomenon in longwall mining, is also investigated.

1. INTRODUCTION

Shale is the most common sedimentary rock that is normally found in coal mines and may be categorised as a clay-bearing rocks. The engineering characteristic of shale has a wide variation, particularly due to of their resistance to contact even with small amount of moisture or to short-term cyclic process by wetting and drying. Shale which contains clay mineral group such as montmorillonite generally has a high slaking potential. When this type of shale contacts with water will have an undesirable properties such as a decreasing of shear strength, and as a consequences it may result in cracking and disintegration of the rock mass. Ichinose (1988) shows that with the increasing of montmorillonite content the strength of the rocks will decrease when wet.

This paper demonstrates the slaking phenomenon of the coal mine shale. The influence of structure and mineralogy in coal mine shale is also discussed. Another evidence to determine the sensitivity of the shale to water is found by examining their water absorption behaviour. The slaking behaviour of the shale is measured by using the ISRM standard method and the observation method. In this experiment coal mine shales which are from Ikeshima and Miike Colliery, Japan and Ombilin Colliery, Indonesia, are used.

2. VERTICAL CLOSURE

In longwall mining shale is normally found as the roof and/or floor of a gateroad. The presence of water i.e. groundwater, mining water or even atmospheric water will tend to cause shale weak and soft due to slaking and swelling. If this phenomenon occurs, under high ground pressure, excessive closure on roadway, such as floor heaving, happens and it may discomfort the stability of the roadway.

Figure 1 shows the vertical closure rate of the roadway during developing stage (Matsui et.al., 1996). The roadways are driven by a roadheader or by drill and blast method at a depth of 400 - 450 m under nearly the same geological and water conditions. The floor is saturated shale. The roadway are supported by the same type of support and spacing. This figure also indicates the general influence of the fault and the distance from the heading face on the overall roadway deformation. It shows that the roadway driven by the roadheader deform and become damage as much as those driven by the drill and blast method. It also shows the different behaviour under wet and dry conditions. Under dry conditions, machine driven roadway exhibits less vertical closure. On the other hand, under wet conditions, no difference can be seen in the both results due to the deterioration of the shale floor by the presence of water.

Figure 2 shows the vertical closure trend under dry and wet conditions at Ikeshima Colliery, Japan.

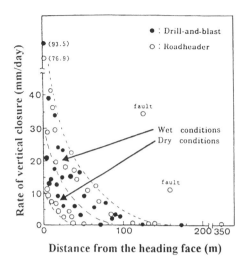

Figure 1. Vertical Closure rate during development

Roadway is driven by the drill-blast method. The seam depth is 450-500 m below see level. The roof and floor rocks are shale. Support in the roadway are set by using 2.8 m x 4.8 m three-piece steel arches (cross section 0.115 m x 0.095 m). The supports are tied and lagged with wood. Under wet conditions the roadway deformation continues with time and non-linear deformational behaviour or time-dependent effects the roadway closure remarkably. The vertical closure in wet condition area reaches 40-60 cm in 30 days after the drivage.

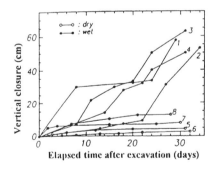

Figure 2. Vertical closure under dry and wet conditions

3. STRUCTURE AND MINERALOGY

Bedding or stratification is the structure that is found in shale and it regards the plane of weakness. Taylor and Spears (1970) in his study of the breakdown of coal measures rocks shows that stratifications absent in only a small portion of the siltstone-sandstone (5-10%). The remainder (about 40% of the total sequence) contains both parallel and cross stratification.

Clay-bearing rocks are the most sensitive to slake deterioration. The important mechanism here is the ion exchange (Veley, 1969; Boswell, 1961). Clay minerals have the ability to take up water, and other liquids, into interlayer structural sites causing swelling, however it also depends on the type of the clay. Montmorillonite is the most common clay mineral that has high potential to adsorb water. Its expansion mainly depends on the nature of the interlayer section. The deterioration of the shale that contains montmorillonite can occur even with very small amount of water such as atmospheric water. The other mechanism of the clay expansion is the enlargement due to capillary action. In this mechanism water meniscus will reduce the capillary tension at grain contacts and at the tip of the cracks. Also water that is drawn into the rock by the action of strong capillary forces may compress air in its path, resulting in disruption of the rock.

X-ray diffraction examination shows that the shale specimen which comes from Ombilin and Ikeshima Colliery contain montmorillonite and smectite mineral.

Anwar (1997) examined the water absorption of some coal measures rocks such as shale, sandyshale and sandstone from Ombilin and Ikeshima Colliery. The result shows that the degree of water absorption of shale is higher than sandyshales.(Figure 3).

4. CHRACTERISTIC OF SLAKING

4.1 ISRM method

The slaking durability of the rock, as described by Franklin (1972), will depend on the permeability and porosity, the action of fluids immediately after penetrating to the rock and the capacity of the rock to resist disruptive forces. The last factor will decide the extent to which weakening, swelling or complete disintegration of the rock will occur. The index of the slaking-durability of the rock is determined by the slake-durability test, as suggested by ISRM (1979). This test is intended to asses the resistance offered by a rock sample to weakening and disintegration when subjected to two standard cycles of drying and wetting.

Two types of specimen i.e. shale and sandyshale were examined. The results, referring to slaking-durability measurement of ISRM method are given in figures 4 and 5. The characteristic of the slaking

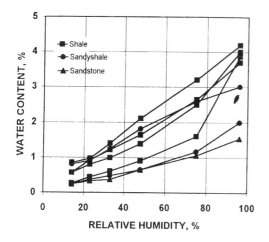

Figure 3. Moisture absorption versus relative humidity.

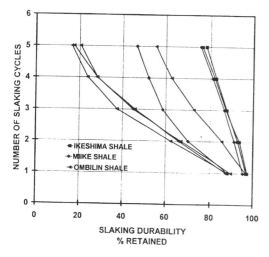

Figure 4. Influence of the number of slaking cycles on slaking durability of shales.

durability may be determined on the basis of the second cycle index (Id_2) as described by Gamble (1971).

From these test, shales from Miike, Ikeshima and Ombilin Colliery are characterised by medium slaking durability (Id_2 between 54,3 and 70.1), which means an accelerated looseness and deterioration after their exposure on the surface.

The describe behaviour of the shale to successive drying and wetting cycles, should be attributed to the

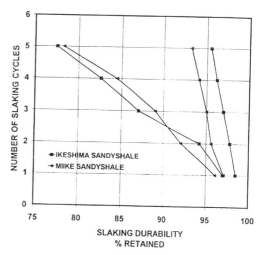

Figure 5. Influence of the number of slaking cycles on slaking durability of sandyshales.

clay mineral group involve. The lowest Id_2 (between 54.3 and 67) are shown by Ombilin shale.

Sandyshale from Ikeshima and Miike Colliery show a fluctuation from very high to extremely high slaking durability (Id_2 92.1 and 97.8).

4.2 Slaking observation method

This test consists of two types of slaking test. One is a visual slaking test when a rock sample is put in water and the other is a dry-wet cycling test. In dry-wet cycling slaking test initially the specimens were dried at 105° C for 24 hours and immersed water for 48 hours after cooling. After wetting or drying, the slaking is observed and estimated by using the visual slake observation index. Five criterion are used to classify the slaking phenomenon in this test :

Index	Criterion
0	No visual change
1	Minor cracking
2	Medium cracking in scales or crack opening
3	Major cracking in scales or minor disintegration
4	Completely disintegration

The result of slaking observation are shown in table 1 and 2. All of the specimen of Ombilin shale present an index of 4 after 48 hours immersed in water. Some specimen even disintegrated within 2 hours in water. The same phenomenon is shown after three cycles in dry-wet slaking test. However shale specimen from Ikeshima presents minor and major cracking. Sandyshale is classified as a very low slaking phenomenon.

Tabel 1. Immersed water slaking observation

Specimen	Hours							
	1	2	4	8	12	24	48	72
I - shale 1	0	1	1	1	2	2	2	2
I - shale 2	1	1	1	2	2	2	2	3
I - shale 3	1	1	1	2	2	2	2	2
O - shale 1	3	4	4	4	4	4	4	4
O - shale 2	3	4	4	4	4	4	4	4
O - shale 3	2	2	3	3	3	3	4	4
O - shale 4	2	3	3	3	3	3	4	4
I - sandyshale 1	0	0	0	1	1	1	1	1
I - sandyshale 2	0	0	1	1	1	1	1	1
I - sandyshale 3	0	1	1	1	1	1	1	1

I : Ikesima Col.; O : Ombilin Col.

Tabel 2. Dry-Wet cyclic slaking observation

Specimen	D-1	W-1	D-2	W-2	D-3	W-3
I - shale 1	0	1	1	2	2	3
I - shale 2	0	2	2	3	3	4
O - shale 1	0	4	4	4	4	4
O - shale 2	0	4	4	4	4	4
I - sandyshale 1	0	0	0	0	0	0
I - samndyshale 1	0	1	1	2	2	3

I : Ikesima Col.; O : Ombilin Col.

CONCLUSION

From the above discussion it is made clear that shales have a high slaking effect soon after they contact with water or in dry-wet cyclic. Although their slake-durability characteristics are classified as medium, however, when contact with water these type of shale will be soon disintegrated. This phenomenon is attributed to the clay mineralogy content. The exception are shown by sandyshale.

The slaking phenomenon of shale are characterised by an accelerated looseness and deterioration immediately after their exposure to the change of moisture content. In longwall mining, special measures have to be taken to eliminate the problem of vertical closure and to make the roadway and the groundwork more stable. Floor reinforcement is one of the method that may be applied to improve the shale strength.

ACKNOWLEDGEMENTS

The authors grateful thank the staff and student of the Rock Mechanics Laboratory, Mining Engineering Division , Department of Earth Resources and Mining Engineering, Kyushu University, the staff of the Geomechanics Laboratory, R&D Centre for Geotechnology-LIPI, the staff of the Geomechanics Laboratory Dept. of Mining Engineering-ITB and also the manager and engineers of the Ikeshima and Ombilin Colliery.

REFERENCES

Anwar, H.Z., Shimada, H., Ichinose, M., Matsui, K., (1997), Deterioration of mechanical properties of coal measures rock due to water, *Proc. of 7th Int. Symp. on Mine Planning and Equipment Selection*, Calgary, Canada, pp.257-261.

Boswell, P.G.H, (1961) *Muddy Sediment*, Heffer, Cambridge.

Franklin, J.A. and Chandra, R. (1972), The slake durability test, *Int. J. Rock Mech. Min. Sci.* Vol. 9, pp. 325-341.

Gamble, G.C., (1971), *Durability-Plasticity classification of shales and other argillaceous rocks*. PhD. Thesis, University of Illinois.

Ichinose, M. and Matsui, (1988), Effect of water on deformation of mine roadways, *Journal of the Japanese Society of Engineering Geology*, 29-3, pp. 1-9.

ISRM, (1979), *Suggested method for determining the slaking, swelling, porosity, density and related rock index properties*, prepared by Commission on Standarization of Laboratory and Field Test.

Matsui, K., Shimada, H., Ichinose, M., (1996), Effect of water on stability of mine roadway, 15th International Conference on Ground Control in Mining, Golder, Colorado, pp. 589 - 598.

Taylor and Spears (1970), The breakdown of British coal measures rocks, *Int. J. Rock Mech. Min. Sci.* Vol. 7, pp. 481-501.

Veley, C.D., (1969), How hydrolizable metal ions react with clays to control water sensitivity, J. Petrol. Technol. 1111-1112.

Mining Science and Technology' 99, Xie & Golosinski (eds) © 1999 Balkema, Rotterdam, ISBN 90 5809 067 1

Dislocation migration analysis of blast-induced block instability in mining

A. Karami & J. Szymanski
School of Mining and Petroleum Engineering, Department of Civil and Environmental Engineering, University of Alberta, Edmonton, Alb., Canada

ABSTRACT: This paper quantifies damage to the rock mass resulting from cyclic loading occurring as a result of blasting. The incurred damage is related to the fatigue of the rock mass and the fatigue life of the rock block is estimated. Results of this study permit determination of the probability of rock failure during the predicted life span of the mine. Damage is studied through the analysis of crack initiation and propagation according to the theory of dislocations. Following the crack initiation at microscopic scale, a macroscopic scale cumulative damage (crack propagation) results leading to local failure of the blocks existing within the rock mass. This paper presents the first part of the related study. It defines the monotonic part of blast cyclic loading of the rock block. The unloading part of the cycle will be presented in another paper.

1. INTRODUCTION

Although substantial progress has been achieved in static design of excavations in jointed rock mass, comparable progress has not been maintained in dynamic design. An underground excavation is produced in stages (or elements), each of that constitutes a small fraction of the complete excavation. As production blasting continues, towards the design geometry (shape and size), the excavation may become unstable due to repeated blast loading. Similarly, in an open pit mine, the stability of the bench and the formed rock blocks in the bench wall is affected by the cyclic loading effect of blasting operations.

Blast damage in its simplest form can be defined as the weakening of the rock mass through fracturing or extension of existing fractures caused by the *near field* blast vibration and entry of explosion gases. *In* the mid - to far - field regions, the dominant mechanism of wall damage or failure is the shaking of wedges, key blocks or pre-conditioned volumes of rock due to cyclic loading of the walls from subsequent mid - and far -field blasts. Repetitive blasting of bench along the ore bodies has been known to severely weaken the excavation and/or bench surfaces by decreasing the joints shear strength and accumulation of shear displacements at the joints.

There have been a number of attempts to find a damage criteria for underground and open pit blasting operations. Most of the research carried out so far considers only the mechanical, physical and the structural properties of the rock mass *(Yu et al., 1996, Chitombo et. al. 1990, Lily 1986).* Few research studies, e.g. Yang et al. (1996), incorporated the blast loading conditions in their model, however, they were only applicable to near field blast damage. To date, no attempt has

been made to model the blast-induced damage to the rock mass in the mid- to far-field region.

The most popular approaches to quantify the blast-induced damage is through determination of the critical peak particle velocities likely to induce damage. These approaches take the form of empirical relationships and generally use a number of parameters, which infer the characteristics of rock mass, the geometry between the explosive charge at the point of interest and the amount of explosive. The main limitations of these approaches are that: i) they do not explicitly take into account actual characteristics of the rock mass; i.e., whether the rock is massive or heavily jointed, the degree of fracturing, joint characteristics or the presence of key blocks; ii) they relate damage to the critical level of the peak vibration velocity alone. These methods do not consider the total energy contained in the vibration, as well as the duration and frequency of the vibration. A realistic model is one which takes into account real structural features and potential key blocks to assess blast damage.

In this paper, the objective is to quantify the damage to the rock mass under blast cyclic loading in mid - to far- field regions. The rock mass is jointed by a set of intersecting discontinuities and bedding planes, which form the blocks of rocks of various shapes and sizes. The fractures surrounding a given rock block may not necessarily be open. The objective here is to determine the damage imposed on rock mass due to ongoing blasting operation in the mine. Then, the incurred damage can be related to the fatigue of the rock mass and the fatigue life of the rock block is estimated. From this study, it would become possible to determine whether or not the block failure would happen in the life span of the mine. Damage is studied through the analysis of

347

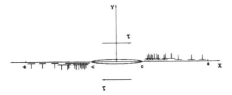

Figure 1: Schematic diagram of the crack, dislocations along the crack line and the associated plastic zone ahead of the crack tip, c<|x|<a

crack initiation and propagation according to the theory of dislocations. Following the crack initiation in microscopic scale, a macroscopic scale cumulative damage (crack propagation) leads to local failure of the formed blocks of rocks in the rock mass.

In this paper the first part of the study involving the monotonic part of blast cyclic loading of the rock block is investigated and unloading part of the cycle will be presented in another paper.

2. DISLOCATION THEORY

The mechanism of elastic-plastic deformation developed in rock masses can be explained through the theory of dislocation. According to this theory, a defect is the irregularity in atomic structure of the material. A *dislocation* can be defined as linear defects, which are oriented arrangements of point defects.

The rock mass is an amorphous body in which atoms are distributed chaotically and their density varies from place to place. When rock mass is loaded in shear, plastic slides occur. These slides then grow and join together, forming a borderline separating the slide region from the region where the slides have decayed. This line is called the *dislocation line* (Gil, 1991). Subjected to shear loading, according to weakest link criterion, material fails at the point where it has the least resistance to failure and that would be the dislocation line. As the applied load increases and exceeds the shear strength of the material, dislocations start to move. If there is a barrier at the crack tip that prevents the dislocation lines to move out of the crack, the induced displacement will be elastic, which is recoverable as soon as the applied load is removed. If there is no barrier at the crack tip (Figure 1), then under the application of external shear load the dislocations move out of the crack provided that the applied load is higher than shear strength of the material. Under sustained loading, dislocations move further into the material due to which the dislocation density ahead of the crack tip increases. Due to high stress concentration at the crack tip, more dislocations glide up at the tip of the crack, which causes higher plastic displacement (damage) at the crack tip. As the plastic displacement accumulates under cyclic loading, the cracks start to grow. If crack grows enough it will be followed by the movement of the pre-defined blocks on the slip plane.

The theory of dislocation was initially used to investigate the plastic behavior of the metals at the notches by Bilby, Cottrell and Swinden (BCS, 1963). Later their model which was for elastic perfectly plastic

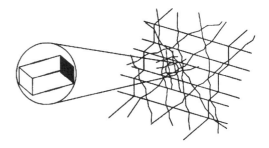

Figure 2: Schematic diagram of the block of rock formed in the bench wall under blast cyclic shear load.

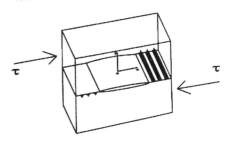

Figure 3: Schematic diagram of the crack at the bottom of the block under blast cyclic shear load

material was further developed to accommodate for linear (BCS, 1964) and nonlinear work hardening (Ellyin et al., 1986) material behavior.

Here, a nonlinear work hardening stress-strain relationship is used to model rock behavior prior to ultimate failure. Under blast cyclic loading, material goes under loading and unloading half cycles. A different constitutive relationship needs to be developed for each case since under unloading material plastic behavior is affected by the Bauschinger effect (memory effect). In this paper only the constitutive relations for the monotonic loading condition is presented and the results are discussed.

3. DISLOCATION MIGRATION ANALYSIS MODEL

3.1 Modeling of monotonic loading stage

Consider a block of rock located in a bench wall under equilibrium conditions. The block geometry is defined by the major joint sets that are known from previous studies and intersect and form a rectangular block (Figure 2). All block sides except the back are partially fractured and separated. The back of the block is assumed to be completely separated and has no resistance to loading. (Figure 3 schematically shows the crack at the bottom of the rock block). The block goes under transient cyclic load that attenuates fast. Depending upon the orientation of the block and the stress wave approaching the block, the applied transient load could have shear and normal components, however only plane shear (mode II shear) component is significant. The applied load in the back of the block creates a mode I opening fracture which can not be analyzed by dislocation theory therefore, in this study it is assumed that the back of the block is already separated and has no resistance to the applied load. As the block goes under cyclic loading and unloading, damage is accumulated at the tip and the very close vicinity of the existing crack. This eventually would lead to crack movement and block instability. The frictional resistance here is the only resisting force against the applied load.

Consider the plane crack at the bottom of the rock block subjected to a shear stress (τ) at infinity. The material contains a distribution of long straight dislocation lines lying parallel to z direction in the xy plane. According to the principles of the fracture mechanics, there is no resistance to the motion of the dislocations inside the crack i.e. crack faces can slip freely and the resistance to the motion ahead of the crack tip ($0<|x|<a$) is governed by the frictional resistance, τ_r. In this region, dislocation movements caused by the shear stress, τ, creates plastic displacement at the tip of the crack, which will extend throughout the plastic zone ahead of the crack tip. As the load is increased from zero to τ these dislocations glide out from the crack tip into the medium, driven by the stress concentrations at the crack tip and opposed by the resisting strength, τ_r. A general power law form of work hardening material behavior was proposed by Ellyin et al. (1986), in which the frictional strength is the only resistance to the dislocation motion. A general power law form work hardening rock behavior can only be applied to rocks prior to ultimate failure. Therefore, this generalized model is represented by:

$$\tau_r(x) = K\varphi(x)^n \tag{1}$$

where $\varphi(x)$ is the total shear displacement and n and K are the work hardening exponents. The total displacement is consisted of plastic displacement $\varphi_p(x)$ and elastic yield intercept, φ_0, i.e. $\varphi(x)=\varphi(x)+\varphi_0$. Therefore, equation (1) can be re-written as:

$$\tau_r(x) = K\varphi_0^n(\frac{\varphi_p(x)}{\varphi_0}+1)^n \tag{2}$$

The post-yield behavior of the material is non-linear of the type of Ramberg-Osgood. For continuously distributed dislocations with Burgers vector, $\mathbf{b}$ and the density function $f(x)$ at each point, the equilibrium condition requires that the sum of all forces, dislocations and external, at each point to be zero (Head et al., *1955*).

$$\int_{-a}^{a} \frac{f(x')dx'}{x-x'} + \frac{P(x)}{A} = 0 \tag{3}$$

where $A=\mu b/2\pi(1-v)$, μ is the shear modulus, v is the Poisson's ratio and $P(x)$ is the applied external load:

$$P(x) = \tau \qquad\qquad |x|<c \tag{4a}$$
$$P(x)=(\tau-\tau_0)-(\tau_r(x)-\tau_0) \quad c\leq|x|\leq a \tag{4b}$$

It is notable that the Cauchy principal value of the singular equation (equation (3)) is to be taken to avoid divergence at x = x'.

The general solution of the equation (3) is given by Muskhelishvili *(1953)* as

$$f(x) = \frac{2(1-v)}{\pi\mu b} \int_{-a}^{a} R(x,y)P(y)dy \tag{5}$$

where,

$$R(x,y) = \frac{1}{x-y}\sqrt{\frac{a^2-x^2}{a^2-y^2}} \tag{6}$$

provided that the P(x) and f(x) satisfy the Holders conditions. If a plastic displacement $d\varphi$ is obtained as $f(x)$ dislocations move from point x to x + dx under the applied stress, knowing the total Burgers vectors in this region (x→x + dx) to be $\mathbf{b}f(x)dx$, the total plastic displacement $\varphi(x)$ can be shown as

$$\varphi_p(x) = b\int_x^a f(x')dx' \tag{7}$$

Substituting equation (5) into (7) and after some manipulations

$$\varphi_p(x) = \varphi_1(x) - B\int_x^a K(x,y)p''(y)dy \tag{8}$$

Equation (8) is a Fredholm Integral Equation of the second kind where $\varphi_1(x)$ is the BCS solution for elastic perfectly plastic material behavior and is known. K(x,y), the kernel and P''(y), the loading function for hardening part of the material behavior are also given by

$$\varphi_1(x) = B\tau_0(x+c)\cosh^{-1}\left|\frac{a^2-x^2}{a(x+c)}+\frac{x}{a}\right|$$
$$- B\tau_0(x-c)\cosh^{-1}\left|\frac{a^2-x^2}{a(x-c)}+\frac{x}{a}\right| \tag{9}$$

$$K(x,y) = \cosh^{-1}\left|\frac{a^2-x^2}{a(x-y)}+\frac{x}{a}\right|+$$
$$\cosh^{-1}\left|\frac{a^2-x^2}{a(x+y)}+\frac{x}{a}\right| - 2\sqrt{\frac{a^2-x^2}{a^2+y^2}} \tag{10}$$

The dislocation density function has to be zero at the end of the plastic zone, meaning that the general solution of the Muskhelishvili should be bounded at the end of this zone (x=±a) i.e. this solution at the boundary reduces to

$$\int_{-a}^{a} \frac{\tau-\tau_r(x')}{(a^2-x'^2)^{1/2}}dx' = 0 \tag{11}$$

To solve (8), using one of expansion methods (Delves, et al., 1974), a perturbation of the elastic perfectly plastic solution of the BCS is introduced and the plastic displacement function, $\varphi_p(x)$, is defined as

$$\varphi_p(x) = \alpha_1(x+c)\cosh^{-1}\left[(1+|\alpha_2|)\left|\frac{a^2-x^2}{a(c+x)}+\frac{x}{a}\right|\right]$$
$$- \alpha_3(x-c)\cosh^{-1}\left[(1+|\alpha_4|)\left|\frac{a^2-x^2}{a(c-x)}+\frac{x}{a}\right|\right]$$
$$+ \alpha_5 e^{-\alpha_6(x/a)} + \alpha_7 e^{-\alpha_8(x^2/a^2)} \tag{12}$$

or simply

$$\varphi_p(x) = \sum_i \alpha_i\varphi_i(\alpha_i,x) \tag{13}$$

349

Substituting $\varphi_p(x)$ into (8) the problem will invert to solving a nonlinear regression problem by minimizing:

$$\min. \quad \sum_{j}^{m} \eta_j^2 \tag{14}$$

where η_j is given by

$$\eta_j = \phi_p(x_j) - \phi_1(x_j) + B\int_0^a K(x_j, y)P''(y)dy \tag{15}$$

The problem now inverts to solving (14) with boundary condition (11). There are eight unknown coefficients in equation (12), therefore, at least 8 equations are needed to solve the system of equations with boundary condition (11). However, by choosing more points an over-determined system of equations is obtained. This system of equations can be solved using one of optimization techniques. In this model, Levenberg-Marquardt's (LM) algorithm (Scales, 1974) is used to find the optimum solution of the system of equations.

In equations (14) and (11), the plastic displacement coefficients, α_i and the extent of the plastic zone, a, are unknown, and the solution process is as followed: For an arbitrary value of plastic zone, 'a', the nonlinear regression problem (14) is solved. Then knowing the coefficients, α_i the boundary condition (11) is tested for this case. Then the extent of plastic zone is modified accordingly and equation (14) is solved iteratively until the boundary condition (11) is satisfied with sufficient accuracy.

Knowing the plastic displacement coefficients, α_i, the plastic displacement function, $\varphi_p(x)$, is obtained as a function of x, distance from the crack tip, using equation (12). Then, the stress field in front of crack tip is found using equation (2).

3.2 crack tip damage modeling

As the rock goes under cyclic shear loading, damage is accumulated ahead of the crack tip, which eventually leads to failure of the rock. Since the failure is due to fatigue of the rock mass, it happens at a lower stress level than the peak shear strength of the rock mass. To quantify the damage, either crack tip stress field or plastic displacement at the crack tip can be used as the damage criterion. A third alternative damage criterion is the energy parameter, which is the amount of energy used to cause the damage. The advantage of the latter over stress and plastic displacement criteria is that energy, by definition, takes into account the effects of both stress field and plastic displacement at the crack tip. J-integral is an appropriate energy parameter, since it provides a means to quantify the damage at the crack tip and characterizes the stress field at the crack tip. For monotonic loading J-integral is used as Rice (1968) initially introduced it for two-dimensional problem as

$$J = \int_\Gamma (W dx_2 - T_m \partial u_m / \partial x_1 ds) \tag{16}$$

where W is strain energy density, $\int \sigma_{mn} d\varepsilon_{mn}$, T_m the surface traction exerted on the material within the contour, u_m is the displacement, Γ is the integration path around the crack tip plastic zone, and Xi5 the coordinate system. For the dislocation model, here, the surface traction is $T_m = \tau_r(x)$, $u_m = \varphi_p(x)$ and Γ is defined as the upper and lower surface of the crack surface, therefore,

$dx_2 = 0$ and the first term in the integral is eliminated. Therefore, for monotonic loading J integral is defined as

$$J = \int_0^a \left(-\tau_r(x) \frac{\partial \varphi_p(x)}{\partial x} dx \right)$$

$$= \int_0^a -K\varphi_0^n \left(\frac{\varphi_p(x)}{\varphi_0} + 1 \right)^n \frac{\partial \varphi_p(x)}{\partial x} dx \tag{17}$$

where from monotonic loading stage the plastic displacement function $\varphi_p(x)$ is determined and then the J-integral can easily be evaluated for this case.

4. RESULTS AND DISCUSSIONS: MONOTONIC LOADING STAGE

A model has been built based on the approach mentioned in previous sections to analyze the plastic displacement at the crack tip and determine the induced-damage in that region. The type of the rock is gneiss with a shear-displacement diagram shown in Figure 4. The material properties are given in Table I. The diagram was digitized and fitted by a power law function. This function will indicate the shear resistance of the material prior to ultimate failure (see equation (1)). The values of fitted function coefficients (K and n) are given in Table 1. The values of the regression coefficients (α_i, i=1. .8), for different applied loads and crack sizes are shown in Table 2.

Table 1: Mechanical properties of the rock type and the function coefficients of the shear strength model

Mechanical properties of Gneiss	
Yield stress (τ_0)	17 MPa
Yield displacement (φ_0)	0.0002058 m
Poison's ratio (v)	0.20
Shear modulus	20 GPa
Peak shear strength	24 MPa
Shear strength function -Model coefficients (see eq. 1)	
K (Pre-Yield region)	372e+6 Pa/m^n
K (Post-Yield region)	27e+9 Pa/m^n
n (Pre-Yield region)	0.881642
n (Post-Yield region)	0.361229

Table 2: Values of plastic displacement function coefficients and the equivalent extent of the plastic zone for different crack sizes under minimum and maximum applied load τ.

2c (mm)	10	10	400	400
τ (MPa)	1	11	1	11
Plastic Displacement Coefficients				
$\alpha 1$	1.0740E-4	1.0485E-4	2.9042E-4	1.4049E-4
$\alpha 2$	7.7417E-9	2.2221E-6	1.2438E-6	2.9746E-2
$\alpha 3$	1.0486E-4	1.0485E-4	1.3057E-4	1.0996E-4
$\alpha 4$	9.2670E-4	-2.518E-5	-3.3037	3.0492E-1
$\alpha 5$	4.7232E-8	2.2179E-9	2.1735E-4	-6.284E-5
$\alpha 6$	9.9888E-1	2.8387	9.9993E-1	2.5027E-1
$\alpha 7$	-4.744E-8	-1.928E-9	-2.174E-4	4.2893E-5
$\alpha 8$	1.0011	2.4676	1.0000	-1.075E-2
Extent of Plastic Zone (m)				
a	5.0206E-3	9.2049E-3	2.0059E-1	3.6367E-1

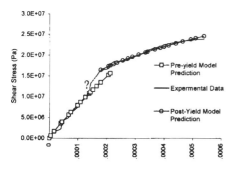

Figure 4: Shear stress-shear displacement diagram of the gneiss (Bertacchi, et. al., 1974)

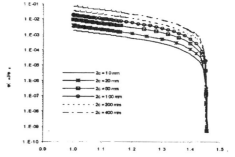

Figure 5: Non-dimensionalized plastic displacement vs. distance from the crack tip

Knowing α_i, one can easily determine the crack tip plastic displacement from equation (12) and the crack tip stress field from equation (2). Figures 5 and 6 show the non-dimensionalized crack tip plastic displacement and crack tip stress field versus distance from the crack tip for the various crack sizes subject to the applied load of 9 MPa. As shown, the plastic displacement is maximized at the crack tip and vanishes at the end of plastic zone, 'a'. The crack tip stress also shows a similar trend. It can be seen from both figures that for a given material under a given applied load, as the crack size increases, the plastic zone also extends proportionally, i.e. end of plastic zone is located at the same a/c ratio. Figures 7 and 8 show the non-dimensionalized crack tip plastic displacement and crack tip stress versus the non-dimensionalized applied stress, τ, for various crack sizes. As seen, the crack tip plastic displacement, $\varphi_p(c)$ increases as the applied load increases. Figure 7 also shows that the plastic displacement at the tip of the longer cracks is larger than the shorter ones and the change in the extent of the plastic displacement is proportional. Unlike the plastic displacement, crack tip stress increases at higher rate when the crack is longer. The higher the stress and the plastic displacement, the higher will be the induced-damage. Damage can either be demonstrated by the stress, plastic displacement or a combination of both i.e. the energy criterion, J-integral. Figure 9 shows the damage that the rock suffered as a result of loading. The damage is determined in terms of J-integral. As discussed earlier, J-integral, which is an energy parameter, is a better choice than the stress or displacement since it takes into account the effects of both parameters and at the same time it is an indicator of the stress field at the crack tip.

5. CONCLUSIONS

The major contribution of this study to the mining science is that in general practice mine openings are designed based on in-situ or laboratory physical and mechanical properties of the rock mass. Therefore, disregarding the energy impact of blasting operations on the rock characteristics. Knowing the amount of damage that the rock mass has suffered due to blasting in vicinity areas, it would become possible to account for the blast-induced damage at the early stages of the mine design and support selection.

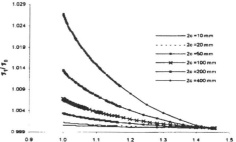

Figure 6: Non-dimensionalized crack tip stress field vs. distance from the crack tip.

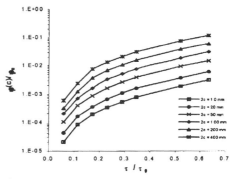

Figure 7: Non-dimensionalized crack tip plastic displacement vs. non-dimensionalized applied stress

Results of this study show that detailed analysis of the crack tip stress and plastic displacement fields provide us with enough information to investigate the crack tip damage and permanent changes imposed on the rock mass by the disturbing loads i.e. blast cyclic load, etc. This information can further be used and combined with results of the unloading part of this study to investigate the stability of the rock block and would provide us with an estimate of the block fatigue life using an appropriate fatigue failure criteria.

351

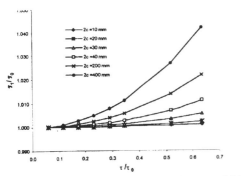

Figure 8: Non-dimensionalized crack tip stress field vs. Non-dimensionalized applied stress

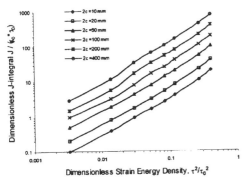

Figure 9: J-integral as damage parameter vs. applied stress ratio for various crack sizes

6. FURTHER WORK

The next step is to solve the constitutive equations for the unloading part of cyclic load that are already developed. The results of this part of the study will be presented near future. Later these results will be combined and a fatigue failure criterion will be introduced through which the fatigue life of the rock block will be predicted. This condition of rock block instability could be used to predict when the bench failure would happen. The limitations are that i) this investigation is purely theoretical, ii) the assumptions involved in this study need to be relaxed with farther study. Among these are the rock block geometry, the boundary conditions, and the loading characteristics.

7. REFERENCES

Bertacchi, P., Sampaolo, A., 1974, Some critical considerations on the deformation and failure of rock samples in the laboratory, Third International Congress on Rock Mechanics, Vol.2, Part A, pp.21-26.

Bilby, B.A., Cottrell, A.H., Swinden, K.H., 1963, The spread of plastic yield from a notch, Proceedings of the Royal Society, A272, pp.304-314.

Bilby, B.A., Cottrell, A.H., Smith, E., Swinden, K.H., 1964, Plastic yielding from sharp notches, Proceedings of the Royal Society, A279, pp.1-9.

Bilby, B.A., Swinden, K.H., 1965, Representation of plasticity at notches by linear dislocation arrays, Proceedings of the Royal Society, A285, pp.22-33.

Chitombo, G., Scott, A., 1990, An approach to the evaluation and control of blast induced damage, Fragblast'90, Brisbane, August 26-31, pp.239-244.

Delves, L.M., Walsh, J., 1974, Numerical Solution of Integral Equations, Claredone Press, Oxford, P.80.

Ellyin, F.~ Fakinlede, C.O.A., 1986, Nonlinear dislocation model of the work-hardening crack-tip field, Int. Journal of Fracture, Vol.32, pp.3-20.

Lardner, R.W., 1974, Mathematical theory of dislocations and fracture, Mathematical Exposition No.17, University of Toronto Press.

Lilly, P.A., An Empirical method of assessing rock mass blastability, The AusIMMIE Aust Newman Combined Group, Large Open Pit Mining Conference, October 1986, pp.89-92.

Muskhelishvili, N.I., 1953, Singular integral equations, (Translated by J.M. Radok) Noordhoff, Groningen, Holland.

Rice, J.R., 1968, A path independent integral and approximate analysis of strain concentration by notches and cracks, Journal of Applied Mechanics, Vol.35, pp.379-386.

Scales, L.E., 1974, Introduction to Non-Linear Optimization, Springer-Verlag New York Inc., P. 115.

Yang, R, Bawden, WF, Katsabanis, PD, New constitutive model for blast damage, International Journal of Rock Mechanics and Mining Sciences and Geomechanics Abstracts, v 33 n 3 April 1996, pp.245-254.

Yu, T.R., Vongpaisal, 5., 1996, New blast damage criteria for underground blasting, CIM Bulletin, Vol.89, No.998, pp.139-145.

Mining Science and Technology'99, Xie & Golosinski (eds) © 1999 Balkema, Rotterdam, ISBN 90 5809 067 1

Mechanism of rock directional fracture when blasting with dynamic caustics

Qing Li, Renshu Yang & Kexue Jin
China University of Mining and Technology, Beijing, People's Republic of China

Qingtian Lang, Ming Zhang & Qiang Tong
Xinwen Bureau of Mining, Shandong, People's Republic of China

Senlin Zhao
Xingtai Bureau of Mining, People's Republic of China

ABSTRACT: In order to make further quantitative analysis on crack propagation under blasting load, the paper first gives a method to research blasting mechanism with dynamic caustics. An experiment was conducted in which transmitted dynamic caustics under blasting load have been traced in a transparent material model, and a series of dynamic caustics photos were obtained. This paper also discusses the general law of crack propagation in slotted cartridge blasting and analyses the stress intensity factor around the crack tip in the progress of crack directional propagation quantitatively. The influence of slot width on the control effect of directional fracture is probed.

1. INTRODUCTION

Although many optical experiment methods have been applied extensively for dynamic mechanics. The determination of characteristic parameters of singular stress fields at the crack tip still renders the difficult solution by conventional experimental methods. In dynamic photoelasticity the interference fringes are used to analysis the singular stress field question. The problem is that the isochromatic fringe patterns have high density nearby a stress singularity field, so the calculation of the intensities in the question is very difficult at the crack tip. In the Moiré method, the fringes representing the deformation are also unable to follow the steep variation of the stresses at the crack tip. The same problems have been encountered in the study of the rapid change in strain at the crack tip when using optical interferometry, holograph as well as strain gages.

The optical method[1] of caustics, introduced by Manogg and Theocaris, has shown to be very effective for the determination of the characteristic

$$\begin{cases} \overline{D}_t = \delta_t (\dfrac{V}{C_s}) \cdot \dfrac{10}{3} \sin \dfrac{2}{5} \pi \\ \overline{D}_l = \delta_l (\dfrac{V}{C_s}) \cdot \dfrac{10}{3} \cos^2 \dfrac{\pi}{10} \end{cases} \quad (1)$$

parameters of singularity stress fields at the crack tip. The difficulty arises mainly from the fact that the high stress region at the singularity is very small and

therefore information gathered by classical experimental methods, as mentioned above, is rather vague and inaccurate. The new method of dynamic caustics has firstly been established to study fracture behavior under blasting load. This method of transmitted dynamic caustics is used for the study of the stress singularity region at the crack tip. The results show the high sensitivity of the experimental method in slotted cartridge blasting.

The above achievements are also applied to driving construction in rock tunnel in several mines. The advantage of directional control blasting technology by using slotted cartridge has been confirmed, which increases the distance of hole in perimeter 50% on present basic, makes the ratio of half hole scar to 80□95% and the uneven degree less than 100mm.

2. THE BASIC OPTICAL THEORY OF TRANSMITTED DYNAMIC CAUSTICS

The experimental method of caustics, which is based on the laws of geometrical optics, transforms the stress singularity field into an optical singularity (i.e. the caustic). This optical singularity field yields much information for evaluation of the singular stress field.

According to the research[2], the relation between dynamic stress factor K_I and D_t/r_0-V/C_s, and D_l/r_0-V/C_s can be expressed as the following.

Where □$_t$(V/C$_s$) and □$_l$(V/C$_s$) are correction factors that are affected by the velocity of a crack propagation.

Then,

$$r_0 = \left(\frac{3}{2\sqrt{2\pi}} \cdot |C_t| \cdot d \cdot Z_0 \cdot K_I^d \right)^{2/5} \qquad (2)$$

Where r_0 is the radius of initial curve of caustics. It depends on optical system and properties of specimen and time.

$$\begin{cases} K_I^d = \dfrac{2\sqrt{2\pi}}{3|C_t|dZ_0} \cdot \left(\dfrac{3}{10\sin\frac{2\pi}{5}} \right)^{5/2} \cdot \left(\dfrac{D_t}{\delta_t} \right)^{5/2} \\[4mm] K_I^d = \dfrac{2\sqrt{2\pi}}{3|C_t|dZ_0} \cdot \left(\dfrac{3}{10\cos^2\frac{\pi}{10}} \right)^{5/2} \cdot \left(\dfrac{D_l}{\delta_t} \right)^{5/2} \end{cases} \qquad (3)$$

From Rels(1) and (2) the dynamic stress factor can be derived as:
Where:
K_I^d: dynamic stress intensity factor for mode I;
Z_0 : distance between the specimen and the reference screen;
d: thickness of specimen;
C_t: the stress optical constant of transmitted caustics,
Where:
C_t=-0.88×10^3 mm^2/kgf;
D_t: the transverse diameter of caustics spot;
D_l: the longitudinal diameter of caustics spot.
In the above formula (3) velocity of a crack propagation of experimental model is less than 400m/s. The values of $\square_t$(V/C_s) and $\square_l$(V/C_s) are about 1.

3. EXPERIMENTAL SYSTEM AND SPECIMEN

The method of dynamic caustics has the advantage of simplicity of equipment. The experimental system is composed of a 16-spark high speed camera and two lenses, as shown in Figure 1. Blasting load is created by the borehole charge Pb(N₃)₂ in slotted cartridge. Two pairs of probes are laid up in charge. Pb(N₃)₂ is fired by a high voltage instrument. The probes transport a short circuit signal to the synchronism instrument. The synchronism instrument then transports a signal to high-speed camera and makes its sparks discharge at the scheduled time. These instruments form so called light-electricity system.

The testing system has advantage of accurate controlling synchronization of blasting load and caustic graphic record. The testing system has also eliminated the effect of blasting fume by and large.

The materials chosen for the present experimental investigation is plexiglas(PMMA) in all cases. Its dynamic caustics parameters are listed in table 1. The specimens are rectangular plates with dimensions 350×250×6mm³ and a borehole of diameter 8mm.

The slot width of the slotted cartridge is chosen as 0.15mm, 0.30mm, 0.50mm, 0.80mm and 1.40mm respectively. The thickness of pipe is 0.9mm. The dynamic caustics characters for different width of slot are studied under blasting load.

4.. EXPERIMENTAL RESULTS AND ANALYSIS

The photos obtained from the dynamic caustics instrument are set up under a projecting apparatus. The position of a crack tip is determined. So the length of a crack propagation and the value of D_t(the transverse diameter)-D_l(the longitudinal diameter) of caustics spots are derived from the respective caustics shapes at each time instant in the test. The

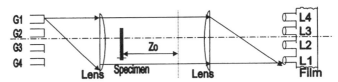

Figure 1 The experimental optical set-up for transmitted dynamic caustics

Table1 The dynamic caustics parameters of experimental material [3]

Material	C_1(m/s)	C_2(m/s)	E_d(N/m²)	$\square$	$|C_t|$(m²/N)
PMMA	2252	1200	4.5×10^9	0.38	0.88×10^{-10}

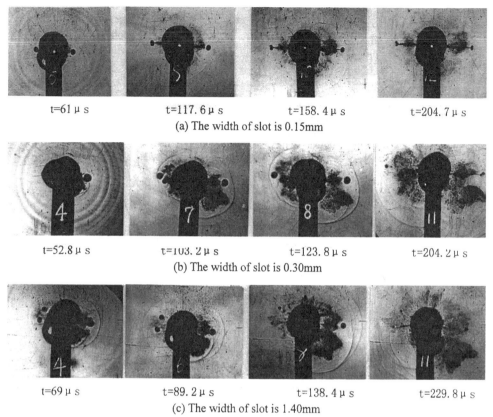

| t=61 μ s | t=117. 6 μ s | t=158. 4 μ s | t=204. 7 μ s |

(a) The width of slot is 0.15mm

| t=52.8 μ s | t=103. 2 μ s | t=123. 8 μ s | t=204. 2 μ s |

(b) The width of slot is 0.30mm

| t=69 μ s | t=89. 2 μ s | t=138. 4 μ s | t=229. 8 μ s |

(c) The width of slot is 1.40mm

Figure 2. Series of photographs showing the crack propagation under blasting load in Plexiglas

change in dimensions of the caustics shape indicates the variation of K_I^d and V. The variation of velocity of a crack propagation and of the dynamic stress factors (K_I^d) versus the crack length for the specimen in Figure 2 are shown in Figure 3 and Figure 4 respectively.

The results have shown that mode of the crack propagation in the initial stages is tensile mode-I and in the later stages is complex mode. Figure 2 shows that the length of crack propagation with the 0.3mm slot width is the maximum comparing to different slot width at the same experimental condition. There are barely cracks at the non-slotted position. The effect of directional broken is best with the 0.3mm slot width. Although some effect of directional broken can be reached with other slot width, many cracks occur at the non-slotted position too.

It is clear in Figure 3 that V of the 0.3mm slot width is the highest of 360m/s comparing with different slot width. The results indicates that the

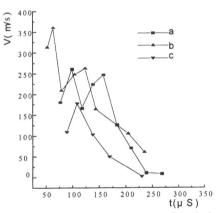

Figure 3 Variation of the crack propagation velocity V versus time under blasting load
(a-the width of slot is 0.15mm; b-the width of slot is 0.30mm; c-the width of slot is 1.40mm)

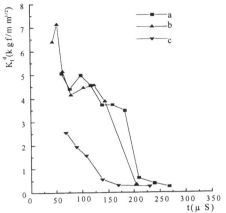

Figure 4 Variation of the dynamic stress intensity factor K_I^d
versus time under blasting load
(a-the width of slot is 0.15mm; b-the width of slot is
0.30mm; c-the width of slot is 1.40mm)

crack propagation is very fast among about
50□100□s. The value of K_I^d is becoming smaller
with decreasing of the slot width as shown in Figure
4. The crack propagation is initiated late when the
slot width is 1.40mm.

A crack at the slotted direction forms
preferentially with the slotted cartridge at fire. The
tip of the crack forms a stress singular field. The
crack has been propagating by the dynamic action of
impact wave and the static action of bursting gas.
The larger slot width makes wider area of the
dynamic action, reduce the power of dynamic action.
Many cracks are formed at the non-slotted direction.
In a word the slotted cartridge leads the dynamic
action in the slotted direction to obtain the crack
directional fracture. The effect is reduced in the non-
slotted direction.

5. CONCLUSIONS

The research shows that optical method of dynamic
caustics is very effective for the determination of the
characteristic parameters of singular stress field at
the crack tip under blasting load. It is very useful for
study the phenomena of initiation and propagation of
the crack under blasting load. The change in
dimensions of the caustics implies the variation of
K_I^d and V. Furthermore the values of K_I^d and V are
more accurate with dynamic caustics.

The results have also shown that mode of the
crack propagation in the initial stages is tensile
mode-I and in the later stages is complex mode in
using slotted cartridge under blasting load. The
effect of directional broken is best with the 0.30mm
slot width. Although some effect of directional

broken can be reached with other slot width, many
cracks occur at the non-slotted position too.

A crack and a stress singular field form
preferentially with the slotted cartridge at the slotted
direction. The crack has been propagating by the
dynamic action of stress wave and static action of
bursting gas. The slotted cartridge leads the dynamic
action in the slotted direction to obtain the crack
directional fracture. The effects are reduced in the
non-slotted direction.

REFERENCES

G.A.Papadopoulos 1993. *Fracture Mechanics—the
Experimental Method of Caustics and the Det.-
Criterion of Fracture*. London.
LiuChen, SuXianji 1988. The effect of velocity of
the crack propagation on caustics lines and
mensuration of the constant of dynamic stress-
optics. *Journal of Experimental Mechanics*.
3(2):111.China: Hefei.
YangRenshu. *Experimental study on the mechanism
of rock shot hole directional fracture by control
blasting with dynamic caustics*. China University
of Mining and Technology of Beijing, Doctor's
degree thesis. 1997:24.

Mining Science and Technology'99, Xie & Golosinski (eds) © 1999 Balkema, Rotterdam, ISBN 90 5809 067 1

Simultaneous extraction of coal and coalbed methane in China

Shugang Li
Department of Mining Engineering, Xi'an Mining Institute, People's Republic of China

Minggao Qian & Jialin Xu
Department of Mining Engineering, China University of Mining and Technology, Xuzhou, People's Republic of China

ABSTRACT: The paper reports on studies of behavior of coal bed methane using the stratum theory. The features related to coalbed methane accumulations and flows are discussed and a technique for simultaneous extraction of coal and coalbed methane is proposed.

1. INTRODUCTION

There is a rich reserves of coal in China, up to now ascertained reserves of raw coal are about $9*10^5$ million tons. China is also one of the richest country in coalbed methane, according to geological survey there are $(3.0-3.5)*10^3$ million m^3 coalbed methane contained in bituminous and anthracite coal field with depth less than 2000m. In view of the sustainable development, the extraction mode which only mines coals but ignores other energy resources such as coalbed methane has to be reviewed again. In fact, for many years, China not only has greatly extracted coal (the output of coal has won the first of the world for several years) to meet the needs of the national economy construction and to export for earning the foreign currency, but also try to greatly drain coalbed methane out in benefit of both assuring the safety of extraction and using of methane (such as Fushun, Yangquan and Furong mining district). But the research and productivity levels still lag well behind advanced coal-mining countries. According to the national statistics, the average rate of drained methane out of 117 coal mines in 42 mining bureaus is 16.5% only (Tu Xigen et al. 1995), the most of them is in Fushun mining bureaus about 30-50%. Recently the field tests of extracting coalbed methane by drilling have be carried out in Liulin, ShanXi, Liujiatun of Fuxing and Daxin of Tiefa, but the result is not ideal. There are many reasons for that, but incomprehension of the inborn nature characteristic of coalbed methane in coal seam is certainly one of them.

2. THE FEATURES OF COALBDE METHANE IN CHINA

Contrasted with USA, Russia, Ukraine and Poland, the coalbed methane of China has the following characteristics:

(1) Plenty of coalbed methane storage. Geological survey made in 1989 by a united research group of China University of Mining and Technology, Huailan Mining Institute and Xi'an Branch of the Central Coal Mining Research Institute showed that the storage of coalbed methane in China is about $3.318*10^3$ million m^3, which is more than 3 times those in USA. There are $2.5*10^3$ million m^3 coalbed methane in 68 poly-coal units locating below surface 300-1500m, which mainly is distributed in the North, Northwest and South of China.(Yang Xilu 1996)

(2) Quite high methane adsorbing ability on coal seam.

(3) Lower pressure of coal seam methane. In China in most of cases the coal seam methane pressure is between 0.5-3.0MPa, while in a few mines the pressure can reach 5.0-8.5MPa below 800-1000m. However, in the Black Warrior Basin and the San Juan Basin of USA, the methane pressure is up to 5.6-8.8MPa at the depth of 600-822m.

(4) Lower scale fissure formed in coal seam by the forced measure such as hydraulic shattering. The half-length of fissure of the Black Warrior Basin is 76-91.44m, but it is only about 30m in China, and in Fushun coal field it can reach 50m.

(5) Lower permeability coefficient of methane in seam. In most of cases the permeability coefficient is less than 0.001md, and the maximum is about 0.54-3.87md in Fushun mine. The permeability of coal seam is the most important parameter in methane extracting, but lower permeability coefficient will result in difficulties in exploration and extraction to some extent.

The permeability has a close relation to the structure of porosity, characteristics of failure, ground pressure, methane content, methane adsorbing and analytic feature, the temperature of in China shows that the modification of coalbed methane moving and the ground pressure caused by mining play a decisive role in the change of coal seam permeability, and in return it is also important for assembling and draining out methane and the distribution of methane pressure.

In China, the annual amount of methane drained out is up to 6 billion m^3. Research and draining practice show that the relations between methane and coal seam feature as "inter-growth" and "coexisting", i.e. coal seam is both generation and reservoir body for gas.

The methane produced and reserved in the coal seam will not immigrate enormously, unless the methane becomes unbalanced due to the change of ground pressure and the deforming, moving, destabilizing of surrounding rocks induced by mining. The methane immigration includes permeation, diffusion, float, emitting to cracked zone or drained out artificially, over assembling, emitting, and even gas burst. The facts have showed that it is not suitable for us to directly extract the coalbed methane on large scale by drilling as USA did. More attentions should be focused on the research of the strata stress distribution, fissure distribution and the immigration of coalbed methane induced by mining in order to extract coal and methane efficiently and safely.

3. THE INSIGHT OF CO-EXTRACTING COAL AND COALBED METHANE FROM KEY STRATUM TERORY

The extraction of underground minerals will lead stress re-distribution, the deformation and fracturing of the surrounding rock, and causes change of the fissure in the surrounding rocks. The movement of stratum around mining fields can cause all kinds of hazards, such as injuries and deaths of miners, collapses of working face and tunnel, and distribution change of water and methane in the coal seams, surface subsidence, even gas and coal burst, or floor water outburst from floor. Recent research has found out that the movement of overlying stratum induced by mining depends dominantly on some of harder thick rock seams, which bear most of mining-induced ground pressure. This one or several harder seams within the overlying stratum are called key stratum (Qian Minggao et al 1996) which play a main control in ground activities. The mechanical behavior of the key stratum, such as deformation, crack, formed structure and movement, will control a large range of strata activities in surrounding rocks of working face, and affect the scope from working face, support system and floor rock mass to the earth's surface. So it can be conclude that the research of stope ground pressure, strata displacement and surface subsidence should be based on the model of key stratum structure integrity.

The key problem lies in that before roof caving the key stratum of above the working face subside down based on Winkler elastic foundation (beam or plate) and while other stratum below the key stratum will deform discontinuously and separation among the stratum will emerge incompatibly. If the sub-key structure existed, the strata can form Voussoir Beam after the local fracturing. In the meantime, incompatible bed-separation under the main key stratum will emerge between the discontinuous and continuous deformation (Qian Minggao et al. 1997). After sub-key strata or key strata fractured and formed Voussoir Beam, discontinuous and disintegrated deformation bed-separation developed around the boundaries above the working face rather than in the middle. The amount of bed-separation depends on the length ratio of fractured rock in the key stratum, the loose coefficient of soft strata and mining depth.

The existence and change of broken fissures and bed-separated fissures of the overlying stratum provide assembly spaces and immigration passages for the immigration and assembly of methane coming from working coal seam and adjacent coal seam during the course of mining. And in this course, the forms of strata structure, disruption and destabilizing of key strata will be greatly influence on the immigration of coalbed methane. Fast or abnormal methane emitting to working face is also a kind of ground behavior caused by the initial fracturing and periodic fracturing of key stratum. The key stratum theory and the following control practice will certainly give new insight to co-extracting coal & coalbed methane. So comprehensive understanding of mechanical behavior of key stratum, features of coalbed methane immigration and assembly, and reasonable and effective measures of draining methane out, will hopefully lead the new technology of co-extracting coal and coalbed methane more safely and economically.

4. THE SPACIAL DISTRIBUTION OF CRACKED STRATUM BEFORE AND AFTER CAVING OF KEY SYRATUM

The recent research has confirmed that the distribution of cracked zone within overlying stratum is not a uniformly bedding distribution as expected by the traditional ground control theory. According to the study of numerical and physical simulation as well as field observation, a prominent bed- separated deformation will develop below the main key strata before floor caving. Before destabilizing of main key strata, the breakthrough of bed-separation fissures and broken fissures forms a banded distribution just like elliptic-paraboloidal zone in space. The bedding cut produces a cracked development elliptical zone, which also is named "O" type circle zone. This cracked space, where coalbed methane will immigrate and assemble, is upper part of collapsed around boundaries over the working face. The middle part of it will re-solidified by the collapsed rockmass, thus the paraboloidal band distribution is formed in the section.

After full mining, the key strata have undergone the initial fracturing and periodic fracturing, and elliptic-paraboloidal crack zone is vanished. But elliptical cracked zone spread on bedding plane still exists. Moreover, the width of cracked zone on the initial mining boundary is equal to the initial fracture distance of key strata, while the width of cracked zone above the working face varies in one or tow times of periodic fractured distance.

Hence, the space distribution of cracked zone of overlying zone, the assembly and immigration methane from working coalbed or adjacent coalbed are in dynamic process, so the methane-drained-out techniques should also follow this dynamic process. During extracting of coal seam, methane coming from working coal seam and adjacent coal seam will assemble and float up because of the density or concentration difference between methane and air (fresh air or leakage air) and diffuse to the cracked zone by the buoyancy. Then methane assembles in the upper developed cracked or bed-separated band. The upward depth of methane floating is directly proportional to the density and emission pressure of methane from working seam and adjacent coalbed. This dynamic process of methane assembly and immigration can be explained by the theory of methane "float-diffuse", which interprets that the mining induced crack zone in overlying stratum is the delivery and assembly zone as well as drainage passages of methane. This provides scientific basis for new technology of methane drainage by drilling.

5. THE KEY TECHNIQUE OF CO-EXTRACTING COAL & COALBED METHANE

By the field measurement and lab experiment, the abutment pressure induced by mining play a key role in the distribution of permeability coefficient. Permeability coefficient of coal seam is rather low in concentrated zone of abutment pressure ahead of working face, hence methane pressure is increased and the outflow of methane is decreased. But in the pressure relieving zone, the outflow of methane is increased and permeability coefficient is increased, sometimes even about 100 times higher than that of concentrated zone, therefore the outflow is increased greatly, this is so called "pressure relieving and outflow increasing effect". So the conclusion is that no matter how low original permeability coefficient

of coal seam is, after the pressure relieving due to mining, the permeability will be increasing greatly, and the seepage velocity of methane as well as the outflow are increasing greatly too. While leakage will lead methane rising, floating and diffusing to cracked zone. So it is more favorable for methane drainage in the cracked zone. All that is the theoretical basis of co-extracting coal and coalbed methane.

According to above discussion, whatever measures would be taken in the drainage of coalbed methane, such as hole suction, roadway drainage and surface drilling, the position of roadway or drilling terminal hole should be selected in the active and enrichment zone of coalbed methane. Practice shows the feasible techniques are as follows:

(1) Rational layout of drilling hole for methane suction should be arranged at zone, where the methane outflow increases while methane pressure relieves, ahead of working face.

(2) In order to extract high density methane, drilling well should be located in the advancing zone of coal face, in stead of goaf zone, according to dynamic process of the cracked zone of overlying stratum during the extracting of coal seam.

(3) In order to extract enrichment methane in the cracked zone, high drainage roadway should be located in cracked zone on strike or dip direction. Alternative is to adopt drilling hole with the major diameter of 200~300mm and long horizontal distance of 500~600m in the mining induced crack zone of the overlying stratum.

(4) To extract methane fully by adopting mining method with protecting layer, such as top slice per-mining method in fully-mechanized top coal caving, or in advance relieving adjacent seam. It not only pre-breaks the hard roof to relieve the ground pressure but also speed-ups the methane delivering, increases the rate of outflow. Through above techniques, the delivering velocity of methane is faster, and floating content is increasing in the range of pressure relieving, and methane can be extracted fully.

6. REALIZABLE GOOD BENEFITS OF CO-EXTRACTING COAL & COALBED METHANE

As well known methane is a harmful gas to threaten mine safety. Since human began to mine coal, easy-to-burn and easy-to-explode methane has resulted in countless vicious accidents. On other hand, however Methane is also a kind of clean and efficient energy. Hence co-extracting coal and coalbed methane must get good economical and social benefits.

In order to explain the above point, we take Luling mine, at Huaibei coalfield in China, as an example. This mine belongs to high methane and easy outburst one with designed capacity of 2.4 million tons per year. The coalbed methane storage is 15 m^3 per ton coal. There are 3.71 billion tons recoverable coal and 64 billion m^3 methane above 1200m depth. At moment this mine produces annual 1.85 million tons coal while the methane emission is about 31.62 millions m^3 each year. According to the features of thicker while softer coal seam as well as lower permeability (0.0067md), the vertical drainage holes were drilled from earth surface. The drainage holes were laid in the elliptic paraboloier cracked zone. The long distance holes along roof and across roof were also drilled horizontally in order to drain out methane. In this technique of co-extracting coal and methane, the methane extracted can provide 4000 household diary usage (Xu Jialin et al, 1997). Suppose if every household uses 3 tons coal each year, that means it can save 12 thousands tons coal and 180 million RMB Yuan every year totally. If the price of methane is 1 RMB Yuan /m^3 and each family is supplied methane 1m^3 everyday, the mine can get 1.46 million Yuan of income in addition. Taking both above into account the total income is 3.26 million Yuan, so the economic benefit is very great. On the other hand, as raw material methane can be used to generate electricity, made for synthetic ammonia, methanol, ethyne and hydrogen and so on. Research shows that 1000 m^3 of methane are equal to 4 tons raw coal if account for equivalent calorific capacity. Taking Luling mine as an example, if only the 80% of methane emission can be used each year, that means 101.2 thousands raw coal will be saved. Moreover the problem of environment pollution caused by methane emission can also be solved.

Co-extracting coal and coalbed methane has the following advantages: reducing amount of emission methane in working environment underground, assuring safety of mining on working face, minimizing the outburst of coal and methane, decreasing the ventilation load and air speed on working face, reducing the coal dust flying up, improving the working condition. is an excellent "green energy"(Xie Zhengyi 1998). It not only can reduce the environment pollution which is caused by draining methane into atmosphere, but also create significant economic benefit. For example, Luling mine economizes on coal 12 thousand tons every year, we can avoid about 96 tons SO_2 and 768 tons smoke draining out if methane was utilized. Therefore the new technique of co-extracting coal and methane can significantly improve the quality of atmosphere environment.

According to geological surveying, the reserves of extractable coal is up to 867 billion tons and coalbed methane is about 9087 billion m^3 within the depth of 2000m in Huainan and Huaibei mine field in China (Liu Huamin 1997). If we can co-extract coal and coalbed methane safely and efficiently in zone of so rich resources. there will be a great strategic

significance for the economic development in East of China. Farther more, there are about 30% of total coalmines with high methane and coal-and-methane outburst in China today. As continuous increase of mining scale and depth, the possibility of accident resulted from methane burst will increase consequently. It is significant to apply and study the theory and technique of co-extracting coal and coalbed methane.

7. CONCLUSIONS

Based on the features of coalbed methane in China, ideal technique of draining methane out seems to be co-extracting coal and coalbed methane in the course of mining, in this way we can drain out the fast moving and plentiful assembled methane efficiently. It implies two aspects: The first is that we can minimize methane-related disaster when extracting coal. Second, we can make full use of mechanical behavior of overlying stratum to study the distribution of delivery and assembly of methane flowing in rock mass or coal seam. If we take effective techniques to exploit and use excellent methane resources, a significant benefit can be obtained in point of good economical and commercial efficiency, the improvement of environment and atmospheric condition, as well as sustainable development of coal industry.

REFERENCES

Liu Huamin 1997. The current situation and prospects of coalbed exploration and use in two Hui coal field of Anhui China Coalbed Methane(2): 21-24.

Qian Minggao, Miao Xiexing & Xu Jialin 1996.Theoretical study of key stratum in ground control Journal of China Coal Society vol 21 (3): 225-230.

Qian Minggao, Miao Xiexing 1997. Mining strata mechanics. Science and Technology Review (3): 29-31.

Tu Xigen,Wang Youan & Wang Zhenyu 1995.The current situation and prospect of controlling coal mine methane in our country Coal Mine Safety (2): 3-7.

Xie Zhengyi 1998. 06. 16 household use "green energy" in Luling coal mine China Science and Technology Daily 6.

Xu Jialin, Liu Huamin 1997 Study on layout of methane suction hole in gob. Coal Science and Technology (4): 28-30.

Yang Xilu 1996. The progress of exploitative exploration of coalbed methane. Coal Geology and Exploration (1): 29-32.

Mining Science and Technology' 99, Xie & Golosinski (eds) © 1999 Balkema, Rotterdam, ISBN 90 5809 067 1

Infrared radiation detection: A potential RS technique for ground control

LiXin Wu & Jinzhuang Wang
Department of Resource Development Engineering, Beijing Campus of CUMT, People's Republic of China

ABSTRACT: The relationships between infrared radiation and rock stress are stated, and the recent research results are presented. Experimental study indicates that coal and sandstone under load have three types of infrared thermal images features, three types of infrared radiation temperature features, as well as the infrared forewarning messages of samples' failure. It was found that the infrared detection is comparable with acoustic emission detection and electrical resistance detection. Stress of around $0.79\,\sigma_c$ could be taken as the dangerous level warning zone both of ground control and ground-pressure disasters.

1. INTRDUCTION

During recent years, many scholars worldwide have conducted quite a few experiments of rock loading in the laboratory and founded some important phenomenon occurred during its fracturing, including electromagnetic radiation from low-frequency to radio-wave, light emission and electron emission. Here one of the most important achievements in Geoscience is the discovery of the meteorological satellite's thermal infrared anomaly just before the coming of an earthquake (Qiang, 1990)and the verification of it both in laboratory and in practical earthquake forecasts (Geng, Cui, Qiang, 1992,1993,1995,1997). A new subject and terminology--*remote sensing rock mechanics (or remote sensing rock physics)* was put forward by Geng and Cui in 1992. Sixty-eight earthquake predications based on satellite thermal infrared anomaly had been made and half of them gave very good results, while the others gave false alarms (Qiang, 1997). And, the famous earthquake of Zhangjiakou, 200Km far away from the north of Beijing in the end of 1996, was fairly exactly foreseen in the satellite infrared images according to abnormal infrared variation.

The physical mechanism of the above should be explained as: when the rock is subjected load and deformation, the mechanical energy accumulates in the rock. Meantime, the accumulated energy transfers into other energy such as acoustic emission energy (micro-fractureing) and thermal energy and electromagnetic radiation energy (the friction between fissures and grains). Thus resulted infrared temperature and its field variation could be detected and could reflect the variation of the thermal and the electromagnetic radiation energy, and could give some forewarning messages (Wu, Wang, 1997,1998).

2. EXPERIMENTAL METHOD

2.1 *Sampling and loading*

In order to observe the features of infrared radiation of rock loading and to study the relationship between the radiation and the rock's mechanical properties, a series experiments had been conducted, including uni-axial loading, bi-axial loading, splitting and three-point bending. Totally more than 100 samples from about 30 lithologies, which were collected both from earth surface and underground mines, had been tested during the past 7 years.

The size of the samples for uni-axial loading experiment was as large as from 5cm×5cm×5cm to 14cm×14cm×28cm, while that for bi-axial experiment was about 4~6cm×8cm×8~10cm. The diameter of the cylindrical samples for splitting and three-point bending experiment was 4~5cm. The loading-bearing ends of sample were grounded evenly and parallel with error within 0.05cm. The other sides of sample were left coarsely except the area where the strain chip attached.

All the samples were loaded until failed. The total

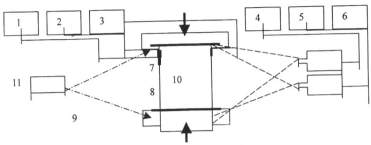

Fig.1 Experiment device and its concept configuration
1 - stress and displacement apparatus
2 - acoustic emission apparatus
3 - electricity resistance apparatus
4 - color monitor
5 - black-and-white monitor
6 - personal computer
7 - IR thermometers ES 2007
8 - IR imaging system
9 - loading plate
10 - sample
11 - plastic plate (for electricity isolate)

loading time of each sample from beginning to failure was controlled to be about 10 minutes, only a few was more than 20 minutes or less than 5 minutes because of the inaccurate estimated strength.

2.2 *Instruments and configuration*

The instruments that had been used in the experiment could also be classified into three parts: loading machine, infrared instruments and universal monitoring devices.

Infrared Radiation Thermometer: Two infrared radiation thermometers of temperature sensitivity 0.1□, FOV 2°and wavelength 8~14□m were selected to continuously monitoring the infrared radiation temperature of the rock surface. The signal was recorded by a 4-pen recorder.

Infrared Imaging System: Four infrared imaging systems had been successively used to observe the rock thermal radiation field and its variation process. Both the detection temperature range and the color scale of these infrared-imaging systems are adjustable. The IR temperature sensibility of these systems is 0.025~0.1□, working band 2.2.~14□m, scanning rate 25~60 images every second. The sample's thermal images were recorded in three manners: by non-standard TV cassettes, by PC disk or on PC diskettes.

The configuration of sample, loading-head and measuring apparatuses is shown in Fig.1. The loading direction was the same as that of the original vertical stress in-situ. The infrared instruments were placed on the same side for comparison. The measuring apparatuses were arranged 2m away from the rock sample to avoid the rock debris, which may hit and damage the instruments when the sample failing and bursting.

3. EXPERIMENTAL RESULTS

3.1 *Infrared thermal image features*

Totally 12 coal and roof-sandstone samples from mine were loading-tested for studying the features of infrared thermal image. The result shows that all the samples' infrared thermal image has reflected the stress conditions and failure features of samples. The results of six samples also show the forewarning messages of its crash. All the results of experimental could be classified into three types:

Type I--Temperature rises gradually but drops later with failure forewarning of low temperature. Its characteristics are: 1) with loading, the infrared radiation temperature rises gradually; 2) as the vertical stress reaches $0.82\sigma_c$ (average), the temperature begins to drop; 3) as the vertical stress reaches $0.94\sigma_c$ (average), the temperature becomes the lowest. When the sample got failure, local high temperature zone, which resulted from stress local concentration or stress friction effect, could be seen on sample's surface through infrared thermal images. The temperature in these zones was $1\sim2^0$ higher than the average temperature on the whole surface. The temperature inside was generally $0.5\sim1^0$ lower than that of surface (coal samples).

Type II--Temperature rises gradually but failure forewarning of rapid rising. Its characteristics are: 1) with loading, the infrared radiation temperature rises gradually; 2) as the vertical stress reaches $0.91\sigma_c$ (average), temperature gets rise rapidly; 3) when the vertical stress reaches $0.99\sigma_c$, the infrared thermal image clearly shows low temperature strip along just generated cracks. It was also found that when the sample began to crash and in the moment of crashing, the infrared thermal image had clearly shown lower temperature inside the samples and higher temperature on the crack surface and local zone of stress concentration (coal samples).

Type III--Temperature drops in the beginning, rises later, drops again in the end and shows low-temperature forewarning (roof sandstone). Its characteristics are: 1)at the beginning of loading, the infrared radiation temperature drops a little; as the vertical stress reaches $0.1\sigma_c$, the average temperature on the whole field of surface drops 0.15^0; 2)then the temperature turns to rise gradually; as the vertical stress reaches $0.67\sigma_c$, the temperature rises nearly to the original temperature before loading; 3)then the temperature drops again, and as the vertical stress reaches $0.96\sigma_c$, the temperature becomes the lowest, and low temperature strip begins to show the site of cracking. These results agree with that of Chui Chengyu and Geng Naiguang (1993,1995,1996).

3.2 *Infrared radiation temperature features*

Totally 8 coal samples from mine were tested for studying the features of infrared radiation temperature. The results of all 8 samples have shown that all variation curves of infrared radiation temperature had given obvious forewarning messages before the sample failure. These forewarning messages could also be divided into three types:

Type I--forewarning of low temperature. Its characteristics are: 1)before the crash of the sample, a low temperature valley point emerges in the fluctuated temperature-variation curve; 2)the moment that this valley point emergence is when vertical stress reaches $0.83\sim0.90\sigma_c$, average $0.86\sigma_c$; 3) the temperature at this moment is $0.1\sim1.1^0$ (average 0.5^0) lower than the original temperature before loading.

Type II--forewarning of high temperature. Its characteristic are: 1)first the radiation temperature curve fluctuates, then a high temperature peak point or zone emerges in this curve just before the crash of the samples; 2)the emergence of this peak point or zone is when the vertical stress reaches $0.82\sim0.95\sigma_c$, average $0.87\sigma_c$; 3)the temperature at this moment is $0.1\sim0.3^0$(average 0.2^0) higher than the original temperature before loading.

Type III--forewarning of continuous high temperature. Its characteristic are: 1)as the vertical stress reaches $0.65\sigma_c$, high temperature emerges in the temperature variation curve; 2)the temperature is $0.2\sim0.3^0$higher than the original temperature before loading, and keeps until the sample suddenly burst, as shown in Fig.2. Among all of the 20 samples tested, only one sample (blind coal of high strength, 82MPa) got burst. When the sample burst, almost all the fragments flied out of the loading platform as far as 5m.

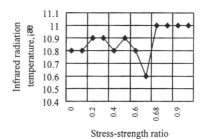

Fig. 2 The features of IR radiation temperature curve

So, it could be preliminary concluded that, the emergence of continuous high infrared radiation temperature is the results of the long-time and abundant accumulation of mechanical energy in the sample. When the stress reaches the samples' strength limit, the accumulated energy reaches its limit and gives out as sudden burst. Detailed explanation waits for further research.

3.3 *The infrared radiation temperature increased with the strength*

Many results shown that the higher the sample's strength, the higher the infrared radiation temperature increased, especially inside the sample during breaking. This includes Mn-steel, iron ore, blind coal, coal and many other rock materials including sandstone, siltstone, granite, limestone and marble. Fig.3 shows the detected relationship-curve between the maximum inside IR radiation temperature during breaking and the samples' strength. The reason maybe lays in that the higher the sample's strength, the more mechanical energy will be accumulated in the sample, and thus more thermal energy will be transferred from the mechanical energy.

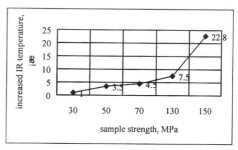

Fig. 3. Maximum IR temperature increase with increasing sample strength

4. THE COMPARISON OF FOREWARNING MESSAGE

The sample's failure-forewarning stress points of infrared thermal image, infrared radiation temperature, acoustic emission and electrical resistance were systematically compared. The statistical results were listed in table 1

Table 1. Comparison of the forewarning stresses of a coal failure

Detection technique	AE Peak valley		Electrical resistance	Infrared image temp.	
Sample sum	9	9	7	6	4
σ_i/σ_c	0.75	0.78	0.79	0.82	0.80
average	0.76		0.79	0.81	
average			0.79		

5. CONCLUSIONS

1. The infrared radiation variation of coal and sandstone could be detected during loading process. And both the variation of infrared thermal image and infrared radiation temperature could be classified into three types;

2. The forewarning stress of infrared remote detection is comparable with that of acoustic emission detection and electrical resistance detection, and the stress nearby $0.79\sigma_c$ should be taken as stress-caution-zone of ground control and its disasters' infrared remote detection, acoustic emission detection and electrical resistance detection.

3. The discoveries have provided the theoretic and experimental basis for the application of the remote sensing technique to the ground control including the monitoring and prediction of rock burst, rock breaking, pillar failure and supporting framework failure.

REFERENCES

Qiang Z.J., et al., Abnormal infrared thermal of satellite--forewarning of earthquake. *Chinese Science Bulletin*, Vol.35 (17), 1324-1327(1990)

Geng N.G., Cui C.Y., et al., Remote sensing detection on rock fracturing experiment and the beginning of remote rock mechanics. *ACTA Seismological SINICA*, Vol.14 (add), 645-652(1992).

Qiang Z.J., et al, Satellite Thermal Infrared Increase Temperature Anomalous with Imminent and Short-term Prediction before Earthquake. *Proc. of IEAS'97 & IWGIS'97*, Beijing, China, 1266-1270(1997)

Cui C.Y., et al.,The Remote Sensing Observation in experiment of Rock Failure and the Beginning of Remote Sensing Rock Mechanics, ACTA Seismological SINICA, 1993(6):4

Cui C.Y., et al., Basic experimental study on infrared remote sensing be applied to earthquake forecast. New Developments of Remote Sensing, Science Publication, Beijing, 151-160(1995)

Wu L.X., *Coal Strength Mechanism And Experimental Basic Study on Rock Pressure Infrared Detection*, Doctor Thesis of Beijing Campus of China University of Mining & Technology, May (1997)

Wu L.X., Wang J.Z., The Features of Infrared Thermal Image And Radiation Temperature of Coal Rocks Loaded□Science in China- Earth Science, 1998,41(2):158-164

Wu L.X., Wang J.Z., Infrared Radiation Features of Coal and Rocks under Loading, Int. J. of R.M. & M.S., 1998, 35(7):969-976

Mining Science and Technology'99, Xie & Golosinski (eds) © 1999 Balkema, Rotterdam, ISBN 90 5809 067 1

Kaiser effect study on determining rock stress in triaxial loading

F.Wang, V.S.Vutukuri & J.O.Watson
School of Mining Engineering, University of New South Wales, Sydney, N.S.W., Australia

ABSTRACT: This paper introduces the experimental investigation on the Kaiser effect in sandstone, which was under different triaxial loading conditions, namely, the vertical stress greater than, equal to or less than the confining stress. The results show that both the vertical stress and the confining stress affect the Kaiser effect. When the previous applied vertical stress is given the stress determined in axial direction from the Kaiser effect take-off point decreases as the confining stress increases if the confining stress is less than the vertical stress. But if the confining stress is greater than the vertical stress the estimated stress increases with the increase of the confining stress. The estimated stress is more or less equal to the previous applied differential stress between the vertical stress and the confining stress if the vertical stress is greater than the confining stress.

1 INTRODUCTION

It has been sought to measure in-situ rock stresses by an easy, fast, economical and reliable method. The discovery of the Kaiser effect in metals (Kaiser 1953), which suggested that previous stress might be detected by stressing the metal to the point where there was a substantial change in acoustic emission rate, brought a promising way to measure stresses.

In 1963, Goodman confirmed from his experiments that the Kaiser effect exists in three rocks. In 1976, Kanagawa et al. first applied it to measure in-situ rock stress. The method involves application of stress uniaxially on a specimen obtained from the site in the compressive testing machine and recording and processing of the acoustic emission data obtained during loading to identify the Kaiser effect take-off point and estimate the corresponding stress. Since this method seems easy, fast and inexpensive it has been paid great attention. A lot of research work has been undertaken to investigate the Kaiser effect in different rocks, under different loading levels. The effects of confining stress, the stress level of the Kaiser effect existence, environment ie water and temperature, the retention time of the Kaiser effect and the method of determining the Kaiser effect take-off point have been studied.

The main achievements on the Kaiser effect study can be summarised as follows: 1. The discovery of the Kaiser effect existence in both metals and non-metals (Hughson & Crawford 1987). 2. The methods of determining the Kaiser effect take-off point, namely, direct observation (Kanagawa et al. 1976), the pivot point method (Hardy & Shen 1992) and the maximum curvature method (Momayez et al. 1992). 3. The system of monitoring acoustic emission is developed, which includes AE piezoelectric transducers, AE signal amplifiers and computer recording and analysis program. 4. Application of the Kaiser effect to determine the in-situ rock stress state (Kanagawa et al. 1976, Momayez & Hassani 1992, Seto et al. 1992, 1996, Jupe et al. 1992).

From literature review (Wang et al. 1999) different opinions exist regarding the effect of confining stress, water and retention time of the Kaiser effect recovery. Especially the conflicting results are reported whether or not the confining stress influences on the determination of the Kaiser effect.

In order to clarify the confining pressure influence on the determination of the Kaiser effect, a different approach has been proposed by Wang et al. (1999). In this approach the testing is immediately conducted following 40 minutes' triaxial loading on a specimen, which is not removed from the triaxial cell after unloading. The obvious advantage is that the AE noises are greatly reduced and the Kaiser effect is clearly seen. The results showed that the stress evaluated by the Kaiser effect decreased as the confining stress increased. However this conclusion is based on the experiments of the differential pre-loaded stresses greater than 5 MPa. This paper will introduce the further experiments in the condition of the differential pre-loaded stress equal to zero, less than zero and as well as greater than zero.

2 EXPERIMENTS

2.1 Procedure

In the present study, the Gosford sandstone is used. The first step is loading and unloading. The cylindrical rock specimen, 45 mm in diameter and 100 mm in height, is loaded in a triaxial cell to the designated values. The vertical stress (σ_1) is applied by the Schenck, a servo-controlled testing machine, at the loading rate of 100 μm/min. The confining stress (σ_3) is applied manually to a pre-set value at the same loading rate as used for the axial stress application. The vertical stress is increased to a pre-set value if the vertical stress is greater than the confining stress, or the confining stress is increased to a pre-set value if the confining stress is greater than the vertical stress. After maintaining the loading for 40 minutes, the specimen is unloaded while maintaining the designated difference between axial and confining stresses. Then the confining stress becomes zero first if the vertical stress is greater than the confining stress. The vertical stress decreases to zero at last. If the confining stress is greater than the vertical stress the vertical stress becomes zero first and then the confining stress decreases to zero. In this step, the rock specimen is still in the triaxial cell, untouched in order to prevent any AE noises arising from any mismatch between the specimen and plates. The specimen is loaded uniaxially in the vertical direction at the rate of 100 μm/min until the substantial acoustic emissions appear.

The MISTRAS 2001 AE detection and analysis computer system was employed with two piezoelectric transducers and two preamplifiers. The two transducers are attached on the spacer that is on the top of the specimen. AE signals and axial stresses were transferred into the computer system. The gain of preamplifier was set at 40 dB. The gain inside the computer system was set at 20 dB. The threshold was set to 45 dB. The frequency filter was set at 20-200 kHz for channel one and 200-1200 kHz for channel two.

Generally, the Kaiser effect point is identified by observing substantial AE increase or the slope change if high resolution data is available. Otherwise, the maximum curvature method (Momayez et al. 1992) is employed.

2.2 Results

The pre-set triaxial stresses applied to the specimens of sandstone are: axial stresses of 10 MPa, 15 MPa and 20 MPa with various confining stresses from zero to 30 MPa. The testing is conducted immediately after unloading. By plotting acoustic emission (hit rate, count rate and accumulative counts) versus time and stress versus time, the Kaiser effect take-off point can be identified. Figures 1, 2

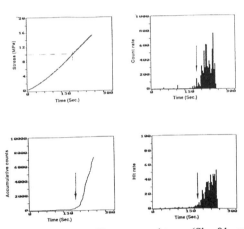

Figure 1 Kaiser effect on sandstone (Sksn81, pre-stress: σ_1 =10 MPa, σ_3=0 MPa, frequency: 200~1200KHz, estimated stress: 10 MPa).

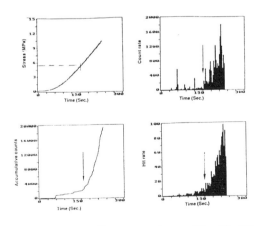

Figure 2 Kaiser effect on sandstone (Sksn33, pre-stress: σ_1=10 MPa, σ_3=10 MPa, frequency: 200~1200KHz, estimated stress: 5.5 MPa).

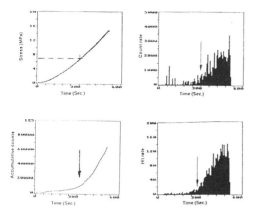

Figure 3 Kaiser effect on sandstone (Sksn62, pre-stress: σ_1=10 MPa, σ_3=20 MPa, frequency: 200~1200KHz, estimated stress: 7 MPa).

and 3 show the estimated stresses by the Kaiser effect, shown by arrows, for three different loading cases, namely, confining stresses 0 MPa, 10 MPa and 20 MPa with the same vertical stress of 10 MPa. Similarly direct observation or the maximum curvature method is used to identify the estimated stresses in other experiments. Basically the Kaiser effect take-off point can be identified whether the confining stress is zero, less than or greater than the vertical stress.

3 DISCUSSION

The results show that the estimated stress is neither the previous vertical stress nor the confining stress except when the confining stress is zero. Then the question is what the estimated stress is? How is the estimated stress interpreted in terms of its previous triaxial stress state?

By analysis of the experimental results the estimated stresses are plotted against the confining stresses in Figure 4.

Obviously the estimated stress is related to the previous applied vertical stress (σ_1) and as well as the confining stress (σ_3). The estimated stress increases with the previous applied vertical stress. In the case of the given previous vertical stress the estimated stress has the polynomial relationships with the confining stress ie:

$$\sigma_e = a\sigma_3^2 + b\sigma_3 + c \qquad (1)$$

where σ_e is the estimated stress by the Kaiser effect; σ_3 is the confining stress; a, b and c are constants of regression, which vary with the previous vertical

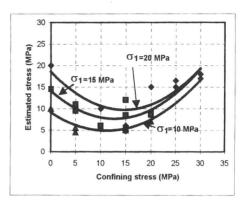

Figure 4 Estimated stress by the Kaiser effect versus the confining stress.

Table 1 Regression constants for Equation 1.

σ_1 (MPa)	x	y	z	R^2
10	0.033	-0.7467	9.07	0.72
15	0.039	-0.9997	14.05	0.53
20	0.041	-1.2084	18.58	0.62

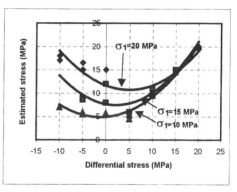

Figure 5 Estimated stress by the Kaiser effect versus the differential stress.

stress. The values of these constants and R-square (R^2) are listed in Table 1.

It can be seen that different combinations of the vertical stress and the confining stress may give the same estimated stress by the Kaiser effect. In other words the Kaiser effect can not determine the previous triaxial stress state unless the previous vertical stress or the confining stress is known.

With the same experimental results the estimated stress is plotted against the differential stress between the vertical stress and the confining stress in Figure 5.

367

It can be seen in Figure 5 that the estimated stress is related to the previous vertical stress and the differential stress (σ_d). The polynomial relationship between the estimated stress and the differential stress is given in Equation 2.

$$\sigma_e = x\sigma_d^2 + y\sigma_d + z \qquad (2)$$

where the x, y and z are constants of the regression, which vary with the previous vertical stress. The values of these constants and R-square (R^2) are listed in Table 2.

Table 2 Regression constants for Equation 2.

σ_1 (MPa)	x	y	z	R^2
10	0.034	0.0837	4.97	0.79
15	0.043	-0.1836	7.68	0.68
20	0.037	-0.3705	11.66	0.62

It can be seen in Table 2 that different previous vertical stresses result in different regression constants x, y and z. When the previous vertical stress and the estimated stress are known the differential stress, which may be either positive or negative, can be determined from Formula 2. If the differential stress is positive the confining stress is less than the vertical stress. If the differential stress is negative the confining stress is greater than the vertical stress. Therefore even if the previous vertical stress is known the confining stress can not be uniquely determined and may have two different values. However when the differential stress is greater than zero, it can be clearly seen that the estimated stress is only related with the differential stress. The estimated stress is more or less equal to the differential stress. If one of the previous vertical stress and the confining stress is known the other can be determined.

4 CONCLUSIONS

In summary, both the previous applied confining stress and the vertical stress have the effect on the stress estimated by the Kaiser effect. The stress estimated by the Kaiser effect has the polynomial relationship with the previous confining stress. The estimated stress decreases with the increase of the confining stress if the confining stress is less than the vertical stress. But if the confining stress is greater than the vertical stress the estimated stress increases with the increase of the confining stress. When the previous applied differential stress is greater than zero the estimated stress is more or less equal to the differential stress.

REFERENCES

Goodman, R. E. 1963. Subaudible Noise during Compression of Rocks. *Geological Society of America Bulletin*. 74: 487-490.

Hardy, H. R. & W. Shen 1992. Recent Kaiser Effect Studies on Rock. *Proc. 11th Int. Acoustic Emission Symp.*, Fukuoka, Japan: 149-157.

Hughson, D. R. & A. M. Crawford 1987. Kaiser Effect Gauging: The Influence of Confining Stress on Its Response. *Proc. 6th Int. Congress on Rock Mechanics*, Montreal, Canada: 981-985.

Jupe, A. J., Barr, S. P. & R. J. Pine 1992. In Situ Stress Measurements Obtained Using Overcoring and the Kaiser Effect of Acoustic Emissions within the Carnmenallis Granite, Cornwall, UK. *Proc. 11th AE Symp.*, Fukuoko, Japan: 167-174.

Kaiser, J. 1953. Erkenntnisse und Folgerungen aus der Messung von Gerauschen bei Zugbeanspruchung von Metallischen Werkstoffen. *Archiv fur das Eisenhuttenwesen*. 24: 43-45.

Kanagawa, T., Hayashi, M. & Y. Kitahara 1976. Estimation of Spatial Geo-stress Components in Rock Samples using the Kaiser Effect of Acoustic Emission. *Proc. 3rd Acoustic Emission Symp.*, Tokyo: 229-248.

Momayez, M. & F. P. Hassani 1992. Application of Kaiser Effect to Measure In-situ Stresses in Underground Mines. Tilerson, J. R. & W. R. Wawersik (eds), *Proc. 33rd U.S. Symp. on Rock Mechanics*: 979-988. Balkema, Rotterdam.

Momayez, M., Hassani, F. P. & H. R. Hardy 1992. Maximum Curvature Method: A Technique to Estimate Kaiser-Effect Load from Acoustic Emission Data. *Journal of Acoustic Emission*. 10(3/4): 61-65.

Seto, M., Utagawa, M. & K. Katsuyama 1992. The Estimation of Pre-stress from AE in Cyclic Loading of Pre-stressed Rock. *Proc. 11th Int. Acoustic Emission Symp.*, Fukuoka, Japan: 159-166.

Seto, M., Nag D. K. & V. S. Vutukuri 1996. Experimental Verification of the Kaiser Effect in Rock under Different Environment Conditions. *Proceedings of Eurock 96*, edited by Barla, G., Torino, Italy: 395-402.

Wang, F., Watson, J. O. & V. S. Vutukuri 1999. Different Approaches to Study the Kaiser Effect in Rocks in Triaxial Loading. The 8th Australia New Zealand Conf. on Geomechanics, Hobart, Australia: 935-941.

Mining Science and Technology' 99, Xie & Golosinski (eds) © 1999 Balkema, Rotterdam, ISBN 90 5809 067 1

Stability of the surrounding rock for openings in fractured rock

Tongyou Liu, Qianli Zhao & Weidong Yin
Jinchuan Non-ferrous Metal Corporation, Jinchang, People's Republic of China

Qian Gao
University of Science and Technology, Beijing, People's Republic of China

ABSTRACT: This paper probes into a method of reliability analysis for support design of large span roadways in fractured stratum with high stress field. First, failure features of fractured rock masses in high original stress field are studied. Then a model of reliability analysis based on shear sliding wedge theory is proposed. Finally Rosenblueth method is used to solve the reliability of sliding wedges and some important conclusions are obtained.

1. DEFORMATION AND FALURE FEATURES OF FREACTURED ROCK IN HIGH STRESS FIELD

Professor Tan Tjongkie (Tan Tjongkie 1986) explored large convergence mechanism of roadways in the fractured stratum with high stress at Jinchuan mine and put forward a mechanics of dilatancy and buckling. He explained that the plastic wedges squeeze into opening is results of rebound of the surrounding rock, release of stress in rock and swelling of rock mass. Summing up researches in Jinchuan and based on authors' experiences of more than thirty years in Jinchuan mine, we consider that structural features and failure mechanism of surrounding rock and bolt-shotcrete system possess several of the following important characteristics (Liu Tongyou & Zhou Chengpu , 1997):

1. Both surrounding rock and bolt-shotcrete support constitute a load bearing system. The stability of the opening not only depends on the load□bearing ability of various components (such as surrounding rock, bolt-shotcrete etc.), but also their interaction between components as well as their coordination of deformation.

2.Interaction between rock and support is a dynamic process. It means that stability of the opening varies with time and space. The reasons for that are sated as follows:

a. Mechanical strength and deformation quality of rockmass are gradually weakening when deformation (convergence) is increasing with time. This not only is owing to creep effect, but also, the most important, it is due to the results of deformation of surrounding rock itself.

b. Owing to adjusting unbalance forces in rock masses with deformation of the rock, values and states of stresses are changing with time. Therefore, acting mechanism of bolts is uncertain. That makes failure models of the system various and multi-typed.

c. During the process of interaction between rock and support system, some local failures will occur at the wall of roadway. The local failures may not lead to instability of the system but will cause the progressive failure of the system and in turn make the structure unstable.

d. Shearing sliding wedges are one of main failure models of surrounding rock of roadways in fractured stratum with high stresses. The destruction model can be analyzed by using the shearing sliding theory proposed by T. L. Rabcewicz .

Based on researches mentioned above, a reliability analysis method was put forward. The method is able to consider these uncertain factors affecting stability of surrounding rock and coordination of deformation.

2. CACLCULATION OF SUPPORT FORCE FOR CIRCLE OPENING IN HIGTH STRESS FIELD

2.1 *Calculation of support resistance without considering deformation coordination of support cells*

Support resistance P_{T1}, for bolt-shotcrete support with a circle section roadway, can be calculated by

"shearing sliding theory":
$$P_{TI} = P_C + P_W + P_G + P_H \qquad (1)$$

where P_C, P_W, P_G, P_H are bear ability of shotcrete layer, metal net, rock bolt and rock bearing arch, respectively. Each term of the above formula can be given as follows:

a. Shotcrete layer

$$P_C = \frac{2d \cdot \tau_B}{b \cdot \sin \alpha} \qquad (2)$$

where α-shearing angle of surrounding rock, $\alpha = 45° - \varphi/2$, φ-interior friction angle of rock mass, b-width of shearing sliding zone on wall of surrounding rock, $b = 2R_0 \cos \alpha$; R_0-radius of circle section. If the section is not circle, then equivalent radius is $R_0 = 1/4$ (H+B), here, H-height of non-circle roadway, B-span, τ_B-resistance shearing strength of shotcrete layer, $\tau_B = 0.2\sigma_c$, σ_c-uniaxial compression strength of shotcrete layer, d-thickness of shotcrete layer.

b. Metal net

$$P_W = \frac{2F_W \cdot R_T}{A \cdot b \cdot \sin \alpha} \qquad (3)$$

where F_W-traversed area of a single metal net; R_T-resistance shear strength of a steel, A-space between metal nets.

c. Rock bolts

$$P_G = \frac{2a \cdot F_G \cdot \sigma_t \cdot \cos \omega}{e \cdot t \cdot b} \qquad (4)$$

where a-an half of the arc length of shearing sliding rock mass, $a = \frac{\pi}{180} R_0(90° - a)$, F_G-section area of a rock bolt; e, t-space between rock bolts along longitudinal and latitudinal directions respectively, ω-mean dip angle of rock bolts, $\omega = \frac{1}{2}(90° - a)$.

d. Rock bearing arch

$$P_H = \frac{2s\tau^R \cos\psi}{b} - \frac{2s\sigma_n^R \sin\psi}{b} \qquad (5)$$

where τ^R-shear stress acted on sliding surface of wedges in surrounding rock; σ_n^R-normal stress acted on sliding surface of wedges surrounding rock; ψ-a mean dip angle of sliding traces of wedges, s-a length of sliding trace; l-length of a rock bolt. There are

$$\tau^R = \frac{\sigma_1 - \sigma_3}{2} \cos\varphi \quad \sigma_3 = P_C + P_W + P_G$$

$$\psi = \frac{1}{2tg\alpha} \ln\left(\frac{R_0 + W}{R_0}\right) \qquad \alpha = 45° - \varphi/2$$

$$\sigma_1 = \sigma_3 + 2(\sigma_3 tg\varphi + c)\frac{1 + \sin\varphi}{\cos\varphi}$$

$$s = \frac{R_0}{\sin\alpha}\left\{\exp\left[(\sigma_0 - \alpha)tg\alpha\right] - 1\right\}$$

$$\sigma_n^R = \frac{\sigma_1 + \sigma_3}{2} - \frac{\sigma_1 - \sigma_3}{2}\sin\varphi$$

$$\theta_0 = \alpha + \frac{1}{tg\alpha}\left(\frac{R_0 + W}{R_0}\right)$$

$$W = (R_0 + l)\left\{\cos\left(\frac{t}{2R_0}\right) + \sin\left(\frac{t}{2R_0}\right)tg\left(\frac{\pi}{4} + \frac{t}{2R_0}\right) - \frac{\sin\left(\frac{t}{2R_0}\right)}{\cos\left(\frac{t}{2R_0} + \frac{\pi}{4}\right)}\right\} - R_0$$

Suppose four cells reach their maximum bearing capability simultaneously, the stable condition of the system of support is given by:

$$P_{TI} > P_{min} \qquad (6)$$

where P_{min} is the minimal support resistance to maintain stability of surrounding rock. It can be calculated based on the solution of elastic and plastic theory.

2.2 Calculation of support resistance considering deformation coordination between support cells

Based on NATM, support and surrounding rock should be regard as the integral and they will interact each other in this system. Generally the bearing capability of each cell of the complex structure can not reach their limit values simultaneously. Therefore the aforementioned calculation of resisting capability of the system is unreasonable. It is necessary to revise formula (1) considering deformation coordination. Then a revised consideration is given as follows:

Suppose metal nets combine tightly with shotcrete and composed of the first subsystem. The stiffness of the subsystem, K_1 has following form:

$$K_1 = \frac{E_c}{R_0} \cdot \frac{R_0^2 - (R_0 - d)^2}{(1 + \mu_c)[(1 - 2\mu_c)R_0^2 + 2(R_0 - d)^2]} \qquad (7)$$

where E_C, μ_C-elastic module and poison ration of

370

the subsystem respectively. The limit bear capability and limit deformation of the first subsystem P_1 and u_1 are given, respectively by:

$$P_1 = P_C + P_W \qquad (8)$$

$$u_1 = P_1 / K_1 \qquad (9)$$

Similarly, bolts and rock bear arch are composed of the second subsystem. The stiffness, limit bearing capability and limit deformation of the second subsystem are respectively:

$$K_2 = \frac{2\pi R_0 F_w E_s}{A \cdot l \cdot e \cdot t \cdot \lambda_c} \qquad (10)$$

$$P_2 = P_G + P_H \qquad (11)$$

$$u_2 = P_2 / K_2 \qquad (12)$$

where K_2, P_2, u_2 are the stiffness, limit bearing capability and limit deformation of the second subsystem respectively.

And E_s is an elastic module of bolt steel; λ_c is an empirical coefficient; The other symbols is the same as mentioned above.

Considering the deformation coordination each subsystem, that is, through the relationship between u_1 and u_2 of the subsystem, we can obtain the limit bearing capability of the system P_{T2} as follows:

□ When $u_1 < u_2$, there are two cases

a. During interaction process of the system, the deformation of first subsystem firstly reaches its limit state, the deformation value of the second subsystem is equal to u_1. At the same time, the limit capability P_{T2} is given by:

$$P_{T2} = K_2 \cdot u_1 + P_1 \qquad (13)$$

b. During interaction process of the system, when the deformation of second subsystem reaches its limit state, the deformation of the first subsystem is over its limit deformation value in the same time and failure has appeared. The limit capability P_{T2} is only the second subsystem, that is,

$$P_{T2} = P_2 \qquad (14)$$

□ When $u_1 > u_2$, there are two cases

a. The second subsystem first reaches its limit state, then the capability of the system is given by:

$$P_{T2} = K_1 \cdot u_2 + P_2 \qquad (15)$$

b. When the first subsystem reaches its limit state, in the same time the second subsystem failure has appeared, the capability of the system is given by:

$$P_{T2} = P_1 \qquad (16)$$

Considering deformation coordination, the stability condition of surrounding rock with critical section is given by:

$$P_{T2} > P_{min} \qquad (17)$$

3. STABILITY RELIABILITY ANALYSES OF SLIDING FAILURE WEDGES FOR ROADWAYS WITH BOLT-SHOTCRETE SUPPORT SYSTEM

A reliability analysis method adopts probability of failure or stability to estimate safety of a structure. For example, a sliding failure of wedges for without considering deformation coordination is defined as follows by formula (6):

$$P_f = P(P_{T1} - P_{min} \le 0) \qquad (18)$$

in which $P_{T1} - P_{min}$ is called safety margin resisting sliding of the wedges. It is found that the safety margin is random variable owing to relation with many random parameters such as shear strength parameters c and φ.

3.1 Limit state equations of reliability analysis wedges sliding considering deformation coordination

Based on analyses mentioned above limit state equations of reliability analysis wedges sliding considering deformation coordination can be given respectively:

□ When $u_1 < u_2$, there are two limit state equations corresponding to each case, that is,

a. When the deformation is up to u_1,, the limit state equation has expression (13):

$$Z_1 = K_2 \cdot u_1 + P_1 - P_{min} = 0 \qquad (19)$$

b. When the deformation is up to u_2, the limit state equation has expression (14)

$$Z_2 = P_2 - P_{min} = 0 \qquad (20)$$

Based on limit state equations (19) and (20) the failure probability P_{f1}, P_{f2} corresponding to above two cases can be found with the help of Rosenblueth method (points estimation method). Considering each failure mentioned above may lead to instability of the support system, then failure probability of the system, when $u_1 < u_2$ $P_{f,u_1<u_2}$ is expressed as

$$P_{f,u_1<u_2} = P(Z_1 \le 0 \ or \ Z_2 \le 0 / u_1 < u_2)$$

$$= 1 - (1 - P_{f1})(1 - P_{f2}) \qquad (21)$$

□ When $u_1 > u_2$, there are two limit state equations

corresponding to cases, that is, the limit state equations when $u_1 > u_2$ can be obtained by expressions (15) and (16) as follows:

$$Z_3 = K_1 \cdot u_2 + P_2 - P_{\min} = 0 \qquad (22)$$

$$Z_4 = P_1 - P_{\min} = 0 \qquad (23)$$

Corresponding to failure probability P_{f3}, P_{f4} can be given. So failure probability of the system, $P_{f,u_1>u_2}$ when $u_1 > u_2$ is expressed by:

$$P_{f,u_1>u_2} = P(Z_3 \leq 0 \ or \ Z_4 \leq 0 / u_1 > u_2)$$

$$= 1 - (1 - P_{f3})(1 - P_{f4}) \qquad (24)$$

When support system (parameters) is determined, the relation between both stiffness of subsystems is invariable, therefore sliding failure probability of wedges of the system, P_f is given by:

$$P_f = \max(P_{f,u_1<u_2}, P_{f,u_1>u_2}) \qquad (25)$$

The reliability of the system is given by:

$$R_r = 1 - P_f \qquad (26)$$

3.2 Calculation of failure probability or reliability of the system and conclusions

In order to solve the reliability of the system Rosenblueth method is used by taking shear strength c and φ as random variables. From the researches, we obtained the following conclusions:

□ Failure probability, P_f of the system is more sensitive for the coefficient of variation of c than φ.

□ Failure probability, P_f of the system considering deformation coordination is larger than not considering deformation coordination.

□ Probability distribution types of random variables have important effect on failure probability values. Results of the analyses show that the probability value taking shear strength as log-normal distribution is larger than taking shear strength as normal distribution.

4. CONCLUSIONS

Based on analyses above, we obtained some important conclusions as follows:

□ It is necessary for us to consider deformation coordination between support cells. If not, it will be over estimating the support action and lead to risk of the structure.

□ It is important for us to carry out researches of the random distribution types of variables. Researches indicate that it is reasonable to adopt

log-normal than normal distribution.

□ Correlation between c and φ is significant for P_f.

The resulting value for the probability of failure of sliding wedges is higher when the correlation between the two strength parameters is taken into account.

REFERENCES

Tan Tjongkie 1986. The mechanical problems on long stability of tunnels. *Research about Engineering Geology and Rock Mechanics for Mining of Jinchuan Nickel Mine, (1), 11- 29.*

Liu, Tongyou & Zhou, Chengpu .1997. Researches on mining rock mech. problems in Jinchuan mine, *Research about engineering Geology and Rock Mechanics for mining of Jinchuan Nickel Mine, (1).*

Mining Science and Technology' 99, Xie & Golosinski (eds)© 1999 Balkema, Rotterdam, ISBN 90 5809 067 1

Modelling of ground stability and environmental impact in mining related problems

D.J. Reddish, T.X. Ren, P.W. Lloyd, G. Swift, M. Lewis & S. Mitchell
School of Chemical, Environmental and Mining Engineering, University of Nottingham, UK

ABSTRACT: Computer modelling has long been a subject of interest and numerous models have been developed for mining and geotechnical applications. The Geotechnics Research Unit at the University of Nottingham has been active in a number of applied research activities, including surface subsidence, Rock Mass Classification, strata behaviour, roadway support design and environmental impact associated with active and abandoned mines. The paper introduces the on-going research using numerical modelling techniques in mining related problems.

1. INTRODUCTION

Applied geotechnics research has been active at Nottingham for over 30 years and funded by mainly European and industrial sponsors. The unit offers a broad range of skills in the areas of rock mechanics, soil mechanics, mine design and environmental engineering. The group operates with a number of core long term study areas intermixed with a wide range of appropriate short term studies. Substantial research work has been carried out in underground excavation design, surface stability, mining method and layout, and mine water. The majority of the research activities are concentrated on computer modelling using both commercial codes and in-house software, and closely supported by laboratory tests (e.g., physical scale models) and fieldwork.

Whilst maintaining its traditional long term research areas, the Unit is engaging in a number of new areas following the dramatic change of the British mining industry in the last a few years. These include rock mass characterisation, numerical modelling/microseismic monitoring of strata behaviour, methane recovery and environmental impact resulting from abandoned mines. This paper briefly introduces these projects involving computer modelling techniques.

2. ROCK MASS CLASSIFICATION

Numerical models are increasingly being applied to a variety of areas related to mining and geotechnical engineering. A major challenge for the modelling work is the acquisition of reliable data for input into the theoretical models. Work is currently being conducted to develop a specific rock mass system for use within coal mine design which is practically applicable to large scale and localised mining excavations in the UK coal mining industry (Lloyd, 1999). The system being developed separates the influences of stiffness from those on strength and produces two ratings that can be rapidly converted into in-situ stiffness, strength and post-failure behaviour parameters. These can then be applied as input parameters for a numerical model that can be used to investigate the stability of underground excavations and their support systems.

Figure 1 shows the concept of Rock Mass Classification System. The process begins with the collection of specially selected field and laboratory data on the site and the processing of these values into an UK Coal Measures Rock Mass Classification rating. All potential influencing parameters have been reviewed and filtered on the bases of measurability and relevance. Gaps in the collected data are filled from a properties database, which is based on geological description and other available parameters. The Rock Mass Rating is further processed into pre and post-failure stiffness parameters and triaxial strength failure envelopes. These data are then fed forward into the chosen numerical model, which is run to provide the prediction.

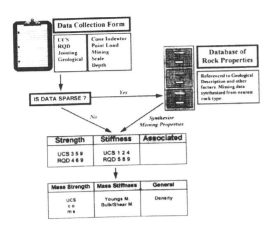

Figure 1 Concept of Rock Mass Classification

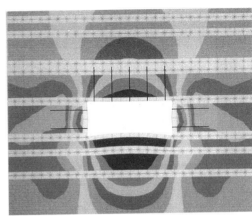

Figure 2 Modelling of roadway rockbolt support

An application of the System is numerical modelling using FLAC (Fast Lagrangian Analysis of Continua, Itasca, 1995)) of UK coal mine roadways using rockbolts for primary support, coal pillar design and caved waste adjacent to roadways (Figure 2). The project aims to determine the most important rock mass parameters concerning the rockbolts interaction with the rock mass. The work initially involved the development of a basic Rock Mass Classification system and followed by fine-tuning of classification scales, importance weightings and parameters by application to varied case study sites.

3. SUBSIDENCE PREDICTION

Mining induced subsidence has traditionally been a key research area in Nottingham University. Over the years numerical models using Influence Function modelling techniques and more recently FLAC code have been developed to predict mining subsidence under various mining and geological regimes (Whittaker and Reddish, 1989; Reddish et al, 1998). One of the most recent activities in this area involves the development of computer models for predicting subsidence associated with weak rock masses (Mohammad, 1998). The project aims to produce a strategy for deriving a balanced set of input parameters for numerical modelling of excavations in weak rock masses, particularly UK Coal Measures. A universal model for subsidence prediction for a range of depth, i.e., 100m to 800m below ground surface, was established. The input values of the pre- and post-failure strength and stiffness parameters were derived using the RMR classification system. The results of the model have been validated using the SEH subsidence prediction

method (NCB,1975). Specific fine-tuning of the input parameters was conducted in order to use the established model for subsidence prediction in each mining condition. This numerical modelling technique has been applied to a number of case histories in Australia, USA, South Africa and India (Reddish et al 1998).

The world-wide trend towards city living ensures that the next 20 years will see increasing redevelopment of underground infrastructure in our cities. Settlement and the use of compensation grouting to mitigate settlement are going to remain a very important and critical issue. The use of Influence function subsidence modelling techniques is being investigated to predict the settlement over shallow tunnels including the simulation of compensation grouting. The project involves the modification of mining based subsidence prediction techniques SWIFT (Reddish, 1989) to simulate the settlements due to face loss in soft ground tunnels. A number of complex case histories from tunnels within central London have been analysed and results compared with monitored data. The technique has a major advantage in terms of its ability to cope with complex 3D tunnel layouts with relatively quick solution times on an ordinary PC. The model has the capability of inserting as well as removing material in any geometry and as such has potential for modelling compensation grouting. Results to date have been excellent with relatively minor recalibration of the model.

4. MODELLING OF STRATA BEHAVIOUR

Seismicity associated with underground mining is probably the most adverse phenomenon in relation

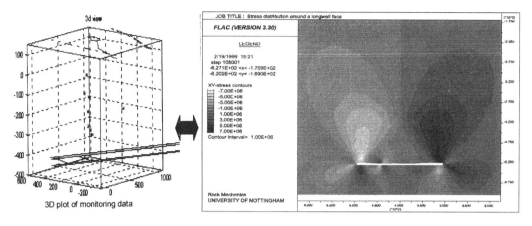

Figure 3 Microseismic monitoring and numerical modelling of strata behaviour around a longwall face

to the safety and productivity of modern longwall mining. The monitoring of microseismic activities associated with longwall mining provides a unique opportunity for studying and analysing the caving mechanisms. Such studies are valuable for providing the knowledge used to improve the stability of mining excavations, the design of roadway support and the prevention of water and gas ingress into coal faces. On the other hand, numerical modelling techniques have the potential of filling in the details of stress distributions and displacements under varying geological regimes. The combination of the two techniques offers major rewards in terms of the level of understanding of the process. The principal objective of the research is therefore to develop a link between the microseismic monitoring and the numerical modelling approach to aid in the understanding of the caving process over longwall panels.

Asfordby Mine was chosen for the study. The microseismic monitoring results obtained by Styles et al (1996) indicated that the bulk of the seismicity occurred within the Sherwood Sandstone aquifer. This activity was generated by fracturing which delineated the margins of a detached block of rock, leading to periodic weightings on the face. Eventually fracturing propagated downwards to the face, generating a hydrogeological pathway for the ingress of water into the longwall face. Floor heave was reported in conjunction with the weightings and was considered to be associated with the microseismic events in the floor strata. Enhanced activity occurred in association with water inflow. The failures within the Sherwood Sandstone were believed to be vertical shear mechanisms.

Numerical models using FLAC have been developed for Asfordby Mine to study the strata behaviour. The modelling results have shown that apart from the ribsides of the longwall face, areas with high stress concentrations were located within the Bunter (Sherwood) Sandstone. The results clearly demonstrated that the failure of the roof strata becomes more extensive and periodic as the face advances, with vertical shear mechanisms being the dominant form. The failure zones propagate through the Permian Basal Sandstones and the Bunter (Sherwood) Sandstones, and well into the strata above. It is likely that water accumulated in the bed separations above will follow the periodic caving and flow into the face through the fractures. The modelling results also demonstrated that extensive yield zones would occur in the floor, which would certainly be associated with the microseismic events observed. In general, a good correlation has been observed between the numerical results and microseismic monitoring data (Ren et al, 1998).

5. METHANE RECOVERY FROM ABANDONED COAL MINES

Exploitation of methane gas from abandoned mines has appeal in environmental, ecological and economic terms. The coal bearing strata surrounding an abandoned coal mine constitutes a naturally fractured, naturally stimulated reservoir with up to 80% of the original gas resource being available for subsequent extraction and utilisation. This very favourable characteristic makes abandoned coal mine reservoirs a very attractive prospect for energy production. However, there are no established techniques to assess the gas potential of an abandoned mine or a mining district so that gas production planning can be made with confidence. As a result, a significant number of abandoned mine gas projects have failed to yield high purity methane. The current challenge facing the industry is the development of reservoir modelling techniques which will enable them to target drainage sectors which would yield and sustain high purity (>70%) methane from abandoned mines. This project addresses this challenge by dealing with the fundamental characterisation of stressed and highly fractured reservoirs and with multi-component flow-gas diffusion modelling at relatively low pore pressures.

The research splits into two main themes: The first element concentrates on stress, strain, displacement, physical fracturing and time dependent consolidation of coal seams and coal measures rocks around abandoned mines. The second element deals with gas release from, and fluid flow through, highly fractured, semi-consolidated media and aims at developing a numerical model which accounts for the multi-component emission, flow and diffusion characteristics of mine gas.

The findings from this research will have a direct contribution to the reduction in greenhouse gas emissions as well as the recovery of a currently wasted resource. There is an additional environmental gain in reducing unwanted emissions of gas in dangerous and hazardous urban situations. There is also an opportunity for the technology to allow enterprising companies to exploit the significant resource for profit.

6. GROUNDWATER RECHARGE IN ABANDONED MINE WORKINGS

The large scale closure of mines as they become exhausted or uneconomic has been a feature of the 1980s and 1990s, and has been apparent throughout Europe. As these mines are abandoned pumping is either reduced or stopped and the local water table can be anticipated to recover. In some cases this is the first time in hundreds of years that pumping has ceased. The older shallower workings are likely to become less stable as they flood as any change generally disturbs these delicate structures. Given that much of the land overlying these workings is developed or required for future development, determining the stability of shallow workings as they

recharge is going to be a regular and important task in the near future.

The primary objective is to develop a PC based numerical modelling system which analyses the stability changes which occur when an abandoned mine floods. The research initiated with an extensive period of field monitoring of some 30 mine outflow sites spread across the UK. This initial phase concentrated upon developing field flow measurement and chemical analysis techniques. These were applied to the 30 sites on a monthly basis for 12 months. Following this initial widespread data collection phase more detailed desk studies of selected sites were undertaken. This involved collection of geological and mining details from national records. These records were then processed and the hydrogeology of the mine assessed in terms off the monitored outflow. Stability models of individual failure mechanisms within abandoned mines were developed and adapted to take into account the influence of recharge water.

Again, the numerical modelling code FLAC, is being ultilised to simulate the development of pore water pressures within the strata surrounding an abandoned mine as recharge takes place, and to quantify the effects of pore pressure development upon the stress distribution. Ultimately, this will provide a greater insight into the stability implicating groundwater recharge on shallow, abandoned mines. The research will deliver a better quantitative understanding of the influencing mechanisms, a document containing detailed case history data, and a computer based package that assesses the stability based on the research findings.

7. ENVIRONMENTAL IMPACT OF MINE WATER AFTER ABANDONMENT

The environmental impact of mine water after abandonment has become a major concern following the large-scale closure of coal mines in the UK. Abandoned mines are a source of poor water quality in specific areas of England and Wales. The situation is a long-standing one in many cases, but as mines are still being abandoned, the situation is continuing to deteriorate. The Coal Authority is currently giving a high priority to preventing polluting mine water from recently abandoned mines reaching sensitive watercourses, as well as dealing as now - with pollution that has already occurred from long-abandoned mines (Coal International,

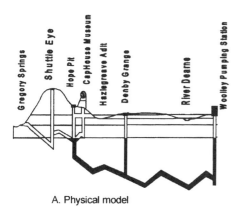 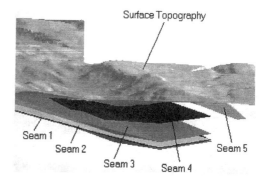

A. Physical model   B. 3D computer model

Figure 4 Environmental impact of mine water after abandonment

1999). Research is needed to evaluate the extent to which existing discharges from mines already abandoned affect water quality, and would therefore need to be tackled if water quality is to be improved, thus ensuring the continuing abandonment of existing mines results in the minimal amount of environmental impact.

Following mine closure, pumping is either reduced or stopped, allowing local water table to rebound, and eventually emerge as a spring, or seepage through the ground or the bed of an existing river or stream. When the water emerges it may well be clear, because the underground water is low in oxygen and the iron is in solution. As this water mix is exposed to air - which may occur before it emerges above ground - the iron rapidly oxidises from the ferrous to the ferric form and precipitates out as an orange deposit. In shallow mines, or in adits set in higher ground, such cycles may be repeated continually as the ground waters fluactuate. In deeper mines connections may be made with underground aquifers. Quite frequently the history and extent of mining is such that neither the hydraulic conditions, nor the chemical nature of the water, can be predicted once the last mining activity ceases.

The Woolley mining region, located in South Yorkshire, comprises a complex group of underground-linked closed collieries. When working these mines discharged their water at a number of locations, but since abandonment discharge has taken place from one centralised pumping station, where various treatment methods are applied. The treatment approach practised at Woolley is passive in nature, using a system of settling pools, cascades,

cloth filters and reed beds, all of which combine to produce an effluent well below the Environment Agency statutory controls.

Work is being conducted to study the hydrogeological regime and environmental impact of mine water following the closure of Woolley and adjacent collieries. The project is concentrated on the questions which arise as to the impact of the current discharge upon the local environment and how it compares in quantity and quality to the situation prior to closure.

The project aims to propose viable options which could be applied in the future, through the development of a flow model of the subsurface network, to other mining regions where data and information is not as abundant. The primary objective of the research is to improve the understanding of the outflow behaviour in terms of quantity and quality and to apply this understanding to coming up with the best environmental treatment strategy for mine abandonment.

A 3D-modelling package Vulcan has been used to model the interconnected multiseam, multi-colliery situation. This allows the visualisation of the entire area at Woolley and therefore provides a better understanding of spatial relationship and the possible behaviour of the mine water under investigation. The model provides a powerful tool for the interpretation and detailed analysis of the current problem and future treatment options for the entire mine complex.

8. CONCLUSIONS

In recent years computer modelling has developed significantly and become an essential tool for many mining-related problems. Improvements in computer software and hardware, especially desktop PCs, have made it possible to build sophisticated models to investigate various ground stability and environmental engineering issues. The paper has briefly summarised some of the research activities using computer modelling at Nottingham University. The majority of the research projects has strong links with and compliments each other.

The mining industry is on the whole undergoing dramatic change as a result of energy requirement readjustment and strict environment protection. Whilst maintaining high quality research in more traditional areas, it is important that research efforts are adjusted to tackle new problems, such as the environmental impact following mine abandonment, through utilising existing computer models, expertise and skills and acquiring new technologies whenever necessary.

9. REFERENCES

Coal International, UK's Coal Authority is to give priority to the prevention of minewater pollution, January 1999.

Itasca 1995. Fast Lagrangian Analysis of Continua, FLAC, Version 3.3, Volume 1-4, Itasca Consulting Group Inc., Minneapolis, Minnesota 55415 USA.

Lloyd, P. W., The Application of Rock Mass Classification Principles to the Design of Mining Systems Utilising Rock Bolts as the Primary Mode of Support. Final Report to the Health & Safety Executive, Bef., 1999,110pp.

Lloyd, P. W., Mohammad, N. and Reddish, D. J., 1997. Surface subsidence prediction techniques for UK coalfields - an innovative numerical modelling approach, Proceeding of the 15th Mining Congress of Turkey, Ankara, May 6-9,

Mohammad N, Reddish D J and Stace L R, The relation between in situ and laboratory rock mass properties used in numerical modelling, Int. J. Rock Mech.. Min. Sci. Vol. 34, No.2, pp289-297, 1997.

Mohammad, N., Reddish, D. J. and Lloyd, P. W., 1997a. Longwall surface subsidence prediction through numerical modelling, 16th International Conference on Ground Control in Mining, Morgantown August 5-7, S. S. Peng (ed) pp 33 - 42.

Mohammad, N., Reddish, D. J. and Stace, L R., 1997b. The relationship between insitu and laboratory rock properties used in numerical modelling, Int. Jr. of Rock Mech. Mining Sc. and Geo. Abstracts, Vol: 34, No. 2. pp. 289 - 297.

NCB 1975. SEH, Subsidence Engineers Handbook, National Coal Board, Mining Department, London. 111 pp.

Reddish, D J The modelling of rock mass behaviour over large excavations using non-linear finite element techniques, Mining Department Magzine, University of Nottingham, Vol.1 No. XLI, pp93-105

Reddish D J, Mohammad, N., and Lloyd, P. W, The Numerical Modelling of International Trends in Subsidence Models, International Conference on Geomechanics/Ground Control in Mining and Underground Construction, 14-17 July 1998 Wollongong NSW, Australia. Vol.2, pp697-708.

Ren, T X, Reddish D. J. and Dunham, R K., A Combined Microseismic Monitoring and Numerical Modelling Strategy to Improve Caving Rock Mass Behavioural Understanding in Longwall Mining, Annual Report for International Mining Consultants Limited, December 1998.

Styles, P. et al, Microseismic Monitoring Work in the Vicinity of Asfordby Colliery Including the Seismic Mapping of the Location of Mechanical Failures in Overburden, Research Contract No 2069/3, Final Report to IMCL, 1996.

Whittaker, B. N. and Reddish, D. J., 1989. Subsidence Occurrence, Prediction and Control,. Developments in Geotechnical Engineering, 56, Elsevier. 528 pp.

Mining Science and Technology'99, Xie & Golosinski (eds)© 1999 Balkema, Rotterdam, ISBN 90 5809 067 1

Geotechnical monitoring and modeling at Cuajone open-pit mine

Ken Rippere, Yonglian Sun & Luis Tejada
Southern Peru Copper Corporation, Tacna, Peru

ABSTRACT: The Cuajone porphyry copper deposit is located on the western slope of the Andes Chain in southern Peru at an elevation of 3500 m. The current 700-meter deep Cuajone open pit will eventually reach a depth well excess of 1000 meters, placing it in the "world class" range in terms of proposed mines. Most of the deep open pits in the world are experiencing some degree of slope instability, as does the Cuajone open pit. The geotechnical issues will become increasingly important as the pit expands.

The displacement of the slope at the Cuajone mine is regularly monitored using various means at a different frequency according to the magnitudes of measured data. Based on the detailed geological information including cell mapping, a structural fabric model was developed in which major structures that have a predominant influence on the stability of the slope were incorporated. A UDEC analysis was then carried out and the slope behavior in the northeast sector of mine was examined. The mechanism of slides occurring at the Cuajone mine was also summarized. The purpose of the project is to ensure that mine production is being carried out in a safe environment.

1 INTRODUCTION

During the past several years, there have been rockfalls and slides in the Cuajone mine. To some extent, the slides are good for the mine, since they indicate that those parts of slope have reached the limit of rock strength. Otherwise, the pit design would be over-conservative and non-economic.

The major slides VII, X and XI, which have occurred at the Cuajone mine, are summarized in Table 1. As can be seen in the table, the primary factor causing the slides is structure. Water is a secondary factor in generating rockfall.

Therefore, efforts should be made in the field to identify major structures, such as faults, joints, tension cracks, etc. This work is tedious but has significant meaning to maintain the slope in a good working condition.

2 FILED MEASUREMENT AND OBSERVATIONS

For the Cuajone mine, the greatest concern is slide VII, located in the northeast section, which has been constantly moving for a long period. Usually, the displacement rate in this region is constant during the dry season, and it accelerates in the rainy season (from late December to early March of the following year). The geology in this area is complicated, including major faults, joints, dikes and tension cracks. As mining goes deeper, all these structures may cause instability problems.

For the northeast section, both extensometers and prisms are utilized for monitoring purposes. As an example, the measured data at three extensometer monitoring points G43, G51 and G52 at elevations 3340 m, 3490 m and 3445 m, respectively, in this cross-section is presented in Figure 1. From November 1997 to March 1998, there was acceleration in the displacement rate, and then the curve increases constantly at a lower velocity. As for the magnitude, G43 has moved 2.8 m and G52 only 10 cm. It is clearly indicated that there is much less horizontal movement in the upper zone at G51 and G52 compared to G43 in the lower zone. This correlates quite well with the modeling results.

Currently, the slope at slide VII is experiencing constant movement with a slight acceleration, which may cause some instability problem in this region. By the time we finish the paper, the movement of the slope in this region accelerated sharply and finally collapsed on late February.

Table 1 Summary of Major Slides at Cuajone Mine

Slide No.	Date	Level (m) from	Level (m) to	Affected Bench No.	Material (tons)	Lithology	Zone	Notes
VII	91/11/20	3475	3535	4	35,000	AL-IA	NE	Structures/lithology/water/blasting
	97/03/05	3385	3550	11	-	IA-BA	NE	Rainwater filtrated in the slide
	98/02/15	3490	3550	4	-	IA	NE	Rainwater filtrated in the slide/cracks opened
X	97/04/29	3205	3340	9	100,000	IA-LP	NE	Structures/mining at toe/previous small slides
	99/02/22	3265	3325	4	8,000	IA-LP	NE	Rainwater/structures
XI	97/10/24	3250	3340	6	145,000	LP	N	Structures/mining at toe
	99/01/16	3280	3325	3	400	LP	N	Structures
	99/02/22	3295	3355	4	1,000	LP	N	Rainwater/structures
NE	99/02/25	3190	3550	24	12,000,000	LP-BA-CB-CT	N/NE	Rainwater/tension cracks/structures

Note: AL for alluvium, IA for intrusive andesite, BA for basaltic andesite, LP for latite porphyry, CB for basal conglomerate, TC for crystal tuff.

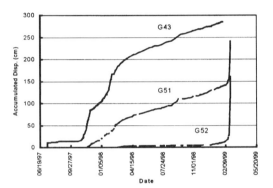

Fig. 1 Extensometer monitoring in the northeast section

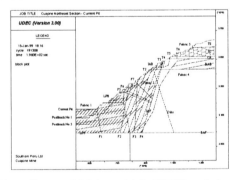

Fig. 2 Structural fabric model for UDEC analysis

3 UDEC MODELING

The two-dimensional distinct element code UDEC (Universal Distinct Element Code) was used to perform the plane stain analysis on the northeast section of the Cuajone mine in an effort to assess the likely behavior of the slope as mining continues at depth. Several models were run in order to evaluate the key factors that affect the slope performance.

3.1 Structural Fabric Model

The structural fabric model that was developed based on the detailed analysis of the geological information, including drill-hole data and field mapping, for UDEC analysis, is shown in Figure 2, where nine types of rock are encountered as following

LPB	latite porphyry broken
LPF	latite porphyry fresh
IAB	intrusive andesite broken
IAF	intrusive andesite fresh
TS	upper tuff
TC	crystal tuff
CB	basal conglomerate
BAB	basaltic andesite broken
BAF	basaltic andesite fresh

• *Near-vertical faults:* Four near-vertical major faults (F1 to F4) were observed in this region, extending from current pit to a level around 3000 m and below. They control the behavior of the slope to a large extent.
• *Eastward dipping faults:* Three major faults (F5 to F7) were measured dipping eastward extending from current pit to somewhere intersecting fault No. 4.
• *Joint structures:* The stereonet technique was applied to construct the fabric structures in the model based on the field cell mapping data. The spacing of those joint sets in the UDEC modeling need not necessarily be equal to those observed in the filed, but the measured ratio of the fabric blocks should be kept the same.

• *Dike:* The dike dipping eastward in this cross section appears to be very strong. The relatively high strength, favorable inclination and location make it less likely exert major detrimental effect on the stability of the pit.

• *Tension cracks:* There were 6 major tension cracks (T1 to T6) near-vertically oriented in the upper benches behind the slope face. Tension cracks that originally may coincide with faults or joints, were developed and opened up as a result of mining at the pit bottom.

3.2 Rock Mass Properties

The properties for all rock types were taken from the report entitled "Cuajone Mine 1994 Slope Design Review" prepared by Call and Nicholas, Inc. in July 1994, except the tensile strength which was assumed to be one tenth of cohesion for the rock types encountered.

3.3 Modeling Procedures

First, cross-section including the geology prior to mining was selected for the analysis. Then elastic properties corresponding to the appropriate rock types and structures were assigned to each element, and the model was allowed to run until reaching elastic equilibrium under gravitational loading for the pre-mining topography. At this point, the displacement for all elements and joints was reset to zero in order to eliminate the influence of the original gravitational loading. Following this the simulation of mining steps and the calibration of the model in order to replicate the observed behavior of the pit were performed.

In the absence of measured pre-mining in-situ stresses, the stress conditions were assumed to be equal for both horizontal and vertical directions. The groundwater table was constructed based on the measured piezometric data.

3.4 Modeling Results

The modeling results for the current pit and pushback No.1 are shown in Figures 3 and 4, respectively. For the current pit, the modeling results show that the major displacement occurs along the near-vertical faults F2, F3 as well as the low angle joints in the latite porphyry. The plot of velocity vectors indicates that for the slope surface, from elevation of 3360 m up, the downward movement is predominant while from that elevation down, the horizontal movement becomes predominant. Also it can be observed that the velocity vectors nearly form a regular pattern, and a sliding failure is most likely to take place in this region with elapsed time. Overall, faults F2 & F3 and the joints that are pervasive throughout the latite porphyry control the movement of the slope.

As mining expands a couple of benches into pushback No.1, it can be seen that more shear displacement occurs along fault 3, joint structures and the upper tension cracks finally tend to interconnect. As such, part of slope in this region is being squeezed out. Also it is seen that the velocity vectors have developed into a regular pattern along the shallow surface. The affected rock mass would be sliding out and slope failure would develop.

The x displacement contours showed that the horizontal displacement near the surface will almost double in pushback No.1 compared to that in the current pit, suggesting that pushback No.1 requires a redesign in this area.

4 DISCUSSION

One of the threats in open pit mining is the potential for the gradual deformation of a large slope to develop into a fast moving slide. This is a poorly understood process. In general, the slope doesn't fail immediately following blasting or the single mining cut. It moves slowly, and may take up to several years before it suddenly fails. The mining at the pit bottom creates more freedom for the newly exposed rock masses, which is more likely to produce tension cracks behind the upper slopes. Those tension cracks develop periodically year round. During the dry season, it moves slowly. When the rainy season comes, the movement usually accelerates. This is a repetitive process with seasons. After several years' development, these tension cracks may develop to such an extent that they connect with the steep faults and near horizontal joints in the lower part of the pit. At that point, a large slide could develop.

5 CONCLUDING REMARKS

Overall, the following concluding remarks can be made:

(1) Based on the detailed mapping, a structural fabric model for the northeast section in the Cuajone mine was developed. This model adequately represents the real case.

(2) The mechanical behavior of the current pit in the northeast section is very much controlled by near vertical faults (F2 & F3) and near horizontal joints that are pervasive throughout the latite porphyry. For the current pit, frequent monitoring of those extensometers and prisms is important to ensure that the mine production is being carried out in a safe environment.

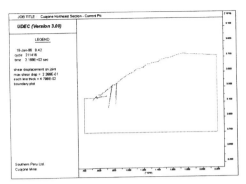

(a) Shear displacement along discontinuities

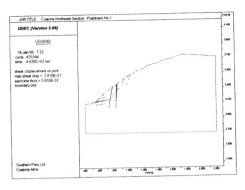

(a) Shear displacement along discontinuities

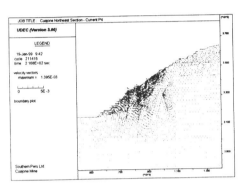

(b) Plot of velocity vectors

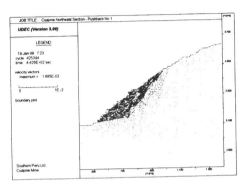

(b) Plot of velocity vectors

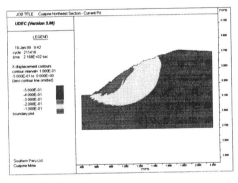

(c) Horizontal displacement contour

Fig.3 UDEC analysis for the current pit

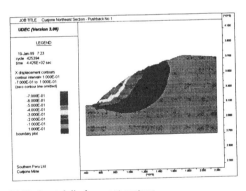

(c) Horizontal displacement contour

Fig.4 UDEC analysis for pushback No.1

(3) The mechanism of major slides occuring at the Cuajone mine is summarized. It illustrates that it is important to identify the faults and fabric joints in the rock masses, since they control the pit behavior to a large extent.
(4) As a case study, UDEC analysis was conducted

for the northeast cross-section. The result indicated that the current pit was deteriorating in a slowly accelerating fashion. The upper part of the slope moved vertically and lower part moved horizontally, which is the same as we currently observe in the pit.
(5) Pushback No.1 is potentially unstable according

to the UDEC modeling, therefore a redesign should be undertaken such as flattening the slope angle or leaving a step-out at the 3145 m level to ensure that the slope remains in a good working condition.

The mechanical behavior of open pit slopes is time-dependent, especially for large-scale pits. Further research focusing on this topic is needed.

6 ACKNOWLEDGEMENT

The authors wish to thank the management of Southern Peru Copper Corporation for the permission to publish the paper. We also acknowledge technical assistance in the UDEC analysis from Loren Lorig, manager of Itasca, SA.

REFERENCE

Call & Nicholas, Inc. July 1994. *Cuajone Mine 1994 Slope Design Review.*
Itasca Consulting Group, Inc. 1996. *UDEC (Universal Distinct Element Code),* Version 3.0.
Hoek, E. & Bray, J.W. 1981. *Rock Slope Engineering.* Published by E & FN Spon. 358.

Mining Science and Technology' 99, Xie & Golosinski (eds) © 1999 Balkema, Rotterdam, ISBN 90 5809 067 1

Underground dimension stone quarrying: Rock mass structure and stability

O. Del Greco, M. Fornaro & G. Oggeri
Dipartimento di Georisorse e Territorio, Politecnico di Torino, Italy

ABSTRACT: The paper examines some cases of dimension stone quarries in different basins in Italy. In these quarries, technical and economic features induced the managers to change the exploitation method from open pit to underground or to begin underground exploitation. These facts are mainly due to the type and origin of the rock and its characteristics, both from the structural point of view and also taking into account the influence of the lithological features on the quarrying technologies and underground structures stability. With the aim of defining the stability conditions of the structures, it is interesting to look for links between the geostructural situations and the planning of quarrying activities, improving both the safety conditions and the productivity of the quarry. The design of the exploitation of underground dimension stone quarries is therefore based on geomechanical, technological, economical and environmental aspects.
The paper examines two significant cases: green ophicalcites (Aostan Alpine Range) and the arabesqued marble (Apuane Alps), both in Italy. The aim of the study is to furnish some design elements in order to define the underground structural pattern and the more appropriate criteria for the stability analyses of the quarry openings.

1 GENERAL ASPECTS

The production of ornamental stones takes on noticeable importance in the sector of natural building materials, both from the economic and technological points of view. These two features are well represented by Italian quarrying activities and both the quality and quantity of the stones are relevant. Furthermore, the technological capabilities of equipment made in Italy used for the manufacturing of finished products (slabs, columns,tiles) are appreciated throughout the world.

As far as the mining activity is concerned, the field of the quarries -both the raw materials for the concrete industry or for aggregates and the dimension stones - has gradually gained on the field of the mining of metallic minerals or industrial minerals (in Italy the main exploited minerals were in the past Pb-Zn and sulphurs, Fe, Mn, Hg, talc, barytes, fluorite, asbestos).

The overall quarry production of ornamental stones in Italy is at present (1999) about 9 million of tons, approximately divided in two equivalent parts for the production of marbles and for all the other stones (granite, gneiss and sedimentary stones). The export percentage is about 40% of the global production, with a peak of more than 50% of the marble production. Exportation consists both of rough blocks and slabs and of finished products.

It is not only the progressive reduction of the number of mines due to the exploitation of the existing orebodies that has determined this focus on quarrying. The particular attention that is increasingly dedicated to the protection and reclamation of the landscape and the environmental constraints which are particularly emphasized in populated areas, such as Italy, and the stability conditions of the slopes in mountaineous areas (e.g. the Northern Alps or the Apuane basin) have induced a growing diffusion of the underground exploitation of quarries of ornamental stones.

Underground exploitation of dimension stones will be increasingly proposed as production methods, not only in the cases where the underground option is imposed by the features of the rock mass, as in the past, but for a number of reasons that have gained weight in recent years, making underground exploitation preferable, even in terms of economy and also in cases where surface operations could be technically feasible.

Five elements should be evaluated when the underground choice has to be considered: 1) good structural conditions of the rock mass (in ornamental stone quarries this aspect is generally satisfied, even though there are very particular rock mass situations, such as the case of stratified rocks with very thin clay filling); 2) technology of the excavation (mainly mechanical cutting to separate the blocks from the mass); 3) commercial features (some properties, for example colour and grain size, are not always the same, due to the limited homogeneity of the rock mass, where change of colour, stains, inclusions can be encountered); 4) economic profitability compared to the costs of open pit exploitation, taking into account the probable lower recovery due to the underground support structures such as pillars but also the savings on overburden removal and muck disposal; 5) safety and environmental reclamation.

In particular, the geostructural conditions determine both the design of the excavations and the methods that should be used to separate the blocks from the faces. The availability of structural data of the rock masses where some underground quarries have been developed, and the technical results after some years of exploitation can allow one to consider the above mentioned elements in a critical way, with an emphasis on the first two elements in particular

2. UNDERGROUND QUARRYING IN ITALY

At the present more than 10 basins have an active quarrying underground in Italy , and the rock masses are principally in sedimentary or metamorphic geological formations. No examples are yet available for igneous rock masses (granite, diorite). This difference is mainly due to the excavation tecniques, that make use of chain saw equipment, diamond wire saw equipment, less detonating cord or conventional blasting. Water jet or heavy duty saw

tecniques in igneous rock are not yet well developed or employed, and the abrasivness of the rocks and the low timely productions are still the negative elements that discourage the choice of the underground for granites.

In the map of Figure 1 it is possible to locate the mentioned quarry basins and the schematic geological structure. In all the cases the underground voids (rooms) are considered to be left in a stable configuration. This criterion arises from the general layout that such exploitations must have in the long period and from the overall stability that must be ensured. The exploitation of adjacent rooms and the influence on external slopes must not compromise the long term stability of the crowns (roof of the rooms) and of the pillars. In general there is not sufficient available detritic waste material to provide a suitable filling of the voids.

Two cases are here reported as significant examples of the features that must be taken into account for an acceptable design and management of underground quarries.

The two cases refers to the green ophicalcites (3 underground quarries examined) of the Aostan Alpine Range and to the arabesqued marble (2 underground quarries examined) in the Apuane Alps.

In the first case, because of the local non homogeneity of the lithotype and the occurrence of frequent structural discontinuities with very large apertures, the development of drifts and caverns and the concomitant positioning of rib pillars should adapt the rentability of the quarry to the stability requirements and the productivity of the stopes, in which bigger and more efficient equipment is used.

In the second case, on the contrary, in the

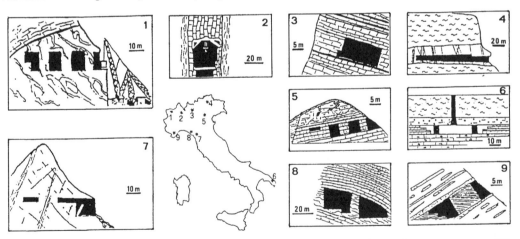

Figure 1. Quarry basins in Italy with underground active exploitations. For each case the quarry recovery (%) and the block recovery (%) are respectively indicated. 1) Green ophicalcite of Aosta Valley (qr 65, br 60); 2) Candoglia marble (qr 95; br 15); 3) Breccia Orobica (qr<80; br 60); 4) Lasa marble (qr 70; br 40); 5) Arenaceous limestone Berica (qr 75; br 70); 6) "Tuff" limestone of Cutrofiano (qr <60; br 70); 7) Apuane marble (qr 70; br 35); 8) Portoro limestone (qr 70; br <50); 9) Ligurian slate (qr <70; br 30).

presence of a relatively homogeneous and lightly fractured rock mass, the shape of the rooms and the design of the pillars must ensure a stable development of the exploitation voids, which nowaday are also of considerable size.

3. GEOSTRUCTURAL FEATURES

A rock body that is exploitable for ornamental products by means of underground excavations should have, in general, more continuity and uniformity than analogous open pit exploitations. This is due to the restriction that underground selection methods are most difficult when change in directions of the openings are requested. There are fewer exposed surfaces than a benched face and the observable exposures are those of blocks that are going to be quarried. At the beginning of the exploitation the observation scale is based on openside surveying and eventual coring. It should also be noticed that many underground quarries are today only the natural or forced development of previous open pit excavations. In these situations the rock mass can already be well known in the main geostructural features. These aspects are of relevant interest because the productivity of the quarry laso depends on the relative orientation of the rooms to the disconinuity pattern.

It can be useful to define the following terms which give a quantitative idea of the quarrying productivity: the quarry recovery is the ratio between the volume of the exploited rock and the volume of available material of the rock body; the block recovery is the ratio between the volume of commercial elements (slabs or small blocks) and the volume of the exploited material; the rentability is the product of the two previously defined quantities.

In general, unless very particular rock body shape, the quarry recovery cannot be more than 65-70 %. These limits are due to the intrinsic resistence of the pillars. In fact, the increasing of the unsupported area determines a stress concentration in the natural supports, in which the rock mass strength should be adequately reduced by the application of a safety factor. The block recovery of the caved blocks presents a wider range of variability, depending on the smoothness of the excavation tecnique and on the state of stress induced on the blocks before quarrying. Block recovery can range from 10% to 60 % in methamorphic rock masses and rise up to 80% in sedimentary homogeneous rock formations.

The three studied rock bodies of the Ophicalcite of the Aosta Valley present similar structural aspects: a rock mass interested by a main discontinuity set of subvertical fractures, with very high persistence, large aperture (some centimeters) and spacing of about 10 to 20 meters. These fractures have a large scale waviness and moderately rough surfaces, with occasional filling of coars and fine material. These main fractures are accompanied by some other subparallel discontinuities, characterized by a lower persistence, closed and with irregular shape. The more evident stability problems are due to the limited spalling rock and to the possible formation of wedges in the crown of the excavations. The thin veins of white calcite can sometime be zones of triggering fracture mechanisms. The size of the rooms is variable: the width range aproximately from 6 to 12 m and the height can reach about 24 m the length can be of some tens of meters if the room is not square shaped (figure 2). The trend is nevertheless toward schemes with more rooms of smaller size instead of few larger rooms.

The here mentioned rock masses of the marble of the Apuane Alps of Carrara (Corchia and Altissimo basin) have a continuous structure, with an orientation of the discontinuities more various and irregular. Sometimes higly fractured bands occur, but they have a significance for the overall stability of the openings. The stability problems are due to the large size of the openings and the related stress concentrations. It is also quite common to have interferences between voids, situated at different levels. The size of the rooms is variable, from about

Figure 2. Rooms and pillars in the Green Ophicalcite of Gressoney.

Figure 3. Exploitation adits at different levels in the Carrara marble basin.

10 m to 20m in width and up to 30 m in height. The length can be of some tens of meters (figure 3).

The common criteria for the studied cases are that the discontinuities must be well localized and that the pillars or the barriers must be left in place simultaneously satisfying the requirements of the stability and of the recovery. As a consequence the planimetric distribution of the pillars is not so regular.

The structural and static conditions that are involved in the global stability assessment deals with the stress concentrations, the resistence of the pillars and of the room crown and the relative orientation of the discontinuities.

If the main set is sub-horizontal the consequences on the stability involve the crown of the room that is, with the subsequent lowering of the base, which becomes more and more unaccessible, and therefore much dangerous. In such a case an active bolting of the roof can be useful (figure 4). The technical design of this depends on the thickness of the rock strata, defined by the discontinuity and the crown itself, on the width of the room and on the eventual occurrence of some other subvertical discontinuities which could form arching structures. The effect on

pillars is negligible except for the case of thin pillars subjected to eccentric loading: this case should be avoided.

The case of inclined discontinuities does not involve important global stability problems for the crown, except for the case of the formation of wedges (key blocks). The influence on the pillar stability is noticeable, as the resistence can be tremendously reduced. In this case a lateral bolting of the pillar could be applied, thus determining a confining pressure that creates a stress state along the discontinuity surface equivalent to the stress condition of the intact pillar (figure 5). The estimation of the shear resistence of the discontinuity can help to reduce the amount of the external support. As a consequence it is advisable to locate the position of at least one dimension of the rooms in such a way that it could be possible to follow the discontinuity strike. The bolts must be accurately protected against corrosion, in the case where their contribution is taken into account in order to ensure long term stability. However, it would be better to adopt only natural support contribution for the after closure stability, if any interference with the surface can involve other

Figure 4. Underground room with sub-horizontal strata and discontinuities. The bolting of the roof is visible. Lasa marble quarry.

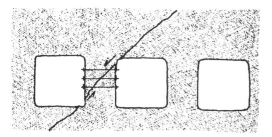

Figure 5. Exploitation scheme with inclined discontinuities across the pillars.

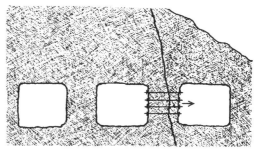

Figure 7. Exploitation scheme with sub vertical discontinities along the pillars.

structures or the overall stability of the slope itself.

The stress state of the pillars in the reported examples is due to rock loading of the overburden thickness that varies from 40 to 100 m or more, which corresponds to a lithostatic vertical stress of 1 to 3 MPa. The induced stress in the pillars increases with the excavation, depending on the dimensions of the pillars, on the size of the rooms and on the external morphology of the slope. The excavations are in fact made in a zone in which the directions of the principal stresses is probably affected by the surface topography of the slope.

The discontinuities with vertical orientation can determine problems in rib pillars when the isolated thin solids could be subjected to flexural loadings, especially when the pillars are next to the external surface and the loads are not exactly vertical (figure 7). It is preferable to cross a vertical discontinuity in a pillar instead of having it parallel to its main longitudinal dimension, as the resistence sections are larger.

Other stability problems are generally of local importance: the relative large spacing of the various sets compared to the nevertheless wide dimensions of the rooms in fact do not allow large instability of the crown, while it is possible to have removable blocks when fractured bands cross the walls of the pillars. An accurate scaling and a local bolting can help avoid safety problems.

4. QUARRYING TECHNOLOGY

The exploitation of the green ophicalcites (serpentinitic breccia with calcite veins) is made in rectangular or square shaped underground rooms and pillar stopes; sometimes rib pillars are left in perpendicular direction to the slope. The cutting of blocks is made by means of diamond wire, after many years of exclusive use of helicoidal wire. The theoreticalc room should be 15-20 m high, with 5-6 m high benches, and the blocks are moved by wheel loaders just to the adit and the external yard, where a derrick is installed (60m, 500 kN). The theoretical width should be of 9 m, the pull length of subsequent cuts is of 1.5 m; by previously removing a lateral slot of 1 m it is therefore possible to immediately use the wire saw for the cut of all the other blocks of the section. The system of the coupled pulleys of the diamond wire technique is particularly useful.

Block squaring is usually performed following the grain plain, that is, the plane along which the main carbonatic veins occur. The blocks are of about 5-6 m3, due to the size of the intact rock mass

Figure 6. A pillar in the Apuane marble crossed by an inclined main discontinuity.

volumes and to the type of equipment. The productivity is of about 700-800 tons/year /people.

In the recent years chain saw equipment has also been widely employed.

In Apuane marble underground exploitation has greatly developed due to the wide dimensions of the rock formations, the high overburden and the good continuity of the rock masses.

The quarrying is based on the known orientation of the weakness planes: discontinuities subparallel to the grain planes have less shear resistence in comparison to the easy way and to the head. In this mode the chinks of the head and the weakness of the grain plane are utilized in order to separate the bench. The blocks have a volume of 8-9 m3.

The underground exploitation starts from an external quarry yard and move from the top to the bottom, by means of lowering horizontal slices. Wide movement areas are thus obtained for the operating equipment. The chain saws are used both for the opening of the development difts and for the lowering of the benches, together with wire saw equipment.

The large rooms allow one to operate like in an open pit quarry, with high benches. The development drifts (6*3 m) allow the production of commercial blocks, using on an advancing length of 2.5 m some splitting hydraulic jacks for the separation of the blocks from the rock mass. The productivity is of about 1000 tons/year/people.

5. STABILITY ANALYSES AND MONITORING

It is very important to perform a correct design and monitoring of this type of underground structure.

The design can be performed on the basis of a careful geostructural survey. The calculation can be made using analytical methods, as the singular structures are generally of simple geometry (pillars and flat crowns). The problems consists of a correct evaluation of the natural state of stress and of the induced state of stress in the natural support. Squared geometries in fact involve local high stress concentrations, and solutions adopting, for example, the tributary area or the linear arch theory criteria are not suitable.

A first consideration can arise from the analyses of the stereographic projection of the discontinuities or by means of more elaborated codes (for example the key block methods by Goodman and Shi). If it were possible, a numerical analyses could also be properly adopted in this case, using, for example, some distinct elements codes (for example the UDEC or 3DEC code from Itasca). The type of rock mass is in fact a discontinuous medium made of big rock elements and some singular discontinuities, in which the behaviour is conditioned by both of these structures. The discontinuities can be generated by means of a deterministic approach when the sets are well recognizable, or by means of a statistical generator when the discontinuities are more ubiquitarious. This last approach presents some more difficulties than the first because it must be locally validated by the surveys. In the case of more fractured rock masses an alternative approach can be performed by means of finite difference codes (for example FLAC from Itasca) or finite elements codes (for example Phase2 by Rocscience). These last approaches are sometimes less suitable for the examined cases because they consider the rock mass as an equivalent continuous medium, and they need a different rock mass characterization in order to individuate the correct geomechanical parameters. The monitoring, in all the cases, of the static conditions of the rock mass around the openings can help both to validate the modelling and to check the evolution of deformations.

6. CONCLUDING REMARKS

The productivity of an underground quarry of dimension stones is linked to the exploitation techniques and to the geometry of the excavations. These features arise from the strength characteristics of the rock bodies and from the structural conditions of the geological formations. It is therefore important to plan the activities on the basis of an accurate study of the geomechanical features of the rock mass, a design of the natural support by means of numerical codes, a monitoring of the stresses and displacements of the underground structures. This procedure has some investment costs and requires a good geomechanical knowledge. The advantages are evident in terms of: reduction of debris waste materials, a reduced environmental impact, the possibility of working independently of the weather, the exploitability of rock bodies at a relevant depth without removal of the overburden, the availability of underground spaces in rock masses with good stability conditions for reuse.

The authors have given equal contribution to this paper.

REFERENCES

Fornaro M. & L. Bosticco 1998. Le cave sotterranee di pietra in Italia. *GEAM Bullettin*, Turin, XXXV, n.1, 21-26.

Del Greco O., M. Fornaro & C. Oggeri 1995. Improvement of dimension stone exploitation using structural analyses. In H.P. Rossmanith (ed.) *Mechanics of jointed and faulted rocks*: 841-845. Rotterdam: Balkema.

Mining Science and Technology' 99, Xie & Golosinski (eds) © 1999 Balkema, Rotterdam, ISBN 90 5809 067 1

Effect of pocketing pillar – A numerical and laboratory study

B.B. Lin
School of Mining Engineering, UNSW, Sydney, N.S.W., Australia

ABSTRACT: Coal pillars are the main support element for underground coalmines. If pillars are under-designed, significant safety, production and operational hazards would be produced. 15 major collapses have occurred in NSW and Queensland in the last 10 years. Over 12 people have been killed and over $10,000,000 of surface damage sustained; or if pillars are over-designed, resources would be wasted and productivity would be jeopardised. Although there are empirical formulae ready to use for regular shaped pillars (square pillars), there are still deficiencies in the methods used to estimate the strength of irregular shaped pillars, such as the irregular pillars caused by pocketing operations, which are sometimes encountered in Australian mining operations. As a result of this, some unplanned pillar failures have occurred. The purpose of this research was to investigate the strength of irregular pillars caused by pocketing in relation to square pillars. It is found that pocketing in pillar produces no noteworthy PCFS reduction when pillar starts yielding. However, it does significantly reduce PRFS accordingly with the pillar mechanical property. It seams that the friction has little influences on PRFS.

1. INTRODUCTION

The strength of a pillar is the maximum load it can support per unit area. Variable pillar strength formulae are available mainly in two forms as the following. Linear form:

$$S_p = k_1 \left[a + b \left(\frac{w}{h} \right) \right] \qquad (1)$$

Power form:

$$S_p = k_2 \frac{w^\alpha}{h^\beta} \qquad (2)$$

Where:

k_1 and k_2 = strength of a cube, MPa
w = pillar width, m
h = pillar height, m
α and β = dimensionless constants derived from field or laboratory
a and b = dimensionless constants derived from field or laboratory satisfying $a + b = 1$

The typical values for a, b and k_1 are 0.64, 0.36 and 6.2 respectively (Bieniawski and van Heerden, 1975), and for k_2, α and β are 7.2, 0.46 and 0.66 respectively (Salamon and Munro, 1967).

Galvin *et al* (1996) proposed the following formulae according to the Australian mining conditions. UNSW Linear Formula:

$$S_p = 5.36 \left[0.64 + 0.36 \left(\frac{w}{h} \right) \right] \qquad (3)$$

UNSW Power Law Formula:

$$S_p = 7.4 \frac{w^{0.46}}{h^{0.66}} \qquad \text{for } \frac{w}{h} < 5 \qquad (4)$$

$$S_p = 19.24 \frac{\left\{ 0.2373 \left[\left(\frac{w}{5h} \right)^{2.5} - 1 \right] + 1 \right\}}{w^{0.1334} h^{0.0667}} \qquad \text{for } \frac{w}{h} \geq 5 \qquad (5)$$

However, the Formulae 3, 4 and 5 are for regular shaped pillars (square pillars) in competent roof and floor strata based on a database of Australian case histories of pillar behaviour. There are some gaps in the methods to estimate the strength of irregular shaped pillar, which have been encountered in Australian mining operations.

To establish the irregular shaped pillar strength, relevant to square pillars, the following irregular shaped pillar formed by pocketing a large, over-designed pillar, as shown in Figure 1, has been investigated in physical model testing.

Figure 1 Sketch of the Irregular Shaped Pillar

In numerical modelling the effect of friction between pillar and roof/floor was also considered as shown in Figure 2.

2. PHYSICAL MODEL TESTING

Materials to be used are sand, plaster and water. The plaster to be used is Boral Hard Finish Plaster. The sand is fine beach sand. The ratio of the mixtures of sand, plaster and water, in weight, is 250:500:250.

The specimens made of these mixtures were cured for at least 28 days in room temperature and moisture. The surfaces of specimen is ground by a grinder with a grinding wheel of 80 to ensure top and bottom surfaces are parallel to each other. Testing is undertaken by using AVERY universal compressive testing machine with a maximum capacity of 360 tonnes. Displacement is measured with a Linear Voltage Displacement Transducer and load is measured with a potential meter. The loading rate applied is 18 tonnes/min.

Ideally the pocketing process should be carried out on the square specimen when it is loaded to a certain stress level. However it is believed that the pillar strength would not change significantly if the irregular shaped specimens is formed before stressing.

3. NUMERICAL MODEL TESTING

Strand7 is a three-dimensional Finite Element Analysis (FEA) package. It makes use of all hardware resources available to it under Windows, including full use of all available RAM (both physical and virtual), use of accelerated graphic display features and printing/plotting to any Windows supported device.

The numerical model is symmetrical to xy, xz and yz of Mohr-Coulomb material, and the ratio of pillar width to height is 2. In the numerical analysis, the tolerance for termination of iterations within a load step was set to 0.0001 for displacement norm and 0.001 for residual norm. The maximum number of iterations was set to 100.

For the clarification of explanation, the terms of Pillar Corner Failure Strength (PCFS) and Pillar Rib Failure Strength (PRFS) are used in the discussion. PCFS is defined as the stress level when a corner of

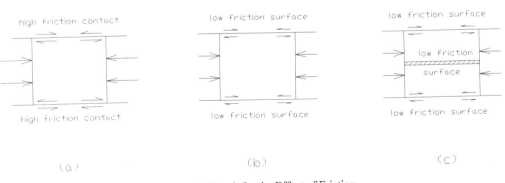

Figure 2 Sketch for the Effect of Friction

pillar starts to yield while PRFS is termed as the stress level when all sides of pillar have yielded.

4. RESULTS AND DISCUSSION

Figures 3, 4 and 5 show that pocketing in pillar does not affect the PCFS for variation of friction level between pillar and roof/floor, cohesion and internal friction angle respectively.

Figure 6 shows that there is no effect of friction on PRFS. However, it should be noted the numerical testing were conducted at strong roof and floor.

As shown in Figures 7 and 8, however, pocketing process does reduce PRFS for various levels of cohesion and internal friction angle. For instance, the reduction of PRFS is 16% for C = 1 MPa at extraction of 24% and about 10% for Fai = 40 deg at extraction of 24%.

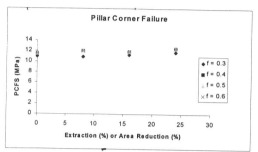

Figure 3. Effect of Friction between Pillar and Roof/Floor on PCFS (E = 1.5 Gb, Fai = 45 deg and C = 3 MPa)

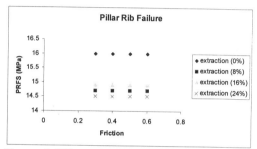

Figure 6 Effect of Friction between Pillar and Roof/Floor on PRFS (E = 1.5 Gb, Fai = 45 deg and C = 3 MPa)

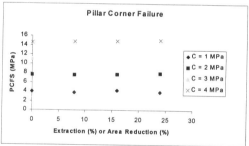

Figure 4 Effect of Cohesion on PCFS (E = 1.5 Gb, Fai = 45 deg and f = 0.3)

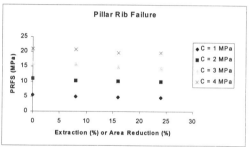

Figure 7 Effect of Cohesion on PRFS (E = 1.5 Gb, Fai = 45 deg and f = 0.3)

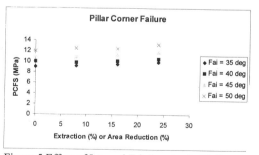

Figure 5 Effect of Internal Friction Angle (Fai) on PCFS (E = 1.5 Gb, C = 3 MPa and f = 0.3)

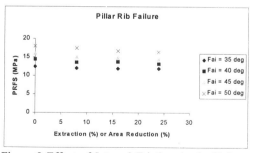

Figure 8 Effect of Internal Friction Angle (Fai) on PRFS (E = 1.5 Gb, C = 3 MPa and f = 0.3)

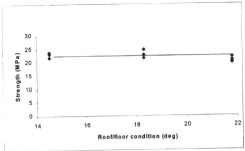

Figure 9 Laboratory Test Results on Effect of Friction

Figure 9 gives the effect of friction on pillar strength from the laboratory testing (tan15 = 0.25 and tan22 = 0.4). There is no significant effect on pillar strength.

Comparison of the testing and numerical results on the effect of pocketing process is presented in Figure 10. They are in general agreement.

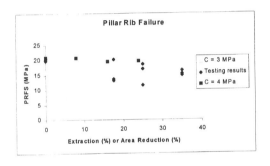

Figure 10 Comparison of Testing and Numerical Results

5. CONCLUSION AND RECOMMENDATION

Pocketing pillar produces no noteworthy PCFS reduction when pillar starts yielding, however, it does significantly reduce PRFS accordingly with the pillar mechanical property.

It was noted that the variation of the testing results is significant. This may be due to the testing limits, such as unparalleled top and bottom surfaces of specimens and platen effects. Therefore, validation of the laboratory and numerical results is essential by underground observation and monitoring.

REFERENCES

Bieniawski, Z.T. and Van Heerden, W.L., 1975. The Significance of in Situ Tests on Large Rock Specimens. *International Journal of Rock Mechanics and Mining Science*, Volume 12, pp101-113.

Galvin, J., Hebblewhite, B. and Salamon, M., 1996. Australian Coal Pillar Performance. *News Journal, International Sociaty for Rock Mechanics*. Volume 4, Number 1, pp33 – 38.

Mark C. and Chase, F.E., 1997. Analysis of Retreat Mining Pillar Stability. *Proceedings: New Technology for Ground Control in Retreat Mining*. IC 9446, US Department of Health and Human Services, USA, pp 17 - 34.

Salamon, M.D.G., Galvin J.M., Hocking G and Anderson I., 1996. Coal Pillar Strength from Back-calculation. *Research report RP1/96*, the Department of Mining Engineering, University of NSW, Australia. ISBN No 0733414893.

Salamon, M.D.G. and Munro, A.H., 1967. A study of the strength of Coal Pillars. *Journal of South African Institute of Mining and Metallurge*. Volume 67.

Mining Science and Technology' 99, Xie & Golosinski (eds) © 1999 Balkema, Rotterdam, ISBN 90 5809 067 1

Hydraulic pressure monitoring system for powered supports at a coal face

Fulian He, Qundi Qu, Zichang Yao & Jianyu Chen
China University of Mining and Technology, Xuzhou, People's Republic of China

Kegong Liu
Luan Coal Mining Administration, Shanxi, People's Republic of China

ABSTRACT: This paper describes the application of 89C52 chip microcomputer in hydraulic pressure information monitoring system of powered supports in coal face, and the details of circuit and software of monitoring system substation. Each substation of the monitoring system possesses the functions of hydraulic pressure information display, memory and communication. It can display the dynamic changing curve of hydraulic pressure information, by connecting with a flameproof portable computer, which is significant to improve the ability of detecting hydraulic system failure and solving powered support failure in longwall coal face.

1. INTRODUCTION

A lot of mining experts (Samir Kumar Das 1991, Robertson 1992 & Smith 1993) thought that condition monitoring is always very important in longwall coal mining. In underground longwall coal face, the adverse environment and complex geological conditions may cause the reduction of supporting force, reliability of powered supports and the severe caving of immediate roof. Therefore, a real time hydraulic pressure information monitoring instrument is urgently needed to detect the failure situation and working state of powered supports in fully mechanised coal face. This monitoring instrument should be functional to accomplish automatic measuring, recording, displaying and transmitting the whole variation process of hydraulic pressure information of powered supports, to realise the field dynamic management of hydraulic pressure information of powered supports, and to perform the field adjustment of support working condition and the further analysis of hydraulic pressure information of powered supports in underground longwall face.

According to the monitoring result, miners can quickly and accurately find out the cause of abnormal working condition of support hydraulic system. By taking effective control measures, the support quality in coal face can be improved; the downtime due to support and surrounding strata can be decreased; the fully mechanised coal face mining with high production and high efficiency can be promoted.

2. CONSTITUTION OF MONITORING SYSTEM

The operating principle diagram of hydraulic pressure information monitoring system of underground powered supports is shown in Figure 1, and the monitoring system is installed on the underground powered supports. The whole monitoring system consists of steel-wire-type hydraulic pressure sensor, monitoring substation, flameproof and intrinsically safe electric power source and flameproof portable computer. One monitoring substation can be connected with 16 sensors and monitor the hydraulic condition of 16 parts of supports such as legs, fore canopy jacks, balance jacks.

Chip microcomputer 89C52 is the main part of monitoring system substation, and the substation also includes clock, program and data memory, sensor gating circuit and liquid crystal display circuit. When the monitoring system is working, chip microcomputer 89C52 controls the gating circuit to excite cyclically 16 sensors (the period is 1 min), and at the same time picks up the vibration signal of the sensor. After the signal is treated by digital filter and calculated, the hydraulic pressure value and sampling time can be displayed on the liquid crystal screen. There are two kinds of data memories, long-term memory and short-term memory. The time between storing operations can be generally set by program as follows: 10 min once for short-term record and 1 h once for long-term record, and the data are stored in the way of cyclic queue. For short-term record, the data amount is 800 groups; for long-term record, the data amount is 400 groups.

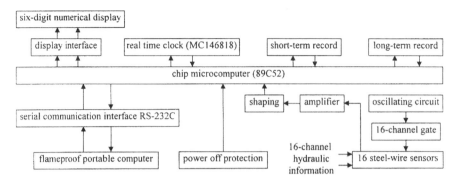

Figure 1. Monitoring system principle structure diagram.

Standby batteries are used for the monitoring system to maintain the operation of clock and stored data when electric power is off, so as to ensure the system operating normally when electric power is on. The monitoring system is equipped with serial communication interface and photo-electric coupler. Then 89C52 chip microcomputer can reliably communicate with flameproof portable computer and transmit the long-term and short-term records stored to the computer system on surface, so that the hydraulic pressure information can be further studied and analysed.

Hydraulic pressure information monitoring system of underground powered supports is a chip microcomputer application system in which main part is 89C52 chip microcomputer. Some special circuit designs of this system are specified as follows.

2.1 *Exciting circuit*

As shown in Figure 2, steel-wire sensor is self-excited continuously in the way of dual coil. The self-excited oscillating circuit is composed of operational amplifiers and two electromagnetic coils with reasonable number of turns and reasonable value of ohms. The amplifier is used not only for amplification but also for compensating the damping energy loss in steel wire vibration, so the continuous frequency signal output of constant amplitude can be got.

In order to eliminate the influence of detuning and drift, the capacitance coupling amplification circuit as well as high precision and low drift operational amplifier ICL7650 are selected. This amplifier operates on dynamic zero setting principle. Thus the intrinsic detuning and drift in MOS circuit can be eliminated, and the drift of low frequency signal transmitted by the steel-wire sensor can be also eliminated.

Around the protected pins 3, 6 of operational amplifier ICL7650, protective rings are installed on the

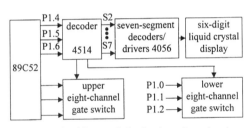

Figure 2. Principle diagram of exciting circuit.

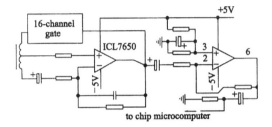

Figure 3. Liquid crystal display interface circuit.

printed circuit board. That is to protect the amplifier pins from the interference due to electric current leakage between input pin and neighbouring pin, as they operate on different voltage levels. Condensers with larger capacitance are adopted in the exciting circuit, so that accurate detection can be got. In addition, condenser is also connected in the feedback circuit, mainly for eliminating high frequency component and high frequency noise.

2.2 *Display interface circuit*

As shown in Figure 3, there are a six-digit liquid crystal static display and seven-segment decoders/drivers 4056 with BCD code, and the above drivers are controlled by decoder 4514. The working principle of this circuit is as follows.

The high three bits of P1 interface of 89C52 chip microcomputer are connected to the three address inputs of decoder 4514 and the rest address pins are connected to ground. Output of decoder 4514 (S2□S7, six states) is connected to the inputs of seven-segment decoders/drivers 4056 with BCD code and makes the decoders/drivers operate in turn. Then the drivers drive the numerical display to operate. The 16-channel analogue switch, which is composed of two chips 4051, is controlled by the signal which comes out by taking inverse logic of states S0 and S1.

2.3 Serial communication interface circuit

RS-232C interface adopts negative logic, and its range of voltage level is -3□-15 V for logic "1" and +3□+15 V for logic "0". Normally, personal computer has standard RS-232C serial interface which can be connected directly. However, for 89C52 chip microcomputer, electric level TTL is used to its serial communication interface instead of standard RS-232C interface, so electric level transformation should be made, which is normally done by using MC1488 output and MC1489 input. If MC1488 and MC1489 are used to perform the electric level transformation, electric power supply of ±12 V is needed. In chip microcomputer system, here is no electric source of ±12 V. To add the electric source of +12 V means costly and troublesome, especially for underground intrinsically safe monitoring instrument. Therefore, MC1488 and MC1489 are replaced with triodes and some other elements which are installed in a circuit shown in Figure 4. This circuit can realise voltage transformation and logic electric level inversion.

There is a lot of interference due to underground adverse and complex environment. In order to suppress the interference, photo-electric isolation is used. According to the feature of RS232C logic electric level, a negative electric potential is connected to input of optically coupled diode, and another negative electric potential is connected through a resistor to its output side. Both of them are come from different voltage converters which make communication signal be transmitted reliably. Since the

impedance of light emitting diode on optically coupled input side is much lower than the internal impedance of interference source, the noise voltage, referred to the optically coupled input, is not high enough to drive light emitting diode. In this way, the

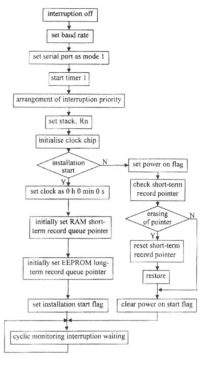

(a)　Main program diagram

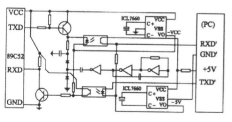

Figure 4. Electric level transformation circuit.

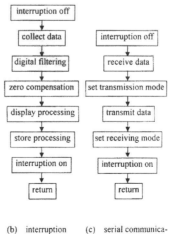

(b)　interruption subroutine diagram　(c)　serial communication subroutine diagram

Figure 5 Monitoring system software diagram

397

noise interference is effectively suppressed, and the ratio of signal to noise in communication transfer process is increased.

3. GENERAL DESIGN OF MONITORING SYSTEM SOFTWARE

The program of monitoring system substation operates according to job priority and sequential scheduling principle. From the beginning of every minute, the monitoring system substation collects 16-channel hydraulic pressure value in pre-set program sequence. Then the monitoring system clock is displayed. The above process is repeated. If the process is interrupted by serial communication, it is prior to execute the serial communication service subroutine. When the subroutine operation comes to end, the monitoring system substation begins to collect 16-channel hydraulic pressure value again, according to pre-set program sequence.

The software of monitoring system includes main program, external interruption subroutine and serial communication service subroutine, as shown in Figures 5a, b, c. Because the methods of zero compensation, digital filtering etc are adopted, the stability and anti-interference of monitoring instrument are improved.

4. APPLICATION

Hydraulic pressure information monitoring system of underground supports has been applied in some Chinese fully mechanised coal faces. As shown in Figure 6, the flameproof portable computer with an expert system communicates with each monitoring system substation in underground longwall face every day. Then twenty-four-hour continuously changing curves of hydraulic pressure of support legs and fore canopy jacks, and different kinds of hydraulic pressure information statistic histograms and tables can be displayed on the flameproof computer screen in fully mechanised coal face. Hydraulic pressure data can be stored in the flameproof portable computer taken to the surface, then printed out, analysed and studied deeply.

By use of hydraulic pressure information monitoring system, more than 2000 support hydraulic pressure changing curves are measured in one fully mechanised coal face. Therefore, dynamic control and adjustment of support working condition are realised in time. This system plays an important role in improving support working condition, raising roof control level, and accurately detecting actual quality of powered supports in fully mechanised coal face.

5. CONCLUSIONS

The hydraulic pressure information real time monitoring of underground supports can be undertaken by a monitoring system installed on powered supports in fully mechanised coal face. A flameproof portable computer is used to make the hydraulic information graphic display and analysis in the coal face. Overall variation process and dynamic variation tendency of hydraulic pressure information of powered supports can be known by miners. This system is advantageous to improve the miner's skill in hydraulic support operation.

Field application results indicate that hydraulic pressure information monitoring system of powered supports is completely suitable to underground environment in longwall coal face, with the feature of low measuring error and high reliability.

Hydraulic information monitoring system of underground supports provides an effective measure to diagnose and improve face support quality in coal face. It is advantageous to decrease the downtime due to supports and surrounding strata, and ensure the reliability of supports and surrounding strata in fully mechanised coal face.

ACKNOWLEDGEMENTS

The authors would like to thank Minggao Qian, Zhiyi Yang & Xiaolei Wang for useful discussions and contributions. Work is supported by the National Natural Science Fund of China (59734090).

REFERENCES

Robertson, B 1992. The changing role for monitoring and communication in underground coal mining. *The Australian Coal Journal*, (38): 21-31.

Samir Kumar Das 1991. Condition monitoring of longwall face machines. *Journal of Mines, Metals & Fuels*, XXXIX(1 & 2): 18-22.

Smith, A 1993. Exploding the myth of condition monitoring and preventative maintenance. *Mining Technology*, 75 (864): 105-111.

Mining Science and Technology' 99, Xie & Golosinski (eds) © 1999 Balkema, Rotterdam, ISBN 90 5809 067 1

Effect of size on the compressive strength of coal

H. Moomivand
Faculty of Engineering, University of Urmia, Iran

ABSTRACT: The compressive strength of coal depends on the distribution, type and condition of discontinuities. In the smaller specimen, the probability of finding *larger* discontinuities is smaller and the compressive strength is thus higher. All groups of laboratory and in situ test results were analyzed by a DataFit computer program separately. It was shown that the effect of size on the compressive strength of 10 different groups of coal is not the same. Compressive strength of different groups of coal specimens had a high scatter for the same size and strength-size relationship had high deviation when all groups of results were mixed. Compressive strength of specimens was divided by the compressive strength of a specimen having a size equal to d (in this analysis 50.8 mm) in any group of test results. Consequently the dimension of strength in all series of tests was omitted and the relationship between the ratio of compressive strengths and size for all groups of results was determined. From extrapolating laboratory and in situ test results, the relationship between compressive strength and size of all cubic coal specimens was derived.

1 INTRODUCTION

The discontinuities of various sizes are present in rock mass. The compressive strength, as a function of discontinuities, increases with a decrease in size of rock specimens. A new definition for size effect on the compressive strength has been given as it can represent the phenomenon. Most of the experimental results available are for coal and are concerned especially with the compressive strength of cubes of various edge dimensions. The effect of size on the compressive strength of coal has been investigated by conducting tests both in the laboratory and in situ. The effect of size on the compressive strength of all laboratory and in situ test results of cubic coal specimens has been analysed using DataFit computer program (1992). From extrapolating laboratory and in situ test results, the relationship between compressive strength and size of all cubic coal specimens has been derived.

2 EFFECT OF SIZE ON THE COMPRESSIVE STRENGTH OF COAL SPECIMENS

2.1 *Laboratory tests*

Coal contains various discontinuities such as cracks, pores, etc. The compressive strength of rock (coal) depends on the distribution, type and condition of discontinuities. In the smaller specimen, the probability of finding *larger* discontinuities is smaller and the compressive strength is thus higher. The effect of size on the compressive strength of different types of rock specimens is not the same (Moomivand, 1993).

The compressive strength increases with a decrease in size of coal specimens (Daniels & Moore 1907, Rice 1929, Lawall & Holland 1937, Steart 1954, Gaddy 1956, Evans et al. 1961). Gaddy (1956) tested the compressive strength of a large number of cubes from 5 different seams including Pittsburgh, Clintwood, Pocahontas No.4, Harlan and Marker having edge dimension from 0.051 m to

Table 1 Values of K and α in Equation (4) with correlation coefficient (r) and standard deviation for ten groups of test results of cubical coal specimens.

K	α	Correlation coefficient (r)	Standard deviation	Group name
25.63	-0.139	0.775	1.01	Pocahontas No.4 (Gaddy 1956)
22.52	-0.176	0.696	4.51	Deep Durffryn (Evans et al. 1961)
27.26	-0.179	0.434	3.29	Clintwood (Gaddy 1956)
33.95	-0.285	0.481	9.46	West Virginia (Lawall & Holland, 1937)
49.99	-0.425	0.908	5.42	South Africa (Bieniawski 1968a)
81.36	-0.436	0.614	7.97	Harlan (Gaddy 1956)
101.78	-0.445	0.994	0.91	Marker (Gaddy 1956)
43.04	-0.463	0.385	5.01	Daniels & Moore (1907)
59.14	-0.467	0.942	2.46	Pittsburgh (Rice 1929, Greenwald et al. 1939, Gaddy 1956)
58.91	-1.311	0.257	19.93	Barnsley Hards (Evans et al. 1961)

1.626 m and he proposed the following relationship between compressive strength and specimen size:

$$\sigma_{cl} = KD^{-0.5} \tag{1}$$

where σ_{cl} = compressive strength of cubical coal specimen having edge dimension D; and K = a coefficient depending upon the chemical and physical properties of the coal.

2.2 *In situ tests*

In the cutting and curing of a rock specimen for laboratory testing, not only are cracks, joints and weaknesses reduced from a large size (rock mass) to a small laboratory size but also with the transporting of specimens from mines to the laboratory and with the cutting of samples to a small size, the environment changes which can affect strength. When the specimens are dried at a constant temperature over a period of a few weeks, a smaller specimen may be drier than a larger one. Therefore, the moisture content of different size specimens cannot be constant. Also, the moisture content of a larger specimen will be less homogeneous from its centre to its surface in comparison to a smaller specimen.

Bieniawski (1968) conducted in situ compression tests of sixty cubical coal specimens

having edge dimension from 0.019 m to 2.012 m underground. He classified the tests into three groups, a small size (up to 0.076 m), a medium size (up to 0.457 m) and a large size (up to 2.012 m). He showed that compressive strength decreases with increasing specimen size and becomes constant when it reaches the critical specimen size (about 1.524 m), and suggested three different equations for different sizes of specimen as follows:

a) Initial constant strength relationship (σ_c=constant). For this case a specific value was not given for the edge dimension and other investigators have found significant variations in strength with sizes in this region (Evans 1970). Different methods were employed to prepare these 3 groups which probably have affected the results obtained (Bieniawski 1977).

b) The subsequent strength reduction relationship:

For W/H < 1 and W < 1.524 m,

$$\sigma_c = 4.772 \frac{W^{0.16}}{H^{0.55}} \tag{2}$$

where σ_c = compressive strength of pillar having width equal to W and height equal to H.

c) The final constant strength relationship:

For W/H$\geq$1 and W$\geq$1.524 m,

$$\sigma_c = 2.758 + 1.517\frac{W}{H} \qquad (3)$$

Pratt et al. (1972) performed in situ tests on quartz diorite and granodiorite specimens ranging

from 0.305 m to 2.743 m in length and laboratory tests ranging from 0.081 m to 0.305 m. They concluded that compressive strength decreases with an increase in the size and asymptotically approaching a constant value for specimens having edge dimension greater than 0.914 m. This critical size for diorite is less than the critical size for coal

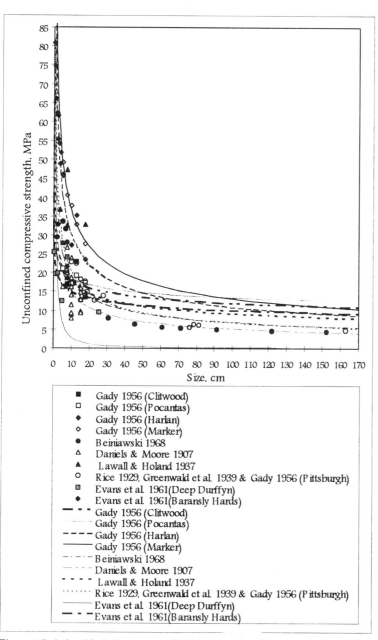

Figure 1 Relationship between unconfined compressive strength and size of all groups of coal specimens.

that was obtained by Bieniawski (1968).

Peng (1993) suggested that there is no difference between the density of cleats in a coal pillar and the density of cleats in a laboratory size specimen, provided the specimen size is larger than the cleat spacing and large fractures or joints are not commonly found in all underground coal pillars. Therefore strength of laboratory size coal specimens is equal to the strength of underground coal pillars. But various experimental results suggest that the size has an effect on the compressive strength of coal specimens. As a matter of fact, the discontinuities of various sizes are present in rock mass. Probably the density of larger discontinuities decreases with a decrease of specimen size. Statistically the number of larger discontinuities present in smaller specimens is smaller and the compressive strength is thus higher. Therefore, the strength, as a function of distribution of discontinuities of different sizes, increases with a decrease of specimen size. In critical size and onward the distribution of discontinuities of different sizes is the same with an increase in the specimen size and the compressive strength approaches a constant value.

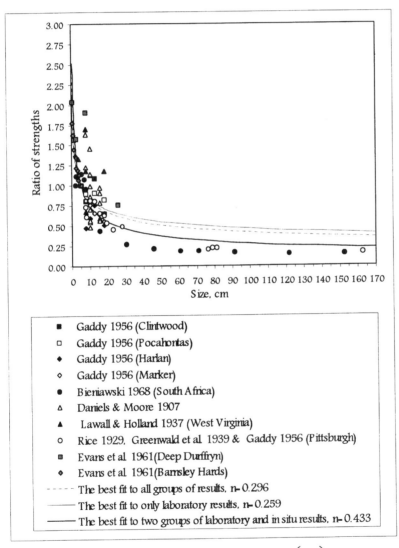

Figure 2 Relationship between ratio of compressive strengths $\left(\dfrac{\sigma_{cl}}{\sigma_d}\right)$ and size of all cubic coal specimens.

3 EXTRAPOLATION OF THE LABORATORY AND IN SITU TEST RESULTS OF COAL

Laboratory and in situ investigations have been carried out by various authors to study the strength-size relationship on different materials, and most of the experiments have been conducted on coal. An analysis of 10 groups of results derived by various authors on coal cubes has been carried out. The best fit function to relate the compressive strength with size has been found to be:

$$\sigma_{cl} = KD^{\alpha} \qquad (4)$$

where σ_{cl} = compressive strength in MPa; D = edge dimensions in cm; and K and α are constants.

Values of K and α in Equation (4) for different groups of test results are given in Table 1 and relationships are according to Figure 1. K varies from 22.52 to 101.78 and α varies from -0.139 to -1.311. Compressive strength of different groups of coal specimens has a high scatter for the same size (compressive strength is from 22.52 MPa to 101.78 MPa for 1 cm cube in different groups of results); so Equation (4) with constant values of K and α to fit the 10 groups of test results together cannot represent the strength-size relationship. The following equation is used to represent strength-size relationship for rocks:

$$\frac{\sigma_{cl}}{\sigma_d} = \left(\frac{d}{D}\right)^n \qquad (5)$$

where σ_{cl} = compressive strength of a cubic rock specimen with edge dimension D; σ_d = compressive strength of a cubic rock specimen with edge dimension d; and n = a constant for a given type of rock ($n \geq 0$; for n=0, size has no effect on the strength).

For extrapolation of the relationship between compressive strength and size of different types of coal, compressive strength of specimens is divided by the compressive strength of a specimen having a size equal to d (in this analysis 50.8 mm) in any group of test results. Consequently the dimension of strength in all series of tests is omitted and the relationship between the ratio of compressive strengths and size for all laboratory and in situ test results has been determined using DataFit computer program. The value of n in Equation (5) has been determined to be 0.296 with correlation coefficient (r) of 0.817 and standard deviation of 0.244.

The number of small scale tests is higher than the number of in situ large scale tests and the best fit is more close to the small scale results. Considering only the results of small specimens with size from 0.32 cm to 25.4 cm, the value of n has been determined to be 0.259 with correlation coefficient of 0.743 and standard deviation of 0.248. If results of some small and large specimens with size from 1.91 cm to 162.56 cm are considered, n becomes 0.433 with correlation coefficient of 0.885 and standard deviation of 0.189. The ratio of strengths versus size of specimens is given in Figure 2.

4 CONCLUSIONS

Coal contains discontinuities of various sizes. In the smaller specimen, the probability of finding larger discontinuities is smaller and the compressive strength is thus higher. The relationship between ratio of strengths and size of all groups including laboratory and in situ test results was determined. The relationship between compressive strength and size of all available cubic coal specimens is expressed in the form of Equation (5); and $0.259 < n \leq 0.433$. The compressive strength decreases with an increase in size and asymptotically approaching a constant value at size equal to 1 m and onward. Size effect is more pronounced in small scale specimens [Equation (5) and Figure 2] and difference in the strength due to increase in the edge dimensions of specimens is negligible at critical size and onward

5 REFERENCES

Bieniawski, Z. T. 1968. The effect of specimen size on compressive strength of coal. *International Journal of Rock Mechanics and Mining Sciences*: Vol. 5, pp. 25 - 335.

Bieniawski, Z. T. 1977. Discussions - A review of pillar strength formulas by W. A. Hustrulid. *Rock Mechanics*: Vol. 10, pp. 107 - 110.

Daniels, J. & L.D. Moore 1907. The ultimate crushing strength of coal. *Engineering and Mining Journal*: Vol. 84, pp. 263 - 268.

DataFit 1992. Data fitting by linear and multiple non-linear regression. P.O.Box 1743, Macquarie Centre, N. S. W. 2113, Australia.

Evans, I. 1970. Discussion of Z. T. Bieniawski's paper -The effect of specimen size on compressive strength of coal. *International Journal of Rock Mechanics and Mining Sciences*: Vol. 7, p. 230.

Evans, I., C.T. Pomeroy & R. Berenbaum 1961. The compressive strength of coal. *Colliery Engineering*: Vol. 38, pp. 75 - 80; 123 - 127; 172 - 178.

Gaddy, F. L. 1956. A study of the ultimate strength of coal as related to the absolute size of the cubical specimens tested. *Engineering Experiment Station*: Bulletin 112, Virginia Polytechnic Institute.

Greenwald, H. P., H.C. Howarth & I. Hartmann 1939. Experiments on strength of small pillars of coal in the Pittsburgh bed. *U. S. Bureau of Mines*: Technical Paper 605.

Lawall, C. E. & C.T. Holland 1937. Some physical characteristics of West Virginia coals. *Engineering Experiment Station*: Research Bulletin 17, West Virginia University.

Moomivand, H. 1993. *Effect of geometry on the unconfined compressive strength of pillars*, M. E. Thesis, University of New South Wales.

Peng, S. S. 1993. Strength of laboratory size coal specimens vs. underground coal pillars. *Mining Engineering*: Vol. 45, pp. 157 - 158.

Pratt, H. R., A.D. Black, W.S. Brown & W.F. Brace 1972. The effect of specimen size on the mechanical properties of unjointed diorite. *International Journal of Rock Mechanics and Mining Sciences*: Vol. 9, pp. 513 - 529.

Rice, G. 1929. Test of the strength of roof supports used in anthracite mines of Pennsylvania. *U. S. Bureau of Mines*: Bulletin 303, 44 p.

Steart, F. A. 1954. Strength and stability of pillars in coal mines. *Journal of Chemical, Metallurgical and Mining Society of South Africa*: Vol. 54, pp. 309 - 325.

Mining Science and Technology' 99, Xie & Golosinski (eds) © 1999 Balkema, Rotterdam, ISBN 90 5809 067 1

Test study on the top cracking process of large coal sample

Zhongming Jin, Jinping Wei, Xuanming Song, Yadong Xue & Yanhua Niu
Taiyuan University of Science and Engineering, People's Republic of China

ABSTRACT: In this paper, the pressing crack effect of the top coal under the abutment pressure in Xinzhouyao Mine is simulated. The characteristics of strength and deformation of top coal are analyzed. It is found out there are six evolution stages during the process of pressing crack of top coal, that are crack compressing, density increasing, expanding, hardening, failure, softening. The changing regularities of pressing crack and lump diameter are given by method of the fractal dimension. According to the damage mechanics theory, the constitutive equations of the top coal pressing crack is obtained. This paper provides scientific foundations for the technical parameter selection and numeral analysis of the lonwall top-coal caving technology.

1 INTRODUCTION

Longwall top-coal caving (LTC) technology is a new mining coal method which has been widely spread to thick seams in China. In LTC technology the top coal is pressed into crack or fissure naturally by the mining induced abutment pressure at the longwall retreating face and is loosened repeatedly by hydraulic powered support sustaining at the face, then the top coal caves from the rear of the support. The practices in China have confirmed that LTC technology is a high-yield, efficient, and economical mining method. In this paper, the phenomenon of the top coal is pressed in to crack or fissure naturally by the mining induced abutment pressure is called the pressing crack effect. It is obvious that the pressing crack effect play a key role in the raising the rate of recovery of the top coal, so evolution of top coal crack under the natural rock pressure should be studied in detail in order to control the broken level of the top coal crack in a expected or favorable way, and this is also the important scientific foundation of the rational selection of the caving technology parameters.

In this paper, by simulating the condition of natural load through triaxial compression test, the crack character and the crack evolution of the top coal are studied. This is a theoritical base of the LTC technology under the double hard conditions (hard roof and hard seam).

2. EXPERIMENT RESULT ANALYSIS

2.1 experiment conditions

The experiment is carried out through the real triaxial compression testing machine with maximum axial compressive force 1500kN and maximum confined compressive force 600kN respectively. The biggest coal sample is 300×300×300mm.

2.2 Coal mass strength and analysis

Experiment shows that the strength depends on the test sample size. Strength reduces non-linearly with increasing size of the test sample size. The standard UCS of 11-12# seam in Xinzhouyao Mine is 30-40Mpa. In the experiment, the average compressive strength of 300×300×300mm test sample is 11Mpa, and that of 150×150×300mm test sample is 15.5Mpa. Fig.1 shows the regression relation between coal mass strength and pressed areas. The average strength is as the following equation:

$$\sigma_c = \frac{3.95}{A^{0.285}} + 50e^{\frac{0.0073}{A}} \qquad (1)$$

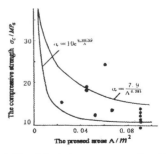

Figure 1. The relation between stress strength and the pressed area

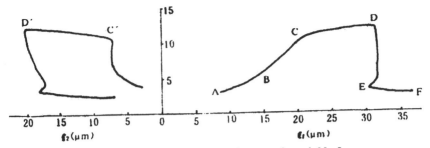

Figure 2. The stress-strain curve of sample No. 2

2.3 The analysis of the characteristic of the coal mass deformation

The typical stress-strain curve of coal test samples is shown in Fig.2, which indicates four comparatively complete zones. In zone AB, the curve is concave, the strain rate is higher than the stress rate, and the primary fissure is compressed. In zone BC, the curve is approximately straight-line, the coal test sample has elastic deformation. In zone CD, the curve is convex, the strain rate higher than the stress rate, this implies that in the sample new fissures emerge, expand and link up, and there is obviously of expansion in the volume. This zone is called the stage of strain hardening, the point D reflects the limit strength. In zone DF, the stress falls and the strain rises, the fissures link up intensively, and this zone is called the stage of plasticity softening. In zone DE, the test sample stress decreases rapidly and the coal lumps flake off.

Table 1. Main mechanics parameters

NO.	$\square_p$ MPa	$\square_p$ mm	E MPa	μ	V cm^3	$V^{'}$ cm^3
1	13.1	0.0294	5871.9	0.74	27000	1.014
2	9.33	0.041	3380.3	0.65	27000	1.012
3	12.45	0.041	5177.7	0.68	27000	1.015
4	11.5	0.0447	3246.5	0.72	27000	1.019
5	13.83	0.0221	26000.0	0.63	27000	1.006
6	5.53	0.024	6000.0	0.52	27000	1.001
7	18.7	0.035	4454.6	1.62	13500	1.078
8	18.44	0.0251	10956.5	0.53	13500	0.976
11	24.55	0.0301	5625.0	0.39	18000	0.976
Av.	14.15	0.0326	7856.0	0.72		1.017

In zone C'D', the increasing rate of lateral stress decreases sharply, the strain increases rapidly, and there is the phenomenon of the expansion of the volume. It is clear that the deformation of the coal mass are closely correlated with the development of the inner crack compression and expansion.

The basic mechanical property of the coal test sample are listed in Table 1.The Poisson ratio is large than 0.5 and the average value is 0.72. The volume of the test sample is:

$$V^{'} = V[1 - \varepsilon_p(1 - 2\mu)] \qquad (2)$$

According to the equation (2), the volume of the damaged coal test sample is increased by 1.37% as shown in Table 1.

2.4 The evolution of crack

In order to characterize the crack's evolution, the fractal dimension D_1 of crack length (Jin, 1994) and the fractal dimension D_2 of crack area (Jin, 1996) are used. Here D_1 represents the change of crack length at different fractal scales while D_2 means the information dimension. The probability of penetrated cracks' distribution is reflected by D_1. D_1 and D_2 can be expressed by

$$D_1 = - \frac{\log N_i}{\log L_i}, N_i = \alpha L_i^{-D_1} \qquad (3)$$

$$D_2 = -\Sigma \, p_i \ln\left[\frac{1}{p_{ii}}\right] \bigg/ \ln L_i \qquad (4)$$

where N_i is the number of the penetrated cracks on the condition of fractal scale L_i, P_i is the probability of area where there are penetrated cracks.

D_1 and D_2 are measured on the two free ends of the coal test sample. To confirm the cracks' evolution inside the test sample, the ultrasonic fissure detector SYC-IIC was used to measure the longitudinal wave velocity, which runs through the two free ends of the sample. Now, we use D_1, D_2 and V to analyze the cracks' evolution of the coal sample under load.

As shown in Fig.3, the changed curve of average crack fractal dimension D_1, D_2 and V appears dropping tendency with the changing of the abutment pressure σ_1. The equations of the average crack fractal dimension and the longitudinal wave velocity are given as follows:

$$\left.\begin{array}{l} D_1 = 1.5882 - 0.024\sigma_1 \\ D_2 = 1.1124 - 0.0048\sigma_1 \\ V = 1852 - 43.9\sigma_1 \end{array}\right\} \qquad (5)$$

The curves showed in Fig 3 reflect the six stages of evolution. In the first zone, named the crack compressing

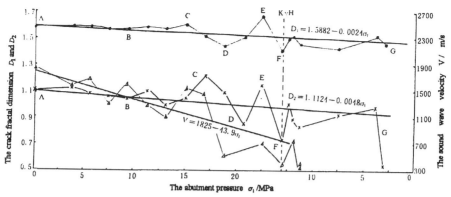

Figure 3 The change curve of D_1, D_2 and V

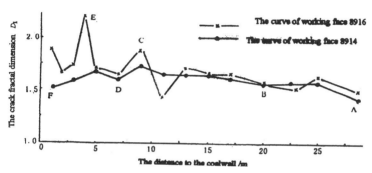

Figure 4. The measured curve of working face 8916 and 8914

stage, D_1, D_2 and V change slightly, $\sigma_1 \leq 1.75\gamma H$ and the average $D_1 = 1.5799$; D_1 is very similar to that measured from the working face No.8914 at Xinzhouyao mine ($D_1 = 1.5507$). In the second zone, named cracks' local density increasing stage, D_1 and D_2 show rising intendance or decrease initially slightly than increase dramatically, this indicates that the short cracks become long first, then intensive; the sound wave velocity declines slightly or drops first, then rises. All those reflect the local evolution of the cracks, at this stage $\sigma_1 = (1.75 \sim 2.0)\gamma H$. In the third zone CD, named the cracks expanding stage, $\sigma_1 = (2.0 \sim 2.75)\gamma H$, the obvious falling of D_1 and D_2 shows that the cracks in coal mass expand everywhere. In the fourth zone DE, named the pressing crack hardening stage, the sharp growing of D_1 and D_2 shows the appearing of lots of new cracks in coal mass makes the sound wave velocity tend to be stable, now the failure limitation of coal mss is achieved and $\sigma_1 = (2.7 \sim 3.0)\gamma H$. In the fifth zone EF named fragmenting stage, D_1, D_2 and V fall rapidly; under the two-dimension load, the coal test sample was destroyed with the gathering of the cracks and the sound wave velocity falls to the lowest figure, now $\sigma_1 = (3.0 \sim 3.25)\gamma H$. In the sixth zone FG named pressing crack softening stage, the main cracks expand, penetrate

and link up inside the test sample, as a result, the sound wave velocity drops to zero and the ultrasonic fissure detector records nothing.

Not all the above six stages can be measured out in working face underground. The peak value of the abutment pressure measured at the working face 8914 at Xinzhouyao mine is $19.8MPa = (2.37 \sim 2.69)\gamma H$, and three stages, i.e., compressing, local density increasing and expanding, can be recognized from the measured results, this implies that the pressing crack hardening stage and the fragmenting stage are not obvious.

According to the field measurement on the coalwall from the face towards the inside of mass at working face 8914 and 8916 at Xinzhouyao mine, the curve of D_1 is shown in Fig 4, which shows the very similar distribution as discussed above. While the hardening of the coal mass is not obvious in working face 8914 and higher D_1 means that the fragmenting degree is not high.

2.5 The analysis of samples' destroy

After the experiment of the sample' pressing crack, the lamps' number and the weight are counted in eight grades of the diameter (150~200, 120~150,

407

80~120, 45~80, 30~45, 10~20, <10mm). Based on the number probability P_i and the average diameter d_c of the first six lump grades, the lump number fractal dimension D_3 of every sample is calculated from the following formula

$$D_3 = -\log P_i \Big/ \log d_c \qquad (6)$$

According to every block grade's weight probability p_i and average lump diameter d_c, the coal lumps' weight

$$D_4 = -\log P_i \Big/ \log d_c \qquad (7)$$

fractal dimension D_4 is

As a result, the degree of the coal sample's pressing crack can be expressed.

2.6 The constitutive equations of the roof-coal pressing crack

According to the Lamaitre's hypothesis, the deformation character of the damaged material may be expressed by the intact material's constitutive equation whose stress has been replaced by the equivalent stress, i.e.

$$\sigma = E\varepsilon = E(1-D)\varepsilon \qquad (8)$$

where D is the damage variable and D can be calculated as follows:

$$D = C\varepsilon^{n-6} \qquad (9)$$

where C is the damage coefficient; n is the material

$$n = \frac{6E_p - 5\sigma_p}{E\varepsilon_p - \sigma_p} \qquad (10)$$

constant. The values of C and N are given in the following formula

$$C = \varepsilon_p^{-(n-5)} \left(\varepsilon_p - \frac{\sigma_p}{E} \right) \qquad (11)$$

where, σ_p is the stress peak value, ε_p is the strain peak value; σ_p and ε_p can be obtained from the experiment. According to σ_p, ε_p and E in Table 1 and the formula (10), (11), N and C of coal seam 11-12# of Xinzhouyao Mine are 7.2 and 431.5 respectively; so D is 0.4476 on the basis of formula (9). As a result, the top coal's consititutive equations is

$$\sigma = 7856.9(1 - 0.4477)\varepsilon \qquad (12)$$

Let σ be $K\gamma H$, the mining depth, on this condition the roof-coal will be crushed. This crucial mining depth is

$$H = 7856.9(1 - 0,4477) \varepsilon \Big/ K\gamma \times 10^3 \qquad (13)$$

3. CONCLUSIONS

1. The cracks' evolution process under the abutment pressure at working face is reappeared through the experiment of the big coal sample's pressing crack, i.e., there are six pressing crack stages: compaction, density increasing, expanding, hardening, failure and softening. The evolution process of roof-coal on the spot has a direct bearing on the abutment pressure $K\gamma H$ and the coal mass strength.

2. There are two forms of failure of the roof-coal: tension fracture and shear breakdown. The former has cracks' expansion mainly and the later has cracks' density increasing mainly. The rational adjustment of the two failure forms is beneficial to the control of the top coal failure and provides a efficient way to raise the roof-coal recovery rate.

3. The formulas (1), (6), (9), (12) and (13) etc. obtained through the experiment can be referred parameter selection and the numerical calculation of LTC technology.

REFERENCES

Xianwei Li, 1993, *Mechanics Features of Rock*, China Coal Industry Publishing House (in Chinese)
Zhongming Jin, *Fractal Features of Coal Fissure and Roof Caving*, Journal of Rock Mechanics and Engineering, 1996, (3):219-228(in Chinese)
Jun Sun, Micro-damage Features of Brittle Rock and Its Time Effect, Journal of Rock Mechanics and Engineering, 1996, (3):219-228(in Chinese)
Zhongming Jin, 1996, *Roof control of Coal Mine in Hard Rock*, China Coal Industry Publishing House (in Chinese)

4 Mine construction and tunneling

Mining Science and Technology' 99, Xie & Golosinski (eds) © 1999 Balkema, Rotterdam, ISBN 90 5809 067 1

Horizontal and inclined tunnels excavated by TBMs: Analysis and comparison of the TBM performances for some cases in Italy

N. Innaurato, R. Mancini & M. Cardu

Dipartimento Georisorse e Territorio, Politecnico di Torino, and CNR, Centro Studi per la Fisica delle Rocce e le Geotecnologie, Torino, Italy

ABSTRACT: The paper, after a brief general discussion of the methods for horizontal, inclined/vertical ways excavation, compares some cases of horizontal and inclined tunnel driving by TBM under the aspect of machine mean performances. The comparison of the cases makes possible to draw some criteria, enabling to forecast the TBM performances in horizontal and in inclined tunnel driving.

1 INTRODUCTION

Horizontal, inclined and vertical ways through the rock are needed for a number of mining and civil (Hydropower, transportation and so on) purposes, and quite often in the same project both horizontal and inclined, or vertical, ways are necessary. These cases offer an opportunity for comparing the performances of the excavation systems in similar geological-geomechanical environments.

The basic mechanism of rock breakage, either cyclical (Drilling and Blasting) or continuous (by TBM or by Raise Borer) are the same for horizontal or inclined ways; in the latter case the fixed motor option (Raise Borer) tends to be preferred for short, small diameter, easily accessible top point cases; as to the D & B system, it tends to be employed only for horizontal ways or very short inclined ways.

In Italy the use of TBM for horizontal and inclined ways became a rather common practice mainly in connection with the hydro-power programs underway by ENEL (National Electricity Board), for new plants and old plants updating, in particular because the general policy is to transfer underground all the water conducts and power houses.

The cases here examined, from 1975 to present time, pertain to hydropower and water supply plants and to the ventilation inclines of a large road tunnel, with diameters in the 2.5 to 3.9 m range (at places, subsequently enlarged).

In Table 1 the main data of the examined cases are collected; to be noticed, the cases are grouped by rock type crossed: cases A pertain to granitic gneisses, B to limestones, C to calcschists and related, D to extremely hard gneiss-pegmatites.

Table 1. General data of the inclined and horizontal tunnel cases discussed in the text.

Case	i	D (m)	L_1 (m)	L_2 (m)	LT
A_1 (1975)	42°	3	3150	2100	GN
A_2 (1977-78)	/	2.57	6215	1900	GRG
B_1 (1985)	43°	3	472	472	CA
B_2 (1989)	/	3,5	3052	2981	DL
B_3 (1992)	47°	3.75	725		DL
C_1 (1978)	46°	3.02	750	715	CCS
C_2-C_3 (1989-91)	/	3,5	14800	14800	CCS, MCS, MTC
D (1997)	44°	3,9	752	450	GN, AMP

D: Tunnel diameter; L_1: Tunnel length; L_2: Length excavated by TBM; i: Tunnel inclination; LT: Lithotype; GN: Gneiss; GRG: Granitic gneiss; CA: calcarenite; DL: Dolomitic limestone; CCS: calcschist; MCS: micaschist; MTC: metaconglomerate; AMP: amphybolite.

2 FURTHER INFORMATION ON THE CASES CONSIDERED

A_1: two inclined shafts for a pumped storage hydropower plant, subsequently enlarged to 6.4 m by D&B. Is the only case, in this report, where roller type tools with carbide buttons have been applied;

A_2: is a pedestrian access tunnel, pertaining to the same plant as A_1, and is quoted to provide a comparison with a horizontal tunnel. Tunnel A_2 has been driven by TBM only for a short stretch (less than 2 km) and completed by D & B due to the excessive consumption of the tools and frequent machine failures;

B_1: two inclined tunnels for a hydropower plant, driven through comparatively soft limestones, subsequently enlarged to 5.8 m by mechanical excavation;

B_2: is a horizontal tunnel driven through dolomitic limestone, harder than case B_1, as a part of an aqueduct;

B_3: inclined tunnel for an hydropower plant, driven through a dolomitic limestone, whose strength is intermediate between B_1 and B_2;

C_1: a couple of inclined tunnels driven for ventilation purposes from a large road tunnel (Fréjus tunnel), through calcschists, subsequently reamed to 5.8 m, again by TBM;

$C_2 - C_3$: two stretches of a hydropower tunnel driven through calcschists (somewhat stronger than C_1):

D: an inclined tunnel for a hydropower plant, driven through the most strong and abrasive rocks amongst the types considered in this report. Excavation by TBM has been discontinued because of a failure, not related to the TBM machine.

3 MACHINES EMPLOYED

Apart from case A_1, machines equipped with disc cutters have been used. The general trend in TBM construction since 1975 has been towards increased disc diameters, which is apparent in Table 2, where the main technical data of the machines are collected.

4 ROCK DATA

The properties of the rocks crossed are not uniform along the alignments of the tunnels, and probably the variability is as important as the average value of the strength in determining the performances of the machines. In Table 3 the range and the average value of the parameter most commonly employed to define the rock strength, which is the uniaxial compression strength C_0, are provided. Other important parameters, for which unfortunately has been not possible to provide reliable data for all the cases examined, are reported too: the tensile strength (Brazilian test, T_0), the RMR (Rock Mass Rating)

and the indicator of the rock abrasivity HK75 (upper quartile of the Knoop microhardness distribution).

Table 2. Machines data

C.	D	S	T_{max}	N	Φ	n	P_{max}	TP
A_1	3		1800	18	(++)	12	480	/
A_2	2.57	0.7	2800	17	300 (*)	12	380	/
B_1	3.0	1.0	4300	30	350	11.4	300	350
B_2	3.5	0.75	7155	27	397	9.57	552	680
B_3	3.75	-1.2	2800	34	350	9.5	300	390
C_1	3.02	0.7	4400	17	300 (*)	12	360	460
C_2-C_3	3.5	1.5	6200	26	412	10.6	600	900
D	3,90	1.6	8000	32	416	11	900	1050

(*): Double disc tools. (++): roller bits with buttons.
C.: Case; D: Head diameter, m; S: Maximum stroke, m; T_{max}: Maximum thrust, kN; N: Number of tools; Φ: Tool diameter, mm; n: Head revolutions per minute, rpm; P_{max}: Maximum power at the head, kW; TP: Total Power, kW.

Table 3. Geomechanical characteristics of the rocks encountered by tunnels.

Case	C_0 (C_{0av}) (MPa)	T_0 (MPa)	RMR	HK75 (MPa)
A_1	100-150 (125)	/	/	/
A_2	70-170 (97)	7-12	70-85	5990
B_1	15-50 (25)	/	/	/
B_2	60-180 (70)	5	60-80	1500
B_3	(40)	2.9	60-85	3600
C_1	40-60 (50)	/	/	/
C_2-C_3	60-130 (83)	5-7	30-70	6300
D	90-200 (112)	10-16	60-80	8200

C_0: Uniaxial strength of the rock; C_{0av}: Average value of the uniaxial strength; T_0: Tensile strength (from Brazilian test); RMR: Rock Mass Rating; HK75: see the text.

Some peculiarities of the rocks affecting the operations are concisely listed:

A_1: quite uniform rock, with scarce joints;

A_2: rock strength progressively increases from the portal: as explained, the machine gave up after approximately 2 km excavated. A major fault has been crossed, with problems;

B_1: calcarenites, softer, alternate with marls, harder, poor rock, faulted, is crossed in the last 50 m of the tunnels;

B_2: distinctly bedded dolomitic limestone, with marl-clay intercalations;

B_3: thick layers; homogeneous, apart from marl intercalations;

C_1: good rock quality, apart from a jointed rock stretch in the initial part of the layout;

C_2, C_3: problems in some shear zones crossed; strength quite variable (calcschists, micaschists and metaconglomerates alternate along the tunnel);

D: hard gneiss alternates with still harder lenses of pegmatites and amphybolites (machine thrust had to be improved after an initial unsuccessful attempt).

5 PERFORMANCES OBSERVED

A very important point in TBM operation, both horizontal and inclined, is thrust. It is seldom possible to exploit the full nominal thrust available from the machine (T_{max} of Table 2) and, consequently, being thrust linked to the power consumed, to exploit the full power available (P_{max} of Table 2). In Table 4 the cases are examined under this aspect: the values, where available, of T_{max}, of the average actual thrust, of P_{max} and of the average power consumed (head only) are collected.

Table 4. Thrust and power data for the cases examined.

Case	T_{max} (kN)	T_{av} (kN)	P_{max} (kW)	P_{av} (kW)	T_{av}/T_{max} (%)	P_{av}/P_{max} (%)
A_1	1800	/	480	340	/	70
A_2	2800	1800-2000	380	160	64-71	42
B_1	4300	1200	300	186	28	62
B_2	7155	4000	552	480	56	86
B_3	2800	2000	300	174	71	58
C_1	4400	/	360	140--180	/	39-50
C_2-C_3	6200	3960	600	504	64	84
D	8000	5800	900	430	72	48

It is possible to conclude that, in the cases examined, the machines were quite reasonably exploited, both in horizontal and inclined excavations, to within, approximately, 60 – 70 % of the capabilities, apart from case B_1, where rock was particularly soft, for the thrust, and to within 50 – 85 % for the power.

The progression and specific consumption data are collected in Table 5.

Referring to the last column of Table 5, it has to be noticed that in most cases the tunnels are not long enough to provide reliable tool consumption data, especially in soft rocks such as limestones: in two cases the same discs set installed on the head at the beginning was still serviceable at the end of the work.

Table 5. Progression and specific consumption data.

Case	N.P.R.	p	U.C.	A.D.P.	Es	S.T.C.
A_1	1.55	2.15	0.24	8.9	115	0.0044
A_2	1.13	1,57	0.31	8.4	97.2	0.021
B_1	5.8--6.5	8.5--9.5	0.18	25-28	15.1	/
B_2	3.7	6.5	0.38 (*)	33.7	48.6	0.0022
B_3	2.5	4.4	0.19	11.4	23	/
C_1	1.1--1.45	1.5	0.22--0.36	5.8--12.5	55.8--63	
C_2-C_3	2.4	3.8	0.27 (**)	15.5	82.8	0.0035
D	2.6	3.9	0.19	11.8	50.4	0.0089

N.P.R.: Net Progression Rate, m/h; p: advance per revolution, mm/rev.; U.C.: Utilisation Coefficient; A.D.P.: Average Daily Progression, m/d; Es.: Specific Power Consumption, MJ/m³; S.T.C.: Specific Tools Consumption, Tools/m³.

(*) Calculated on the worked days.

(**) Rock reinforcement time is not taken into account.

6 DISCUSSION

The cases examined refer to the use of different machines and, more important, in different years, in different rocks, which makes difficult to answer to the simple question: how much more difficult is to drive by TBM an inclined tunnel with respect to an horizontal tunnel?

In order to minimise the disturbance, at least, of the rock factor in a comparison, in the Figures 1 to 4 some important features of the operations are plotted against the rock strength indicator C_0: penetration per revolution, specific energy, thrust per disc, machine utilisation coefficient (thrust comparison is meaningful because the cases do not refer to operations widely differing in bore diameter and tools number).

It can be seen that penetration tends to be lower, but not dramatically lower, than in horizontal excavation; the difference is more apparent in soft rocks (Figure 1); the specific energy Vs. C_0 correlation tends to remain the same (Figure 2) and the same applies to the thrust Vs. C_0 correlation (Figure 3); the machine utilisation coefficient tends to be lower in inclined tunnels. Cases examined, however, refer only to inclinations in a 40° – 50° range.

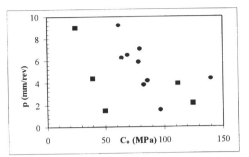

Figure 1. Penetration per revolution Vs. rock compression strength. Circles refer to horizontal tunnels, squares to inclined tunnels.

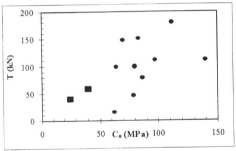

Figure 2. Specific energy consumption Vs. rock compression strength. Same symbols as in Figure 1.

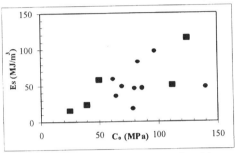

Figure 3. Thrust per disc Vs. rock compression strength. Same symbols as in Figure 1.

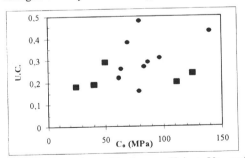

Figure 4. Machine utilisation coefficient Vs. rock compression strength. Same symbols as in Figure 1.

In the Figures have been also added points representing horizontal tunnels, not quoted in this report.

7 CONCLUDING REMARKS

TBM have been, and are, used in Italy in a number of projects, mainly hydropower, and it does not seem that tunnel inclination necessarily leads to much lower performances that in the horizontal case; another point is, however, important: inclined tunnels are often driven by drilling and blasting, which probably is a cheaper method, but tends to be abandoned: one reason is, plainly, the scarcity of well trained and tolerant workers, but another, important, is the excavation speed, which undoubtedly favours the TBM.

Coming to our cases, the tunnel of case C_1 is served, in addition to the ventilation inclines driven by TBM, by other ventilation inclines, driven by D&B (Alimak method), practically of the same length. The excavation time by D&B has been 22 % longer, though considering, for the TBM case, the machine launching time. Both data are referred to the pilot bore.

REFERENCES

Arnoldi, L., E. Roncada, G. Salvini, F. Toso & L. Vezzoli 1994. L'impianto idroelettrico ENEL di Riva del Garda. *Quarry and Construction* 1: 90-99.

Fantoma, D., D. Bogo & L. Arnoldi 1986. Scavo di due condotte forzate con fresa a piena sezione. Impianto idroelettrico di pompaggio dell'Anapo (Siracusa, Italia). *Proc. Int. Congress Large Underground Openings*: 141-151. Firenze.

Garrone L., S. Pelizza, L. Stragiotti & T. Viaro 1980. Scavo con frese dei pozzi di ventilazione nel traforo autostradale del Fréjus, lato Italia. *Boll. Ass. Min. Subalpina* XVII, 1: 77-97.

Hardin, P. 1980. Fresh air for Frejus: vent shaft excavation in Europe's tunnel. *Tunnels & Tunnelling* 5: 25-27.

Innaurato, N., R. Mancini & M. Cardu 1997. An Analysis of tools consumption in TBM operations: Italian cases examined. *Proc. Int. Mining Tech. 97 Symp.*: 757-766. Shanghai.

Mining Science and Technology' 99, Xie & Golosinski (eds) © 1999 Balkema, Rotterdam, ISBN 90 5809 067 1

Modeling the properties of sliding layer of shaft lining

Weihao Yang, Henglin Lu, Jiahui Huang & Yongjun Jin
China University of Mining and Technology, Xuzhou, People's Republic of China

ABSTRACT: Based on analysis of forces acting on inner lining of a vertical shaft the best characteristics of sliding layer between outer and inner shaft lining were determined. Secondly, a series of simulation tests were carried at different temperatures and pressures in order to study the mechanical features of selected materials of sliding layer, and the relations among the long-term shear strength, temperature and pressure were defined. The results provide guidelines for design of inner shaft lining. The investigations indicate that when suitable material with reasonable thickness of sliding layer is used, the horizontal pressure is a key factor in the determination of the thickness of inner shaft lining.

1. INTRODUCTION

About 50 shaft linings lying in deep overburden have fractured in China since 1987. The fracture mechanism of the shaft lining is: if aquifer in alluvium is directly covered over, or has an obvious contact with the coal measure strata, water pressure in aquifer drops due to mining activities. The effective stress of aquifer increases, resulting in soil consolidation and subsidence. The vertical

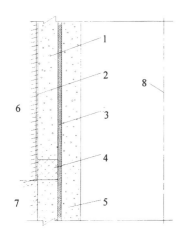

Figure 1 schematic diagram
for sliding and yieldable shaft lining

1-outer lining 2-foamed plastic layer 3-sliding layer
4-yieldable segment 5-inner lining 6-overburden
7-rock 8-center line of shaft

additional force acts on the outside of the shaft lining by the settling soil. Because this additional force is very large, the fracture of the shaft lining occurred.

Since 1988, a new shaft lining structure called "vertically sliding and yieldable shaft lining" was applied to shaft lining by freezing to guard against the fracture of lining. Vertically yieldable segments are inserted in the outer lining and a sliding layer (hereafter "SL") is set up between inner and outer lining (fig.1). So the outer lining can slide along inner lining. The inner lining of vertically sliding & yieldable shaft lining was mainly acted by gravity, horizontal pressure of SL, and the vertical additional force transferred from outer lining. The vertical additional force depends on component, temperature, pressure, thickness of SL, and approximately equals long-term shear strength of SL. The relationships among the long-term shear strength of SL, temperature and pressure were studied by simulation test in this paper.

2. SIMULATING DESIGN

On the basis of many experiments, the suitable component of SL was selected. Specific gravity of SL is $1.16g/cm^3$, softening point is 49.5^0, ductility is 61mm~65mm, and penetration is 63~65 @ 1/10mm. The thickness of SL should be greater than 100mm, but when it is greater than 140mm, its effect on decreasing vertical additional force no longer increases obviously(Cui, Guangxin et al 1996).

Therefore, the thickness of prototype of SL was assumed to be 120mm in tests.

When the thickness and component of SL are known, parameters which influence the long-term shear strength of SL are as follows:

$$F(r_2, r_1, z, \delta, T, p, V, T_0, \gamma, \tau, \mu, E, \upsilon, g, U, t) = 0 \quad (1)$$

where r_2, r_1=inside radius of outer shaft lining and outside radius of inner shaft lining; δ=thickness of SL, z=depth; T=temperature; p=horizontal pressure; V=loading rate; T_0=softening point of SL, γ=bulk density of SL, t=time, μ=Poisson's ratio; υ=viscosity factor of SL, τ=the shear strength of SL, E=deformation modulus of SL, g=gravity acceleration; U=displacement.

From Eq.1, we can get following similitude criteria (Cui Guangxin 1990):

$$\pi_1 = \frac{z}{r_1}, \quad \pi_2 = \frac{r_2}{r_1}, \quad \pi_3 = \frac{\delta}{r_1}, \quad \pi_4 = \frac{U}{r_1}, \quad \pi_5 = \frac{T}{T_0},$$

$$\pi_6 = \frac{r_1\gamma}{p}, \quad \pi_7 = \frac{Vt}{pr_1^2}, \quad \pi_8 = \frac{\tau}{p}, \quad \pi_9 = \frac{E}{p}, \quad \pi_{10} = \frac{pt}{v},$$

$$\pi_{11} = \frac{gt^2}{r_1}, \quad \pi_{12} = \mu.$$

According to the specification of the test rig, the accuracy of sensor and the size of prototype, C_L=31 is determined (fig.1). Where C=symbol of the proportion constant for the parameters between the prototype and the model. It shows that the geometric dimensions of the model are equal to 1/31 of that of the prototype.

The equivalent material of SL and reinforced concrete are the same as those of prototype in tests, so that $C_{T_0} = 1$, $C_E = 1$ and π_{12} is satisfied. From criterion π_5, we get $C_T=1$. That is to say, the temperature of SL in the model should be equal to that of the prototype. The temperatures of SL in tests were 10, 15, 200 respectively.

Table 1 geometric character of model shaft lining

terms	r_1 /mm	r_2 /mm	δ/mm	Height of sliding layer /m
prototype	6200	6440	120	62
model	200	208	4	2

From criterion π_9., we get C_p=1. It means that the pressure acting on SL in tests was the same as that in prototype. The pressures in tests were 0MPa, 1MPa, 2MPa. respectively. From π_8, we get C_τ=1. That is, the shear strength of SL in tests is equal to that in the prototype.

From π_6, C_γ=1/C_L can be determined, i.e. The bulk density of the equivalent material should be 31 times of that of the prototype. But these kinds of materials are difficult to find out. In tests, the

Table 2. Test plan

Test no.	p /MPa	T /□
1	0	20
2	1	20
3	2	20
4	0	10
5	1	10
6	2	10
7	0	15
8	1	15
9	2	15

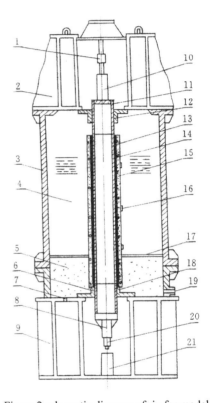

Figure 2 schematic diagram of rig for model test
1-load sensor 2-upper lid 3-bearing steel cylinder 4-pressured water 5-soil 6, 11-supporting steel ring 7-drainage hole 8-support 9-pedestal 10-jack 12, 18-steel guide ring 13-inner lining 14-sliding layer 15-outer lining 16-AD590 temperature sensor 17-water resistance layer 19-drainage hole 20-displacement meter 21-bearing set

method of adding water pressure for simulating was adopted.

From π_{10} and π_{11}, we know: C_t=1 and $C_t=C_L^{1/2}$. Obviously, they conflict with each other. According to π_7, we obtain $C_V=C_tC_L^2$.

In tests, the method for increasing loading rate for simulating was adopted. The loading rate in tests is equal to $C_L^{5/2}$. Because the shear strength of SL under rapid loading condition is more than that under slow loading condition, if we use the shear strength of SL in tests to design shaft lining, shaft lining will be safer. So, the method is allowable for engineering.

3. TESTS

The tests were carried out in the large-scale simulation test rig at Underground Laboratory in China University of Mining & Technology. The test plan is listed in table 2.

Test equipment is shown in fig.2.In each test, water pressure and temperature are stable, outer lining is static, but inner lining acted by vertical load moved along outer lining. Vertical load was acted on the top of the model of inner shaft lining by hydraulic jack and iron weight at different stage.

Considering that the shear strength of SL is small, creep displacement of inner lining due to the gravity of the model was measured from beginning to 4.5h, then each 50N load was added on the top of inner shaft lining in every 2h. The test did not end until total displacement of inner lining reached 5mm.

4. TEST RESULTS

Fig.3 shows curves of load-displacement (p=2MPa,T=10, 15 and 20°C respectively). When non-damping creep occurred in tests, the shear stress of SL is regarded as the long-term shear strength, see table 3.

5. DESIGN EXAMPLE

Horizontal pressure of SL p_L is:

$p_L = \gamma H = 0.012 \times 293 = 3.516 \text{MPa}$

where H=height of SL,H=293m, γ=bulk density of SL, γ=0.012MN/m³.

From table 3, we can see that long-term shear strength of SL is less than 5kPa in all tests.

For grade C50 reinforced concrete, its allowable compression strength can be calculated:

$[R_z] = (R_a + \mu_{min} R_g)/k = 16.65 \text{MPa}$

where R_a=design compression strength of concrete, equal to 23.5 Mpa, μ_{min}=minimum percentage of reinforcement, equal to 0.15%, R_g=design tensile strength of steel bar, equal to 310 Mpa, k=safety factor, equal to 1.44.

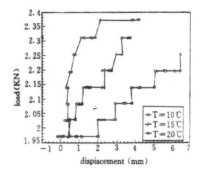

Figure 3 load-displacement curves (p=2MPa)

Table 3. Test results

Test no.	p/MPa	T/□	U/mm	τ/kPa
1	0	20	3.4	3.18
2	1	23	4.5	2.9
3	2	19.7	6.2	2.25
4	0	11	5.8	1.88
5	1	10	4.7	2.4
6	2	10	4.3	2.38
7	0	15	5.5	2.64
8	1	15	2.2	4.2
9	2	15.5	3.8	2.4

The thickness of inner shaft lining e can be computed:

$$e = r\left(\sqrt{\frac{[R_z]}{[R_z] - \sqrt{3}\, p_L}} - 1\right) = 0.831\text{m} \approx 0.850\text{m}$$

Where r=3.25m.

Tangential stress at inner edge of inner shaft lining is:

$$\sigma_t = \frac{2(r+e)^2 p}{(r+e)^2 - r^2} = 18.92 \text{MPa}$$

Gravity stress in inner shaft lining at the interface between overburden and rock is:

$$\sigma_{zz} = \gamma_h H = 7.032 \text{ MPa}$$

where γ_h=bulk density of reinforced concrete, equal to 0.024 MN/m³.

Because the long-term shear strength of SL is greater than or equal to the vertical additional force acted on the inner lining by outer lining, if we use the long-term shear strength of SL in tests to design shaft lining, shaft lining will be safe. So, according to the test results, we can assume:

$f' = \tau = 10\text{kPa}$

Thus vertical additional stress in inner shaft lining at the interface between overburden and rock is:

417

$$\sigma_f = \frac{2(r+e)\int_0^{tl} \tau \, dz}{(r+e)^2 - r^2} = 3.845 \text{MPa}$$

Total vertical stress σ_z is:

$$\sigma_z = \sigma_{zz} + \sigma_f = 10.877 \text{MPa}$$

According to the fourth strength theory, computing stress $\square_0$ is:

$$\sigma_0 = \sqrt{\sigma_t^2 + \sigma_z^2 - \sigma_z \sigma_t} = 16.446 \text{Mpa}, [R_z]$$

Therefore, inner shaft lining is safe.

6. CONCLUSIONS

1) Sliding layer is the key portion of the vertically sliding & yieldable shaft lining. In order to reduce the vertical additional force acted on inner lining, the long-term shear strength of the sliding layer should be as low as possible, and meet the need of construction.

2) Three sorts of force acting on the inner lining of vertically sliding & yieldable shaft lining should be taken into consideration when designing shaft lining. They are gravity, horizontal pressure of sliding layer, and the vertical additional force transferred from outer lining which can be taken as the long-term shear strength of sliding layer

3) When suitable material with reasonable thickness of sliding layer was used, loads which determine the thickness of inner shaft lining is still horizontal pressure.

4) It is suggested that the thickness of sliding layer should be equal to 100mm~150mm, and the bulk density of sliding layer should be 11kN/m^3~12kN/m^3.

REFERENCES

Cui, Guangxin.1990.The similitude theory and the model test. Xuzhou: Press of China University of Mining & Technology.

Cui, Gangxin et al.1996.Study on the composite shaft lining with an asphalt layer lying in thick overburden. Journal of China University of Mining & Technology. 24(4):11~17.

Mining Science and Technology' 99, Xie & Golosinski (eds) © 1999 Balkema, Rotterdam, ISBN 90 5809 067 1

Estimation of thrust in the construction of small-diameter tunnels using pipe-jacking

H. Shimada, S. Kubota & K. Matsui
Kyushu University, Fukuoka, Japan

ABSTRACT : The major purpose of this study is to find a way to predict the thrusts in using pipe-jacking. In pipe-jacking, the performance of the mud slurry plays important roles in the pushing process. The thrusts in pipe-jacking can be predicted accurately by using the initial thrusts and the resistance between the mud slurry and the concrete pipes.

1 INTRODUCTION

Pipe-jacking is an often-used method for the non-disruptive construction of underground pipelines such as water, electricity, gas, and so on. Pipe-jacking, in its traditional form, has occasionally been used for short railways, roads, rivers, and other projects (Hunt, 1978). It involves the pushing or thrusting of a drivage machine through concrete pipes ahead of jacks. It utilizes the mud slurry, which is formed around the pipes in order to stabilize the surrounding soil (Shimada and Matsui, 1995, 1996).

From these perspectives, the major purpose of this study is to find a way to predict the thrusts when using pipe-jacking.

2 CONSTRUCTION OUTLINE

Pipe-jacking, like many other below-ground construction methods, should be used in stable, water-free soil conditions. Unfortunately, with the demands on available space and the need to provide more services, it is not always possible to select stable strata, which means that contractors have to contend with unstable ground below the water table (Cole, 1977, Hough, 1978). The pipe-jacking system can be used in the above situations. Figure 1 shows the pipe-jacking system. It is particularly suited to both cohesive and sandy soil, and can be used to construct pipe tunnels that are up to 2,000mm in diameter.

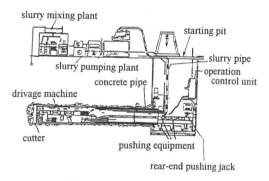

Figure 1. Pipe-jacking scheme.

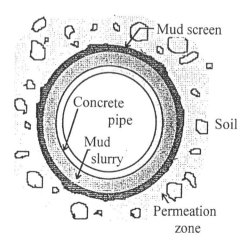

Figure 2. Illustration of a concrete pipe and soil.

Recent technological developments have led to successful methods that can stabilize unstable strata by excluding water from excavations by means of the mud slurry around the pipes. This system meets modern environmental standards. This type of mud slurry is formed by water, decomposed granite soil, carboxyl methyl cellulose, lost circulation control material, bentonite, fine aggregate, and other materials. The proportional mixes vary the viscosity and the seepage capacity of the mud slurry and will be determined by whether the mud slurry is required to strengthen the soil.

During the pushing processes, the mud slurry is injected into the face and the over-cutting area that is between the concrete pipe and the soil in Figure 2. After the slurry fills voids in the soil, it stabilizes due to slurry pressure. To minimize ground deformation in the pushing process, it is necessary to maintain the slurry pressure that was kept on underground water pressure at over $0.5kgf/cm^2$ (Shimada and Matsui, 1994). The mud slurry plays an important part in reducing the friction resistance in pushing the pipes.

3 PREDICTION EQUATION OF THRUST IN LINEAR JACKING

3.1 Outline of the prediction equation of thrust

The general form of the fundamental prediction equation of thrust for pipe-jacking consists of the initial thrust and the frictional resistance between the soil and the pipes as follows:

$$F = F_0 + \delta \tau_0 L \qquad (1)$$

where F=thrust, tf; F_0=initial thrust, tf; δ =contact length between soil and the pipe, m; τ_0=friction resistance between soil and the pipe, tf/m^2; and L= pushing distance, m.

3.2 Fundamental concept of the prediction equation of thrust

In the pipe-jacking method that uses slurry, the performance of the mud slurry is important in the pushing process (Shimada and Matsui, 1997). Hence, it is more significant to consider several parameters, which are dependent on the resistance between the mud slurry and the pipes in order to predict the thrust and/or the pushing rate. Industry experts generally believe that the overburden pressure provides the most useful parameter for predicting machine performance. For this reason, the Japan Sewage Association has proposed several empirical prediction equations of thrust when using the pipe-jacking method. These equations are given by theoretically considering the frictional resistance between the soil and the pipes.

In the pipe-jacking method that uses slurry, the mud slurry is injected into the face, which is located ahead of the drivage machine. The pipe is advanced after spoke cutters stir the face and the cutting soil is removed from the face by the discharge pipes. The soil is kept stable at the face by the injected mud slurry. The initial thrust, therefore, is dependent on the slurry pressure.

Second, the mud slurry is injected into the face and over-cutting area that is between the soil and the pipes during the pushing process. After the slurry fills the voids in the soil, the soil stabilizes due to the slurry pressure. Hence, as the slurry pressure prevents the soil from collapsing into the over-cutting area, the soil does not come into contact with the pipes in the pushing process. It is clear that an increase in the thrust is dependent on the resistance between the pipe and the mud slurry.

3.3 Prediction equation in linear jacking

It is necessary to establish a precise prediction equation in order to develop a suitable design. As stated previously, it is clear that the drivage performance of the pipe-jacking method that uses slurry is different from the concept of drivage for the previously proposed predicting equations (Shimada and Matsui, 1998). In regard to predicting the thrust, it is more important to analyze the mechanism that stabilizes the soil in the pushing process.

First, it was concluded that the initial thrust was dependent on the slurry pressure. This is why the soil is kept stable at the face by injecting the mud slurry. So, the first term of the predicting equation, which is expressed as the initial thrust, can be shown as follows:

$$F_0 = P_w (B_c / 2)^2 \pi \qquad (2)$$

where P_w=slurry pressure, tf/m^2; B_c=outer diameter of the pipe, m.

Second, the frictional resistance around the pipes, τ_0 which is expressed as the second term of the predicting equation was determined by the effects of the overburden pressure on the pipe by keeping the pipe in contact with the surrounding soil.

$$\tau_0 = T \mu + C \qquad (3)$$

where T=load acts on the surface of the pipe due to the overburden pressure, the weight of the pipe, etc., tf/m^2; μ =coefficient of kinematic friction between soil and the pipe; C=cohesion between soil and the pipe, tf/m².

As mentioned above, the mud slurry is injected into the face and the over-cutting area, which is between the concrete pipe and the soil during the pushing processes. The uniform load of the slurry pressure always acts on the surface of the pipes. In the pipe-jacking that uses slurry, the soil does not come into contact with the pipes in the pushing process. Therefore, the frictional resistance around the pipes should be used as the value for the resistance between the mud slurry and the pipes. Consequently, equation (3) should be applied to the following equation:

$$\tau_a = P_w \mu' + C' \qquad (4)$$

where τ_a = friction resistance between the mud slurry and the pipe, tf/m²; P_w=slurry pressure, tf/m²; μ'=coefficient of kinematic friction between the mud slurry and the pipe; C'=cohesion between the mud slurry and the pipe, tf/m².

In light of the above, the following prediction equation for the above pipe-jacking method can be derived:

$$F = P_w (B_c/2)^2 \pi + \pi B_c (P_w \mu' + C')L \qquad (5)$$

4 PREDICTION EQUATION OF THRUST IN CURVED JACKING

The difference of the thrust between linear jacking and curved jacking is due to the frictional force of outside components of the curved area of thrust. The general form of thrust in curved jacking is as follows;

$$F_1 = (F_0 + F_1' + T_1 \tan\phi')\sec\alpha \qquad (6)$$

where F_1 = the thrust when the first concrete pipe is pushing by the second pipe in the curved area, tf; F_0 = initial thrust, tf; F_1'= thrust if the first pipe is pushing in the linear area, tf; T_1 = the exterior component in the curved area of F_1, tf; ϕ'= internal friction angle, deg; α = deflection angle of pipes; deg, as shown in Figure 3.

The Japan Sewage Association has proposed an empirical prediction equation of thrust in curved jacking. This equation is given by theoretically considering the increased ratio of the thrust λ in the curved jacking toward the linear jacking as follows;

$$\lambda = \frac{k^{n+1} - k}{n(k-1)} \qquad (7)$$

where n = the number of pipes in the curved area and k is expressed by equation (8).

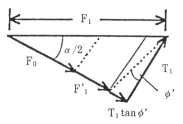

Figure 3. Forces of outside components of the curved area of thrust.

$$k = \frac{\sec\alpha}{1 - \sin\alpha \tan\phi' \sec\alpha} \qquad (8)$$

The coefficient of friction between the pipes and the soil, $\tan\phi'$, is always used as 0.5 by the Japan Sewage Association.

In the pipe-jacking method that uses slurry, the mud slurry is injected into the face and over-cutting area that is between the soil and the pipes during the pushing process, as stated previously. Hence, an increase in the thrust is not dependent on the resistance between the pipe and the soil but the resistance between the pipe and the mud slurry. For this pipe-jacking method, it is better that the coefficient of friction, $\tan\phi'$, is less than 0.5.

Figure 4 shows the relationship between μ/μ' and λ'/λ in two cases of the radius of curvature, 50m and 100m, respectively. The number of the pipes N in the curved jacking is 20. Where μ = the coefficient of the friction between the pipes and the soil=$\tan\phi'$=0.5, μ'= the coefficient of the friction between the pipes and the mud slurry, λ = the increased ratio of the thrust in curved jacking toward linear jacking in contact with the pipes and the soil, and λ'= the increased ratio of the thrust in the curved jacking toward the linear jacking in contact with the pipes and the mud slurry. λ and λ' are given by equation (7). μ/μ' is the dimensionless parameter which is larger, the coefficient of the friction between the pipes and the mud slurry is smaller. Also, it means the smaller the λ'/λ, the smaller the increased ratio of the thrust that has been obtained. This graph shows that higher μ/μ' reduces the increased ratio of the thrust in the curved jacking. From this point of view, the mud slurry plays an important part in reducing the friction resistance in the curved jacking of the pipes.

Figure 5 shows the relationship between μ/μ' and λ'/λ in three cases. The number of pipes in each case is 20, 40, and 60, respectively. This graph shows that the increased ratio of the thrust decreases

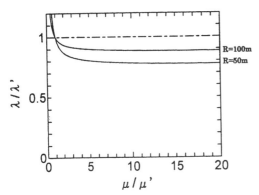

Figure 4. Relationship between μ/μ' and λ'/λ in two cases of the radius of curvature.

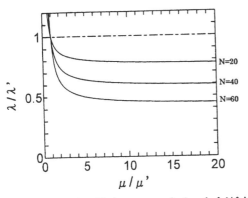

Figure 5. Relationship between μ/μ' and λ'/λ in three cases of the number of the pipes.

with an increasing number of pipes. It means the effect of a lower coefficient of the friction between the pipes and the mud slurry is significant when the distance of curved jacking is larger.

Consequently, it is considered that if the use of mud slurry on the face and over-cutting area is successful, lower increased ratio of the thrust is obtained in the curved jacking. In order to establish a more useful prediction equation in the curved jacking, more study, experience, and field data must be obtained.

5 CONCLUSIONS

This study analyzed the prediction equation in order to explain the characteristics of the thrust and the friction resistance.

It is clear that friction resistance can be continuously kept at a low value for various types of soil if the effect of mud slurry on soil can be maintained under optimum conditions, both in linear jacking and curved jacking. However, in order to fully understand the behavior of pipe-jacking, more research and field data are needed.

REFERENCES

Cole,J.M. 1977. Pipe-jacking Case Histories. *Tunnels and Tunneling*, July : 91-93.

Hough, C.M. 1978. Pipe-jacking Case Histories. *Tunnels and Tunneling*, April : 51-52.

Hunt, M. 1978. Pipe-jacking the Harburg Sewer. *Tunnels and Tunneling*, July : 19-21.

Katano, S. and Ogawa,T. 1994. Effect of Slurry Shield Tunneling in Soft Alluvial Clay on an Adjacent Underground Subway Structure. *Proc. of Int. Congress on Tunneling and Ground Conditions* : 151-156.

Shimada, H. and Matsui,K. 1995. Shallow Tunnel Drivage by Using Long Distance Curve Method. *Proc. of the 26th Sympo. on Rock Mechanics* : 216-220 (in Japanese).

Shimada,H. and Matsui,K. 1996. Stability of Soil around Shallow Tunnel Using Pipe-jacking. *Proc. of Mining Science and Technology* : 287-292.

Shimada,H. and Matsui,K. 1997. Effect of Mud Slurry on Surrounding Soil in Using Pipe Jacking. *Proc. of the 1st Asian Rock Mechanics Sympo.* 1 : 221-226.

Shimada, H. and Matsui, K. 1998. A New Method for the Construction of Small-diameter Tuinnels Using Pipe-jacking. *Proc. of Resional Symposium on Sedimentary Rock Engineering* : 34-239.

Mining Science and Technology' 99, Xie & Golosinski (eds) © 1999 Balkema, Rotterdam, ISBN 90 5809 067 1

Stereoscopic monitoring and prediction model of slope engineering system in complicated conditions

Z.Q.Wang

Institute of Geotechnical Engineering, College of Architecture and Civil Engineering, China University of Mining and Technology, Xuzhou, People's Republic of China

X.Q.Wang

San Hekou Mine, Jining, People's Republic of China

ABSTRACT: Under the direction of modern and dynamic design thought of advance prediction, monitoring in time, feedback design and synthetic decision, taking a large and complicated slope engineering system as a example, a stereoscopic-dynamic monitoring method of slope engineering system safety is put forward in this paper. The minimum state variables of slope engineering system stability are determined by chaotic attractor according to information surveyed in-situ. Grey system prediction models of GM(1,n) are built up, which can more exactly forecast dynamic process and stability of slope with minimum state variables which are related, relied on and exchanged each other.

1 INTRODUCTION

Natural slope or artificial slope is one of the most common circumstances of natural geology in process of man life and engineering action. With the expansion of engineering scale and field limit, all kinds of rock slope or soil slope and related underground engineering are excavated on complicated geology and circumstances condition. Sometimes, for the purpose of engineering demand and safety assurance during construction and service period, some complicated methods of excavation and support are used. Complicated slope engineering system is formed in which its projects are influenced, crossed and limited in time and space, and behave as the combination of surface engineering with underground engineering, of temporary engineering with permanent engineering, and of stable engineering with unstable engineering. This slope engineering system is called slope engineering system on complicated conditions in this paper. Obviously, it is very important for construction safety to reinforce the study of advance prediction and forecast of the slope engineering system on complicated conditions in time.

The high slope engineering system of factory in Geheyan Hydraulic Power Station is shown in Fig1, the main excavation project is high slope engineering system of factory in which its height is 180m and its length is 350m. Because of the need of water diversion and factory construction, four water-diversion tunnels are excavated and passed through the high slope engineering system of factory, the diameter of each tunnel is 12m, and the tunnels are

parallel to each other. At the same time, in the high slope feet, a deep slope of factory foundation is excavated whose height is 50m and length is 200m. In addition, for the purpose of improvement of slope stability, five large chambers were formed in the 201 soft layer between limestone and shale before lower shale slope was excavated, then the chambers were poured with concrete.

The geology conditions are also very complex. The hard and fissured limestone is located in the soft shale, a rock stratum combination of upper hard and lower soft is formed, 201 soft layer between limestone and shale has obvious characteristics of creep and small shear strength, so it is the main weak part in the slope.

2 STEREOSCOPIC-DYNAMIC MONITORING METHOD OF SLOPE ENGINEERING SYSTEM ON COMPLICATED CONDITIONS

All kinds of geology conditions and excavation elements of slope engineering system are influenced and restricted each other, so it is needed to hold the relation and transforming conditions between overall stability and part stability. The slope engineering stability can be improved greatly by the method discussed in this paper, i.e., advance prediction, monitoring in time, dynamic information feedback and design optimization and synthetic decision.

Based on rock mass quality difference and the extent of excavation influence as well as the height of slope and excavation quality control, the monitoring method of slope engineering system safety in

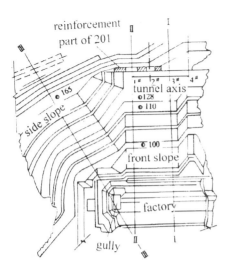

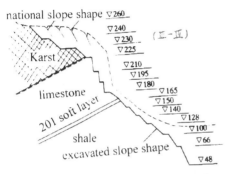

Fig1 Slope engineering system in Geheyan project

Geheyan project has been put forward. This method has the characteristics of three line and one plane, stereoscopic crossing, combination of dynamic and static method, verification of many variables each other. Firstly, the front slope is the main monitoring part according to the rock mass quality difference and the extent of excavation influence, taking the fact of many engineering items and the existing of power factory under the front slope into account, and monitoring lines were installed in the front slope (Fig1). considering the fact of maximum height of slope and the serious influence of unstable gully slope engineering construction, monitoring line was installed in the side slope. Three monitoring lines can basically grasp the overall stability of the slope. Based on the complex and changeable geology engineering as well as excavation conditions, many monitoring points are installed in the face of overall slope (including front and side slope, especially the parts having serious changes in excavation and the parts having movable block bodies). This method shows the combination of overall with part. In addition, the combination method of static with

dynamic method was also adopted in order to consider the influences of explosive dynamic load. In other words, it is not only to monitor stresses and displacements, but also to survey blasting vibration intensity and fracture circle change. The outer and inner displacement was surveyed at the same time.

According to above demands, slope displacement (outer and inner), anchor stresses, crack width, blasting vibration velocity and the fracture circle sizes were chosen to monitor the stability state of slope excavation and 201 soft layer replacement chambers, forecasting the possible time of loss of chambers stability and gully slope during excavation and reinforcement. Efficient monitoring and information feedback was also conducted to improve the influences of part project on slope engineering system, ideal results were obtained and the correctness of monitoring method was verified.

3 DETERMINATION OF MINIMUM STATE VARIABLES OF SLOPE ENGINEERING ON COMPLICATED CONDITIONS

Generally speaking, many state variables can be used to evaluate the stability state, such as stress, strain(displacement), underground water level and rock mass parameters. Different models can be built up based on individual variable or multiple variables. But there should exist optimum variables in question, in other words, which variables mainly control the stability of slope?

If $\{X_0(t)\}$ is the displacement-time series, n dimension phase space was built up based on displacement -time series, time interval $t=m\Delta t$(m is integer, Δt is time interval of collection) is used to increase $\{X_0(t)\}$ by integer times. N points were drawn out from above data, then dispersed variables were shown below:

$$X_0 : X_0(t_1), X_0(t_2), \ldots\ldots X_0(t_n)$$

$$X_1 : X_0(t_1 + \tau), X_0(t_2 + \tau), \ldots\ldots X_0(t_n + \tau)$$

$$\ldots\ldots \quad \ldots\ldots$$

$$X_{n-1} : X_0(t_1 + (n-1)\tau), X_0(t_2 + (n-1)\tau), \ldots\ldots$$
$$X_0(t_n + (n-1)\tau)$$

X_i (: $X_0(t_i), X_0(t_i+), \ldots, X_0(t_i+(n-1))$) is any point of phase space, where X_i is Vector symbol, the distance $|X_i-X_j|$ between X_i and other (n-1) point X_j is calculated. Data point in phase space is achieved when the X_i is center with radius r. The process is repeated for all i, then we can get follow equation:

$$c(r) = \frac{1}{n^2} \sum_{i,j=1}^{n} \theta(r - |\vec{X}_i - \vec{X}_j|) \qquad (1)$$

Where υ is Heaviside function, when x= 0, Δ (x)=1, when x>0, Δ(x)=0; Δx_i-x_j is the distance between two points in any n points of time series;

424

$c(r)$ is integration related dimension of attractor.

A trace amount is taken for a given size and used to survey the attractor results. The results have shown that the number of point is proportional to $(r/c(r))^D$ if attractor has D dimension manifold, and if r is small enough, then:

$$D = -\frac{Lnc(r)}{Lnr} \qquad (2)$$

Where D is the fractal dimension of manifold evolution in phase space.

After calculating different D for different n, we can find that D becomes stable when n exceeds some value, here D is the dimension of attractor, and n corresponding to D is the minimum state variables describing complicated dynamic system (Fig2).

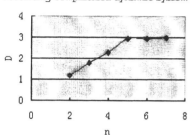

Fig 2.· D-n relation curve

Taking the key variables as well as limited datum of monitoring into account, follow three variables are used to forecast the stability of slope engineering system, namely displacement, Stress and underground water level.

4 GREY SYSTEM MODEL OF SLOPE DYNAMIC STABILITY FORECAST

Because grey theory has the characteristics of good function of dynamic forecast and less information needed, it is widely used to forecast geo-engineering stability at present, and generally single GM(1,1) model is used widely. From above research, we know that the state variables determining stability of slope engineering system on complicated conditions are not single, and the important extent of these variables may be exchanged greatly in process of dynamic evolution of slope engineering stability. Moreover, there are influences and reactions among variables, so we should use multiple variable models of grey system GM(1,n) to forecast the stability state of slope engineering system on complicated conditions.

For slope engineering system described by n state variable, time-series are:

$$x_1^{(0)} = \{x_1^{(0)}(1), x_1^{(0)}(2),......,x_1^{(0)}(N)$$

$$x_2^{(0)} = \{x_2^{(0)}(1), x_2^{(0)}(2),......,x_2^{(0)}(N)$$

....

$$x_n^{(0)} = \{x_n^{(0)}(1), x_n^{(0)}(2),......,\quad x_n^{(0)}(N)$$

Where N is the number of time-series data, every x_i represents an original state, n is the number of minimum state variables determined by above research. Because there are some related actions between variables, so GM(1,n) model should be established for every variable. If:

$$x_i = \sum_{i=1}^{k} x_i^{(0)}(i) \qquad (3)$$

Follow equations can be achieved for i=1,2, ...n:

$$\dot{x}_1 = a_{11}x_1 + a_{12}x_2 + ... + a_{1n}x_n + u_1$$

$$\dot{x}_2 = a_{21}x_1 + a_{22}x_2 + ... + a_{2n}x_n + u_2 \qquad (4)$$

....

$$\dot{x}_n = a_{n1}x_1 + a_{n2}x_2 + ... + a_{nn}x_n + u_n$$

The state equation form is:

$$\hat{x} = AX + BU \qquad (5)$$

Where:

$$A = \begin{bmatrix} a_{11} & a_{12} & ... & a_{1n} \\ a_{21} & a_{22} & ... & a_{2n} \\ ... & ... & ... & ... \\ a_{n1} & a_{n2} & ... & a_{nn} \end{bmatrix}, B = \begin{bmatrix} 1 & 0 & ... & 0 \\ 0 & 1 & ... & 0 \\ ... & ... & ... & ... \\ 0 & 0 & ... & 1 \end{bmatrix}$$

$$\hat{x} = [\dot{x}_1, \dot{x}_2, ... \dot{x}_n]^T, X = [x_1, x_2, ... x_n]^T$$

$$U = [u_1, u_2, ..., u_n]^T$$

$$\hat{x}(t) = e^{A(t-t_0)}x_0 + \int e^{A(t-\tau)}B(\tau)u(\tau)d\tau \qquad (6)$$

The solution of state equation is:

Where A, B can be determined by least-squares method. Then forecast model system of n variables is achieved, and can be used to forecast dynamic stability of slope engineering system in process of excavation, support or service period.

5 CASE APPLICATION OF FORECAST MODEL

Fig3 shows the changing curve of displacement, stress and underground water level in front slope monitored in-situ, the number of minimum variables is three by analysis of D-n relation, and D-n curve is shown in Fig2.

The prediction models of GM(1,n) grey system are built up:

Where $x^{(1)}, y^{(1)}, z^{(1)}$ are the weighed accumulation results of displacement, underground water level and stress series in one step, through the treatment of deceleration in one step, the prediction results can be

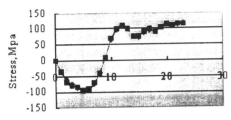

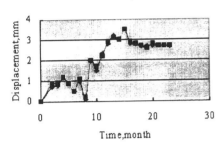

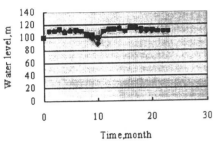

Fig3 Displacement, water level and stress curve

2) Multiple grey system models can exactly embody the complicated influences among variables compared with single grey model.
3) A new method is provided for slope engineering stability prediction by these models.

7 ACKNOWLEDGEMENTS

The research described in this paper was supported by National Natural Science Foundation of China under grant No 49802026.

REFERENCES

Qing Siqing etal. An introduction to nonlinear engineering geology. Chengdu: Jiaotong University of Westsouth Press, 1992,104-106.

Zhou Cuiying etal. Study on nonlinear dynamics of sliding disaster system. Journal of Changchun institute of geology, 1995, No3.310-316

Dong Julong. Grey prediction and decision. Wuhan: Huazhong University of technology Press,1988.185-187

Wang Zaiquan. Study on dynamic stability prediction of slope and its application on one project. Journal of Rock Mechanics and Engineering, 1998,No2,117-122.

given . From Fig3 and above prediction results of GM(1,n) models. The prediction results have a good accuracy compared with in-situ.

From above prediction equations, we can find that the influences of underground water level on displacement and stress are limited (coefficient is 10^{-4} or 10^{-3}), this is because the changing scope of underground water level is very small (Fig3(c)). Moreover these models can reflect influences of one factor on another, and can forecast possible loss time of slope according to prediction result of displacement combining with Verhulst model of landslide forecast (Wang zaiquan, 1998).

6 CONCLUSIONS

1) The stereoscopic-dynamic monitoring method of slope engineering system can embody the modern and dynamic design thought of advance prediction, monitoring in time, feedback design and synthetic decision.

Mining Science and Technology'99, Xie & Golosinski (eds)© 1999 Balkema, Rotterdam, ISBN 90 5809 067 1

Lateral large deformation of a bolt in discontinuous rock

Bo Liu, Xianwei Li, Longguang Tao & Songshan Yang
China University of Mining and Technology, Beijing, People's Republic of China

ABSTRACT: Based on the new large deformation theory (i.e. Stocks-Chen Strain-Rotation Decomposition Theorem, Chen Zhida, 1979), a rational lateral large deformation model of bolt due to shear-tensile loads has been proposed in this paper. The uniform strain (& stress) theoretical formula has been derived. A new method of determining bolt deformation and internal force based on displacement measurement has also been presented. Theoretical complete curves between strength of bolted joints and bolts' deformation have been set up. The validity of theory has been confirmed by experimental results.

1. INTRODUCTION

The phenomena of bolts' failure with tensile-shearing large deformation can be frequently seen in-situ in collapsed tunnels and slopes supported by bolts. A typical example is that bolts have been used to reinforce the stratified rock in coal mine (Liu, 1997). Snyder (1983), Holmberg (1992) and Ferreo (1995) pointed out that displacement and strain of bolts nearby joints' intersection are usually large. Compared with axial reinforcing action of bolt, the lateral behavior of bolt almost plays the same important role when it is used to support jointed rock. Because any failure of bolt always begin with local position, the research on bolt's lateral behavior is important to bolt-reinforcement design in the joint rock.

2. ANALYSIS ON THE LATERAL LOCAL LARGE DEFORMATION OF BOLT

2.1 *Strain tensor analysis on the large deformation of bolt*

The analysis on local deformation mode of bolt indicates that the performance of bolt in jointed rock shows a typical synthetic deformation due to extension, shear strain and local rotation (Liu, 1998). In order to determine large deformation behavior of bolt correctly, the analysis should be based on actual configuration (after deformation) instead of original one. In the local deformation zone of bolt intersected with joint plane, we chose $\{X^i\}$ as fixed orthogonal

coordinate (see Figure.1). Its gauge is unit vector, $\{\dot{x}^i\}$ is original co-moving coordinate and it has the same embargo with $\{x^i\}$, where $\{x^i\}$ is actual co-moving coordinate and it deforms with bolt.

Considering bolts deformation mode subjected to tensile-contract, lateral shearing and local rotation, displacement components of bolt can be written as follows

$$u^1 = x^1(\cos\theta\cos\hat{A}-1)+x^2 tg\theta$$

$$u^2 = -x^1\cos\theta\sin\hat{A} \tag{1}$$

Where β is local rotation angle, γ is shearing angle, deformation angle $\theta=\beta+\gamma$.

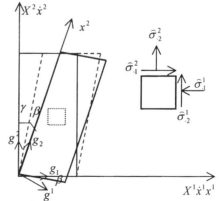

Figure 1. Model of bolt local-lateral large deformation based on co-moving coordinate

Length formula of line element are $(d\dot{s})^2 = \dot{g}_{ij}dx^i dx^j$, $(ds)^2 = g_{ij}dx^i dx^j$, Where $\dot{g}_{ij}$ and g_{ij} are metric tensors of initial and actual configuration respectively.

Covariant derivative of vector u^i to x^j is

$$u^i\Big|_j = \frac{\partial u^i}{\partial x^j} + \Gamma^i_{jt}u^t \qquad (2)$$

where $\Gamma^i_{jt} = \frac{1}{2}\dot{g}^{ik}[\dot{g}_{kj,l} + \dot{g}_{kl,j} - \dot{g}_{jl,k}]$. Because $\dot{g}_{ij}$ are constant, $\Gamma^i_{jl} = 0(i,j,l=1,2)$.

Physics components of displacement gradient $\hat{u}^i\Big|_j = \left[\sqrt{g_{(ii)}/g_{(jj)}}\,u^i\Big|_j\right]$, Mean rotation angle Θ can be calculated according equation

$$\sin\Theta = \frac{1}{2}(\hat{u}^2\Big|_1 - \hat{u}^1\Big|_2). \qquad (3)$$

Local large deformation strain tensor of bolt can be expressed

$$\hat{S}^i_j = \frac{1}{2}(\hat{u}^i\Big|_j + \hat{u}^j\Big|_i) + \delta^i_j(1 - \cos\Theta) \qquad (4)$$

where δ^i_j is Kronecker symbol,

$$\delta^i_j = \begin{cases} 1 & i=j \\ 0 & i \neq j \end{cases}, \cos\Theta = \sqrt{1 - \frac{1}{4}(-\sin\beta - \sin\theta)^2}.$$

According to formula (1)□(4), strain tensors can be derived as follows

$$[\hat{S}^i_j] = \begin{bmatrix} \dfrac{\cos\theta\cos\beta - 1}{\cos\theta} + 1 - \cos\Theta & \dfrac{1}{2}(\sin\theta - \sin\beta) \\ \dfrac{1}{2}(\sin\theta - \sin\beta) & 1 - \cos\Theta \end{bmatrix} \qquad (5)$$

Because the ratio of β to θ is different, components of tensile and shearing strain can be calculated directly. The experiments show that the values of bolts' deformation and stress can be calculated by using $\beta = k \cdot \theta$ ($k \leq 1 \square k$ determined by experiments) easily. When $k=1 \square$it indicates that bolt is in a state of tensile deformation. The less is the value of k, the more is local shearing strain of bolt. Theoretical curve of strain components under tensile-shearing large deformation is shown in Figure 2. Each strain component increases with deformation angle's increase. Where shearing strain

$\hat{S}^1_2$ is the most sensitive to the change of k value.

2.2 Theoretical method of determining deformation angle θ of bolt's large deformation

It is difficult to measure bolt's deformation on joint plane, When bolted rock serves as a system to bearing extern load, it is easy to measure its tangential and normal displacement of joint. By measuring its displacement of bolted joint rock, we can get deformed state of bolt on joint plane, thus the stress of bolt can be calculated easily. The research is just based on the reversing analysis.

$$\theta = tg^{-1}\left(\frac{u\sin\alpha}{l_{ef}\cos\psi}\right) \qquad (6)$$

Where l_{ef} is effective length of tensile-shearing deformation of bolt, Ψ is shearing dilation angle of rock, α is setting angle of bolt (acute angle with joint plane), u is tangential displacement along bolted joint plane.

Parameters l_{ef} and k which stands for the relation between deformation angle and rotation angle can be determined by experiments on different rock and bolts' materials. By measuring tangential (or normal) displacement of rock joint, the deformation angle of bolt can be derived. Strain rules of bolt can also be obtained. If the allowable displacement is given, it provides theory basis for bolt design to reinforce joint rock.

Because the elasto-plasticity stress solution had been derived (Liu, 1998), the plastic large deformation stress solution can also be obtained.

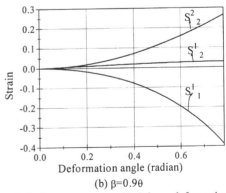

Figure 2 Strain components of bolt large deformation

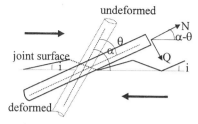

undeformed

joint surface

deformed

Figure 3. Analysis on bolted rough joints

$$\hat{\sigma}_1^1 = \frac{2\sqrt{2}\sigma_s}{3\gamma_8}\left[\frac{2\cos\theta\cos\beta - 2}{\cos\theta} + 3(1-\cos\Theta)\right]$$

$$\hat{\sigma}_2^2 = \frac{2\sqrt{2}\sigma_s}{3\gamma_8}\left[\frac{\cos\theta\cos\beta - 1}{\cos\theta} + 3(1-\cos\Theta)\right] \quad (7)$$

$$\hat{\sigma}_1^2 = \hat{\sigma}_2^1 = \frac{\sqrt{2}\sigma_s}{6\gamma_8}(\sin\theta - \sin\beta)$$

Where γ_8 is octahedral shear strain， Θ is mean rotation angle, σ_S is tensile yield limit of bolt.

3. GLOBAL RESISTANCE OF BOLT AND BOLTED JOINT IN ROUGHNESS DISCONTINU-OUS ROCK

Accurate description of mechanical behavior should be based on the deformed system (see Figure 3). Under the load of shearing-tensile, bolt emerge local deformation of θ angle. Then the bolt's axial force N and shear force Q are shown as follows respectively.

$$N = \hat{\sigma}_2^2 \cdot A_b$$
$$Q = \hat{\sigma}_2^1 \cdot A_b \quad (8)$$

Where $\hat{\sigma}_2^2, \hat{\sigma}_2^1$ are bolt's elasto-plasticity tension and shear stress of bolt, A_b is section area of bolt.
The combination of axial tension and shearing dowel effect leads to the raising of the resistance along joint interface. If P_{ts} is defined as bolt's global resistance along joint surface, then we have

$$P_{ts} = N\cos(\alpha - \theta) + Q\sin(\alpha - \theta) -$$
$$[Q\cos(\alpha - \theta) - N\sin(\alpha - \theta)]tg\varphi_j \quad (9)$$

Where φ_j is joint friction angle.

The global shearing strength τ_{jb} of bolted joint includes joint itself shear strength τ_j and the

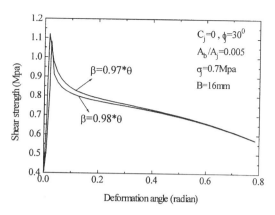

Figure 4. Theory curve of shear strength of flat joint reinforced by steel bolt (limestone)

increasing strength τ_b on joint surface by bolt, i.e. $\tau_{jb} = \tau_j + \tau_b$. Where $\tau_j = c_j + \sigma_j tg\varphi_j$, c_j is joint cohesion ,σ_j is normal stress on joint surface, and $\tau_b = P_{ts}/A_j$,(A_j is joint area bolted by single bolt). Then the global shear strength of bolted flat joint is

$$\tau_{jb} = c_j + \sigma_j tg\varphi_j + \{\hat{\sigma}_2^2[\cos(\alpha - \theta) + \sin(\alpha - \theta)tg\varphi_j]$$
$$+ \hat{\sigma}_2^1[\sin(\alpha - \theta) - \cos(\alpha - \theta)tg\varphi_j]\}A_b/A_j$$
$$(10)$$

For rough joint face of undulate angle i as shown in Figure 3, if normal stress is not high (before the convex destroyed), φ_j in equation (9)， (10) should be substituted by joint's remain friction angle. Joint friction angle φ_j should change to $(\varphi_j + i)$ while σ_j approaches to uniaxial compressive strength.

According to the above, it is easy to obtain the whole curve between shear strength of bolted joints and bolt's deformation angle. Theory curve of shear strength of flat joint reinforced by steel bolt is shown in Figure 4.

In elastic deformation phase, shearing strength of bolted joint increases with the raising of angle θ. The less is the k value in equation $\beta = k \cdot \theta$, the more is shear strength .As the bolt goes through a smaller deformation procedure, its lateral resistance reach peak value (see Figure 4).

In case of smaller k value, when bolt deforms little, the strength emerge maximum value quickly, and the peak value strength occurs long before the failure of bolted joint. Shear strength drops gradually after the

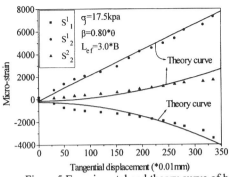 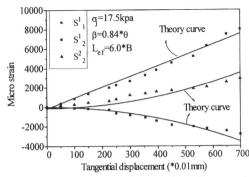

Figure 5 Experimental and theory curve of bar strain components on bolted joint (granite and concrete)

peak value. However, when deformation of blot is larger, the value of k has no obvious effect on shear strength. The strength after peak value would trend to a constant.

4. SIMULATED EXPERIMENT AND ANALYSIS

Two prismatic rock blocks with a square of 200 side and a height of 100mm have been used for simulated specimen. In each block center, bolt's hole of Φ20mm has been drilled. In order to eliminate the influence of samples non-even, the joint plane has been processed into flat and perfectly smooth surface. There is no cohesion between joints. Many ways have been adopted for simulating bolt, but it is difficult to guarantee the demands of setting strain rosettes. In order to make sure the close contact between bar and drill hole, resin and steel bars with square of 14×14mm serve as bolt, which are covered with Φ20mm plastic pipe. For the sake of measuring the complete state of stress, the rosettes are placed on the bolt's axial-center of bolt nearby shear plane, which composed of three gauges at $0°$, $45°$, $90°$. Tensile instrument is adopted at the end of bolt. Twelve groups of direct shear tests have been performed. Shear forces, normal stress, shearing displacement and bolts' strains have been recorded during the tests.

Bolt's deformation angle and joint displacement can be calculated from each other, and the displacement of bolted joint can be measured easily. Some strain results of shear test on bolted granite and concrete are shown in Figure 5. The rosettes are set on the intersection of bolt's axial-center and joint plane. Experimental results show good agreement with theoretical model.

5. CONCLUSIONS

The research shows that the finite deformation mechanics based on co-moving coordinates can describe the bolt's shear-tensile deformation behavior in discontinuous rock more correctly. The lateral larger deformation behavior of bolt can be predicted by the theoretical formula proposed in this paper. The study also indicates the bolt deformation and internal force can be calculated on joint displacement measurement. According to the presented equations of global resistance of bolted joint in roughness discontinuous rock, theoretical complete curves between strength of bolted joints and bolts' deformation have been set up. Tests results on different rock joints show good agreement with analytical model.

REFERENCE

Chen Zhida. 1994. *Large Deformation theory of rod, plate and shell*, Beijing: Science press.
Ferrero A M. 1995. The shear strength of reinforced rock joints, *Int. Journal. Rock Mech Min.& Geo Abs.tr* 6. 595-605
Holmberg M et al. 1992. The mechanical behavior of a single grouted bolt, *Rock support in mining and underground construction*, 1992: 473-481
Liu Bo et al. 1997. Study on two typical geotechnical problem by using large deformation geometry theory, *Journal of China University of Mining & Technology* Vol.7 No.2: 30-34
Liu Bo. 1998. Study on lateral behavior of bolts and their global resistance, Doctor thesis, *China University of Mining & Technology*.
Syder V M. 1983. Analysis of beam building using fully grouted roof bolts, *Proc. of the Inter Symp. on Rock Bolting* : 187-194

Mining Science and Technology'99, Xie & Golosinski (eds) © 1999 Balkema, Rotterdam, ISBN 90 5809 067 1

The design of and experimental study on lateral behavior of truss-bolt system

Xianwei Li, Bo Liu, Longguang Tao & Yuqi Zhou
China University of Mining and Technology, Beijing, People's Republic of China

ABSTRACT: Lateral forces acting in truss bolts used in a roadway support was studied. The arching influence on lateral behavior of bolt reinforcing breaking roof has been analyzed. The formula involving lateral force of bolt, tension of rod in truss and the reinforcement behavior have been derived. Simulated experiment test on the stratified roof supported by truss and its parameters optimization have been performed. The test results show that the lateral action of bolts provides an important contribution to roof stability. The theory formula can be used in the design of truss. The reliability of theory has been verified by application examples.

1. INTRODUCTION

Truss, an advanced and reliable support system of bolt, is becoming an effective means in reinforcing fracture roof. As its simple structure (two bolts, one rod and jointing device), easy drilling and setting, truss is popularized in application. Because of the tension of rod, the mechanical behavior of bolt in truss is different from the traditional viewpoints of arching and suspension. The lateral force as well as the axial force play the same important role in reinforcing broken roof. Ensuring the essential tension of rod becomes the key to the setting of truss system successfully.

2. THEORETICAL ANALYSIS OF BOLT LATERAL FORCE

Applications show that the lateral resistance of bolt is related to the integrity of surrounding rock. It is necessary to distinguish the counter-force of plastic rock (broken zone and plastic zone is not equal) and elastic rock. Because the diameter of bolt is small and it is located in higher rock stress field, the bolt should be taken as flexibility rod.

With the help of horizontal tension, the rock can always provides a certain lateral counter-force to bolt. Because the bolt is flexibility rod, the counter-

force of rock to bolt will reduce quickly from surface to deep of surrounding rock. The counter-force may be lower than that of perfectly elastic rock. So the counter force can be considered as linear distribution (see Figure 1).

The horizontal force H_t at the end of bolt is given, so the counter-force of plastic zone is

$$q(z) = -(\frac{p_{u2} - p_{u1}}{a}z + p_{u1})B \qquad (1)$$

Where p_{u1}, p_{u2} are the counter-forces at the place of z=0 and the interface between the elastic and plastic zone respectively; a is the thickness of plastic zone in surrounding rock; B is the diameter of bolt.

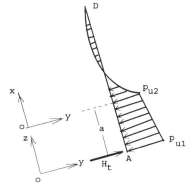

Figure 1 Analysis diagram on bolt lateral behavior

The bending moment M_z (M_x) and shearing force S_z (S_x) of the bolt in the plastic (elastic) zone had been derived (Liu & Li, 1997).

$$M_z = -H_t \cdot z + B\left(\frac{p_{u2} - p_{u1}}{a} \cdot \frac{z^3}{6} + p_{u1} \cdot \frac{z^2}{2}\right) \quad (2)$$

$$S_z = -H_t + B\left(\frac{p_{u2} - p_{u1}}{a} \cdot \frac{z^2}{2} + p_{u1} \cdot z\right) \quad (3)$$

$$M_x = -\frac{H_t + Q(a)}{\beta} e^{-\beta x}[(1 + \beta h_0)\sin\beta x + \beta h_0 \cos\beta x] \quad (4)$$

$$S_x = -[H_t + Q(a)]e^{-\beta x}[\cos\beta x + (1 + 2\beta h_0)\sin\beta x] \quad (5)$$

Where $\beta = \sqrt[4]{\dfrac{Bk_h}{4EI}}$, $h_0 = \dfrac{H_t \cdot a + R(a)}{H_t + Q(a)}$,

$Q(a) = -\dfrac{B}{2}(p_{u2} + p_{u1})$, $R(a) = -\dfrac{Ba^2}{6}(p_{u2} + 2p_{u1})$

Where EI is the flexural rigidity of bolt, k_h is the coefficient of rock counter-force.

3. DESIGN THEORY OF TRUSS BASED ON LATERAL BEHAVIOR OF BOLT

When cracked rock beam arched, the stability of abutment depends on necessary arch thrust. There is a pressure-bearing arch of truss and cracked rock beam interaction (Figure 2). The lateral force of bolt can enhance the ability of anti-sliding at the abutment and apex of arch and improve the state of stress in roof.
Horizontal thrust T at the abutment is:

$$T = qs^2 / 8e \quad (6)$$

where S is the span of rectangle tunnel; q is the uniform load at the crack beam, ($q = q_0 + \gamma bt$, where γ is the unit weight of rock, b is the array pitch of truss, t is the height of cracked rock beam); e is the tension arm of horizontal thrust, $e = t - 2a/3$, (a is the imposed height of arch pressure) it can be taken as the thickness of equivalent plastic zone. We have

$$e = \frac{\sqrt{s^2 + f^2}}{6}\left[\frac{(\gamma h + c \cdot ctg\varphi)(1 - \sin\varphi)}{c \cdot ctg\varphi}\right]^{\frac{1 - \sin\varphi}{2\sin\varphi}}$$
$$-\frac{f}{2} + \frac{\sqrt{s^2 + f^2}}{3} \quad (7)$$

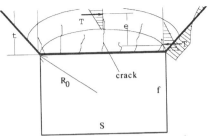

Figure 2. Arching action of truss to crack roof

Where f is the side wall height of roadway; γh is the stress of primary rock (we can get it by calculating γh, where h is the depth of roadway); φ, c are the internal friction angle and cohesion of broken rock. Considering T and thrust p which provided by the counter force of bolt. The safety coefficient of anti-sliding F_s can be defined as:

$$F_s = \frac{2(p + T)tg\varphi}{\gamma st \cdot b} \quad (8)$$

Considering the difference between coefficient of counter force and the state of rock damage, we can determine $P_{u1} = P_{u2} = P_u$, so the horizontal thrust provided by the lateral resistance of bolt can be written as:

$$p = p_u aB \sin\alpha \quad (9)$$

Where α is the setting angle of bolt (the acute angle with the horizontal direction). According to the formula (6) to formula (9), we have

$$p_u = \frac{F_s \gamma stb}{2tg\varphi \cdot aB \sin\alpha} - \frac{qs^2}{8eaB \sin\alpha} \quad (10)$$

To measure the coefficients k_h and β is very difficult. Due to the influence of horizontal force H_t, the counter-force p_{u1} is generated.

$$p_{u1} = k_h \cdot \frac{H_t}{2EI\beta^3} \quad (11)$$

Where $k_h = 4EI\beta^4 / B$.
According to equations (10) and (11), it is easy to derive

$$H_t = \frac{p_{u1} \cdot B}{2\beta} = \frac{B}{2\beta}\left(\frac{F_s \gamma stb}{2tg\varphi \cdot aB \sin\alpha} - \frac{qs^2}{8eaB \sin\alpha}\right) \quad (12)$$

At the interface of elasto-plastic zone,

$$p_{u2} = K_h \frac{1+\beta h_0}{2EI\beta^3}[H_t + Q(a)] \tag{13}$$

From formula (11) and (13), we have

$$p_{u2} = \frac{2\beta(1+\beta h_0)}{B}[H_t + Q(a)] \tag{14}$$

The plastic zone is exist before supporting, so formula (14) can be substituted by the formula (12) and $h_0, Q(a)$, according to $p_{u1} = p_{u2}$ we can determine

$$\beta = 2/a \tag{15}$$

where

$$a = \frac{\sqrt{s^2 + f^2}}{2}\{[\frac{(\gamma h + c \cdot ctg\varphi)(1 - \sin\varphi)}{c \cdot ctg\varphi}]^{\frac{1-\sin\varphi}{2\sin\varphi}} - 1\} \tag{16}$$

From formula (11) (15), k_h can be shown as follows:

$$K_h = \frac{64EI}{Ba^4} \tag{17}$$

So when the dimensions of section is determined, the stress of primary rock and the parameters c, φ and F_s then the pre-tightening force of rod can be calculated easily. The internal force of truss system can be designed and verified too. If a and t have been measured, then the two parameters can be used for design directly.

4. EXPERIMENT ANALYSIS

4.1 Experiment system and its parameters

In order to study the lateral behavior of bolt in friable roof, the experiments on dichotomous laminated beams supported by trusses have been performed. By choosing the geometric proportion 1/10, the beams with scale of 760×200×200mm have been made. The adding ratio of cement to aggregate is 1 to 4.5. The thickness of single layer is 100mm. When poured, the bolt's hole with diameter φ16mm is kept in advance. According to the setting angle of 30°,45°,60°,75° and 90°, no less than two laminated beams were made at each angle respectively, which have been cured for 28 days before the experiments.

In the experiment, the simulated steel bolts with the diameter of φ16mm were adopted. Strain rosettes with 0°,45° and 90° gauges have been set along the bolt center near the interface. At the axial center of rod, foil strain gauges have also been set to measure the deformation of rod. The span of simulated beam is 340mm. The length of rod is 300mm. Two trusses have been instrumented in each model. At the head of each bolt, the tension bridles were used for pre-tightening. The strains of bolt and rod have been measured by 7V07 automatic data logger continually.

4.2 Experiment result analysis

The comparison of average bearing capacity at different setting angles trusses shows that the initial breaking load is smaller (see Figure 3). The influence of arch between truss and crack beam result in the bearing capacity of truss system

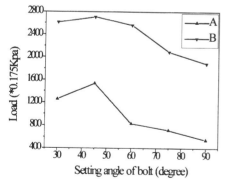

Figure 3. The comparison of bearing capacity

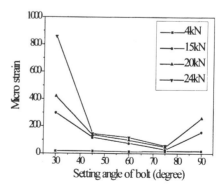

Figure 4. Comparison of rod strain under different load

increase continually. The figure 3 indicates that when the setting angle of bolt is 45°, both initial and final breaking loads are large. When the setting angle is 30°, 60° and 75°, the final breaking load have no obvious difference. However, both initial and final breaking loads are small when the dip angle is 90°.

Figure 4 shows that even at the same load, the strain of rod is different due to the different setting angle. When the setting angle is from 45° to 75°, the deformation of rod is small. So from the viewpoint of interaction between truss and surrounding rock, the rational setting parameters of truss should make the system bearing capacity larger and internal force of truss smaller. So the setting angle with 45° to 75° is suitable.

The small anchor force and short length bolts are widely used in coal mine. Experimental result indicates that the effective parameter of truss design should avoid the vertical setting of bolt and ensure to develop the lateral and axial global counter-force of bolt.

5. ENGINEERING APPLICATIONS

The in-situ tests have been performed in a coal mine in Shanxi .The average thickness of coal seam is 6.13m. The dip angle is 6°. The false roof is less than 0.3m, which is made of carbon mudstone, and fall with mining. The direct roof is argillaceous rock. The thickness is 1.2-1.3m. The test result of compressive strength of argillaceous rock is $S_c = 17.3$ MPa. $c = 5.1$ MPa. $\varphi = 30°$. The unit weight of direct roof is 27kN/m³. There is a thick fine-sandstone above the direct roof. The primary stress of rock is 15MPa. The span of roadway is 3.4m. The heights of walls are 2.6m and 2.3m respectively. The roadway is excavated along the roof.

From theoretical computation and analysis of in situ, the thickness of equivalent broken zone (i.e. a) is 0.58m. The height of crack rock beam t=1.38m. The tension arm of horizontal thrust is e=0.99m, and =3.45m⁻¹. Referring to Zhu's research (Zhu, 1993), to determine $q_0 = 30$ kN/m is suitable. If the given safety coefficient $F_s = 2.5$, from formula (12) we have derived p_u =4.8Mpa and result of rod's tension H_t =20.8kN. According to theoretical design, the tube-split bolt is used in the test roadway of 200m length, and the length and diameter of bolt are 1.8m and 0.03m respectively. In order to enhance the anchor force, the pretreatment measure on bolts is used. The setting angle of bolt is

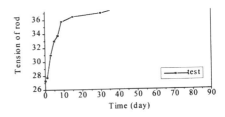

Figure 5. The variation curve of tension of rod

$\alpha = 60°$, and the array pitch of truss b=0.7m. The tension rod is made from ϕ18mm-steel bar, and the length of rod is 3.2m. Patent technology is adopted

to design the joint and setting method (Li,1993), in which the rod is close to the surface of rock. Diamond steel nets are used to maintain the roof. The further check on internal force of bolt and rod shows that they can satisfy the requirement of safety.

Tension measurement of rod has been completed in underground. The variation curve is shown in Figure 5. Few literatures reported the related result. The tensions of rod are 28.3 38.1kN from 1day to 90 days. Test results are larger than 20.8kN. So it satisfies the requirement of safety coefficient 2.5. The anchor force of bolt, at the early stage, is 30kN, and reach 49 - 70kN at later period. From July 1995 excavation to June 1996, the tunnel has been constructed safely and undertaken the mining dynamic pressure successfully.

REFERENCES

Brady, B H G & E.T.Brown.1985. Rock Mechanics for Underground Mining. George Allen and Unwin (Publishers):214-222.

Li Xianwei. 1990. *Mechanics properties of rock body.* Beijing: Publisher of coal industry of China

Liu Bo, Li Xianwei et al 1997, Study on reinforcement of mine excavation based on latetral resistance of bolts, *The 4th Int. Geotech. Conf. & Exhi., Beijing Sep.* 1997:232-237

Liu Bo.1998, *Study on lateral behavior of bolts and their global resistance,* Doctor thesis, China University of Mining & Technology, Beijing.

Zhu Fusheng & Zhen Yutian. 1993. The mechanism of full column archoring bolts and roof truss system. *Chinese Journal of Rock and Engineering, Vol.12 No.3* :249-252

Mining Science and Technology' 99, Xie & Golosinski (eds) © 1999 Balkema, Rotterdam, ISBN 90 5809 067 1

Analysis of large deformation and the numerical simulation of the slope sliding process

Manchao He, Zhida Chen, Han Zhang & Yanxiang Jiang
China University of Mining and Technology, Beijing, People's Republic of China

ABSTRACT: The large deformation theory and the finite element modeling methods were used for analysis of the process of the large slope deformation in Fushun open-pit mine. The results fit well with those obtained using the large deformation theory created by Professor Chen Zhida.

1. INTRODUCTION

Up to now, there are three theoretical defects in large deformation analysis of the slope sliding: (1) Continuum problem, i.e. rock mass is usually believed as a kind of medium of high discontinuity. And on the other hand, the continuum mechanics theory is almost used to every analysis of rock masses stability. (2) Constitutive equation problem, is hardly determined in theory or in practice, but it must be used for all analysis of deformation mechanics of rock masses. (3) Large deformation problem is usually quite large, however, the analysis theory used at present belongs to the mini-deformation theory[1~3].

The full review of the problems mentioned above is beyond scope of this paper, and only the large deformation in question is discussed here. Some concepts concerning about the continuity or constitutive equation of engineering rock masses can be found in reference [3] and [4].

2. NON-RATIONALITIES OF CLASSIC DEFORMATION THEORY IN SLIDING ANALYSIS

2.1 *The error of the classic deformation theory*

While adopting the classic deformation theory to analyses typical slide in Fushun open-pit, the error occurred which proves the unrationalities of it. As shown in Fig. 1~2, it is a typical circular slide and its curvature of sliding plane is constant. When block A is sliding, it can be believed as a rigid body moving from A to A', the rigid rotation angle is (φ'-φ).

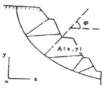

Fig. 1. A circular slide before sliding (above)

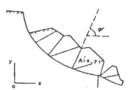

Fig. 2. A circular slide after sliding

So the displacement equation of each point within the block A is

$$\left.\begin{array}{l} u = (\cos\theta - 1)\cdot x - \sin\theta\cdot y + a \\ v = \cos\theta\cdot x + (\cos\theta - 1)\cdot y + b \end{array}\right\} \quad (2.1)$$

where: u, v = Components of displacement vector along x, y axis direction; a, b = Components of translation of particle center $\theta = (\varphi'-\varphi)$.

According to classic theory, the strain components ε_{ij}

$$\left.\begin{array}{l} \varepsilon_{xx} = \dfrac{\partial u}{\partial x} \\[2mm] \varepsilon_{yy} = \dfrac{\partial v}{\partial y} \\[2mm] \varepsilon_{xy} = \dfrac{1}{2}(\dfrac{\partial v}{\partial x} + \dfrac{\partial u}{\partial y}) \end{array}\right\} \quad (2.2)$$

Mini-rotation angle w_x

$$w_x = \frac{1}{2}(\frac{\partial v}{\partial x} - \frac{\partial u}{\partial y}) \qquad (2.3)$$

Depending upon the formula above, ε_{ij} can be expressed as:

$$\begin{cases} \varepsilon_{xx} = \cos\theta - 1 \\ \varepsilon_{yy} = \cos\theta - 1 \\ \varepsilon_{xy} = 0 \\ w_x = \sin\theta \end{cases}$$

If $\theta = \varphi' - \varphi = 10°$, then,

$$\begin{cases} \varepsilon_{xx} = -0.015 \\ \varepsilon_{yy} = -0.015 \\ \varepsilon_{xy} = 0 \\ w_x = 0.1737 \end{cases}$$

If the movement is rigid movement, ε_{ij} should be zero. It is clearly that the results are quite ridiculous, since there isn't any strain in block A. So strictly speaking, it is unrational that the body movement is described by using the classic theory of deformation mechanics.

2.2 The defects of the classic quantitative deformation mechanics

In real sliding movement (as shown in Fig. 3), every element in sliding masses has a large deformation. For solving this problem, the FEM method depending upon the classic deformation mechanics is unrational in geometrical equations, and it goes against the law of energy conservation. As shown in Fig. 3, the classic FEM analysis is carried out by a fixed coordinate system. When sliding masses are deformed, a shadow elements in Fig. 3 is changed into a inclined one, if it is, as usually, divided into elements by a fixed coordinates, the element

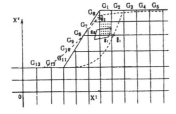

Fig. 3. Theoretic model of large deformation slope

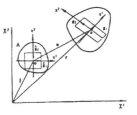

Fig. 4. The co-moving coordinate system

boundary must be straight. Obviously, the element is very different from a real deformed one. So the particles in the element have also some different from a real deformed one. These facts are against the law of energy conservation according to the mass-energy relation.

3. NONLINEAR LARGE DEFORMATION THEORY

3.1 Co-moving coordinate system

As shown in Fig. 4, we chose two reference systems, a co-moving system x^i and a fixed system X^i. For large deformation theory, the method of co-moving system developed by Professor Chen Zhida is used to describing the motion of a deformed body. Fig. 4 shows a transformation of continuum body A_0 in large deformation. In a time process sequence, A_0 is continuously transformed, i.e., $A_0(t_0) \rightarrow A(t)$. At the moment t_0 the embedding coordinate system $\{x^i\}$ is initially isomorphic with a fixed coordinate system $\{X^i\}$, i.e.:

$$\overset{o}{X}{}^i = X^i(x^i, t_0) = \overset{o}{x}{}^i$$

At the moment t:

$$X^i = X^i(x^i, t)$$

The local basic vectors:

$$\overset{o}{\vec{g}}_i = \frac{\partial \overset{o}{\vec{r}}}{\partial x^i} \qquad \vec{g}_i = \frac{\partial \vec{r}}{\partial x^i} \qquad (3.1)$$

where $\overset{o}{\vec{r}}$, $\vec{r}$ is the local vectors of any point of the deformed body A_0 at the moment t_0 or t.

3.2 S-R decomposition theorem [2]

The basic vectors $\overset{o}{\vec{g}}_i$ of undeformed state is changed to $\vec{g}_i$, in deformed state, the transformation tensor F_i^j gives:

436

$$\bar{g}_i = F_i^j \overset{o}{g}_i \qquad (3.2)$$

By S-R decomposition theorem [2], where:

$$F_j^i = S_j^i + R_i^j = Strain\ tensor + rotation\ tensor \quad (3.3)$$

Finite strain tensor:

$$S_i^j = \frac{1}{2}(u^i|_j + u^j|_i^T) - (1 - \cos\theta)L_k^i L_j^k \qquad (3.4)$$

Finite mean local rotation:

$$R_i^j = \delta_j^i + L_j^i \sin\theta + (1 - \cos\theta)L_k^i L_j^k \qquad (3.5)$$

Mean angle of rotation θ:

$$\theta = \arcsin\left\{ \frac{1}{2} \left[\begin{array}{c} (u^1|_2 - u^2|_1)^2 \\ + (u^2|_3 - u^3|_2)^2 \\ + (u^3|_1 - u^1|_3)^2 \end{array} \right]^{\frac{1}{2}} \right\} \qquad (3.6)$$

Axis of local rotation:

$$L = L^i g_j$$

$$L_i^j = \left(\frac{1}{2\sin\theta} \right)(u_i|^j - u_i|_j^T) \qquad (3.7)$$

3.3 Verification of the rationality

For large deformation of slide in Fig. 1~2 we shall use the large deformation theorem (equation (3.3) ~ (3.7)) to calculate the strain of every point in sliding masses, can get:

$$|S_j^i| = \begin{bmatrix} S_1^1 & S_1^2 \\ S_2^1 & S_2^2 \end{bmatrix}$$

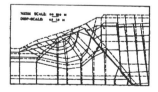

Fig.5 Slope displacement of w800section in 1984

$$= \begin{bmatrix} \dfrac{\partial u}{\partial S_1} + (1 - \cos\theta) & \dfrac{1}{2}\left(\dfrac{\partial u}{\partial S_1} + \dfrac{\partial v}{\partial S_2}\right) \\ \dfrac{1}{2}\left(\dfrac{\partial u}{\partial S_2} + \dfrac{\partial v}{\partial S_1}\right) & \dfrac{\partial v}{\partial S_1} + (1 - \cos\theta) \end{bmatrix}$$

$$= \begin{bmatrix} \dfrac{\partial u}{\partial x} + (1 - \cos\theta) & \dfrac{1}{2}\left(\dfrac{\partial u}{\partial x} + \dfrac{\partial v}{\partial y}\right) \\ \dfrac{1}{2}\left(\dfrac{\partial v}{\partial x} + \dfrac{\partial u}{\partial y}\right) & \dfrac{\partial v}{\partial y} + (1 - \cos\theta) \end{bmatrix}$$

$$= \begin{bmatrix} 0 & 0 \\ 0 & 0 \end{bmatrix}$$

$$w_x = \frac{1}{2}\left(\frac{\partial v}{\partial S_1} - \frac{\partial u}{\partial S_2} \right)$$

$$= \frac{1}{2}\left(\frac{\partial v}{\partial x} - \frac{\partial u}{\partial y} \right)$$

$$= \sin\theta$$

For the movement of circle slide in Fig. 1~2, the strain components within the sliding masses are zero, but the rotation takes place along the circle sliding plane. These results are clearly rational.

Considering the general case of slope large deformation shown in Fig. 3, two coordinate systems, namely a fixed system and a co-moving system, are used according to the FEM equations of non-linear large deformation geometrical field. The deformed element is described by the co-moving coordinate system. So the boundary of the element is the same as the real deformed one, and the energy after or before deforming keep a constant. Obviously, this FEM technique of nonlinear large deformation is quite rational for describing the large deformation body.

4. THE NUMERICAL SIMULATION OF THE PROCESS OF SLIDE DEFORMATION-INSTABILITY

According to the historical data, fourteen landslides have occurred due to cutting the weak layer near the region of station w800 of Fushun open-pit since 1960. The reasons are that this station locates in the nucleus of the downfold, and the weak intercalations are very developed. In order to study on the sliding mechanism of this station, We used the nonlinear large deformation FEM method to analyze and simulate the process of slide deformation-instability from 1984 to 1993.

Fig. 5 shows the deformation state of the rock masses in station w800 in 1984. The dotted line shows un-excavated state of the rock masses, the solid line shows the deformation state of the rock masses excavated to the slope in Fig. 5. We should pay attention that the two faults produced obvious bench malposition slump.

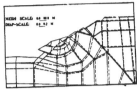

Fig. 6. Incremental displacement of w800 during the period of 1984~1990

Fig. 6 and Fig. 8 simulated the whole sliding process of the deformation instability due to the excavating of the open-pit during 1984~1993. With excavation, the front position of the slide produced large displacements. The front part of the sliding plane produced compress-shearing deformation, the middle part of the sliding plane produced tension deformation, the back part of the sliding plane produced tension-shearing deformation. The slide is getting to form, but it is not developed. The whole sliding plane showed in Fig. 8 is shearing deformation, the slide is formed. Fig. 7 and Fig. 9 show the whole process of the shearing plane of the slide deformation.

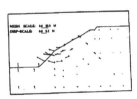

Fig. 7. Incremental displacement vector of w800 during the period of 1984~1990

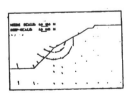

Fig. 8. Incremental displacement of w800 during the period of 1991~1993

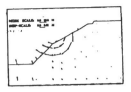

Fig. 9. Incremental displacement vector of w800 during the period of 1991~1993

5. CONCLUSIONS

An unrational result is got when we analyze a large deformation sliding by using the classic theory of the deformation mechanics. The calculation process of the FEM method that based on this theory is against the law of energy conservation. So it is quite necessary using the large deformation theory to analyze the large deformation sliding. The example gives a good fit to the large deformation theory whose nucleus is the "S-R" decomposition theorem in this paper.

REFERENCES

A Cemal Eringen, 1980. *Mechanics of Continua.* New York: Kricger Publishing Company.

Chen Zhida, 1988. *Rational Mechanics* (in Chinese with English abstract). Xuzhou, China: China Mining University Press.

He Manchao, 1991. *High Slope Geomechanics* (in Chinese). Beijing, China: Coal Science Press.

He Manchao, 1991,8. *Continuum of the rock masses.* Beijing, China: A collection of thesis for the Doctorate in China.

He Manchao, Sep. 13, 1990. *Constitutive Relation of Plastic Dilantancy due to Weak Intercalation in Rock Masses.* Proceedings of 26th Annual Conference of the Engineering Geology of Weak Rock. Rotterdam: A. A, Balkema Press.

Mining Science and Technology' 99, Xie & Golosinski (eds) © 1999 Balkema, Rotterdam, ISBN 90 5809 067 1

Development and application of a testing device for study of frost heave and thawing settlement of artificially frozen soil

Jianping Wang, Wenshun Wang & Tianshen Shi
China University of Mining and Technology, Xuzhou, People's Republic of China

ABSTRACT: This paper analyses differences between artificially and naturally frozen soils, and comparing the related testing devices worldwide. A new testing device is introduced for study of frost heave and thawing settlement of artificially frozen soil. The device was applied to test the frost heave and thawing settlement of normal clay in Xuzhou, which proves the device is reliable and useful. The study also shows that the frost heave in vertical direction is by far smaller than the thawing settlement in the same direction.

1. INTRODUCTION

More than 400 projects have been constructed with the frozen method, during last 40 years in China, which range from mine construction to urban civil engineering, railway and water pipeline construction etc.

The frost damage of buildings in cold regions is a common engineering question. The frost heave mechanism has been studied in detail worldwide and a lot of conclusions have been drawn.

For naturally frozen soil, for example, the law of the distribution of the frost heave along the depth, the experiential formulas of some influential factors, the estimate formulas of the frost heave and frost strength based on the different water migrate theory, the relationship between the frost heave and the water content and the relationship between the frost heave and the load have been studied in detail.

But for artificially frozen soil, the study has been focused on the load-carrying ability and the deformation-carrying characteristic of the frozen-soil-wall in the process of mine-shaft-construction. It is little known about the law of the frost heave and thawing settlement in the vertical direction while the cold front extends in the horizontal direction, which is the key problem when constructing projects near or underneath buildings.

This paper mainly introduces the development and application of a new testing device for studying the frost heave and thawing settlement of artificially frozen soil. By using advanced semiconductor thermoelectric cooling technology, the cooling temperature can be adjusted to the minimum of -35 with the accuracy of ±0.2 . The cooling power is adjustable with maximum 2000 watt. There is no noise, vibration, or pollution, which will occur in the mechanical cooling system.

2. FREEZING-THAWING CHARACTERISTICS OF ARTIFICIALLY FROZEN SOILS

The artificially frozen method is to remove heat from the ground, causing the pore water to freeze and act as a bonding agent named as a frozen-wall. The heat removal is accomplished by use of coolants circulating through freezing pipes, which are buried in the zone of ground to be frozen. Then, construct the underground works safely under the protection of the frozen-wall.

The naturally frozen soil is formed from top to bottom step by step with the air temperature becoming colder in the cold season, and it is thawed simultaneously both from top and from bottom in the warm season.

3. THE DEVELOPMENT OF THE DEVICE

Several testing devices for studying the frost heave and the thawing settlement have been developed worldwide. But most of them are used for studying the permafrost and the seasonally frozen soil. The main characteristics of those devices are:

1. They are applied to study the one dimension frost heave and thawing settlement of the naturally frozen soil.

2. The cold source is probably the cold circulating fluid.

3. The samples are frozen entirely in some direction in the process of the freezing.

4. The samples are so small that the diameters are approximately 10cm and the ratios of the length to the diameter are 1.5~2.0.

5. The aims of their studies are mainly the physics and the strength of the naturally frozen soil instead of the relationships between the frozen soils and the buildings.

The frost heave in-situ is presented little by little after condensing the unfrozen soil in the process of the freezing. The thawing settlement is presented due to the loss of underground water, the structural damage and the dropping of the load-carrying ability in the process of the thawing. So it is very difficult to simulate truly the frost heave and the thawing settlement on the spot engineering by using a smaller sample in spite of that it is reasonable to study the physics and the strength of the frozen soil.

The device, based on the simulation theory, has been developed in order to simulate truly the freezing and thawing process of the artificially frozen soil. Its functions are:

1. It can be used to simulate the one and three-dimensional freezing and thawing process of the artificially frozen soil.

2. Adopt the advanced thermoelectric technology, no vibration, no noise, no pollution.

3. It can be used to freeze the samples separately in the horizontal direction, from top to bottom or from bottom to top according to the design. It can simulate the water levels and the flow of the underground water.

4. Both big and small samples can be frozen simultaneously.

5. The aim of the device is to study the frost heave and the thawing settlement and the relationship between the artificially frozen soil and the ambient buildings.

The device (figure 1) mainly consists of the testing boxes, the cooling system, the electric power supply system, the temperature control system, the loading system, the control system of the underground water level, the thermal insulation system, the data logging and processing system.

3.1 The testing boxes

The testing boxes are made up of a big and a small mental box. Set a steel plane with lots of holes on the bottom of the big box to make a filter. The hard and the soft thermal insulation materials adhered separately on the internal and external wall to reduce the energy loss and to keep the boundary conditions. The load is applied on the top of the soil on the basis of the lever principle. The control system of the underground water can simulate the level and the flow of the underground water.

The dimensions of the big box is 100×80×120cm.

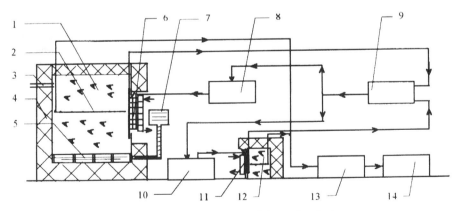

Fig.1. The simple drawing of the testing device

1- the big sample 2-temperature sensors 3-the hole inputting water 4-the thermal insulation 5-undergroud water 6-1# cooler 7-the device for adjusting water level 8-1# current electric power 9-the control circuit 10-2# current electric power 11-2# cooler 12-the small sample 13-the data logging system 14-the data processing system

There is a big hole (15×50cm) in one side of the box to install the freezer. Some soils are put into the box by layers to test the frost heave and thawing settlement in the open system. But the dimensions of the smaller box is 20×20×40cm with a hole (15×35cm). It is used to freeze the soils in the close system.

3.2 The cooling system

The cooling system supplies the cold energy with thermoelectric coolers that utilize the Pelter effect discovered in 1834 (figure 2 The principle of thermoelectric cooler). During operation, DC current flows through the thermoelectric coolers causing heat to be transferred from one side to the other, which creates a cold and hot side. At the cold junction, energy is absorbed by electron as they pass from a low energy level in the p-type semiconductor element, to a higher energy level in the n-type semiconductor element. The electric power supply system provides the energy to move the electrons through the system. At the hot junction, energy is expelled to a heat sink as electrons move from a high energy level (n-type) to a lower energy level element (p-type).

The cooling plane is a copper plane, on which a few thermoelectric coolers are combined. The soil is frozen through the cooling plane. The heat generated in the process of the freezing is taken away by the circulating water.

The thermoelectric coolers are solid state heat pumps without moving parts, fluids or gasses. They have some benefits:

1. Reduced space, size and weight
2. Quick cooling to standard the designed
3. Precision temperature control capability
4. The difference between the cold and hot surface can reach 100^0.
5. Reliable solid-state operation, with no sound and vibration.

3.3 The temperature control system

The temperature control is the key problem to maintaining the testing precision. The cold surface temperature confines the freezing velocity, the temperature gradient and the amount of water immigration. So it affects the distribution and amount of ice lens. The temperature control is accurate to within plus or minus 0.2 , which is available for the testing.

A precise platinum thermometer is installed in the center of the cooling copper plane, which sends the information to the control circuit continuously. Then

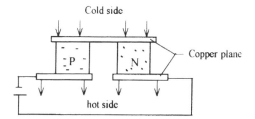

Fig.2. The principle of thermoelectric cooler

the control circuit takes different actions according to the information.

3.3 The data logging and processing system

The system is responsible for sampling and processing the data on time. It is made up of computers, Data Taker 100 and series of sensors, by which the frost heave, the thawing settlement and the temperature distribution are to be observed in time.

4. THE APPLICATION OF THE DEVICE

The samples were normal clay in Xuzhou. The water content was 23%. The unit weight was 1.95g/cm³. The void ratio was approximately 41.5% and the liquid and plasticity limits were 32.3% and 21.2%. 106 temperature sensors and 10 load sensors were installed in the soils in the process of filling the boxes. 15 displacement sensors were put on the top of the soils to get the frost heave and thawing settlement.

After the cold front extended to the desired extent, on one hand let the freezing go on to keep the thickness of the frozen wall and on the other hand shut down the supply power to let the frozen wall thaw naturally according to the design.

In the process of the freezing and thawing the data logging and processing system were on duty unless the temperature of the soils resumed to the initials.

In figure 3 (The temperature of the cold copper plane and the freezing time), the temperature of the cold copper plane dropped quickly and then kept at -19^0.

In figure 4 (The boundary temperature and the testing time), the change of the boundary temperature was small in the process of the freezing and thawing.

In figure 5 (The temperature of the center point in

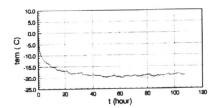

Fig.3. The temperature of the cold copper plane and the freezing time

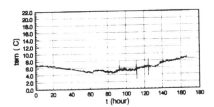

Fig.4. The boundary temperature and the testing time

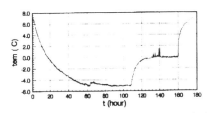

Fig.5. The temperature of the center point in the sample and the testing time

the sample and the testing time), the freezing period was from the start to the 108th hour and the thawing was from the 108th hour to the end. The temperature dropped slowly and then kept about at -5^0 in the process of the freezing and it went up quickly on turning off the electric power for the thawing. Once the temperature went up to the freezing point the frozen soil began to thaw and absorb so much heat that the temperature kept stable for 30 hours (from the 130th hour to the 160th hour). After that, the temperature began to go up rapidly.

At the same time the frost heave in the vertical direction was by far smaller than the thawing settlement in the same direction.

5. CONCLUSIONS

1. Adopting the advanced thermoelectric technology, the device runs safely and reliably without vibration, noise and pollution. It is available for freezing the soil in lab.

2. It can be used to simulate simultaneously the one and the three dimensional freezing and thawing process of the artificially frozen soil in the open or close system.

3. The frost heave in the vertical direction was by far smaller than the thawing settlement in the same direction. Meanwhile, much attention should be paid to the thawing settlement on the spot engineering.

The project supported by National Natural Science Foundation of China.

REFERENCES

J.P. Wang et. al.1997. Analysis of shaft sinking with concrete hoist tower through artificially frozen strata. *Mining Science and Technology*.

W.S. Wang et. al. 1996.The analysis on the frost heave and thawing settlement laws of the artificially frozen soil. *Shanxi Mining Institute Learned Journal*.Vol.14.

Mining Science and Technology' 99, Xie & Golosinski (eds) © 1999 Balkema, Rotterdam, ISBN 90 5809 067 1

A micromechanical study on the flow rule for granular materials

Jianxun Wu
China University of Mining and Technology, Xuzhou, People's Republic of China

Yuji Kishino
Tohoku University, Sendai, Japan

Hideo Asanuma
Nishimatsu Constructing Company, Japan

ABSTRACT: The authors have performed a set of numerical probe tests in terms of the Granular Element Method. These test results help explaining the internal mechanism hidden behind the macroscopic behaviors. This paper deals with the plastic behavior whose mechanism has its main origin on the slipping between grains. A theoretical yielding condition, which explains the result obtained by the simulation, is derived from non-slipping condition for stable assemblies of rigid particles. Performing eigen-value analyses of the instantaneous stiffness matrix utilized in the Granular Element Method, the authors demonstrate the similarity of displacement fields accompanied by plastic deformations for different loading directions.

1 INTRODUCTION

To investigate elastoplastic character of materials experimentally, we need performing a set of probe tests. However, it is extremely difficult to prepare samples with the identical properties and the same loading history. Bardet (1994) carried out numerical probe tests by means of Distinct Element Method (DEM). The authors also carried out numerical probe tests and discussed the elastoplastic character of granular materials by means of Granular Element Method (GEM) (1998) and found that the results obtained by GEM are far more accurate than those obtained by DEM. It is due to the reason that GEM utilizes a stiffness method which enables us to perform stress control tests exactly. In this paper, GEM is adopted to investigate the flow rule for granular materials micromecanically.

2. LOADING AND STRESS PROBE TESTS

The main difference between GEM and DEM is that GEM utilizes a stiffness method (Kishino 1990). A global stiffness relationship of an assembly of disks is obtained as

$$\Delta F = S\Delta U \qquad (1)$$

where ΔF and ΔU are the global force and movement vectors and S is the global instantaneous stiffness matrix.

The disk assembly used in numerical tests consists of 395 disks with diameters of 0.5 to 1.0 cm in a circular region as shown in Figure 1. The spring constants at all contact points are $k_n = 1000$, $k_t = 700$ (kN/m) in normal and tangential directions respectively, and the interparticle friction angle is $\phi = 25°$. The disk assembly was compacted until the isotropic stress become 0.2 MPa (the point A in Fig. 2). Then a shear loading test was performed by increasing σ_2 while σ_1 being kept constant. The loading curve is shown in Figure 3. At points B, C, D the stress probe tests were performed. The results obtained from the point B is shown in this paper.

The stress probe test consists of a set of loading and unloading tests in 8 directions with the same magnitude of incremental stress(Fig. 4). By these tests the incremental strain can be divided into the elastic part $\Delta\varepsilon^e$ and the plastic part $\Delta\varepsilon^p$.

Plots of the incremental plastic strain vectors in the principal strain space for the loading direction (0, 1, 6, 7) lay almost on a unique direction (flow direction) as shown by black squares in Figure 5, while those for other directions (2, 3, 4, 5) are zero.

3. FLOW RULE

The flow rule for the stress probe direction where the plastic deformation takes place is expressed as

$$\Delta\varepsilon^p = h^{-1}(n : \Delta\sigma)m \qquad (2)$$

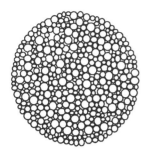

Figure 1 Granular model with 395 disks.

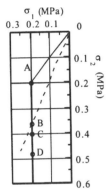

Figure 2 Loading path.

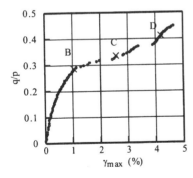

Figure 3 Loading curve.

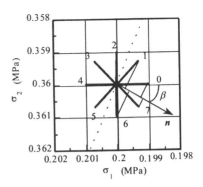

Figure 4 Stress probe directions(B).

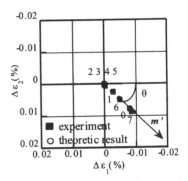

Figure 5 The incremental plastic strains(B).

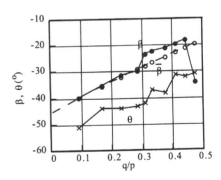

Figure 6 Changes in direction angles.

where n and m are the unit normal vector of yield surface and the flow direction, and h is a hardenning factor. As shown in Figure 4, the unit normal vector n is determined as the direction angle β, which can be calculated by the least-squares method (Wu & Kishino, 1997)

Theoretical values obtained by (2) are plotted by circles in Figure 5. It is found that the experimental

and theoretical values are in good agreement. The normal direction angle β is plotted in Figure 6 as well as the flow direction angle θ (Fig.5). As β and θ are not identical, the flow rule becomes non-associative.

4. CONSIDERATION ON YIELD SURFACE

The plastic behavior obtained by the previous numerical tests is well explained by the flow rule. Thus the flow rule may have its origin on the micromechanics. The proportionality of the incremental plastic strain to projection of the incremental stress can be explained from micromechanical point of view as follows.

Let the disk assembly consist of rigid particles in a statically stable state. Then every disk in the assembly must fulfill the following equilibrium conditions

$$\Sigma_c f_c = 0 \qquad (3)$$

$$\Sigma_c f_c \times r_c = 0 \qquad (4)$$

where f_c and r_c are the contact force vector and the radius vector connecting the contact point, and Σ_c means to take sum over all contact points.

The granular assembly also needs satisfying the following conditions at every contact point

$$p_n \geq 0 \qquad (5)$$

$$|p_t| \leq p_n \tan \phi \qquad (6)$$

where Δp_n and Δp_t are the normal and tangential components of f_c at contact point.

Let us consider another state where the every contact force f_c is changed to αf_c, then the conditions (3-6) are still satisfied. Thus no slippage and no plastic strain are expected in the loading path from (σ_1,σ_2) to $(\alpha\sigma_1,\alpha\sigma_2)$. Thus the yield lines are taken to be straight lines passing through the origin of stress space, as shown by the broken line in Figure 2. The normal direction angle $\bar{\beta}$ determined by the above consideration is plotted in Figure 4.

The above results are valid for a particle assembly with any combination of arbitrary shapes and particle sizes, as far as the particles are rigid. These results may be applied approximately to the assembly of elastic particles, as far as the deformation is small enough.

5. PLASTIC DISPLACEMENT FIELDS

To check the validity of the flow rule, a new type of equi-projection probe test was carried out (Fig.7). In this probe test, the probes are performed only on the probe directions 0, 1, 6, 7, and the incremental stresses are chosen so that they have a common magnitude of projection onto the normal direction of the theoretical yield line. If the assumptions for the yield line is adequate, the incremental plastic strains

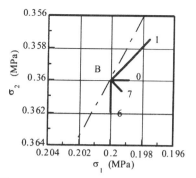

Figure 7 Equi-projection test(B).

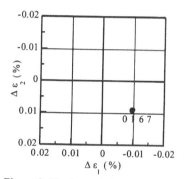

Figure 8 Plastic strain by equi-projection test(B).

for the 4 probe directions will be the same. As shown in Figure 8, four incremental plastic strains for the point B are almost the same as predicted. The micromechanical origin of this result may be explained by observing unrecoverable displacement patterns. As shown in Figures 9 and 10, the unrecoverable displacement patterns for two probe directions (1, 7) at point B are almost the same.

This result is verified quantitatively by eigenvalue analysis. Let the n-th eigenvalue and normalized eigenvector of the global stiffness matrix S of the equation (1) of rank N be λ_n and U_n, then the following equations holds.

$$SU_n = \lambda_n U_n \qquad (n = 1, 2, \cdots N) \qquad (7)$$

Utilizing the eigenvectors, the movement vector which consists of displacement components and rotations of all disks in the assembly can be expanded as

$$U = \Sigma_{n=1}^N \alpha_n U_n \quad (\lambda_n < \lambda_m \quad when \quad n < m) \qquad (8)$$

In Figures 11 and 12, the distributions of expansion coefficients α_n $(100 < n < 200)$ for the

445

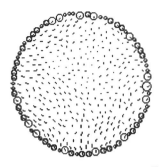

Figure 9 Displacement pattern (B-1, ×300).

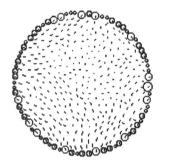

Figure 10 Displacement pattern (B-7, ×300).

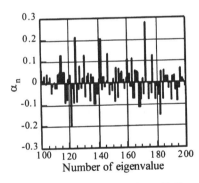

Figure 11 Expansion coefficients (B-1).

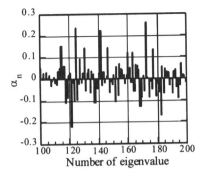

Figure 12 Expansion coefficients(B-7).

6. CONCLUSIONS

The numerical probe tests carried out by means of the granular element method give both macroscopical and microscopical results. The main results obtained in this paper are the following: 1) The plots of incremental plastic strains are in good agreement with the predicted values determined by the non-associative flow rule. 2) The normal direction of the yield surface coincides with the direction determined from a micromechanical consideration. 3) The existence of common flow direction is qualitatively explained in terms of eigenvalue analysis.

The more detailed discussion will be expected in terms of the similar methodology.

REFERENCE

Bardet, J.P.(1994). "Numerical simulations of the incremental responses of idealized granular materials",. *Int.J. Plasticity.*, 10(8):879-908.

Kishino Y. (1990). "Quasi-Static Simulation of Liquefaction phenomena in Granular Materials". *Proc. 2nd International Symposium for Science on Form*, Tokyo.

Wu, J. & Kishino, Y. & Kyoya, T. (1998) "A micromechanical study on the flow rule for granular materials", *Journal of applied mechanics* (JSCE), Vol.1, 497-506.

Wu, J. (1998):On the micromechanics of Granular material, *PhD thesis*, Tohoku University, Sendai, Japan

probe directions 1 and 7 of the point B are plotted. The distribution characteristics are almost the same for the two directions. This fact explains why the incremental plastic strains have a common direction irrespective of the stress probe directions.

It is very interesting to note that the limited number of eigenvectors dominate the real displacement fields. In the examples above, the number of freedom is N=972, and mainly 10% of all eigenvectors determine the 80% of real displacement field.

Mining Science and Technology' 99, Xie & Golosinski (eds) © 1999 Balkema, Rotterdam, ISBN 90 5809 067 1

Development of bolting technology for coal roadway in China

Chaojiong Hou, Nong Zhang & Jianbiao Bai
China University of Mining and Technology, Xuzhou, People's Republic of China

ABSTRACT: This paper introduces the newly popular enhancement theory on the strength of rock surrounding roadway by bolting. As for bolting design, special attention is paid to the introduction of engineering analogy method based on the stability classification of rock surrounding roadway and a systematic design method with computer numerical stimulation.

1. INTRODCTION

Bolting is a kind of efficient, economical and safe supporting method of roadway. It has two obvious advantages: 1) It belongs to "active" support, thus can provide support resistance as soon as bolts are installed. 2) The deformation of rock surrounding bolted roadway can be decreased by more than a half as that of roadway supported by sheds under similar conditions. It also can keep roadway unblocked, decrease cost, lighten the labor of workers and improve operating environments. So bolting has become a main support technology for coal roadway all over the world.

China began to use bolting technology in 1956. Just several years ago its development was still rather slow because of soft and loose rock surrounding roadway, complex construction, impact of mining activity, and inadequate study of bolting theory, design methods, bolting materials, working machines and tools, monitoring means and so on. Of newly driven coal and semi-coal roadways only 15% and 30% respectively were bolted in 1995. It limited increase of production and efficiency [1]. However in the latest three years, China University of Mining and Technology, Central Coal Mining Research institute, Xintai Mining Administration, Xinwen Mining Administration, and others have worked on solving key problems, and made significant progress. In 1997 the bolting proportions in coal and semi-coal roadway raised to 20% and 33% respectively.

2. ENHANCEMENT THEORY [2]

Besides the action of support, the stability of surrounding rock is mainly decided by rock strength and stress condition. In recent years China has conducted study on how to get better mechanic properties of rockmass after being bolted. These researches mainly aim at such as water conservancies, tunnels, slopes and other shadow-burying and underground projects without mining influence. Meanwhile the improvement of C φ E in pre-peak strength zone of bolted rockmass has been discussed in varying degrees. Soft and loose surrounding rock, deep shaft, mining-induced stress and tectonic stress cause strong strata pressure in coal roadway; and then lead to serious destruction, and appearance of broken zone, plastic zone and elastic zone. Bolted rockmass may be found in the broken zone, even in two or three zones as mentioned above, consequently, the strength of bolted rockmass is post-peak strength or residual strength. The main contents of the enhancement theory on the strength of rock surrounding roadway by bolting are: Bolts can provide lateral pressure to bolted rock, enhance the strength of rock surrounding roadway, especially in post-peak zone, and increase peak strength and residual strength. And mechanic properties of post-peak rockmass are improved by bolting.

Now take a coal roadway of medium hard rock(uniaxial compressive strength is about 15 -20 MPa) as an example, and use laboratory simulation tests to explain strength enhancement of rock surrounding roadway. The original bolted rock is $2.0 \times 2.0 \times 2.0 m^3$,and the model is $20 \times 20 \times 20 cm^3$, bolts are laid out only on one of the surfaces. Load tests are performed on unbolted rockmass and on bolted ones with various numbers. The results are shown in table. 1. It can be seen that K_j and K_c are both larger than 1 after bolts are installed, which explains that the strength of bolted rockmass is always larger than that of unbolted rockmass. K_j and K_c rise with the increase of the number of bolts installed. Meanwhile C' and ϕ' become larger too. So the mechanic properties of rock get improved obviously.

3. BOLTING DESIGN METHODS

Table.1 The strength enhancement of bolted rockmass

Number of bolts	K_j	K_e	C'(MPa)	ϕ'(^{0})
0	1	1	0.0168	31.51
2	1.03	1.08	0.0182	31.53
3	1.09	1.13	0.0183	33.51
4	1.16	1.19	0.0184	35.57
5	1.21	1.25	0.0186	37.14
6	1.27	1.35	0.0194	38.80
8	1.35	1.48	0.0221	40.40

Where:

K_j- enhancement coefficient of limit strength of bolted rockmass, $k_j=\sigma_1/\sigma_c$

σ_1- limit strength of unbolted rockmass ;

σ_c- limit strength of bolted rockmass;

K_c-- enhancement coefficient of residual strength of bolted rockmass, $K_c =\sigma'_1/\sigma'_c$;

σ'_c- residual strength of unbolted rockmass ;

σ'_1- residual strength of bolted rockmass ;

C'- corresponding cohesion;

ϕ'- corresponding internal friction angle.

China University of Mining and Technology, together with China Central Coal Mining Research Institute, adopts a method that combines theoretical study with extensive field monitoring and numerical analysis, and puts forward bolting pattern and selecting scheme of main parameters of coal roadway. It firstly uses fuzzy cluster analysis to classify the stability of surrounding rock of roadway into 5 categories according to 7 parameters, which are rock strength of roof, two sides and bottom, burying depth of roadway, width of pillar, coefficient of mining activity and integrated coefficient of rock surrounding roadway. The five categories are: (1) extremely stable, (2) stable, (3) medium stable, (4) unstable, (5) extremely unstable. Bolting pattern and parameters vary from category to category.

A systematic bolting design method characterized by computer numerical simulation has been successfully used in Xinwen, Xintai, Kainan, Yanzhou, Jinchen, Luan and other mining districts.[3] Evaluation of geological mechanic, design, construction, monitoring and information feedback are taken as an integral system in this method, and design is closely connected with practice, meanwhile it tests, revises and perfects design through practice. A set of software for bolting design of coal roadway, which includes pre- and post- handling programs, has drawn up. It can be used to input and transform original parameters, form and compare schemes, select best scheme, output results and so on. Main parameters of roadway are input in practical use, various possible schemes will be automatically formed and further analyzed via a computer. The best scheme will be selected according to the principle of low cost and small displacement, as shown in Figure1. Information will be fed back from construction and field monitoring. And five parameters (relative displacement of two sides, incoherent deformations of in- and out-bolting rock zone, the number of yield points of testing bolts and anchorage power of bolts) are input into the computer in order to decide whether design should be revised and how.

4. HIGH & SUPER HIGH STRENGTH BOLTS[4]

In the past, most metal bolts used in China were made of low carbon steel, its yield strength is 235MPa, and diameter is $\phi14 \sim \phi16mm$. The bolts were usually broken in the thread part because its section area is 20% smaller than that of rod. But once the thread part is heat-treated, its strength will exceed that of the rod, and it will have enough elongation percentage. Then the drawing breakage of bolt occurs in the rod instead of at the end. This kind of bolt is called extensile bolt, whose strength and elongation percentage both gets enhanced. The comparisons of mechanic properties with common bolts are shown in Table 3.

High and super high strength bolts can be made if 20SiMn-twist steel of construction and left-handed rotation is selected. Their mechanic properties are shown in Table 4 and Table 5. These bolts have been widely used with excellent results.

5. BOLTING OF COAL ROADWAY UNDER DIFFICULT AND COMPLEX CONDITIONS

The so-called difficult and complex conditions, which occupy about 40% of all roadway, are "three soft" surrounding rock (soft roof, soft sides and soft floor), deep shaft, compound roof, gob edge roadway, mining roadway for fully mechanized mining face with top-coal caving and so on. Maintenance of roadway under these conditions is rather difficult.

Table 3 Test mechanic properties of bolt with extensible rod and general bolt

Bolting pattern	Rod diameter (mm)	Free rod Length (mm)	Yield load (kN)	Limit load (kN)	Extension range (mm)	Yield and breakage point	Enhancement percentage(%)		
							Yield load	Limit load	Extension range
General bolt	14	1300	31.4	49.3	30-40	bolt end			
	16	1500	44.9	66.6	30-40				
Bolt with extensible rod	14	1200	44	69	200-260	bolt rod	40.1	40.0	400-500
	16	1400	60	89	260-330		33.6	33.6	500

Table 4 Mechanic properties of 20MnSi-twist steel

wist steel diameter	bolt end	thread	test time s	yield load (kN)	Limit load (kN)	elongation percentage(%)	breakage point
φ18	unheat-treated	M16	3	53.7	87	13.8	bolt end
	heat-treated	M16	3	92.0	137	20.0	bolt rod
φ20	unheat-treated	M18	3	65.9	102	16.4	bolt end
	heat-treated	M18	3	114.0	171	21.3	bolt rod
φ22	unheat-treated	M20	3	83.0	136	16.6	bolt end
	heat-treated	M20	3	141.5	216	23.3	bolt rod

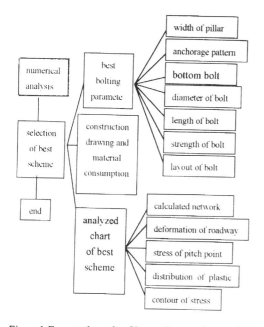

Figure1 Exported result of best scheme of numerical analysis

Table5 Mechanic properties of super high strength in practice

Bolt diameter	Yield load (kN)	Limit load (kN)	Elongation percentage(%)
φ18	175	206	17
φ20	206	257	17
φ22	254	314	17.7

Firstly, it is vertically important to keep two sides of roadway stable. Bolts should provide strong support resistance in order to control the deformation of coal sides, meanwhile they should have proper elongation for rock deformation. Enhanced extensible bolts made of Q235steel or 20SiMn steel can satisfy these demands, whose anchorage power is more then 40KN. The bolts should exceed 1.8m in length. Secondly, the integral sinking of two sides is bound to cause the floor heave and incoherent deformations in and out of roof anchorage zone, so it is necessary to keep roof stable. Supporting with full-length resin grouted bolts can decrease deformation of roof and pressure of bolt head obviously, and few bolts lose effectiveness. These factors justify the choice of high and super high strength bolts with full-length resin grouted. There are two basic bolting patterns: one is bolt-steel reinforcement-mesh, the other is truss bolting. When coal or immediate roof is too thick, cable anchor of small diameter may be applied to keep roof stable. The cable anchor is 28mm in diameter and 6 9m in length, which should be installed in stable strata. Thirdly, bolting should reinforce stress concentrated area in two bottom angles.

Grouting reinforcement may be applied if the surrounding rock is unstable and the fracture and joint develop. Grouting can greatly enhance strength of broken rock and improve the stability of roadway. Chemical and cement materials are commonly used for grouting. High-water-content, quick-hardening material is a new popular reinforced material, whose advantages are quick-hardening and adjustable, 100%-consolidation- coefficient and slight swelling property under high water-content, high ratio of water and cement, fine permeability, nice plastic property of consolidation, low cost [5]. Comparisons of liquid properties of high water-content material and cement material are shown in Figure 2.

Fig.2 Comparisons of liquid properties of high water-content material and cement material

Engineering instance: stating cut of No.2 coal seam in Xiandewang coal mine, Xintai administration

The testing roadway is 1725 starting cut lying in No.2 coal seam, which is a 2.5m-thick extremely soft seam, a 20^0 seam pitch and 400-450m working depth. Its uniaxial compressive strength is 1.5-2.0MPa. Its false roof is 0 0.4m-thick carbonaceous mudstone. Its immediate roof is 0.5-2.5m-thick siltstone, which contents three lines of seams, it is broken and easily to fall off. Its main roof is sandstone whose stratification is developed. Its floor is 2.5-3.0m-thick sandy shale, which will swelling when meeting water. The 1725 starting cut is 5.0m wide and 2.6m high, it is affected by mining activity of 1721 mining face.

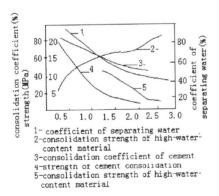

1- coefficient of separating water
2-consolidation strength of high-water-content material
3-consolidation coefficient of cement
4-strength of cement consolidation
5-consolidation strength of high-water-content material

Technology scheme: 1) High strength bolt with full-length resin grouted(ϕ 22×2400mm,made of 20SiMn), whose breaking power is 216KN, was adopted in the roof,. Advance grouting was used to deal with fracture zone of roof. Small-diameter cable anchor is used to reinforce the thicker immediate roof. The anchor is 22mm in diameter and 9.0m in depth. 2) Enhanced flexible bolts (ϕ16×1800mm,made of Q235) were adopted to reinforce the bearing ability of two sides.

Support effect: in comparison with I-steel support, relative displacement of two sides is decreased by 51.7%, the advance rate of roadway is increased by 35%, the cost of support and maintenance is deceased by 36.3%.

CONCLUSIONS

(1) After a rather slow beginning, bolting has become an important support pattern of roadway in coal mine in China because of its technological and economic superiority.

(2) Enhancement theory on the strength of rock surrounding roadway by bolting suggests: mechanism of bolting is to strengthen rock surrounding roadway, especially in post-peak zone, enhance its peak strength and residual strength, improve mechanic properties of rock in post-peak zone.

(3) The stability of coal roadway is divided into five categories by using the method of fuzzy cluster analysis. Basic bolting pattern and main parameters are designed for every category of roadway.

(4) A systematic bolting design method characterized by computer numerical analogue has been successfully studied, meanwhile a set of corresponding software for bolting design of coal roadway has been drawn up.

(5) Enhanced flexible bolt high strength bolt or super high strength bolts can be made by heat-treating the thread part of bolts end or the whole rod of bolts. These high-property bolts have successfully applied in some mines.

(6) Great progress has been made in bolting under difficult and complex roadway conditions.

REFERENCES

Hou Chaojiong et al.1996.Development of bolting of coal roadway in China. Journal of China Coal Society(2)113-118.

Hou Chaojiong et al.1998.Study on key bolting of gob edge roadway for fully mechanized mining face with top-coal caving. Ground Pressure and Strata Control (sup.)

Ma Lianjie et al. 1997.Design method based ground stress for bolting of coal roadway. Ground Pressure and Strata Control (3,4):195-197.

Hou Chaojiong et al.1997.High strength bolt. Ground Pressure and Strata Control (3,4):176-179.

Mining Science and Technology' 99, Xie & Golosinski (eds) © 1999 Balkema, Rotterdam, ISBN 90 5809 067 1

Cable bolting for soft rock support

C. Wang
Department of Mining Engineering and Mine Surveying, Western Australian School of Mines, Kalgoorlie, W.A., Australia

Wending Ma
China University of Mining and Technology, Xuzhou, Jiangsu, People's Republic of China

ABSTRACT: Cable bolting application for soft rock support is introduced in this paper. Factors influencing the generation of bond strength of grouted cable bolts installed in soft rocks and approaches to the achievement of a high cable bolting system capability in soft rock circumstances are highlighted. The feasibility of using cable bolting as an alternate soft rock support under certain conditions is discussed.

1 INTRODUCTION

Rock bolts have been playing a significant role in rock reinforcement and ground support. However, it is not uncommon that rock bolting failures occur in underground openings or roadways excavated in soft strata. The following photographs (Figure. 1) show some sever collapses of underground roadways within soft rocks and initially supported by rock bolts.

Investigation into the failures of rock bolt supported soft rocks indicated the insufficient bond between the rock bolts and the surrounding rocks of the roadways was the main reason of the failure. Further study revealed that the insufficient bond can be attributed to the poor anchorage and low load bearing capability of rock bolts when installed in soft rocks.

For rehabilitating these roadways, enclosed yieldable steel sets are usually the first option. In most cases, steel sets, some times in conjunction with backfill between steel sets and the surface of the surrounding rock of a roadway work very effectively due to the yieldability and high support resistance of the steel sets.

However, due to the high cost of steel sets and the complicated operational procedure for installation, steel set is not an economically preferred option for mining practice. Therefore, the onus is on engineers to search for a cost effective and competitive alternative to steel sets. In view of that, discussion on the capabilities of cable bolts and their applications for soft rock support was undertaken and is presented in this paper.

Figure 1. Failure of split sets in roadways within soft rocks.

2 STATE-OF-THE-ART CABLE BOLTING

Cable bolt was initially developed for cut-and-fill mining in underground metal mines. Its application for ground support dates back to early 1960s when degreased hoist ropes were used as cable bolts. The poor load transfer characteristics of the plain wire and the limited availability of hoist rope for mine-wide cable bolting instigated the development of specific cables for cable bolting. Since the early 1970's, rock reinforcement in metal mines using long, flexible cables has been under continual development with major contributions being made by mining groups in Australia, Canada and South Africa. The development of cable bolts has been focusing on efforts to increase the cable's pull out resistance by altering its configuration through the addition of buttons, coatings, and unwinding the individual strands of the cable. To date, a variety of configuration of cable blots have been developed and in use in different mines world wide. Table 1, after Windsor (1992) and Hutchinson (1996), gives a summary of some of the cable bolt configurations which are offered for the achievement of a maximum bond strength at the cable/grout contact.

Table 1. Some of cable bolt configurations and their capacities (After Windsor 1992 and Hutchinson 1996).

Type	Longitudinal Section and Cross Section		Pullout capacity (t)
Plain strand			20~25 t/m Rupture=25 t
Epoxy coated			30% higher than plain
Birdcaged strand			35~80% higher than plain strand
Buttoned strand			150% higher than plain
Nutcaged strand			100~200% greater than plain strand
Bulbed stand			Bond strength=3 times plain
Duble plain strand			1.5~2.0 times plain strand

During the early 1990s, glass fibre cable bolts were developed. One of the first prototypes of high strength glass fibre cable bolt was constructed in Canada using Polystal or DAPPAM. It was a result of requirement for a cuttable cable bolt facilitating use of continuous mining (Pakalnis, 1994). The ultimate tensile strength of Polystal cable bolt (consisting of four Polystal tendons of 9mm diameter) was 243kN. Whereas, a cable bolt composed of 10 straight tendons of 6.4mm DAPPAM produced an ultimate tensile strength of 270kN.

To achieve good bonding of cable bolt applied to rock reinforcement, grouting is usually employed. Therefore, cable bolts with fully capsulated grouting are typical. Three methods of grouting are used. The traditional grouting method as shown in Figure 2(a) is called "Breather tube method". The grout, usually having a water/cement ratio≥0.4, is injected into the bottom of the hole through a large diameter tube. The breather tube, which is typically 13~19mm diameter and allows the air to bleed through is attached to the cable bolt element and extends to the end the hole before it is installed into a hole. Both the tube and cable bolt are sealed into the bottom of the hole by means of a plug of cotton waste or quick setting mortar. The direction of grout travel is upwards in the hole. The second method called "Grout tube method" is illustrated in Figure 2(b). In this case a large diameter grout injection tube extends to the end of the hole and is taped onto the cable bolt. The cable bolt and tube are held in place by a wooden wedge inserted into the hole collar. Grout is pumped to the top of the hole and is injected down the hole until it appears at the hole collar. The third method is called "Self-Retracting grout tube method"(Figure 2(c)) which has been successfully developed and practiced in Australia since the early 1990's (Villaescusa, 1999). By this method, a cable bolt is inserted into a hole and is held in place. Subsequently, a grout tube is inserted into the toe of the hole. The grout paste which usually has a w/c ratio of 0.3 to 0.35 is then pumped through this tube and pushes the grouting tube to retract down the hole as the grouting process proceeds until the grout paste appears at the collar of the hole. The optimal grouting rate is such that the self-retracting tube should be in minimal contact with the advancing grout paste inside the hole. Empirically, for minimizing voids at the collar of the hole when the tube is pulled out of the hole, only 30cm of hose should be placed in the grout past. Obviously, this method requires the operators to be well trained.

Plates can be attached to the exposed ends of cable bolts using a barrel and wedge anchorage system. This is particularly significant in weak rock support using cable bolts in conjunction with mesh and shotcrete. In an underground excavation, the closer to the rock surface, the lower is the confining stress applied to the grout by rocks surrounding a borehole. The voids between the cable bolt and the

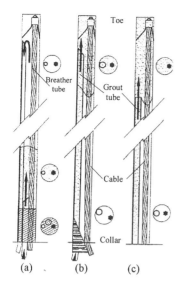

Figure 2. Cable bolt installation and grouting operation.

grout near the collar of a borehole are inevitable due to the breather tube installation or the pull-out of a retracting tube.

According to Kaiser (1992), stress change and modulus reduction with the rock deformation will reduce bond strength. Therefore, the use of face plate and anchor can secure a holistic rock and bolting system to be created rather than only the inner part of the surrounding rock of the excavation being reinforced.

Steel barrel and multiple component tapered wedges are used as the surface anchors for multiple wires cable bolts. A typical barrel and wedge anchor assembly is shown in Figure 3. The mechanism of cable bolts surface anchorage and the installation procedure were extensively addressed by Thompson (1992)

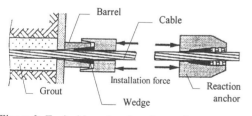

Figure 3. Typical barrel and wedge anchor assembly and their installation for tensioning a cable bolt (After Thompson 1992).

3 PERFORMANCE OF CABLE BOLTS IN SOFT ROCK

A cable bolting system capacity is a function of the component strengths of the steel strand, the bond (load transfer), the deformation response and retention characteristics of the system. There are five factors that influence the performance of a cable bolting system. They are as follows:
- Bond strength at the cable/grout interface;
- Bond strength at the grout/rock interface;
- Strength of the cable strand;
- Mechanical key at the cable/grout interface; and
- Mechanical key at the grout/borehole interface.

3.1 Bond strength at the cable/grout interface

The bond strength of a bolt refers to the shear resistance per unit contact surface area and is obtained by dividing the pullout resistance force by the surface area of the bolt. According to Yazici and Kaiser (1992), Young's modulus of rock and grout, strength of grout, borehole diameter and the frictional coefficient of the bolt-grout interface are four parameters affecting the bond strength at the grout/cable interface. Generally, higher bond strength can be obtained for stiffer rock, higher Young's modulus and compressive strength of grout and smaller diameter of borehole.

Yazici and Kaiser (1992) discussed details of influence of rock stiffness on the bond strength. These are presented in Figure 4. These results were based on an assumption of a single 15.2mm plain cable bolt grouted into a hole of approximately 53mm diameter. From Figure 4, it can be found that, for rock of Young's modulus less than 5GPa, bond

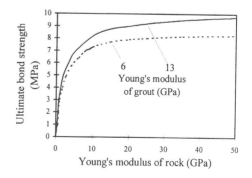

Figure 4. Influence of Young's modulus of rock on the bond strength at grout/cable interface (After Yazici and Kaiser 1992).

453

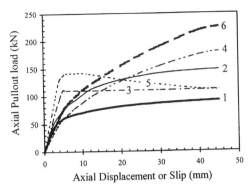

1-Plain strand, w/c=0.45; 2-Plain strand, w/c=0.35;
3-Epoxy coated strand, w/c=0.45;
4-Strand with button at 150mm depth, w/c=0.45;
5-Single birdcaged strand, w/c=0.45;
6-Bulbed strand, w/c=0.45

Figure 5. Influence of cable configuration on cablebolt pullout load capacities (Based on Villaescusa, 1992 and Goris, 1990).

strength reduces tremendously with the reduction of Young's modulus. For Young's modulus over 5GPa, the influence of rock stiffness on the bond strength is not significant. Compared with high stiffness rock, reduction of bond strength caused by rock stiffness for a typical soft rock with a Young's modulus ranging from 5GPa to 30GPa is about 20%.

As mentioned previously, a modification of cable configuration can greatly increase the interlock action at the cable/grout interface. In this case, the mechanical key at this interface plays a significant role in the creation of bond strength at the grout/cable interface. Figure 5 presents increase of pullout load for different configuration of cable bolt at an embedded length of 300mm.

3.2 Bond strength at the grout/rock interface

For a grouted cable bolt, pullout load is also determined by the bond strength at the grout/rock interface. This is vital to an application of cable bolt in soft ground. With the adoption of bulbed, caged or buttoned strand, the bond strength at the grout/rock interface becomes the predominant factor dictating the actual working capability of a grouted cable bolt. For a grouted cable bolt with an embedded length of L, the pullout capacity P can be calculated from the following equation.

$P=\pi D L \tau_{ult}$

Where, τ_{ult} = ultimate bond strength at the rock/grout interface;
D=diameter of the borehole.

This approach is used in many countries such as France, Italy, Switzerland, UK, Australia, Canada and USA (Littlejohn, 1993). In practice, a safety factor is commonly incorporated into this equation. Therefore, $\tau_{working}$ instead of τ_{ult} is actually used for cable bolt design. Table 2 presents some of the empirical data for $\tau_{working}$.

Table 2 Rock/grout bond strength value (After Daws, 1991 and Littlejohn, 1977).

Rock type	Grout type	Working bond strength (MPa)	Factor of safety
Mudstone	Cement	0.63	2
Sandstone	Cement	1.44	2
Limestone	Cement	1.0	2.8
Slate and shale	Cement	0.42~0.69	2
Siltstone	Resin	4	2

For soft rocks, a rule of thumb is that the maximum average working bond stress at the rock/grout interface should not exceed the minimum shear strength divided by the relevant safety factor (normally not less than 2). Where shear strength data are not available, one-tenth of the uniaxial compressive strength of massive rock up to a maximum value τ_{ult} of 4.2 MPa is often taken as ultimate bond stress. However, field anchorage tests are strongly recommended to confirm bond values in cable bolting design as the estimation of bond magnitude and bond distribution is complicated and the precision of the estimation is unsecured.

4 IMPROVING THE PERFORMANCE OF CABLE BOLTING IN SOFT ROCKS

Cable strand, grout and rock and their interactions compose a cable bolting system. In this system, the steel strength is dominated by steel quality and cable diameter. Where a high tensile strength is required for rock support, twin strand cable can be used. For a given strand, its tensile strength can be fully used only when an excellent match between the strand strength and the pullout load exits. As previously discussed that the pullout load of a cable bolt is a function of borehole diameter (D), embedded length (L) of the cable bolt and ultimate bond strength (τ_{ult}). Where the τ_{ult} should be the smaller value of the bond strength at grout/cable interface and that at rock/grout interface.

The influence of cable bolt embedded length on the pullout load capacity was extensively studied by the Spokane Research Center, US Bureau of Mines.

A linear relation between embedded length and load capacity was reported by Goris (1991) and is presented in Figure 6.

Similar results were reported by Benmokrane (1992) who confirmed Goris' observatios on this particular issue. These results indicate cable bolt grout or full length cable bolt grout is an effective approach to achieving a higher pullout load capacity. Changing the borehole diameter is another option for realizing a high cable bolt pullout load capacity. This might be a challenge to the traditional drilling equipment. However, this also hints the future orientation of drilling technique development in soft rock excavation and support.

Optimization of cable bolt array and using cable bolt in conjunction with conventional rock bolts in a staggered pattern are two approaches to an achievement of satisfactory cable bolting results in soft rock support application. Cable bolting optimization can be realized by changing either embedded length or borehole diameter or both to achieve a good match between cable bolt pullout load capacity and cable strand strength. A joint use of cable bolt and conventional rock bolt may be applied in such a way that rock bolts installation immediately follows the heading face using a small spacing and cable bolts with a large spacing are employed with a certain distance away from the heading face.

5 APPLICATION OF CABLE BOLTS IN ROADWAYS WITHIN SOFT ROCK IN UNDERGROUND COAL MINES

Both successful and unsuccessful applications of conventional bolts, such as split sets and rebars, have been reported for roadway support in underground coal mines with soft strata. Poor design, inappropriate installation and a lack of post installation quality monitoring all contribute to the failures of bolting in supporting soft ground. However, in most cases, using conventional bolts beyond their capacities in soft rock support is the main reason for subsequent roadway collapse. As mentioned previously yieldable steel sets are often chosen to rehabilitate a failed bolted roadway in engineering practice. In some circumstance, steel set is the first consideration in designing the support pattern for a roadway within soft rocks. Truly, yieldable steel sets is a technically proved approach to soft rock support. Taking into account of the operation cost and the productivity of using steel sets, alternatives to steel sets are economically and technically feasible.

Requirements for controlling a roadway within typical soft rocks were defined by Lu (1994). The critical parameter for soft rock support design is a support resistance of 0.3 to 0.5 MPa, which a support system should provide to the surrounding rock of a roadway. Reasonably, if a bolting system could offer

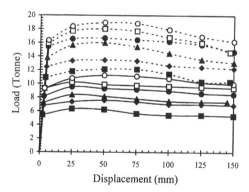

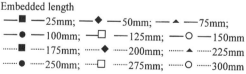

Embedded length
—■—25mm; —♦—50mm; —▲—75mm;
—●—100mm; —□—125mm; —○—150mm
······■······175mm; ······♦······200mm; ······▲······225mm
······●······250mm; ······□······275mm; ······○······300mm

Figure 6. Effect of embedded length on cable bolt load capacity (after Goris 1991).

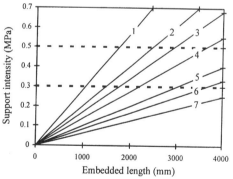

1-70mm bore hole and 700mm×700mm spacing;
2-50mm bore hole and 700mm×700mm spacing;
3-42mm bore hole and 700mm×700mm spacing;
4-70mm bore hole and 1000mm×1000mm spacing;
5-50mm bore hole and 1000mm×1000mm spacing;
6-42mm bore hole and 1000mm×1000mm spacing;
7-70mm bore hole and 1500mm×1500mm spacing

Figure 7. Support intensity with cable bolting parameters.

a supporting resistance equivalent to this value, a satisfactory soft rock support practice would be achievable. Figure 7 gives an insight into the feasibility of cable bolts of plain strand for supporting roadways within mudstone (bond strength at grout/rock interface is 0.63 MPa as presented in Table 4).

Figure 7 shows that a supporting intensity equivalent to 0.3 to 0.5 MPa can be achieved by optimizing either bore hole diameter or embedded length of cable bolt or both. Apparently, most strata encountered in mining practice are stronger than mudstone. Optimization of cable bolting design for excavations in these stronger strata would be expected to be easier.

6 CONCLUSION

Cable bolting for ground support has been widely used in underground cut-and-fill, open stope mining and hard rock support. Its introduction into soft rock support shows both opportunity and challenge.

In terms of load bearing capacity, cable bolts with a variety of configurations are capable of supporting soft rock.

The weak part of a cable bolting system in soft rock support application is the grout/rock interface. Using large diameter boreholes and long embedded cable bolts are two options for achieving a high cable bolt system capacity. This may put a challenge on to the conventional drilling equipment.

ACKNOWLEDGEMENTS

The authors acknowledge the Visiting Research Fellowship granted to one of them by the Department of Mining Engineering and Mine Surveying at the Western Australian School of Mines. It facilitated conduct of studies presented in this paper. Permission of China University of Mining and Technology to accept this scholarship is very much appreciated.

REFERENCES

Lu S.& Chen Y. 1994. *Strata control around coal mine roadways in China.* pp447-462. Publishing house of China University of Mining and Technology.

Hutchinson D.J. & Diederichs. M.S. 1996. *Cablebolting in underground mines.* pp126-135. BiTech Publisher.

Windsor C.R. Cable bolting for underground and surface excavations. *Proceedings of the International Symposium on Rock Support in Mining and Underground Construction.* pp349-366. Ontario, Canada, June, 1992.

Hoek E., Kaiser P.K & Bawden W.F. 1995. *Support of underground excavation in hard rock.* pp165-174. Balkema, Rotterdam.

Villaescusa E. An Australian perspective to grouting for cablebolt systems. *Proceedings of Rock support and reinforcement practice in mining.* pp83-90. Kalgoorlie, Australia, March, 1999. Balkema, Rotterdam.

Pakalnis R., Peterson D.A. & Mah G.P. Glass fibre cable bolts-an alternative. *CIM Bulletin.* January, 1994. pp53-57.

Daws G. Cable bolting. *The Mining Engineer.* February, 1991. Pp261-266.

Littlejohn S. Overview of rock anchorage. *Comprehensive Rock Engineering.* Volume 4. pp414-450.

Choque P. & Hadjigeorgiou J. The design of support for underground excavations. *Comprehensive Rock Engineering.* Volume 4. pp34014-345.

Fuller P.G. Cable support in mining. *Proceedings of the International Symposium on Rock Bolting.* pp-511-522. University of Lulå, Sweden, September, 1983.

Yazici S. & Kaiser P.K. Bond strength of grouted cable bolts. *International Journal of Rock Mechanics and Mining Sciences.* pp293-306. Vol. 29, No. 3, 1992.

Kaiser P.K., Yazici S. & Nose J. Effect of stress change on the bond strength of fully grouted cables. *International Journal of Rock Mechanics and Mining Sciences.* pp279-292. Vol. 29, No. 3, 1992.

Wang C. Joint support system for roadway within extremely soft rock in underground coal mines. *Proceedings of the Second International Symposium on Hard Soil-Soft Rock.* Naples, Italy, Oct. 1998. pp1021-1026.

Thompson A.G. Tensioning reinforcing cables. *Proceedings of the International Symposium on Rock Support in Mining and Underground Construction.* pp285-291. Ontario, Canada, June, 1992.

Hutchinson D.J. & Diederich M.S. The cablebolting cycle-Underground support engineering. *CIM Bulletin.* June, 1996. pp117-122.

Mining Science and Technology' 99, Xie & Golosinski (eds) © 1999 Balkema, Rotterdam, ISBN 90 5809 067 1

Rock support systems design and application in underground mines particular reference to Dolphin and Bold head mines of King Island Scheelite

D. K. Nag

Faculty of Engineering, Monash University, Melbourne, Vic., Australia

ABSTRACT: The paper describes the philosophy used in the design of support systems for Dolphin and Boldhead, two underground mines near the southern end of King Island at the western approach to Bass Strait, Australia. In 1991 when a combination of high processing and surface site costs coupled with the low mineral prices led to cessation of operation. Design consideration for the reinforcement of mine driveways and stopes in two mine were based on a quantification of the rock response together with a reinforcement strategy to enhance structural integrity of the roof and side walls. A different type of support systems was used, and tested their effectiveness in the mines. Effectiveness and economic factors were main consideration in the final selection of the support system. Successful roof support was achieved with fully column grouted rock bolts, cable bolts, swellex bolts with systematic timely installation of bolts in proper designed grid pattern.

INTRODUCTION

The Scheelite ore body is situated on the south east of King Island. The deposit is a mineralised skarn which outcrops on the shore of the island and continues below the seabed. Two underground scheelite mines _ Dolphin and Boldhead operated by King Island Scheelite, at Grassy, King Island (Fig 1). The original discovery of scheelite on the island was made in 1911, and opencut mining commenced soon after. The deposits were worked intermittently up to 1960. Since that time production was continuous until the 1991 closure when long term low world demand for tungsten and high processing costs lead to the closure of the operation. Up until 1972 all mining was by open cut methods, obtained from an open cut mine. After that time all production came from the Dolphin and Boldhead underground mines. The Boldhead mine had a lower grade and resource tonnage than the Dolphin mine, and the low prices of the metal mid to late 1980 led to its closure in 1985. The Dolphin mine with higher grade and resources but associated with poor ground conditions continued to operate largely as a result of improved mining productivity and cost levels.

GEOLOGY

King Island consists of a basal metamorphic complex unconformably overlain by trough and shelf sequences (shales, sandstone, carbonates) which have

been subject to varying degrees of deformation, metamorphism and acid and basic igneous activity. Subsequent intrusion of granite metamorphosed and metasomatised the local sedimentary pile resulting in a mineralised skarn consists of biotite and pyroxene rich hornfelses with granitiferous zone . The Dolphin deposit is located in an anticline fold which plunges form 10 to 45 degrees south east and consists of five main ore bodies, which are bound by major faults which dislocate the deposit. The principal ore zone consists of two mineralised bands separated by a barren marble sequence. The orebody is massive in nature, thickness varies 1-50metre, lay below sea level and adjacent to the original seashore. The

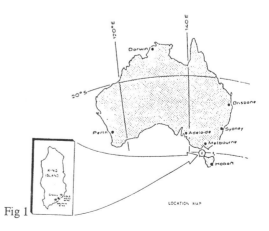

Fig 1

457

Table 1.

ROCK MASS INDEX		DOLPHIN MINE	BOLDHEAD MINE
1.	**ROCK MATERIAL PROPERTIES**		
1.1	Rock type	Hornfels(hornblende facies)	Hornfels
1.2	Uniaxial Compressive Strength	High; range 100-200 MPa	High range 120-180 MPa
1.3	Young's Modulus	Uniaxial tests: High, 145 GPa Biaxial tests: Med., range 15-125 GPa	Same Same
1.4	Poisson's Ratio	0.22	0.24
1.5	Tensile Strength	8 –15 MPa	9-15 MPa
2.	**DISCONTINUITY PROPERTIES**		
2.1	Number of sets	Five; 1 Bedding & 4 major joint sets	Four; 1 Bedding & 3 Major Joint sets
2.2	Orientation	Bedding: 3.5^0 to $40^0 \rightarrow 140^0$ Joint set 1: $82^0 \rightarrow 195^0$ Joint set II $85^0 \rightarrow 258^0$ Joint set III $72^0 \rightarrow 308^0$ Joint set IV: $84^0 \rightarrow 225^0$	Bedding 20^0 to $50^0 \rightarrow 140^0$ Joint sets I $62^0 \rightarrow 195^0$ Joint sets II $82^0 \rightarrow 252^0$ Joint set III $84^0 \rightarrow 225^0$
2.3	Spacing	Bedding: range, few cm to few metres Joint set 1. 0.54m Joint set II, 0.32m Joint set III, 0.44m Joint set IV, 0.87m /9minor faults : 10-18m)	Same (estimates) 0.49m 0.32m 0.29m 0.52m
2.4	Persistence	Typically exceeding room & pillar dimensions	Same
2.5	Roughness	Typically planar smooth or undulating smooth	Same
2.6	Infilling	Tight & clean or thinly infilled with carbonate & gauge. Typically welded	Same
3.	**GROUNDWATER**	Minimal	Minimal /same location moist
4.	**DRILL CORE QUALITY**	Predominantly bedded strata: 63 (joints/metre : 10) Massive strata 82	Bedded strata: 67 (joints/metre : 9) Massive strata:82 (joints/metre 6)
5.	**BLOCK SIZE**	Medium; volumetric joint count : 8	Same; Vol. Joint count: 10
6.	**CLASSIFICATION**	Class III, fair rock (Bieniawski, 1979)	Class III fair rock

Boldhead deposit is 3km away comparatively small consists of a separate lenses, faulted and a single ore horizon. All zones are continuous with a shallow 25 degree dip, vary in thickness between 0.3 – 10 metre. The mine series rocks are approximately 100 metre thick, fair to good condition within which majority of mine openings had been excavated.

The Dolphin orebody is wide and tabular extended upto 450 metres along strike 350 metres down dip, thickness varies between 5-50 metres, average dip 32 degrees lay below the sea bed and adjacent to the original sea shore. On any one horizon the ore body is divided into four principal stoping blocks by six major faults. The main ore zone lies within mine series rocks are approximately 200 metres thick and are relatively soft giving rise to generally fair condition for mine opening Extremely poor ground

Table 2			
Virgin Principal Stresses – Dolphin Mine			
Mine Level	Maximum Horizontal, N/S (MPa)	Intermediate Horizontal, E/W (MPa)	Minimum, Vertical (MPa)
-- 75m	4.5	3.1	2.0
-130m	6.0	4.6	3.5
-195m	7.2	6.0	5.2

condition are found in fault zones, adjacent to other orebody and at the change of orebody dip due to a significant increase in the frequency of jointing. The summary of rock, mass indices are given in Table 1.

The main source of instability in any mine, mainly due to high stress conditions, weak ground, low strength stratigraphic contacts and joints and faults interaction. The virgin principal stresses measured at different depths of the mine were low and given in Table 2.

The amount of information which was available during the exploration ,and preliminary design stages limited to that obtained from regional geology maps, surface mapping, open cut mine geology and exploration boreholes. During the early stages of the exploration work, detailed information including RQD, joint spacing, rock type ,ground water etc. was collected to build up a database on characteristics of each rock type and respective geotechnical domains. Later on data collection focussed on specific mining needs, stability analysis at the stope or development back. Consequently it was only possible to construct very general rock mass classification upon which rock support requirements were based for both mines (Terzaghi1946 & Deere1964). More detailed information was gathered gradually and boreholes were positioned judiciously at different depths mainly for rock mechanics studies, logging, CSIR's rock mass rating (Bieniawski1979) and Q method (Barton1974) were used. The rock mass classification for each area was established and support system was considered as per suggested support chart. Later on back analysis was carried out to assess the local ground condition and support requirements were predicted on the basis of observation and checking geological information for each block to conform local and regional interpretations, and the final support system was as per table 3.

A number of different types of rock reinforcement was tried in both mines to judge their effectiveness. Both were finally selected for their area on the basis stiffeners parameters, time for installation to full load bearing capacity corrosion resistance and cost effectiveness.

Table 3		
Final Rock Support Systems – Dolphin & Boldhead Mine (as per Rock mass classification system)		
Q – 15-50	2.5 – 7.5	03 – 0.1
RMR: 70-95	40 – 60	10 - 25
Support: unsupported	Untensioned grouted rebar/swellez @1.5mx1.5mx2.4m long	Untensioned grouted rebar/swellez @1.5mx1.5mx2.4m long + 50 –100mm shotcrete with mesh

Development of the Support System

The highly faulted nature of the ground, the low angle of dips of the rock strata and inherent weakness of some of the rock units, drive and stope supports had always been important consideration in the underground operation. The primary objective of support practice was to mobilize and conserve the inherent strength of the rock mass so that it became self supporting. Primary support was applied during or immediately after excavation to ensure safe working conditions during subsequent excavation.

The first underground opening undertaken at each mine was the decline which was to be the life of mine access to the orebodies, steel sets were used in very difficult ground and in other places expansion shell type of rock bolts were used. In practice, a support system provided that was quick to install and required little maintenance.

Expansion shell bolts had been used continuously to support local instability in the sidewall and back. Their installation cost was comparable to other available system but their effectiveness as a support systems was subject to a number of constraints. The results of anchorage tests for freshly installed bolts., and tests on bolts installed for some time show a diminished result, and used for temporary support. only.

A number of different types of rock reinforcement was tried in both mines to judge their effectiveness. Bolts were finally selected for the area on the basis of stiffness parameters, time required for installation, full load bearing capacity, corrosion resistance and cost effectiveness.

Chemical anchor bolts using resin cartridges marketed by several companies had been tried successfully, the anchorage strength achieved within design requirements. Due to practical difficulties in drilling bits, operation skill and cost structure, the system was not used extensively in the mines.

Full column untensioned cement grouted rebar tend to slip less in poor ground ,was introduced and tested for anchorage tests, and results showed the system was very effective and economical and suitable for both mines.

Frictional type of rock bolts (Rock stabiliser) both split set and swellex type tried in the mines and tested. In all conditions (dry and wet) they provided very effective support system. It was also found that

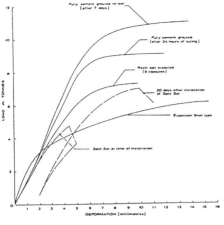

Fig 2 TYPICAL ANCHORAGE TEST RESULTS OF VARIOUS ROCK BOLT TYPES

Pillars Support

In case of Dolphin and Boldhead mine, the mining system relied completely on pillars which were designed to bear the total weight of the overburden and effectively control the displacement of the hanging wall. Pillars were mainly in joined rock and low stress environment, deformations were expected to take place through sliding and rotation of blocks, and the failure load of pillars would vary markedly with joints initiation . Because of the highly discontinuous nature of the rock mass almost all pillar continued either or all of joints bedding and fault structures capable of causing pillar failure. A variety of techniques researched and tried to restrict rock deformation ,to maintain pillars integrity and to improve pillars load bearing capacity prior to sand filling.

over a period of time, anchorage became greater due lock up of the system due to ground movement. Strength was also retained during nearby blasting. The bolt gave immediate support over the full length without use of chemicals or cement. The anchorage test results for all types of bolts tested in the mines are shown in Fig 2.

A number of different mining methods were used in both mines (Fudge and Nag 1995), and the majority of stope working areas were supported by cable bolts with systematic design patterns. The design was based on site geological mapping, use of statistical method to generate fabric data from joint survey information and use of block theory to determine potential unstable block in the stope area. The selection of appropriate cable support type was made to control block movements, and economic assessment and observation of performance were recorded. In some areas in the stope cable bolting system was used in conjunction with grouted rockbolts/swellex bolt. The shorter bolts provided immediate local support against minor spalling. The cable bolts were used to reinforce entire rock mass and provide long term support against major spalling. The most common configuration of cable bolts used in the mine was two cables per hole, 20 metre long, 15.2mm diameter, 7 strand, high strength steel, and untensioned. Displacements in cable bolts supported area were monitored by multiple extensometers measurements. Results showed that the bolts were effective and reduced ground displacement by 40-50%. The cable bolt support provided greatly increase personal safety and stoping productivity.

Resin Injection: A trial using epoxy resin injected at high pressure into the pillar was initiated to achieve significant bond between rock blocks and improvement in joint cohesion. Several pillars were injected with resin at different areas of the mine, successful results were achieved , drilled core showed higher penetration of resin through joints than cement grout at similar pressure and adhesion properties were also improved. But the system had some problems — on site mixing of epoxy resin, and to maintain quality control was extremely difficult in underground mining environment. The materials used were of some degree of toxic and were liable to produce skin infections if carelessly handled, personal required to use barrier skin cream and adequate protective clothing. All machinery used to pump the resin must be thoroughly cleaned on conclusion of the work for the day. These restrictions necessitate the use of trained and properly supervised personnel for the work. This system was not seen as a standard support system.

Cement Grouted Re-bar: The different cement grouted rockbolts were tried and monitored, the most successful technique had been used in pillars to arrest deterioration, were horizontally grouted rockbolts plain rebar 20mm diameter 2.4m long, 6 numbers in each pillar face. This grouted rockbolting system was used systematically as the pillar face was available, and provided good load deformation characteristics and high shear resistance.

GROUND CONTROL MANAGEMENT

At the early stage of mining, the mine management recognized that a well managed ground control plan was necessary component for successful mining operation in very difficult ground conditions.. The ground control management plan was produced using a combination of in house, outside consultant and CSIRO division of applied geomechanics (Nag 1980). The ground control plan was divided into two parts i). Local scale ii) Global or Regional scale, and was revised once a month and critically revised half yearly to correct areas of deficiency.

Regular updates of local geological structure plan, and, study their adverse influence on ground control and geotechnical logging and drilled core in the area, the water inflow were maintained. Significant damage to rock at the perimeter of an excavation, pillars and stope back rock block failure were photographed and recorded in systematic folders. Ground support and reinforcement based on appropriate methods and the timing of installation of ground support was considered as an integral part of the design to limit the potential of unraveling of the rock mass. It was recognized that in certain critical areas of the mine, the support was required to be installed within 24 hours of the face development.

All stope faces and development ends were examined each shift to examine conditions and make recommendations for ground support.

Systematic rock bolting was established method of roof support. The rock bolts were installed after blasting, scaling ,before a new cut was made ,and bolts placed as close to the face as possible to prevent the movement of newly exposed rock face.

CONCLUSION

The King Island Scheelite mining operation, geotechnical engineering was seen as an integral part of the total mining process. Rock stability and instability issues from the economic extraction ore always received priority. The combination of sound mining experience and judgement with geotechnical design and analyses, the operation became very cost efficient and successful mining operation in Australia. Systematic rock bolting using rock mass classification system and timely installation, proved to be successful in reducing accident rates and assisted in raising productivity in most difficult ground conditions.

REFERENCES

1. Barton,N., R. Lein & J. Lunde, (1974). Engineering Classification of Rock Masses for the design of Tunnel Support. Rock Mechanics Vol 6.5 pp 189 – 236
2. Bieniawski, Z.T. (1979) –The geomechanics classification in rock engineering applications Proceedings 4th International Conference on Rock Mechanics ISRM, Montreux. Vol.2. pp.41-48
3. Deere,D.U. (1964). Technical Description of Rock Cores for Engineering Purposes, Rock Mechanics and Engineering Geology. Vol 1. No. !.
4. .Fudge,A. & D. K Nag (1995) – Experience with different underground mining methods at Dolphin Mine of King Island Scheelite. Proceedings. Underground Operators Conference, Kalgoorlie, WA. pp 129-136
5. Nag,D.K. (1980) – Ground Control Management Plan of Dolphin & Boldhead Mine King Island Scheelite. Internal Reports
6. Terzaghi,K. (1946). Rock Defects and Load on Tunnel Supports. Publication 418 – Soil Mechanics, Series 25 , Harvard University.

Mining Science and Technology' 99, Xie & Golosinski (eds) © 1999 Balkema, Rotterdam, ISBN 90 5809 067 1

Rational match of drill hole, bolt and resin diameters in a coal roadway

Nianjie Ma
China University of Mining and Technology, Beijing, People's Republic of China

Shanyue Guan
Journal of China Coal Society, Beijing, People's Republic of China

ABSTRACT: Based on the result of the laboratory and field tests, a theory is put forward about the rational match among drill hole and bolt diameter, and the resin., whether the bolt has longitudinal reinforcement or not. The theory is important for selection of the design parameters for coal roadway support.

1. INTRODUCTION

Bolt supporting is an important technique in coal roadway. Now, it has become one of the main supporting forms of sectional roadways in principal coal production countries. It is spreaded and applied fast throughout the world due to the advantages of the bolt supporting's safety, economy and high speed. Bolt supporting has also become one of the important supporting methods in Chinese coal mine developing entry, and more than 15 of roadways are supported by this way in 1997. From 1995, bolt supporting is defined as a development aim of coal roadway supporting in Chinese coal mines, and 50% of rock and coal roadways in coal mines will be supported by bolt by 2000. Most of bolts in road supporting are resin bolts at home and abroad. For example, more than 60% of bolts in America are resin bolts; In Britain and Australia, most bolts in active mining roadways are resin bolts; Most bolts in roadway support are also resin bolt in China.

When using resin bolt, the rational match of bolt's diameter, drill hole's diameter and stick resin's diameter is the key for bolt to get the biggest bolt anchor force and the best economic effect. Australia, British and other countries pay more attention to this research, they consider that the bolt anchor force reaches the greatest point when the difference between drill hole's diameter and bolt's diameter is 4 10mm for left-hand twist steel bolt without longitudinal reinforcement. So the 28mm diameter drill hole matches 22mm diameter of left-hand twist steel bolt without longitudinal reinforcement. But there are no research reports on the diameter match

relationship between drill hole and twist steel bolt with longitudinal reinforcement. The rational match among bolt's diameter, drill hole's diameter and stick resin's diameter has never been studied in China. These parameters are often determined by experience, sometimes they are so irrational that the bolt anchor force is too low to give full play to bolting and even caving in roof on a large scale in roadway. Therefore the research of the three diameter's rational match is very important for theory and practice.

2. DRILL HOLE DIAMETER

The diameters of drill hole are mainly 28mm 33mm and 43mm in China coal mines. The diameter of drill hole should be determined mainly by anchorage cost. It was indicated that bigger the drill hole's diameter, and the more expensive the cost, because bigger drill hole need more anchor resin. The cost of the 33mm drill hole is 25% more than that of the 28mm drill hole; and the cost of the 43mm drill hole is about 120% more; and the cost of the 26mm drill hole will be 10% less than that of the 28mm drill hole. Additionally, greater diameter drill hole also increase the cost and time of drilling. Therefore, it is a trend that the drill hole's diameter gradually becomes smaller in Chinese coal mines.

In addition, there are some shortcomings as follows when the diameter of drill hole is too large: First, the bolt can't stab resin effectively and mix up the anchor resin evenly, which will result in low anchorage capacity; Second, large drill hole needs more drilling time; Third, large drill hole needs more anchor resin which results higher anchorage cost. Therefore, the 28mm drill hole's diameter is the ordinary choice in roadway bolting, and the diameter over 33mm is unfavorable.

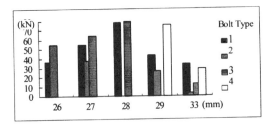

1—22 no longitude   2—18 with longitude
3—20 with longitude   4—with longitude
Figure 1. The bolt anchor force of 100mm length threaded steel bolt in different diameter of drill hole

3. DIAMETER OF BOLT

Figure 1 shows the bolt anchor force along 100mm length of threaded steel bolt with longitudinal reinforcement when the diameters are 18mm, 20mm and 22mm, and the 22mm diameter of threaded steel bolt without longitudinal reinforcement in siltstone roadway. the figure shows :

(1) When the 22mm threaded steel bolt without longitudinal reinforcement is used in drill holes varying from 26 to 33mm respectively, the bolt anchor force varies obviously. At first the anchor force increases gradually with the increase of diameter of drill hole, and reaches its greatest point (79kN) in 28mm diameter drill hole, then it decreases with the increase of diameter of drill hole. Thus it can be seen that bolt can be anchored effectively to surrounding rock when the diameter difference between drill hole and bolt ranges from 4mm to 10mm. The bolt anchor force of the 22mm left-hand twist threaded steel bolt without longitudinal reinforcement gets its greatest point when the drill hole's diameter is from 27mm to 28mm.

(2) As the figure shows that 22mm diameter threaded steel bolt with longitudinal reinforcement has big anchorage capacity in 29mm diameter drill hole, and can be effectively anchored to 33mm diameter drill hole. It can be seen that the difference between the diameter of drill hole and nominal diameter of bolt should be 6mm or more when threaded steel bolt with longitudinal reinforcement is used.

(3) The bolt anchor force of the 20mm nominal diameter threaded steel bolt with longitudinal reinforcement used in drill holes with vary from 26mm to 33mm is similar to that of the 22mm nominal diameter threaded steel bolt without longitudinal reinforcement used in the same size of drill hole. The bolt anchor force also reaches its maximum (80kN) in 28mm diameter drill hole and bolt can be anchored effectively to surrounding rock in 27mm and 29mm diameter drill hole, but the bolt anchor force is small in the 33mm diameter drill hole.

(4) When the 18mm diameter threaded steel bolt with longitudinal reinforcement is used in drill hole varying from 26 to 33mm, the bolt anchor force decreases gradually with the increase of diameter of drill hole and nearly becomes zero in 33mm diameter drill hole. It indicates that the difference of the nominal diameter of drill hole and bolt diameter should be less than 13mm when threaded steel bolt with longitudinal reinforcement is used.

Therefore, the diameter difference between drill hole and bolt should be in the rang from 6mm to 12mm, and 7mm or 8mm would be better. This rational match relationship differs from the threaded steel bolt without longitudinal reinforcement. The main reason is that the maximum diameter of threaded steel bolt with longitudinal reinforcement is 2mm or 3mm larger than that of threaded steel bolt without longitudinal reinforcement.

The above analysis shows that the bolt anchor force is very small when the difference between the diameters of drill and bolt is too large or too small, only the rational match can gain maximum bolt anchor force. The main reason is that the different diameters will be the key to fully stirring resin thus evenly mixing up resin adhesive and firming agent. If the diameter of bolt is too large, and the diameter difference between drill hole and bolt is only 1mm to 3mm, there would be no enough space for resin adhesive and firming agent to be mixed up evenly and solidified, that results in lower bolt anchor force. On the contrary, smaller bolt diameter, larger difference between diameter of drill hole and maximum bolt diameter will result in two harmful consequences: First, the resin can't fill up drill hole after stirring, thus resulting in lower bolt anchor force; Second, when stirred, the resin can't be fully blended, thus results in incompletely solidified resin and too low bolt anchor force.

4. THE DIAMETER OF RESIN

Under the situation of ensuring fix successfully, the diameter of stick resin should be big enough. There are two reasons: First, the rational stick resin's diameter should make resin adhesive and firming agent be mixed up successfully and solidified, and then bolt would get bigger anchor force. Second, when resin stick's diameter is big, the original length of resin stick is short. It will give more space to fix bolt when roadway's height is low.

According to above principles, resin stick's diameter D can be:

$$D'_{max} \leq D - 3 \quad D'_{min} \geq \sqrt{(D^2 - d^2)L/l}$$

Where, D'_{max} is resin stick's maximum diameter; D is drill hole's diameter; D'_{min} is resin stick's minimum diameter; d is bolt's diameter; L is bolt anchor length; l is resin stick's length.

5. ANALYSIS OF ANCHORAGE DESTROY

Based on the analysis of lab simulation tests and field tests, bolt anchorage destroy form can be got as the following.

5.1 *The anchorage slides following the bore wall of the drill hole*

When bolt being pulled out, resins adhere firmly with bolt, sliding along the hole, and the anchorage is very smooth. It is because the resin was mixed up and solidified, and the adhesive force is big enough to achieve design intensity. On the condition of strong rock surrounding, the weak point is the part where resin adheres to the hole. The adhesive force is lower than both resin strength after solidifying and drill hole's strength. So, this part is destroyed first. It is a normal destroy form. The high anchor force indicates that the bolt diameter match very well to the drill hole's diameter.

5.2 *The bolt's diameter is too big to have enough space to mix up resin*

When bolt being pulled out, there is little solidified resin around the bolt. This will happen when the size of bolt reinforcement is too big, and only 1 3mm less than the hole's diameter. When the difference between bolt maximum diameter and the hole's diameter is only 1 3mm, this not only makes fixing bolt difficult, but also has no enough space to mix the resin (form with resin adhesive and firming agent) and solidify. It fails to anchor for so low anchor force.

5.3 *The bolt's diameter is too small to stir the resin evenly*

When bolt being pulled out, there is a lot of solidified and unconsolidated resin around the bolt. The main reason is that nominal diameter and maximum diameter is small, or the drill hole's diameter is big, and the difference between the maximum diameter with the hole's diameter is more than 7mm. These will result in two harmful consequence: First, the resin can't fill up drill hole after being stirred, resulting in too low bolt anchor force; Second, when being stirred, the resin can't be fully blended, thus resulting in incompletely solidified resin and too low bolt anchor force.

5.4 *Bolt's diameter is so small that resin stick can't be stabbed*

The main reason is that the bolt nominal diameter and maximum diameter is too small, or drill hole's diameter is too big, the difference between maximum diameter and drill hole's diameter is more than 10mm. For example, when bolt whose diameter is 18mm (maximum diameter is 21 21.5mm) is used in 43mm or 33mm drill hole, it will happen, so the difference between the bolt's diameter and the hole's diameter is never more than 10mm when threaded steel bolt with longitudinal reinforcement is used.

6. PRACTICAL EXAMPLE

As Figure 2 shows is real stress state of whole long anchorage roof bolt in the gate of 2707 face in Dongpang coal mine in Xingtai coal mine administration (left-hand twist threaded steel bolt without longitudinal reinforcement is used, the diameter is 22mm, the drill hole's diameter is 28mm). From the figure, when the rational match of roof bolt was used, the anchor force would reach 140kN on the day, reach 200kN after 14 days, and reach bolt's yield load or ultimate load when the bolt work was normal. As Figure 2 shows that the bolt's anchor force is the biggest in the middle of bolt, and least around the collar, the stress to surrounding rock is even, the bolt and the surrounding rock's stress state is good, and real anchor force is big.

7. MAIN CONCLUSIONS

The parameters design such as like bolt's diameter, drill hole's diameter, resin stick's diameter and anchor form have important meaning for coal

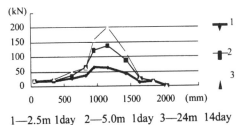

1—2.5m 1day 2—5.0m 1day 3—24m 14day

Figure 2. Real stress state of roof bolt under whole
long anchorage

roadway bolt supporting. Adopting three diameter
rational match would improve the anchor force very
much, thus improve the effect of bolt supporting and
stability of coal roadway. It would not only validly
control deformation or roadway, decrease or expire
roof caving in bolt supporting coal roadway, but
also reduce supporting cost and remaintance cost
very much. There would be reached good technical
and economical results.

(1) The hole's diameter will affect greatly to bolt
supporting cost and anchor force. On this condition,
choosing 28mm drill hole may get the best drill hole
efficiency, high anchor force and better technical,
economical results.

(2) The anchor force is the biggest when 22mm
threaded steel bolt without longitudinal
reinforcement or 20mm threaded steel bolt with
longitudinal reinforcement are used in 28mm drill
hole.

(3) The size of the bolt's diameter is the key factor to
stir the resin evenly, get big anchor force and fix the
bolt successfully. When the threaded steel bolt
without longitudinal reinforcement is used, the
difference between drill hole's diameter and bolt's
diameter should be 4~10mm, and the best is 5~6mm;
When the threaded steel bolt with longitudinal
reinforcement is used, the difference between drill
hole's diameter and bolt's diameter should be
6~12mm, and the best is 7~8mm.

(4) For whole long anchorage, the improvement of
bolt's diameter will not raise much anchorage cost.
The technical economic result will be outstanding if
the diameter is increased properly.

Mining Science and Technology '99, Xie & Golosinski (eds) © 1999 Balkema, Rotterdam, ISBN 90 5809 067 1

Experimental study of marble constitutive properties under uniaxial impact[1]

Renliang Shan, Quanchen Gao & Wenjiao Gao
China University of Mining and Technology, Beijing, People's Republic of China

ABSTRACT: This paper presents the impact constitutive relationships of series tests on marble samples using split Hopkinson pressure bar (SHPB). It is found that the stress-strain curves of marble generally have three stages. The first stage goes up along an approximate straight line with positive slope, and usually the higher the strain rate is, the larger the slope of line is. The second stage represents the idea plastic character of marble after being stricken. The third stage goes down along an approximate straight line with negative slop, and is not capable of rebound.

1. INTRODUCTION

Rock impact fragmentation is widely used in the engineering of geology, mining, petroleum, civilian etc. Many problems therein have become popular in rock mechanics and rock engineering, and have received growing attention of the experts in geotechnics, geology and mining etc.

Rock impact constitutive relationships are key points to recognize the stress wave propagation laws in rock and the dynamic failure laws of rock, meanwhile they are also the bases of numerical modeling of rock dynamic fragmentation such as drilling and blasting, and of design of earthquake-resistant or explosion-protection engineering.

Because the costly equipment such as SHPB, light gas gun are needed in measuring rock impact constitutive characteristics. So they are rarely known compare to static constitutive characteristics.

There are four stages of a typical static complete stress-strain curve as shown in figure 1(Jaeger & Cook 1979).

In the first portion (OA), the curve is concave upwards for the compressed closing of microcracks within the rock specimen.

In the second stage (AB), the curve is almost straight, which shows the essential linear behaviour of rock.

In the third stage (BC), usually starting from a stress level of about two third of the compressive strength, the curve is concave downwards, that is to say the curve's slope becomes smaller till zero with the stress increasing. The local failure inside of sample gradually develops till peak point of C. When unloading at any point P in this stage, the curve PQ and the forever strain ε_0 will be formed. If reloading, the curve QR will be formed.

In the fourth stage (BC), the stress decrease, the strain increase, the stress-strain line slope is minus. The rock failure continually develops till to split to several blocks finally.

Elastic modulus and compressive strength have been focused in the study of rock dynamic properties for along time, but it is inadequate about stress-strain relationship itself that represents the failure progress really. The experimental rocks are mostly granite, sandstone, basalt and limestone. With SHPB, a lot of marble constitutive curves under uniaxial impact were studied in this paper.

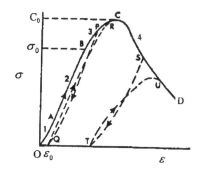

Figure 1. Rock static complete stress-strain curve

---

[1] The project is supported by National Sceince Found of China (49502041)

2. TEST SYSTEM AND TEST PRINCIPLE

2.1 *Test system*

SHPB equipment is the core of the test system (figure 2). It includes: *dynamic system*, which is made up of barrel and nitrogen bottle, can accelerate impact bar; *load generation and transmission system* that is made up of impact bar, rock sample, input bar and output bar that compactly contact with the two extremity of rock sample; *impact velocity test system* that is made up of condensing bulb, photo-diode, *magnifier and counter; strain test system* that is made up of strain gauge, ultra-dynamic strain equipment, the dynamic test and analysis equipment of CS2092.

All the diameters of impact bar, input bar and output bar are 30 cm, but their lengths are 500mm, 1500mm and 1400mm respectively. Fifty marble specimens are used in the experiments. Both their diameter and length both are 30 cm, densities are 2.67~2.68g/cm^3, static uniaxial comprssive strength are 35~45Mpa, the Poisson's ratios are 0.19~0.22, and the sound wave velocities are 5625~6013m/s.

2.2 *Test principle*

The expanding of highly pressured gas provided by nitrogen bottle in the barrel accelerates impact bar and makes it strike input bar at a certain velocity, in which a rightward input wave ε_I is generated. When ε_I gets to the interface 1-1, a part of which rebound to the input bar, generates reflection wave ε_R, the other part of passes through sample and delivered to output bar, generates output wave ε_T (figure3). Strain wave signals are accepted by strain gauges and then transferred to electrical signals by ultra-dynamic strain equipment, further are transferred to dispersed signals and stored by CS2092.

According to the theory of wave transmission and the uniform hypothesis, rock sample's stress, strain, and strain rate can be determined from input wave, reflected wave and output wave measured in experiments by following 3 formula:

$$\sigma = \frac{A_0}{A} E_0 \varepsilon_T \qquad (1)$$

$$\varepsilon = -\frac{2C_0}{L} \int_0^t \varepsilon_R dt' \qquad (2)$$

$$\varepsilon = -\frac{2C_0}{L} \varepsilon_R \qquad (3)$$

Real rock impact constitutive curves should be acquired by eliminating time variable in strain rate-time, strain-time and stress-time curves. While material's constitutive property are closely related to the magnitude of strain rates, the history of strains and strain rates, we simultaneously inspect the properties of strain rate, strain and stress changing with time, and the corresponding constitutive property of rock sample in the ensuing paragraphs.

3. TESTED CURVES AND ANALYSIS FOR MARBLE

Because the samples of marble can be heavily broken (even broken to pieces) at the impact velocity of 8~9m/s, maximum impact velocity for marble ought to be lower then 10m/s.

On the other hand, du to some samples may be broken to several big blocks at the velocity of 6~7m/s, and the others also appear some cracks on their surfaces although kept entirety. Experience also shows that it is difficult to control the impact system if the velocity is lower then 6m/s. The minimized impact velocity of marble test is set over 6m/s.

Typical strain rate-time curves, strain-time curves, stress-time curves, stress-strain curves of marble samples at the different impact velocities are shown in figures 7-10.

From figures 4-7 we can conclude that:

Strain rate-time curves have two big up and down fluctuations and many tiny surges in amplitude. The higher the impact velocity is, the larger the amplitude of fluctuation is. The first fluctuations keep only 20-30 microseconds and their amplitudes are not sensitively dependent upon impact velocities. The second fluctuations need a longer time to go up,

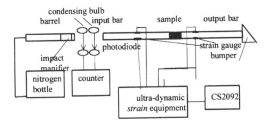

Figure 2. Uniaxial SHPB experimental system

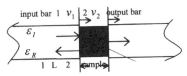

Figure 3. The principle sketch of SHPB

468

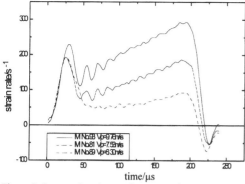

Figure 4. Impacted strain rate-time curves for marble

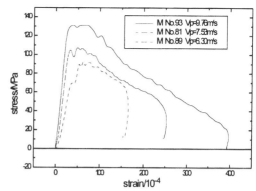

Figure 7. Impacted stress-strain curves for marble

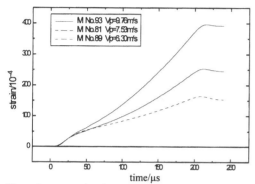

Figure 5. Impacted strain-time curves for marble

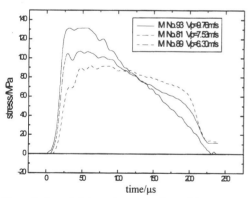

Figure 6. Impacted stress-time curves for marble

but a shorter time to go down, and their amplitudes are sensitively dependent upon impact velocities. When the velocity rises a few, the amplitude rises a lot. The above characteristics of strain rate are determined by the input rectangle stress waveform (Li, Xibing & Gu, Desheng 1994).

The strain almost goes up all the time, but there is an inflexion obviously. The slopes before the inflexions are relatively larger, but they are smaller after inflexions. While impact velocity is greater; the discrepancy between the two slopes of the former part and the later is smaller.

The stress go up quickly but go down slowly with time, the declined parts intersect one another, the drop rate is correlated with impact velocity. The higher the velocity of impact bar is, the faster the unloading velocity of specimen is. It implies that the higher the impact velocity is, the quicker the failure velocity of specimen is. There is still a stable period of a few ten microseconds with some small waves between the up and down parts.

Impact constitutive curves of marble often ascend in line along positive slope before peak. Their linear slopes (modulus of elasticity) are closely correlative with strike velocities or loaded strain rates, usually the bigger the strain rates are, the bigger the moduli are. Dynamic modulus of elasticity of marble are about $1.8 \sim 5.0 \times 10^4$ MPa in the test. This stage represents the elastic phase of the marble. Constitutive curves of marble after peak mostly descend in line along negative slope then suddenly descend along vertical line, that means marble only have the first type of unloading (negative slope) and strong unloading (unlimited slope), seldom have the second type of unloading (positive slope). This stage delegates the brittleness characteristic of marble. There is a short approximate horizontal line between two stages of loading and unloading, which represent yieldability or flowability of marble after stricken.

All the impact constitutive curves of marble shown in figure 7 and the rock static constitutive curve shown in figure 1 obviously have elasticity before peak and brittleness after peak, but there exit obviously difference on the flowing two aspects.

1. Under the static load there exists an obvious

initial compressive portion on the stress-strain curve, but under impact load there exists elastic stage.

2. Marble has an obvious stage of yieldability or flowability under impact load, but when under static load, such stage will occur only when the surrounding compression gets more then 800 Mpa (Li , Xianwei 1983).

4. CONCLUSION

Stress-strain curves of marble generally have three stages. The first stage goes up along an approximate straight line with positive slope, which (modulus of elasticity) has some correlation with strain rate, and usually the higher strain rate is, the higher modulus of elasticity is. The second stage goes along an approximate horizontal line, which represent the plastic character of marble after being stricken. The third stage goes down along an approximate straight line with negative slop, and is not capable of rebound.

REFERENCES

J.C.Jaeger & N.G.W.Cook 1979. Fundamentals of Rock Mechanics. Third Edition. London: Chapman and Hall.

Li, Xibing & Gu Desheng 1994. Rock Impact Dynamics. Changsha: Central South University of Technology Press.

Li, Xianwei. 1983. Mechanical Properties of Rock. Beijing: China Coal Industry Publishing House.

Mining Science and Technology' 99, Xie & Golosinski (eds) © 1999 Balkema, Rotterdam, ISBN 90 5809 067 1

Real time data processing system for photoelastical mechanics

Liqian An & Wenying Yu
China University of Mining and Technology, Beijing, People's Republic of China

ABSTRACT: This paper introduces a new real time data processing system for photoelastical mechanics, in which two or more isoclinics can be recorded together, isochromatics can be erased directly, and isoclinics can be abstracted just from one image in the experiment. Compared with the traditional measurement in the photoelastics, this new system greatly shortens time of abstracting the whole-field isoclinics, and also considerably enhances the detection precision of the experiment.

1. INTRODUCTION

From the beginning of the photoelastical mechanics, how to abstract the isoclinics fringe from the whole field has been a difficult problem. Traditionally just one isoclinical fringe, which represents the one principle stress direction, can be detected at once during the experiments, so the abstracting of whole isoclinics field is very time-consuming and laboursome, the precision also can not be guaranteed. In order to overcome this problem, some auto-photoelastical system has been developed recently by using modern photoelectric technology and computer image processing methods, detection precision has been increased considerably, but the problem is that in these systems only one fringe of isoclinics is detected at one time. The methods are also less visualized.

The new method introduced in this paper is to generate light impulse on the photoelastic plane polarization system, in which both polarizer and analyzer rotate in synchronism. Through the computer controlling, each light impulse can correspond to one angle of the polarization axis, which also correspond to one isolinicis primary stress direction angle. By controlling the light impulse and recording time appropriately, we can obtain the image which includes the information of two fringes of isoclinics along with different primary stress direction angle.

The photoelastic stress image taken by this way contains both information of the isochromatics and isoclinics. So we have to separate them. The separating of the isoclinics and isochromatics has

been focused by many researchers worldwide, and a lot of methods have been put forward, such as image division, image differential, images gray value comparing method and so on. But all these method ask for two image which are taken in the same situation (one includes only isoclinics and the other include isoclinics and isochromatics). After a serial of image processing, the isoclinics can be abstracted. By the method introduced in this paper we can conduct the procedure mentioned above by using only one image, thus we will shorten the time of capturing image and data processing. This leads a great improvement of the process precision.

2. EXPERIMENT PRINCIPLE

When we place a plane-stressed model on plane polarized light field, according to photoelectric effect, the light intensity from the polarized photoelastic system is expressed as:

$$I = E^2 Sin^2 2\alpha Sin^2 R / 2$$

where:

E——plane polarized light amplitude

α——included angle of the direction primary stress $\sigma 1$ of the stressed model

R——phase difference of two plane polarized light which transmit out on the direction of primary stress $\sigma 1$ and $\sigma 2$

In terms of above discussion, when one dot is on the primary stress direction which is concordant to polarization analyzer axis ($\sigma = 0$), no light will eject from the system. All these dots will comprise

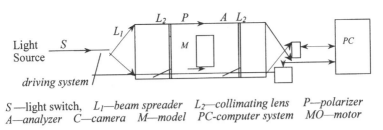

S —light switch, L_1—beam spreader L_2—collimating lens P—polarizer
A—analyzer C—camera M—model PC-computer system MO—motor

Fig 1 Experiment system sketch

one black fringe, all dots on this fringe have the same primary stress direction. This fringe is called isoclinics. Because isoclinics will occur when the primary stress direction is concordant to polarized axis, we can only observe one isoclinics of primary stress direction when we rotate isoclinics and isochromatics once at the same time. So we can't detect whole-field isoclinics at the same time but detect them one by one. While using the new method in this paper, we can get two or more pieces of isoclinics at the same time. The diagram of experiment structure is shown in Fig 1.

This new experiment system includes high-speed light switch, both polariscope and polarization analyzer rotated with high speed in synchronism, image capturing and processing system.

The experiment system is controlled by computer. At the beginning, after starting dynamo drives the polariscope and polarization analyzer rotate with high speed, a special switch makes each light impulse to correspond to one isoclinics primary stress direction, CCD (Charge Coupled Device) camera will record those isoclinics. CCD camera uses the way of line interlacing to record information. One picture has two fields: the first field is called odd field, the second field is called even field. During the odd field optical integration time, by controlling the signal concerned, working-storage section and horizontal shift register output the information of the last field (the second field of last image) line by line to the end. Then it is a turn of field blackout period. In this period, the sensitive region of pattern and register quickly transmit the charge in the sensitive region of pattern to working-storage register until the field blackout period ends. The second light integration begin, at the same time, the first field signal output. Controlling the light integration signal of CCD makes each field light integration time just to correspond to one primary stress direction angle, thus one frame picture can record two direction isoclinics at the same time. Because

two fields record two different direction isoclinics, wherever the isoclinics exists one the stress_optic image is flashing. The reason is that the gray value of two adjacent lines is greatly different. There are two stable zones between two flashing zones. These two zones correspond to another two pieces of isoclinics with two different primary stress direction angle. This image can gain four pieces of isoclinics at the same time by the computer image processing.

3. ISOCLINICS DISPLAYING AND THE SEPARATING OF ISOCHROMATICS

The image obtained according to the above method includes not only two or more isoclinics but also isochromatics. On the base of this image's characters, it's easy to separate isoclinics and isochromatics from one image. At the flashing zone, where lies the isoclinics zone, the gray value of elements which located in adjacent line has a great difference. But in the stable zone, they are almost equal to each other. Based on the above discussion we can decide whether the dot is on the isoclinics fringe or not. The dots, which match the isoclinics's condition, will give the gray value. On the other hand the dots located on the last line have written the gray value to a date file. The above processing procedure is controlled automatically by a new developed software which is written in C language. The above processing flow chat is as follows:

Setting two file: file one and file two and setting three arrays: buff1 [N], buff2 [N], buff3 [N]

Read one line information to buff1

1.read one line information to buff2

2.comparing dates in buff1 and buff2

If match the isoclinics characters: buff1=0 and buff3 0

If not match isoclinics characters: buff1=0 and buff3=0

3.buff3 =file one, buff1 =file two, buff2 =buff1

Fig 2 Isoclinics and isochromatics fringe picture

Fig 3 Isoclinics fringe centerline of 0 and 20

Repeat step 1,2,3 until the end of image. Display the information in file one and file two separately. They are the isoclinics picture and isochromatics pictures. Fig2 is the image of two kinds fringe.

4. FRINGE THINNING AND WHOLE-FIELD ISOCLINICS DISPLAYING

In this paper we use "detecting gray peak-valley value method" to abstract the fringe center line (photo centerline). At first, we build up an image element gray value display window whose center is cursor. From this window, we can know the gray value of dots which is in the window, so we find the dots whose gray value is smallest on the isoclinics fringe along width direction, then trace the smallest gray value dot and connect them. In this way we will get the isoclinics fringe center line. This method also can be used in abstracting isochromatics fringe centerline. Fig 3 is the isoclinics fringe centerline of 0 and 20 degree.

Put all direction isoclinics information of fringe centerline taken by above mentioned method to one date file and display them, we will get the fringe centerline of whole-field isoclinics. Fig 4 is the image.

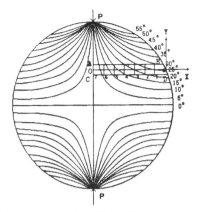

Fig 4 Image of whole-field isoclinics fringe center line

5. CONCLUSIONS

The method introduced in this paper give a way to gain two to four pieces of isoclinics with different primary stress direction angle. This method improves the speed of abstracting isoclinics greatly. It's more coincident to real experiment situation. It can acquire the whole-field isoclinics image more accurately by interpolation method.

By using image subtraction method to separate or abstract isoclinics, we have to take two pictures, which exactly correspond to each other dot by dot. While using the method introduced in this paper, only one picture which includes both information of isoclinics and isochromatics is needed. From this picture, we can separate isoclinics and isochromatics, and gain two image that include only one kind of fringe. This will decrease the errors made by the two times photos and the change of gray. This way also reduces the experiment procedure, improve the abstracting precision and speed considerably.

REFERENCES

Zhao Qingcheng photomechanics Shanghai Science and Technology Press, 10,1988

An Liqian "the Method of Photoelastic Isoclinics Displaying Together", Mechanics and Experience, 17(1), 1995

Zhang Dongsheng, Dong Jingwei, Li Hongqi, "Image Processing Technology of Photoelastrics Isoclinics Abstracting", Exp.Mech., 8(2), 1993

Tan Haoqiang, "C Language program design", Tsinghua University Press, 1995

Mining Science and Technology' 99, Xie & Golosinski (eds) © 1999 Balkema, Rotterdam, ISBN 90 5809 067 1

Application of a natural co-moving coordinate system for description of rigid body plane rolling

Yanqi Song & Zhida Chen
China University of Mining and Technology, Beijing, People's Republic of China

ABSTRACT: A new S-R finite deformation theory, first proposed in 1979 by Chen, has significantly developed in recent years and helped address problems related to the large deformation problem in the mining industry, geology, surveying and other fields. When a body undergoes a large deformation, the coordinate line embedded in the strained body transforms and initial linear orthogonal coordinate system becomes an arbitrary curvilinear coordinate system. In order to get better understanding of this new theory applied in the curvilinear coordinate system, the tensor method of describing the plane rolling of rigid object is emphasized in this paper.

1. INTRODUCTION

In underground mining, excavations and strata movements give rise to large deformations of rock mass. Complications of deformation geometry and the difference in engineering properties of the earth's material(rock) usually cause its mathematical solutions with highly nonlinear characters. Under such circumstance, the analyses of classical small deformation theory most likely leads to great errors, sometimes even no result can be obtained.

The classical nonlinear theory basing on Green's strain tensor lacks the definition of finite rotation compatible with the strain, while finite rotation is potentially important to large deformation. The Polar decomposition theorem of the deformation gradient loses its utility for nonuniqueness of decomposition due to the noncommutative property of matrix products.

On the basis of SS theorem (Green Love) and SR-RS theorem (Finger-Truesdell), the new S-R decomposition theorem (Chen, 1979) which is compatible in strain and rotation has been established, it is a revision of Stoke's S(strain)-R(rotation) decomposition theorem. In this theory, the moving coordinate method of Euler is extended to deformed body and become the rational natural coordinate (or called embedding co-moving coordinate method). This is an important progress for deformed body mechanics. The most advantage of this new S-R decomposition theorem is fit to solve both small and large deformation problems and to prevent unnecessary error which will occur in the classical finite deformation theory. In order to get better understanding of this new theory applied in the curvilinear coordinate system, the tensor method of describing the plane rolling of rigid object is emphasized in this paper.

2. THE CO-MOVING COORDINATE DESCRIPTION METHOD AND THE S-R THEOREM

In order to determine the deforming state, we select two reference frames:

(1) fixed reference system $\{X^1, X^2\}$,

(2) natural co-moving coordinate system $\{x^1, x^2\}$

embedding in the strained body.

In general cases, the initial reference frame is chosen as a rectilinear or curvilinear orthogonal system. As the body is deformed, the natural frame embedding in the body changes, consequently allowing the deformation to form a new curvilinear system.

During the transformation, We define the basic vectors before and after deformation as $\overset{0}{g_i}$ and g_i.

The transformation $\overset{0}{g_i} \to g_i$ is realized through

$$g_i = F_i^{\,j}\, \overset{0}{g_j} = \frac{\partial X^j}{\partial x^i}\, \overset{0}{g_j},$$

$$F_i^{\,j} = \delta_i^{\,j} + u^j|_i \qquad (i, j = 1,2,3)$$

where F_i^j is the tensor of displacement gradient and $u^j|_i$ is the covariant derivation of displacement component u^i with respect to co-moving coordinate x^j.

The S-R decomposition theorem proved that: *for a physically possible transformation induced by a deformable body point set, F_i^j can be decomposed into a representation of symmetrical transformation and an orthogonal transformation,* i.e.

$$F_i^j = S_i^j + R_i^j$$

where S_i^j and R_i^j are the components of strain tensor and rotation tensor respectively, which are determined as

$$S_i^j = \frac{1}{2}\left(u^i\mid_j + u^i\mid_j^T\right) - L_k^i L_j^k (1 - \cos\theta)$$

$$R_i^j = \delta_i^j + L_j^i \sin\theta + L_k^i L_j^k (1 - \cos\theta)$$

In the above δ_i^j is Kronecker identity tensor and L_j^i is the unit vector of the rotation axis.

The mean rotation angle θ is determined by the following formula

$$\sin\theta = \pm\frac{1}{2}\left[\left(u^1\mid_2 - u^1\mid_2^T\right)^2 + \left(u^2\mid_3 - u^2\mid_3^T\right)^2 \right.$$
$$\left. + \left(u^3\mid_1 - u^3\mid_1^T\right)^2\right]^{1/2}$$

and

$$L_j^i = \frac{1}{2\sin\theta}\left(u^i\mid_j - u^i\mid_j^T\right)$$

For practical computation, the tensor components must be changed into physical components.

3. PLANE POLAR COORDINATE SYSTEM

In the finite deformation co-moving coordinate system, the linear orthogonal reference has the same basic vector everywhere which is called uniform gauge geometry field. But the curvilinear reference system has different basic vectors for every point, we call it nonuniform gauge geometry field. Now we take the plane polar coordinate system as an example to explain the measure properties of curvilinear coordinate system

As the following paragraph, we define

$$x^1 = r, x^2 = \theta$$

$\mathbf{r}$ is the position vector of a point, and

$$d\mathbf{r} = \frac{\partial \mathbf{r}}{\partial x^i}dx^i = \frac{\partial \mathbf{r}}{\partial r}dr + \frac{\partial \mathbf{r}}{\partial \theta}d\theta = \mathbf{g}_1 dr + \mathbf{g}_2 d\theta$$

where $\mathbf{g}_1$ has no means of dimension, but $\mathbf{g}_2$ has the physical dimension of length.

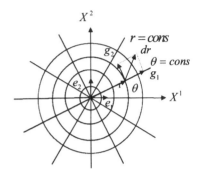

Figure 1. Plane polar coordinate system

$\mathbf{r}$ is expressed by the components of linear coordinate system as following

$$\mathbf{r} = \mathbf{e}_1 r\cos\theta + \mathbf{e}_2 r\sin\theta$$

thus

$$\mathbf{g}_1 = \frac{\partial \mathbf{r}}{\partial x^1} = \frac{\partial \mathbf{r}}{\partial r} = \mathbf{e}_1\cos\theta + \mathbf{e}_2\sin\theta,$$

$$\mathbf{g}_2 = \frac{\partial \mathbf{r}}{\partial x^2} = \frac{\partial \mathbf{r}}{\partial \theta} = \mathbf{e}_1(-r\sin\theta) + \mathbf{e}_2(r\cos\theta)$$

$$|\mathbf{g}_1| = 1, \quad |\mathbf{g}_2| = r$$

$$g_{11} = \mathbf{g}_1\cdot\mathbf{g}_1 = 1, \ g_{22} = \mathbf{g}_2\cdot\mathbf{g}_2 = r^2,$$

$$g_{12} = \mathbf{g}_1\cdot\mathbf{g}_2 = g_{21} = 0$$

$$\mathbf{g}^1 = \mathbf{g}_1, \ \mathbf{g}^2 = \mathbf{g}_2, \ \mathbf{g}^3 = \mathbf{g}_3$$

$$\left[g_{ij}\right] = \begin{bmatrix} 1 & 0 \\ 0 & r^2 \end{bmatrix}, \ g = |g_{ij}| = \begin{bmatrix} 1 & 0 \\ 0 & r^2 \end{bmatrix} = r^2, g^{ij} = \frac{G_{ji}}{g}$$

In the above formula, G_{ji} is the algebraic cofactor of determinant g. So we get

$$\left[g^{ij}\right] = \begin{bmatrix} 1 & 0 \\ 0 & \dfrac{1}{r^2} \end{bmatrix}$$

Differential quadratic expression is

$$(ds)^2 = g_{ij}dx^i dx^j = g_{11}(dr)^2 + 2g_{12}(dr)(d\theta) + g_{22}(d\theta)^2$$
$$= (dr)^2 + r^2(d\theta)^2$$

476

4. APPLICATION

When a rigid body moves in a plane, the distance between two arbitrary points remain unchanged. If a point is taken as the main point, the other point will revolve around the main point in addition to the same translation.

In the following we take an example to explain that the natural co-moving coordinate system can also be applied in the calculation of plane motion of a rigid body.

Suppose a rolling wheel moves on a rigid plane without deformation. As shown in Figure 2, the initial fixed coordinate system and co-moving coordinate system are chosen as the plane polar coordinate system.

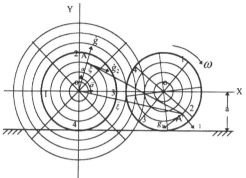

Figure 2. Polar co-moving frame

Let $\{\xi,\alpha\}$ be co-moving coordinate system, then the rolling motion makes a transformation

$$T:\{\xi,\alpha\}\rightarrow\{\bar{\xi},\bar{\alpha}\}$$

Where $\{\bar{\xi},\bar{\alpha}\}$ is coordinate of a point after motion referred to the fixed frame, the transformation is

$$\bar{\xi}=\left[\left(a\omega t+\xi\sin(\omega t+\alpha)\right)^2+\xi^2\cos^2(\omega t+\alpha)\right]^{1/2}\Big\}$$

$$\bar{\alpha}=\alpha+\omega t$$

The position vector of an arbitrary point in the wheel is

$$\mathbf{r}=\mathbf{e}_1\xi\cos\alpha+\mathbf{e}_2\xi\sin\alpha$$

After motion the vector is

$$\mathbf{R}=\mathbf{e}_1\bar{\xi}\cos\bar{\alpha}+\mathbf{e}_2\bar{\xi}\sin\bar{\alpha}$$

as

$$\overset{0}{\mathbf{g}}_i=\frac{\partial\mathbf{r}}{\partial x^i},\quad\mathbf{g}_i=\frac{\partial\mathbf{R}}{\partial x^i}$$

we obtain

$$\frac{\partial\mathbf{r}}{\partial\xi}=\overset{0}{\mathbf{g}}_1=\mathbf{e}_1\cos\alpha+\mathbf{e}_2\sin\alpha$$

$$\frac{\partial\mathbf{r}}{\partial\alpha}=\overset{0}{\mathbf{g}}_2=\mathbf{e}_1\left(-\xi\sin\alpha\right)+\mathbf{e}_2\xi\cos\alpha$$

moreover

$$\mathbf{g}_1=\frac{\partial\mathbf{R}}{\partial\xi}\ ,\mathbf{g}_2=\frac{\partial\mathbf{R}}{\partial\alpha}$$

$$\overset{0}{g}_{ij}=\overset{0}{\mathbf{g}}_i\cdot\overset{0}{\mathbf{g}}_j\qquad g_{ij}=\mathbf{g}_i\cdot\mathbf{g}_j$$

So the metric tensor in the initial frame and the moving frame are as following

$$\left[\overset{0}{g}_{ij}\right]=\begin{bmatrix}\overset{0}{g}_{11}&\overset{0}{g}_{12}\\\overset{0}{g}_{21}&\overset{0}{g}_{22}\end{bmatrix}=\begin{bmatrix}1&0\\0&\xi^2\end{bmatrix}$$

Because of rigid motion, no metric changes. We have

$$g_{ij}=\overset{0}{g}_{ij}$$

So we get

$$[g_{ij}]=\begin{bmatrix}1&0\\0&\xi^2\end{bmatrix}$$

The velocity of a point A in the rolling wheel is

$$\mathbf{V}=\mathbf{g}_1 a\omega\sin(\omega t+\alpha)+\mathbf{g}_2\left(\omega+\frac{a}{\xi}\omega\cos(\omega t+\alpha)\right)$$

$$=V^i\mathbf{g}_i\qquad\qquad(i=1,2)$$

The covariant derivatives of V^j with respect to $x_i\left(x^1\equiv\xi,x^2\equiv\alpha\right)$ in the real time co-moving system are $V^j\|_i$

$$\frac{\partial\mathbf{V}}{\partial x^i}=V^j\|_i\,\mathbf{g}_j,\quad V^j\|_i=\frac{\partial V^j}{\partial x^i}+\Gamma_{ik}^j V^k$$

with $V^j\|_i$ being the covariant derivative of velocity component V^j with respect to the co-moving coordinate x^i in real time local system.
Where

$$\Gamma_{jk}^i=\frac{1}{2}g^{il}\left[\frac{\partial g_{li}}{\partial x^k}+\frac{\partial g_{lk}}{\partial x^j}-\frac{\partial g_{jk}}{\partial x^l}\right]$$

is the Christoffel symbol of the second kind. We obtain

$$\Gamma_{11}^1=\Gamma_{12}^1=\Gamma_{21}^1=\Gamma_{11}^2=\Gamma_{22}^2=0$$

$$\Gamma_{22}^1=-r,\quad\Gamma_{12}^2=\Gamma_{21}^2=\frac{1}{r}$$

477

hence

$$\left[V^{j}\|_{i}\right]=\begin{bmatrix}V^{1}\|_{1} & V^{1}\|_{2}\\ V^{2}\|_{1} & V^{2}\|_{2}\end{bmatrix}=\begin{bmatrix}0 & -\xi\omega\\ \dfrac{\xi}{\omega} & 0\end{bmatrix}$$

and the physical component of $V^{j}\|_{i}$ is

$$\hat{V}^{j}\|_{i}=\sqrt{\frac{g_{(jj)}}{g_{(ii)}}}V^{j}\|_{i}$$

where $(ii),(jj)$ indicate no sum over the double index. We have

$$\left[\hat{V}^{j}\|_{i}\right]=\begin{bmatrix}\hat{V}^{1}\|_{1} & \hat{V}^{1}\|_{2}\\ \hat{V}^{2}\|_{1} & \hat{V}^{2}\|_{2}\end{bmatrix}=\begin{bmatrix}0 & -\omega\\ \omega & 0\end{bmatrix}$$

By the kinematics equation:

$$\hat{V}^{i}\|_{j}=\hat{L}^{i}_{j}\dot{\theta}+\hat{S}^{i}_{j}$$

$$\hat{L}^{i}_{j}\dot{\theta}=\frac{1}{2}\left(\hat{V}^{i}\|_{j}-\hat{V}^{j}\|_{i}\right)\quad \hat{S}=\frac{1}{2}\left(\hat{V}^{i}\|_{j}+\hat{V}^{j}\|_{i}\right)$$

The superscript $\wedge$ means physical component. For plane rotation, the axis direction is:

$$L^{2}_{1}:L^{1}_{3}:L^{3}_{2}=1:0:0$$

Therefore the angular velocity $\dot{\theta}$ and the strain rate $\hat{S}^{i}_{j}$ are

$$\dot{\theta}=\omega,\quad \hat{S}^{1}_{1}=\hat{S}^{2}_{2}=\hat{S}^{1}_{2}=\hat{S}^{2}_{1}=0$$

i.e. strain everywhere in the wheel is equal to zero. At the contact point

$$\xi=a,\quad \alpha=180^{0}-\omega t,\quad V=0$$

For large deformation as shown in figure 3, we can obtain the results by similar operation process. At the contact region of a circular cylinder, we have

$$\sin\theta=-\frac{1}{2}\left(1-\frac{v_{0}}{a}\right)\frac{\sin\alpha}{\cos^{2}\alpha}$$

$$\hat{S}^{r}_{r}=\cos^{2}\alpha\left[\frac{1-v_{0}/a}{\cos\alpha}+\frac{\sin^{2}\alpha(1-v_{0}/a)}{\cos\alpha}-1\right]$$
$$+\left(1-\cos\theta\right)$$

$$\hat{S}^{\alpha}_{\alpha}=-\sin^{2}\alpha+\left(1-\cos\theta\right)$$

$$\hat{S}^{\alpha}_{r}=-\frac{1}{2}\sin2\alpha\left[1-\frac{1-v_{0}/a}{2\cos^{2}\alpha}\right]$$

where v_{0} is maximum vertical displacement. At the point $r=a,\alpha=0,\theta=0$, we have

$$\begin{bmatrix}\hat{S}^{r}_{r}\\ \hat{S}^{\alpha}_{\alpha}\\ \hat{S}^{\alpha}_{r}\end{bmatrix}=\begin{bmatrix}-v_{0}/a\\ 0\\ 0\end{bmatrix}$$

From the above, we know co-moving coordinates also apply for large deformation.. The detailed discussion of this issue will be presented in another paper.

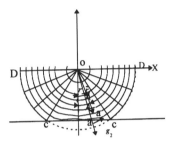

figure 3 $T:a(r,\alpha)\rightarrow a'(\bar{r},\bar{\alpha})$

5. CONCLUSIONS

Under the condition of orthogonal transformation, the formula for rigid body rotation in classical mechanics can also be derived from the S-R decomposition theorem. So the new S-R finite deformation theory is suited for solving both small and large deformation problems.

REFERENCE

Chen Zhida, 1988. Rational Mechanics (in Chinese). Xuzhou, China: China Mining University Press.
Xie,H., 1993. Fractal in Rock Mechanics, A. A. Balkema Publishers, Rotterdam
Chen Zhida, 1986. On the representation of finite rotation in nonlinear field theory of continuum mechanics, *J. of Applied Mathematics and Mechanics*, Vol7, No.11, 1017-1026.
Chen Zhida, 1992 The mathematical theory of finite deformation applied to contact mechanics-Nonlinear theory of contact mechanics, *International Symposium Of Contact Mechanics, Lausann.*,
Chong Wang,Antnio Airton Bortolucci & Tarcsio Barreto Celestino 1998. Local rotation gradient applied to determine fracture locations in rock masses. *In proceeding of international congress of Computational Mechanics, Spain.*

5 Coal processing and clean coal technology

Mining Science and Technology '99, Xie & Golosinski (eds) © 1999 Balkema, Rotterdam, ISBN 90 5809 067 1

Research on triboelectric beneficiation of ultrafine coal

Qingru Chen, Xinxi Zhang, Zengqiang Chen & Zhenlian An
China University of Mining and Technology, Xuzhou, People's Republic of China

Jicai Qian
Qitaihe Beidou Chemical Plant, Heilongjiang, People's Republic of China

ABSTRACT: The tribocharging relation between mineral components in coal and stainless steel and the effect of the environment humidity and friction time on electrification are researched and analyzed in this paper. Tests were conducted in both bench-scale and pilot system based on basic study. Test results demonstrated the feasibility of deep removal of ash and pyrite from fine coal. Scale-up design of the system was successful.

1. INTRODUCTION

The deep removal of pyrite and ash to produce clean coal is one of the development trends in coal preparation. For deep removal of pyrite and ash minerals, coal must be ground to liberate minerals from coal, so the preparation of ultra-fine coal is the key link for producing superclean coal.

Wet processing is currently the main method of fine coal treatment, which requires a considerable investment in both processing and dewatering treatment. On the contrary, triboelectric processing hasn't the problems associated with wet method.

2. THE BASIC RESEARCH OF TRIBO-ELECTRIFICATION

2.1 *The mineral components of coal*

The mineral components of coal usually consist of kaolinite, pyrite, siderite, quartz, calcareous spars, gypsum, etc. Research on the charging process of minerals and coal is the presupposition of triboelectrification beneficiation.

The mineral components in coal are classified into semiconductor and insulator according to their electric conductivity. Therefore, the contact charging between the mineral components and a metal is classified into semiconductor contact and insulator contact charging.

2.2 *Triboelectrification of mineral components of coal*

2.2.1 *The study of minerals' original charge*

The original charge of a mineral particle is defined as its equilibrium charge, which lays statically on natural condition in air. Amount of tribocharge of a particle is equal to the difference of its original charge and that after contact with another material. So, the measurement of original charge of minerals is the presupposition of measuring tribocharge. Measurement apparatus used composed of an EST111 digital static charge meter and Faraday cages. Original charge of three coals with different coalification rank (ash <3%) and six minerals were firstly measured. The measure results are shown in Table 1.

It can be seen from Table 1 that the measure results are very small under different temperature and relative humidity. Considering the measurement error, the original charge of mineral components are considered to be zero, since minerals are composed of electrical neutrality atoms in general. Even some ions exist in minerals, they are neutralized by charged particles in air adsorbed on them.

The study of triboeletrification between mineral components in coal and metal is one of presupposition for understanding the unlike charged mechanism of coal and mineral in tribo-electrification apparatus. Conducted had been the experiments of triboelectrification between three coals and stainless steel and between six minerals and stainless steel.

Table 1. The original charge of mineral components under different temperature and humidity (PC*)

Temperature(°C)	8.5	26.5	26.5	27	28	29
Relative humidity(%)	43.2	56	53.2	54	55	58.5
Quartz	-1	0	-2	0	0	-1
Siderite	0	1	0	1	0	0
Gypsum	-1	2	-1	0	0	-1
Calcareous spar	1	2	-1	-1	0	-1
Kao linite	-2	0	-1	-1	1	1
Pyrite	1	0	-1	0	1	-1
Longflame coal	2	1	0	1	0	2
Meagre-lean coal	-2	1	-1	0	0	1
Anthracite	0	2	1	-1	0	1

* 1PC=10^{-12} C

Table 2. The triboelectrification of mineral component (PC)

Relative humidity(%)	39	41	44.5	51.5	55	57	58
Quartz	-367	-321	-177	-57	-5	4	20
Siderite	-357	-337	-263	-174	-145	-105	-86
Gypsum	-1207	-863	-626	-225	-143	-56	-40
Calcareous spar	-822	-687	-664	-519	-353	-284	-207
Kaolinite	-103	-85	-58	-39	-35	-27	-22
Pyrite	-30	-24	-22	-15	-13	-14	-10
Longflame coal	-5260	-4600	-49.70	-4520	-3340	-2995	-2250
Meagre-lean coal	574	533	483	390	327	302	226
Anthracite	-2200	-1934	-1557	-1210	-877	-800	-780

Temperature: 20.5°C   Friction time:15sec.

Table 3. The charges of minerals under different friction time

Friction time (sec.)	15			30			45		
Relative humidity (%)	48	50	52	48	50.5	61	56	58	59.5
Quartz	-70	-63	-19	-70	-46	18	8	15	26
Siderite	-283	-230	-203	-290	-209	-51	-177	-133	-72
Gypsum	-566	-376	-252	-537	-355	-22	-97	-50	-25
Calcareous spar	-505	-500	-375	-403	-383	-97	-245	-186	-103
Langflame coal	-4250	-4260	-3150	-4235	-3250	-1830	-2470	-2050	-2040
Meagre-lean coal	262	253	170	276	260	176	208	170	162

Temperature: 20.5°C

Table 4. Test results for coal of different rank

Coal rank	Raw coal ash %	Cleaned coal		Tailings	
		Yield (%)	Ash (%)	Yield (%)	Ash (%)
Langflame coal	4.08	66.04	1.98	33.96	8.16
1/3 caking coal	9.84	58.64	3.78	41.36	18.43
Bottle coal	3.22	42.15	1.85	57.85	4.22
Meagre coal	12.50	56.26	4.98	43.74	22.17
Anthracite	14.06	60.31	6.85	39.69	25.02

2.2.2 Triboelectrification between mineral components in coal and metal.

2.2.2.1 Influence of the relative humidity of air on triboeletrification

In the process of experimental study for coal and mineral particles, stainless steel cup was used as the friction surface. The tribocharging between particles and surface was obtained by pouring particles into a Faraday cage from stainless steel cup directly or after a period of friction time. In order to avoid the influence of conductive induction charge, the stainless steel cup was hung and laid in the air by clamping apparatus and insulation pole, in which the particles are stated with original charge. When measuring charge, hands only get in touch with the end of insulation pole. The measure results of triboelectrification of above mineral components are shown in Table 2.

It has been shown that the charges of coal and mineral components indicated obviously variation except quartz, that the magnitude of charge is decreased monotonously with the increasing of environmental relative humidity and the charged polarity does not change. The charged polarity of quartz varied from negative to positive while the relative humidity was over 55%

2.2.2.2 *Influence of friction time on triboelectrification*

In order to research the influence of friction time on triboelectrification, the charges measure of six components is conducted under different friction time. The measure results are shown in Table 3. Because of that friction between particles and stainless steel cup have a certain random and crushing phenomenon, the results of many times measuring repeatedly for the some mineral aren't completely alike. Table 3 shows that the charges have little obvious variation, and it indicates that the charged of mineral would reach equilibrum charges after passing a certain friction time[1].

3. TEST RESULTS

3.1 *Laboratory tests*

On the basic of theoretical study, triboelectric separation tests of ultra-fine coal in a model triboelectric separator had been done. Test parameters included factors of tribocharger structure, rate of air flow, feed density and electric field strength.

Tribocharger is the key part of the separator which should allow the particles fully charged. The influence of tribocharger length on separation is shown in Figure 1. It can been seen that a long tribocharger will improve the separation.

High air velocity gave a good charging effect on particle and then a good separation result. The test result is shown in Figure 2.

Test results given in Table 4 show that triboelectric separation is adaptive for cleaning of many coals.

3.2 *Result of pilot test*

A pilot system of triboelectric separation was built and experiments conducted for verification of separator scale-up. Capacity of the pilot separator was 50kg/h. Figure 3 is one of the test results.

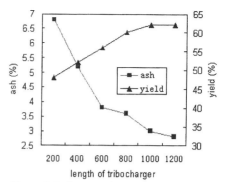

Figure 1 Influence of tribocharger length on separation

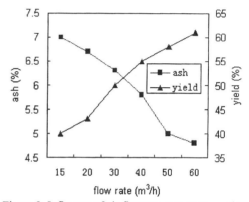

Figure 2 Influence of air flow rate test on separation

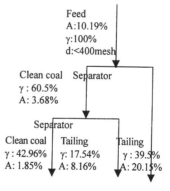

Figure 3 Flowsheet test of triboelectric separation

Pilot test results denoted that the scale-up of separator did not affect the separation efficiency.

4. CONCLUSIONS

1. The equilibrium charges of mineral components in static atmospheric condition are zero.

2. The triboelectric charging of different particles shows that: meagre coal attains positive charge in the testing range of humidity and temperature; quartz attins negative charge in lower humidity and , in higher humidity, postive charge; other particles attain negative charge.

3. With the adsorption of water molcules on particle surface, their surface state and electric conductivity are changed. Triboelectric charging of each mineral component varys with the variation of environment humidity, that the charges decrease with the increase of environment humidity.

4. Increasing friction time could generally increases the quantity charge change obviously, however while the quantity of charge approaches to equilibrium point, the charges are hardly changed.

5. The laboratory test result shows it is feasibility to remove pyrite and ash from ultra-fine coal use triboelectric separator.

6. The pilot test result shows the scale-up of separator was successful and provides data for further industrial tests.

REFERENCES

An Zhenlian, Chen Qingru and Zhang Xinxi, Journal of China University of Mining & Technology, 1997 Vol.2 No.1:71~74

Mining Science and Technology' 99, Xie & Golosinski (eds) © 1999 Balkema, Rotterdam, ISBN 90 5809 067 1

On-line forecast of raw coal washability curves

Zhenchong Wang & Maixi Lu
China University of Mining and Technology, Beijing, People's Republic of China

Jing Liu
Tangshan Branch, Central Coal Mining Research Institute, People's Republic of China

ABSTRACT: To specific coal preparation plant, the ash content of raw coal is related to its density. The density, in turn, is related to elementary ash content. Based on these relations an on-line forecast system of raw coal washability curves was developed. It uses Windows with C++. The system can define the washability curves automatically and in real-time. The experiments demonstrate that for cumulative yield, forecast error is less than 0.4 and for cumulative ash, forecast error is less than 0.7.

1. INTRODUCTION

The implementation of process automatic control is important to raising the separation efficiency of coal preparation plant. Gravity separation is about 80 % of processing capacity of coal preparation plant in China, the automatic control of which is key to the automation of the whole plant. On-line forecast of the washability of raw coal is an important part in the automatic control of gravity separation process.

On the forecast of washability of raw coal, authors studied in detail. Analyzing the data of raw coal of the same mine and plant indicates that the cumulative float yield of specific gravity fraction is correlative to the raw coal ash, and the elementary ash of coal is correlative to its density fraction. Further more, they analyzed 75 groups of raw coal data of 6 mines, which have different kind of washability, the conclusion was confirmed (Lu 1996).

On the base of the result, an on-line forecast system of raw coal washability is developed under Windows.

2. MATHEMATICAL MODELS FOR FORECAST

2.1. The correlative relationship between the ash content of raw coal and the density consists of it

The research result (Lu 1996, Liu etc al 1998) indicates that for the same coal body, there exists linear correlation between the ash content of raw coal and its density fractions <1.8, <1.6, <1.5, and <1.4:

$$R_i = a_i + b_i A \tag{1}$$

where: R_I = cumulative float yield of density fraction i; A = ash content of the raw coal; a_i, b_I = regressive coefficients of density fraction i.

2.2. The correlative relationship between elementary ash of raw coal and the density consists of it

The correlative relationship between elementary ash of raw coal and its density consists is as follows

$$\lambda = a_0 + a_1 g + a_2 g^2 \tag{2}$$

where: λ = elementary ash; g = density; a_0, a_1, a_2 = correlative coefficients.

2.3. Fitting models for density curve and cumulative curve

The following models are used for the fitting of density curve and of cumulative float curve (Lu 1990):

Arctangent:

$$Y = 100[t_2 - arctan(k(x-c))]/(t_2-t_1) \tag{3}$$

Hyperbolic tangent:

$$Y = 100[a + c*th(k(x-x_0))] \tag{4}$$

Modified hyperbolic tangent:

$$Y = 100[a + bx + c*th(k(x-x_0))] \tag{5}$$

$$Where:\ Th(u) = (e^u - e^{-u})/(e^u + e^{-u})$$

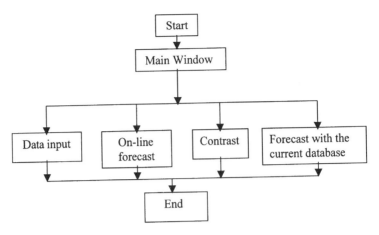

Figure 1. Diagram of the program

where: Y = cumulative float yield of density i or cumulative float yield of certain cumulative ash; x = density or cumulative ash; others are the parameters of models.

3. ALGORITHM OF ON-LINE FORECAST

3.1 *Obtain on-line forecast models*

The forecast models are obtained with the following steps:

1. Make regressive analysis and fitting by using raw coal database, form the mathematical models for forecast and store them in forecast information base.

2. Check the accurate of the models with given data.

3. Considering the changes of the characteristics of raw coal with time, the data in raw coal database will be scrolled up and refreshed when the forecast error is out of the permitted value, and establish forecast models again.

3.2 *The forecast of the washability curves of raw coal*

According to the source of raw coal ash, there are three types of forecast.

1. Key input ash. Forecast the washability of raw coal at given ash.

2. Auto ash-scan interface. Through the communication port of computer read in the ash data with given interval and make real-time forecast.

3. Ash generator. Determine the range of raw coal ash by the raw coal database and produce ash data by random generator, shows washability curves dynamically. It is mainly for demonstration.

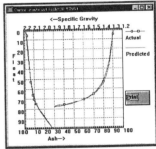

Figure 2. Contrast

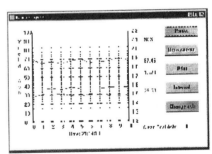

Figure 3. Tendency

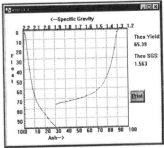

Figure 4. Forecast

4. THE DESIGN AND FUNCTIONS OF THE FORECAST SYSTEM

According to the discussion in the last section, the program is designed, shown in Figure 1.

The forecast system of the washability curves of raw coal is written with C++ under Windows. It consists of four parts.

1. Establish and maintain raw coal database. Input groups of practical float-sink data, establish raw coal database, which is for making forecast models later. The database can be modified, appended and deleted dynamically. For the accurate of forecast, there must be 8 or more groups of data in the raw coal database.

2. Set up forecast models. According to the raw coal database, calculate the parameters of model (1) and (2) in section 1. Then, the system will select automatically the optimum models for the density curve/cumulative float curve by least square method. The parameters of the models and the relative information will be stored in forecast information database for use later.

3. Check the accurate of the models. For checking the accurate of the models, the function of forecast of historic data is provided in the forecast system. The program will draw the practical and forecast curves at the same screen and print out the data (see Figure 2)

4. Display and refresh the washability curves dynamically: shows the tendency figure of raw coal ash, the tendency figure of theoretical yield, theoretical separate density and washability curves in real-time (see Figure 3, 4). The sample interval can be changed at any time manually. The forecast models come from forecast information database.

When the forecast error is over the permitted value, the system will scroll up the data in the raw coal database and set up the forecast models again.

The forecast system is Windows MDI style.

5. EXPERIMENTS

With the data of Ling Xi coal preparation plant, Kai Luan mining bureau (Jan.-Oct., 1990), we use the forecast system to set up the models and forecast the washability of November and December. The results are shown in Table1, 2.

Table 1. The practical and forecast results of cumulative float yield of density fractions

	-1.4		-1.5		-1.6		-1.8	
	practical	forecast	practical	forecast	practical	forecast	practical	forecast
1	30.88	30.82	38.97	37.42	42.95	41.79	48.03	47.21
2	30.83	31.16	38.65	37.78	42.52	42.16	48.60	47.59
3	32.64	32.05	39.57	38.74	44.19	43.14	49.16	48.59
4	32.22	34.75	41.28	41.61	45.94	46.10	51.86	51.61
5	36.19	36.34	44.84	43.32	48.59	47.86	54.13	53.40
6	31.57	31.51	37.96	38.16	42.40	42.54	48.11	47.98
7	34.55	35.79	42.40	42.73	46.93	47.25	53.34	52.78
8	31.80	31.37	37.46	38.00	41.26	42.39	48.04	47.82
9	31.69	31.57	37.41	38.22	40.59	42.61	47.93	48.04
10	36.66	36.32	44.92	43.29	48.35	47.83	52.76	53.37
11	34.10	34.93	40.65	41.81	45.57	46.31	50.48	51.82
12	36.77	36.81	43.77	43.82	48.03	48.37	54.76	53.93

Table 2. The practical and forecast results of cumulative ash content

	-1.4		-1.5		-1.6		-1.8	
	practical	forecast	practical	forecast	practical	forecast	practical	forecast
1	8.94	8.92	11.61	10.84	13.32	12.68	16.41	16.16
2	8.70	8.43	11.17	10.83	12.94	12.65	16.46	16.12
3	8.96	8.57	11.15	10.80	13.30	12.58	16.36	16.01
4	9.15	8.82	11.81	10.71	13.68	12.38	16.85	15.70
5	8.83	8.52	11.27	10.66	12.74	12.28	15.55	15.54
6	9.45	8.93	11.46	10.82	13.38	12.63	16.70	16.07
7	8.23	8.50	10.62	10.68	12.51	12.31	16.08	15.60
8	8.31	8.14	10.18	10.82	11.91	12.64	16.24	16.09
9	8.18	8.30	10.21	10.81	11.68	12.62	15.94	16.08
10	8.08	8.52	9.77	10.66	11.32	12.28	14.91	15.54
11	7.56	8.10	9.60	10.70	11.90	12.37	15.81	15.68
12	7.89	8.21	9.98	10.65	11.78	12.25	15.56	15.50

The calculation indicates that for cumulative yield, the forecast error is less than 0.4 and for cumulative ash, the forecast error is less than 0.7.

6. CONCLUSIONS

On the base of the forecast models of raw coal washability the corresponding forecast system is studied and developed. The forecast system is examined with practical data from Ling Xi coal preparation plant. The forecast system can be used for on-line forecast and computer simulation of raw coal washability.

REFERENCES

Cai Mingzhi 1995. Programming of Borland C++ 4.0. *Tsinghua university press.*

Liu Jing, Lu Maixi etc al 1998. Predicting the washability of raw coal from its total ash content. *XIII International Coal Preparation Congress.* Volume II, Brisbane: Australia.M. Lu 1990. Computer application in coal preparation industry in China. *XXII International Symposium APCOM.* Berlin.

Lu Maixi, Liu Jing 1996. Study on the relationship of raw coal ash and raw coal float consist. *Coal preparation technology.* Vol.5.

Mining Science and Technology'99, Xie & Golosinski (eds) © 1999 Balkema, Rotterdam, ISBN 90 5809 067 1

Superimposed-pulsation jig: Development and performance tests

Kunliang Yao, Jing Liu, Kang Yang, Dahai Yang, Yun Lin, Xuejun Yan, Jie Lu, Xing Xu & Xuexin Gao
China Coal Research Institute, Tangshan Branch, People's Republic of China

Abstract: Jig is the most widely used piece of coal cleaning equipment in China. In order to optimize the effect of pulsating current on jigging bed, to allow the particles to undergo perfect stratification according to density, the SKTFZ superimposed-pulsation jig has been developed in China. It differs from similar equipment in that it has a twin air-supply system supplied from one air source. Through rational regulation of air pressure and air quantity of the main and auxiliary pulsation cycles, and operating with frequency and cycle values defined in numerous tests, a remarkable improvement in jig performance has been obtained.

1. INTRODUCTION

A suitably loosened jigging bed constitutes a prerequisite condition for the particles to undergo stratification and transposition. A ideal pulsation curve within a pulsating cycle can be obtained in the following manner: to attain a greater acceleration of flow of water current at the initial air intake period, so that to enable the bed to rapidly lift up from the bed-plate through overcoming the inertia between particles, and after obtaining an appropriate bed mobility, a certain velocity of upward water current should be maintained to keep the desired bed mobility. The longer the time for maintaining the bed mobility within a cycle, the better is the stratifying result and the faster the stratifying speed.

Use of the numerical-controlled air valves with fast opening speed can cater to the energy requirement at the initial stage through gaining a rapid pressure rise in the air chamber, making therefore the air admission time possible to be flexibly and conveniently adjusted. However after air admission the bed tends to expand, causing a faster attenuation of the velocity of upward water current. To tackle this problem, several small impulses are superimposed onto the numerical-controlled jigging cycle. This is aimed at providing some additional energy to maintain the bed mobility after the bed is rapidly lifted. The essence of superimposed-pulsation lies in solving the contradiction of energy requirements in different phases. The energy in this case is released in a timed manner, and a general energy supply system is changed to one featuring supply of limited energies at several local places. This is designed to meet the different energy requirements in different phases of a jigging cycle. Through superimposed pulsation, the loosening time of material in the jig can be relatively prolonged, extending consequently the time required for the material to undergo stratification, and hence effectively enhancing the stratification precision and capacity of the jig.

Coal is stratified in a jig under the effect of pulsation of water current powered by compressed air in the air chamber. The stratification effect is directly governed by water current pulsation characteristic and the law of variation of bed mobility.

2. OUTLINE OF SKTFZ JIG AND INDUSTRIAL TEST RESULT

The SKTFZ superimposed-pulsation jig can be used for treating 100-0mm raw coal with a capacity as high as 13-20t/ m² per hour.

The jig is provided with a single air source, unique superimposed-pulsation air valves and air supply system. The jig features a U-shape air chamber beneath the screen-plate, supply of water across the total width of wash-box, modular structure, hopper type quietly operating impeller discharge mechanism and automatic discharge control system. The jig is highly adaptable and can be incorporated

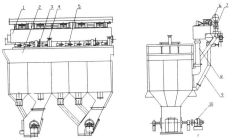

1. Air box   2. Washbox   3. Float   4. Follow-up overflow weir   5. Butterfly valve   6. Auxiliary air pipe
7. Auxiliary air valve   8. Air pipe   9. Water pipe
10. Discharge mechanism

Figure 1 Schematic diagram of the SKTFZ Jig

into the existing jigging circuit without any major modification. Therefore the jig is highly suitable for updating and renewal of existing jigs.

SKTFZ Jig has passed the strict inspection on all aspects made by China National Quality Supervision and Testing Center for Coal Preparation Plants and has also passed the technical appraisal made by the former Ministry of Coal Industry. As evidenced by field operation over a year, the jig is rational in design, advanced in performance and reliable in operation, and is up to the expected standards.

Industrial test of the jig was conducted in 1997 at the Liangzhuang coal preparation plant, Xinwen Mining Administration. The plant originally used the LTX 14 m² jig. In 1996 the jig was replaced by a SKTFZ Jig.

After determination of the optimum air valve operating parameters through numerous routine observations, industrial test runs of the jig were made. The results are listed in Table 1.

Table 1. Result of industrial test.

Jig type	Refuse (1st stage)		Middling (2nd stage)		Organic efficiency
	I	Ep	I	Ep	□ (%)
LTX 14	0.13	0.11	0.20	0.11	87.67
SKTFZ	0.06	0.07	0.16	0.07	93.93

When used for treating extremely difficult-to-wash coals with a content of ±0.1 near-density material of 40%, the organic efficiency, yield of clean coal are respectively increased 6.26% and 3.38%. The content of coal in refuse is reduced to 2.23%. The capacity of SKTFZ is 80% higher than the original jig. Other advantages include a reduction of 50% of water. The results gained, if converted to economic benefit, are equivalent to 3.18 million Yuan a year for Liangzhuang plant.

3. MEASUREMENT AND ANALYSIS OF MAIN OPERATING PARAMETERS

In order to measure the effect of the jig and to analyse the operating principle, the main parameters, especially the water and air related parameters have been measured and checked with the jigging parameter monitor developed through our own effort. The monitored parameters are shown in Figure 2.

The monitor used is based on the STD V40 System II, which is compatible with IBM PC, and provided with clock and watchdog circuit. The system structure of the monitor is schematically shown in Figure 3.

After the industrial controller is power on, and upon turning on the start/stop switch, the monitor will start to collect the measurement signals and store them into a semi-conductor disk. At the same time the monitor will give a real-time display. The sampling rate is 30 times per second, and analog signals from a maximum of 36 points can be simultaneously collected. The monitor can be operated independently of the keyboard. After completion of signal collection, the sampled signals will automatically transfer from the semi-conductor disk into the diskette. The stored data can be displayed on PC for observation, analysis, editing and printing-out of waveforms of various parameters.

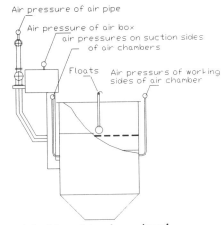

Figure 2 Position of signals monitored

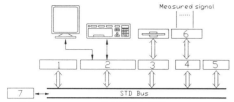

1. V40 system card 2. VGA/KB support card
3. Diskette drive support card 4. A/D card
5. Semi-conductor diskette 6. Interface card 7.
Monitoring and control software

Figure 3 Schematic diagram of STD jig parameter monitor.

For facilitating analysis, the sampled signals are processed. After shifting the signal values in each group, the values can be printed on unified coordinate paper on which the values are grouped together.

The results measured of the jig working with and without superimposed pulsation are shown respectively in Figure 4 and Figure 5. The measurements were made on October 20, 1997, with the numerical-controlled air valves to provide the main pulsation with a frequency of 50 and the rotary air valve to provide auxiliary pulsation with a frequency of 150. The sampling rate was 70/cycle. The signal represented by each curve is as follows:

1-sampled signal from the 2nd-stage float;
2-sampled signal from the 1st-stage float;
3-air pressure signal from air pipe;
4-air pressure signal from air box;
5-air pressure signal from the air chamber (air-intake side) for the 1st cell;
6-air pressure signal from the air chamber (working side) for the 1st cell;
7-air pressure signal from the air chamber (working side) for the 3rd cell.

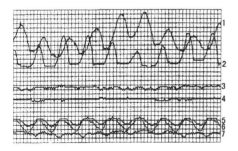

Figure 4. Curves of parameters measured (without superimposed pulsation)

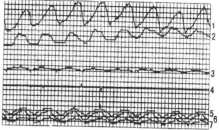

Figure 5. Curves of parameters measured (with superimposed pulsation)

It can be seen from Figure 4 and 5 that:

1) When superimposed pulsation does not existed, the motions of the 1st and 2nd-stage floats assume a sine pattern. The 1st-stage float lowers rapidly, and for a certain period of time, it remains at the lower position. When auxiliary pulsation is superimposed, both the 1st and 2nd-stage floats will move upward with reduced fluctuations. The bed is found, in this case, to move forward in a loose state. Moreover the uprising period of the bed in the first stage is prolonged and the falling of the bed is quickened while for the bed in the second stage, there exists evidently a quiescing period after the bed is fully loosed, with noticeably lengthened loose time.

2) The air pressures in the air pipe and air box remain in a non-fluctuated state, with the air pressure of the former slightly higher than that of the latter. It indicates that to provide a high and low-working pressure with one air source is feasible and can be successfully effected.

3) The intensity of the signal from the air-intake side of the air chamber for cell one is higher than that from the working side of the same chamber, and a leading phase can be observed. It is evident that the air pressure of the air intake side is higher, and a certain period of time is required for the pressure to transfer to the other side. In case of a larger pressure difference and a longer delay time, non-uniform pulsation of the bed would occur across the width of the jig, jeopardizing as a result the stratification and separation result in severe cases. This is a problem that should be avoided in design and start-up of a jig. However as viewed from the motion of the bed of a superimposed-pulsation jig, there is no distinct difference in pressure and delay time. Generally in jigging operation, less air with lower pressure is required for stage 2 than for stage 1. Hence the intensity of the signal from the working side of the air chamber for the 2nd stage should be lower than that from the 1st stage. Nevertheless with superimposed pulsation, the air

pressure in the air chambers tends to vary acutely. This can be reflected by the bed where pulsation with small amplitude yet high frequency can be observed along with the upward and downward motion of the float -- a phenomenon helpful to weaken the active forces between particles and water and between particles for quicker stratification and separation.

4) The result of measurement is in good agreement with that of industrial test, indicating that the SKTFZ Jig works with a pulsation waveform fit for separation of coal. The jig proves to be able to attain the expected performance.

4. CONCLUSIONS

1. The SKTFZ jig is high in separation precision, organic efficiency and capacity, and low in water consumption and noise level.

2. The measurement and testing of jigging parameters offer an important means for making investigations into the separating performance, principle of operation and theory of jigging separation of the novel jig. Sample analysis shows the SKTFZ jig can produce an ideal pulsating waveform, conducive to stratification and separation of coal.

3. The SKTFZ jig is designed according to the characteristics of coal preparation plants in China. As it uses only a single air source, the system is simple in design, suitable for updating and technical renovation of jigs used in various kinds of coal preparation plants. There is no need to make reinvestment on system remoulding or add any other equipment. The Jig has a broad prospect for widespread application.

REFERENCES

J. Onodera et al. 1994 Aspects on the Plant-Scale Experiments of "Variwave" air-pulsated jig at Mitsui Miike Coal Preparation Plant. *The 12th Int. Coal Prep. Congr., Cracow. Poland.*, May 23—27.

Zhang Yongzheng. 1984 Oscillometric technology for jig and preliminary result analysis. *Proc. of the 1st Gravity Separation Technology Seminar.*

Mining Science and Technology '99, Xie & Golosinski (eds) © 1999 Balkema, Rotterdam, ISBN 90 5809 067 1

The fractal characteristic of particle size distribution in coal grinding process[1]

Fangui Zeng
Department of Geoscience, Taiyuan University of Technology, People's Republic of China

Shuquan Zhu & Zuna Wang
China University of Mining and Technology, Beijing, People's Republic of China

Abstract: Fractal particle size distribution (FPSD) is a powerful tool for characterizing particle population. Researchers have used it to characterize the particle population of bulk material in nature and crushed form. In our work a coal has been comminuted in an experimental ball mill and the ground product on different grinding time are characterized by FPSD. It was discovered that the fractal dimension (D) of FPSD of the product at different grinding time was similar, which denoted that the fractal dimension of FPSD can be used to characterize the self-similarity of coal comminution. For same maceral constituent the D value of FPSD was almost unchanged with the grinding time, but distinct for different marcral particles. The mineral particle has smallest D and that of inertinite is larger than vitrinite's. It reflects the distribution of macerals in coal.

1. INTRODUCTION

Coal comminution is an important operating unit in coal beneficiation. But the comminution is a operation with unsolved problems (1), of which energy consumption is the major one. Researchers have focused on this problem. A solvation way is to characterize the particle size distribution (PSD) and analyze its correlation with the energy consumption (2,3). Thus, characterization of PSD is an important task in comminution process. For coal, as many researchers (4-8) have indicated, the PSD is an important technical parameter. It has been pointed out (4-8) that the PSD plays very important role in preparation of coal-water slurry fuel and affects the combustion of coal fine (8). But for coal of various rank and component, their PSD differ. Zhu Shuquan et al (7) pointed out that the powder product from coal of different rank has different PSD characteristics and some of which cannot be fitted very well by Rosin--Rammler equation. For coal comminution, an important way is to search for a suitable PSD model.Turcorette (9) indicated that fractal particle size distribution (FPSD) can characterize very well the particle population of material in nature form and that from a particulate process, and he utilized the data published by Bennet

(1936) to calculate the fractal dimension (D) of FPSD of the crushed coal with a result of D = 2.5. It must be realized that coal of various rank and subjected to different grinding condition will give different FPSD characteristics.

But little attention has been focused on this area. Another question is whether or not the FPSD characteristics is the same or similar for ground product of same coal at different grinding time. In our work, the D value of Huainan coal powder at different grinding time had been investigated in detail, and their distribution characteristics of particle population composed of different macerals was found.

2. EXPERIMENT

Taken from Huainan Coal Mine was a bituminous coal sample with properties showed in Table 1. Coal was reduced in a jaw crusher to smaller than 3mm. The crushed coal was then ground in a batch ball mill with interior dimension of Φ20x123mm. The top size of steel balls as grinding medium was 30mm and the smallest size, 8mm. Total mass of ball load was about 13 kg. For each test, 1 kg of crushed coal was put into the mill operating at 72 rpm. At

---

[1]* Supported by National Fund for Doctoral Advisory, #9529006

predetermined grinding time the coal powder was taken out and its PSD for different maceral constituents measured under MPV-3 microscope. Particles were divided into three types, mono-maceral, bi-maceral and tri-maceral. Line scanning was used to measure the particle size. The size is the random diameter of particle and the size distribution was transformed into mass distribution.

Table 1 Proximate and Ultimate Analyses of Coal Sample, %

Mad	Ad	Vdaf	St,d	Cdaf	Hdaf	Odaf	Sdaf	Ndaf
1.52	13.37	40.21	0.25	83.73	5.76	8.68	0.29	1.540

3. RESULTS AND DISCUSSION

3.1. Model of fpsd and calculation of its fractal dimension D

Turcorette (9) proposed a fractal particle size distribution model as follows:

$$N_{(>d)} = Cd^{-D} \qquad (1)$$

Where N - number of particles larger than d;
d - particle diameter;
D - the fractal dimension of the particle size distribution;
C - constant.

With equation (1), the fractal dimension D of its FPSD can be obtained. However, N in equation (1) is difficult to obtain from experiment, especially for powder. The equation can be transformed into the form with mass distribution and rewritten as:

$$\frac{W_{(>d_i)}}{W_0} = 1 - (\frac{d_i}{d_{max}})^{3-D} \qquad (2)$$

$$\frac{W_{(<d_i)}}{W_0} = (\frac{d_i}{d_{max}})^{3-D} \qquad (3)$$

where W - mass of particles larger or finer than d_l in Equation (2) or (3) respectively;
W_0 - total mass of the particle population;
d_i - diameter of the ith particle d, micron;
d_{max} - diameter of the largest particle in the particle population;
D - fractal dimension of FPSD of particle population.

Eq. (2) and (3) can be rewritten in logarithmic form as:

$$\log \frac{W_{(>d)}}{W_0} = (D-3)\log \frac{d_i}{d_{max}} \qquad (4)$$

$$\text{or } \log \frac{W_{(<d)}}{W_0} = (3-D)\log \frac{d_i}{d_{max}} \qquad (5)$$

$\log \frac{W_{(>d)}}{W_0}$ or $\log \frac{W_{(<d)}}{W_0} - -\log \frac{d_i}{d_{max}}$ diagram must be a straight line whose slope of is equal to (D - 3) or (3 - D) and D can then be calculated as shown in Table 2.

3.2. Fractal dimension characteris-tics of mono-maceral particle size distribution

Table 2 shows the fractal dimension D and correlation coefficient R calculated by Eq. (4) for particle population of vitrinite, inertinite and mineral. All correlation coefficients are larger than 0.92; that is to say, the D values can reflect the distribution characteristics of particle population. For same component particle, fractal dimensions D of the particle size distribution at different grinding time are almost unchanged. The D values of vitrinite and the inertinite particles are larger than that of mineral. The larger the D value is, the more concentrated the distribution of particle population is. Under microscope it can be seen that mineral particles distribute in a wide size range, while the vitrinite and inertinite particles mainly distribute in fine population. It can be explained as follows: minerals enbedded in coal distribute in a extensive extent, and the vitrinite and inertinte, being more brittle than the minerals, were ground to finer size.

3.3. Fractal dimension characteristics of bi- and tri-maceral particle size distribution

Tab. 3 gives the fractal dimension of bi- and tri-maceral's PSD and its corresponding correlation coefficient calculated by Equation (4). All relative coefficients R are larger than 0.92. As the mono-maceral particle does, D of bi- and tri-maceral ones are almost unchanged with grinding time. But D of V+M, I+M andV+I+M particle population is smaller than vitrinite and inertinitie ones. The D of V+M particles is larger than that of I+M and V+I+M particles, approximately equal to that of mineral particles. D value of I+M is the smallest.

Table 2. The fractal dimension of mono-maceral particle size distribution

Grinding time,min	Vintrinte		Inertinite		Mineral	
	D	R	D	R	D	R
10	2.147	0.966	2.170	0.923	2.000	0.985
20	2.137	0.955	2.160	0.965	2.007	0.963
50	2.144	0.943	2.162	0.938	2.006	0.974
80	2.148	0.950	2.166	0.984	2.014	0.985

Table 3 The fractal dimension of bi- and trimaceral praticle size distribution

Grinding time,min	V+M		I+M		V+I+M	
	D	R	D	R	D	R
10	2.053	0.985	1.643	0.945	1.816	0.954
20	2.052	0.957	1.662	0.932	1.815	0.921
50	2.051	0.987	1.607	0.958	1.820	0.939
80	2.047	0.982	1.618	0.920	1.814	0.947

3.4. The character of fractal dimension of particle size distribution in coal grinding

According to previous results, the fractal dimension D of same type of particles are almost unchanged with grinding time. Gutsche, O. et al (3) pointed out the unchanged character of PSD with the energy input and pertained it to the self-similarity of micro-crack distribution in particle; but they just investigated the grinding of single mineral particle. For coal, it composes of macerals and minerals, the micro-crack distribution in each maceral and mineral is different, and conseqently, their behaviors are different in grinding process. Thence, fractal dimension D of different type of particles are different. The D of vitrinite and inertinite particles are larger because microcracks in them distribute densely, and they are easy to be ground finer. Generally, the minerals enbedded in coal have smaller size but distribute in wide size range. Besides, micro-cracks in minerals are less developed than both vitrinite and inertinite. Therefore, minerals in coal are uneasy to be ground finer and D value of maceral-mineral particles are smaller. It could be considered that complete liberation of minerals from coal is generally impossible.

4. CONCLUSIONS

From previous analysis, the main conclusions can be obtained as:
1. The fractal dimension D of particle size distribution of same material is unchanged in grinding proces.
2. The fractal dimension D of particle size distribution of different material are distinct.

REFERENCES

King, R.P., Powder Technology, 81 (1994), 217-234

Li, G., Xu, X., Minerals Enginering. 6(1992), 163-172

Gutsche,O., Kapur, P.C., Fuerstenau, D.W., Powder Technology 78 (1994) 263-270

Zhang, W., Zhu, S., Wang, Z., in the Proceedings of the 21st International Technical Conference on Coal Utilization and Fuels System, 251-257, March 18-21, Clearwater, Florida, USA

Danier, M., Proceedings, 21st International Technical Conference on Coal Utilization and Fuels System, 543-548, March 18-21, Clearwater, Florida, USA

Wu, C., ChineseCoal, 5 (1996), 12-14, (in Chinese)

Zhu, S., Zhang, W., Wang, Z., in the Proceedings of the 21st International Technical Conference on Coal Utilization and Fuels System, 245-250, March 18-21, Clearwater, Florida, USA

Chow, O.K., in the Proceedings of the 21st International Technical Conference on Coal Utilization and Fuels System, 327-337, March 18-21, Clearwater, Florida, USA

Turcottle, D.L., J. Geophys. Res., 91(1986), 187-190

Mining Science and Technology' 99, Xie & Golosinski (eds) © 1999 Balkema, Rotterdam, ISBN 90 5809 067 1

Separation behavior of binary-density fluidized beds

Lubin Wei
China University of Mining and Technology, Beijing, People's Republic of China

Jicai Qian
Qitaihe Beidou Chemical Plant, Heilongjiang, People's Republic of China

Qingru Chen & Yaomin Zhao
China University of Mining and Technology, Xuzhou, People's Republic of China

ABSTRACT: A binary-density-fluidized bed has been formed by specially designed bed structure. It can turn out three products in a fluidized cascade at the same time. The bed densities are uniform in the same density area and largely different between two distinct density areas. Three-product separation experiments were carried out with both artificial tracer and coal particles. The experimental results showed that the partition density was around 1.52 g/cm^3 and the probable error E_p, about 0.08 in the separating area of low density. In the separating area of high density, the partition density was around 1.87 g/cm^3, and the E_p value, about 0.11.

1. INTRODUCTION

A gas-fluidized bed has many properties similar to that of a liquid. It is natural to try to use fluidized beds for minerals dry beneficiation. The first process for coal was described by Fraser and Yancey (1926), and many studies have been carried out since then (Eveson 1966, Douglas et al. 1973, Dong & Beeckmans 1990). This technology has been available to separate coal of 50-6mm size fraction and a commercial coal dry preparation plant built in China (Chen et al. 1996a). However, the current separator with air-dense medium fluidized bed can only turn out two products at the same time with single separating density. Two separators must be employed in the way of serial connection to turn out three products, which leads to the complication of technical system and the increase of construction investment and operation cost. Moreover, it is difficult to operate continuously owing to the problems of preparation and recovery of dense media (Chen et al. 1996b). To improve further this technology, research on the three-product separator was carried out. The formation of binary-density fluidized bed is the key to three-product separation. It means that two separating areas with different densities are formed in a fluidized cascade. In each separating area, the bed density is uniform and meets the technical needs of coal preparation respectively. Therefore, the processed materials can be separated into three products according to density.

2. EXPERIMENTS

2.1 *Bed structure design and experimental system*

The exploratory study carried out in conventional bed with rectangular cross-section structure showed that it was impossible to turn out three products by conventional fluidized bed because the ordinary-structure bed can only form a multiple-density bed (Wei et al. 1996, Wei 1997). The multiple-density bed made the separation materials easily accumulated in bed and the quality of products greatly affected by the cut points along the bed height.

The exploratory study indicates that the great difference of minimum fluidization velocity between the magnetite powder and magnetic pearls asks the operation condition under high gas velocity in order to fluidize uniformly the dense media particles mainly composed of magnetite powder in the lower bed. At high velocity of gas, the mixture of magnetite powder with magnetic pearls caused by bubbles will be enhanced, which results in the formation of density gradient along the total height of fluidized bed.

On the basis of exploratory study, a special structure of bed was exploited. The schematic diagram of model system for three-product separation is shown in Figure 1. The compressed air first enters air buffer (8) for its pressure stabilization. The air pressure is measured by pressure manometer (7). The gas flow rate can be adjusted by valve (6)

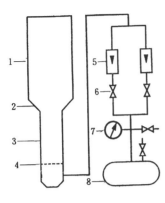

Figure 1. The schematic diagram of model system for three—product separation

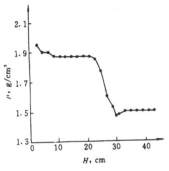

Figure 2. Density distribution of the binary-density bed vs. height

and measured by rotameter (5). The model separator is composed of a distributor (4) and three parts: the lower part (3) with smaller cross-section area of rectangular prism, upper part (1) with larger cross-section of rectangular prism and middle part (2) with pyramidal shape, connecting the upper and lower parts. In this special-structure fluidized bed, dense media in upper part, mainly composed of magnetic pearls, can be fluidized at lower superficial velocity of gas, while dense media in lower part, mainly composed of magnetite powder, can be fluidized at higher superficial velocity of gas. Besides, the gas velocity decreases along the height of pyramidal part, which strengthens the segregation between two media. Therefore, a binary-density fluidized bed is formed. Figure 2 shows the distribution of bed density along the bed height.

2.2 Materials and experimental procedure

Considering the preparation and recovery of dense

media, two match sizes of magnetic powder and magnetic pearls were chosen as dense media. Their physical properties such as density ρ_p, mean size d_p and minimum fluidization velocity U_{mf} are tabulated in Table 1.

Table 1 Physical properties of dense media

Mat-	Magnetite powder			Magnetic pearl		
ch	ρ_p (g/cm)	d_p (μm)	U_{mf} (cm/s)	ρ_p (g/cm)	d_p (μm)	U_{mf} (cm/s)
A	4.6	176	13.0	3.6	51	1.5
B	4.6	136	7.9	3.6	64	2.2

Artificial tracer and coal particles were used as separated materials in the experiments. Two sizes of tracer particles with mean diameter 13mm and 6mm were made. For each size of tracer particle, there were nine grades of density distinguished by different color.

During each operation, the magnetite powder and magnetic pearls with a certain ratio were fluidized by compressed air at a given gas flow rate. When the distribution of bed bulk density measured by bed pressure was kept constant, indicating the state of steady fluidization, separated materials were scattered into the fluidized bed. After 70 s separating, the air supply was suddenly cut off and bed would collapse. When the bed completely rested, three layer samples were taken from the bed. The thickness of layer were 14.5 cm, 21 cm and 7 cm respectively.

3. RESULTS AND DISCUSSION

3.1 Preliminary separation experiments

The preliminary experiments were carried out with dense media of match size A at gas flow rate ranging from 7.5 m³/h to 8.5 m³/h, and density of upper bed around 1.43 g/cm³ and that of lower bed around 1.90 g/cm³. Considering only from the point of density distribution, the fluidized beds seemed applicable for three-product separation. However, the separation results with artificial particles of 13 mm were unsatisfactory. The main reason was too big the difference of minimum fluidization velocity between magnetite powder and magnetic pearls. In order to make the media particles in lower bed, which were lmainly composed of magnetite powder, fluidized homogeneously, the operation condition had to be at higher gas flow rate. High gas flow rate leaded to heterogeneous radial distribution of gas velocity after gas entering pyramid, so resulting in considerable back mixing of dense media particles in the upper bed. The back mixing leaded to bad separation in the lighter area in the upper part. From the observation on fluidization state in lighter area, it can be deduced that the misplacement was mainly caused by misplacing effect by motion of dense media particles (Chen et al. 1996a). For reduction of misplacing effect internal parts were placed in the upper beds. The geometry of the internal part was a rectangular prism grid made of screen mesh. The geometrical properties of the mesh grid are given in Table 2. The experimental results showed that the separation was improved with the increase of mesh size. As mesh size was larger than 2 mm, the separation was improved greatly. The grids with smaller mesh size made media particles hardly passing through the apertures so the circular motion of dense media reformed within it. The good effect of the mesh grit was due to that they kept the separated materials from entering the area close to bed wall, where the medium particles moved considerably.

Table 2 Geometrical parameters of mesh grids

No.	Mesh size (mm)	Grid dimension (mm)
1	0.5	150×150
2	1	150×150
3	1	130×130
4	2	140×140
5	5	150×150

The mesh grid did improve separation effect, but it made the arrangement of discharge apparatus difficult.

3.2 Three-product separation

The preliminary experiments showed that the difference of minimum fluidization velocity between magnetite powder and magnetic pearls is too big. Therefore, the size match of media was changed to match B, so the fluidization performance of lighter area was well improved and satisfactory separation results obtained.

In a wide range of gas flow rates, the experiments of three—product separation were carried out in the binary-density fluidized bed. The comprehensive evaluation to results of separation is shown in Table 3 and Table 4.

The results of separation showed that for 13 mm tracer particles, the partition density ρ_{50} equaled to the bed density ρ_b approximately. With the increase of gas flow rate Q, both separation density ρ_{50} and probable error E_p value increased in lighter density area, but decreased in denser density area. The results of separation for 6mm size tracer particles were somewhat arbitrary and worse than that for 13 mm particles. At too low gas flow rates, the activity of dense media particles was small, so the separation

Table 3 Comprehensive evaluation to separation of tracer particles in lighter area

Q (m^3/h)	ρ_b (g/cm^3)	13 mm tracers		6 mm tracers	
		ρ_{50} (g/cm^3)	E_p	ρ_{50} (g/cm^3)	E_p
4.0	1.50	1.50	0.03		
4.2	1.50	1.49	0.04	1.53	0.09
4.4	1.50	1.48	0.06	1.52	0.13
4.5	1.50			1.51	0.06
4.6	1.50	1.50	0.06	1.52	0.11
4.8	1.56	1.62	0.07	1.75	0.08

efficiency was low in the denser density area. At too high flow rates, the difference of separation density between two density areas was too small to meet the technical needs for coal preparation. The gas flow rate should be chosen in the range of 4.4-4.6 m^3/h.

Table 4 Comprehensive evaluation to separation of tracer particles in denser area

Q (m^3/h)	ρ_b (g/cm^3)	13 mm artificial particles		6 mm artificial particles	
		ρ_{50} (g/cm^3)	E_p	ρ_{50} (g/cm^3)	E_p
4.0	1.95	2.02	bad		
4.2	1.94	2.01	bad	1.94	bad
4.4	1.88	1.85	0.07	1.91	0.11
4.5	1.88			1.91	0.07
4.6	1.88	1.85	0.06	1.97	0.08
4.8	1.74	1.76	0.07	1.83	0.07

Three-product separation experiments were also carried out with size of 25-6 mm coal particles. Some results are listed in Table 5. The results showed that the relationship between separation result and gas flow rate was concordant to some conclusions with tracer particles. In addition, the E_p value increased with increasing feed mass, because the more the feed mass, the worse the fluidization performance. The separation results of 25-6 mm coal were not obviously improved in comparison with that of 6mm tracer particles. The possible reason was the effect of irregular shape of coal.

4. CONCLUSIONS

1. It is impossible to obtain a three-product separation in an air-dense fluidized bed separator of current model. By means of specially designed bed structure with a pyramidal part, a binary-density fluidized bed is formed. In suitable range of gas flow rates, satisfactory results of three-product separation are reached.

Table 5 Comprehensive evaluation three-product separation of coal

Q (m³/h)	Feed mass (g)	Ligher area ρ_{50} (g/cm³)	E_p	Denser area ρ_{50} (g/cm³)	E_p
4.5	150	1.52	0.06	1.85	0.09
4.4	350	1.50	0.09	1.90	0.11
4.5	350	1.54	0.09	1.87	0.11
4.6	350	1.53	0.08	1.84	0.09
4.5	550	1.49	0.13	1.91	0.14

2. The separation can be well improved with application of a suitable internal mesh grids.

3. With the increase of gas flow rate, both separation density ρ_{50} and probable error E_p value increase in lighter density area, but decrease in denser density area.

4. With the decrease of feed size or increase of feed mass, the separation performance become bad.

5. For 25-6 mm coal separation, the partition density is around 1.52 g/cm³ and the probable error E_p value is about 0.08 in the separating area of low density. In the separating area of high density, the partition density is around 1.87 g/cm³, and the E_p value is about 0.11.

ACKNOWLEDGEMENT

The authors appreciate the financial support of the Multiphase Reaction Laboratory of Chinese Academy of Sciences, the Nation Natural Science Foundation of China, and the Coal Science Foundation of China.

REFERENCES

Chen, Q., L. Wei & Z. Luo 1996a. Theory and practice of coal dry beneficiation with air-dense medium fluidized bed. In Y.G. Guo & T.S. Golosinski (eds). *Proc. '96 Int. Symp. Mining Sci. & Technology*, Xuzhou, 16-18 October 1996: 767-770. Rotterdam: Balkema.

Chen, Q., L. Wei & C. Liang 1996b. Three-product separator with binary-density-air-dense media fluidized beds. *Chinese Patent ZL 96231636.9*, April 10, 1996.

Dong X. & J.M. Beeckmans 1990. Separation of particulate solids in pneumatically driven counter-current fluidized cascade. *Powder Technol.*, 62: 261-267.

Douglas E., A.S. Joy & T. Walsh 1973. Development of equipment for the dry concentration of minerals. *Filtration & Separation*, 9: 532-538.

Eveson G.F. 1966. Pneumatic process used in coal cleaning. *Coal Preparation*, (July/August): 135-139.

Fraser T. & H.F. Yancey 1926. Artificial storm of air-sand floats coal on its upper surface, leaving refuse to sink. *Coal Age*, (March): 325-327.

Wei L., Q. Chen & Y. Zhao 1996. Study on the binary—density fluidized bed. J. Basic Sci. & Eng. 4: 275-279 (in Chinese).

Wei L. 1997. Segregation and mixing of two types of particles in air—dense media fluidized beds. *The Chinese J. of Nonferrous Metals*, 7(suppl.3): 7-9 (in Chinese).

Mining Science and Technology' 99, Xie & Golosinski (eds) © 1999 Balkema, Rotterdam, ISBN 90 5809 067 1

The influence of hydrophilic groups on slurry-ability of Chinese coals

Shuquan Zhu, Qiaowen Yang, Xinguo Wang, Xianhua Zhi, Fengqi Li, Zuna Wang & Rongzeng Zhang
China University of Mining and Technology, Beijing, People's Republic of China

ABSTRACT: Two groups of representative coal samples have been characterized, and their slurry-ability has been evaluated under strictly controlled conditions in laboratory. Through computer program analyses, the Oxygen content of coal on dry-ash-free basis is found to be the most relevant coal parameters governing all the coals' slurry-ability. In order to understand the behavior of different oxygen functional groups, FTIR analysis have been made for 10 of those coals with narrow coalification range. Study on the FTIR adsorption peak area corresponding to hydroxyl and carboxyl groups leads to a definition of modified hydroxyl content. The result shows that the influence of carboxyl groups to slurry-ability is 13 time stronger than the hydroxyl group.

1. INTRODUCTION

Coal water slurry is a mixture of mainly coal powders and water. It has attract much attention as a substitute fuel for oil and as a more convenient and clean form of coal for combustion and transportation since it was brought about in the 70's of this century. Therefore, numerous studies (Kaji 1983, Wu 1986, Warchol 1985, Kanamori 1990) have been carried out on the influence of coal properties to slurry's characteristics. Slurry-ability is a widely accepted term of characterizing a coal's slurry concentration at a fixed viscosity. In the area of coal slurry-ability study, our work could be divided into two periods. At the beginning of our research, a simple comparison of coal slurry concentration was made for the coals selected in the coal slurry preparation work. The other preparation conditions, such as additive formulation and agitation strength, were not kept the same. In the 80's, a systematic study of coal slurry-ability was carried out. That is, coal sample collection, additive dosage, particle size distribution and agitation time and strength were all controlled or kept constant. This study is intent to summarize our result at the second period, which will reflect more surface properties than the earlier.

There are two groups of coal samples involved in this study. One of the groups is ranging from brown coal to anthracite collected across China. The other ten coals mainly collected in Shenmu coal basin, which is similar in coalification range and deposition conditions.

2. EXPERIMENTS

2.1 *Coal samples*

Representative coal samples (0x12mm) from different parts of China were collected according to Chines National Standard Sampling Procedure. Detailed characterization of these samples had been reported earlier (Zhu 1995). Table 1 shows the oxygen content of the coal samples.

Table 1. The oxygen content of coals (daf).

Coal	PZ	DS	YM	FX	JY	XHY	HN	BS	ZZ
Odaf%	20.3	13.93	17.76	10.17	11.52	13.49	8.68	7.95	7.1
Coal	TS	BY	SZ	ST	HB	LA	XC	YQ	SDL
Odaf%	6.99	7.62	5.85	3.88	2.4	3.84	3.37	1.83	12.74
Coal	WJL	DXM	LJP	MJT	HJT	NTT	YJL	GJL	MYK
Odaf%	12.95	11.05	10.03	9.49	10.93	11.05	10.4	11.56	9.55

The saturated inherent moisture content (MHC) of these coals ranges from 2.3% to 22.3%, ash content (Ad) ranges from 3% to 26%, volatile matter content (Vdaf) ranges from 43.7% to 9.10%, oxygen content (Odaf) ranges from 1.83% to 20% and carbon content (Cdaf) 71.1% to 92.6% on dry ash free basis.

2.2 Experimental procedure

The coal samples were firstly reduced to minus 3 mm and then ground with a laboratory ball mill in two stages. The average diameter of fine products and croaser products were controlled by milling time. A computer program was used to calculate the ratio of the two products for each coal to attain the constant packing efficiency for the final mixture. Laboratory standard coal water slurry preparation procedure was followed.

Rheological properties of each fresh-prepared slurry were measured by Haake RV 12 Viscometer at a constant ambient temperature. Plot of apparent viscosity at shear rate 30 s^{-1} against the coal concentration was drawn, from which the slurry-ability index(C_{FV}) was obtained, shown in table 1.The index is the solid concentration of the slurry with viscosity of 1000 mPa.s.

Table 1. The slurry-ability index of coals.

Coal	PZ	DS	YM	FX	JY	XHY	HN	BS	ZZ
C_{FV}%	52.3	59.2	60.2	63.5	65.7	64.7	68.3	65.5	66.7
Coal	TS	BY	SZ	ST	HB	LA	XC	YQ	SDL
C_{FV}%	71.1	69.8	71.8	73.6	74.9	73.0	72.2	70.8	56.7
Coal	WJL	DXM	LJP	MJT	HJT	NTT	YJL	GJL	MYK
C_{FV}%	59.2	59.5	60.2	60.4	61.2	62.0	63.2	64.3	66.3

2.3 Hydrophilic group analysis

Nicolet 710 FTIR Spectrometer with multi-reflectance accessory was used in measuring the functional groups in coal. The adsorption spectra of the 10 samples with similar oxygen content are shown in figure 1.

Since the FTIR peak area analysis can only provides the semi-quantitative information of oxygen functional groups, so the chemical reagent analysis was also applied to some of the coals as a remediation.

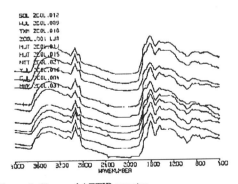

Figure 1. Ten coals' FTIR spectra

3. RESULTS

3.1 Oxygen content

The oxygen content of coal is related to the coal rank. It is an indication of coal's slurry-ability as well. Figure 2 shows that there exists a proximate lineal relations between them.

Using multi-element regressive analysis program, taking the critical F equal to 10, a optimal correlation can be obtained from all the coal parameters determined (Zhu et al 1998).

$$C_{FV} = 76.36 + 1.067 O_{daf} \qquad (1)$$

The coefficient of this equation is 0.9343, standard error 2.1459.

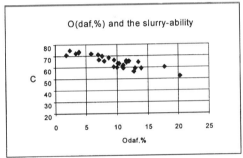

Figure 2. The relationship of oxygen content to the slurry-ability index

3.2 Modified hydroxyl groups

Although figure 2 shows that a proximate lineal relation exits between oxygen content and slurry-ability of coal, there is an apparent deviation for the coals with similar oxygen content. Probably, it is the form of oxygen lead to the difference. From this point of view, the FTIR spectra peak areas corresponding to hydroxyl and carboxyl groups was plotted in figure 3 against slurry-ability index for coals with similar rank and deposition conditions. Obviously, the relation of both -OH peak area (white circles) and -COOH peak area (black triangles) is not good. But, if points of -OH and -COOH are connected with a line for the same coal, the centers (black circle) of these lines are in good correlation with slurry-ability index. From figure 3, equation (2) can be obtained:

$$A(OH)' = 1/2[A(OH) + 25A(COOH)] \qquad (2)$$

Here: $A(OH)'$ = a modified hydroxyl peak area,
$A(OH)$ = peak area at 3400 cm^{-1},
$A(COOH)$ = peak area at 1710 cm^{-1}

Existence of the relation indicate that coal slurry-ability depends on the combination of hydroxyl and carboxyl groups

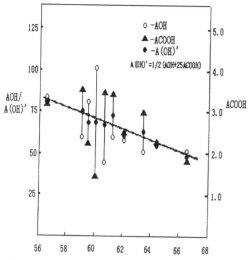

Figure 3. Relations of oxygen functional groups to slurry-ability index

$A(OH)$ and $A(COOH)$ in equation (2) is FTIR spectra corresponding to the hydroxyl and carboxyl adsorption respectively. According to Beer's Law, mole adsorption coefficient needs to be known to convert the area into real mole content.

Two of the coals were analyzed for their hydroxyl content and carboxyl content using selective agents chemically. From these, the average coefficients can be calculated as follow:

$$A_{OH} = 27.37\,g/meq$$

$$A_{COOH} = 13.94\,g/meq$$

Hence, equation (2) can be converted into equation (3) – (5):

$$A(OH)' = 1/2\,[\,27.37\,C(OH) + 25*13.94\,C(COOH)\,] \quad (3)$$
$$A(OH)' = 27.37*1/2\,[\,C(OH) + 12.7\,C(COOH)\,] \quad (4)$$
$$C(OH)' = 1/2\,[\,C(OH) + 12.7\,C(COOH)\,] \quad (5)$$
or
$$2C(OH)' = [\,C(OH) + 12.7\,C(COOH)\,] \quad (6)$$

Here, $C(OH)' = A(OH)'/27.37$, is defined as the modified hydroxyl content. It is clear that the

influence to slurry- ability of one carboxyl group is about 13 times larger than that of one hydroxyl group. The definition of modified hydroxyl group illustrate that coals with more carboxyl group is difficult to achieve high solid concentration in making coal water slurry. It explains reasonably why low rank coal and weathered coal has poor slurry-ability.

3.3 Improvement of coal slurry-ability

In order to verify the influence of carboxyl to coal slurry-ability. A test of de-carboxyl experiment had been made.

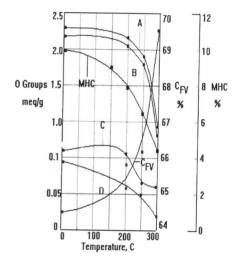

A -- total acidic groups   B -- phenolic hydroxyl
C -- carboxylic salts   D -- Carboxylic acid

Figure 4 The properties and slurry-ability index variation with heating temperature.

The NTT coal particles under 3 mm were put into a steel tube with nitrogen gas flowing through. The temperature of the tube was raised in a rate of 10 C/min to the maximum temperature of 200, 250 and 300 degree Centigrade and kept the temperatures for 30 minutes. Then, the heat-upgraded coals were analyzed and used as a feed to make the slurry. The results were plotted in figure 4. It can be seen that the hydroxyl and carboxyl groups decrease in coal surface lead to an increase of slurry-ability (C_{FV}) and a decrease in saturated inherent moisture content (MHC)

4. CONCLUSIONS

1. Oxygen content of coal has a linear relationship with slurry-ability index from brown coal to anthracite.

2. The definition of modified hydroxyl content is a well define parameter that combine the influence to slurry-ability of hydroxyl and carboxyl contents, which demonstrate that the influence of one carboxyl group is about 13 times larger than that of one hydroxyl group.

3. Both hydroxyl and carboxyl group decreases upon heating under inert atmosphere. As a result, the slurry-ability of coal improved.

REFERENCES

Kaji, R. et al 1983. Effects of coal type, surfactant and coal cleaning on the rheological properties os coal water mixture. *Proceedings of the 5th international symposium on coal slurry combustion and technology*: 151. Tampa: Florida.

Kanamori, S. et al 1990. Studies on the relationship between coal properties and characteristics of CWM. *Proceedings of 15th international conference on coal and slurry technologies*: 433. Florida.

Wu, J. et al 1986. The surface properties of coal and their effects on behavior of CWS. *Proceedings of the 8th international symposium on coal slurry combustion and technology*: 10. Orlando: Florida.

Warchaol, J. et al 1985. The effects of coal properties on slurry quality. *Proceedings of the 7th international symposium on coal slurry fuels preparation and utilization*: New Orleans : Louisiana

Zhu, S. et al 1995.The slurryability of Chinese coals and the role of macerals. *Proceedings of the 1st UBC-McGILL bi-annual international symposium on fundamentals of mineral processing*:491. British Columbia.

Zhu, S. & Zhan, L. 1998. Study on the slurryability of Chinese coals. *Journal of coal sci. & techno.* Vol.23(2):198

Mining Science and Technology' 99, Xie & Golosinski (eds) © 1999 Balkema, Rotterdam, ISBN 90 5809 067 1

Application of M-COL to Chinese coals

T. Murata
Technical Research and Development Headquarters, Mitsui Engineering and Shipbuilding Company Limited, Chiba, Japan

K. Abe
Plant and Energy Engineering Headquarters, Mitsui Engineering and Shipbuilding Company Limited, Chiba, Japan

Y. Katoh
Chiba Technology Center, Mitsui Engineering and Shipbuilding Company Limited, Japan

ABSTRACT: M-COL Process is improved coal preparation technology which can achieve high de-ashing and pyritic sulfur removal for low grade coal by lower cost. The M-COL process improve flotability and selectivity of fine coal through the alteration of coal surface property. Therefore, by applying the M-COL process to low flotability fine coal , a significant enhancement of recovery rate of fine clean coal also can be achieved. In this presentation , the advantage and rough F.S. by applying the M-COL process to low grade Chinese coal is introduced on the basis of verification test results using several Chinese coal samples.

1 PREFACE

Coal accounts for about 75% of the production and consumption of primary energy in China. In addition, its coal industry is positioned as an important domestic industry and coal's superiority as an energy source is expected to be held for a while. However, low ratio preparation plant feed can be pointed out as one of the problems of mass coal consumption in China. Currently, the ratio is estimated at 23% and most distributed coals are low grade, which causes to expanded air pollution and lowered energy efficiency. Considering the gravity of this problem, authorities concerned are now making efforts to increase the ratio of preparation plant feed to 30% or more.

On the other hand, the conbination of a JIG for coarse coal preparation and a flotation machine for fine coal preparation is being used as a common coal preparation process in China. However, feeding coal with low flotability into this process causes new problems, such as coal leakage for tailings due to lower flotation velocity and fine coal mixing into over flow water in thickener due to increased load on treatment of waste water. To propose a solution for these problems, this paper will review the application of M-COL process to Chinese coals as a improved preparation for the coal preparation process.

2 PROPERTIES OF COAL

2.1 *Properties of Chinese coal*

China has huge coal deposits, which proved reserves is estimated at about 1 trillion ton. Coal seams are widely distributed in the country with wide variety of coals. Fig.1 shows the characteristics of Chinese coal by [H]-[O] two dimensional coordinate, each of which represents the atomic ratio to Carbon. In Fig.1 evaluating coals which [H] are over 80 or which [O] are nearly below 7 as "coals with excellent flotability" , there are many Chinese coals with extremely bad flotability, which [H] are below 60 or which [O] are over 10.

2.2 *Properties of test coals*

Table 1 shows proximate, ultimate analysis and [H], [O] values of the two kinds of coals used for M-COL test. Both of the test coals low grade coals containing more than 20% of ash content and nearly

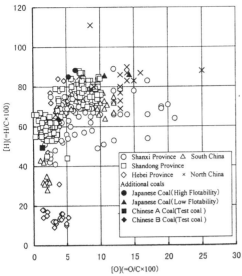

Figure1. Characteristic of Chinese coal
by [H]-[O] two-dimensional coordinate

Table 1. Proximate ,Ultimate Analysis and [H],[O] value of the
Coal used for M-COL application Test

	Coal		A	B	Note
Item					
Proximate Analysis	Moisture	%	1.0	0.9	JIS M 8812
	Ash	%	25.0	27.4	
	Volatile matter	%	8.9	15.5	
	Fixed carbon	%	65.1	56.2	
Ultimate Analysis	Ash	%	25.28	27.69	JIS M 8813
	C	%	65.3	62.0	
	H	%	2.69	3.28	
	N	%	1.04	1.08	
	O	%	1.15	3.13	
	S	%	4.54	2.82	
	[H]		49.4	60.6	
	[O]		1.3	3.6	

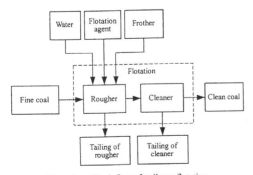

Figure 2.   Block flow of ordinary flotation

over 3% of sulfur. By calculating [H] and [O] of A
and B coals and evaluating them by Fig. 1 showing
the characteristic of Chinese coal, we can find the
properties of low flotability due to higher coal
rank in both of them, each of which [O] is below 5
and [H] is nearly below 60.

3   M-COL APPLICATION TEST METHOD

3.1 Ordinary flotation test process

Fig.2 shows a typical block flow of ordinary
flotation process. Under this flotation, flotation
agent and frother are added in a conditioner. If fine
coal fed into flotation process has low flotability,
coal is not well adhered to the flotation agent,
which may causes shortage in flotation time, by
flowing into flotation machine. The flotation time
shortage can lead to lower clean coal yield and
tailing ash to eventually harm the following
processes. Adding more flotation agent may be
considered as a solution to this problem, however, it
will cause another problem such as more difficulty
in treatment of waste water. These problem become
more serious with smaller particle diameter of fine
powder coal processed by the flotation machine as
well as lower flotability.

3.2 M-COL application test process

Fig.3 shows a block flow of M-COL application to
ordinary flotation process to alter coal surface
property into more lipo-philic, which is especially
effective to coals with faulty flotability. Just as
shown in Fig.3, M-COL process alternates coal
surface property by adding little flotation agent,
below 0.5wt% per coal, to target coal slurry and
mixing them in field of high shear force of a high
shear mixer. In M-COL application test to Chinese
coals, we used fine coal, which particle diameter
was 0.5mm and under, and simulated each process
of the block flow in Fig.3 with bench scale-sized
batch equipment.

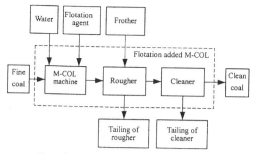

Figure 3. Block flow of M-COL process application

4 TEST RESULTS AND CONSIDERATIONS

4.1 Correlation between flotation time and yield

Fig.4 shows the correlation between cleaner flotation time and yield under M-COL process in comparison with yield between M-COL process and ordinary flotation. The Fig.4 proves that M-COL process can quadruple A coal's yield of ordinary flotation and double B coal's yield of ordinary flotation. We must pay special attention to the fact that M-COL ensure almost same level of yield even for A coal with extremely low flotation as that B coal which flotation is relatively higher.

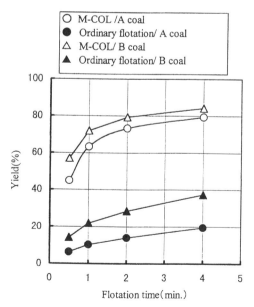

Figure 4. Comparison of yield between M-COL process and ordinary flotation

Fig.4 also shows that M-COL process can substantially reduce flotation time through the comparison of flotation times necessary to reach the same yield between M-COL and ordinary flotation, which means a substantial improvement in flotation velocity of M-COL applied coals.

4.2 Evaluation depend on material balance

Fig.5 and Fig.6 show material balance of M-COL and ordinary flotation test to A coal and B coal respectively. Both figures prove that MCOL can substantially increase recovery ratio of combustible of raw coals in comparison with ordinary flotation.

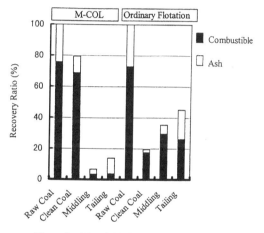

Figure 5. Material ballance of M-COL and ordinary flotation test to A coal

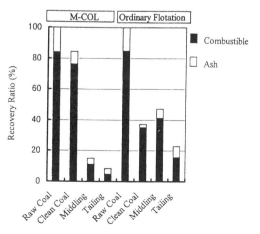

Figure 6. Material ballance of M-COL and ordinary flotation test to B Coal

We must also pay attention to tailing ash content. While tailing ash content in ordinary flotation is 42% for A coal and 60% for B coal, that in M-COL is 74% for A coal and 76% for B coal. Especially, the 42% ash content in A coal under ordinary flotation says that is mixed with the coal supposed to be recovered as clean coal in tailing, which lead to various problems of the treatment of waste water in the following processes. On the other hand, M-COL ensure above 70% ash content. This ash content of tailing is enough higher and also improves the problems regarding clean coal under tailing of ordinary flotation, which eventually smoothes the concentration of thickener in the following processes and the utilization of circulating water of concentration tailing's dehydration and over flow water.

Table 2.   Result of preliminary feasibility study for M-COL process application

Annual operation time	Hour/Year	6,000	
The amount of clean coal product (increase)	Ton/Year	36,000	
Price of clean coal	Yuan/ton	230	
The amount of annual sold	Yuan/year	8,280,000	
Plant investment	Yuan	8,000,000	
Depreciation expenses of plant	Yuan/year	2,000,000	4 years pay
Electric power	Kwh	120	
Electric power cost	Yuan/year	302,400	0.42yuan/kwh
The amount of added oil	Ton/year	600	
Annual cost for oil	Yuan/year	840,000	1,400yuan/ton
Annual expenses	Yuan/year	3,142,400	
Annual income	Yuan/year	5,137,600	

5 M-COL APPLICATION TO FLOTATION PROCESS

The test results in the previous chapter verify that M-COL process can be possibly applied to the flotation process, which processes coals with low flotability. Fig.7 outlines the P&ID of M-COL process application to ordinary flotation. In this figure, M-COL process is placed above the flotation conditioner and the main equipment consist of slurry tank to adjust concentration, the M-COL machine for surface property alternation, a pump to provide slurry and flotation agent, and frother volume feeder. Each component equipment is connected by piping another. Therefore, equipment

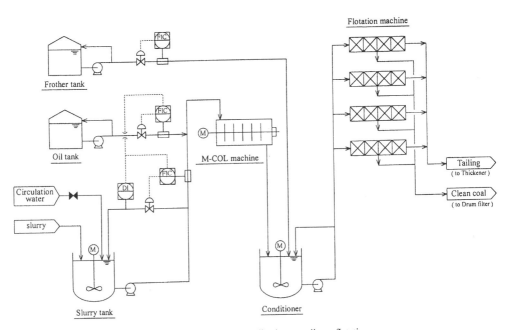

Figure 7.   P&ID of M-COL process application to ordinary flotation

for M-COL application can be easily integrated into pre-installed flotation process.

Table2 outlines the result preliminary feasibility study for M-COL process application to pre-installed flotation process. This study is carried out based on the premise that tonnage of flotation supply fine coal is 20ton/h and recovery volume with M-COL process is increased by 30% in comparison with ordinary flotation based on the test result of chapter 4. According to this table, if plant investment is depreciated in four years, profits can be expected from starting year of the operation. In other words, M-COL can be also expected to bring economic merits.

6 CONCLUSION

1. Experiments verified that M-COL process can relieve the problems caused by coals with low flotability in flotation process.
2. M-COL can be easily applied to existing flotation process and the preliminary feasibility study on target coals proved M-COL process's high economics.

REFERENCES

Sadakata, Murata & Ishida (1997), *China Environment Handbook*, Science Forum Co..

Mining Science and Technology' 99, Xie & Golosinski (eds) © 1999 Balkema, Rotterdam, ISBN 90 5809 067 1

The study and practice of cyclonic microbubble flotation column in ash and pyritic sulfur rejection from coals

Guangyuan Xie & Zeshen Ou
China University of Mining and Technology, Xuzhou, People's Republic of China

ABSTRACT: The operation principle and the structural feature of a new type of cyclonic micro-bubble flotation column are introduced. The results of related laboratory research and industrial investigations are also presented in this paper. It has been proven that this flotation column permits coal desulfurization and ash reduction. It has the potential to substitute the flotation equipment used at present.

1. INTRODUCTION

China is a great country of coal production taking coal as a principal energy source. Along with rapid development of Chinese economy, the coal consumption has been increased rapidly. Most Chinese coals are difficult to be separated, and have high inherent ash. The deposit of high- and medium-sulphur coals with sulphur content more than 2.0% account for about 14% of the total coal reserves in China.

The flotation is one of the important methods for desulfurization and ash reduction of high-sulphur coals, because the grain size for flotation is fine and the liberation of ash-forming minerals and intergrowth constituents of pyrite is relatively sufficient. But the selectivity of the conventional flotation is not high, the hydrophobicity of pyrite, the high-ash slime as well as pyrite entrained seriously to the cleaned coals, so that the efficiency of desulfurization and ash reduction is not obvious. This paper will introduce the structural principle of a new type of cyclonic micro-bubble flotation column, the laboratory research and the commercial applied results.

2. ANALYSIS RESEARCH ON MECHANISM OF MICRO-BUBBLE FLOTATION COLUMN AND CONVENTIONAL FLOTATOR

In regard to the flotation operation, the amount of float coal accounts often for 60~80% of the feeds. Thus, the sufficient gas-liquid interface is favorable to the separation. But the size of bubble of the micro-bubble flotation column is smaller than that of the conventional flotator. In conditions of the aeration quantity of the same pulp, the gas-liquid interface for separating the coal is greater. In addition, because the fluid state of the small bubble and the moving state within the flotation column are relatively more stable than that of the conventional flotation, it is more favorable to recovering both the fine grain and the coarse grain. The mechanism is shown in Figure 1.

From the water line in the flotator and the flotation column, it can be seen that the micro-grain of fine slime and the intergrowth micro-grain mingled in the cleaned coals are decreased to the

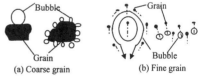

(a) Coarse grain   (b) Fine grain

Figure 1 Effect of bubble size on catching coal grain.

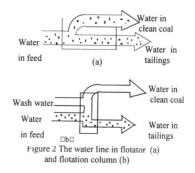

Figure 2 The water line in flotator (a) and flotation column (b)

lower limit. Thereby, the micro-bubble flotation column has possessed a higher selectivity and cleaned coal recovery. The mechanism is shown in Figure 2.

3. STRUCTURAL PRINCIPLE OF A NEW TYPE OF CYCLONIC MICRO-BUBBLE FLOTATION CLOUMN

The cyclonic micro-bubble flotation column was developed successfully by Research Center of Mineral Processing Engineering of China University of Mining & Technology for over ten years. The structure and principle are shown in Figure 3.

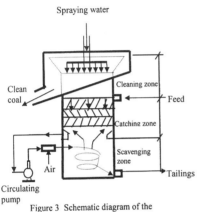

Figure 3 Schematic diagram of the cyclonic microbubble flotation column.

The cyclonic microbubble flotation column consisted of flotation segment, cyclonic segment and microbubble segment. The flotation segment can also be divided into two zones, i.e. the catching zone between the feed point and the cyclonic segment, the cleaning zone between the overflow weir and the feed point. The collecting groove of froth concentrate and the water sprayer are set at the top of the flotation segment. The feed pipe is located at about 2/3 of the column height, and the separated final tailings are discharged from the flow outlet of the cyclonic end bottom. The microbubble generator is in the outside of the column body, joining the cyclonic segment along the tangential direction. The additive pipe for the frother and the air suction pipe are set on the microbubble generator. The microbubble generator is a key part for realizing the separation. Through utilizing the recycle slurry to be sprayed with pressure, it can suck air to be mixed with the frother agent and crash the bubbles, and a great amount of microbubbles are separated by the pressure release, and then the

microbubbles enter into cyclonic segment along the tangential line. When the bubble generator creates the suitable bubbles, the cyclonic force field is also provided for the cyclonic segment.

After the gaseous, solid and liquid recycle slurry entered into the cyclonic segment at a high speed along the tangential line, the cyclonic movement of slurry occurs in the cyclonic segment under the action of the centrifugal force and the flotation force. The bubbles and the mineralized air pellet flocs, moving into cyclonic centre, can rapidly enter the flotation segment. The bubbles with the slurry been feed from upper part are mineralized by the backward collision, so that the rate of adhesion and contact would be increased, and the separation with a high selectivity in the flotation segment would also be realized. The function of cyclonic segment is to recover those fine coals, which are not separated in time, to raise the recovery of fine coals and to increase the ash of the flotation tailings. The bubbles in the column body can be mineralized upwards and continuously cleaned by water. And the carried matters with high ash content can be removed. The upper thick froth layer and the spraying action of rinsing water should raise the concentrate grade greatly. The cyclonic microbubble column has a high recovery of clean coal and has a good selectivity.

4. SEMI–INDUSTRIAL EXPERIMENTAL RESEARCH ON DESULFURIZATION AND ASH REDUCTION

Figure 4 and Table1 have shown the experimental results of the coal slurry from Datun coal preparation plant in the flotation column with diameter of 200mm and height of 3.5m. And the obvious gradients of sulphur content and ash content are created along the height in the flotation column. The ash content has reduced by 27.55% from point 9 to point 1 at the overflow weir of the cleaned coals. In the reverse, the ash content has increased by

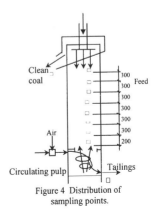

Figure 4 Distribution of sampling points.

35.11% from point 9 to point 10 at tailings outlet. It is obvious that the separation effect of desulfurization and ash reduction is in both the cyclonic segment and the flotation segment of the flotation column.

Table 1. Ash and sulfur of samples.

Sample	$A_{d(\%)}$	$S_{td(\%)}$
Feed slurry	20.37	0.74
Clean coal□	6.81	0.51
□	16.67	0.68
Feed point□	18.72	0.76
□	20.51	0.81
□	20.38	0.79
□	22.68	0.85
□	24.37	0.85
□	30.48	1.10
□	34.36	1.00
Tailings□	69.47	1.55

Table 2 Separation result of flotation column on Jiangzhuang coal.

Size	Feed		Clean coal		Tailings	
mm	$A_d\%$	$S_{td}\%$	$A_d\%$	$S_{td}\%$	$A_d\%$	$S_{td}\%$
>0.50	11.62	0.96	5.01	0.654	38.54	1.895
0.25□0.50	13.99	1.215	6.22	0.678	59.45	1.608
0.125□0.25	19.25	1.343	5.51	0.710	62.88	2.59
0.075□0.125	19.35	1.622	6.42	0.770	62.97	3.684
<0.075	26.91	1.605	7.18	1.019	65.75	3.044
Total	22.53	1.491	6.65	0.878	65.16	3.014

The separation results of the coal slurry from Jiangzhuang in the flotation column are listed in Table 2. The ash content has decreased from 22.53% to 6.65%, and the sulphur content has reduced from 1.491% to 0.878%. It may be seen that the result of the flotation column for desulfurization and ash reduction is distinct.

5. COMMERCIAL PROSPECTS

On the basis of the research on the laboratory flotation column with diameter of 50mm and on the modified semi-industrial flotation columns with diameters of 150mm and 200mm, the industrial flotation columns with diameters of 1m, 1.5m and 3m etc. have found a broad commercial application.
The commercial applied results of the flotation columns with diameters of 1m, 1.5m and 3m from Zhongliangshan coal preparation plant, Mowo coal preparation plant in Xingtai and Datun coal preparation plant are listed in Tables 3, 4, and 5 separately.

Table 3 Applied results of flotation column with 1m diameter in Zhongliangshan.

Feed		Clean coal		Tailings	
Ad%	Std%	Ad%	Std%	Ad%	Std%
15.32	1.45	9.50	1.25	40.51	2.03
16.08	1.41	9.75	1.28	42.04	1.98
16.07	1.39	9.56	1.27	40.11	1.92
15.39	1.42	9.87	1.25	39.03	1.95

From Table 3, it can be seen that the flotation column has possessed not only a remarkable function as ash reduction, but also a notable result of desulfurization.

Table 4 Applied results of flotation column with 1.5m diameter in Mowo.

Year	Month	Flotation			Washed fine coals
		Feed	Clean coal	Tailings	Ad%
		Ad%	Ad%	Ad%	
95	9	29.70	9.96	50.17	9.17
95	12	27.06	9.81	51.03	10.51
96	4	22.19	9.53	46.49	10.34
96	6	26.35	10.52	51.74	10.39
96	7	27.22	9.97	57.86	10.09
96	9	24.83	9.07	50.42	10.02
96	12	25.60	9.15	52.31	9.19
Average		26.14	9.71	51.43	9.98

As seen from Table 4, the flotation column has possessed an obvious function as ash reduction. Additionally, the flotation column may make the ash content of the flotation concentrate to be equal to the ash content of washed clean coals. The cleaned coal recovery of the coal preparation plant may be greatly raised and the greater economic efficiency would be gained.

Table 5 Applied results of flotation column with 3m diameter in Datun.

Year 97		Feed	Flotation column		Flotator Clean coal
Month	Day		Clean Coal	Tailings	
		Ad%	Ad%	Ad%	Ad%
6	17	19.85	7.34	50.06	8.50
6	18	19.18	7.26	42.59	8.94
6	19	19.64	7.21	53.20	8.79
6	21	19.00	7.21	46.19	8.13
6	22	21.86	7.54	52.44	8.35

The ash content of cleaned coals of the flotation column can decrease by 1~1.5% in comparison to the flotator as shown in Table 5. Thus, it ensured that the ash content of cleaned coals is smaller than 8.0%.

The processing capacity and the disposed motor power of different specifications of the flotation columns are listed in Table 6.

Table 6 Processing capacity and disposed motor power of the different specifications of flotation columns.

diameter, mm		1000	1500	2000	3000
Capacity	Pulp flow, m³/h	20□30	50□8 0	100□1 50	180□2 50
	Dry slime, t/h	2	5	10	15
Motor power, kW		13	37	55	75

From Table 6, the flotation column saves the energy by 1/2~1/3 in comparison with the flotator with equal processing capacity.

6. CONCLUSIONS

The semi-industrial experimental research and the commercial applied results indicated that the cyclonic micro-bubble flotation column has a high efficiency of the coal desulfurization and ash reduction, and its technological index is advanced. Owing to such advantages as unique novel structural design, short column body, small power consumption, obvious saving on energy, small maintenance and stable reliable operation etc., the cyclonic micro-bubble flotation column has been effective and economized technological equipment for desulfurization and ash reduction of the fine coal. And it will have a broad popularizing and applying prospects.

REFERENCES

Olson, T. J. and Aplan F.F. 1984. The floatability of locked particles in a coal flotation system. *Applied Mineralogy, Proc. 2nd Int. Cong. on Applied Mineralogy in the Mineral Industry,* W. C. Park, D. M. Hausen and R. D Hagni(Eds). The Metallurgical Society of AIME, Warrendale.

Ou Zeshen et al. 1996. Coal desulfurization by column flotation. *Proc. of the'96 Int. Symp. on Mining Science and Tech.* Guo Y. & T.Golosinski (Eds).

Mining Science and Technology' 99, Xie & Golosinski (eds) © 1999 Balkema, Rotterdam, ISBN 90 5809 067 1

Performance of Climax_x Magnetic Separators for high efficiency, high capacity magnetite recovery

A.J. Dynys & J.V. Ghelarducci
CLI Corporation, Canonsburg, Pa., USA

ABSTRACT: Climax_x Magnetic Separators, developed and patented by CLI Corporation, are the dominant magnetic separators in North America. They were developed to allow high capacity operation with excellent magnetite recovery. These separators have been installed in a number of plants, including the Greenrup Preparation Plant, Lady Dunn Preparation Plant and Homer City Coal Preparation Plant. This paper will discuss the advantages of the Climax_x design and present data to substantiate its superior performance.

1 INTRODUCTION

Heavy media processing (finely ground magnetite suspended in water) has been proven as the most efficient means of cleaning high ash coals. Statistics show that the price of coal has been steadily decreasing for the past few years. However, magnetite is very expensive, approximately US$100 per ton ($/T). For heavy media processing to be economically viable, the magnetite must be recovered and reused. This is usually done by taking advantage of the highly magnetic properties of magnetite versus the relatively non-magnetic properties of coal. The most common device used for recovery is a wet-drum magnetic separator.

Magnetite consumption (or loss of magnetite) has historically been around 1.75 pound per ton (lb./T) which equates to adding approximately $0.10 per ton of clean coal. Engineers at CLI Corporation noticed this and began to examine traditional magnetite recovery circuits to determine where improvements could be made. It was determined that the biggest improvement could be made in redesigning the magnetic separators specifically for coal processing. This paper will discuss the improvements made in the design of magnetic separators and the corresponding improvements in magnetite recovery.

2 DESIGN FEATURES

Figures 1 and 2 show cross sections of a traditional magnetic separator and a Climax_x Magnetic Separator, respectively. For discussion, the separator will be divided into two parts, the drum and the tank. The advantages of each are discussed in the following paragraphs.

2.1 Advantages of Drum Design

Drums have been designed at both 30-inch and 36-inch diameters, offering more flexibility in design and installation. The drum consists of a 750-gauss interpole magnetic element incorporating 6 main poles and 5 interpoles for an overall arc length of 135°. This extended arc length provides additional scavenging capacity, resulting in a very high level of magnetite recovery. This drum was designed specifically to take advantage of the improvements employed by the Climax_x tank design.

2.2 Advantages of Tank Design

In traditional magnetic separators, the gap between the drum and the tank is fairly small, which causes an increase in the velocity of the slurry as it passes through the gap. The increased velocity creates higher drag forces, which leads to misplaced magnetite in the tailings and the need for secondary magnetic recovery equipment. The Climax_x tank design addresses the detrimental effects of slurry velocity by providing a larger gap between the tank and the drum. This leads to less misplaced magnetite and helps eliminate the need for

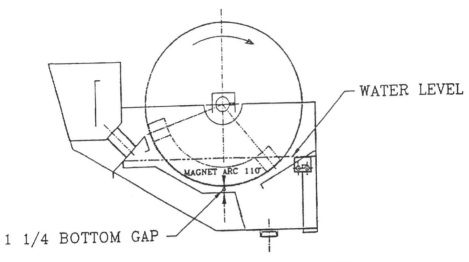

Figure 1 – Cross Section of a Traditional Magnetic Separator

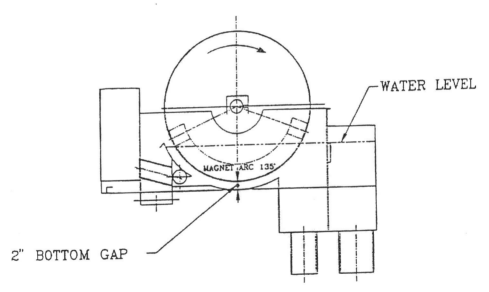

Figure 2 – Cross Section of A Climax$_x$ Magnetic Separator

secondary recovery equipment.

The drive is designed to rotate the drum at approximately one half the speed of a conventional separator. This feature reduces the eroding effect of the magnetite at the face of the drum and also helps minimize misplaced magnetite by reducing the velocity in a similar manner as described above.

Another beneficial feature of the Climax$_x$ tank design is the higher operating water level. In addition to providing higher capacities, this exposes the slurry to more of the magnetic element, reducing the probability of losing magnetite. The higher water level also requires the discharge point to be raised, minimizing "short-circuiting" of the feed to the magnetic concentrate.

The Climax$_x$ tank design has also addressed another problem with the traditional magnetic separator design. The patented tailings removal

Homer City Coal Preparation Plant

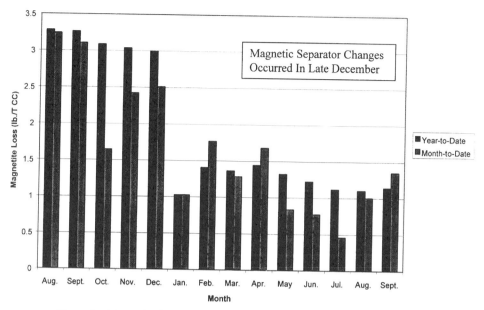

Figure 3 – Magnetite Consumption of Homer City Coal Preparation Plant

system prevents plugging of the tailings spigot. Level is controlled by an overflow weir and does not require operator attention.

It should be noted that the $Climax_x$ separator is set up in the counter-rotation style, instead of the concurrent style. Historically, counter-rotation separators have provided a "cleaner" concentrate due to the "scrubbing" that occurs as a result of the concentrate moving in the opposite direction of the flow. In other words, it is harder for non-magnetic material to become entrained in the concentrate.

3 SIZING OF MAGNETIC SEPARATORS

$Climax_x$ Magnetic Separators are sized based on the volume and magnetic loading of the feed slurry. As previously stated, the tank design permits higher capacities, normally in the range of 120 to 140 gallons per minute slurry per foot of drum width (GPM/ft.). All magnetic separators are limited by the amount of magnetite they can carry over the discharge lip. The $Climax_x$ drum is capable of removing 6.0 tons per hour of magnetite per foot of drum width (TPH/ft.).

Other design constraints include percent solids in the feed slurry (10% - 20%) and percent

nonmagnetics (10% - 60%) in the feed. The high intensity of the magnetic field is less affected by solids in the feed slurry, allowing larger tolerances in control of the feed slurry.

4 PERFORMANCE RESULTS

The superior performance of the $Climax_x$ magnetic separator has been reported previously by Chedgy et al. (1997) and Norrgran (1997). These authors reported magnetite losses of less than 1.0 gram per liter of tailings, corresponding to magnetite recoveries greater than 99.8% for complete $Climax_x$ separators, and 99.5% for traditional drums mounted on $Climax_x$ tanks.

More recently, the Homer City Coal Preparation Plant has been converting its traditional style magnetic separators to the $Climax_x$-style design. They are reconditioning old drums with salvaged parts to have 6 main poles and no interpoles, providing them with an overall arc length of about 120°. These drums are then mounted on new $Climax_x$ tanks and placed in service. As a result of these modifications, their magnetite consumption has been reduced by more than 50% (Figure 3).

5 SUMMARY

The Climax$_x$ Magnetic Separator was designed for high efficiency, high capacity recovery of magnetite in coal processing operations, without plugging. The advantages of its drum and tank designs have been presented, along with information on sizing of separators and test data supporting its superior performance.

6 REFERENCES

Chedgy, D.C., J.V. Ghelarducci, M.A. Sharpe and S. Yu, "Climax$_x$ High Capacity Processor - A Novel Design in Coal Pre-Combustion Cleaning," SME Preprint 97-153, Feb. 1997.

Norrgran, D.A., "Wet Drum Magnetic Separators - Heavy Media Duty," presented at Coal Prep 97 Operators Workshop, 1997.

Mining Science and Technology' 99, Xie & Golosinski (eds) © 1999 Balkema, Rotterdam, ISBN 90 5809 067 1

Design and operation of large diameter heavy media cyclone circuits in coal preparation

A.J. Dynys, J.V. Ghelarducci & A. Lambert
CLI Corporation, Canonsburg, Pa., USA

K.E. Harrison
Harrison Consulting Services, Pittsburgh, Pa., USA

ABSTRACT: In the past decade or so, coal preparation equipment (particularly heavy media cyclones, screens and magnetic separators) has undergone significant changes. With coal prices trending to lower costs per ton, traditional coal washing circuitry is now growing obsolete due to higher processing costs. New circuits focus on fewer pieces of equipment processing higher tonnage at higher efficiencies. This paper discusses the selection of equipment, circuit design considerations and operating performance of large diameter heavy media cyclone circuits.

1 INTRODUCTION

The Climax$_x$ High Capacity Processor (HCP), developed and patented by CLI Corporation, is a novel design in coal washing (Chedgy et al. 1997). As seen in Figure 1, the core part of this design is a large diameter (typically one meter) heavy media cyclone circuit. The HCP also features a single-line design concept, in which the required number of units is minimized. This paper discusses the authors' experience in selecting heavy media cyclones, the design of heavy media circuitry and the selection of screens and magnetic separators with emphasis on important operating principles.

2 CYCLONE SELECTION

The first step in circuit design is to select the correct cleaning device. In the past decade or so, heavy media cyclone (HMC) manufacturers have made great strides in increasing capacities such that heavy media cyclones are now available in sizes from 660 mm up to 1300 mm. The largest of these cyclones can handle lumps of coal up to 105 mm in diameter, which means that HMC's can now be used to process both coarse (6mm and larger) and intermediate (6mm x 0.5 mm) size coal fractions. The HCP design takes advantage of this fact, eliminating the need for a typical coarse coal circuit (jig or heavy media vessel).

Cyclone selection is based upon tons of coal to be processed and the volumetric flow rate to the cyclone (a function of the media-to-coal ratio and the operating pressure). There are numerous mechanical variations that need to be considered during the selection process. Most important are inlet area, apex diameter, length and diameter of the vortex finder and the length of the cylindrical or "barrel" section of the cyclone. Inlet area sets the flow that the cyclone can handle at normal feed pressure and the vortex finder and apex diameters determine how much flow will exit each end of the cyclone. The length of the "barrel" section of the cyclone determines residence time in the cyclone, an important consideration for coals with a large amount of near-gravity material (±0.1 specific gravity units). CLI Corporation has a wealth of experience in selecting the proper heavy media cyclone for a variety of different coal cleaning installations.

3 SELECTION OF FEED SYSTEM

The next step in circuit design is to determine how to deliver the coal to the cyclone. There are two common systems that are used, gravity and pumping. Each system has its own advantages and disadvantages, as discussed below.

Pump-fed cyclones require the minimum amount of space and ancillary equipment (piping, platework, etc.) and facilitate plant layout. Pump-fed cyclone circuits typically require lower capital

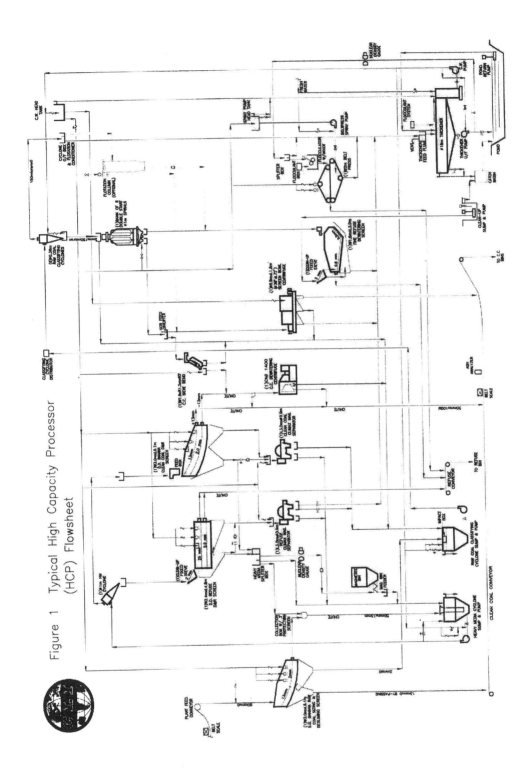

Figure 1 Typical High Capacity Processor (HCP) Flowsheet

costs and offer the added advantage of being somewhat flexible (i.e., it is possible to change the pressure and flowrate by changing the mechanicals on the pump). However, this arrangement is limited by the top size of particle the pump can handle. Fortunately, there are different types of pump and impellers that allow larger particles to pass. Proper pump selection is critical in this arrangement.

Two of the potential problems with this type of arrangement are maintaining accurate density readings on the media return pipe and fluctuations in pump performance. Pump-fed cyclones require that the media gravity be controlled by measuring the gravity of the return line from the drain and rinse screens. If not properly designed, the "S-tubes" can plug or the flow in a straight length of pipe can vortex near a restriction, both of which will cause erroneous readings in the gravity control systems. Fluctuations in pump performance affect the separating performance of the cyclone, so that the optimum separating conditions are not held constant.

The other option is to gravity-feed the cyclones. This arrangement generally requires plants to be taller, making plants larger and thus more expensive (Figure 2). Expanding plant capacity becomes more difficult too, since there may not be enough head to install larger cyclones. However, a properly sized gravity-feed system offers several advantages.

There is less variability in the flow of the feed and there is constant head on the cyclone feed inlet. Gravity control is more accurate since the gravity can be measured at the pump discharge pipe, which is more representative of what is feeding the cyclone. This arrangement also removes the restriction of particle top size that the pump can handle, since it is only necessary to pump media.

4 SCREEN SELECTION

There are two types of screens generally associated with a heavy media cyclone circuit, the raw coal/deslime screen and the drain and rinse (D&R) screens for the clean coal and refuse products. Each of these screens can be either the traditional horizontal screen with a sieve bend or the newer multi-slope "banana" screen. Multi-slope screens have become the preferred type in many applications due to their higher capacity per

unit area. The advantages of multi-slope screens have been well documented, so it is not necessary to describe them again (Anon. 1993).

Raw coal/deslime screens are the most important from a performance standpoint. For the cyclone to perform optimally, the ultrafines (less than 10 micrometers) must be removed prior to treatment. The ultrafines adversely affect slurry viscosity, hinder the performance of the magnetic separators and can build up in the circuit if the operator is not attentive. The drain and rinse screens are important from an economic standpoint. The duty of the D&R screen is to recover magnetite while minimizing the loss of clean coal.

Screens are sized on the basis of tons passing through the screen opening, with a number of correction factors for screen duty, type and application. CLI Corporation has developed its own empirical screen sizing information, based on work with screen manufacturers, so that it can quickly select the proper screen for any coal cleaning application.

5 MAGNETIC SEPARATORS

Since it was first patented in 1990, the $Climax_x$ Magnetic Separator has become the dominant magnetic separator in the U.S. coal industry. Due to its advanced design, the $Climax_x$ Magnetic Separator offers:

- Superior magnetics recovery at higher capacities than traditional separators.
- The ability to recover magnetite in one pass (no secondary recovery circuit).
- The ability to handle higher concentrations of nonmagnetics in the feed.

Dynys & Ghelarducci (1999) have published a paper detailing the design and performance benefits of the $Climax_x$ Magnetic Separator.

6 OVERALL CIRCUIT PERFORMANCE

CLI Corporation has installed several large diameter heavy media cyclone circuits. Most recently, a HCP with a one-meter heavy media cyclone was constructed and commissioned in Bilaspur, India. This washery is cleaning raw coal with an average of 44% ash to a clean coal product of 30% ash. Initial performance data indicate that

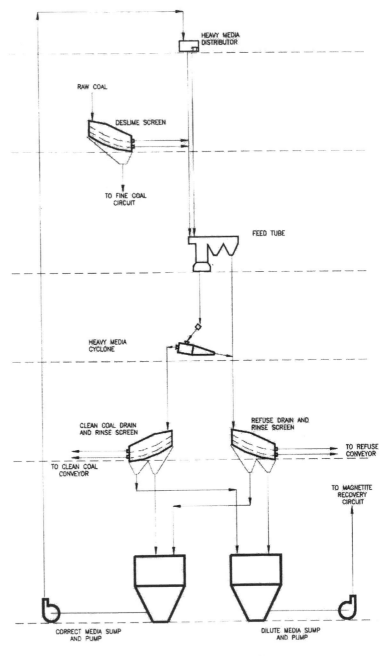

FIGURE 2   TYPICAL GRAVITY−FED HEAVY MEDIA CYCLONE CIRCUIT

the cyclone is performing at an organic efficiency greater than 96% (Organic efficiency is defined as actual yield divided by theoretical yield at the same clean coal ash).

Another one-meter heavy media cyclone circuit was installed at Pittston's UK #1 Preparation Plant. Plant testing results showed that organic efficiency was greater than 99% and probable error (E_p) was less than 0.04. Probable error is a measure of misplaced material based on the distribution curve. Chedgy et al. (1996) have previously published results from this work.

Another one-meter cyclone, along with two 800-mm cyclones, was installed at Cannelton Coal Company's Lady Dunn Preparation Plant. Additionally, four 800-mm cyclones were installed at US Steel Mining's Concorde Preparation Plant. Results from these installations have met or exceeded quality and performance expectations.

7   CONCLUSIONS

Large diameter heavy media cyclone circuits offer a cost-efficient method to process both coarse and intermediate size coals. The sizing of cyclones, design of feed systems and selection of screens and magnetic separators has been discussed. Overall circuit performance has proven to be highly efficient.

8   REFERENCES

Anonymous, "Banana Screen in Sizing Application," NSW Coal Preparation Society, July 1993.

Chedgy, D.C., J.V. Ghelarducci, M.A. Sharpe and S. Yu, "Climax$_x$ High Capacity Processor - A Novel Design in Coal Pre-Combustion Cleaning, " SME Preprint 97-153, Feb. 1997.

Chedgy, D. C. F. Addison, F. Stanley, and S.Yu, "High Capacity Heavy Media Processing," Coal Prep '96, Lexington, KY pp. 72-86, 1996.

Dynys, A.J. and J.V. Ghelarducci, "Performance of Climax$_x$ Magnetic Separators for High Efficiency, High Capacity Magnetite Recovery." 1999 International Symposium on Mining Science and Technology, Beijing, China, August 1999.

Mining Science and Technology '99, Xie & Golosinski (eds) © 1999 Balkema, Rotterdam, ISBN 90 5809 067 1

Use of computerized pattern recognition techniques for separation of waste rock from lump coal

Fuqiang Liu, Rong Shen, Xinhong Wang & Jianshen Qian
China University of Mining and Technology, Xuzhou, People's Republic of China

ABSTRACT: This paper describes application of advanced, computerized pattern recognition techniques to identification of coal and waste rock, and for automatic separation of the two. Industrial application of this method is expected to improve working conditions of workers, reduce the labor intensity of coal cleaning process, and greatly decrease the pollution of environment.

1. INTRODUCTION

The waste rock selection, separation of waste rock from lump coal is an important link in the production of mining. At present in our country there are two methods to separate waste rocks from lump coal. One is separating by hand, not only the intensity of labor is very high and the efficiency of manufacturing is low, but also the circumstance of working is bad, which do harm to the physical and mental health of workers heavily. The other is separating by machine, such as using oscillating screen. But if adopting this method, generally the lump coal needs to be fragmented in advance. So we will encounter two problems: one is the serious pollution of environment, the other is that partly float coal can't be concentrated.

In order to improve working environment and increase production efficiency and automation level of ore concentration, some researchers at home and abroad have carried out initial study. For example, in Spain R.Manana and J.J.Artieda analyzed the visual sense characteristics of mineral; In Australia Bullisban Asin private limited company developed a mineral color scanning system, using laser scanning technology to separate high-ash material form low-ash commercial coal. But there is still no efficient approach to separate massive waste rock form lump coal. These are sketchy researches in theory, and further work is needed. In application the test study has still not been in progress in our country.

In this paper we plan to use the technology of image process and pattern recognition to make theory and application study of automated separating waste rock. The application of image process and pattern recognition is mature in medical science, remote sense and military. But it is seldom used in mine. In this paper we will use these theory and technology to establish waste rock and lump coal's mathematics model and their recognition.

Therefore, based on advanced technology of computer, this paper adopts the technology of image process and pattern recognition to achieve the goal of automatic selection of waste rock. It can fill in the gaps in the fields of research at home and abroad. Once the contents of research are transformed to product and being spread and applied, it will improve working conditions of workers, reduce the labor intensity of workers and decrease the pollution of environment greatly, so as to increase the labor productivity by a big margin. It will have a good benefit of society and economy.

2. THE STRUCTURE OF SYSTEM

In the process we capture the original images from working face of selecting waste rock, then these

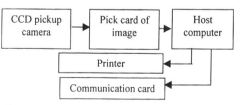

Figure 1

images are sent to our computer and disposed. The hardware system includes: (1) CCD pickup camera; (2) pick card of image; (3) host computer: computer upon 486, no less than 16M RAM. Operating system is Windows 95 and development language is Visual C++. The structure of system is shown by Figure 1.

3. THE RESEARCH OF OPERATOR

In reality the workers' judgement upon lump coal and waste rock is mainly according to the differences between density and texture. From this angle this paper discusses the problem on the separation of lump coal and waste rock according to their distribution situation. Histogram indicates the appearance frequency of some gray level in an image. It points out the range of density and approximate distribution in this image. The shape of histogram itself can also express the specialty of an image or the area of the image. If the peak value of histogram leans to low lightness, we can draw a conclusion that the average lightness is low; if the density range of histogram is narrow, its contrast can not have a good effect. From the angle of statistic histogram represents probability density function of the area. Therefore, its statistical texture such as mean and variance can be used to express the characteristic differences between mutual sorts.

In the images we want to deal with lump coal and waste rock. From the distribution of histogram both of them have evident specialty of double pear. Lump coal itself is dark and its density is low. Only when lump coal reflects light, the parts reflect light appear bright point and the gray level of these parts will be high. In other words, most parts of lump coal concentrate on "dark area", while the gray level of waste rock generally will be high. So in the whole image the lightness are the parts that waste rock and lump coal reflect light. We take the mean and variance of density distribution as characteristic value to distinguish with lump coal and waste rock . The flowchart of system process is shown by Figure 2.

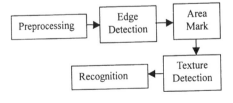

Figure 2.The flowchart of process.

3.1 *Image preprocess*

There is much coal dust on work face of selecting waste rock and the light intensity of illumination is not in good condition. The purpose of image preprocess is to make image clear and edge prominent. Image preprocessing includes two steps: adjusting the lightness and contrast of original image and smoothing filter. It will make the characteristic of image obvious and make recognition easy to go on if lightness and contrast adjust well. The purpose of smoothing filter is to remove sharp and discontinuous noise point and prepare for edge detection. We adopt mean value filter.

Mean value filter can effectively restrain interfering pulse and point noise, at the same time it can keep the information of edge. Mean value filter is a slip window containing odd pixel. After sorting median is used as the output for the center pixel of window. Scanning image $f(i,j)$ in use of window w and supposing the center pixel of window $g(m,n)$, then

$$g(m,n)=Median\{f(m-k,,n-l),(k,l) \in w\} \quad (1)$$

3.2 *Edge detection*

Due to analyzing the density distribution of each coal or waste rock in an image, it is necessary to find out its area position. We adopt the operator of edge detection to detect edge, so as to find out the position of each coal or waste rock.

Edge is the reflection of discontinuity of local characteristic in an image. It can draw the outline of object, which means end of an area and start of the other area. Presently edge detection often adopts edge operator, surface fitting operator, template matching operator, and threshold processing etc. Here we adopt Sobel operator. Edge can correspond with the comparative change between areas. Sobel operator is a kind of gradient operator and it has a good effect of edge sensing.

As for digital image $f(i,j)$

$$S(i,j)=|[f(i-1,j+1)+2f(i,j+1)+f(i+1,j+1)]- [f(i-1,j-1)+2f(i,j-1)+f(i+1,j-1)]|+|[f(i-1,j-1)+2f(i,j-1)+f(i,j+1)]- [f(i+1,j-1)+2f(i+1,j)+f(i+1,j+1)]| \quad (2)$$

Adopting suitable threshold *THs*, if $S(i,j)>THs$, then *(i, j)* will be edge point.

3.3 *Area mark*

If the edge of each coal or waste rock in an image is found, we ought to track edge and mark in allusion

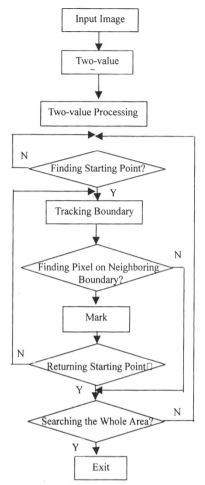

Figure 3.The flowchart of area mark program.

to every close area. Boundary tracking adopts the left-hand rule. After scanning binary edge image and searching pixel 1 unmarked (assume it $f(0)$), we start the tracking of one boundary. Searching pixel 1 in anticlockwise direction in NO.8 neighbor area of $f(0)$, the first found is called $f(1)$. Then go on searching pixel 1 in anticlockwise direction in NO.8 neighbor area of $f(0)$, the first found is called $f(2)$. Repeat this tracking until returning starting point $f(0)$ and the tracking of one boundary ends. During the process of tracking each $f(i)$ is endued a tracking mark respectively. Then we start the tracking of another boundary and endue different tracking mark for this boundary.

3.4 *Density analysis*

According to the coordinate of each marked area, if it is reflected to original image, we can analyze density distribution of each coal or waste rock respectively. Comparing with the histogram of a lump of coal and waste rock, we can know that the curves of their density distribution are obviously different. Lump coal's curve of density distribution is sharp, while waste rock's is flat. In other words, the density of lump coal mostly concentrates on a narrow range, while the density distribution of waste rock is well-proportioned. Though analyzing mean and variance of these two kinds of image, we can discriminate them.

Assume an image ($M \times N$):

Mean: $\quad \mu = \sum\limits_{i=1}^{M} \sum\limits_{j=1}^{N} f(i,j) \Big/ M \times N \quad$ (3)

Variance: $\quad \sigma^2 = \sum\limits_{i=1}^{M} \sum\limits_{j=1}^{N} f(i,j) - \mu \Big/ M \times N \quad$ (4)

where density $= f(i,j)$, $i=1,2,\ldots,M; j=1,2,\ldots,N$.

4 RESULTS

Figure 4a is an original image of lump coal and

| a. Original image | b. Preprocessed image | c. Edge sensing |

Figure 4:The progress of image processing

$\mu = 154.2$ $\mu = 169.61$
$\sigma = 91.91$ $\sigma = 77.49$

a. The image of lump coal
b. The image of waste rock

Figure 5.The balance of lump coal 's histogram and
waste rock's histogram.

waste rock we used during the test process. Figure
4b is a preprocessed image. The edge image
adopted Sobel edge sensing operator is shown in
Figure 4c. Figure 5 involves an original image of a
lump of coal and waste rock as well as their
respective histogram distribution, mean and
variance. We can see there are great differences
between their mean and variance, so mean and
variance can be used as characteristic value to
identify lump coal and waste rock.

REFERENCES

Kenneth. R. Castleman. 1998. *Digital imag
processing.* Prentice-Hall International, Inc.
Liu .1997. Research on the multimedia monitoring
system in coal mine and its key technologies.
Journal of China Coal Society: 412-416.
Liu.&Yu.&Ma.1997. Study on the management
system of mine's multimedia monitoring in
dispatching room. *Journal of China University
of Mining & Technology*: 23-26.

Mining Science and Technology' 99, Xie & Golosinski (eds) © 1999 Balkema, Rotterdam, ISBN 90 5809 067 1

MZ series of coal flotation agents

Hong Zhu, Xia Chen & Zeshen Ou
China University of Mining and Technology, Xuzhou, People's Republic of China

ABSTRACT: This paper discusses three factors that must be considered in designing a flotation agent: floatability of coal, collectivity and selectivity of the agent, and economic effect of its use. The chemical and petroleum industry by-products are used to develop the MZ series of highly efficient coal flotation agents suitable for processing of various coals. These agents were subject to an experimental study, which confirmed that they have high coal collection ability and good selectivity, and their use has a beneficial economic effect.

1. INTRODUCTION

In China most coal-preparation plants have flotation processes, in which the coal slurry accounts for 14% of the total amount of coal, that is, every year more than 25,000,000 tons of coal slurry have been treated. At present, the coal collectors used are kerosene and diesel oil mixed with small amount of chemical products. Those agents are low in efficiency and high in consumption. In recent years, most countries have been developing chemical by-products as flotation agents to substitute for petroleum products, in order to lower cost and raise the yield of cleaned coal(R.H.Yoon 1991).

Sufficiently utilizing chemical by-products, and changing waste materials into things of value are practical in China at present. In recent years, we have preliminarily explored and experimented on low price by-products from petrochemical works, coking plants, wood processing and other chemical processing plants. We have made up a series of MZ agents as coal collectors agent, with the function of producing foam, and good flotation effect. .

2. BASES FOR DESIGNING FLOTATION AGENT

In design of coal collector the following respects should be considered:
(1) flotability of coal.
(2) collectivity and selectivity of agent.
(3) the ratio of property to price of the agent.

2.1 *The factors which influence the floatability of coal*

The floatability is a synthetic result, which refers to hydrophobic property of coal surface, density, size of component and ash content.

The whole particle surface of pure coal is hydrophobic. But the degree of hydrophobic is related to that of metamorphic grade of coal. The coal with medium metamorphic grade, such as fat coal and coking coal, has the best hydrophobicity. When the coal is stored for a long time, its surface would be oxidized and weathered, then the hydrophobicity is lowered, because there are more hydrophilic functional groups produced, which contain oxygen atom. Even the humic acid and similar compounds would form.

Coal slurry is composed of many components with different density. The low density components contain few mineral substances, which produce ash, and have good hydrophobicity. The high density components contain more mineral substances, and has poor hydrophobicity. If more components have medium density, this coal is difficult to be separated, and impossible to obtain two products: the low ash cleaned coal and the high ash tailing .

The size of coal slurry component has extreme influence on the consumption of flotation agent and selectivity of flotation process. Every kind of coal has their best size range, usually the size is 0.1~0.3mm. The fine particle has huge specific surface, which would adsorb a large amount of flotation agent, would affect the float of aiming particle. The content of super fine particles with high ash in coal slurry would evidently influence the flotation. Because they expend large quantity of agent, and usually cover the surface of fine coal, the decreasing hydrophobicity is resulted and the recovery of cleaned coal decreases either. If the coal contains some wastes, which would easily become muddy, the coal separating process will produce a great deal of very fine (from few micron to tens micron) high ash mire. Some mire covered the surface of cleaned coal, some mire existed in the water layers between the bubbles carried to cleaned coal, which leads to staining the cleaned coal.

It's evident that the flotable property is not only the function of hydrophobicity, but also is influenced by the coal density, the size of component, the ash content and the mineral composition of waste. Obviously, when the coal slurry has different floatability, the flotation agent used must have different characters. When the slurry has poor hydrophobicity and contain a little high ash fine mire, agent with strong collective ability must be used; while coal slurry has good hydrophobicity, but contain more high ash fine mire, the agent with high selectivity must be used. Select right flotation agent for every kind of coal slurry must be determined by the experiments.

2.2 *The factors to influence the character of flotation agent (Guo MengXiong 1989)*

The collective quality of non-polar hydrocarbon oil mainly influenced by three factors: hydrophobicity, viscosity and solubility in water.

The longer carbon chain of the hydrocarbon molecule, the higher distillation temperature and the better hydrophobicity it has. But the long chain molecules bears higher viscosity, lower solubility and worse dispersion ability in coal slurry. For this reason, the sequence of flotation activity of each component of hydrocarbon oil are: aromatic hydrocarbon > unsaturated hydrocarbon > isomerized alkane > cyclic alkane > normal alkane. This order had been proved by experiment.

The existence of complex polar molecules gives strong affects on collective ability of hydrocarbon oil. First of all, the polar terminal of the complex polar molecule is adsorbed on polar area of coal surface and the non-polar terminal points to water, it improves the hydrophobicity of coal surface and furthermore the adsorption of hydrocarbon oil on coal surface. Next, polar molecules possess dispersion ability for hydrocarbon oil, they use the non-polar terminal to adsorb hydrocarbon, and the polar terminal point to water, so that it decreases the surface tension of oil-water interface, promotes the oil drop to disperse, and obstruct the oil drops to join together. At last, the complex polar substances can

produce peptization, because they are adsorbed on surface of fine mire, which changes the surface electricity, and then gets off from coal surface. The intensity of above three functions is related to the ratio of polar terminal to non-polar terminal in the complex polar molecule. So that some polar molecules can raise the collective character of hydrocarbon, and some can raise the selectivity. Therefore, according to the coal property we can make up suitable floatation agents.

2.3 *Economic factor*

In order to modify the effect of hydrocarbon oil, we can treat the kerosene or diesel oil by catalytic oxidation to produce or made up complex polar substances by adding selected appropriate surface-active agent and solubilizer. But these are expensive ways. In processes of petro-chemistry, wine-making, fertilizer production, detergent processing, there are a number of by-products containing the complex polar substances, we can select the appropriate by-products to make up flotation agents. This method has a weakness, that is, the composition of raw material lacks stability, so that every batch must be analyzed.

By experiment, we select some by-products of petrochemical plants and other chemical works to analyze and experiment with, and make up a series of coal collectors to suitable for different coals.

(1) MZ-101 is suitable for coal of hard to float, containing low high-ash-mire.

(2) MZ-102 is suitable for coal of easy to float, containing more high-ash-mire.

(3) MZ-103 is suitable for coal of hard to float, containing medium high-ash-mire.

(4) MZ-104 is suitable for coal of hard to float, containing more high-ash-mire.

3. EXPERIMENT RESULTS AND DISCUSSION

3.1 *Contrast Tests*

Synthetic MZ series high effective collectors for coal flotation was used in experiment with different kinds of coals, and comparison with ordinary collectors (kerosene or diesel oil) was made either. The results of experiment shown in Tab.1, bothe Jiulishan and Zhaozhuang coal are difficult to separate. When ordinary collector was used, the recovery of cleaned coal is low. For instance, the kerosene is double of MZ-101, producing the Jiulishan cleaned coal with nearly same ash content. Meanwhile, the recovery of cleaned coal using kerosene is 24% lower than that using MZ-101. The coal-preparation plants want the ash content in

Table 1. The Results of Coal Flotation with MZ-series Agents for Different Coals

Coals	Agent Name	Dosage g/t	Froth		Tailing		Feed
			A%	r%	A%	r%	A%
Jiulishan (Anthracite)	MZ-101 GF	700 200	11.93	85.40	78.84	14.60	21.70
	Kerosene GF	1520 83.0	11.55	61.50	40.82	38.50	22.78
Zaozhuang (Coking coal)	MZ-101 GF	1020 81.0	15.85	89.26	73.65	10.74	22.06
	Kerosene GF	1020 120	16.91	47.30	26.05	52.70	22.24
Xinglong Zhuang (Fat coal)	MZ-102 GF	500 100	9.85	72.30	55.28	27.70	22.24
	Diesel oil GF	800 150	9.57	70.20	50.07	29.80	21.62
Baguanhe (Coking coal)	MZ-103 GF	500 100	11.12	58.20	33.20	41.80	20.35
	Diesel oil GF	500 100	11.24	49.50	29.48	50.50	20.45
Shitai (Coking coal)	MZ-104 GF	326 72.0	12.38	83.51	65.51	16.49	21.14
	Diesel oil GF	650 150	12.60	59.70	42.73	40.30	24.74

tailing higher than 75%, then discard the tailing. But by using kerosene or diesel oil, the ash content is far behind this desire.

It is shown in Tab.1, when collector is kerosene, the dosage used is 1520 g/t, the ash content in the tailing coal just is 40.82%, and use MZ-101 as collector only 700 g/t, the tailing ash is 78.84%, exceeding the desired level., Zhaozhuang coal also get good result.

3.2 MZ-101 Agent

MZ-101 agent is specially synthesized agent based upon the coal character and technical needs. First, from many petrochemical by-products is found out some oils, one of which is named floating oil. The result of analysis shows floating oil is a mixture of many components. In addition to a great deal of alkanes, it contains fatty amines (17.95%), various fatty acids (12.8%) and small amounts of aldehydes, ketones, alkenes and phenols (6.4%), in which the fatty amines and fatty acids separately belong to cationic and anionic surface-active agents. All would strengthen the collective ability for coal. Known from experiment, the collective ability of fatty amines is more powerful. The floating oil contains aldehydes, ketones and olefins also, they were all efficient compositions, so that it is possible to use floating oil to synthesize flotation agents, but the defect is that it contains some phenols, which is poisonous and has bad smell. Therefore, it is

necessary to take off phenols from floating oil, then the floating oil can be used as main raw material to synthesize the high efficient collector (MZ-101). This is an ideal collector for coal hard to separate.

3.3 MZ-102, 103, 104 Agents

Collectors MZ-102, 103 and 104 are made up by using other petrochemical and chemical by-products. They have appropriate collective ability and good selectivity. Shown in Tab.1, one can see that they also get good effect. For Xinglongzhuang coal, when the cleaning has similar ash content and recovery, the consumption of collector decreases 60%; and for flotation of Baguanhe coal, when same amount of oil is consumed and the clean coal has similar ash content, the recovery using MZ-103 as collector is 9% higher than that of using diesel oil; for Shitai coal, when the cleaning has similar ash content, using of MZ- 104 decreases half the amount of collector, at the same time, the recovery of cleaning increases 23.8%. All these facts shown MZ series of collectors have attained high technical and economical effect.

CONCLUSION

1. Only considering the floatability of coal, collective ability and selectivity of agent, and economic factor at the same time, the design of

flotation agent would get reliable bases. The floatable property of coal is the comprehensive result of hydrophobicity of coal surface, density, size component and ash content.

2. Adding of appreciate amount of surface-active agent into hydrocarbon oil can raise the collective ability of reagent, at the same time it can promote the dispersion of hydrocarbon oil and produce the peptization to the mire in the coal slurry. It can modify the flotation effect.

3. The developed MZ coal collector can be made up according to the coal property, so it is adaptable to different coals. The flotation process that uses MZ-agents is less consumption, low cost and good effect. So this process will get distinctly economic effect.

REFERENCES

R.H.Yoon 1991. Hydrodynamic and Surface Forces in Bubble Particle Interactions. Dept. of Mining and Minerals Engineering, Virginia Polytechnic Institute and State University Blacksburg, Virginia 24061. 1991.

Guo MengXiong 1989. Flotation, Publishing House of CUMT, 1989 ,141.

Mining Science and Technology' 99, Xie & Golosinski (eds) © 1999 Balkema, Rotterdam, ISBN 90 5809 067 1

Flotometric hydrophilicity of coal particles

Jan Drzymala

Technical University of Wroclaw (I-II), Poland

ABSTRACT: Experimental contact angles of coals depend on the procedure of measurement. The angles are high on polished and dry surfaces measured with a sessile water drop, smaller on wet and polished surfaces measured with a captive bubble, still smaller on particles attached to bubbles, and very small on coal particles submerged in water and measured by dynamic flotation methods. It was shown that the flotometric hydrophilicity and small hydrophobicity of coal particles most likely result from very low receding water-contact angles of coals. Possible contribution of other factors to the hydrophilicity of coal was also discussed.

1 INTRODUCTION

Coal is an important industrial material and source of energy. Run of mine coals and selected products of coal preparation contain unwanted ash-forming minerals which can be separated from the carbonaceous matter by various methods including gravity, flotation, and oil agglomeration techniques. Therefore, a knowledge of surface properties of coal is very important. It is widely believed that coal is hydrophobic. Its hydrophobicity has been measured by numerous researchers, for instance by Brady & Gauger (1940), Horsley & Smith (1951), Gutierrez-Rodriguez et al. (1984), and recently by Holysz (1996). The contact angle of coal in the absence of any reagents has been measured by various techniques. It is now well established that the contact angle depends not only on degree of coalification but also on the method of measurement. When the contact angle is measured by the sessile water drop method, the bituminous coals appear to be very hydrophobic with contact angles between $40°$ and $80°$. However, it is evident from microflotation tests that bituminous coals in the absence of any reagents either float very little or do not float at all. It can be seen in the paper of Klassen & Laskowski (1964) that the microflotation recovery of the 0.385-0.49 mm size fraction of a high volatile bituminous coal is only 5 percent. Similarly negligible collectorless and frotherless flotation of bituminous coals was reported by Holysz (1996). Numerous other flotation tests also point to a negligible or lack of flotation of various coals while their sessile water drop contact angles are very high.

In this paper an attempt was undertaken to examine closer the apparently high hydrophobicity and exceptionally low floatability of coals in the absence of frothers and collectores.

2 EXPERIMENTAL

2.1 *Materials*

Polish coals were used in the flotation tests. Their rank and corresponding carbon content are given in Table 1. Other solids used in the flotometric tests were graphite from Sri Lanka and gilsonite. Gilsonite is a natural hydrocarbon also known as grahamite with density of 1.35 g/cm^3 and carbon content about 40% (Lewis, 1994).

Table 1. Characteristics of coals used in flotometric tests. Sulfur content, heat of combustion, and ash content can be found elsewhere (Sablik 1986).

Coal, Polish classification	H$_2$O %	C daf, %	H$_2$ daf, %	O$_2$ daf,%
31.1	13.18	78.6	4.70	13.5
32.1	6.33	81.4	4.84	12.2
33	2.07	85.8	5.12	6.6
34	1.74	85.5	5.26	6.7
35	1.46	87.4	5.01	5.1
37	1.43	89.7	4.38	3.0
42(anthracite)	0.9	92.0	3.04	1.3

2.2 *Flotometric tests*

Flotation tests were carried out according to a procedure called flotometry (Drzymala & Lekki,

1989). In a typical test a sample of 0.2 cm³ of coal was introduced into 120 cm³ of distilled water in a beaker and agitated for 10 minutes. Then, the mixture was transferred to a Hallimond tube and floated with air at a constant flow rate equal to 0.625 cm³/sec. The time of flotation was 30 minutes. The flotation recovery was measured as a function of time by reading the volume occupied by particles reporting to the receiver of the Hallimond tube. The flotation tests were carried out at 20°±2°C for different narrow size fractions of each coal at natural pH.

When hydrophobicity of particles is greater than about 25° there is an interaction between particles which reduces the recovery of floating particles in the Hallimond tube (Drzymala 1999a). Then, it is more accurate to perform flotation tests with individual particles and such tests were carried out for most coals. For the one particle-flotation test, ten particles were selected from each needed for testing size fraction and equilibrated in distilled water for 10 minutes. Then, one particle and the solution were transferred to the Hallimond tube. If within 30 minutes of bubbling the air through the tube particle reported in the calibrated receiver of the Hallimond tube, it was counted as floatable. Subsequently, the particle was removed from the solution, and a second particle introduced for flotation. In this way all ten particles were subjected individually to flotation. The flotation recovery was calculated as a ratio of the number of floating particles to all 10 tested particles multiplied by 100%.

3 RESULTS

Flotometric tests provide relationships between recovery of particles subjected to flotation and time of flotation (Fig. 1). The curves, representing the kinetics of flotation, initially increase with time of flotation and then level off. This occurs, depending on the solid and its size fraction, after 10 to 15 minutes of flotation (Fig.1a). Therefore, the recovery after 30 minutes of flotation provide the maximum recovery for a given size fraction of particles floated in the Hallimond tube. When the maximum recovery is plotted as a function of particle size, we obtain the so-called separation curve (Fig.1b) with D_{50} in the middle of the curve. The relationship between maximum recovery and particle size should be vertical while in practice (Fig. 1b) it is bell-shaped due to fluctuations in the chemical composition of the grains, variations in particle geometry within a given fraction used for flotation, and also from the imperfect performance of the flotation cell as a

separation device. D_{50} well represents the flotation system because each particle having a size equal to D_{50} possesses the same odds for floating as well as for sinking. D_{50} is therefore the maximum particle size which can be successfully floated ($D_{50} = D_{max}$), though the recovery of the whole size fraction having the mean diameter equal to D_{50} is only 50%.

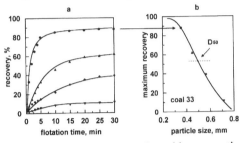

Fig.1. Flotometric tests (a) and resulting separation curves for coal 33 (b). D_{50} from separation curve is used to calculate detachment contact angle. Size fractions used in the tests: ● - 0.315-0.40, ▲ - 0.40 - 0.50, ◆ - 0.50-0.63, ▼ - 0.63- 0.80 mm.

D_{50} together with the density of the particle can be used to determine the hydrophobicity of the investigated material in terms of contact angle, utilizing an equation which relates the maximum size of floating spherical particle and its density with detachment contact angle (Drzymala 1994b):

$$\pi r_{max}\sigma(1-\cos\theta_d)-[(4/3)\pi r^3{}_{max} \rho_p g - \pi r^3{}_{max}\rho_w$$

$$g \{2/3+\cos (\theta_d/2) - (1/3)\cos^3(\theta_d/2)\} - \pi r^2{}_{max}$$

$$(1 -\cos \theta_d) (\sigma/R - R\rho_w g) = 0 \qquad (1)$$

where r_{max} is the maximum radius of floating particle ($r_{max} = D_{max}/2$), θ_d is the detachment angle, σ surface tension of water, ρ_p density of particle, ρ_w density of water, R radius of bubble (1.55 mm), g acceleration due to gravity, π is equal to 3.14. It should be noted that the hydrodynamic, inertial, and particle-particle interaction forces were not taken into account in Equation 1.

When only capillary and gravity forces as taken into account, Equation 1 reduces to the so-called Scheludko equation (Scheludko et al. 1976):

$$D_{max}{}^2 \rho' = 6\sigma g^{-1} \sin^2(\theta_s/2) \qquad (2)$$

where θ_s is the approximate detachment contact angle and ρ' stands for ρ_p-ρ_w. Equation 2 is accurate for the detachment contact angles smaller than about 30°. Equations 1-2 are valid for the detachment contact angles between 0° and 90°. Above 90° there is a rupture of the bubble rather than the bubble-solid interface, and therefore the flotometric method is not able to distinguish between contact angles equal to $90°$ and greater. This is a theoretical upper limit of detachment contact angle which can be measured by flotometry.

Having the detachment contact angle θ_d one can calculate the equilibrium contact angle θ related to the gravity-free system as it is required by the thermodynamic definition of the contact angle in the Young equation. It results from the geometry of the system (Drzymala, 1994b) that:

$$\theta = \arcsin\left[(r_{max}/R)\sin(\theta_d/2)\right] + \theta_d/2 \quad (3)$$

Equations 1-3 describe flotation systems when one particle is attached to one bubble. Since in the flotometric tests the hydrophobic interactions of particles can be neglected for $\theta < 20°$ (Drzymala 1999a), the determination of small detachment contact angles can be performed using individual particles as well as many particles at once. For more hydrophobic particles the flotometric tests should be carried out with individual particles and the contact angles calculated from Equations 1-3. It is also possible to run flotometric test with a collection of particles and use another formula for the detachment contact angle (Drzymala, 1999a).

$D_{50} = D_{max}$ and the density of particles can be used for calculating hydrophobicity provided that the entrainment of particles, which always occurs in flotation cells, is negligible. The entrainment of particles in our flotation Hallimond tube is govern by the equation (Drzymala and Lekki 1989, Drzymala 1994a):

$$a_{max}(\rho_p - \rho_w)/\rho_w = L_H = 0.023 \pm 0.002 \text{ cm} \quad (4)$$

for particles with density greater than about 2.0 g/cm^3. For particles with density smaller than 2.0 g/cm^3 the following equation should be applied (Drzymala 1994a, 1999b):

$$a_{max}((\rho_p - \rho_w)/\rho_w)^{0.75} = L_L = 0.0225 \pm 0.0025 \text{ (cm)} \quad (5)$$

In Equations 4-5 the term a_{max} stands for the maximum size of entrained particles, ρ_p for particle density, ρ_w denotes density of water, while L_H and L_L are constants. It was shown in previous papers (Drzymala 1994a, 1999b) that Equation 4 results from Newton's turbulent mode of settling of particles behind levitating bubble while Equation 5 from intermediate laminar/turbulent setting mode. In this work the validity of Equation 5 for the delineation of entrainment of hydrophilic materials in our Hallimond tube was checked against gypsum mixed with oak wood dust.

The results of entrainment measurements, flotometric tests with coals and other solids, and calculated contact angles are shown in Table 2.

Table 2. Results of flotation. pH of flotation was 6.5-6.9 except for coal 37 for which the pH was pH=6.1. D^1_{max} - results of one-particle flotometric tests, D_{max} - results of multi-particle flotometric tests.

Sample	ρ	D^1_{max}	D_{max}	L_L	θ_d	θ
	g/cm^3	mm	mm	cm	deg	deg
coal 31.1	1.30	0.485	0.450	0.020	4.5	2.6
coal 32.1	1.20	0.490	0.490	0.019	4.0	2.3
coal 33	1.26	0.530	0.530	0.020	5.0	2.9
coal 34	1.26	0.565	0.550	0.020	5.2	3.0
coal 35	1.20	0.775	0.715	0.022	6.0	3.7
coal 37	1.45	0.475	0.460	0.025	5.6	3.2
anthracite	1.36	0.420	0.410	0.021	4.5	2.5
graphite	2.28	-	0.315	0.038	6.4	3.5
*gypsum +wood(1:1)	1.32	-	0.515	0.022	-	-
*gypsum +wood(3:1)	1.64	-	0.325	0.023	-	-
coal 42 +3MNaCl	1.36	-	0.575	0.027	6.4	3.8
coal 35+ Na₂PO₄**	1.20	-	0.775	-	6.0	3.7
paraffin on coal 32.1	1.20	6.65	-	-	>90	-
gilsonite	1.37	-	1.7	0.081	23	18

*hydrophilic samples used as standards, **10^{-3}M

4 DISCUSSION

It results from the flotometric tests that the investigated coals do not float well. Their floatability, characterized by parameter L_L, is similar to the mechanical carryover of hydrophilic standard materials. Thus, from the flotation point of view the studied here coals are hydrophilic, except for coal 35 and graphite, which are weakly hydrophobic. If one assumes that the observed recovery of the coal particles is not due to entrainment but due to flotation, the calculated contact angles would be very small and equal to $4°$ - $6°$ (Table 2 and Fig. 2). Flotation tests carried out by Li & Somasundaran (1993) by the so-called levitation technique, which relays on gentle lifting wet 0.1-0.4 mm coal particles by ascending water-air interface, point to a slightly hydrophobic nature of bituminous coal particles. This observation indicates that not only coal 35 but other coals may be slightly hydrophobic with their flotometric hydrophobicity masked by entrainment.

It can be seen in Figure 2 that the greatest contact angles $(40°-75°)$ coals is observed when hydrophobicity is measured by the sessile water drop method on polished dry coal surfaces. Smaller hydrophobicity of high rank coals is observed when the contact angle is measured on polished coal surfaces by the captive bubble method, especially for low rank coals with moisture content above 2 percent. The reduced hydrophobicity very likely results from the presence of a water film on the surface of coal formed when this heterogeneous material is immersed in water. It is not clear whether or not polishing itself influences contact angle. On one hand polishing may orient the graphite-like structures present in high rank coals (Mahajan 1984) and carbon blacks (Marsh & Menendez 1989) parallel to the surface. Since (0001) planes of graphite interact with water by the van der Waals forces only, this surface is very hydrophobicity while other planes, due to broken bonds, are hydrophilic or only weakly hydrophobic (Cooper et al. 1985). On the other hand it was demonstrated by Vargha-Butler et al. (1986) that contact angles on freshly cut coal without polishing, roughly polished surface, and on highly polished coal surfaces did not change in a noticeable pattern.

When hydrophobicity is measured by the static bubble attachment method developed by Hanning & Rutter (1989) on particles instead on polished surfaces, the contact angle of coal is still lower and lies between $12°$ to $29°$ (Fig.2). Similar numbers were reported by Good & Keller for receding contact angles measured by the sessile water drop method on polished coals.

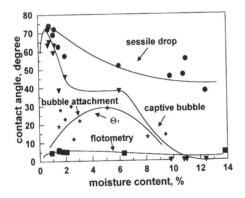

Fig.2. Comparison of contact angles of coals measured by different methods as a function of coal moisture: sessile drop and captive bubble - Gutierrez-Rodriguez et al. (1984), bubble attachment (★)-Hanning & Rutter (1989), sessile drop water receding Θ_r - Good & Keller (1989), flotometric contact angles - this work.

Consequently, the smallest values of contact angles should occur for water receding from coal particles pressed with a captive air bubble as it happens in the first stage of flotation process. Thus, the flotometric contact angles of coals are small or none because the water receding contact angles on coal particles are either very small or zero. The small water receding contact angles of coal particles immersed in water, in turns, are caused by the presence of the water films, resulting from a mosaic of hydrophilic

/hydrophobic sites on the surface of coal. It is also known (Good et al. 1990) that water receding contact angle indicates the population of hydrophilic sites while the water advancing contact angle the concentration of the hydrophobic sites. Since the receding contact angle for bituminous coals is small and the advancing angle is high, the amounts of hydrophilic and hydrophobic sites in these coals are of the same order.

There are other factors which can reduce the flotometric contact angles of coals. Among them are slime coating (Jowett et al. 1956), adsorption of humic acids released from coals (Laskowski 1992), water molecules forming bridges anchored to the hydrophilic sites of coal (Laskowski 1995), pores filled with water (Laskowski 1995), low density and low inertia of coal particles, and dynamic nature of flotation tests. These factor play rather negligible role because additional flotometric tests carried out with graphite, gilsonite, and coals in the presence of Na_2HPO_4 (slimes and humic acids disspersant, Jowett et al. 1956) and NaCl (surface water bridges destroyer, Laskowski 1995) showed no much improvement of the flotometric hydrophobicity. Since low density gilsonite and coal particles coated with paraffin floated well, the influence of low inertia of coal particles on flotation can also be ruled out. Only the dynamic nature of flotation may somehow reduce the flotometric contact angles.

5 CONCLUSIONS

It results from this work that the contact angle of coals significantly depends on the method of measurement. The highest contact angles are observed by the sessile drop method, smaller by captive bubble, still smaller by bubble attachment and lowest by flotometry. Thus, the discussed methods are sensitive to different contact angles ranging from water advancing to water receding. Most likely zero or small water receding contact angles observed in the coal particle-air bubble-water system are responsible for the flotometric hydrophilicity or only weak flotometric hydrophobicity of coals. Some other factors such as dynamics of flotation process, water in pores of coal, lack of graphitic layers oriented parallel to the surface as in polished coals, slime particles, humic acids adsorption, and low inertia of particles may also slightly contribute to the flotometric hydrophilicity of coals.

ACKNOWLEDGEMENTS

Flotation tests were carried out by Ms.A. Szulmanowicz and Mr. A.Kawaler. The author wish to thank prof. J.Sablik for supplying coal samples.

REFERENCES

Brady, G.A. & A.W. Gauger 1940. Properties of coal surfaces. *Industrial and Eng. Chem.* 32(12): 1599-1604.

Cooper, H., M.C. Fuerstenau, C.C. Harris, M.C. Kuhn, J.D. Miller & R.F. Yap 1985. Flotation. In: N.L. Weiss (ed.), *SME Mineral Processing Handbook*: v 1, Ch.5, 1-109. New York: SME.

Drzymala, J. & J. Lekki 1989. Flotometry another way of characterizing flotation. *J. Colloid Interface Sci.* 130: 205-210.

Drzymala, J. 1994a. Characterization of materials by Hallimond tube flotation. Part 1: maximum size of entrained particles. *Int. J. Miner. Process.* 42: 139-152.

Drzymala, J. 1994b. Characterization of materials by Hallimond tube flotation. Part 2: maximum size of floating particles and contact angle. *Int. J. Miner. Process.* 42: 153 -167, and Erratum 1995, 43:135.

Drzymala, J. 1999a. Characterization of materials by Hallimond tube flotation. Part 3: maximum size of floating and interacting particles. *Int. J. Miner. Process*, in press.

Drzymala, J. 1999b. Entrainment of particles with density between 1.01 and 1.10 g/cm^3 in a monobubble Hallimond flotation tube. *Minerals Engineering*, in press.

Good, R. & D.V. Keller 1989. Fundamental research on surface science of coal in support of physical beneficiation of coal. *Quarterly report no. 7.* State University of New York at Buffalo, p.25.

Good, R.J., N.R. Srivasta, M. Islam, H.T.L. Huang & C.J. van Oss 1990. Theory of the acid - base hydrogen bonding interactions, contact angles, and the hysteresis of wetting: application to coal and graphite surfaces. *J. Adhesion Sci. Tech.* 4:607-617.

Gutierrez-Rodriguez, J.A., R.J., Purcell & F.F. Aplan 1984. Estimating the hydrophobicity of coal. *Colloids and Surfaces* 12: 1-25.

Hanning, R.N. & P.R. Rutter 1989. A simple method of determining contact angles on particles and their relevance to flotation. *Int. J. Mineral Process.* 27: 133-146.

Holysz, L. 1996. Surface free energy and floatability of low rank coal. *Fuel* 75(6): 737-742.

Horsley, R.M. & H.G. Smith 1951. Principle of coal flotation. *Fuel* 30: 54-63.

Jowett, A., H. El-Sinbawy & H.G. Smith 1956. *Fuel*, 35: 303-308.

Klassen, W.I. & J. Laskowski 1964. Relationship between zeta potential, surface potential and flotation, *Przemysl chem.* 43(1):12-16, in Polish.

Laskowski J. S. 1992. Oil assisted fine particle processing. In J. Laskowski & J. Ralston (ed.), *Colloid Chemistry in Mineral Processing*: 361-394. New York: Elsevier.

Laskowski, J.S 1995. Coal surface chemistry and its effect on fine coal processing. In S.K. Kawatra (ed.), *High Efficiency Coal Preparation*: 163-176. Littleton, SME.

Lewis, D.H. 1994. Gilsonite. In D.D. Carr (senior ed.), *Industrial Minerals and Rocks*: 535-541. Littleton: SMME Co.

Li, C., P. Somasundaran & C.C. Harris 1993. A levitation technique for determining particle hydrophobicity. *Colloids and Surfaces, A. Physicochemical and Engineering Aspects* 70: 229-232.

Mahajan, O.P. 1984. Physical characterization of coal. *Powder Technology* 40: 1-15.

Marsh, H. & R. Menendez 1989. Mechanism of formation of isotropic and anisotropic carbons. In M. Harsh (ed.), *Introduction to Carbon Science* : 197-228, London: Butterworths.

Sablik, J. 1986. Surface properties of coals and promoters of their flotation, *Prace Glownego Instytutu Gornictwa (seria dodatkowa)*, in Polish.

Scheludko, A., B.V. Toshev & D.T. Bojadjiev 1976. Attachment of particles to a liquid surface (capillary theory of flotation). *J. Chem. Soc., Faraday Trans. I*, 12: 2815-2828.

Vargha-Butler, E.I., M. Kashi, H.A. Hamza & A.W. Neuman 1986. Direct contact angle measurements on polished sections of coal, *Coal Preparation*: 3: 53-75.

Mining Science and Technology' 99, Xie & Golosinski (eds) © 1999 Balkema, Rotterdam, ISBN 90 5809 067 1

Research on acceleration of elastic flip-flow screen surface

Yaomin Zhao, Chusheng Liu & Maoming Fan
China University of Mining and Technology, Xuzhou, People's Republic of China

ABSTRACT: This paper established the mechanics model of flip-flow screen. Based on the theory analysis, the authors obtained the differential equation of screen surface, the distribution formula of acceleration along the surface of screen, the acceleration formula and the numerical solution. All the formulas were tested by experiments. Based on the above, the authors designed a model flip-flow screen and did the pilot experiments. All the results obtained will instruct us in designing and applying the flip-flow screen in industry.

1. INTRODUCTION

In recent years, the surface moisture of raw coal delivered to the preparation plants is usually over seven percent. Because of the rapid blinding of screen apertures, conventional dry screening of such moist material at mesh sizes of 6mm is very difficult.

There are now various types of equipment for dry screening of raw coal in China. The apparatus of probability sizing screen, rotating probability screen, banana screen, spiral screen, piano-wire screen and so on in early 1980's have been widely spread in coal industry. Satisfactory results are usually obtained when over 13mm aperture is used. But the problems for 6 mm coal sizing, especially the dry screening of sticky and moist material, have not been solved.

The flip-flow screen is efficient for screening moist fine coal (Dietz 1994). The acceleration of the surface of screen is high, so that the apertures will not be blinded (Hirsch 1992). However, the distribution of acceleration of the surface of flip-flow screen has not been studied in detail. This paper introduced a calculation formula and an acceleration distribution of the surface of flip-flow screen based on the studies of theory and experiments. This provides the basis for the designing and applying the flip-flow screen.

2. THE DYNAMIC MODEL OF THE SURFACE OF FLIP-FLOW SCREEN

For the sake of studying the acceleration of the surface of flip-flow screen, a model flip-flow screen was designed as shown in Figure1. The surface of screen is considered as an elastic press rod with two movable ends to establish the dynamic model (Zhao&Liu 1998). We also presumed that:

1. Under the condition of no-load, the surface of screen moves at elastic moving displacement;

2. The length of the neutral plane of the surface of screen is constant during the elastic moving displacement.

As shown in Figure 2, the modulus of elasticity of the screen surface's material is E(polyurethane).

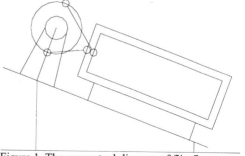

Figure 1. The conceptual diagram of flip-flow screen

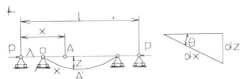

Figure 2. The dynamic model of the surface of flip-flow screen.

The moment of inertia is I. The loads of both sides are p. The length of screen surface is l. Taking the arc length $X=OA'$ as abscissa and the vertical displacement as Z, we have the differential equation of large deflection.

$$Z^{(4)}[1-(Z')^2]^{-1}+4Z^{(3)}Z''Z'[1-(Z')^2]^{-2}$$
$$+(Z'')^3[1+3(Z')^2]\times[1-(Z')^2]^{-3} \tag{1}$$
$$+\frac{2P}{EI}Z''\times[1-(Z')^2]^{-\frac{3}{2}}=0$$

where

$$Z^{(4)}=\frac{\partial^4 Z}{\partial X^4};\quad Z^{(3)}=\frac{\partial^3 Z}{\partial X^3};\quad Z''=\frac{\partial^2 Z}{\partial X^2};\quad Z'=\frac{\partial Z}{\partial X}.$$

boundary condition:

$$Z(0,t)=Z''(0,t)=0;\ Z(l,t)=Z''(l,t)=0.$$

In order to analyzing the deflection curve equation, we found the solution by perturbation method (Minde 1988). Taking the deflection f of the middle point of screen surface as a perturbing parameter, we developed the solution of equation (1) and the load p in the form of power series of f:

$$Z=\overline{Z}_1(X)f+\overline{Z}_2(X)f^2+\cdots$$
$$P=P^0+P^{(1)}f+\frac{1}{2!}f^2P^{(2)}+\cdots \tag{2}$$

boundary condition:

$$\overline{Z}_i(0)=\overline{Z}_i'(0)=0;\ \overline{Z}_i(l)=\overline{Z}_i'(l)=0.$$

where $\overline{Z}_i''=\dfrac{d^2\overline{Z}_i}{dX^2}$

auxiliary boundary condition:

$$\overline{Z}_1(\tfrac{1}{2})=1;\qquad \overline{Z}_r(\tfrac{1}{2})=1\ \ (r\neq1)$$

We could get the first-order equilibrium differential equation from the equation (1) and the equation (2):

$$\frac{d^4\overline{Z}_1}{dX^4}+\frac{2P^0}{EI}\frac{d^2\overline{Z}_1}{dX^2}=0$$

the solution of above equation is

$$\overline{Z}_1(X)=\sin\frac{\pi X}{l} \tag{3}$$

We also could get the second-order equilibrium differential equation by comparing the coefficient of f^2:

$$\frac{d^4}{dX^4}\overline{Z}_2+(\frac{\pi}{l})^2\frac{d^2}{dX^2}\overline{Z}_2+\frac{2P^{(1)}}{EI}\frac{d^2}{dX^2}\overline{Z}_1=0$$

$$\overline{Z}_2(X)=0$$

So, the approximate solution of the equation (1) is

$$Z(X,l)=Z_1(X)f+Z_2(X)f^2=f\sin\frac{\pi X}{l} \tag{4}$$

because,

$$2\lambda=l-\int_0^l[1-(Z')^2]^{\frac{1}{2}}dX=\frac{1}{2}\int_0^l(Z')^2dX$$

$$f=\frac{1}{\pi}\sqrt{8\lambda_l} \tag{5}$$

$$\lambda=e(1+\cos\omega t) \tag{6}$$

where $x=$ the displacement of slide block in the crank and slide block mechanism; $e=$ the eccentricity of transmission system; $\lambda=$ the circular frequency of the crank.

The vertical displacement of screen surface is

$$Z(X,t)=\overline{Z}_1(X)f(t)=\sin\frac{\pi X}{l}\sqrt{\frac{8el}{\pi^2}}\sqrt{1+\cos\omega t}$$

The vertical acceleration of screen surface is

$$\ddot{Z}(X,t)=\frac{\partial^2 Z}{\partial t^2}$$

$$=\frac{1}{\pi}\sqrt{8el}\left[-\frac{\omega^2\cos\omega t}{2\sqrt{1+\cos\omega t}}+\frac{1}{4}\frac{\omega^2\sin^2\omega t}{(1+\cos\omega t)^{\frac{3}{2}}}\right]\sin\frac{\pi X}{l} \tag{7}$$

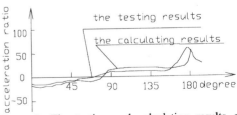

Figure 3. The testing and calculating results of acceleration in middle point of the surface of screen.

We can learn from the equation (7) that the distribution of acceleration of screen surface is a sinusoidal distribution.

3. EXAMPLE OF REAL COMPUTATION

The parameters of model machine were as follows: length of screen surface l=200mm, eccentricity e=8mm, circular frequency λ =41.87 (rad/s), screen aperture is 6mm×15mm. The calculating and the testing results of acceleration in the middle point of screen surface are shown in Figure 3. We find that the testing results correspond to calculating results very well.

The maximum acceleration of testing result is 53g. However, the maximum acceleration of calculating result is infinite. In reality, when the screen surface extended near horizontal state, the tensile plays a main role and makes the real acceleration decrease greatly.

Figure 4 is the illustrator of the test points of the system. Compared results of calculated and tested accelerations for the points in Figure 4 are listed in Table 1. It is clear that the calculating results are corresponding to the test results very well.

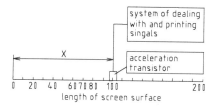

Figure 4. The illustration of the test points of the system.

Table 1. The acceleration of test and calculation results.

Test point	1	2	3	4
Distance (mm)	0	20	40	60
$\ddot{Z}$max/ g(testing)	0	17.64	29.49	36.00
$\ddot{Z}$ max/ g(calculating)	0	16.30	28.50	39.48
Test point	5	6	7	
Distance (mm)	70	80	100	
$\ddot{Z}$max/ g(testing)	46.87	47.63	53.00	
$\ddot{Z}$ max/ g(calculating)	46.90	50.4	57.40	

4. SCREENING TEST OF THE FLIP-FLOW SCREEN IN THE PILOT PLANT

The optimization parameters of the flip-flow screen have been studied. According to the characteristics of raw coal, the kinetic parameters (inclined angle, tension of screen surface and frequency) of the flip-flow screen have been changed. The tests of 6 mm aperture for raw coal have been carried out and the favorable operating status have been found out. The test system consists of three parts: raw coal screening, moisture examination and size distribution analysis. The pilot test system is shown in Figure 5.

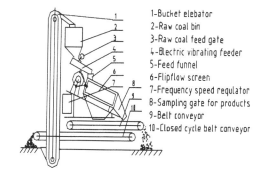

Figure 5. The pilot test system.

The test procedure is: blending the water and raw coal (50~0mm) evenly, conveying it to the bucket elevator to go through the raw coal bin and electric vibrating feeder which feeds it into the model flip-flow screen. The over- and the under-products (0~6mm) were collected. Then the size analysis and surface moisture examination on the samples of the screen products are carried out.

Table 2. Test results of flip-flow screen.

Surface Moisture of feed (%)	<1mm weight of feed (%)	Aperture size (mm)	Length of screen surface (m)	Screening efficiency (%)
0	30	6	1.4	90
4.5	33.3	6	2.1	90
7.5	35	6	2.1	85
8.3	36.7	6	2.8	86
10	36.7	6	3.5	85
12	35	6	3.5	84
14	35	6	3.5	85

Table 2 presents the sizing result of the flip-flow screen in the pilot plant. The conclusion is that when the surface moisture of feeding material changed from 0-14%, the screening efficiency of flip-flow screen is over 84% at 6 mm aperture sizing.

CONCLUSIONS

1. The acceleration of flip-flow screen surface is very high and the maximum acceleration of the middle point of screen can reach 53g. The high acceleration benefits loosing and demixing of materials.

2. The acceleration distribution of the surface of flip-flow screen is corresponding to a sinusoidal distribution.

3. The screening efficiency of the flip-flow screen is over 84% at 6 mm aperture sizing.

REFERENCES

Dietz. G. 1994, 25 years of polyurethane screen planets in mineral processing, *Aufbereitungs Technik*, 35(8): 404--412.

Hirsch. W.1992, Flip-flow screens of the third generation, *Aufbereitungs Technik*, 33(12):686--691.

Minde. W. 1988, Stability theory of elastic stick, Beijing: Academic Education Press.

Zhao, Yaomin & Liu, Chusheng. 1998, Research on polytechnic parameters of flip-flow screen, *14th Annual International Pittsburgh Coal Conference*, CD-ROM. Taiyuan, October.

Mining Science and Technology' 99, Xie & Golosinski (eds) © 1999 Balkema, Rotterdam, ISBN 90 5809 067 1

Performance characteristic curve of heavy-medium separator

Feng Liu & Yumei Xing
The Tangshan Branch of China Coal Research Institute, People's Republic of China

Xiaoyun Shan
Hebei Institute of Technology, People's Republic of China

ABSTRACT: Based on the theory of terminal settling velocity in gravity separation of minerals, coupled with overall consideration of the structural and operating parameters of a separator, a method for calculating the effective performance of heavy medium (HM) separators is proposed. By using this method the performance curve of the JLT (vertical lifting wheel separator) series HM separator can be drawn. This provides a simple and practical tool for appropriate selection of separator and precise determination of the rated capacity of a separator selected for use under different process requirement.

1. INTRODUCTION

Determination of the effective capacities is of important significance for study, design and application of HM separators. In practical operation, the capacity is generally determined in terms of the separator's mechanical capacity (float and sink materials discharge capacities) or based on an empirical value. Based on the state of motion of particles in the separator, the theory of terminal settling velocity in gravity separation and by taking overall consideration of such factors as structural parameters, feed property and separation performance, a method for calculating a effective separation capacity under various operating conditions is proposed. By this method, a further study has been made on the characteristic performance curve of the JLT series separators. The method proves to be a simple and practical tool for appropriate selection of HM separator and correct determination of selected saparator's nominal capacity.

2. CALCULATION OF EFFECTIVE CAPACITY

There exist three factors affecting the HM separator's capacity, i.e. feed property, structural parameters of the separator and operating parameters. The terminal settling velocity of a particle is governed by properties of feed and suspension while the critical terminal settling velocity in the separator

depends on the separator's structural parameters, and the separation performance is determined by operating parameters. Therefore the calculation method proposed is a feasible and scientific method because an overall consideration is made of the hindered terminal velocity of particles, critical terminal velocity in the separator and the probable error (Ep) of the separator.

2.1 Calculation of terminal settling velocity of particles

The free-falling terminal velocity of a particle in a static suspension v_0 is expressed by:

$$v_0' = \xi \sqrt{\frac{\pi d(\delta - \rho)g}{6\psi\rho}} \qquad (2.1)$$

where ξ = coefficient, it is 0.5 (density $>1.8\times10^3$ kg/m³) and 0.5-0.65 (low-density separation) for coal, refer to references for other materials; δ = density of particle; ρ = density of (media) suspension; g = gravity acceleration; ψ = coefficient of resistance of motion of particle; φ = particle size (Li & Zhang 1992). Following is a detailed discussion of the individual parameters.

1) Density of particles: The Ep value is an important factor for evaluating the separating performance. The property of the particles in our study should be in compliance with the required Ep value and has ξ higher than that of the separation

density δP. Hence, the density of particles is chosen as:

$$\delta = \delta_p + E_p \qquad (2.2)$$

2) Particle size: The smaller the particle, the slower the separating rate, to enable all the particles to be separated, a lower size limit of the feed d_{min} is taken.

3) Coefficient of resistance of motion of particle $R_e^2\psi$ is first calculated by:

$$R_e^2\psi = \frac{\pi \cdot d^3}{6}(\delta - P) \cdot g \frac{\rho}{\mu^2} \qquad (2.3)$$

where μ = dynamic viscosity of medium; R_e = Reynolds' number. Then calculation are made of R_e by $R_e^2\psi$ vs R_e curve, volume concentration of suspension λ based on apparent viscosity of suspension, and finally the ψ value by R_e vs ψ (Sun 1982) for spherical particles in suspensions with varying volume concentrations.

4) Calculation of terminal settling velocity of particle: v_0 can be derived according to equation (2.1). Then the terminal velocity of such particles in dynamic suspension is calculated by using the following expression:

$$v_0 = \begin{cases} v_0' - v_1 & \text{Separation in horizontal plus} \\ & \quad \text{upward currents} \\ v_0' + v_2 & \text{Separation in horizontal plus} \\ & \quad \text{downward currents} \end{cases}$$
$$(2.4)$$

where v_1, v_2 = velocity of upward current and downward current in separator respectively.

2.2 *Critical terminal velocity in separator* v_{HL}

The velocity v_{HL} refers to the minimum terminal velocity required for all the particles in the feed to be separated.

As illustrated in Figure1, when a critical particles

Figure 1. Schematic diagram of v_{HL} calculation.

(with the required density and size as discussed in 2.1) moves from point A to point B, its distance to the liquid surface is H, and L to point A. In its motion from point A to point B, the velocity in the horizontal direction is equal to that of the horizontal current v_s. While in its motion from point C to B, the particle moves uniformly with a hindered terminal settling velocity v_{HL} if the time required for the particle to reach its terminal settling velocity is neglected. Thus at point B, we have:

$$\frac{L}{v_s} = \frac{H}{v_{HL}}$$

$$v_{HL} = \frac{H}{L}v_s \qquad (2.5)$$

where H = working depth of scraper, L = effective length of separator, v_s = velocity of horizontal current of medium.

2.3 *Critical volume concentration in the bath's separating zone*

During separation, the terminal settling velocity of a particle is, in fact, its hindered terminal settling velocity. The difference between the latter and free-falling terminal velocity is related to the volume concentration of the particles in the separating zone i. Restricted by critical terminal velocity in separator v_{HL} the hindered terminal settling velocity of critical particles is equal to v_{HL}. In other words, the particles with a settling velocity smaller than v_{HL} are light product, the reverse is true.

$$v_{HL} = (1-i)^n v_0$$

After transformation and substitution of, it becomes:

$$i = 1 - \sqrt[n]{\frac{v_{HL}}{v_0}} = 1 - \sqrt[n]{\frac{Hv_s}{Lv_0}} \qquad (2.6)$$

where i = critical volume concentration of particles in bath; n = function of Reynolds' number of free-falling particles (Sun 1982).

When I value is known, the effective capacity Q of the separator can be determined.

2.4 *Calculation of effective capacity Q*

The volume concentration is defined by:

$$i = \frac{V_1}{V} \qquad (2.7)$$

The volume of the particles in separating zone (m³), can be expressed by:

$$V_1 = \frac{L \cdot Q}{3.6 v_s \delta} \qquad (2.8)$$

In equation (2.7), V denotes the volume of separating zone (m³). For a trough separator with a shape as illustrated in Figure 2, the V value is defined by:

$$V = \frac{h}{3}\Big[BL + (L - 2hctg\alpha)(B - 2hctg\alpha) +$$
$$\sqrt{BL(L - 2hctg\alpha)(B - 2hctg\alpha)}\ \Big]$$

where B = width of trough; h = depth of effective separation zone (generally taken as $H \square H + 0.1m$); a = inclination angle of the trough .
By relating to equations (2.6) to (2.8), we have :

$$\frac{QL}{\delta v_s 3.6} = (1 - \sqrt[n]{\frac{H v_s}{L v_0}})V$$

By referring to equation (2.2), the formula for calculating the capacity Q can be obtained:

$$Q = 3.6\frac{(\delta_p + 2E_p)v_s V}{L}(1 - \sqrt[n]{\frac{H v_s}{L v_0}}) \qquad (2.9)$$

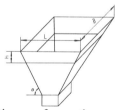

Figure 2. Diagram of separating zone of trough

Although the equation (2.9) is derived based on the model of a trough separator, it can however be applied to different forms of HP separators.

3 STUDY OF THE CHARACTERISTIC PERFORMANCE CURVE

Though the relationship between the various characteristic performance parameters is well depicted in equation (2.9), yet its practical use is inconvenient in many as pests. It is imperative to have a characteristic curve that can well reflect the relationship between the above mentioned parameters. Thus a multi-variable coordinate system similar to a homograph has been developed, based on which the characteristic curve of the performance of a HM separator can be obtained. This may serve as a simple and practical tool for optimum selection of separator and determination of the nominal capacity of a separator.

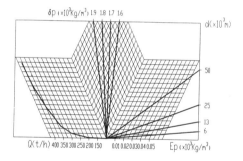

Figure 4. Performance Characteristic Curve of the JLT 2040 H.M. separator.

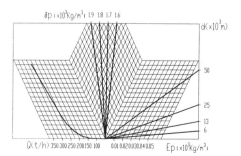

Figure3. Performance Characteristic Curve of the JLT 1630 H.M. separator.

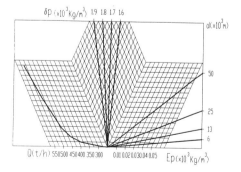

Figure 5. Performance Characteristic Curve of the JLT 2545 H.M. separator.

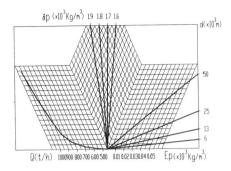

Figure 6. Performance Characteristic Curve of the JLT 4565 H.M. separator.

The performance characteristic parameters involved have been calculated for the JLT1630, 2040, 2545 and 4565 separators (see Figure 3□6). The parameters include lower size limit of feed (6□50 mm), separating density (1.6□1.9) and *Ep* value (0.01□0.05). By reference to the curves, the capacity of any types of HM separators with given parameters can be obtained.

4 DISCUSSION

1) The capacity of a separator depends not only on structural parameters, but also on other parameters such as property of feed, separating performance, property of suspension and operating parameters. In selection of a separator, great importance should be attached to the influence produced on capacity Q by non-structural parameters.

2) The appropriate separation length and working depth of the scraper under given operating conditions can be calculated through equation (2.9).

Equation (2.9) also can be applied to determine the operating parameters involved under conditions of given structural parameters, such as velocities of horizontal, upward and downward currents, etc. or to predict the separating performance of a separator (δ_p, E_p) with given structural and operating parameters. It is of great reference significance for study and optimum design of a HM separator.

3) The relationship between the main performance parameters can be directly perceived through the performance characteristic curve. Though the curve, the capacity of a separator operating with different operating parameters and lower size limit of feed can be readily determined, facilitating the selection of separators.

4) The operating parameters and velocities of horizontal, upward and downward currents of suspension all have a great bearing on separating performance and capacity of a separator. By using the Equation (2.9), operating curve can be drawn to provide reference data for optimum operation of the separator.

5) Due to fluctuation of feed rate and the composition of the suspension, the volume concentration of particles in the separating zone is a random variable. Likewise the size composition of feed exerts a pronounced influence on the hindered terminal settling velocity of the critical particles. Further investigations should be made for its improvement.

REFERENCES

Li Xianguo, Zhang Rongzeng 1992. *Principle of Gravity Concentration of Mineral.* Beijing Coal Industry Publishing House.

Liu Feng, Xing Yumei. 1997. Study of Effect of State of Flow of Suspension on Performance of HM Separator; *Coal Science And Technology*; 25(4):31□34.

Sun yubo 1982. Gravity Concentration of Mineral. *Metallurgical Industry Publishing House.*

Mining Science and Technology' 99, Xie & Golosinski (eds) © 1999 Balkema, Rotterdam, ISBN 90 5809 067 1

Study on floatability of maceral

Chengfeng Cai
China Coal Research Institute, Tangshan Branch, People's Republic of China

ABSTRACT: As evidenced by result of study on physio-chemical property and flotation characteristic of maceral, the maceral shows a remarked size segragation tendency, with the content of fusinoid in the −0.043 mm fraction about 20-30 percentage points higher than those in the +0.043 mm fractions. Fractionated flotation of +0.074 mm can produce a concentrate with the contents of vitrinite and fusinoid groups being 85.9% and 11.6% respectively, resulting in a removal rate of the latter as high as 58.48%. Whereas flotation of the −0.074 mm fraction produces no remarkable result in enrichment of vitrinite group. It is demonstrated by maceral analysis that the monomeric fusain can be preferentially floated up due to its better floatability.

1 INTRODUCTION

Dongsheng-Shenfu Coalfield rich in reserve of good-quality coal is one of the preferred coal liquefaction bases. However due to its specific coal-forming condition, the coal is high in inertinite content, with the fusinoid group being 25-70% in content. Coal liquefaction experiments conducted by China Coal Research Institute (CCRI) demonstrate that, through removal of fusain, the content of fusinoid group can be reduced from 28.7% to 11.0% while the conversion rate of THF in clean coal is enhanced to 4.3% with a rate of increase of oil yield from 59.7% to 66.8%. Therefore it is necessary to develop an inertinite group separation method with industralization potential for adding to the technological reserve for effecting direct liquefaction of coal.

It has been proposed by senior engineer Xu Jianping et al to separate, based on the principle of density difference, the fusain with a small-diameter H.M. cyclone using ultrafine medium. With this method separation can be made with a lower size limit down to 0.043 mm, an Ep value of 0.085 and a fusinoid group removal rate of 56.4% (Xu Jianping,1997), exhibiting a certain commercial prospect. However for the -0.043 mm fraction the maceral components should be separated with

flotation process. It has been demonstrated that enrichment of vitrinite and removal of fusinoid through flotation process may have some uncertainties.

2 EXPERIMENTAL STUDY

The study was made by the author with samples collected from Ningtiaota 2-^2 seam of Shenfu mining area. Whereas the macropetrographic components were sorted through handpicking for determining the effect on floatability of maceral components produced by the physio-chemical properties, such as size, density, electrokinetic potential (EP), elementary composition and maceral components. Study on floatability characteristics of maceral was made with batch flotation in combination with maceral analysis.

Electrokinetic potential: The handpicked vitrain and fusain were crushed to −0.5 mm for measuring the streaming potential with the ZP-100B potentiometer of Japan.

Floatability: Use was made of the laboratory batch flotation method in accordance with the standard GB 4757-84 and maceral analysis.

Maceral analysis: □20 mm micro-sections were prepared for measurement with 50x more oil-immersion objective and 10x eye piece.

3 PHYSIO-CHEMICAL PROPERTIES OF PETROGRAPHIC COMPONENTS

1) Petrographic characteristic
The four maceral components in the sample differ in property: a) The vitrain is the lowest in ash content. The durain being mineralized has a higher ash, yet with the increase of ash content, the durain shows an enhanced hydrophilicity; b) The vitrain has a higher contents of hydrogen and oxygen, and move side chains and cyclanes in its organic structure, presenting a natural hydrophobicity. However the oxygen in vitrinite belongs essentially to the hydroxyl radicals; c) The fusain is high in moisture content because of the presence of well-developed fissures. However it has a noticeably higher content of volatile-matter, lower contents of hydrogen and oxygen and much higher content of carbon as compared with the other components. This indicates that in the inertinite, the aromatic nucleus has the highest degree of condensation(Chen Peng et al.,1988), and there are less side chains and oxygen-containing functional groups on the multi-ring aromatic nuclues.

2) Charcteristic of distribution of maceral in different density fractions
The raw coal in full size range was first liberated after crushing to –0.5 mm. The characteristic of distribution of maceral in different density fractions are shown in Figs 1 & 2. The vitrinite and fusinoid groups tend to concentrate each in a specific density fraction, and the peak value, scope and density of each group vary with degree of liberation. The peak value of each of the abovementioned groups will vary along with the enhancement of level of liberation of maceral. The curve of content of mineral matter as a whole tends to go upward with increase of density and a small crest appears in the 1.4 density fraction, indicating that the cell cavities of fusain are filled up with fine clay.

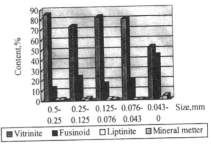

Fig. 2. Distribution of Maceral in Various Size Fractions

3) EP of maceral
The result of experimental study made by the author is shown in Fig. 3. When neutral ion exchange water is used as wetting medium, both vitrain and fusain will have negative EP values of being □= -27.34 mv and □= -9.32 mv respectively. The fusain is smaller than vitrain in absolute EP value□, indicating that the fusain has a better hydrophobic property because of its weaker surface hydration effect and thinner hydration layer. This is beneficial to its preferential flotation due to preferential absorption of reagent with the better action of non-polar collector. With the increase of dosage of cationic conditioning agent, the Eps of both vitrain and fusain will turn from negative to positive. Test demonstrates that after soaking in water for a certain period of time, the fusain's negative EP will increase, indicating that the surface of fusain is liable to be oxidized.

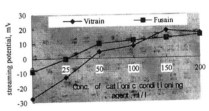

Fig. 3. EP values of Vitrain and fusain vs concentration of cationic conditioning agent

4 FLOATABILITY OF MACERAL

Comparative tests on floatability of all macromaceral constituents were made under identical conditions, with the exception of fusain because of the limited amount of handpicked fusain. The test result is tabulated in Table 1, from which it can be seen that:

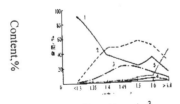

Density, g/m³

Fig. 1. Distribution of Maceral in Different Density Fractions. 1.Vitrinite, 2.Semi-vitrinite 3.Fusinite 4.Liptinite 5. Mineral matter

1). Through flotation process only a part of fusain and limited amount of fusinite can be removed.

2). Under identical conditions, the floatability of vitrain is the best, which is followed by clarain and durain. With the increase of fusinoid, the flotation rate drops, resulting in decrease of final concentrate yield and noticeably higher concentrate ash. The ash of concentrate is closely related to content of fusinoid.

In order to avoid contamination of concentrate by ultra-fine fusain, fractionated flotation process was applied. According to the test result, the curves showing the flotation rates of different size fractions and the flotation characteristic of fusinoid are drawn as illustrated in Fig. 4 and Fig. 5.

With a vitrinite content of > 70%, a better result can be obtained in a flotation time of 3 minutes: a higher concentrate yield of 68.5%, a lower fusinoid content of 11.6% and the ash of the concentrate of the +0.073 mm fraction is evidently lower than that of the −0.073 mm fraction. The flotation rate of the −0.073 mm

size fraction for fusain enrichment is noticeably reduced. As can be seen from the curve showing the content of fusinoid group in the concentrates of the −0.073 mm size fractions (Fig. 5), the

flotation time for concentrate I is only 0.5 minutes, and its fusinoid content is remarkably higher than that of concentrate II. However the vitrinite content of concentrate II is evidently higher than those of concentrate I and concentrate III. It can be seen from microscopic analysis of flotation concentrates that the concentrate I includes individual fusains, pyrofusinite, semi-fusinite and intergrown vitrain and fusain. It shows that preferential floatation is made by fusain which is then followed by vitrain and finally by the mineralized fusain and the unliberated coarser intergrown vitrinite and fusinoid groups. The result of study is in good agreement with the study of physio-chemical properties of macromaceral constituents.

5 CONCLUSIONS

1) Maceral analysis of flotation concentrates indicates that during conventional flotation process, the monometric fusains show a good floatability and will float up first, followed by vitrain and mineralized fusain, and finally by unliberated coarser intergrown vitrinite and fusinoid;

2) The EP value of fusain is found to be somewhat different from the result of study made by other researchers. According to the present study, the EP value of fusain in neutral water solution is $\square$= -9.32 mv, and that of vitrain is $\square$= -27.34 mv, with the absolute value of the former smaller than that of the latter;

3) With the use of fractionated flotation process, the contamination of +0.073 mm concentrate caused by monomeric fusain can be reduced. With a flotation time of 3 minutes, a better result can be obtained: a concentrate yield of 68.5% with a content of vitrinite of 85.9%, a content of fusinoid of 11.6% and a removal rate of fusinoid of 58.48%.

The method for suppressing the fusain awaits further investigation.

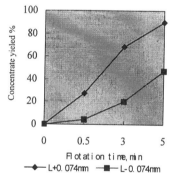

Fig.4 Effect of Size on Flotation Rate

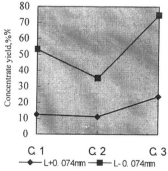

Fig.5 Flotation Characteristic of Fusinoid

Table 1. Result of Comparative Flotation Test Made with Coal Samples with Different Contents of Fusinoid

	Product	Float. time	Result				Maceral						η_f
			r %	Aad %	Σr %	Aad %	V %	SV %	SI %	I %	L %	M %	
Vit-rain	C.1*	3	88.32	2.35	88.32	2.35	82.60	1.60	7.00	7.80	0.60	0.40	6.89
	C.2	1.5	6.36	3.52	94.68	2.43	77.90	2.10	6.80	12.00	0.20	1.00	
	C.3	1.5	1.89	4.05	96.57	2.46	79.20	0.40	9.00	10.20	0.20	1.00	
	T.		3.43	10.91	100.0	2.75	57.70	2.60	12.90	18.80	1.00	7.00	
	Vitrain		100.0	2.75			80.80	2.80	7.90	7.30	0.20	1.00	
Cla-rain	C.1	3	83.54	2.72	83.54	2.72	75.80	2.20	13.00	7.60	1.00	0.40	33.18
	C.2	1.5	8.41	3.22	91.95	2.77	69.20	2.10	15.60	11.40	0.70	1.00	
	C.3	1.5	3.50	3.43	95.45	2.79	71.90	2.30	15.10	9.70	0.80	0.20	
	T.		4.55	9.37	100.0		61.20	4.10	12.10	18.30	0.90	3.40	
	Clarain		100.0	3.33			66.70	2.40	9.90	16.60	1.40	3.00	
Du-rain	C.1	3	71.93	5.55	71.93	5.55	53.80	3.60	26.50	13.60	1.40	1.10	32.7
	C.2	2	10.48	6.94	82.41	5.73	57.10	4.70	16.70	16.90	1.30	3.30	
	C.3	2	4.09	6.56	86.50	5.77	47.20	4.70	14.00	31.80	1.10	1.20	
	T.		13.50	15.20	100.0	7.04	45.00	8.00	26.40	11.20	3.30	6.10	
	Durain		100.0	7.69			57.00	1.60	16.30	21.70	0.60	2.80	

Note: rpm—1800 Aeration quantity—0.25m3/m2.min Dihexyl—1000g/t MIBC—100g/t
C.= Concentrate T.=Tailings η_f=Removal rate of Fusinoid

REFERENCES

Xu Jianping. 1997. Study on separation of fusain using a small-diameter H.M. Cyclone operating with ultra-fine medium. *Proc. 14th Int. Coal Conference*, 1997.

Chen Peng et al. Study of coal maceral of Yanzhou coal using Fourier infra-red spectrometry. *Journ. of Fuel Chem*. 1998: Vol 16, No. 3, 3.

Mining Science and Technology' 99, Xie & Golosinski (eds) © 1999 Balkema, Rotterdam, ISBN 90 5809 067 1

A highly selective flotation cell with oscillating separator

Hongzhi Cheng, Changfeng Cai, Xiaojun Zhang & Jing Zhang
China Coal Research Institute, Tangshan Branch, People's Republic of China

ABSTRACT: It is assumed that the mechanism of entrainment of ultra-fine slimes, a source causing contamination of flotation concentrate, is based on the coating and lifting up action of bubble aggregate for hydrophilic particles. The highly selective oscillating separator developed by authors has been proven to be effective for lowering the ash of concentrate. The FJG-S8 high-selective flotation cell has also been developed based on the principle of similarity, the main structural and kinetic parameters has been determined scientifically. Field operation of the cell for treating typical difficult-to-float fine coal demonstrates that the cell offers a good selectivity. As compared with the XJM-S8 cell originally used, with identical ash of tailings, the ash of concentrate is 1.39% lower with a floatation perfection index 3.54 % higher.

1. INTRODUCTION

Diversified flotation cells are currently used in China's coal preparation plants, yet none offers a idea selectivity. When the feed pulp is high in content of ultra-fine slimes, a higher ash content of concentrate can be observed due to contamination caused by mechanical entrainment – a bottle-neck incapable of being tackled over years. With regard to the various types of flotation columns, resort is essentially made of a large amount of wash-water for reduction of contamination of concentrate.

The FJG-S8 flotation cell operating with oscillating separator passed the technical appraisal made by the former Ministry of Coal Industry in 1997. The novel separator featuring high selectivity is developed on the basis of study of mechanism of slime contamination and the principle of oscillation separation (Patent application No. 98216764.4) for further reduction of slime contamination due to mechanical entrainment. The successful developed high-selective FJG-S8 cell operating with oscillating separator has changed the passive situation of only relying on wash-water for reducing the concentrate ash – a conventional practice applied for a long period of time. The cell proves to be a high-efficiency equipment for use in coal preparation plants treating high-ash and hard-to-float fine coal for reducing the contamination as described above.

2. PRINCIPLE OF DESIGN

For the design, development and operation of the FJG-S8 cell, the following basic principles are applied:

2.1 *Mechanism of slime contamination*

The contamination of concentrate by high-ash fine slimes is subject to a series of mechanisms of actions, including mechanical entrainment, covering up with fine slimes and coagulation of particles. Among them the mechanical entrainment is the predominant form of action.

Mechanical entrainment refers to the report of hydrophilic gangue particles to the concentrate product via mechanical conveyance. As a main cause for concentrate contamination, the mechanism of mechanical entrainment involves the following aspects:

1) Coating and lifting up action of bubble aggregate for hydrophilic particles

The majority of the mineralized bubbles in flotation process usually float up in form of bubble clusters. Namely the air bubbles will group together to form aero-flocs with voids between bubbles filled up with pulp. The particles adhered onto the bubbles are thus brought up to the forth layer under the effect of bubble buoyancy. During this process the pulp filling up the voids between bubbles assumes a dynamically balanced state. Under the effect of mutual coherence between water films and turning of bubbles, filter flocs with leaching action can be formed to enable the pulp to undergo filtration during floating up of bubble aggregates. This allows a portion of the undesirable hydrophilic impurities to retain in the bubble aggregate and float up with bubbles. Hence the high-ash particles report to the froth layer via mechanical entrainment as described above.

2)   Non-selective adhesion of particles

Because of the larger specific surface area of ultra-fine particles, a surface force will go into action. Based on the EDLVO theory some particles will lift up to the froth layer through non-selective adhesion onto air bubbles (Yoon 1993).

3)   Effect of secondary enrichment of froth layer weakened due to unsteadiness of liquid level

If the liquid level in the cell remains unstable and particularly when it is in a turbulent state, the upward current will bring the high-ash particles in the pulp directly into the froth layer, weakening, as a result, the effect of secondary enrichment.

2.2 *Principle of improvement of selectivity with oscillating separator*

The oscillating separator is applied with an aim to: exciting the pulp, mineralized bubbles and bubble aggregates in the separating zone to undergo oscillation with the aid of mechanical vibration. This method may break up the bubble aggregates, permit the bubbles rising individually. Hence the entrained hydrophilic particles are detached from the bubble aggregates under the effect of vibration-induced inertial force. In addition it causes the hydrated films of air bubbles to undergo oscillation-induced deformation. So that the detachment energy is provide for the hydrophilic particles adhered to bubbles in non-contact manner (Dai et al. 1997). Furthermore under the effect of intensive high-frequency oscillation the horizontal current is caused to have an alternating velocity much greater than that of upward current in the cell. The greater alternating velocity plus the much greater alternating acceleration serves as a buffer against the upward current, stabilizing the pulp level and froth layer. The above-mentioned factors may lead to reduce the mechanical entrainment and hence improve the flotation selectivity.

2.3 *Principle of simulation scale-up design*

The main parameters of the flotation cell are determined with simulation scale-up method based on the following similarity criteria (Cheng et al. 1998):

$$D/L = constant \tag{2.1}$$

$$H_t = kD + H_0 \tag{2.2}$$

$$N_{Qa} = Qa/(ND^3) = constant \tag{2.3}$$

$$N_p = p/(\rho N^3 D^5) = constant \tag{2.4}$$

$$ND*(D/L) = constant \tag{2.5}$$

Where H_t = depth of cell (m); H_0 = total height of the separating and foam zone (m) (H_0 is a fixed value for same series of flotation cells); D = diameter of impeller (m); k = dimensionless coefficient of agitating depth in trough; L = characteristic dimension of trough width(m); P = Agitating power, kw; N == impeller speed (rpm); N_p = power factor; ρ =pulp density (kg/m³); N_{Qa} = number of air-flows (dimensionless); Qa = aeration quantity (m³/s).

When design is made based on above criteria, the flotation cells in the same series can be assured to have similar agitating intensity, aeration rate and solids-liquid suspension state.

3. CELL DESIGN

The FJG-S8 cell is designed to have 4 compartments each bank, with a unit volume of 8 m³ and a total volume of 32 m³.

3.1 *Structural parameters*

With volume of 8 m³ flotation cell, it can be derived from the above mentioned equations and similarity criteria that $D = 0.6985$ (m), $L = 2.728$ (m) and $H_t = 1.368$ (m). The above value of D is rounded off to 700 mm.

3.2 *Kinetic parameters*

The main kinetic parameters involved for the design of the FJG-S8 cell are listed in Table 1.

3.3 Vibration parameters

The oscillating separator uses a horizontal high-frequency and low-amplitude mechanical vibration generator as oscillating source. The vibration generator consists of vibration exciters and a vibration body. The exciters are driven by two self-

Table 1. Main kinetic parameters of the FJG-S8 flotation cell.

Main parameters	Designed	Actually-measured	Deviation
SIMILARITY			
Fixed linear velocity, m/s	8.90	8.94	+0.04
Fixed number of air flows	0.085	0.086	+0.01
Fixed power factor	1.42	1.48	+0.06
KINETICS			
Aeration rate, $m^3/m^2/min$	1.20	1.22	+0.02
Agitating power, kw	15.8	16.7	+0.90
Uniformity of aeration, %	> 80	85.44	---

synchronous vibration motors running in a direction opposite to each other to produce a horizontal vibration excitation force varying according to the sine function.

The vibration body is composed of motor support, vibration framework, screen mesh, connecting links and main vibration springs. Through the lateral vibration of the screen mesh, the mineralized air bubbles and pulp in the vicinity of the mesh are directly brought into oscillation. The experimentally determined vibration intensity is $k = 2g – 4.8g$ (g – gravity acceleration, m/s^2), and the maximum exciting force required is 38621 (N).

Two VB20114-W vibrating motors are used, each with a maximum exciting force of 20000 N, a rated speed of 1430 rpm and a rated power of 1.1 kw.

Through adjusting the separation angles of eccentric blocks of motors, stepless regulation of the exciting force and hence the vibration amplitude can be made. The vibration frequency is regulated through adjusting the frequency of power source by means of a frequency converter. The vibration amplitude can be infinitely regulated in a range of 0 – 2 mm. The operating vibration amplitude is 1 – 2 mm. While for the vibration frequency, regulation can be made in a range of 0 – 25 Hz. The working frequency is 12 – 20 Hz.

4. PERFORMANCE

Since the initial operation of the FJG-S8 flotation cell in 1997 at Jiahe coal preparation plant, Xuzhou mining administration, a remarked result has been obtained. When used for treating a feed ash of 26 – 34 %, in which the content of – 0.043 mm fraction is 68.13 % with high-ash (39.88 %), the cell can produce concentrate with average ash of 12.80 %

and tailing with ash of Ad >50 %. The pulp throughput of the cell is up to 315-330 m^3/h. As compared with the XJM-S8 cell previously used, with the same ash of tailing, The ash of concentrate is reduced by 1-2%, or 1.39% on average while the flotation perfection index improved by 3.54%, bringing forth a favorable economic result.

5. CONCLUSIONS

As revealed by study on mechanism of contamination of concentrate by fine slimes, the contamination caused may be mainly attributed to mechanical entrainment which is induced essentially through the coating and carrying actions of bubble aggregates for hydrophilic particles. Based on result of study of the theory of oscillation separation, an oscillating separator, the first of its kind, featuring high selectivity, has been developed. Through exciting the pulp, mineralized bubbles and bubble clusters in the separating zone to undergo oscillation via mechanical vibration means, the flotation selectivity can be improved due to effective elimination of the hydrophilic particles mechanically entrained. Furthermore the FJG-S8 flotation cell with high selectivity, developed on the basis of main structural and kinetic parameters scientifically determined by applying similarity theory, proves to be an effective equipment for treating the hard-to-float high-ash coal fines.

REFERENCES

Cheng Hongzhi et al. 1998 On similarity criterion of mechanical agitation machines. *Coal Preparation Technology*. (5): 10—13.

Dai Qiang et al. 1997 Bubble deformation and oscillation flotation method – A high-efficient ultra-fine mineral flotation method. *Metallic Ore Dressing Abroad*. (1): 28—32.

R. H. Yoon. 1993 Hydrodynamics and surface force in particle-bubble interactions. *Mettallic Ore Dressing Abroad*. (6): 5—11.

Mining Science and Technology' 99, Xie & Golosinski (eds) © 1999 Balkema, Rotterdam, ISBN 90 5809 067 1

The study of FXZ static column and its industrial application

Maixi Lu, Fan Wang & Wenli Liu
China University of Mining and Technology, Beijing, People's Republic of China

Erkang Xu & Qindong Xu
Chuandong Mechanical Design and Manufacturing Company, People's Republic of China

ABSTRACT: Laboratory flotation tests have been done for five coals. The results show that the construct of bubble generator and the aeration rate are important for the performance of column. One m diameter by 6 m high FXZ-1 static column has been used in four coal preparation plants for more than one year. Industrial application is very successful. The FXZ column has the advantage of low power consumption, high separate efficiency and high throughput. It has past the identification and is suggested to be spread out in coal preparation plants of China.

1. INTRODUCTION

Column, as an advanced fine coal cleaning equipment, has been developed for more than 20 years. Microcel, developed by Virginia Tech., has been successfully used in several countries, the biggest one is 4.2 M diameter (Yoon 1998). Jameson cell has been widely employed in western country (Sablik 1994). Turboflotation, developed by CSIR, has been put to use in the industry (Ofori&Firth 1998).

Column was adopted in 1960' in China. But the usage of column was failed quickly because of the bad operating results. Since 1990, the cyclonic micro-bubble flotation column has been developed (OU 1996). The test of 1.5M packed column is executing in Shitai coal preparation plant, Huaibei coal mine. Several other columns are in developing. The characteristic of Chinese fine coal is quite different compared with the coal of western country. Its native ash content is high and the floatability is difficult. It is necessary to study the column in detail under Chinese conditions.

2. LABORATORY TEST RESULTS

2.1 *Raw material*

Five samples from Taixi, Datong, Linghuan and Qitaihe Coal Mines were collected. Among them, Linghuan is bituminous coal, others are anthracite. There are two Taixi coal samples: Taixi-1 and Taixi-2.

The size distribution of Taixi-1 is relative coarser and easier to separate. For Taixi-2 coal, the ash content is 23.13%, the weight of minus 45 micro fine particles is 78% with ash content of 27.88%. Obviously the Taixi-2 coal is difficult to separate.

2.2 *Test system*

The diameter of column for laboratory test is 45mm, the height of column can be adjusted, static mixer is in the bottom of column as the bubble generator, high pressure air from pump is sucked before static mixer. Kerosene was the collector and JF oil was the frother. The concentrate of flotation feed is 5%. Feed was conditioned with kerosene for 3 minutes before separation. After 10 minutes running, the clean coal and tailings were collected for 10-20 minutes as samples. In order to compare with conventional flotation cell, the conventional flotation tests were done in a one-liter flotation cell.

2.3 *Aeration rate is important for the performance of column*

Producing enough fine air bobbles is the most important factor for flotation. Aeration rate is a symbol of the number of the air bobbles. The limitation of the aeration rate should be non-turbulent occurrence in the column. Under this limitation, the higher is the aeration, the higher is the combustion recovery. The test results of superficial gas rate with the ash content and combustible recovery of clean coal for Taixi-1 coal is showed in Figure 1. It is clear that the combustible recovery

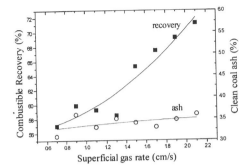

Figure 1. Superficial gas rate with the ash content and combustible recovery of clean coal for Taixi-1 coal.

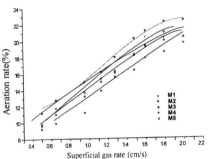

Figure 2. The relationship of aeration rate under superficial gas rate.

Table 1. Parameters of the static mixers.

Name	Unit type	No. Of mixer unit	L/D	Total length (cm)
M1	A	6	1.17	14
M2	A	12	1.06	25.5
M3	B	9	1.42	25.5
M4	B	5	1.40	14
M5	C	1		25.5

Table 2. Separate results of five static mixers.

Raw coal	Static mixer	Combus. recovery %	Ash of clean coal %	Ash of tailings %
Taixi-1	M1	55.91	2.22	25.55
	M2	78.67	2.67	40.13
	M3	86.66	2.34	48.17
	M4	75.95	2.64	39.10
	M5	75.38	4.10	34.09
Taixi-2	M1	61.24	2.63	43.32
	M2	66.48	3.23	47.52
	M3	74.47	3.59	54.97
	M4	73.71	3.37	53.77
	M5	71.38	4.92	46.32

increased from 57% to 71% with the ash content nearly unchanged while the superficial gas rate increased from 8 to 20 cm/s. The limitation of the aeration can be changed with different bubble generator. The bubble generator should meat following requirements: high aeration rate; producing small enough bubble; low energy consumption; no clean water needed; low floatation reagent consumption; high reliability and easy maintenance. Five types of static mixers were used in the tests. The parameters of the static mixers are listed in Table 1. Compared tests were done for five static mixers while keeping other conditions constant. Figure 2 is the relationship of aeration rate with superficial gas rate. Table 2 is the separation results of five static mixers for Taixi-1 and Taixi-2. Obviously M3 is the best one. Its aeration rate is always the highest one under different superficial gas rates. For Taixi-1 and Taixi-2 raw coals when other conditions unchanged, M3 gets much higher combustion recovery. M1 with M2, M3 with M4 have the same type of mixer unit, but the length or the number of the mixer unit is different. The results show that the more is the number of unit, the higher is the aeretion rate. The type of mixer unit is very important for the quality of aeration, although M5 has the same length with M3, the aeration quality is much worse than M3.

2.4 *The separate results of column are satisfactory*

Figure 3 shows the separation results of column, conventional flotation cell with the curve of release analyze for Taixi-2 raw coal. Evidently, the separation results of column for different conditions are much close to the release analyze curve. When the ash of clean coal is nearly unchanged, the yield of column is much higher than the yield of conventional flotation cell.

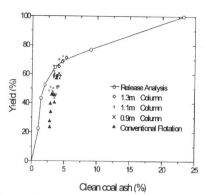

Figure 3. The separation results of column, conventional flotation cell with the curve of release analyze for Taixi-2 raw coal.

3. INDUSTRIAL APPLICATION

One miter diameter by 6 m high FXZ-1 static column has been used in four coal preparation plants since 1997. Figure 4 is the picture of FXZ-1 in Yalai Coal and Cock Limitation, Shichuang province. It is installed outside of the building because the weather is not cold in Shichuang. It has been successfully operated since August 1998. All of the control valves and air inlets are above the liquid lever on the top of the column. Therefore it makes the operating easier. Furthermore, the column does not need any high-pressure air. The power consumption of the recycle pump, which produces fine air bubbles, is only 11 kw. A froth scraper with 0.75 kw power is mounted on the top of the column, the scrubber makes the clean coal moving quickly and reduces the entrainment of high ash particles. The total reagent dosage (collector and frother) is 1.15 kg/ton.

Table 3 and Table 4 are a part of industrial operating results for FXZ-1 static column in Yalai Coal and Cock Limitation. It is certain that the separation results are satisfactory for industrial using, although the size of particles are very fine and the ash content of fine particles are very high. When the size distribution of raw coal is very fine (61% minus 45 micro with ash content of 51%), the ash content

Table 3. Industrial operating results for FXZ-1 static column.

Date	Ash content (%)			Clean coal yield(%)
	Feed	Clean coal	Tailings	
8/6/98	34.63	8.74	57.78	47.21
8.7/98	27.76	11.14	55.55	62.58

Figure 4. The picture of FXZ-1 in Yalai Coal and Cock Limitation.

Table 4. Size distribution of feed for Yalai Coal and Cock Limitation.

date	8/6/1998		8/7/1998	
Size (mm)	Mass (%)	Ash (%)	Mass (%)	Ash (%)
0.5-0.25	0.90	16.97	6.83	6.65
0.25-0.125	15.08	7.84	26.66	8.87
0.125-0.074	12.07	10.95	11.72	13.74
0.074-0.045	10.76	16.63	9.38	21.88
-0.04	61.19	51.01	45.41	48.12
Total	100.00	35.65	100.0	28.33

of raw coal is as high as 35%, the separation result is satisfied. The ash contents of clean coal and tailings are 8.74% and 57.78% respectively. The throughput of FXZ-1 is 60M³/h, 6-7 tons dry raw coal per hour.

4. CONCLUSIONS

1) When no turbulent is occurrence, the higher is the aeration, the higher is the combustible

557

material on recovery for column.

2) The construct of bubble generator is important for the aeration rate and the size of bubble, therefore the bubble generator is important for the performance of column.

3) The separation results of column for different conditions are much better than conventional flotation cell. The test results of column are close to the release analyze curve.

4) It is possible for column to produce super low ash coal.

5) FXZ-1 static column has the advantages of high output, low power consumption and high separation efficiency. It is easy to operate and maintain. It has past the identification and is suggested to be spread out in coal preparation plants of China.

REFERENCES

Ofori, P K. & Firth, B. 1998. Turboflotation for fine coal. *XIII International Coal Preparation Congress.* Brisbane: Australia.

Ou, Zeshen et al. 1996. Coal desulfurization by column flotation. *Proc. of the'96 Int. Symp. on Mining Science and Tech.* Guo Yuguang and Tad S. Golosinski (Eds).

Sablik, Jerzy. & Nicol, Stuart K. 1994. Recent scientific and technological advances in coal flotation. *12th International Coal Preparation Congress.* Cracow: Poland.

Yoon R.H. Pillips D I. & Luttrell G. H. 1998. Industrial application of advanced fine coal cleaning and dewatering technologies. *XIII International Coal Preparation Congress.* Brisbane: Australia.

6 Mine environment

Mining Science and Technology' 99, Xie & Golosinski (eds) © 1999 Balkema, Rotterdam, ISBN 90 5809 067 1

Mine subsidence control in P.R. China: Engineering practice

Yuzhuo Zhang
China Coal Research Institute, Beijing, People's Republic of China

ABSTRACT: Underground coal mining causes twenty thousand hm^2 of land to subside annually, which affects a quarter of million of Chinese farmers. This paper reviews the methods to predict subsidence and measures to reduce and mitigate the damage. The profile functions, influence functions and numerical models are developed and improved with acceptable accuracy. The control approaches including coal pillars, backfilling of gobs, ground grouting etc. have been utilized. The future trends are to take a comprehensive approach to extract as much coal as possible within the environmental acceptable extent.

1. INTRODUCTION

Coal production in China reaches 1.3 billion tons annually. Coal mining and utilization cause environmental problems, among them the surface subsidence and deformation is one of the major concerns. Twenty thousands hm^2 of land subside each year due to coal mining. Totally half million hm^2 were damaged and this makes six million people lose their living land. Many villages have to be moved far away. Farmers are not insured enough to compensate their loss, as that in developed countries such as USA. Moreover, protection of surface buildings, railroads and water bodies during mining activities is necessary, which calls for less subsidence, deformation and fewer disturbances on the surface.

Reliable prediction and effective control of surface subsidence are the two fundamental issues. Hundreds of prediction models have been developed in major coal mining countries such as Poland, Russian, Germany, USA and South Africa. Advances in China include that the negative power function was proposed and probability integral method was improved [Liu, 1982, Zhang,1992]. More than two hundreds stations of field subsidence observation were installed and data collected to analyze model parameters. The purpose of prediction is to plan and control the surface damage. With regard to subsidence control strategies, in addition to conventional methods such as leaving pillars and backfilling gob area, some new approaches as ground grouting and reinforcement of pillars have been developed.

2. PREDICTION OF THE SUBSIDENCE IN CHINESE COAL FIELDS

Principles of overburden failure, strata movements and surface subsidence due to mining of tabular coal seams are correlated and predictable with certainty when longwall mining is used. The subsidence profile is usually trough shaped. It is essential that prediction of subsidence be developed with good accuracy. In China, most widely used prediction methods are profile functions, influence functions and mechanical methods. As an empirical method, the negative power function as one of profile

$$W(x) = W_{cm}e^{-a(c-x/n)^n}$$

functions is applied, which has the following form:

Where W(x) is the subsidence value of any point in main section of profile; W_{cm} is the maximum subsidence value; H is the mining depth; and a,n,c are the constants in a certain case.

Another widely used method is called probability-integral method, which is evolved from geometrical method developed by Polish scientists.

$$W(x) = \frac{W_{cm}}{\sqrt{\pi}} \int_{-\sqrt{\pi}\frac{x}{r}}^{0} e^{-\lambda^2} d\lambda$$

in which W_{cm}=m cosα; c is subsidence factor; m,x, and r are geometrical parameters.

Based on the study of Berry (1961), the author developed a model using transversely isotropic medium theory of elasticity with back analysis. It possesses the following forms:

$$W(x) = \frac{-M}{2\pi(\alpha_1 - \alpha_2)}[\alpha_1 \sum_{n=1}^{4} \arctan\frac{\alpha_1^2 X_n^2}{r_n^{'}}$$
$$- \alpha_2 \sum_{n=1}^{4} \arctan\frac{\alpha_2^2 X_n^2}{r_n^{''}}]$$

where M, X and r are geometrical parameters and $\alpha_1\alpha_2$ are mechanical parameters, which are functions of five elastic constants E_1, E_2, $\square_1, \square_2$ and G, where E_1, E_2 are the horizontal and vertical elastic modulus; λ_1, λ_2 are the Possion ratio in two directions and G is shear modulus.
Based on field data, the relationship has been established between the parameter of probability-integral method and those of the elastic model as follows:
The reason is that parameters of probability integral

$$\frac{E_1}{E_2} = \frac{1}{2\eta^2} + 0.1$$

$$\frac{E_1}{G} = 1 + (\frac{1}{2\eta^2} - 0.1)\tan\beta$$

$$\frac{\mu_1}{\mu_2} = \frac{3b\tan\beta}{\eta^2}$$

model were back analyzed from hundreds of field observation stations.

These functions are the simplified relationship reflecting the major factors in subsidence calculation. It is, however, powerful when combined with numerical methods such as FEM, DEM and BEM.

3. CONTROL OF SURFACE SUBSIDENCE USING SUPPORT PILLARS

The principle of this method is to choose a reasonable way to leave coal pillars supporting the overburden permanently. The maximum resources are mined for allowable surface displacements and deformation. As a traditional approach, practices in China were mainly use of long pillar mining or strip mining. Some examples are given in Table 1. It can be seen from Table 1 that engineered cases using long pillar mining are usually with extraction ratio of 40 to 60 % and subsidence factors are between 0.05 to 0.3. The difficulties to extend the use of this method are two folds: low extraction ratio and incapability of equipping modern mining machinery.
Room and pillar mining is not prevalent in large mines in China, because longwall mining is regulated as the major method for a relatively higher extraction ratio. The utilization of modern continuous mining system in room and pillar mining is gradually accepted in China for its high efficiency. Table 2 gives several examples in China.
Research is under way for artificial reinforcement of coal pillars. Both bolting and chemical method are being studied in laboratory. By using these methods,

Table 1   Engineering cases of strip mining (long pillar mining)

Site	Depth (m)	Seam thickness (m)	Panel width (m)	Pillar width (m)	Roof support method	Subsidence factor
Pingan Mine, Fuxin	144	1.4	30-50	20	Caving	0.15
Jiaohe Mine, Jilin	85	1.0	12-20	10	Caving	0.03
Hebi Mine No.9	164	1.0	30-40	16	Caving	0.164
Shengli Mine, Fushun	505	16.6	28	38	Hydraulic sand backfilling	0.040
Dayaogou Mine, Nanpiao	355	5.7	110	30	Caving	0.029
Fengfeng Mine No.1	119	5.1	25-35	25-48	Caving	0.073
Fengfeng Mine No.2	134	1.5	17-18	12-13	Caving	0.069
Fengfeng Mine No.3	135	1.4	15	17	Caving	0.130
Nantong Mine, Sichuan	280	1.45	12	12	Caving	0.056

Table 2   Examples of Room and Pillar Mining

Site	Depth (m)	Seam thickness (m)	Pillar size (m X m)	Entry width (m)	Extraction ratio (%)	Note
Shenfu	30-50	6.0	7.5X7.5	5.0	64	
Xiaohengshan	100-400	1.3	33X8	4.5-6.0	80	Retreat mining
Jixi Xiqu,xishan	150-176	2.2	15X13.5	5	44	Protect power tower

Table 3   Examples of ground grouting to reduce surface subsidence

Site	Mining depth (m)	Seam thickness (m)	Seam angle (°)	Panel width (m)	Panel length (m)	Subsidence velocity reduction (%)	Subsidence reduction (%)	Mining method
Laohutai mine, Fushun	602	36	26			65	67	Sand filling
Xuzhuang Mine, Datun	529	2.6	20-23	110	960	42.8	35	Caving
Tangshan Mine, Kailuan	764	2.8	15	135	425			Caving
Huanfeng, Xinwen	787	2.2	30	292	1100	2.9		Caving
Tongtan14307W,Yanzhou	545	5.4	2-3	179	854	66	36	LTCC*
Dongtan 14307 E,Yanzhou	560	6.0	1-4	179	926			LTCC

* Longwall Top Coal Caving

it is expected that the recovery ratio can be increased by more than 5%. An innovative idea was proposed to backfill a long pillar with high strength materials and then to mine all the other coal resources.

4. CONTROL OF SURFACE SUBSIDENCE USING GOB BACKFILLING

Using backfilling materials to maintain coal roof has a long history. As backfilling keeps timely compact, roof caving can be avoided, disturbed volume is limited and therefore surface subsidence dramatically reduced. In different backfilling methods, hydraulic filling of sands is the most effective one. From 1950 to 1970's, hydraulic filling was used in Xinwen, Fushun and other mines for protection of surface buildings and rivers. This is, however, not cost-effective in most cases. Supply of sand sources is more and more expensive. Study shows that more than 15% of the cost will be consumed in backfilling, and this makes the coal companies less competitive. Pneumatic filling was also used in several mines. Recent trends in backfilling are to develop highly efficient filling materials and technologies such as DMT technique, strip-substitutes, and boundary hardening of pillars using filling.

5. CONTROL OF SURFACE SUBSIDENCE USING GROUND GROUTING

The technique of ground grouting is also called Filling Overburden-separation to Reduce Subsidence (FORS)[Zhang,1998]. In 1985 Qi and Fan performed the first industrial test in Fushun Coal Mine with success. More than seven mines in China have used this method to reduce land subsidence, as shown in Table 3. This includes Yanzhou, Datun, Xuzhou, Kailuan, Xinwen, etc.. The results reported that 30 to 60% of predicted subsidence is reduced due to use of FORS.

The philosophy of this technique is as follows. When the overburden contains such a rock group that mechanical properties of the immediate strata is greatly different and the upper layer is stronger mechanically, discontinuity or separation occurs between the layers, both in time and space. A borehole is drilled and grouting system installed prior to influence of mining. Once the advance of working face influences the strata where the borehole located and a separation occurs, backfilling materials are pumped into the room. The materials take the rule of support upper strata, which reduces the development of separation to overburden and finally reduce surface subsidence. This can be shown in Figure 1.

Three major questions need to be concerned. Under what conditions separation occurs and how to effectively support ground is a major concern. Study shows that mining area should be big enough to allow the growth of separation for filling materials. It is obvious that separation also exists in fractured zone. These rooms, however, can not be utilized as the filling rooms because water can flow into the working face. Therefore, only those rooms which are above fractured zone are significant when use the technology.

Secondly, a long-term stable structure in overburden should be maintained under the support of the filled materials. It is proposed that ground grouting has four effects, namely occupying rooms, supporting posts, compressing the underlying strata, and expansion. The last question is to predict how the ground is controlled and how the subsidence reduced.

6 TRENDS OF SUBSIDENCE PREDICTION AND CONTROL

Even though the surface subsidence prediction is relatively solved for simple mining and geological conditions, there are still some difficulties need to be tangled. Three major directions are proposed by the author [4]. One is subsidence forecasting in complex geological conditions such as encountering large scale faults. The second direction is to predict subsidence evolution in Longwall Top Coal Caving

(LTCC) method. LTCC method disturbs strata and the surface more intensively. The third direction is to distinguish the effects of mining and dewatering on subsidence. In many areas of China, degradation of water table also causes surface subsidence. The coupling effects have not clearly been understood.

As matter of fact, the control of subsidence is an economic problem. The research should focus on comprehensive means and evaluation of the total gain and loss to mine a certain volume of coal. Pillars as supports to hold the entire overburden are truly effective when carefully designed. In many places such as urban area, room and pillar mining, long pillar mining, and seam-thickness-limited longwall mining can still be an alternative. The gob backfilling method is not eliminated but rarely used because of its high cost. It is expected that some cost-effective materials can be invented and utilized in backfilling. For FORS method, intensive studies should be made to extend the test conditions. Geological conditions fitting to this technique needs to be identified for major coalfields in China. On the other hand, the technology itself should be improved in the material mix, injection procedures, multiple level grouting, quality inspection, etc.

7. CONCLUSIONS

China's economy requires sustainable development. This means energy utilization has to satisfy environmental capacity. Large volume of coal extraction induces surface subsidence which endangers farmer's living space, especially in the east and north parts of China. Therefore, surface subsidence control has to be carefully planned. The quantity and extent of mining influences should be minimized at the highest possible extraction of coal.

Surface subsidence has its own mechanisms, which are complicated, site specific and dependant on geological and mining conditions. To meet the requirements of accurately predicting the surface subsidence and deformation, several models were developed based on field observation data obtained from hundreds of mines in China. Under normal geological conditions, usually a good accuracy can be obtained.

Reduction of surface subsidence is the key issue to protect buildings and other objects to be protected. Conventional approaches are used at the expense of wasting some resources or backfilling materials. Newly developed ground-grouting method, if it is carefully designed and implemented, can nicely form structurally stable strata and reduce surface subsidence using relatively small portion of grouting materials. The final goal would be combining different methods to minimize the surface damage with the lowest cost that should include compensation of environmental damage.

REFERENCES

Berry, D.S. & Sales, T. W.. 1961, An elastic treatment of ground movement duo to mining—transversely isotropic ground. J. Mech. Phy. Solids. 9:1-10

Brady, B. H. G. & Brown, E. T. 1985. Rock mechanics for underground mining, London: George Allen & Unwin

Fan, Xueli. 1998. Advances and trends of overburden separation filling to control surface subsidence in coal mines in China, New Technology on Surface Subsidence Control, (Zhang Yuzhuo, Xu Naizhong Eds.), Publishing House of China University of Mining and Technology. (in Chinese)

Farmer, I. 1985. Coal mine structures, London: Chapman and Hall.

Zhang, Yuzhuo. 1992. Principles and programs for surface subsidence prediction, Publishing House of Coal Industry. (in Chinese)

Zhang, Yuzhuo. 1997. Two approaches for mining thick coal seam under buildings. Journal of China Coal Society. 22(7): 48-51

Zhang, Yuzhuo and Chen Liliang.1996. Conditions for bed separation of overlying burden in longwall mining, Journal of China Coal Society. 21(6):576-581.

Zhang, Yuzhuo, Chugh, Y.P. 1996. Similitude of strata movements due to underground mining using equivalent material modeling. Journal of Coal Science and Technology (China), 1(1):1-8

Mining Science and Technology'99, Xie & Golosinski (eds) © 1999 Balkema, Rotterdam, ISBN 90 5809 067 1

Minimising the environmental impact of mining operations

D. Schofield, G. Brooks & A. Tucker
SChEME, University of Nottingham, UK

ABSTRACT: The landscape is a vital part of the natural and man-made environment. Often mine planners fail to adequately predict, recognise or deal with the visual impact of a new development or the pollution that results from that development. The result has been a serious erosion of the character and quality of many of the Chinese urban and rural landscapes.

Visibility mapping and visualisation techniques are central to the effective prediction and communication of landscape, pollutant and visual impacts. The techniques must be carefully chosen and rigorously applied, as they will be subject to close scrutiny and, in the case of contentious developments, may need to be explained and substantiated at a public or government enquiry (The Landscape Institute, 1995).

Computer Graphics (CG) and Virtual Reality (VR) techniques have been used for a number of environmental visualisations. The exponential development of this technology has left the development of application methodologies behind; resulting in an under-utilisation of the advanced visualisation techniques and technology on offer.

1 INTRODUCTION

World population is growing at an extraordinary rate, even if 100% recycling could be achieved it would not meet the future demand for raw materials. Emerging new markets based in the developing industrialised nations (for example China, India, the Far East 'Tiger' economies) continue to demand more consumer goods for their growing middle classes. Populations in developed countries continue to demand ever increasing quantities of material goods. Western governments are relying more and more on tax and export income derived from the resource sector. We may not be pleased with the ever-increasing demand on non-renewable resources but it would take fundamental changes in the attitude of the majority of the population to bring about an ethos of sustainability. Therefore it becomes clear that mining will not only continue in the future, but also continue to expand on a worldwide scale (Earth Science Australia, 1998).

Mining has a large impact on the environment in the process or moving and processing cubic kilometres of material. Fortunately, many mining areas are remote and unsuitable for most other purposes. Mining methods, which have an environmental impact, are determined by economic considerations, particularly the taxation structure.

Most governments see mining as a steady source of income despite the fact that mining resource prices are often volatile. In order to provide a steady tax flow, mines are often levied per tonne of ore. Charging per tonne of ore means that in tough times ore of higher grade must be mined and simultaneously ore of lower grade left behind. This is a short-term strategy and means that mining communities often must suffer the hardship associated with premature closures. This also leads to problems of the mining companies being unable afford to comply with strict environmental legislation or fund restoration of the land affected by the mine workings. This has led to governments in many countries formulating policy to ensure environmental regulations are met.

2 AGENDA 21

The Chinese Government has responded actively to the United Nations Conference on Environment and Development (UNCED) by taking the lead in formulating the first national Agenda 21 in the world, titled China's Agenda 21 - White Paper on China's Population, Environment and Development in the 21st Century (ACCA21a, 1997). This document integrates the principles of UNCED with China's particular conditions and needs.

Chapter 13 of the Agenda 21 white paper is entitled "Sustainable Energy Production and Consumption". This states that the energy industry is fundamental to China's national economy, is of critical importance to socio-economic development and the improvement of people's living standards. In a rapidly expanding economy, China's energy industry is confronted with dual pressures for economic development and environmental protection (ACCA2b, 1997).

China has a coal-based energy structure, with coal consumption amounting to 75% of total energy consumption. Cleaner energy sources constitute only a small proportion of the total energy supply. Because of this, China emits large quantities of pollutants, resulting in serious atmospheric and water pollution.

If the present pattern of energy production and consumption is maintained, China would be hard pressed to meet future energy demands. This is due to the shortage of resources, low financial inputs and inadequate transportation, combined with the need for environmental protection. Therefore, major components of China's strategy for sustainable development include changing present energy production and consumption patterns, diversifying energy sources and the structure of power production, and establishing an energy structure that is less or not at all harmful to the environment (ACCA2b, 1997).

China has rich coal resources. The foreseeable future, coal will continue to be China's main source of energy. Both the output of coal and the sulphur content in coal resources will increase. Environmental problems are associated with the entire process of coal extraction, processing and consumption.

In comparison with developed nations, China still has a long way to go in terms of developing low-polluting techniques for coal extraction and for high efficiency, clean utilisation. One of the main areas for attention indentified by the white paper was land reclamation. In 1991, the land reclamation rate at the coal extraction sites was only 16% and the coal pit water utilization rate was 15% (ACCA2b, 1997).

3 PROPOSED ACTIONS

While some work has already been done on China's Agenda 21 since its inception, much remains to be done. It is proposed that all departmental and regional government agencies should integrate sustainable development thinking into their economic and social plans.

Efforts to raise public awareness about and participation in China's Agenda 21 will continue.

This will include publishing the national sustainable development strategy in newspapers, magazines, and booklets, and broadcasting it on television, radio, and other mass media. Computer representations of proposed developments have been used in the west, to introduce the public to the impact of proposed mineral developments. Photo-realistic renderings of mine sites or landfill tips, at various phases of development, give the public a better understanding than plans and maps.

China will continue to seek international cooperation both in the form of financial support and information exchange. Opportunities for technology transfer and foreign investment should be sought out and considered as China's Agenda 21 continues. In particular, China will try to establish relevant financial mechanisms such as joint ventures. These mechanisms should help finance environmentally sound technology, recycling, cleaner production, improvement of industrial products, reclamation of secondary resources, protection of ecologically fragile regions, and other projects which can attract foreign expertise and capital for China's sustainable development. International experiences in sustainable development management will also be valuable to China (ACCA2a, 1997).

The transfer of computer technology has been highlighted as an important part of Agenda 21. The exponential development of the technology has created a range of new environmental applications. CG and VR technology are being widely applied in the west for environmental visualisation.

4 COMPUTER GRAPHICS (CG) AND VIRTUAL REALITY (VR) TECHNOLOGY

CG and VR are continuously evolving technologies, and this evolution seems to be accelerating at an ever increasing rate. It is true to say that anything in hard print is probably out of date and most of the real information about the current state of the art can actually be found on the Internet (Denby and Schofield, 1998).

The term 'computer graphics' refers to a set of computer applications which can be used to produce images and animations which would have been impossible with the technology available only a few years ago. To produce high resolution computer graphics, a three-dimensional geometry is defined using conventional CAD software. A range of texture maps are then applied to the solid three-dimensional CAD objects (texture maps are the computer graphics equivalent of applying patterned wallpaper over an object). Lighting conditions are then defined and objects are viewed from a range of different camera positions (Denby et al, 1998).

The term 'virtual reality', as used in this paper,

can be described as the science of integrating man with information. It consists of three-dimensional, interactive computer generated environments. These environments can be models of real or imaginary worlds. Conceptualisation of complex or abstract systems is made possible by representing their components as symbols that give powerful sensory cues, related to their meaning (Warwick et al, 1993).

In contrast with static three-dimensional images from CAD systems, virtual reality models allow the user to interact with the world. A virtual reality user can, for example, sit in a vehicle and drive it. The vehicle will have responsive behavior associated with it, and other vehicles in the world may have predefined actions and responses.

The rapid developments in PC technology in recent years and the huge potential market for desktop VR in a wide range of sectors means that this is the area where some of the fastest developments are occurring. Whilst much of the development is for the leisure industry there are real engineering based opportunities that can be exploited. Desktop PCs with the 3D graphics acceleration necessary for VR training are now available. Such systems can produce high levels of realism with smooth motion and stereo/surround sound capabilities. As this machine will also be usable for other applications this dilutes the argument that the hardware costs are excessive.

At the same time that hardware costs and performance have reached accessible levels a growing market for VR software is resulting in continuous reductions in software development costs. As the market for VR applications in mining increases this will result in major reductions in unit costs of complete systems. Hence, the argument that VR is an expensive technology no longer holds and the opportunity for mass use of VR opens up as unit costs drop dramatically (Denby and Schofield, 1998).

5. ENVIRONMENTAL APPLICATIONS OF CG AND VR TECHNOLOGY

Many large mine planning packages (such as Surpac and Vulcan) have the ability to translate complex data from different sources into highly visual, 3D graphics, this makes them useful for environmental management. The software's proven ability to evaluate and present possible scenarios during planning and remediation exercises, facilitates communication, ensuring optimum decision making.

The AIMS Research Unit at the University of Nottingham has taken this further, applying advanced CG and VR technology to a range of environmental applications. The rest of this paper will describe some of these applications.

5.1 Visual Impact Studies

Understanding the environmental impact of any engineering project is an integral aspect of the assessment process. Large amounts of multi-source information must be evaluated and incorporated into an acceptable model, readily understandable by all participants in the decision making process.

Figure 1. A CG representation of a proposed mine

AIMS Research have used CG and VR software to provide an interactive, intuitive approach to meeting planning requirements and raising community awareness. The land surfaces are often modelled using mine planning\landform manipulation software such as Vulcan or Surpac. The wireframe models created are then exported into 3D software, where textures are draped over natural or rehabilitated terrain and realistic lighting is added. Users may then view the world from any position.

Figure 2. Example of a field of visual influence

Once built the model can be used many ways:
- To help engineers visualise the design and proposed after-use of the land.
- To provide high-resolution still images of the proposed changes from key viewpoints,
- To provide high-resolution animations – either 'fly-by' or 'walkthroughs' of the world.
- To provide 'zones of visual influence' calculated from key viewpoints.
- To provide real time VR software, enabling users to interact with the model.

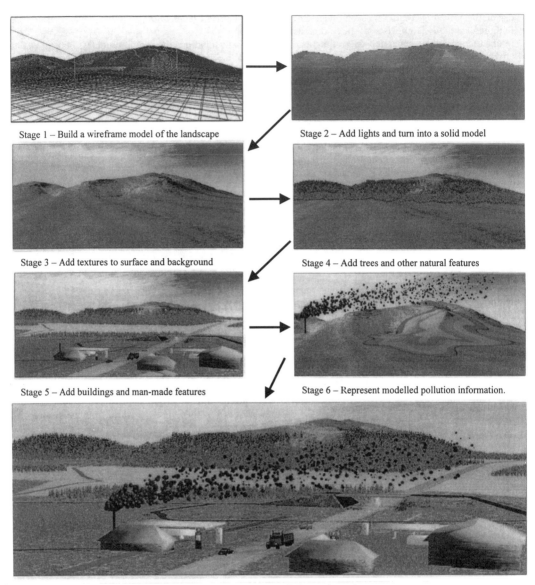

Stage 1 – Build a wireframe model of the landscape

Stage 2 – Add lights and turn into a solid model

Stage 3 – Add textures to surface and background

Stage 4 – Add trees and other natural features

Stage 5 – Add buildings and man-made features

Stage 6 – Represent modelled pollution information.

Stage 7 – Combine 3D model with pollution data to show effect of pollution on the environment.

Figure 3. Modelling pollution in a 3D environment

CG and VR visual impact studies raise public awareness of and participation in the planning process of any proposed impact to their environment.

5.2 Pollution Modelling

Resource assessment, hydrogeological studies, waste disposal and mining\civil engineering projects benefit from accurate 3D interpretation, providing successful environmental solutions. Interactive modelling and visualisation capabilities allow the effect of pollutants to be easily assessed.

A true three-dimensional pollution modelling system allows the generation of 3D models, the calculation of volumes, averages and totals of attributes to provide statistically significant and accurate results which simulate the real conditions encountered.

Figure 3 shows the step-by-step creation of a virtual world used to display pollution data. The landscape is first created in a wireframe form, turned into a solid model with lighting and textures. Both natural and man-made features are then added –

Figure 4. An overview of a landfill site

giving a realistic landscape. The result of the numerical pollution modelling are then represented graphically and combined with the 3D model. This allows the engineers and planners to see the effect of the pollution in the environment.

This technique may also be extended to visualise a range of other information in the virtual world. Other work at Nottingham includes visualising the results of groundwater modelling, noise pollution and an examination of the impact of introducing different flora and fauna to a landscape.

5.3 *Site Assessment*

Site Assessments are an internationally recognized investigative processes conducted through a systematic, phased process to determine the likelihood of a site being contaminated and establishing the type and extent of contamination. These assessments are necessary to assess short and long-term liability potential and to assess estimates of site remediation volumes and costs (Canadian Environmental Solutions, 1999).

Depending on the nature and magnitude of the environmental impact, and of the remediation technology employed, remediation of contaminated subsurface soil and surface/groundwater bodies may require a period from a few weeks, to several years to complete. It is imperative that prior to initiating a remediation plan, a clean-up criteria be developed which considers factors such as proven effectiveness of a particular technology, site applicability, the remediation time frame, and the probable costs associated with available alternatives.

The importance of training personnel to perform site assessments of mineral operation, during both the operating phase and the restoration phases, has been noted. The AIMS Research Unit at Nottingham University is recognised as one of the leading developers of VR based training systems for the mining industry. The University of Nottingham has developed a number of applications (both research and commercial systems). AIMS are concentrating

on studies into the use of VR training technology for the minerals industry.

SAFE-VR, developed by AIMS, has been designed to be a tool to allow the rapid creation of a great variety of hazard spotting and training applications. One of the initial applications developed by the AIMS unit using SAFE-VR is an environmental assessment appraisal system for training personnel to perform site assessments.

Figure 4 shows a site used to train site inspectors to perform an environmental audit on a landfill site. Landfill and waste disposal is a popular after-use of surface mine workings. Within the virtual environment the user has to walk around the site spotting and correcting a range of problems and hazards. These hazards include toxic waste dumping, broken fences, windblown litter, unsafe material handling and damaged fencing. The user is scored not only for how many items they identify correctly, but also for selecting the correct remedial action.

Figure 5. Windblown litter attached to the site fence

Other systems developed using SAFE-VR include (Denby et al, 1999):
- A surface mining truck pre-shift inspection
- A full operational land drilling rig.
- Pumping systems within a chemical plant.
- Oil rig production modules.

6 CONCLUSIONS AND PREDICTIONS

The environmental problems of China are self evident, they are not hidden away in remote mines and hidden landfill sites. The environmental problems affect the local population every day and any visitor to China will experience pollution and the eroded urban and rural landscapes first hand. A change in way the planning and design of new projects is undertaken is needed.

Agenda 21 starts to address this change, trying to place the responsibility on the planners to design their new projects with the environment in mind, to plan for restoration and after-care of the land, to minimise pollution and to reduce the visual impact of their designs.

CG and VR technology offers a range of environmental solutions. The ability to visualise the engineering design, assess the impact, modify the design, and most importantly communicate that design to a wide audience in a simple visual format. The real time VR experience can be extended to produce a range of interactive systems to train staff to spot and correct environmental hazards in their workplace.

The AIMS Research Unit at the University of Nottingham has been recognised as one of the leading developers of CG and VR tools for the minerals industry. We believe that the extension of this technology into all areas of environmental assessment and design offers a great potential to solve some of China's environmental problems.

7 REFERENCES

ACCA21a, 1997. Introduction to China's Agenda 21, government web page at URL: www.acca21.edu.cn/ca21ht.html, (version current 29th March 1999).

ACCA21b, 1997. China's Agenda 21, chapter 13 – sustainable energy production and consumption, government web page at URL: www.acca21.edu.cn/chnwp13.html, (version current 29th March 1999).

Canadian Environmental Solutions, 1999. Remediation of soil and water contamination: strategis.ic.gc.ca/SSG/es30567e.html (version current 29th March 1999).

Denby, B. and Schofield, D., 1998. The role of virtual reality in the safety training of mine personnel, proceedings of SME annual meeting and exhibit, Orlando, Florida, pre-print 98-134, 9th – 11th March 1998.

Denby, B., Schofield, D., Hollands, R., Ren, T. X., Walsha, T. and Williams, M., 1998. Using advanced computer modelling techniques to improve the safety of mining operations, proceedings of CCRI - international mining technology symposium, coal mining safety and health, Chongquing, China, 14th-16th October 1998.

Denby, B., Hollands, R. and Schofield, D., 1999, Virtual reality based training technology for the mining industry, 28th international conference on safety in mines research institutes, June 7th-10th, Sinaia, Romania, awaiting publication.

Earth Science Australia, 1998. Mining a perspective, educational web page at URL: www.bushnet.qld.edu.au/schools/msb/enet/eres/rr es/depfile/minper.htm,

(version current 29th March 1999).

The Landscape Institute and The Institution of Environmental Assessment, 1995. Guidelines for landscape and visual impact assessment, London: Chapman and Hall.

Warwick, K., Gray, J. and Roberts, D., 1993. Virtual reality in engineering. London: IEE, 1993.

Mining Science and Technology'99, Xie & Golosinski (eds) © 1999 Balkema, Rotterdam, ISBN 90 5809 067 1

Solidifying tailings slurry with new binder at Meishan Iron Mine

Wenyong Liu, Henghu Sun, Baogui Yang & Xiaoyang Liu
China University of Mining and Technology, Beijing, People's Republic of China

Qilin Shi, Bohua Liu & Kanglin Ning
Shanghai Meishan Mining Industrial Limited, People's Republic of China

ABSTRACT: Meishan Iron Mine produces 2300 tones of tailings every day. The tailings slurry contains a large amount of clay minerals such as kaolinite, montmorillonite and the like. These tailings are very difficult to concentrate and store using traditional methods. This paper presents a new technology developed to dispose of and make use of these taiilings. A solidifying material has been used as binder to solidify the tailing slurry into lumps, solidifying agent added when the tailings slurry is concentrated to about 70% of solids by weight. The tailing lumps can be used for filling, to fill in exhausted underground mines or for other purposes.

1. THE PROPERTIES OF MEISHAN TAILINGS

In order to solidify tailings effectively, many researches on properties of Meishan tailings have been done. The properties include ingredient, size, viscosity, free subside and filtrate of compound slurry, which are described as follows:

1.1 *Chemical composition and ingredients*

Table 1. Chemical composition of Meishan tailings.

Composition	SiO$_2$	Al$_2$O$_3$	Fe$_2$O$_3$	MnO$_2$
Content %	26.03	14.59	31.14	0.29
Composition	TiO$_2$	CaO	MgO	loss
Content %	0.49	9.19	2.79	14.13

Table 2. Ingredients of tailing sands.

Mineral	magnetite		siderite	hematite
Content(%)	10.3		16.03	5.6
Mineral	pyrite		calcite	kaolinite
Content(%)	5.2		13.42	13.72
Mineral	quartz	feldspar	apatite	dolomite
Content(%)	15.43	7.0	1.0	9.7

From above two tables, the main chemical compositions of Meishan tailing sands are SiO$_2$, Al$_2$O$_3$ and Fe$_2$O$_3$. These are main chemical compositions of clay. The mass of them is more than 70 percent. In the tailing, there are many clay minerals, such as kaolinite, apatite, etc., which make tailings be difficult to dispose.

1.2 *Size composition of tailing sands*

Size composition of tailing sands was measured by Maxter-sizer. The result is shown in figure 1.

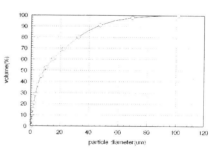

Figure 1. Size composition of tailing sands

It is found that high clay content is a remarkable property of its size composition. About 10% is less than 3.93 m, 30% less than 34 m, and nearly 49% is less than 37 m. High clay content causes a series of problems in concentration. If being stored at a little density, it needs more area of land. If being solidified directly, it is very difficult using normal solidifying materials and costs much more money.

1.3 *Free subside test*

After a series of experiments, we found that it's easy to concentrate tailing slurry from 25% to 40%. Then, for the high clay content, it is very difficult to concentrate tailing slurry by free subside method. Test was conducted at a 1000ml vessel. Original density is 10%, 20%, 30%. Results as follows:

Table 3. Free subside test results

Pulp density (%)	subside speed (cm/min)	2hours density(%)	30hours density (%)
10	0.164	22.43	34.95
20	0.083	31.04	44.73
30	0.021	33.82	46.47

We could find that original density has great influence on original speed of subsiding. When the slurry is getting dense, its original subside speed becomes less sharply. When slurry of 20% has been free subsided for 30 hours, its density could reach 40%. The fact is beneficial to slurry disposing.

1.4 *Viscosity test*

Fann six-speed viscosimeter was used to test slurry's shearing stress at different shearing strain and its viscosity was obtained. All test samples were disposed from original sample of 58% in density. Results as table 4:

Table 4. Viscosity test results

N(r/min)	3	6	100	200	300	600
20%	2	2	4	5	6	10
25%	3	4	6	8	9.5	14
30%	6	7	10	12	15	21
35%	13	15	21	25	27	34
40%	15	21	32	36	40	47
45%	25	40	57	65	70	85
50%	43	49	112	120	130	155
55%	47	66	145	160	164	181

From the above table, the slurry viscosity can be calculated. We find that the viscosity increases quickly while the density is above 40%. The higher viscosity is not good for adding solidifying materials.

2. THE SOLIDIFYING EXPERIMENT OF MEISHAN IRON TAILING SANDS SLURRY

2.1 *Solidifying material*

A type of solidifying materials was developed by Beijing Gaoshui Institute of Mining Engineering and Material. It mainly has the following properties:
(1) Strong water solidifying capability. In the solidifying body, the total volume ratio of water can be up to 90 percent and its mass ratio of water to solid is about 3:1. That's why this sort of materials has strong water solidifying capability[4].

(2) Rapid solidification and strength performance: When the material and water are mixed in a 1:9 volume ratio, the resultant pulp will solidify into a type of strong artificial stone within 5-30 minutes. Its strength increases very quickly and will reach 3.0MPa in one day and 5.0MPa in seven days[4].

(3) Strong aggregate solidifying capability. This material can solidify many kinds of aggregates, even if mud. From the following strength experiment data (table 6), we can find it's very effective for the material to solidify Meishan iron tailings[4].

2.2 *Technique of solidifying*

Considering with the properties of Meishan tailings, the solidifying technique has been designed. The disposing processes can be divided into three steps[1][2][3].

First, 25% tailing sands slurry should be subsided and concentrated to 40%. This step is relatively easy.

Then add the solidifying material into it and mix it. This is the best time to add the solidifying material. As the density becomes higher, the viscosity becomes higher too, and adding and mixing of the material becomes more difficult.

At last, press and filtrate the slurry to 60-75% in 0.5-0.8 MPa pressure. The purpose of this step is to reduce water and save material as it can as possible. For the existing of solidifying material, solidification has taken place partly, so it is not too difficult to filtrate the slurry. Finally shape it to product. The test results are as follows (table 5):

Table 5. Press experiment results

No.	Original density (%)	final density (%)	time (min)
1	40	60	15
2	40	65	23
3	40	70	30
4	40	75	38

2.3 *Experiment data*

The solidifying body must meet the following two requires. It must be ensure that the packing body of tailing sands can't be changed into mud when meeting water. The curing time is short enough for the convenient of transportation. The second require can be met easily because of the material's property of fast strength performance.

Table 6. Experiment data of solidifying body strength

No.	Filter density	Ingredient radio	specific gravity	mixing time	strength of body before dipping in water				Strength of body after dipping in water			
					1d	3d	7d	28d	1d	3d	7d	28d
1	60	7	1.42	10'	0.04	0.04	0.05	0.06	0.03	0.035	0.06	0.05
2	65	7	1.42	10'	0.14	0.12	0.2	0.25	0.15	0.12	0.18	0.15
3	70	7	1.42	10'	0.25	0.25	0.4	0.45	0.25	0.25	0.4	0.35
4	75	7	1.42	10'	0.5	0.6	0.8	0.9	0.5	0.35	0.65	0.71
5	60	8	1.43	10'	0.12	0.12	0.15	0.15	0.12	0.1	0.15	0.1
6	65	8	1.43	10'	0.28	0.3	0.5	0.55	0.25	0.25	0.35	0.3
7	70	8	1.43	10'	0.4	0.4	0.65	0.7	0.3	0.45	0.65	0.65
8	75	8	1.43	10'	0.95	1.1	1.45	1.5	0.75	0.78	1.1	1.2
9	60	9	1.44	10'	0.13	0.25	0.3	0.2	0.15	0.2	0.2	0.15
10	65	9	1.44	10'	0.3	0.5	0.65	0.5	0.3	0.35	0.65	0.65
11	70	9	1.44	10'	0.8	0.85	1.15	1.35	0.7	0.75	1.0	1.1
12	75	9	1.44	10'	1.3	1.6	2.1	2.6	1.2	1.4	1.9	2.1
13	60	10	1.45	10'	0.26	0.35	0.3	0.35	0.25	0.35	0.4	0.52
14	65	10	1.45	10'	0.4	0.45	0.65	0.7	0.4	0.45	0.75	0.8
15	70	10	1.45	10'	0.9	0.95	1.0	1.1	0.8	0.95	1.25	1.35
16	75	10	1.45	10'	1.55	1.8	2.5	2.8	1.3	1.8	2.65	2.9

In order to know whether the packing body could change into mud, we put some product into sealed container and the others into water. The curing temperature is 16-18°.Then we compared the compressive strength of them in different phase. If the compressive strength in sealed container is same as that in water, we could say the packing body hasn't been changed into mud. If the compressive strength in water is lower than that in sealed container obviously, we could say that the body has been changed into mud. That is, the solidifying of tailing sands slurry couldn't meet the requirement. Experiment data are shown in table 6.

3. ANALYSIS OF RESULTS

3.1 *We could discover some regularities from the above results of experiment:*

With the increase of quantity of the solidifying material, the compressive strength of tailing sands packing body could become higher and higher and packing body couldn't change into mud. The relation between ratio of solidifying material and compressive strength is shown in figure 2.

With the increase of the concentration, the compressive strength of packing body is increased. The relation between compressive strength and concentration of body is shown as fig.3.

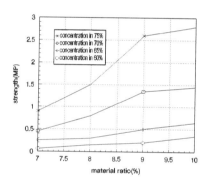

Fig 2. The relation between strength and ingredient radio in different concentration after 28 days

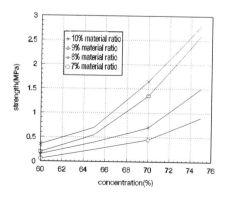

Fig 3. The relation between strength and concentration in different ingredient after 28 days

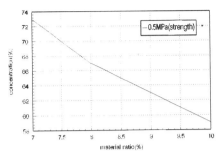

Fig 4. The relation between concentration and ingredient radio in 0.5 MPa pressure

Table 7. The cost of solidifying material

No.	ingredient radio(%)	Unit consumption (Kg/T)	unit price (y/T)	cost of material (y/T)
1	7	70	450	31.5
2	8	80	450	36.0
3	9	90	450	40.5
4	10	100	450	45.0
Ave	8.5	85	450	38.25

In the same strength and condition for changing into mud, the amount of solidifying material could be decreased with the increase of concentration.

We need to explain the data in the table 6: In the experiment, in order to shape the mixture body, we must smash it to pieces and shape it. This action could break its internal structure and make the strength of packing body low. So in reality, the strength of production is probably higher than that in the experiment.

The results of experiment show that: In the condition of concentration and content of solidifying material, packing body of tailing sands didn't change into mud when its strength reached over 0.6MPa. This shows that using the solidifying material to dispose tailing sands is completely advisable in technology.

3.2 Economical analysis

The aim of solidifying tailing sands is easy to stack the tailing sands and let it not to change into mud. So meeting the requirements of technology, the less the quantity of the used solidifying material is, the better it is. That is, when the ingredient radio is 7%, the cost of material is 31.5yuan per ton. If we think of the condition of the body's breaking in experiment and strength's decreasing, the cost of material is also lower when meeting the same requirement of strength on the spot. At present,

though the cost of condensing and filtrating is difficult to determine, the cost of comprehensive disposal could be controlled in 45 yuan per ton.

4. CONCLUSIONS

After the study of solidifying tailing sands in Meishan mine, the following conclusions are obtained.

1. Meishan iron tailings' content have a large amount of clay and it's difficult to be concentrated. But it is feasible to dispose the tailing sands by using solidifying materials developed by Beijing Gaoshui Institute of Mining Engineering and Material. When the packing body's compressive strength reaches 0.6MPa, it couldn't become mud again.

2. The process of the new technique is: concentrate slurry from 25% to 40% at first, then compound with adhesive. The high density slurry was filtrated to 60-75% and put into model and stacked in ground subsiding area or crammed into excavated area.

3. Using this new technology, the tailing storage capacity is greatly reduced and large area of land will be saved. The environmental pollution caused by tailings storage will greatly be decreased.

4. The ground subsiding area can be disposed by using tailing sands packing body. This is beneficial to the land recovering. We can also fill tailing sands solidifying body into excavated area and it can decrease ground pressure and sinking of the ground surface. In rainy season, the solidifying body can protect the ground subsiding area and under ground stopes against the damage of rain water.

REFERENCES

[1] Henghu,S., Wenyong,L., Yucheng,H. & Baogui,Y. 1998. *The Use of High-water Rapid-solidifying Material as Backfill Binder and Its Application in Metal Mines*. Brisbane, Australia: Sixth International Symposium on Mining with Backfill.

[2] H Sun, G Baiden & M Grossi 1998. *New Backfill Binder and Method*. Brisbane, Australia: Sixth International Symposium on Mining with Backfill.

[3] Henghu,S., Qinglin,L. & Wenyong,L. 1991. *Study on A New Technology of High-water Rapid-Solidifying Backfill in Metal Mines*. Beijing: Nouferrous Metal.

[4] Henghu,S. 1994. *High-water Rapid Solidifying Material and Its Application*. Beijing: The press of CUMT.

Mining Science and Technology' 99, Xie & Golosinski (eds) © 1999 Balkema, Rotterdam, ISBN 90 5809 067 1

Partial backfilling with high-water rapid setting material to control mining subsidence

Ximin Cui, Baogui Yang & Liqun Shen
Department of Resource Exploitation Engineering, China University of Mining and Technology, Beijing, People's Republic of China

Jinan Wang
Institute of Civil Engineering, Beijing Science and Technology University, People's Republic of China

ABTRACT: This paper presents a new method for controlling mining subsidence and damage in room and pillar mining by partial backfill of mined rooms with high-water rapid setting material. Stress ratios in filled pillars are investigated using the geometric method for predicting stresses in inclusions,. The results of numerical simulations show that the new method is effective in controlling mining subsidence and simple in operation.

1. INTRODUCTION

Research on ground movement and mining damage can be traced back to 19th century and have attracted considerable attention. Many theoretical and experimental investigations have been carried out. It is well known that mining damage can be reduced or prevented by structural and underground precaution. The latter includes measures such as leaving pillars, filling, partial extraction, rapid extraction, and rational design of workings for reducing ground movement and deformation.

The design and construction of surface structures with the object of reducing mining damage is an engineering field. Most structural precautions take one of the following treatments or the combination of several treatments:

1. Strengthening foundations or structures of buildings by rigid design.

2. Flexible design, enable a structure to adapt itself to the transmitted deformation without losing its strength.

3. Reducing the forces transmitted from ground deformation. This can be accomplished by designing small plan areas of houses, providing gaps between adjacent units of large buildings, reducing the bond between ground and structure, cutting trenches around buildings in order to absorb compressive ground strain.

4. Orientation of new buildings with regard to the mining conditions: Outcrops of faults or other prospective sites of discontinuities should be avoided, and when the major lateral axis of a building is parallel to the lines of equal subsidence, the strain will have the least effect.

The observations of subsidence may be summarized as follows:

1. Total extraction without stowing, causing caving or continuous subsidence of the immediate roof, induces a subsidence factor between 0.60 and 0.95.

2. Partial stowing, so-called strip packing, has little or no influence, which may results a subsidence factor ranging from 0.60 to 0.90.

3. Pneumatic, slusher, and hand stowing has a subsidence factor between 0.30 and 0.70, the most common value being 0.50.

4. Hydraulic stowing induces subsidence factors ranging from 0.10 to 0.30. Smaller values are obtained when packing under pressure: 0.08 is quoted for Upper Silesia.

5. When partial extraction is combined with hydraulic stowing of the panels, the subsidence factors can be less than 0.03 or even smaller than measurable.

However, leaving pillars and partial extraction must leave permanent pillars and it will be more difficult or impossible to extract in future. Filling, mainly in the form of pneumatic and hydraulic stowing, is too expensive and complex to apply in practice in China's coal-field. The main purpose in this paper is to introduce a new method for carrying out total extraction and ensure the final and temporary deformation not to exceed the allowable values by partial back filling with high-water rapid setting material.

2. STRESSES PREDICTION METHOD

When an elastic leaving pillar or filling body of any

shape is loaded in a condition of plane stress or plane strain, the interface between the pillar or filling body and rocks around them has the deformation in a complicated way under the stress action along this interface. Exact mathematical results are very complex even for few geometric shapes that their solutions are known as circle or elliptic ore-body. Exact solutions are not known for even such simple shapes as rectangular or oval ore-body, and exact solutions for irregular or complex shapes are nearly hopeless to find and impossible to use.

In order to analyze the stresses and stability of filling body, an imaginary effective inclusion is adopted (Babcock 1970,1974), shown in Figure 1.

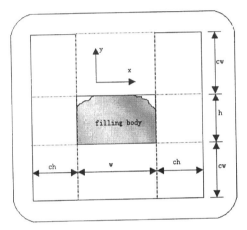

Figure 1. Imaginary rectangle used to estimate average filling body stresses.

The effective Young modulus can be expressed as

$$\overline{E}_c = \overline{E}_{1y} = \frac{\sum_{i=1}^{n} h_i}{\sum_{i=1}^{n} \frac{h_i}{E_i}} = \frac{2cw + h}{2cw / E_p + h / E_1} \quad (1)$$

Where $\overline{E}_{1y}$ is the effective Young modulus for the column in loading direction, E_p is the Young modulus for the plate, E_1 is the Young modulus for filling body, and the element heights h_1, h_2, h_3 are cw, h, and cw, respectively, c is a constant to be determined and w is the width of the filling body. So we have the average filling body stress $\overline{\sigma}_{yy}$ produced by the applied stress S_y as

$$\overline{\sigma}_{yy} = \frac{2cw + h}{2cw + h / k} S_y \quad (2)$$

3. BACK FILLING DESIGN AND STABILITY ANALYSIS

The high-water rapid setting material (HWRSM) has the characteristics of rapid solidification, strength regeneration and no water lost. The material is solidified in half or an hour after filling and 24 hours latter the strength is up to 1.0-2.5Mpa. No water lost can make the filling body contact very well with the roof and get the efficient support. When the density of mortar is 65% and the compounding ratio is 12%, the properties of high-water rapid setting material are shown in table 1.

Table 1 Properties of high-water rapid setting material.

Properties	Value
Elastic modulus	0.273Gpa
Poisson's ratio	0.248
Uniaxial compressive strength	3.82Mpa
Cohesion	0.80Mpa
Internal friction angle	16.7degree

The partial backfill process is filling by HWRSM with the extraction of coal shown in Fig.2.

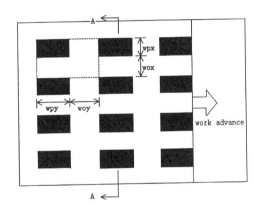

Figure 2 Room and pillar filling with high-water rapid setting material.

Figure 3 Section A-A

576

The effective Young modulus of a single level room and pillar back filling system in the pillar length was given as

$$\overline{E}_y = \frac{A''}{A'}E \tag{3}$$

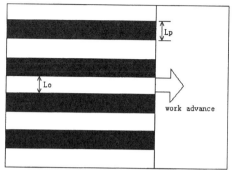

Figure 4 Partial filling with high-water rapid setting material.

$$A' = (w_{oy} + w_{py})(w_{ox} + w_{px}) \tag{4}$$

$$A'' = w_{px}w_{py} \tag{5}$$

Where A'' is the single filling pillar area, and A' is the total area enclosed by the dotted line in figure 2. The variable E is the Young modulus of high-water rapid setting material in the direction of filling pillar length. The total mined-out region w_1, w_2, h can be treated as an an-isotropic soft inclusion for which the average stress ratio in the y direction is determined from the smallest of w_1 and w_2 dimensions and the h dimension. Because of $w_1 < w_2$, the average stress ratio for the y direction over area A' produced by the weight of overlaying rock S_y is

$$\overline{\sigma}_{yy} = \frac{2w + h}{2w + h/k}S_y \quad , \quad k = \frac{A''}{A'} \tag{6}$$

The average pillar stress ratio is

$$\frac{\overline{\sigma}_{py}}{S_y} = \frac{2w_1 + h}{2w_1 + h/k} \frac{E}{\overline{E}_y} \tag{7}$$

If the percentage filling is defined as $\delta = A''/A'$, we have

$$\frac{\overline{\sigma}_{py}}{S_y} = \frac{2w_1/h + 1}{2w_1\delta/h + 1} \tag{8}$$

For partial back filling with high-water rapid setting material the percentage filling δ can be expressed as

$$\delta = \frac{L_o}{L_o + L_p} \tag{9}$$

The ratio $\overline{\sigma}_{py}/S_y$ is a function of the w_1/h ratio and the percentage filling as shown in figure 5.

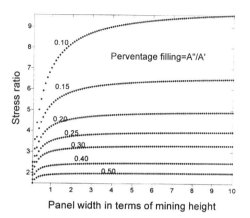

Figure 5. Filling pillar stress ratio $\overline{\sigma}_{py}/S_y$ as a function of the w_1/h ratio.

In some coal-fields, the mining height is a constant, if the panel width is very large and the percentage filling is less than 30%, it is impossible to maintain nearly a uniform average filling pillar stress. That is to say the percentage filling must be larger than 30% for ensuring the stability of filling body.

4. NUMERICAL SIMULATION

In order to illustrate the efficient of subsidence control, the Fast Lagrangian Analysis of Continua program developed by Itasca Consulting Group. Inc. is adopted to simulate the deformation. The rock mass properties and the panel characteristics are shown in table 2 and table 3, respectively. See Figure 6 - 7.

Table 2 Properties of rock mass.

Rock	Compress strength	Elastic modulus	Poisson's ratio
Coal	17.83MPa	10.0Gpa	0.40
Sandy mud	26.64MPa	16.34Gpa	0.31
Sandstone	88.4MPa	25.57Gpa	0.26
Silt-stone	52.5MPa	19.3Gpa	0.29

When the filling percentage is 40%, and the width of single filling body is 20m, the ground subsidence value is 215mm and the subsidence factor is 0.1075. If the filling percentage is 30%, and the filling width is 30m, the ground subsidence value is 665mm and the subsidence factor may reach 0.3325. However, neither the final nor the temporary deformations exceeds the allowable values.

Table 3 Characteristics of panel

Characteristics	Value
Mining depth	182m
Mining height	2.0m
Panel length	400m
Panel width	180m
Dip angle	0°

5. CONCLUSIONS

The high-water rapid setting material has the characteristics of rapid solidification, strength regeneration and contact with the roof very well. Its natural properties determine that it has great uses in practice in future. The research results in this paper show that it can be used in mines to filling mined areas, to control the subsidence and deformation, to ensure the stability of mining rock. When the filling percentage is greater than 30%, it can be sure that neither the final nor the temporary deformations exceed the allowable values. The new backfill method could control the ground subsidence effectively while all the underground natural resource is excavated. Thus, it is different from the traditional pneumatic and hydraulic stowing, and very importance to sustainable development of the natural resource mining.

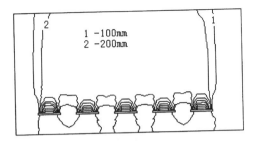

Figure 7 Subsidence contours for 40 filling percentage

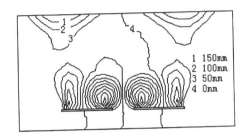

Figure 8 X-displacement for 30 filling percentage

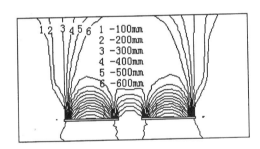

Figure 9 Subsidence contours for 30 filling percentage

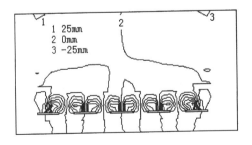

Figure 6 X-displacement contours for 40 filling percentage

REFERENCES

Babcock, C. O. 1970. *Plates with rows of holes considered an an-isotropic soft inclusion models of three-dimensional room-and-pillar mining systems.* BuMines RI 7436:28-29.

Babcock, C. O. 1974. *A geometric method for the prediction of stresses in inclusions, ore-bodies, and mining systems.* BuMines RI 7838:19-21.

Brady, B. H. G. & Brown, E. T.1985. *Underground mining rock mechanics.* Beijing: Coal industry press.

Mining Science and Technology' 99, Xie & Golosinski (eds) © 1999 Balkema, Rotterdam, ISBN 90 5809 067 1

Studies of high water and quick setting (HWQS) cementitious mine backfill

T.S.Golosinski
University of Missouri-Rolla, Mo., USA

Caigen Wang & R.Ganeswaran
Western Australian School of Mines, Kalgoorlie, W.A., Australia

ABSTRACT: High water and quick-setting (HWQS) cementitious material is used in several underground mines to produce relatively strong backfill with considerable early strength. This paper presents the investigations undertaken to define properties of HWQS based backfill made with highly saline mine water and with highly saline gold mill tailings.

1 INTRODUCTION

Many mines use their mill tailings or their part to produce backfill. In most applications such backfill is placed as a hydraulic or as a paste backfill. In both cases only part of tailings slurry is used. Hydraulic backfill requires removal of majority of fines from tailings, especially so if the mill grind is very fine. Paste backfill allows using more fines but the unused portion needs to be disposed of by other means in case of fines of minus 15μm being less than 15 to20% in the total tailings solids by weight. In addition preparation and transport of the paste backfill are costly.

For both the hydraulic and paste backfill, the bulk of tailings water needs to be disposed of, often at a significant cost. Efforts are being made to develop a backfill that would utilize a larger portion of tailings thus minimizing the cost of tailings disposal. As an example Amaratunga et all (1997) have recently proposed modifications of paste fill whereby higher portion of fines is used (total paste). While promising, this modification does not address the tailings water disposal problem. Furthermore, the cost of *total paste* preparation and transport appears to remain high.

In some situations all tailings, both its solid and liquid component, can be disposed of as an underground backfill. This is facilitated by use of a cementitious binder called further the HWQS binder (an acronym for High Water content Quick Setting binder). Properties of backfill based on this binder include high water absorption rate, quick solidification when mixed with water and early development of strength, and re-solidification after

cracking (Hou et al, 1996 and Wang et al 1995). These properties together with the ability to dispose of mill tailings in their entirety, and in an environmentally friendly manner, make the HWQS mine backfill of interest to the underground mining industry.

Unlike the mines that use the HWQS fill at present, most of the Western Australian mines operate in highly saline environment. As a result mine water and tailings have unusually high salt content and their use for production of backfill may significantly change HWQS backfill properties.

Consequently, the ability of the HWQS backfill to develop significant strength in a highly saline water environment needs to be confirmed before any industrial applications are undertaken. In addition, ability of the binder to bind the very finely ground gold mill needs to be confirmed and quantified. This paper presents the laboratory investigations undertaken to provide this information and in particular to quantify the properties of HWQS backfill made of representative gold mill tailings and of highly saline water as present in Goldfields area mines.

2. PROPERTIES OF HWQS MATERIAL

The backfill binder reported on in this paper consists of two components. These are called further *component A* and *component B*. Grinding and mixing sulphoaminous cement clinker, a suspension agent and a super retarder, makes component A. Component B is made by grinding and mixing calcium, sulfate, lime, a small amount of suspension

Table 1. Properties of HWQS backfill at different W/C ratio (a mixture of components A, B and industrial water)

Water ratio W/C	Water content in volume %	Initial setting time minutes	Initial compressive strength, MPa					Compressive strength 28 days after cracking			
			2h	24h	3d	7d	28d	2h	24h	3d	7d
2.00:1	85.20	14	3.33	6.26	7.27	7.92	8.70	9.84	9.36	8.72	9.28
2.25:1	86.60	15	2.42	4.74	5.38	6.19	7.08	7.08	6.82	6.74	7.12
2.50:1	88.00	15	1.78	3.93	4.64	4.74	5.32	5.20	5.53	5.50	5.21
2.75:1	88.80	14	1.30	3.08	3.57	3.84	4.22	n/a	n/a	n/a	n/a
3.00:1	89.70	15	0.92	2.54	2.99	3.22	3.60	n/a	n/a	n/a	n/a

Table 2 Properties of HWQS samples produced with salty water at the W/C ratio of 2.25

NaCl concentration %	Initial setting time minutes	Sample strength for different sample curing times, MPa				
		2 h	24 h	3 d	7 d	28 d
0	15	2.29	4.20	5.24	6.25	7.11
4	15	2.21	4.01	4.82	5.73	6.84
8	15~20	2.09	3.34	3.69	5.27	6.31
12	20~25	1.96	2.78	3.00	3.73	4.81
16	20-25	1.46	2.51	2.81	2.94	3.33

Table 3. Properties of HWQS samples produced with saline mine water at three different W/C ratios. Water with 46,600 mg/L TDS.

W/C ratio	Initial setting time minutes	Sample strength for different curing times MPa				
		2 h	24 h	3 d	7 d	28 d
2.0:1	12~15	2.70	3.54	4.47	4.83	5.05
2.25:1	15~20	1.82	2.38	3.55	3.91	4.17
2.5:1	20~25	1.52	2.34	2.85	3.43	3.66

agent and small quantities of other agents. When mixed together in water the components react and produce ettringite which contains a large amount of water. Other products of the reaction include calcium silicate gel, unhydrated calcium hydroxide and calcium sulfate. The final product contains from 85% to 90% water. Details of chemical reactions taking place when the components A and B are mixed with water are reported by Hou et al (1996).

Properties of the HWQS fill obtained by mixing components A and B with industrial water were originally researched by Wang et al (1995). These were later confirmed by tests conducted at Western Australian School of Mines (WASM), the results of the latter presented in Table 1. The ratio of water to solids by weight (W/C) shown in the first column can be adjusted in a fairly wide range of 2.0 to 3.0 to fit the requirements of the application.

The volumetric water content in backfill, shown in the second column, depends on W/C ratio and can reach 90%. The initial setting time of the backfill, shown in the third column, is an order of magnitude shorter than that of traditional cement based backfill. Early strength development is of particular importance in some mine backfill applications where it may simplify both backfill application and the mining methods.

Backfill strength measured for various sample curing times is quantified in columns 4 to 8 of Table 1. Interestingly, no significant strength increase takes place after 7 days of sample curing so for the practical purposes the 7-day strength may be considered to be the final one. For most of the mixtures the strength of 1 MPa or higher is developed as early as 2 hours after components A and B are mixed.

Finally, the HWQS backfill re-consolidates and develops strength after cracking. As shown in the last columns of Table 1, the backfill reaches the 28-day strength close to the original one at a re-curing age of 28 days after cracking. This property is of importance to mine backfill applications as it means that HWQS backfill is able to provide continuing support even when temporarily cracked by blasting or rock movements.

3. SALINE WATER BASED HWQS BACKFILL

As the mine water in many WA mines is highly saline, effect of salt contained in mine water on the properties of HWQS fill was investigated in detail. A series of backfill samples were cast using both salty and saline mine water and their properties were defined in laboratory tests. Several representative

curing times were used, with wet sample curing intended to reflect prevalent mine conditions.

3.1. Salty water

In the first series of tests the HWQS samples were produced using industrial water to which a varying content of dissolved salt (NaCl) was added. The purpose of these tests was to define a threshold salt content, if any, at which the properties of samples deteriorate to an unacceptable level. The salt concentrations varied from 0% to 16% by weight at the W/C ratio of 2.25. Four different curing times were used, varying from 2 hours to 28 days.

The results of the tests are shown in Table 2. Following observations can be made:

- The initial setting time of samples increases with the salt content. For the highest salt content it is almost 50% longer than the corresponding time for water only based samples. Overall, the setting time remains below ½ hour for all investigated conditions, including water with 16% of NaCl (by weight).
- Backfill strength decreases with increasing content of salt content. At the highest salt content this strength decreases to approximately 50% of the original strength as obtained for backfill samples based on water with no salt.

3.2. Saline mine water

The next series of tests investigated the properties of the HWQS samples produced with saline mine water obtained from an underground gold mine located in Eastern Goldfields. Results of water analysis indicated content of 46,600 mg/L of total dissolved solids (TDS) with sodium chloride accounting for 70% to 85% of the total. Other properties of the mine water used for production of backfill samples for this series of investigations are listed in Table 4 below. Three different W/C ratios were used in preparation of the samples: 2.0, 2.25 and 2.5. Measured properties of backfill samples are shown in Table 3.

The results of the tests do not differ significantly from those obtained for salty water based backfills at the same salinity level. Sample strengths are somewhat lower than those measured for samples based on water only (Table 1). The change of W/C ratio does not alter this trend significantly, with a small drop of strength evident throughout the whole range of tested W/C ratios.

While the backfill samples produced with saline mine water show loss of strength, the retained strength is still significant and at 3 MPa or more. It

Table 4. Properties of slurrys used to produce HWQS samples.

Source of water	pH	Concentration of cations in mg/L				Concentration of anions in mg/L				TDS	Sodium to total cation ratio
		Ca	Mg	Na	K	HCO₃	SO₄	Cl	NO₃	mg/L	%
Tailings liquid	7.6	4,260	1,410	44,900	424	36.4	2,660	77,400	24.9	131,000	85
Saline mine water	7.4	1,050	2,100	14,100	134	328	5,780	23,300	1.4	46,600	73

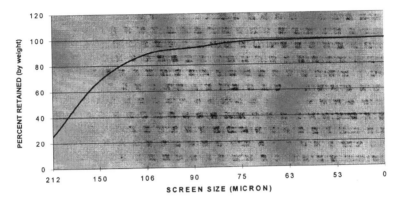

Figure 1. Particle size distribution of gold tailings.

Table 5. Properties of HWQS backfill samples produced from gold tailings.

W/C	Sample setting time	Uniaxial compressive Strength, MPa				
	Minutes	2h	24h	3d	7d	28d
2.25:1	20~25	1.38	1.78	2.43	3.05	3.25
2.0:1	20~25	1.96	2.31	3.21	3.77	4.13

is sufficient to meet most mine backfill applications (Bloss, 1996). In addition to retaining significant strength, the saline water based HSQW sets very quickly (see Table 3) and re-solidifies after cracking.

4 HWQS BACKFILL MADE OF GOLD MILL TAILINGS

The next series of experiments was intended to: (1) define feasibility of producing HWQS backfill from the fine ground tailings produced by a gold mill, (2) the feasibility of utilizing such tailings in their entirety (both the solid and the liquid content), and (3) to define the properties of HWQS backfills based on such tailings. A supply of run-of-mill tailings from a Goldfields area mill was secured and used to produce backfill samples, which subsequently were tested for strength and other relevant properties.

Two groups of samples were produced, one with W/C ratio of 2.25 and the other with W/C ratio of 2.0 (tailings to HWQS product by weight). The properties of tailings liquid and the particle size distribution of its solids are listed in Table 4 and shown in Fig. 1 respectively. In addition, Table 4 shows typical properties of saline mine water from an adjacent underground gold mine which was used in the tests reported on earlier in this paper.

Properties of the tested fill samples are shown in Table 5 below. The strength of the samples is much lower than the strength of the samples made with industrial water. The tailings slurry contained exceptionally high concentrations of salts with TDS at 131,000 mg/L. As a result, the strength of the HWQS backfill samples made of tailings is somewhat lower than the strength of samples made with saline and salty water, although this drop is judged not to be significant. At 3.21 MPa and 3.77 MPa after 3-day and 7-day curing period respectively, and at the W/C ratio of 2.0 by weight, the strength of the tailings based HWQS backfill is sufficient for most mine backfill applications.

5 DISCUSSION

The results of investigations presented above indicate that the strength of HWQS backfill depends, between other factors, on salinity of water used in its production. For HWQS backfill cured for 7 days and at W/C ratio of 2.5 (88%

contained water by volume) the strength of samples made of industrial water exceeds 6 MPa, while that of samples made of highly saline water with 12 % NaCl content reaches roughly 60 % of that. Similar strength differences were defined to exist for other W/C ratios and curing times. It is noted, however, that the backfill strength of 2.5 is sufficient in most mine backfill applications (Bloss, 1996).

While the HWQS strength decreases with increasing salinity of water, at higher salt concentrations this decrease moderates. HWQS samples made of very highly saline tailings (see Table 5) show the strength of 3.05 MPa if cured for 7 days and made with W/C ratio of 2.25. Similar samples made of saline mine water (see Table 3) had shown the strength of 3.91 MPa. This indicates the loss of strength of approximately 20%. In comparison the loss of strength in samples made of highly saline water, compared to the samples made of industrial water (traces of salt) is in the range of 60%.

Finally, HWQS backfill can be produced from finely ground tailings as discharged by a gold mill. In the reported tests the HWQS samples made of gold mill tailings had the strength of approximately 2MPa after 24 hrs curing and in excess of 3 MPa after seven days curing. It is likely that the strength of backfill derived from gold mill tailings may be higher if the salt content of the tailings decreases.

6 FURTHER INVESTIGATIONS

The reported investigations quantify several key properties of HWQS backfill as related to its possible application in WA mines. High salinity ground water environment and very high salt content of tailings characterize many of these mines. However, to fully assess the feasibility of using HWQS backfill several of its other properties need to be investigated, defined and quantified where appropriate. These include:

Effects of highly saline mine water, and other factors specific to underground mine environment on performance of the backfill over time. The ability of the backfill to retain its strength for prolonged periods of time is of importance to most backfill applications.

Feasibility of adding coarse aggregate and mine waste rock to the backfill to reduce the requirement for HWQS components and to facilitate mine waste rock disposal.

Ability of the backfill to withstand and resist the blast damage which may result from blasting in adjacent stopes need to be confirmed and quantified

7 CONCLUSIONS

The HWQS backfill as described in this paper offers a potential for significant reduction of cost of mine backfilling operations. It also offers an opportunity for cost efficient disposal of mill tailings in their entirety, thus to reduce tailings related environmental hazards and to lower overall cost of mining.

Thanks to its short setting time the fill has the ability to provide early support of mine voids. It develops significant strength with time. This strength is somewhat reduced if water used in production of fill is highly saline. However, even if very highly saline mill tailings are used for production of the fill, it achieves strengths in excess of 3 MPa, which are sufficient for most mining backfill applications.

Several other properties of HWQS fills need to be investigated and quantified to allow for full assessment of their applicability in Australian underground mines. These include: retention of strength properties with time, ability to resist impact of blasting adjacent rock and feasibility of adding coarse rock or aggregate to the fill.

8 ACKNOWLEDGMENTS

The authors thank Mr. Jianping Li for his help and assistance with the laboratory investigations of backfill properties. Tailing and saline mine water used in the tests was supplied by Kalgoorlie Consolidated Gold Mines. Special thanks go to Professor Chaojun Hou of the China University of Mining and Technology for supply of HWQS materials and advice on conduct of tests..

REFERENCES

Amaratunga L.M., Hein G.G. and Yachyshyn D.N. 1997. Utilization of gold mill tailings as a secondary resource in the production of a high strength total tailings paste fill. CIM Bulletin, vol. 90. No. 102, pp. 83-88.

Bloss M. L. 1996. Evolution of cemented rock fill at Mount Isa Mines Limited. Mineral Resources Engineering, vol. 5, no. 1, pp. 23-42.

Hou C., Zhang L., Feng G. and Yan Z. 1996. *Research on ZKD quick-setting and high-water-content material.* Proceedings, Guo and Golosinski (eds), Mining Science and Technology '96. A.A. Balkema. Rotterdam/Brookfield, pp. 37-40.

Wang Y., Wang C. and Zhou, H. 1995. *Backfill technology used in roadway support* (in Chinese). Mining Industry Publisher, Beijing.

Mining Science and Technology' 99, Xie & Golosinski (eds)© 1999 Balkema, Rotterdam, ISBN 90 5809 067 1

Event-based modeling for land management system in mining area

Jie Jiang & Jinzhuang Wang
China University of Mining and Technology, Beijing, People's Republic of China

Jun Chen
National Geomatics Center of China, Beijing, People's Republic of China

ABSTRACT: Both structural and behavioral aspects need to be modeled in some spatio-temporal databases. In this paper, an event-based approach is proposed for modeling the system's behavior of building reviewing. The spatio-temporal process is divided into three kind of components, that is agent, event and state, among them event is the key factor. All of the three components are taken into account while modeling the process. The relations among agent, event and state are analyzed, and special consideration is given to the event. Some problems for further investigations are also discussed in this paper.

1 INTRODUCTION

Land recultivation in mining area is very essential for sustainable development. A GIS based computer supported system may be useful for land management. Such a CSCW (Computer Supported Collaborative Work) system should provide the end user with an environment in which they can handle their routing review work easily, as well as manage various spatio-temporal data dynamically. Traditional GIS data models focus on structural aspects of the spatial systems but take little consideration on behavioral aspects. Although someone try to extend the traditional vector and raster data model to involve the time factor (Langran 1992, 1993; Raafat 1994; Chen 1996), they cannot describe the relations between land reviewing actions and data which are results of these actions. Some people use GIS and OA to build enterprise GIS that can perform both data management and decision supporting (Chen 1998), but they still cannot integrate the structural and behavioral aspects really due to the limitation of inherent data model of these commercial systems. Thus a number of complex problems remain unsolved, such as the efficient storage and access of the spatio-temporal data dynamically, the behavior based query, spatio-temporal reasoning and the construction of a suitable query language (Worboy 1994). There exist some initial efforts to represent behavioral aspects in the spatial data model (Peuquet 1995; Claramunt 1995; Allen 1995),

however, these models are not well suited for representing the hierarchical behavior in the land management process, as well as describing various relations between behaviors and objects affected by the behaviors.

2 COMPONENTS IN LAND SUBDIVIDING

Land subdividing is a collaborative work by a group of urban planning staff and land managers to review land use applications submitted by public agencies or private citizens. An application would be reviewed in a legally approved and sequential steps, requiring a quite long processing period from the submission to issuing the legal permits (Chen 1998). The results of the reviewing actions are different states of spatio-temporal objects, such as site location, permitted building area or registered property. If we call the organizations or persons who execute these review actions by the name of agent, and the actions event. Then this spatio-temporal process of land subdividing can be viewed as a composite of agent, event and states. Various agents execute a sequence of events. Occurrence of events cause spatial objects change in both the geometric and thematic states, such as creating a new land parcel at a vacant space, decomposing an existing land parcel into smaller parcels (Chen, et al. 1996), as shown in Fig.1. Among the three components, the event is more important. That is because the intentions of agents have to achieve through events, and the different

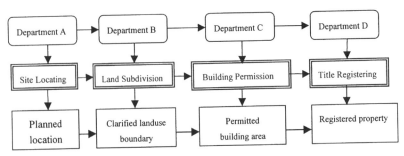

Fig.1 Agents, Events and States in Land Subdividing

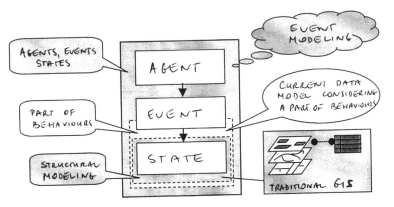

Fig.2 Components to be modeled in event-based modeling

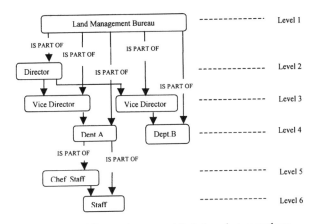

Fig.3 Different Agents and Relations between them

states of spatio-temporal objects are results of events. These three components interact through various connections between them, then construct the overall behavior of the land subdividing.

The main functions of land subdividing system are data management, decision supporting and information query & analyzing. They are also corresponded to these three components. The data management handles the states. The decision supporting is related to events and agents. And the query and analyzing depend on the relations between agents, events and states.

3 EVENT BASED MODELING FOR LAND SUBDIVIDING SYSTEM

We have discussed in section 1 that traditional GIS data models, though extended to involve the time factor, cannot describe the relations between land reviewing actions and data which are results of these actions. And the method of using GIS together with OA still cannot integrate the structural and behavioral aspects really. A number of complex problems remain unsolved. There exist some initial efforts to represent behavioral aspects in the spatial

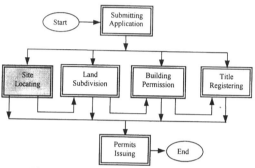

Fig.4 First-level events in land subdividing

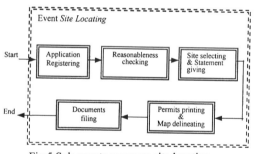

Fig.5 Sub-events compose site locating

data model. Peuquet et al (1995) proposed an event-based spatio-temporal data model where the sequence of events through time was organized in increasing order along a time-line. Another event-oriented approach was discussed by Claramount et al. (1995) to model changes among a set of entities. Spatial entities and their temporal versions were associated through intermediary logical tables (Past events, Present events and Future events) that permit description of complex succession, production, reproduction and transmission processes. The time was treated as a complementary facet of spatial and thematic domains that are separated into distinct structures and unified by domain links. Allen et al. (1995) tried to develop a generic model for explicitly representing casual links within a spatio-temporal GIS. A small number of elements were presented in that model via an extended Entity-Relationship formalism, including objects and their states, events, agents and conditions, as well as the relations (Produces, Is Part Of, Conditions). However, the priority in these models was given to some local or partial behaviors of the applications rather than an overview of the system's behavior. So we need to design a model which is more close to the real world system of land subdividing. From a viewpoint of system theory, we should study the overall system rather than any part of it. And while describing the land subdividing process, we should represent all the three components, as well as various relations between them, as fig.2 shown. Moreover, we should take special consideration on the events as they are the key factor in the system behavior.

4 AGENTS, EVENTS AND STATES IN BUILDING REVIEWING

There are various kinds of relations among agents, events and states. Five of them need to be studied for modeling the land subdividing process. That is

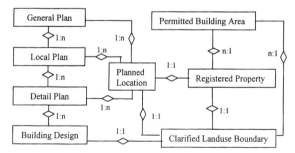

Fig. 6 A causal entity-relation diagram of different states

relations between agents & agents, events & events, states & states, agents & events and events & states. Organizations and individual persons is treated as agents, they are hierarchical and have relations between themselves (fig.3). Events are various actions. Each of these events may be caused by an agent or a previous event. They are also hierarchical. Any event in a upper level may be decomposed into a sequence of sub-events (see fig.4 and fig.5). As to different states of spatial objects, there exist causal link between present state and previous states. It is also important to maintain the relations between historical states and current states (see fig.6).

It should be noticed that occurrences of events may cause geometric changes and different events might bring about different geometric states. It is essential to explicitly preserve known relations between events and their consequences.

5 FURTHER INVESTIGATIONS

A behavior-based approach is proposed in this paper. This approach divides the land subdividing process into three components, that is agent, event and state. By modeling the three components and relations among them, it is possible to integrate the structural aspects and behavior aspects and construct a framework for such kind of systems that is more close to the real world system. However, there remains many other problems and issues to be solved. Further investigation includes:

A. Formal definition and specification of agents, events and states, as well as their relations;

B. Event-based spatio-temporal data organization and query;

C. Role and effect of agents. There exist causal relations between agents and events that reflect the human-human interactions in the collaborative reviewing process. A better understanding of the relations between agents and events will be important for representing the dynamic behavior of the system. This will help the designers to classify and define the system functions properly, which will enable the end users to concentrate on problem solving and decision making tasks at the semantic level.

ACKNOWLEDGEMENT

This research was supported by Natural Science Foundation of China (NSFC) under grant number 49671061.

REFERENCES

Allen, A., Edwards, G. and Bedard, Y., 1995, Qualitative causal modeling in temporal GIS, *Spatial Information Theory- A theoretic basis for GIS*, Springer-Verlag, No. 988, pp.397-412

Claramunt, C. and M. Theriault, 1995, Managing time in GIS: an event-oriented approach, *Recent Advances in Temporal Databases*, Springer, pp.23-43

Claramunt, C. and M. Theriault, 1996,Toward semantics for modelling spatio-temporal processes within GIS, SDH'96, pp.2.27-2.43

CHEN, Jun, Yanfen LE, 1996, Defining and representing temporal objects for describing the spatio-temporal process of land Subdivision, in *Intenational Archieves of ISPRS* , Part B2, pp. 49-56, Vienna, 1996

CHEN, Jun, Jie JIANG, Ronghua YAN, Shuping JIN, Lilin XU, 1998, Towards collaborative building reviewing by integrating GIS with office automation, Proceedings of *International Conference on Modeling Geographical and Environmental Systems with GISs*, Hong Kong, June, 1998, pp.357-362

Langran, G. 1992. *Time in GISs,* Talor&Francis, 1992.

Langran, G., 1993, Issues of implementing a spatiotemporal system, *Int. J. Geographic Information Systems,*Vol.7, No.4, 305-314

Raafat, H., Zhongshen Yang and D. Gauthier. 1994. Relational spatial topologies for historical geographical information, *Int. J. Geographic Information Systems*, Vol.8, No.2, pp.163-173.

Peuquet, D. J. and N. Duan, 1995, an event-based spatiotemporal data moadl (ESTDM) for temporal analysis of geographical data, *Int. J. Geographic information Systems,* Vol.9, No.1, pp./-24

Worboy, M.F., 1994, A uniform model for spatial and temporal information, *The Computer Journal 37(1)*:26-34

Mining Science and Technology' 99, Xie & Golosinski (eds) © 1999 Balkema, Rotterdam, ISBN 90 5809 067 1

Farmland damage due to coal mining subsidence and its remediation in Eastern China

Zhenqi Hu & Hehe Gu
China University of Mining and Technology, Beijing, People's Republic of China

Dehui Liu & Hu Feng
Nanjing Agriculture University, People's Republic of China

ABSTRACT: Subsidence damage on farmland due to coal mining can be classified into two types: landscape damage and soil damage. The water accumulation, land sloping and fissures are typical damage characteristics of the subsidence landscape in Eastern China. Based on the dynamic tests of soil properties in subsided land, the rules governing soil changes were defined. It is believed that changing soil properties are the main reason of lower farmland productivity. Soil physical properties are sensitive to mining subsidence, but soil chemical properties are not sensitive, except for soil electrical conductivity. Soil salinisation is the most important change resulting from mining subsidence. The soil analysis of subsided lands also indicates soil erosion problem due to subsidence, most severe in the middle of the affected area. To remedy the subsidence damage, several reclamation treatments are discussed in this paper, such as "Digging deep to fill shallow", appropriate drainage and backfilling.

1. INTRODUCTION

Coal accounts for approximately 75% of the annual energy consumption of China. Underground mines in China account for 96% of the coal output. The underground coal mining has caused large amount of land subsidence, which has led to farmland losses and caused severe conflicts between farming and mining. According to statistics, the subsidence land due to coal mining exceeds more than 400,000 hectares at present and is increasing at the rate of more than 20,000 hectares each year. Half of the subsidence land is located in the plain area of northern and eastern China, which is prime farmland. This situation makes the reclamation of subsided farmland due to coal mining become an urgent task for our country.

2. SUBSIDENCE DAMAGE ON FARMLAND DUE TO COAL MINING IN EASTERN CHINA

2.1 *Landscape Damage of Farmland Due to Coal Mining Subsidence*

Land subsidence has a long history and has typical damage characteristics in China. Because of multi-seam excavation in eastern China, the subsidence areas usually subside more than one times. Thus the land damage is severe and the resulting subsidence is often very deep. Usually, the subsidence depth ranges from 1.5m to 6.0m. As the subsurface water level is high in the eastern China, the parts of the subsidence prone area may impound the subsurface and surface water. Therefore, accumulation of water and prone (downward sloping) land without water accumulation are two typical damage characteristics of subsided landscape in eastern China (see Figure 1). In some areas, the water accumulated in the trough may be polluted by acid mine water, which will induce a series of environmental pollution problems. The prone land without water accumulation still has the capability of cultivation, but it is erodible and its productivity is low. The crack or fissure is another damage feature of subsided landscape. Generally, cracks or fissures exist in all subsidence prone areas because of the ground movement caused by the mining subsidence (He et al 1991). They are open or invisible cracks ranging from a few centimetres to several meters wide. This sort of landscape disturbance not only causes the damage to surface facilities such as buildings, railroads, highways, etc., but also leads to the losses of soils and the necessary nutrients for plant growth, which is harmful for agricultural production. Thus surface cracks or fissures is another damage characteristic of subsided landscape.

2.2 *Subsidence Damage on Soil Properties of Farmland*

To explore the subsidence damage on soil

Table 1. Impact of mining subsidence on soil properties in Eastern China

soil Properties	subsiding land					old subsided land					newly subsided land			
	CR	I	II	III	IV	CR	I	II	III	IV	I	II	III	IV
bulk density (g/cm³)	1.18	1.22	1.29	1.33	1.27	1.15	1.17	1.25	1.36	1.35	1.08	1.18	1.35	1.35
porosity (%)	55.0	56.7	56.5	56.8	54.5	56.0	56.0	51.2	46.7	48.7	57.1	54.0	43.9	43.9
macro-porosity (%)	2.9	2.2	2.6	2.6	3.0	2.7	2.8	1.6	1.9	1.8	4.2	2.7	1.8	1.8
moisture content(g /kg)	150	129	126	124	150	160	162	168	325	373	163	180	286	304
hydraulic conductivity (cm h⁻¹)	1.53	1.63	0.48	0.43	0.02	2.50	0.09	0.28	0.01	0.01	1.57	1.03	0.10	0.02
Organic Matter(g /kg)	17.7	17.4	16.5	17.2	17.2	16.5	14.6	13.4	15.2	14.5	18.7	17.6	18.1	18.9
Total N (g/kg)	1.08	1.02	0.96	0.76	0.96	1.04	1.04	0.90	1.18	1.07	1.18	1.10	1.07	1.09
Plant available P (ppm)	10.8	11.5	14.7	17.9	13.5	10.2	12.6	13.5	2.9	2.9	15.4	9.3	8.2	8.2
Plant available K(ppm)	180	137	162	189	138	184	132	126	174	211	240	261	280	
pH	7.90	7.93	7.94	7.89	7.90	7.88	7.96	7.93	7.93	8.01	7.94	7.96	7.92	
Electrical Conductivity (ds/m)	1.9	1.9	1.8	2.0	2.5	1.6	1.7	1.9	6.0	5.2	1.4	1.6	2.4	2.2
Biomass C (ppm)	157	254	320	246	242	248	255	296	230	145	307	228	120	107

Note: CR ---- undisturbed farmland; I, II, II, IV ---- soil sampling position along the subsided prone land

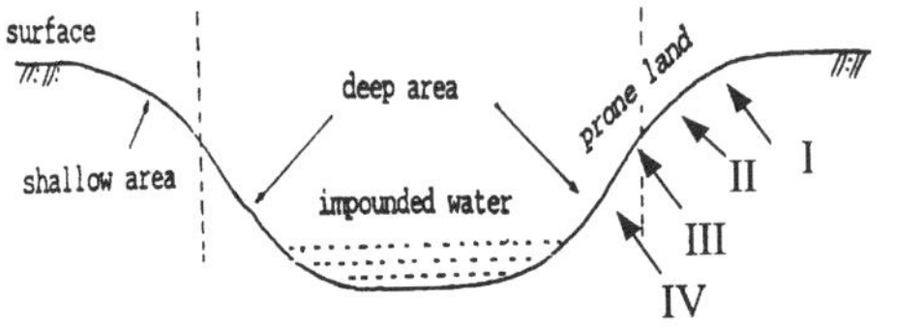

I~IV soil sampling positions, I,II,III – top, middle, bottom of the prone land, IV – centre of the trough

Figure 1. Profile of subsidence trough

productivity of farmland, three plots of subsided farmland were selected from Liuqiao and Jiahe mines in Eastern China. One was old subsided land in Liuqiao mine, which was 10 years' old. The second plot was newly subsided land in Liuqiao mine, which was about 1km from plot one. The third plot was an on-going subsiding farmland in Jiahe mine. Two undisturbed farmland plots were also selected from the two mines for comparison. The maximum subsidence was about 3.0 m in plot one, 2.0 m in plot two and 1.4 m in plot three at the sampling time. Soil sampling was undertaken in different positions along the subsided prone land (see Figure 1): Top, Middle, Bottom and Centre. The centre of subsidence trough was filled with water partially. The tested soil properties were: soil bulk density, soil moisture content, organic matter, hydraulic conductivity, total N, plant available P, plant available K, electrical conductivity, pH and soil Biomass C. The results are listed in Table 1.

The test results (see Table 1) showed that from the top of subsided prone land to the centre of subsidence trough, soil bulk density is increasing, This is mainly due to soil being compacted while the surface settles down with compression deformation. Soil moisture contents have the same tendency to increase from top to centre. Therefore, subsidence changes the water distribution in soils. The subsurface water may have risen by soil capillary, leading to the higher water content in the lower position of the subsidence land and water accumulation in parts of the subsidence trough. The value of hydraulic conductivity in the three subsided plots was decreased from top to centre (see Table 1). Therefore, the soil physical properties are sensitive to mining subsidence and generally they become worse from the top of the subsidence prone land to the centre of subsidence trough.

Soil organic matter (OM) is a very important factor for plant growth. The tested results (see table 1) indicated that: (1) The content of organic matter decreased from the top of the subsided prone land to the middle; (2) The middle of the prone land with the maximum slope had a relatively low content of organic matter compared to the other sampling positions, which reflected the influence of the land slope due to surface subsidence. This slope led to major run-off and erosion of soils. (3) The bottom of the prone land with higher water content and naturally growing grass had a maximum content of organic matter, which indicted that the compression deformation with a concave subsidence curvature in this area might lead to an accumulation of soil organic matter eroded from the top and middle.

The electrical conductivity (EC) is an important parameter for measuring total salt content (salinity) of soils (Rowell 1994). In nearly every case the electrical conductivity of subsided soils were higher than that of farmland soils (see Table 1). The salt content of soils was increasing from the top to the centre of the subsidence trough with the maximum salt content at the bottom of the prone land. The old subsidence land had a higher salt content than that of the newly subsided and subsiding land. The results showed that: (1) the mining subsidence led to tremendous increase of salt content in the soil; (2) the bottom of the prone land generally accumulated maximum salt content because of compression deformation with concave subsidence curvature and erosion from the top and middle of the prone land; (3) the salt accumulation of soils in subsided land varied with time, the older the subsidence land the higher the amount of salt accumulated. Therefore, soil salinisation resulted in the subsided land was the most important effect caused by mining subsidence, having an affect on the germination and growth of crops and subsequently leading to poor yield in subsidence areas in Eastern China.

A direct measurement of the C and nutrients contained in the microbial biomass is essential for studies of the role of the soil microbial biomass in organic matter dynamics and nutrient cycling (Carter, 1993). The present study only tested the biomass C by the fumigation-extraction method for assessing the impact of mining subsidence on soil biological properties. The soil biomass C in newly subsided land showed a significant decreasing tendency from the top to the centre of subsidence trough. In the old subsided and subsiding land, soil biomass C also showed a decreasing tendency from top to the centre of subsidence trough exception of the middle of subsided prone land (see Table 1). The result indicated that mining subsidence affected soil biological properties, but the extent depended on the subsidence time. The impact of mining subsidence on soil biological properties in subsiding land might not be the key problem because of the effects of many other factors over the time period. At the beginning of subsidence and with light subsidence (smaller settlement and smaller land slope), the impact of mining subsidence might be insignificant. , which was proven in Beishu mine (Shandong Province) with coal seam thickness of about 0.8 m and a maximum subsidence of 0.6 m. The impact of mining subsidence on soil biological properties could have occurred in a certain period after the subsidence process. Studies are underway to attempt to establish the exact time of these biological changes.

3. RECLAMATION TREATMENTS FOR SUBSIDENCE LAND IN EASTERN CHINA

There are four types of reclamation treatments in eastern China:

Drainage: In some coal mines, the subsidence depth is not very deep and the depth of impounded water in the subsidence trough is shallow. For this case, establishment of a system of drains such as a trench can make the impounded water drain away and lower the subsurface water level so that the subsidence land is relatively raised. After the establishment of the system of drains, levelling the subsidence land is needed if the topography affects agricultural production very much. This technical approaches is simple and cost-effective, which is mainly for reclaiming farmlands. But it requires the drainage trench is deep and wide enough to drain impounded water away and lower the subsurface water to a available level for plant growth. In Pingdingshang coal mine, the drainage trench has 3.0 m of wide and 2.0 m of depth.

Digging Deep to Fill Shallow (Hu, 1994) : This is a popular, simple and practical method for reclaiming the subsidence prone areas in China. It usually divides the subsidence prone area into tow parts: deep and shallow (see Figure 1), then makes the deep area deeper by a excavator such as Hydraulic Dredge Pump (HDP) for digging a fish pond. The soil excavated from the deep area by excavators is used for filling the shallow area so that the shallow area can reach a desire elevation for crop growth. Obviously, this technical approach for developing the subsidence area is used to construct both lands and fish ponds. Therefore, the effectiveness of this reclamation method is very high, which has become the most encouraging treatment in eastern China.

Filling: This is a type of landfill method ---- filling subsidence land with some filling materials such as coal wastes and fly ash. Coal wastes are generated in the process of coal mining and processing, which do not occupy lands, but also pollute environments. About 3.0 billion tonnes of old coal mining and processing wastes lie scattered about China; 150-200 million tonnes are generated each year. In addition to comprehensive utilisation of coal wastes, the coal wastes are mainly used as materials for filling the subsidence prone areas in eastern China. Usually, two types of coal wastes are used as filling materials. They are new coal wastes from underground and old coal waste piles. The use of coal wastes as filling materials into the subsidence area can restore the disturbed lands and reduce land losses due to coal wastes piles, and the environmental pollution due to coal wastes could also be alleviated. This sort of reclaimed land is generally used for construction and agricultural purposes in China (Yan 1987). If the reclaimed land is for agricultural purpose, soil cover (topsoil, more than 0.5m) should be placed on the filled coal wastes.

Coal ash generated from coal-fired power-plant is another kind of filling material, which is also widely used for filling the subsidence prone areas in eastern China. The procedures are: (1) removing the topsoil from the subsidence land for forming a circle bank, which results a large pit called ash stockyard; (2) filling the ash stockyard with the coal ash using hydraulic transportation; (3) draining the water and settling the coal ash; (4) replacing the topsoil (0.2m-0.5m) on the coal ash.

If subsidence areas near a lake or river, the lake mud or fluvial mud can be as the filling materials. The large amount of garbage could also be the filling materials.

Directly reconditioning : If there are not any water in some shallow subsidence areas and the subsurface water level is not very high, the method of directly reconditioning the subsided land can be used. Usually, levelling of the subsided land by bulldozers or manual work is often used. If the slope of the subsided land is large, the terraces should be used. This sort of reclamation method is usually used in the mining area with low subsurface water level. Thus, it is widely used in northern China. However, there are still some subsidence land in eastern China that can use this method, for example, 13.6 ha. of subsidence prone areas in Feicheng, Shangdong province were directly graded.

4. CONCLUSIONS

The conclusions based on this study are summarised as follows:
1. Mining subsidence destroyed original landscape. The water accumulation, prone land and fissures are typical damage characteristics of the subsidence landscape in eastern China, which are also the main factors affecting farm production. Soil erosion was a severe problem with mining subsidence, and the middle of the prone land generally had the maximum erosion.
2. Soil physical properties were sensitive to mining subsidence and they became worse from the top to the centre of the subsidence trough. Except for electrical conductivity, tested soil chemical properties were not so sensitive to mining subsidence and might be changed after subsidence process. Most of chemical properties showed a worsening change from top to the centre of the trough. The bottom of prone land accumulated

nutrients and salt. Thus, the most important impact of mining subsidence on soil chemical properties was seen in soil electrical conductivity reflecting high salt content, which might be the vital change resulting in poor productivity. The impact of mining subsidence on soil biological properties might occur after the subsidence process. The soil biomass C in newly subsided land showed a significant decreasing tendency from the top to the centre of subsidence trough, but no significant tendency in the old subsided land and subsiding land.
3. Drainage, "Digging deep to fill shallow", Filling and Direct Reconditioning are main treatments for reclaiming subsided land in eastern China. Of these, "Digging deep to fill shallow" and "filling" are most popular and encouraging treatments in eastern China. Since land shortage is much more serious in eastern China than that in the rest of China, the restoration of damaged farmland due to coal mining subsidence has become a focus of research activities in China.

ACKNOWLEDGEMENT

This study was supported by National Natural Science Foundation of China (NSFC) under grant number 49401007.

REFERENCES

Carter, Martin R. 1993. *Soil Sampling and Methods of Analysis*. Lewis Publishers. USA. 823pp.

He, Guoqin, L. Yang, G. Lin, F. Jia and D. Hong. 1991. *Mining Subsidence*. Printing House of the China University of Mining & Technology. 375pp. (in Chinese)

Hu, Zhenqi. 1994. The technique of reclaiming subsidence areas by use of a hydraulic dredge pump in Chinese coal mines, *International Journal of Surface Mining, Reclamation and Environment* 8(4):137-140, 1994. Rotterdam: Balkema

Rowell, D.L. 1994. *Soil Science: Methods & Applications*. Englan: Longman Scientific & Technical

Yan, Zhicai. 1987. Technical approaches and related policies for land reclamation and treatment in Chinese coal mines. *Proceedings of the second international symposium on the reclamation, treatment and utilization of coal mining wastes*: 123-131. Rotterdam: Balkema

Mining Science and Technology' 99, Xie & Golosinski (eds) © 1999 Balkema, Rotterdam, ISBN 90 5809 067 1

Surface subsidence rules and parameters of Yanzhou mining district

Kan Wu, Juanle Wang & Jianming Jin
Department of Mining Engineering, CUMT, Xuzhou, People's Republic of China

Jiaxin Ge & Zhenxiu Huang
Yanzhou Mining Bureau, Zhoucheng, People's Republic of China

ABSTRACT: Field data collected in subsidence area related to a thin seam excavation, slicing and longwall top-coal caving in Yanzhou mining district, China, were collected. Analysis of this data allowed construction of a model that allows prediction of the surface subsidence under variety of geological and mining conditions. The model inputs vary with mining conditions and the size of working areas, with the model validated for a variety of site conditions.

1. INTRODUCTION

To solve the problems of mining under railways, buildings, and water bodies in Yanzhou mining district, China, we need determine a prediction model of surface movement and deformation which adapts to this district's geological condition and mining system. In existed prediction models, the probability integral predict model is used widely (He et al. 1991). But, its accuracy used in Yanzhou mining district and reliability used in varieties of mining system need further study.

This paper's main work is studying the adaptability of probability integral predict model in Yanzhou mining district. Our work is based on the field's observation data. So we collect some observation data of surface subsidence at three coal mines of Yanzhou mining district. These data are the representative of thin coal seam's excavation, slicing and longwall top-coal caving in Yanzhou mining district.

2. SITUATION OF OBSERVATION STATION

The 4314 Observation Station: The working face's width along the coal seam's dip direction is 160m; its length along the coal seam's strike is 1579.2m; its area is 0.25 km^2. The extracting coal seam is No.3. The depth between working face and ground is between 331 and 319m (the absolute elevation of ground is 46m). The structure of the coal seam is simple and steady. The thickness of the coal seam is between 8.05 and 8.42m, and the average is 8.22m. Its average slope angle is 4.17°. Mining systems used in this working area are longwall, fully mechanized and top-coal caving method along the strike. Both designed and actual height of working face is 2.8m, caving height is 5.42m. The average advanced velocity of the working face is between 2.0 and 4.5m every day.

The 3303 Observation Station: The extracting coal seam is No.3, its mining system is layer mining (having two slices). Every slice's extracting thickness is about 3m.

Table 1. The fitting parameters of probability integral predict method and its precision

Observation	q	tgβ	b	θ	S	ma$_q$	ma$_b$
4314	0.85	2.15	0.24	88	0.05~0.15h*	2.2	160
3303-1	0.49	1.90	0.33	90	0.0~0.05h	2.1	7.1
3303-2	1.00	1.30	0.25	82	0.0	4.7	12.2
Sixth mining area along strike	0.82	1.70	0.32	89	0.1~0.15h	2.5	29.6
Sixth mining area along dip direction	0.82	1.70	0.32	89	0.01h	1.9	4.8

* h represent the depth of extraction (unit: m)
Note: q = subsidence factor; tgβ = tangent of main effect angle; b = displacement factor; θ= angle of maximum subsidence; S = displacement distance of inflection point; ma$_q$ = the relative mean error of subsidence fitting; ma$_b$ = the relative mean error of horizontal movement fitting

Observation stations in the sixth working area of Beisu coal mine: In this working area, we extract two coal seams, No. 17 and No. 16. The thickness of the coal seams is about 2m in sum. Coal seams' strike is near to east-west and has a flat dipping structure towards northwest. Its average slope angle is 5°; the mining depth is between 260 and 350m; the average thickness of the alluvial of Quatermary system is 55m. There are many buildings over this working area. To solve the problem of extracting coal under these buildings, this coal mine takes a measure to advance 4 working face (the length is 440m) at the same time, and lay out two observation lines (one along the strike of coal seams, the other along the dip).

3. THE FITTING EFFECTIVENESS OF PROBABILITY INTEGRAL PREDICT MODEL AND ITS PARAMETERS

3.1 The method of determining parameters

To eliminate problems with the existing methods of determining parameters and to solve the problem of determining parameters by using of the field observation data with arbitrary shape's working faces and dynamic field observation data, we adopt a new method to determine the parameters (Pattern Search) by deep study and comparison again and again.

Pattern Search is a good method to solve the problem of unconstrained extremum. It was provided by Hooke and Jeeves in 1961, as a direct method. This method has two advantages: (1) programing by using it is easier; (2) determination of parameters by using the field observation data under the condition of arbitrary shape's working faces or dynamic surface subsidence is easier.

3.2 The fitting effectiveness of parameters

Using probability integral predict model and determining parameters' principle of Pattern Search, we determined the steady state's fitting parameter and fitting precision for three observation stations are involved in this paper. The fitting parameter and its precision are listed in table 1.

From table 1, we can get some views as follows.

1. Subsidence factors are smaller than the normal tremendously when small working faces are excavated. (eg. working face 3303-1);

2. In the small working face situation, the fitting error of displacement is small (eg. working face 3303-1). But after reached the supercritical mining, the fitting error of displacement is obviously increased along the strike (eg. The observation line along the strike over sixth mining district of Beisu coal mine). Whereas it is irrelevant with the degree

of mining along the dip direction (eg. The observations line along the dip direction over sixth mining district of Beisu coal mine).

3. Tangent of main effect angle is related to mining methods. The effecting order from small to big is fully mechanized top-coal caving, slicing (first slice), thin seam mining and slicing (second slice).

4. Exception of these rules, the rule of ground subsidence and prediction parameters have not significant differences each other.

5. It is reliability to predict ground subsidence by using probability Integral method in Yanzhou mining district. Generally, it does not need to develop a new model. But we must study the changing rule of parameters according to the degree of mining. In some special situation, we need also correct the model partially.

4. CORRECTION OF PARAMETER

From the principle of probability integral method, when we predict subsidence, the premise condition is the degree of mining should reach supercritical mining or near (Wu 1998). So, we need correct the prediction parameter when we predict subsidence with extra subcritical extraction or subcritical (small working face). The prediction parameters change with the size of working areas (the degree of full extraction). This series of parameters should be the function of factor of full extraction.

Through the study of many case studies of fitting of subsidence curve and displacement curve, we find that other parameters are steady excepting the subsidence factor and displacement factor. So, we only need to study the changing law of subsidence factor and displacement factor varying with degree of mining when we predict subsidence of small working areas. The correction of parameters can use two methods: (1) setting up a correction equation; (2) using time series analysis.

4.1 Correction equation

For Yanzhou mining district, we can get the representative function between the correction factor of subsidence factor and the coefficient of mining degree by using a sectional function:

$$y_w = q_d \Big/ q$$
$$= \begin{cases} 0.97n^2 - 0.07n + 0.39 & (0.1 < n <= 0.83) \\ 1.00 & (n > 0.83) \end{cases} \quad (1)$$

Where y_w = the correction factor of subsidence factor; q_d = the subsidence factor for small working face extraction; q = the subsidence factor when supercritical mining; n = factor of full extraction, and we can use a formula to determine its value as follow:

$$n = \sqrt{n_1 \bullet n_2} \qquad (2)$$

where

$$n_1 = \frac{D_1}{(2r)} \qquad (3)$$

$$n_2 = \frac{D_2}{(2r)} \qquad (4)$$

if $n_1 > 1$ then $n_1 = 1$; if $n_2 > 1$ then $n_2 = 1$. Where D_1 = the actual length of working area along the strike; D_2 = the actual length of working area along the dip direction; r = the main effect radius.
The subsidence factor determined from 3301-1 observation station data is q_d, but not q.

4.2 The method of the time series analysis

Time series modeling is building the relationship between the past value and future value of data series (Yang 1990).
To a smooth and steady random process $\{x_t\}$, we can build a differential equation as follow.

$$x_t - \varphi_1 x_{t-1} - \varphi_2 x_{t-2} - \cdots - \varphi_n x_{t-n}$$
$$= a_t - \theta_1 a_{t-1} - \theta_2 a_{t-2} - \cdots - \theta_{m1} a_{t-m} \qquad (5)$$

This model is called ARMA(n,m). When the value of θ makes 0, it turns into another widely used model, AR(p). We can portray it by using a equation as follows, too.

$$x_t - \varphi_1 x_{t-1} - \varphi_2 x_{t-2} - \cdots - \varphi_n x_{t-n} = a_t \qquad (6)$$

By this model, we can regard the mathematics expectation of this series at time of *t* as the end value of next time. For the model AR(p), we can get its prediction equation

$$x_t(1) = E(x_{t+1})$$
$$= E(\varphi_1 x_t + \varphi_2 x_{t-1} + \cdots + \varphi_p x_{t-p+1} + a_t) \qquad (7)$$

Using dynamic field observation data, we can determine the probability integration parameters every observation moment; and a sequence of parameters is formed. By analyzing and studying, the dynamic Probability Integration parameters don't meet the demand of time series modeling. So, we should draw their trend terms at first. We can suppose:

$$x_t = q_t + p_t \qquad (8)$$

In this formula, q_t represents certainty part and p_t represents the surplus after drawing the trend terms.
We could predict parameters by this model based on a group of new observation data. A new model could be built immediately. Then we could use the new model to predict the next set of parameters and use the predicted parameters to predict the subsidence further.

5. CORRECTION OF PREDICTION MODEL

The main aim of this correction is to correct the displacement curve of the main section of strike. From table 1, we can find that when extra super-critical mining, the fitting error of displacement has a trend of increasing. The main cause is that the value of displacement dose not equal 0 in the area of supercritical extraction (by the principle of Probability Integration Method, the value should be 0).
The original expression of unit horizontal movement basin is:

$$U_e(x) = -\frac{2\pi b x}{r^2} e^{-\pi \frac{x^2}{r^2}} \qquad (9)$$

When reached supercritical extraction, the unit horizontal movement basin is:

$$U'_e(x) = U_e(x) - \frac{\pi b}{2}\left(\frac{0.237}{r^2} + \frac{0.687x}{r} + 0.231\right) \qquad (10)$$

6. CONCLUSIONS

We can use the Probability Integration Model to predict the ground subsidence in Yanzhou mining district. To get higher precise in predicting ground movement and deformation, it is needed to choose rational parameters or to do some necessary correction for the parameters or the expression of unit horizontal movement based on the results obtained in this paper.

ACKNOWLEDGEMENT

This study was supported by National Natural Science Foundation of China under grant number 59634030.

595

REFERENCES

He Guoqing, Yang Lun et al. 1991. *Mining Subsidence*. Xuzhou: Publishing House of China University of Mining & Technology.

Wu Kan, Ge Jiaxin et al. 1998. *Integration Method of Mining Subsidence Prediction*. Xuzhou: Publishing House of China University of Mining & Technology.

Yang Shuzi & Wu Ya 1990. *Engineering Application of Time Series Analysis*. Wuhan: Publishing House of Science University of Central China.

Mining Science and Technology'99, Xie & Golosinski (eds) © 1999 Balkema, Rotterdam, ISBN 90 5809 067 1

Comprehensive treatment for mining subsidence area in East China

Guangli Guo, Guoliang Zhang, Guoqing He & Zhengfu Bian
Department of Mining Engineering, China University of Mining and Technology, Beijing, People's Republic of China

ABSTRACT: The mining subsidence has serious influence on environment in coal mining areas of East China. This paper proposes a comprehensive treatment of this problem. It is proposed that affected areas are filled with coal waste to restore the village buildings and to reclaim the land. Restoration of groundwater table is also required. By adopting these measures, the reclaimed land can be used again, lots of coal reserves can be freed from the pillars left under villages, and the cost of reclamation can be minimized.

1. ENVIRONMENTAL IMPACT OF MINING SUBSIDENCE IN EAST CHINA

There are a lot of villages and productive farmland in the plain mining area with the high ground-water table in East China. In 1995, the coal output of East China was 25.9 per cent of total national output. There are more coal seams with larger minable thickness in this district, so the ground subsidence and deformation values caused by underground mining are larger, which has serious influence on environment. In 1996, the mining subsidence and waste dump area in East China were 5 710 ha. By the end of 1996, the total area of land subsidence in the plain mining area of East China had been up to about 52 430 ha of which the farmland was about 70 percent, the extinct farmland 10 467 ha, and the farmland area of accumulated-water 8 667 ha. The environmental problems caused by mining subsidence mainly include:

(1) Most subsidence land was the fertile farmland, and the soil quality degenerates seriously in subsidence area. The large area of farmland becomes relatively low-lying land in which drainage is difficulty in raining seasons, then the surface water flows to subsidence area, which makes the farmland become the seasonal or everlasting accumulated-water area.

(2) The ground-water table controlled by regional hydrogeological condition goes relatively higher, and is close to or even exceeds the ground surface in subsidence area, so the farmland becomes moist land or swamp or lake, and the ecological environment takes a turn for the worse.

(3) Mining subsidence damages the village buildings over the working face. Since the build-ings are soaked in the accumulated-water area for a long time, their service lives are great threatened.

(4) Mining subsidence damages the former water conservancy facilities, which worsens the environment of accumulated-water area.

(5) Large accumulated-water area also gives rise to serious land utilization contradiction because of dense population and well-developed economy.

How to resolve the problems of mining under villages, treatment of accumulated-water area and land reclamation is very important for the sustainable development of coal mine district.

2. THE MEASURES FOR MINING UNDER VILLAGE AND TREATMENT FOR ACCUMULATED-WATER AREA

The main measures for mining under village and treatment for accumulated-water area include underground measures and surface measures both.

The main aim of underground mining and strata control measures is to reduce the ground subsidence and deformation and control it within acceptable extent. These measures include stowing mining, thick-limit mining, partial mining (partial strip extraction method, room-and-pillar method, etc.), grout filling in layer-separated zone of overburden to reduce subsidence method, etc. However, these methods are not very suitable or hard to carry out in this district because of large losses of coal resources, technical cause and cost problem.

The surface measures mainly include village removal, on-situ reconstruction anti-deformation buildings, nearby filling subsidence area in order to rebuild anti-deformation building, land reclamation and comprehensive treatment for the accumulated-water area, etc.

Coal mining after village removal is that the village is removed outside of coal-field to avoid subsidence influent and re-laying on the virgin field. This method is suitable to the villages near the coal-field boundary. But it is very hard to carry out for the villages inside coal-field because the new site is far from the original site. On-situ reconstruction is that the old buildings in the village are replaced with the new anti-deformation buildings and then coal seam is extracted directly under the village. The inhabitant needn't remove over long distance and leave their land when this method is adopted, but there is great cost of anti-deformation construction and difficulty in management and technology in condition of mining thick seam. The method of nearby filling subsidence area to rebuild anti-deformation new village can refrain from the long distant removal, lower the cost of anti-deformation measures, and effectively utilize waste subsidence land. Experience shows that is a better method to solve the problem of mining under villages in East China (Guo et al. 1998).

Large accumulated-water subsidence area caused by underground mining is the main environmental problem in East China mining district. The subsidence treatment would be designed based on the condition of regional topography and landforms as well as the distribution features of rivers, ditches, railways, roads and residential areas.

3. COMPREHENSIVE TREATMENT AND LAND RECLAMATION FOR ACCUMULATED-WATER SUBSIDENCE AREA

The comprehensive treatment of regional ground water net should be utilized together with land reclamation in accumulated-water area.

(1) It is to dredge the former rivers and ditches, repair, reinforce and heighten dikes and dams, and set up flood-control dams or water gates on the upper and lower reaches of the river near or through the subsidence area, on which the pumps with big flow and lower-lift are installed for draining in raining seasons.

The larger subsidence area should be divided into relatively smaller treatment units with roads, railways, rivers or ditches as delimitation. Based on the size and distribution of subsidence in a treatment unit, dig deeper to form deepwater fishpond in the deeper accumulated-water area. The shallow water area of depth less than 0.5 m can be used to plant aquatic economic crops such as lotus, arrow-head and wild rice stem etc.. The boundaries of shallow water area should be cultivated as vegetable plot. Every deepwater fishpond is about 50 m in width, 100 m in length and 4 ~ 5 m in depth with the shelter

belt and fodder grass. The soil dug out of the fishpond is used to backfill the shallow subsidence area reclaimed as farmland. This reclamation method is named *digging depths and backfill shallows*. Figure 1 shows the sectional sketch map of accumulated-water area treatment.

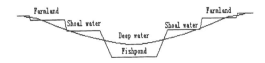

Fig.1. The sectional sketch map of accumulated-water area treatment

(3) Dig trenches (or ditches) at certain intervals to dewater or drain out water for lowering ground-water table in farmland reclamation area. The ground-water table should be kept lower than 0.8 m below the farm surface. The interval L between two trenches can be calculated according to the formula of soil vadose (Xue 1986).

The subsidence area treatment should be carried out in batches and by stages because the mining subsidence is a long-time and dynamic progress. The methods of manual digging, back filling and leveling reclamation of farmland are preferable in the section with no or little accumulated-water. Although these methods are inefficient, it will do less harm to topsoil structure and result in less loss of soil fertility, so the soil production capacity can recover quickly. In the deep accumulated-water section, the waterpower machinery can be adopted to complete the work of excavating intensive fishpond, transporting mud and filling. Namely, the high-pressure pump system is used to break up and excavate soil and the slurry pump system is used to transport mud from deep-water section to the farmland reclamation section to precipitate. The method is of high efficiency and low cost, but the original topsoil structure will be damaged seriously, the losses of soil fertility will be very great, and the topsoil aeration and permeability will be reduced. In general, after farmland reclamation it takes 2 ~ 3 years to improve the fertility of the topsoil to recover its production capacity.

4. CONSTRUCTION ANTI-DEFORMATION BUILDING ON THE BACK FILLING COAL WASTES GROUND

Construction reclamation section should be selected at the relative stable subsidence area near roads. Usually coal wastes are adopted as the main back filling material and transported by truck from refuse-disposal site to the reclamation section (Guo et al. 1998). Two filling methods can be used. One is slice filling wastes method, in which every slice is 30 cm thick and is rolled four times round by roller with 60 ~ 80 KN in weight. Another method is filling wastes up to the design level once and the heavy rammer is used to ram down the filling wastes for foundation. A layer of loess of 30 cm thick should be laid over the wastes foundation and be rammed down. The elevation of backfill ground must be higher than the local highest flood level. If there are some other virgin seams the design elevation should also be above the highest flood level after mining subsidence in future.

Because of the potential uneven settlement and deformation existed in the wastes foundation and the subsidence impacts of other deep seams extracted in future, new buildings should be designed with the anti-deformation structures based on the prediction reports of wastes foundation settlement and mining subsidence. The main measures adopted in anti-deformation design include as follows:

(1) Large structures should be divided into independent units by means of deformation joints through the structure and foundation. Deformation buffer grooves should be dug around the building to absorb ground compression deformation.

(2) Adopt reinforced concrete slab of 100 mm thick as foundation bedding in order to avoid excessive uneven settlement which may occur in the wastes foundation.

(3) Set up the horizontal slip bedding between foundation and ground frame in order to reduce additional stress transmitted upwards and to facilitate relative horizontal movements between the building and the ground caused by subsidence.

(4) Set up reinforced concrete ground frame, eaves frame, structure column and belt below the window board of first floor of single-story house or storied building.

5. COMPREHENSIVE TREATMENT PROJECT FOR ACCUMULATED-WATER SUBSIDENCE AREA IN XUZHOU MINE AREA

5.1 Introduction of the experimental plot

Xuzhou is one of the main coal districts in East China. Large area of mining subsidence has formed because of under-ground mining over 50 years. The depth of ground subsidence area is generally 1.5 m ~ 5 m. The comprehensive treatment experimental spot of accumulated-water subsidence area is situated in the eastern part of Xuzhou mine area, where there is a development network of rivers. There are two rivers, one canal and some pitches linked each other, namely the Grand Canal in south,

Tuntou river in north and Bulao river oblique crossing the area as shown in Figure 2.

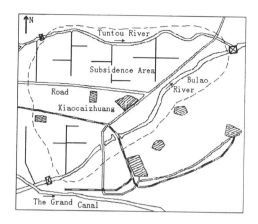

Fig.2. The sketch map of network rivers in the eastern part of Xuzhou mine area

The coal-bearing strata are Permo-Carboniferous system. 4 minable seams of coal (seam-3, 17, 20, 21) present from the top to bottom in the area. Seam 3 is 4 ~ 4.8 m thick, distributing stably. It is the main seam to be mined. Other coal seams are thin seams of 0.6 ~ 0.8 m thick, buried 245 ~ 278 m below seam 3, and will not be extracted in resent years. Dip angle of seams is 7 ~15°. Seam 3 is mined by descending inclined-slicing longwall mining along strike with bulk caving. Quaternary alluvium has an average thickness of 29 m, in which the upper part is silt bed of 8 m and the ground-water table is between +28.5 and +29.0 m.

Xiaocaizhunag in the experimental plot is a typical village with 312 families, 1215 residents in Xuzhou mine area. All houses in the village are single-story house which can not stand ground deformation caused by underground mining because of lower quality.

The ground elevation will decrease from +30.5 ~ +32m to +27.3 ~ +28.8m after underground mining, which is close to or even lower than the ground-water table. Subsurface water will reveal on surface in parts of subsidence area for a long time so that large area of farmland will become swamp and can not be cultivated. Thus the land lies waste and the ecological environment takes a turn for the worse. The village houses will be damaged seriously because of mining subsidence and deformation. Meanwhile they are soaked in accumulated-water for a long time so can not be used any more.

5.2 Reclamation and comprehensive treatment measures for accumulated-water area

Based on the conditions of natural, social and economic environment and the prediction reports of mining subsidence, the comprehensive treatment measures had been determined as follows by means of comparing many versions in technical and economic analysis.

(1) Due to the conditions of Bulao river's natural irrigation and drainage capacity and ground irrigation and drainage system, the measure of lowering ground-water table was carried out, which includes repairing and constructing several dams or water gates across Bulao river and setting up drainage pump station on it. Keep the water level in Bulao river about +27m in experimental area. Drain accumulated-water in farmland by means of pump station in raining seasons and open the water gate for irrigating farmland in dry seasons.

(2) The subsidence area near the road reclaim with coal wastes become the new site of village after removal and new anti-deformation houses were built on it.

(3) Land reclamation were carried out in two steps. For the dynamic subsidence area of one slice, the main treating measure was leveling sloped land and keeping the ground-water table 0.8 m below the surface through dewatering and draining measures, which ensures the farmland is cultivable. The steady subsidence area where all slices have been mined was developed into the deep water fishpond, the shallow water aquatic economic crops section and farmland reclamation section through *digging depths and backfill shallows*. The elevation of reclaimed farmland should be kept between 28 m and 28.5 m, together with the measures of digging ditches to dewater subsurface water and pump draining water in raining seasons, which can ensure the reclaimed farmland can be cultivated normally. Based on the calculation and analysis results the interval near two drain ditches should be less than 310 m in order to meet the design requirement of dewatering and draining accumulated-water and lowering ground-water table. In fact, the average interval of original ditches is only 200 m, so widening and deepening the original ditches can only meet the requirement of dewatering the ground-water table.

Meanwhile, adopting the comprehensive measures of dewatering and draining water as well as reclamation measures suit to local conditions greatly reduced the essential elevation of reclaiming farmland, cut down the cost.

5.3 Coal wastes backfill and construction new anti-deformation village

The section of construction reclamation by means of coal wastes backfilling is 1.5 kilometers south from old village site near the highway of Xuzhou to Jiawang town, south of mine railway line. The new village site is convenient for transportation. Seam 3 has been mined and other thin seams will not be mined in resent years.

Based on the subsidence area condition and the predicted results of mining subsidence of other seams in future, the ground surface elevation should be +32 m, which can ensure no cumulated water and no blocks to drainage easily. The average thick-ness of backfilled wastes is about 2.5m. The anti-deformation houses can resist the uneven subsidence impacts caused by the artificial wastes foundation and deep seams mining in future. The new village, with 415 anti-deformation single-story and two-story houses and construction area of 62 750 m^2, were built five years ago. Up to now, all the houses are in good conditions, and no walls or floors of the houses were found damaged.

5.4 Effect of comprehensive treatment project in experimental plot

Along with underground mining, the comprehensive treatment project for accumulated-water subsidence area were carried out by stages. New village were built in 1993, peasants removed from the old and bad quality houses to the bright, commodious and safe new houses, the average housing area per person reaches 37 m^2, and living environment has been greatly improved. Total recoverable reserves of 4 865 400 tones under the former village were emancipated completely, which released the tension of working face replacement in the coal mine.

By the end of 1997, over 60 ha subsidence land had been treated and reclaimed comprehensively, land productivity had been improved. A lot of peasants losing their farmland settled down again. The peasants' incomes had been notably raised, and the ecological environment had been improved and recovered.

REFERENCES

Guo Guangli, Zhang Guoliang, et al., 1998, Study on comprehensive treatment of subsidence area in Xuzhou mining area. In Guo Yuguang (ed.), *Environmental Monitoring and Control in Mine Area*: 286-291. Xuzhou: The Publishing House of China Uni. of Min. and Tech..

Xue Yuqun 1986. *Dynamics Principle of Ground-water*. Beijing: Geology Publishing House.

Mining Science and Technology' 99, Xie & Golosinski (eds)© 1999 Balkema, Rotterdam, ISBN 90 5809 067 1

Environmental applications of mine surveying technology in mining areas

Jingxiang Gao, Peijun Du & Shubi Zhang

Department of Mining Engineering, China University of Mining and Technology, Xuzhou, People's Republic of China

ABSTRACT: Environmental treatment and protection, ecological restoration and rebuilding are the key issues to sustainable development of mining areas. As an important technical support, mine surveying technology plays important role in environmental protection and treatment of mining areas. The typical example of modern mine surveying technology is the application of GIS, GPS and RS, which includes the following aspects: (1) environment monitoring and management. (2) the theories and techniques of ecology rebuilding and land reclamation. (3) mining subsidence monitoring and studying on its mechanism. (4) the rational exploitation and utilization of resources.

1. INTRODUCTION

It is well known that coal is China's main energy resource. By statistics, coal accounts for above 70% in the structure of energy resource consumption and this proportion will be little changed in a rather long time. Mining leads to environmental pollution and ecological damage in mining areas while bringing economic benefit to society. Resources and environment are the core of sustainable development for national economics, and advanced and new technology must be applied to solve those increasing problems caused by mining, which is vital to sustainable development strategy. In this paper what is reviewed and discussed in detail is the new development in controlling environmental pollution caused by mining in China through modern mine surveying technology based on spatial information technologies.

2. ENVIRONMENTAL POLLUTION AND TREATMENT IN MINING AREAS

2.1 *Treatment of solid waste in mining areas*

Debris and living garbage are the main solid wastes in mining areas, which are also the main polluting sources in mining. The influences of wastes on environment mainly include occupying farmlands, blocking up rivers and affecting ecology.

Treatment on waste rock mountains is launching in the whole state now, and other solid wastes such as slime, coal ash and boiler ash are being given high attention and effective treatment. Some mining areas have done much work in synthetically utilizing and harnessing the waste rock mountains and produced good social, economic and ecological benefit. The main measures to process solid wastes are as follows:

1. Eliminate the material base of spontaneous combustion of waste rock mountains. It is forbidden for washing debris to be driven to waste rock mountains. To those in spontaneous combustion, effective measures are adopted to extinguish the fire.

2. Distil useful material from solid waste.

3. Take waste rock as the raw material of producing building products, packing the subsiding holes and packing material of roadbed.

4. Set up power plant in order to use waste rock as raw material;

5. Afforest on the waste rock mountains and set up site of entertainment around it.

6. Try to use some wastes to produce microorganism fertilizer and improve soil.

2.2 *Treatment of waste water in mining areas*

Mine waste water mainly includes mine water, coal cleaning waste water, living waste water, waste water of miners' lamp plant, and so on.

Over 2.2 billion m^3 mine water is produced each year in China. The features of mine water are large pumping capacity and low toxicity. Suspending substances are the main pollutants in mine water, which usually doesn't contain poisonous material, and is neutral or alkalescent, sometimes acidic because of its high sulfur content. The treatment of mine water has attracted more attention and made some improvement. Some mines such as Quantai mine of Xuzhou mine bureau and Jiangzhuang mine of Zaozhuang mine bureau have already adopted some effective methods to process mine water and produced better benefit.

The pumping capacity of coal cleaning waste water is about 30 million tons per year. The pollution of coal cleaning water is much more serious than mine water, and it has great toxicity because it contains medicaments and PAM which are used to clean coal and treat water.

The Mine pharmacy plants drain over 1.2 million tons per year. It will bring great harm if not processed properly. The draining capacity of this kind water is small, so it needs to be harnessed effectively.

According to relative research results and practical measures, the treatment of mine waste water can be processed in the following four aspects:

1. Processing waste water by physical and chemical methods.

2. Improving waster water by biological techniques.

3. Purifying waster water.

4. Searching for new equipment to depurate water.

2.3 *Treatment of waste gas and dust pollution in mining areas*

Coal smoke pollution caused by combustion is the main source of atmospheric pollution. 80% of the smoke and dust, 90% of SO_2 of the whole state comes from coal combustion. Causing acid rain, SO_2 is the main polluting material. The suspending particles in the atmosphere of some northern cities are over 800 $\Box g/Nm^3$, which is beyond the standard excessively resulting in serious atmospheric pollution. By statistics, the waste gas emitted only by coal combustion is over 170 billion m^3, smoke and dust over 370,000 tons, SO_2 over 560,000 tons. Furthermore, There is a kind of pollution with small quantity but great harm, that is, plumbum powder and dust in coal mining industry.

Up to now, China has taken some steps on coal combustion pollution, for example, cleaning and processing coal, developing cycling sulfur bed boiler and smoke gas purifying technology. Those steps have had waste gas controlled partly in mining areas, but not solved problems essentially yet.

2.4 *Treatment of Ground Subsiding Areas*

Coal is exploited mainly by underground mining, which covers 97% of all output. Because many mines in China adopt the method of long-wall mining method and top-coal caving (LTC) technology, ground subsidence is rather serious. The subsidence makes ground structures damaged, causes large areas of farmlands destroyed, leads to large areas of water accumulated in subsiding area and also causes land salinization. By statistics, 0.2 ha land will subside if mining every 10, 000 tons coal. By 1994, the area of subsiding land is about 400,000 ha. At present, the reclamation of subsiding land has already caused the state's higher attention. But the work started untimely, so there are many old problems and great difficulties at hand. Moreover, the development among different regions is not balanced. By statistics, the ratio of land reclamation in the whole state is less than 1%, and it is no more than 13% in coal mining industry, while fortunately it reaches 40% in TongShan county, JiangSu Province.

Development of land reclamation technique is fairly rapid, and many effective techniques such as filling method, digging the deeper and filling up the lower areas, afforesting waste rock mountains have been widely applied in some mining areas. Ecological reclamation and biological reclamation techniques have been transferred from experiment to practice.

3. APPLICATION OF MINE SURVEYING TECHNOLOGY TO ENVIRONMENTAL TREATMENT IN MINING AREAS

Mining area is a kind of special geographical area. When mineral resources are exploited, ecological and environmental protection should be considered, and regional resources, environment and development should be taken as a whole, then the technical system of monitoring, analyzing, appraising should be set dynamically and synthetically up in mining areas. The theories and techniques of multi-disciplines including environment, mining, surveying, geology, information and others can be used to monitor environment and disaster phenomena in mining areas, and then effective treatment of regional ecological environment and sustainable development of social economy can be realized. Mine surveying technology is one of the important technologies.

3.1 *Mine surveying technology and its Development*

Mine surveying technology is a multi-disciplinary integrating science of surveying and mapping science, mining engineering and other disciplines. Traditional mine surveying technology was represented by theodolite, level, steel tape and other instruments, and it played important roles in mining safely and effectively. With the scientific development and progress of surveying instrumentation, modern surveying instruments such as Total Station, EDM, gyro have got wide use in mining. The latest development of mine surveying is the applications of "3S" technologies, i.e., GIS (geographic Information System), GPS (Global Positioning System) and RS (Remote Sensing). GIS is the technical system applying computer to process and manage spatial geographic data and attribute information. It can realize the input and output of multi-source data and have great data processing and analyzing ability. GPS is a real-time spatial positioning system by satellite information. It has many advantages, such as high-precision, money and time saving, automation, global overlay, full-time work. RS can receive ground information of resources, environment and other aspects. It has the advantages of large information capacity, high speed, real-time and others. Mine GIS(MGIS) is the typical embodiment of "3S" technologies applying to mines.

3.2 Application of modern spatial information technology to environment monitoring and management

Modern spatial information technology mainly includes RS, GIS, GPS integrated technology ("3S" technology). Its influences are permanent and far-reaching in the applied fields of earth science, especially in resources and environment. Spatial information technology has obvious advantages in mineral resources exploitation, environmental protection and restoration, disaster monitoring, prevention and reduction and other fields.

RS can collect real-time and synthetical information of multi-phase, multi-field and multi-spectrum about atmosphere, ground, globe and region. After processed, the necessary thematic and synthetical information. can be provided. When all information is input and processed by GIS, environmental information of different periods and different regions can be compared and analyzed in order to realize environmental treatment and management. The processed information can be output by text, figure and charts. GPS can realize the connection of environmental information and its spatial position. With the development of ES (Expert System), GIS can be added with ES to provide decision support according to the practical situation. "3S" integrated technology can realize the integrative automatic system of environmental information collection, storage, management, analysis, processing and output, which can provide all necessary information with rapid, effective, and accurate decision-making.

It has already obtained gratifying achievements to apply 3S technologies to environment monitoring, analysis, appraisal, renovation, planning and decision-making and other respects in JinCheng, KaiLuan, TongChuan and other mining areas in China. It is just the beginning, however, there are

many problems should be explored, such as the interpreting, reversed deduction and pattern recognition of RS image, the method of applying RS environmental information to assess environment quality of mining areas, expert system of environmental problems, theories and techniques of 3S technology integration, and so on. All those issues should be focused in the future.

3.3 Theories and techniques of ecological environment rebuilding and land reclamation

Surveying and mapping technology is the primary technology of land reclamation. It can be applied in the reconnaissance, marking, monitoring and executing of reclamation area from begin to end. The following is the main surveying tasks: setting up control network on reclamation area, surveying topographically before and after reclamation, drawing all charts related to reclamation planning and executing, providing original surveying information for current and long-term plan of land reclamation, marking geometric elements according to design, monitoring the subsidence of reclaimed lands, and so on. Land reclamation is the base of ecology rebuilding, surveying technology can be used to direct ecological environment rebuilding in mining areas.

Besides the conventional surveying work, modern mine surveying technology has been widely applied in the following fields: assessment of influences of mining on environment, planning and designing of land reclamation, studies on technologies of land reclamation, economic assessment caused by mining, technical assessment of the productivity of reclaimed land in mining areas, and so on. Especially 3S technology has a bright prospect in those aspects. RS information integrated with GPS, GIS can realize monitoring, comparing, assessing dynamically and synthetically, find problems and direct production.

3.4 Mining subsidence monitoring and study on its mechanism

In the past, leveling was used to capture original data of mining subsidence. With the development of GPS, the traditional leveling now can be replaced by real-time, dynamic and automatic survey of ground displacement. GPS information can be input, processed and analyzed by GIS to provide useful information.

But the applying fields should be expanded and improved further. The forecasting theories and methods of mining subsidence should be perfected further, and the forecasting precision should be improved. Conventional monitoring technologies should be connected with modern spatial information technology to monitor the subsidence and disaster dynamically and synthetically. Furthermore, the following issues should be studied in detail: the mutual operating mechanism of mine engineering and ecological environment in mining areas, the theories and techniques of dynamic monitoring, of visual design, virtual reality(VR) and dynamic simulation, the decision support system(DSS). Then optimized exploitation of mineral resources and ecological environmental synthetical protection and treatment in mining areas can be realized. Also the unfavorable effects of mining subsidence on environment could be diminished and avoided, so that the ecological environment in mining areas develops towards good direction.

3.5 Optimized exploitation and appraisal of mineral resources

The mining speed of coal resource is very rapid, but the utilization ratio is very low. So waste and loss of energy are rather serious. As one-off energy source, coal is the base of economic development, and the exploitation must be optimized. Mine surveyors have been using surveying technology to do reserves managing work. Henceforth, it is better for them to use 3S technology to serve optimized exploitation and sufficient use of coal resources. The main methods are to have conventional reserves work well managed, and then the RS information should be analyzed comprehensively in time and the resource information and its dynamic changing laws should be directly put into production. GIS can also be used in assessment of mining resources, providing enough original information for MGIS with other information of mining areas.

CONCLUSIONS

When analyzing and appraising the damage degree of mining to environment and studying rational use of resources and environmental treatment, the advantages and uses of mine surveying technology should be exerted adequately. With information of geological, surveying, mining and managing, varied methods and knowledge can be used to study the effects of mining factors on environment from technological, economic, social and environmental aspects in terms of GIS, GPS, RS and their integrated technology. As a result, various environment protecting measures can be put forward and the feasibility and rationality comparison can be given from technological and economic point of view, then the optimized environment protecting policies can be decided. This is of great importance to achieve optimized exploitation of mineral resources, synthetical treatment and sustainable development of mining areas. With the development of environmental science, mine surveying technology and correlative disciplines, mine surveying technology will play more important roles in environmental treatment and provide forceful technical support for sustainable development of mining areas.

REFERENCES

Guo Dazhi 1998. Application of spatial information technologies and modern tasks of mine surveying. *Mine Surveying:* 1998(1)
Xu Guanhua 1996. Some issues about developing the application of spatial technology. *Remote Sensing Information*: 1996(2)
Du Peijun, Gao Jingxiang & Zhang Shubi. 1998 Discussion on the application of modern surveying and mapping technologies in solving resources and environmental problems in mining areas. *Environmental Engineering*: 1998(3)
Wei Chaoyang, Zhang Licheng & He Shujin 1997. A discussion on Eco-environmental conditions in the coal mining areas in China. *Acta Geographica Sinica*: 1997(4)
Ministry of Coal Industry 1997. Ninth-five Plan of environmental protection of coal industry. *Coal Mine Environmental Protection*: 1997(1)

Mining Science and Technology' 99, Xie & Golosinski (eds) © 1999 Balkema, Rotterdam, ISBN 90 5809 067 1

Land deterioration caused by mining and optimum land reclamation technique

Xiaoting Wang & Guanghua Lu
Taiyuan University of Technology, People's Republic of China

Jianqiang Wang
Shigejie Colliery, Lu'an Bureau of Coal Industry, People's Republic of China

ABSTRACT: The paper analyzes the main cause and forms of land deterioration in coal mining areas , and the related hazards. The model for prediction of subsidence is presented. It can be used to analyze and control subsidence and to assess the suitability of various subsidence control measures.

1. INTRODUCTION

In China, underground mining is the main way of coal mining. When bringing economic benefit the mining, specially underground coal mining, also leads to serious land deterioration such as ground subsidence, overburden removal of land and waste accumulation According to the statistics, by the end of 1994, the total area of land collapse was 281,000 hm² □taking over 10 mm of subsidence as a criterion□and the land collapse was 0.2 hm² per 10,000 t of coal. By 2000 year, the total area of land deterioration is predicted to be up to 500,000 hm².

In Shanxi province, its coal-bearing area covers 36.5% of nationwide and the coal mining spreads over 85 counties.

In 1997, the coal production in this province was up to 330 million t. Such high-intensity and large-scale of coal mining must lead to large-area land deterioration. According to the statistics made by the former Ministry of Coal Industry of PRC in 1996, gob caving, open-pit spoil bank, waste rock dumping yard, water and soil loss and so on caused by mining exploitation, covered an area of 6015hm². However, reclamation and recovery land was only 1380 hm². Therefore, it's of great importance to protect the present cultivated land and reclaim damaged one for Shanxi so as to keep its economy prosperity and stability.

2. THE MAIN FEATURE OF COAL INDUSTRY IN SHANXI PROVINCE

1. With wide range of coal field and large scale mining, state-owned mines have adopted longwall mining system and roof caving method. Although they are safe and efficient, high yield and low consumption, the earth surface collapse is serious, leaving great influence on environment.

2. Most coal seams with multi-level are thick or medium-thick gently inclined and lap-over. The extraction of them does have a serious influence on earth surface.

3. Coalfield lie in Loess Plateau, which has complicated topography, developed structure and a crisscross network of gullies. Because of a long-term influence from external geological forces and new tectonic movement, the ecological environment of this region is fragile, and water and soil loss is serious.

4. From northern to southern and from west to east, the railway and high way form a crisscross network in Shanxi province. Because of cultivated farmland and dense population, any surface subsidence and collapse will result in serious negative problems in Shanxi.

5. There are many scattered mines in Shanxi with multi-level and various types of coal production configuration. Besides six major State-owned coal bureaus, many small coal mines spread all over the province. There are 4458 mines whose production capacity is more than 30,000 t and 1116 mines whose production capacity is less than 10,000 t. The average minable area of small coal mine is 0.9 km². However, due to the differences of management principles and manager's concepts, the plundered mining has led to destruction of earth surface. 98 of 118 counties in Shanxi province have being exploited. Such wide scope and area of mining is rare in China. Therefore, Shanxi is a province suffering the most serious land deterioration in China.

3. MINING SUBSIDENCE AND LAND DETERIORATION

3.1 The cause of land deterioration

Earth surface collapse□Land destruction mainly includes surface subsidence, overburden removal of land and waste stacking caused by coal mining. The main reason of surface subsidence is that underground mining destroys original mechanical balance of overlying rock mass, so lead to displacement, deformation and damage of rock mass. Meanwhile, surface subsidence has brought about many bad results. For example, engineering facilities such as various pipelines, railways, roads and bridges are deformed or even damaged. It also makes farmland uneven and irrigation facilities out of order, influences agricultural production, changes the type of water body and pollutes water source, etc. To some extent, it also returns to threaten underground working and worker's safety. The wider the mining area and more mined layers are, the more serious problems are rising

Overburden removing: In open-pit mining, because overburden removal covers a lot of land, meadow and farmland, geomorphy will be changed, vegetation will be destroyed and ecological balance will be broken. For example, exploitation of Pingshuo open-pit mine in Shanxi covers 20,000 mu of cultivated land and 18,000 mu of forest land. In addition, open-pit mining also aggravates weathering and erosion of soil in mining district.

The feature of land deterioration in open-pit mining is that before starting production a lot of land should be stripped, whereas after starting production land loss quantity per 10,000 t coal is less than that of underground mining. At Antaibao open-pit mine, the average stripping ratio is 5.5, the average thickness of overlying loess and strata is 40 m, and coal production every year is 15 million t, so there will be more than 70 million m^3 stripping waste to be dumped. According to the design and planning, the whole area of the mine is 2152 hm^2 which includes external waste disposal dump of 756 hm^2 and internal waste dump and pit of 1028 hm^2, coal preparation plant, mine yard and road of 368 hm^2. The maximum stacking height of waste disposal dump is 135 m, and the maximum depth of pit is 200 m. In mining area, six villages have to be relocated. It indicates that in open-pit area land form features will be changed, and original land structure will be completely destroyed. With the coal mining, it may be more serious and go into a vicious circle.

Waste accumulation: According to statistics, there are 1000 waste rock dumps in state-run key mines in China, and their amount of accumulation is 3000 million t. These waste rock dumps cover about 5500 hm^2. In the future, there will still be millions of waste to be discharged every year. Waste rock dump not only covers land, but also pollutes atmosphere by spontaneous combustion. In addition, filter and flowing water from waste rock dump presents acidic or contains poisonous and hazardous elements, which pollutes the surrounding soil and water body, and influences the grow of crops and fish. Therefore, waste accumulation is one of the main problems of environmental pollution.

3.2 the Type of land deterioration

There are many factors which may lead to land deterioration, such as depth and thickness of coal seam, coal mining method, overburden property, roof control, pillar size and strength, gob area size and continuity or discontinuity of gob areas, topography and geomorphy, climate and under ground water level. According to these factors, the types of land deterioration caused by mining subsidence are as follows:

1. digging: This kind of land deterioration is caused by direct digging and mining of overburden land and strata in open-pit mine. For example, at Yuanbaoshan open-pit mine managed by Pingzhuang Coal Mining Administration, land loss of stripping covered 30 hm^2 before starting production.

2. collapse: This type of land deterioration is caused by underground mining and includes earth surface collapse, fracture, rock dislocation. In Shanxi Province, the land area of collapse reached 450 km^2 because of fracture. By the end of this century, it will reach 730 km^2. Mountainous slip areas with a certain scale are more than 30, most of which are found in such mining districts as Datong, Yangquan, Xishan and Xuangang, etc.

3. Covered land: This refers to the deterioration caused by solid waste dump and covered land discharged from open-pit mining and under ground mining. There are 186 waste rock dumps in Shanxi province and they covers 2000 hm^2.

The mining-induced environmental damage is very grave. It makes land depletion and dry. For this reason, the large area of land is deserted and unproductive, while the loss and leakage of surface and underground water may bring about great difficulties to people's living and production.

4. THE TECHNICAL WAYS OF COMPREHENSIVE LAND TREATMENT

4.1 collapse control technique
Shanxi lies in Loess Plateau, where vegetation is rare, topography is complex, mountainous area is much more than farm land, water resource is scarce, and water and soil loss and sand cover are grave. There are many mining collapse regions. Natural environment and geological condition and climate are different all over the province. Therefore, measures suited to local conditions should be taken when dealing with the treatment of earth surface collapse caused by coal mining. According to the principle of protecting ecological environment and decreasing water and soil loss, we take valid measures as follows to help with the comprehensive treatment.

4.1.1 Adopt the protective mining techniques
The primary and valid way of decreasing land resource deterioration and collapsed loss from coal mining is to layout pillars and use backfilling method in order to protect ecological environment of district. The methods such as pillar system, backfilling system, partial extraction, thickness-limited mining, thick or multi-seam coordinated mining can all be used to support load of overlying strata, reduce and control the movement of overlying strata and control the movement and deformation of earth surface so as to protect buildings, topography, geomorgraphy and underground structures. As the cradle of Yellow River Culture, Shanxi has many historic spot and cultural relicts within her area, among which the spots like Jin Temple and Mount WuTai deserve special protection in this way.

4.1.2 Technology of keeping waste in mine
In driving of alternate heading, wide heading face should be used, so ripping and dinting waste can be backfilled into two sides of the roadway, or used to backfill gob area. As for coal seam with thick dirt band, coal and dirt can be mined separately and the picked-up waste can be directly discharged in gob area, which not only reduces both the amount of labor and land area covered by waste disposal on surface, but also supports overlying strata above gob area. Thus movement and deformation of surface are reduced and controlled.

4.1.3 Backfilling gob with pulverized coal ash
Shanxi province is the base of energy resources and heavy chemical industry in China. Especially after carrying out the strategy of changing from coal export to both coal and power export, the number of power plants and installed capacity are greatly increased, therefore, the coal ash coming from power plants is also increased. According to statistics, power plant costs 11 to 15 □to discharge 1 t pulverized coal ash. Moreover, the pulverized coal ash from power plants covers a large area of land and pollutes the air and water around. Therefore coordinating relation between power plants and coal mines, using pipe to transmit pulverized coal ash from power plants by machines or water power to fills underground gob, can not only eliminates the pollution from pulverized coal ash, but also cuts down the cost for placing and discharging coal ash. This way has been put into use in some mining district of China and proves effective.

4.2 land reclamation technique
Land reclamation is a branch of environmental science. Reclaiming of land not only recovers damaged or deteriorated land resource, solves a series of contradiction caused by land shortage, but also is of important technique to protect environment and keep ecological balance. Reclaiming damaged or deteriorated land can greatly lighten or eliminate the environmental and visual impact.

4.2.1 Reclamation technique of building terraced field
The side slope of no water accretion or water accretion, the waste dumping of tripping in open cast mine and waste rock dumps in underground mines are fit to adopt the method of trimming land and building into terraced field or green terraced field to rehabilitate.

Antaibao open-pit mine in Pingshuo has adopted the technique mentioned above, and 408 hectares land has been reclaimed in waste disposal dump of Er'pu, where grass, herbs, industrial crops shrub and arbor have been planted. Good economic and environmental benefits have obtained. After this technique was adopted in Wangzhuang mine in Lu'an administration, 100 t and 4 hectares of covered waste rock dump have become green. At the same time it has become a place for miners and residents to tour and rest. A new road of greening and afforestation in waste rock dump has been explored in China.

4.2.2 Filling reclamation technique of waste rock dump
Filling mining collapse area and open-cast site with

coal mine dirt, can not only reduce land deterioration caused by mining and solid waste, but also utilize the filled land, and eliminate the environment pollution caused by vast rock dump. Jincheng Coal Mining Administration has built up parks, sports ground and living building by filling 17 hm² of land with waste rock, which has improved the environment of mining districts and perfect worker's living quality.

4.2.3 *The reclamation technique digging deeper pound*

In the damage districts which have steady, partial or seasonal water accretion, damage districts should be dug deeply under permitted hydrogeological condition and change it into fishpond or water storage which can fight a drought.

Water resource is lack in Shanxi province, together with high and steep mountain slope, scarce vegetation and weak regulation of pondage, leading to the situation of flood in rainy season and water lacking in drought season. The reclamation technique of digging deeper pound will improve ecological environment.

Near some mining areas in Shanxi, there are villagers who have made damaged districts into water storage for irrigation by deepening them to 1.5 ~ 2.0 m.

4.2.4 *reclamation technique of ecological agriculture and microorganism*

According to ecology and ecosystem economics, best economic and environmental benefits will be made by applying land reclamation technique and ecological engineering technique combined with many-sides plantation and breed of agronomic, animal and microorganism after filling and reclaiming of mining damage district.

5. GREY FORECASTING MODEL FOR MINING SUBSIDENCE

The expression of GM(1,1) model is:

$$X^{(1)}(k+1) = (X^{(0)}(1) - \frac{b}{a})e^{-ak} + \frac{b}{a}$$

GM(1,1) is a dynamic model of imitation differential equation which needs not a great many sample data. GM(1,1) model is the basis of gray forecasting. It can be used for data forecasting, catastrophe forecasting, topology forecasting and so on.

According to the measuring values of the observation stations certain mine at surface in May, June, July and August in 1997 □table 1□, using the model, we forecasted the its values of subsidence in September, October.

Table 1. The measured values in May □ August
units: mm

point / time	May	June	July	August
I	18	22	25	34

Building G□1,1□model, making once accumulative product from raw data into resultant data:

$$X^{(1)}(k) = (18,40,65,99)$$

After calculating, we obtain the forecasting model:

$$X^{(1)}(K+1) = 75.2e^{0.20935k} - 57.2$$

The calculation result is shown as table 2. From table 2, we know that in September and October the average relative error between the measure value and the forecasting value is 4.05%, i.e., the calculating precision is 95.95%, and meets the demand of forecasting.

Table 2. The calculation result units: mm

time	measure values	forecasting values	relative error □%□
May	18.0	18.0	0
June	22.0	21.7	1.4
July	25.0	26.8	-7.0
Aug.	34.0	33.0	3.0
Sept.	42.2	40.7	3.6
Oct.	48.05	50.2	-4.5

6. CONCLUSIONS

The technique proposed in this paper could strengthen the study of theory and technique for tackling the subsidence area and land reclamation.

The Grey Model introduced in this paper can be used to forecast the mining-induced subsidence with such advantages as simple calculation, higher precision and common use.

Mining Science and Technology' 99, Xie & Golosinski (eds) © 1999 Balkema, Rotterdam, ISBN 90 5809 067 1

Reclamation of lignite open pits in Poland

Z. Kozlowski
Technical University, Wroclaw, Poland

W. Koziol & R. Uberman
University of Mining and Metallurgy, Krakow, Poland

ABSTRACT: The paper discusses problems connected with reclamation of big open pits left after exploitation of lignite. The pits are up to 300 m deep and cover area of up to 2000 ha. Complicated geological, hydrogeological and geotechnical conditions make the reclamation difficult. A series of research studies defined the criteria for the choice of the most profitable ways of reclamation. The paper addresses technical and economic aspects related to closure of lignite mines, as well as utilization and reclamation of pits and other areas disturbed by mining.

1. INTRODUCTION

Construction and exploitation of bigger and bigger lignite open pits, with the area exceeding 2000 ha and depth of up to 300 m, complicates considerably restoration and reclamation works after exploitation is completed.

Nowadays in Poland there are two big operating lignite open pits, namely: Belchatow with capability of mining 38 million t of lignite annually and Turow with designed output capacity of about 16-12 million t of lignite per year. The third strip mine - Szczercow, located next to Belchatow has the designed production capacity of about 28 mln t of lignite per year and is currently under construction (figure 1). The above mentioned opencast strip mines in a final exploitation phase are planned to achieve the pit dimensions listed in Table 1. It has been planned to develop these final pits into water reservoirs, a first venture of this type in the history of Polish mining.

2. RECLAMATION METHOD

Unfortunately, subsequent modifications of the mining plans resulted in considerably smaller volumes of the excavated pits. In addition the geological and hydrogeological conditions have proven to be unfavorable. Basic difficulties for development of pits into water reservoirs resulted from limited subterranean water resources available at the sites as well as limited alternate surface water resources located close to the mines.

In case of Belchatow and Szczercow additional difficulties are connected with:
- localization - two huge pits are located next to each other (Figure 2), requiring nearly 4000 mln m³ of water for their flooding,
- necessity of isolation of saline trench that separates both pits to eliminate the danger of water pollution by chlorides;
- extension of mine life, and resulting delayed closure of both mines requires that Belchatow

Table 1. The dimension of lignite pits in the final exploitation phase.

Strip mine	Year of exploitation's end	Parameters of final pits		
		Area [km²]	Depth [m]	Capacity [mln m³]
Belchatow	2020	17,5	250-280	1910
Turow	2026	19,0	300	2000
Szczercow	2038	15,0	280	1862

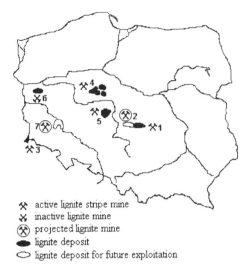

☓ active lignite stripe mine
☓ inactive lignite mine
⊗ projected lignite mine
● lignite deposit
◌ lignite deposit for future exploitation

Figure1. Lignite Mining Regions in Poland
1-Lignite Mine Belchatow 2-Projected
LM Szczercow 3-LM Turow, 4-L M Region Konin
5-LM Region Adamow, 6-LM Sieniawa in
liquidation, 7-Lignite Deposit Legnica

dewatering systems remains operational and maintained until exploitation in Szczercow open pit is completed.

Even though the remaining mine life is significant the studies of the most efficient and environmentally friendly ways of mine closure need to be studied well in advance. Such studies are conducted at the University of Mining and Metallurgy in Krakow as well as by Poltegor-Projekt Ltd. in Wroclaw and the Technical University in Wroclaw. As a first step the hydrogeological research was undertaken to get reliable prognosis of groundwater behavior and flows. Using a software package called MODFLOW the groundwater inflow into Belchatow and Szczercow pits were determined. The study indicated that it would take over 100 years to fully fill the both pits with water, assuming that the pit depth remains unchanged.

Similar studies were done for Turow lignite mine. By flooding of the final pit using water from the self-recharging ground water aquifer located close by, with some additional surface run-off water, filling the pit up would take approximately 65 years. These evaluations assumed that water from the near-by Nysa Luzycka River would not be diverted into to pit.

That long time required to fill the pits with water after they are deactivated creates numerous technical, geotechnical, organizational and economic problems. Reducing the length of flood periods constitutes the first condition to facilitate this form of pit reclamation that, incidentally, is preferred and desired by local population.

The most realistic way to achieve this goal appears to be decreasing of the pit depth using mined overburden or other mined waste materials. Advanced feasibility studies on reduction of remnant pit depth have already begun for Belchatow and Szczercow mine.

These studies indicate that Belchatow pit should receive 350 mln m^3 of backfill materials from Szczercow mine, beginning in year 1998 and ending in year 2020. Furthermore, after closure of Belchatow pit, the Sczercow mine would have to provide further 376 mln m^3 of backfill material. Furthermore, according to these studies about 180 mln m^3 of ashes from Belchatow powerstation would be stored in Belchatow final pit. All this would allow reduction of the pit depth from presently planned 250-280 m to 105-115 m, for 1435 mln m^3 pit volume difference. The bottom of water reservoir will be situated 45-55 m above the ceiling of salin trench (figure 3).

In case of Szczercow pit, it would be required to relocate about 850 mln m^3 of earth materials from temporary outer cover dumping into the pit. After the pit backfill its depth would be reduced from 280 m to 95-100 m, with pit bottom some 55-60 m above the ceiling of saline trench. The estimated schedule of pit flooding process has been shown in figure 4.

The time of pit filling with water could be reduced by 10-20 years if the surface water could be diverted to the pits from outside, in volumes of 2-4 m^3/s.

3. DISCUSSION

The proposed scheme of abandoned pit backfilling and subsequent flooding sets the general trend for reclamation of large lignite pits in Poland. It allows to satisfy the wishes of the population while addressing environmental issues. These include protection of water reservoirs against a salinity that could arise if the water would get into the salt deposit, as well as disposal of surplus waste material.

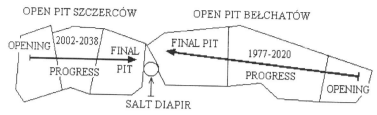

OPEN PIT SZCZERCÓW   OPEN PIT BEŁCHATÓW

Figure 2. Directions of opening and progress in strip mines Belchatow and Szczercow

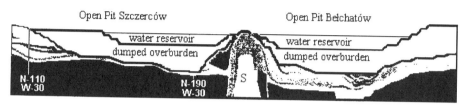

Figure 3. Scheme of planned reclamation of final pits Belchatow and Szczercow. S-salt diapir

Water protective function against the salt leaching process is now being fulfilled by a special well barrier surrounding described salt diapir. It will work until the process of depth decreasing will be finished i.e. bottom in both of pits (Belchatow and Szczercow) will reach altitude above ceiling of salt deposit.

The long time required to flood the pits implies additional difficulties and problems. To the most sensitive of these is maintenance of stability of pit slopes. The slope will have to remain stable 5-6 times longer than the present slopes created in the Belchatow's pit.

Present experience indicates that no significant adjustment in pit shape will be required. The only planed works are works that will lead to flattening of highest slope benches, currently 20 m high, to target inclinations of 3-5 degrees. This flat slope would fulfill the role of shelf braking the erosive action of water waves. It has been estimated, that on the surface of the planned reservoirs (3 km wide and 5 km long) the water waves may arise to the height of 1.5-2.0 m, which would disadvantageously affect shore stability and preservation. Currently there are no plans to shape or grade the slopes at deeper benches. It is assumed that the functions of benches would meet the slope stability requirements, with the current bench widths of 40-80m. The above water slopes will be shored by dense planting of plants and vegetation with firm root systems.

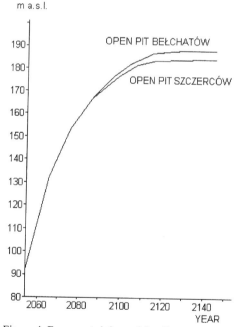

Figure 4. Forecoasted time of flooding process.

The research and design work conducted so far have shown that reclamation of large pits remaining after mining out of lignite will be technically difficult, time-consuming and very expensive. It is estimated that reclamation of a large lignite pit by creating a recreational lake in its place may cost as much as construction of a lignite pit of the similar dimensions.

In case of Belchatow and Szczercow the most expensive would be long term of dewatering, ensuring isolation of the saline trench, as well as excavation and transportation of pit backfill from overburden dump. For these reasons scientists are searching for optimal ways of reclamation and utilization of such open pits, particularly by looking for an additional possibilities of filling pits with bulk earth volumes and waste from outside of lignite mines area.

It has been proposed to transport waste rock remaining after coal washing in bituminous coal mines of Upper Silesia. Also given consideration was use as the pit backfill of power station ash accumulated in victinity of several power stations in the area. Flattening of the pit slopes of the final pit and using thus created earth mass for filling the pit to achieve the depth of water reservoirs circa 50 m is also given consideration.

4. CONCLUSIONS

Based on current research and design work on restoration activities related to closure of lignite mines in Poland following conclusions are drawn:

1. Reclamation and use of big lignite pits after mine closure is associated with a number of technical difficulties; its time and cost are comparable to that required for construction of a new lignite mine.
2. Main problems in reclaiming the lignite pits into water reservoirs are that of long time required and availability of huge volumes of backfill material.
3. In the analyzed cases of Belchatow and Szczercow pits additional difficulty is presented by possibility of water becoming saline through its contact with the trench present in the victinity.
4. The research revealed that for achievement of fair results the design work and work on choice of most feasible restoration ways should begin well in advance.
5. The work on shaping of the should start during pit exploitation, well before mine closure.

REFERENCES

Kozlowski, Z. 1998. *Analyze of technical possibilities of depth decreasing of final open pit Belchatow*. Krakow: Uczelniane Wydawnictwo Naukowe Akademii Górniczo-Hutniczej w Krakowie.

Uberman, R. 1994. *Utilization of post exploitation open pits in lignite mining*. Wroclaw: Oficyna Wydawnicza Politechniki Wroclawskiej. Prace Naukowe Instytutu Gornictwa. Zeszyt 74.

Seweryn, L. Kozlowski, Z. 1996. *Utilization of the post exploitation final pit in lignite mine Belchatów*. Wroclaw. Oficyna Wydawnicza

Mining Science and Technology' 99, Xie & Golosinski (eds)© 1999 Balkema, Rotterdam, ISBN 90 5809 067 1

Quality and treatment of coal mine drainage water in China

Jiguang Guo & Zhongjian Shan
China University of Mining and Technology, Beijing, People's Republic of China

ABSTRACT: About 71% of collieries in China are short of water resources. 2.2 billion m³ of coal mine drainage are discharged annually. Making full use of mine water is the efficient approach for solving the shortage of water resources in coal mine areas. The authors studied field and statistical data to assess the status of discharging and utilizing coal mine water in China, mine water quality, and its distribution. In this paper the Chinese coal mine drainage systems were classified into six types. The mine water treatment technology is also summarized.

1. INTRODUCTION

The total amount of freshwater resource in China is 2800 billions m³, ranging the sixth in the world. But the amount of water resource per person is 2700 m³, only accounting for 1/4 of the world's average. Thus China has serious problem of freshwater shortage. The distribution of both water resource and coal resource is not equalization in China. The coal resource is 75% in north China, but the water resource is only 19%. Therefore, the area having more abundant coal resources usually has much more serious problem of water shortage. It is surveyed that 71% of 86 main state-owned mines are short of water, 40% of them are very severe shortage in water supplies (Chen 1997). Coal is the main energy resource in China. With the increase of coal production and scales of mines, the problem of water shortage in mining areas will become more serious.

With the excavation of coal resources from underground, a large quantity of coal mine drainage is discharged. Now the annual total amount of the mine drainage is 2.26 billions m³, but the utilization ratio is less than 20% of the total amount. A large amount of untreated mine drainage containing coal and rock powder is discharged into rivers directly, which not only wastes the valuable water resource, but also makes different effects on the area around the coal mine. To solve the problem of water shortage and assure the sustainable development of China's coal industry, utilization of coal mine drainage has become an urgent task in China.

2. THE DISCHARGE AND UTILIZATION OF COAL MINE DRAINAGE IN CHINA

The total amount of mine drainage in China was 2.26 billions m³ in 1996, which meant that 1.8 m³ of mine drainage per tons of coal was produced, 19.5% of it was utilized by coal mines.

Main characteristics of discharge and utilization of coal mine drainage in our country were (see Figure 1)(The ministry of coal industry 1996):

(1) Some provinces such as HeBei, HeNan, ShanDong, HuNan, HeiLongJiang and LiaoNing generated much more mine drainage than other provinces. Each of them produced mine drainage over 1 billion m³ per year. But NingXia, ShanXi and NeiMengGu produced less amount of mine drainage than that of other provinces.

(2) Mine drainage per tons of coal ranged from 0.6 m³ to 20.1 m³. HuNan, HeBei, HeNan and ShanDong provinces had the bigger value of it than that of other provinces, but ShanXi, NeiMengGu and NingXia provinces had the smaller.

(3) ShanXi province had the biggest value of utilization ratio of mine drainage, which was 56.2%. But HuNan had the smallest, 5.2%. In general, the utilization ratio of mine drainage in water abundant districts is lower than that in water shortage districts.

3. QUALITY OF MINE DRAINAGE

3.1. The source of coal mine drainage

During processes of coal excavation, underground water runs into opening and coal surface because of

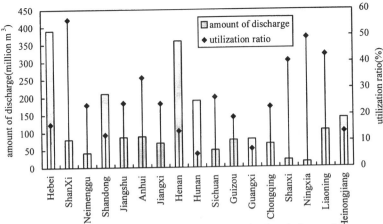

Fig.1 The discharge and utilization of mine drainage of coal producing provinces in China

excavation and gob area falling, and flow into mine water storehouse, then pump out of underground, this is mine drainage. It comes from five kinds of sources: atmosphere precipitation, surface water, water-bearing stratum water, abandoned-goaf water and fault water. Water-bearing stratum is often the main source of water in most of Chinese underground mines. Karst stratum has the largest water capacity and cranny water-bearing stratum has the smallest. So the karst water has the most effects on mine production. For example, very large amount of mine drainage mainly come from karst water-bearing stratum in Fengfeng and Jiaozuo collieries,.

3.2 The quality of mine drainage in China

Mining depth of most of coal mines in China is 200~600 meter, few of them up to 1000 meters. At this depth, water alternates vehemently, and the chemistry elements transfer by filtrating and dissolving. So the quality of mine drainage is affected by the condition of geology, water kinetics, geochemistry, mineral-bed geological construction, mining conditions and underground worker's activities and so on. The main indexes of quality of mine drainage are discussed as follows:

Suspended solid The main suspended solids (SS) in mine drainage are coal and rock powder produced by extracting opening and coal and transportation of coal and rocks. So the color of mine drainage is black. The index of suspended solid in mine drainage varies from 10 mg/l to 1000 mg/l, majority of it ranges 100~500 mg/l.

pH value Majority of coal mine drainage is neutral, few of it is acid. The coal mines with acid mine drainage are accounting for 5~8 percent of the total in China, which usually locate in the mining

areas such as northwest, southwest and south of Yanzhi river with high content of sulfur in coal. In general, when coal seam contains 5% sulfur (Guo 1991), acid mine drainage will be generated. (Tab.1)

Salinity and hardness The salinity and hardness of coal mine drainage have a close relationship with hydro-geological conditions of water-bearing strata and the depth of coal seams. The salinity and hardness of mine drainage is high in the coal mines with poor mineral-bad and discharging conditions. For example, in the northwest China, the salinity and hardness are very high. The salinity and hardness have also a relationship with the mining depth of coal mines. Coal mines with deeper mining depth usually have a higher degree of salinity and hardness. The salinity of mine drainage in China is around 1000 mg/l, the highest of it can be up to 4000mg/l. The minimum of its hardness is 50 mg/l, the maximum is 5000 g/l.

Tab.1 relation of pH value of acid mine drainage with sulfur content of coal seam.

Sulfur content of coal seam %	pH
5~7	6.0~5.5
7~9	5.5~3.5
9~11	3.5~2.5
>11	<2.5

COD and BOD5 COD value of coal mine drainage in China generally ranges 20~1000 mg/l and the BOD_5 value is within 0~10 mg/l. COD value of mine drainage is high, but BOD_5 value is low. COD value has a great relation with the SS and it is an important character of coal mine drainage. The reason is that most of organic substance in mine drainage is coal powder and the others that consuming oxygen is relatively few.

Petroleum class Petroleum class substance of mine drainage mainly comes from working oil used by miner. The range of oil content in mine drainage varies between 1.0~20.0 mg/l. Because of the oil content in mine drainage, it is difficult to be purified into drink water.

Bacteriology index The range of the total amount of bacterium in mine drainage varies between 500~2000 per liter, the range of coliform group varies between 20~2000 per liter. The bacterium in mine drainage comes from the providing water source, the workers excrement and urine and everyday living water.

Fluoride Fluorin is one of elements contained steadily in the coal. So the most of coal mine drainage in China contains fluoride. The fluoride index of mine drainage varies between 0.4~15 mg/l.

Radioactivity Coal excavation damages rocks' structure and underground water running through rock rifts into coal seam. Mine drainage have some radioactivity because of the underground rock contains radioactive elements, i.e. uranium, carbon-14 and so on. The radioactivity of mine drainage in most of coal mines in China aren't high. Total α-radioactivity is less than 0.6 Bq/l, and total β-radioactivity is not more then 1.5 Bq/l. For low radioactive mine drainage, coagulation and sedimentation methods are needed, but we can't disregard of the danger of mine drainage with high radioactivity, and special treatment are needed.

3.3 Classifications of mine drainage in China

From the view of water treatment and utilization, mine drainage can be classified into six types based on its quality in China:

Clean mine drainage All kinds of index of water quality fundamentally reach the potable water standards of China, only a little of treatments and detoxification is needed for reaching the standard. This type of mine drainage is clean water section of the underground mine water which clean and dirty water flow separately.

General mine drainage The SS in coal mine drainage is the only one index not to meet the standards, the others meet with the standards by

normal treatment. Most of mine drainage in China belong to this type.

High salinity and hardness mine drainage The total solubility solid is more than 1000 mg/l, the total hardness is more than 450 mg/l and SO_4^{2-}, Cl^- more than 250 mg/l.

Acid mine drainage The pH value of water is less than 6.5. Although its quantity is small, its danger is severe. The pH of mine drainage in few of coal mines may be up to 3~2, which can't be used as potable water and industry utilization water even though after treatment.

High fluorine Fluoride in water is more than 1.5 mg/l. If it is used as potable water, the treatment of deflourine must be needed.

Other mine drainage This type of mine drainage is the one containing oil, heavy metal elements and radioactive elements. Its distribution is very little.

4. THE TECHNOLOGY OF PURIFYING TREATMENTS OF MINE DRAINAGE IN CHINA

Selection of clarifying treatments should be based on the local environment, quality of mine drainage and planed use after treatment. The different uses of mine drainage also decide its degree of treatment. Now there are several uses of mine drainage after treatment in China: (1) potable water. (2) boiler water. (3) bathing water. (4) coal preparation water. (5) dustproof and sprinkling water. (6) irrigation water (7) standard discharge.

4.1 The technique of purifying treatment of general mine drainage

The emphasis of purifying treatment for general mine drainage is removing the SS (most of them are coal and rock powder). Its treatment way is the normal technology of treatment, namely coagulation and sediment, filter, disinfection (Fig.2).

In the above technical flowchart, reaction tanks include three types: separable board, U-turn and rotational flow, among which separable board type is often used. The type of slanting pipe with honeycomb shape is the best of the sedimentation

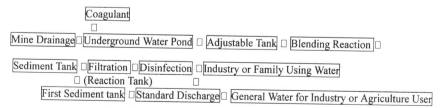

Fig.2 Treatment flow chart of general mine drainage

tank, which has high sediment efficiency. The general types of filtration tank used in mine drainage purifying center are siphonate type and gravitational no-value type, which has the advantages such as automatic wash and convenient control. There are two layers of filter bed in the filtration tank, the upper usually is anthracite with 1.0~1.5 mm granular length, the lower is gravel sand with 0.5~1.0 mm granular length.

If the quantity of mine drainage is not large and water quality of it is relatively good, all-in-one purification equipment with small volume, rational structure and integrating aggregation, sedimentation filtration into whole should be selected. This equipment is installed conveniently and its investment is relatively less.

4.2 The treatment of mine drainage with high salinity and hardness

For high hardness mine drainage with high carbonate and low soluble solid, the technology of lime softened and aggregation sediment simultaneously are usually adopted. The coagulant is often $FeSO_4 \cdot 7H_2O$

The technology of electrodialysis has been adopted for treatment of high salinity mine drainage with high soluble solid and low carbonate since 1980s in China. It requires raw water to meet with following conditions: turbidity is less than 3 degree, ferrous is less than 0.3 mg/l, Mn is less than 0.5 mg/l and water temperature is 5~40 ^{0}C. The cost of it is high because the electrodialysis consumes more electricity.

4.3 The treatment of acid mine drainage(AMD)

Now in China, the method of integrating with biology and chemistry neutralization is adopted to dispose acid mine drainage, the neutralizer is limestone with low price.

The purpose of using biological disk is to oxidize Fe^{2+} in the acid mine drainage, a reaction is:
$$4FeSO_4 + 2H_2SO_4 + O_2 \rightarrow 2Fe(SO_4)_3 + 2H_2O$$
Neutralizing reaction by limestone is below:
$$CaCO_3 + H_2SO_4 \rightarrow CaSO_4 + H_2O + CO_2\uparrow$$
$$Fe_2(SO_4)_3 + 6H_2O \rightarrow 2Fe(OH)_3\downarrow + 3H_2SO_4$$

4.4 The treatment of high fluorine mine drainage

Using active alumina as de-fluorine agent is the most usual way that turns mine drainage into potable water by getting rid of fluorine at present in China. The active alumina usually is treated with 5% of $Al_2(So_4)_3$ solution at first. The reaction is below:
$$Al_2O_3 + nH_2O + Al_2(So_4)_3 \Leftrightarrow Al_2O_3 \cdot Al_2(So_4)_3 \cdot nH_2O$$
The active alumina after treatment reacts with fluorine ion:
$$Al_2O_3 \cdot Al_2(So_4)_3 \cdot nH_2O + F^-$$
$$\Leftrightarrow Al_2O_3 \cdot 2AlF_3 \cdot nH_2O + SO^{2-}$$

When absorbing fluorine ion is up to saturation, we can use $Al_2(SO_4)_3$ solution dip washing and make them regenerated.

The characteristics of using active alumina as de-fluorine agent are strong selectiveness, large exchange capacity to fluorine ion, no effect by the density of SO_4^{2-} and Cl- in the mine drainage and low price.

Except the method of using active alumina to remove fluorine, the methods to remove fluorine in mine drainage include hydroxyl-phosphate-calcium, Electro-condensation and electrodialysis.

5. CONCLUSION

Exploitation and utilization of mine drainage resource is the most effective solution to resolve the problem of water shortage in the collieries of China. To explore and utilize coal drainage, we mainly consider following factors: (1) water capacity (the source of mine drainage, capacity and steady condition). (2) water quality (mainly test the suspending solid in mine drainage, pH value, salinity hardness, petroleum class, fluoride and the index of radioactivity). (3) Selection of feasible purification treatments based on the quality and capacity of mine drainage. These factors were studied deeply in this paper. Current successful experience in use of mine drainage in China was summarized. The results obtained in this paper are helpful in further utilizing mine drainage resources.

REFERENCES

Chen, Minzhi 1997. The protection, exploitation and utilization of water resource of mine in China, *Coal Mine Environmental Protection*, 1997, (1).

The ministry of coal industry 1996. The 'ninth-five' plan of coal industry for environmental protection, (1996~2000.).

Guo, Jiguang, 1991. The investigation and analysis of discharge condition and treatment & utilizing for coal mine drainage in China, *The master's degree thesis in Beijing graduate school of CUMT*, in 1991.

Mining Science and Technology '99, Xie & Golosinski (eds) © 1999 Balkema, Rotterdam, ISBN 90 5809 067 1

Models of energy and materials flow for reclaimed land in mined areas and their application*

Zhengfu Bian, Xikuan Hu & Guoliang Zhang
China University of Mining and Technology, Beijing, People's Republic of China

ABSTRACT: By means of the compartment model in systems ecology, the state of energy and materials' flow of Weizhuang reclaimed land use system is analyzed. From the example, we find that a compartment model is an effective model for designing the input level of reclaimed land use system and adjusting the scale of each land use unit. To overcome shortcomings of the existed compartment model, variable parameters' compartment model (VPCM) is developed. It is proved that VPCM can reflect the parameters of the system varying with time and reveal the nonlinear relationship between input and output of the system.

1. INTRODUCTION

In systems ecology, compartment model is a universal model to evaluate the e&m flow in an ecological system (Niu Wenyuan 1984, Odum 1988). Agronomists in our country ever applied compartment model to analyze the evolution laws of soil quality and the fertilizing efficiency at a farm located in the suburb of Beijing (Liu Tiebing & Han Cunlu 1988). Foreign ecologists ever used the model to discuss the material recycling efficiency in a lake ecological system (Odum 1988). Reclaimed land use system also include the process of e&m flow, so the compartment model provide a ready-made model for analyzing the evolution laws of reclaimed land quality. But those applications of compartment model stated above usually deal with a simple ecological system or a simple land use system with a hypothesis that the input and output of a system keep linear relationship. In this paper, existed compartment model is used for analyzing the state of e&m flow in Weizhuang reclaimed land use system. To overcome the shortcomings of existed compartment model, variable parameters' compartment model (VPCM) is developed which takes the parameters of the system as variable with time and considers the relationship between input and output of the system as nonlinear.

2. EXISTING COMPARTMENT MODELS

Take reclaimed land use system at Weizhuang near Xuzhou City as an example, land use types consist of pig farm, fishery, cattle farm and crop land (see table 1). The process of e&m flow of Weizhuang reclaimed land use system can be summarized as figure 1.

In figure 1, $x_1 \sim x_5$ represent the state variables or stocks of each compartment, $Z_{10} \sim Z_{50}$ represent the input from outside of the system, $f_{01} \sim f_{05}$ represent the output of each compartment, f_{ij} represents the e&m flow from j compartment to i compartment in the system. Then, variation of stock of each compartment with time can be figured out by compartment model as series of equation (1). If we introduce the ratio of e&m flow a_{ij} as the flow capacity per unit stock from j compartment to i compartment, which is determined by formula (2), then, series of equation (1) change into (3).

$$
\begin{cases}
\dfrac{dx_1}{dt} = f_{15} - f_{41} - f_{01} + Z_{10} \\
\dfrac{dx_2}{dt} = f_{25} - f_{42} - f_{02} + Z_{20} \\
\dfrac{dx_3}{dt} = f_{35} - f_{43} - f_{03} + Z_{30} \\
\dfrac{dx_4}{dt} = f_{41} + f_{42} + f_{43} + f_{45} - f_{54} + Z_{40} \\
\dfrac{dx_5}{dt} = f_{54} - f_{15} - f_{25} - f_{35} - f_{45} - f_{05} + Z_{50}
\end{cases}
\quad (1)
$$

$$
a_{ij} =
\begin{cases}
\dfrac{f_{ij}}{x_j} & i \neq j \\[2mm]
-\displaystyle\sum_{k=0}^{n} \dfrac{f_{ki}}{x_i} & i = j, n
\end{cases}
\quad (2)
$$

* Supported by National Natural Science Fund of China (Approved No. 49771040)

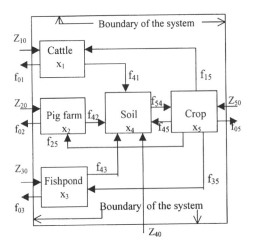

Figure 1. The e&m flows in the system of Weizhuang reclaimed land use system

$$
\begin{cases}
\dfrac{dx_1}{dt}=a_{15}x_5-a_{11}x_1+Z_{10}\\[2mm]
\dfrac{dx_2}{dt}=a_{25}x_5-a_{22}x_2+Z_{20}\\[2mm]
\dfrac{dx_3}{dt}=a_{35}x_5-a_{33}x_3+Z_{30}\\[2mm]
\dfrac{dx_4}{dt}=a_{41}x_1+a_{42}x_2+a_{43}x_3+a_{45}x_5-a_{44}x_4+Z_{40}\\[2mm]
\dfrac{dx}{dt}
\end{cases}
\tag{3}
$$

Where, $X=(x_1,x_2,\cdots x_5)^T$, $Z=(z_{10},z_{20},\cdots z_{50})$,

$$
A=\begin{bmatrix}
-a_{11} & 0 & 0 & 0 & a_{15}\\
0 & -a_{22} & 0 & 0 & a_{25}\\
0 & 0 & -a_{33} & 0 & a_{35}\\
a_{41} & a_{42} & a_{43} & -a_{44} & a_{45}\\
0 & 0 & 0 & a_{54} & -a_{55}
\end{bmatrix}
$$

It is clear to find that provided the initial stock of each compartment, the e&m flow between each two compartments, the input and output of each compartment are given, the dynamic laws of each variable with time can be derived by means of series of equations (3). We can also adjust some related

Table 1. Reclaimed land utilization structure Unit: ha

Land use type	Crop land	Cattle farm	Fishery	Pig farm
Area	60	2	14	3.5

parameters to make the reclaimed land use system run in desirable pattern and achieve desirable state. Moreover, according to the existed compartment model, the output of each compartment can be calculated by formula (5).

$$
\begin{aligned}
Y=(y_1,y_2,\cdots y_n)&=(a_{01},a_{02},\cdots a_{0n})(x_1,x_2,\cdots x_n)^T\\
&=CX^T
\end{aligned}
\tag{5}
$$

Recycling index, which is defined as the percentage of a certain element returning to original compartment from which it flows out, can be determined by formula (6).

$$
R_i=\frac{n_{ii}-1}{n_{ii}}
\tag{6}
$$

Where, n_{ii} is the element of matrix N which is defined as $N=[I-G]^{-1}$, $G=(g_{ij})_{n\times n}$, and

$$
g_{ij}=\begin{cases}
0 & i=j\\[2mm]
\dfrac{f_{ij}}{\displaystyle\sum_{k=0}^{n}f_{kj}} & i\neq j
\end{cases}
\tag{7}
$$

Total recycling flow of the system is defined as:

$$
T=\sum_{i=1}^{n}\left[(\sum_{k=0}^{n}f_{ki})\times R_i\right]
\tag{8}
$$

Total recycling index R of the system can be calculated by formula (9):

$$
R=\frac{T}{\displaystyle\sum_{i=1}^{n}(\sum_{k=0}^{n}f_{ki})}
\tag{9}
$$

By the compartment model, the stock and yield of each compartment in different years are predicted (Bian 1997). We find that the compartment model can be used to design appropriate input level for achieving desirable yield. Thus the existed compartment model is effective in the application to the reclaimed land use system.

3. VARIABLE PARAMETERS' COMPARTMENT MODEL (VPCM) AND ITS APPLICATION
3.1 Development of VPCM

The vital shortcomings of existed compartment model is that the parameters, such as input and the e&m flow between each two compartments, keep a constant as time goes on, although the stock and output of each compartment in the system have changed with the lapse of time. Because of the vital shortcoming, the predicted results do not coincide with the reality.

Based on existed compartment model, time variable is introduced into the model, then, the stock x_i, input z_i, e&m flow f_{ij} would be written as $x(i, t)$, $z(i, t)$, $f(i, j, t)$, and existed compartment model would change into VPCM as following series of equations:

$$\frac{dx(i,t)}{dt} = \sum_{j=1}^{n} a(i,j,t)x(j,t) + z(i,t) \quad (i=1\sim n) \quad (10)$$

$$y(i,t) = f(0,i,t) = \sum_{i=1}^{n} a(0,i,t) \times x(i,t) \quad i=(1\sim n) \quad (11)$$

$$a(i,j,t) = \begin{cases} \dfrac{f(i,j,t)}{x(j,t)} & (i \neq j) \\ -\sum_{k=0}^{n} f(k,i,t)/x(i,t) & (i = j) \end{cases} \quad (12)$$

$$g(i,j,t) = \begin{cases} 0 & (i=j) \\ \dfrac{f(i,j,t)}{\sum\limits_{k=0}^{n} f(k,j,t)} & (i \neq j) \end{cases} \quad (13)$$

3.2 *Application of VPCM*

Also take Weizhuang reclaimed land use system as an example, we assume:

$$z(i,t) = k_z(i)x(i,t) \quad (14)$$

$$f(i,j,t) = k_f(i)x(i,t) \text{ or } k_f(i)x(j,t) \quad (15)$$

$$f(0,i,t) = \begin{cases} k_{of}(i)x(i,t) & (i=1\sim 4) \\ x(5,t) - \sum_{k=1}^{4} f(k,5,t) & (i=5) \end{cases} \quad (16)$$

Where, $k_z(i)$, $k_f(i)$ and $k_{of}(i)$ are coefficients, which vary with the stock of each compartment. When calculating the e&m flow from j to i compartment, if

f (i, j) depends on x (i, t), then, $f(i,j,t) = k_f(i)x(i,t)$, Otherwise, $f(i,j,t) = k_f(i)x(j,t)$.

From formula (10), following formula can be derived:

$$x(i,t+1) = x(i,t) + \int_t^{t+1} \left[\sum_{j=1}^{n} a(i,j,t)x(j,t) + z(i,t) \right] dt \quad (17)$$

VPCM is feasible for analyzing the relationship between stock and output and the relationship between output and input. Here the results related to cattle and crop compartments are listed in table 2 by using VPCM. The parameters of the system such as the input and output of each compartment, the e&m flow between each two compartments and so on, are the same as those in the application of existed compartment model.

Table 2 shows: 1) Input is not a constant, but varies with the value of stock. 2) In crop compartment, there exists a nonlinear relationship between input and output because the hypothesis is taken that there exist two nonlinear relationships, one between input and stock and another between stock and output.

As an example, we design the input level and the state of e&m flow of Weizhuang reclaimed land use system by means of VPCM. The study indicates that: 1) If the input level and the parameters of e&m flow in the system are kept original state, the scale of fishery, cattle farm and pig farm should be cut down appropriately. 2) Because pig farm, fishery and cattle farms have higher incomes, farmers in Weizhuang are not willing to cut them down. Under this circumstance, the input level of some compartments and the e&m flow among them must be changed in order to keep the development of the system sustainable. The adjusted parameters in initial year of the system are as table 3. The output of each compartment in the system keep stable increase and the recycling ratios of nitrogen, phosphorous and potassium of the whole system reach 46.3%, 45.8% and 51.4% respectively.

Table 2. The stock, input and yield of cattle and crop compartment calculated by VPCM. Unit: kg/ha

Time	Element	Stock		Yield		Input	
		x_1	x_5	Y_1	y_5	Z_{10}	Z_{50}
0	Nitrogen	166.7	415	67.7	51.9	47.0	3.8
	Phosphorous	33.3	192.5	26.7	72.3	15.8	1.6
	Potassium	100.0	405.0	53.3	113.5	39.0	3.7
15	Nitrogen	183.6	1008.3	74.6	656.7	51.8	9.2
	Phosphorous	32.7	401.6	26.2	270.1	15.5	3.3
	Potassium	119.1	899.1	63.5	602.3	46.4	8.2
30	Nitrogen	202.2	1669.7	82.1	1229.8	57.0	15.4
	Phosphorous	32.1	640.0	25.7	480.5	15.2	5.3
	Potassium	141.8	1462.9	75.6	1095.8	55.3	13.3

Table 3. The adjusted parameters of Weizhuang reclaimed land use system in initial year.   Unit: kg/ha

Element	Initial input					Materials flow between each two compartments							
	Z_{10}	Z_{20}	Z_{30}	Z_{40}	Z_{50}	f_{41}	f_{42}	f_{43}	f_{54}	f_{15}	f_{25}	f_{35}	f_{45}
Nitrogen	141	1970	569	350	7.6	113.3	1080	541.7	415	62.6	586.3	85.6	75.7
Phosphorous	32	659	129	71	3.2	53.3	456	130.1	92.5	21	196.3	288	24.0
Potassium	78	1640	444	35	7.4	133.3	1440	451.7	405	52	488.3	71.2	52.4

4. CONCLUSIONS

1. Either existed compartment model or VPCM can be used for designing reasonable input level of each compartment, i.e., reclaimed land use unit, and adjusting the scale of each compartment according to the fixed input level and parameters of e&m flow. However, we must pay more attention to that the existed compartment model can not describe the actual state of the system and the result is qualitative.

2. VPCM is an improved model on the basis of the existed compartment model. In contrast to the existed compartment model, VPCM can reflect the actual state of e&m flows; for example, e&m flow between each two compartments and input of each compartment is variable with time. As a result, VPCM can reveal the nonlinear relationship between input and output of the system.

3. The results by applying VPCM to design Weizhuang reclaimed land use system are useful for instructing farmers' production in reclaimed land.

REFERENCES

Bian, Zhengfu 1997. Study on succession laws of essential factors of mined land reclamation interface and monitoring methods. Dissertation of Ph.D. degree submitted to CUMT.1997.10 (in Chinese)

Odum, H T 1983. *Systematic ecology*. John wiley & sons. (Translated from English into Chinese by Jiang Youxu, 1993 . Beijing: Science publisher.)

Liu, Tiebing & Han Cunlu 1988. Compartment model for materials recycling in rural ecosystem. *Journal of Rural Ecoenvironment*. 1988(4): 52-58 (in Chinese)

Niu Wenyuan 1984. Latest progress in physical geography. Beijing: Science publisher.

Mining Science and Technology' 99, Xie & Golosinski (eds) © 1999 Balkema, Rotterdam, ISBN 90 5809 067 1

Reclamation of surface coal and oil sands mines in Western Canada

Raj K. Singhal & Kostas Fytas
Department of Mining and Metallurgy, Laval University, Quebec City, Que., Canada

ABSTRACT: A sizeable surface mining industry exists in Western Canada. There are the surface coal mines in the Provinces of Alberta, British Columbia, and Saskatchewan producing the bulk of Canadian Coal. The two oil sands plants in Fort McMurray, Alberta producing synthetic crude oil are surface mining operations. The Western Canadian surface oil sands and coal mines have many special features due to the geology and location of mineral deposits. This paper discusses (1) the current surface mining methods in both of these industries, (2) the status of abandonment and reclamation of surface mines in both of these industries.

1. CANADIAN COAL: RESERVES, PRODUCTION & EXPORTS

Canada's total coal resources are estimated at approximately 320 billion tonnes. Of this, about 240 billion tonnes are considered to be unmineable by current technology at prevailing prices. The remaining 80 billion tonnes are defined as resources of immediate interest which can be exploited by using current technology. About 10 per cent of these 80 billion tonnes would typically be considered as proven reserves. One recent estimate places Canada's proven Coal resources at about 7 billion tonnes. There are 28 coal mines in Canada with combined thermal and metallurgical production of 78 million tonnes.

2. MINING METHODS

Mining methods used on all locations are governed by the prevailing geology at the mine site and as a result a brief mention of the overall geological structure is appropriate. Several Canadian provinces have coal resources but Alberta is estimated to have the largest reserves of bituminous and sub-bituminous coal. The province of Alberta is divided into three regions each containing deposits of similar age and characteristics. There are the Mountain, Foothills and Plains Regions.

The Mountain Region extends along the eastern flank of the Rocky Mountains. It contains mostly medium and low volatile bituminous coal deposits mainly of Upper Jurassic and Lower Cretaceous age. The coal of these deposits is usually friable. The coal bearing strata are highly folded and faulted which has caused considerable tectonic thickening and thinning of coal seams in certain locations.

The Foothills Region adjoins the eastern edge of the Mountain Region and contains mainly highly volatile bituminous coals of Upper Cretaceous and Lower Tertiary ages. Like their counterparts in the Mountain Region these coalfields have been sheared, faulted and folded, and in certain locations show considerable thickening and thinning of coal seams.

The Plains Region coal is mostly sub-bituminous in rank mainly of Upper Cretaceous and Lower Tertiary age. Most of the seams of this region are flat-lying except those close to the surface that have been glacially disturbed.

The coal deposits of south-eastern British Columbia (Crowsnest Coalfield) are Upper Jurassic in age and are part of the southern end of the Mountain Region as previously described for Alberta, with the notable difference that they contain considerably more seams of greater thickness than those in adjoining areas of Alberta. The coal deposits of north-eastern British Columbia are however Cretaceous in age and are the northern extension of the coals that occur in the Mountain Region of Alberta. Several small basins of Tertiary coal occur in the south-central part of the province including Tulameen Princeton, Merritt and Hat Creek. Most of these basins contain coal seams, which are intensely folded, faulted and irregular in thickness.

In the provinces of British Columbia and Alberta most surface coal mines are open-pit terrace operations. The operations use typical truck-shovel methods extracting multiple coal seams from several multi-bench work faces. In the Foothills and Plains Regions draglines are also used.

3. MINING OF OIL SANDS IN ALBERTA

All Canadian oil sands operations are located in the province of Alberta. In Canada the first commercial plant to extract bitumen from Oil Sands was installed in 1967 by Great Canadian Oil Sands Ltd. (now Suncor Inc.). The second commercial plant managed by Syncrude Canada Ltd. came on stream in 1978. Both plants are located in the city of Fort McMurray, Alberta. The bituminous sands that constitute the orebody at Suncor and Syncrude mine sites are contained within the Fort McMurray formation. Although both operating oil sands plants use the hot water process for extracting bitumen from the oil sands, different types of mining equipment are used (Singhal et al., 1999).

One of the most interesting recent developments in Canadian oil sands mining is the replacement of continuous mining systems – bucketwheel excavators/ conveyors with shovels/ trucks and crushers. Oils sands companies believe that recent technological advances in shovels and trucks make this the most economical mining system. Oil sands operations have recently introduced the largest truck (Komatsu Haulpack 930E) and the world's largest hydraulic shovel (O&K RH 400). Oil sands operations have also introduced the largest semi-mobile crusher (7,500 tph) in the industry. It incorporates a diamond shaped hopper that can handle two 320-t trucks dumping simultaneously. It is the largest double-roll crusher in the world and it is designed to accept lumps as large as 3.5 m. The new Syncrude's North Mine, that was officially opened in October 97, will be operating three of these crusher stations by the year 2000, when it will be at full capacity (Fair, Coward & Lipsett).

It is in the North Mine where Syncrude uses the hydro-transport system. Syncrude first tested the hydrotransport concept back in 1989. Hydrotransport is the process of combining oil sand and hot water to create slurry, which can be screened and subsequently pumped through centrifugal pumps and pipeline. In 1997 Syncrude succeeded with the start-up of the first North Mine Hydrotransport system. The crushed oil sand is fed to a surge facility where two apron feeders distribute the oil sand to the cyclofeeders. The cyclofeeders mix oil sand with hot water to create the required bitumen slurry. This slurry is screened and pumped through centrifugal pumps and a pipeline. This hydrotransport system is designed to average 6,200 tph with a yearly target of 42 Mt. Slurry is pumped over 5 km by four 2,500 hp slurry pumps to the extraction plant. New monitoring technology is continuously improving the way process variables are measured.

The Aurora Mine is another significant evolutionary step in this direction for Syncrude. Aurora will require an investment of $500 million to bring the first production train into operation in 2001. Aurora will have two open pits and four production trains with a production target of over 35 million barrels/year of bitumen. Within the Aurora site, 15 km of pipeline will be used to hydro-transport oil sand to the new Aurora extraction plant delivering tailings and returning water to the process. Aurora's development and progress with Suncor's Steepbank Mine are sure to stimulate further technological advances (Mining Magazine, 1998).

4. ENVIRONMENTAL GUIDELINES AND REGULATIONS

Several provinces in Canada have recently issued environmental guidelines and regulations relative to mine development. A few related publications are:

- Rehabilitation of Mines Guidelines for Proponents, Ontario Ministry of Northern Development and Mines (July 1992).
- A guide to the Review and Certification of Mine Developments, B.C. Ministry of Energy, Mines and Petroleum Resources and Ministry of Environment Lands and Parks (January 1992).
- Guidelines for Mineral Exploration: Environmental, Reclamation and Approval Requirements, B.C. Ministry of Energy, Mines and petroleum Resources (January 1992).
- Alberta Environmental Protection and Enhancement Act (September 1993).
- Bill 29 - Environmental Assessment Act Ministry of Environment, Lands and Parks, British Columbia (1994).

Similar acts and regulations have been

enacted in several other Canadian provinces. Considerable emphasis is on Environmental Assessment Process and Environmental Impact Assessment (EIA). The purpose of the Environment Assessment process as given in the Alberta's Environmental Protection and Enhancement Act is:

a) to support the goals of environmental protection and sustainable development.

b) to integrate environmental protection and economic decisions at the earliest stages of planning and activity.

c) to predict the environmental, social, economic and cultural consequences of a proposed activity and to assess plans to mitigate any adverse impacts resulting from the proposed activity, and

d) to provide for the involvement of the public, proponents, departments of the Government and Government agencies in the review of proposed activities.

In Alberta an Environmental Impact Assessment is mandatory for coal mines producing more than 45 000 tonnes per year; any oil sands mine regardless of production; any coal processing plant; and any oil sands producing more then 2 000 m^3 of bitumen or derivatives per day. Furthermore there are no exemptions for coal mines, oil sands mine, coal processing plants or oil sands plants.

The content required in an Environmental Impact Assessment report must be as follows:

a) a description of the proposed activity and an analysis of the need for the activity;

b) an analysis of the site selection procedure for the proposed activity including a statement of the reasons why the proposed site was chosen and a consideration of alternative sites;

c) an identification of existing baseline environmental conditions and areas of major concern that should be considered;

d) a description of potential positive and negative environmental, social economic and cultural impacts of the proposed activity, including cumulative, regional, temporal and spatial considerations;

e) an analysis of the significance of the potential impacts identified under clause (d);

f) the plans that have been or will be developed to mitigate the potential negative impacts identified under clause (d);

g) an identification of issues related to human health that should be considered;

h) a consideration of the alternatives to the proposed activity, including the alternative of not proceeding with the proposed activity;

i) the plans that have been or will be developed to monitor environmental impacts that are predicted to occur and the plans that have been or will be developed to monitor proposed mitigation measures;

j) the contingency plans that have been or will be developed in order to respond to unpredicted negative impacts;

k) the plans that have been or will be developed for waste minimisation and recycling;

l) the manner in which the proponent intends to implement a program of public consultation in respect of the undertaking of the proposed activity and to present the results of that program;

m) the plans that have been or will be developed to minimise the production or the release into the environment of substances that may have an adverse effect.

5 RECLAMATION SUCCESSES IN COAL AND OIL SANDS MINING

According to the Coal Association of Canada, Alberta's sub-bituminous coal mines have an outstanding record in achieving successful reclamation. The goal of Alberta's coal producers is to restore all land that has been mined to a state "equal to or better than" that which existed before mining began. To date, over 62 per cent of the land that has been disturbed by sub-bituminous coal mining has been completely reclaimed, with a large proportion of the remaining land currently in the process of being reclaimed. Alberta's coal producers return mined areas to productive uses as quickly as possible; in most cases this involves agricultural uses. This rapid restoration of land means that less than *one-tenth of one per cent* of Alberta's agricultural land base is out of production due to coal mining at any one time, and mined lands are normally out of agricultural production for less than four years.

Similar observations apply to the reclamation at Canadian Oil sands operations (Tuttle et al., 1999). For example at Syncrude Canada where land reclamation efforts have traditionally focused on reforestation several innovations have been introduced. For example:

- For thousand of years the Wood Bison was an integral part of the Northern Alberta landscape and the lives of the region's Native peoples. Until they all but vanished almost a century ago. In 1993, a small herd of these still scarce animals were reintroduced to the area as part of a research project to determine the feasibility of a Bison grazing range on mine land reclaimed by Syncrude (Pauls,1999). This five year research project by Syncrude and the Fort McKay First Nations has demonstrated the feasibility of raising bison on land reclaimed after disturbance by oil sands mining.

- During the construction phase of Syncrude's development a special team of environmental scientists, engineers and construction specialists was formed to oversee the diversion of an existing creek away from the mine area and the creation of a new "natural area" to replace it. The diversion of Beaver Creek cost approximately $ 35 million and created a new water habitat that quickly attracted aquatic plant, insect, fish, and animal species normally found in northern Alberta lakes.

- As a part of its comprehensive land reclamation program, Syncrude plants a wide variety of trees, grasses and shrubs each year. Most of the tree species Syncrude replants including white spruce, jackpine and aspen are indigenous to the region. However, exotic species such as Siberian larch are also planted. These trees are both hardy and well adapted to the harder northern climate.

On the hand Suncor's reclamation program has been focusing on the following objective: " Disturbed lands shall be reclaimed with gentle slopes to primarily a forest use compatible with predisturbed terrain, providing a range of end uses including wildlife, traditional use and recreation; and dyke slopes shall be revegetated primarily for erosion control providing natural-end-use possibilities" (Nix, 1997). Here are two examples of Suncor's reclamation innovations at the Steepbank mine:

- Refugia or "islands of intact natural ecosystems are introduced within the larger development areas in order to increase the recolonisation rate of reclaimed landscapes. These refugia serve as sources of seed for native plant establishment and assist in speeding recolonisation of reclaimed areas by amphibians, birds, small mammals and hundreds of species of invertebrates that exist in forest soils.

- Wildlife corridors are maintained between the mining area and the neighbouring Athabasca River in order to mitigate concerns regarding wildlife movements.

Prior to opening a mine in most provinces in Canada a formal application is to be filed with appropriate authorities. A publication entitled "Health Safety and Reclamation Code for Mines in British Columbia" issued by the Ministry of Energy, Mines and Petroleum Resources deals with the procedure in British Columbia. The province of British Columbia in this regard provides a role model.

CONCLUSIONS

The concept of sustainable development dictates that environment is not to be sacrificed simply to achieve economic benefits. Canadian mining industry recognises this and has taken various steps to fulfil its obligations. At the mine-site the mining engineers have had to revise their work on mine planning by keeping abreast of legislation.

On the technical front, Canadian mining industry particularly the open pits have managed their environmental affairs very well. Reclamation programs, rehabilitation of abandoned mines, stability of geotechnical structures, water quality, have all received their due share of attention. Environmental Impact Assessments encourage mine developers to make environmental concerns a part of the process of their planning.

In Canada, an excellent consultative process is in place between the federal and provincial governments and the industry. Industry has participated in discussions, which have resulted in various pieces of legislation mentioned in this paper including those related to reclamation.

REFERENCES

Singhal, R.K. & Fytas, K. 1999, Developments in Canadian Open Pit Mining, *Mining Engineering*, February, Vol. 51, no. 2, pp. 26-30.

Fair, A., Oxenford J & Lipsett M. 1998, Technology visions for mining at Syncrude, *Proceedings of the Mine Planning & Equipment Selection (MPES) '98*, Raj K. Singhal, editor, Balkema Publisher, pp. 453-467.

Mining Alberta's oil sands, *Mining Magazine*, August 1998, Volume 179, No.2, pp. 58-73.

Tuttle,S. & Sisson, R. 1999, Closure Plan for the proposed Millenium Project, *CIM Bulletin*, Volume 92, No. 1026, pp.95-100.

Pauls, R.W. 1999, Bison and the oil sands industry, *CIM Bulletin*, Volume 92, No. 1026, pp. 91-94.

Nix, P. & Gulley, J. 1997, Benefits of Suncor's Conservation and Reclamation Plan for the proposed Steepbank Mine, *Proceeding of the 99th CIM Annual General Meeting*, Vancouver, CD-Rom .

Mining Science and Technology' 99, Xie & Golosinski (eds)© 1999 Balkema, Rotterdam, ISBN 90 5809 067 1

Comparison of different soil reconstruction methods for reclaiming contaminated lands in Geevor mine

K. Atkinson & P.H. Whitbread-Abrutat
Cambourne School of Mines, University of Exeter, UK

Zhenqi Hu
China University of Mining and Technology, Beijing, People's Republic of China

ABSTRACT: The presence of land contamination is an inevitable legacy of an industrial past. There are about 100 000 and 220 000 ha of land is contaminated in the UK. To reuse the contaminated land, some treatments such as soil reconstruction are useful. This paper evaluated three treatments of soil reconstruction in Geevor mine based on soil tests. The results revealed that copper pollution was the most serious in this mine site. The plot with the treatment of topsoil removal was still of high contaminants and low pH. Remedied plot by industrial mineral amendments could removal some of the contaminants and buffer pH effectively, but it still had high level of metals. Subsoil cover plot showed the lowest content of contaminants, which was less than the threshold value. The remedied plot with industrial mineral amendments and subsoil cover plot have higher macro nutrients for plant growth. Therefore, subsoil cover and mineral amendments are effective methods in reclaiming contaminated lands.

1 INTRODUCTION

The UK has a long industrial history and many sites have been damaged as a result of their former use. The presence of land contamination is an inevitable legacy of an industrial past. There are about 100 000 and 220 000 ha of land is contaminated in the UK, representing between 0.4 and 0.8% of the total UK land area (POST 1993). Cornwall in particular shows the legacy of past mining activity, with 3871 ha of land associated with metalliferous mining and processing. High contents of heavy metals always exist in the land, which may pose significant risks to human health or the environment. Areas of contaminated land are normally visually unattractive, ranging from totally barren at one extreme to nearly complete vegetative cover at the other. Most areas, however, lie somewhere between. Therefore, reclamation of these contaminated lands has become an important task in Cornwall.

The general characteristics of contaminated site are however fairly easy to predict acidic pH values, high toxic metal availability, poor drainage qualities, lack of nutrients. Removing the contaminated soil layer to somewhere and use of uncontaminated soil from other place to cover the contaminated land are two typical treatments for remediating contaminated land. They are called treatments "Topsoil removal "and "soil cover". Use of industrial minerals as

amendments is a new method to remediate contaminated land, having being done by CSM (Camborne School of Mines) for several years (Mitchell and Atkinson 1993). These methods have been applied to the reclamation in Geevor mine, Cornwall in recent years. But the comparison of these methods have not been done so far.

2 DESCRIPTION OF EXPERIMENTAL SITE AND METHODS

Geevor mine is situated between the villages of Pendeen and Trewallard on the Lands End, Peninsula, south-western corner of the county of Cornwall in England. It is 2 miles from St. Just and 7 miles from Penzance, the nearest commercial and industrial centre of the area. It was a tin mine closed in 1990, the last survivor of an industry that had worked hereabouts for well over two-thousand years. Though the exact date when mining started is still unknown, it is thought to have reached its peak in the 1860's, producing tin and copper and a lesser extent of arsenic, lead and zinc. The mine site lies at about 60m above sea level in a gently sloping ground, along the incised and rugged coastline. The coastline is exposed to tremendous amount of erosion due to the beating action of the sea, and the land is swept by strong winds blowing from the sea.

Three remediated plots and one comparison plot as reference were selected, the details of each plot are described as followings based on field survey:

Plot A: unremedied contaminated soil as reference
(1) about 20~25 degree slope
(2) surface description: dark (high organic matter), many dead roots and fibrils, loose
(3) no vegetation

Plot B: Topsoil removed
(1) 15cm contaminated topsoil was removed in 1994
(2) surface description: bare, stone, dark (high organic matter content), some dead roots
(3) 20~25 degree slope
(4) no vegetation

Plot C: Remedied soil (Montmorillonite 1:30)
(1) remedied by using of industrial minerals (Montmorillonite) as amendment in September 1995
(2) about 20~25 degree slope
(3) surface description: dark (high organic matter), many dead roots and fibrils, loos
(4) some vegetation with about 50% of coverage

Plot D: subsoil cover
(1) subsoil covered on the contaminated land, the subsoil was taken from the Penzance

(2) about 20 degree slope
(3) surface description: yellow, soil and stone, less organic matter, more clay than plot B and C, few vegetation

To compare the effectiveness of these remediation treatments, the main physical and chemical properties of soils were tested. The toxic metal contents and pH were the focus of this study.

3 RESULTS AND DISCUSSIONS

3.1 Stone content and soil texture

The particle size distribution was obtained by using of sieving method (see table 1 and figure 1). The results show that Plot D (subsoil cover) has the highest stone content and Plot A (reference plot) has the lowest stone content (using a 10 cm sieve (Rowell 1994)).

3.2 Soil Moisture Content

Soil moisture content is an important parameter for plant growth. The tested results (see figure 2) revealed that the plot D (subsoil cover) has the lowest moisture content and plot A (contaminated soil as reference) has the highest moisture content, which is coincidence to the stone content.

Table 1. Particle Size (PS) Distribution and stone content: (Unit: %)

	plot A	plot B	plot C	plot D
PS≥10mm	2.91	5.44	6.96	30.28
4mm≤PS<10mm	5.77	8.37	4.26	18.47
2mm≤PS<4mm	9.42	12.70	12.91	9.36
0.4mm≤PS<2mm	38.34	33.63	40.65	30.26
0.075mm≤PS<0.4mm	28.23	25.01	23.93	11.62
PS<0.075mm	15.33	14.85	11.28	7.43
Stone content: (≥10mm)	2.91	5.44	6.96	30.28

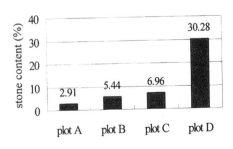

Figure 1. Comparison of stone content among the selected plots

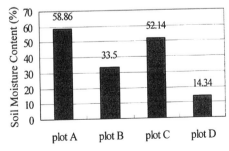

Figure 2. The mean soil moisture content

3.3 Chemical properties and contaminants

3.3.1 pH and EC

Soil pH and electrical conductivity are important chemical properties of soils. The tested results (see table 2 and figure 3) showed that contaminated soils were very acid, the pH value were about 3 and the reference plot (Plot A) had the lowest pH value. Topsoil removal plot still had the lower pH value, which revealed that this remediation method is not effective. Subsoil covered plot had the best pH value for plant growth and remedied plot by using of industrial minerals (plot C) had also better pH values, which shows the beneficial from the remediation.

Reference plot (plot A) had the highest soluble salt content. But all the EC values are suitable for plant growth.

3.3.2 heavy metal elements

Many of contaminated lands are polluted by heavy metals. Heave metal elements in soils of the selected plots were analysed by AAS (see table 2). The results showed that the copper pollution in contaminated soil (plot A) is very serious (see figure 4), which is more than nine times of the threshold value (ICRCL 1987). The subsoil cover plot was the only one without contamination. Remedied plot by using of industrial minerals could remove Fe and Pb distinctly while it removed Cu and Zn slightly, but it still contained high levels of Cu and Fe.

Iron in soils is an important indicator of soil acidity. High level of iron usually results acid soils. Untreated plot (plot A) had the highest content of iron (see figure 3), which induced the most acid soil. Removal of contaminated topsoil could reduce the amount of iron tremendously, about half of the original amount of iron, but it still contained very high level of iron and had lower pH value. Subsoil cover could let the contaminated soil become normal effectively (with normal level of iron and pH value). Using the industrial minerals as amendments (plot C) could removal the iron from contaminated soil, even slight better than the topsoil removal treatment, but it still had higher level of iron. Normally, this plot (plot C) was also acid, but the tested pH value was 5.26. This is mainly because that the amendments decreased ability of removing metal cations from more acid solutions and increasing the competition for exchange sites by H^+ ions, which revealed that the use of minerals amendments could decontaminate soil pore water and buffer the pH to more neutral values. The minerals containing high level of Ca (see table 2 in plot C) might be the key reason to buffering the pH.

Table 2. Chemical properties and heave metal element contents

	plot A	plot B	Plot C	plot D	Threshold
Cu (μg/g)	1088.675	838.83	1021.93	37.74	130
Zn (μg/g)	14.79	3.39	12.10	7.25	300
Fe (μg/g)	4186.46	2160.41	1910.54	174.54	
Mn (μg/g)	3.89	8.06	46.36	166.33	
Pb (μg/g)	166.52	47.24	68.47	28.62	500
Cd (μg/g)	0	0	0.05	0.1	3
Ca (μg/g)	80.15	45.42	2921.25	1514.65	
Mg (μg/g)	35.63	29.65	117.31	167.48	
K (μg/g)	78.16	85.17	142.15	200.04	
pH (μg/g)	3.34	3.4	5.26	6.94	Soil:Water =1:5
EC (μscm^{-1})	227	122	97	164	

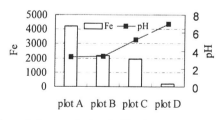

Figure 3. Iron content and pH value

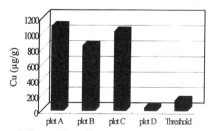

Figure 4. Copper in soils

Subsoil cover plot and the plot remedied by minerals had higher macro-nutrients for plant growth and much more vegetation on the surface than the other plots. The plot with mineral amendments had the largest vegetation coverage while the reference plot and topsoil removal plot almost had no vegetation. This result showed that the mineral amendments could accelerate natural revegetaion of metal contaminated sites.

4 CONCLUSIONS

Based on the comparison of different remediation treatments, some conclusions could be summarized as followings:

(1) Copper pollution was the major problem in Geevor mine sites and the Cu value was more than nine times of the threshold value.

(2) Topsoil removal method could removal some metals, but it still had high level of metals and low pH. Thus this treatment is not effective. Furthermore, the treatment of removed contaminated topsoil is another difficult work.

(3) Subsoil cover treatment was the only one reclaiming a desired land without contamination, containing low level of metals, neutral pH value and higher macro nutrients. High stone content is the problem in this treatment.

(4) Remediation with industrial minerals could remove Fe and Pb distinctly while it removed Cu and Zn slightly, but it still contained high levels of Cu and Fe. Some good results such as nearly neutral pH value, higher macro nutrients and higher degree of vegetation coverage have been achieved. Thus, the major benefits from this remediation were buffering the pH value and accelerating natural revegetation.

(5) Different soil reconstruction treatments have shown different properties of soils. Subsoil cover and mineral amendment are effective methods in reclaiming contaminated lands.

ACKNOWLEDGEMENT

This paper is supported by National Natural Science Foundation of China and partial support by the UK Royal Society Fellowship.

REFERENCE

International Committee on The Redevelopment of Contaminated Land. 1987. *Guidance on the Assessment and Redevelopmet of Contaminated Land: ICRCL 59/83*. ICRCL, 2nd end.

Mitchell, P.B & K. Atkinson. 1993. The possible role of industrial minerals and mineral processing in contaminated land rehabilitation. *Land contamination and reclamation*, 1(2):111-119

Parliamentary Office of Science and Technology (POST), 1993. *Contaminated Land*, London:POST

Rowell, D.L. 1994. *Soil Science:Methods & Application*. England: Longman Scientific & Technical

Mining Science and Technology' 99, Xie & Golosinski (eds) © 1999 Balkema, Rotterdam, ISBN 90 5809 067 1

Decision making process concerning the mine closing from the chronological point of view

J. Dvořáček

Institute of Economics and Control Systems, VŠB, Technical University of Ostrava, Czech Republic

ABSTRACT: Mine closure is caused by reserves depletion or by economic conditions what is the case of the Czech Republic in last years. The decline of demand for coal let to closure of many mines and production damping. The closure was performed in two ways: 1. rapid variant: without utilization of prepared coal reserves, with fast finishing of underground workplaces liquidation (disassembly of some machines only for ecological reasons), with rapid liquidation or sale of surface objects. 2. slow variant: with prepared coal reserves utilization, with extensive dismantling of machines and equipment, with withdrawal of support, with gradual liquidation of separated parts of mine and surface buildings and their alternative utilization. Each of both variants has its advantages and disadvantages. The contribution deals with the method of mining activity damping variant on the basis of economic criteria.

1 INTRODUCTION

The orientation of the Czech Republic in last years towards the market economy was connected with many substantial changes that affected considerably also the mining industry. In the course of some years there came to the extinction of home ore mining, to substantial limitation of uranium production with the perspective of closing of all mining capacities and also to limitation of coal mining.

This development, in the main, corresponds to long-term cycle of entrepreneurial activity in mining industry - development and damping of mining activity accompanied the mining industry during the whole historical development. The limitation of mining was the consequence of two groups of impacts acting:

1. Natural conditions that could not be got through by the mining and coal preparation technology of that time (including the reserves depletion).

2. Economic conditions when the deposit exploitation was ineffective with the respect of costs and revenues of mining activity. This situation concerns also the underground coal mining in the greatest hard coal deposit of the Czech Republic - in the Ostrava-Karviná district.

2 DAMPING OF UNDERGROUND COAL MINES IN THE CZECH REPUBLIC

The Ostrava-Karviná district was the greatest producer of hard coal mined in underground way in former Czechoslovakia and remains the most important deposit also nowadays, although the coal output has decreased substantially. The greatest coal output in the district history was reached in the year 1979, and namely in the level of 24,8 mil. tones per year. The economic activity of the organizational unit - the Ostravsko-karvinské doly (OKD) /Ostrava-Karviná Mines (OKM)/ - mining coal in the Ostrava-Karviná district has come through, after the year 1989, the important restructuralization changes. These changes have led, in the production sphere, to relatively considerable decline of coal output because of the fact that the base of restructuralization was damping of ineffective mines and mining concentration into more productive deposit parts.

As the mining damping the organized mining decreasing or stopping connected with possible following liquidation of ineffective mine or its determined ineffective part is identified. The final decision concerning the concrete mine damping makes the government of the Czech Republic because the substantial part of costs connected with the mining damping resp. with mine closure is settled from the so called non-investment appropriation from the state budget. In addition to the appropriation from the state budget (there is no legal claim for this appropriation) the damping of mining activity can be financed from the reserve created according to the Mining Law, from own means of the mining organization or yields of mining damping (e.g. yields of scrap sale, of coal sale after the mining damping announcement etc.).

From the chronological point of view the damping of mining activity runs in the following stages:

Preparatory stage of mining damping: it includes the working out of projects and appurtenant documents and their approval.

The 1st stage of mining damping: it takes the time since mining damping announcement till mining activity completion.

The 2nd stage of mining damping: it takes the time since mining activity completion till completion of liquidation or securing of main workings.

The 3rd stage of damping: it takes the time since completion of liquidation or securing the main workings mouthing on surface, its end depends on surface behavior after the completed liquidation or securing of a mine and at its possible flooding.

Still in the preparatory stage of mining damping two marginal variants of damping can be taken into account:

The rapid variant: without utilization of coal reserves technically prepared for extraction, with rapid completion of liquidation or securing works, it means - without workings robbing, with clearance of the most necessary machines, equipment and material only for ecological reasons (e.g. due to oil substances content), without alternative utilization of workings for other purposes (e.g. depositing of flue ash in underground), with rapid liquidation or sale of surface buildings.

The slow variant: with utilization of coal reserves technically prepared for mining, with larger robbing of workings and clearance of machines, equipment and material, with gradual liquidation or securing of single areas of a mine, with the construction of greater number of closing barriers (stoppings), resp. with alternative utilization of workings for other purposes and with gradual liquidation of surface buildings (objects).

In conditions of the Ostrava-Karviná district the following mines were damped: the Jan Šverma Mine, where the damping was announced in October 1991 and the mining was stopped in December 1991, the Ostrava Mine with the damping announcement in

November 1991 and production finishing in December .992, the Heřmanice Mine where the damping was announced in January 1993 and the mining was stopped in June 1993, the Odra Mine with the damping announcement in January 1994 and mining finishing in June 1994, the Paskov Mine where the damping was announced in January 1998 and the production was finished in December 1998. The specific case is the Julius Fučík Mine where the partial damping was announced already in the year 1992, the full damping in January 1995 but it came to the stopping of mining in March 1998.The duration of the 1st stage of mining damping then fluctuated from 3 up to 39 months.

The analysis of statistic indicators leads to the conclusion (Dvořáček 1998) that very short duration of the 1st stage of mining damping has led to the inefficiently driven meters (footage) of preparatory workings, to substantial decrease of important indicators of mining concentration which are the average daily output from one working face and average mined area in m^2 per day. In the contrary, the extension of the duration of the mining damping 1st stage caused the improvement of concentration indicators, especially the average daily output from one working place and further, the increase of seams thickness mined, i.e. putting through the selective mining method - mining in geological conditions enabling the reaching of relatively best economic results.

In addition, it can be presupposed a priori that the rapid variant of the mining damping 1st stage decreases the costs for operation of costly power consumers (i.e. compressors, ventilator fans, pumps etc.), however, it can be the source of social problems for the reason of great number of employees releasing. The disadvantage is also the impossibility of property relations solving in the sphere of pieces of land and other mining enterprise assets. The solution of these relation have to be transferred on other juridical subject.

It can be stated then that the duration of the mining damping 1st stage has significant economic consequences. These consequences should be taken into account at projecting of mining damping in further mines.

3 ECONOMIC CRITERION AT SELECTION OF THE MINING DAMPING VARIANT

At the decision making concerning the selection of mining damping variant from the enterprise point of view let us base on the following variants:

The 0th variant: It is the question of basic mine state with negative economic result, the mining would run then as up to now. The economic result following from this activity can be characterized:

$$HV_{(0)} = T_{(0)}.z_{(0)} \tag{1}$$

where $HV_{(0)}$ = annual economic result of the mine in the framework of the 0th variant, $T_{(0)}$ = annual mine output in the framework of the 0th variant, $z_{(0)}$ = economic result falling to one ton of output share in the framework of the 0th variant.

Taking into account the analyzed problems it is evident that the economic result per one ton of output and the economic result of the whole mine as well is negative one.

The 1st variant: it is the question of the rapid damping variant, i.e. very quick completion of mining with non-utilization of all coal reserves prepared for mining, with clearance of machines and equipment only for ecological reasons without yields of mining damping. In addition, it can be presupposed that one part of costs for damping must be settled by enterprise from its own financial means. The economic result of this variant can characterized as follows:

$$HV_{(1)} = -N \tag{2}$$

where $HV_{(1)}$ = annual economic result of the mine in the framework of the 1st variant, N = costs for damping settled from mining enterprise sources.

The 2nd variant: it is the question of slow damping variant, i.e. longer period of decreasing coal output the sale of which creates the most important yield of damping, with the possibility of machines clearance and sale, with robbing of long workings and scrap sale. In the framework of this variant the decreasing of variable costs per 1 ton of output can be presupposed (especially due to stopping of preparatory workings driving) and decreasing of fixed costs as well (decreasing of main ventilator fan capacity, stopping of unnecessary property depreciation, decreasing of workers number etc.). The economic result of this variant can be characterized as follows:

$$HV_{(2)} = T_{(2)}.z_{(2)} + HV_{\acute{u}} = T_{(2)} \left(c - nv_{(2)} - \frac{NF_{(2)}}{T_{(2)}} \right) + HV_{\acute{u}} \tag{3}$$

where $HV_{(2)}$ = annual economic result of the mine in the framework of the 2nd variant, $T_{(2)}$ = annual output of the mine in the framework of the 2nd variant, z_2 = economic result per one ton of output in the framework of the 2nd variant, $HV_{\acute{u}}$ = economic result following from the yields of mining damping, c = price of one ton of coal, $nv_{(2)}$ = variable costs per one ton of coal, $NF_{(2)}$ = total fixed costs in the framework of the 2nd variant.

In case we compare both possible variants, i.e. the rapid or slow one, with the basic state, we can set the conditions when, at realization of certain variant, the mining enterprise reaches the economic effect compared with the basic state, then, compared with the 0th variant.

The comparison of the 1st and 0th variant: economic effect arises for the mine, if:

$$HV_{(1)} > HV_{(0)} \tag{4}$$

$$-N > T_{(0)} \cdot z_{(0)} \tag{5}$$

The rapid variant will mean benefit for the enterprise if the costs for damping which are settled by the mine from its own sources are, in absolute value, lower than the absolute value of loss following from the mining activity going on. The experience up to now from the mining damping in conditions of the Ostrava-Karviná district where more than 90 % of costs for damping were settled from the appropriation from the state budget show that the relation (5), at the realization of the rapid variant, is valid. However, the problems arise in the social sphere, at solving of property relations, in ecological sphere.

The comparison of the 2nd and 0th variant: economic effect arises for the enterprise, if:

$$HV_{(2)} > HV_{(0)} \tag{6}$$

$$T_{(2)} \left(c - nv_{(2)} - \frac{NF_{(2)}}{T_{(2)}} \right) + HV_{\acute{u}} > T_{(0)}.z_{(0)} \tag{7}$$

If the presuppose the knowledge of price, variable costs per one tone of output, total fixed costs and economic result from the damping, it all in the framework of the 2nd variant, we can calculate, from the relation (7), the output height which will ensure the economic effect to the mining enterprise during the slow damping variant realization.

For conditions of one unprofitable mine in the Ostrava-Karviná district the simulation calculation was carried out where it was presupposed as follows: the same coal price in the framework of the 0th and 2nd variant too, decrease of variable costs per 1 ton of output by 22 % in the framework of the 2nd variant and decrease of total fixed costs by 6 % in the framework of the 2nd variant. The calculation showed that in case of such costs decrease the mine overcame the loss at the output which was even under the output level in the basic state, it means, in the 0th variant. The presumption for it is represented, however, by sufficiently prepared coal reserves and the possibility of fixed costs and also variable costs decreasing.

It can be presupposed then that the closing of an unprofitable mine will be economically effective for the mining enterprise both at realization of rapid and slow variant as well. The selection of the concrete variant must be supported by another economic criterion, and namely, by attainability of appropriation from the state budget for covering of mining damping costs and by the amount of this appropriation.

The allotment of appropriation for damping of the concrete mine demands the decision on the level of the Czech republic government, in addition, the amount of this appropriation does not cover all costs for mining damping. Under this situation the importance of yields of mining damping increases: the receipts for sale of coal, unnecessary machines, scrap, receipts for services sale etc. The practical experience shows, however, that the majority of these activities is ineffective (costs for scrap obtaining are higher than its market price makes), realization of these activities is connected with problems (small number of interested persons in purchase of mining machines), these activities are limited by ecological barriers (yields of waste depositing in underground of mines) or are not substantial from the quantitative point of view. The only important benefit in the time period of the 1st mining damping stage which can decrease the necessary amount of appropriation from the state budget are the yields of coal sale which is mined in the course of the 1st mining damping stage.

4 CONCLUSION

On the basis of previous considerations and practical experience in the sphere of mines closing in the Ostrava-Karviná district for selection of the concrete variant in the framework of the 1st mining damping stage the following can be recommended:

1. The selection of the rapid variant: under the situation of the appropriation from the state budget attainability and its sufficient height, in case that the possibility of selective mining is limited, the sufficient quantity of coal reserves is not prepared, it is relatively easy to employ the released workers from the mine being damped, it is not possible to decrease variable and fixed costs connected with mining process sufficiently.

2. The selection of slow variant: under the situation when the appropriation from the state budget allotment is not certain or its height is not sufficient, the mining damping can be prepared with sufficient time advance, it is possible to ensure the sufficient quantity of prepared coal reserves, it is possible to realize the selective mining, it is possible to decrease the variable and fixed production costs. The term „sufficiently" can be concretized with the using of equations mentioned in previous chapter.

Financial means expended for the mining damping in the Ostrava-Karviná district are very high and are permanently increasing because the costs for covering of mining activity consequences and social costs must be expended still long time after mine closure and its technical liquidation. It can be presupposed then that in the future we can await the realisation of the mining damping slow variant. It indicates also the duration of the 1st mining damping stage of closed mines in chronological succession from the year 1991 till the year 1998 that have the increasing trend.

This contribution was realised thanks to the support of grant from GAČR No. 105/99/0080.

REFERENCES

Dvořáček, J. 1998. K problematice útlumu uhelných dolů v Ostravsko-karvinském revíru. *Acta Montanistica Slovaca*, vol. 3, No. 1: 85-88

Dvořáček, J. 1998. Management of underground mine economics in the period of production decrease. *Proceedings of the Seventh International Symposium on Mine Planning and Equipment Selection, Calgary, Canada, October 6-9, 1998, A.A. Balkema, Rotterdam, 373-376*

Šnapka, P., Dvořáček, J. 1989. K ekonomické efektivnosti exploatace zásob vázaných pod zvodněnými horizonty. *Uhlí, 37: 394-400*

7 Mine control, automation and mechanization

Mining Science and Technology '99, Xie & Golosinski (eds) © 1999 Balkema, Rotterdam, ISBN 90 5809 067 1

Advanced computer techniques: Developments for the minerals industry towards the new millennium

B. Denby & D. Schofield
SChEME, University of Nottingham, UK

ABSTRACT: This paper provides a critical review of the developments and the state of the art of advanced computer techniques, including:
- Information Technology (IT)
- The Internet
- Computer Aided Design (CAD)
- Geographical Information Systems (GIS)
- Simulation
- Artificial Intelligence (AI)
- Computer Graphics (CG) and Virtual Reality (VR)

This paper is not intended to cover all technology of relevance to the mining industry, but to provide a brief snapshot of current technology and insight into the future in these areas.

1 INFORMATION TECHNOLOGY (IT)

Computer development is driven by the needs of the user. Computer hardware has been doubling in speed approximately every eighteen months for decades now. This increase in speed has led to a radical change in software development with modern software expanding to include an never ending set of 'add on' features. A word processor may have some image manipulation and spreadsheet capabilities, while the spreadsheet itself may offer word processing and also provide database functionality.

As more computer capacity becomes available, users attempt to extract ever greater performance from their existing systems. Expectations grow and this creates pressures to improve hardware performance to meet the increased expectations. As hardware performance grows software develops to use the increased power and the upward performance spiral continues. This drive for computer power thus fuels the development of better and faster systems.

It would be foolish to quote the specifications of a modern personal computer and speculate about the future, as in years to come anyone reading this will smile at the predictions and look back nostalgically at the technology described. However, two recent developments are worth mentioning Silicon Graphics, who now own Cray, have announced the

T3E series of supercomputers which use up to 2048, liquid cooled, 600Mhz, processors with up to 2 GB of memory per processor (Silicon Graphics, 1999). Also Princeton University last year announced a 500 gigabyte 'CD ROM' which is the size of a five cent piece (Whittle, 1998), this is based on Stephen Chou's 49-square-nanometer "dot" gate technology which uses a single electron to switch it on and off (InSCIght, 1997)

Figure 1. A Cray TE3-1200 system

The mining engineer at head office was quick to embrace the digital computer as it became available, but it was not until the 1980s when PCs came into their own, that mining engineers started to make daily use of computers. As computer technology developed, each advancement would be greeted with a headline or sales pitch that assured the engineer that "the following problems can now be solved.... ". All that happened was that the engineers posed new, more complex problems that needed a new generation of more powerful computers in order to obtain solutions (Sturgul, 1999). This is still the case, as the computers get more powerful, the users redefine their problems to use the capabilities of the technology.

The mining engineer's demands for new hardware and software facilities go beyond simple advancements of what already exists. Computer technology has been introduced which has altered the fundamental way certain parts of the mine planning and operation phases are undertaken.

Computerised databases store exploration data, which was itself often recorded using electronic surveying and monitoring equipment. Computer interpretation of exploration data, such as geophysical logs, is necessary due to the enormous amounts of data being processed. Most orebody modelling is now undertaken using advanced geological software capable of manipulating the three dimensional data sets to give realistic representations of the position of the mineral and associated material. Financial software, primarily spreadsheet based, is used to assess the viability of a proposed extraction and many smaller financial decisions will revolve around the spreadsheet calculated cost implications. When the mine is finally operating some equipment will be computer controlled, most will have components which rely on silicon based technology to optimise their operation and ensure safe working. Automatic or tele-remote operation of mining equipment is becoming more commonplace in the industry.

The future of computing in the minerals industry is not always led by the demands of the mining engineer, but sometimes by the academics and commercial software developers who come up with new ideas and products which may change the way certain traditional operations are performed.

2 THE INTERNET

In 1969, the *Internet* was originally founded as a government project to network military bases and missile silos. This suite, or agreed upon language of communication, allowed thousands of diverse computer networks to co-operate and collectively form a larger computer network. Currently, the Internet is a rapidly expanding global communications network with over 30 million daily users in over 120 countries. It has a world wide growth rate of over 1,000 new users per day. Commercial activity now accounts for nearly two-thirds of Internet traffic. This network is available to

any of its users, and the majority of the information on the Internet is available to *all* of its users. In a very short space of time, the Internet has affected the lives of millions of people. This network of networks is still in its infancy however.

WWW stands for "World Wide Web". The WWW project, started at CERN (the European Laboratory for Particle Physics), tried to build a "distributed hypermedia system". In practice, the web has become a vast collection of interconnected documents, spanning the world. Searching the Internet for these documents can be *very* frustrating! Information overload is here to stay and even the search engines may not help you to find the information you really need. Manual or undirected searches for specific items can use up a lot of time. A number of intelligent search mechanisms are being developed to assist the directed searcher to look for specific information.

Alan Sugar, the chairman of Amstrad Corporation, in a recent article gave a very good example of the hype surrounding the Internet. In his example he asked us to consider two students, one with a PC connected to the Internet and one in a university library, both were asked to obtain information for their chemistry examination on copper, who would get information first? Alan Sugar states "the bookworm would win hands down. The information on copper is in the library, always has been there, always will be" (Sugar, 1999).

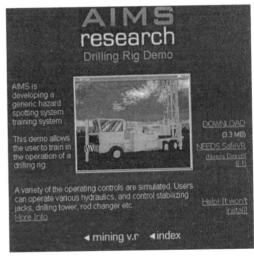

Figure 2. AIMS site (www.nottingham.ac.uk/aims)

In the mining industry the use of the Internet as a means of communication has been as widespread as in every other industry. The Internet and e-mail has fundamentally altered the way engineers interact, communicate and hence work with each other. Web pages are used for publicity and information, a page from the AIMS web site at Nottingham is shown in figure 2. In fact much of the information for this paper was found on the Internet.

In the future we are likely to see more training and testing facilities being delivered over the Internet. An example of this is the Mine Emergency Response Interactive Training Simulation (MERITS), being developed by NIOSH in Pittsburgh. This will be an interactive multimedia computer simulation, delivered via the Internet, of an underground mine that is undergoing some type of emergency. It will simulate both underground and surface activities at the mine site, and provide a means to inform users (command center trainees) of those events and allow them to attempt to resolve the emergency.

MERITS will run on two computers simultaneously - the host PC and the local PC. These two computers will communicate by a local area network (LAN) or Internet connection. The host PC will perform the simulation of the underground and surface activities. The host PC (at a training centre perhaps) will pass relevant information concerning those activities to the local PC (at a remote mine site perhaps) which will communicate with the trainees (NIOSH, 1997).

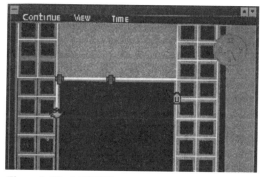

Figure 3. The MERITS system

The Internet has also altered the dissemination of academic information. Conferences now take place in cyberspace, with the papers being 'published' as web pages with active content. In 1996, the first international mining symposium on mine simulation, MINESIM '96, was held on the Internet. This was co-sponsored by the National Technical University of Athens, Greece, and the University of Idaho, USA. There were about 200 papers presented and over 2,500 people visited the WEB site that was maintained in Greece. Such a format allowed people in remote areas from around the world to "attend" the symposium and visit various other web sites that were linked to the main site (Panagiotou and Sturgul, 1996).

3 COMPUTER AIDED DESIGN (CAD)

CAD technology allows for the drafting of engineering design using computer equipment. Drawings are represented in the computer by point coordinates and vertices. More advanced CAD applications will allow for the construction of three-dimensional models of nearly any object.

Those who have watched the CAD industry since the introduction of AutoCAD in the early 1980s know that as we approach the last few years of the 20th century, the future for CAD software is very uncertain. New vendors enter the market almost daily, introducing CAD for the masses and dropping prices to boost the "CAD is a commodity" mentality (Lurins, 1995).

Mining planning software began in the large mining companies, who developed in-house software during the 1960s and 1970s to help with the planning of large open-pit mining projects. In the 1980s most of the present mining CAD companies were formed: these included such companies as KRJA/Maptek, Datamine, Mincom, Geomin, Gemcom, Surpac, Geomath, Geostat and Minemap.

The development of this software can be viewed as a succession of technological advances. Initially orebody modelling and ultimate pit design were the only objectives. The availability of graphics workstations in the early 1980s meant that variable sized block models and interactive graphical pit design applications became available. In 1986, Lynx applied CAD technology and released the first 'solid modelling' products which were to revolutionise the geometry of orebody modelling techniques. Advances using this technology rapidly spread to other products to the extent that 3D wireframe and rendered models are now an essential part of most mining software vendors' marketing (Henley, 1998). Originally available for large Unix workstations most mining software is now being ported to PCs, to run under a Windows NT operating system.

Modern mining software suites provide advanced 3D spatial information, modelling, visualisation and analysis systems for the mining industry, covering: exploration, geological modelling, mine design and planning, mine production, scheduling and surveying, resource extraction, geostatistics, groundwater modelling and rehabilitation.

Most of the larger mine planning now have advanced graphical capabilities, for example Vulcan software, from Maptek, can map photographic, satellite, airborne, survey, texture, seismic and other data over 2D or 3D triangulated surfaces, producing much more effective model presentation than that offered by solid shading. Surface features and topography can be draped over a model, allowing details (houses, vegetation stands and infrastructure) to be seen. Pasminco Ltd recently contracted Maptek to produce a corporate video to announce the Century Zinc project. The video uses Vulcan to "fly" through various aspects of the planned project, a still is shown in figure 4 (Maptek, 1998).

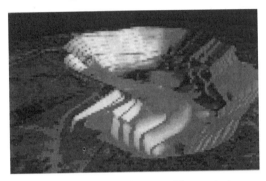

Figure 4. Vulcan image of Century Zinc project

Following the recession of 1991-2 the mining industry software market is nowhere near as buoyant as it was ten years ago. Most of the large and medium mining companies already have all the software they want and need. This situation of saturated markets has been seen for the last few years, short term solutions have been found in less obvious markets such as eastern Europe, India and South America. Long term solutions such as broadening the product ranges into new markets are now being seen as an alternative growth strategy.

Henley (1988) stated some general points about the future of mining (CAD) software:

- There is an increasing need and demand for real integration of more varied applications, for example geotechnical analysis with mine planning to help 'real' mine design.
- There will be more emphasis on communication: the Internet, Intranets and real time transfer of information.
- Data capture and database validation will become increasingly a problem as the data volume increases.

The authors would add to this list the need for improved graphical realism, taking the technology from mainstream CAD packages, and hardware advances, to create realistic, interactive 3D models.

4 GEOGRAPHICAL INFORMATION SYSTEMS

On the walls of caves near Lascaux, France, Cro-Magnon hunters drew pictures of the animals they hunted 35,000 years ago. Associated with the animal drawings are track lines and tallies thought to depict migration routes. These early records followed the two-element structure of modern Geographical Information Systems (GIS): a graphic file linked to an attribute data base.

Today, biologists use collar transmitters and satellite receivers to track the migration routes of caribou and polar bears to help design programs to protect the animals. In a GIS, the migration routes were indicated by different colors for each month for 21 months. Researchers then use GIS to superimpose the migration routes on maps of mineral development plans to determine the potential for interference with the animals (USGS, 1997).

Traditional maps are abstractions of the real world, a sampling of important elements portrayed on a sheet of paper with symbols to represent physical objects. People who use maps must interpret these symbols. Topographic maps show the shape of land surface with contour lines. The actual shape of the land can be seen only in the mind's eye. Graphic display techniques in GIS's make relationships among map elements visible, heightening one's ability to extract and analyse information (USGS, 1997).

In the strictest sense, a GIS is a computer system capable of assembling, storing, manipulating, and displaying geographically referenced information, i.e. data identified according to their locations. GIS technology can be used for scientific investigations, resource management, and development planning.

At present, the use of GIS is not widespread within the mining industry but the reduction in the price of the software, increased hardware performance and the need to handle larger amounts of georeferenced data means that GIS are becoming necessary many aspects of modern mining projects. GIS can also be applied in a number of applications associated with mining, which are not core mine design or extraction activities.

An example of this is the extensive environmental data collected to describe groundwater flow, the extent of contamination and its environmental consequences, and the evaluation of remedial measures. HSI GeoTrans have developed a variety of computerised systems for data management, predictive simulation and innovative presentation formats. HSI GeoTrans also uses GIS to perform site investigations to better understand the nature and extent of contamination and potential environmental liabilities (HIS GeoTrans, 1999).

The U.S. geological Survey have applied GIS to many mining related problems, including mapmaking, natural resources site selection, emergency response planning and site assessment. In a particularly interesting project, the National Forest Service was offered a land swap by a mining company seeking development rights to a mineral deposit in the Prescott National Forest of Arizona.

Using a GIS and a variety of digital maps, the USGS and the Forest Service created perspective views of the area to depict the terrain before and after mining. GIS was used to combine map types and display them in realistic 3D perspective views that conveyed information more effectively and to a wider audience than traditional, 2D maps (USGS, 1997).

Figure 5. Prescott National Forest

Walsh Environmental has conducted GIS projects for oil and gas exploration and development, mining companies and state and local governments, clients include: Mobil, Shell, ASARCO, Amoco, and Premier Oil. Recently, Walsh provided historic image analysis interpretation as part of a legal case. Historic images were examined to calculate the impacts of a mine during a 40 year period. Area calculations were used to identify the level of impact during different time periods (Walsh Environmental, 1999).

GIS technology can be applied to mine design, reclamation and the relocation of public infrastructure and settlements. At the University of Nottingham GIS is being used to aid in the resettlement of villagers in Ghana who live on the site of a new surface mine development. The representation of the resettlement areas allows assessment by the Ghana Environmental Protection Agency before the project begins (Mensa, 1999)

5 SIMULATION

The formal definition of discrete system simulation is a system where, at any instant in time, a countable number of changes can take place. Trucks being loaded, travelling to a dump, dumping, being refueled, returning to the mine, etc., are classic examples of such systems (Sturgul, 1999). The CAD models used to represent a mine can be considered as the "static" representation of the mine. A simulation tries to mimic the mine in operation and can be considered as a "dynamic" representation of the mine.

The first task of a simulation is to accurately reproduce the actual operation of the mine. When the simulation behaves exactly as the real mine does,

producing the same production tonnage, the same travel times for items of equipment and the same downtimes, the simulation is then used for sensitivity analysis. What if a new truck is added to the mine? What if the shaft hoists for an extra hour each day? What if the repair crews were 2% more efficient? This allows the mine management to make strategic decisions about a mining operation, often before planned production has started.

Sturgal (1999) believes that the most important advance in recent years has been the introduction of animation into simulation software. This shows the results of the simulation in a two dimensional "cartoon" fashion. Sturgal justifies this statement with two reasons:

- The first, and probably the most important, reason was that it shows that the logic used in creating the simulation computer program is correct. Often, simulation programs can contain insidious errors that do not show up when the program is being run and can be detected best when viewing the animation.
- The other main reason for the animation is that management readily can appreciate the work that has been done. Prior to animation, the results of the simulation study were often presented as a set of numbers obtained from the simulation study.

Most modern simulation programming languages and software have animation components embedded within them.

Simulation is now taught at many mining education institutions and specialist consultancies offer simulation expertise to the industry. From 1961 until 1995, there were approximately 125 published papers on the subject of discrete mine system simulation, showing that although becoming more important this is still a relatively undervalued field of work. The software and techniques used to obtain the results in these papers have been presented by Sturgul (1996).

One of the largest mining projects undertaken recently was Western Mining Corporation's proposed expansion of its Olympic Dam mining operation. The expansion plans required the development of the mine, materials handling and metallurgical plant to allow for the treatment of the planned annual production. In order to achieve the target production, various mining strategies were to be evaluated using a large scale simulation system. The model was also used to determine the most efficient operator shift rotations, calculate the number and capacity of load haul dump units (LHDs) required, formulate an effective rail operating philosophy and optimise maintenance scheduling (Fluor Daniel Simulation, 1997)

Knowledge Based Engineering (KBE) are using a complex expert system \ simulation system called

G2 to build very complex simulations for a number of large mining operations. The G2 mine planning system is in the form of a simulation model which simulates the entire business of a mine over a selected time period in a fraction of the real time. For example, a five-year simulation of the mine's entire business activities can be simulated in one hour. The speed of the simulation is merely dependent on the computing power at hand. The model essentially comprises four components, namely the mining model, the financial model, the capacity model and the logistics model (Varendorff and Hatten-Jones, 1999).

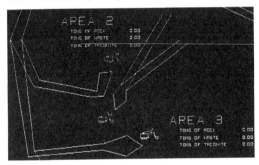

Figure 6. Hibbing taconite simulation

Sturgul uses the GPSS simulation programming language, in association with the Proof animation software. An example of one of these simulations is the work done at the Hibbing Taconite Co., Hibbing, Minnesota. The mine is a surface mine that uses three types of loaders and two types of trucks. The mine was installing a dispatch system and wanted to be able to predict production for a variety of combinations of loaders and trucks. The simulation set up allowed for up to 28,000 different combinations of truck/shovel configurations. Using this simulation the mine planners were able to select an optimum configuration of loaders and trucks (Sturgul, 1998).

6 ARTIFICIAL INTELLIGENCE (AI)

Artificial Intelligence (AI) is an umbrella term for a wide range of computing methods that mimic various aspects of human intelligence. The diversity of the methodologies that are utilised in AI often makes the subject difficult for the non-specialist to comprehend, and for this reason it is too often dismissed as an area for research only, which in reality it is not.

In the last two decades improvements in the power of computers and advanced software have moved AI from the laboratory into the industrial world, although some sectors have been quicker to realize the potential than others. AI still has strong roots in academia, but many commercial systems are now available (Schofield and Denby, 1994).

A large number of methodologies can be included under the banner of AI, the most important being expert systems, neural networks, genetic algorithms and case-based reasoning. Associated topics include intelligent databases, fuzzy logic, knowledge induction, artificial life, cellular automata, fractals and chaos theory. In fact, the boundaries of AI and conventional computing are becoming difficult to define distinctly.

EXPERT SYSTEMS: An expert system can be defined as '...an intelligent computer program that uses knowledge and inference procedures to solve problems that are difficult enough to require significant human expertise for their solution' (Barr and Feigenbaum, 1981). Expert or knowledge-based systems are perhaps the most widely applied type of AI technique and, indeed, an incorrect perception of many is that AI is concerned solely with expert systems.

These systems have been applied in a number of areas within the mining sector, including instruction, technology transfer, monitoring and control, diagnosis, prediction, design and planning and interpretation. They are primarily useful in situations where decision making is undertaken.

For many years, the University of Nottingham has been involved in the development of a number of AI systems for the minerals sector. Expert systems developed at Nottingham include tools to aid surface mining equipment selection, spontaneous combustion prediction, slope stability, subsidence prediction, aggregate petrography, contaminated land reclamation and mineral process plant control. Expert systems are now used as a standard software development technique rather than an AI research area.

NEURAL NETWORKS: Humans often learn by trial and error and neural networks operate in an analogous manner. A network must be trained by repeatedly being fed input data together with corresponding target outcomes. After a sufficient number of training iterations the network learns to recognise patterns in the data and, in effect, creates an internal model of the process that governs the data. The network can then use the internal model to make predictions for new input conditions. Neural networks thus have the important ability to learn from their own experience without explicit programming for each new input.

Neural networks have been applied in a number of mining areas particularly in pattern recognition, data compression, system modelling and financial modelling. They are primarily applicable where data are complex and large in quantity and/or where a

640

large number of case histories is available. Neural network systems developed at the University of Nottingham include systems for intelligent drilling control, orebody grade estimation, control of froth floatation plants and on line size analysis.

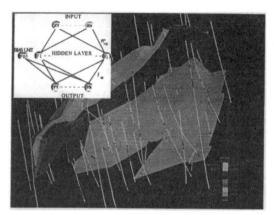

Figure 7. Grade estimation using neural networks.

GENETIC ALGORITHMS: Genetic algorithms provide an approach to programming that mimics the reliance of natural evolutionary processes on the mechanics of genetics and natural selection. Populations of organisms evolve as a result of selective pressures, mating between individuals and alterations, such as those arising from genetic mutation. In genetic algorithms similar effects are simulated by operators that duplicate, recombine and change solutions in the current population to create a new population.

Genetic algorithms by their very nature consider a wide area of the search space for a particular problem. By using simple operations genetic algorithms are able to optimise parameters rapidly after examining only a small fraction of the search space.

Genetic algorithms yield good results for many practical problems and have proved to be successful in a variety of fields. Successful applications in engineering include pipeline optimization, structural optimization and the calculation of natural frequencies. At the University of Nottingham genetic algorithms have been applied to the problems of open pit scheduling (Denby et al 1998), ventilation network optimisation, mineral process plant control and product blending.

7 COMPUTER GRAPHICS (CG) AND VIRTUAL REALITY (VR)

Virtual reality is a technology which allows the creation of computer-generated three dimensional worlds. The technology is similar to that which creates the computer animated, CG, feature films, such as "Toy Story" or "Bug's Life", but with one important difference. Whereas computer animation can take days to create a few minutes worth of film, VR must generate its view of the virtual world in real time. This is similar to the technology used in many modern computer game products where a player controls movement within a three-dimensional, VR environment in real time. This ability means that the computer can instantly respond to human interaction, allowing the user to roam around the virtual environment, view it from any angle and position, and interact with any virtual objects within it.

The way a virtual world is experienced can vary widely. The ultimate interface is known as 'full immersion' where the user views the virtual world through a head-mounted-display (HMD). The HMD presents an image directly in front of each eye and magnifies it so that it fill a wide field of view, creating the impression of actually being in a world, rather than looking at a screen. As the user moves their head, the HMD monitors the movement and the computer generates a new view of the virtual world as seen from the new head position. However, HMD technology is fairly crude, and rather expensive, and has a number of disadvantages including encumbrance, isolating experience and occasional simulator-sickness.

Many VR systems opt to use a standard computer screen as an interface into the virtual world. Although not as immersive as an HMD, the computer monitor still displays a view of the virtual world. Using this system, standard computer peripherals: joysticks, mice etc. can be used to navigate around the world and interact with it. Larger monitors make the experience more immersive, and the use of projectors to display very large images allows a number of people to be involved at the same time.

A number of research groups around the world have been developing prototype systems to evaluate the potential for applications using VR for mining personnel. These systems indicate the current areas that are being considered for primary implementation of the technology. The AIMS Research Unit at Nottingham University is a leading developer of VR based systems for the mining industry. The University of Nottingham has developed a number of applications (both research and commercially based).

The VR applications developed by the AIMS include environmental visualisation, safety awareness training systems, accident reconstruction, data visualisation, ergonomic design, simulation systems, driving simulators, training systems and hazard awareness systems (Schofield, 1997).

Figure 8. An AIMS accident reconstruction image.

Large government research institutions around the world are also involved in developing VR applications for the mining industry:

- CSIR in South Africa have developed a training systems for the gold mine stope environment.
- CSIRO in Australia have a number of VR projects underway including a large mine data visualisation project.
- NIOSH in the U.S.A. are currently developing a hazard awareness system for underground coal environments.

AIMS are currently concentrating on studies into the use of VR training technology for the minerals industry. SAFE-VR, developed by AIMS, has been designed to be a tool to allow the rapid creation of a variety of hazard spotting and training applications.

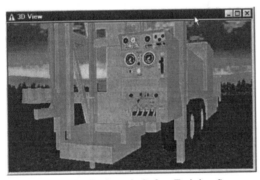

Figure 9. SAFE-VR Generic Safety Training System

The trainee must investigate the virtual world, spotting any hazards present and classifying them correctly. At the beginning of each session the actual hazards present can be selected randomly, so that the experience is different each time. This ideally means that the user stays interested, but at the very least insures that they must complete a full inspection each session, instead of just learning the package once, as with some computer based training tools.

SAFE-VR also includes a logic-based simulation engine. This can be used to simulate procedural tasks, such as equipment operation, and to model more complex hazard situations using the equivalent of fault trees. The number of possible applications is endless, and existing systems include a haulage truck pre-shift inspection, chemical plant pump operation, a drilling rig simulation, and an environmental assessment appraisal (Denby et al, 1999).

8 CONCLUSIONS AND PREDICTIONS

What can the mining engineer expect in the next few years? To finish the authors will briefly predict some future advances in each of the topic areas mentioned.

IT: Advances in hardware will inevitably lead to more exploration computing being done in the field, probably with verbal input / output. Software will take over most, if not all, of the data interpretation tasks. In day to day operations we can expect better control and management of personnel and equipment and remote operation of equipment will become more widespread (Whittle, 1998). This is assuming any computer equipment continues to work after the year 2000!

INTERNET: It goes without saying that the Internet is going to have a large impact on most mining companies. Day by day the Internet is changing the way companies communicate and perform their business. As more users connect and the speed and bandwidth increases new applications are going to be available, remote sales and commerce, remote training, remote monitoring of mine sites and remote teleoperation of equipment.

CAD: As new technologies appear that revolutionise 3D usability and performance, there will be a dramatic and eager shift from 2D design to 3D design. The new CAD systems will have integrated high-end 3D graphics solids modelling capability and 3D animation functionality. The average CAD user will be able to work with texture-mapped, shaded renderings much like they work with vector images today.

GIS: Many disciplines can benefit from GIS techniques. An active GIS market has resulted in lower costs and continual improvements in the hardware and software components of GIS. These developments will, result in a much wider application of the technology throughout the mining industry. GIS will help greatly in the management and analysis of large volumes of data, allowing an understanding of processes and better management of mining activities to maintain economic vitality and environmental quality (USGS, 1997).

SIMULATION: Further advances in software will result in better interaction between the simulation and the CAD models and animations. In 1973, Alan Bauer and Peter Calder recognized the extreme power and usefulness of simulation models for mine design and commented "When a simulation is based on operating data and procedures that truly represent those in the real situation, then it becomes a true working model on which major decisions can be based with confidence. As such it should play an essential part in every open pit planning operation." Some argue that simulation models are already reaching this level of representation.

AI: The 'de-skilling' of tasks through AI technology, may enable decisions to be made by those with considerably less expertise than an expert. This may be highly beneficial as the skills of leading professionals could be made available in the form of an expert system to the remotest (or poorest) parts of the world.

Despite the increased power of hardware and software, computers still do not encompass the richness and depth of human reasoning and will not do so for the foreseeable future. This is not a technology to fear; it offers an opportunity to improve performance through the careful selection of application areas and sensible implementation of the correct techniques.

VR: CG and VR based training systems will form an important element of the training systems of the future. The technology is now well developed and reducing in cost to levels whereby even small mining organisations can consider its introduction. Acceptance by the workforce is a key issue and implementation must be considered carefully.

Increasing interest in this technology by a range of mining organisations suggests that the methodology is starting to become accepted as a valid enhancement to conventional training. As understanding of the techniques grows and further systems and examples develop, acceptance will increase further. It is only a matter of time before VR training for the mining industry develops to reach its full potential.

In general there will be a drive towards convergence of many areas of IT to produce higher levels of integration. Much of this will revolve around using VR as an interface medium. Improved simulation and AI techniques will make the VR world behaviour better mimic the real world. Use of the Internet and real time data integration will provide better and timely data and information for engineers and managers to base their decisions on. Improved graphics will lead to increased realism and better visualisation of our simulated environments.

However it is impossible to predict the future with any real degree of accuracy. The upward spiral of performance improvements will probably meet another step change in technology (e.g. Internet2) that will change the way we all must work. The future is far from certain!

9 REFERENCES

Barr, A. and Feigenbaum, E. A., 1981. The handbook of artificial intelligence, volume 1, Los Altos, CA: William Kaufman.

Bauer, A. and Calder, P., 1973. Planning open pit mining operations using simulation, Proceedings of APCOM '73, South African Institute of Mining and Metalurgy, Johannesburg, South Africa, pp. 273 – 278.

Denby, B., Schofield, D. and Surme, T. 1998 Genetic algorithms for flexible scheduling of open pit operations, proceedings of APCOM '98, Institution of Mining and Metallurgy, London pp 605-616

Denby, B., Hollands, R. and Schofield, D., 1999, Virtual reality based training technology for the mining industry, 28th international conference on safety in mines research institutes, June 7th-10th, Sinaia, Romania, awaiting publication.

Fluor Daniel Simulation, 1997. Olympic Dam expansion underground mine simulation – Western Mining Corporation, commercial web page at URL: http://www.angelfire.com/biz/fdsim/od1.html, (version current 25th March 1999).

Henley, S., 1998. Mining software and computing into the next millennium, proceedings of 27th APCOM conference, London, pp 31–46.

HSI GeoTrans, 1999. HIS GeoTrans on the web, commercial web page at URL: www.hsigeotrans.com/, (version current 25th March 1999).

InSCIght, 1977. Tiniest transistor, Academic Press daily inSCIght web page at URL: www.apnet.com/inscight/01301997/grqapha.htm, (version current 23rd March 1999).

Lurins, S., 1995. Face the future: Provocative visions and reliable predictions about CAD in the year 2000, CADALYST article, web page at URL: www.cadonline.com/src/future1.htm, (version current 24th March 1999).

MAPTEK, 1998. Maptek online, commercial web page at URL: www.maptek.com/, (version current 24th March 1999).

Mensa, A., 1999. Application of software systems to mine design in Ghana, M.Phil. thesis, awaiting publication, University of Nottingham.

NIOSH, 1997. Mine emergency response interactive training simulation (MERITS), Government web page at URL:

www.cdc.gov/niosh/pit/mer_merits.html,
(version current 23rd March 1999).

Panagiotou, G. N. and Sturgul, J. R. eds., 1996.
Mine simulation, proceedings of MINESIM '96,
1st international symposium on mine simulation,
Balkema, Rotterdam.

Silicon Graphics, 1999. Supercomputers,
commercial web page at URL:
www.sgi.com/products/supercomputers.html,
(version current 24th March 1999).

Schofield, D., 1997. Virtual reality associated with
FSV's quarries and open cast vehicles - training,
risk assessment and practical improvements,
workshop on risks associated with free steered
vehicles, safety and health commission for the
mining and other extractive industries, European
Commission, Luxembourg, 11th-12th November,
1997.

Schofield, D. and Denby, B., 1994. Artificial
intelligence in the minerals sector, transactions of
the Institution of Mining and Metallurgy, volume
103: A1-54, pp A1-A2.

Sturgul, J. R., 1996. Annotated bibliography of mine
system simulation, proceedings of MINESIM '96,
1st international symposium on mine simulation,
Panagiotou, G. N. and Sturgul, J. R. eds.,
Balkema, Rotterdam.

Sturgul, J. R., 1998. Personal web page at URL:
http://www.uidaho.edu/~sturgul/,
(version current 25th March 1999).

Sturgul, J. R., 1999. Personal communication.

Sugar, A., 1999. Show us the money, Computer
Weekly, 34-36, 4th March 1999.

USGS, 1997. Geographical information systems,
government web page at URL:
info.er.usgs.gov/research/gis/title.html,
(version current 25th March 1999).

Varendorff, R. V. and Hatten-Jones, T., 1999. World
class mine planning using expert system
simulation, commercial web page at URL:
http://www.kbe.co.za/main.htm,
(version current 25th March 1999).

Walsh Environmental, 1999. GIS and remote
sensing page, commercial web page at URL:
www.walshenv.com/gisweb/,
(version current 25th March 1999).

Whittle, G., 1988. The future of mining computing,
keynote address, proceedings of third regional
APCOM conference, Kalgoolie, Australia, 3–6.

Mining Science and Technology' 99, Xie & Golosinski (eds) © 1999 Balkema, Rotterdam, ISBN 90 5809 067 1

Mining robotization: Past, present and future

V. L. Konyukh
Institute of Coal and Coal Chemistry, Kemerovo, Russia

ABSTRACT: Some preconditions of robotization in mining are analyzed. Mining robotics has a lot of differences from factory robotics, because it must work in unpredictable mine environment. Mine robotization moves through the following stages: robotization of the simplest objects (1967-80), world discussion about the great potentiality for robotics in mining (1980-85), mine robotization is not simple problem (1985-90), mine robotics as a basis for intelligent mine (1990-present). There are three shapes of mine robotization: remotely- controlled manipulators, multifunctional robots for robotics-based mining, mine rescue robots. Both direct and indirect sources of robotics influence on mining are enumerated.

1 INTRODUCTION

Expenditures of manual labor are formed by both mining-and-geological conditions and a level of mine mechanization. Mostly expenditures of manual labor are increased because a deterioration and changeability of mining conditions can not be compensated by traditional mechanization. Besides, a cost of miner's labor is increased every year. Employers in mining must pay a lot of money for non-productive human factors, such as safety, training, insurance, ventilation, idle times, movement to working place. At the same time a cost of automatic devices is reduced.

People wouldn't like to work under dangerous mining conditions. Recently the big groups of miners

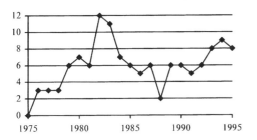

Figure 1. Dynamics of world publications in mining robotics

were lost owing to accidents in Russia (Kuzbass-67 men), China (Yunnan-38 men, Zhejiang-32 men). That is why, robotics is topical for mining as a way to remove a miner out of face. Since 1980 there are more than 160 world publications about possible applications of robotics in mining (Konyukh 1995). The first boom of mining robotics took place in 1983-84 (Figure1). It was expected to use factory robotics for mining. However, mining robotics is a specific scientific field, owing to important differences from traditional robotics.

2 PECULIARITY OF MINING ROBOTS

There are two opposite views on a mining robot: some experts say- one must have at least a manipulator with a gripper, but other experts apply mining machines with a traditional automation to robots. Let me determine a mining robot as an autonomous mining machine with a flexible control that provides the multifunctional use of working head during mining. As a distinct from factory robots, mining robot must work in unpredictable agressive environment with super-high loadings on a working head in a limited working space. It needs a programming at a working place and repairs on a surface. Besides, mining robot use must give a good cost-effect.

Program-controlled robots will find very limited

application in mining. It may be a manipulator for a mounting of drill string or a rail transport robot. Adaptive robots with sensors can be used for drilling and charging of blastholes, shotcreting of workings, loading/haulage of rock masse. Intellectual robots are very intricate and can recognize unknown objects in dynamic mine environment. There are a lot of these tasks during mining: aimed breaking, supporting, face inspection, loading/dumping of equipment, communications mounting, equipment maintenance. However, it is impossible to introduce reference pattern for such recognition. That is why, mining must be robot-oriented to introduce adaptive robots.

Some ideas of factory robotics can be used for mining, such as new kinematics, ways of positioning, reprogrammable control, master-slave control, some sensors (Konyukh 1997 a).

If a roadheader has the kinematics SCARA with rotary movement of manirulator links, a working head has a maximum service space (Figure2).

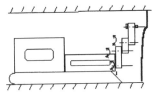

Figure 2. Application of SCARA-kinematics for roadheader

The specific scientific problems of mine robotization are arisen: terminology, evaluation of mining technologies from a view of robotization, navigation of mine robot in a workings, visibility of working head, synthesis and modeling of mining robotics systems for concrete technologies, safety control in changing environment, limited sensors set, non-traditional sources of efficiency. Perhaps, robotization is not high -level automation, because one uses the other basis, namely multifunctional working head, reprogrammable control, and decision-making in a changing environment.

3 SHAPES AND APPLICATIONS OF ROBOTICS IN MINING

Replace a miner by remotely-controlled manipulator is the obvious shape of robotization. A technology is not changed but a miner has to keep a remote participation for every working cycle. The main problem is how to ensure the telepresence of an operator in a working space. A manipulator must be equipped by TV- cameras and force sensors.

However, high-speed actions of manipulator, work in invisible environment, rotary and telescopic movements can not be realized because an operator limits wide possibilities of manipulator.

Multifunctional technological robots for robotics-based mining can be considered as the second shape of mining robotics. It is about dangerous for men technologies. If an underground mine with robots is filled by methane, expenses on ventilation, driving, and equipment will be reduced to a great extent. The main problems are to create non-traditional mining technologies, remove manual actions, and recognize technological situations during mining.

As the third shape of mining robotics a mine rescue robot can be considered. This remotely-controlled robot must have self-propelled movement, accumulator drive, TV-camera, and a lot of sensors for gas kinds, temperature, fire, pressure, water invasion. One must transmit or write down an information about conditions in dangerous spaces for mine rescue team.

4 ROBOTIZATION OF THE SIMPLEST OBJECTS (1967-80)

In 1967 the first unmanned rail haulage was introduced at coal mine "General Blumenthal" (Germany). Then the same systems were developed for ore mines "Tashtagol" (Russia), "Kiruna" (Sweden), "Henderson" (USA), and coal mine "Stashice" (Poland). Later the locomotives without drivers were classified as rail robocars. A drill carriage with reprogrammable positioning of grilling boom was created in France, Japan, Norway. As a result, a time of movement between blastholes was reduced in two times. At the same time a remote control of LHD-machines in dangerous chambers was realized. In spite of waste of time because of remote control these systems at ore mine "Vihanti" (Finnland) were repaid after 400 working cycles owing to extraction of ore losses in chambers (Laatio 1980). These objects with reiteration of standard operations were good suitable for robotization.

In 1976 Prof. M.Thring (Great Britain) had constructed the model of autonomous remotely-controlled manipulator for extraction of thin coal seams (Thring 1977). So called telechir consists of front and back bodies with three jacks, two manipulators, two TV-cameras, and bi-directional connection with 4 operators on a surface. The autonomous manipulator "Hunter SA100" was developed by "Quasar Industries" (USA) for charging of blastholes with the help of elastic gripper.

5 WORLD DISCUSSION ABOUT THE GREAT POTENTIALITY FOR ROBOTICS IN MINING (1980-85)

In 1980 some Soviet experts suggested to use multifunctionality, reprogrammability, and manoeuvrability of robotics for mining operations such as loading / dumping, mounting, auxilliary works, supporting, railway laying. The national program was composed to make various robots and two full- robotized underground mines. The program was not excecuted because mining engineers were not prepared to ideas of robotics. Harold (1980) analyzed the preferable areas of robotization in mining such as rail haulage and rockbolting. The Japanese Association of Industrial Robots recommended to use robots for tubbing erection, rock-bolting, shotcreting, drilling and charging, loading/dumping, work after accidents. A lot of experts thought it is possible to use factory robots for mining and forecast a wide use of robots in American coal industry since 1995. The national programs of mine robotization are formed in Great Britain (NGB), Canada (CCARM), Australia, France (RAM).

6 MINE ROBOTIZATION IS NOT SIMPLE PROBLEM (1985-90)

Remote control of a shearer from surface was realized by CERCHAR (France). An operator can see a full information from 15 on-board sensors on a screen. The robotized drilling rig "Robofore" (Montabert, France) enables to point two borers on 2400 drilling points with precision 10 mm. The same machine "Robot Boomer" was suggested by "Atlas Copco". US Bureau of Mines developed the robotized continuous miner and rock-bolter. Some applications of articifial intelligence and technical vision in mining robotics are analyzed.

At the same time there were some vain endeavours to make a remotely-controlled manipulator for auxilliary operations in a face. It was difficult, to sense any efforts on a manipulator and observe the actions of a gripper in a working zone. Actions of manipulator during man-oriented mining were not effective, because a lot of intermediate actions were inaccessible for robotics.

The experience has shown a single robot can not give a good cost-effect in mining. To avoid a lot of mistakes of factory robotization it was necessary to create an all-round scientific base of mining robotics.

7 MINING ROBOTICS AS A BASIS FOR INTELLIGENT MINE (1990-PRESENT)

Some theoretical principles of mine robotization include (Konyukh 1997 a):
- Evaluation of mining technologies from viewpoint of changeability of working environment, visibility of working zone, stability of machine transfer functions, steadiness of control algorithms, functional flexibility of machine;
- Synthesis and dynamic simulation of mining robotics systems by Petri nets;
- Principles of flexible control of mining robots;
- Evaluation of both direct and indirect efficiency of mining robotics;
- Expediency of robotization by comparison of materialized and present labor expenditures.

At that time a guidance of LHD-machine along reflected belt (Canada), inspection robot NUMBAT (Hainsworth 1990), shotcreting manipulator "Stabilator" (Sweden), automatic manipulator "Scanska" for a mounting of drill string (Sweden), roof - bolter "Robolt" (Finnland), and learning control of powered support "Gullick Dobson" (Great Britain) were developed. An improvement of mining equipment with the aid of cheap microelectronics use becomes more profitable than a modernization of mechanical parts. Equipping of mining machines by technical vision systems made it possible a selective extraction of multi-layer seams by roadheader (Spain), automatic scooping from any stockpile (Japan), automatic shotcreting and charging of blastholes (Australia), recognition of coal interface (USA). The control of a few underground LHD-machines from a long distance on a surface was the great advance in mining robotics (Baiden 1994).

These results were good base to join underground equipment by mine-wide information network, that is connected with surface computer. The Scandinavian technology program "Intelligent Mine" is aimed to increase a working time use by information change in real-time, robotization of autonomous mining machines, applications of high technologies in mining, forming of mining machinery for 21-th century (Seppänen 1993). One consists of 28 projects such as drilling, navigation of autonomous machines, high-speed information network, computer control of mining in real-time, robotized LHD-machines, automatic charging, shotcreting, and drilling. The same program "KUJ 2000" is developed for the greatest ore mine "Kiruna" in Sweden (Nilsson 1993). Germany develops the conception "Partially Automated Mine", that consists of five levels: on-board automation of machine, dispatching of several automated machines from surface, computer planning of mining in full, analysis of mining (Czwalinna 1993). The Canadian "Mining Automated Plan" is aimed to replace of miners by automatics and robotics (Chadwick 1996). Video-, data-, and sound-information is transmitted by mine-wide telecommunication network.

8 PROFITABILITY OF MINING ROBOTICS

Large primary investments and a deep influence on mining are typical for robotics. That is why, a simulation of miners' actions by robot in traditional mining is not promising. However, intelligent mining on a base of robotics will be very profitable. Direct efficiency of mining robotics can be evaluated as stoppages, increase of working speeds, economy of labor cost, shortening of roadways and equipment, and reduce of expenditures on miners' safety, energy, and materials (Konyukh 1997). Indirect influence on mining is more problematical. It may be improvement of working conditions, multifunctionality of robotics, stabilizing of working actions, flexibility of a control, work under inaccessible for a man conditions, group service of equipment units by a robot, reduce of wastes for miners' know-how. Let me forecast the following stages of mine robotization:
1. Telemanipulators for mechanization of manual works are developed;
2. Operator alternates remote control and automatic control of telemanipulator under direct visibility, then control is realized from surface;
3. Coordinated actions of a few robots in some technologies are realized;
4. Control systems of robots are joined by mine-wide information network;
5. Computer Aided Design (CAD) and Computer Aided Mining (CAM) are put into practice to realize mine planning, monitoring, and control in real-time. Mining robots change trajectories of working heads, movement in space, step and speed of both extraction and supporting depending on mining conditions.

CONCLUSION

There are essential preconditions for robotization in mining, such as increase of labor cost, dangerous for miners work, limited resources of both traditional equipment and miners. After "the boom of early 80's" mining robotics is considered as a basis for a future intelligent mining. In spite of the recent success with long distance remote control of underground LHD's a lot of scientific problems of mine robotization have not been decided jet. First of all it is about direct and indirect sources of robotizatin effeciency. The deeper an integration of robotics in mining is, the more its profitability is , because an expensive hardware is replaced by a cheap software.

REFERENCES

Baiden, G.R. 1994. Combining teleoperation with vehicle guidance for improving LHD productivity an INCO Limited. *CIM Bulletin,* June, vol. 87, № 981, pp. 36-39.

Chadwick, J. 1996. Advanced mine automation. Mining Magazine. Nov., pp. 258-264.

Czwalinna, J. 1993 Novel development in mechanization and automation in German coal mining. *Mine Mechanization and Automation.* A.A. Balkema, Netherland, pp.91-97.

Hainsworth, D.W., Mallett, C.W., Stacey, M.R. 1990. Project NUMBAT an emergency mine survey vehicle. *Proc. of the 3rd Intern. Conf.on Robotics.* Melbourne, June, pp. 91-98.

Harold, R. 1980. Continuous Miner Manufacturers: Each Offers Something Special. *Coal Age,* February.

Konyukh, V. 1995. Applications of a flexible automation and robotics in mining: database for designers. *Proc. of the 3-rd Int. Symp. of Mine Planning and Equipment Selection.* Canada, Calgary, pp.605 - 609.

Konyukh, V.L. 1997 a. Mining robotics: the theoretical fundamentals. *Proc. of the Int. Mining Tech'97 Symp. "Mining Equipment and Technology towards 21 Century-Review and Prospect".* Shanghai, China, pp.603-610.

Konyukh, V. 1997 b. Mining robotics-a profitability of the technological applications. *Proc. of the 4-th International Symposium on Mine Mechanization and Automation.* Australia, Brisbane, 3-6 June.

Laatio, E., Hursti, H. 1980. Pemote control loading at Vihanti base metal mine. *World Mining.* v.33, N12, pp.32-34.

Nilsson, J.O. 1993. Future production system - KUJ 2000. A multi-million construction project. *Proc. of the 2-nd International Symposium on Mine Mechanization and Automation.* Sweden, Lulea, pp. 55-60.

Seppänen, P and Pukkila, J. 1993. Intelligent mine technology programme. *Proc. of the 2-nd Intern. Symp. On Mine Mechanization and Automation.* Lulea, Sweden, June 7-10, pp.79-82.

Thring, M. 1977. UK study: remote controlled miner, *Coal Age,* N6, pp.16-22.

Mining Science and Technology' 99, Xie & Golosinski (eds)© 1999 Balkema, Rotterdam, ISBN 90 5809 067 1

Intelligent surface mining systems

Y.D.Jiang
China University of Mining and Technology, Beijing Campus, People's Republic of China

Tad S.Golosinski
University of Missouri-Rolla, Mo., USA

ABSTRACT: The paper presents the current status of surface mine automation. It discusses the current technological advances related to mining applications of advanced computer, communications, GPS and other technologies. Following this the recent philosophies guiding the development of intelligent surface mining systems are outlined and the current status of these systems is reviewed..

1 INTRODUCTION

Over the several last decades a number of attempts were made to automate and robotize mining operations. Automated longwall operations were tried in Britain, in Poland and in the former USSR. Autonomous equipment operation in surface mines was introduced and is used in a Japanese quarry. While several of these systems have worked for periods of time, the applications failed to prove their economic or technical viability, or both. Some of the main problems encountered was the lack of and unreliability of suitable sensors used in the harsh mining environment, the lack of sufficient data transmission and processing power, and lack of full understanding of how various parts of the system interact.

Recent advances in computer, communications and GPS technology have spurred renewed effort to develop an intelligent mine in which a real time control and management of the operation is possible with the objective to optimize the mine performance. At present development of such system is receiving lots of attention from the leading mining equipment manufacturers and suppliers, their interest stimulated by a fiercely competitive market.

The latter is a result of consolidation of numerous mining equipment suppliers into much fewer but stronger groups. It also stems from the low commodity prices, an environment in which many mining companies are forced to look for ways to cut costs and improve performance. Intelligent mining systems are seen by many as facilitating significant mine performance improvement.

2 RECENT OPEN PIT MINE DEVELOPMENTS

Over the past decade the cyclical surface mining has been developing and changing rapidly. Relatively mobile, large hydraulic shovels have become available, able to develop huge digging forces, mine selectively and move fast between mining faces. Similarly, several families of large wheel loaders have been developed able to significantly lower the cost of excavation in many mines. However, the most spectacular developments have been related to off-highway truck developments. These are briefly discussed below (Golosinski, 1999).

2.1. Truck size

Facilitated by the development of new diesel engines, new types of truck tires and drive transmission trains, the size of mining trucks has been increasing continuously. While in the early 1980's the largest truck was designed to carry 170 st (short tons) of material, some 150 trucks with capacity of 320 st are in operation in mid-1999, the number expected to reach 400 units within the next year.

Most recently 360 st trucks have become available from at least two manufacturers. Doubling of the truck payload, therefore doubling of its productivity, has lowered the unit cost of truck transport by up to 50%. Moreover, trucks with payloads in excess 400 st are on the drawing boards at present.

2.2. Efficiency of drive trains

In their desire to outperform competition the suppliers of both the electrical and mechanical drive trucks have improved significantly the efficiency of the truck drives and speeds at which the trucks move. Powered by efficient AC electric drive the 320 st truck moves faster than the 170 st truck of mid-1980's. It is also able to stop at a much shorter distance due to characteristic features of AC retarding system, thus able to travel faster while going downhill.

Comparison of the corresponding truck rimpull charts indicates that today's truck speeds are up to 30% higher, especially so on uphill and downhill ramps. This allows for significant reduction in truck cycle time. The related productivity increase is in the range of 20% or more. Even the same model trucks perform significantly better, a result of incremental improvements of design, manufacture, drive train efficiency, more efficient controls and the like. As an example the Caterpillar Model 777D truck manufactured in 1999 has twice the productivity of the 777 model manufactures in early 1980's (Golosinski, 1999).

Wider introduction of trolley assist systems, pioneered in South Africa in early 1980's, offers the potential to further increase the truck speed and its productivity.

2.3. Availability and reliability

In early 1980's the typical availability of truck fleets was in the range of 70%, and sometimes less. Improvements in truck design and manufacture, reduced maintenance requirements, more effective maintenance, and easier truck operation have led to significant increase of its availability. Truck purchase or lease agreements are being signed today which guarantee truck availability in excess of 90%, with 85% to 90% availability commonly achievable by well run fleets. This high availability translates directly into lower unit cost of mining.

Following improvements in truck availability, their reliability has improved as well. Many mines work towards and are close to achieving the target of eliminating any truck failures between the consecutive PM (preventive maintenance) service. The industry standard for MTBR (mean time between repairs) stands currently at 125 operating hours.

2.4. Fleet management, dispatch and control

The computerized fleet management and dispatch systems are a common feature in most mines today. These allow for optimization of truck assignments and achievement of production objectives at significantly lower overall cost. Mines reported productivity increases of over 20% after introduction of those systems (Zoschke and Vesterdal, 1998).

2.5. The results

As the result of all the described developments the unit cost of cyclical mining has decreased significantly. In well run, modern surface mines using the new generation of the cyclical mining equipment the unit cost is often less than one-third of that in early 1980's. This lead to significant change in mine preferences, with the cyclical systems preferred under most circumstances over the continuous mining. In addition introduction of newest technology into cyclical surface mining created a firm basis for further advances in development of intelligent mining systems.

3. CURRENT ADVANCES IN MINE CONTROL

3.1. Background

More recently a number of new technologies became available that are applicable to open pit mining operations. These include not only advanced digital computing equipment but also such technologies as reliable and accurate GPS (Global Positioning Systems), high-speed, high-band, bi-directional wireless data communication, flat panel displays and the like.

Spearheading the introduction of these new technologies to mining are the largest surface mining equipment manufacturers, Caterpillar and Komatsu. Locked in fierce competitive battle for command of the market. While originally lacking the expertise in control systems and mine communications both have recently formed alliances with providers of this expertise. Caterpillar has formed a number of alliances with such companies as Mincom (mining software and computing), Trimble Navigation (GPS technology), and Aquila Mining System (equipment performance control and management). Komatsu took a more direct route by outright acquisition of Modular Mining Systems (mine control systems and equipment dispatch), after rounding out the in-house equipment manufacturing capacity by acquisitions Demag of Germany (hydraulic shovels) and Dresser of USA (off-highway trucks).

3.2. METS - Mining and Earthmoving Technology System

The new mine control strategies and systems available at present incorporate collection and processing of real time information on equipment status and location. This allows computerized generation of optimum real time equipment assignments. In the new development the information provided by these systems is being tied in to the digitized terrain and orebody models to facilitate reliable separation of ore and waste, or separation of various types of mined materials (Greene, 1999, Harrod and Sahm, 1998).

The most widely publicized, advanced real-time mine information management systems are CAES (Computer Aided Earthmoving System, see fig. 1) and METS (Mining and Earthmoving Technology System) developed by Caterpillar led consortium. The METS system combines several crucial technologies into one package, namely:

- Digital orebody model created with Mincom's Minescape software; this module provides the system with knowledge of spatial position of ore, waste and their respective properties
- Centimeter accuracy GPS developed by Trimble that provides real time information on position of the mining equipment and allows tracking progress of mining
- VIMS (Vital Information Management System) of Caterpillar that provides real-time information on status of equipment, both in terms of its performance and health monitoring

The tasks imposed by the mine plan, such as production of certain tonnage of material with defined grade and within prescribed time are entered into METS system Manager Module. Likewise entered into this module is the equipment available to do the job, its specifications and performance characteristics.

METS Manager processes this information, together with that gathered by the sub-systems named before, ensuring that all mining related tasks are performed in an optimum way. Several of these systems are now installed in US mines, all mines reporting significant improvements in performance as a result. Typical of the applications is that reported on by Phelps, 1998.

Competing systems developed by other consortia has recently become available. Most notable between these is the system developed by the group led by Modular Mining, a subsidiary of Komatsu group of companies.

3.3. Autonomous truck operation

While work on development of economically feasible autonomous truck operation has begun almost two decades ago, it has been largely unsuccessful so far. It is only recently that availability of the RTK (Real Time Kinematic) GPS systems, that allows centimeter accuracy truck positioning, provided added impetus to this work. At present the autonomous truck technology is proven to be technically feasible. To confirm its economic feasibility a fleet of four autonomous Komatsu trucks operates on a trial basis in an Australian coal mine. The initial indications are that this technology is superior in all respects to the manually operated trucks.

4 AN INTELLIGENT MINING SYSTEM

The discussion above indicates that most of the basic technologies required to develop a fully automated open pit mine are currently in place, or will become available soon. These include:

- Data acquisition systems that provide accurate, real-time information on location and status of all pieces of mining equipment, and facilities
- Digital orebody models with ability to interact with the mining equipment and control its advances on one side, and interacting with the computerized mine plans on the other
- Vital signs monitoring systems that allow collection of reliable equipment performance data on one side, and provide equipment health diagnostics on the other
- Wide band, high speed, bi-directional communication systems that facilitate fast transmission of voluminous data

The availability includes both the

- hardware, sufficiently durable to withstand harsh mining environment and
- software with ability to fulfil the intended function under a variety of site-specific conditions.

Continuing strong competition between major technology suppliers to mining make further rapid progress of these technologies feasible.

The final hurdle that must be overcame to

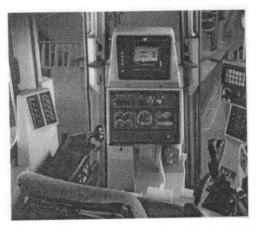

Figure 1. Loader operator's cabin with the CAES system installed in fornt of the operator.

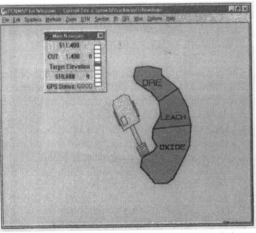

Figure 2. OreNav system of Thunderbird Mining System - equipment position monitoring screen.

achieve full automation of open-pit mining operation is development of an intelligent supervisory / managerial system able to coordinate and control the performance of all the other systems named above. The first step towards development of such system, the METS is already in place. The experience gathered with it will likely prove of great value in further work on development of autonomous mine.

5. CONCLUSIONS

Significant progress has been taking place in development of component technologies of a fully autonomous open pit mine. These technologies include hardware and software related to: real time definition of equipment status and position, digital modeling of orebodies, equipment monitoring and control, and ability to efficiently transmit voluminous data. To facilitate autonomous mine operation all these technologies need to be made compatible, and a system manager needs to be developed to coordinate their outputs and inputs. Work on development of such system continues at present.

REFERENCES

Golosinski, T. S. 1999. Mining system selection and applications: continuous vs. cyclical mining. Braunkohle - Surface Mining, vol. 51, no. 1, 33-38. January.

Greene, D. 1999. Computer aided earthmoving systems. Mining Engineering, vol. 51, no.2, pp. 49-53. February.

Harrod, G.R. and Sahm, W.C. 1998. Mining information goes real-time. Preprint 98-113. SME Annual General Meeting, Orlando, Florida, March 9-11.

Phelphs, R.W. 1998. Morenci improves ore control. E&MJ, vol. no. pp.34-36, July.

Zoschke, L.T. 1998. Mine optimization in the north: truck dispatch in Canada. CIM Annual General Meeting, Montreal, Canada

Mining Science and Technology' 99, Xie & Golosinski (eds) © 1999 Balkema, Rotterdam, ISBN 90 5809 067 1

Preliminary design of conveyor section for belt conveyor with cable-suspended idlers

Yinglin Liu
College of Mechanical Engineering, Tai Yuan University of Technology, People's Republic of China

ABSTRACT: Traditional design of belt conveyor sections for cable type conveyor with suspended idlers may lead to sub-optimum results and poor conveyor performance. Through analysis of the forces acting in conveyor frame a new, preliminary design is proposed. It includes the design reference for selection of cable frame of most appropriate size.

INTRODUCTION

Because cable frame pendent-type belt conveyor runs steadily and has small dynamic load, also it is easy to mount and remount, so it has been applied widely. As to working face of coal mine, the bulging or heave phenomenon often appears because the floor pressure is unsteady. Under this condition, cable frame pendent-type belt conveyor has distinctive adaptability.

Frame cable is the main part of cable frame pendent-type belt conveyor. Loads, belt, cable separating frame and all supporting rolls are hanged by cable frame. The frame cable bears heavy load force. However, the selection of frame cable lacks of theoretical basis and is simply determined according to experience and analogy in China. The result incurs wide discrepancy. Through analyzing the forces on the frame cable, this paper derives the related formulas and provides reference for correct selection of frame cable.

1. STRUCTURE OF CABLE FRAME PENDENT-TYPE BELT CONVEYOR

The structure diagram of cable frame pendent-type belt conveyor is shown in Fig.1. Its framework is mainly composed of two parallel frame cables 4 and many pendent cables 3. A pair of floor frames 2 are posed every 60~100m unevenly along the belt conveyor to set cable tensioner 1 for tightening the frame cable. In order to reduce the forces frame cables, parallel rolls 7 are posed under or near the pendent cables ordinarily. Two trough-type rolls 6 are put between every two pair of pendent cables. There is a cable separating frame 5 between two trough-type rolls which keeps two frame cables

from approaching. Parallel rolls also have the effect on separating cables.

2. THE DERIVATION OF SIZE SELECTION FORMULAS

2.1 *Tension of frame cables when belt conveyor is horizontal and stationary*

Call the distance between two pendent cables a cable span. Fig.1. only represents one cable span. The equivalent model of one cable-span is shown in Fig.2 .

In order to ensure the normal running, the maximum deflection y of frame cable catenary is limited to 1% of unit cable–span as convention. Because the defection of frame cable is very small, while tension of frame cable is very large, regard frame cable as a smooth catenary. The load of cable frame is symmetrical as to point A and B, so $R_{Ax}=R_{Bx}$, $R_{Ay}=R_{By}$. According to laws of mechanics, if the maximum deflection of frame cable is within the range 1.5-2.0% deflection per cable-span, and

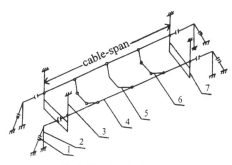

Fig.1 The sketch of cable frame
pendent-type belt conveyor

the calculation error caused by elasticity deflection is less than 1%, the elastic deflection of frame cable can be negligible.

All loads embodying to frame cable are three concentration loads $P1$, $P2$, $P3$, which act on trough-type rolls and at the hanging points of cable separating frame respectively. Therein:

$$P_1 = [(q_1 + q_2 + 2q_3)l + G_1]g \qquad (N) \qquad (1)$$

$$P_2 = G_2 g \qquad (N) \qquad (2)$$

Where,

q_1---loading mass per unit length, kg/m;
q_2---belt mass per unit length, kg/m;
q_3---frame cable mass per unit length, kg/m
l---span of trough-type rolls (per cable-span equals to 2l), m;
G_1, G_2---mass of trough-type rolls and cable separating frame, kg, respectively.

Before the frame cable be chosen, let q_3=0.6~1.2 kg/m. Because this value is much smaller than q_1 and q_2, its effect on calculation result is very small. If frame cable has been determined, q_3 equals to its actual value, and the formulas this paper derived become the checking formulas.

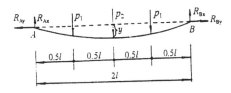

Fig.2 The mechanics model of a cable span

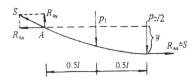

Fig.3 The mechanics model of
the left half cable-span

Analyze the left half cable span as Fig.3. In Fig 3, the resultant of R_{Ax} and R_{Ay} is S. S is the tension of frame cable. Because the maximum deflection of frame cable is very small (It is exaggerated in Fig.3), so $S \approx R_{Ax}$. Take moment equation at pendent point A:

$$S \cdot y - 0.5l \cdot p_1 - \frac{1}{2}l \cdot p_2 = 0$$

Substitute $y=2l\times1\%$ into above equation, we get
$$S=25(P_1+P_2) \qquad (N)$$
Tension in each frame cable is:

$$P = \frac{1}{2}S = 12.5(p_1 + p_2) \qquad (N) \qquad (3)$$

2.2 Additional tension of frame cable when belt conveyor is inclined and stationary

If the belt conveyor is posed inclined along angle β, trough-type rolls, parallel rolls, cable separating frame and weight of frame cable itself will increase the tension of frame cable. For this reason, the tension increased in unit cable-span is:

$$\Delta S_1 = (2G_1 + G_2 + G_3 + 2l \cdot 2q_3)g \cdot Sin\beta \qquad (N)$$

The additional tension of each frame cable in unit cable span is:

$$\Delta P_1 = \frac{1}{2}\Delta S_1 = (G_1 + 0.5G_2 + 0.5G_3 2lq_3)g \cdot \sin\beta$$
$$(N) \qquad (4)$$

Where, G_3---mass of each parallel roll (including parallel roll frame), kg.

2.3 Additional tension of frame cable when belt conveyor is running

Running resistance of belt conveyor can make frame cable to produce additional tension in the course of running. The directions of additional tensions of frame cable produced by top and bottom belts are opposite.

The additional resistance produced by top belt in unit cable-span is:

$$W_1 = \pm(2lq_1 + 2lq_2 + 2G_1')g\omega_1 \cos\beta$$
$$= \pm2[(q_1 + q_2)l + G_1']g\omega_1 \cos\beta \qquad (N) \qquad (5)$$

Where,
G_1'---mass of rotating part in each trough-type roll, kg;
ω_1---running resistance coefficient of each trough-type roll.

If the belt runs upward, the formula (5) uses "-" sign, else uses "+" sign.

The additional resistance produced by bottom belt in each cable-span is:
Where,

$$W_2 = \mp(2lq_2 + G_3')g \cdot \omega_2 \cos\beta \qquad (N) \qquad (6)$$

G_3'---mass of rotating part in each parallel roll, kg;
ω_2---running resistance coefficient of each parallel roll.

If the belt runs downward, the formula (6) uses "+" sign, else uses "-" sign.

The additional tension of frame cable subjected to unit cable-span is:

$$\Delta S_2 = W_1 + W_2 \qquad (N) \qquad (7)$$

The additional tension of each frame cable is:

$$\Delta p_2 = \frac{1}{2}\Delta S_2 \qquad (N) \qquad (8)$$

Because the absolute value of W_1 is much larger than W_2, ΔP_2 can be simplified as:

$$\Delta p_2 = \frac{1}{2}\Delta S_2 \approx \frac{1}{2}W_1 = \pm[(q_1 + q_2)l + G_1']g \cdot \omega_1$$
$$(N) \qquad (9)$$

Belt conveyor running horizontally is a special case of running inclinedly. In formulas (5),(6) and (9), let $\cos\beta=1$ when the belt conveyor runs horizontally.

2.4 *Effect of ambient temperature on the tension of frame cable*

In the different ambient temperature, frame cable expands with heat and contracts with cold, which makes the tension of frame cable change. If the deflection of frame cable is within the range of one percent per cable span, use temperature coefficient K_1 to take the ambient temperature effect on frame cable tension into account. If the temperature θ is above -10, (like coal mining district), let $K_1=1.0$; If $\theta=-10\sim-15$, let $K_1=1.2$; If $\theta=-15\sim25$, let $K_1=1.5$; If $\theta=-25\sim40$, let $K_1=1.8$. Cable frame pendent-type belt conveyor is rarely used in low temperature in China.

2.5 *Effect of two approaching frame cables on frame cable tension*

Some belt conveyors have no special cable separating frame. This makes the two frame cables have the tendency of approaching together which leads to additional tension. Use coefficient K_2 to take this tension into account. Use trough-type rolls (each group trough-type rolls are composed of a horizontal middle roll and two inclined rolls). If the angles of inclined rolls are 20°, K_2 can not be less than 2; if $\alpha=30°$, $K_2=1.5$. If the structure of trough-type rolls can avoid the case that inclined angle of trough-type rolls is larger than the given value, then $K_2=1$.

2.6 *Total tension of frame cable*

If belt conveyor is arranged inclinedly, every point at frame cable has unequal tension. The preliminary design of frame cable ought to base on maximum tension. The maximum tension usually locates on the supporting point of the top floor frame. Suppose the length of each frame cable is L, that is to say, the distance between the two floor frames is L. Suppose

$$F = K_1K_2[P+(\Delta P_1 + \Delta P_2)n] \qquad (N)$$

each L includes n cable span ($n=L/2l$). Total tension of each frame cable is:

In addition, the start of belt conveyor also affects the tension of frame cable. Reckon this factor to safety coefficient m. The safety coefficient usually is taken as 4~5. As a result, the maximum tension in frame cable is:

$$F = mK_1K_2[P+(\Delta P_1 + \Delta P_2)n] \qquad (N) \qquad (10)$$

Consult appropriate specification table of steel cable according to the maximum tension F, then we can determine the parameters of frame cable such as the nominal tensile strength $6(N/mm^2)$, diameter $d(mm)$, and so on.

3. LIVING ANALYSIS AND RESULT

The author calculated the domestic DSP-1063 cable frame pendent-type belt conveyor. The technical parameters are: $q_1=92.6kg/m$, $q_2=14.86kg/m$, $q_3=0.8kg/m$, $G_1=27.9kg$, $G_1'=22kg$, $G_2=5kg$, $G_3=35.1kg$, $G_3'=17kg$, $\omega_1=\omega_2=0.03$, $l=1.5m$, $L=100m$ (length of conveyor is 300m), n=34, $\beta=0$, $K_1=1$, $K_2=1$, m=5. Calculated results are as follows.

$$P_1 = [(q_1+q_2+2q_3)l+G_1]g$$
$$= [(92.6+14.86+2\times0.8)\times1.5+27.9]\times10$$
$$= 19149 \qquad (N)$$
$$P_2 = G_2g = 5\times10 = 50 \qquad (N)$$
$$P = 12.5(p_1+p_2) = 12.5(19149+50)$$
$$= 2456.125 \qquad (N)$$
$$\Delta P_1 = (G_1+0.5G_2+0.5G_3+2lq_3)g\sin\beta = 0(\beta=0)$$
$$\Delta P_2 = [(q_1+q_2)l+G_1']g \cdot \omega_1\cos\beta$$
$$= [(92.6+14.86)\times1.5+22]\times10\times0.03\times1$$
$$= 55 \qquad (N)$$
$$F = mK_1K_2[P+(\Delta P_1+\Delta P_2)n]$$
$$= 5\times1\times1\times[2456.125+(0+55)\times34]$$
$$= 13215.625 \qquad (N)$$

According to the value of F, choose $6X(19)$-15-1550 steel cable. Its total breaking force $S_p=41000N$. By checking computation, the actual safety coefficient is 5.3. By means of living example analysis, we can draw the conclusion as following:

(1) Compared with the specification of a Russian

belt conveyor of the same type with the following parameters: belt width B=1000mm, productivity Q=420t/h, and belt velocity v=1.6m/s. Choose the steel cable as frame cable, its diameter d=14mm. The results calculated with the formulas offered by this paper is quite consistent with Russian ones.

(2) The diameter of domestic frame cable is larger and safety coefficient are also higher. Frame cables of DSP-1063 belt conveyor are $6X(19)$-21.5-1550 steel cable. Calculating by the recommended formulas, its safety coefficient m=11.75 and approaches to the safety coefficient of belt.

(3) Apply the formulas in this paper to choose frame cable, the diameter is smaller than current frame cable. The author made calculation on this case, and the elastic elongation effect on maximum deflection of the cable is not go beyond 1%, when frame cable chosen is subjected to maximum tension. Therefore, it is unnecessary to worry about the diameter of frame cable chosen is too small.

(4) under same tensile stress σ, the belt conveyor with different width belt ought to use frame cable of different diameter. However, belt with B=800mm or B=1000mm, the diameters of frame cables are roughly equal to (20-22) mm, even though the phenomenon of "small belt width with big cable diameter" will appear. For this reason, the author suggests the design department of belt conveyor must carry out the force analysis to ensure the selected size of the frame cable is correct and rational.

REFERENCES

Fuxing Yang. 1983 *Construction of belt conveyor*. Mining Industry Publishing House. (In Chinese)

Ronghai Xia et al. 1987 *Mining hoist machinery equipment*. Publishing House of China University of Mining and Technology. (In Chinese)

Xueqian Yu et al. 1998 *Mining conveyance machinery*, Publishing House of China University of Mining and Technology. (In Chines

Mining Science and Technology' 99, Xie & Golosinski (eds)© 1999 Balkema, Rotterdam, ISBN 90 5809 067 1

Development of a steep angle conveyor for surface mining applications

Tad S. Golosinski
University of Missouri-Rolla, Mo., USA

Mahinda D. Kuruppu
Curtin University of Technology, Kalgoorlie, W.A., Australia

Songnian Zhao
Maanshan Institute of Mining Research, People's Republic of China

ABSTRACT: The steep angle conveying concepts are analyzed with the purpose of modifying them for open pit mining applications. A pilot test rig suitable to transport blocky ore in hard rock mining is designed, constructed and tested. The test results will reveal the operating parameters and provide guidelines for industrial scale installation of steep angle conveyors.

1 INTRODUCTION

The efficiency of open pit mines depends to a large extent on the volume of waste removed. Therefore, a mining or material transport method that allows for reduction of the volume of waste removed would have a major impact on efficiency of mining. One such method may be steep angle conveying which can be employed to lift ore at a steeper angle than the ordinary belt conveyor is capable of transporting. If feasible, this may allow for significant steeping of pit walls by elimination of various restrictions imposed by in-pit haulage ramps, thus reducing the volume of stripped waste and enhancing economics of mining operations. The objective of this study is to determine an appropriate design of a steep angle conveyor, which can be successfully used for transporting blocky ore from open pit mines, and to develop the operating parameters for a conveyor suitable for industrial applications.

In 1995 the first named author (Golosinski) initiated a research project on the application of steep angle conveyors to open pit mining in collaboration with the last named author (Zhao) and his research team. The proposal was well received by the governments of Western Australia and China and consequently the Western Australian Government provided financial support. The Australian team completed the first phase, which involved the literature survey of existing design in 1996 (Kuruppu and Golosinski 1998).

2 SURVEY OF EXISTING STEEP ANGLE CONVEYORS

Several concepts of steep angle conveying have been tried in surface mines over the past decades. The conveying angle of standard belt conveyors is restricted by the instability of the conveyed material to less than 18-21 degrees. However, a steep angle conveyor can move material up any slope, including vertical lift although an angle up to 60 degrees is more practical with many mine designs. The two belt conveying systems that appear to offer the most promise of successful open pit mine application were sandwich belts and formed belts. Sandwich belt type steep angle conveyors developed by the Continental Conveyor Company of the United States shown schematically in Fig. 1 employs two endless rubber belts one of which applies pressure on the other containing the conveyed material. The cover belt makes operative proximate contact with the carrying belt via a system of hugging idlers except in the material loading zone and the discharge zone where the belts are moved out of contact. The additional force on the belts provides hugging pressure to the conveyed material in order to develop sufficient friction at the material-to-belt and material-to-material interfaces to prevent sliding back at the design conveyance angle (Dos Santos 1989, 1990). The formed belt conveyor shown in Fig. 2 was developed by Conrad Scholtz Company of Germany and is available commercially. The belt consists of three components; a flat cross-rigid base belt, corrugated side walls, and cross cheats to prevent slide back of material. The belt is designed with the sufficient strength and flexibility to hold and transport the conveyed material at any angle (Paelke and Germany 1988, 1990).

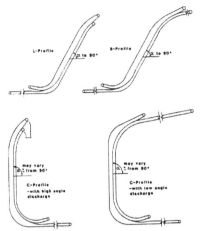

Figure 1. Profiles of Continental High Angle Conveyor

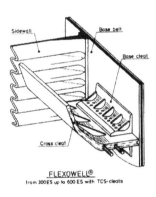

FLEXOWELL®
from 300ES up to 600ES with TCS-cleats

Figure 2. Close up view of FLEXOWELL belt conveyor

However, the relatively large lump size of ore in hard rock mining is a drawback to use the above type of conveyors. Alternatively, bucket type conveyors with well-designed buckets can be used for hard rock mining applications producing relatively large lumps of ore. The use of an in-pit crusher cannot in general be eliminated altogether in order to reduce the size of the blasted rock before loading on to the conveyor.

After consulting with each other the two research teams agreed to develop a pilot bucket conveyor. It is now under construction at the Maanshan Institute of Mining Research in China. Subsequently it will be tested to determine the reliability, the operating characteristics and the general performance

3 PILOT CONVEYOR DESIGN

The overall design of the pilot steep angle conveyor was done by Chinese research team in consultation with the Australian team based at the Curtin University of Technology, Kalgoorlie, Western Australia and the detailed design was done in China. Subsequently, the construction of the pilot plant was undertaken by the engineers at the Maanshan Institute of Mining Research in China. They will also be responsible for instrumentation and testing. The tests are to define the technical and economic feasibility of the concept, the performance and operational characteristics of the installation for various material types, and to provide other information required for future industrial installations.

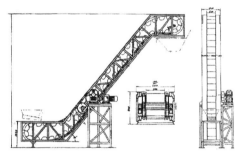

Figure 3. Pilot conveyor design

The pilot conveyor shown in Fig. 3 consists of about 56 buckets, each of which can carry chunks of ore up to 0.5 m in size. It will be installed in the laboratory and tested to determine its performance. The specifications of the pilot conveyor are as follows.

conveyor angle	50^0
height of conveying	5 m
bucket width	0.5 m
bucket volume	0.025 m^3
bucket spacing	0.47 m
driving speed	0~2 m/s

The main features are:

- The proposed design is a continuous cycle bucket conveyor. A centrally located drive gear drives it and the speed can be varied up to a maximum value of 2 m/s.
- The buckets are made of welded steel. They are designed for strength and resistance to cutting and abrasion caused by conveying rock. The size of the bucket is adequate to convey bulk ore up to 0.5 m lumps. The shape is designed to be compatible with the high conveying angle and to have a high filling ratio.
- Spillage protection devices have been implemented between buckets. They prevent leakage of ore particles during conveying.
- The central positioning of the drive gear promotes stable haulage at high speed while reducing vibration. All natural frequencies have been avoided in the range of operating speed. This helps smooth operation and also minimizes maintenance requirements.
- The overall design and in particular the central transmission features linking the conveyor with other similar units without leaving transfer points. This facilitates the extension of the conveyor length as the mining pit extends to higher depths. The ability to eliminate transfer points improves the efficiency of operation and reduces costs.
- Whenever the application demands bucket clearing devices can be installed at the material unloading end.

The conveyor design has been patented in the Peoples Republic of China. Patenting in other countries is under consideration.

4 MATHEMATICAL MODELING OF THE PERFORMANCE

It is intended to use the MATLAB mathematical modeling software to develop a mathematical model simulating the pilot conveyor (MATLAB 1992). This will comprise of simulating the mechanics of the conveyor, and the analysis of its dynamic performance. The software can be used to answer 'what if', scenarios and will be useful for optimizing the performance at almost no cost in comparison with that required for experimenting with a mechanical plant. Correctness of the model will be validated against the performance of the pilot plant. Used to simulate operation of the pilot installation under a variety of conditions, the model will allow for optimization of various test run scenarios. The model will be of critical importance for selection of the design parameters for the industrial steep angle conveying installations. The essential features of the MATLAB program that will be used for this purpose are:

- MATLAB gives a programming environment and a number of different library functions for developing the user's own programs.
- SIMULINK is a subprogram of MATLAB. It is a graphically oriented software for modeling, analyzing and simulating a variety of physical and mathematical systems. This software is especially good for simulating dynamic systems such as the steep angle conveyors.
- SIMULINK has two phases of use; model definition and model analysis. Model definition consists of building the model up from basic or previously constructed elements such as gain blocks and integrators. Model analysis is the problem solution by iterative simulation as the model converges to the desired behavior.
- User may build up the mathematical model by making use of the basic block functions and suitably combining them to simulate the intended application.

The steep angle conveyors do not present many problems during the normal operation. However, if a sudden power failure occurs under load the consequences can be quite serious unless the conveyance is equipped with an efficient and effective braking system. In extreme cases the dynamic transient loading resulting from the sudden loss of drive power and the subsequent braking under load can lead to catastrophic failures. Mathematical modeling helps design the conveyor to minimize such damage.

5 CONCLUSIONS

Steep angle conveying is a viable and an attractive alternative to truck haulage in large open pit mining operations. Selection of an appropriate design and the determination of correct operating parameters by monitoring the performance of a pilot plant enable the optimization of the design. The results of this research will help to build industry scale steep angle conveyors applicable for large open pit hard rock mining operations typically found in China, Western Australia and elsewhere. The collaborative research effort of Chinese Mining research institutions and overseas participants help the technical drive, the technological transfer and to give the open pit mine operators an attractive alternative to costly truck transportation.

ACKNOWLEDGEMENTS

This paper results from an on-going research program supported by the Department of Resources Development of the Government of Western Australia. Mr David Bachman is the Executive Officer of the program.

REFERENCES

Dos Santos, J.A. and J.R.McGaha 1989. Modern Continuous haulage Systems and Equipment. *Bulk Solids Handling.* 9(3):361-365.

Dos Santos, J.A. 1990. High angle conveyors – HAC. *Bulk Solids Handling.* 10(3):357-360.

Kuruppu, M.D. and T.S. Golosinski 1998. Maintenance practices of mining machinery – A Western Australian perspective. *Proc. Mine Planning and Equipment Selection 1998*:607-612. Balkema:Rotterdam.

MATLAB User Manual 1992. *The MathWorks Inc..* Natick, MA 01760, USA.

Paelke, J.W. and F.R.Germany 1988. FLEXOWELL vertical lift systems in underground mining and construction industries. *Bulk Solids Handling.* 8(3):327:342.

Paelke, J.W. and F.R. Germany 1990. 25 years of experience in steep angle and vertical conveying. *Bulk Solids Handling.* 10(3):319:326.

Mining Science and Technology' 99, Xie & Golosinski (eds)© 1999 Balkema, Rotterdam, ISBN 90 5809 067 1

Strength of conveyor belt splices

M. Hardygóra & M. Madziarz
Wrocław University of Technology, Poland

M. Wojtylak
Silesian University, Katowice, Poland

ABSTRACT: The durability of the conveyor belt- the most expensive component of the conveyor- depends to a large measure on the strength of the splice, which is lower than that of the solid belt. Splices in the belt are the areas in which the continuity of plies or steel cords has been disrupted. In such an area, a zone is formed in which the stresses in adjacent plies or cords differ markedly, which results in their different elongation and thus in additional shearing stresses in the layer of rubber. The strength of the splices depends on several factors such as the parameters of the joined belt, on the connecting layer and the technology of splicing, as well as on the materials used to make the splice. The strength of the splice constitutes a criterion for the selection of a belt suitable for the operating conditions. This paper presents the investigations on belts splices which were conducted in the Institute of Mining Engineering of the Wrocław University of Technology. The investigations determined the strength of the full-length splices of belts and show the effect of geometric and physical parameters of the joined components on the ultimate strength of the splices.

1. INTRODUCTION

The proper functioning of belt conveyors is determined by many factors, among which the proper work of the conveyor belt is of primary importance. Splices are the weakest links in the conveyor belt loop. The strength of these joints, and thus their structure and the method and quality of splicing, determine the strength of the whole conveyor belt loop. The high strength of the conveyor belts used in mining, the longer conveyor routes, the ever higher power of the drives and the growing reliability requirements which transport systems must meet demand high-strength splices. Despite the long experience in the manufacture and use of proven belt designs, splices between multiply fabric belts are still a serious obstacle to the full exploitation of the belt strength. Strength tests conducted in the Conveyor Belt Transport Laboratory of the Institute of Mining of Wrocław University of Technology on conveyor belt splices made in the field have shown that these joints have significantly lower strength than that which could be obtained for the same belt designs and bonding materials. It turns out that the causes are basic errors

in execution and a mismatch between the spliced belts. Since different conveyor belt manufacturers use different raw materials and materials (e.g. ply fabrics), belts having the same nominal strength and an identical number of plies can differ markedly in their properties, as a result of which the strength of the belt loop will be diminished. Thus it becomes necessary to conduct research, both theoretical and experimental, to determine what effect the difference in the mechanical properties of the spliced belts has on the obtained strength of the splices.

2. THEORETICAL MODEL OF MULTIPLY BELT SPLICE

To analyze the stress pattern in the conveyor belt and its splice, a suitable model is needed. The problem of modelling the multiply belt splice has been tackled many times. Taking into account these researches, a belt model based on the following assumptions:
- the splices are treated as elastic elements subjected to Hooke's law;

- normal stresses in the interply layers and in the belt covers are neglected because of the low modulus of elasticity of the rubber, incomparable to the rigidity of the plies;
- stresses in the plies even out in the belt cross-section situated at a finite distance from the damage;
- in the load disturbance zone the interply layer is subject to non-dilatational strains;
- the interply rubber is regarded as a linearly elastic body subjected to Hooke's law;

was adopted for the analysis of the effect of the mechanical properties of the plies on the distribution of stresses in the splice. In addition, it was assumed that:
- the values of Young's moduli of the plies of the spliced belts are different;
- the difference in the values of Young's moduli of the particular plies in the same belt is negligibly small in comparison to that between the spliced belts;
- when the undamaged core of the belt is subjected to tension, the deformations of the particular plies are equal.

The state of stress in the plies in any longitudinal section of the belt for the above assumptions is shown in fig. 1.

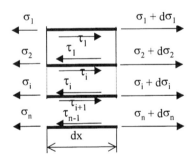

Figure 1. State of stress in elementary cross-sectional segment of belt core.

Considering the conditions of equilibrium of the forces acting on the elements of the successive plies in the cross-section of a belt element of length dx (fig. 1), we get the following equations of equilibrium:
- for the i-th ply, where $1 \leq i < n$, "n" denotes the number of plies in the belt:

$$\sigma_i + \tau_i dx = \tau_{(i-1)} dx + \sigma_i + d\sigma_i$$

- for the n-th ply:

$$\sigma_n = \tau_{(n-1)} dx + \sigma_n + d\sigma_n$$

The adjacent plies in the crosssection of the belt core are bonded together by an interply rubber layer (or a glue layer). Taking into account this strain-shear stress relation for the interply rubber layer:

$$\tau_i = \frac{G}{d} \Delta U$$

we get the following differential equations which describe the state of stress in the particular plies:
- for the first ply (i =1):

$$\frac{d^2\sigma_1}{dx^2} = \frac{G}{d}\left[\frac{1}{E_2}\sigma_2 - \frac{1}{E_1}\sigma_1\right]$$

- for the second ply (i =2) to the last but one ply (i = n-1):

$$\frac{d^2\sigma_i}{dx^2} = \frac{G}{d}\left[\frac{1}{E_{(i+1)}}\sigma_{(i+1)} - \frac{2}{E_i}\sigma_i + \frac{1}{E_{(i-1)}}\sigma_{(i-1)}\right]$$

- for the last ply (the exterior ply) in the cross-section (in = n):

$$\frac{d^2\sigma_n}{dx^2} = \frac{G}{d}\left[\frac{1}{E_{(n-1)}}\sigma_{(n-1)} - \frac{1}{E_n}\sigma_n\right]$$

The above relations form a system of "n" differential equations ("n" stands for the number of plies in the belt) which describe the state of stress in the plies of the belt subjected to uni-axial tension. Furthermore, the following equation of equilibrium, which can replace any of the above equations, holds for any cross-section of the belt:

$$\sum_{i=1}^{n}\sigma_i = const$$

where:
σ_i – the stress in the i-th ply,
n – the number of plies in the belt.

If it is assumed that the strains in the plies beyond the bounds of the notch-affected zone are identical, then the value of stress in a ply depends on the value of the modulus of elasticity of the ply, and it is given by the following relation:

$$\sigma_i = \sigma_T \cdot \frac{E_i}{\sum_{i=1}^{n} E_i}$$

where:
σ_T – the stress in the conveyor belt,
E_i – the modulus of elasticity of the i-th ply.

The derived system of differential equations enables the analysis of the stress pattern in the multiply belt splice area in which the continuity of the successive plies is broken at the contact points between the particular steps. The characteristic geometric and material parameters of the splice, such as the location of the damage in the cross-section of the core, the length of the step, the mechanical properties of the belts and the bonding materials, are de-fined by appropriate boundary conditions of the solution.

3. METHOD OF SOLVING THE SYSTEM OF EQUATIONS

The computer mathematical modelling of physical or technological processes includes four repeatable elements:

1. the adaptation of the model to the specificity of the modelled object;
2. the simplification of the model to a form ensuring a sufficient accuracy of modelling;
3. the discretization of the model, enabling its computer representation by a finite number of procedures, variables and parameters;
4. the formal description of the model in a language which the computer can accept.

The multiply conveyor belt splice model is a system of „n" 2nd order normal differential equations, where „n" stands for the number of plies in the conveyor belt. If this model is to be applied to the calculation of the distribution of stresses in the splice, the number of equations, the values of the parameters in the equations and the initial conditions must be provided. The calculation consists in the numerical solution of the system of differential equations by substituting difference quotients for the derivatives in the equations and solving the resultant system of linear algebraic equations.

The aim of converting the system of differential equations into the difference problem is to obtain a system of linear algebraic equations. The difference (network) method is the replacement of a derivative by a difference quotient. Let us take a system of equations describing the state of stress in the splice of a four-ply belt. After the transformation to a differential problem we get the following system:

$$\sigma_1(x_q) = \frac{1}{2 + agh^2}\left[\sigma_1(x_{q-1}) + \sigma_1(x_{q+1}) + bgh^2\sigma_2(x)\right]$$

$$\sigma_2(x_q) = \frac{1}{2 + 2bgh^2}\left[\sigma_2(x_{q-1}) + \sigma_{21}(x_{q+1}) + agh^2\sigma_1(x_q) + cgh^2\sigma_3(x_q)\right]$$

$$\sigma_3(x_q) = \frac{1}{2 + 2cgh^2}\left[\sigma_3(x_{q-1}) + \sigma_3(x_{q+1}) + bgh^2\sigma_2(x_q) + dgh^2\sigma_4(x_q)\right]$$

$$\sigma_4(x_q) = \frac{1}{2 + dgh^2}\left[\sigma_4(x_{q-1}) + \sigma_4(x_{q+1}) + cgh^2\sigma_3(x_q)\right]$$

where: h – a step of the difference method network.

$$a = \frac{1}{E_1}, \quad b = \frac{1}{E_2}, \quad c = \frac{1}{E_3}, \quad d = \frac{1}{E_4}, \quad g = \frac{G}{d}$$

The system is solved by solving the corresponding algebraic equation for each node of the network. The equations are interdependent, e.g. the value of function σ_2 at point x_{ih} depends on the value of function σ_2 at point $x_{(i-1)h}$, point $x_{(i+1)h}$ and σ_1 at point x_{ih}. Such systems can be solved numerically for given initial and boundary conditions. The matrices of the systems of algebraic linear equations formed as a result of the discretization of differential problems are unique, i.e. most of their coefficients equal zero. Nonzero coefficients are located on several or a dozen or so diagonals. It is practically impossible to solve such systems of equations by exact methods as, for example, Gaussian elimination. The size of the matrix may be in the order of several thousands and it depends on the calculation step for which the difference problem is to be solved. Exact methods necessitate the transformation of the whole matrix, and consequently the entire matrix must be available. For this, in the order of 10^6 real numbers must be stored in the available memory and about 10^9 multiplication and addition operations must be performed. A substantial proportion of these operations would have zeros as their operands. Furthermore, the round-off error generated during the performance of such a number of operations will, most probably, distort seriously the solution of the system.

Most iterative methods do not have such drawbacks. They require only that a procedure for multiplying the matrix by an assigned vector should be specified. An algorithm exploiting the advantages of the Czebyshev method and the gradient method was used for the solution. A block diagram of the algorithm is shown in figure 2.

The advantage of the Czebyshev method is numerical stability. This means that the round-off error for this method is in the order of the data and calculation accuracy. The drawback of this method is slow convergence. Gradient methods are characterized by fast convergence, but they are numerically unstable. This means that even after a large number of steps the next term in the iterative sequence is far from the solution because of the cumulation of round-off errors. The algorithm was used to calculate the stress patterns in the splice area for a four-ply belt in the following cases:

- identical mechanical properties of the plies of the spliced belts,
- different mechanical properties of the plies of the spliced belts.

Series of computations were performed for different assumed values of the difference between Young's moduli of the spliced belts. A diagram showing a typical overlap splice between four-ply belts is shown in Figure 3. It follows from the splice structure that only two cases of the location of the damage in the cross-section of the core need to be considered:

- the outer ply is slit (beginning of step 1-end of step 3),
- the ply directly beneath the outer ply is slit (step 1-step 2 contact and step 2-step 3 contact).

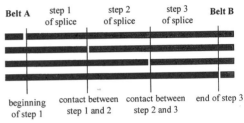

Figure 3. Diagram of splice between four-ply belts.

4. RESULTS OF CALCULATIONS

Taking into account the strength systems existing in the splice between belts with different mechanical properties (Fig. 3), the magnitude of the stress concentrations at the contacts between the particular steps in the splice was calculated. The results of the calculations have been compiled in tables 1-4. The following values of constants: $E = 2000$ kN/m, $G = 750$ kN/m², $d = 0.001$m were used in the calculations.

The splice is layer structure in which, depending on the considered cross-section, there are different combinations of the plies of the spliced belts. If the materials of the plies have different mechanical properties, the particular plies carry uneven loads. As a result of damage induced disturbance, stress concentrations appear at the contacts between the plies of the spliced belts. As the load is transferred from the slit ply onto the adjacent plies, the stresses in the fabric layers closest to the damage increase. Consequently, the most unfavourable distribution of stress appears in the system in which the rupture is located directly beneath the single outer ply, characterized by a high modulus of elasticity, with which the other plies with much lower values of this parameter collaborate. A less elastic ply takes over most of the

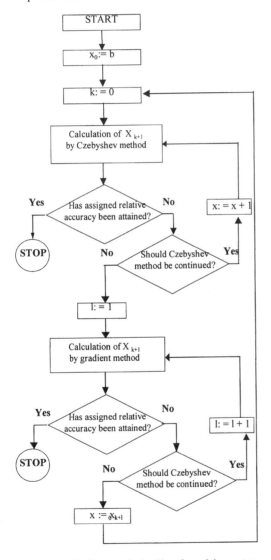

Figure 2. Block diagram of algorithm for solving system of linear equations.

load which used to be carried within the undamaged belt by the slit ply. In this case the maximum stress concentration, among all the contacts between the steps in the splice, occurs in the ply whose modulus of elasticity is high. As a result of this stress, which tensions the ply, this

Table 1. Calculated stress concentration values for beginning of step 1.

Difference between moduli of elasticity of plies $E_1 < E_2$	Maximum stress concentration in comparison to:							
	stress in splice between identical belts				stress in undamaged belt			
	σ_1/σ_0	σ_2/σ_0	σ_3/σ_0	σ_4/σ_0	σ_1/σ_{01}	σ_2/σ_{02}	σ_3/σ_{03}	σ_4/σ_{04}
50 %	0,83		1,09	1,08	1,24		1,59	1,12
40 %	0,86		1,07	1,06	1,28		1,56	1,1
30 %	0,89		1,05	1,05	1,33		1,53	1,09
20 %	0,93		1,03	1,03	1,39		1,5	1,07
10 %	0,96		1,02	1,02	1,43		1,49	1,06

Table 2. Calculated stress concentration values for step 1-step 2 contact.

Difference between moduli of elasticity of plies $E_1 < E_2$	Maximum stress concentration in comparison to:							
	stress in splice between identical belts				stress in undamaged belt			
	σ_1/σ_0	σ_2/σ_0	σ_3/σ_0	σ_4/σ_0	σ_1/σ_{01}	σ_2/σ_{02}	σ_3/σ_{03}	σ_4/σ_{04}
50 %	0,79		1,11	1,1	1,18		1,62	1,14
40 %	0,83		1,09	1,08	1,24		1,59	1,12
30 %	0,85		1,08	1,07	1,27		1,58	1,11
20 %	0,9		1,05	1,05	1,35		1,53	1,09
10 %	0,94		1,03	1,03	1,4		1,5	1,07

Table 3. Calculated stress concentration values for step 2-step 3 contact

Difference between moduli of elasticity of plies $E_1 < E_2$	Maximum stress concentration in comparison to:							
	stress in splice between identical belts				stress in undamaged belt			
	σ_1/σ_0	σ_2/σ_0	σ_3/σ_0	σ_4/σ_0	σ_1/σ_{01}	σ_2/σ_{02}	σ_3/σ_{03}	σ_4/σ_{04}
50 %	0,88	0,89		1,22	0,92	1,3		1,82
40 %	0,9	0,91		1,19	0,94	1,33		1,78
30 %	0,92	0,94		1,14	0,96	1,37		1,7
20 %	0,94	0,96		1,1	0,98	1,4		1,64
10 %	0,97	0,99		1,04	1,01	1,45		1,55

Table 4. Calculated stress concentration values for end of step 3.

Difference between moduli of elasticity of plies $E_1 < E_2$	Maximum stress concentration in comparison to:							
	stress in splice between identical belts				stress in undamaged belt			
	σ_1/σ_0	σ_2/σ_0	σ_3/σ_0	σ_4/σ_0	σ_1/σ_{01}	σ_2/σ_{02}	σ_3/σ_{03}	σ_4/σ_{04}
50 %	0,88	0,89		1,24	0,92	1,28		1,85
40 %	0,89	0,91		1,2	0,93	1,33		1,79
30 %	0,91	0,93		1,16	0,95	1,36		1,73
20 %	0,94	0,95		1,11	0,98	1,39		1,66
10 %	0,96	0,96		1,07	0,99	1,42		1,6

element of the splice cross-section is most likely to fail. The magnitude of the increment in the load acting on the ply with the higher modulus of elasticity and the resultant decrease in the strength of the splice are dependent on the difference between the values of Young's moduli characterizing the plies of the two belts. The results of the analysis have been verified and corroborated by experimental studies carried out on specimens of four-ply belts. The relationship between the increase in stress concentration in the plies, at the contacts between the particular steps in the splice, and the magnitude of the difference between Young's moduli of the plies of the spliced belts has been determined through a theoretical analysis based on the verified model of the multiply conveyor belt. The solution of the problem by the numerical method has opened up wide possibilities of the analysis of the effect of several strength and geometric parameters on the distribution of stress in the belt and its splice. It has been shown that the decrease in the strength of the splice between belts having different mechanical properties occurs because of the additional (in comparison to the splice between identical belts) increase in the concentration of tensile stress in the plies and the increase in shear stress in the interply rubber (adhesive) layer. If two different belts are spliced, the rupture strength of the obtained splice is much lower than that of a splice between two identical belts. The heterogeneity in the strength properties of the spliced belts results in the nonuniform loading of the particular plies and of the glue bond, which leads to the gradual rupture of the overloaded plies and to the unsticking of the splice. The results of tests on splices made in laboratory conditions have confirmed the conclusions drawn from the tests carried out on belt splices made in the field and the results of the theoretical calculations of the distribution of stress in the splice area.

ACKNOWLEDGEMENTS

Thanks are due to FTT „Stomil" Wolbrom Ltd. For having supported preparation of the belt samples.

REFERENCES

Hardygóra, M. *Methods of testing joints in conveyor belts*. 5-th International Symposium on Mine Planing and Equipment Selection. Sao Paulo. 1996.

Madziarz, M. *The influence of the layout and technology of multiply belt splices on their strength*. In Polish. Wrocław University of Technology. Dissertation. 1998.

Hardygóra, M., Gładysiewicz, L. *Procedure for calculation of stress patern in conveyor belt plies as a result of cutting one of them.* In Polish. Górnictwo Odkrywkowe 1979. No.1/2.

Hardygóra, M. Properrties of conveyor belts.In Polish. Górnictwo Odkrywkowe. 1995. No 5.

Mining Science and Technology'99, Xie & Golosinski (eds)© 1999 Balkema, Rotterdam, ISBN 90 5809 067 1

Development of technology for conveyor belt rope condition assessment in China

YuLin Gao & Li Cai
China University of Mining and Technology, Beijing, People's Republic of China

ABSTRACT. This paper reports on development and application of the conveyor belt condition assessment technology. The technology has been under development for over ten years in China. The main steps are: visual defect detection, electromagnetic induction detection, and X-ray detection.

1 INTRODUCTION

Because of the large transportability and the long conveyer distance, the conveyer belt with wire ropes has got the favor with more and more people. It has become one of the primary equipment in modernized mine carrying coal system in China. But when the belt is breakdown, especially in slope carrying, it may cause serious accidents of equipment damage and personal casualty, and also interfere greatly in production. Besides sudden impact of external force, the main reasons of transverse failure of belt are the following two. One is that water flows along the wire ropes in belt, and causes rope rusting and breakdown that result in strength reduction of original belt. The other is that the cohesive reduction between wire ropes and rubber at belt causes the elongation that result in strength reduction or breakdown. In the last ten years, scientists and technicians have done a lot of work to preventing these accidents and made encouraging progress.

2 MANUAL DETECTION

According to the above reasons of belt fracture, maintenance workers observe directly the running condition of belt at production every day. They watch whether there are abnormal phenomena such as blistering on belt and estimate whether wire ropes in belt are broken down or rusted. They make signs on two ends of belt joint, and then measure the distance between the two ends manually every day to estimate whether there is joint elongation.

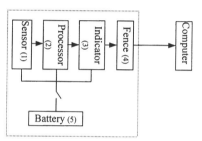

Figure 1. The block diagram of nondestructive defectoscope for the conveyer belt with wire ropes type GXD.

3 ELECTROMAGNETIC INDUCTION DETECTION

In the 1980's, an Australian scientist A.Harrison developed the conveyer belt monitor (CBM) according to the principle of electromagnetic induction. In practice, some scientists have also developed various kinds of similar defectoscope in the world. Take one of them as an example, the block diagram is shown as Figure 1.

Sensor(1): According to the principle of electromagnetic induction, when the primary coil of sensor is applied with a voltage, the magnetic field can be formed. The magnetic circuit consists of iron core, airspace (including rubber coating of belt) and wire ropes in belt .The magnitude of magnetic flux has relation to the magnetic-resistance of the three objects. When wire ropes in belt are rusted or broken down, magnetic-resistance of the circuit becomes larger, so the magnetic flux through magnetic elements becomes smaller, thus rusting or breakdown of wire ropes in belt can be detected. Signal processor(2): It is made of printed wiring board with integrated circuits with explosion-proof. Indicator(3): Detection signal can be displayed with luminotrons. At the same time, the signal through safety fence can be transmitted onto ground by multicore signal cable, and then processed and displayed by computer. Safety fence(4): It is used for explosion-proof requirement. Battery(5): It is used for the power supple by nickel-cadmium accumulators with the current-limiting resistance, which accords with the request in GB3836.4 that 'the service power of the current-limiting resistance is no more than 2/3 of its nominal value under normal working state, and no more than its nominal value under fault state'. It is known from the block diagram (Figure 1.) that sensor (1), processor (2), indicator (3) and battery (5) are set up in a box. The protective performance accords with manufacture method standard prescribed in GB4208 "The classification of box protection grades". The insulation between the casing box and the sparkles circuit in defectoscope, and between the casing box and internal wires can bear the voltage test of 50Hz and 500V alternating current lasting for one minute, but there is not breakdown phenomenon. The performance of explosion-proof box accords with the request of the item IP54 in GB4208.

4 X-RAY DETECTION

4.1 Direct observation or photographing with portable X-ray defectoscope

This type of equipment is manufactured by a few factories in China. For example, the product of ShangHai Defectoscope Factory is used for film photographing. The largest output is 200KVp and 5mA; the power supple is 220V and 6.5A; the control temperature is 60°C; the beam angle is 40°C; the focus is 3×3mm; the distance from lens to belt is 600mm; the film sensitive time is 1.5min once; break for 5min after one exposure.

4.2 X-ray real-time penetrometer for full cross section of belt

4.2.1 The principle of the system

On the basis of the electric drive and control theory, the optical theory, the communication system theory and the digital image processing and pattern recognition theory, the equipment can detect the faults of breakdown, rusting and joint elongation of wire ropes in belt. The principle is the following.

Through the running belt, X-ray irradiates on X-ray receiving board (board(1)). There is a photo-diode array of one dimension covered by crystal which can change the projected picture of wire ropes in belt into current signal. After amplification and collection, the current signal turns into discrete analog pixel signals, and then delivered to image control board (board(2)). In this board, image pixel analog signals are amplified and changed into eight-bit digital signal by A/D converter and delivered to image receiving board (board(3)) in PC where the signals are processed and displayed. Initially, digital image signals are delivered to image processing board (board(4)). Then image signals are stored in frame memory in board(4) to form into one frame which is 512×512×8 bits. After that, image signals are changed into analog signals by D/A converter again and displayed on image display unit (5) under the control of PC. On the other side, a frame of image can be stored in magnetic disk by man-machine interaction. If necessary, the stored image signals can be written into frame memory, and then displayed on image display unit (5). At the same time, the processing such as compensation, amplification and roam can be realized; the calculations such as joint elongation and original belt strength can also be done. The block diagram is shown as Figure 2 and the detection result of original belt and joint is shown as Figure 3.

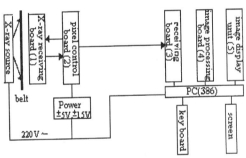

Figure 2. The block diagram of the nondestructive defectoscope

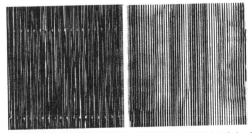

Figure 3. The detection result of joint and original belt.

668

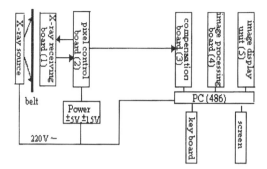

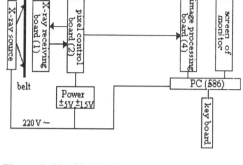

Figure 4. The block diagram of the nondestructive defectoscope after improvement.

Figure 6. The block diagram of the nondestructive defectoscope after the second improvement .

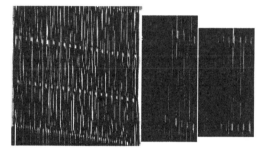

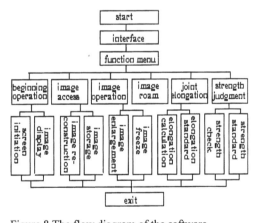

Figure 5. The detection result of slant joint with half-screen display after software flattening.

Figure 7. The detection result of joint and original belt with break wire ropes

Later, in order to improve image definition, board (3) is replaced by compensating board. After real-time compensation, the image is displayed on image display unit (5) more clearly. The block diagram is shown as Figure 4.

At the same time, in order to ensure the calculation accuracy of joint elongation under high belt speed, half-screen display function is added, and slant joint is processed by software flattening. The detection result is shown as Figure 5.

After the second improvement, board(2) delivers digital image signals to a new image processing board(4) directly. Real-time compensation and displaying image on the PC monitor are realized by using software, which makes image compensating board(3) and analog image display unit(5) omitted. The block diagram is shown as Figure 6. The detection result of joint and original belt with break wire ropes is shown as Figure 7.

4.2.2 The software of the system

The software is programmed all with TURBO C, and used to operate and manage image frame memory in image processing board (4). The

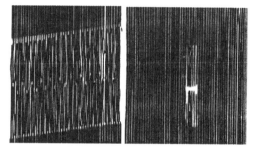

Figure 8.The flow diagram of the software

software includes the interface and the function menu. The flow diagram is shown as Figure 8.

After improvement, the software is programmed with TC and Borland C++ for Windows. The part of TC is run under DOS to control real-time image display and storage. Borland C++ for Windows realizes the calculations of strength and joint elongation of belt, and manages all memory information.

5 CONCLUSIONS

5.1 *Manual detection:*

The labour intensity is large to the Manual method. Detection accuracy relates to the experience and responsibility of worker, so the reliability is poor. But because of the low investment, a lot of users still adopt this method today.

5.2 *Electromagnetic induction detection:*

The result of the detection is a few curves to this method. It only can detect the location and the degree of wire rope breakdown approximately. Its accurate location and the detail of condition must be judged by other apparatus (such as X-ray defectoscope), so the detection period is longer. But relatively, the cost of the equipment is cheaper, and it can be easily made into explosion-proof type, so some users adopt it.

5.3 *Direct observation or photographing with portable X-ray defectoscope:*

The portable X-ray defectoscope used at present has strong X-ray radiation and small detection area. It can't realize continuous detection. What it is used for direct observation, labour intensity is large, and X-ray radiation is serious, so the safety of worker can't be ensured, specially, in the long detection period of time. If it is used for photographing, it costs more than 300 yuan to photograph a joint of belt. The cost is too much, so detection must be simplified and is not accurate. But because of little initial investment, many users adopt this method.

5.4 *X-ray real-time penetrometer of full belt cross section*

Under the principle that production is not influenced, the equipment can quickly and clearly detect the real-time and overall images of wire ropes in belt. It also can access the images by frames and then calculate them. Using the calculation, when the strength of original belt has a more than 10% reduction or the value of the joint elongation is more than 10.24mm, there are alarm and prompt on the screen of computer. The X-ray source used in the equipment can continuously work for a few hours, and has little radiation that is 0.2mR/h at most outside 5cm near the section of detection, so the safety of worker can be reliably ensured. Detection

report can be got in time with little labour intensity. When image transmission is realized by optical cable, (the technique has already been mature,) the operation of computer can be done on ground, and the detection part can be made into explosion-proof type. But because of high initial investment, only four users have used the equipment today. These users give it a high evaluation.

REFERENCES

Gao Yulin & Cheng Hong 1994. Autodetection and predicting system for across splitting of reinforced belts. *Coal Mine Automation.* 2: 32-34.

Gao Yulin, Cheng Hong & Song Wei 1995. Principles of a system for prediction of transverse fracture of conveyer belt. *J.of China Coal Society.* 3: 288-291.

Tan Jiwen & Huang Fengqi 1996. The study of quantitative detection system for the conveyer belt with wire ropes. *Coal Mine Safety.* 1:34-36.

Gao Yulin, Yue hui & He zhiqang 1996. Application of penetrometer used for real time monitoring powerful belt in coal mine. *Coal.* 5:45-47.

Gao Yulin et al 1996. Application of restoration and enhancement inspecting the equipment for malfunction in coal industry. *Proceedings NCIG'96.* WuHan.

Zhou Pengcheng et al 1987. The study of penetrometer for prediction of transverse fracture of conveyer belt in explosion-proof type. *Supplementary Issue of Coal Mine Automation*: 43-46.

Liu Zhihe et al 1998. On-line and real-time observation system for the conveyer belt with wire ropes. *Coal Science and Technology.* 5:38-40.

Zhao Shujiang et al 1998. Spectrum analysis of serial image signals from a powerful belt conveyer. *J.of China Coal Society.* 1:97-101.

He Zhiqing & Gao Yulin 1998. The collecting and displaying images of NDT. *Proceedings NCIG'98.* XiAn: 98.5.

Mining Science and Technology' 99, Xie & Golosinski (eds)© 1999 Balkema, Rotterdam, ISBN 90 5809 067 1

Test rig for quality control of conveyor belt idlers

W. Bartelmus, L. Gładysiewicz & W. Sawicki
Wroclaw University of Technology, Poland

ABSTRACT: The paper deals with description of a rig for quality control of conveyor belt idlers. Quality of conveyor belt idlers is given by vibration measurements. The paper shows the results of investigations, which gives vibration properties of an idler and a quality rig. Vibration measurements are presented in the form of water fall spectrums and spectrums of vibration for an idler rotated at its rated rotation speed.

1. INTRODUCTION

Conveyor belt idlers are elements, which have vital influence to reliability of a belt conveyor, so is generated noise by conveyors. To fulfil the proper reliability and proper leval low noise suitable quality of idlers is needed. The proper quality of idlers is fulfilled by quality control of components, which give final products/idlers, and the quality control of final products. The final product quality may be tested by investigation of its dynamic behaviour when an idler rotates with its rated rpm. Some remarks on idlers dynamic behaviour are given in (Bartelmus et al. 1996) and (Bartelmus 1997). Dynamic properties of idlers are reviled by vibration generated by an idler during its rotation. On the discussion given in (Bartelmus 1992, 1997) vibration generated machinery components is caused by four groups of factors: design, production, technology, operation, change of condition. Design factors include specified flexibility/stiffness and shape of idler components, specified machining tolerance of components. Production technology factors include deviations from specified design factors given during machining and assembly of an idler. Operational factors include rotational speed of an idler and its load. Changes of condition factors are not discussed in the paper. Computer based investigation presented in (Bartelmus et al. 1996) and (Bartelmus 1997) leads to conclusion that for quality control of idlers there is a need to design a special test rig and vibration signal has to be a measure of idlers quality. During the quality test of an idlers it is rotated with its rated rotation speed - rpm and is loaded by a flexible belt.

2 RIG DISCRIBTION

The scheme of the rig is given in Figure 1. An idler (1) is driven by flexible belts (2). The idler (1) is supported in two points (A) and (B) to the revolving/swing-able frame (3). The circular velocity of the flexible belts is controlled by an electric frequency inventor which controls an electric motor (4). The swing-able frame is not connected to the driven frame (5). During quality control of an idler constant rated rotation is kept. Several examination test for the rig has been done at the condition of a rig deceleration. The rotating system has possibility of linear changing of rotating speed with time.

3. DYNAMIC PROPERTIES OF IDLER AND RIG

As it was mentioned the quality of an idler are assessed on the base of vibration measurements. Any

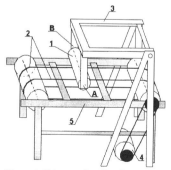

Figure 1. Scheme of quality rig

mechanical structure has its mass and stiffness so has ability to vibrate. Vibration can be characterised by displacement, velocity, and acceleration for a given frequency of vibration. The physical values like: d-displacement [m.], v-velocity [m/s], a-acceleration [m/s^2] can be presented in dB according to ISO 1682 for reference values of (d,v,a). Physical values (d,v,a) can be plotted as function of frequency [Hz]. Using impulse excitation for a mechanical system causes its vibration with its so called natural frequencies. Response to impulse excitation of freely supported idler is given in Figure 2. Figure 2 shows the spectrum of an idler natural frequencies. Response to impulse excitation of the swing-able frame (3) together with the idler (1) (the idler is supported on elastic belts 2) is given in Figure 3.

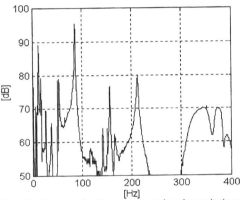

Figure 2. Response vibration spectrum to impulse excitation of idler

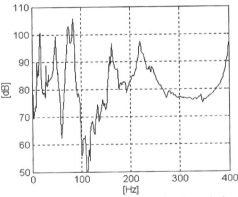

Figure 3. Response vibration spectrum to impulse excitation of system (swing-able frame (3) together with the idler (1) (the idler is supported on elastic belts 2))

Figures 2 and 3 give vibration properties of an idler and vibration properties of the system (swing-able frame (3) together with the idler (1) (the idler is supported on elastic belts 2)).

4. VIBRATION GENERATED BY IDLER DURING OPERATION

The vibration generated during operation is investigated by the rig (Figure 1) at condition of system deceleration and at condition of constant rated rotation. During operation of an idler all factors like: design, production, technology, operation, change of condition are revealed. Figure 2 reveals design factors of an idler. To these design factors has influence shape/dimension of an idler and material properties. More detailed consideration on factors having influence to vibration generated by machinery systems is given in (Bartelmus 1998). Sources of vibration generated by an idler can be divided. First source depends of rotation speed of an idler second doesn't depends of rotation speed of an idler. Signal of vibration revealing condition of an idler are received from two points A and B (Figure 1). Vibration signals are measured in vertical directions. To investigate relation between two sources vibration signals are presented in the form of water fall spectrums. The water fall spectrums are

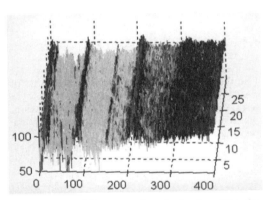

Figure 4a - Water fall spectrum of acceleration for point A, the idler in bad condition

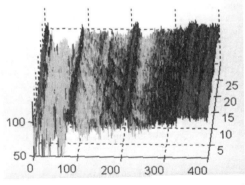

Figure 4b - Water fall spectrum of acceleration for point B, the idler in bad condition

672

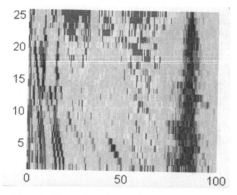

Figure 5 - Upper view of water fall spectrum for point A, idler in bad condition

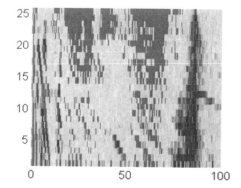

Figure 6 - Upper view of water fall spectrum for point B, idler in bad condition

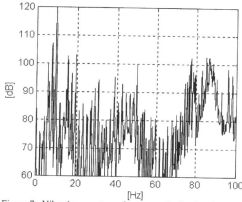

Figure 7a Vibration spectrum for scope of vibration 0-100 Hz for point A, idler in bad condition

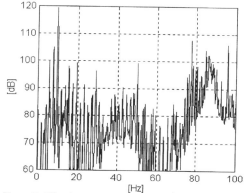

Figure 7b Vibration spectrum for scope of vibration 0-100 Hz for point B, idler in bad condition

obtained during deceleration of the system/quality rig. During the period (100s) of deceleration 30 successive spectrums at different rotation speed is obtained. The results of vibration measurements in the form of water fall spectrums is given in Figure 4. The idler is in bad condition is so classified by a maintenance worker on the base of subjective assessment. It is expected that to vibration may have influence two mentioned sources. In Figure 4a is presented signal received from point A, In Figure 4b is presented signal from point B. The results show constant vibration frequencies and changeable frequencies. One can see the constant frequencies revealing properties of the system with natural frequencies given in Figure 3. One can see that changeable frequencies are better seen Figure 4a. These frequencies are caused by change of condition of the idler in comparing to new one.

One of very frequent drawback of the idlers the is unbalance and run-out of circumference shape. Rated frequency rotation for investigated idlers is approximately 10Hz. The unbalance and run out may course vibration with 10 Hz and its harmonics. Results of measurements are given in Figure 5 and 6. The scope of frequencies 100 Hz. The Figures show water fall spectrums and its upper view. Frequencies of 10 Hz end harmonics are seen.

Spectrums of acceleration for different scopes of frequencies are given in Figure 7 to 9 for points A and B. The Figures shows that bearings of an idler are in different condition because there are different levels of vibrations.

Similar investigation have been done for an idler in good condition. Figure 10 shows water fall spectrums for vibration signals received from points A and B.

Figure 10 shows only frequencies connected with natural vibration, that is approximately 80 and 220 Hz, compare to Figure 3. Figure 10 doesn't show rotation dependably frequencies as it is given Figure 4 and lower intensities in the scope of frequencies

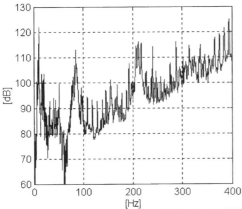

Figure 8a Vibration spectrum for scope of vibration 0-400 Hz for point A, idler in bad condition

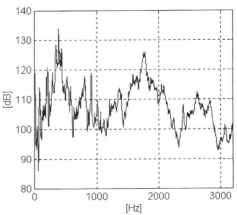

Figure 9b Vibration spectrum for scope of vibration 0-3200 Hz for point B, idler in bad condition

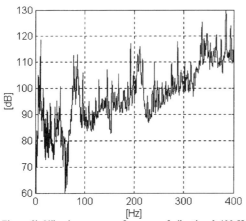

Figure 8b Vibration spectrum for scope of vibration 0-400 Hz for point B, idler in bad condition

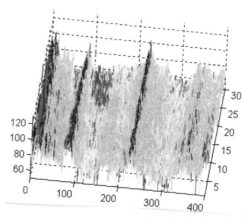

Figure 10a - Water fall spectrum of acceleration for point A, idler in good condition

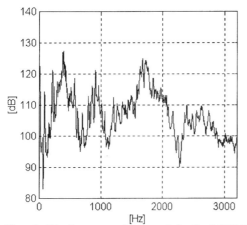

Figure 9a Vibration spectrum for scope of vibration 0-3200 Hz for point A, idler in bad condition

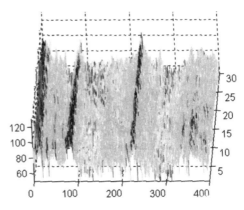

Figure 10b - Water fall spectrum of acceleration for point B, idler in good condition

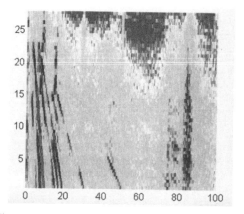

Figure 11 - Upper view of water fall spectrum for point A, idler in good condition

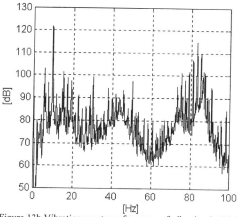

Figure 13b Vibration spectrum for scope of vibration 0-100 Hz for point B, idler in bad condition

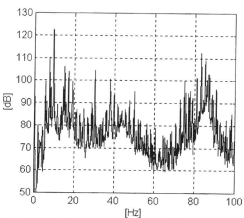

Figure 12 - Upper view of water fall spectrum for point B, idler in bad condition

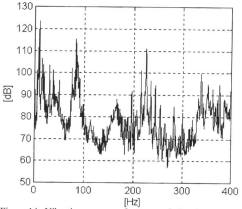

Figure 14a Vibration spectrum for scope of vibration 0-400 Hz for point A, idler in bad condition

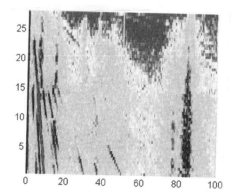

Figure 13a Vibration spectrum for scope of vibration 0-100 Hz for point A, idler in bad condition

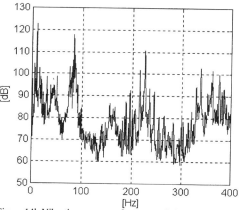

Figure 14b Vibration spectrum for scope of vibration 0-400 Hz for point B, idler in bad condition

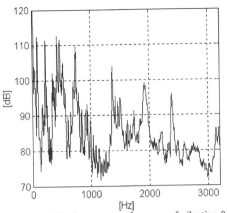

Figure 15a Vibration spectrum for scope of vibration 0-3200 Hz for points A, idler in bad condition

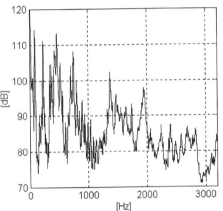

Figure 15b Vibration spectrum for scope of vibration 0-3200 Hz for points B, idler in bad condition

between vibration signals and condition/quality of conveyor belt idlers. The results show that vibration generated by an idler has double nature. There are frequencies connected with natural vibration of an idler or system in which an idler is mounted and vibration components which are connected with rotation speed of an idler. Components connected with unbalance and run-out of circumference shape of an idler are well seen in an idler in bad and in good condition. These are frequencies connected with rotation of an idler and its harmonics. For an idler in bad condition one can see an increase of the vibration intensity in natural frequencies of an idler and creation of rotation dependably frequencies.

REFERENCES

Bartelmus, W. 1992. Vibration condition monitoring of gearboxes. *Machine Vibration;* 1: 178-189. Springer-Verlag London Limited

Bartelmus, W. Gładysiewicz, L. Sawicki, W. 1996. Noise causes of conveyor belt (in Polish). *Scientific papers of the Institute of Mining of Wrocław University of Technology, no 79.*

Bartelmus, W. Gładysiewicz, L. Sawicki, W. 1996. Structure of noise spectrum of conveyor belt (in Polish). *Scientific papers of the Institute of Mining of Wrocław University of Technology, no 80.*

Bartelmus, W. Mathematical modelling and computer simulation of conveyor belts mechanical parts co-operation for supporting diagnostic inference. *Proceedings of the sixth International Symposium on Mine Planning and Equipment Selection,* Ostrava, A.A. BALKEMA/ROTTERDAM/ BROOKFIELD

Bartelmus, W. 1998 Mining Machinery Diagnostics-Open cast Mining (in Polish) Published by Śląsk Katowice

300 to 400 Hz. Water fall spectrums for the scope of frequencies 0 - 100 Hz is given in Figures 11 and 12 for the idler in good condition. Figures show rotation dependably frequencies of 10, 20, 30, 50 Hz. That is caused by unbalance and run-out of circumference shape of a new idler.

Acceleration spectrums for an idler in good condition for different scopes of frequencies are given in Figures 13 to 15.

If you compare the acceleration spectrums Figure 7 -9 to Figures 13 -15 the biggest increase of the vibration leval is in the scope 300 - 400 Hz. In this scope of frequencies you can obtained information on change of an idler condition.

5. CONCLUSIONS

The paper gives first trial for finding relation

Mining Science and Technology' 99, Xie & Golosinski (eds) © 1999 Balkema, Rotterdam, ISBN 90 5809 067 1

Development of the pin-tooth mine car handling technology

Youfeng Qin
Nanjing Design and Research Institute of Coal Industry, People's Republic of China

ABSTRACT: The main contents, characteristics and application of the pin-tooth mine car handling technology are briefly stated in this thesis, and the operating process arrangement of the system is described. Its impacts on the correlated fields in China and aboard together with future development are also introduced.

INTRODUCTION

An innovation of the pin-tooth mine car handling technology gives overall consideration to the operating process, equipage, motive power and controls. It simplifies the process systems and brings about the complete set equipment for mechano-hydraulic integrated mine car handling system and improves efficiency and security.

1. OPERATING PROCESS OF PIN-TOOTH MINE CAR HANDLING SYSTEM

The characteristics of the operating process of pin-tooth mine car handling system is that in each rail there are one multifunctional pin-tooth mine car handling machine and two mine car arrestors of which the front one limits the position of mine car caging and the back one limits the train to be used, together with the swing deck, safe gate, electric control system and full hydraulic drive system. (See Fig.1). There are two parallel rails in cage-in side of cage shaft in this system. The locomotive pushes train to the back arrestor and then the first group mine cars is conveyed to the front arrestor by the back pull pawl. Exchange method is used in the mine-car's cage-in and cage-out in general, when it is needed to dump cage alone, handling machine will move in the cage and push mine car out. Using the handling system in the whole marshalling yard and pitbottom of cage shafts ,the mine cars are operated mechanically and the self-slide travel of mine cars is canceled, consecutive operations can be realized and the requirements about handling mine cars under various conditions can also be met.
The pin-tooth mine car handling system is simple in structure, reliable in ability and multifunctional in

application. It needs no deep trench base and drive cave. The whole system is relatively simple because there are fewer devices, more functions and higher efficiency for the system. When it works in the marshalling yard of cage shafts, the pin-tooth mine car handling machine of which working stroke can be selected willingly. It is flexible in combination according to the process and the jobs such as caging in, caging out, marshalling of mine cars can be carried out The mechanization of handling mine cars can be realized and the handling machine is easy in producing and lower in cost, and convenient in mounting and maintaining.
In the marshalling yard of cage shafts using the pin-tooth mine car handling technology, the types of rails are unified into one standard. Rails in the cage-in side are laid out horizontally or with a little slope, mine cars are forced to move full-mechanically and fierce impact between the mine car and equipment in self-slide handling can be decreased, thus the life-span of the system can be lengthened.
The pin-tooth mine car handling system consists of pin-tooth mine car handling machine, two wheel mine car arrestors, swing deck, safe gates, central hydraulic system and electric control equipment. Also, the safety interlocking of car handling under various hoisting modes can be realized. The process system is simple, reliable and advanced.

2. PIN-TOOTH MINE CAR HANDLING MACHINE

Pin-tooth mine car handling machine comprises a train of pin-cars run on the rail, under the cars, pins are disposed in equal distance throughout the whole length. The sprocket gear driven directly by hydraulic motor. It drives the pin-cars by pin-sprocket gearing.

677

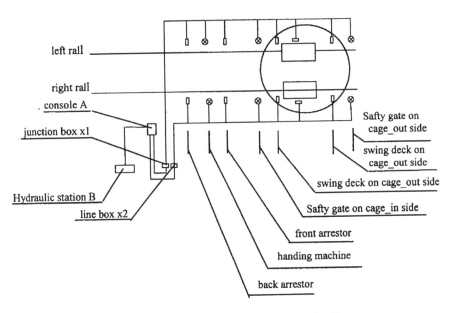

left rail

right rail

console A

junction box x1

Hydraulic station B

line box x2

Safty gate on cage_out side

swing deck on cage_out side

swing deck on cage_out side

Safty gate on cage_in side

front arrestor

handing machine

back arrestor

Figure 1 Schematic arrangement of the pin-tooth mine car handling system

On the pin-cars push and pull pawl used in handling are mounted. The push pawl rises up to push the dog of the mine car to cage-in. When marshalling, the pull pawl rises up to pull the apron or axle of mine car.The push and pull pawls are kicked back when mine cars move along the cage-in direction.The pin cars pass through under the mine cars when return.

Rail of the handling machine is mounted beside the basic rail, occupyong the altitude of 38kg/m rail. U shaped steel beams are used as the rail in ground, while angle steel beams are used as the rail in swing deck and cage, and so the mine cars are protected from being dropped off when move on such rail.

The wheel type car arrestor matching with the pin-tooth handling machine has a pair of arrest pawls and spring dampers. When the wheel of mine car strike the pawl, compressing the spring mounted in the back end of the shaft, the impact energy are absorbed by the spring. The end of the pawl connects with the operating mechanism by bar linkage. The two pawls are opened or closed when the piston of hydraulic cylinder move back and forth. Mine cars are arrested while the pawl is closed and mine cars pass through while the pawl is opened.

The arrestors can be divided into two types as front arrestors and back arrestors according to their positions on line. The front arrestor limits the position of the mine car waiting to cage-in, the back one limits the train. The back arrestor interlocking with the mechanical device for raising pull pawl is permitted to marshall mine cars.

The amelioration of the swing deck should permit the handling machine work in cage, and the side localization should be ensured reliably by the structure of the swing deck's head and rail liner in cage when the arm of swing deck connects with the cage. The accidents that the mine car drops off the rail or the handling machine can not move in back cage smoothly should be avoided, this is duo to the malpositioning between the basic rail and the handling machine rail in and out cage.

Stop dog for hindering the pin cars is mounted on the cage-in side of the swing deck. The stop pawl of the dog is lifted when swing deck rises up and the pin-cars can not pass through. In the other hand, the stop pawl is laid down for the pin cars pass through when the swing deck goes down. The whole mechanism interlocks with the swing deck preventing the pin cars from dropping into the shafts.

The Coal Mine Safe Regulations of China demands that the safe gate should be interlocked with the signal of cage position and the hoisting signals. The safe gate can not be opened when cage does not arrive, only the signals can be only given out for adjusting position of cage or changing cage deck when the safe gate is not closed. The chain mechanism of safe gate matching with the pin-tooth mine car handling machine is driven by hydraulic motor, the open and close of the safe gate can be realized when the hydraulic motor rotates in positive or reverse direction. With the hydraulic system without any other mechanism, the interlock required by the Safe Regulations can be realized and the speed of safe gate can be easily adjusted and controlled.

3. CENTRAL HYDRAULIC SYSTEM OF PIN-TOOTH MIN E CAR HANDLING MACHINE

Central hydraulic system is especially matched with the handling machines to control them on or under the cage shafts centrally. Compared with the power system of the old handling machine, the new hydraulic system has advantages as follows:

(1) All the motive power source of the pin-tooth mine car handling machine is full hydraulically driven, the motive source is simple and easy to control.

(2) Double hydraulic sources are used, one works while the other one is taken as the spare station, and when the handling machine does not work, the motor of the hydraulic station is idling and electric power are greatly economized, so the operating cost are cut down.

(3) The load and speed of handling machine can be controlled by adjusting the pressure and flow of the hydraulic system which has subsystem and automatic protect function avoiding over-pressure and over-tempture,the speed of the system can be set at the optimum .

(4) Low speed and high torque motor is used in handling machine and deceleration mechanism is no longer needed, so the system is easy to install and arranged. Because the mass movement inertia of the drive system is small, the machine can be started in full load and shock can be damped. The motor needs not to stop when the turning direction is changed.

(5)The executive mechanism of handling machine is hydraulically driven, which has the advantages such as compact structure, dampproof, explosion proof, convenient maintenance, safety and reliability because there is no electric drive equipment under the rail.

In the two rail marshaling yard using pin-tooth mine car handling machine central hydraulic system, the devices hydraulically controlled on each rail includes: a set of swing deck having two hydraulic motor on the cage-in and cage-out side, a front arrestor equipped a hydraulic cylinder, a set of pin-tooth mine car handling machine equipped a hydraulic motor, a back arrestor mounted a hydraulic cylinder. In this type arrangement, there are fourteen devices centrally controlled by the hydraulic system.

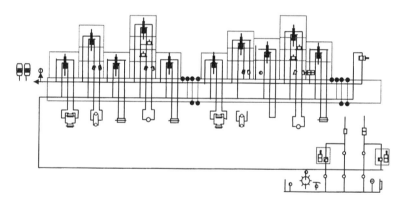

Figure 2 Principle of the central hydraulic control system for pin-tooth handling machine

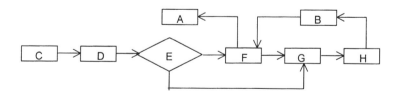

A:display unit   B:monitoring unit   C:operating unit
D:safety interlock   E:selection of the control types
F:logic unit   G:output unit   H:control devices

Figure 3   Operating Processing Chart

The principle of central hydraulic control system for the pin-tooth mine car handling machine see fig.2.

4. ELECTRIC CONTROL EQUIPMENT OF PIN-TOOTH MINE CAR HANDLING MACHINE

Electric control system (Fig. 3) of pin-tooth mine car handling machine has the characteristics as follows:

(1) The whole set of handling system is coordinately controlled by a device selecting switch and a group of buttons named as Obverse, Inverse and Stop.

(2) Multifunctional travel limiting device is used by the monitor unit, only one limiting device are need for one handling device. Passive and static monitor is now turn into initiative and dynamic monitor, in the past, one handling machine needs several travel switch to monitor.

(3) Object oriented programming method is used in the design of the handling control program, and the travel control for the car pusher can be adjusted and set by the user through operate unit, the system is convenient to shakedown and the requirement for the control adapting to different work process can also be met.

(4) Main electric cells are all from American A-B Company and Japan OMRON Company of which the products are in keep with the international standard such as IEC□NEMA□DIN and etc.. The quality is reliable and the capability is stable.

The electric control equipment is installed in the single control room of the mine marshaling yard. The basic structure of the console is fully closed indoors.

The man-machine intellectual interface of the process control can be adjusted and set by users through operate unit, the intellectualization of the mine car handling work and the digitalization of dynamically tracking can be realized, the operating method like fool camera is used in monitoring .the electric control cells, and power supply devices are reduced because of the simplification of the mechanical drive system. The whole control system is simple and reliable.

5. COMPARISON BETWEEN HOME AND ABROAD

The pin-tooth mine car handling system brings forth creative concept in design. It is quite different from the conventional types, the efficiency and safety in work site are greatly enhanced.

Compared with the alike equipment in China and aboard such as type B74, T81 and the equipment using in the Qianjiaying coal mine which was imported from German, the pin-tooth mine car handling system adopts the central hydraulic and computer control technology. The process arrangement is simplified and the complete set of equipment is machine-electric-hydraulic integrated, there are fewer constructional components, more functions and higher efficiency in the system.

Compared with the chain and air-driven car pushers, 20% weight of the mechanical equipment are decreased, 50% base construction work are reduced, 40% cost for the civil engineering, coal mine building and equipment installing are saved and two-thirds of the installing time are shortened. It consumes only one-fifth of power consuming by the B74and T81, one-fiftieth of power consuming by the linear motor pusher and the handling machine used in the Jining No.2 coal mine. The pin-tooth mine car handling system hold a lead position in the field of coal mine, metal mine and chemical mine marshalling.

The complete set of pin-tooth mine car handling system is quite different from the conventional product .The unitary power source of the whole set of machine is hydraulic. The handling system has also the function of reliable logic interlock by using the programmable control system. The control method for operating all devices of the system is one selecting switch matching with a group buttons, so the handling system is convenient for maintenance and installation.

6. APPLICATION AND PROSPECTS

The pin-tooth mine car handling machine obtained Chinese patent of invention in 1992 and Germany patent of invention in June of 1998. The mine car handling system and the relevant units was appraised by the Ministry of Coal Industry in 1993 and 1995.

In recent years, 52 sets, about 43.398 million Yuan RMB, have been used in coal mining district, namely, Zaozhuang, Xinji, Yongxia, Jibei etc., Lingbao Sulphur Mine of Chemical Industry Ministry and Daye Iron Mine of Metallurgy Ministry. Contracts which are worthy 1.798 million and 1.836 million have just been signed with the Daizhuang coal mine and Jianlinshan Iron Mine respectively. 21 sets have been selected in the design of Getting, Xutuan, No.5 mine of Tengbei coal mine districts and Chengjiao coal mine. In the past three years the machine having been improved is appreciated by the users and has obtained more social and economic benefits.

It can be concluded that the pin-tooth mine car handling technology is not only suitable for different new mines in constructing, but also convenient for the reform of the handling system now in use.

Mining Science and Technology' 99, Xie & Golosinski (eds) © 1999 Balkema, Rotterdam, ISBN 90 5809 067 1

Skip loading and unloading system and equipment in Jining No. 2 Coal Mine

Xiaoqun Liu
Nanjing Design and Research Institute of Coal Industry, People's Republic of China

ABSTRACT: Jining No.2 Coal Mine is equipped with a 34t multi-rope coal-hoisting skip and 34t vertical measuring hopper, the largest in China. A study has been carried out on the skip loading and unloading system and equipment to maximize its production. As a result the design production level was successfully achieved.

1. INTRODUCTION

In the production of coal mines, skips are generally used to hoist coal in the main shaft where vertical shaft development method is adopted. In 1970s and 1980s, the largest tonnage of the skip designed and used in China is 16t. Therefore, for the large-scale mines designed at that time, two pairs of 12t or 16t skips and hoisting equipment should be equipped in one shaft to meet the hoisting demand of the main shafts. But this would make not only the shaft section larger but also the investment high. Therefore it is obviously not economical and reasonable. In order to raise the skip-hoisting efficiency and the economic benefit of mines and meet the requirement of large-scale mine construction in China, it is necessary and important to develop as quickly as possible the heavy-duty coal-hoisting skips and loading and unloading equipment which are suitable for the practical service conditions in China.

Located in the east of Jining, Shandong Province and subordinated to Yanzhou Mining (Group) Ltd. Corp., Jining No. 2 Coal Mine is a extra large mine with the designed productive capability of 4Mt/a. Service life is 67.5 years. The mine is developed with vertical shaft, and designed with main, auxiliary shafts and central airshaft. The main shaft is used for hoisting the raw coal of the whole mine. In order to meet the hoisting capacity requirement of this mine, it is preliminarily designed and considered to install one pair of 34t multi-rope coal-hoisting skips for vertical shaft, with fixed weight loading, pneumatic unloading and single-stage hoisting adopted. The skip hoisting height is 631 m.

According to the practical service conditions of this mine, the following technical option has been set forth. The coal hoisting skip is designed as 34 t multi-rope type for vertical shaft; the loading equipment is designed as 34t vertical measuring hopper; the unloading equipment and its loading and unloading system are designed as horizontal movement type. The industrial operation result of this skip loading and unloading system and equipment in Jining No. 2 Coal Mine has shown that the system and equipment can meet the production demand . Both are very important for raising the skip hoisting efficiency and the economic benefit of this coal mine.

2. DESIGN AND STUDY OF SKIP LOADING & UNLOADING SYSTEM

2.1 *Loading system underground*

The loading system of the main shaft is composed of shaft bottom hopper, loading conveyor roadway and skip loading chamber. The loading system is arranged in half-rising mode, i.e. the loading conveyor roadway is arranged at the yard level. The loading system equipment is mainly composed of K-4 reciprocating feeder, belt conveyor, 34 t skip loading equipment and etc. The net diameter of the shaft bottom hopper is 10 m and the capacity is 1750 t. There are four mouths designed under the hopper, and each of them is installed with one reciprocating feeder. And there is one loading belt conveyor designed under the feeders. According to the underground wall rock conditions of this mine, after investigation, analysis and comprehensive technical and economical comparison, the equipment of vertical measuring hopper loading equipment using four-point (load cell) -adding method is for the first time put forward and designed. The arrangement of "One-to-two" is adopted for the skip loading equipment, i.e. one 34 t loading equipment (with breeches chute designed under) is designed for two skips respectively. Both the switching of the loading equipment sector gate and the reversing of breeches chute deflecting plate are operated pneumatically. The loading system underground is shown in attached figure. Compared with the plate conveyor loading equipment, this kind of arrangement can not only save the capital investment around 50%, but also greatly reduce the operation cost of equipment.

Skip loading procedure: coal from the shaft hopper / reciprocating feeder / the loading belt conveyor/ loading equipment/skip. When the coal quantity in the measuring hopper reaches 34t, both the reciprocating feeder and the loading belt conveyor shall be stopped, and the breeches chute deflecting plate of the loading equipment shall be reversed to the corresponding position where skip is to reach. When the skip reaches in place, the sector gate of the loading equipment opens and the coal is filled into the skips through the breeches chute. Then the heavy-duty skips are hoisted, the sector gate of the loading equipment is closed, the blending plate of the breeches chute transverses, and the abovementioned procedure will be repeated.

2.2 Pitmouth unloading system

The pitmouth unloading system is relatively simple, which is mainly composed of the skip unloading equipment, the heavy-duty reciprocating feeder and etc. The unloading equipment includes the two parts of skip gate switching device and coal-receiving movable tongue plate, of which switching action is operated pneumatically. In order to meet with the hoisting capability of the main shaft, two heavy-duty reciprocating feeders with the production capacity of 800 – 1000 t/h are arranged under the shaft mouth receiving hopper.

After the skip reaches in place, the coal-receiving movable tongue plate of the unloading equipment is lowered and put over under the skip unloading mouth, the switching device moves horizontally and the skip gate is opened to unload. After being unloaded into the shaft mouth receiving hopper through the moving tongue plate, the coal is loaded onto the belt conveyor by the heavy-duty reciprocating coal feeder under the hopper to be transported to the coal preparation plant for treatment.

3. THE DESIGN & RESEARCH OF SKIP AND LOADING AND UNLOADING EQUIPMENT

3.1 34t multi-rope coal-hoisting skips for shaft

The diameter of the main shaft of this mine is 6 m, equipped with one pair of 34t multi-rope coal-hoisting skips for vertical shaft. The structure of fixed skip box and bottom-discharge is adopted for the skips. Being the most important part in skip structure design, the gate structure demands safe and reliable action and fast unloading speed, of which the safe and reliable action is the most important, esp. for heavy-duty skip. Absorbing the concerned foreign data, and considering the practical working conditions of Chinese coal mines, the structure of outer-roller-type bottom-sector gate is determined to be used for the skips in the mine. With simple structure, the gate has the outstanding virtues of safe and reliable operation. After being closed, even under the pressure of self-weight and from the coal, the gate will not be automatically opened. The gate cannot project beyond the skip outline when it is opened to unload. Therefore, the accident will not happen to damage the shaft equipment and loading and unloading machines. The open area of the gate is large and the coal-unloading speed is fast.

In order to simplify the structure of skip, so as to make the process and fabricate convenient and reduce the self-weight of skip, the bearing rods on the upper plate of the skip are all made of steel plate hot-bent into trough-shaped girder structure; the skip box is divided into two parts, adopting the two-layer structure of face plate and lining plate, with U-steel girth fixed around the skip box to reinforce; flat steel is used for the upright column. When skips are overhauled, the braking operation is started from the upper plate. There are wedge-shaped cage shoes for braking are designed at the both sides of the upper plate of skip, so as to improve the bearing conditions of the main rods in case of the skip is overwound. The main technical parameters of this kind of skip is as the following: the nominal coal-carrying capacity of skip is 34 t; the valid volume of skip box is 35 m^3; the section dimensions of skip box is 3470×1900 mm^2; the skip height is 16 m; the max. load permitted on the head rope hanging plate is 1098 kN; the max. hoisting height is 800 m; the self-weight of skip (including the head and tail rope hanging device) is 40.2 t.

3.2 Vertical measuring hopper loading equipment using four-point-adding method

According to the wall rock conditions of the shaft

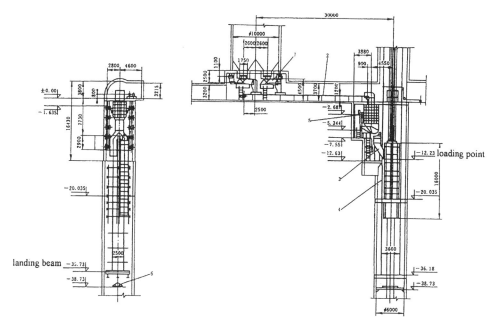

Shaft Bottom Loading System of Jining No. 2 Mine

1. reciprocating feeder
2. loading belt conveyor
3. 34t skip loading equipment
4. metal support structure wedge-shaped guide and landing beam
5. pneumatic device of skip loading equipment
6. shaft bottom tail rope buffer beam

bottom, the managing and application level in China at present, the option of vertical measuring hopper loading equipment using four-point-adding method is adopted in this design. With the nominal coal-loading capacity 34t, and effective volumetric capacity 35 m³, this vertical measuring hopper loading equipment is the largest tonnage one developed and used in China at present. This kind of loading equipment is mainly composed of hopper box, sector gate, load cell and breeches chute.

In the past, measuring hopper loading device was supported and measured in single point. This not only makes the hopper box less stable, but also makes the measuring accuracy easy to be influenced by the loading gravity center, esp. for measuring hoppers with large capacity. In order to ensure the safe operation and measuring accuracy of 34t measuring hopper, four load cells are arranged on the fixed beam at the two sides of the middle of the measuring hopper, the load of the whole hopper can be supported by the four load cells and then added together.

3.3 Horizontal movement unloading equipment

The skip unloading equipment is composed of two parts, i.e. switching device of skip gate and coal-receiving movable tongue plate. The gate switching device is used to open the gate of heavy-duty skip to unload when skip reaches in place and being stable, and to close the gate after the coal is unloaded. The coal-receiving movable tongue plate is installed between the skip unloading mouth and the pitmouth coal receiving bunker to function as transit chute.

In foreign countries, there are two kinds of outer-roller type sector gate switching device commonly used, one is swing link type and the other is horizontal movement type. The former requires the skip to reach accurately in place, otherwise, the open dimension of skip gate will be influenced, simultaneously the swing link is huge and heavy with large space occupied by swinging during the operation. Similar with the car arrester switching device inside the 1t cage produced in China, the later is simple in structure, convenient in installation and maintenance. In addition, it is not strictly required for the skips to reach accurately in place. Therefore, the skip gate switching device is designed as horizontal moving type.

4. INDUSTRIAL EXPERIMENT RESULT

This skip loading and unloading system was put into use in Aug. 1996 in Jining No. 2 Coal Mine and the total coal hoisting load reached 100 Mt by June, 1997. From the industrial operation, it can be seen that the operation of the skip is stable, safe and reliable. The gate can open or close swiftly with quick unloading speed. The loading equipment can be volume-measured accurately and reliably, with low operation cost and little maintenance work. The unloading equipment is safe and reliable in operation, with its gate being opened and closed fast and operated easily. This system was technically appraised by the former Coal Ministry Science and Education Bureau on June 23, 1997.

The actual test result shows that the unloading time at the skip mouth in average is 12.4 s, the loading time at the shaft bottom in average is 21 s and the stopping time after one cycle of loading and unloading operation is only 22s, it means 8s less than that specified in *Mine Design Code of Coal Industry*

5. CONCLUSIONS

In the above mentioned study, combining the coal mine design and fully considering the Chinese actual conditions, we have done some creation and improvement works in improving the skip loading and unloading condition, ensuring safe and reliable operation of skip and loading and unloading equipment and facilitating the replacement of skip, and developed some special large type coal mine equipment. The successful use of the above-mentioned equipment has filled a gap of large tonnage coal hoisting skip and loading and unloading equipment in China, which plays an important role in the development and construction of large scale mines in China. This technical achievement has reached the international advanced level and the national leading level and has important popularization and application value. This has reference meanings for the design and application of large scale mine skip loading and unloading system henceforth.

Since the relevant information from China and abroad is limited and the design experience of heavy-duty skip loading and unloading system is insufficient, there are still some aspects in the above-mentioned study need to be improved and perfected, such as the overwind protection measures and etc. For all these, further study and research should be carried out in the design practice in future.

Mining Science and Technology '99, Xie & Golosinski (eds) © 1999 Balkema, Rotterdam, ISBN 90 5809 067 1

Study on the identification of load spectra in cepstrum domain*

Miao Wu, Chong Chen, Dunyong Lu & Renzhi Wei
China University of Mining and Technology, Beijing, People's Republic of China

ABSTRACT: The load spectra are important data in designing and improving the machines. In some cases it is difficult to measure them directly, so indirect methods have to be adopted. A new method based on cepstrum analysis is proposed to identify the load spectra. This method can be used to eliminate the filtering influence of transfer function on the response signals so that the load spectra can be determined indirectly. It can also be used to identify torque spectrum. An example of engineering application on a road header was introduced in the end of this paper.

1. INTRODUCTION

Load spectra have important meaning in designing and improving machines. It is especially useful to design a machine by modern design methods (Zhifang, Fu 1990). The basic method to obtain load spectra is to measure load directly from the machine. But for many engineering structures, especially for mining machines, it is difficult to measure the load directly for their bad operating environment and high cost of measurements. In order to meet engineering needs, the cepstrum analysis method was developed to determine load spectra from response signals in the paper (Miao, Wu 1992). This method has the advantage that the calculation is simple, and the measurement is easy. It seems to be more suitable for the feature of structures and working circumstance of mining machines.

2. THEORICAL BASIS OF CEPSTRUM ANALYSIS TO IDENTIFY LOAD SPECTRA

It is known that load identification is the inverse problem of structure dynamics. The principle of load identification can be shown in Figure 1: $x(t)$ is the input signal (i.e. dynamic load) of a machine system, $h(t)$ is the impulse response function, $y(t)$ is the response signal.

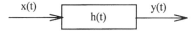

Figure 1. Input and output signal of a machine system

The relation of $x(t)$, $y(t)$ and $h(t)$ is

$$y(t) = \int_0^\infty x(\tau) \cdot h(t-\tau) = h(t) * x(t) \qquad (1)$$

The Fourier transform of Equation (1) is

$$Y(f) = H(f) \cdot X(f) \qquad (2)$$

Equation (2) is rewritten with the form of amplitude and phase

$$|Y(f)|e^{j\varphi_y} = |H(f)|e^{j\varphi_h} \cdot |X(f)|e^{j\varphi_x} \qquad (3)$$

The complex cepstrum of the Equation (3) can be obtained by inverse Fourier transform and logarithm calculation.

$$\overline{C}_y(\tau) = F^{-1}\{\ln|Y| + j\varphi_y\}$$
$$= F^{-1}\{\ln|H| + j\varphi_h\} + F^{-1}\{\ln|X| + j\varphi_x\}$$

*Supported by National Natural Science Foundation of China, approved number: 59775004

$$= \overline{C}_h(\tau) + \overline{C}_x(\tau) \tag{4}$$

where, F^{-1} denotes the inverse Fourier transform and real cepstrum can be written as:

$$C_y(\tau) = F^{-1}\{\ln|Y|\}$$
$$= F^{-1}\{\ln|H|\} + F^{-1}\{\ln|X|\}$$
$$= C_h(\tau) + C_x(\tau) \tag{5}$$

The phase spectrum is

$$C\varphi_y(\tau) = F^{-1}(\varphi_y)$$
$$= F^{-1}(\varphi_h) + F^{-1}(\varphi_x)$$
$$= C\varphi_h(\tau) + C\varphi_x(\tau) \tag{6}$$

From formula above, once the system dynamic characteristic $h(t)$ is known and $C_h(\tau)$ is subtracted from $C_y(\tau)$, then by inverse Fourier and antilogarithm transformation of $C_y(\tau)$, the load spectrum $|X(f)|$ can be obtained. Using the similar transformation to Equation (6), φ_x is obtained from $C\varphi_y$. Since $X(f)$ includes both amplitude and phase spectrum, $x(t)$ can be obtained through the inverse Fourier transform of $X(f)$. Then dynamic load can be identified in the time domain after calibration.

For some cases there is no reliable method to get system transfer function $H(f)$, the system characteristics and the dynamic load have to be identified directly from the response signals. In general the dynamic load $X(f)$ shows dense harmonic, and system transform function $H(f)$ shows a slow-changing wave in the frequency spectrum. Correspondingly in the cepstrum domain, $C_x(\tau)$ usually locates at high cepstrum frequency and $C_h(\tau)$ at low cepstrum frequency. The system dynamic characteristics can be obtained through adding a window function $W(\tau)$ at low cepstrum frequency. There is

$$C_h(\tau) = C_y(\tau)W(\tau) \tag{7}$$

Dynamic load can be obtained as:

$$C_x(\tau) = C_y(\tau)[1 - W(\tau)] \tag{8}$$

3. DETERMINATION OF WINDOW WIDTH IN CEPSTRUM DOMAIN

The accuracy of the load spectrum depends on whether the cepstrum window can effectively filter the system dynamic characteristics or not, so the determining of window width is of importance for identifying the load spectra. The Green function of time series was adopted here to determine the width of the window.

The model can be obtained by a time series of output signal y_t. The transfer function of its equivalent model includes the system dynamic characteristics. Input load signal x_t is not the noise signal, but it can be regarded as the signal produced after white noise signal a_t through a forming filter. There is the model of ARMA (n, m)

$$y_t - \sum_{i=1}^{n} \varphi_i y_{t-i} = a_t - \sum_{i=1}^{m} \theta_j a_{t-j} \tag{9}$$

or

$$y_t = \frac{(1 - \sum_{j=1}^{m} \theta_j B^j)}{(1 - \sum_{i=1}^{n} \varphi_i B^j)}, \quad B^j P_t = \theta P_{t-j} \tag{10}$$

where B is a delay operator. Then z transform of the equation above is

$$H(z) = \sum_{j=0}^{m} \theta_j z^{-j} / \sum_{i=0}^{n} \varphi_i z^{-i} \tag{11}$$

or

$$H(z) = A \frac{\prod_{i=1}^{n}(1 - \eta_i z^{-1})}{\prod_{i=1}^{m}(1 - \lambda_i z^{-1})} \tag{12}$$

where A is an arbitrary constant. From the property of the z transform of complex cepstrum, it is known that this time series is a stable exponential series. The complex cepstrum of $H(z)$ decreases as the value of $|n|$ increases and its decaying speed is n times faster than that of exponential series $\alpha(n)$, there is

$$|\overline{C}_H(n)| < C \left| \frac{\alpha^2(n)}{n} \right|, \quad -\infty < n < +\infty \tag{13}$$

where C is an arbitrary constant, and α is defined as:

$$\alpha = \max[|a_k|, |b_k|, |c_k|, |d_k|] \tag{14}$$

where, $z = a_k$, $z = c_k$ are the pole and zero points inside of the identical circle and $z = 1/b_k$, $z = 1/d_k$ are the pole and zero point outside the identical circle respectively. Let $a_k = \eta_j$, $c_k = \lambda_j$, so when $n = j$, there is

$$\alpha_j = C\eta_i^j \quad \text{or} \quad \alpha_i = C\lambda_i^j \qquad (15)$$

In addition, Green function of *ARMA (m, n)* can be written as:

$$G_j = \sum_{i=1}^{n} g_i \lambda_i, \quad |\lambda_i| < 1 \qquad (16)$$

g_i can be determined and regarded as a constant after the model *ARMA (m, n)* has been determined. From equations above it is known that α_j is one term of G_j and is also the slowest decaying term. Thus, $\alpha(n)$ can be replaced by G_j because $G_j \to 0$, there is $\alpha_j \to 0$. Now it is feasible to estimate the decaying speed of the exponent series and the width of system characteristic in cepstrum domain from the decaying speed of G_j.

When n>m, the recursive algorithm of Green function can be given as:

$$G_0 = 1$$
$$G_1 = \varphi_1 G_0 = \theta_1$$
$$G_m = \varphi_1 G_m + \varphi_2 G_{m-1} + \cdots + \varphi_m G_1 - \theta_m$$
$$G_{m+1} = \varphi_1 G_m + \varphi_2 G_{m-1} + \cdots + \varphi_m G_1 + \varphi_{m+1} G_0 \qquad (17)$$
$$\vdots$$
$$G_j = \varphi_1 G_{j-1} + \varphi_2 G_{j-2} + \varphi_n G_{j-n}$$

Green function of the system represents the repulse response function, so the window function estimated from G_j includes the width region where the system characteristics locate in the cepstrum domain.

The method above is synthesized as follows

(1). Construct *ARMA (m, n)* model from response signal y_i and evaluate the value of parameters φ_i 、θ_j.

(2). Calculate G_j from φ_i, θ_j.

(3). When $G_j \to 0$, let $j = n_1$.

(4). Let $W(\tau) \le n_1$. ($W(\tau)$) denotes the window function width).

4. THEORETICAL BASIS OF TORQUE IDENTIFICATION USING ENERGY METHOD

The basic idea of the method is to identify torque spectrum of the machine from electric motor parameters (Chong, Chen 1995). The distinctive advantage of the method is that the electric motor parameters are easily acquired through measurements, such as parameters of electric current, voltage and power. Because of the inertia moment of the electric motor rotor and mechano-electrical coupling, some transformation is necessary to determine the torque spectrum from the electric motor parameters. This is the task of the torque spectrum identification.

According to the basic theory of electric motor engineering, the instantaneous power of a three-phase alternating current motor is given as:

$$p = \sqrt{3}V_l I_l \cos\varphi - 2V_l I_l \cos(\varphi + 60°) \cdot \sin 2\omega t \quad (18)$$

where V_l, I_l are line voltage and line current respectively, φ is the phase angle between the line voltage and line current. The first term in equation (18) is the active power and the second term is sinusoidal term with frequency 2ω. The information of the torque spectrum involved in the instantaneous power can be revealed through proper transformation.

According to some experimental results, the changes of low frequency components of the instantaneous power reflect the changing tendency of the torque, and the changes of high frequency components involve the information of sinusoidal quantity with frequency 2ω. The spectra of instantaneous power have dominant energy at working frequency and its harmonic frequency. From state above, two problems have to be solved to identify torque spectrum from instantaneous power. Firstly, in order to eliminate the influence of dominant working frequency and its harmonic frequencies on torque spectra, adaptive point-blocking filter should be used. Secondly, in order to eliminate the filtering effects of transfer function of the machine system on the input torque, cepstrum analysis technique introduced above should be adopted.

5. APPLICATION EXAMPLES

Load spectra of a road header in three rectangular axes were identified from acceleration response signals. It is difficult to obtain the transfer functions between measurement location and each cutter because the construction of the cutting head shown in Figure 2 is very complex. As comparison, the curves in Figure 3 were the measured load of three

Figure 2. The construction of the cutting head of a road header

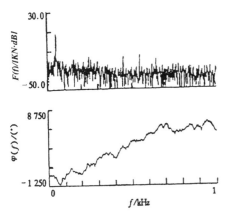

The measured load spectrum and its phase

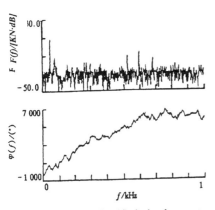

Figure 4. The identified load spectrum using cepstrum

rectangular axes directly from strains. The curves in Figure 4 were the identified load at the same working condition. From the results above, it can be known that it is feasible and effective to indirectly determine the load spectra by cepstrum analysis method.

6 CONCLUSIONS

1. The load identification is the process to eliminate the filtering effects of the transfer function from response signals. The identification quality depends on whether the method adopted can effectively eliminate the filtering effects or not.

2. Cepstrum techniques can be used to separate system characteristics or load spectra from the response output by windowing, so filtering effects of the system can be eliminated and the load spectra can be identified. To reduce the blindness in deciding the window width, a method based on Green function of a time series model is necessary.

3. The method to identify load spectra by cepstrum analysis is easier than other methods to be used in actual situation. Especially it is more suitable to mining machines such as road headers, for between cutting teeth and measuring points there are many transfer functions acting on the response signals.

REFERENCES

Chong, Chen. 1995. Research of torque spectrum identification by energy method: 17-27. Dissertation of China University of Mining & Technology. (In Chinese)

Daxian, Zhang. 1995, Modern signals analysis: 154-158. Tsinghua University Press. (In Chinese)

Miao, Wu. 1992. Study on the method of working load spectrum identification of mining machines: 28-51. Dissertation of China University of Mining & Technology. (In Chinese)

Zhifang, Fu. 1990. Modal analysis and parameter identification: 254-272. China Machine Press. (In Chinese)

The non-linear dynamic analysis of inner flow in centrifugal impeller

Y.M. Li & H. Xie
China University of Mining and Technology, Xuzhou, People's Republic of China

ABSTRACT. Based on experiments complex flow in the centrifugal machines is studied. The non-liner dynamics are used to define the phenomenon of inner fluid flow separation of centrifugal machines at points away from the design operating point. The approach allows quantifying complex three-dimensional flow in centrifugal machines, explaining the reasons for swirl flow, mechanisms of flow separation and stall, and their effect on pump performance.

INTRODUCTION

The calculation of the flowing boundary properties and characteristics in the rotor of centrifugal machines is mostly based on the data of experiments. The complicated inner fluid flow in the centrifugal machines makes it difficult to clarify completely the structure of the flow field. So it is significant to study the qualitative theory of differential equation and understand the flow principles of the inner fluid flow in centrifugal impellers. The theory analysis is justified by the experiment of oil flow.

1. PHYSICAL AND MAHTEMATICAL MODEL OF THE OIL FILM ON THE SURFACE OF BLADES

The curvature of the blades is supposed to be zero, shown in Fig.1. On the pressure side of the blades, the oil film moves slowly and elastically in the

boundary. The upper one is the separation boundary of oil film and gas flow, on which the speed and shear stress of the oil flow are same. Considering a tiny oil film group, shown in Fig.2, the Coriolis-force and the centrifugal force are taken into account, the control equations are as follows:

$$
\frac{\partial u_2}{\partial t} + (u_2 \frac{\partial u_2}{\partial x} + v_2 \frac{\partial u_2}{\partial y} + w_2 \frac{\partial u_2}{\partial z}) \mp 2\Omega v_2 = -\frac{1}{\rho_2} \frac{\partial \overset{*}{p}_2}{\partial x} + v_2 (\frac{\partial^2 u_2}{\partial x^2} + \frac{\partial^2 u_2}{\partial y^2} + \frac{\partial^2 u_2}{\partial z^2})
$$

$$
\frac{\partial v_2}{\partial t} + (u_2 \frac{\partial v_2}{\partial x} + v_2 \frac{\partial v_2}{\partial y} + w_2 \frac{\partial v_2}{\partial z}) = -\frac{1}{\rho_2} \frac{\partial \overset{*}{p}_2}{\partial x} + v_2 (\frac{\partial^2 v_2}{\partial x^2} + \frac{\partial^2 v_2}{\partial y^2} + \frac{\partial^2 v_2}{\partial z^2}) \quad (1)
$$

$$
\frac{\partial w_2}{\partial t} + (u_2 \frac{\partial w_2}{\partial x} + v_2 \frac{\partial w_2}{\partial y} + w_2 \frac{\partial w_2}{\partial z}) \pm 2\Omega u_2 = -\frac{1}{\rho_2} \frac{\partial \overset{*}{p}_2}{\partial x} + v_2 (\frac{\partial^2 w_2}{\partial x^2} + \frac{\partial^2 w_2}{\partial y^2} + \frac{\partial^2 w_2}{\partial z^2})
$$

$$
\frac{\partial u_2}{\partial t} + \frac{\partial v_2}{\partial y} + \frac{\partial w_2}{\partial z} = 0
$$

The boundary conditions are given as:

$$z = 0 \pounds \quad u_2 = v_2 = w_2 = 0 \quad (2)$$

$$z = h \pounds \quad u_2 = u_1 \pounds \quad v_2 = v_1 \pounds \quad w_2 = w_1 \quad (3)$$

$$\mu_2 \frac{\partial u_2}{\partial z} = \mu_1 \frac{\partial u_1}{\partial z} \pounds \quad \mu_2 \frac{\partial v_2}{\partial z} = \mu_1 \frac{\partial v_1}{\partial z} \quad (4)$$

Where, h represents the thickness of the oil film. The subscripts "1" and "2" represent the gas flow and oil film, $\overset{*}{p}$ represents the static pressure.

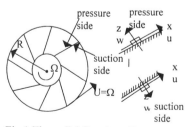

Fig.1 The radial flow impeller with straight blade

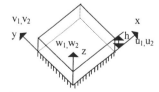

Fig.2 The coordinate of oil film and gas flow on the blade

$$p^{\cdot} = p - \frac{\rho}{2} r^2 \omega^2 \qquad (5)$$

The boundary condition at $z = h$ is the matching condition of the motion equation of the oil flow and gas flow on the interface. Under the assumption of boundary layer, the gas flow governing equations near the oil film, according to Brount's idea, can be reduced to:

$$u_1 \frac{\partial u_1}{\partial x} + v_1 \frac{\partial u_1}{\partial y} + w_1 \frac{\partial u_1}{\partial z} \mp 2\Omega w_2 = -\frac{1}{\rho} \frac{\partial p^{\cdot}_1}{\partial x} + v_1 \frac{\partial^2 u_1}{\partial z^2}$$

$$u_1 \frac{\partial v_1}{\partial x} + v_1 \frac{\partial v_1}{\partial y} + w_1 \frac{\partial v_1}{\partial z} = -\frac{1}{\rho} \frac{\partial p^{\cdot}_1}{\partial y} + v_1 \frac{\partial^2 v_1}{\partial z^2} \qquad (6)$$

$$\frac{\partial p^{\cdot}_1}{\partial z} = \pm 2\Omega u_2$$

$$\frac{\partial u_1}{\partial t} + \frac{\partial v_1}{\partial y} + \frac{\partial w_1}{\partial z} = 0$$

In order to establish the relationship between the oil film flow and gas flow, the governing equations of the oil film flow are simplified by the analysis of order of magnitude. In this experiment, the characteristic coefficient is much bigger than that of viscosity of air. So $\lambda = \frac{\mu_1}{\mu_2}$ is a very small term (in fact $\lambda = 10^{-3}$). Order analysis of each term of the oil film governing equation can be conducted. The boundary conditions at $z = h$ are

$$\mu_1 \frac{\partial u_1}{\partial z} = \mu_2 \frac{\partial u_2}{\partial z} \quad and \quad \mu_1 \frac{\partial v_1}{\partial z} = \mu_2 \frac{\partial v_2}{\partial z}, \qquad (7)$$

and the derivatives of velocity to time can be written as:

$$\frac{\partial u_2}{\partial t} = \frac{\partial u_2}{\partial h} \frac{dh}{dt} \pounds \frac{\partial v_2}{\partial t} = \frac{\partial v_2}{\partial h} \frac{dh}{dt} \pounds \frac{\partial v_2}{\partial t} = \frac{\partial v_2}{\partial h} \frac{dh}{dt} \qquad (8)$$

In this way, the orders of derivatives of velocity to time and location are deduced in the governing equations of oil film flow. Since Coriolis-force exerts on it, and $\frac{\partial p^{\cdot}_1}{\partial z} \neq 0$, according to equation (6), the pressure varies in the direction of z in the boundary layer of gas flow and the oil. Comparing each term of the expression, neglecting the tern of high degree, the governing equations can be reduced to

$$v_2 \frac{\partial^2 u_2}{\partial z^2} = \frac{1}{\rho_2} \frac{\partial p^{\cdot}_2}{\partial x} \qquad (9)$$

$$v_2 \frac{\partial^2 v_2}{\partial z^2} = \frac{1}{\rho_2} \frac{\partial p^{\cdot}_2}{\partial y}$$

The pressure term of the equation above can be derived from the non-viscous flow outside the boundary layer, so the equations above are the two degree ordinary differential equation of u_2 and υ_2. Through direct integration and boundary conditions

(2) and (4), the expression of u_2 and υ_2 can be obtained as:

$$u_2 = \frac{\mu_1}{\mu_2} \left\{ (\frac{1}{\mu_1} \frac{\partial p_2}{\partial x})(\frac{z^2}{2} - hz) + z(\frac{\partial u_1}{\partial z})_{z=h} \right\} \qquad (10)$$

$$v_2 = \frac{\mu_1}{\mu_2} \left\{ (\frac{1}{\mu_1} \frac{\partial p_2}{\partial y})(\frac{z^2}{2} - hz) + z(\frac{\partial v_1}{\partial z})_{z=h} \right\}$$

On the interface of gas and oil, $z = h$

$$(u_2)_{z=h} = \frac{\mu_1}{\mu_2} \left\{ (\frac{\partial u_1}{\partial z})_{z=h} h - \frac{1}{\mu_1} \frac{\partial p^{\cdot}_2}{\partial x} (\frac{h^2}{2}) \right\} = (u_1)_{z=h}$$

$$(v_2)_{z=h} = \frac{\mu_1}{\mu_2} \left\{ (\frac{\partial v_1}{\partial z})_{z=h} h - \frac{1}{\mu_1} \frac{\partial p^{\cdot}_2}{\partial y} (\frac{h^2}{2}) \right\} = (v_1)_{z=h} \qquad (11)$$

From the above two equations, it is understood that in the experiment the boundary condition in the gas flow around a body is different from that in true flow without oil film. First, the existence of the thickness of oil film transforms the effective shape of the gas flow around a body ($h \neq 0$). But the oil film is very thin, so its influence is very small (in the experiment, $h=0.1$ mm). Second, the gas flow on the surface of the body should match the no-slip condition in the real flow around a body. But, because of the existence of oil flow, u_2, v_2 on the surface of the body is usually not zero (the thickness of oil film is neglected), only when $(u_2)_{z=h}$ and $(v_2)_{z=h}$ are very small, the boundary condition of the gas flow on the surface of a body can be approximated as the no-slip condition of the true flow around a body. The two equations above show that if h is very small and $\frac{\mu_1}{\mu_2} \ll 1$, $(u_2)_{z=h}$, $(v_2)_{z=h}$ are high-degree small quantity, so the no-slip condition can be similarly met. Therefore, if the oil film is thin enough and $\frac{\mu_1}{\mu_2} \ll 1$, the influence of the oil flow on the gas flow around a body is very little.

The flow line equation of the oil film is

$$\frac{dy}{dx} = \frac{v_2}{u_2} = \frac{\frac{1}{\mu_1} \frac{\partial p_2}{\partial y} (\frac{z}{2} - h) + (\frac{\partial v_1}{\partial z})_{z=h}}{\frac{1}{\mu_1} \frac{\partial p_2}{\partial x} (\frac{z}{2} - h) + (\frac{\partial u_1}{\partial z})_{z=h}} \qquad (12)$$

The thickness of oil film is very small, meanwhile, the pressure gradient $\frac{\partial p^{\cdot}_2}{\partial x}$ and $\frac{\partial p^{\cdot}_2}{\partial y}$ is not very big.

Although $p^{\cdot}_2$ is affected by the centrifugal pressure, as shown in expression (5), the centrifugal pressure doesn't work in the field of velocity without free surface. Therefore equation (12) can be expressed as the pressure term.

$$\frac{dy}{dx} = \frac{v_2}{u_2} \approx \frac{(\frac{\partial v_1}{\partial z})_{z=0}}{(\frac{\partial u_1}{\partial z})_{z=0}} = \frac{\tau_{wy}}{\tau_{wx}} \qquad (13)$$

Where, τ_{wx} and τ_{wy} represent the components of the frictional force in the x and y direction of the gas flow on the surface of blade. The equation above indicates that the oil flow line approximately demonstrate the limit flow line of gas around airfoil blades.

2 THE DYNAMIC ANALYSIS OF THE FLOW SPECTRUM ON THE BLADES.

It is significant in theory and application to expound the principles of the development and break down by the study of the principles of the development and variation of the frictional force line. Here the principles of the flow state and its development are studied by dynamic analysis and experimental method.

To profile flow, every point on it has its certain vector of the frictional force. The frictional force lines exist in the vector field of friction stress too. Suppose the surface of airfoil blades is plane, the differential equation of the force line of friction is as follows:

$$\frac{dy}{dx} = \frac{\tau_{wy}(x,y)}{\tau_{wx}(x,y)} \qquad (14)$$

According to the qualitative theory of differential equation, the equation above can be reduced to

$$\begin{cases} \dot{x} = \tau_{wx}(x,y) \\ \dot{y} = \tau_{wy}(x,y) \end{cases} \qquad (15)$$

Let the flow on the blades be in a certain state, the force vector of friction at point $O(x_0, y_0)$ on the blades is zero, that is

$$\tau_{wx}(x_0, y_0) = \tau_{wy}(x_0, y_0) = 0 \qquad (16)$$

where point $O(x_0, y_0)$ is the singular point of expression (14). Take point $O(x_0, y_0)$ as origin of co-ordinates and spread τ_{wx} and τ_{wy} at point $O(x_0, y_0)$ according to Tailor-series, and take one-degree term, then

$$\begin{cases} \dot{x} = (\frac{\partial \tau_{wx}}{\partial x})_0 x + (\frac{\partial \tau_{wx}}{\partial y})_0 y \\ \dot{y} = (\frac{\partial \tau_{wy}}{\partial x})_0 x + (\frac{\partial \tau_{wy}}{\partial y})_0 y \end{cases} \qquad (17)$$

The coefficient matrix is

$$B = \begin{bmatrix} \frac{\partial \tau_{wx}}{\partial x} & \frac{\partial \tau_{wx}}{\partial y} \\ \frac{\partial \tau_{wy}}{\partial x} & \frac{\partial \tau_{wy}}{\partial y} \end{bmatrix} \qquad (18)$$

The characteristic equation of B is

$$D(\lambda) = \lambda^2 + p\lambda + q = 0 \qquad (19)$$

where

$$p = -(\frac{\partial \tau_{wx}}{\partial x})_0 - (\frac{\partial \tau_{wy}}{\partial y})_0 \qquad (20)$$

$$q = (\frac{\partial \tau_{wx}}{\partial x})_0 (\frac{\partial \tau_{wy}}{\partial y})_0 - (\frac{\partial \tau_{wx}}{\partial y})_0 (\frac{\partial \tau_{wy}}{\partial x})_0 \qquad (21)$$

The characteristic roots are

$$\lambda_{1,2} = \frac{-p \pm \sqrt{p^2 - 4q}}{2} \qquad (22)$$

The characteristics of λ are completely determined by p and q. The flow condition in the impeller decides the condition of flow around a body and the distribution of the friction force line on the surface of the impeller. So different voiles of p and q can be obtained at the singular point $O(x_0, y_0)$. According to the qualitative theory of differential equation, there are seven types of singular point in the p-q phase plane. The pathes of three types of singular point combining with the study of oil flow experiment are discussed in this paper.

2.1 $q < 0$, λ_1 and λ_2 are real roots of different sign

According to the qualitative theory of differential equation, the track lines can be drawn near the saddle point, shown as Fig.3. It can be seen that there are two special frictional force lines which extend to or against the saddle point, the others all extend away from the saddle. In the experiment, the oil flow spectrum of the inner profile of centrifugal impeller is got, shown as fig.4, when relative rate of flow $\overline{Q} = 0.44$, the properties of partial flow of saddle point type appear near the entry of the suction side of the blades.

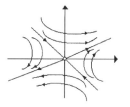 

Fig.3 The track lines of the saddle point.

Fig.4 The oil flow spectrum on the blade.

2.2 $q > 0$, $p^2 \square 4q > 0$, λ_1 and λ_2 are real roots of different sign

Singular point $O(x_0, y_0)$ is a node. The same goes for its track line system. The four semi-coordinate line and system of parabolic are shown in fig.5. If $p > 0$,

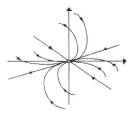

Fig.5 Singular point O(0, 0) is a node.

Fig.6 The oil flow spectrum of node on the blade.

all the track lines extending to node are called adhesion node. If $p < 0$, all the track lines against the node are called separation node. From the experiment of oil flow, this shape of track lines also exists in the impeller. Fig.6 shows the oil flow spectrum on the surface of blade in centrifugal impeller when the relative flow rate is $\overline{Q} = 0.44$.

2.3 $q > 0$, $p^2 \square 4q > 0$ and $p \neq 0$

Under this condition focus forms, shown as Fig.7. It is found that the integral properties of the singular point exist in the oil flow fluid spectrum according to properties of topology. In the flow state that airfoil has angle of attack, besides one or several saddle points, a force line of friction from these saddle point come into the focus in terms of spiral line. Hence the flow analyses from the angle of topology, focus is bound to exist in the separation flow of the centrifugal impeller. The spiral fluid-spectrum of the oil flow was discovered in experiment shown as Fig.8. A separation spiral point is a common type of singular point in the oil flow spectrum on the surface of a body in the case of angle attack. In the centrifugal impeller separation spiral point occurs near the front cap on suction side of the blades when the flow rate is small.

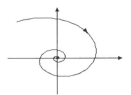

Fig.7 Spiral line (focus)

Fig.8 The spiral fluid-spectrum of the oil flow on the blade

3 CONCLUSIONS

1. In the oil film flow on the surface of the blades of centrifugal impeller the force line of fiction can be approximately taken as the limited flow line.
2. The analysis and discussion above show clearly that there are many singular points in the oil flow fluid spectrum on the surface of a body in any certain condition. Those singular points compose the fluid spectrum according to certain distribution and combination principles.
3. Although the type, number and distribution of singular points in the fluid spectrum on the surface of a body vary and present different oil flow fluid spectrum, the local shape near every singular point is restricted to the several types of the plane diagram of p-q.

REFERENCES

Y.M.Li, H.Xie & H.W.Zhou 1996A. Bifurcation of inner fluid flow of mining-turbomachine at the off-designed operating point. Guo Yuguang & Tad S. Golosinski (eds), *Mining Science and technology*. Rotterdam: A. Balkema.
Li Yimin, Xie Heping & Li Jianglin et al 1988. Dynamic Description of Flow in Centrifugal Impeller Inlet. *Journal of China University of Mining & technology*. 27(2).
Liu Shida & liu Shikuo 1994. Advanced Series in Nonlinear Science. *Solitary Wave and Turbulence*. Shanghai Scientific and Technological Education Publishing House:

Mining Science and Technology' 99, Xie & Golosinski (eds) © 1999 Balkema, Rotterdam, ISBN 90 5809 067 1

High property electronic ceramics humidity transducer used in mine

Shuyan Leng, Aiguo Kang, Qi Wang & Ping Huang
ShanXi Mining Institute, People's Republic of China

ABSTRACT: The principle and the technology of new electronic ceramics humidity transducer with fast response time, good-linear function and widen testing region is presented in this paper. The relation between ceramics surface potential barrier and the humidity has been developed.

1 INTRODUCTION

The humidity transducer, using the humidity ceramic as the sensing elements, is a new kind of equipment detecting humidity in mine. In order to control the density of donor and the number of fill atom and to control the stability of surface barrier, a new technology of doping and decorating electrode surface are advanced based on the general technique of ZnO-Cr_2O_3 humidity ceramics. As a result of above improvement, a new humidity transducer with better stability, better repeatability, easy operation, better linearity, longer operation life, higher accuracy, faster response, no-striking point, no-spark, and no-special demand with regard to environment has been made. It is an ideal instrument used to measure humidity in the poor environment of mine.

The relation of humidity ceramics resistance versus humidity has been explained by some people (Shimizu 1985, Cewen Nan 1987, Hongtao Sun 1988) based on the model of series shunt and composite dielectric. But the humidity ceramics is a complex system. The property of ceramics resistance versus humidity is related to many factors. In all those factors, the surface barrier is the chief. According to above viewpoint a new law relating to the grain surface barrier and humidity is proposed, which explained the property of $ZnCr_2O_4$ ceramics resistance versus humidity well.

2 THE IMPROVEMENT OF MAKING TECHNIQUE ON $ZnCr_2O_4$ HUMIDITY CERAMICS

The chemical composition of powder is composed according to the molecular formula: $(ZnO)x$ $+(Cr_2O_3)y$ $+(Al_2O_3)p$ $+(Li_2CO_3)q$ $+(CaO)r$, where, x, y, p, q and r are component quantity; and x:y=1:$10 \sim 10$:1; $p,q,r = 1\% \sim 5\%$wt. The humidity element is made on the base of general $ZnCr_2O_4$ ceramics technique improved by

some means such as to add a doping process, to change a sintering process, to decorate a surface electrode.

2.1 The affection on the characteristic of $ZnCr_2O_4$ ceramics by different ratio between ZnO and Cr_2O_3

The curve of the characteristic of humidity ceramics is showed in Fig.1, where the ratio between ZnO and Cr_2O_3 is different while other techniques are the same. It is indicated that the characteristic of humidity ceramics is very bad either more or less the ZnO is. The semi-conductivity of the sample can be improved by defining the optimum amount of ZnO.

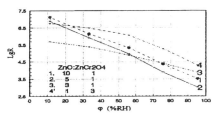

Fig. 1 The affection of different ratio between ZnO and Cr_2O_3

2.2 The affection of Ca-doping on humidity ceramics characteristic

By Ca-doping, the sintering temperature of humidity ceramics can be reduced and the mechanic strength can be increased. The linearity of resistance versus humidity is also improved. It is shown in Fig. 2.

2.3 The affection of Li-doping on humidity ceramics characteristic

The resistivety of ceramics can be lowed by adding the ion of Li that become the fill atom in grains. The

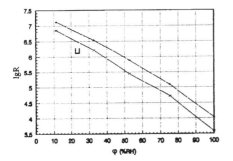

Fig.2 The affection of Ca-doping on humidity
ceramics characteristic

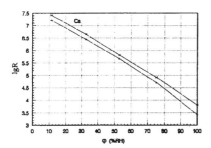

Fig.3 The affection of Li-doping on humidity
ceramics characteristic

result of experiment is shown in Fig. 3.

*2.4 The affection of Al-doping on humidity ceramics
characteristic*
The aim of adding Al-ion is to overcome the
shortcoming of the intrinsic semiconductor and to
improve the sensing property, the stability and the
linearity of humidity ceramics (Fig. 4).

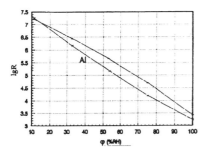

Fig.4 affection of Al-doping on humidity ceramics
characteristic

*2.5 The decoration of surface electrode on humidity
ceramics*
Because the sinter temperature of ceramics is high,

the electrode of RuO_2 can be adopted. After adding
some materials in electrode serosa and decorating
the surface, the response velocity is increased
largely. The hygroscope time is less than 3s and the
dehydration time is less than 10s.

3   THE   MECHANISM   OF   MEASURING
HUMIDITY BY $ZnCr_2O_4$ CERAMICS

The humidity ceramics, which is made by right
amount of $ZnCr_2O_4$ and excessive amount of ZnO,
has a structure of polycrystalline and multi-hole.
This structure is the base of the humidity-sensing
characteristic. The electrical property change of
surface with the amount of absorbing water is the
base of the resistance- humidity property.

*3.1 The surface states and the surface conductivity
of humidity ceramics*

In $ZnCr_2O_4$ humidity ceramics, the replacement of
Cr^{3+} to Zn^{2+} forms the donor impurity. So the
humidity ceramics become negative semiconductor.
The conductivity of ceramics can be adjusted in
certain range. The aim of improving ceramics
property by doping is to form the stable surface
states and interface states, and to restrain the
dependent connection between the intrinsic semi-
conductivity and other physic parameter, such as
temperature et al. It is sure that the conducting
process of humidity ceramics is related to only the
humidity of environment. The structure of surface
energy band of ion-crystal and the surface potential
barrier of semiconductor are shown in Fig.5 and
Fig.6 respectively.

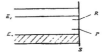

Fig.5 The structure of surface energy band of ion-
crystal

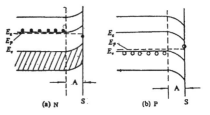

Fig.6   The   surface   potential   barrier   of
semiconductor

*3.2 The influence of absorbing water on conduction
of humidity ceramics*

In $ZnCr_2O_4$ humidity ceramics, the tiny hole appear

between the same gains which have different crystal direction and between the gains which have different crystal shape. The tiny hole can also be formed by volatilizing of additives (PVA) at high temperature sintering process. The volume of holes makes up to 30%□40% of humidity ceramics. The holes in ceramics take effect in forming capillary of water and in increasing the ratio of areas of ceramics that is good to sensing humidity.

Water is one kind of strong polar dielectric. The electric dipole moment of water molecular is 1.9× 10⁻¹⁸ Deby. The water molecular will take the precedence of being attracted by the surface of ceramics. The electrical conduction and the ionic conduction all act on the process of sensing humidity. Under the condition of low humidity, electrical conduction is the chief. But under high humidity, the ionic conduction is the chief.

3.3 *The model of humidity versus resistance about ceramics*

The humidity ceramics is a complex system not only in composition but also in its structure. The property of resistance versus humidity relate to many factors such as the category of ceramics, the semi-conductivity of grains, the grain surface barrier, the surface area, the surface states, the scale of hole, the shape of hole and the distribution of the hole. Among these factors, the surface barrier, the scale of the hole and the distribution of the hole are the main.

According to the principle of surface physics and the theory of semi-classic, if the ceramics surface can be regarded as a model of even one dimension, the distribution of potential can be demonstrated by Poisson equation (Zhu lubing. et al. 1992)

$$\frac{d^2\Phi(z)}{dz^2} = -\frac{\rho(z)}{\varepsilon\varepsilon_0} \qquad (1)$$

where ρ is the charge density.

If the density of donor energy level shown Nd is a constant, then:

$$\rho(z) = eN_d \qquad (2)$$

After integral, Poisson equation is changed to:

$$\Phi(z) = -\frac{e}{2\varepsilon\varepsilon_0} N_d^+ (z-d)^2 \qquad (3)$$

In above equation, $z = d, \Phi(z) = 0$. Where d is the thickness of surface charge layer. When $z = 0$, the surface potential is:

$$\Phi_s = -\frac{e}{2\varepsilon\varepsilon_0} N_D^+ d^2 \qquad (4)$$

Because the resistance formed by surface potential is larger than resistance of ceramics, the conductivity of humidity ceramics can be indicated approximately as (Heywang. 1961)

$$\rho = \rho_0 \exp(\frac{-e\Phi_s}{kT}) \qquad (5)$$

Under the condition of water absorbtion, the surface potential will be affected. If the surface defects, the scattering, the complexation of absorption and the change of surface can all be ignored, according to the simple Henry law, it follows:

$$\Phi_s = b\varphi(T) \qquad (6)$$

where φ is the relative humidity of enviroment. The relation between the conductivity of humidity ceramics and the relative humidity of enviroment is logarithmic linearity,

$$\ln \rho = -\frac{eb(T)}{kT} \cdot \phi + c \qquad (7)$$

4 THE TECHNIQUE PARAMETER OF $ZnCr_2O_4$ HUMIDITY CERAMICS

4.1 *The structure, the shape and the scale of humidity ceramics*

The shape of humidity ceramics is a rectangle thin. Its length is 8.4mm and the width is 6.1mm.The error of length and width is no larger than 0.05mm. Its thickness is 0.4mm,whose error is no larger than 0.02mm.The two surfaces of sample is smeared by RuO_2 thick liquid.The Pt slivers is used as electrode lead.

4.2 *The microstructure of $ZnCr_2O_4$ humidity ceramics*

The photograph of electronics microscope indicates the microstructure of $ZnCr_2O_4$ ceramics (Fig. 7).

Fig. 7 The electron microscope photograph of microstructure

4.3 *The property of resistance versus humidity of humidity ceramics*

The measured temperature of environment is 22.5□ and the relative humidity is 54%RH. The voltage of power is 2.0v, the frequency of power is 20Hz and the standard resistance is 5000Ω. The test curve of resistance versus humidity is shown in Fig.8.

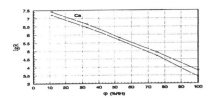

Fig.8. The LgR-φ curve of humidity ceramics

Table 1. The main technical parameters of humidity ceramics

Property	Value
Measuring Range (RH)	11%—100%
Linearity	0.02
Sensitivity	0.05
Precision	0.01
Temperature Coefficie	0.25%
Hygroscope time(s)	2.90
Dehydration time(s)	8.52

4.4 The main technical parameters of humidity ceramics

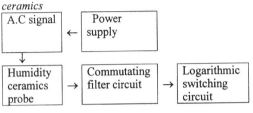

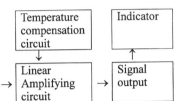

Fig. 9 The principle frame of humidity transducer

5. THE PRINCIPLE AND THE TECHNIQUE OF HUMIDITY TRNSDUCER

5.1 The principle of humidity transducer

The principal frame is shown in Figure 9.

5.2 The circulate characteristic of humidity transducer

1. The temperature compensation circuit

Because the temperature property of humidity ceramics presents a negative temperature coefficient (NTC), a thermal ceramics with positive coefficient (PTC) can be series connected in the circuit to make the temperature compensation.

2. The linear circuit

Because the resistance-humidity property of humidity ceramics presents a logarithms linearity, the antilogarithm amplifier is adopted to make the action of linearizing and amplifing.

5.3 The protection of humidity transducer

The flameproof of humidity transducer include: the power used by humidity transducer is a intensive safe low voltage D.C. power. The largest output current is less than 180mA.The mark of flame proof is Exi_bI. The grade of dustproof and waterproof is IP54 (wang Qi.1996).

5.4 The chief technical parameters of humidity transducer

Table.2 The chief technical parameters of humidity transducer

Property	Value
Humidity Range(RH)	11%-100%
Linearity	0.02
Linearity	0.05
Temperature Coefficient	0.25%
Hygroscope time(s)	3
Dehydration time (s)	10

REFERENCES

Heywang W. 1961. *Solid State Electron.* 3(1):51.
Nan Cewen 1987. *Physics Journal.* 36(10):1298-1305 .
Sun Hongtao & Wu Mingtang 1988. *Materials Science Progress.* 2(6):P66.
Wang, Qi 1996. *Jnl of China Coal Society.* 21(2): 216.
Y. Shimizu, H. Arai & T. Seiyama 1985. *Sensors and Actuators.* 7(11).
Zhu Lubing, Bao Xing 1992. *Physics of Surface and interface.* Tianjing University Publishers.

Mining Science and Technology' 99, Xie & Golosinski (eds) © 1999 Balkema, Rotterdam, ISBN 90 5809 067 1

Development of a very low frequency magnetic sonde to measure distance through rock

P.M.Stothard & W.J.Birch
Department of Mining and Minerals Engineering, University of Leeds, UK

ABSTRACT: The move towards room and pillar extraction methods with roof bolting for support has resulted in stopes and roadways needing to remain open for considerable periods. Monitoring of rock mass movements is important not only from a safety aspect but also from an economic standpoint. Rock mass behaviour is a valuable indicator of roadway life and allows accurate mine planning to be achieved. Manual data collection is expensive and often impractical. Automated methods are currently available in the form of single and multipoint extensometers, strain gauges, and stress meters. These systems require precise installation, are expensive and often have a limited range of measurement. Instrumentation of this type eventually becomes inaccessible and this can prove expensive. This paper discusses the development of a low cost alternative that can monitor rock mass movements over a range of 1-20m through rock and other solid material.

1 INTRODUCTION

The magnetic sonde prototype discussed here has been developed in the Department of Mining and Minerals Engineering at the University of Leeds.

The aim of developing the prototype was to discover whether distance and hence movement could be measured using low frequency magnetic fields. From the literature available on geophysical induction methods and other electromagnetic applications it appeared that magnetic fields passed through solid materials. It was found that these systems exploit the relationships that exist as a result of the propagation of electromagnetic waves and it was proposed that with some modification, a magnetic induction system could provide a means of short-range distance measurement through rock or other solid material. However, the relationships that exist as a result of the propagation of electromagnetic fields for anything other than very simple geometric arrangements are complicated. For this reason, an empirical approach was adopted that provided an initial data set from which distance could be calculated. The data set was obtained under laboratory conditions and demonstrated that distance can be measured using very low frequency EM magnetic fields through free-space (air). Field trials were performed with the system transmitting through solid rock and the results demonstrate that distance can also be measured through rock over the intended range of operation.

2. MAGNETIC SONDE PRINCIPLES

The magnetic sonde prototype operates via a low power transmitter-receiver arrangement similar to other induction and ground conductivity methods described by McNeill (1997). However, unlike these methods, it is the primary and not the secondary magnetic field that is of interest. The basic components of the system are a transmitter (Tx) coil and receiver (Rx) coil aligned co-axially (Figure 1). When current flows in the windings of Tx, a primary electromagnetic (EM) field is generated. Via inductive coupling between the two coils, the primary EM field present at Rx induces a voltage in Rx that is a function of distance and angle between Tx and Rx. The induced voltage at Rx is also a function of 'out of phase' secondary EM fields generated by the primary EM field in conductive materials adjacent to the system. However, the amplitude of the secondary EM field is generally small when compared to the primary EM field.

Phase differences between the primary and the secondary EM field depend on any adjacent conductor's properties as an electrical circuit (Telford 1995).

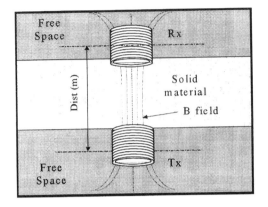

Figure 1. Co-axially aligned Tx and Rx.

The mathematical relationships that exist between the primary field and the many secondary fields present at Rx are very complex for any systems other the most simplistic (Keller et al 1966, Telford 1995). Therefore, an empirical approach was adopted to investigate the relationships that applied to the prototype thus allowing a set of calibration data to be established. The data provide a set of constants that are used to indicate the position of Rx with respect to Tx for a known angle θ along the axis of Tx.

3 ACQUISITION OF CALIBRATION DATA

Tx and Rx are essentially a series of coils that form a pair of solenoids. The primary EM field along the axis of a coil can be modelled by equation 1.

$$H_x = I \frac{a^2}{2(a^2 + x^2)^{3/2}} \qquad (1)$$

Where, a = radius of coil, x = distance along axis from centre of coil, I = current flowing in coil, H = B*μ_o, B = magnetic field, and μ_o = permeability of free space (Hammond 1993).

Using equation 1, 'H' was modelled along the axis of a coil and plotted in figure 2. The model demonstrated that 'H' decreases rapidly with distance and also that to establish the relationship between Tx and Rx a signal analysis method that could detect the very small Tx signal present at Rx above background noise was required.

Tx and Rx were placed on a calibration bed that

positioned them co-axially. Tx was mounted so that it could be turned through 90° in 5° increments (Figure 3). The bed was marked in 0.25m increments.

Tx was supplied with a low frequency alternating current to generate a primary EM field. The current was generated as a 10 Hz sine wave so that the received signal was easily recognisable and any transmission line or other noise would be carried on the 10Hz signal. The Tx and Rx signals were connected to a 12-bit A/D card installed in a Personal Computer. Rx was amplified by a gain of 500.

Signal processing software was used to capture and process the receiver and reference signals via the A/D card. The software incorporates many digital functions in module form that if approached electronically would produce complex circuitry. A schematic of the system is shown in figure 4.

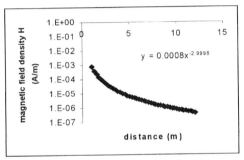

Figure 2. B field along the axis of a solenoid.

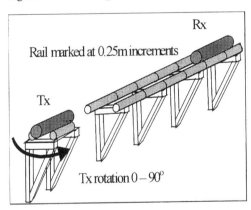

Figure 3. Calibration bed.

The most significant module used in the signal processing is the Fast Fourier Transform (F.F.T.) module. The F.F.T. is an algorithm that allows the frequencies present in a signal to be broken down into their individual components (Lynn 1994).

Several other modules were also used in the processing of the signal. The low pass Butterworth filter is a commonly used anti-aliasing filter and was set to 'second' order with a cut off frequency of 20Hz. The Data Window module allows selection of the signal sampling rate and type of data window that should be applied to the captured signal data. A Hanning data window was used with a sampling rate of 512Hz i.e. Above the Nyquist frequency. A blocklength of 2048 was used. The outputs of each module were displayed directly to the screen of a P.C. as a chart or digital meter.

However, despite using a highly accurate signal processing method, it was found that using VRx alone introduced errors. If the magnitude of the transmitted signal at Tx varied, there was a corresponding change in the voltage on Rx and the calculated distance would vary introducing a considerable error. To compensate for any fluctuations in the signal power the ratio VTx/VRx was used that provided a set of calibration data valid for all values of VTx. VTx being a reference value.

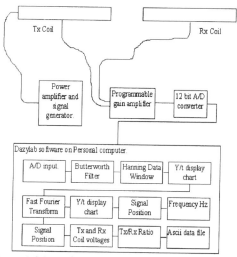

Figure 4. Magnetic sonde data capture.

To obtain the calibration data Tx was energised and the reference and receiver voltages recorded at each 0.25m increment from 12m to 1m along the calibration bed (figure 3). The process was repeated for angles from 0 to 90°. The data of the 0° line are plotted in figure 5. The plot of VRx against distance shows that the VRx follows the form of figure 2. Data was plotted to obtain a curve equation for each Tx-Rx angle and the constants relating to these curves were used to give a calculated distance for any point on the axis of Tx for a known angle

between Tx and Rx. A three-dimensional plot showing the effect of distance and angle on the receiver volts is plotted in figure 6.

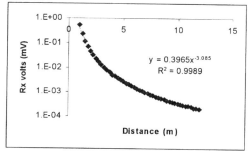

Figure 5. Plot of VRx against distance.

4 DISTANCE FROM CALIBRATION DATA

The worksheet of figure 4 was modified to incorporate the curve equation constants gained from the data so that a sonde distance could be calculated directly from the reference and receiver signal ratio for a known angle.

To examine the system accuracy under laboratory conditions Tx and Rx were placed on the calibration bed and the system powered. The angle between Tx and Rx was maintained at 0° and the calculated distance given by the sonde recorded at known distances from 1 to 12m. The sonde distance under laboratory conditions at 0° is plotted in figure 7 against measured distance.

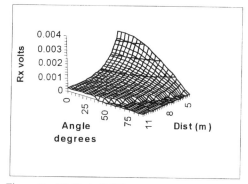

Figure 6a. 3D Plot of VRx.

During the tests it was noted that the distance given by the sonde fluctuated slightly despite of the use of the VTx/VRx ratio. Further inspection revealed that the reference and receiver coil voltages were fluctuating slightly causing sonde distance to differ after each block of data (Figure 8). To investigate whether the fluctuations were periodic the receiver was placed at 6m and the experiment run over several days. The data obtained over this period is shown in figure 8 and show that generally, the receiver voltage has the same curve as the reference voltage, however, there are several clearly visible troughs on the Rx curve that are not present in the Tx curve. An experiment was run over a week to investigate this further and the results are shown in figure 9 where variations in sonde distance are clearly visible.

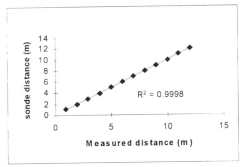

Figure 7. Sonde distance v measured distance.

The variations were found to be due to secondary magnetic fields that result from vehicles moving about outside the laboratory. The peaks in sonde distance represent the presence of vehicles.

During the investigation into the signal fluctuations the effects of temperature were also considered and several platinum thermistors were placed around the system, one on each coil and one in the centre of the calibration rail to monitor air temperature. The temperature data obtained in the experiment are plotted in figure 10. Sonde distance is affected by temperature. An increase of approximately 5°C produces a decrease in sonde distance of approximately 0.004m at a 6m separation.

5 PROTOTYPE FIELD TRIALS

Electro magnetic waves travelling through free space are unattenuated and those travelling through other media are attenuated to varying degrees depending on the conductivity and permeabilty of the material (Telford 1995). The very low frequency magnetic field used by the magnetic sonde has a wavelength much longer than the linear distance involved in monitoring with the result that little attenuation or phase change occurs during propagation through rock and air.

The depth of penetration of an EM field can be estimated using the 'skin depth', the distance over which the signal is reduced by 1/e (i.e. 37%) (Telford 1995).

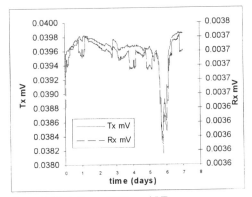

Figure 8. Variation of VTx and VRx.

In the Slingram method described by McNeill (1997), the transmitted signal frequency is kept low enough so that the electrical skin depth in the ground is always significantly greater than the inter-coil spacing provided that ground conductivity is also low. Operation of the sonde during field trials was at 10 Hz through the Millstone grit series where common accepted values for the resistivity of the rocks vary from $1.5*10^4 \Omega m$ for siltstones containing 0.54% water to $4.2*10^2 \Omega m$ for coarse grained sandstones containing 1.0% water (Telford 1995).

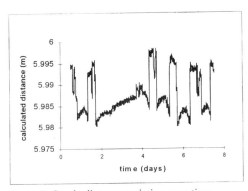

Figure 9. Sonde distance variation over time.

Using these values, and the formula (equation 2) .....

$$z_s \approx 500(\rho/f)^{1/2} \qquad (2)$$

.... where z_s = skin depth, ρ = Ωm and f = Hz, the skin depth was estimated to be in the region of 12 to

77m and therefore the EM field generated by the sonde should penetrate the rock easily over the distances involved and be detected at the receiver via inductive coupling as an Rx voltage similar to that obtained in the laboratory.

To demonstrate that the field would penetrate the ground and that distance can be measured through solid rock using the calibration data obtained from laboratory, a site was chosen that allowed easy access and positioning of survey stations so that measured distances and sonde distances could be compared. The Departmental Mine in the Greenhow Mining Area in North Yorkshire proved an ideal site, the Gilfield level is straight for a considerable distance allowing survey points to be accurately positioned using a Total Station (figure 11).

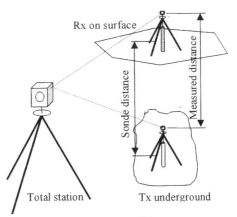

Figure 11. Positioning Tx and Rx.

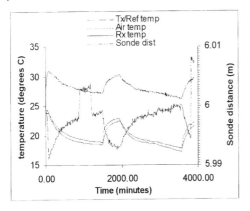

Figure 10: Temperature and sonde distance.

The Tx and Rx solenoids were mounted below survey tripods as shown in figure 11. To provide an initial position the receiver was positioned at the mine entrance and its position recorded by the total station with reference to G1 (a known survey point). The transmitter was then placed exactly above the receiver and using the Total Station in tracking mode. Data provided by the Total Station for Tx and Rx were recorded and the necessary corrections made. The sonde was energised and the sonde distance recorded after one minute. i.e. the mean of 15 data blocks. This procedure was repeated at regular intervals up to a distance of 130m from G1. The results are shown in figure 12 where the profile of the hillside and adit are clearly visible. Sonde distance against measured distance are plotted in figure 13. The R^2 for the data indicates a good fit of the data to the line suggesting that the distance given by the sonde be very close to that given by the Total station.

6 CONCLUSIONS

The relationship between Tx and Rx has been established using the calibration bed. The relationship proves that the voltage induced in Rx resulting from the primary EM field generated by a low frequency alternating current circulating in Tx decreases rapidly with distance from Tx as indicated by equation 1. Amplification of VRx is essential if the full-scale of the A/D card is to be utilised and 10Hz signal extracted from background noise and processed successfully. Without amplification and accurate signal processing of the Rx signal via the Fast Fourier Transform, the system would not function as a distance-measuring device.

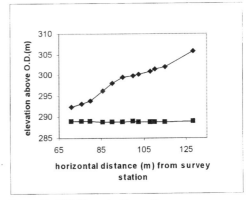

Figure 12. Hillside and adit profile.

Using VRx alone to provide a sonde distance measurement is unsatisfactory, the calculated sonde distance is affected by transmitted signal amplitude

701

variation that introduces errors. The ratio of VTx/ VRx is a more suitable method and compensates for variations in VTx.

The constants obtained from the calibration data can be used to determine distance by comparing the voltage on Tx to the voltage on Rx for a known angle along the axis of Tx.

The distances given by the sonde compare well with measured distances under laboratory and field conditions.

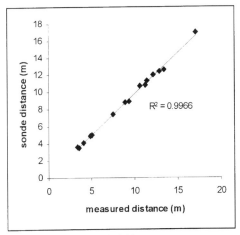

Figure 13. Accuracy of Field Data.

Varying the quantity of conductive material adjacent to the sonde affects the sonde distance reading. Large variations would have to be compensated for and a method of directly measuring secondary fields incorporated if highly accurate measurement was required.

Sonde distance readings are affected by temperature and will change if the coils do not maintain a constant or near constant temperature. If there is a temperature gradient between Tx and Rx this will introduce error in the sonde distance. However, the effect of variation of adjacent conductive material and temperature on sonde is small and the accuracy required would have to be agreed depending on the application. If these variables were kept constant or minimised, their impact on the system would be negligible.

Measuring distance using low frequency EM fields through air and rock was successful. The ability of the sonde to transmit through 16m of rock with little reduction in signal strength suggests that the sonde was operating within the 'skin depth' described by McNeill (1997) and Telford (1995). The profile of the hillside at Gilfield mine given by the magnetic sonde demonstrates that if the sonde was buried or placed in a borehole differential settlement could be monitored through rock or other solid media in real time.

The inclusion of an inclinometer to measure angle on Tx and Rx would also allow ride to be calculated.

7 FURTHER WORK

The field trials on the magnetic sonde prototype have proved successful and demonstrate that distance can be measured through rock using magnetic fields. The next stage of development of the sonde will be to incorporate angle measurement in to the device. This will allow position in three-dimensional space to be calculated using several Tx and Rx coils sets arranged orthogonally.

The next step in development of the one-dimensional model is to install the system in a working environment where any practical or logistics problems will be revealed.

REFERENCES

Hammond, P. 1993, Electromagnetism for Engineers, Pergammon Press, pp 70–100.

Keller, G.V. and Frischknecht, F.C., 1966. Electrical methods in Geophysical Prospecting: New York: Pergammon Press.

Lynn, P.A. 1994. An introduction to the Analysis and Processing of Signals. Macmillan.

McFarlane, J. 1984. Leeds University Mining Association Journal..

McNeill, J.D., 1997. The application of electromagnetic techniques to environmental geophysical surveys. In McCann, D.M., Eddleston, M., Fenning, P.J., & Reeves, G.M. (eds.). Modern Geophysics in Engineering Geology. Geological Society Engineering Special Publication No.12, pp. 103-112.

Telford, W.M., Geldart, L.P., Sheriff, R.E., 1995 Applied Geophysics. Cambridge University Press, pp 306-364.

Mining Science and Technology' 99, Xie & Golosinski (eds) © *1999 Balkema, Rotterdam, ISBN 90 5809 067 1*

Application of infrared spectrum in mining environmental inspection

RuLin Wang, Xia Wang & YongPin Liu
China University of Mining and Technology, Beijing, People's Republic of China

ABSTRACT: Compared with the widely used catalysis methane sensor in mining, the analysis principle of infrared gas absorption spectrum offers a number of advantages and is becoming an important research subject in mining environmental inspection and control. The analysis of infrared gas absorption spectrum, the design of optics system, the elimination of environmental noise, the signal processing and development of the instrument are reported on in this paper. An advanced design of optics system based on non-dispersed infrared absorption is also described as developed recently in China. Finally the further developments related to application and development of infrared spectrum technique in mining environmental inspection are discussed.

1 INTRODUCTION

With the development of mining industry, the safety problem becomes more and more important. The number of personal injuries and deaths caused by the methane explosion made up to more than 50 percent of all the serious injury in the coal mining accidents in China, so it is very important to measure the methane concentration in mining process.

Compared with other measuring methods, the system-inherent advantages of infrared absorption spectrum technique in mining environmental parameter inspection are:

1. The same inspecting method can be used to the measurement of different gases such as CH_4, CO_2, CO, NO and so on. This is good for the unified production, overhaul and maintenance in the environment control in the future.

2. The high sensitivity and wide measurement range make it possible to detect gas concentration from several PPM to 100% by Vol..

3. The response is rapid. It needs only 0.1 sec. from the measured gas coming into analyzer to gas concentration displayed.

4. The disturbance of background gases can be avoided, and different gases can be detected precisely and selectively.

5. The life span of the infrared absorption spectrum analyzer is over five years. In comparison with this, the life of generally used catalysis methane sensor or electrochemical CO meter is shorter because of the limit of their inspection principle.

With the wide application of infrared spectrum in different field, it can be expected that the infrared gas analyzers will be substituted for the catalysis instruments in the measurement of flammable gases and vapors, since the reasonable price and low running costs speak in favor of this technique.

2 THE PRINCIPLE OF INFRARED ABSORPTION SPECTRUM INSPECTION

The infrared radiation can be transformed into another form of energy, the vibration and rotation energy of molecules, when the material was illuminated by infrared radiation. The molecule oscillation increases as a function of energy uptake during the absorption process. The frequencies of these oscillations are the characteristics of the molecule. Radiation is only absorbed at the wavelengths that correspond to these frequencies. Basically, absorption can be easily measured.

Most gases have infrared absorption bands at the characteristic wavelengths. In non-dispersed infrared absorption, the transmission of infrared light through an absorbing gas follows the Beer-Lambert relation (Heng, Cheng 1985),

$$\tau_\lambda(c,l) = \frac{I(\lambda)}{I_0} = \exp[-k(\lambda)cl] \qquad (1)$$

Where $\tau_\lambda(c, l)$ is the transmissivity at wavelength λ, I_0 is the incident intensity of the light, $I(\lambda)$ is the intensity at wavelength λ observed after propagation through a length l of the absorbing medium, and $k(\lambda)$ is the spectrum absorption coefficient.

The design of infrared spectrum gas analyzer is based on the selective absorption of infrared light by gas molecules. Its working wavelength and the measured object are fixed in advance, and it can be used for continuous quantitative analysis.

3 THE STRUCTURE OF INFRARED SPECTRUM GAS ANALYZER

3.1 *The structure of light path*

According to the difference of the light path structure, the infrared spectrum gas analyzer can be divided into non-dispersed infrared (NDIR), certain-dispersed infrared (CDIR) and gas filter correlation three types. The design principle of these three kinds of analyzer is different, but they share the same working theory, main technique index and application field. Taking CDIR analyzer as an example, the measurement principle is described in the following(YunTing, Ling 1994).

The simple structure is the main advantage of CDIR. There is no gas leak problem and get the life-span by using solid detector. This kind of analyzer has clear superiority at instrument design in measure noxious, strong corrosive or unstable gas. Small size and low-cost are also merits of this instrument.

Photoconductive and pyroelectric infrared detectors, such as InSb and PbS, are in common used. One or more wavelengths must be determined according to absorption band of measured gas and detector. An additional interference filter makes the system work at the selected wavelength.

CDIR can be divided into geometrical single light path and dual light paths by the light path structure, and can be divided into time dual light beam and alternate light beam by signal processing.

1. Geometrical dual light paths

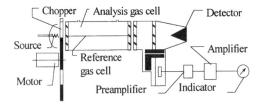

Figure1 Geometrical dual light paths

Figure1 shows the typical system of geometrical dual light paths. The optics system is consisted by a light source, two gas chambers, infrared interference filter and pyroelectric detector. The infrared beam, modulated by the rotating chopper, is focused through a narrow band-pass filter and then alternately transmitted through the measuring cell and reference cell, and the output of the analyzer is proportional to the difference of absorption between the two cells.

2. Geometrical single light path

Figure2 shows the typical system of geometrical single light path. The optic components comprise only a light source, a gas chamber and a detector. Two kinds of filter are equipped on chopper, a measurement filter and a reference filter, which select the appropriate wavelengths for the analysis. The infrared source radiation, modulated by the rotating chopper, result in time dual beams, reference and measurement light beam correspondingly. The detector receives two alternate signals of reference and measurement light beam respectively. Then at the effect of synchronous signal, the output of the analyzer is proportional to the gas concentration.

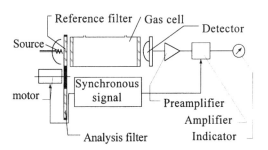

Figure2 Geometrical single light path

The analyzer of geometrical single light path is a simple, small, low-cost instrument, and is easy to be integrated. So it is especially fit for small portable instrument.

3.2 *Infrared source*

A current pulsed infrared LED can be selected as infrared source, with suitable filters two narrow-band infrared beams will be available, and the analyzer will has an expected working life of more than five years(S.F. Johnston 1992, J. Saffel & S. Johnston 1995).

The relation between the infrared radiation through measured gas and the output current of the detector can be expressed as following:

$$I = P\tau_\lambda(c,l)m\Omega R \qquad (2)$$

Where c is the gas concentration, λ is the selected wavelength, m is the transmission coefficient of filter, Ω is the detector solid angle of receiving radiation, P is the light source power, R is the responsibility of the detector, $\tau_\lambda(c,l)$ is the transmissivity of infrared light through an absorbing gas at wavelength λ.

The minimum radiation power of the source can be obtained from Eqs.(2).

$$P_{min} = \frac{I_{min}}{\tau_\lambda(c_0,l)m\Omega R} \qquad (3)$$

Where I_{min} is minimum detector output current, c_0 is the maximum measured gas concentration.

Thus infrared source can be selected and designed in accordance with Eqs.(3).

3.3 Light path length

When the change of gas concentration is Δc, the change of detector output ΔI can be obtained from Eqs.(2)

$$\Delta I = P_{min}m\Omega R\Delta\tau \qquad (4)$$

if $k(\lambda)\, c\, l \ll 1$, in accordance with Eqs.(1), Eqs.(4) can be expressed as:

$$\Delta I = P_{min}\, m\Omega Rl\Delta c \qquad (5)$$

Δc is the sensitivity of measurement, take 0.01% for example to CH_4 measurement, so the detector sensitivity ΔI can be obtained from Eqs.(5). Under the same principle, when P_{min}, ΔI and Δc are fixed in advance, the best path length l_{best} can be determined according to Eqs.(5). It is noted that the signal noise ratio problem of the measurement system should be considered fully, when P_{min}, l_{best} and ΔI are determined.

3.4 The structure of sensor

The key of the infrared spectrum gas anlyzer design is the optic system design and signal processing. The optic system can be designed as geometrical single light path (time dual light beams) structure, and use an open-path cell (Figure3), which is drawed from the design of existed infrared spectrum analyzers (WenKai, Cheng & YongPing, Liu 1997, F. Gume 1994). The gas pump and complex gas tube can be omitted in this open-path cell design scheme, so the whole instrument has the characteristics of small size, low power supply and is suitable to be a portable instrument used in mining.

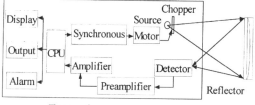

Figure3 The frame of analysis structure

A microprocesor which provides signal sample, calibration and processing is used in the signal processing system. This simplifies circuit design, easifies the correction of equipment, and increases the reliability of circuit. The functions of signal processing system are: (1) sample, comparation and average of the singnals of dual light beams; (2) signal linearization; (3) periodical auto-calibration; (4) over-level and fault alarm; (5) display and signal output.

It is easy to realize intelligent and intrinsic safe by processing data with microprocessor. The frame of analyzer structure is shown in Fig 3.

4 THE APPLICATION PERSPECTIVE OF MINING INFRARED SENSOR

Originally, infrared measurement technique was used as the analysis instrument in laboratory. However, they are less suitable for use in occupational protection and workplace monitoring, e.g. in mining. In recent years, with the development of optoelectronic technique in the world, it has been possible that an infrared spectrum gas analyzer used in mining will be developed soon by adopting the merit of the infrared spectrum gas analyzer in laboratory and considering coal mining condition.

4.1 The main problems of the infrared spectrum gas analyzer used in mining

1. The environment condition in mining is poor, water, dust and so on affect seriously the infrared spectrum gas analyzer used in mining. So the instrument should be considered specially to be equipped with dust-proof and moisture-proof. If open-path gas cell is adopted, the problem of glass dew-proof should also be considered under high humidity condition.

2. The Problem of flameproof must be considered in the infrared spectrum gas analyzer. The traditional method of sealed gas chamber, and the using of gas pump and pipe make the power and size of instrument large, it is not suitable to be used under mining. If open-path gas cell is adopted, the special flameproof must be arranged, it is a good ideal to adopt a compound structure by combining flameproof partition and intrinsic-safe partition.

3. Because the cost of producing the infrared spectrum gas analyzer is high, it is not suitable to used as portable instrument. Thus, at present, the main target is to develop a long-term continuous used sensor. The sensor should have a series function of indication and alarm, and can be connect with monitoring system used in mining by output suitable signal.

4.2 Background gas interference

The relative humidity is very high in most mining well, almost in the range of 80% □95%, and the carbon dioxide, CO_2, content is higher than on ground, especially in some cases CO_2 content can reach 0.5%. In different infrared wavelength, water vapor and CO_2 have different transmission coefficient. The cross sensitivity with respect to water vapor, carbon dioxide and other interfering gases is not wanted. So the correct selection of infrared waveband for CH_4 measurement can avoid the direct absorption influence of water vapor and CO_2.

According to the infrared absorption spectrum of H_2O, CO_2 and CH_4, it is clear that the transmission coefficient of H_2O and CO_2 at 3.39μm and 3.9μm are very small. So the influence of direct infrared absorption by water vapor and carbon dioxide is virtually eliminated in measuring CH_4 by using narrow band-pass optical filters if 3.39μm and 3.9μm are selected as measurement and reference light beam respectively.

4.3 The compensation of zero-point and sensitivity drift

In order to decrease zero-point drift and eliminate the effect of dirt, the structure of geometrical single light path and time dual light beams is adopted. The interference filter of the measuring light beam is set to wavelength of 3.39m. This wavelength is perfectly suited to detect methane. Gas mixtures in the open-path cell, that contains methane, absorb part of the radiation at this wavelength so that the measuring signal decreased. The signal of the reference light path (wavelength 3.9μm) is not effected in this process. The change of the output of the infrared light source, dirt on mirrors and windows as well as dust or aerosols contained in the air, have an identical effect on both light paths and are almost completely compensated by formation of the two alternate signals. The alternating electrical signals of the infrared sensor are amplified and transmitted to the microprocessor that performs all processing steps of calibration, linearization and temperature compensation. The concentration value obtained is finally displayed on the indicator.

5 RESULTS

1. By the study of theory and application, it can be induced that mining environmental inspecting sensors using infrared absorption principle have a series of merits, and it is the main developing direction in the future.

2. The suitable optics structure should be adopted in infrared gas sensors used in mining. It is an ideal scheme that the optics structure is geometrical single light path, time dual light beams and open-path gas cell with narrow-band filter.

3. The influence of the change of water vapor and CO_2 content in measuring of methane concentration can be avoided by selecting 3.39μm and 3.9μm narrow-band spectrum as the measurement and reference light respectively.

4. In order to increase the reliability and accuracy of the sensor in long-term continuous working condition, the intelligent auto-calibration functions of zero-point and sensitivity should be adopted.

REFERENCES

WenKai Cheng & YongPing Liu 1997. The analysis of France flame-proof infrared CO_2 analyzer. *Mining science and technology*. 25(3):7-9.

F. Gume 1994. Methane Monitoring. *Journal of the Mine Ventilation Society of South Africa*. 47(12):257-258.

S. F. Johnston 1992. Gas monitors using infrared luminescence diodes. *Meas. Sci. & Tech*. 3(2):191-195.

J. Saffel & S. Johnston 1995. Mid-IR LEDs for broadband optical gas detection systems. *Proc. SPIE - Int. Soc. Opt. Eng*. 2366:77-89.

YunTing Ling 1994. The choice guide of infrared products in China (eds). *Publishing house of electronics industry*:92-100.

Heng Cheng 1985. Infrared physics. *Publishing house of national defense industry*:18-21.

Mining Science and Technology' 99, Xie & Golosinski (eds) © 1999 Balkema, Rotterdam, ISBN 90 5809 067 1

Condition monitoring of machines in mining industry using vibration analysis

Bernd Bauer, Jianfeng Shan & Paul Burgwinkel
Aachen University of Technology, Institut für Bergwerks- und Hüttenmaschinenkunde (IBH), Germany

ABSTRACT: While vibration analysis is a useful tool for condition monitoring, reliable early fault detection in large and complex drive systems that are used in mining industry remains problematic. On the one hand the large size of such transmission parts requires multiple sensors and on the other hand common vibration diagnosis techniques do not work at extremely low rotating speed. This paper describes the monitoring and diagnosis of power transmission systems in mining industry using a patented sensor technology that is able to cope with these problems.

1 MACHINE DIAGNOSIS IN COAL MINING

In order to increase the economic efficiency of coal mining great efforts have been made to decrease the production cost. This aim can be achieved by reasonable maintenance of machines used in coal mining processes. Worldwide the maintenance strategies in all industries are changing towards a condition based predictive maintenance strategy. Only with this strategy a planned maintenance is possible because faults and failures of machines can be detected in a very early stage. This reduces unforeseen stillstands and production loss. But the condition-based maintenance requires powerful diagnostic methods to receive reliable information about the condition of machines and systems. Vibration analysis is certainly one of the most powerful methods for machine diagnosis.

The technique of vibration analysis has been widely used for many years as a tool for the detection of machine faults and has shown its advantages in many aspects. For example with the help of the vibration analysis it is possible to detect a fault of a machine at an early stage before the fault develops to an eventual failure and interrupts the production process. Its application saves a large amount of time for machine maintenance and reduces the production losses greatly.

But the application of the vibration diagnosis to the reliable fault detection of large and complex power transmission system in coal mining remains a great challenge. First of all, the complexity and size of those drive systems call for the use of many permanently mounted sensors. This is, of course, expensive. Secondly, extremely different speed and load ratios occur and the power take-off from the shift gear runs at very low speed which is a big problem for common vibration analysis techniques. Furthermore the environment in mining industry is very rough. Therefore the measurement equipment must be very robust. In underground mining the equipment even has to be flame and explosion proofed.

To deal with these problems two monitoring and diagnosis systems for machines in coal mining industry have been developed in close cooperation with the company ACIDA GmbH:

- a system for monitoring and diagnosis of the power transmission gearbox of a bucket-wheel excavator as used for open coal pit mining.
- a system for monitoring and diagnosis of machines in longwall mining for underground hard coal production

These two systems are in use for several years now and have successfully shown their advantages for the machine condition based, predictive maintenance strategy.

2 VIBRATION DIAGNOSIS FOR MACHINE MONITORING IN MINING INDUSTRY

The vibration diagnosis is normally carried out in the following main steps: signal measurement, signal analysis, diagnosis and strategic decision.

2.1 Signal measurement
To cope with the above mentioned problems in mining a low-cost sensor technology is developed which is called dtect. The applied sensors are knock sensors. Usually these accelerometers are used in cars. There they act as a part of the control device which regulates the combustion process of the car engine. These sensors are offered for a very low price. Unfortunately the frequency response characteristic of these sensors is not constant over the frequency. But a signal processing using the envelope analysis makes it possible to use these inexpensive and robust sensors even for precise fault diagnosis.

2.2 Signal analysis and diagnostics
For the signal analysis a method called envelope analysis was investigated. The envelope analysis technique is used for detecting even early faults on machines running with very low speed.

- *Diagnosis of antifriction bearings*
The defect of an antifriction bearing, e.g. a pitting on the outer race, is periodically run down by the rollers. Each of these run downs excites resonances of machine parts adjoining the damaged bearing. This may be considered as a kind of amplitude modulation with the resonance frequencies as the carrier and the shock impulses as the modulation signal. In general, enveloping separates the amplitude modulation from the carrier frequencies. In the case of bearing defects the shock impulses are separated from resonance frequencies. The shock impulse frequency is specific for the bearing defect and may be calculated with the geometric dimensions and the revolution speed of the bearing. Even in very complex machines like gear boxes an early

detection and isolation of bearing defects is clearly possible with the envelope analysis.

- *Diagnosis of tooth meshing*
Frequency analysis and Cepstrum analysis are mostly used for diagnosis of the gearing of wheels. Often wear on the teeth of gear wheels does not change the tooth meshing amplitudes significantly but produces sidebands around the tooth meshing frequencies and its harmonics. These amplitude modulations of the tooth meshing frequencies can be examined with the envelope analysis. Up to now the amplitudes of the rotary frequencies of the gear wheels and their harmonics are evaluated. Increased amplitudes may be a indication for increasing wear, but unfortunately sidebands also occur if imbalance or misalignment arises. That is why checking the amplitudes of these frequencies is not sufficient for detection of wear.

The envelope analysis was also investigated concerning parameters for detecting tooth wear. Research showed, that the amplitude of the hunting tooth frequency may be an appropriate parameter. Its amplitude in the envelope spectrum seems to increase if the wear of the teeth is increasing.

3 EXCAVATOR DIAGNOSIS

The monitored excavator has a length of about 200 m and a weight of 7760 tons. It conveys 100.000 tons of coal per day. The bucket wheel needs a very high driving torque at an extremely low speed. The drive consists of two 520 kW separate power trains, setting the bucket wheel (Fig. 1) going with four gear wheels. Sixteen vibration sensors are mounted at the gear box of the bucket wheel drive.

The vibration signal is transmitted through the multiplexer (Fig. 2) to the on-board computer in the excavator's control room. Additionally, signals of the load and the speed of the bucket wheel are measured together with the vibration signals. The software automatically sets the filter and amplification parameters of the vibration signal channels inside the multiplexer.

Channels of the multiplexer can be adjusted separately concerning sample rate, measuring time and low pass filter parameters. Moreover,

Fig. 1: The monitored Bucket wheel gearbox.

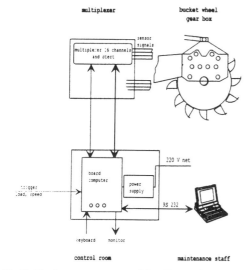

Fig.2: Hardware structure of the diagnosis system

the optimum amplification factor for the signal level is automatically chosen. This guarantees the best signal/noise ratio. The vibration signals depend on speed and load of the bucket wheel. The signals of these parameters are used for triggering to guarantee that only vibration signals under similar conditions are compared to each other. The dtect-signal which is an envelope signal is transformed into the frequency range. Characteristic frequencies in the power spectrum are extracted and evaluated. If a predefined alarm limit of a characteristic frequency is crossed an alarm message appears in the control room and is also sent to the handy of the responsible maintenance staff vice SMS (Short Message Service). On-board computer can carry detailed

data analysis . This is important because checking the amplitudes of characteristic bearing frequencies is sometimes not enough for a reliable diagnosis. Therefore, normally human experts compare the pattern of the vibration spectrum with typical pattern of the vibration spectra of bearing defects. The on board software uses a fuzzy-logic module which imitates this action of human experts. In this way the software is able to automatically recognize typical bearing defects as well as unbalance, beating of disengaged machine parts and other kind of machine failures.

The stored data records and classified fields can be read by the maintenance staff at any time by connecting a notebook to the serial interface link. It is also possible to supply a central maintenance computer with measured data and diagnosis results via radio transmission. Thus, continous online access to the information about the condition of various machines could be realized.

Figure 3 shows one example of the diagnosis results with the monitoring system in which a bearing defect in the gear box of the Bucket wheel excavator is detected. The envelope power spectrum is obtained from an enveloped vibration signal measured with a sensor mounted close to the defect bearing which runs at 1490 rpm. The roller defect frequency is about 62 Hz and the cage frequency is 10 Hz. The spectrum shows the typical pattern of a roller defect consisting of the roller defect frequency and its harmonics. Furthermore, sidebands with cage frequency and its harmonics occure around the roller defect frequency and its harmonics. The automatic diagnosis system detected this defect in a very early stage of damage. A visual inspection of the

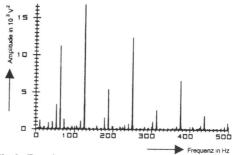

Fig.3: Envelope power spectrum

bearing after removal confirmed a small pitting on one of the rollers. It is considered that all excavators will be supplied with the presented monitoring system.

4 MONITORING OF MACHINES IN LONGWALL MINING

For monitoring and diagnosis of machines in longwall mining areas a special diagnosis system has been developed, which is different from the system in open pit mining. In regards to the conditions and environment underground the sensors and measure equipment in the longwall mining area must -be flame protected and explosion proofed. In addition, measurement equipment attached to the machines must be very robust. The less measure and analysis equipment is located in the longwall mining area the better. Therefore only the sensors and measure modules are in this area. The analysis, diagnosis and storing of data is performed by an online diagnostic system (Fig. 4).

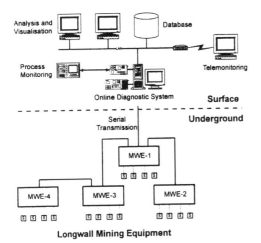

Fig. 4: Longwall monitoring system

To satisfy the requirements of longwall mining environment knock sensors are further developed. One or two of these sensors (flame proofed) are attached to each of the machines to be monitored. The vibration signals are measured with a data aquisition module (MWE), which was developed especially for the use in underground mining. Depending on the monitored machine additional trigger-signals, like speed or load, can be

measured simultaneously. Up to 8 vibration sensors may be connected to one MWE module. Each module has a microprocessor for the communication with the other modules and/or the communication with the surface online diagnostic system.

The measured data is stored in a database from which each analysis computer can call the data at every time. The data can also be transmitted from the coal mine to an external company vice internet, where the diagnosis experts can analyze the vibration signals more detailed. Thus tele-diagnosis and teleservice are possible worldwide.

The online diagnostic system performs an automated monitoring of the machines which are presented by colored symbols. The color describes the condition of the machine. When the machine is in good condition, the symbol is green. It becomes red when the thresholds (LdZ and RMS values) of the machine are crossed.

On the analysis computer the following analysis possibilities are available: the deep FFT analysis, the trend analysis, the time related FFT analysis and week analysis. Therefore it is possible to detect defects by different ways.

Since mid 1997 this monitoring system is used in one longwall coal mining face. From Nov. 1997 to Dec. 1998 many faults were detected and failures were prevented. Until now this system has been already applied to some other long mining faces and performs well.

REFERENCES

Bauer, B.; Ziegler, B. (1998). Verzahnungs-diagnose Neue Ansätze am Beispiel eines Voith-Turbogetriebes. In: Proceedings of AKIDA 98 in Aachen

Kempkes, A, Burgwinkel, P, Geropp, B (1995). Pattern Recognition with Fuzzy-Logic for Diagnosis of Antifriction Bearings. In: Proceedings of EUFIT'95 in Aachen

Kessler, H-W, van den Heuvel, B, Bauer, B, Geropp, B (1995). On-Line Fault Diagnosis of a Bucket Wheel Gear Box in a German Open Cast Coal Mine. In: Proceedings of the 4th International Symposium on Mine Planning & Equipment Selection in Calgary

Mining Science and Technology' 99, Xie & Golosinski (eds) © 1999 Balkema, Rotterdam, ISBN 90 5809 067 1

Error compensation method for light scattering dust sensor

GuoZheng Tian
China Agricultural University, Beijing, People's Republic of China

JiPing Sun & JianMing Zhu
China University of Mining and Technology, Beijing, People's Republic of China

ABSTRACT: In this paper, the principle of dust monitoring using light scattering method is introduced. Several factors influencing light scattering dust sensor and occurrence of errors are analyzed. The methods of reducing errors resulting from falling dust pollution on glass window, non-stability of light source, interference of background light and circuit property drift are presented. These result in better accuracy of measurements.

1 INTRUDUCTION

Using light scattering method to monitor dust concentration is a thought of better dust sensing technology. When dust is monitored on-line, pre-collector is not used, the particle's floating state can be preserved in air. The physical and chemical character of dusty air is not changed. Automatic continuous monitoring can be realized easily. But the response of light scattering dust sensor is influenced by dispersion of particles, light wavelength, receiving angle and optic-character of particles. Some errors are exist caused by dust pollution on glass window, unstability of light source, background light, circuit characteristics and selected devices.

2 THE PRINCIPLE OF MONITORING DUST USING LIGHT SCATTERING METHOD

When action takes place between light wave and measured field, some detected information can be sensed by changing of light wave's characters such as light intensity, phase, frequency or spectrum, polarization direction, spread direction, spread time, etc.

The technique of monitoring dust using light scattering is that; when light wave go through dusty air, light intensity in some direction is changed because of scattering of particles, so the information of dust concentration is obtained. This changed optical parameter is converted into electronic signal by opto-detector. The basic structure scheme introduced optoelectronic sensor of monitoring dust concentration is shown as Fig.1.

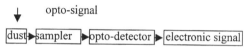

opto-signal

Fig.1 Principle scheme of opto-electronic dust sensor

In Fig.1, the opto-signal has the function of sensing component. Opto-detector and relative circuit has function of transducer. The prominent specialty of optoelectronic sensor is un-touch. It is not suffer from interference of electromagnetism (optical part). Sensitivity and accuracy are good. Resolution ratio of time and space is high.

According to scattering theory of Mie, the scattering light intensity of dust particle is related to wavelength(λ) of incident light, scattering angle(θ), radius(α) of dust particle and complex coefficient of refraction. Multi-factors should be considered when light source wavelength and receiving angle of dust sensor are selected. First, Mie.theory is based on scattering of global particle. The scattering light distribution of non-global particle is not related with Mie.theory completely. The mine dust particles have different shape. So the strict solution can not be obtained in theory. Second, Mie.theory is only suitable for conditions of single particle scattering and single dispersion independent scattering. The mine dusty air is multi-dispersion. When dust concentration is up to specific value, the light intensity illuminated to particles is weaker than incident light. The multi-scattering between dust particles will happen. In this situation, Mie.theory is not suitable. Even for single global particle, the relation among scattering light intensity, particle radius, scattering angle and concentration is very complex. The accurate theoretical calculation is very difficult and experimental method is often used. When ensuring optimum light source wavelength and receiving angle, the light scattering character curve of dust must be first measured at first.

3 THE METHODS OF REDUCING THE INFLUENCE OF DUST POLLUTION ON GLASS WINDOW

In light scattering dust monitoring system, the glass window can be polluted by falling dust. After a period of time, dust can be accumulated on glass window. The light penetrating ability of glass window is decreased and it influences the monitoring effect. In order to avoid measuring error caused by glass pollution, the following steps can be taken:
1. Using clean air to blow the glass window to avoid dust falling on glass surface. This method is used only when measurement is stopped timely. If cleaning and measuring are carried out at same time, the distribution of dusty air in sampling area will be changed, the measuring effect will be influenced. Pressing air equipment and clean air must be provided in this method. This will make equipment structure complex and expensive.
2. Protecting the glass of dust sensor from dust pollution, such as heating, adding protection board. These methods will make equipment structure complex and expensive. Dust monitoring effect will be influenced.

A compensating method was studied in this paper. The compensation method is introducing adjustment of dust pollution to measuring result. It includes fixed compensation with unknown dust pollution extent and auto-compensation according to dust pollution extent monitored continuously.

Fixed compensation is using software to eliminate dust pollution error according to dust pollution rule. Falling dust quantity on glass has a fixed rule along with monitoring time. Homologous math model can be found. Dust pollution error can be eliminated by computer software.

Auto-compensation is directly monitoring falling dust on glass window and eliminating dust pollution error. This paper uses light scattering method, designes propriety sampling cavity structure and monitoring angle. A detector is installed outside of the glass window to monitor scattering light intensity of falling dust. The influence of falling dust is eliminated. A light path structure of dust pollution auto-compensated system is designed shown in Fig.2.

In Fig.2, 1 is emitting light source, 2 is monitoring receiver,3 is compensating receiver,4 is light trap used for reducing reflection light. These four windows are round and the centers are at a plane. Because window 2 and 3 is very closed. The falling dust pollution of which is nearly same. The dust pollution error of window 2 can be eliminated by monitoring falling dust of window 3, the dust pollution error of window 1 can be also eliminated. When dust pollution is increased, the emitting light is decreased by falling dust on window 1 ,the scattering light intensity of floating dust to receiver 2 is also decreased by falling dust. Receiver 3 receives scattering light from floating dust and falling dust. When the dust concentration in air is increased, the scattering light intensity of floating dust is increased. At the same time, the scattering light intensity of falling dust is also increased along with falling dust. This shows that monitoring receiver 2 receives dust scattering light only from A area. The output signal is

$$V_M = K_1 C_F - K_2 C_A \qquad (1)$$

Compensating receiver 3 receives floating dust scattering light from B area and falling dust scattering light from glass window 3. The output signal is

$$V_c = K_1 K_3 C_F + K_4 C_A \qquad (2)$$

Constants $K_1 K_2 K_3 K_4$ can be obtained by test. The output signals of receiver 2 and 3 are sent to micro-controller system. The floating dust concentration in air C_F and falling dust concentration on glass C_A can be obtained. The measuring error of dust pollution is eliminated.

4 STABILITY OF LIGHT SOURCE AND CANCELING OF STOCHASTIC ERROR SIGNAL

When the reaction between light and dust happens, the scattering light intensity of dust is changed along with dust concentration. The scattering light is received by opto-detector and converted into electronic signal representing dust concentration. The principle is shown as Fig.3.

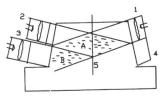

Fig.2 Light path structure of dust pollution auto-compensation system

light beam scattering light

Fig.3 The block diagram of light intensity
monitoring system

The output electronic signal V_0 is

$$V_0=(I_L+I_0+I_L')K_C+V_0'$$
$$=[\square_0Kf(Q)+I_0+I_L']K_C + V_0' \qquad (3)$$

where I_L= light electric current produced by signal light; I_0=dark current of opto-detector; $I_L^\square$=light electric current produced by background light; Q=dust concentration; $\square_0$=light source power; K_C=gain of amplifier; K=sensitivity of opto-detector; $V_0^\square$=zero drift of amplifier.

This shows that output signal is related not only to dust concentration but also to other factors such as background light, environmental temperature, circuit parameter etc. Any change of these factors can influence measuring accuracy by light source power, dark current, opto-detector character and circuit parameters. When monitoring dust concentration directly from scattering light intensity, above errors must be analyzed and compensated for necessary accuracy.

4.1 The stability of light source

The stability of light source is very important to monitor directly light intensity. It is seen from formula (3),the output has a close relation to light source power $\square_0$. The stability of light source is thought to be the first important in measurement. The emitting power of light source is mainly related to voltage or current of driving power supply. A stable driving power supply is very necessary for the stability of light source. There are many types of light source used in light measurement technology. In this paper, the LED is used to produce near infrared light. The emitting power of LED is nearly proportional to driving current in some range. A constant current driving circuit is used as Fig.4. The circuit used near infrared LED HG505 as

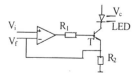

Fig.4 LED constant current driving circuit

scattering light source. It is a pulse light source modulated by synchro pulse signal (frequency is 34HZ).Light power can be changed by adjusting restrictive resistance R_2.

4.2 The canceling of background light, dark current, and zero drift of amplifier

Background light can not only bring error to monitor, but also reduce dynamic range of monitoring system. In order to eliminate the interfere of background light, a series measures are taken in this paper. Band pass light filter is installed in receiver to allow available light in the monitoring range to go through. Light trap is set up in sample cavity structure to eliminate reflection light right in front of monitoring receiver. Optic black paint is spread in sample cavity to decrease reflection light of cavity wall. Some relative measures are taken in detecting circuit.

The error caused by background light is usually a normal direct current. It is a slowly changing free signal. Errors caused by dark current of opto-detector and zero drift of amplifier are also direct current. A effective method to eliminate direct current background error is modulating light source to a fixed frequency pulse light. The modulated light source is a amplitude modulated carrier wave alternating current signal. Normal direct current or random interfere error signal can be cancelled by adding a narrow band filter in behind processing circuit. This filter has same frequency as modulated light source. The detecting circuit used in this paper is shown as Fig.5.

The scattering light signal of dust particles is converted into voltage signal in proportional to dust concentration by the circuit. After received by opto-detector, scattering light of dustparticles is converted into pulse voltage. The signal is very small. It is transmitted to pre-amplifier by capacitor. Interference signal is eliminated by following band pass filter. Phase demodulating is conducted by synchro-demodulator. Random noise is cancelled. Valuable periodic signal is picked up. At last, a direct voltage signal is exported by low pass filtering amplifier. It is proportional to scattering light intensity (or dust concentration).

5 CONCLUSIONS

1. Experiment shows that, it is a better dust sensing technology using light scattering method to monitor floating dust concentration. Floating dust concentration can be measured by monitoring scattering light intensity of dust particles in air. Automatic continuous on-line monitoring can be realized.

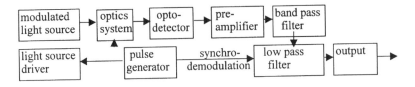

Fig.5 Diagram of light source modulating and detecting circuit

2. A light scattering compensating method is first put forward for eliminating error caused by falling dust on glass window. The relative light path system is designed. The test proved that the sensor's accuracy and uninterrupted monitoring time are increased.

3. In this paper, interference of background light, dull current and zero drift is cancelled by light source modulating and synchro-monitoring. The interference of light out of band is eliminated by interfere filter. The stability of light source is ensured by constant current driving method.

4. Sensing cavity structure is designed to sample natural airflow. The light reflection interference is avoided by light trap and light absorbing paint.

5. Many error compensating measures are taken. The performance of light scattering dust sensor is greatly improved. The accuracy is advanced.

REFERENCES

GuoZheng Tian, JiPing Sun & JianMing Zhu 1997. Study on compensation of dust pollution and light intensity distribution in the light scattering dust sensor. *Journal of china society*. 6:632-636.

GuoZheng Tian 1996. A study using a near infrared light scattering method to monitor coal mine respirable dust concentration. *A.A.BALKEMA*: 245-249.

Born.Max & Wolf Emil 1980. *Principles of Optics*. Oxford pergamen.

Anzhi He & Dapeng Yan 1995. *Prnciples of modern sensors and applications*. BeiJing.

Mining Science and Technology' 99, Xie & Golosinski (eds) © 1999 Balkema, Rotterdam, ISBN 90 5809 067 1

Dynamic simulation of driving technologies

V.L. Konyukh & V.V. Sinoviev
Institute of Coal and Coal Chemistry, Kemerovo, Russia

ABSTRACT: Dynamic simulation of driving by Monte-Carlo technique, specialized languages, problem-oriented simulator, animation, and object-oriented simulators is analyzed. Method of simulation consists of 9 stages. Road heading and drilling-and-blasting technology are simulated in Petri nets, GPSS/H, and PROOF Animation. Some results of simulation are adduced.

1 INTRODUCTION

Let's introduce the term "dynamic simulation" as a mathematical modeling of systems that consists of a lot of discrete elements changing in both space and time. There are both analytic and simulation methods of modeling. A simulation is more effective for compound dynamic systems that can not be described by differential equations.

It is not easy, to choose the best technology of driving for mining-and-geological conditions because there are a lot of stochastic equipment interactions in space and time.

After a creation of model a designer carries out so called "What if?" - experiments to find the best variant of driving. The aim of this work is to develop methods of driving design with the aim of computer simulation.

2 METHODS OF DYNAMIC SIMULATION

2.1 Monte-Carlo technique (1960-70)

At first a process was described by consecutive equations in engineering languages such as Fortran or Pascal (Rist 1961). Some coefficients in equations were changed randomly. Behaviour of a model was analyzed "step-to-step". The step of simulation was chosen for the quickest element. As a result, a designer had to analyze a lot of unnecessary information after simulation. Besides, a simulation was very difficult and unaccessible for mining engineers.

2.2 Specialized languages of simulation (1970-present)

These languages, such as SIMAN, SLAM, MODSIM, GPSS consists of special operators "create", "move to", "give to", "wait for" that are written in engineering languages (Sturgul 1997). As a distinct from the first method a process is described as an interaction of single dynamic models E1, ..., En (Figure 1).

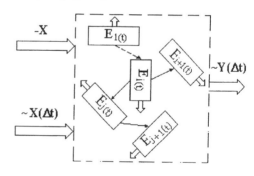

Figure 1. Method of simulation by specialized languages

Behaviour of a common model is analyzed "event-to-event" where an event means beginning or completion of a technological operation. The intermediate steps of modeling time are not shown. It enables to simplify a creation of model and change initial data quickly. During experiments a user evaluates output $\sim Y(\Delta t)$ for various combinations of both random $\sim X(\Delta t)$ and determine X inputs. The

General Purpose Simulation System (GPSS) is good applicable for mining (Sturgul 1997b). However it was not used for driving. GPSS is based on a queueing theory, where technology is simulated by random movement of demands through service instruments. The simulation by service system answers there questions: a queue of demands before instrument, a common time of demand service, an utilization of instruments. We regard any heading machine as a service instrument, that has to service demands, such as breaking, haulage, support setting. Common service system is the closed one because a service begins if a previous demand had been serviced by system. Perhaps, a queue of demands can be determined as $m=L/l$, where L=length of working before a service system; l=advancement for a single working cycle.

Let's imagine standart driving operations as sets of GPSS/H-blocks (Figure 2).

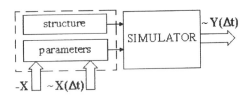

Figure 3. Method of modeling by simulator

Only Petry nets-simulator was created for mining in 1989 to simulate mining robotics systems (Konyukh 1997). It was written in Fortran, Turbo-Pascal 6.0, and Visual Basic. Petri net simulates a process as a movement of so-called tokens across transitions and places. Tokens linger in intermediate places for a duration that corresponds to given time of operations. Let's simulate a road heading by Petri net (Figure 4).

a)

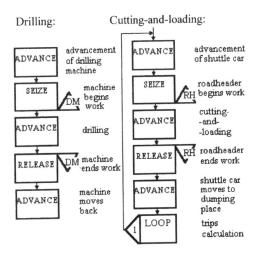

Figure 2. Standard GPSS-sets for driving

As you see, any operation is simulated by "SEIZE-ADVANCE-RELEASE"-blocks with random duration. A common technologies of driving were collected from these totalities.

2.3 Problem-oriented simulators (1989-present)

These simulators have standard forms where a non-programmer describes but structure and parameters of object (Figure 3). Simulator reflects a process as a change of object states for steps of modeling time. There are simulators MAST, MAP/1, SIMDIS that were written in Fortran for mechanical engineering.

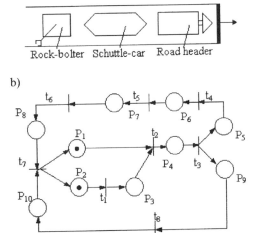

Figure 4. Road heading: a-technology; b-Petri net

At first an extraction by a roadheader (P1) and loading of a shuttle car (P2) take place. After a back filling (P3) a shuttle car moves to road mouth (P4). Then its unloading (P5), loading by support units (P6), return to heading (P7), and unloading of bolters at rock-bolter (P8) are fulfilled. At the same time inspection of heading (P9) and rock-bolting (P10) take place. Finally, all tokes go back in initial places P1, P2.

Changing time delays of tokens in any places of net a designer can choose an optimum set of equipment, determine "bottlenecks", evaluate an equipment utilization, simulate failures in any places of technology.

2.4 Animation and object-oriented simulation (1988-present)

Dynamic simulation must be clear proof for mining designers ang engineers. Lately new languages of computer animation such as CINEMA, ARENA, PROOF Animation are created (Sturgul 1995). Being connected with specialized languages of simulation, an animation enables to reflect movement of equipment on a computer schema of process according to simulation experiments. User can change any parameters of equipment, speed up a simulation, enlarge any part of process, collect statistics data, jump in time and space to analyze and forecast any technological situations.

Sometimes it is possible to use only animation for simulation of mining. There are object-oriented languages of animation TAYLOR II and WITNESS, that enable to construct and analyze technologies straight on a computer screen. WITNESS was used in Canada to simulate pillar mining (Vagenas 1995).

3 CONSISTENCY OF DRIVINC SIMULATION

Dynamic simulation consists of the following stages:
- description of technology;
- idea of driving as a service system;
- transformation of technological parameters, such as cross-section, rock hardness, length of blasthole, support increment into time of technological operations;
- programming in GPSS/H or in Petri nets;
- testing of accordance between a model and reality;
- computer imagination of technology;
- simulation experiments with random time of operations;
- animation;
- analysis of results.

For this simulation we used the complex GPSS/H-PROOF Animation under the license of Wolverine Software Corp., USA.

4 ROAD HEADING

The compound model consists of 31 GPSS/H-blocks that were connected with files of PROOF Animation. Besides, we developed the software to connect problem-oriented simulator with PROOF Animation. Dynamics of Petri net for roadheading was reflected as animation also. Driving equipment moves on a computer screen according to simulation experiment (Figure 5).

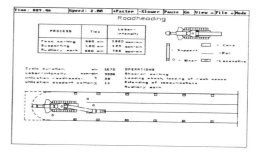

Figure 5. Still of road heading animation

A user can change any conditions of driving as a time of operations and get final indices of driving variant for any operation: duration of cycle, labor-intensity, utilization of machines, As an example, influence of shuttle car capacity V on the both roadheader utilization k and the duration of working cycle t was estimated (Figure 6).

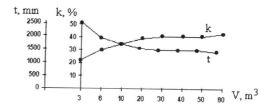

Figure 6. Cycle duration t and road header utilization k depending on a shuttle car capacity V

5 DRILLING-AND-BLASTING TECHNOLOGY

At first, two autonomous drilling machines drill blastholes for 180-260 min. Then, a charging and blasting (50-86 min), ventilation (30 min), and barring-down (8-12 min) are realized. After that, a scooptram removes rock masse to cars (110-170 min).

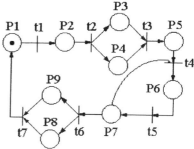

Figure 7. Petri net for drilling-and-blasting

Finally, support setter drives into a heading face to set support for 150-170 min. At the same time, auxilliary operations such as blastholes marking, communications advancing, and drainage extraction are fulfilled. This process was simulated by Petri nets (Figure 7) where P1 is movement of drilling machines to heading, P2-drilling of blastholes, P3-charging, P4-departure of machine, P5-blasting and ventilation, P6-work of a scooptram, P7-counter of n trips, P8-support setting, P9-auxilliary operations.

The other model consists of 49 GPSS/H-blocks including ADVANCE (drilling, movement, charging, ventilation, marking of blastholes, loading), ASSIGN (random working duration, number or trips), LOOP (cycles calculation). It is difficult, to learn a distribution law for every duration of operation by practice. We introduced the uniform, normal, and exponential distribution of operations duration into the model. A difference of cycle duration, labor-intensity, utilization of machines was about 5-7% for these laws. A user can change total length of a working, heading increment after a working cycle, and a random duration of any operation. After simulation he can see a total cycle duration, labor-intensity, and machines utilization on a computer screen (Figure 8).

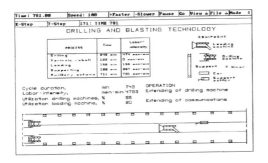

Figure 8. Still of animation for drilling-and-blasting technology

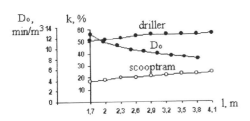

Figure 9. Utilization of machines k and specific extraction duration D_0 depending on blasthole length l

During experiments an utilization of equipment k and duration D_0 of one cubic metre extraction depending on blasthole length l were evaluated (Figure 9).

6 CONCLUSIONS

Monte-Carlo technique is difficult and not perspective for driving simulation. Models of road heading and drilling-and-blasting driving with the aid of GPSS/H, Petri nets were developed. Driving models were animated by PROOF Animation. This method enables to answer a lot of "What if"-questions such as choice of technology, co-ordination of machines work, discovery of bottlenecks. Using simulation and animation for driving designer or engineer can economize on investments before mining.

ACKNOWLEDGMENT

Thank to Dr.John Sturgul (Univ. of Idaho, USA) for the friendly advices during this work that was executed thanks to the grant N OUTR.CRG 960 628 "Simulation and Animation of Russian Coal Mines" of the Scientific and Environmental Affairs Division of NATO and modern software from Wolverine Software Corp. (USA).

REFERENCES

Rist, K. 1961. The solution of a transportation problem by use of a Monte Carlo technique. *Mining World, Nov.*

Sturgul, J.R. 1997 a. Simulation languages for mining engineers. *Mine simulation*: 29, incl. CD-ROM. Rotterdam: Balkema.

Sturgul, J.R. 1997 b. Annotated bibliography of mine system simulation (1961-1995). *Mine simulation*: 28, incl. CD-ROM. Rotterdam: Balkema.

Konyukh, V. & O. Tailakov 1995. Dynamic simulation of mining robotics systems. *Proc. of the 3-rd Intern. Symp. on Mine Mechanization and Automation*: Golden, v.2., 17.25-17.36. USA.

Sturgul, J.R. 1995. Simulation and animation: come of age in mining. *Engineering and Mining Journal*, Oct.

Vagenas N., M. Scoble & G. Baiden 1995. Simulation for design, planning and control in the automated mine. *Mine Planning and Equipment Selection: 271-276*. Rotterdam: Balkema.

Mining Science and Technology' 99, Xie & Golosinski (eds)© 1999 Balkema, Rotterdam, ISBN 90 5809 067 1

Technological progress in mechanization of lignite surface mining as a condition to maintain strategic role of this fuel in Polish power industry

Zbigniew Kozłowski

Mining Engineering Faculty, Technical University of Wrocław, Poltegor-Engineering Limited, Poland

ABSTRACT: In Poland lignite which constitutes some 40 percent of fuel for production of electric energy achived this strategie position mainly owing to the integral mechanization of mining operations. Introduction of technological progress in lignite mining systems is dated to the 60ties when one of world's first integrally mechanized Turów Lignite Mine was constructed and then, in the 80ties when one of world's largest Bełchatów Surface Mine with production capacity of 38.5 million tons per year was constructed. The use of comprehensive mechanization of mining operations coused that lignite in Poland makes it possible to produce electric energy cheaper by approx. 35% as compared to the production of electric energy based on the hard coal. A condition to maintain current role of lignite in Polish power industry is further implementation of technological progress in newly designed lignite open pits, such as Szczerców and Legnica.

1 INTRODUCTION

Although the history of lignite deposits exploitation is dated to the 18th century its considerable economic importance falls to the second half of 20th century, which is shown in fig.1.

Lignite in Poland is generally used for production of electric energy (almost 99 percent in 1998), and there are no real prospects for its substantial utilization in an other way in the near future (e.g. in horticulture or for manufacture of filters).

At present, in domestic lignite-fired power plants, there are installed 8703 MW, which is about 27 percent of generating capacity at industrial power plants. In the 90ties, production of electric energy from lignite varies between 37 percent and 43 percent of electric energy from industrial thermal power plants. An evidence for its strategic role in national economy is that lignite as a fuel fired in the industrial thermal power plants has been since many years remaining at a level of approx. 40 percent.

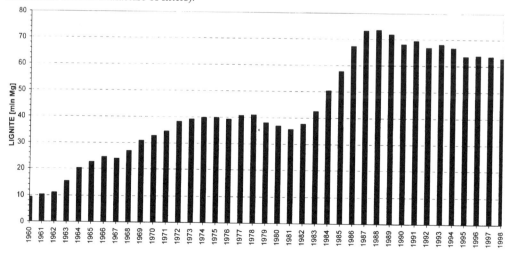

Fig. 1. Lignite production in Poland in the years 1960-1998 (mln Mg)

It is also manifested by a substantial portion of lignite in the national balance of primary energy recovery, where it is placed second after the hard coal. Costs of electric energy production based on the lignite are by approx. 35 percent lower than while using hard coal as a fuel.

A strong trump card in hands of Polish lignite producers is abundance of workable reserves of this mineral enabling them to be exploited at a current level during more than 300 years.

It also results from long-term economic forecasts that till 2020 electric energy requirements in Poland shall increase relative to the current level by more than 60 percent.

It is also known that about 97 percent of electric energy in Poland is generated based on the solid fuels, i.e. hard coal and lignite, and that during next 20 years, it will remain to be substantially unchanged.

This provides chances for lignite sector in Poland to maintain all operative mining basins until exhaustion of reserves and to undertake exploitation of prospective lignite deposits as f.ex. Legnica deposit area with reserves of more than 3 milliard Mg.

An essential condition to take adwantage of this opportunity is, due to more and more distinct conditions of market economy, to maintain competitive prices of lignite as compared to other fuels. In turn, complying with this requirement will only be possible by modernization of production systems to enable exploitation of lignite deposits with less favourable geological and mining conditions without considerable increase in production costs.

2 TECHNOLOGICAL PROGRESS IN POLISH LIGNITE MINES - EXPERIENCE AND PRESENT STATE OF THE ART.

A specific feature of Polish Lignite mines has been and still is their adaptation during construction to the up-to-date technological trends in the worldwide techniques of operation with due consideration of individual geological and mining characteristics of particular managed deposits.

Therefore, in particular lignite mining areas in Poland, there are employed considerably differing techniques of operation depending on the time-period when they were introduced, and local mining conditions as well.

In the Great Poland basin, two mines are operated, i.e. Konin Mine with production of about 12 million Mg/year and Adamów Mine with production of about 4.8 million Mg/year, in a multi-open pit system where depth of mining, except for one open pit (Lubstów) does not exceed 80 m. They are characterized by the use of multi-bucket excavators with low and medium capacity of the order of 2-10 million m^3/year and by a mixed belt conveying and rail mounted haulage.

The technological progress in these mines during particular time-periods was manifested by a gradual modernization of excavators by eliminating single bucket excavators and by more and more common employment of bucket wheel excavators instead of bucket chain ones, with a simultatioes expulsion of rail mounted haulage by belt conveying systems where successful options were introduced, among others, such as bench lift conveyors reducing transport distances of worked rocks.

The Turów Mine extended intensively at the beginning of 60ties was at that time an example for the integral mechanization of mining operations by using at about 150 deep Turów II Open Pit fully mechanized ECS complexes (Excavator-Conveyors-Spreader). There were employed BWEs with average capacities of the order of 30-40 thons. m^3/day and 1400-1800 mm and 2000 mm belt conveyors (collecting ones) with 320 and 630 kW main drive units.

The successful implementation of continuous mining systems in the 60ties has paved the way for Polish lignite industry to the economic management of deep open pits down to 250 m.

This experience enabled to construct at the turn of 70ties and 80ties the Bełchatów Mine with production capacity of 38.5 million Mg per year with use of a next generation of main machinery. In the overburden, there were employed so-called one hundred capacity excavators with production capacity of the order of 30 million m^3/year operating in conjunction with 2250 mm steel-cord belt conveyor systems and 1000 kW drives. In lignite, due to selective working, there were employed excavators with efficiency of the order of 8-10 million m^3/year and 1800 mm belt conveyors and 630 kW drive units. The next time-period of importance for the development of mining technology in lignite mines was the beginning of 80ties when mobile head stations of conveyors B-1800 mm were introduced on lignite benches of Bełchatów Mine and at the turn of 80ties and 90ties when, in connection with transition to the internal dumping and liquidation of stationary overburden and lignite distribution station, a necessity has occurred to adapt equipment complexes to the reversion of mined material on successively reconstructed inclined planes (distributing conveyors) and to the haulage of overburden downhill the open pit to the internal dumping site (decelerating conveyors).

The above solutions were forced by the necessity to adapt handling systems to the varying technological options different during the period of deposits opening, exploitation and also, for external and internal dumping.

The present-day indices for production in Polish lignite mines are given in table 1.

Table 1. Lignite production and overburden stripping volume in 1998.

Surface Mine	Lignite production mill. Mg	Overburden stripping mill. m³	O:L (overburden :lignite)
Adamów	7,8	34,9	7,35
Bełchatów	35,4	132,1	3,73
Konin	12.7	60,4	4,77
Turów	9,9	47,8	4,82
Total	62,8	275,2	4,38

In 1998, the achieved indices enabled all lignite mines to achieve favourable financial effects in electric energy production based on this fuel. In the future, however, there is to be prepared for necessity to undertake exploitation of lignite deposits with less favourable conditions at an average O:L ratio of some 6-7 and also-due to the necessity to collect funds for final management and reclamation of more and more deep and large, as regards their volume, final excavations.

As yet, the modernization of equipment complexes in Polish surface lignite mines has mainly consisted in the employment of more and more efficient and reliable main machinery and common introduction of continous mining techniques of operation. To maintain favourable indices in the future under worse and worse operating conditions it will be necessary, however, to investigate possibilities to improve the effects in the substantially wider range.

3   NECESSARY   DIRECTIONS   OF TECHNOLOGICAL PROGRESS IN LIGNITE MINING

More than fifty-year practice of lignite exploitation in Poland has shown a number of less or more individual features of this branch of industry, which have to be taken into consideration to achieve best possible both technical and economic effects.

These features comprise necessity to:
– achieve high concentration of lignite extraction and overburden stripping,
– adapt equipment complexes to a quick mining face advance and to different   stages of

industrial deposit management starting from opening phase through the exploitation period up to the period of management and reclamation of final excavations,
– become proficient in the mining technology of associated minerals occurring in the overburden,
– create possibility at mine-site disposals to store power plant waste,
– provide during lignite deposits exploitation suitable conditions for management of final excavations.

To ensure high concentration of lignite production and overburden stripping under conditions of Polish lignite mines modernization of continuous mining systems shall be expected. The excavators and spreaders in all the mines, due to the assumed extension of their service life by 20-40 years, are subject to their profound modernization or replaced by new equipment.

It is similar situation with haulage dominated by belt conveying. In planned to be constructed new Legnica Mine with production capacity of minimum 25 million Mg/year and overburden stripping of about 200 million m³/year, it will be necessary to use next generation of so-called two thousend capacity excavators with yearly efficiency of about 50 million m³. The employment of this machinery will bring about necessity to use other components of equipment complex, such as conveyors and spreaders of adequately higher technical parameters.

The quick mining face advance in mines of high production concentration amounting to several hundred metres per year brings about the necessity of permanent reconstructions of equipment complexes, especially conveying lines. These reconstructions are also caused by variable technological assignments at particular stages of exploitation (opening up, mining operations with external dumping, mining operations with internal dumping, reclamation of post-mining open pits).

The solution of this problem may be expected by using equipment   easy shiftable or easy to be transported at working faces and also, easy to be relocated to the successive haulage ramps.

The technological progress in this field has been initiated and will be continued in lignite mines by eliminating mobile head stations within conveying lines on working levels and by replacing them by the stations on pontoons capable of being hauled. Also, stationary conveyor drive stations on inclined planes are practically constructed in a way enabling them to be easily relocated. To ensure an efficient organization of constructions and reconstructions of handling systems it will be required, apart from adaptation of handling systems components for haulage, to have the disposal of suitable equipment to carry the whole head stations and components of belt conveyors. In this area, Polish mines are well equipped having the disposal of suitable carrying units with 500 Mg load capacity and also, equipment to carry conveyor components of weight up to 60 Mg.

A next important element which requires to use different types of equipment within mining systems is necessity to incorporate in equipment complexes handling overburden and lignite special systems making possible selective working and storage of associated minerals and also, systems to transport power plant waste to the mine disposal site. An exemplary solution of this problem in the Bełchatów Mine is illustrated in fig. 2 showing layout of belt conveyors in this mine as of 1997.

In such a multi-task handling system, the extremely important element is a suitable reversible system making it possible to freely control excavated material to the power plant, overburden disposal area, storage yards of associated minerals and also, wastes to the selected dump levels.

The movement of excavated material takes place here bench by bench up and down and therefore, it is necessary to use extremely different elements of conveying routes. The future for such multi-task equipment complexes, apart from the differences between conveyors, is also optimization of technological process control systems based on dynamically developing techniques.

The next problem analyzed in Polish lignite mines since a short time only is necessity to adapt working processes to the planned future programme of final reclamation, especially very deep open pits. It is connected with the appropriate management of stockpiled masses of overburden and wastes in order to shallow mine excavations, and not only as regards particular open pits but more and more frequently, in respect of adjacent mines.

An important and frequently technically quite complicated is the problem of geotechnical protections which require to deliver definite portions of stripped overburden to the dump benches endangered by slides, or to deliver appropriate overburden into different areas of open pit in order to exchange the soil.

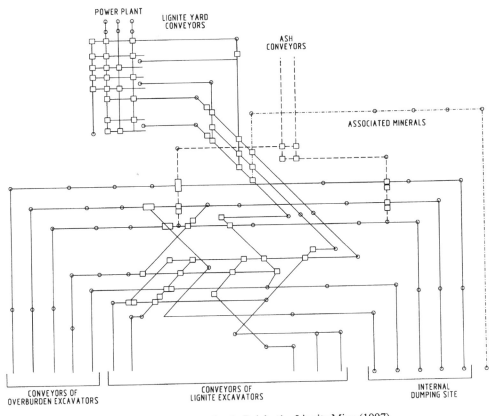

Fig. 2. Equipment complex in Bełchatów Lignite Mine (1997)

The solution of these problems is achieved by making appropriate revisions within handling systems, or by using equipment to unload transported material along conveying routes. Nevertheless, options used till now should be improved.

4 CONCLUSIONS

The present-day state of the art. in Polish lignite mines was achieved gradually along with world's technological trends in this field.
The achieved technological progress in surface mining enabled to undertake economic exploitation of lignite deposits despite worse and worse geological conditions of their working.
During the nearest tens of years, the lignite in Poland should still be a strategic fuel for production of electric energy. However, a condition for this is to maintain the rate of technological progress implementation while undertaking exploitation of deposits with less favourable mode of occurrence considering higher and higher requirements of environmental protection. Specialists involved in lignite exploitation in Poland are entirely convinced, however, that they shall meet assignments imposed to them owing to a substantial reserves of lignite deposits, proved technology, proper scientific-design and manufacturing background for machinery and equipment and also, with regard to the social acceptance to continue intensive working lignite deposits for purposes of power industry.

REFERENCES

Kozłowski Z. *Developmental Trends of Continuous Mining Systems in Polish Lignite Mining.* 5th International Symposium on Continuous Surface Mining. Wrocław, May 26-29, 1998

Mining Science and Technology'99, Xie & Golosinski (eds) © 1999 Balkema, Rotterdam, ISBN 90 5809 067 1

Management and planning in the sphere of production equipment maintenance

J.Černý
Chemopetrol (JSC), Power Production and Distribution Litvínov, Czech Republic

J.Dvořáček
VŠB, Technical University of Ostrava, Institute of Economics and Control Systems, Czech Republic

ABSTRACT: The authors present the importance and way of financial means planning for the creation of financial reserves covering the equipment repairs - costs items deductible from tax base of the manufacturing enterprise. These means are strictly conditional ones and the consequence of their erroneous statement presents losses of available net profit of the manufacturing enterprise due to excessive tax load or, alternatively, due to additional tax load from the next year profit. The way of legal reserves amount assessment was applied in production division Energetika where the production equipment is closely connected with the production equipment of Mostecká uhelná společnost (Most Coal Company), and in the whole series of cases, the machinery is identical with equipment of this largest brown coal mining company in the Czech Republic.

1 INTRODUCTION

The development of tax system and making it more accurate result in consecutive limiting of possibilities for legal reserves application for equipment repair - costs items deductible from tax base. The tax systems regulations settle these reserves as strictly specific, conditional and nontransferable ones as to the time and purpose are concerned. The consequence of their excessive assessment and eventual non-drawing-down are additionally taxed items (and, from that, resulting decrease of disposable profit of production organization) or, in case of overdrawing of these means, by creating of insufficient reserve as the interference into direct costs increase with the same result - i.e. decrease of profit volume. Factor of legal reserves creation is important both in the part of new production capacities putting into operation when the application of standard estimations for presupposed costs for equipment maintenance does not give sufficiently accurate data for assessment of legal reserves amount, with all consequences resulting from that. It is possible, with using of statistic methods of costs planning in the sphere of machines and equipment maintenance and

transformation results, to increase substantially the accuracy of planning of legal reserves actual costs, and, by that, to optimize consecutively the tax load in the relation to production organization profit. The analysis of costs for operation and maintenance of production capacity newly being put into operation, course of costs for maintenance during operation and consecutive decision making concerning the equipment reconstruction or renewal, as well as the prognosis determination of costs course in single operation periods can be applied in dependence on the scope of equipment considered.

2 PREPARATION OF DATA FOR EVALUATION

For data creation and their consecutive processing there is necessary to define spheres of production equipment where the greatest presuppositions for necessity of legal reserves creation can be met. In most cases it is the question of production complexes or single parts of machine equipment which are subject to permanent wear and have corresponding technological character. In further justified cases also single production equipment elements or their groups can be determined.

The following inputs are necessary for these groups:
1. Purchase price of tangible investment property.
2. Percentage of annual depreciation.
3. Structure of annual depreciation (depreciation cycle, depreciation percentage).
4. Equipment operation period.
5. Annual costs amount for similar equipment maintenance.

The determination of accurate data is necessary for obtaining of the presupposed costs corresponding characteristics.

3 SELECTION OF GROUPS BEING COMPARED

For this contribution the production equipment elements were selected which differ, in their characteristics, with the respect of usage duration of these equipment elements, variance of depreciation rates, as well as the different amounts of actual maintenance costs:
1. Machine equipment, new one, in operation up to 5 years.
2. Measuring equipment, new one, in operation up to 5 years.
3. Machine equipment, in operation for 5-10 years.
4. Measuring equipment, in operation for 5-10 years.

Number of defined groups depends, in real using, on itemization according to the point 2.

4 EVALUATION OF DATA INPUTS DEPENDENCIES

4.1 Production capacity being newly put in operation, event. in operation up to 5 years (actual costs transferred in percentage values).

In this table output the actual purchase price of tangible investment property was decreased by depreciation and this net book value was given in the second column of table No.1 in percentage from original purchase price. The actual maintenance costs were related both to net book value and to purchase price. These data are included in the third and fourth column of table No. 1. From table values the graphic output was worked out - see Fig. No.1. From graphic output the relation of gradually increasing difference between maintenance costs proportion in purchase price and maintenance costs proportion in net book value is evident, and, especially the cyclicality of course of maintenance costs proportions in ensuring of equipment operation capability. For definition of legal reserve amount, on the basis of operation experience gained, the value of maintenance costs proportion can be used, and namely of costs decreased by the difference of maintenance costs proportion in net book value and maintenance costs proportion in purchase price - we have obtained with the help of equipment characteristic technically similar, event. quite identical one. This difference ensures, with sufficient accuracy, the determination of legal reserve amount for repairs but it came to its non-

Production capacity being newly put in operation (machine equipment)

Table No. 1

Equipment operation duration	Equipment net book value	Maintenance costs . 100 Net book value	Maintenance costs . 100 Purchase price	Technical reason
[year]	[%]	[%]	[%]	
1	100,00	9,700	9,7	technical adjustments of operating parameters
2	90,00	3,556	3,2	costs for repairs according to reality
3	81,00	4,815	3,9	costs for repairs according to reality
4	72,900	8,505	6,2	costs for repairs according to reality
5	65,610	4,268	2,8	costs for repairs according to reality
>5	59,049	5,419	3,2	costs for repairs according to reality

consecutively

further repeating cycle according to using of tangible investment property up to physical and moral depreciation

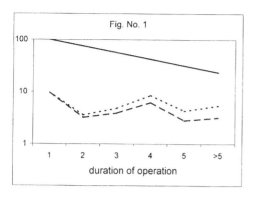

Fig. No. 1

duration of operation

Production capacity being newly put in operation (measuring equipment).

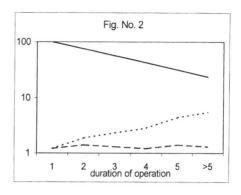

Fig. No. 2

duration of operation

drawing-down, and, at the same time, it decreases the actual amount of costs means expended for equipment maintenance in given year, and, by that, it optimizes the consecutive tax load of production organization. The maintenance costs amount can't be used in the level gained in case of similar equipment in the past because the presupposition of non- drawing-down as a result of higher equipment quality arises there. The value of legal reserve decrease ensures, to a sufficient extent, the covering of maintenance costs and minimizes the part of costs means which has to be split-off from the costs of the given year for covering of eventual maintenance costs increase in this year. By that, the state is reached, when the majority of costs is covered from tax-free legal reserve, and only a small part of maintenance costs is covered from costs of given year. In this way the dependence on profit creation in given year can be influenced because a part of maintenance costs is covered by tax-free created legal reserve of previous year. At the same time, the risk of additional tax load from non- drawn-down legal reserve means is being minimized.

4.2 Production capacity being newly put in operation, event. in operation up to 5 years (actual costs transferred in percentage values)

From the graphic output the relation of gradually speeded-up increasing difference between the maintenance costs proportion in purchase price and maintenance costs proportion in net book value and partial approaching to linearity of course relation of costs for equipment operation capability is evident. The substantial difference is evident there, and namely compared with machine equipment when

the influence of shorted depreciation period occurs in the relation to maintenance costs proportion by speeded-up decrease of tangible investment property value at observing of maintenance costs. In a final view then, there is evident that, compared with machine equipment, in this case the maintenance costs proportion in net book value is being increased, and, this fact, compared with development of machine equipment maintenance costs proportion explicitly defines the state where it is possible, for planning of legal reserves for new equipment, to use only equipment identical or similar from the viewpoint of creation and selection of groups being compared. For definition of legal reserves amount the values can't be used which are decreased by difference of maintenance costs proportion in net book value and maintenance costs proportion in purchase price, but it is necessary, with the respect of costs proportion linearity, to determine them in a such way that we'll use the value of maintenance costs proportion in purchase price decreased by the value of standard deviation (root mean square deviation) of the maintenance costs proportion file (set). On the basis of practical experience we take into account the maintenance costs fluctuation in time. The cause of legal reserve amount decrease is identical with the point 4.1 This difference ensures, with sufficient accuracy, the determination of legal reserve amount for repairs but it came to its non-drawing-down, and, at the same time, it decreases the actual amount of costs means expended for equipment maintenance in given year and, by that, it optimizes the consecutive tax load of the organization in the connections as mentioned in point 4.1.

4.3 Production capacity in operation for 5 - 10 years (actual costs transferred in percentage values)

Production capacity in permanent operation (machine equipment)

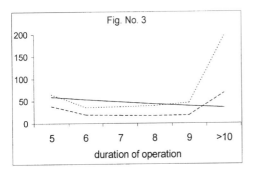

Fig. No. 3

From graphic output the relation of minimizing value of tangible investment property is evident when there was necessary to carry out the major overhaul for further using that increased only the utility value and, it is the signal to preparation if tangible investment property replacement (penetration of equipment net book value curve with the curve of maintenance costs proportion in net book value). For further period it is possible, for definition of legal reserve amount, to use the value of maintenance costs proportion in purchase price decreased by deviation value of the file maintenance costs proportion in purchase price - and namely - for the reasons mentioned in point 4.1. This difference ensures, with sufficient accuracy, the determination of legal reserve amount for repairs but it came to its non-drawing-down. From further course it is evident that it comes to penetration of equipment net book value curve with the curve of maintenance costs proportion in net book value, in short time horizon to further penetration with the curve of maintenance costs proportion in purchase price, and then, it is explicitly defined that further expending of financial means would not increase the equipment value but it is necessary to commence the total equipment reconstruction or to put it out of operation.

4.4 Production capacity in operation for 5 - 10 years (actual costs transferred in percentage values)

From the graphic output the relation of quick decrease of equipment net book value, nearly linear maintenance costs proportion in purchase price and increasing maintenance costs proportion in net book value is evident. For definition of legal reserve amount it is necessary, with the respect of

Production capacity in permanent operation (measuring equipment)

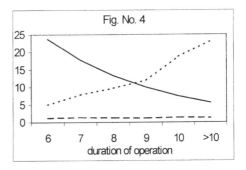

Fig. No. 4

maintenance costs proportion in purchase price linearity, to determine them in a such way that we'll use the value of maintenance costs proportion in purchase price of identical or similar equipment decreased by the value of standard deviation (root mean square deviation) of the file. The procedure is applied on the basis of practical experience with the respect of maintenance costs fluctuation (variation) in time. This difference ensures, with the sufficient accuracy, the determination of legal reserve amount for repairs but it came to its non-drawing-down. Further, from graphic output is evident that from the moment of penetration of equipment net book value curve with the curve of maintenance costs proportion in net book value the prolonged operation time period of the equipment remains without increasing of book value, however, from penetration moment the equipment can be characterized as morally obsolete and its replacement can be considered.

5 OUTPUTS EVALUATION

From analyses carried out and characteristics worked out the following conclusions result:

1. Creation of legal reserves for repairs and maintenance of tangible investment property from profit of existing calendar year for the following year makes possible the including of this legal reserve in costs of existing year and, by that, the decreasing of tax load of the existing year.
2. The legal reserve is useable strictly conditionally and only in the following year.
3. In case of legal reserve non-drawing-down it comes to additional tax load in the height of tax rates.

4. Correct determination of legal reserve amount makes possible the covering of presupposed costs for maintenance of tangible investment property of the following year, and, by that, the ensuring of these costs also in case of adverse development of profit creation in the following year.
5. It is possible, with the help of statistic method, with a sufficient accuracy, to define the creation of legal reserves amount for equipment repair.
6. With the correct definition of legal reserve amount it is possible to decrease the tax load substantially and dependence on profit in given year.
7. For single equipment types and groups it is necessary to work out separate outputs.
8. The defining of legal reserves amount for new equipment can be carried out with the help of outputs processed for equipment types identical or similar to equipment in operation.
9. Graphic outputs draw attention to presupposed terms of reconstruction or putting the corresponding equipment out of further operation.
10. The outputs inform sufficiently about equipment utilizability period without further costs increasing.
11. This way of analysis is applicable and verified for equipment with operation period longer than 5 years.
12. Equipment net book value depends on valid regulations for single equipment types (annual depreciation rate set by law).

In all figures:

————————— equipment net book value

- - - - - - - - - maintenance costs proportion in net book value

— — — — — maintenance costs proportion in purchase price

Mining Science and Technology' 99, Xie & Golosinski (eds) © 1999 Balkema, Rotterdam, ISBN 90 5809 067 1

Influence of cutting conditions and assistance of high-pressure water jet in the process of rock cutting upon the wear, cutting force and dustiness of the shearer pick edge

A. Klich & K. Kotwica
University of Mining and Metallurgy, Kraków, Poland

A. Meder
Mining Mechanisation Centre, Gliwice, Poland

ABSTRACT: Paper presents the results of research received on laboratory stand of the University of Mining and Metallurgy, Kraków during cutting of artificial samples of rock with uniaxial compressive strength from up to 105 MPa with and without cutting process assistance of high-pressure water jet. During the researches the pick edge wear, pressure, tangential and side force and dustiness were measured under several selected mining parameters. For the research there were used shearer adjacent *Alpine* and *Boart* picks. The cutting with picks was assisted with high-pressure water jet from the front and back. For cutting with assistance of high-pressure water jet the *Saphintec* type nozzles of d = 0,3 to 0,8 mm diameter and water pressure of p. = 1 to 60 MPa were used.

1. INTRODUCTION

The problems of mechanical hard rock cutting have become more intensive in Polish mining in recent years due to deeper coal seams and driving workings in harder rocks containing more highly abrasive inclusions. It was connected mainly with small-cutting advance rate and very high wear of the shearer picks, so the cutting process became not effective economically. The attempts to use the cutting tools of new types, mainly the tangential tools with sintered carbide inserts of increased diameter and disc cutters with small diameter, did not lead to successful results in hard rock cutting. Therefore, the steps towards new rock cutting technologies have been undertaken. An effect of pick edge wear on tool loading as well as an impact of high-pressure water jet assistance on the rear of the shearer pick edge and in addition on dust concentration resulting from rock cutting have been investigated on unique test stand for single cutting tool testing at the Department of Mining Machines and Waste Utilisation Equipment, University of Mining and Metallurgy in Kraków. The test stand used during research studies as well as the measuring method and results obtained during cutting the rocks of about 100 MPa in uniaxial compression strength are presented below.

2. TEST STAND AND MEASURING METHOD

The measurements were taken on the test stand for single picks which have been subjected to some modifications in order to create similar to real pick edge working conditions. The stand (Figure.1) consists of a frame along which a traverse is moved vertically. A slide support with a cutting tool-holder of L shape is mounted on the traverse. The support can move along the traverse. The measuring head enabling an independent measurement of cutting force components: tangential P_s, radial P_d and lateral P_b is mounted on the tool-holder, on the lower part of the longer arm. The replaceable *Boart Longyer* RH8 standard tool-holder on which a cutting tool is directly mounted is used. The holder is installed in such a way that the picks can cut in the layout which is close to that occurring on the cylindrical part of traverse organs, i.e. at setting angle $\kappa = 45°$ and lateral angle $\rho = 8°$.

There is also a clamp for fitting the nozzles to assist the cutting with high-pressure water jet at 75 MPa in front or at the rear of the pick. To generate water jets the *Saphintec* saphire nozzles of $\varnothing 0.30 \div 0.80$ mm have been used.

The concrete rings modelling a rock have been placed on the rotary table of continuously adjusted speed within the range from 0 up to 20 rpm. The samples have been made as concrete rings of outer diameter $\varnothing$ 1200 mm and 450 mm in height, reinforced with steel rings from the inside. The rings have been made of special concrete of guaranteed

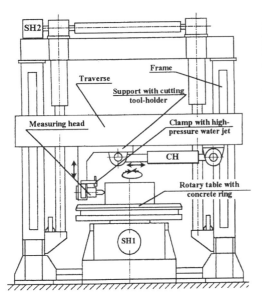

Figure 1. Scheme of the test stand for single picks

uniaxial compression strength R_c = 105 MPa. Large arenaceous shale pieces of R_c = 119.2 MPa taken from the "Halemba" coal mine in Ruda Śląska have been installed in the rings. This material has been characterised by relatively very high abrasive resistance.

The tests were carried out with three types of tangential picks: *RM8 V5-25* and *RM8-520* manufactured by *Boart Longyear* and equipped with sintered carbides of ∅25 mm and ∅22 mm and *Alpine* rotary pick of carbide diameter ∅18 mm. The picks have been especially chosen for test purposes in order to obtain a relationship between the insert diameter and the pick wear which could be achieved by selecting the picks of similar profile and the same edge angle guaranteed by the manufacturer, but of increased insert diameter.

The aim of the tests was to compare the wear and load for cutting tools of different sintered carbide insert diameters by mechanical cutting of a concrete sample being an imitation of hard rock. In addition, an effect of high-pressure water jet assistance on pick wear and dustiness has been investigated.

3. PICK WEAR MEASUREMENTS

The wear measurements were made for "dry" cutting by using all three tool types described above as well as for "wet" cutting by using two *RM8 V5-25* picks. The measurements were taken both for mechanical and "wet" hydromechanical cutting at the same working parameters: cutting depth g = 8 mm, pitch t =10 mm and cutting velocity v = 1 m/s. Water jets at pressure

p = 20 MPa were applied in front or at the rear of the pick. The jets have been generated by the *Saphintec* nozzles of 0.40 mm in diameter. In the case of the front assistance a nozzle was placed at a distance of about 55 mm from the pick corner and the water jet was directed tangentially to the plane of shear at a distance of 2 mm before the pick corner. In the case of rear assistance a nozzle was placed at a distance of about 95 mm from the pick corner and the water jet was directed tangentially to the plane of shear at a distance of 2 mm behind the pick corner.

The measurement was made before cutting and then every 500 m of pick run over the sample circumference until the total cutting length of 2500 m was obtained. The differences in total length for individual measurements did not exceed ± 40 m. Then the pick was removed from the tool-holder and after allowing its cooling down to ambient temperature the actual pick height h_n from the pick shank base surface was measured. In addition the pick edge profile was taken in special moulds filled with *Duracryl Plus* - a dentistry material which reproduces the impressed shapes perfectly. The linear edge shortening h_n was determined by measurements made with the height optimeter. Then a tin casting of imprint was made. Knowing the mass of edge casting, its volume was determined. The edge volume wear was determined by comparing the consecutive volume measurements.

The results of tool shortening and wear measurements are presented in Figure 2 and Figure 3. It is clearly evident that the edge volume wear decreases with the increasing diameter of sintered carbide insert. Comparing to the *Alpine* pick of smallest sintered carbide insert of ∅ = 18 mm the volume wear decreased by 68% for *RM8-520* (sintered carbide insert diameter ∅ = 20 mm) and by 84% for *RM8-V5-25* (sintered carbide insert diameter ∅ = 25 mm), while the edge shortening was reduced by 28% and 36% respectively. When comparing the wear values obtained for the high-pressure water jet assistance, the tool wear decreased by 82% was observed for the front application and by 18% only for the rear application. Similar figures were obtained for edge shortening which was decreased by 48% and 8 % respectively.

4. CUTTING RESISTANCE TESTING

In the case of "dry" cutting the load measurements were made for the same cutting tools as above and for the same cutting parameters, within 15 seconds approximately every 500 m of cutting length at the half of sample height after the cutting process was stabilised.

On the base of recorded force measurements the maximum and mean values were determined for the force components: tangent P_s, radial P_d and lateral P_b. Mean values of cutting force for three picks under investigation are presented in Figure 4. By comparing

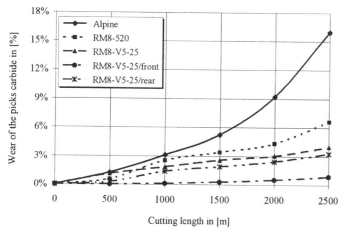

Figure 2. Wear of the picks carbide in the function of the cutting length

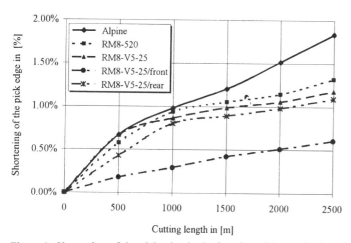

Figure 3. Shortening of the pick edge in the function of the cutting length

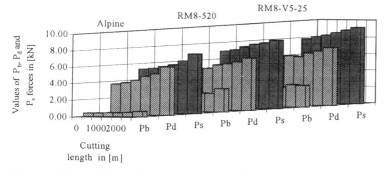

Figure 4. Values of P_b, P_d, P_s forces in function of cutting length

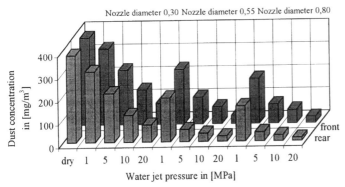

Figure 5. Dust concentration in function of nozzle diameter and water jet pressure

the obtained results one can find that values of the tangent P_s, radial P_d and lateral P_b force components increase with an increasing wear due to cutting. Also, an increase in force occurs for an increased diameter of sintered carbide insert. It should be noted that the increased tool wear leads to more rapid increase in the radial force than in the other forces. In addition, it was found that the *Alpine* picks showed the most rapid increase of force due to most rapid insert wear. For this pick the values of P_s and P_d increased by 36% and 57% respectively, while for the picks with inserts of $\varnothing =20$ and $\varnothing = 25$ mm the same figures were 20%, 28% and 18%, 24% respectively. When cutting a concrete sample with the picks mentioned above, it was found that the lateral force for the *Alpine* pick was very small (about 4 times lower than those of other picks). It reduced the pick rotary motion in tool-holder, thus affecting the pick wear characteristics. The changes of force components and observations of the cutting process might also indicate a lower cutting dynamics for the picks with inserts of $\varnothing =20$ and $\varnothing 25$ mm comparing to the pick of $\varnothing 18$ mm insert.

The force measurements were also made for the *Alpine* pick both for the front and rear water jet assistance at pressure of 20, 40 and 60 MPa. No significant changes of cutting forces were observed for the rock sample under testing. For the nozzle diameter $\varnothing 0.80$ mm and the highest pressure value the decrease of the mean tangent P_s and radial P_d components was 12% and 9 % respectively for the front assistance and 4% and 3.5% for the rear assistance.

5. DUSTINESS MEASUREMENTS

The dust concentration was measured during application of "dry" and "wet" cutting methods. In the latter case the nozzles of three diameters $\varnothing$ 0.30, 0.55 and 0.80 mm were used at four pressure values of high-pressure jet p = 1, 5, 10 and 20 MPa for the rear and the front

assistance. The dust concentration was measured with the aerosol suction apparatus of *Barbara 3a* type designed for measurements of respirable dust which is most hazardous for a worker. The measurements were carried out for 3 minute continuous operation at fixed distance from the pick of 1.2 m and for the same cutting parameters as above, while the *Alpine* pick was used.

The obtained results are presented in Figure 5. It is clearly evident that an implementation of high-pressure water jet leads to considerably lower dust concentration. The best results have been obtained for highest pressure values and nozzle diameters. The dustiness was reduced by 97% for the rear and the front assistance respectively. A decrease of nozzle diameter and pressure reduces this effect, but the results still remain satisfactory (dust concentration drop by about 80%) for the nozzle of $\varnothing 0.55$ mm and p = 5 MPa (rear assistance). In the case of front assistance the same effect was obtained for the same nozzle at pressure p = 10 MPa.

6. CONCLUSIONS

1. Rotary picks with large diameter of sintered carbide edge are most useful for mechanical cutting of hard rocks. Both their wear and tool load increase less rapidly than in the case of those of smaller edge diameters. In addition, these picks show lower operational dynamics.

2. An application of high-pressure water jet assistance in front of the cutting tool improves its life time, while keeping the same cutting force. An indirect effect of reduced force is achieved due to lowered tool wear.

3. The best dust concentration reduction is obtained for a high-pressure nozzle placed at the back of the cutting tool.

4. Due to energy consumption reasons, the diameter of jets used for assistance and pressure should not exceed 0.50 mm and 20 MPa respectively.

Mining Science and Technology '99, Xie & Golosinski (eds) © 1999 Balkema, Rotterdam, ISBN 90 5809 067 1

Investigation of the per phase double-switch power converter for four-phase switched reluctance motor drive*

Hao Chen, Jianguo Jiang & Guilin Xie
China University of Mining and Technology, Xuzhou, People's Republic of China

ABSTRACT: The main circuit topology of the per phase double-switch power converter for 4-phase Switched Reluctance motor, and the adjustable-speed control strategy are introduced in the paper. The mathematical model of the power converter main circuit is given. The comparison of the per phase double-switch converter and the split supply converter for 4-phase Switched Reluctance motor, and the comparison of the 4-phase per phase double-switch converter and the 3-phase per phase double-switch converter, are made. Those show that the 4-phase per phase double-switch converter is suitable for the high-power or high-current Switched Reluctance motor.

1 INTRODUCTION

The Switched Reluctance motor drive is a type of adjustable-speed motor drive system, which consists of the Switched Reluctance motor (SRM), the power converter and the controller (Lawrenson et al. 1980). Switched Reluctance motor has the simple and robust structure. There are no squirrel cage fault of the induction motor and there is not a brush in the rotor, which lead to maintenance easy, flammable and explosive environmental adaptability, so the SRMs have wide applicable prospects for the drives in coal mine, such as conveyor, the electric drive of mining machine, locomotive drive (Chen & Xie 1998), local fan and pump. For the large power requirement of the Switched Reluctance motor drive on that equipment, the scheme should be advanced in accordance with the large power applications. The power converter is one of the major components of the Switched Reluctance motor drive, the performance and the cost of the drive depend to a large extent on the topology of the power converter main circuit.

2 THE MAIN CIRCUIT TOPOLOGY AND THE ADJUSTABLE-SPEED CONTROL STRATEGY

A per phase double-switch power converter main-circuit for 4-phase Switched Reluctance motor is shown in Figure1.

There are two main switches and two flywheel diodes in the per phase circuit. The PWM adjustable-speed control strategy (Chen & Xie 1997) is adopted, such as the triggering pulse of the main switches, $S_5 \sim S_8$, are modulated by the pulse width modulation (PWM) signal, the speed control of the drive can be achieved by regulating the duty ratio of the PWM signal. Taking A phase for example,

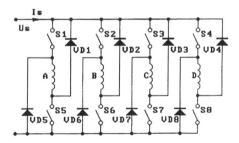

Figure 1. Main circuit of the per phase double-switch power converter for 4-phase SRM

1) At the stage of supply, the main switches, S_1 and S_5, are turned on, the supply source U_S supply to A phase winding of the motor.
2) At the stage of the PWM signal chopping, the main switche, S_1 is turned off, A phase current flows

*Project Supported by China Coal Science Foundation (97-E-10104), China Coal College & University Youth Science Foundation (97-012) and China University of Mining & Technology Science & Technology Foundation (97-A-8).

through the main switch, S_5 and the flywheel diode, VD_5.

3) At the stage of commutation, the main switches, S_1 and S_5 are turned off, A phase current is forced to flow through the two flywheel diodes, VD_1 and VD_5, the electric energy is returned from the motor to the supply.

3 MATHEMATICAL MODEL

Neglecting the mutual inductance, the mathematical model of the 4-phase per phase double-switch power converter is as follows:

$$
\begin{bmatrix} U_{AL} \\ U_{BL} \\ U_{CL} \\ U_{DL} \end{bmatrix} =
\begin{bmatrix}
R_A + \dfrac{\partial \psi_A}{\partial_A} \cdot D & 0 & 0 & 0 \\
0 & R_B + \dfrac{\partial \psi_B}{\partial_B} \cdot D & 0 & 0 \\
0 & 0 & R_C + \dfrac{\partial \psi_C}{\partial_C} \cdot D & 0 \\
0 & 0 & 0 & R_D + \dfrac{\partial \psi_D}{\partial_D} \cdot D
\end{bmatrix}
\begin{bmatrix} i_A \\ i_B \\ i_C \\ i_D \end{bmatrix}
$$

$$
+ \frac{2\pi n}{60} \cdot \frac{\partial}{\partial \theta}
\begin{bmatrix}
\psi_A & 0 & 0 & 0 \\
0 & \psi_B & 0 & 0 \\
0 & 0 & \psi_C & 0 \\
0 & 0 & 0 & \psi_D
\end{bmatrix}
\tag{1}
$$

Here $DL = 1,2,3 (U_{k1}) U_{k2}$ and U_{k3} $(k = A,B,C,D)$ is the phase winding voltage at the stage of supply, the PWM signal chopping and commutation, respectively, and

$$
\begin{bmatrix} U_{k1} \\ U_{k2} \\ U_{k3} \end{bmatrix} =
\begin{bmatrix}
1 & -2 & 0 \\
0 & -1 & -1 \\
-1 & 0 & -2
\end{bmatrix}
\begin{bmatrix} U_S \\ U_T \\ U_D \end{bmatrix}
\tag{2}
$$

Where U_S is DC supply voltage, U_T and U_D is the on-state drop of the main switches and the flywheel diodes, respectively, $Di = \dfrac{di}{dt} \Box R_k$ is the phase winding resistance and ψ_k is the flux linkage, n is the rotor speed of the motor.

4 COMPARISON

The 4-phase split supply power converter is generally used for the 4-phase 8/6 structure Switched Reluctance motor at present, the main circuit is shown in Figure2.

Two series capacitors constitute the midpoint of the power source in the power converter. While A phase or C phase is supplied, the return circuit consists of the DC supply source, capacitor, C_2, and

A phase winding or C phase winding. While A phase or C phase is at the period of the continuous-current, the capacitor, C_2, is charged by A phase current or C phase current through the diode VD_A or VD_C. While B phase or D phase is supplied, the return circuit consists of the DC supply source, capacitor C_1, and B phase winding or D phase winding. While B phase or D phase is at the period of the continuous-current, the capacitor, C_1, is charged by B phase current or D phase current through the diode VD_B or VD_D. The midpoint potential is floated because the capacitors, C_1 and C_2, are charged and discharged. Especially on the condition of the starting or single-phase conducting, the midpoint potential is floated so seriously that cause the trouble of the drive because of the asymmetry in the manufacture and the control parameters. So the midpoint potential should be regulated. The voltage rating of the main switches and the diodes is equal to the DC supply voltage, U_S, but the voltage rating of the phase winding is equal to the half of the DC supply voltage.

4.1 Compared with the 4-phase split supply converter

In view of the same 4-phase 8/6 structure Switched Reluctance motor, the motor is supplied by the 4-phase per phase double-switch converter and by the 4-phase split supply converter respectively. At the range of 0 °~30 °, the phase windings of the motor are supplied by the power converter with the square wave , and the DC supply voltage is U_S , the DC supply current is I_S. The two power converters for four-phase 8/6 structure Switched Reluctance motor, such as the per phase double-switch power converter and the split supply power converter are compared while the input is same, the results are shown in Table 1.

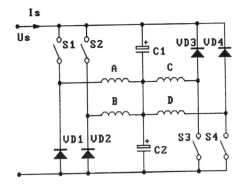

Figure 2. Main circuit of the split supply power converter for 4-phase SRM

736

Table 1. Comparison of the two 4-phase power converters.

Items	double-switch	split supply
Main switch numbers	8	4
Flywheel diode numbers	8	4
Gate driver numbers	8	4
Auxiliary power source numbers	5	3
Phase voltage	U_S	$U_S/2$
Phase current	$I_S/2$	I_S
Rated voltage of the main switch	U_S	U_S
Rated current of the main switch	$I_S/2$	I_S
Total volt-ampere ratings of the main switch	$4\,U_S\,I_S$	$4\,U_S\,I_S$

Table 2. Comparison of the two per phase double-switch power converters.

Items	4-phase	3-phase
Main switch numbers	8	6
Flywheel diode numbers	8	6
Gate driver numbers	8	6
Auxiliary power source Numbers	5	4
Phase voltage	U_S	U_S
Phase current	$I_S/2$	$2\,I_S/3$
Rated voltage of the main Switch	U_S	U_S
Rated current of the main Switch	$I_S/2$	$2\,I_S/3$
Total volt-ampere ratings of the main switch	$4U_S\,I_S$	$4\,U_S\,I_S$

The costs of the main switches occupy large proportion of the power converter cost. Although the number of the main switches, the flywheel diodes, the gate drive circuit and the auxiliary power source in the 4-phase per phase double-switch converter is more than that in the 4-phase split supply converter, the total volt-ampere ratings of the main switches in the two power converter is same, so the costs of the two power converters are same. While the capacity of the two power converter is same, the voltage rating of the phase windings in the per phase double-switch converter is higher than that in the split supply converter, so the former has the low current rating, the low loss and high efficiency, it is suitable for the high-power or the high-current Switched Reluctance motor. Moreover, the each phase of the per phase double-switch converter is independent, there is no trouble of the midpoint potential floated in the split supply converter.

4.2 *Compared with the 3-phase per phase double-switch converter*

The 4-phase 8/6 structure Switched Reluctance motor is supplied with the square wave by the 4-phase per phase double-switch power converter at the range of $0° \sim 30°$, and the 3- phase 6/4 structure Switched Reluctance motor is supplied with the square wave by the 3-phase per phase double-switch power converter at the range of $0° \sim 45°$. Table 2 gives the comparison while the input is same, such as the *DC* supply voltage is U_S, the *DC* supply current is I_S.

The total volt-ampere ratings of the two per phase double-switch power converter are same, so the costs of the two power converters are same. While the input capacity of the two power converters is same, the current rating of the main switches in the 4-phase per phase double-switch converter is only 75% as the same as that in the 3-phase per phase double-switch converter. So the 4-phase 8/6 structure Switched Reluctance motor with 4-phase per phase double-switch power converter has the advantage in the choice of the main switches for the high-power drive.

5 EXPERIMENT

A 4-phase 8/6 structure Switched Reluctance motor prototype with the 4-phase per phase double-switch power converter is tested. Figure3 shows the phase current waveform while it operated at the speed $n = 800r / min$, and the output torque $T_2 = 1.6 N \cdot m$

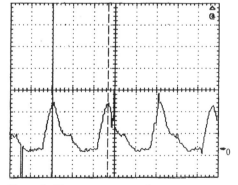

Figure 3. Phase current waveform
Scale: Abscissa: 5.0 ms/div. Ordinate: 4.0 A/div.

with the DC supply voltage $U_S = 264V$, the frequency of the PWM signal is 6KHz

6 CONCLUSIONS

The main circuit topology of the per phase double-switch power converter for 4-phase Switched Reluctance motor and the adjustable-speed control strategy are introduced in the paper. The comparison of the per phase double-switch converter and the split supply converter for 4-phase Switched Reluctance motor, and the comparison of the 4-phase per phase double-switch converter and the 3-phase per phase double-switch converter, are made. Those show that the 4-phase per phase double-switch converter is suitable for the high-power or high-current Switched Reluctance motor in coal mine.

REFERENCES

Lawrenson, P. J., J. M. stephenson & P. T. Blenkinsop 1980. Variable-speed switched reluctance motors. *IEE Proceedings of Pt.* B(127):253-265.

Chen, H. & G. Xie 1998. A Switched Reluctance Motor Drive System for Storage Battery Electric Vehicle in Coal Mine. *Preprints of the 5th IFAC Symposium on Low Cost Automation*: Ts4-1-Ts4-5. Shenyang: China.

Chen, H. and G. Xie 1997. Theory and practice of PWM control for switched reluctance motors. *Journal of China University of Mining & Technology.* 26: 23-27.

Mining Science and Technology' 99, Xie & Golosinski (eds) © 1999 Balkema, Rotterdam, ISBN 90 5809 067 1

Application of Computer Integrated Manufacturing System (CIMS) methodology to design of coal preparation plant

Shifan Xu, Dunwei Gong & Xuesong Wang
China University of Mining and Technology, Xuzhou, People's Republic of China

ABSTRACT: The Computer Integrated Manufacturing System (CIMS) for coal preparation plant has been developed in China, the first known application in the world. This paper presents the optimization aim and the architecture of coal preparation plant CIMS which consists of three levels of promotive control structure. In the presented example the business environment and production condition of coal preparation plant, the structures of function subsystem and the support subsystem are given. The function subsystem consists of management information subsystem, production schedule subsystem and workshop automation subsystem. The system has been successfully applied to coal preparation plant of Nantun Coal Mine with very encouraging result.

1. INTRODUCTION

CIM was put forward by Dr. Joseph Harrington in 1974, whose kernel points are system and information (Wenjian Liu 1994). System point indicates that each link in the production chain is an indiscerptible entirety and should be considered entirely. Information point indicates that the essence of the process of production management is the process of information collection, transmission and procession and products are the material presentation of information. CIMS is a system based on the theory of CIM.

Our country drew up the "863 plan" that aims to high technology development in 1986. CIMS is one of the 17 main research subjects and is subordinate to the field of automation (Shifan Xu 1998). Experts and scholars not only tackled key techniques of CIMS but also began to implement CIMS exemplary project under the rule of "benefit driving, planing for the whole, breaking through key projects, carrying out by steps" in 1989. Many enterprises, which implemented CIMS, have acquired striking economic benefit and social benefit.

The production process of coal preparation plant is short, small batch and continuous process, moreover, the process is comparatively simple, but its automation and mechanization are in comparatively high level. There are varieties of products in coal preparation plant, the main-product is cleaning coal and by-products are reiteration and coal slime. Manufacture, maintenance, storage and transportation move alternatively and at the same time the balance, stabilization and high responsive speed of production are highly required. The quality of the staff is comparatively low (especially in China). For all of the above reasons, we should carefully consider some questions as follows when implement CIMS in coal preparation plant: propaganda, implementation, application and maintenance. The CIMS implementation should improve cleaning coal grade, low the cost of manufacture, improve the staff's quality of performance and skill, then the overall benefit will be improved.

2. OPTIMIZATION AIM AND ARCHITECTURE OF COAL PREPARATION PLANT CIMS

2.1 *The optimization aim of coal preparation plant CIMS*

The optimization aim of discrete enterprise CIMS can be summarized to T, Q, C, S and E. T represents the time of new products to the market, which reflects the capability of plant responding to the market. Q represents the quality of product. C represents the cost of management, which reflects the potentiality of plant participant in the market competition. S represents the quality of service, excellent quality of service will attract more and more customers in the market competition. E represents the elimination of pollution of the environment. Considering that coal preparation plant has the following problems: the first thing is that its main economic benefit comes from cleaning coal, but cleaning coal is difficult to rewash, when it can't meet the customer's requirement once it was washed. The second is the disposing of slurry water. The third is that the cleaning coal is classified by ash content, so there doesn't exist the problem of new products and services after sale. Based on all of the above, we can summarize the optimization aim to P, Q, C, E. Here P represents improving of cleaning coal's productivity. Q represents guaranteeing of cleaning coal quality. C represents the cost of manufacture. E represents eliminating of the pollution of environment.

2.2 *The architecture of coal preparation plant CIMS*

The coal preparation plant CIMS is a complex multi-layers and multi-structures system that integrates production and management as a whole. Coal preparation plant isn't like discrete plant and it has several unique characteristics comparing with the other continuous industries, so it is essential to establish architecture of coal preparation plant CIMS according to the features of coal preparation plant CIMS and production management.

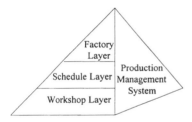

Figure 1.Promotive control structure of coal preparation plant CIMS

In the view of comparative central scale, simple process, small layers of production management, the present organization and the function requirement in coal preparation plant, the architecture of coal preparation plant CIMS can be divided into three layers with promotive control structure, factory, schedule and workshop layers. The equipment layer is included into the workshop layer (Figure 1).

The schedule layer forms a connecting link between the proceeding and the following and plays an essential role in such an architecture that embodies the characteristics of production management.

3. COMPOSITION OF COAL PREPARATIION PLANT CIMS

Coal preparation plant CIMS consists of function subsystem and support subsystem. Function subsystem includes management information subsystem (MIS), production schedule subsystem (PSS) and workshop automation subsystem (WAS) (Shifan Xu 1997). Support subsystem includes network subsystem (NS) and database subsystem (DS). The function trees are demonstrated as Figure 2, Figure 3 and Figure 4. In fact the functions of function subsystem can be extended or compressed according to requirement of different coal preparation plants.

The establishment of support subsystem should be determined by function subsystem. In the coal preparation plant CIMS established by the authors, the servers and the clients adapts to Sun workstation, HP microcomputer and other 586 microcomputers. The network topology structure adapts stack heap tree-type, the prime net adapts 100M and 10M optical cable, the switchers adapt CISCO3200 and CISCO1900, the background databases adapt Oracle for Sun, Solaris and Oracle for Windows NT, and the front ground databases adapt Access.

4. INTEGRATED TECHNOLOGY ROUTINE OF COAL PREPARATION PLANT CIMS

The integrated system of coal preparation plant CIMS can be achieved through three steps: 1) Physical integration. It can be accomplished through establishing Windows 95, Windows NT and Unix heterogeneous net operation console, Oracle and Access heterogeneous data console, integrated programming environment composed of application development tools such as Office 97 / Oracle / VC++/VB/FIX and TCP/IP protocol. 2) Information integration. It can be accomplished through interoperation of application program interface of heterogeneous database carried out by integrated platform environment and loaded module, adaptation of united code and centralized database. 3) Function integration. Function integration is carried out around three main lines that are production management, assurance of quality and cost management. The integrated line of production management is embodied by three subsystems that are production management, schedule management and workshop management. Each line consists of some subsystems

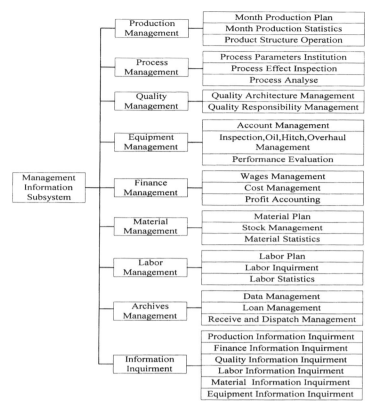

Figure 2. The function tree of MIS

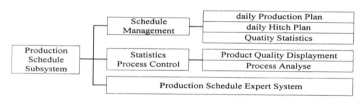

Figure 3. The function tree of PSS

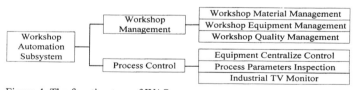

Figure 4. The function tree of WAS

5 .PROBLEMS TO RESEARCH OF COAL PREPARATION PLANT CIMS

There has been no precedent of applying CIMS to coal preparation plant so far. In order to popularize and apply the theory and the technology of CIMS well, there are many theories and application problems worth researching further, which include establishing the structure of ERP in and applying the technology of virtual reality and quality management and quality control to coal preparation plant.

5.1 Establishing the structure of ERP in coal preparation plant

Enterprise Resource Planning (ERP) is a advanced and amateur model of business management in the present world, which was first put forward by the America Garter Group Inc.(Yimin Shen 1998). ERP considers enterprise as an organic whole and starts from the point of entirety optimization and uses scientific method, makes an overall balance to various resources of the enterprise including labor, capital, materiel, equipment, method (production technique), information and time. Considering relative great difference in the process of production between the continuous and discrete enterprise, the coal preparation plant ERP should be established. The aim and function of ERP should be defined, and the function and information of MIS and PSS of present coal preparation plant CIMS should be in full use.

5.2 The application of virtual reality applied to coal preparation plant

The technology of virtual reality is a new technology that was appeared recently and became hot point quickly. The technology of virtual reality can imitate and recur the realism realistically and response to users' operation timely and provide a platform for users that can interact with three-dimensional images produced by the computer (Shujun lv etc 1998). The production system of coal washing is a complex and huge system, so there has been no adaptive model so far. The usage of the technology of virtual reality can simulate real production process through the immersion and interaction of man or retrospect the production process through real production data, so the production process can be optimized. The application of the technology of virtual reality in coal preparation plant needs tackling some problems such as the establishment of virtual production environment and the acquirement of information through the production system of coal washing.

5.3 Quality management and quality control

Although the process of quality management has been developed into total quality management, there exist many problems in quality management of coal preparation plant (especially in China) because of several kinds of factors. The quality management and quality control in coal preparation plant CIMS should import advanced management idea, which include systematic idea, feedback idea and total quality management idea. Total quality management idea includes the idea of customer-focus, team, empowerment, the deployment of policy and so on (T.Thiagarajan & M.Zairi 1997). On the base of these advanced management ideas the integrated quality architecture can be established that should emphasize quality management and quality control in the production process and those functions should embody the modern management ideas of the above(Joseph Sarkis & Michael Reimann 1996).

6. CONCLUSIONS

This paper presents not only the whole optimization aim and architecture of coal preparation CIMS but also the corresponding application system that has been first applied to a coal preparation plant in all over the world. The implementation has received striking economic benefit and social benefit. CIMS provides powerful support of the capability of the market competition for coal preparation plant.

REFERENCES

Wenjian Liu 1994. The Introduction Of Computer Integrated Manufacturing. Haerbing: The press of haerbing university of industry: 5-12.

Shifan Xu 1998. Computer Integrated Manufacturing System / Computer Integrated Processing System (CIMS/CIPS). The automation of coal mining. 3:3-6.

Shifan Xu. 1997.The introduction of coal mining CIMS. Xuzhou: The press of China university of mining and technology.42-43.

Yimin Shen 1998. Distributed business-oriented ERP. Proceeding of computer integrated manufacturing system of China. 3:84-87.

Shujun lv etc 1998. The application of virtual reality to industrial process control, design and train. Proceeding of computer integrated manufacturing system of China. 2:28-32.

T.Thiagarajan & M.Zairi 1997. A review of total quality management in practice: understanding the fundamentals through examples of best practice applications-part I. The TQM Magazine. 9(4):270—286.

T.Thiagarajan & M.Zairi 1997. A review of total quality management in practice: understanding the fundamentals through examples of best practice applications-part II. The TQM Magazine. 9(5):334-356.

Joseph Sarkis & Michael Reimann 1996. Quality information systems in advanced manufacturing environments. Quality Engineering. 8(3):419-431.

Mining Science and Technology'99, Xie & Golosinski (eds) © 1999 Balkema, Rotterdam, ISBN 90 5809 067 1

Analysis and identification of leakage failure in hydraulic system of hydraulic powered support

Zhichang Yao & Zhiyi Yang
China University of Mining and Technology, Xuzhou, People's Republic of China

Bin Lin
Department of Mining Engineering, University of New South Wales, Sydney, N.S.W., Australia

ABSTRACT: Diagnosing the faults of hydraulic support system, used in underground longwall mining, is very complex and time consuming as sometimes the relation between the cause and the result of the faults is vague. In addition, the definitions of cause and result are vague as well. This paper details the application of the fuzzy set theory to diagnosis of hydraulic support system.. The development of a practical and workable method and instruments to diagnose the leakage of hydraulic support system and the emulsion pump set has been introduced in this paper.

1. INTRODUCTION

Today hydraulic roof supports are widely used for underground longwall mining in China. However, diagnosing breakdowns of the supports is always time-consuming, as the relation between the cause and the result of the faults is sometimes vague. In addition, the definitions of cause and result are generally vague. In most cases there is no clear boundary between the normal and abnormal working status of hydraulic supports. This study investigates the application of fuzzy set theory to diagnosing the hydraulic faults. It has been found that the application of fuzzy concepts together with proper instrumentation could speed up diagnosis. As a result, maintenance time of breakdown can be reduced substantially.

2. FUZZY SETS TO THE DIAGNOSIS OF HYDRAULIC SUPPORTS

Hydraulic supports are comprised of pipelines, valves and legs. Valves control the movement of the support system. The faults of the system are characterized by a set of working parameters, that is

$$U = \{u_1, u_2, ..., u_m\} \tag{1}$$

Where u_1 = the system abnormal pressure, which is function of valve fault (u_{11}), valve seating worn out (u_{12}), mechanical fault (u_{13}) and seal failure (u_{14}). u_2 = the system abnormal flow quantity, which is consisted of the pump fault (u_{21}), the pipeline leaks (u_{22}).

u_3 = the noise level
u_4 = age of the system
u_5 = others rather than above

The subset of (u_m), therefore, is generally in the form of:

$$u_m = \{u_{m1}, u_{m2}, ..., u_{mt}\} \tag{2}$$

Weights have been assigned to all elements as shown Figure 1. They are:

$$A_1^{(1)} = (a_{11}, a_{12}, ... a_{1k})$$
$$A_2^{(1)} = (a_{21}, a_{22}, ... a_{2s}) \tag{3}$$

$$......$$

$$A_m^{(1)} = (a_{m1}, a_{m2}, ... a_{mt})$$

$$A^{(2)} = (a_1, a_2, ... a_m) \tag{4}$$

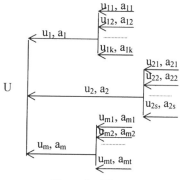

Figure 1   The Tree for U set

In order to pinpoint the fault of the system and find the cause, a set of locations is proposed as follows:

$$V = \{v_1, v_2, \ldots v_n\} \quad (5)$$

Where v_1 = pipelines
v_2 = safety valves
v_3 = operation valves
v_4 = one-way valves
v_5 = legs
v_6 = hydraulic jack

Relations to the sets U and V are presented in Equation 6:

$$R_1^{(1)} = \left[r1_{ij}^{(1)}\right] \quad i=1,2,\ldots,k \; and \; j=1,2,\ldots,n$$

$$R_2^{(1)} = \left[r2_{ij}^{(1)}\right] \quad i=1,2,\ldots,k \; and \; j=1,2,\ldots,n$$

......

$$R_m^{(1)} = \left[rm_{ij}^{(1)}\right] \quad i=1,2,\ldots,k \; and \; j=1,2,\ldots,n$$

$$(6)$$

Then we have the primary compositions:

$$B_1^{(1)} = A_1^{(1)} \circ R_1^{(1)} = \left[b_{1j}^{(1)}\right] \quad j=1,\ldots,n$$

$$B_2^{(1)} = A_2^{(1)} \circ R_2^{(1)} = \left[b_{2j}^{(1)}\right] \quad j=1,\ldots,n$$

......

$$B_m^{(1)} = A_m^{(1)} \circ R_m^{(1)} = \left[b_{mj}^{(1)}\right] \quad j=1,\ldots,n$$

$$(7)$$

Let

$$R^{(2)} = B^{(1)} \quad (8)$$

Then we have the secondary composition for all elements of U:

$$B^{(2)} = A^{(2)} \circ R^{(2)} = \left(a_1, a_2, \ldots, a_m\right) \circ R^{(2)}$$
$$= \left(b_1, b_2, \ldots, b_n\right) \quad (9)$$

By comparing the magnitudes of all elements (b_i) we can pinpoint the system faults quickly.

From the database we have collected in China and experience, we decide the magnitudes for weights:

$$A_1^{(1)} = \left(a_{11}, a_{12}, a_{13}, a_{14}\right) = \left(0.2, 0.3, 0.1, 0.4\right)$$
$$A_2^{(1)} = \left(a_{21}, a_{22}\right) = \left(0.4, 0.6\right)$$
$$A^{(2)} = \left(a_1, a_2, a_3, a_4\right) = \left(0.4, 0.3, 0.2, 0.1\right)$$

The relations from the pressure (u_1) and flow quantity (u_2) to the set V are given in Table 1 and 2.

Table 1 The Relation $(R_1^{(1)})$ from Pressure (u_1) to Set V

	v_1	v_2	v_3	v_4	v_5	v_6
u_{11}	0.0	0.4	0.4	0.2	0.0	0.0
u_{12}	0.0	0.3	0.5	0.2	0.0	0.0
u_{13}	0.1	0.2	0.3	0.1	0.2	0.1
u_{14}	0.1	0.2	0.3	0.2	0.2	0.0

Or

$$R_1^{(1)} = \begin{bmatrix} 0.0 & 0.4 & 0.4 & 0.2 & 0.0 & 0.0 \\ 0.0 & 0.3 & 0.5 & 0.2 & 0.0 & 0.0 \\ 0.1 & 0.2 & 0.3 & 0.1 & 0.2 & 0.1 \\ 0.1 & 0.2 & 0.3 & 0.2 & 0.2 & 0.0 \end{bmatrix}$$

And

Table 2 The Relation $(R_2^{(1)})$ from Flow Quantity (u_2) to Set V

	v_1	v_2	v_3	v_4	v_5	v_6
u_{21}	0.7	0.1	0.1	0.1	0.0	0.0
u_{22}	0.2	0.3	0.3	0.1	0.1	0.0

Or

$$R_2^{(1)} = \begin{bmatrix} 0.7 & 0.1 & 0.1 & 0.1 & 0.0 & 0.0 \\ 0.2 & 0.3 & 0.3 & 0.1 & 0.1 & 0.0 \end{bmatrix}$$

By applying the following formula, suggested by Yang (1995) for better accuracy

$$b_j = \sum_{i=1}^{n} a_j r_{ij} \quad j = 1,2,\ldots m \quad (10)$$

We can get the primary composition for the pressure fault

$$B_1^{(1)} = A_1^{(1)} \circ R_1^{(1)} = (0.05, 0.27, 0.38, 0.19, 0.10, 0.01)$$

And the primary composition for the shortage of flow quantity

$$B_2^{(1)} = A_2^{(1)} \circ R_2^{(1)} = (0.40, 0.22, 0.22, 0.10, 0.06, 0.00)$$

Then we can build up the relation from Set U to Set V as follows.

Table 3 The Relation $(R^{(2)})$ from Set U to Set V

	v_1	v_2	v_3	v_4	v_5	v_6
u_1	0.05	0.27	0.38	0.19	0.10	0.01
u_2	0.40	0.22	0.22	0.10	0.06	0.00
u_3	0.10	0.20	0.20	0.20	0.20	0.10
u_4	0.10	0.25	0.25	0.10	0.20	0.10

Or

$$R^{(2)} = \begin{bmatrix} 0.05 & 0.27 & 0.38 & 0.19 & 0.10 & 0.01 \\ 0.40 & 0.22 & 0.22 & 0.10 & 0.06 & 0.00 \\ 0.10 & 0.20 & 0.20 & 0.20 & 0.20 & 0.10 \\ 0.10 & 0.25 & 0.25 & 0.10 & 0.20 & 0.10 \end{bmatrix}$$

Finally we obtain the secondary composition for the weak support capacity of the system.

$$B^{(2)} = A^{(2)} \circ R^{(2)} = (0.170, 0.239, 0.283, 0.132, 0.118, 0.034)$$

From the above calculations, it is found that the possibilities of the locations for the abnormal pressure are 5%, 27%, 38%, 19%, 10% and 1% for pipelines, safety valves, operation valves, support legs and hydraulic jack respectively; For the abnormal flow quantity, the most possible locations is pipelines and the most unlikely location is hydraulic jack. For the weak support capacity, the most likely locations are operation valves and safety valves (28.3% and 23.9%).

3. ANALYZE AND DIAGNOSE THE LEAKAGE OF HYDRAULIC SYSTEM

There are many state parameters to diagnose the hydraulic support leakage, such as pressure, noise, flow and so on. Because of high internal pressure of the hydraulic system, it can cause high pressure fluid leaking and flowing when leakage arise, while the leaking fluid will cause wider band noise signals, its frequency bands vary from audio frequency to supersonic. Taking pressure, leakage noise and flow as the characteristic parameters of diagnosing, they have strong coherence to the hydraulic system state, high sensitive reactivity, and the character that they can also be analyzed and determined according to the measured value. Based on these characters we adopt appropriate instruments and methods to detect the hydraulic system leakage comprehensively and to enhance the accuracy of fault diagnosis.

By combination of the three diagnoses any leak in a hydraulic system can be quickly located. An expert system program is in consideration to identify a hydraulic system by computer.

This method, including the instruments, is one of the best approaches in China to diagnosing the leakage of the hydraulic system.

3.1 The real-time and on-line working pressure monitoring of hydraulic supports

Because the working pressure of the hydraulic system is strongly influenced by the leak fault, it is necessary that we should take the real-time and on-line monitoring to each branch pressure of the hydraulic system as the first step when diagnose the leak fault, and we take KJH4 strata pressure check device as the collecting unit to gather and store each branch pressure, then from the underground strata pressure diagnosing and measuring system together with an explosion-proof portable computer to check the dynamic variable trend of the pressure. At last we can use graphics/images to show the variety of each branch pressure.

The branch pressure of supports should correspond to the load, otherwise it's an abnormal value. The abnormal behavior of the pressure is that the setting pressure can not be enhanced, or is that the yield pressure decrease and the pressure fluctuation become wider. Therefore, select the measured position, where the setting pressure can not be enhanced and analyze the pressure-time variety trends graph, we can determine the position and reasons of the leak fault accurately.

3.2 Detect the leak fault of hydraulic supports

When we use KJH4 strata pressure diagnosing measuring device detecting each branch pressure of supports underground coal mine, the abnormal value of the pressure can be reflected. But in fact, the reasons of the abnormal pressure are not all caused by the leakage of support themselves. It's necessary that we should use the device, which is designed to find out the leakage directly, to identify comprehensively. Based on analyzing the characteristic parameter to the leak fault of supports, we develop KBF2 fluid leak detector which together with KJH4 strata pressure diagnosing and measuring device provides a quick and accurate analyzing method to determine the leak fault of the supports. Fig 2 shows the principle of detection.

When external leaks on the line of a hydraulic system arise, only a ultrasonic sensor can detect the signals. While inside the control valve of supports, at the time a failure seal loop causes emulsion to flow from the high pressure cavity to the low one, the crossflow of the leak fluid will arise, it needs a high sensitive crystal acceleration sensor to detect the leaking noise signals caused by the high pressure fluid leaking at the clearance. The signals display on the LCD panel after they have finished impedance matching, electricity amplifying, filtering and demodulating.

3.3 *Diagnosing and measuring emulsion pump sets*

The internal leaks of emulsion pump set occur when a cylinder, a piston, sealing or internal valves were wearing down. As a result of this the volumetric efficiency decreases and the output of the pump can not reach its rated capacity. For detecting this type of leaks we have developed the BGC emulsion pump set leak finder. This instrument, the principle diagram of the BGC emulsion pump set leak finder as shown in Figure 3, can separate the supports and pipelines of the hydraulic system from the emulsion pump set by the cut-off valve 1 and 2 and can exclude the leak effect of the supports and pipelines. In this way we need not detect the hydraulic system one point by one point, and will save much time, labor and material.

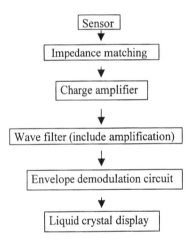

Fig 2.The principle diagram of the fluid leak detector

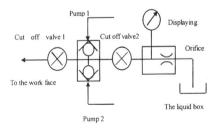

Fig 3. The principle diagram of the BGC emulsion pump set leak finder

Table 4 The working state of emulsion pump set

Working condition	normal	alert	fault
The efficiency of the emulsion pump set	>85%	70-85%	<70%

With this instrument we can measure the flow-rate through the orifice and the pressure difference between the front cavity and the rear cavity of the orifice. Based on the relationship between flow-rate and pressure, we can diagnose the leaks inside the emulsion pump set itself by determining its volumetric efficiency with the measured data. Both laboratory study and field practice has indicated the following relationship between pump efficiency and leak level, as listed in Table 4.

These three parameters have strong coherence to the hydraulic system state and are sensitive to system leakage. Based on these characteristics we can adopt appropriate instruments and methods to detect the hydraulic system leakage comprehensively, then enhance the accuracy of diagnosis.

4. CONCLUSIONS

(1) Application of fuzzy theory is an alternative to reduce the diagnosis time of hydraulic faults and to enhance the breakdown maintenance efficiency with combination of proper instrumentation. It must be noted, however, that the reliability of this method depends on the experience to assign the weights and the quality of the database to decide the relations.

(2) Hydraulic information monitoring instruments and methods of hydraulic supports provides an effective measure to diagnose and improve face support quality in coal face. It is advantageous to decrease the downtime due to hydraulic system fault in underground longwall mining.

REFERENCES

Yang L. and Gao E., 1995. *Fuzzy Theory and Its Application.* Publishing House of University of Technology South China. PRC. (In Chinese)

Nguyen, V. U., 1985. *Some Fuzzy Set Applications in Mining Geomechanics.* International Journal of Rock Mechanics and Mining Sciences and Geomechnics abstracts, Vol 22(6), pp 369 – 379.

Samir Kumar Das, 1991. Condition monitoring of longwall face machines, *Journal of Mines, Metals & Fuels*, XXXIX (1 & 2): 18-22.

Mining Science and Technology' 99, Xie & Golosinski (eds) © 1999 Balkema, Rotterdam, ISBN 90 5809 067 1

A theoretical model for lubrication with o/w emulsions

Shengming Yan
China University of Mining, Beijing, People's Republic of China

Shigeaki Kuroda
University of Electro-Communications, Tokyo, Japan

ABSTRACT   A theoretical model based on the continuum mechanics of mixtures is proposed for lubrication with o/w emulsions. It contains a set of governmental equations and two sets of formulas calculating viscosity coefficients. The results of numerical analysis agree qualitatively with the experimental observations.

1 INTRODUCTION

As is well known, Oil-in-water (o/w) emulsions are widely used in hydraulic systems of mining machine, because of their favorable properties such as good lubrication ability, excellent cooling efficiency, nonflammability and low coast. In fact, when the emulsion is entrained into the gap between two surfaces of connected parts, the concentration of oil phase will increase, so that the phase inversion occurs sometimes and an oil pool may be formed in the inlet zone (Fig.1). In order to explain this curious phenomenon, Wilson et al.[1-3] and Kimura et al.[4] developed theoretical models respectively. Both these models assumed that the formation of the oil pool depended geometrically on the average droplet size and the oil concentration of emulsion. However, the experiments of Nakahara et al.[5,6] and Zhu et al.[7] showed that the formation of oil pool was a strong function of operating conditions such as the entraining speed and the like, the model used by Zhu et al. cannot explain the mechanism for causing the oil pool to disappear at high speeds.

On the basis of continuum mechanics of mixtures, Al-Sharif et al.[8] developed a new theoretical model which relates the size of oil pool to operating conditions. However, if the gap of solid surfaces is smaller than the diameter of droplets, this model becomes invalid, although this case is usually encountered under operating conditions of low entraining speed.

A theoretical model developed by Yan et al.[9,10] is based also on continuum mechanics of mixtures, but its field of application is not restricted by the size of gap of solid surfaces. In this paper, we will further discuss this model and present some numerical results.

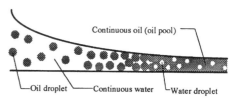

Fig.1 Schematic diagram of phase inversion

2 THE GOVERNMENTAL EQUATIONS

For lubrication with emulsion, we considered the motion of emulsion as a superposition of motions of water and oil, and obtained the governmental equations as follows:
for oil (disperse phase),

$$\frac{\partial}{\partial x}\left(\frac{\phi_d h^3}{12\xi_d}\frac{\partial p}{\partial x}\right)+\frac{\partial}{\partial y}\left(\frac{\phi_d h^3}{12\xi_d}\frac{\partial p}{\partial y}\right)=$$
$$\frac{\partial}{\partial y}\left[\frac{\phi_d(u_A+u_B)h}{2}\right]+\frac{\partial(\phi_d h)}{\partial t} \qquad (2.1)$$

for water (continuous phase),

$$\frac{\partial}{\partial x}\left(\frac{\phi_c h^3}{12\xi_c}\frac{\partial p}{\partial x}\right)+\frac{\partial}{\partial y}\left(\frac{\phi_c h^3}{12\xi_c}\frac{\partial p}{\partial y}\right)=$$

$$\frac{\partial}{\partial y}\left[\frac{\phi_c(u_A+u_B)h}{2}\right]+\frac{\partial(\phi_c h)}{\partial t} \quad (2.2)$$

and for emulsion (superposition of two phases),

$$\frac{\partial}{\partial x}\left(\frac{h^3}{12\xi}\frac{\partial p}{\partial x}\right)+\frac{\partial}{\partial y}\left(\frac{h^3}{12\xi}\frac{\partial p}{\partial y}\right)=$$

$$\frac{\partial}{\partial y}\left[\frac{(u_A+u_B)h}{2}\right]+\frac{\partial h}{\partial t}$$

(2.3)

where

$$\xi_d=\frac{\mu_d\mu_c-\mu_{dc}\mu_{cd}}{\phi_d\mu_c-\phi_c\mu_{dc}}$$

$$\xi_c=\frac{\mu_d\mu_c-\mu_{dc}\mu_{cd}}{\phi_c\mu_d-\phi_d\mu_{cd}} \quad (2.4)$$

$$\xi=\frac{\mu_d\mu_c-\mu_{dc}\mu_{cd}}{\phi_c^2\mu_d-\phi_d\phi_c(\mu_{dc}+\mu_{cd})+\phi_d^2\mu_c} \quad (2.5)$$

If we used the conventional lubrication theory for the lubrication with emulsions, its governmental equation would be Reynolds equation

$$\frac{\partial}{\partial x}\left(\frac{h^3}{12\mu}\frac{\partial p}{\partial x}\right)+\frac{\partial}{\partial y}\left(\frac{h^3}{12\mu}\frac{\partial p}{\partial y}\right)=$$

$$\frac{\partial}{\partial y}\left[\frac{(u_A+u_B)h}{2}\right]+\frac{\partial h}{\partial t} \quad (2.6)$$

where μ is the effective viscosity of emulsion and has the following relation with the four viscosity coefficients in Eqs. (2.1)~(2.3):

$$\mu=\mu_d+\mu_c+\mu_{dc}+\mu_{cd} \quad (2.7)$$

Although Eq.(2.3) has the same form as Eq.(2.6), they are not identical because $\xi\neq\mu$. If the volume fraction of oil or water became 1.0, ξ would becomes μ and Eqs.(2.1)~(2.3) would be reduced to Eq.(2.6). Therefore, we can consider the conventional Reynolds equation (2.6) as a special case of our governmental equations.

3 THE FOUR VISCOSITY COEFFIENTS OF EMULSIONS

An emulsion has two possible states in the gap between two surfaces as in Fig.2. It has a normal state in the thick film zone, where the oil droplets suspend in water, but has an abnormal state in the thin film zone where the oil droplets are deformed and sandwiched between two solid surfaces.

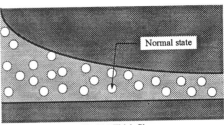

Fig.2(a) Thick film

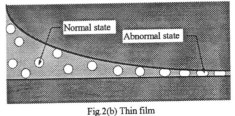

Fig.2(b) Thin film

Fig.2 Two states of emulsion in film

For abnormal state, the four viscosity coefficients of emulsions can be calculated by

$$\mu_d=\phi_d\eta_d$$
$$\mu_c=\phi_c\eta_c$$
$$\mu_{dc}=\mu_{cd}=0 \quad (3.1)$$

and for normal state, they are calculated by

$$\mu_d=\phi_d^m\eta_d+\mu_m$$
$$\mu_c=\phi_c^n\eta_c+\mu_m$$
$$\mu_{dc}=\mu_{cd}=\mu_m$$
$$\mu_m=\phi_d\phi_c\exp(a\phi_d+b) \quad (3.2)$$

where m,n,a,b are constants that related with fabrication method of emulsion and vary from emulsion to emulsion. These constants can be obtain with the aid of measuring the effective viscosity of emulsion.

4 AN EXAMPLE OF NUMERICAL ANALYSIS

By solving Eqs.(2.1)~(2.3), we can obtain the pressure distribution, the oil concentration distribution, the size of oil pool and the film thickness. For example, the governmental equations of steady line contact problems will be written as

$$\frac{d}{dx}\left(\frac{h^3}{12\xi}\frac{dp}{dx}\right)=u_m\frac{dh}{dx} \quad (4.1)$$

$$\frac{d}{dx}\left(\frac{\phi_d h^3}{12\xi_d}\frac{dp}{dx}\right)=u_m\frac{d(\phi_d h)}{dx} \quad (4.2)$$

From these equations we can obtain some results

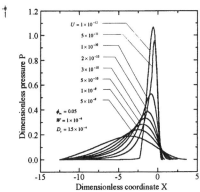

Fig.3 The pressure distribution in contact area

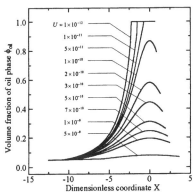

Fig.4 The oil concentration distribution

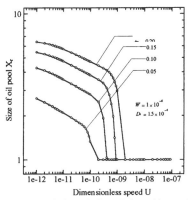

Fig.5 The variation of oil pool size with speed

of numerical analysis as Fig.3~Fig.5.

Fig.4 and 5 show that at low entraining speeds the volume fraction of oil will increases rapidly and the oil pool will be formed, but at high entraining speeds the increment of oil concentration will become slow and the oil pool will become small, even not appear.

5 CONCLUSIONS

A theoretical model based on the continuum mechanics of mixtures has been proposed for lubrication with o/w emulsions. By solving its governmental equations, we can obtain the pressure distribution, the oil concentration distribution, the size of oil pool and the film thickness.

For two different states of emulsion, two set of formulas are suggested to determine four viscosity coefficients in these governmental equations.

The numerical analysis that is not restricted by the size of gap of solid surfaces has been done. Its results agree qualitatively with the experimental observations.

NOMENCLATURE

h	film thickness
p	film pressure
t	time
u_A	speed of surface A
u_B	speed of surface B
u_m	entraining speed
x	coordinate of entraining speed direction
y	coordinate of film width direction
ϕ_c, ϕ_d	volume fraction of water and oil
η_c, η_d	viscosity of water and oil in pure states

REFERENCES

1. Sakaguchi, W.R.D. Wilson, Proceedings of the 5th Conference on Plastic Working, Tokyo, 1984, pp. 457-460
2. W.R.D. Wilson, Y. Sakaguchi, et al. Wear 161 (1993) 207-212
3. W.R.D. Wilson, Y. Sakaguchi, et al. STLE 37 (3) (1994) 543-551
4. Y. Kimura, K. Okada, Proceedings of the JSLE International Tribology Conference, 8-10 July Tokyo, 1985, pp. 937-942
5. T. Nakahara, T. Makino, et al. ASME Journal of Tribology 110 (2) (1988) pp. 348-353
6. T. Nakahara, S. Shiokawa, et al. Japanese Journal of Tribology 35 (12) (1990) pp. 1435-1446
7. Zhu, G. Biresaw, et al. ASME Journal of Tribology 116 (1994) pp. 310-320
8. A. Al-Sharif, K. Chamniprasart, et al. ASME Journal of Tribology 115 (1993) pp. 46-55
9. S. Yan, S. Kuroda, Wear 206 (1997) pp. 230-237
10. S. Yan, S. Kuroda, Wear 206 (1997) pp. 238-243

Mining Science and Technology'99, Xie & Golosinski (eds) © 1999 Balkema, Rotterdam, ISBN 90 5809 067 1

A method of multi-information fusion for fault diagnosis of large gearboxes in mining

Yuwei Yang
Beijing University of Aeronautics and Astronautics, People's Republic of China

YuanChang Jing
China University of Mining and Technology, Beijing, People's Republic of China

ABSTRACT: The method of multi-information fusion and its application in monitoring and fault diagnosis of large mining gearboxes are studied in this paper. In this application the structural characteristics of the belief function allow for its simplification and development of application specific functions. Based on simple belief function and using statistical method to create evidence model, a new simplified combination algorithm is inferred, different from the traditional Dempster's rule. It has been proven that this new algorithm is effective and practical for oil analysis in the field of fault diagnosis.

1. INTRODUCTION

The Dempster's rule is an important tool for the combination and inference of evidence in the Dempster-Shafer theory. With the increase of focal elements in the frame of discernment, the number of elements in 2^Θ increases exponentially, thereby the computing process complexity increases exponentially. Several ways have been proposed to solve this problem. One of them is to create an effective algorithm in the light of some bodies of evidence with special structure. The other is to study the approximation algorithm. The following are the achievements in these two aspects. Barnett dealt with cases where bodies of evidence focus on atomic hypotheses and their complement (Barnett, J.A. 1981). Gordon and Shortliffe proposed an algorithm for approximating the results of Dempster's rule of combination for the case where the combined evidence arranged in a hierarchical or tree-like structure (Gordon, J. & E.H. Shortliffe 1985). Shafer and Logan gave an algorithm for an exact implementation that is linear in its computational complexity (Shafer, G. & R. Logan 1987). Dubois and Prade made a study on a computational approximation approach (Dubois, D. & H. Prade 1987). Voorbraak employed random coded messages to clarify the requirements for the use of evaluating Dempster's rule (Voorbraak, F. 1990). In this paper a method that is a fast-simplified fusion algorithm of evidence is presented; it is based on simple belief function and can create evidence model by using statistical model.

2. BASIC CONCEPTS IDENTIFICATION

2.1 *Frame of Discernment (FD)*

In the theory of evidence, let θ_i be a variable, Θ be the set-universe of all possible values of θ_i. Θ is called the frame of discernment that may be expressed as:

$$\Theta = \{\theta_i, \ i = 1, 2, \cdots, n\}. \tag{1}$$

If $I \neq j$, $\theta_I \cap \theta_j = \phi$. So θ_i and θ_j express the different answers for a problem. The set 2^Θ consists of all subsets of Θ.

2.2 *Basic Probability Assignment (BPA)*

BPA is a function defined on universe 2^Θ. It is named m: $2^\Theta \rightarrow [0,1]$. The function satisfies the following conditions:

(1) $m(\phi) = 0$

(2) $\sum_{A \subseteq \Theta} m(A) = 1$,

where $m(A)$ is the belief for a certain sub-set A of Θ in the case of the current evidence, but doesn't include the belief for any subset of A, that is, if $B \subset A$, $m(B)$ has no relation to $m(A)$. $m(A)$ expresses such a meaning: on the condition of the information provided by the current evidence, the belief of the answer for a problem in A can be got, but the belief for a specific answer in A is unknown.

2.3 *Focal Element and Body of Evidence*

For $A \subseteq \Theta$, if $m(A) \neq 0$, A is called a focal element. A set, kernel F, consists of all focal elements. All the answers for the problem provided by an evidence are set in a dually group (m, F) called a body of evidence.

2.4 Belief Function (Bel) and Plausibility function (Pl)

For a given basic probability assignment m, the belief of a subset A of $2^{\circ} \rightarrow [0,1]$ can be expressed as

$$Bel(A) = \sum_{B \subseteq A} m(B), \qquad (2)$$

where $Bel(A)$ represents the belief of the answer in A on the condition of the current evidence.

For special note, $Bel(A)=0$ doesn't mean that the anti-answer is in A, but means the lack of evidence of the answer in A. The belief of anti-answer in A can be expressed by $Bel(\overline{A})$ as

$$Bel(\overline{A}) + Bel(A) \leq 1. \qquad (3)$$

According to the above definition, BPA function m is one-to-one correspondent with the Bel function described as follows:

$$m(A) = \sum_{B \subseteq A} (-1)^{|A-B|} Bel(B). \qquad (4)$$

Use BPA function m to define Pl as

$$Pl(A) = \sum_{A \cap B \neq \phi} m(B) = 1 - Bel(\overline{A}) \quad \text{when } A \subseteq \Theta, \quad (5)$$

$Pl(A)$ expresses how much we should believe in A if all current evidence support A. $Pl(A)=1$ indicates that the answer may be in A, but doesn't mean the answer must be in A because at this moment $Bel(A)$ may be zero. Thus it can be seen that Pl reflects the probability, while Bel reflects the certainty.

3. SIMPLIFIED FUSION ALGORITHM OF EVIDENCE BASED ON THE SIMPLE BELIEF FUNCTION

If a Bel is described as follows:

$$Bel(A) = \begin{cases} m(B), & B \subseteq A, \ A \neq \Theta \\ 1, & A = \Theta \\ 0, & \text{others} \end{cases}, \qquad (6)$$

it is called a simplified belief function. Its BPA function m is defined as

(1) $m(A) > 0$
(2) $m(\Theta) = 1 - m(A)$
(3) $m(B) = 0$, for any other $B \subseteq \Theta$.

Actually if the evidence is provided by experts, it generally emphasizes a certain subset of A, that is, only one focal element is provided besides Θ.

At this moment the Bel provided by experts is a simplified belief function. Therefore it is very necessary to study on the simplified combination algorithm of evidence based on simple belief function.

3.1 A simplified fusion algorithm of evidence based on simple belief function with same focal elements

When $m_1, m_2, \ldots, m_n$ are assumed to be the simple Bel with same element $A(A \subseteq \Theta)$ and

$$m_i(A) = \alpha_i, \ 0 \leq \alpha_i \leq 1, i = 1,2,\cdots,n, \qquad (7)$$

the BPA function m achieved after combining the evidence according to Dempster's rule is:

$$\begin{cases} m = \bigoplus_{i=1}^{n} m_i & \text{£¬} \\ m(A) = 1 - \prod_{i=1}^{n} (1 - \alpha_i) & \text{£} \qquad (8) \\ m(\Theta) = \prod_{i=1}^{n} (1 - \alpha_i) \end{cases}$$

3.2 A simplified fusion algorithm of evidence based on simple belief function with different n focal elements

When $m_1, m_2, \ldots, m_n$ are assumed to be the simple Bel with different focal elements $A_1, A_2, \ldots, A_n$ and

$$m_i(A_i) = \alpha_i \ \text{£¬} 0 \leq \alpha_i \leq 1 \ \text{£¬} i = 1,2,\cdots,n, \qquad (9)$$

where for any given $I \neq j$, $i, j = 1,2, \ldots, n$, there is $A_i \cap A_j = \phi$, then

$$m = \bigoplus_{i=1}^{n} m_i \qquad (10)$$

$$m(A_i) = \frac{\alpha_i \cdot \prod_{\substack{j=1 \\ j \neq i}}^{n} (1 - \alpha_j)}{\sum_{i=1}^{n} \alpha_i \cdot \prod_{\substack{j=1 \\ j \neq i}}^{n} (1 - \alpha_j) + \prod_{i=1}^{n} (1 - \alpha_i)}. \qquad (11)$$

3.3 A simplified combination algorithm of evidence based on simple belief function with embedded focal elements

When $m_1, m_2, \ldots, m_n$ are assumed to be the simple Bel with different focal elements $A_1, A_2, \ldots, A_n$, and $m_i(A_i) = \alpha_i$, $0 \leq \alpha_i \leq 1$, if $A_1 \subset A_2 \subset \ldots \subset A_n$, the BPA function,

$$m = \bigoplus_{i=1}^{n} m_i, \qquad (12)$$

can be changed to

$$m(A_i) = \alpha_i \cdot \prod_{j=1}^{i-1}(1-\alpha_j). \qquad (13)$$

4. A SIMPLIFIED FUSION ALGORITHM OF EVIDENCE BASED ON USING A STATISTICAL MODEL TO CREATE AN EVIDENCE MODEL

According to the Bayes statistical model, an evidence model with a similar form can be set up and the correspondent BPA function m can be got which is represented by:

$$m_1(A) = \frac{\prod_{\theta_i \in A} p_i \cdot \prod_{\theta_j \in A}(1-p_j)}{1 - \prod_{i=1}^{n}(1-p_i)}, \qquad (14)$$

where $\theta_i \in \Theta$, $i=1,2,\ldots,n$, $A \subseteq \Theta$, f denotes the value of a certain feature, $p_i = p(f|\theta_i)$ is the conditional probability by the statistical model.

For general p_i, if $\forall A \subseteq \Theta$, $m(A) \neq 0$, this means that the created evidence model has $|2^\circ|$ focal elements at most. Another evidence can be set up as:

$$m_2(A) = \frac{\prod_{\theta_i \in A} g_i \cdot \prod_{\theta_j \in A}(1-g_j)}{1 - \prod_{i=1}^{n}(1-g_i)}, \qquad (15)$$

according to s—the value of another feature different from f and conditional probability $g_i = p(s|\theta_i)$, the computing complexity of the combination of two evidence by Dempster's rule is $O(|2^\circ| \cdot |2^\circ|)$ that can't be dealt with any other algorithm.

According to the above BPA function, a simplified combination algorithm of evidence using Dempster's rule is inferred as follows::

For any two evidence (m_1, F_1) and (m_2, F_2) belong to frame of discernment $\Theta = \{\theta_i, i=1,2,\ldots,n\}$, when $A \subseteq \Theta$, the following equations hold:

$$m_1(A) = \frac{\prod_{\theta_i \in A} p_i \cdot \prod_{\theta_j \in A}(1-p_j)}{1 - \prod_{i=1}^{n}(1-p_i)} \qquad (16)$$

$$m_2(A) = \frac{\prod_{\theta_i \in A} g_i \cdot \prod_{\theta_j \in A}(1-g_j)}{1 - \prod_{i=1}^{n}(1-g_i)}, \qquad (17)$$

the combination result, $m = m_1 \oplus m_2$, $\quad (18)$ can be expressed as

$$m(A) = \frac{\prod_{\theta_i \in A} p_i g_i \cdot \prod_{\theta_j \in A}(1-p_j g_j)}{1 - \prod_{i=1}^{n}(1-p_i g_i)}. \qquad (19)$$

5. CONCLUSIONS

Consider the following equation,

$$\sum_{\theta_i \in A} m_1(A) = \frac{p_i}{M_1} = Pl_1(\theta_i)$$

$$\sum_{\theta_i \in A} m_2(A) = \frac{g_i}{M_2} = Pl_2(\theta_i) \qquad (20)$$

for each θ, p_i is in direct ratio with the possibility of θ_i in the created model when evidence model is set up. Since in the created BPA function, the value of the BPA function for the different focal elements have a certain relationship, for example,

$$\frac{m_1(\{\theta_1\})}{m_1(\{\theta_1, \theta_2\})} = \frac{1-p_2}{p_2}, \qquad (21)$$

it is a special BPA function. If there isn't the above relationship in $m_1(A)$, it is impossible to rebuild the initial BPA function m from $Pl_1(\theta_i)$ according to the created formulae. For example, when

$$\Theta = \{\theta_1, \neg \theta_2\} \, \text{£¬} \, m(\{\theta_1\}) = \frac{1}{3} \, \text{£¬} \, m(\{\Theta\}) = \frac{2}{3}$$

There is,

$$Pl(\theta_1) = \frac{1}{3} \quad \text{£¬} \quad Pl(\theta_2) = \frac{2}{3}$$

$$M = 1 - [1 - Pl(\theta_1)] \cdot [1 - Pl(\theta_2)] = 1$$

The BPA function created by the above formulae is:

$$m'(\{\theta_1\}) = 0 \quad , \quad m'(\{\theta_2\}) = \frac{2}{3} \quad , \quad m'(\Theta) = \frac{1}{3},$$

which is different from the initial function. Therefore, the simplified algorithm proposed above is only applicable for the evidence model we created. For the general model, because the BPA function can't always be expressed in the demand form, it is impossible for the simplified algorithm to make an evidence combination.

According to the Dempster's rule, as the evidence increases, the computing complexity increases exponentially, thus it is very impractical for application. The simplified algorithm proposed above is very effective for our evidence model. This algorithm is applied to GOAFDS and the accuracy of diagnosis is near 90%.

For example, here is the analysis and diagnosis of oil samples of gearbox for a certain belt conveyor on 11, July. Several forms of diagnosis about gearbox

753

are listed in Table2. The detection results of these oil samples are listed in Table 1. According to the record of oil samples, it is noted that the oil brand is L-AN46 engine oil and the type of gearbox is NGW. After comparison with the limit value, the detecting result of various bodies of evidence has exceeded standard in varying degrees. It shows that the equipment perhaps has problems.

Table 1. The detecting result of oil sample

Spectro(ppm)				
Fe	Cu	Cr	Si	Na
68	23	1	7	12
FT-IR(relative value)				
OXI		NIT		Water
4		4		7
Ferrography (grade value)				
Metal-oxide		Fatigue particle		Oxide
5		7		2
Sphere particle		Laminar		Cut
5		5		2
Viscosity(mm²/s)				VI
V40°		V100°		VI
39.15		6.52		111

Table 2. The several forms of diagnosis about gearbox

	Gear
A	Case crush.
B	There is scratch on the sliding direction.
C	There are crackles on the tooth face or point or gear mass.
D	There are undulate ridges on tooth face.
E	The tooth face changes coarse and has some press trucks. The side of gear has wear and
F	The corrosion of the tooth face causes rust and caves.
G	The gear mass changes color and softening.
H	The tooth face has scratch, cut and concave.
	Roller bearing
A	The surface of roller and rollaway nest peals
B	The roller and rollaway nest have pitting and caves.
C	The rollaway nest has long scratch or has elliptical concave caused by rub and
D	The roller holder wears heavily or cracks.
E	The variation of bearing increases and the degree of coarse increases.
F	The bearing has burn or coherence.
G	The roller and roller race have crackle.
H	The roller and rollaway nest have dint and concave.
I	The fitting surface corrodes and has fretting.
J	The side of roller bearing wears heavily.

By using the above algorithm to make diagnosis, the following results are received:

For gear:
$\max[Bel]=Bel(\{A,B,E,F\})=0.98345$
$\max[Pl]=Pl(\{A,B,E,F\})=0.98368$
Judgement: the fault type is group (A,B,E,F).
Explanation:
(1) The viscosity of oil samples is much lower than the standard value.
(2) Due to the high oxidability and nitriability of oil samples, it is difficult to form an oil film.
(3) The face of the gear is corroded and heavily worn so as to damage the gear.
For roller bearing:
$\max[Bel]=Bel(\{A,C,E,I,J\})=0.95082$
$\max[Pl]=Pl(\{A,C,E,I,J\})=0.95111$
Judgement: the fault type is group (A,C,E,I,J).
Explanation:
(1) There is too much difference between the value of viscosity and the corresponding standard value.
(2) The value of viscosity is so low as to make lubrication insufficient and to case the bearing to wear heavily.
Suggestion: change the oil. We propose to use the lubricating oil of the branding N220.

The researchers on the spot adopted the suggestion of the experimental center and replaced the oil with N220. According to the feedback information, new detection results of the oil samples are normal. The diagnosis results are correct, which help to avoid further wear and tear of the equipment, assure that the equipment will work properly, reduce the frequency and times of maintenance and save nearly 10,0000 RMB in maintenance expenses.

REFERENCES

Barnett & J.A. 1981. Computational Methods for Mathematical Theory of Evidence. *Proc. 7th IJCAI*: 868-875.

Gordon, J. & E.H. Shortliffe 1985. A Method for Managing Evidential Reasoning in a Hierarchical Hypothesis Space. *AI*. 26: 323-357.

Shafer, G. & R. Logan 1987. Implementing Dempster's Rule for Hierarchical Evidence. *AI*. 33: 271-298.

Dubois, D. & H. Prade 1987. An Approach to Approximate Reasoning Based on the Dempster's Rule of Combination. *Int J. Expert Systems*. 1(1): 67-85.

Voorbraak, F. 1990. On Justification of Dempster's Rule of Combination. *AI*. 48: 171197.

Mining Science and Technology' 99, Xie & Golosinski (eds) © *1999 Balkema, Rotterdam, ISBN 90 5809 067 1*

Load capacity of cycloid-involute composite tooth profile gears in open gear transmission

Xun Fan, Hongbing Qiao, Junmei Zhang & Huirong Meng
China University of Mining and Technology, Beijing, People's Republic of China

ABSTRACT: This paper presents the load capacity of a new type of cycloid-involute composite tooth profile gear used in open gear transmissions. The wear resistance and strength of the gears are described in detail, and the calculation and test results are discussed and compared with those of involute gear. It is shown that the load capacity of the composite gears is superior to that of involute gears.

1. INTRODUCTION

Open gears operated under a low speed and heavy load condition are widely used in the mining mechanical engineering, with a very poor working environment generally. Because of the accumulation of abrasive particles in lubricants, the original tooth profile surface is easy to be worn away and the tooth cracks are usually occurring. The service life of this sort gears is very short, with the minimum only about one or two month. It is economically significant to improve the load capacity of open gears and prolong their service life. One way to solve this problem is to optimize the tooth profile of gears, so that the load capacity of the tooth surface can be improved. At present, most open gears are adopted involute as the tooth profile curves. Although the inviolate has many advanced characteristics, it is not the most optimal profile for the open gears.

In addition to the involute tooth profile, it is already known that cycloid can also be used as the tooth profile. As compared with involute gears, the relative radius of curvature of the cycloid gears is large at the contact point, and its tooth thickness on the critical section at the tooth root is thick, and its specific surface sliding at both the addendum and the dedendum holds constant values. The contact strength, the bending strength and the wear resistance of the cycloid gear may be superior to those of the involute one. However, the cycloid gear can not be used for transmitting loads, because its radius of curvature of the tooth surface is zero at the pitch point.

In order to make good use of the characteristics of the cycloid gear and avoid its demerits, a composite tooth profile gear which consists of the involute tooth profile near the pitch point and the cycloid tooth profile at the addendum and dedendum was designed (see Fig.1). The load capacity of this type of composite tooth profile gear will be discussed in this paper.

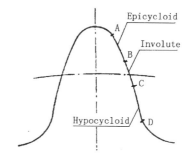

Fig. 1 Generation of the composite tooth profile

2. PROFILE EQUATION OF COMPOSIS GEAR

The tooth profile of a composite gear consists of three parts of curves shown in Fig.1. They are epicycloid AB, hypocycloid CD and involute CB. The equations of the cycloid-involute composite profile are expressed as follows.

2.1 *Cycloid profiles*

$$\begin{Bmatrix} x \\ y \end{Bmatrix} = \begin{bmatrix} \cos\dfrac{\pi}{2z} & -\sin\dfrac{\pi}{2z} \\ \sin\dfrac{\pi}{2z} & \cos\dfrac{\pi}{2z} \end{bmatrix} \begin{Bmatrix} x_1 \\ y_1 \end{Bmatrix} \qquad (1)$$

in which epicycloid in section AB:

$$x_1 = (a+r)\sin\varphi + a\sin(\theta+\varphi)$$
$$y_1 = (a+r)\cos\varphi - a\cos(\theta+\varphi) \qquad (2)$$

hypocycloid in section CD:

$$x_1 = -(r-a)\sin\varphi - a\sin(\theta-\varphi)$$
$$y_1 = (r-a)\cos\varphi - a\cos(\theta-\varphi) \qquad (3)$$

where, r is the radius of pitch circle;
a is the radius of rolling circle;
θ is the rotational angle of rolling circle;
φ is the rotational :

$$\varphi = \frac{a}{r}(\theta + z\operatorname{inv}\alpha_o)$$

α_o is the angle of the gear
The other symbols are same as those of familiarized involute gears,

2.2 Involute profile

$$\begin{Bmatrix} x \\ y \end{Bmatrix} = \begin{bmatrix} \cos(\frac{\pi}{2z}+\varphi_0) & -\sin(\frac{\pi}{2z}+\varphi_0) \\ \sin(\frac{\pi}{2z}+\varphi_0) & \cos(\frac{\pi}{2z}+\varphi_0) \end{bmatrix} \begin{Bmatrix} x_2 \\ y_2 \end{Bmatrix} \qquad (4)$$

in which

$$x_2 = r_b\sin\theta_v - r_b\theta_v\cos\theta_v$$
$$y_1 = r_b\cos\theta_v + r_b\theta_v\sin\theta_v \qquad (5)$$
$$\varphi_0 = \operatorname{inv}\alpha_0$$

The method to calculate the geometry sizes of the composite gear is similar to that of involute one.

3. WEAR RESISTANCE ANALYSIS

3.1 Wear resistance test

In order to make the analysis and comparison more clear, three kinds of spur cylindrical gears, which are standard involute, modified involute and cycloid-involute composite gears, are adopted as the test gears in wear resistance tests. Geometry parameters, the materials and their heat-treatments are identical basically □See Table.1□. The test gears with composite tooth profile are cut by a special self-designed hob.

The wear resistance tests were conducted on a closed power loop type of gear test rig. The test gears were lubricated by gear grease and tested in two different conditions with and without the presence of abrasive particles in the lubricant. In order to imitate the practical operating environment in mines, ore powder is used as the abrasive particles. Grade of the ore is 32.5%; density is 3.4 t/m^3; Vickers diamond hardness is 12-18; and the particle-size is 180-200 mesh.

3.2 Wearing capacity

Two methods were used to measure the wearing capacity. One is the weighting method and the other is choral thickness method with which the distribution law of the wearing capacity on the surface can be obtained through measuring the chordal thickness at any point along the tooth height.

Wearing capacity as measured by the weighting method is shown in Fig.2. In Cartesian coordinates, x axis stands for number of circulation and y axis represents accumulative total wearing capacity. With the help of Fig.2, it can be observed that the wearing capacity of composite gears is much less than that of involute gears. While modified involute gears' wearing capacity is between that of standard involute gear and composite gear at the beginning, but after a period of wearing, the tooth profile lost its original curve, and the wearing capacity increased. In order to make the comparison more clear, the mean square-root values of the specific wear ability was calculated□See Table 2□. From Table.2 it can be found that in the absence of abrasive particles, the specific wear ability of composite gears dropped by 23.7%, compared to standard involute gears, and by 25.1%, compared to modified involute gears; and they dropped by 15% and 22.8% respectively in the condition of abrasive wear.

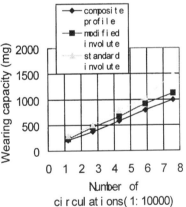

Fig. 2 Wearing capacity of the test gears

4 BENDING STRENGTH ANALYSIS

4.1 Calculating modeling

For the analysis of the bending strength of the composite tooth profile gears mentioned above, the finite element method is used. In order to make a

Table.1 Main parameters of the test gears

Contents	Involute Gear	Modified Involute Gear	Composite Gear
Center distance (mm)	140	140	140
Number of tooth (z_1/z_2)	21/49	21/49	21/49
Gear ratio (i)	2.3	2.3	2.3
Module (mm)	4	4	4
Face width (mm)	14	14	14
Modified factor (x_1/x_2)	0/0	+0.28/-0.28	0/0
Pressure angle (α)	20°	20°	20°

Table 2 Wearing capacity of test gears

Profile	Gear	Non-abrasive Wear （mg/h）	Abrasive Wear （mg/h）
Standard involute	Pinion	4.832	183.52
	Gear	6.135	212.52
Modified involute	Pinion	4.920	202.13
	Gear	6.740	219.48
Cycloid-Involute	Pinion	3.685	156.13
Composite	Gear	5.400	153.50

Table 3 Maximum stresses of composite gear and involute gear

Contents	Number of Element	Composite Gear	Involute Gear
Finite Element Method	155	92.8 MPa	114.6 Mpa
Boundary Element Method	57	95.6 MPa	118.5 Mpa

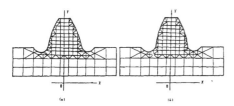

Fig. 3 Mesh generation of gear models

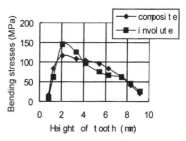

Fig. 4 Distributions of bending stresses of gears

comparison between composite and involute gears, both composite and involute gear are discussed. The mesh generation of the tooth in two-dimensional model of the two gears is shown in Fig.3, in which the composite gear is on the left, and the involute one is on the right. The basic parameters of the gears are the same as those given in Table 1. From the Fig.3, it can be found that the tooth thickness at the

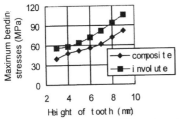

Fig.5 Maximum stresses at different mesh point

dedendum of the composite gear is thicker than that of the involute one, so the former is superior in the bending strength.

4.2 *Calculating results*

The results of the calculation by SAP5 and BEM2D programs are shown in Fig.4 and Table 3. It shows obviously that the bending strength of the composite gears is larger than that of the involute ones. The maximum bending stresses vary with the change of the mesh points when gears mesh with each other from the addendum to the dedendum as shown in Fig.5.

5 CONCLUSIONS

Based on the above research, the conclusions can be drawn as follows:

The load capacity including wear resistance and the bending strength of cycloid-involute composite tooth profile gears developed in the present study are superior to that of involute gears.

The composite tooth profile gear is especially useful in the case of open gears with low-speeds, high-load and a large gear ratio.

REFERENCES

Yoshio TERAUCHI, Bulletin of JSME, Vol.25, No.199, 1982, 118-126.

AGMA, Method for Evaluating the Risk of Scuffing & Wear, ANSI/AGMA 2001-B88, WG-6 Nov. Zhengzhou, China, 1988.

Raymond. J. Drago, Fundamentals of Gear Design, Butterworth, USA, 1988.

Fan Xun, Meng Huirong, et al., Study on the Wear Resistance of Cycloid-Involute Composite Tooth Profile Gears, Proceedings of the 2nd China/Japan International Symposium on Machine Elements, China, 1996, 176~180

8 Mine economics, management and administration

Mining Science and Technology' 99, Xie & Golosinski (eds) © 1999 Balkema, Rotterdam, ISBN 90 5809 067 1

Modelling the interests of the stakeholders at Chinese state coal sector

Rick Tamaschke & Trevor Grigg
University of Queensland, Brisbane, Qld, Australia

Xiubao Yu
Tongji University, Shanghai, People's Republic of China

ABSTRACT: This paper analyses the relationship between the performance of organizations within the CSC and the interests of stakeholders, the trade-offs between these interests, as well as how these interests are affected by the industry practice. Pooled cross-section and time series data are employed and a simultaneous equation regression model is developed for the analysis. The results indicate strong quantitative relationships between organizational performance and the interests of stakeholders in the Chinese State Coal (CSC) sector. Executives' interests, employees' interests government's interests and other stakeholders' interests are found to be significantly related to organizational performances. Results relating to important issues in the industry are discussed and suggestions both to the government and organizations within the CSC sector are outlined in this paper.

1. INTRODUCTION

According to Freeman, the term stakeholder refers to "any group or individual who can affect or is affected by the achievement of an organisation's purpose" (Freeman, 1984, p.52). Stakeholders include shareholders, managers, employees, customers, suppliers, government etc.

From a strategic point of view, stakeholders are important. Stakeholders have gained increasingly more attention both from academics and industries since Freeman's work in 1984. There have been a number of articles analysing the relationships between the various stakeholders of an organisation and strategies in dealing with their interests. Unfortunately, the concepts stakeholder, stakeholder model, stakeholder management and stakeholder theory are explained and used by various authors in very different ways and supported with diverse and often contradictory evidence and arguments. Moreover, this diversity and its implications are rarely discussed (Donaldson, 1995). Few studies have systematically assessed or analysed the relationships between stakeholder interests and organisation performance at a quantitative level.

Stakeholders have various interests and expectations of an organisation. The organisation may be thought of as an axis, which unites and reconciles the various interests of all stakeholders. The reconciliation of various interests of stakeholders mostly happens when the interests of the organisation are satisfied. The approach of reconciliation differs according to the different management practice and different environment the organisation is in. In other words, a specific management practice and the environment mainly determine a specific approach of reconciliation in which the organisation operates. This paper analyses quantitatively the relationship between the performance of organisations within the CSC sector and the interests of stakeholders, the trade-offs between these interests, as well as how these interests are affected by the industry practice.

2. THE SIMULTANEOUS EQUATION MODEL

Eight dependent variables were selected to represent the interests of various stakeholders for the study. The performance variables are profit, Total Factor Productivity (TFP), coal output, executive income, employees' average income, employment at the non-coal section, farmland compensation and the demand for materials used in coal production. The sample selected for the modelling consists of all organisations in the Shandong State Coal (SDSC) sector. The period selected is 1990 to 1996 as during this period, the structure and other conditions in SDSC were fairly stable. The variables are analysed using a simultaneous equation regression performance model (SERM). The SERM seems particularly useful because of the interactions between the various stakeholder performance variables and trade-offs between stakeholders' interests. The model is summarised in Table 1. Tables 2 and 3 provide the definitions of the variables.

The model consists of twelve equations, including

seven identities. The identities define profit (equation 1), Total Factor Productivity and its components (equations 2 and 3), the capital- labour ratio in the coal sector (equation 5), employment in the coal section (equation 6), the components of total cost (equation 8), and material cost (equation 12). To be estimated are equations 4, 7, 9, 10 and 11. Equation 4 postulates that current output is positively related to lagged profits, the capital labour ratio and coal section employment. Equation 7 postulates that employment in the non-coal section is positively determined by employment in the coal section and profits, and inversely by changing employment in the coal section. Equation 9 postulates that executive incomes are positively related to employment income, Total factor Productivity and firm size. Equation 10 postulates that employment income is positively related to profits, and negatively to resource category. And

finally equation 11 postulates that farm compensation is positively related to coal output and the capital labour ratio.

To provide an adequate sample size, the (annual) time series data for the eight organisations were pooled, and the equations estimated using a pooled time series/cross-section approach (Kmenta, 1990). The model has as many equations (including identities) as endogenous variables, and the 5 non-identity equations to be estimated are over-identified. Hence, the models were estimated by two-stage least square (2SLS). The equations were analysed for multicollinearity and these tests gave no undue cause for concern. Inspection of the residuals about the fitted equations indicated no obvious patterns. The results of the estimations for equations 4, 7, 9, 10 and 11 are given in Table 4.

TABLE 1: The simultaneous Equation Stakeholder Performance Model

(1) $\text{profit}_{it} = \text{output}_{it} \times \text{coalpric}_{it} - \text{totlcost}_{it}$	*(identity)*
(2) $\text{TFPdf}_{it} = (\text{outpindx}_{it} / \text{inptindx}_{it})^* \text{TFPdf}_{it-1}$	*(identity)*
(3) $\text{outputindx}_{it} = \text{output}_{it} / \text{output}_{it-1}$	*(identity)*
(4) $\text{output}_{it} = \alpha_0 + \alpha_1 \text{profit}_{it-1} + \alpha_2 \text{clratio}_{it} + \alpha_3 \text{emplcoal}_{it} + \varepsilon_{it}$	
(5) $\text{clratio}_{it} = \text{capital}_{it} / \text{emplcoal}_{it}$	*(identity)*
(6) $\text{emplcoal}_{it} = \text{emplt}_{it} - \text{emplnc}_{it}$	*(identity)*
(7) $\text{emplnc}_{it} = \beta_0 + \beta_1 \text{emplcoal}_{it} + \beta_2 \text{profit}_{it} - \beta_3 \text{emplcchg}_{it-1} + \varepsilon_{it}$	
(8) $\text{totlcost}_{it} = \text{exeinco}_{it} + \text{empinco}_{it} \times \text{numbempl}_{it} + \text{matcost}_{it} + \text{farmcomp}_{it} + \text{othecost}_{it}$	*(identity)*
(9) $\text{exeinco}_{it} = \delta_0 + \delta_1 \text{empinco}_{it} + \delta_2 \text{TFP}_{it} - \delta_3 \text{firmsize}_{it} + \varepsilon_{it}$	
(10) $\text{emplinco}_{it} = \phi_0 + \phi_1 \text{profit}_{it} - \phi_2 \text{rescatg}_{it} + \varepsilon_{it}$	
(11) $\text{farmcomp}_{it} = \gamma_0 + \gamma_1 \text{output}_{it} + \gamma_2 \text{clratio}_{it} + \varepsilon_{it}$	
(12) $\text{matcost}_{it} = \text{output}_{it} \times \text{unituse}_{it}$	*(identity)*

Note: The ε_{it} are the random error terms.

TABLE 2: Definitions of Endogenous Variables of the Model

Variables	Descriptions
profit_{it}	Total profit after tax earned from sales of coal in period t (10000 yuan).
TFP_{it}	Value of Total Factor Productivity deflator in year t.
outpindx_{it}	Coal output index. The index is the coal output in year t divided by the output in year t-1.
output_{it}	Annual coal output (tones).
clratio_{it}	Capital - labour ratio (1000 yuan/employee). The value in year t is the production capital in year t divided by the employment at coal section in year t.
emplcoal_{it}	Employment at coal section in year t (persons).
emplnc_{it}	Employment at non-coal section (persons).
totlcost_{it}	Total coal cost (10000 yuan) in year t.
exeinco_{it}	Executive annual income in year t (yuan). The income is the sum of basic salary plus organisational bonus plus contract bonus.
emplinco_{it}	Employees' average income in year t (yuan/person). The income includes basic salary plus all types of bonuses. All employees at the coal section are included: production workers, workers at auxilliary sections and office staff.
farmcomp_{it}	Total compensation (10000 yuan) to farmers in year t due to the farmland damages caused by underground mining.
matcost_{it}	Total cost of all materials used at the coal section in year t in constant price (10000 yuan). The cost reflects the quantities used or the demands for all materials in constant price.

Note: The subscript *it* refers to organisation *i* in period *t*.

TABLE 3: Definitions of the Predetermined Variables of the Model

Variables	Descriptions
coalpric$_{it}$	Average coal price over year t for all types of coal sold (yuan/tone).
inptindx$_{it}$	Input index. The index is the input in year t divided by the input in year t-1.
TFP$_{it-1}$	Value of Total Factor Productivity deflator in year t-1.
output$_{it-1}$	Total coals produced and sold in year t-1 (tones).
profit$_{it-1}$	Total profit after tax at year t-1 generated from sales of coal (10000 yuan).
capital$_{it}$	Total capital input used in coal production in year t (1000 yuan).
emplt$_{it}$	Total employment both at coal section and non-coal section in year t (persons)
emplcchg$_{it-1}$	Lagged employment change at the coal section (persons). The value is the difference of coal employment in year t-1 minus coal employment in year t-2.
othecost$_{it}$	All other coal costs (total) excluding the costs of labour, material and farmland compensation in year t (10000 yuan).
firmsize$_{it}$	Dummy variablei generated from the scale of output in year t.
rescatg$_i$	Resource category. A dummy variable for resource category. The numbers for the categories vary from best (1) to worst (5).
unituse$_{it}$	Unit cost of materials in year t (yuan / tone).

TABLE 4: Results of the Estimations

(4) output$_{it}$= -3021922 + 198.17 profit$_{it-1}$ +104024.9 clratio$_{it}$ +339.42emplcoal$_{it}$
        (-3.41#)   (6.82*)   (6.00*)   (8.76*)   (F = 119.8*; R^2 .91)

(7) emplnc$_{it}$ = 463.19 + 1.17emplcoal$_{it}$ + .23profit$_{it}$ – 1.4 emplcchg$_{it-1}$
        (.324)   (8.45*)   (3.1*)   (-2.48*)   (F = 44.8*, R^2 .79)

(9) exeinco$_{it}$ = -13767 + 5.91 empinco$_{it}$ + 30.30 TFP$_{it}$ – 162.47firmsize$_{it}$
        (-5.37#) (6.65*)   (2.00*)   (-.73)   (F = 40*, R^2 .77)

(10) emplinco$_{it}$ = 3928 + .026 profit$_{it}$ – 88.12 rescatg$_{it}$
        (13.85#) (5.3*)   (-1.12)   (F = 37.5*, R^2 .68)

(11) farmcomp$_{it}$ = - 749.96 + 7.89 *10^{-5} output$_{it}$ + 52.18 clratio$_{it}$
        (-5.13#)   (5.00*)   (11.88*)   (F = 169*, R^2 .90)

Values in parenthesis are 't' statistics.
* Denotes significant at at least the 5 per cent level, one-tailed test.
# Denotes significant at at least 5 per cent level, two tailed-test.

DISCUSSION OF RESULTS

The results of the model suggest that:
1) Lagged profits, capital-labour ratio and coal employment, are positively related to the coal output.
2) Employees' income and TFP performance are related to executive income, which is consistent with what the paper expected. However, firm size is found not to be related to executive pay at CSC, which conflicts with the (namely western) literature. The results suggest that social comparison theory plays the most important role in the formation of executive pay at CSC. Organisational (economic) performance plays an important role as well.
3) Profit is significantly related to employees' average income. Resource category does not have an important impact on the dependent variable, which indicates that there is not much income difference between employees at organisations with good resource conditions and with poor resource conditions.
4) Employment at the coal section, profit, and changes of employment at the coal section are important factors that explain the employment at the non-coal sector. The employment at the coal section provides the basis for the employment at the non-coal section (to support and provide the necessary services). The linkage between employment at the non-coal section and profit suggests that part of the profit is used in providing more support services for the coal section by the non-coal section. The results do suggest that there is a relationship between employment at the non-coal section and employment change at the coal section. The decrease in employment at the coal section (negative sign for change of coal employment) will result in the increase of employment at the non-coal section. This reflects the practice of the industry that redundant employees of the coal section are transferred to the non-coal section.
5) The results of Equation 11 lend support to the view that both output and capital-labour ratio are important predictors for farmland compensation.

Overall the results suggests that most of the stakeholders can benefit from the improvement of the performance of the CSC. For example, as incomes for both executives and employees at the coal section are

763

performance related, improvements of economic performance such as profit and productivity will benefit both groups of stakeholders. To protect their own interests, executives not only need to improve organisational performance (profit, safety, state assets, productivity), but also have to look after the interests of their employees at the coal sector as executives' income are importantly related to employees' income. Employees also need to look after the interests of their organisations (especially profit) as their income is affected by the profit performance of their organisations.

To improve the productivity performance of the coal sector, the government sets policies for state coal organisations to reduce employment at the coal section. According to government policy, the redundant employees are transferred out of the coal section and re-deployed to the non-coal section. The results of the simultaneous equation regression model illustrate that the employment capacity of the non-coal section at state coal organisations is mainly determined by the total employment at the coal section. The results also suggest that there are limitations within the non-coal section to accommodate the redundant employees from the coal section. In contrast to current industry practice, the model implies that the employment at the non-coal section should be reduced, along with downsizing at the coal section. Apparently, the practice of downsizing can improve the productivity performance of the coal sector, but will cause difficulties for the non-coal sector business. To keep sustainable development for all the sections, strategies have to be worked out for the government or the industry to fundamentally address this issue. One possible approach is to widen the focus of the non-coal more towards servicing the community at large, so that non-coal business becomes progressively less dependent on the coal section.

Output growth was considered to be in the interests of executives (power) and organisations (stability and security of production and market share). The model shows that the growth of coal output is strongly influenced by lagged profits, the capital labour ratio and coal section employment. This indicates more capital and relatively less input of labour are required to stimulate output growth. Ouput growth is obviously very important to the government's objective to rationalise the coal sector, and the government needs to emphasis the importance of improving profit performance and inputting capital at CSC.

The results also show that the greater the output of coal and the thicker the coal seam (measured as capital-labour ratio in the study), the more compensation there will be for farmlands. It was also seen that suppliers benefit from the growth of coal output. The improvement of Total Factor Productivity reduces the cost of coal production and hence has a negative impact on the demand for coal production materials. Although suppliers can not get the benefit of the TFP improvement, many other stakeholders can enjoy the benefits. Coal organisations can reduce coal costs and reap more profits. Both employees' and executives' income can be increased by the increase in profit. Customers can also enjoy a lower market price with the decrease of coal cost if the TFP improvement is significantly widespread throughout the industry.

CONCLUSIONS

This paper has attempted to apply Freeman's (1984) stakeholder model to analyse the performance of the CSC sector. A simultaneous equation performance model was constructed. The model suggests that most stakeholders will benefit from improved performance at CSC, but that trade-offs are necessary between the interests of the various groups. The model should be useful the Chinese government as it moves to improve the efficiency of the coal sector.

REFERENCES

Conyon, M. and Leech, D. (1994), *Top pay, company performance and corporate governance*, Oxford Bulletin of Economics and Statistics, Aug. Vol.56, 3, p.229 – 247.

Donaldson, T. and Preston, L. (1995), *The Stakeholder Theory of the Corporation: Concepts, Evidence, and Implication*, Academy of Management Review, Vol. 20, 1, p.65-91.

Freeman, R.E. (1984), *Strategy Management: A Stakeholder Approach*, Boston, Pittman.

Frost, F. A. (1995), *The use of stakeholder analysis to understand ethical and moral issues in the primary resource sector*, Journal of Business Ethics, Vol. 14, 8, p:653-661, Aug.

Griner, E. H. (1996), *Corporate Performance, Stock Option Compensation, and Measurement Error In CEO pay*, Journal of Applied Business Research, Vol. 12, 1, Winter 1995 – 1996.

Jensen, M. C. and Murphy, K. J. (1990), *Performance pay and Top Management Incentives*, Journal of Political Economy. Vol. 98, 2, p.225 – 264.

Kmenta, J. (1990), *Elements of Economics*, 2nd edition, Maxwell MachMillan International, New York.

Labour Department, Shandong State Coal Administration Bureau (1995), *An application for the distribution methods for top managers' income*, Internal industry profiles, June

Labour Department, Shandong State Coal Administration Bureau (1989-1997), *Applications for distribution methods for top managers' income*, Internal industry profiles.

Mallette, P. Middlemist, R. and Hopkins, W. (1995), *Social, Political and Economic Determinants of ZC Chief Executive Compensation*, Journal of Managerial Issues, Vol. 7, 3, p.253-276, Fall.

Mining Science and Technology '99, Xie & Golosinski (eds) © 1999 Balkema, Rotterdam, ISBN 90 5809 067 1

US policy, law and regulation on surface mining control and reclamation

Paul H.Chen
International Management Institute, Pacific Palisades, Calif., USA

ABSTRACT: This paper discusses the U.S. federal and state mining policies, law, and regulations. It studies how the federal government, mainly the OSM, and the state agencies conduct the oversight of reclamation activities and enforcing and regulating the surface mining law. The financial responsibilities and management of federal, states and coal mining industries are analyzed. Special focus is placed on how the coal agencies cooperate with other environmental agencies and citizen groups in safe guard the environment.

1 BACKGROUD

The North American Continent is rich in coal deposits especially in the Appalachia Plateaus regions. The typical coal deposit is shallow, about 30 to 90 feet beneath the ground surface. Hence the most popular technique for coal extraction is surface mining, which is cost effective and relatively safe in comparison to others such as underground mining.

There were no coal mining regulations required by law before the 1930s. West Virginia enacted the first mining law in 1939. It was followed by other coal states such as Indiana, Illinois, and Pennsylvania. During World War II, the demand for coal energy heightened. It superceded the worry of environmental protection. This resulted in massive land erosions and water pollution in the coal mining areas.

In the 1960s coal demands continued to rise. Individual coal states enacted laws and regulations to require permits and inspections for coal mining. The United States Congress enacted the major coal mining law, the Surface Mining Control and Reclamation Act in 1977 to combat this problem.

2 THE U.S. SURFACE MINING LAW

The surface mining control and reclamation Act (SMCRA), signed into law in 1977, is the main

law established by the federal government for protecting people and the environment from the effects of coal mining. It contains five regulatory provisions. Among them, Title IV and V are the most important ones. Title IV regulates abandoned mine reclamation and Title V specifies the terms for controlling the environmental impacts of surface coal mining.

Title IV establishes the Abandoned Mine Land (AML) program which provides for the restoration of eligible lands and waters, mined and abandoned or left inadequately restored before passing the law in 1977. (An amendment, the 1990 Abandoned Reclamation Act, provides for the extension of reclamation the abandoned mines created after 1977.) The following are the key provisions,

- Title IV creates an Abandoned Mine Reclamation Fund for use in the reclamation and restoration of land and water resources adversely affected by coal mining activities. The AMR Fund is drawn from the reclamation fee collected from mining operators based on the amount of coal produced. The money is then allocated to fund various AML projects.

- Besides to pay reclamation fee, all operators are required to report the amount mined, coal removal method, permit and safety and health ID number. Making knowingly false statements or failure to make accurate statements are punishable by fine and/or imprisonment.

- Each State with an approved regulatory program may submit a State Reclamation Plan and annual projects to carry out reclamation . The Surface Mining Office is to define eligible lands and water for reclamation by prioritizing all eligible sites in the AML inventory data base.
- Provision on the reclamation of rural land and land adversely affected by past coal mining are regulated in Title IV. They give power to the Secretary of Agriculture and the Secretary of Interior or their agents to protect public health, safety, and general welfare.
- The law provides regulation for other important details such as how to handling the liens on reclaimed land, carry out filling voids, sealing tunnels etc. and emergency events dangerous to public health.
- Finally, the Law provides power to the Secretary of Interior to approve certification upon satisfactory completion of reclamation.

Title V provides law and regulations for environmental protection standards in ongoing surface coal mining. It governs mining from the beginning of the mining process to completion of the reclamation and restoration of the land. It holds the operators fully accountable for the entire life cycle of coal mining including restoration of land by imposing performance bonds. The key provisions of Title V are summarized as below.

- Title V establishes the environmental protection standards. It defines procedures for carrying out the regulations which starts with promulgation, publication, public hearing and commenting, and concurring with other environmental laws.
- The law regulates the state surface coal mining programs by requiring that states meet the Federal capability requirement for carrying out the Title V provisions. All State mining laws are superseded by the provisions of this Act except if it proved that they provide more stringent regulations than the Federal law requirements.
- Title V provides strengthen regulations for the coal mining permits. It gives detailed procedures for permit application requirements. It defines the terms and conditions for approval, denial and renewal of the permits.
- The law requires mining operators to post a performance bond, an amount sufficient to assure the coverage of reclamation plan to safe guarded the environment.

- Environmental protection standards, inspection and monitoring procedures as well as enforcement and penalties are written specifically in this provision.
- Any citizen can file a civil action against the U.S. Government and other agencies if his/her property legal rights are invaded. Any approval or disapproval of state program by Federal agencies is required by judicial review.

3 US MINING POLICY

The U.S. policy toward coal mining development is based on the need to meet the nation's energy needs therefore protecting national security, economic well being, and general welfare of the Nation. U.S. mining policy attempts to create a balance between sufficient coal supply (which relates to national economy well-being) and sustainable development of coal resource. The following approaches are adopted by the Surface Mining Law.

- The law creates a uniform standard among all mining states to insure the competition in interstate commerce among sellers of coal will not undermine the ability to improve and maintain standard of mining operations.
- To streamline the administer function the Act specifies the States being the primary agencies responsible for developing, authorizing, issuing, and enforcing regulations. The federal agency, mainly the Office of Surface Mining, is responsible for administration, oversight and enforcement the law, as well as managing the AML projects for federal lands.
- The law creates an accountable and responsible financial approach. It uses two financial steps; the reclamation fund and the performance bonds. The reclamation fund provides financial mean for reclamation and restoration of abandoned mine lands and water, while performance bonds provides sufficient fund to cover the cost of reclaiming the ongoing mine sites in the event the operator does not complete reclamation.
- The Act calls for the public participation in the development, revision and enforcement of regulation, standards, reclamation planes and programs. Using matching fund, joint programs and other methods, the local citizen groups are encouraged to raise private funds, manage the AML reclamation projects

through local initiatives and involve in safe guarding environment.

- The law requires the cooperative effort be established among government agencies. It includes joining programs, sharing resources, and facilitating interagency communication in management. The OSM is to identify and eliminate of duplication and overlap programs among agencies.

- Congress specifies in the Act to protect unique cultural and natural resources. It prohibits mining within the boundaries of national parks, forests, wildlife refuges, trials, wild and scenic rivers, wilderness and recreations areas and historic places. Its general policy is to assure that surface coal mining operations are not conduct where reclamation as required by this Act is not feasible.

4 OSM AND STATE MINING AGENCIES

Title II of SMCRA creates the Office of the Surface Mining Reclamation and Enforcement (OSM) within the Department of the Interior. OSM is responsible for implementation and enforcing the law, upgrading regulations, oversight environmental restoration and environmental protection programs.

OSM provides financial grant-in-aid and technical assistance to state programs and manage AML programs on the federal lands. The agency upgrades and enacts new regulations through a sequences of public actions. It first, publish and promulgate rule and regulation, then solicit comments and hold public hearings, and finally, implement the law.

The states are the primary governmental responsibility for mining operations. OSM grants the states "primacy status" through enacting programs at the states level that demonstrates their capability to carry out the provisions of the law. With the primacy status, the states are then authorized to establish regulations, to enforce the law and to operate the programs with adequate funding and manpower resources on private land within the state boundary.

The OSM 1997 Revision Oversight Directive (REG-8) establish policies procedures and responsibilities for conducting oversight of state programs. The newly proposed regulations of Permit Eligibility; Definitions of Ownership and Control, and nationwide computer database of the

Applicant Violator System (AVS) enhances the permit and application regulations.

5 FINANCIAL RESPONSIBILITY AND MANAGEMENT

The Surface Mining Law establishes an Abandoned Mine Reclamation Fund to provide for reclamation and restoration of land and water resources adversely affected by past mining. Title IV section 402 details the reclamation fee that is to be collected and deposited to the fund. The fee is collected from the mining operators based on the tonnage of coal produced (35 cents per ton of surface mining and 15 cents of underground mining). The law holds the Secretary of Interior accountable for fund management and fee collection. The Fund provides resources for state reclamation programs, emergency projects and high-priority projects, among others. Between 1978 and 1996, over $4.5 billion reclamation fee were collected. In the 1998 fiscal year, OSM provides near $200 million in grants to the state programs. The authority to collect reclamation fee will expire in the year 2004. The OSM is amending its rules concerning the incidental extraction of coal. An innovative way, called "enhancing reclamation program", to maximize available AML fund is under study. This may allow the contractor to sale incidental extracted coal that would reduce the governmental financial burden.

For environmental protection of the ongoing surface mining, the law holds the mining operator financial accountable. Before a mining permit can be granted, an operator must post a performance bond sufficient to cover the cost of reclaiming the site in the event the operator does not complete reclamation. The bond is not fully released until performance standards have been met and full reclamation of the site has been successfully completed. This policy guarantees for the environment protection of the site minded.

6 COOPERATION OF AGENCIES AND CITIZEN GROUPS

The SMCRA Act requires cooperative efforts among government agencies. The policy includes joining programs and sharing resources (information, manpower, techniques, equipment and material.) and facilitating interagency communication in management. The OSM is to cooperate with other Federal agencies and state regulatory authorities to minimize duplication in inspections, enforcement and administration of the Act.

The Act calls appropriate procedures be provided for the public participation in the development, revision and enforcement of regulation, standards, reclamation planes or programs. The AML fund and other government resources would not have enough revenues to address every potential eligible sites. Addition funding is actively solicited from private sources. Using matching fund, joint programs and other innovative procedures, local citizen groups are encouraged to raise private funds and manage the AML reclamation projects through local initiatives (such as Clean Stream Initiative.) The reclamation and restoration work is then carried through by environmental, industrial, local agencies and/or other citizen groups.

On January 4, 1999 a new program to assist non-profit organizations undertake local acid mine drainage reclamation projects was announced. As part of the 1999 Appropriations Act, $750,000 was provided for cooperative agreements through the Appalachian Clean Streams Initiative. The cooperative agreements will be in the $5,000 - $80,000 range in order to assist as many groups as possible complete projects to clean streams impacted by acid mine drainage. Information and applications forms are available on line.

7 DISCUSSION

Coal is a major natural resource for both industrial and developing countries. It provides energy and chemical material for the society. The byproduct of coal mining is environmental degradation. How to develop coal resources and the same time protect the environment is a major challenge for the industrial country such as the United States and also the developing countries like China. The US Surface Mining Law, the Surface Mining Control and Reclamation Act of 1977, is the major legal system developed to achieve this goal. The Law enacted in 1977 provides fundamental blue print and strategy to deal with the environment impact from the abandoned mines and the environmental protection of the ongoing mining in the United States. For the last 20 years it has served its purpose well.

Environment protection is not just a local issue. It becomes a global problem. Air and water pollution is not limited within the boundaries of an individual country. The global warming problem is the best example. The effective way to protect the environment of the earth is therefore, dependent upon the cooperation of international communities. The success in US environmental protection of coal mining provides a good model for other coal mining countries.

A joint partnership program should be developed in the international level to deal with the environmental protection of coal mining. Coal mining techniques, legal system, management experience and information should be shared in the international community. An international mining regulation or law should be developed to provide environmental protection and natural resources conservation such as rainforests, wildlife, fish, and vegetation. An international coal organization or association can provide the forum for all important coal mining and environment related issues.

8 REFERENCES

Abandoned Mine Land (AML) Reclamation Program: Enhancing AML Reclamation , 7483 Federal Register Vol.64, No.29, Friday, February 12, 1999, Rules and Regulations.

Twentieth anniversary, surface mining control and reclamation act. A report on the protection and restoration of the nation's land and water resources under the surface mining law, United States Department of the Interior, Office of Surface Mining, Washington D.C. 20240, August 3, 1997.

US Department of the Interior, Office of Surface Mining Reclamation and Enforcement Directive system, REG-8, "Oversight of the State Regulatory programs," dated 06.20.1996.

Compilation of selected laws concerning minerals and mining, with amendments through the 102D congress, prepared for use of the committee on natural resources of the 103rd Congress, first session, January 1993, US government printing office, Washington DC.

Mining Science and Technology' 99, Xie & Golosinski (eds) © 1999 Balkema, Rotterdam, ISBN 90 5809 067 1

Township-owned coal mines in China: Managing sustainable development*

Jianhua Ai, Hongliang Zuo & Jingbiao Xu
College Business Administration, China University of Mining and Technology, Beijing, People's Republic of China

ABSTRACT: This paper presents a new model for evaluating the overall benefit of China's township-owned coal mines. The benefit of coal mines should be evaluated from the perspective of various partners concerned, including investors, local government, local residents, local enterprises and the society as a whole. In order to find ways to manage effectively township-owned coal mines, we also discuss about such measures as technological policy, resource development planning, entry and withdrawal of barriers, tax and fee collection policies, resource pricing policy and management system as well.

1. INTRODUCTION

Adequate supply of energy has contributed greatly to the rapid development of China's economy ever since 1978 when reform and opening policy was initiated. At present more than 70% of China's energy supply comes from coal. Between 1978 and 1997, China's output of coal has increased at an annual rate of 4.12%, while that of township-owned coal mines has increased at an average annual rate of 10.51%. Township-owned coal mines are playing an important role now. In 1997 township-owned coal mines produced 581 million tons of coal, which represents 43.68% of China's total coal output. The number of township-owned coal mines reached as many as 58,000.

However, fast development of the township-owned coal mines has also led to some serious problems that can not be ignored. (1) Low recovery rate. Recovery rate in advanced countries reaches as high as 70%, and that of China's key state-owned coal mines and local state-owned coal mines is respectively 50% and 30%, but that of township-owned coal mines is only between 15% and 20%. (2) Heavy pollution of the environment. (3) Poor safety conditions. In 1997, death rate per million tons of coal output in township-owned coal mines reached 7.9 persons, while that of key state-owned coal

mines and local state-owned coal mines was respectively 1.2 and 4.3 persons. (4) The fact that people of various economic status are engaged in coal mining has resulted in great disorder and frequent conflicts over coal mining right. Some small coal mines often dig cross their boundaries, which has brought about many serious accidents and heavy losses to big coal mines. The conflict between big and small coal mines is the most serious problem China's coal mining industry is now facing. (5) The overall quality of China's coal mining has been declining. Though total output of township-owned coal mines accounts for 43.68% of China's total, each small coal mine produces only an average of 10,000 tons per year, which implies that nearly half of China's coal mines is using primitive methods. (6) Vicious competition is rampant. Small coal mines have been driving their state-owned counterparts out of the market by taking advantage of their lower cost. Because of unfair competition and sluggish market, almost all key state-owned coal mines have been suffering heavy losses.

As sustainable development is the theme of economic development for the 21st century, it is urgent and important for the Chinese government to exercise effective management over township-owned coal mines. China's mining industry is in a disadvantage in management compared with its

* Supported by National Science Foundation of China (NSFC)

foreign counterparts, so research in this respect is of great practical value (Hou Zhencai 1995, Lu Chaodong 1989). By referring to research achievements(Wallace, N 1992, Gow. L. J. A 1993, Hogan, L 1992, Siddayao, C.M & Griffin, L. A 1993), we hereby design a new model for evaluating the overall benefit of China's township-owned coal mines by taking into account the interests of various parts, including investors, local government, local residents, local enterprises and the society as a whole. In order to find ways to manage effectively township-owned coal mines, we consider it necessary to establish technological policy, resource development planning, entry and withdrawal of barriers, tax and fee collection policies, resource pricing policy and management system as well.

2. THE MODEL

2.1 Evaluation of investors' benefits

Benefits of investors can be evaluated with formula (1)

$$NR_i = R_i - (IC_i + RC_i) - (DC_{ig} + T_{ig}) - (WC_{ip} + RSC_{ip} + EC_{ip} + SC_{ip}) - (RSC_{im} + EC_{im} + SC_{im}) - RSC_{ir} \qquad (1)$$

NR_i stands for the net benefit of investor(s). R_i stands for the benefit of investor(s). IC_i and RC_i respectively stand for investment cost and operating cost, which shouldered by the investor(s). DC_{ig} and T_{ig} respectively stand for drainage charges and various taxes and fees paid to the government. WC_{ip}, RSC_{ip}, EC_{ip} and SC_{ip} respectively stand for wages, resource compensation cost, environment compensation cost and safety compensation cost paid to local residents. RSC_{im}, EC_{im} and SC_{im} respectively stand for resource compensation cost, environment compensation cost and safety compensation cost paid to local enterprises. RSC_{ir} stands for resource cost paid to the owner(representative of the central government). Because mineral resources have not yet commercialized, RSC_{ir} only covers compensation fee for resource prospecting.

Though township-owned coal mines usually sell their products at lower prices, they can still obtain sound economic benefit by taking advantage of their lower investment cost, operating cost, resource cost, environment cost and safety cost.

2.2 Evaluation of local government's benefits

The benefits of local government can be evaluated with formula (2):

$$NR_g = (DC_{ig} + T_{ig}) - MC_g \qquad (2)$$

NR_g stands for the net revenue of the local government. ($DC_{ig} + T_{ig}$) stands for revenues of the local government. MC_g stands for the management cost shouldered by the local government.

Direct benefits of the local government include drainage charge, taxes and other fees, while indirect benefits of the local government include creation of job opportunities and contribution to the development of the local economy. Generally speaking, net revenue of the local government is positive, representing the local government's tendency to develop township-owned coal mines.

2.3 Evaluation of local resides' benefits

Local residents' benefits can be evaluated with Formula (3):

$$NR_p = (WC_{ip} + RSC_{ip} + EC_{ip} + SC_{ip}) - (RSC_p + EC_p + SC_p) \qquad (3)$$

NR_p stands for the net revenue of local residents. WC_{ip}, RSC_{ip}, EC_{ip} and SC_{ip} stand for wages, resource compensation fee, environment compensation fee and safety compensation fees paid to local residents by township-owned coal mines. RSC_p, EC_p and SC_p refer to the actual resource cost, environment cost and safety cost incurred by local residents.

The wage level of township-owned coal mines is usually low. However, because of serious land subsidence, environmental pollution and other safety problems, various compensation fees paid to local residents are usually less than actual losses. That's why the net revenue of local residents in some regions is negative.

2.4 Evaluation of local enterprises' benefits

Local enterprises' benefits can be evaluated with Formula (4):

$$NR_m = (RSC_{im} + EC_{im} + SC_{im}) - (RSC_m + EC_m + SC_m) \quad (4)$$

NR_m stands for the net revenue of local enterprises. RSC_{im}, EC_{im} and SC_{im} respectively stand for resource compensation fee, environment compensation fee and safety compensation fee obtained by local enterprises. RSC_m, EC_m and SC_m stand for losses of resource, environment and safety incurred by local enterprises.

Township-owned coal mines tend to break rules and regulations by digging beyond their borders, which often leads to such serious underground accidents as flooding and gas explosion in key state-owned coal mines. Hence the net revenue of local key state-owned coal mines in most regions is negative, while the net revenue of local coal-using enterprises is positive because of township-owned coal mines' supply of cheaper coal and reduced transportation cost.

2.5 Evaluation of the resource owner's benefits

Benefits of the resource owner (representative of the central government) can be evaluated with Formula (5):

$$NR_r = RSC_{ir} - RV_r \quad (5)$$

NR_r stands for the net revenue of the owner of the resource. RSC_{ir} stands for resource revenue obtained by the owner. RV_r stands for the actual value of coal resource and its associated resources.

At present coal enterprises are only charged for resource prospecting, so in a sense they have been using coal resource free of charge. Moreover, almost all the compensation fee for resource prospecting is retained by local governments, so the net revenue of the resource owner is negative.

2.6 Evaluation of social benefits

Social benefits can be evaluated with Formula (6):

$$NR_s = NR_i + NR_g + NR_p + NR_m + NR_r \quad (6)$$

NR_s stands for the net social benefits. NR_i stands for the net revenue of the investor(s). NR_g stands for the net revenue of the local government. NR_p

stands for the net revenue of local residents. NR_m stands for the net revenue of local enterprises. NR_r stands for the net revenue of the owner of the resource.

At present NR_i and NR_g are positive☐while NR_p, NR_m and NR_r are negative, so we can roughly conclude that NR_s in some regions is negative, which indicates that the operation of township-owned coal mines is inefficient.

In theory we may presume the following conditions as sufficient and essential for efficient operation of township-owned coal mines: (1). $NR_s >= 0$; (2). $NR_i > 0$; (3). $NR_g >= 0$; (4). $NR_p >= 0$; (5). $NR_m >= 0$; (6). $NR_r = 0$.

Then the system for evaluating the benefits of township-owned coal mines can be described with Figure 1.

3. EFFECTIVE MANAGEMENT OF TOWNSHIP-OWNED COAL MINES

1. *Technological policies.* The government should draw up technology policies that require township-owned coal mines to produce in an economic scale, adopt reasonable technology and equipment, secure a high recovery rate, avoid environment pollution and ensure safety.

2. *Resource development plan.* The government should work out scientifically resource development plans, and then request bids for the exploitation of coal resources. It is also advised that the government sell out complete coal mines when it finishes constructing them.

3. *Policy barriers.* Overall benefits of all coal mines should be evaluated. Those that have failed or will not be able to bring about sound overall benefits should be either suspended or closed down.

4. *Taxation and fee policies.* Drainage fee and compensation charges for property, environment and safety should all be raised, while taxes and other fees levied by the local government should all be cut down.

5. *Policies for pricing resources.* A perfect system should be established to evaluate the value of mineral resource, on the basis of which the prices of mineral resources can be reasonably determined.

6. *Management system.* The government should reform the existing resource ownership so as to establish resource property market. It is also important to divide reasonably power and function

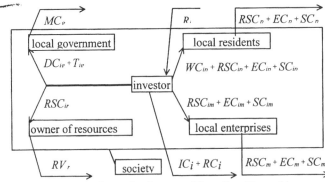

Figure 1 A system for evaluating overall benefits of township-owned coal mines

of resource management between the central government and local governments.

4. CONCLUSIONS

1. The benefits of township-owned coal mines should be considered and evaluated from the prospective of all partners concerned, including investors, local government, local residents, local enterprises and the society as a whole.

2. For the purpose of implementing the sustainable development strategy and managing township-owned coal mines effectively, the Chinese government should have to carry out strategic renovations in its technological policy, resource development planning, entrance and withdrawal barriers, taxation and fee policies, resource pricing policies and management system related to township coal mining.

REFERENCES

Gow. L. J. A. 1993. The use of economic instruments for resource management, Coal Research Association of New Zealand Inc.

Hogan, L.1992, The efficiency of Australia's mineral taxation arrangements, Australian Bureau of agricultural and Resource Economics, Canberra, ACT (Australia).

Hou Zhencai 1995 Supervision and management of mineral resource development in Australia. *Mineral Protection & Utilization:* 1995(2):10-11

Lu Chaodong 1989. A number of issues regarding American mining policy, *Mineral Protection and Utilization* 1989 (1):3-6

Siddayao, C. M. & Griffin, L. A. 1993. Energy investments and the environment, International Bank for Reconstruction and Development. Washington D.C(USA).

Wallace, N, 1992. Natural resource management: an economic perspective, Australian Bureau of agricultural and Resource Economics, Canberra, ACT (Australia)

Mining Science and Technology' 99, Xie & Golosinski (eds) © 1999 Balkema, Rotterdam, ISBN 90 5809 067 1

Comparison analysis between China and Japan for small coal mines

N. Horii

Institute of Developing Economies-JETRO, Tokyo, Japan

ABSTRACT: This paper presents the current status of coal production in China, especially focusing on dominant share of small coal mines. Through comparison with evolution of Japan's coal industry, those factors can be pointed out for growth of small coal mines in China, such as 1) better condition of mining resource, 2) small and dispersed market structure, 3) low level of labor cost and lack of capital. Lastly, recent policies for China's small coal mines is considered and evaluated.

1 INTRODUCTION

The outstanding characteristic of China's coal industry is that Township Owned Mines (TOMs), most of which are small, producing only less than 7,000 tons a year, have dominant shares in the market. Those small coal mines have some advantages over State Owned Mines (SOMs) in the competition in China's coal market, such as their flexible and efficient management system, employment creation in rural areas and so on. After some reforms in 1980s, which were liberalization of TOMs' coal production and price reforms, TOMs' production increased dramatically and solved energy shortage problem even in the rapid economic growth period. Compared with oil industry, reforms in China's coal industry towards market economy were successful at least from increase of coal production.

However, TOMs' coal supply increase caused many other problems. We can point out, for example, air pollution caused by unprocessed high sulfur coal, increase of miner's fatality incidence rates and resource degradation caused by room and pillar mining. In 1998, the State Bureau of China Coal Industry decided to shut down above 25,800 small coal mines, most of which were TOMs, for the reason of its environmental pollution and resource degradation. China's government policies are also changing from liberalization to regulation of TOMs again.

Thinking from international comparison, the growth of TOMs is very interesting phenomenon. There are no other countries, in which small coal mines could accomplish such a remarkable growth as China's TOMs. In the United States, Russia and Japan, coal industry have been made more concentrated into small number of large mines and production from small coal mines continues decreasing.

The purpose of this paper is to verify what factors have influenced the growth of TOMs through the comparison with Japan's small coal mines. Those factors analyzed here are 1) condition of mining resource, 2) market structure, 3) labor cost and capital. Through comparison of those above factors with Japan's small coal mines, it can be understood that China's transformation to market economy caused surplus entry of coal producing firms and scattered investment. As a result of that, China's coal industry is facing environmental and resource degradation, and surplus supply problems.

2 ENERGY SYSTEM CHANGE AFTER 1980s

2.1. Economic growth and energy source shift

The contribution of coal to satisfying growth in energy demand during economic reform period is reflected in the contribution rate of each energy sources to growth in total energy consumption, shown in Table 1.

By this measure, coal displayed an overwhelmingly high contribution rate of 93.8% to energy growth from 1953 until 1960, when the Daqing oilfield was discovered,. In contrast, a shift away from coal continued from 1960 to 1975, with oil at 54.7% and natural gas at 6.6% displaying extremely high contribution rates, as also shown in the Table 1. However, from 1979, when the policy of reform and openness was established, until 1996, the contribution rate of coal again rose, reaching 77.2%, while oil declined to 15.2%, and natural gas to 0.7%.

In addition, contribution rates before oil imports were approved in the 1990s, i.e., from 1979 to 1990, show that the share of coal had climbed further to 83.3%, while oil had actually declined to 9.0%, and natural gas to 0.3%. These facts clearly show that burgeoning energy demand during the economic reform period was regulated by additional coal production.

Table 1. Contribution rates of each energy sources to total energy consumption growth (%)

contribution rates	coal	oil	natural gas	water power
1953-1960	93.8	4.2	0.5	1.5
1960-1975	28.2	54.7	6.6	10.6
1979-1996	77.2	15.2	0.7	6.9
1979-1990	83.3	9.0	0.3	7.3

Source: Calculated by author

2.2. Shift towards market economy and growth of small coal mines

As shown in Table 1, after 1979 coal has played major role in China's energy supply again. The shares coal occupied in total energy consumption, increased to 76.2% in 1990 again, from 69.9%, lowest level in 1976. The reason for increase of coal share, is that entry into coal industry by new enterprises was easier than other energy industries. Most of coal production increase after 1979 was supplied by TOMs. Because of deregulation, TOMs's entry into coal market was strongly promoted in 1980s. As a result of those deregulation policy, the number of TOMs increased surprisingly, from 18,000 in 1980 to 63,000 in 1985. Even in 1995 the number of TOMs was still numerous, 73,000. (Yan, 1997)

The policies, influencing entry activities by TOMs, are deregulation of selling surplus coal above planned target and price liberalization started in 1981. Both policy reforms also succeeded in raising economic incentives for starting production of coal. Consequently such huge number of TOMs rushed into the market. The shares of TOMs to total coal production increased dramatically from 14.1% in 1978 to 46.4% in 1995. Even though most of TOMs are small scaled, whose average production per year are less than 7,000 tons. In the next section, the factors influencing the growth of TOMs are verified

through comparison with Japan's coal industry.(Chinese Ministry of Coal Industry, 1997; Yan, 1997)

3 COMPARISON BETWEEN JAPAN AND CHINA FOR SMALL COAL MINES

3.1. Condition of mining resource

Coal production in Japan already started in the 19th century. At that time, in Japan there were feudal lords ruling in their regions. Feudal lords monopolized and controlled coal production. After political revolution of Meiji period in the middle of 19th century, feudal lords lost their power to control coal market and many farmers began their coal production without order.

However, the new government of Meiji periods, made the policy at once to prohibit private coal mining, and all coal mining firms were transformed to state owned firms. After several years later, those state owned companies were sold off to large companies which had large amount of capital. Under rich endowments of coal resources, those large companies increased their scale of coal production through large investment.

On the other hand in Japan there were also certain amount of small scaled coal mines. Most of them mined poor conditioned coal seams because of government's policy to allocate rich conditioned to large companies and poor ones to new entry. Consequently, Japan's small coal mines, which mined poor resource endowments, were not competitive with large coal mines. (Sumiya, 1968; Yata, 1975)

As for China, whose coal resource endowments are much better than Japan's, small coal mines are very competitive in the market. It is because TOMs mines rich conditioned coal seams. Coal industry policy prior to 1979, investments in coal industry were not towards "Energy Bases", which is the present center of coal production in China, i.e., Shanxi and Neimenggu. In 1960s and 70s, for reasons of security and transportation, the south-western part and eastern coastal areas received large part of investments for development of coal resources. Investments to "Energy Bases" had been stayed at low level before economic reform period, only 12.4% of total national investments were allocated to those areas, prior to 1979. After 1980, the proportion of investments allocated to "Energy Bases" increased to 30.0% in the Sixth, 32.9% in the Seventh, and 37.1% in the Eighth Five-Year Plan.(Yan, 1997)

When the policies for deregulation of TOMs' coal mining were applied, huge resource was still underdevelopment in "Energy Bases" area. During last 20 years, the increase of coal production in Shanxi province was the most outstanding. The fact is notable that in Shanxi province, the proportion of TOMs' production is 50.3%, much higher than northern provinces, such as Liaonin, and Heilongjiang (Table 2). The reason for that difference is that northern provinces has been mined for long history, and most of resource endowments had already been mined by large scaled SOMs, as a result of which there were less room for TOMs to entry into.

Table 2. Coal production and proportion of TOMs' in each major coal producing provinces

	Year 1996	
	Total (10^4tons)	Proportion of TOMs' production ()
Hebei	7,409	33.2
Shanxi	34,946	50.3
Neimenggu	7,317	42.9
Liaonin	6,041	26.0
Heilongjiang	8,147	29.4
Anhui	4,642	13.5
Shandong	8,949	21.0
Henan	10,780	40.3
Hunan	5,093	72.9
Sichuang	9,567	60.9
Guizhou	6,143	79.7
Yunnan	3,072	65.7
Shaanxi	4,613	49.4
Total	130,799	47.0

Source: *China Coal Industry Yearbook 1997*

3.2. Market structure

At its first stages in the 19th century, main users for Japan's coal industry were salt pan. Then demands for steamship fuels, rail fuels and chemical materials, increased. After that, the share for the use for cokes and electricity fuels got dominant. In a process of user shift from small and dispersed to large and concentrated ones, the share of small coal decreased in Japan. (Sumiya, 1968)

The more concentrated and the larger scaled the users changed, the more advantages large scaled coal mines acquired in the market competition. For large users, the stability of coal supply becomes important element to decide the supplier. To satisfy large demand of large users, large coal mines has more

advantages. As for transportation cost, supplying to large and concentrated users can decrease their transportation cost because of scale economies.

In China, coal users are extremely small scaled and dispersed. Here power industry, the largest sector in terms of coal consumption, should be taken as an example. China's power industry accounted for 32.3% of total coal consumption in 1995. Table 3 shows installed capacity by size of each power generator. According to these figures, in 1995 generator of 200MW or more accounted for no more than over 40% of total installed capacity, while the proportion of generator of 25MW or less exceeded 30%. In addition, 6MW or less generator accounted for about 12% of total. It shows that China's coal market is extremely small scaled and dispersed. From these facts, it is understood that SOMs, most of which are large coal mines, lose their advantages over TOMs in China's coal market.

Table 3. Installed capacity by size of each power generator (Year 1996)

capacity of power generator	umber of units	installed capacity	percentage
over 300MW	175	5925.7	25.1%
200-300MW	215	4380.5	18.5%
100-200MW	337	3843.7	16.2%
50-25MW	423	2332.8	9.9%
25-12MW	608	1731.8	7.3%
12-25MW	1034	1364.1	5.8%
6-12MW	1727	1258.5	5.3%
under 12MW	n.a.	2817.1	11.9%

Source: *China Power Yearbook 1996-97*

Accompanied with the conversion towards market economy, former planned economic system almost collapsed. Before, SOMs supplied their coal products to state owned firms, under this system SOMs had advantages with economies of scale, as discussed above. But at present, state owned firms, former SOMs' users, faces decline of their economic activities. Instead, township owned firms have grown up and occupied dominant shares in national economy. Most of those township owned firms are small scaled and dispersed. Consequently, SOMs are losing their competitive advantages at now.

3.3. Labor cost and capital

In Japan, there were two major coal producing regions, i.e., Hokkaido and Kyusyu area. Both areas also had rich resource endowments, but the applied mining systems were completely different between two areas. In Kyusyu area, where labor mobility was

775

high from near rural areas, labor cost stayed at lower level. On the other hand, in Hokkaido area, which was newly developing area, and in which labor mobility was quite low, labor cost was very high. Because of that difference, in Hokkaido there were much fewer small coal mines and the mining system transformed from room-and-pillar mining to long-wall mining much earlier than in Kyusyu area. (Shouda, 1987 and Ogino, 1993)

In China, there exists abundance of labor, especially in rural areas. As shown in Table 2, major coal producing provinces are located in rural area. Consequently, in such provinces coal mines can make use of low labor cost. Not only low labor cost but also lack of capital gives TOMs no incentives to invest capital for mechanization. The transformation towards market economy worsened lack of capital in coal producing firms. As discussed before, those policies for coal industry to transform into market economy were to promote new entry by TOMs, which is managed by much lower level of political system than former SOMs. The success of those policies made the number of TOMs dramatically but the capital for each coal mine to invest was scattered. As a result of that outcome, most of TOMs are mining with room-pillar mining system.

Exploiting China's rich resource endowments, TOMs raise their efficiency to mine shallow and thick coal seams with room-and-pillar mining system. After reaching a certain depth, at which room-and-pillar mining system is not able to continue mining any more, then TOMs ,especially in Shanxi, abandon their mining coal seam and remove other new coal seam. This resource degradation occurs because of low labor cost and lack of capital.(Cui, 1995)

4 CHINA'S RECENT POLICY CHANGE

In recent years, China's government changed their policy emphasis to reducing TOMs' production, which means drastic reform of energy system supporting China's rapid economic growth during economic reform period. The most outstanding example for policy change is to shut down 25,800 small coal mines, based on "China Coal Law". This policy's targets are following:

- shutting down 25,800 TOMs operating without license
- reducing total coal production by 250 million tons by the end of 1999
- controlling TOM's coal production between 360-400 million tons by the end of 1999

These policy reflects seriousness of those problems caused by TOMs' production, such as environmental and resource degradation.

5 CONCLUSION

The discussion in the section 3 indicates that growth of TOMs can be explained by three factors. These include:

- 1) better condition of mining resource,
- 2) small and dispersed market structure,
- 3) low level of labor cost and lack of capital

The above analysis also shows that those three factors were influenced by changes caused by conversion towards market economy after 1980s.

We also discussed China's recent policy to solve problems brought about TOM's coal production. But recent policy would be only temporarily effective without change of above three factors.

REFERENCES

Chinese Ministry of Coal Industry, 1997. *China Coal Industry Yearbook*, October. (in Chinese)

Chinese Ministry of Power Industry, 1997. *China Power Industry Yearbook*, December. (in Chinese)

Cui, Y. 1995. *History of China Coal: Shanxi Province*, Meitangongyechubenshe, October. (in Chinese)

Ogino, Y. 1993. *History of Labor and Capital in Chikuhou Area*, Kyusyu University Press, February. (in Japanese)

Sumiya, M. 1968. *Analysis of Japan's Coal Industry*, Iwanami Press, February. (in Japanese)

Shouda, S. 1987. *History of Coal Industry in Kyusyu Areas*, Kyusyu University Press, April. (in Japanese)

Yata, T. 1975. *Japan's Coal Industry after World War II*, ShinHyoulon, September. (in Japanese)

Yan, C. (ed.) 1997. *China's Energy Development Report*, Jingjiguanlichubenshe, June. (in Chinese)

Mining Science and Technology '99, Xie & Golosinski (eds) © 1999 Balkema, Rotterdam, ISBN 90 5809 067 1

Mining industry reorganisation in Jiu Valley, Romania

R. Sarbu, E. Traista, G. Madear, Cl. Ionescu, M. Rebrisoreanu & M. Popescu
Petrosani University, Romania

ABSTRACT: In the past the Romanian Government was managing the coal industry through the state plans. The last one covers the period of up to year 2010. This plan required that Romania mines enough coal and metals to cover its domestic needs. After 1990 the Government policies have changes. The new Government stresses preservation of mineral resources with the resulting downsizing of the mining industry. Many uncertainties and difficulties are associated with this new program and it is difficult to predict its final outcome. The paper presents reorganization of the mining activities in Romania's Jiu Valley coal field, with its socio-economics and environmental impacts.

1. INTRODUCTION

Jiu Valley is a mountain area placed in the central part of Romania, surrounded by four main groups of mountains, two of them higher then 2500m. This fact made climate to be cold and rainy and limited the economical development. The traditional occupation was for a long period of time shepherd. In this period in the area live 12.000 people.

In the second part of 19th century, coal was discovered and begun a fast development of are. Figure 1 shows the coal extraction evolution in time and the increasing of population.

Coal production increased with a low ratio till 1960, when a forced production increasing begun. The inhabitants increasing followed the coal production's evolution. After 1960 the area have a

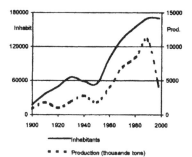

Figure 1. Evolution of coal production and inhabitants in Jiu Valley

forced development that exceed the local supplies. These facts decreased the standard of life and affect strongly the environment

After 1990 the coal production was reduced to 4.0 millions tons, but all mine kept the employers. The coal production may be sustained in this condition only by subsidies. In 1996 was obvious that the mining industry had to be re-organized because the Romanian economy cannot sustain it further. The govern planning for mining industry is not well seen by the Parliament, because a mining reorganization will unemployed hundreds of thousands' miners and will decrease all economical activities in respective areas. By this reason, mining industry is too large developed yet, which need very large subsidies from govern. Now for Govern is very difficult to leave poor mine because social pressure is very great. All mining areas from Romania are developed only to serve mining and closing a mine will affect all economically activities. By this reason resources management is neglected for policy benefit. Practically, all poor mine must be closed, and their resources preserved till further mining technologies will allow to extract minerals in profitable conditions. All financial resources should be given to the rich ores mines that can be made profitable. The most important issues that will appear are mineral extraction is developed in some area in which other industries are not developed. By this reason, the employers have not alternatives to find another job. By the other way the Unions are very strong in the mining industry and they are fighting for keeping the job, get salary and a lot of another right. These facts made very expensive mineral extraction's and also very difficult reorganization or closing the mine. All mining areas are developed in mountain area with very precious landscape resources. These resources are strongly affected by mining buildings and dumps and in same cases when the extraction is made in quarries are completely destroyed. The reorganization has to be made keeping the efficient mine. The unemployers which was expected to appear may create very serious social problems. By this reason the Romanian Govern decided to pay all unemployers with sum of money in values of 20 salaries. With this money was expected that they will begin their own business or they leave this area. The main business is expected to be in tourism due to geographical conditions, but this involve the environmental improvement. in this paper are shown the results of this reorganization.

2. SOCIO ECONOMIC IMPACT

After 1990 because of necessity in under going transformation from a controlled economy to the market economy, Romanian Govern is drawing up a special program to hurry the structural reforms and to create appropriate conditions of market economy. A government program proposes to achieve some actions as follows:
- Economy demonopolisation and it diversification;
- Transition to the privatization process and a free trade;
- Reduce government subsidies and create a real concurential environment in market economy.

This new economic environment will have an impact on coal mining industry restricted it. To be in accordance with these restrictions Coal Company have initiated a restructuring program regarding its own activity which have to take into account:
- Structure reorganization;
- Financial recovery and costs control (in order to reduce costs);
- Reconsideration of a legislative framework;
- Changes in the exploitation methods;
- Adapting the coal output taking into consideration coal demand which decrease with 30% in 1997 in comparison with 1989;
- Analyze mines performance (there are 3 categories: viable, stationary and, unlivable - closing gradually those with weak

performances);
- Reduce number of employees.

Restructuring mining sector in Jiu Valky has the aim to achieved optimal conditions to a sustainable development in the area. These conditions could be achieved only if the effects of restructuring process can be tolerable for all those which are affected by the process. Otherwise it is possible to appear rejected reactions, which could degenerate in strikes and demonstrations of popular protest, and usually their consequences are very hard to estimate. So that the workforce and mining communities are not accepted restructuring willingly.

The most important and also, strong impact during the restructuring process due to measures which have to be applied is on human factor (inhabitants and communities). An important measure to achieve of activity takes into consideration the necessity of adapting the existing number of the workforce in the company.

The employee before beginning of restructuring process was *54000* (1989) and decreased to 20300 at the end of 1998.

In comparison with 1989, into a first phase of restructuring in 1992 the employee number decreased with 13.76%; in the second stage (till 1994) with *15.6%*, in 1997 employees was reduced with more then a half - *56,5%* and in 1998 with 62%.

In the future it is estimated the same decreasing trend of employment till 2005 with an average of 300-500 employee every year.

The major decreasing was registered in 1997 as a result of applying the Govern policy which was a policy of no compulsory redundancies.

The mineworkers were therefore offered a financial compensation to have the coal industry and to accept redundancy compensation terms. Each mineworker who was agreeing to leave his job in the coal mining industry was paid with an amount of money (compensation) depending on years of service till twenty monthly wages, with the condition that the worker not to wish to get his job back in the mining sector.

Coal Company in the first phase of restructuring calculated carefully the terms of the compensation packages offered to employees in order to encourage takes up by older workers in the initial phases of the restructuring. The package was designed to attract the older workers into early retirement.

The money that is given to the employers is used in another way then the governance hoped. The most important part of this money is spent to buy consumer goods. The sales of every shops increased

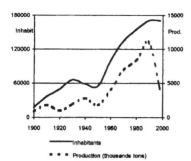

Figure 2. Sales evolution in re-organization period

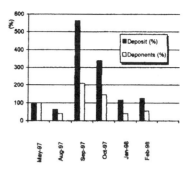

Figure 3 Deponents number and deposits evolution in reorganization period

very much in reorganization period (Fig. 2).

This chart show that in reorganization period sales rise very much exceeding sales in Christmas period. The number of deponents in banks and the deposits increase in reorganization period (Figure 3).

All this data show that the money is not used in the way in which they are given. One reason for this evolution is the fact that the money given is too reduced to allow a private business. Another reason is the fact that the Jiu Valley have a reduced supply of utilities like drinkable water or gas and the pollution is too great

The impact of the redundancies has to be quantified not only on local level because it is well known that Jiu Valley population increased in time as a result of mining activity development or because of a migratory spore.

The concern is not really justified because only a few mineworkers (7-8 percent from the number of workforce) wanted to turn back. So that, the main attention has to be given to Jin Valley, area really affected by this restructuring process.

The number of immigrant unemployed which leave this is was less then was expected. Only 1200 returned to their native counties and moreover approximate one-quarter of them come back in Jin Valley. The reason of immigrant employer remaining in this area is the fact that their native areas are undeveloped and they have not any chances to find another jobs.

Nevertheless, the Govern decide to encourage the tourism development. The landscape of area offers a lot of places suitable for this purpose but the mining industry damaged it in many places. Also, the pollution in area is too great for tourist feature.

3. ENVIRONMENTAL IMPACTS

Air is polluted by a power station and by mining and

Table 1. Air pollutants

No.	Pollutant	U.M.	Legal	Value
1	HC1	mg/m^3	0.100	0.000
2	NO_2	mg/m^3	0.100	0.023
3	SO_2	mg/m^3	0.250	0.018
4	Suspension particles	mg/m^3	0.150	0.101

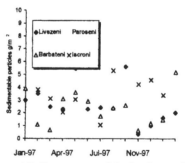

Figure 4 Sedimentary particles in air near Paroseni power station and in surrounding areas

coal processing activity with dust. Another pollutants Sedimentable particles in air in the reorganization period are presented in Figure 4. It shows that air pollution was not reduced by mining reorganization.

To assess what are the pollution sources with dust is interesting to show the daily evolution of suspension particles in air (Fig. 5.).

In mining companies Saturday and Sunday are free days so that it was expected that suspension particles in air to be reduced in these days. The fact that Suspension particles have not a weekly variation show that air is polluted in a large measure by other companies like power station.

In Figure 6 is shown a comparison between air content in suspension particles before mining reorganization and after re-organization. This figure show that air quality decrease after re-organization.

Water quality is also strongly affected by the mining activity due to coal processing activity, especially.

Jiu Valley has a very developed hydrological network compound mainly by low flow rivulets. The total length of hydrological network is 380 km covering an area of 1010 km^2 and a flow of 18 m^3/s. From all this length, just 20 km are polluted, but they are the main receiver area so that all rivulets are affected. River Jiu is not polluted chemically, just with organic substances due to social activities and with slurries due to coal processing activities. Jiu pollution damaged strongly the aquatic life. The Jiu pollution effects were transmitted in area that is not polluted, reducing trout, molluscs and other aquatic fauna effective. The daily evolution of pollution with organic substances are shown in Figure 7.

The daily evolution on slurries content is shown in Figure 8.

Figure 7 and Figure 8 show that water impact decrease little after mining re-organisation.

Carrier mining activity development and the great number of sterile and burned coal ash dump affected soil quality on large areas (Tab. 2).

Table 2 show that only one mine have 5 dumps placed in different places and some of them are very small.

These dumps together with the deforestation activities provoke the soil erosion phenomena.

There are 41 waste dumps of which 23 operational and 18 preserved. The amount of waste stocked in this dumps is 37 million m^3 and they are spread on surface of 219 ha. The existence of this dumps and carriers have a strong negative impact on landscape that was damaged in condition of a great tour potential existence.

The soil and landscape recovery have to take into account in order to create conditions for tourism developing.

4. REDEVELOPMENT MEASURES IN THE AREA

Restructuring programs from Coal Company (taking into account the dimension of implications) and

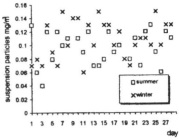

Figure 5 Daily evolution of suspension particles content

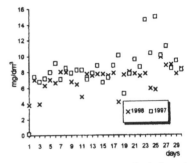

Figure 6 Comparison between air quality before mining re-organization and after re-organization

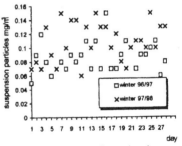

Figure 7 Daily evolution of organic substances content

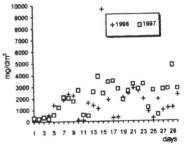

Figure 8 Daily slurries content

Table 2. Lonea mine dumps

Dump name	Volume [m3]	Dump surface [ha]
(2)	(3)	(4)
Jiet	24,080	0. 57
Lonea 1	125,085	1.49
Defor	2,149,970	12.65
Valea lui Ciort	982,472	7.19
Valea Arsului	200,00	2.1

Table 3 Programs in field of public interest

Programs which could be applied in the fields of public interest	Estimated funding -million- ECU	Estimated number of jobs created
Rehabilitation of transport in Jiu valley		
-bus version	11.5	250
-trolley version	15.8	250
Rehabilitation of waste collecting and dumping system	12.6	150
Rehabilitation of the cleaning residual water system	9	50
Modernization of the road Petrosani-Herculane	100	1100
Modernization and development of the loca road network	17.5	70
Ecologsation of Jiu reiver	28.1	150
Rehabilitation of the thermal network in Jiu valley	45.2	150

social programs which help to achieve reconstruction and reconversion of the area are strongly interconditions, the final aim being to create viable and attractive alternatives to ensure in time a sustainable development in Jiu Valley.

The main attention of a managerial team in the mining company is to find and to apply some instruments and methods to minimize the social consequences - more important being the methods of social protection and reconversion of workforce.

Few instruments were already applied in the initial phase of restructuring process, such as:
- Early retirement of older workers;
- Financial compensations to workers who accepted to leave the coal mining sector in the previous conditions. These leads to conclusion that the workers themselves increasingly perceived that remaining in a contracting industry will be not in their long term interest;
- Training and reconversion programs for redundancy workers in accordance with the existent opportunities at the moment.

All these measures will have a positive effect only on a short term if the basic aspects of initiating and applying some regional development programs will not become operational.

Researches and analyses elaborated till now in the area suggest the priority of such programs which have to put into practice in Jiu Valley in the first phase. These programs require substantial financial funding but in the same time create new jobs which could be considered like a first step in absorption of mining workforce.

These programs will have benefit effects on infrastructure modernization and will contribute to increase the attractiveness of Jiu Valley in business field. So that, could be mentioned the programs presented in table3.

Programs and projects mentioned together with the others in this field which will be identified in the future, have to be supported with the public financial funding, and it is possible with foreign support.

The financial effort necessarily is substantially, but not only the direct results - creations of new certain jobs - but also the indirect effects - increase area attractiveness to stimulate business - lead to the conclusions that the investments will be quickly recouped and will generate profits.

These are also, identified projects which could be applied in short terms and could be applied creating new jobs and new business which are shown in table 4.

Regarding all these and programs of local economic development have to specific the fact that was analyzed only the jobs directly created in that activities but will be jobs indirectly created in concordance with the new business opportunities, the estimated number of indirectly jobs being of 3 to 1 from basic activity.

Another alternative in the are take into consideration the higher tour potential of Jiu Valley which was no explored till now, but in the future could become one of the most serious resources to ensure an opportunity for further development. This tourism industry in Jiu Valley can create many secure jobs in specific activities like:

Table 4 The projects which could be applied

Programs which could be applied in the fields of public interest	Estimated funding -million- ECU	Estimated number of jobs created
Reconstruction of mine yards	131	1150
Restoration of old furniture factory at Petrila	4	80
Establish a new timber processing factory	6	40
A factory of construction materials (brick prefabs)	8	60
Exploration and processing limestone	70	150
Electronic industry (optic fibers, silicon plates)	60	80
Optics industry (lens of higher quality)	50	100

accommodation, services, tours objective service, and can generate profits, and also, in time we could talk about a tour industry in Jiu Valley.

The major impediment in implementation of these programs is the lack of funding., but we could say if the local development programs are consequently applied there is the possibility to create in the following 20 years, almost 20000 to 25000 new jobs in extra mining activities. The new opportunities will lead to improve substantially life quality for inhabitants in Jiu Valley, and to increase the level for economic activities generating profits on long term.

A financial estimation in Jiu Valley in order to create about 15000 jobs in the next 8 years is 450 million $ (30,000 $ for each job created).

5. CONCLUSIONS

Mining reorganization have a strongly negative socio-economic impact because cannot create new jobs or developments.

Air and water impacts are positive but are very small.

Landscape impact is expected to appear after some years when the environment will be rebuilt.

The damaging consequences of coal sector restructuring upon local economic activity and social life may be mitigate by resettlement and social protection measures and regional reconversion programs.

With appropriate action and public and private initiative the social impact of coal sector restructuring can be minimized and a more secure economic future for the coal fields can be foreseen.

In spite of these programs at the beginning of 1999 there was in Jiu Valley a very strong strike when miners start to Bucharest and then entire country was destabilized. By this reason, Jiu Valley became a national issue that shows there is a more complex problem than governance believe.

REFERENCES

Sarbu R., Georgescu M., Ciocan V., Badulescu C., Traista E., Polluting sources and their impact on the environment in the Jiu Valley coal field, A.A.Balkema/RotterdamlBrookfield/l 997, p 957;

Glasson J., Therivel R., Introduction in environmental impact assessment, Oxford University Press, 1995;

Morris P., Therivel R., Method of environmental impact assessment, Oxford University Press, 1995;

Traista E., Improvement of wastewater provide from coal mining activity cleaning, PhD. Thesis, Petrosani, 1998.

Coal Company, Study regarding restructuring process in Jiu Valley, Technical Report, Restructuring Department Petrosani, Romania, 1997

European Bank Of Reconstruction and Development, Jiu Valley. Regional Water and Environment Project, Technical Report, Petrosani, Romania, 1997.

Mining Science and Technology' 99, Xie & Golosinski (eds) © 1999 Balkema, Rotterdam, ISBN 90 5809 067 1

Research of the sustainable development of Yanzhou coalmine area

Fuchang Huang
Yanzhou Mining Group Corporation Limited, Shandong, People's Republic of China

ABSTRACT: The reasonable use of coal resource is an important issue for the sustainable development of Yanzhou mine area. Coal mining under buildings, railroads and water-bodies (three under reserves) is the main way to realize the coal mine area's sustainable development. This paper presents new achievement of the coal mining technology under buildings, railroads and water-bodies in the Yanzhou mine area, and introduces the important research topic concerning sustainable development.

1. INTRODUCTION

As a resource intensive industry, development of coal mining enterprises is dominantly dependent on the situation of resource reserves. So compared with other industries, the research of the sustainable development of coal industry is more important.

Yanzhou coal mine area is located in the southwest of Shandong Province, where a large population live on fertile land divided by many rivers and lakes. The forth measure alluvium is thick, and coal under buildings, railroads and water-bodies covers a big part of the total resource. By the end of 1995, industrial coal reserves under buildings, railroads and water-bodies is as high as 1432Mt, which is 65.1% of the total; the mineable reserves under buildings, railroads and water-bodies is 650Mt, which is 51.2% of the total (as shown in Tab.1).

The research of mining under buildings, railroads and water-bodies is an important project for the sustainable development of Yanzhou coalmine area. Therefore, since 1970s Yanzhou Mining Group Corporation Ltd. (YMGC) has conducted industrial experiments to safely mine coal under railroads bridges, villages and the forth measure water-bearing bed. In order to decrease surface subsidence, the new technology of grouting overburden-separation is also developed at YMGC. In the past two decades, the total coal output under buildings, railroads and water-bodies of the coal mine area reached 23.6173Mt (as shown in Tab.2). The enterprise has received good profit (as shown in Tab.3) and made outbreak in technology. Meanwhile it has furthered mining under buildings, railroads and water-bodies and ensured the sustainable development of Yanzhou coal mine area.

2. MAIN ACHIEVEMENTS OF MINING UNDER BUILDINGS, RAILROADS AND WATER-BODIES AT *YMGC*

2.1 Minimizing the anti-water coal pillar below No.4 water-bearing bed

Yanzhou coal mine area belongs to hidden coal fields. The forth measure covers the seams. The anti-water coal pillar with the vertical height of 80m is left in the mine designing to prevent *No.4* bearing water from getting into the coal mine.

Through tens years of research, Xinglongzhuang Mine colliery, which belongs to YMGC, has successfully decreased the anti-water coal pillar from 80m to 53m. Up to now, the mine produces 4.05Mt coal under No.4 water-bearing bed through this way.

By using of fully mechanized slicing method with anti-water coal pillar to water-bearing coal seam, YMGC has successfully mined out 37Mt of coal under No.4 water-bearing bed. While in the industrial experiment of fully mechanized longwall top-coal caving, YMGC has safely produced 1.971Mt coal under No.4 water-bearing bed through longwall face No. 2304 and No. 2303.

Table.1 Coal reserve under buildings, railroads and water-bodies of Yanzhou coal mine area (Mt)

	industrial reserves	mineable reserves	three-under reserves	permanent coal pillar	three-under reserves	
					industrial	mineable
total	2200	1270	1432	613	819	650

Tab.2 Coal output under buildings, railroads and water-bodies of Yanzhou coal mine area (Mt)

total	under villages	under water-bodies	under railroads
23.6173	8.7178	6.9180	7.9815

Tab.3 Three-under mining economic profit table

	output (Mt)	output-value (million yuan)	profits-tax (million yuan)	social profit (billion yuan)
lifting upper limit in Xinglongzhuang Mine	4.0502	244.5602	82.989	15.325
under railroads in Xinglongzhuang Mine	2.1025	346.900	164.700	7.956
under Wuguan Village in Beishu Mine	0.6758	104.8233	1.4259	2.557
total	6.8285	696.2835	249.1149	25.838
average per ton coal (yuan per ton)		101.97	36.48	3783.85

2.2 *Mining technology under railroads (bridges)*

The thick seam mining under the specified railways of YMGC was begun in 1984. Initiating the longwall slicing under railroads in China, GMGC accumulated a set of effective experience for predicting deformation of railroads and its maintenance.

In 1993, the top-coal caving method was adopted under railroads. However, this resulted in the increase of surface subsidence, rise of subsidence rate and violent deformation. Cracks of 300—500mm width and settlement benches of 600mm were also found on the surface. A lot of time has to be spent on maintaining railroad and the cost of maintenance increases greatly.

Based on the observation of subsidence caused by longwall top- coal caving, a quick surveying system of the trackage deformation was developed. As the trackage base was widened and raised before mining. Even if the maximum subsidence velocity is as high as 218 *mm/d*, the maximum subsidence is 5200mm without limiting the advance rate of the longwall top-coal caving face, the railways can keep operating as usual. Up to now, YMGC has produced 2.1025Mt coal by this technology.

2.3 *Thin-seam mining under villages*

Beisu Mine colliery of YMGC made the mining experiment under Wuguan Village to realize mining thin seam without removing villages. According to the geometrical shape of the village and computer simulation, the layout of longwall face was optimized and the width of the face was increased to 450m. The purpose is to have the village located in the subsidence basin center caused by mining. So the village only suffered the dynamic but homogeneous deformation during the mining process, and avoids of inhomogeneous and repeated deformation.

According to observation of the in-situ, slight damaged buildings of Wuguan Village was 37.3% of the total; moderate damaged houses was only 4.9%. The effect of protecting buildings is very obvious. Meanwhile, the mine successfully produced 0.6758Mt of No.16 upper seam under the village.

2.4 Technology of grouting overburden separation to minimize surface subsidence

The coal seam of Yanzhou coal mine area is high, which results in large surface deformation after mining. The maximum of surface subsidence reaches about 7m. Therefore, that makes gathering water over large space in rainy season and damaging rich soil, and villages have to be relocated. In order to control the surface subsidence and realize the mining of high seam without removing villages finally, YMGC has initiated the research project of grouting overburden separation to minimize surface subsidence.

The experiment shows: by using this new technology, the total subsidence is reduced over 50%. At present, Dongtan Mine Colliery of YMGC is making industrial experiments at two longwall top-coal caving faces.

In summery, YMGC has made significant achievement on coal mining under buildings, railroads and water-bodies, which makes full reasonable use of the coal resource and insure the sustainable development of the coal mine area. YMGC also provides certain experience for the sustainable development of China Coal Industry.

3. FUTURE RESEARCH AND DEVELOPMENT OF YMGC

3.1 Improving the mining under buildings, railroads and water-bodies

3.1.1 Further minimizing the anti-water coal pillar

YMGC plans to further minimize the anti-water coal pillar, and improve it into anti-sand coal pillar and anti-subsidence coal pillar; the anti-water coal pillar below No.4 water-bearing bed is minimized to 30m step by step. According to the preliminary estimates, 50Mt *coal* below No.4 water-bearing bed will be finally mined in this way.

3.1.2 Improving the mining of the thick seams under buildings

Considering the relationships of recovery ratio, surface deformation, the layout of working face and assistant technological systems, YMGC will carry out experiments of the thick seam mining under buildings to realize mining without moving villages. In the coming years, grouting overburden separation to minimize the surface subsidence is the research emphasis.

3.1.3 Searching possibility of coal mining under state-owned railroads

The coal reserve of YMGC under state-owned railroads is as much as 153Mt. According to some relative regulations of China, this part coal is preserved, which is an immense waste of source. YMGC intends to confer with the supervisory department of the railroads and tries to temperately mine coal under state-owned railroads on the basis of government ratification.

3.1.4 Improving and developing other technology of mining under buildings, railroads and water-bodies

The technologies that need to be improved and popularized include thin and medium seam mining under the railroads in the mining field, mining under rivers (lakes) and thin seam mining without removing villages.

With the development of industry, more buildings of various kinds on the surface will be set up in the mining district. So, YMGC will develop new mining methods, aiming at mining under high-voltage lines and highways.

3.2 Develop retreat mining of special seams

Because the thin seams of YMGC contain hard streak and pyritic nodule, mechanized coal breaking can not be effectively realized for a long time, which has greatly limited profit of thin seam mining. Coal at irregular-shaped zones in thick seam mining district should be exploited. The above mentioned problems block the sustainable development of YMGC. The relative technologies need to be developed as soon as possible.

3.2.1 Develop the thick and medium seam mining technology at irregular districts

YMGC plans to mine such coal with the following technologies:
---Shortwall room and pillar mining will be used to mine coal under buildings, railroads and water-bodies and below No.4 water-bearing bed.
---Special top-coal caving technology will be

adopted to mine certain coal seams with more complicated geological conditions exiting in No.3 seam.

3.2.2 Overcoming the technological difficulties to realize the mechanized mining of thin seam containing hard streak

YMGC will take several steps to solve the difficulties. Now, the working emphasis is to realize mechanized loading of thin seam faces, which is troublesome and YMGC are going to invite experts worldwide to overcome this technical problem of the mining science.

3.3 Reasonable mining of parags mixed with coal

Some other useful minerals were mixed with coal in the coal-bearing series of YMGC. For example, the No.18 floor at Beishu and Tangcun Mine collieries is made of kaolin of better quality. With the joint efforts of YMGC and other research institutions, iron and titanium are removed, which establishes the base for the further exploration of kaolin. Now the exploration of kaolin is being undertaken.

3.4 Unified management of the surface subsidence

The surface subsidence caused by mining severely damages the environment. YMGC tries its best to develop such technologies as follows:
---The surface subsidence of important districts was eased by using grouting overburden separation.
---The subsidence field was refilled with gangue and the fine dust of its own power plants.

4. CONCLUSIONS

Yanzhou Mining Group Corporation Ltd. has contributed its efforts to innovate new coal mining techniques under buildings, railroads and water-bodies, and made this advanced technique used extensively in China. However, the further research need to be carried out in order to keep the sustainable development of YMGC in the future.

Mining Science and Technology' 99, Xie & Golosinski (eds) © 1999 Balkema, Rotterdam, ISBN 90 5809 067 1

On the optimal way of coal transportation

Shuren Tao, Wuyuan Ren, Kunming Sun & Liping Xing
China University of Mining and Technology, Beijing, People's Republic of China

ABSTRACT: According to the characteristic of coal supply from western to eastern China, there are many ways of coal transportation to be chosen: railway, transferring to electricity with high voltage, and pipeline. In this paper we will discuss in detail the feasibilities for the different ways of coal transportation from the economic point of view.

1. INEVITABILITY OF STRATEGIC WEST-TRANSFER OF COAL INDUSTRY

Due to the nature, traffic and other reasons, during the long history in China, the social development standard of the coastal provinces is high, and the quantity of coal consumed is great, accounting for 72 percent of countrywide. The quantity of net coal transferred is two hundred million tons or so, and the circumstance is unlikely to change in a short time.

According to the countrywide coal-field forecast data of the year 1981, middle and west China accounts for 89.2 percent of China's total.

It will be an inevitable trend that emphasis of coal exploitation is to transfer to the west step by step. Because the transportation capacity by railroad can not satisfy the need of coal transportation, the joint ways of coal transportation and electricity transmission will replace coal transportation solely by railroad mainly.

2. DESIGN OF WAYS OF COAL TRANSPORTATION

Three ways are discussed in the article: railroad, electricity with high voltage, and pipeline, as shown in Graph 1. The way of electricity transmission includes DC (direct current) with high voltage (500KV) or AC (alternative current) with high voltage (500KV).

The basic enactment is that the starting is pithead crude coal (not taking mining way into account), and that the ending is 500kv electric net (not taking

tideway distribution of electric net account). Furthermore, the same part in system is not participated in the comparative analysis.

3 SELECTION AND CALCULATION OF ECONOMIC INDEXES

At present, there are two problems in the research on coal transportation technology selection.

The first one is that the terminal is not comparable with each other. In many researches, economic comparison is based on the output of coal transported by different transportation ways. Here, users get different products (electricity, coal, coal water fuel, etc.). But natural conditions and technical factors influence the conversions between these matters, from which great difference is produced. So the comparable character disappears. Furthermore, the energy which users get is different because the energy consumption caused by different transportation is different. Even if compensatory computation is adopted, the conclusion differs because different parameters are selected. The second problem is that the phase "the same problem, many conclusions " come into being because indexes of transportation scheme and selections of technical criterion are different, such as voltage level, circuitry wire, construction type (build or rebuild), draught way of locomotive, etc.

To overcome the above shortcomings, technical criteria of each technical scheme are put forward, and the long-run average cost ($LRAC$) is selected so as to reflect correctly the economic characteristic of each technical scheme.

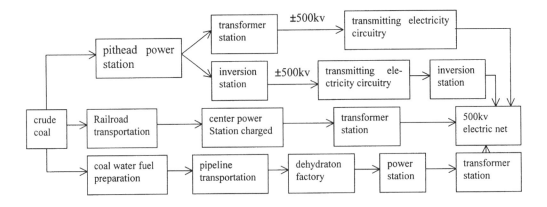

Graph 1 Economic analytic system of coal transportation technology

$$LRAC_j = [C_j + (A/P, I, T_j)]/Q_j \quad (1)$$

Where C_j— annual operation and maintenance expense of scheme j, which includes operation expense and equipment maintenance expense (including salary, material, welfare, etc.) and others, if any; I_j— total investment or total construction cost of scheme j; $(A/P, I, T_j)$ — capital recovery factor of technical scheme j, and rate of return on capital is I when its life-span is T_j; and Q_j— design transportation capacity of technical scheme j in normal production year.

4. ESTABLISHMENT OF GENERAL ECONOMIC ANALYTICAL MODEL

According to Formula 1 and the concrete circumstances of each scheme, long-run cost calculation models of each transportation scheme are established.

4.1 Estimating the cost of construction

4.1.1 *Estimating the cost of circuitry construction*

The investment of railroad circuitry construction is influenced by all kinds of natural condition factors, such as landform, climate, etc. owing to different landform. We suppose that railroad—the per unit cost is UC_{r1}(10,000 yuan /km) when the length of plain region is L_{r1}; UC_{r2} when the length of hilly region is L_{r2}; Uc_{r3} when the length of mountainous region is L_{r3};so the average unit cost of the whole circuitry TUC_{r1} is:

$$TUC_{r1} = (L_{r1} \times UC_{r1} + L_{r2} \times UC_{r2} + L_{r3} \times UC_{r3})/L_r \quad (2)$$

Where L_r — the total length of circuitry,
$$L_r = L_{r1} + L_{r2} + L_{r3} \text{ (km)}$$

In addition, we suppose that time limit for the construction is T_r (year), distribution coefficient of each year's investment is $\beta_{1(r)}$, $\beta_{2(r)}, \cdots \beta_{Tr(r)}$, serving period of the circuitry is n_{r1} (year), then recovery cost of circuitry construction cost TUC_{r1} is:

$$TUC_{r1} = [\sum_{t=1}^{T_r} TUC'_{r1} \times (1+mc)' \times \beta_{t(r)} \times (1+i)^{T_r - 1}]$$
$$\times (A/P, I, n_{r1}) \quad (3)$$

Where mc — rate of escalation for engineering cost.

4.1.2 *Investment expenses estimation of locomotives and vehicles*

If serving time of locomotive and vehicle is n_{r2} (year), the capital recovery cost TUC_{r2} of locomotive and vehicle is:

$$TUC_{r1} = (UI_{r1} + UI_{r2}) \times (A/P, I, n_{r2}) \quad (4)$$

UI_{r1}—per unit investment of locomotive , 10,000 yuan/kilometer; and UI_{r2}— per unit investment of vehicle , 10,000 yuan/kilometer;

4.2 *Estimating the operation cost*

Operation cost includes management cost, maintenance cost and loss cost of coal transported, and so on.

4.2.1 *Cost evaluation of operation and maintenance. Its formula is:*

$$C_{r1} = C_w \times Q_r (1+\mu) \times L_r + C_f \times L_r \quad (5)$$

788

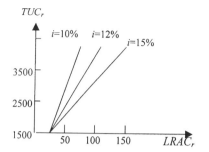

Graph 1. Relation between $LRAC_r$ and TUC_r
(L_r =500km , Q_r=6000)

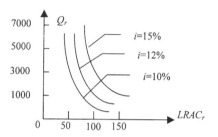

Graph 2. Relation between $LRAC_r$ and Q_r
(TUC_{r1} =1754, L_r=500)

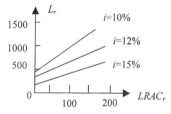

Graph. 3 relation between $LRAC_r$ and L_r
(TUC_{r1} =1754,Q_r=6000)

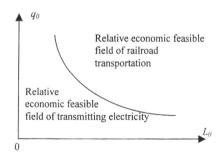

Graph 4. L_0-q_0 curve sketch

C_{r1}—operation cost and maintenance cost; 10,000 yuan/year; C_w— ton·km running cost, 10,000 yuan/ton·km; λ —conversion coefficient of empty car's running cost, λ =0.57; and C_f—per unit maintenance cost of railroad, 10,000 yuan/km;

4.2.2 Loss cost of coal transported

According to detailed implementation rules of coal-freight means promulgated by Ministry of Coal Industry, Ministry of Railway and Ministry of Communication, we assume two percent of total quantity transported as loss quantity transported by railroad. So loss cost arisen by transportation is:

$$C_{r2} = Q_r \times 2\% \times P_m \tag{6}$$

P_m— coal price, yuan /ton;
Then total management coat of railroad transportation C_r (10,000 yuan/km) is:

$$C_r = (C_{r1} + C_{r2}) / L_r \tag{7}$$

So the all-purpose model of the long-run average cost of railroad transportation technique scheme $LRAC_r$ is:

$$LRAC_r = (TUC_{r2} + TUC_{r2} + C_r) / Q_r \tag{8}$$

The relation between $LRAC$ and per unit cost, transportation distance and quantity transported are shown in Graph 1-3. The relative economic feasible region of each scheme is gained on the basis of comparison between different schemes: for example, the feasible region of railway transportation scheme and 500kv transmit electricity AC (see Graph 4). The rest may be deduced by analogy. The economic feasible regions of 500kv transmit electricity AC and DC, railway transportation and pipeline transportation, pipeline transportation and transmit electricity AC are gained, as shown in table 1.

REFERENCECS

Huabei electric power design institute. 1982. *The technical economic comparison between transmit electricity and transportation coal in huazhong area(e,yu section) electricity net (research report).*
Shipping department of Science research institute of railway ministry etc.1980. *The technical economic comparison between transportation coal and transmit electricity (research report) from Lianghuai to south of jiangsu province.*

Table 1 Economic feasibility transportation distance among different transportation means

No	item		condition		Feasible economic transportation
			Kcal/kg	MJ/kg	
1	Transportation way of railroad and transmit electricity AC	q_0	2000	8.36	□1733, railway excels transmit electricity
			2500	10.45	□1386, railway excels transmit electricity
			3000	12.54	□1155, railroad excels transmit electricity
			3500	14.63	□990, railroad excels transmit electricity
			4000	16.72	□866, railroad excels transmit electricity
			4500	18.81	□769, railroad excels transmit electricity
			5000	20.90	□692, railroad excels transmit electricity
			5500	22.99	□629, railroad excels transmit electricity
			6000	25.08	□576, railroad excels transmit electricity
			6500	27.17	□532, railroad excels transmit electricity
			7000	29.26	□493, railroad excels transmit electricity
2	Transportation way of AC and DC		500kv level		□450, transmit electricity DC excels AC
3	Transportation way of railway and pipeline		Flat hill		□150, pipeline excels railway
			mountainous region		□100, pipeline excels railway
4	Transportation way of pipeline and transmit electricity AC	q_0	2000	8.36	□1152, pipeline excels transmit electricity
			2500	10.45	□944, pipeline excels transmit electricity
			3000	12.54	□804, pipeline excels transmit electricity
			3500	14.63	□703, pipeline excels transmit electricity
			4000	16.72	□627, pipeline excels transmit electricity
			4500	18.81	□568, pipeline excels transmit electricity
			5000	20.90	□520, pipeline excels transmit electricity
			5500	22.99	□481, pipeline excels transmit electricity
			6000	25.08	□448, pipeline excels transmit electricity
			6500	27.17	□420, pipeline excels transmit electricity
			7000	29.26	□396, pipeline excels transmit electricity

Wuhan coal mine design institute. 1994. *Pre-feasibility research report on pipeline transportation from Yu county to Weifang.*
Shuren Tao. 1991. *Technical economic evaluation.* Higher educational publishing house.

Mining Science and Technology' 99, Xie & Golosinski (eds)© 1999 Balkema, Rotterdam, ISBN 90 5809 067 1

Modeling optimal exploration and extraction of mineral resource

Lijie Wang, Jianwei Rui & Rijia Ding
China University of Mining and Technology, Beijing, People's Republic of China

ABSTRACT: With consideration of the mineral resource owner's attempt to maximize his benefits from the total resources, this paper puts forward a dynamic model on the optimal exploration and extraction rates of resources. The paper points out that the optimal exploration and extraction rates calculating model is as same as the maximum benefits calculating model. The necessary conditions for the maximum benefits or the optimal rates of exploration & extraction are discussed. The optimal rates of exploration and extraction at any different time are presented.

1. INTRODUCTION

Mineral resource is a kind of resource property (here the mineral resource refers to the resource worthy of extracting or further exploiting economically). The mineral resource is self-valued and should be developed on the basis of rights-charged. All these opinions have been widely accepted by most scholars. In the process of conducting the propertied management, the benefits of the resource's owner must be protected effectively. It means that the resource's owner should recover the value of the resource. The total profits can only result from the sale of the mineral products. Thus, the corresponding exploiting and extracting activities are necessary to get some mineral products to sell in the market. Then different arrangements on extraction and exploration activities could influence the output of mineral products and further influence the total profits from developing the mineral resource. At the same time, the extraction and exploration need magnitudes of labor and capital inputs. All the input should get returns and the returns also come from the profits of selling the mineral products. From the above analysis, it could be believed that the total sale profit of the mineral products not only includes the value of mineral resource per se, but also returns to the labor and capital inputs in the extraction and exploration. Apparently, the mineral resource owner always hopes to maximize total profits form

developing the resource (here the resource owner is supposed to be the extractor & the exploiter). In order to maximize his total profits, the resource owner must arrange the extraction and exploration rates properly. The proper rates will be the rates that can maximize the profits from developing the mineral resources.

2. THE MODEL

As described above, the problem the mineral resource owner must face is twofold. First, for a given (known) stock of the resource, what extraction over time will maximize the net present value of the resource; second, when should exploration be undertaken to increase the level of (known) resource.

In order to solve the problem, we suppose that, at any time of t, the total deposit available for extraction is N, the price of the output in the market is p. Because of the exploration and the change of the market, the value of N or p is different from time to time. So, N and p are respectively a function of time t, marked as N (t) and p (t). At any given time t, the owner of the resource's costs of extraction from T^h deposits depend on the rate of extraction from that deposit q (k, t) and the amount of the deposit remaining R (k, t). The costs will be expressed as C (q, R, k). It should be pointed that C, q and R can continuously change with time t and they only change their function style if concerning with different deposit. The owner also incurs the costs of exploration. The exploration is expressed, as V. It should be the function of w (t), where w (t) is exploratory efforts over time.

Thus, at any point of time, the total profits obtained by the owner of the resource could be calculated by summing net revenues from each deposit, and subtracting total exploration costs:

$$TR= \sum_{k=0}^{N(t)} \{p(t)q(t) -C [q (k, t), R (k, t), k]\}$$

$$-V [w (t)] \qquad (1)$$

Over the whole time, the total benefits of the resource is equal to a present value derived by summing each period's profits and discounted with rate□.

Here we let the benefits be discounted completely continuously. So the actual discounted rate i will be as the following:

$$i=(1+\delta/m)^{m} ;$$

i—the actual discounted rate at the nominate rate□;

m—the compound frequency.

δ—the averaged social discounted rate(with risk concerned).

While m→∞, i will be equal to e^{δ}.

Thus the net present value of the total benefit will be:

$$NPV = \int_{0}^{\infty} e^{-\delta t} \{\sum_{k=0}^{N(t)} \{p(t)q(t)$$
$$- C[q(k,t), R(k,t), k]\} \qquad (2)$$
$$- V[w(t)\}dt$$

It should be pointed out that when the profits are not discounted to a present value, they include not only the returns to the inputs of extraction and exploration but also the value of the depleted. mineral resources. If they are discounted at a rate□, the net present value will not include the returns to inputs of extractions and explorations any more, because all the inputs have received returns by the discounted rate. If the□is a risk contained rate, the returns to risk taken be excluded from NPV. Thus the NPV is only the value of the mineral resources. If we sum up the profits from all the deposits of every period and discount them at proper rate to generate a present value, the final value of NPV which we get is equal to the value of the whole

mineral resources in the tract. It is also the calculating method of the mineral resource's value (See Wang Lijie & Rui Jianwei *Study on the Mechanics of How to Transfer the mineral Rights.* Beijing: Journal of Coal Science & Engineering (China), No.1996 (2), p65-68.).

When the owner maximizes the NPV, the following constrains must be met:

(1) $R_t'(k)=-q$ (k ,t). It is because that existing deposit is depleted on the rate of extraction from that deposit k. R_t' refers to a time derivative.

(2) $N_t'(t)=f[w(t),N(t)]$. It is because that the number of discoveries depends on the rate of exploration w(t) and the cumulative number of discoveries already made N(t). N_t' refers to a time derivative.

Thus the problem of maximizing the total benefit changes into the problem of solving maximum equation with respect to 2 costarring. To the owner, the objective is to choose optimal q (k, t) and w (t) so as to maximize the present value of profits. If the present value of profits is maximized, the optimal q and w will be derived

3. THE OPPTIMAL RATES OF EXTRACTION & EXPLORATION

In order to use those model equations, equation (1), equation (2), equation (1) and equation (2) are combined as:

$$L= e^{-\delta t} \{\sum_{k=0}^{N(t)} \{p(t)q(t)$$

$$-C [q (k, t), R (k, t), k]\}$$

$$-V [w (t)]\}-\lambda_1(k, t)q(k ,t)$$

$$+ \lambda_2(t)f[w(t),N(t)] \qquad (3)$$

Apparently, λ_1 is the shadow price associated with the reserves remaining in the kth deposit at time t . Since mineral resources are non-renewable, what is extracted in one period is not available for production in future time periods. This is the opportunity cost of current extraction. Depletion of the resource must generate not only returns to the labor and capital inputs, but also to the owner of the resource as compensation for depletion of his resource assets. In a perfectly competitive equilibrium, this return to the resource owner should equal to the return that could be earned on the next best capital in the margin. and λ_2 is the shadow price, or marginal benefit, associated with making a new discovery at time t.

Maximizing equation (3) with respect to q (k, t) and w (t), the first order conditions for a profit maximum can be derived:

$$C_q[q(i,t),R(i,t),i]+\lambda_1(i,t)\,e^{\delta t}$$
$$=C_q[q(j,t),R(j,t),j]+\lambda_1(j,t)\,e^{\delta t}$$
$$(i,j=(0,N(t)))\qquad\qquad(4)$$

Thus the sum of marginal extraction cost plus the shadow price on the remaining reserves must be same for all deposits; otherwise profit could be increased by shifting production to lower-cost deposits.

$$v_w'/f_w'=e^{\delta t}\lambda_2(t)\qquad\qquad(5)$$

V_w and f_w are, respectively, the additional cost and additional discoveries associated with one more unit of exploratory effort. Hence v_w'/f_w' is the marginal discovery cost. The returns to exploration activity are maximized when this benefit of an additional discovery.

$$\lambda_1'=-\partial H/\partial R\qquad\qquad(6)$$

$$\lambda_2'=-\partial H/\partial N\qquad\qquad(7)$$

Combining equation (4), equation (5), equation (6) and equation (7), the equilibrium (optimal) rates of extraction and exploration over time can be solved for as the following:

$$q=\frac{1}{C_{qq}}[C_{qR}q+(f/N-\delta)(p-C_q)-C_R]\quad(8)$$

$$w=\frac{V_w[(f_{wN}/f_w)f+\delta-f_N]}{V_{ww}-V_wf_{ww}/f_w}$$
$$\frac{-f_w\{(pq-C)+N(t)[q_N(p-C_q)-C_RR_N-C_N]\}}{V_{ww}-V_wf_{ww}/f_w}$$
$$(9)$$

In equation (8) and (9), the single subscripts refer to first order partial derivatives, and double subscripts refer to second order partial derivatives.
From the equation (8) and equation (9), we can find that initially (before time zero), exploration maybe

high and extraction zero. As soon as the first deposit is found, extraction rates rise and exploration will fall down because extraction costs are low. As extraction costs start to rise with depletion, exploration will start to pick up again until the costs of finding and extracting new deposits are high and both exploration and extraction decline to zero, where the mineral resource just comes to the end of its economic life.

4. CONCLUSIONS

(1) The total profits of developing mineral resource include two main parts. One is the value of the mineral resource; the other is the return to the activities of extraction & exploration. If the profits are discounted at a proper rate, the final NPV will only includes the value of the mineral resources.

(2) The total profits from developing the mineral resource are influenced not only by the extraction rate but also by the exploration rate. The optimal rates of the extractions and explorations are the rates, which can maximize the total profits from the resources, and with the proper rates, both the profits from developing the resources and the value of resources could be maximized.

(3) The optimal extraction rate and the exploration rate at any given time can be expressed as the above equation (8) and (9) through maximizing the total profits from developing the mineral resource by stimulating the processes of extractions and explorations under some special assumes.

REFERENCE

1.Wang Lijie. 1996.*Study on Theory and Method of Coal Resource Evaluation.* Beijing: China Coal Industry Press.
2.Wang Lijie, Rui Jianwei.1998.*Study On the Mechanics of How to Transfer the mineral Rights.* Beijing: JOURNAL OF COAL SCIENCE & ENGINEERING (CHINA), NO.1996 (2), p65-68.

Mining Science and Technology' 99, Xie & Golosinski (eds) © 1999 Balkema, Rotterdam, ISBN 90 5809 067 1

Method of mineral resources accounting

L.A. Puchkov & V.M. Shek

Moscow State Mining University, Russia

The concept "mineral resources" has explanated. This terminus as a rule is connected to profitability of their improvement such as the concept "reserves". We give the another notion, which is not connected to economic indexes and used the geologic concept "Clark". In this paper it is show the suitability of such explanation.

The concept " mineral resources " has some explanations. Thus resources of mineral resources also, as well as their reserves, are measured by mass or volumetric units. And if it is possible to differentiate reserves, which definition is connected to profitability of their improvement, with enclosing "poor" ores (coal, not metallic minerals), the contouring of resources is in most cases problematic.

We give another explanation. It's purpose is determination of a not economic estimation of a limiting content of mineral in local formation in a certain place of earth crust for reference of this formation to resources category. Such notion, which is not connected to economic indexes, and allows to carry out mineral resources separation in rock masses, is the geologic concept "Clark" representing a prevalence of a chemical element in a nature (earth crust).

Thus, we can give the following definition: " the Resources of minerals are natural phenomenon, when in some limited space of earth crust (or both of water or gaseous envelope of a planet) there is any useful element in an amount exceeding its Clark.

The Clarks characterize only average distribution of elements in earth crust on mass. We are interested in index of the element content in concrete elementary volume of the upper lays of earth crust. If all these elementary volumes would be formed under identical conditions, each of them would contents any elements in a quantity, equaled to their Clarks. But due to a diversity of forming conditions the concentration of separate elements in these volumes varies from zero up to rather significant values. If the unitary factor's influence would be approximately equal, the distribution of elements would correspond to the normal law. However in most cases some group of the factors have prevailing influence on intensity of concrete element inflow into some elementary volume. For example, the intrusion of magma into some elementary volume of earth crust can reduce, when it is hardening, forming groups of minerals with various masses containing of non-ferrous metals. Therefore mass distribution of each element in elementary volumes most likely should corresponded to the logarithmically normal law

$$y = \frac{1}{x\sigma\sqrt{2\pi}} e^{-\frac{(\ln x - \mu)^2}{2\sigma^2}} \quad \text{For x > 0.}$$

(1)

It is accepted expectation μ for any element $\check{O}$ is accepted equal to its Clark C, i.e. $\mu = C(\check{O})$.

We are interested in a part of a distribution curve, where is that minimum content of an element (mineral) in elementary volume, from which this volume should be referred to resource. It is possible to describe this data area by the Pareto distribution

$$P(\xi > x) = \left(\frac{x_c}{x}\right)^\alpha, (x > x_c, \alpha > 0),$$

(2)

where α is statistic index, X_c – parameter, which have to be set earlier, before realization of statistical accounts.

The problem is too definite a magnitude X_c. The beginning of 1989 [1] reduces the mineral resources, in particular copper, in the world. Indexes of distribution of the confirmed reserve copper on the countries, the content of metal are given as dot estimation (rate). For definition of logarithmically normal distribution of mineral we have to count the natural logarithms of minimum (threshold) content for each country. According to square of the considered countries and, accepting the maximum depth of exploration and mining of deposits as 2000 m, we determine the mass an earth crust site for each country, in which the mineral resources can place.

The limited depth of mining is accepted to those reasons, that on depth equal 2000-m temperature of rocks exceeds 65° C, and on depth 3000 m is exceeds 100 °. The modern mining engineering can not supply operation of the mineworkers in such conditions. Smaller value of temperature is inexpedient for elimination of "missing" ("passing") of a part of mineral resources developed by super-deep ore mines (for example, in SAR).

For all countries we find a long of this mass per studied mineral copper for example. It is possible to tell that in the case of a rectangular distribution of minerals on continents, each country can get one's amount proportional to the square. But this is a rather rough supposition. For a strict regularity of reasoning we should investigate not sites of continents in geographical boundaries of the states, but areas of geologic provinces (i. e. regions with similar conditions of certain type deposit shaping).

However it is possible to show applicability of the offered mode for the mineral resources deposits estimation on the base of available statistical data.

Let's assume, that in each country, included in account, all mineral reserves with a content above threshold are found. Then it is possible to determine a quintile of investigated distribution as a difference between a complete amount mineral and their common reserves. Further we determine a dispersion of sample D and standard deviation s. It is estimated, that the average value of a statistician is equal to a Clark (for copper - 0,01 %). We determine values of quintiles for a minimum content of mineral for each country. Their comparison with accounted data shows, that amount of the mineral referenced to deposits with a content, which is above the minimum minerals amount according to a parent distribution, is higher than it is fixed in calculations of reserves. It can be explained firstly by insufficient studding of rock masses (especially of strata belong lower two-kilometers of earth crust). Therefore received values are necessary to perceive now as a limit which is a criterion of complex geologic studding of certain stratum of earth crust. So the ration of available common reserves to assume has received a title "degree of an extent of exploration" of earth crust K_{cp}. The histograms of distribution of common copper reserve and degree of an extent of exploration of copper ores for the considered countries and continents were constructed.

The account of the statistic index, according the data in [1], show, that the copper deposits of Panama ($K_{cp}=41,65$ % from probable reserves), Chile (24,56 %), Fiji (15,57 %) are better explored. Besides, these data can be interpreted as presence of very favorable geological conditions of copper deposits shaping of in these countries.

Further there are Philippines (6,38 %), Zambia (6,27 %), Peru (3,37 %), Israel (2,57 %), Cyprus (2,55 %), Papua New Guinea (2,36 %), Jordan (2,15 %), Afghanistan (1,88 %), Zaire (1,76 %) and Iran (1,48 %). And only then there are USA with 119600 thousand tons (1,34 %) and Mexico with 22290 thousand tons of common reserves (1,18 %).

The common reserves of metal as a whole on the world exceed authorized in 1,5 times, and in the separate countries - in 1,1 - 7,0 times. The minimum contents of metal in accounted blocks in other countries axe as fallow: in Australia - 0,11 %, in Finland - 0,14 %, in Greece - 0,2 %, in USA - 0,25 %, in Mexico - 0,25 %. The minimum contents are higher in enveloped countries: in ARE - 1,12 %, in Congo and Yemen - 1,2 %, Oman, Jordan - 1,36 %, in Uganda - 1,46 %, Afghanistan - 1,5 %, Mozambique - 1,7 %, Mauritania- 1,83 %, Bolivia - 1,99 %, Zaire - 2,8 %.

For the last countries an acceptance of lower values of a minimum contents gives mineral reserves much above accepted on the base of present accounts. Thus reference of resources to a reserve category is concept rather conditional, which depend on method of their calculations and economic situation at the moment of accounting. One of confirmations is the fact of acceptance low values of a minimum content of metal (0,4 % and 0,45 %) in Papua New Guinea and Fiji according the Australian method of reserve calculation.

For an approximate estimation of potential resources at the countries, we shall assume that the minimum copper content for reference to a category of resources is equal to value accepted at calculation of reserves for deposits with the most favorable conditions of getting the metal. In Australia it is 0,11 % (Australia). This index shows a relation of common reserves in the country to probable resources. For the countries with the accepted minimum copper content > 1,0 % increasing of such resources is 32 - 43 %.

Coming back to equation (2) we can conclude that $X_c \approx 0,1$ % and it is quintile of square equal to 0,787133. Taking into account, that in the most explored countries (Chile) K_{cp} equal to 0,25, it is possible assume a possibility to reach of common world copper reserves $23,1 \cdot 10^9$ ton. It is more then 35 times greater, then common reserves in 1989 (622,17 million tons). It is supplied by exploration of deposits without the missing.

Using an inequality (3) we can more precisely to determine magnitude s.

$$P\left(\left|x - \mu\right| > k\sigma\right) < \frac{1}{k^2} \text{ For } k \geq . \qquad (3)$$

The definition of parameter m and s allows noting the equation (1) for any element (mineral).

REFERENCE

1. Mineral resources of developed and developing countries. Observe, VNIIzarubeggeology, 1990.

Mining Science and Technology '99, Xie & Golosinski (eds) © 1999 Balkema, Rotterdam, ISBN 90 5809 067 1

Reserve management for production capacity maintenance in Konin mine, Poland

S. Jarecki, M. Jaruzel & Z. Kasztelewicz
Lignite Mine Konin, Poland

R. Uberman
University of Mining and Metallurgy, Krakow, Poland

ABSTRACT: The paper presents problems related to management of lignite deposits with complicated geological structure and located in an urban and industrialized region. The new technological solutions related to removal, transport and disposal of overburden are illustrated. The problems involved in management of lignite quality coming from a new mines to two power plants are discussed as are the problems involved in utilization of the abandoned lignite pits for industrial and municipal purposes.

1. KONIN MINING COMPLEX.

Lignite surface mining and energetic complex Konin is situated in Central Poland, about 200 km to the west of the Polish capital, Warszawa. The deposit consists of two mineable seams with the thickness ranging from several meters to the maximum of 12 meters. The lignite is covered by oberburden the maximum thickness of which reaches 50 meters and which acts as aquifer. As a consequence lignite requires dewatering. The structure of the deposit is very complicated. Overall the reserves are not big averaging the average pit containing around 50 mln tonnes of lignite. Only in two pits do the reserves exceed 100 mln tonnes. The indicated resources of lignite in Konin region are sufficient to secure operation of the mine till at least year 2040. Total developed and mineable reserves in Konin region amount to about 568 million tonnes, of which some 125 mln tonnes exploited at present. A pit containing additional 60 mln tonnes is being developed at present. The plans call for development of further pits in the future. Although most of the lignite is of poor quality and is used for power generation, some of it can be used for production of briquettes. Its heating value fluctuates from 7.5 MJ/kg to 9.1 MJ/kg, ash content ranges from 8.2% to 18.0%, and the sulfur content is in the range of 0.33 - 0.88%. The lignite here is exploited for about 50 years. Currently there are 4 active pits: Patnow, Jozwin I and IIA, Lubstow, Kazimierz North (Figure 1). The total lignite production capacity fluctuates from 12 to 15 million tonnes per year. Each pit individually mines about 2.0-5.0 million tonnes of lignite annually. For each ton of mined lignite and average of 5.6 bm³ of overburden needs to be mined and 8.1 m³ of water pumped out. Between 97 and 98% of lignite is burned in two power stations: Konin with power output of 583 MW and Patnow with power output of 1200 MW. The remainder of lignite is used for briquettes production, organic fertilizers production and also for retail sales. Bucket-wheel excavators as well as chain-bucket excavators are used for lignite and overburden excavation. The overburden, interburden and lignite excavated in the pits is transported by belt conveyors. Stackers are used to construct waste dumps. Lignite is delivered to the power plants by the mine railway.

The exploitation of Konin lignite differs significantly from that in two other Polish lignite mines, Belchatow and Turow. There are differences in:

- lignite quality, which is lower in Turow,
- in deposit structure, which is highly irregular in Turow (erosion effects, number of seams exploited at the same time,
- in mining technology, and particularly so in material conveyance
- in seam location in relation to the existing industrial and other infrastructure (power lines, railways, pipelines, canals, roads etc.).

Mine activity in such difficult geological conditions and in complicated terrain required development of many unique technical and organizational solutions. Some of these were innovative firsts on the world scale, and without precedents in Polish lignite mining.

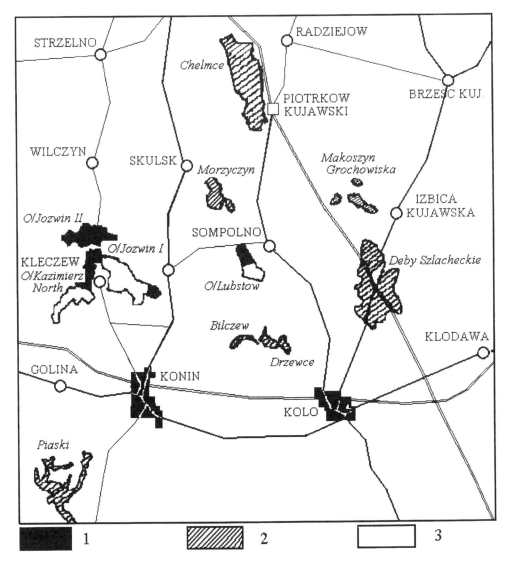

Figure 1. Lignite Mining Region. 1-Exploited deposit 2-Perspective deposit 3-Deposit in preparation for exploitation

2. MINING TECHNOLOGY

Limited accessibility of terrain subject to mining activities, high costs of land buying and also high land tax related to its mining use required a change in conventional mine design thinking. The external dumps have been eliminated with all waste material being disposed into the existing pits. This applies to pit Kazimierz North which is being developed at present with the waste material transported to the final pit of Kazimierz South. Similar procedure will

be used to develop the Jozwin IIB pit, the overburden from which will be dumped in Jozwin I and IIA pits. The same procedures will be used to develop further pits in the future, namely: Deby Szlacheckie, Drzewce and Morzyczyn - overburden will be filled partly in the deep pit Lubstow. This would help with reclaiming the pit into a water reservoir by reducing the required water requirements.

Some of prospective lignite deposits are located far away from the area worked currently. It could reduce the ability to use the present pits for overburden disposal by making it very expensive. As a result current plans call for construction of temporary out-of-pit dumps in the area. These would eventually be re-excavated and disposed into the pits during the final stages of their exploitation. The technological and economical analyses done for this scenario confirms its economic feasibility. The described solution alongside of economical advantages brings other benefit, namely it would result in filling up of the abandoned pits, thus to make easier their later utilization (in case the final land use is a water reservoir).

Characteristic for mine complex Konin is simultaneous exploitation of several seams. Due to small reserves the time of exploitation of individual seams is relatively short - often much shorter in relation to amortization of excavators and conveyors. Mining in the majority of pits lasts not longer than 20 years. This requires frequent relocation of excavators and other mining machinery to newly developed pits to take full advantage of undepreciated mining equipment. The relocations are normally done without disassembly of the equipment, a major technical task indeed. Up to now several relocation have taken place of both the excavators and the conveyors (SRs 1200, Rs560, A2RsB5000, A2RsB8800) for distances from several to 27 km. Relocations maybe made difficult by unfavorable terrain conditions (existence of power lines, roads, canals, pipelines etc.).

The precedent was set with a transport of excavator SmRs315 on the special pontoon across the lake for a distance of 3 km. Mastering of this technique has allowed to use better the technical potential of the equipment at hand and could lower the costs of new opencast mine construction.

3. RESERVE MANAGEMENT

Varying lignite quality cased against the demands of receiving power stations require intricate and complex mine planning and reserve management techniques. The objective of which is to provide the required quantity of lignite with the predetermined quality. A software package INTEGRAPH is used for design and planning work. It was found to be well suited to the needs of the situation, and particularly the ease with which it allows manipulation of volumes as well as geological and mining data. To meet the requirements of this systems a special database has been developed that stores the geological data for exploited and perspective beds. To optimize the mine performance a proprietary scheduling software package was developed by the mine in cooperation with the University of Mining and Metallurgy in Krakow. The package allows inclusion of economical information in the scheduling process in addition to the geological and other parameters and properties. Taking into consideration these parameters allowed optimizing the mining schedules for middle and long-term periods.

Current planning and quality control of supplied lignite has been made up to now using a laboratory sampling of lignite made by mining and dispatcher services. In the nearest future the exploited seams of lignite will have higher variability of lignite quality. In addition the lignite purchasers requirements, concerning of quality fluctuation, will raise considerably. As a result the mine prepares itself to initiate of continuous lignite quality control system, with lignite quality monitored directly on the belt conveyors. Lignie blending will also be initiated. Combination of extraction-planning system with continuously control of lignite quality should help to achieve of qualitative and quantitative stabilization of lignite supplying process.

Taking into consideration necessity of complex usage of the resources in place. The mine has begun the exploitation of minerals associated with lignite and occurring in overburden of lignite deposit. It has selectively exploited loam and sand for ceramic and building industries. The extracted loam and sand have been dumped on the special heaps forming so-called „anthropogenic beds", which could be the base of stock for ceramic and building plants for a large space of time.

Mineralized mine water is also used for consumption purpose. It is pumped by the drainage system (from the cretaceous well 100Z on the Lubstow outcrop).

4. RECLAMATION

Approximately 10,000 ha of agricultural land was disturbed by mining activities so far. Of this 5,100 ha is still an active mining area. About 3900 ha has been already reclaimed, the remainder of mined out land used for industrial purposes, such as disposal of power plant ashes. Land reclamation is an on-going process. Because the mine occupies the mostly agricultural land with good developed agriculture the final land use is planned to be agriculture. Thanks to a manner of fertilizing of overburden materials, successful rape and wheat growing takes place on the reclaimed area. Slopes of dumps are reclaimed for forest purpose.

Exploited pits are used in the first instance for mining-industrial purposes and after that, reclaimed

Tab. 1. Characteristic of post exploitation open pits.

Open pit	Year of end of exploitation	Area of final pit [ha]	Depth of pit [m]	Manner of utilization
Goslawice	1974	320	40	ash dumping
Kazimierz South	1997	100	54	overburden dumping→for agriculture (forestation)
Patnow	2000	500	65	ash dumping
Jozwin I, IIA	2003	700	68	overburden dumping→for agriculture (forestation)
Lubstow	2010	340	150	water reservoir

to the final land use (Table 1). In the open pits of mine Goslawice and Patnow ashes from power plan have been stored. Open pits Kazimierz South and Jozwin I and IIA will be used for overburden disposal from Kazimierz North and Jozwin IIB, what has been described above.

REFERENCES

Mazurek, S. 1997. *Price of mineral as the main bed's parameter.* Krakow. Gospodarka surowcami mineralnymi. T. XIII. Z.1,

Kasztelewicz, Z. Mazurek, S. Rosiak, T. Uberman, R. 1995. *Building of strip mines for lignite exploitation concerning the aspects of new formal standards.* Konin. Materialy VI Krajowego Zjazdu Gornictwa Odkrywkowego. Wyd SITG.

Uberman, R. Mazurek, S. 1997. *Estimation of recoverable resources of lignite concerning economical criterion.* Wegiel brunatny Nr 1 (18)

Mining Science and Technology' 99, Xie & Golosinski (eds) © 1999 Balkema, Rotterdam, ISBN 90 5809 067 1

Comprehensive evaluation and analysis of coal enterprises in China

Jingwen An, Xingping Zhang & Pu Jin
China University of Mining and Technology, Beijing, People's Republic of China

ABSTRACT: The comprehensive indexes of evaluation are proposed in this paper. On the basis of the Grey System theory, the paper gives the Grey Cluster Model. 93 major state-owned coal enterprises are evaluated and clustered by the model. The results can be used to find the problems existing of the enterprises, and to help make decision.

1. INDEX OF A COMPREHENSIVE EVALUATION

In a market economy, enterprise should be an independent market and benefit entity. Its goal is to maximize its benefit. It will get different conclusions if the enterprises are evaluated from different goals and angles. Therefore the basic principle is that the index system should reflect various aspects of enterprises. The index system is put forward through analyzing the overall of coal enterprise, referring to the comprehensive indexes of evaluation of the Economy and Trade Ministry of China, considering the actual situation of coal enterprises, and consulting with experts and managers of coal enterprises. The comprehensive indexes for evaluating coal mine enterprises are based on the analysis of their general status, consisting of 14 indexes in five categories: (1) Profitability includes three indexes, *rate of return on assets, rate of return on capital,* and *return on sales.* (2) Operational covers *receivable accounts turnover, current assets turnover,* and *inventory turnover.* (3) Development comprises *assets accumulation, science and technology level,* and *capital accumulation.* (4) Solvency comprises *the ratio of liabilities to assets,* and *quick ratio.* (5) Environment and social aspects involve three indexes, *social contribution, death rates per million tons,* and *satisfactory disposal of pollutant.*

The index system overcomes the shortage of single index, and stresses the comprehensibility. It emphasizes the indexes that can reflect the quality and benefit but not the output growth speed and extensive operation ones. So the attention of coal enterprises is paid to the increase of the benefit and equilibrium development. Moreover, the new national economic accounting system, the financial regulations, and the enterprise reform are considered. For the index system, the data applied should be receivable and comparable.

2. GREY CLUSTER MODEL

2.1 *Whitened matrix*[1]

Before *whitened matrix* is defined, the index data should be standardized. Suppose that the variable range of index i is $\left[f_k^1, f_k^2\right]$, then

(1) If index i is the beneficial one, the better the result, let

$$c_k^i = \left(f_k^i - f_k^1\right)\Big/\left(f_k^2 - f_k^1\right) \tag{1}$$

(2) If index i is the cost one, let

$$c_k^i = \left(f_k^2 - f_k^i\right)\Big/\left(f_k^2 - f_k^1\right) \tag{2}$$

(3) If index i is fixed, let

$$c_k^i = \begin{cases} 1 \\ 1 - \dfrac{\left|f_k^i - f_k^*\right|}{\max\left|f_k^i - f_k^*\right|} \end{cases} \tag{3}$$

Where, f_k^* is the optimum value of index i.

(4) If index i is the internal one, let

$$c_k^i = \begin{cases} 1 - \dfrac{q_1 - f_k^i}{\max\left(q_1 - f_k^1, f_k^2 - q_2\right)} & f_k^i \prec q_1 \\ 1 - \dfrac{f_k^i - q_2}{\max\left(q_1 - f_k^1, f_k^2 - q_2\right)} & f_k^i \succ q_2 \end{cases} \quad (4)$$

Where, $[q_1, q_2]$ is optimum internal of index i.
Then

The *whitened matrix* $c = \begin{bmatrix} c_{11} & c_{12} & \cdots & c_{1n} \\ c_{21} & c_{22} & \cdots & c_{2n} \\ \vdots & \vdots & \ddots & \vdots \\ c_{m1} & c_{m2} & \cdots & c_{mn} \end{bmatrix}$

Where, c_{ij} is the *whitened Cluster number*, i presents the number of enterprises, and j presents the number of indexes.

2.2 Whiten function

Suppose the value of indexes can be classified into three levels, i.e. the high, middle, and low.

2.2.1 benefit indexes

Suppose the *grey number* of the high level is:
$\otimes_{i1} \in [G_{i1}, \infty]$□
Where, $\otimes$ presents *grey number*.
Then, the *whitened function* is

$$f_{i1} = \begin{cases} \dfrac{1}{G_{i1}} x & 0 \le x \le G_{i1} \\ 1 & x \ge G_{i1} \end{cases} \quad (5)$$

Suppose the *grey number* of the middle level is:
$\otimes_{i2} \in [G_{i2} - \Sigma, G_{i2} + \Sigma]$,
Where, □ is a *fuzzy number*.
Then the *whitened function* is

$$f_{i2} = \begin{cases} \dfrac{1}{G_{i2}} x & 0 \le x \le G_{i2} \\ \dfrac{1}{G_{i2}}(2G_{i2} - x) & G_{i2} \le x \le 2G_{i2} \end{cases} \quad (6)$$

Suppose the *grey number* of the low level is:
$\otimes_{i3} \in [0, G_{i3}]$
Then the *whitened function* is

$$f_{i3} = \begin{cases} 1 & x \le G_{i3} \\ \dfrac{1}{G_{i3}}(2G_{i3} - x) & G_{i3} \le x \le 2G_{i3} \\ 0 & x \ge 2G_{i3} \end{cases} \quad (7)$$

2.2.2 Fixed indexes

Suppose the optimum *grey number* of indexes is:
$\otimes_{i1} \in [G_{i1} - \Sigma, G_{i1} + \Sigma]$
Then the *whitened function* is

$$f_{i1} = \begin{cases} \dfrac{1}{G_{i1}} x & 0 \le x \le G_{i1} \\ \dfrac{1}{G_{i1}}(2G_{i1} - x) & G_{i1} \le x \le 2G_{i1} \\ 0 & x \ge 2G_{i1} \end{cases} \quad (8)$$

2.2.3 Cost indexes

Suppose the *grey number* of the high level is:
$\otimes_{i1} \in [0, G_{i1}]$
Then the *whitened function* is

$$f_{i1} = \begin{cases} 1 & x \le G_{i1} \\ \dfrac{1}{G_{i1}}(2G_{i1} - x) & G_{i1} \le x \le 2G_{i1} \\ 0 & x \ge 2G_{i1} \end{cases} \quad (9)$$

Suppose the *grey number* of the middle level is:
$\otimes_{i2} \in [G_{i2} - \Sigma, G_{i2} + \Sigma]$
Then the *whitened function* is

$$f_{i2} = \begin{cases} \dfrac{1}{G_{i2}} x & 0 \le x \le G_{i2} \\ \dfrac{1}{G_{i2}}(2G_{i2} - x) & G_{i2} \le x \le 2G_{i2} \end{cases} \quad (10)$$

Suppose the *grey number of the low level* is:
$\otimes_{i3} \in [G_{i3}, \infty]$
Then the *whitened function* is

$$f_{i3} = \begin{cases} 1 & x \ge G_{i3} \\ \dfrac{1}{G_{i3}} x & 0 \le x \le G_{i3} \end{cases} \quad (11)$$

2.2.4 internal indexes

Suppose the *grey number* is:
$\otimes_i \in [G_{i1} - \Sigma, G_{i2} + \Sigma]$
Then the *whitened function* is

804

$$f_{i2} = \begin{cases} \dfrac{1}{G_{i1}} x & 0 \le x \le G_{i1} \\ 1 & G_{i1} \le x \le G_{i2} \\ \dfrac{1}{G_{i1}} \left(G_{i1} + G_{i2} - x \right) & G_{i2} \le x \le G_{i1} + G_{i2} \\ 0 & x \ge G_{i1} + G_{i2} \end{cases} \qquad (12)$$

2.3 Critical matrix:

$$\lambda = \begin{bmatrix} \lambda_{11} & \lambda_{12} & \cdots & \lambda_{1t} \\ \lambda_{21} & \lambda_{22} & \cdots & \lambda_{2t} \\ \vdots & \vdots & \ddots & \vdots \\ \lambda_{s1} & \lambda_{s2} & \cdots & \lambda_{st} \end{bmatrix} \qquad (13)$$

Where, $\square_{ik}$ is the intersection of $f_{jk}=1$ and $f_{jk}\square 1$.

2.4 Cluster weight matrix:

$$\eta = \begin{bmatrix} \eta_{11} & \eta_{12} & \cdots & \eta_{1t} \\ \eta_{21} & \eta_{22} & \cdots & \eta_{2t} \\ \vdots & \vdots & \ddots & \vdots \\ \eta_{s1} & \eta_{s2} & \cdots & \eta_{st} \end{bmatrix} \qquad (14)$$

Where, $\eta_{kj} = \dfrac{\lambda_{kj}}{\displaystyle\sum_{j=1}^{n} \lambda_{kj}}$

2.5 Cluster coefficient matrix :

$$\delta = \begin{bmatrix} \delta_{11} & \delta_{12} & \cdots & \delta_{1t} \\ \delta_{21} & \delta_{22} & \cdots & \delta_{2t} \\ \vdots & \vdots & \ddots & \vdots \\ \delta_{s1} & \delta_{s2} & \cdots & \delta_{st} \end{bmatrix} \qquad (15)$$

Where, $\delta_{ij} = \displaystyle\sum_{k=1}^{t} f_{kj} c_{ik} \eta_{kj}$

2.6 Judgement and classify

If $\delta_{ik}^{*} = \max_{k} \{ \delta_{ik} \}$ $\qquad (16)$

Then cluster elements belong to category k.

3. THE EXAMPLE OF EVALUATING THE COAL ENTERPRISES

Based on the *grey cluster model*, the 93 state-owned coal enterprises of China are evaluated in this paper. In evaluation, different criteria have been determined according to different goal. In the process of evaluating coal enterprises, the criteria are determined as follows: refer to the state average level, the *grey number* of high level is adjusted according to the actual situation of coal enterprises. Refer to the average level of coal industry, the *grey number* of middle level is adjusted forward, but the low is adjusted backward. Moreover, the theoretic optimum values are considered for some indexes. For example, in theory, the optimum value of *current ratio* is close to or bigger than 2, the *quick ratio's* is bigger than or equal to 1, and the *death rates per million tons'* is 0, etc. Considered the above situations, the *grey numbers* of the high, middle and low levels are determined as follow after discussing with experts.

$G_{i1}=(4,3.5,3,4,2,6,110,40,50,3,30,30,0.4,0.9)$
$G_{i2}=(2,1.9,1.3,2.5,1.3,3.9,90,50,35,1.3,23,25,1.5,0.6)$
$G_{i3}=(0.9,0.05,0.06,1,0.95,1.24,65,60,20,0.2,16,20,2,0.35)$

Data of samples are calculated by formula (1)-(16), and the results are as follows:

Table one: the results of evaluation

categories	High benefit	Middle benefit	Low benefit
Quantity	11	39	43
percentage	12%	42%	46%

The results reflect that the overall situation of coal enterprises is not good: of the all major state-owned coal enterprises, only 12 percent are better, but 42 percent are middle, and 46 percent are low.

The results of evaluating and the values of indexes are analyzed thoughtfully, we find main problems in coal enterprises and suggest taking the measures as following

(1) The overall level of enterprises is low. And coal enterprise lack for competition. Because the coal enterprises had been operated in the planed economy system, they lack for the competitive idea and the ability of operation. In the recent years, the macro-economy of China landed softly, and the world economy slumped, so the coal enterprises of China have supported the great market pressure, that is, faced to the fact that the supply of coal is bigger than the demand. Coal enterprises have not been free from the old system, the old opinion and custom restrict them. So there are many problems in transforming operation system, in opening up the new product market, and in finding the new economic growth point. Consequently, coal enterprises should speed up 'the two basic transforms', lead themselves to the actual market entity, in order to increase their power of the market competence.

(2) The level of profitability is low, and the operation of capital lacks of efficiency. Many coal enterprises pay more attention to the management of assets in kind, than the efficiency of capital; so the capital turnover are very low. It was calculated that the average return on sale of coal enterprises is 3.4% in contrast to 5.2% of that of the state, and the average Rate of Return on Asset of coal enterprise is 3.1% in contract to 7.6% of that of the state. So the measures that they should adopt are as follows: improve the capital's efficiency; prompt capital flows rationally and rapidly, and enhances the profitability.

(3) Resources were not be used rationally, there are many enterprises whose scale and technology are invalid. Invalid scale means the scale of enterprise is not economic. Invalid technology means the technology situation and the inputs or combined inputs are not rational, management falls behind, and by which enterprise can't obtain the productivity that they should. Applying the data envelopment analyses method, we have measured that enterprises whose scale is invalid account for 58 percent, and the invalid technology account for 75 percent, the valid technology and scale account for 15.6 percent only. So management falls behind, resource is not made the most and wasted greatly in more than half of all the coal enterprises, overlook the function of the existing assets, and the enterprises don't get the scale benefit which they should obtain. So advanced management methods, organization form, and advanced technology must be adopted in order to make full use, which is one of the most important tasks of the coal enterprises.

(4) The number of payable and receivable is very large, which restrict the product and hinder the development. The average account receivable turnover of coal enterprises is 1.94 in contrast to the national level (4.56). Enterprises each other greatly debit, which hinder the normal production and operation. So they should clear up debts and accelerate the account receivable turnover.

(5) The productivity is low, the burden about labor is heavy. In China, the number of labors per ten thousands tons coal, are seventyfold as much as that of American, tenfold as much as that of German, and twice as much as that of Poland. Although low technology is the main cause, heavy labors burden is also an important one. So they should decrease the labors largely, which is one of the gaps of changing economic growth mode.

In brief, the management administration should be formulated by the appropriate measures according to the different situations of coal enterprises. The high quality enterprises should be sustained by various policies in order to prompt their development. The middle and low ones should make full use of the existing resource, and increase the efficiency of capital utilizing. The decrepit, scrapped and deficit ones should bankrupt or transform their products.

4. CONCLUSION

The indexes of comprehensive evaluation and the Grey cluster model are efficient in evaluating coal enterprises. The results of evaluating can reflect objectively the whole operating condition of coal enterprises, and the problems that exist in the operation. They will offer more effective macro-guidance to the administrators and enhance the management level of coal enterprises. The successful application confirms that the Grey cluster model is an effective way by which to evaluate enterprises.

REFERENCES

Julong Deng. 1985. *Grey system theory*. Huazhong science and Technology University publishing house.

Coal enterprise accounting statements.1997.

Jingwen An, xingping Zhang, Xiangyang Xu.1996.*The evaluation on technology and scale validity of coal enterprise (research report)*.

Mining Science and Technology' 99, Xie & Golosinski (eds) © 1999 Balkema, Rotterdam, ISBN 90 5809 067 1

Effectiveness of organizational structure of a mining enterprise

Gabriela Paszkowska

Institute of Mining Engineering, Wroclaw University of Technology, Poland

ABSTRACT: The paper presents results of a study on effectiveness of the organization structure of a Polish mining company. An attempt has been made to assess the complexity, centralization and formalization i.e. three dimensions of the organization structure of an existing mine. A questionnaire survey performed among the employees of the mine has revealed their attitudes towards the organization structure and their evaluation of the three dimensions. In both the objective and the subjective approaches the organization structure came out to be complex, very formalized and centralized. This type of organization structure traditionally dominates in Polish mining industry. Is it still effective in the dynamically changing economic, political and technological environment?

1. INTRODUCTION

Organizational design involves selecting the combination of organizational structure and control systems that lets a company pursue its strategy most effectively — that lets it create and sustain a competitive advantage. The patterns of information flow and the lines of authority and responsibility constitute the structure of an organization.

The primary role of organizational structure and control is twofold:
- to coordinate employees' activities so that they work together to most effectively implement a strategy that increases competitive advantage
- to motivate employees and provide them with incentives to achieve superior efficiency, quality, innovation, or customer responsiveness.

Organizational structure and control shape the way people behave and determine how they will act in the organizational setting. Good organizational design allows an organization to improve its ability to create value and obtain a competitive advantage. The role of organizational structure is to provide the vehicle through which managers can coordinate the activities of various functions or divisions to fully exploit their skills and capabilities

2. DIMENSIONS OF ORGANIZATION STRUCTURE

Most authors in organizational theory recognize three major variables of organizational structure (Robbins 1987):

- complexity
- formalization
- centralization

2.1 Complexity

Complexity can be defined as the degree of differentiation that exists within an organization (Robbins, 1987). *Horizontal differentiation* considers the degree of separation between units of the same level. *Vertical differentiation* refers to the depth of the organizational hierarchy, and is usually inversely related to the span of control.

Spatial differentiation encompasses the degree to which the location of an organization's facilities and personnel are dispersed geographically. Horizontal differentiation is usually achieved by *specialization*. There are two ways of specialization. The first is division of labor, also called *functional specialization*, in which jobs are broken down into simple and repetitive tasks. If individuals are specialized, rather then their work, we have *social specialization*. Social specialization is achieved by hiring professionals who hold skills that cannot be readily routinized. Generally, the greater the number of different functions or divisions in an organization and the more skilled and specialized they are, the higher is the level of differentiation.

Implementing a structure to coordinate and motivate task activities is very expensive. The costs of operating an organizational structure and control system are called bureaucratic costs. The more complex the structure — that is, the higher the level of differentiation — the higher are the bureaucratic costs of managing it.

2.2 Formalization

Formalization refers to the degree to which jobs within the organization are standardized. If a job is highly formalized, there are explicit job descriptions, numerous organizational rules, and clearly defined procedures. Where formalization is low, employees' behavior would be relatively nonprogrammed.

2.3 Centralization

Centralization is defined by most theorists as the degree to which decision making is concentrated at a single point in the organization. Authority is centralized when managers at the upper levels of the organizational hierarchy retain the authority to make the most important decisions. When authority is decentralized, it is delegated to divisions, functional departments, and managers at lower levels in the organization. By delegating authority in this fashion, management can economize on bureaucratic costs and avoid communication and coordination problems because information does not have to be constantly sent to the top of the organization for decisions to be made.

3. AUTHORITY DELEGATION

When strategic managers delegate operational decision-making responsibility to middle managers, this reduces information overload, and strategic managers can spend more time on strategic decision making. Consequently, they can make more effective decisions and economize on their time, which reduces bureaucratic costs. When the managers in the bottom layer of the organization become responsible for adapting the organization to suit local conditions, their motivation and accountability increases. The result is that decentralization promotes organizational flexibility and reduces bureaucratic costs because lower-level managers are authorized to make on-the-spot decisions.

4. ORGANIZATIONAL STRUCTURE AND THE TASK ENVIRONMENT

In order to have fit between a unit's design and its tasks, there must be a fit between the various design elements and the activities of the unit itself. When the unit fits its tasks, employees know to perform particular tasks in effective and efficient ways. According to Lawrence & Lorch (1967), the nature of the tasks and the design depends on many factors. If a unit's tasks are certain, predictable and clear, the most effective design is the one that is formally structured — the one with clear rules, procedures, and job descriptions. Managers can identify the best

ways to perform a task. These can be written into a job description. If on the other hand, tasks are uncertain and unpredictable, then a relatively unstructured design may be best, because formal delineations of ways to perform tasks may be impossible.

The number of people reporting to a manager (span of control) can vary with the characteristics of a manager's job. The more complex the task, the more attention the manager has to devote to hiring, training, and reviewing subordinates, thus the span of control should be small. The same small span of control is important for tasks that are highly interdependent; the interdependency is relatively easy to manage if fewer people are in the group. Organizations that are „delayering" and increasing the number of direct reports a manager has in order to cut costs need to be careful to access the effects that the enlarged span of control has on effective job performance. While on the one hand, people may feel free to make decisions and take initiative, it is difficult for a manager to maintain communication and manage the complex interdependencies between tasks with a large span of control. This situation also makes it more difficult for managers to have the type of relationships necessary to coach and develop all their subordinates

5. ORGANIZATION'S DESIGN AND EMPLOYEES

Compatibility must exist between the organization's design and its employees. The structure, measurements, reward system, and selection and development systems must be all compatible with employees' needs, abilities and expectations. Managers may make inappropriate assumptions about the job's requirements and the abilities and skills of people asked to do the jobs. Factory jobs that are designed to be narrow, routine, and predictable no longer fit the characteristics of today's better-educated, affluent worker who has his or her own expectations about challenge and job variety. Some organizations counteract this misfit by changing the job designs to enrich or enlarge the jobs. Job enlargement combines elements of several routine jobs to increase the variety of the work. Job enrichment changes the job design to allow for more responsibility, autonomy, and challenge. These techniques fail when job designs are altered without the analysis necessary to understand the required tasks or employee needs, and further mismatches occur.

As organizations grow and add more people to perform more tasks, it becomes necessary to increase the number of organizational units, which adds levels to the hierarchy. This in turn increases individual job specialization, increases the formal rules and procedures, and restricts decision-making authority. These changes can be very demoralizing for employees who are used to being relatively autonomous and informal. Effective organizational design would try to minimize these mismatches before they occur.

6. RESEARCH METHOD

In the social sciences the term "scaling" is applied to the procedures for attempting to determine quantitative measures of subjective abstract concepts. Rating scales are used to judge properties of objects without reference to other similar objects. One of the most frequently used form is Likert scale. With this scale the respondent is asked to respond to each statement in terms of five degrees of agreement. Scale values are assigned to each possible answer. Such scales may be unidimensional or multidimensional. Multiple dimension scaling recognizes that an object might be better described by the concept of an attribute space of n dimensions rather than a single-dimension continuum.

The object of research was to measure attitudes (opinions) of managers in a chosen mine towards the organization structure of the mine. A multidimensional Likert scale has been developed for this research. The scale was meant to measure the three dimensions of organization structure, i.e.: complexity, formalization and centralization in terms of subjective opinions of managers. Several questions have been asked to measure one dimension. Some general questions have been asked additionally. This part of the questionnaire consisted of 20 statements. Respondents have been asked to evaluate them by assigning one value to each statement, where „1" stood for „strongly disagree", 2 — "disagree", 3 —"undecided", 4 — "agree", 5 — "strongly agree".

The questionnaire contained also a set of extra questions asked in order to allocate the respondent to pre-defined manager groups (management levels and functions).

One of the largest (in terms of output and mining area) Polish mines was chosen for the survey. 50 questionnaires were distributed among managers. Three levels of managers: high, middle and low and three functional groups: mining line-people, supportive function representatives and administrative staff have been included. The response

rate was 52%. Structure of manager groups that responded to the survey is presented in Figure 1.

7. RESULTS OF THE SURVEY

7.1 Complexity

Managers of the high hierarchy level admit that their span of control is too large. On the contrary, managers of the middle level declare that the number of people directly reporting to them is to small.

Information flow between departments is not effective according to the answers of the higher level managers. The answers of representatives of other management levels are inconsistent. Generally, 30.8% of respondents agree that the information flow in their company is effective, while 42.3% of respondents disagree.

The number of departments is too big in the opinion of high level. Managers of administrative function and of supportive function as well as middle level managers disagree that all departments in the currant structure are necessary. In general, 42.3% of respondents regard the number of departments as to big (while 23.1% disagree) and 38.5% are of the opinion that not all departments are necessary (while 15.4% disagree). Administrative managers agree that the duties of different departments overlap. Generally, 46,2% of respondents share this opinion and only 23,1% disagree.

7.2 Formalization

On average, all respondent groups except middle level managers, agree that numerous rules and regulations make their work more difficult. 80.8% of all respondents share this opinion and only 11.5% disagree. 57.7% of all respondents are of the opinion that bureaucratic regulations hinder inter-division communication. Especially high level managers, on average, share this opinion. 30.8% of all respondents disagree.

7.3 Centralization

65.4% of respondents declare to have too many duties on their posts. On average, high level managers support this opinion. 26.9% of all respondents disagree. All respondents agree that their superiors take into account their suggestions and remarks, they also rely on their opinion and reckon with them (80.8%).

Majority (76.9%) of respondents is of the opinion that their authority is sufficient to exact abiding their orders by subordinates and to ensure introduction of changes and improvements (80.8%).

Lower level managers and administrative managers agree that high level managers have the authority to make too many decisions. 50% of all

Table 1. Structure of manager groups that responded to the survey

	High level	Middle level	Low level
Administrative	46.2%	20.0%	12.5%
Supporting	23.1%	40.0%	25.0%
Mining	30.8%	40.0%	62.5%

respondents support this opinion while only 19.2% disagree. On average, lower level managers do not agree that they have enough freedom to make decisions, on the contrary, high level managers feel free to make decisions. In general, 53.8% of respondents claim to be independent in decision-making. Generally, 65.4% of respondents would like to be more independent in decision making. Majority (65.4%) of respondents regards their decisions as very important for the company and only 15.4% do not agree.

7.4 Other issues

Most of respondents (92.3%) are satisfied by their jobs and consider them adequate to their professional qualification. As far as restructuring of the company is considered, there is no generally shared opinion whether the changes go in the right direction. Only higher level managers (on average) support this thesis. Generally 46.2% of all respondents agree and only 3.8% disagree with the opinion

8. DISCUSSION OF THE RESULTS

Although the response rate of the survey and the final number of responses was not high, some conclusions may be drawn. The attitudes shown by respondents i.e. managers of the chosen mine (subjective approach) can be verified by observation of the organization design and analysis of used procedures.

The survey has revealed that managers working in the mine are aware of high degree of formalization and complexity of the organization. Especially high level managers admit that the number of departments is too large and the information flow among them is ineffective. Most managers agree that the functions of different departments overlap and not all departments are necessary. Very high percentage of respondents shares the opinion that bureaucratic regulations hinder inter-divisional communication and interferes with the job.

As far as centralization is concerned the answers do not clearly prove that managers are aware of it. Most respondents are satisfied with the amount of authority and decision making power they are given. They also feel appreciated by their superiors. The rationale for such answers might be reasoning that more authority means more duties. Most respondents admit to have too many duties already and they probably do not want to assume extra responsibility. However, majority of respondents would like to have more freedom in decision making. This might mean the need to reduce the degree of formalization but also might indicate the desire for authority delegation. Statements, on too much decision-making power in the hands of high level man-

agers, supported by lower level and administrative managers, seem to confirm the latter thesis.

9. CONCLUSIONS

The business environment in which mining companies operate in Poland is dynamically changing, volatile and complex. It requires flexibility — ability to change in order to adjust to environmental changes. High complexity and formalization of mining organization, reveled by both objective approach and attitude testing, hinder the ability to adjust to environmental changes. Although high centralization of power can be proved by means of observation and organization structure analysis, managers' answers did not confirm this characteristic. Assuming more power is probably feared because of more duties expected. This leads to a conclusion that reduction of excess bureaucracy and complexity in the organization needs to be the first step of reingeneering. After that, probably the need for authority delegation and decentralization will become more visible and desired.

REFERENCES

Bonoma, T., Zaltman G. 1981. *Psychology for Management.* Kent Publishing Company, Boston.
Emory, C.W. 1980. *Business Research Methods.* Irvin, Homewood, Illinois.
Lowrence, P.& Lorsch, J. 1967. *Organization and Environment.* Harvard Business School Division of Research, Boston.
Robbins, S.P. 1987. *Organization Theory: Structure, Design, and Applications.* Prentice-Hall, Inc., Englewood Cliffs, New Jersey.

Mining Science and Technology' 99, Xie & Golosinski (eds) © 1999 Balkema, Rotterdam, ISBN 90 5809 067 1

Optimizing development-extension plan: The case of a Chinese mine

Rijia Ding, Lijie Wang & Jianwei Rui
China University of Mining and Technology, Beijing, People's Republic of China

Jinrong Ma
Beijing Mining Research Institute of Central Coal Research Institute, People's Republic of China

ABSTRACT: With simulation and optimization methods, the optimal computer model of development-extension of mine is defined in this paper. This model fully utilizes the advantage of simulation and optimization technique, it can not only determine optimal development-extension plans reliably but also supply using effects and all kinds of parameters. This model was applied at a mine in China and proved that this model is practical and reliable.

1. INTRODUCTION

From the view of systematic engineering, every stage of coal mining is dependent and inheriting. Thus, in order to determine the optimal plan of development-extension, we should take the advantages of present producing experiences as well as equipment ability and should avoid their disadvantages under the present condition, and it is an indispensable part of development-extension designing to analyze present producing level. In order to combine optimization method with simulation method and to perform both methods' advantages, the process of lifting, transporting and mining of present producing level are simulated and analyzed by simulating. On the basis of simulation result and supplied technical data, the muster of shaft development-extension plan is determined, which means to avoid the "explosion" of plan amount. Then the optimal plan is selected from plan muster by optimization method. In addition, many technical parameters are determined and possible effects and problems are forecasted. The idea of modeling can be described as "Simulation-Optimization-Simulation", which makes development-extension plan more reliable and practical.

2. MODELING

Mining process is a dynamic, ploy-dimension, complicated system. The producing and variation of its many facts obviously have random characteristics. Simulation is the most suitable method to analyze such dynamic system. Based on the modeling idea of "Simulation-Optimization-Simulation", the computer simulation and optimization models are established respectively, both of which are combined organically at last.

2.1 *Computer Simulation Model*

In order to find an optimal design plan of development-extension, every possible plan should be analyzed and compared from the viewpoints of technology and economy. Furthermore it is necessary to analyze and simulate the producing system of each plan, so as to determine its reliability and technical parameters of every producing link. However, because there are too many plans, the simulation program should run quickly under the condition of reality if every plan is simulated. Therefore, considering the characteristics of development-extension plans, we modify and improve previous simulation model of producing system [2].

(1) Simulating coal streams source

The simulating of the mining plan can only provide coal output and working face number of every mining block and can not determine detailed technical parameters. Thus, when development-extension plans were simulated, it is necessary to determine coal stream distributing law of every mining block by recording the variation of coal streams with time. Researches show that the variation of out-transported coal streams of mine block conforms to Normal *Distributing Law*. Using the average and the standard aviation of out-transported coal supplied by sub-simulating-block of mining plan at different time, coal stream simulating block automatically forms Normal Distributing model of coal output stream of every mine block. This model can basically display random characteristics of coal transporting stream. Used as input stream and loaded to main-road transporting stream, it can simulate the producing system dynamically.

(2) Simulating The Railway Transportation

The road transporting net of every development-extension plan is different from each other. When plans are designed, it is difficult to give their detailed transporting network explicitly, so transporting net can not be established with the help of *natural node method*. In order to simulate every development-extension plan, we take advantage of some main nodes of mine transporting stream to establish a simplified transporting net. These main nodes (such as on-hook & off-hook place and sending car place of pit-bottom, on-hook and off-hook place of loading station of mine block, main cross point of railway, etc.) form a circled transporting net. The simulating program can automatically form a transporting net for every plan, and the only difference of these nets is the difference of running time.

After establishing the railway net, we used the scanning method of clock of dominated object to simulate coal transportation. Railed cars are regarded as dominated objects. According to their activity characteristics, the movements are divided into seven statuses. Every train is assigned with two dimensions (simulation sub-clock and status), through which train's moving process is simulated.

(3) Calculating the optimal transporting and managing fees

In order to determine an optimal extension plan, the optimal transporting and managing fees should be calculated under the condition of different transporting facilities. The optimal specifications and amount of facility of every plan are determined by simulating. On the basis of simulation, "five-kind-fee" is calculated and optimal transporting and managing fees are determined. Using out-transported rate as the criterion, simulation program provided optimal transporting fees and some parameters automatically. By this way, transporting and managing fees of different transporting facility can be also compared and optimized. So the optimal transportation plan could be determined.

With the improvement stated above, running time (5□6 min) of each plan is saved 45 min. A reliable and practical simulation model of designing extension-plan is established, which can run in coincidence with the optimization model.

2.2 Establishing the optimization model

At present, the methods to assess the technical and economical effect of coal mine design plan are mainly Net Present Value Method, Internal Rate of Return Method, Dynamic Payback Period Method, etc. But all these methods have their disadvantages. Thus, A new criterion, that is ACDCPT (Average Cost of Discounted Cash Per Ton), was put forward in this paper. We can use this criterion to assess which plan is optimal. Its calculating equation is as following:

$$TM = \frac{1}{A}[\sum_{t=1}^{t_p}(C_i + K_i)(1.0 + E)^{-i}]\frac{E(1.0+E)^{t_p}}{(1.0+E)^{t_p}-1}$$

Where, TM—Average Cost of Discounted Cash PerTon, *RMB* per ton
A—output capacity of mine (level) million ton/yr.
t_p—total years of every level servicing duration of mine, year
C_i—level producing fees of No i year, million *RMB*/a
K_i—facility fees and tunneling fees of No i year, million *RMB*/yr.
E—basic discounted rate.

ACDCPT is calculated in term of the way of apportioning discounted cash of total fees into every year of optimal period and then dividing every year's average cost by annual output. It is also called average cost per ton. This criterion comprehensively displays the condition of producing and managing costs and all kinds of input costs. It takes into account of time value of capital and can assess extension plan of different producing scale. It is a kind of practical and comprehensive criterion.

3. EXAMPLE OF APPLYING THIS MODEL

3.1 The m*ine condition*

The mine's development pattern is vertical-shaft and multi-level. Its mining field is 5 km in length and 5km in width. Its mining coal seams are inclined and have 19 coal seams (total thickness: 71.5m). Its previous output is 1.5 million ton/yr. after three times extension and building, its present output capacity is 4 million ton/yr.. Currently, both second and third producing levels are in works. The second level's developing pattern is public main road of every coal seam group and rock-tunnel-connection at every mine block, and the third level's is public main road of every seam group and rock-tunnel-connection at every mine block. Now next level (deep elevation: -300m~-700m) is prepared to be developed.

Table 1 Analysis of transporting system capacity

Index	Possible output (0.01million Ton/yr.)		Output from work-face (0.01million Ton /yr.)		Lifting output 0.01million Ton/yr.		Duration of injecting main coal barn min/shift		Rate of lifting %		Time of waiting for sending min/shift/train		Time of waiting for unloading min/shift/train	
level	2	3	2	3	2	3	2	3	2	3	2	3	2	3
plan☐	252	136	249	129	228	114	408	312	90	83	164	176	81	77
plan☐	239	133	237	129	228	114	214	298	95	85	173	179	74	75
plan☐	228	125	227	122	222	114	19.8	294	98	89	179	199	10	70
plan☐	220	115	220	115	218	114	254	254	99	99	196	205	8.6	56.2

Table 2 Lifting capacity of main lifting shaft

Name	No	Elevation	Maximum Lifting Capacity 0.01million Ton/a	Recycle time of lifting min	Name	No	Elevation	Maximum lifting capacity 0.01million Ton/yr.	Recycle time of lifting, min
Mixed shaft	1	-100	228.22	85.00	Old main shaft	2	-300	114.86	143.00
		-140	218.53	89.44			-320	112.70	146.01
		-180	208.95	93.89			-340	110.03	149.02
		-220	198.32	98.33			-360	108.64	152.02
		-260	190.53	102.78			-380	106.72	155.03
		-300	182.47	107.22			-400	104.88	158.04
		-340	175.05	111.67			-420	103.11	161.05
		-380	168.19	116.11			-440	101.40	164.05
		-420	162.84	120.56			-460	98.76	167.06
		-460	157.95	125.00			-480	95.17	170.07
		-500	151.46	129.44			-500	92.64	173.08

3.2 Simulation of current producing level

3.2.1 Maximum capacity of transporting system

In order to determine maximum capacity and weak link of current transporting system, many plans are simulated to supply reliable information for extending design. Main simulating results are listed in Table 1.

Table 1 shows that the maximum transporting capacity of second and third levels are 2.28 million Ton/yr. and 1.14 million Ton/yr. respectively. It also shows that maximum lifting capacities are 2.28 million Ton/yr. and 1.14 million Ton/yr. respectively. The weak link of Transporting System is the inadequate capacity of lifting system.

3.2.2 Forecasting the lifting system's capacity

Based on the survey data of current lifting system, the lifting capacity of two main lifting shaft are simulated under the condition of different elevation, results are listed in Table 2.

Table 2 shows that the lifting capacity of current second level of mixed shaft is 2.28 million Ton/yr. and is a little inadequate (because second level annual output is 2.20 million-Ton). If mixed shafts are extended directly and do not change lifting facility, its third level lifting ability should be 1.8247 Million Ton/yr.. Similarly, the lifting capacity of third level of Old Main Shaft (1.14 Million Ton/yr.) is inadequate. For this reason, Old Main Shaft can not be extended directly while the Mixed Shaft can be extended from −100m level to −300m level. In order to solve the problem of inadequate main lifting capacity, we put forward the plan of developing a new shaft to ~300 level from ground surface.

3.3 Optimizing Extension Plan

Based on the simulating result, three reserved plans (3.60 Million Ton/yr. 4.00 Million Ton/yr. and 4.50 Million Ton/yr.) are put forward. Considering the condition of coal seam group, pattern of level, pattern of shaft and pattern of main road, we establish reserved plan muster. For different Plan, simulating model is used to calculate and we find that, of all plans, the extension cost of inexplicit incline shaft is the lowest, and also find that the pattern of main road of all plans should be arranging two main centralized roads in coal seam group. The results are listed in Table 3.

Table 3 result of simulating extension plan

Plan (million Ton/yr.)	Pattern of extension	Arrange pattern of main road	Fee of Total discounted cash, million RMB	cash of producing fees million RMB	Cash of project fees million RMB	Cash of facility fees□milli on RMB□	Discount ed cost per ton million RMB/ton
3.60	Inexplicit inclined shaft	Two main road	36919.67	22305.15	3475.89	11138.63	10.27
4.00	Inexplicit inclined shaft	Two main road	37941.32	22800.58	3805.08	11335.88	9.5
4.50	Inexplicit inclined shaft	Two main road	43119.63	25069.52	5198.51	12851.6	9.62

Table 4. Fees of every transport type

Transport type	Optimal train number		Number of main transport ing	Total car number	Five kinds fees(million RMB/a)					Total transport fees(million RMB/yr)
	Level third	Level forth			salary	Electricity	Discount	Maintain and repair	other	
A	11	10	408	25	47.6	6.9	19.1	10.3	11.4	95.0
B	11	6	298	21	39.6	5.8	20.4	11.6	9.5	86.3
C	3 belts (1000mm in width) or 4 belts (800mm in width)				30.1	29.0	73.8	155	7.2	295.0

Note: Plan A - both levels using 3-ton railed car
Plan B - the third level using 3-ton railed car, the forth level using 3-ton bottom-unload railed car
Plan C - both levels using belt.

Table 3 shows that the plan of 4 Million Ton/yr. is optimal because its average cost of discounted cash per ton is the lowest, It also shows that the forth and the fifth level should be arranged with the pattern of inexplicit inclined shaft and two centralized main roads.

3.4 Simulating the optimal extension plan

After being extended, third level and forth level were in works at the same time. For 3-ton railed cars (gauge: 900mm) are mainly used in third level, so its possible transporting facilities are 3-ton railed cars and belt. For the transporting pattern of third level and the fact that the forth level was new, forth level's possible transport facilities could be 3-ton railed car, 3-ton bottom-unload railed-car and belt. Computer simulation model is utilized to simulate different combination of transporting type mentioned above for four days. Through repeatedly automatic changing the number of train, simulating model could give optimal transporting parameters and transporting fees all at once under the condition of different transporting type. See Table 4.

Form Table 4, main road transporting pattern of optimal plan is 3-ton railed car in level third and 3-ton bottom-unload railed car in level forth. Both levels' optimal train numbers are 11 and 6 respectively.

4. CONCLUSIONS

1. By using Modeling method of simulating-optimizing-simulating, the problem of plan number "explosion" during determining extension plan is effectively avoided, which makes the reserved plan reliable.

2. By using simulation and optimization method comprehensively, the two problems (optimal result is hard to be got by simulation method and optimization method can not be applied in complex system) are solved.

3. The way of using simulation method to select transporting pattern primly overcomes the obstacle that it is hard to calculate cost while taking into account of producing random variation, which ensures the reliability of carrying out plan.

4. Optimal extension pattern and optimal transporting pattern of example mine are determined. Compared with previous plan, the new one cuts down input cost by more than 30 million RMB. The economic benefit is obvious.

REFERENCES

Research Report on Optimal Plan of Mine Extension, Liaoning Technical and Engineering Uni. 1993

Nai Yue, Research on Computer Simulation of Main Producing System of mine, Journal of Fuxin Mining Institute, 1988

Mining Science and Technology' 99, Xie & Golosinski (eds) © 1999 Balkema, Rotterdam, ISBN 90 5809 067 1

Higher education in mining engineering in 21st century China

Yongchang Xing, Weiya Chang & Liquan Wang
China University of Mining and Technology, Xuzhou, People's Republic of China

ABSTACT: This paper introduces the preliminary research on teaching program and curriculum reform in the undergraduate education in mining engineering in China to order to meet the needs for versatile engineering staff in the new century. It focuses on the adjustment of specialty direction, education of train objectives and layout of the curriculum.

1. INTRODUCTION

It's known that higher education in the mining engineering has a long history in China. After 1949, a considerable number of superior mining engineers and management have graduated from universities and colleges subordinate to the related ministries such as Mining Ministry and Metallurgical Ministry, which ensures an army of well-educated and trained engineering staff in the economic construction in new China.

Now we are standing at the threshold of the new millenium, when science and technology develop faster than ever before and economy in China is on its way to market-oriented economy, which severely challenges the current higher education in the mining engineering as well as in the other fields. Therefore, a research program on training of the qualified mining undergraduates and changing of subject contents and curriculum is launched by China University of Mining and other relative universities. To meet the requirement of knowledge economics for the qualified mining engineering staff in the 21st century, the present situation of higher education in the mining engineering in China and around the world are carefully studied before the speciality adjustment and conversion. A teaching program is drawn up in the concerned speciality of mining engineering, geology, petroleum engineering and mineral resources. The reform is initiated in curriculum and courses with the help of new textbooks and syllabus for major subjects. All these frame the program reform in the undergraduate education in mining engineering in China.

2. ADJUSTMENT OF SPECIALITY

According to the past-ten-year statistical data, industries in developed countries have covered their reflections on the traditional higher education as follows:

1. The quality of the engineering graduates could not meet the needs for engineering staff in the 21st century, especially for those who are good at dealing with interdisciplinary problems, carrying out projects simultaneously and cooperatively, and keeping good interpersonal relationships. These abilities are extremely important for the industries under the increasingly sharp competition. Engineering graduates can not start to work as qualified engineers (Christiansen, D. 1992, & Waston, G. 1992).

2. Modern industry is suffering from the serious consequences of highly complicated and narrow fields by much-specialized education, which leads to the exchange difficulty, both internal and external; and it is a long period of time before the industry can gain the advantage over the competition and the beneficial results. Industry does not work the same way as departments of the universities do. Industry needs staffs who are experts in cooperation (Crookall, J. R. 1994).

3. Education of science and engineering must offer the graduates the knowledge and abilities, with which they will efficiently work under the changing circumstances of many factors such as the re-arrangement of multi-functional team-work and resources, and the use of developing science and technology. (Lacocca Institute Leheigh University 1991)

Since 1950s, new and high technology has been developed in the form of cooperation with the main characteristics of the interdisciplines of many fields, the mixture of various techniques, the high concentration of information and technology and the leading position of comprehensive technology. Any great inventions of new technology did not result from the accumulation and perfection of simple experiences, but from the systematical and comprehensive research with both good basis of theories and strong influence and guide of theories.

It is evident that the qualified engineering staffs for the 21st century can not be those who will only be able to work with one specialized technique of a narrow field for long or even for the whole life, but those who should master the knowledge and techniques of many fields or specialities with broadened view and active mentality. Under the changing situations and with more than one task to do at one time, they should be good at cooperation and comprehensive consideration or analysis on problems of interdisciplines to obtain the best solutions. Therefore, to cultivate talents to meet the needs of the 21st-century development of economy, science and society, reform in teaching materials and curriculum layout must be undertaken to broaden and adjust specialities.

In the earlier time, there was only one speciality in mining engineering where subjects were provided in coal and non-coal mining, strip and underground mining, mine construction and mine ventilation & safety. Later in order to meet the needs of planning economy, the mining engineering was disintegrated into four sections, that is, underground mining, open-pit strip mining, mine construction and mine ventilation & safety. Meanwhile the coal mining departed from the non-coal mining. This disintegration, now, is obviously unadaptable to the market-oriented economy. After careful adjustment and conversion the narrow fours are merged into a new broader mining engineering so that graduates can widely pursue career in the relevant fields of Mining planning and development, mine designing (both underground and strip), mine safety engineering, and even in mine ground pressure & rock mass engineering. They can also join the management or further advanced studies in the mining engineering.

3. TRAINEES AND THEIR ADAPTABILITY

Education should meet the times and society, and the same goes for the higher education in China. We are living in a time when intellectual economy challenges the traditional industrial economy. Science and technology develop fast, characterized by high disintegration and integration, with the later being a current trend. So modern higher education focuses more on knowledge synthesis, basic and engineering courses, and broader speciality which makes it possible for students to be educated in their capacities especially in professional competence and originality.

The successful career pursuit of graduates and their high qualification in work can, to some degree, embody the fulfillment of education theory and concept in a university. The needs of the society and times squarely influence education code, speciality conversion and curriculum layout. In order to face the future and market, the graduates in mining engineering must qualify themselves in the following aspects:

1. They should bear a strong responsibility to develop mining industry.

2. They should be quite cultivated and mentally healthy. They should acquire military knowledge; learn laws and humanities. Diligence and team-work spirit are necessary qualities for the future engineers. The capacity to do well in public relationship will help to their advantages.

3. They should systematically learn theories of basic sciences, acquire broad knowledge in their own field and related sciences. They should be able to do drawing, perform experiments and tests, and know how to use computer to help themselves work more efficiently.

4. They should have a general understanding of market economy and management. They should have an acquisition of knowledge of technological quality and efficiency concerned with the relevant project.

5. They should learn a foreign language with ability to understand relevant materials in this language. They should be good not only at reading and writing but also at listening and speaking.

6. They should be able to further self-education and develop a habit of independent thinking. With a preliminary knowledge of Creation & Invention they should be capable of analyzing and finding out solutions to the problems in the future work.

7. They should possess basic knowledge of modern sports and develop a good habit of exercises to obtain good health for the hard jobs and difficult circumstances in the future.

4. REFORM IN TEACHING PROGRAM

Only a scientific and reasonable teaching program or curriculum will ensure the educating and training of qualified graduates. The reform in the teaching program is based on the following considerations:

4.1 Integration

The students are assured of a broader speciality. After integration, the mining engineering involves the three previous specialities of underground mining, strip mining and mine ventilation & safety. Non-coal mining is also listed in the curriculum. The new program is characterized by "5+3", which consists of sessioned teaching program and leveled curriculum.

"5+3" means that in the first five semesters a uniform program is provided to all the students in the same college or department. In the last threes the students in different specialities follow different curricula.

A curriculum is arranged in four levels. The first involves basic courses accounting for 46% of the total (1208 hours). The second are special basic courses intended to broaden students' educational backgrounds, which account for 864 hours. The third are those of professional interests (360 hours). The last but important are the 96 hours of special and interdepartmental electives for the purpose of enlarging their knowledge and enhancing personal interests.

4.2 *Versatility*

Students are provided more basic courses to meet the requirement of fast development of science and technology for versatile and qualified engineers. Besides humanities, mathematics and other relative natural sciences, foreign languages, computer science, creation & invention and basic courses for engineering are added in the new curriculum.

4.3 *Practice*

Students are specially educated in their practice capacities. The goal is achieved by practice sessions in the curriculum with 13 weeks arranged for cognition, production and graduation field work, 4 weeks and a half for curriculum design. 200 hours are set aside for operation and practice in computer science.

4.4 *Mental Health*

Mental health is another concern in the new program. Philosophical and moral courses are listed in the required courses. A Grand Moral Engineering is framed by two terrains (classroom teaching and out-of-class activities) and three field practices (military training, volunteer work and social practice). At the same time courses in humanities and arts are provided to cultivate students' minds.

4.5 *Flexibility*

A feature of flexibility is reflected in the new program, where a full account is taken of personal interests and abilities. Redistribution has been arranged in the required courses and electives. The

former are divided into the common and the special which occupy 78.7% of the entire teaching hours involving basic and special knowledge, moral and behavioral codes. The later are electives composed of common electives, departmental electives and special electives. Meanwhile interdepartmental electives are opened to all the undergraduates.

4.6 *Auxiliary*

The new program gives an increasing attention to out-of-class activities. Through years education has been centered on classroom teaching and hence auxiliary courses are not included in the curriculum, but the research carried out by authors shows that these auxiliary courses are exercising invisible and formative influence on the students' characters and capacities. On that score, they are listed in new curriculum to ensure that graduates are not only physically but also mentally strong enough to shoulder tough tasks in their future careers.

4.7 *circumstances for talents*

The unity of systemization and flexibility of the new program helps a lot in the development of the students' character and provides a fine circumstance for the versatile and the talent. The curriculum is so sessioned and leveled that students can select courses according to their own abilities and interests. All the students are permitted to attend common electives grouped in philosophy and psychology, history and culture, literature and arts, law and sociology, economy and management, science and technology. Students are allowed to choose interdepartmental courses.

5. REFORM IN THE SPECIAL COURSES FOR MINING ENGINEERING

A new curriculum is designed for mining engineering to broaden students' knowledge and provide them more chances in their future job-hunting. The curriculum consist of three platforms . The first platform is common basic cause platform, its main causes are as follows:

Higher Mathematics(1)	5 credit/ 80 hours
Higher Mathematics(2)	5.5 credit/ 88 hours
Linear Algebra	2.5 credit/ 40 hours
Probability & Mathematical Statistics	3 credit/ 40 hours
Higher Physics (1)	3.5 credit/ 56 hours
Higher Physics (2)	3.5 credit/ 56 hours
Higher Chemistry	2.5 credit/ 40 hours

The second platform consists of special basic courses, its main causes are as follows:

Theoretical Mechanics	4.5 credit/ 72 hours
Material Mechanics	5 credit/ 80 hours
Mining Rock Mass Mechanics	2 credit/ 32 hours
Electrical & Electronic Technology	5 credit/ 80 hours
Fundament Of Mechanical Design	3 credit/ 48 hours
Hydro-mechanics & Hydro-Mechanica Engineering	3 credit/ 48 hours
Openings & Development Engineering	3.5 credit/ 56 hours

The third are special course composes of three directions of underground mining, open-pit strip mining and computer application in mining engineering. The main causes of underground mining direction are as follows,

Mining Engineering	11.5 credit/ 184 hours
Mine Ventilation & Safety	5 credit/ 80 hours
Excavation Equipment Technology of Mine	4 credit/ 64 hours
Lifting &Conveying	2 credit/ 32 hours

A Striking feature in the curriculum is integration. Here, the coal mining merges into non-coal mining; underground mining is integrated with open-pit strip mining; mining engineering is united by mechanical and electronic engineering; computer science is applied in the mining engineering; mining engineering and mining rock mass mechanics are interrelated; mining engineering is joined with management.

CONCLUSION

We like to communicate to the experts of the other countries and study their advanced experience by presenting this paper to the conference It is believed that with the perfection and effectuation of the new teaching program and curriculum the higher education in mining engineering will supply more and more qualified and versatile graduates for the 21st century.

REFERENCES

Christiansen, Donald 1992. New Curricula. *IEEE Spectrum*.

Crookall, John R. 1994. Engineering the Integration of Manufacturing Education and Industry. *Journal of Manufacturing System*. V13. No. 1.

Lacocca Institute Leheigh University 1991. *21st Century Manufacturing Enterprise Strategy*.

Shen, Shituan 1998 Deepening the Reform in High Engineering Education for 21st Century. *Researches in High Education of Engineering*. No. 4.

Wastion, George F. 1990. Refreshing Curricula. *IEEE Spectrum*.

Wu, Minsheng 1998. Personnel training and Educational Innovation in the 21st Century. *Research on Education Tsinghua University*.

Mining Science and Technology' 99, Xie & Golosinski (eds)© 1999 Balkema, Rotterdam, ISBN 90 5809 067 1

Economic analysis of internal competition of coal industry in China

Min Zhou & Hongliang Zuo
China University of Mining and Technology, Xuzhou, People's Republic of China

ABSTRACT: In this paper game theory and informative economics are both applied to analysis of the current situation of China's coal industry, where over competition is rampant. By analyzing business decisions of two coal enterprises and then n coal enterprises under the condition of games, the conclusion is that coal enterprises can achieve maximum profits only through cooperation among themselves. While attempts by any individual coal enterprise to decide its own output and exceed its quota, which results in Nash equilibrium output, will not be able to bring about maximum profit. Finally, causes for the ineffectiveness of macro policies by China's coal industry are probed into, and countermeasures are thereafter put forward.

1. INTRODUCTION

Similar to the situation of the international coal market, China's coal market has been continually sluggish in the 1990s. Such problems as oversupply, falling price, idle production capacity and declining economic benefit has seriously hindered the healthy development of China's coal industry. Though the management system, policies, special features of the coal industry itself and some other factors are all responsible for the current situation, there's no denying that most important factors include the long existing over competition between key state-owned coal enterprises and small local coal enterprises for precious coal resources and markets, and low barriers to enter the coal industry, which inevitably results in over competition featuring too many enterprises, low efficiency and elimination of few inefficient enterprises. The government has for many times tried to limit coal output, but hasn't improved the situation very much. Therefore, in this paper we intend to describe quantitatively business decisions of coal enterprises under the condition of competition by applying analytical methods of game theory (Zhang 1995) and from the perspective of microeconomics. The purpose is to reveal the laws lying behind those business decisions by coal enterprises so as to provide scientific bases for making macro policies for China's coal industry.

2. QUANTITATIVE ANALYSES OF BUSINESS DECISIONS BY COAL ENTERPRISES UNDER THE CONDITION OF COMPETITION

2.1 *Analyses of business decisions by two coal enterprises in the game*

Let's make the following assumptions first. There exist only two coal enterprises. They produce coal of the same quality, and sell the same product on the same market. Coal outputs of the two enterprises are respectively q_1 and q_2, and the total market supply $Q=q_1+q_2$. The price at which all the produced coal can be sold out is p. Then p becomes function of the whole market supply, i.e., $p=p(Q)=320-Q$. We may further assume that fixed costs of the two coal enterprises are not considered, their marginal costs are the same, i.e., $c_1=c_2=80$, then their total cost is respectively $80q_1$ and $80q_2$. Finally we assume the two coal enterprises decide their own coal outputs, and neither coal enterprise knows how much coal the other enterprise produces. The above assumptions constitute a game with coal enterprise 1 as one party and coal enterprise 2 as the other party, and their strategy space is made up of various outputs. Due to the limitation of production capacity, the outputs can be regarded as continual and divisible and have an upper limit in theory. Hence the two enterprises each has many alternative strategies and many possible corresponding profits (income minus cost). With the above assumptions, profits of the two coal enterprises can be expressed as the following

$$u_1=q_1p(Q)-c_1q_1=q_1[320-(q_1+q_2)]-80q_1$$
$$=240q_1-q_1q_2-q_1^2 \qquad (1)$$

and $u_2=240q_2-q_1q_2-q_2^2 \qquad (2)$

The two formulas indicate that the profit of each party depends on the strategies of both parties, i.e., their outputs.

If we assume $(q_1{}^*, q_2{}^*)$ is the only Nash equilibrium of the game, then the sufficient prerequisites for solving the Nash equilibrium are actually maximization of $q_1{}^*$ and $q_2{}^*$.

$$\begin{cases} \max_{q_1}(240q_1-q_1\dot{q_2}-q_1^2) \\ \max_{q_2}(240q_2-\dot{q_1}q_2-q_2^2) \end{cases}$$

Then the maximum value can be worked out by making derivatives of $q_1{}^*$ and $q_2{}^*$ respectively corresponding to q_1 and q_2 become 0, i.e.,

$$\begin{cases} 240-\dot{q_2}-2\dot{q_1}=0 \\ 240-\dot{q_1}-2\dot{q_2}=0 \end{cases}$$

The sole solution to the game is then $q_1{}^*=q_2{}^*=80$. In other words, (80,80) is the only strategy mix for Nash equilibrium, the profit of each party is $u_1=u_2=80\times(320-160)-80\times80=6400$, the total output $Q_1{}^*=160$, the market price $p=320-160=160$. The total profits of both parties $u_1{}^*=u_1+u_2=12800$. The result is based on the output decision respectively made by each of the two enterprises in line with the principle of maximizing its profit, but in this case the two coal enterprises can not actually maximize their profits.

Now we regard the total profits of the two enterprises as the basis for deciding the optimum output for the market. The total output corresponding to the maximum profit under the market condition is Q, total profit $U=Qp(Q)-cQ$ $=Q(320-Q)-80Q=240Q-Q^2$. The solution is $Q_2{}^*=120$. The total profit of the two enterprise is $u_2{}^*=14400$, and $q_1'=q_2'=60$.

By comparison, we can observe $Q_2{}^*<Q_1{}^*$, $q_1{}^*=q_2{}^*$ $>q_1'=q_2'$, but $u_2{}^*>u_1{}^*$(12.5% higher), which indicates that if the two coal enterprise could cooperate with

each other to find out the output that can bring about maximum profits to them as a whole, and then share the total output equally (60), they each would be able to get a higher profit than otherwise (7200>6400). In order to make this idea more convincing, we can also apply the principle of game theory "prisoner's predicament" for further analysis (refer to Table1). Strategies of Party 1and Party 2 are abiding by the quota and exceeding the quota, Table 1 shows the game competition between Party 1and Party 2. In the game, as far as Party 1 is concerned, if the strategy of Party 2 is to abide by the quota, Party 1 can achieve a profit of 7200 if it abides by the quota, but achieve a profit of 8000 if it does not abide by the quota, the optimal strategy is not to abide by the quota strategy (8000,6000). For the same reason, if the strategy of Party 2 is to exceed the quota, Party 1 can achieve a profit of 6000 if it does not exceed the quota, but a profit of 6400 if it does exceed the quota, in this case the best strategy for Party 1 is to exceed the quota. The other way round, for Party 2, whatever strategy Party 1 chooses, the best strategy for Party 1 to choose is to exceed the quota. This is the reason for attempts by the two parties to exceed the quota and break the equilibrium.

Table 1. Prisoners' dilemma

	Coal enterprise 2 abiding by the quota	(Party 2) exceeding the quota
coal enterprise 1 abiding by the quota	7200, 7200	6000, 8000
(Party 1) exceeding the quota	8000, 6000	6400, 6400

2.2 Analyses of business decisions by n coal enterprises in the game

The above model can be extended to cover output decision of n coal enterprises, whose respective outputs are $q_1,\ldots, q_n$, $p=p(Q)=k-Q(Q>Q)$, marginal cost $c_1=\ldots=c_n=c$. When n coal enterprises each decide their respective output, their total output is

$$Q_1^{\cdot}=\frac{n}{n+1}(k-c)$$

The total profit of n coal enterprises is

$$u_1^{\cdot}=\frac{b(k-c)^2}{(n+1)^2}$$

If n coal enterprises cooperate with one another in

order to decide an output that can generate maximum overall profit for the enterprises as a whole, the optimum output of n coal enterprises is

$$Q_2^* = \frac{k-c}{2}$$

The total profit is $u_2^* = (k-c)^2/4$. When $n>2$, we have

$$Q_1^* = \frac{n}{n+1}(k-c) > \frac{k-c}{2} = Q_2^*,$$

$$u_1^* = \frac{n(k-c)^2}{(n+1)^2} < \frac{(k-c)}{4} = u_2^*$$

To sum up, the output mix decided by each individual coal enterprise to maximize its profit shall not be able to result in maximum profit of the coal enterprises as a whole, while cooperation among all the coal enterprises to decide their respective output can greatly enhance their overall profit.

3. POSITIVE ANALYSIS

The best example to illustrate the above analyses and models is the government's effort to limit coal output VS coal enterprises' attempts to exceed quotas (Zhou,1998). Each of the state-owned coal enterprises, local and township-owned coal enterprises has its own interest, and each tries to maximize its profit. The game between the coal enterprises will surely result in declining coal prices, decreased profit and even loss of the whole coal industry, therefore coal enterprises all share the intention to limit their output so as to maintain reasonable price. However, if the government makes policies to limit coal output and stipulate quotas for individual coal enterprises, each individual coal enterprise takes for granted that it achieve more profit and does not affect the market price much if it exceeds the set quota while all the other coal enterprises abide by their quotas. Motivated by its own interest, each coal enterprise tries to produce more, and not one coal enterprise abides by the quota. The final result is that coal prices fall down sharply, and each coal enterprise can only get the unsatisfactory profit of Nash equilibrium.

Moreover, we admit that the government's effort to limit coal output has some positive effects, particularly on key state-owned coal enterprises. In this case, key state-owned coal enterprises limit their output, leave their production capacity unused, and

lose their original markets. However, non-state-owned coal enterprises take the advantage to increase their output and seize the market that used to belong to state-owned coal enterprises. Please refer to Table 2. In the end, total coal output has not been effectively controlled, coal prices have fallen down. By the end of 1997, the ratio of coal market concentration c_4=8%. 64000 coal enterprises produce 1330 million tons of coal. The number and output of various small coal enterprises respectively account for 93.8% and 46.6% of the total. Structure of the coal industry is becoming even more diversified and unreasonable.

Table 2. Comparison between the outputs of state-owned coal enterprises and non-state-owned enterprises*

Unit: 10,000 tons

Type of coal enterprise	1980	1985	1990
State-owned	50654	58904	68531
Township and private	11364	28324	38911

1992	1994	annual increase rate of raw coal
68536	67463	2.37%
42548	54822	27.32%

* For Yearbook of China's Coal Industry 1980~1994

4. POLICIES AND MEASURES

1. Close down small coal enterprises and maintain big ones. Legal measures and administrative methods should be adopted to close down, or stop, or merge, or transfer some small local township and private coal mines, particularly those coal enterprises that compete with state-owned coal mines for resources. The purpose is to cut down the output of non-state-owned coal mines, maintain normal coal production order, and protect the effective market for profitable key state-owned coal mines.

2. Keep supply falling short of demand. The government needs to exercise effective macro regulation of coal production so as to ensure that demand for coal basically equals to market supply of coal. It is suggested here that the supply is slightly be short of demand, so that the coal enterprises can be in a better position in price bargains.

3. Make common commitment. Through guidance and regulation by coal management society,

commitment and self-discipline by all individual coal enterprises, optimum coal output and most reasonable price are expected to be determined. Perfect coal industrial management system is to be established so as to increase the marginal cost of those coal enterprises that deliberately exceeds quota, and severely penalize them whenever they are discovered to have done so.

REFERENCES

Zhang Weiying. 1997. *Game theory and informative economics*, Shanghai People's Press.

Liu Wei, Zhou Min. 1998. *Regulate the market structure of China's coal industry from the perspective of exit and entry barriers*. Coal Economics Research, 9:13-15.

Yearbook of China's Coal Industry 1980~1994

Mining Science and Technology' 99, Xie & Golosinski (eds) © 1999 Balkema, Rotterdam, ISBN 90 5809 067 1

New technology of underground coal gasification and its economic evaluation

Jianpei Xu, Xuejian Liu, Zhaoxiang Zhang & Xianling Liao
Management Department, China Coal Economic College, Yantai, People's Republic of China

ABSTRACT: The new technology of underground coal gasification (UCG) has been developed on the bases of coal mining technology, especially on the traditional methods in China. With the features of "long tunnel, large section and two stage", this new technology would provide a potential solution to the problems faced by coal industry worldwide, such as numerous abandoned resources, surface subsidence, serious environmental pollution and low economic benefits if it is industrialized. In this paper the engineering feasibility and economic evaluations of UCG are analyzed.

1. INTRODUCTION

Being rich in natural resources, coal makes up a substantial of approximately 70% China's energy structure. But as the non-regenerated resources, the part of coal resources is held up or abandoned underground by traditional mining methods which have changed the valuable coal resources into "the dead coal resources". How to transform "the dead coal resources" into usable coal resources for the benefits of mankind?

The new technology of underground coal gasification (UCG) has been developed in abandoned mine in China by the UCG Research Centre at China University of Mining & Technology. The semi-industry trial of UCG at 2nd shaft of Xinhe coal mine in Xuzhou was first finished in 1994. Then the trial of UCG at Liuzhuang mine in Tangshan has been conducted since 1996. In the trial the long tunnel, large section, advancing gasifier was designed, and the air successive gasification process and its moving velocity of gasification face were researched

This new technology of UCG has been made breakthrough in China because it can be applied to make the best use of abandoned coal resources. At present there are 30-million-kiloton "dead coal resources"—abandoned coal resources in China, so this technology has practical significance.

In comparison with the traditional technology of coal gasification, this technology has its specific characteristics—extracting useful components from coal and leaving the noxious or the useless underground.

Industrialization of the technology of UCG is to reach a certain scale industrial groups of electricity, chemistry and gasification on the bases of high-quality, low-cost gas processed by the technology; to transform the traditional technologies of coal industry. It will be good for developing coal resources, increasing the economic and social benefits, as well as improving ecological environment.

2. TARGET LOCATION AND APPLIED AREAS OF INDUSTRIALIZING THE TECHNOLOGY OF UCG

2.1 *Target Location of Industrializing the Technology of UCG*

There are 30-thousand-million-ton abandoned resource and a great many of declined coal enterprises that have to face the pressure for self-development in China. Over 70% of total electric energy are generated by burning coal way of low efficiency and serious pollution, so we have an urgent need to look for a new, clean resource.

Gas is an important raw material for chemical industry and a fuel for civil use. The costs of ground gasificatian are usually 3-6 times as many as those of UCG, meanwhile some problems for instance controlling environmental pollution, fully utilizing existing mining equipment, protecting cultivated land are demanded to settle fundamentally.

The new technology of UCG and its industrialization are the vital way to solve the problems mentioned above, and consequently arrive

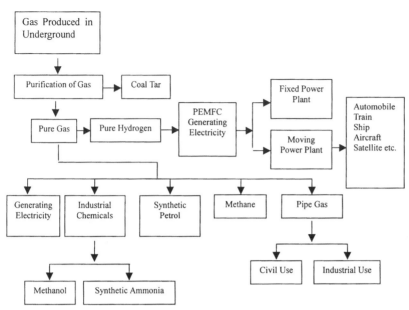

Figure 1 A product chain and a series of products produced by the technology of underground gasification

at the conclusion that in the middle-run, the target location of its industrialization in China should be to choose the optimal target and way for the industrializing according to the internal conditions, external environment and the reserve of abandoned coal resources, and set up the certain scale industry or industrial garden areas including electricity, coal, fuel hydrogen and coal chemistry etc. On bases of high quality, low-cost produced gas will make contribution to increasing the quantity of employment, controlling environmental pollution and for the local economic development.

2.2 *Applied Areas of Industrializing the Technology of UCG*

The new technology of UCG will produce a series of products and product chain, as illustrated in figure 1. Moreover industrializing the technology would bring the high-added value and the high rate of return, as well as lay a real foundation for different countries and different regions to agilely choose industrial targets and ways.

Some important applied areas of industrializing the technology of UCG should be coal industry and producing hydrogen industry in the light of developing trend of the advanced technology and China's special conditions.

2.2.1 *Coal Industry*

The most important applied area of industrializing the technology is coal industry. Because it really realizes the "no person, no equipment" goal on working face, and changes traditional ways of mining coal and will replace it in future. At the same time, application of this technology will cut down productive cost, reduce accident rate, and give declined mining enterprises the vitality and full vigor.

There are two advantages of the technology applied in coal industry besides what is mentioned above; it is unnecessary to wash coal, therefore a quantity of investments will be cut down, as pure gas produced is extracted from useful component of coal; secondly; the problem of surface subsidence that exists in traditional mining coal will be solved and protect cultivated land is protected.

2.2.2 *Generating Hydrogen Industry*

It is well known that hydrogen is a new clean resource with broad prospects, and it makes up nearly 60 percent of water gas generated by the technology of UCG.

To extract hydrogen from water gas is possible not only in terms of technology but will also get economic benefits. (The rate of input-output is 1:6)

2.2.3 Electricity Industry

By applying the technology of UCG, electrical energy will be generated by burning gas instead of coal, so the burning rate will raise and the cost will be cut down, and environmental benefits will be improved greatly since no waste slag and no dust are produced during the production.

2.2.4 Industry of Coal Chemistry

The application of the technology is also closely related to the industry of coal chemistry. The cost of affiliated products, for instance chemical fertilizer and methanol and so on will decrease by using the higher-quality, lower-cost gas which is an important raw materials for the industry of coal chemistry. South African has been a good example for developing coal chemistry to produce liquid fuel (artificial petroleum) with synthetic water gas.

3. THE WAYS AND COUNTERMEASURES FOR INDUSTRIALIZING THE TECHNOLOGY OF UCG

3.1 The key to apply the technology of UCG is to keep the higher-quality, lower-cost gas to be generated continuously.

To obtain stable and controllable gas, (if coal layer is complicated) we should take the complex system theory as our guide, enforce such means as experimenting, modeling tests and monitory to generate gas steadily and continuously.

3.2 The important way to cut down the cost is to make full use of abandoned resources.

At present it is a necessity for the industrialization to apply the technology to the gasification of the abandoned coal resources.

3.3 Selecting industrial Programs

Different countries and a variety of regions should select an optimum program according to self-conditions from a lot of programs for industrializing the technology of UCG. These include:
◆ underground gasification
◆ underground gasification—generating electricity
◆ underground gasification—processing chemical fertilizer
◆ underground gasification —producing methanol
◆ underground gasification—generating pure hydrogen

3.4 Building an Example Project and Setting up an Industrial Garden Zone

Chinese President Jiang Zemin said at the Fourteenth APEC Summit: "Setting up industrial garden zone is the greatest deed of science and technology on industrialization in the century. It combines scientific researches with industrial development, therefore inventions can be transferred smoothly into industrial areas to reach their economic and social benefits."

We should firstly set up an example project for the sake of reducing the risk of market and investment, and then promote gradually the technology.

The technology of UCG is industrialized in the form of industrial garden zones, and generally the industrial garden zone has generally the following features

3.4.1 Promotion features

Industrial garden zones are a comprehensive collection of technologies of UCG, processing hydrogen and intellectual faculty management, and they are good examples for "developing advanced sciences and technologies and reaching their industrialization".

3.4.2 Applicability

To set up the industrial garden zone has tremendous influence on promoting continuous development of coal resources owing to changing traditional production methods and using clean energy-hydrogen and electric.

3.4.3 Market Prospects

Gasified products (gas, electric power, hydrogen, methanol and other related products) have the obvious potential market on the bases of first analyses of market forecasting.

3.4.4 Strategic Signification

In China The area between Baize Coal Mine and LiangJia Coal Mine, Longkou, Mining Bureau is ideal for setting up an example industrial garden zone. It is bounded by the Yellow Sea to the north, Yantai City to the east (About 100 Km), Qingdao to south and WeiFang to the west. There are a 1000 kkw coal-burning electric plant; a 12 kkw electric plant in it. LiuHai Coal Mining and farmland in the east, and an express highroad in the south, so this region is a ideal zone for building industrial garden zone on account of scenic beauty, transport facilities, pleasant weather and a variety of institution.

The total thinking for setting up the industrial

garden zones is to stress the essentials, highlight cruxes and emphasize innovation under the guiding ideology of "developing high sciences and technologies, reaching their industrialization " and China's macro economic policies.

A management committee composed of the management departments of government, companies belonged to the industrial garden zones, research institutes of universities and related banks should be set up in order to build Longkou industrial garden zone as a good example of industrializing the technology of UCG in China or even in the world.

There should be outstanding traits in Longkou demonstrative industrial garden zone, for example, full utilization of abandoned coal resources and existing equipment, high rate of retune on investment, clean and high-added value products, transformation of traditional management and productive ways and driving power for developing local economy and self-circulation.

The successful demonstration of industrializing the technology of UCG will certainly give impetus to its development and speed up its pace.

4. THE PROSPECTS RESEARCH ON APPLYING THE TECHNOLOGY OF UCG

The technological value is determined by whether a technology has a vast range of prospects, so the system of economic evaluation should also be considered in terms of the following aspects:

4.1 *Prospects Research on Making Full Use of Coal Resources and Its Economic Benefits*

(1) Prospects Research on Applying the Technology to Abandoned Coalmines

The rate of extraction of underground mining is low, about 50% in China's state-owned coal mines, 10—30% in local and villages and township coal mines. By the end of 1993, there have been 60.2 million kiloton industrial reserves that are surplus in state-owned coal miners, and 16.7 millions kiloton of them are "the dead coal" that can't be mined forever, in addition to the non-state-owned coal miners', there will be 30 million kiloton "dead coal" in China.

In addition, 163 important coalmines in China will have been abandoned or closed by the end of the century. And about 1/3 producing level of 630 important coal-miners have be abandoned according to *China coal Industry Yearbook*. Abandoned coal reserve of various permanent pillars will have been increasing. Therefore spreading the application of the technology of UCG to abandoned coalmines is of great immediate significance.

(2) Prospects Research on Applying the Technology of US to Productive Coalmines

Just as we discussed above, only can 41 million kiloton coal of 60.2 million kiloton coal industrial reserves that are surplus in state-owned important coal mines be mined, while the technology of UCG will make the full use of the surplus 60.2 million kiloton industrial reserves and non-state-owned coal mines' coal reserves, and it will open up a broad prospects for using of surplus industrial reserves at the moment.

(3) Prospects Research on Applying the Technology to Non-underground Mining

There are 114.5 million kiloton demonstrated mineable reserves in China on the base of reference materials of the 16th World Energy Meeting that accounts for 11.19% of the total resources and ranks the third in the world. In accordance with present yield, the total coal reserves will be only mined for eighty-five years.

Comparing with average mineable years of 224 in the world; 249 in US; 286 in Germany, 364 in Australia, 500 in the Commonwealth of Independent States, it shows that we are called for making full use of coal resources from the angle of strategy so as to keep continuous development of economy.

REFERENCES:

Liang Jie. 1997. Study on the stability and controlling technology in steep incline seam□dissertation □ Xuzhou:China Univ. of Mining & Techn.
Yang Lanhe et al.1998. Industrial experiment of underground coal gasification. *Journal of China University of Mining & Technology.* 23(3):254~258.

Mining Science and Technology' 99, Xie & Golosinski (eds) © 1999 Balkema, Rotterdam, ISBN 90 5809 067 1

Author index